D1001651

mechanics, waves, and thermodynamics

physics:

VOLUME ONE

HOLDEN-DAY
San Francisco

mechanics, waves, and thermodynamics

DUANE E. ROLLER
Late Professor of Physics
Harvey Mudd College

RONALD BLUM
Academic Coordinator
Maryland State Universities and Colleges

with the special editorial assistance of

SUMNER DAVIS *Professor of Physics*
University of California, Berkeley

ANTHONY J. BUFFA *Professor of Physics*
California Polytechnic State University, San Luis Obispo

THIS BOOK IS IN THE HOLDEN-DAY SERIES IN PHYSICS

Duane E. Roller *received his Ph.D. in physics from the California Institute of Technology. During his career he held professorships at the University of Oklahoma, Hunter College, Wabash College, and Harvey Mudd College. At each of these institutions, as well as at Harvard University and the California Institute of Technology, he taught introductory physics, and in 1949 he was awarded the Oersted Medal in recognition of his distinguished contribution to the teaching of physics. From 1933–1948 he was Editor of the* American Journal of Physics, *and later became Editor of* Science *and* Scientific Monthly. *Dr. Roller was author or coauthor of a half-dozen books, one the widely used text* Mechanics, Molecular Physics, Heat and Sound *with R. A. Millikan and E. C. Watson. He also served as President of the American Association of Physics Teachers.*

Ronald Blum *is currently Academic Coordinator for the Maryland State Colleges Information Center. He received his Ph.D. in physics from Stanford University and has done research at Stanford University, NASA Institute for Space Studies, and the University of Chicago. At the University of Chicago he taught physics and physical science to physics majors, as well as to nonscientists. He served for three years on the Commission on College Physics, pioneering in the development of computer-related curriculum materials for physics.*

Text design Nancy Clark, cover design Edward Riley
Illustrations David A. Strassman, Evanell Towne, John Foster, J & R Services
Composition Allservice Phototypesetting

Photographs on pages 496, 500, and 506 reproduced by permission from PSSC Physics, *2d ed., 1965, D. C. Heath & Co. with Educational Development Center, Newton, MA*

physics volume one
mechanics, waves, and thermodynamics

Library of Congress Catalog Card Number: 81-81011

ISBN: 0-8162-7284-0

Printed in the United States of America

9 8 7 6 5 4 3 2 1

To my parents, Sol and Helen Blum, for love that never failed
and to my children, Elana, Tamara, and Don, with love that never will

foreword

This book has as its foundation the highly respected classic text *Mechanics, Molecular Physics, Heat, and Sound* by Millikan, Roller, and Watson, first published in 1938, as well as a set of class notes on electricity, magnetism, and light written by the same authors in collaboration with Carl Anderson and Wolfgang Panofsky. The principal author of this early version was Professor Duane E. Roller who later undertook—as a major ambition—the revision and completion of the project for publication during his lifetime. To facilitate this, he accepted a faculty position in 1957 at the invitation of Joseph Platt, President of the recently established Harvey Mudd College in Claremont, California, where he taught physics and worked on the manuscript. It was during this period that Professor Roller brought the book to my attention, stirring my own enthusiasm for its revision and publication. At the time of his death in 1965, however, the project was still in the developmental stage.

At the suggestion of Professor David Saxon several years later, I asked Dr. Ronald Blum, Professor of Physics at the University of Chicago, to complete the manuscript. It turned out to be a sizeable undertaking which Dr. Blum pursued while with the Commission on College Physics at the University of Maryland, where he directed a pioneering effort toward exploring, developing and encouraging the use of the computer in the teaching of introductory physics. The present book, then, represents a classic treatment of introductory physics made up-to-date by a complete revision, with the addition of new chapters, abundant new examples, applications, problems, and illustrations. Moreover, it features calculator and computer usage within the context of introductory physics. Dr. Blum has fused the scholarly treatment and time-respected content of the earlier work with a modern approach, to offer the student a unique introductory physics text.

The publication of this work is the fulfillment of a longstanding ambition of both authors. Professor Roller was one of the most respected and inspirational teachers of physics of his era, and in recognition of this was awarded the Oersted Medal in 1949 by the American Association of Physics Teachers. During his lifetime he spent considerable time in the preparation of the earlier version and, later, with its revision. Dr. Blum has devoted well over a decade to the expansion, further development, and modernizing of the material. Few would have given such effort, care and dedication to the completion of this project, and to him go my personal thanks and sincere appreciation for his accomplishment. In a personal sense, the publication of this book has meant a very great deal to us both.

I am indebted to the many reviewers who read the manuscript and provided valuable comments and creative suggestions that resulted in the text's overall improvement. Richard Olson and Norma Campbell were very helpful in the earlier stages. Special thanks go to Professor Anthony Buffa for his meticulous effort in the book's editorial development, and to Professor Sumner Davis who was more than generous with his time and advice in a consulting capacity. Many thanks also go to those involved in the actual production of the book—in particular to Nancy Clark who designed the book and edited the manuscript, and to Edward Riley, Managing Editor.

You, the student, will be the beneficiary of the combined efforts of these dedicated individuals. Our hope is that you will come to understand and appreciate the thoughts and ideas set forth in this book, and that your dedication to its study will reflect the dedication that made possible its creation. Much of what you learn from this book will remain with you all your life and prove most valuable in your future studies. Regardless of your field of endeavor, perhaps you too will pass on some of what you learn from the study of this text to future generations for their better understanding and deeper appreciation of the profound and beautiful science that is physics.

San Francisco
January, 1981

Frederick H. Murphy
Publisher

preface

Einstein once remarked that the most astonishing single fact about the universe is its very reasonableness. This book is an introduction to physics, the most reasonable of the natural sciences. The goal has been to bring together the sources, theory, and practice of each topic to form a unified narrative in which the unfolding stages in the development of physics issue logically, reasonably, one from the other, without mystifying gaps or acrobatic leaps of intuition. Where it has not been possible within the compass of the text to proceed with inexorable logic from one step to the next, the student is at least led to an appreciation of the reasonableness of the result by heuristic arguments, clearly labeled as such. Whatever shortcomings are to be found in these pages, it is sincerely hoped that hand-waving is not among them.

This book is the first of a two-volume text on introductory physics intended for serious students of the physical sciences, life sciences, and engineering. Those taking this course should have studied geometry, algebra, and trigonometry as prerequisites. Calculus and analytic geometry are also required, but may be taken concurrently. The first volume covers mechanics and thermodynamics; the second treats electricity and magnetism, light and optics, and the foundations of quantum mechanics. The organization of the material is essentially classical, with three important exceptions.

First, we have written of concepts rather than isolated phenomena. Thus we begin with four chapters whose central theme is description. The next five chapters introduce particle mechanics, and Chapters 10, 11, and 12 discuss rotations and the mechanics of rigid bodies. Chapter 13 on gravity and central forces serves as a bridge between the study of rotation and the properties of matter, of which gravity is the most fundamental apart from mass itself. Chapters 14 and 15 continue the discussion of matter and introduce the first ideas of disturbances in material media and their propagation.

Chapters 16, 17, and 18 discuss periodic phenomena in matter. Here the emphasis is on the unifying concepts of oscillatory disturbances, their propagation as waves, and superposition effects which give rise—using Huygens' construction—to the basic laws of reflection, refraction, and interference. Several types of waves in material media are considered, and their conceptual similarities are stressed. Thus acoustics and music, which usually appear in a separate chapter at the back of the bus, are part of the mainstream discussion of waves and interference. Chapter 20, which covers relativity, is the longest chapter and capstone of our treatment of mechanics.

The last part of this volume, Chapters 20 to 25, covers a broad range of thermodynamic phenomena, not just temperature and heat. The emphasis is on the first law of thermodynamics and equations of state for various substances, not just those for an ideal gas. Elementary statistical concepts are introduced in Chapter 22 and further developed in Chapter 24 with the Maxwell-Boltzmann distribution and kinetic transport theory. Chapter 23 deals with phases of matter and concludes the treatment of the first law of thermodynamics with a discussion of intermolecular forces. It also provides a breather before the relatively difficult Chapter 24. The final chapter covers the second law of thermodynamics, the definition of entropy, and some statistical and philosophical considerations of its significance.

The second exceptional feature for a book of this nature is the inclusion of as much historic material as practicable, not out of mere deference to the past or scientific patriotism but so that today's student of physics can trace the organic growth of our knowledge. Physics did not spring full blown from the brow of genius, but was the result of many varied experiences and experiments and the contention of many different theories. Our purpose is served if the student can see the rationality behind such growth and perhaps think "Given such facts might I not have drawn these conclusions myself?" This does not disparage genius, but rather, illustrates the eminent reasonableness of physics.

We come to the third unusual feature of this text. Given the historic development of physics, what are the tools needed for its continued unfolding? One very important tool—so important that it becomes a conceptual factor in itself—is the computer and more recently the hand-held programmable calculator. Physics as a body of knowledge must draw to itself the power of electronic computation, just as in an earlier age it assimilated algebra, calculus, higher mathematics, and technology. Modern computation techniques should be made available to students within the context of physics at the earliest possible stage in their intellectual development. Hence, this book contains a course-within-a-course that develops the principles of numerical analysis and computing as applied to the study and understanding of physics.

Instead of presenting many short sections of varying length I have elected to use a modular approach, keeping the length of each section roughly equal. There are five or six sections per chapter, and a typical lecture would generally cover one or two sections. Any chapter can be completed in a week's time, with the exceptions of Chapters 19 and 24, and the entire volume can be taught in one regular academic year or less. Each section with its problems may be treated as a separate study unit, making it easier for students and teachers to know where they stand and to pace themselves accordingly. Modularization not only lends itself to personalized self-instruction (PSI) but also has enabled the development of a unique three-track system. The tracks, which are illustrated in flowchart form on pp. xii to xv, are (*A*) basic introductory physics, (*B*) computer-oriented methods in physics, and (*C*) advanced introductory physics. Each track is self-consistent and assumes the preceding track, with the more advanced material generally occurring at the end of each chapter. A more detailed discussion of the track system is given in the *Teacher's Guide*.

There are 187 worked examples in Volume One, and they should be considered an integral part of the text. The problems are also of great importance. Students cannot learn to think like physicists unless they do interesting and challenging problems, and do them in quantity. There are 916 problems in this volume and most of them have two or more parts; some of these are thought problems rather than computations. Almost every answer is given at the end of each chapter where it is readily available without being immediately visible. Problems are divided according to sections in the text, with the more difficult ones at the end of each section. The more advanced problems lead the student through a number of steps to arrive at some interesting or valuable conclusion. It is suggested that the instructor scan the problems at the end of each chapter and assign some of them as recommended reading, if there is not enough time to do them.

The appendices in this volume have been carefully selected to provide students with a valuable reference work that they can continue to use for many years to come. The definition of any symbol can be located quickly in the glossary of symbols (Appendix A); tables of the International System of Units (SI) comprise Appendix B and give the definitions and abbreviations of the standard mksa units, as well as the system of prefixes. The conversion tables (Appendix C) are extensive and easy to use, and the periodic table (Appendix G) is practically a small handbook of physics. The mathematical and statistical tables (Appendices D, E, F, and J) include all the formulas needed in the text and more; the BASIC computer language, treated in Appendix I, provides a handy reference to the most popular and widespread language for instructional and personal computing.

Computer topics are clearly marked with an asterisk, and computer-oriented problems are also flagged, although many of them do not actually require programming but are primarily concerned with methods. The computer programs in the text are written in the universal computer language—the flowchart—so that the presentation is language-free. Furthermore, *no prior experience of computing is necessary.* In fact, the computational methods can be studied profitably without even going near a computer, although hands-on experience is recommended.

As one understands more about computers and computer methods, one becomes less dependent on them, psychologically and practically. A knowledge of numerical analysis, approximations, and errors can often lead to quick solutions on a hand-held calculator, and the student is also aided in avoiding the pitfalls of substituting number-crunching for insight. The obvious advantages of computing are ease of computation (more physics with less mathematics) and the treatment of more realistic situations. It also encourages the study of errors and algorithm construction, the use of recursive procedures and scaling, and the simulation of experiments. Computer use compels precision and discipline, leads to a greater awareness of our own processes of abstraction and analysis, and even raises fundamental questions about the nature of our description of physical reality. The computer can introduce a fresh sense of discovery into even the most well worn problems and makes possible an "open-endedness" in the physics curriculum, homework, and laboratory.

The writing of this book has taken more than twelve years. Numerous individuals provided advice, meaningful suggestions, criticism and guidance, and certainly they merit special recognition here. First and foremost is the highly respected physicist, teacher, writer, and scholar, Duane Roller, who provided the initial manuscript for Volume One, as well as an example of scholarly purpose and dedication that has been a guide and inspiration to me. Without that foundation this book would not have been written. My only regret is that I was not privileged to know Professor Roller.

Alfred Bork suggested the integration of computers into the text. Julius Brandstatter provided valuable and timely encouragement. Anthony Buffa did a marvelously devoted and painstaking job in his detailed reviews of the manuscript in its several drafts, as well as in proof. Others who reviewed the manuscript and made many helpful comments and suggestions were Sumner Davis, Charles Bordner, Douglas Shawhan, Robert Leighton, Harold D. Rorschac, Walter D. Wales, David Cook, Robert March, Stanley Williams, Charles Whitten, Jr., R. Wayne Crews, Don Martin, Malcolm Smith, Warren Blaker, Jerry Pine, Anthony Leitner, David McDaniels, D. Murray Alexander, J. Gordon Stipe, Jr., H. R. Brewer, M. E. Oakes, John L. Powell, Herbert D. Peckham, Susan Schwartz, Harry Bates, Peter Sturrock, Philip DeLavore, David Saxon, Kenneth Jolles, Mark Zemansky, Burton Fried, Arthur Leuhrmann, John Robson, Timothy Kelley, and Louis Deegan, Jr.

Frederick H. Murphy, President of Holden-Day, fortified my resolve when the going was rough by his own commitment to the book; it would not have been written without his faith. Others who made the production of this book possible were Nancy Clark, Edward Riley, and Edward Millman. To each of them I want to express my deep and sincere appreciation for their outstanding and valuable effort.

Baltimore *Ronald Blum*
January, 1981

course guide

VOLUME ONE

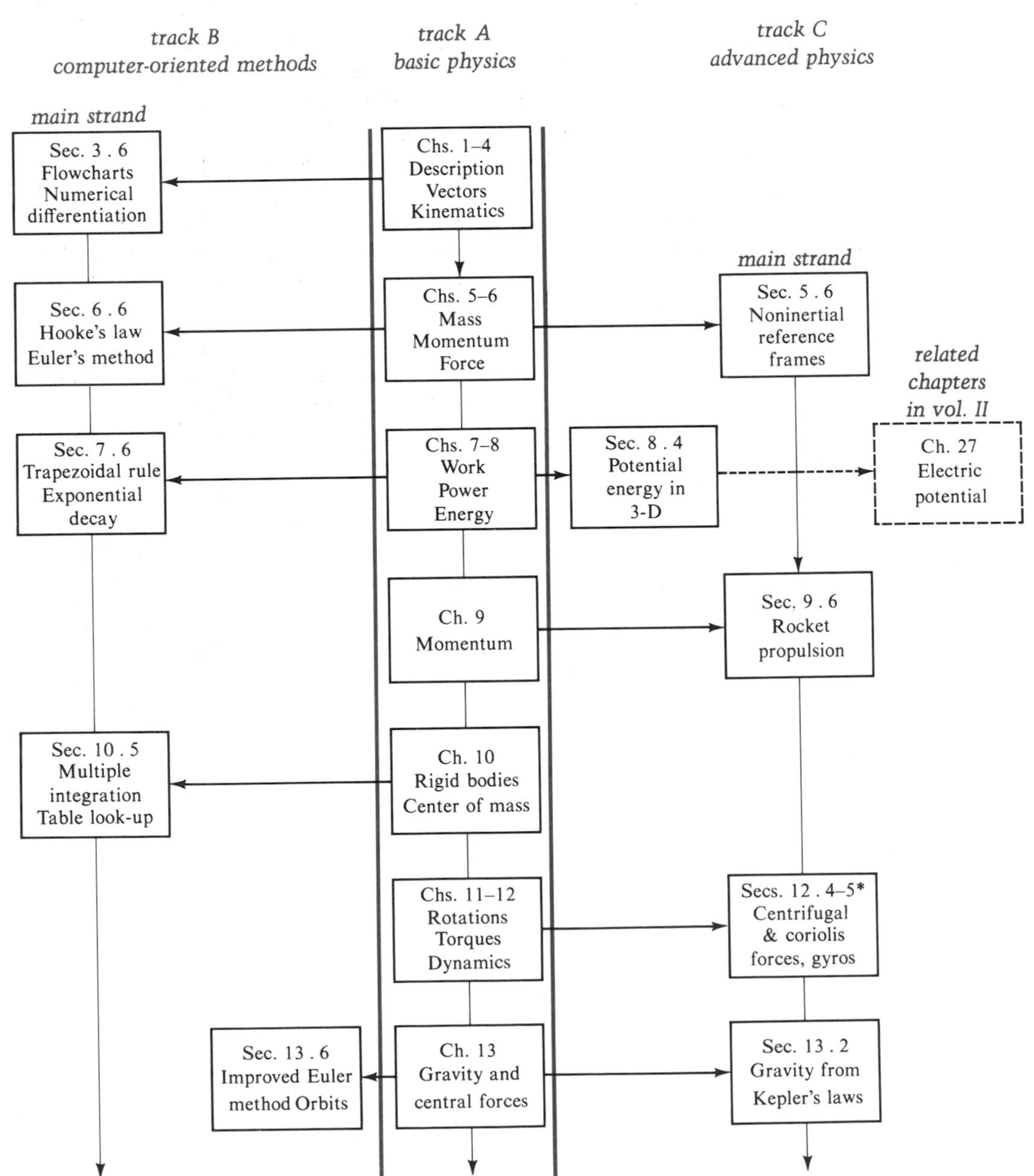

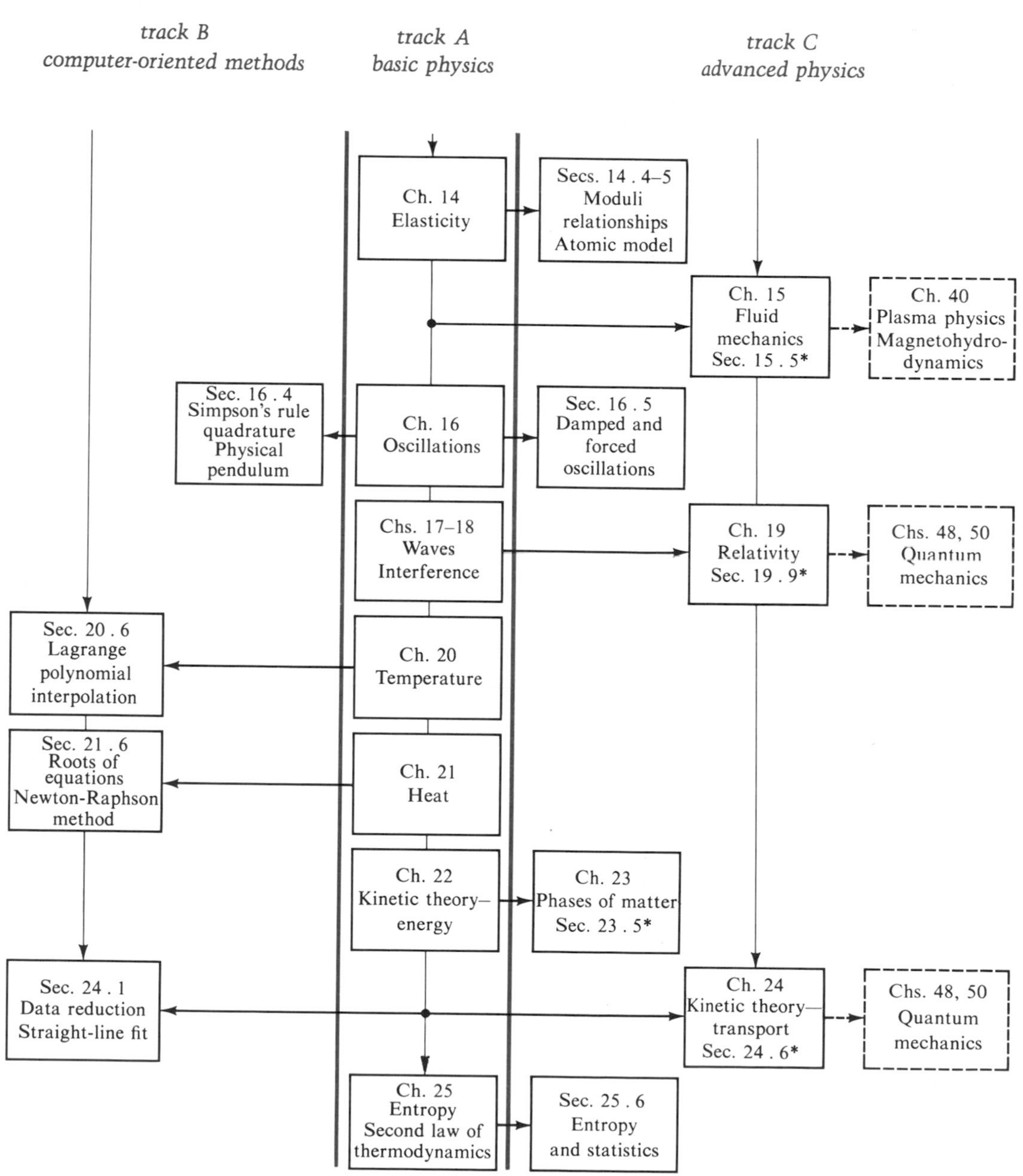

*Optional section.

VOLUME TWO

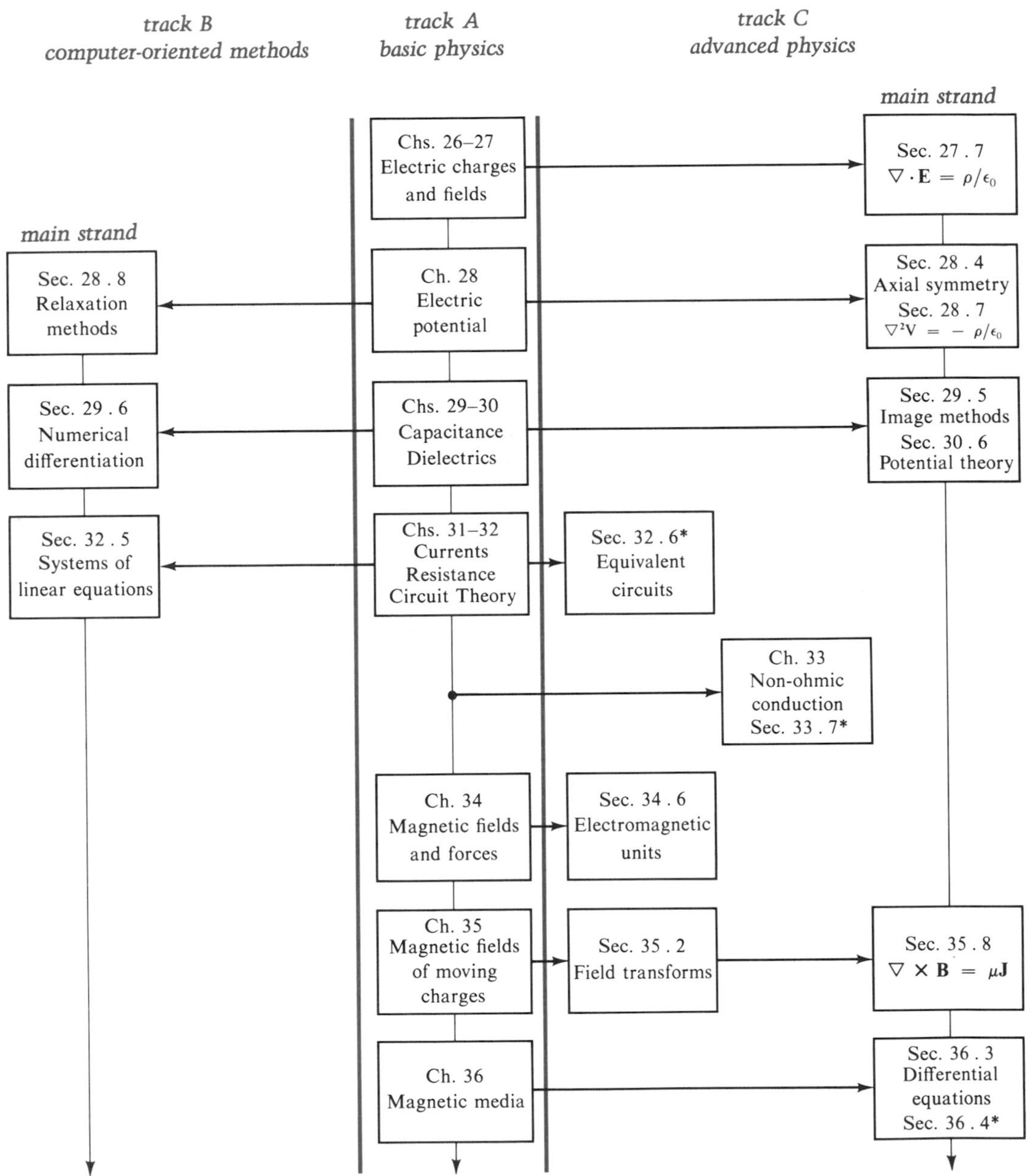
track B
computer-oriented methods
track A
basic physics
track C
advanced physics
main strand
Chs. 26–27
Electric charges
and fields
Sec. 27 . 7
$\nabla \cdot \mathbf{E} = \rho/\epsilon_0$
main strand
Sec. 28 . 8
Relaxation
methods
Ch. 28
Electric
potential
Sec. 28 . 4
Axial symmetry
Sec. 28 . 7
$\nabla^2 V = -\rho/\epsilon_0$
Sec. 29 . 6
Numerical
differentiation
Chs. 29–30
Capacitance
Dielectrics
Sec. 29 . 5
Image methods
Sec. 30 . 6
Potential theory
Sec. 32 . 5
Systems of
linear equations
Chs. 31–32
Currents
Resistance
Circuit Theory
Sec. 32 . 6*
Equivalent
circuits
Ch. 33
Non-ohmic
conduction
Sec. 33 . 7*
Ch. 34
Magnetic fields
and forces
Sec. 34 . 6
Electromagnetic
units
Ch. 35
Magnetic fields
of moving
charges
Sec. 35 . 2
Field transforms
Sec. 35 . 8
$\nabla \times \mathbf{B} = \mu \mathbf{J}$
Ch. 36
Magnetic media
Sec. 36 . 3
Differential
equations
Sec. 36 . 4*

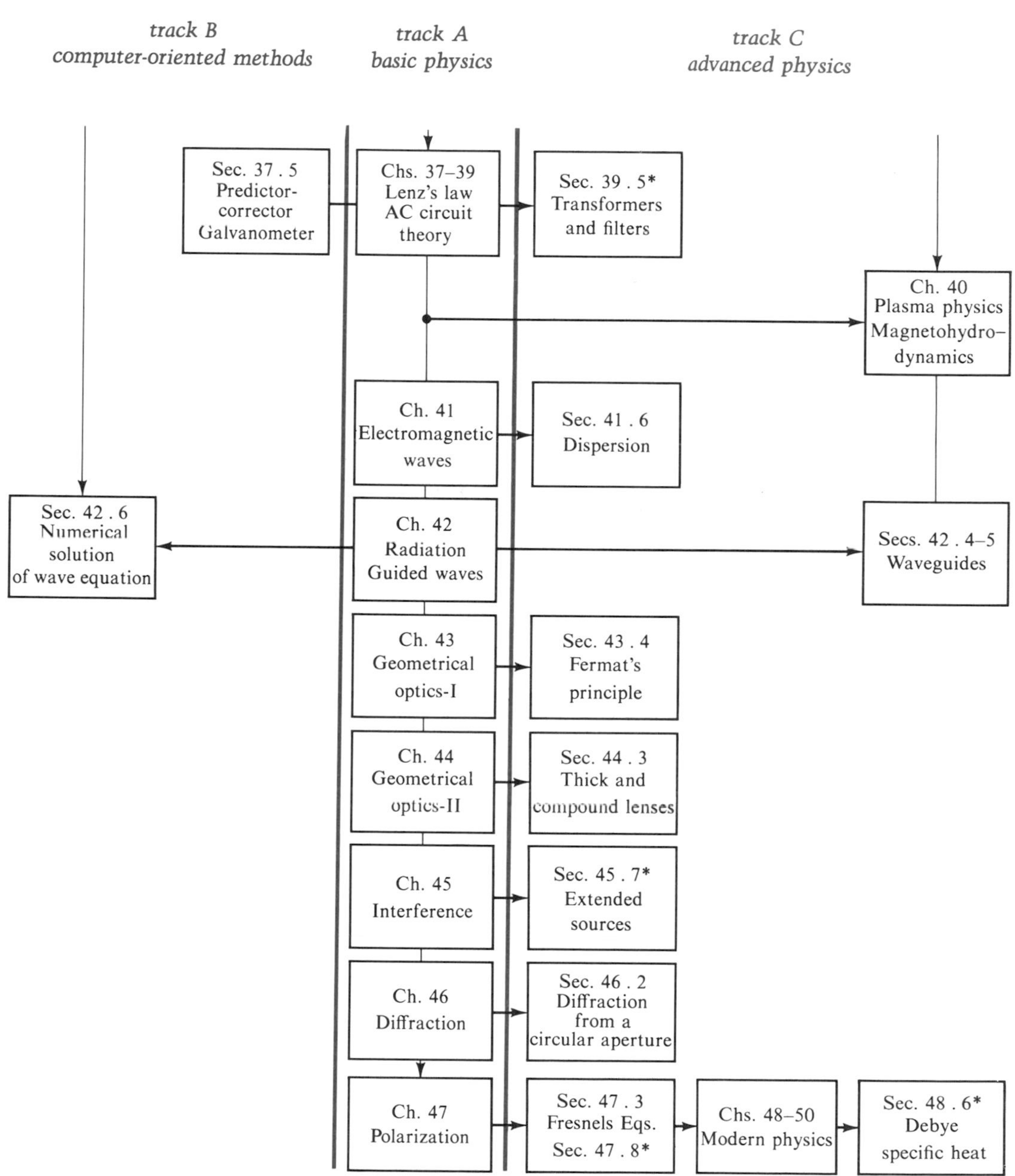

*Optional section.

contents

CHAPTER FIFTEEN *mechanics of fluids* 380

CHAPTER SIXTEEN *oscillations* 414

CHAPTER SEVENTEEN *waves* 450

CHAPTER EIGHTEEN *interference* 480

mechanics, waves, and thermodynamics

CHAPTER ONE

the description of physical reality

Yet . . . it is certain that mathematics generally, and particularly geometry, owes its existence to the need which was felt of learning something about the relations of real things to one another. . . . [Therefore] geometry must be stripped of its merely logical-formal character by the coordination of real objects of experience with the empty conceptual framework of axiomatic geometry. To accomplish this, we need only add the proposition: solid bodies are related, with respect to their possible dispositions, as are bodies in Euclidean geometry of three dimensions. . . .

Geometry thus completed is evidently a natural science; we may in fact regard it as the most ancient branch of physics. Its affirmations rest essentially on induction from experience, but not on logical inferences only. We will call this completed geometry "practical geometry." . . . All linear measurement in physics is practical geometry in this sense, so too is geodetic and astronomical physical measurement, if we call to our help the law of experience that light is propagated in a straight line, and indeed in a straight line in the sense of practical geometry.

Albert Einstein, Geometry and Experience,
in Sidelights on Relativity, *Methuen, 1936*

The physical phenomena that first attract our attention are the motions of objects about us. Motions and changes of motion, and the conditions that determine them, are the concern of the science of *mechanics*. Mechanics provides the foundation on which physics is built, and its fundamental concepts and generalizations will be found to permeate all the basic physical and engineering sciences. The ultimate aim of mechanics is to describe and predict motions and changes of motion. First, however, we must be able to describe the positions that physical bodies and their parts occupy in space, relative to one another, and the changes of position that may occur as time progresses. Thus there are three stages in the development of mechanics: *physical geometry,* concerned with spatial properties, their measurements, and their relationships; *kinematics,* which differs from physical geometry only in that it introduces the concept of time, and hence of motion; and *dynamics,* which describes and predicts how a body will move under any particular set of physical circumstances.

1 · 1 the role of physical geometry

The first notions of spatial properties arose with the surveying and construction measurements of the ancient Egyptians and Babylonians and the early Greeks. They devised rough rules for finding areas and volumes in terms of lengths, but these rules were applied mainly to specific physical objects. A *point* was an actual mark on the object, and a *line* was a measuring stick or a stretched string. An *area* was a field, a floor, or the surface of a monument or other structure; a volume was the capacity of a particular structure, such as the amount of water held by a given container. The areas and volumes of various *shapes* were also visualized in terms of physical objects, but apparently not as abstract entities. There seems to have been little concept, say, of a triangle as the abstraction representing all triangular objects. Nevertheless, this accumulating store of practical knowledge was the origin of *empirical-inductive geometry*—the observation and description of the spatial characteristics of physical objects, their classification into idealized shapes, and the formulation of empirical rules or "laws" to embody the relationships among them. In its mature form, empirical-inductive geometry is a branch of *experimental physics,* and the most elementary branch.

Indications are that Greek geometry owed a considerable debt to Babylon. Beginning around the sixth century B.C., the Greeks developed geometry into a logical discipline, a *postulational-deductive system* based on empirical knowledge (observed facts). By means of a few relatively simple definitions and postulates, they were now able to deduce empirical results theoretically. This development, beginning probably with Thales of Miletus about 600 B.C., culminated in *Euclidean geometry,* so named because its methods were codified in Euclid's *Elements* about 300 B.C.

Euclidean geometry is of particular interest to us here, not only because it played a dominant role in the development of mechanics, but also because it illustrates most of the important characteristics of the postulational-deductive stage of inquiry, which is widely and effectively employed in theoretical physics. Euclidean geometry is, in fact, the most elementary branch of *theoretical physics.*

The postulates or axioms of a theory are statements assumed to be true for the purpose of the theory in light of experience. They are not "self-evident truths"; rather, they represent idealizations or abstractions, such as "perfect" points, lines, areas, and volumes, of which there are no exact counterparts in nature. Each postulate of a given theory should provide an implicit definition of some one of the concepts appearing in it, and the postulates must be consistent and independent of one another. However, since postulates are often dredged up from intuition, in many cases it is extremely difficult or impossible to *prove* that the members of a set of postulates are mutually independent or even self-consistent.

Euclid's theory has been found to have many logical defects. Some of his definitions have postulates hidden in them; others are partly theorems. Often the reasoning involves tacit assumptions. Thus the assumption that "figures may be freely moved in space without change of shape or size," which provides the basis for the important physical concept of rigid bodies, is used in several proofs of theorems, but it is not stated explicitly as a postulate. Although none of this detracts from the great value of Euclid's pioneer work, it does illustrate a crucial fact: no physical theory of any breadth has ever been developed that was initially free of logical flaws. Only slowly and through the effort of numerous workers do theories acquire the relatively faultless and elegant formulations that eventually find their way into treatises and textbooks.

1 • 2 physical quantities

In order to express relationships between *physical quantities*—concepts which lead to measurable values—it is first necessary to define the quantities. The link between a theoretical abstraction and a physical quantity is an *operational definition,* which specifies the way in which the quantity can be measured. An operational definition will always include some direct or indirect reference to an arbitrary standard and a set of manipulations for measuring or comparing the quantity in question against that standard. A definition cannot be considered operational unless it encompasses a manual or experimental procedure which can in fact be carried out.

The concept of length, for example, might be defined in theoretical terms as "spatial extent." The operational definition of length, however, specifies a counting process. In operational terms, *length* is the number of times, including fractions, that a designated unit length, denoted by $[l]$, fits into the particular linear distance l to be measured. An operational definition may also include certain carefully specified physical or mathematical concepts. For example, in addition to the operational procedure of counting, the definition of length includes the concept of *rigid bodies,* objects in which all points have a fixed relationship to each other which remains constant under all conditions. A perfectly rigid body is a theoretical ideal, but for all practical purposes we can consider solids to be rigid bodies unless otherwise stated.

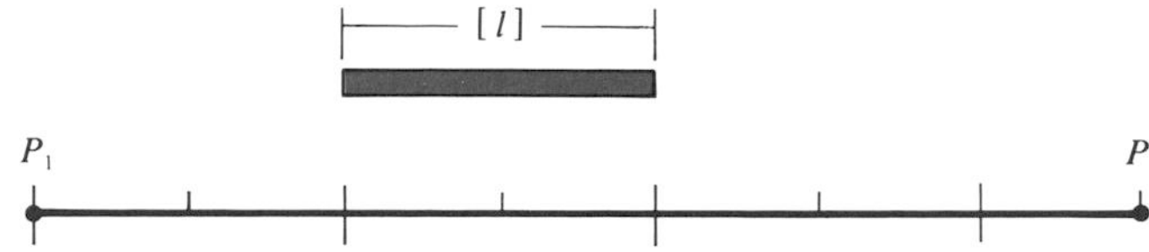

figure 1 • 1 *Measurement of length;* $l = \overline{P_1P_2} = 3.5[l]$

The immediate result of the *direct* measurement of a physical quantity is a pure number and a unit of measure. In Fig. 1 . 1 the numerical value of the linear distance from P_1 to P_2 is a pure number $\{l\}$, where

$$\{l\} = \frac{l}{[l]} = 3.5$$

However, the length l itself consists of two terms, the unit $[l]$ and the numerical value $\{l\}$ with respect to that unit. If $[l] = 1$ meter, then in this case

$$l = \{l\}\,[l] = 3.5 \text{ meters} \qquad [1 \cdot 1]$$

It is important to remember that the numerical value alone does not specify a physical quantity. The distance l between two fixed points remains constant regardless of our choice of $[l]$; hence the numerical value $\{l\}$ has meaning only in relation to the units $[l]$ in which it is expressed.

Except for simple counting, successive measurements of a given object exhibit discrepancies due to *random errors* in measurement. Therefore, if the true length of a particular rod is l_0, the arithmetic mean of a large number of successive measurements will be some number that represents an average length $\bar{l}$. Any individual measurement l will deviate from the average by an amount ϵ; that is,

$$l = \bar{l} \pm \epsilon \qquad [1 \cdot 2]$$

If we square each such value of ϵ and take the average of all ϵ^2, we obtain a quantity $\overline{\epsilon^2}$ known as the *variance* of the set of measurements. The square root of this average is the statistical quantity known as the root-mean-square (rms) error, or *standard deviation* σ:

$$\sigma = \sqrt{\overline{\epsilon^2}} \qquad [1 \cdot 3]$$

The larger the number n of measurements, the smaller the difference between their average $\bar{l}$ and the true length l_0—that is, the smaller the *standard error of the mean,* $\sigma/\sqrt{n}$. Therefore the best *estimate* of l_0 is the quantity

$$l = \bar{l} \pm \frac{\sigma}{\sqrt{n}} = \bar{l} \pm \Delta l \qquad [1 \cdot 4]$$

where the *uncertainty* Δl determined from n measurements is such that for truly random errors, 68 percent of the time the average $\bar{l}$ will fall within a distance Δl of the true, but unknown, value l_0.

What does this mean in practical terms? If a physicist measures the length of a bar and finds the average of these measurements to be $\bar{l} = 0.832$ meters, with a standard deviation of $\sigma = 3$ centimeters, the uncertainty in this average is $\Delta l = \sigma/\sqrt{100} = 3$ millimeters. On this basis, he

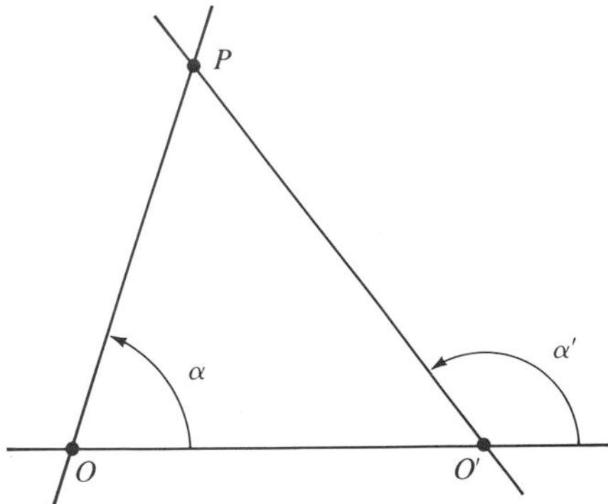

figure 1•2 *Location of a point P by triangulation*

would report the length of the bar to be $l = 0.832 \pm 0.003$ meters. In other words, there is a 68 percent probability that its true length l_0 lies in the range $0.829 \leq l_0 \leq 0.835$ meters. Furthermore, since l_0 is known only to three significant figures, any formula containing this element will be correct only to three significant figures; anything more would be pure conjecture. The statement of a physical quantity in terms of its average value and uncertainty is the accepted standard, and we shall use it when we deal with exact values of important physical constants.

Sometimes a direct measurement is impossible and an indirect measurement must be used. For example, from the rules of Euclidean geometry, we can compute the distance to a remote or inaccessible object by the indirect method of *triangulation* (see Fig. 1 . 2). This entails the selection and direct measurement of a base line OO' of some convenient length, the direct measurement of two angles α and α', and the assumption that PO and PO' are straight lines. All other unknown quantities can then be calculated from geometry. However, we can also test the principle of triangulation by selecting some length interval on the earth's surface that can be measured both directly and by triangulation; that is, we can obtain the distance l in two different ways. When this is done, the two measurements must correspond, after an allowance for errors in measurement. In order to say that we are dealing with a single physical quantity, we must be sure that any two operational definitions with overlapping ranges of application give rise to the same physical results.

A *constitutive definition* states the measurable properties of a physical quantity in terms of other physical quantities that have been defined operationally. All spatial properties, for example, can be defined constitutively in terms of the single quantity length. We can define the area of one face of a cube as $A = L^2$ and its volume as $V = L^3$. Similarly, for a spherical object, $A = 4\pi R^2$ and $V = \frac{4}{3}\pi R^3$, where R is radial length. Clearly, if we had only an operational definition of each quantity, there would be no means of relating physical quantities to each other, and hence no means of developing theories. A physical quantity does not lose its operational significance when use is made of one of its constitutive definitions; both definitions can be utilized consistently in a physical theory. Ideally, every physical quantity should be so formulated that it can be defined both operationally and constitutively.

1 · 3 dimensions

Physical quantities that are fundamental to an entire set of constitutive definitions are called *dimensions*. The simplest example is length. We shall later develop operational definitions of other physical dimensions, such as time and mass, as well as constitutive definitions of various quantities which can be derived from them, such as velocity, acceleration, force, momentum, and energy. To some extent the designation of a physical quantity as fundamental or derived is arbitrary and may depend on the problem. For instance, instead of choosing length as fundamental and defining area as $A = L^2$, we could just as easily choose area as the primary quantity and treat length as a derived quantity, $L = \sqrt{A}$. To be classified as fundamental, however, a physical quantity must be entirely independent of the units of all other dimensions; that is, it must be measurable in units of itself, which are specified in its operational definition.

Although we can *define* a physical quantity in terms of its dimensions, when we want to specify its measured value, we must also include the units in which that value is expressed. Two quantities can be equated only if they are dimensionally consistent. We can equate 100 cm = 1 m or 1 ft = 0.3048 m, since centimeters, meters, and feet are all units of length, but we cannot equate a length and a speed, since the dimensions of speed are length/time. This requirement for dimensional consistency is a useful check on calculations. For example, a well-known formula for the vertical distance y traveled by a falling body in some time t is $y = \frac{1}{2}gt^2$, where g is the acceleration due to gravity. Both sides of any equality must have the same dimensions. Thus if the dimension of y is length, then the dimension of $\frac{1}{2}gt^2$ must also be length, and dimensional consistency requires that

$$\dim(y) = \dim(g) \times \dim(t^2) \qquad \text{or} \qquad L = \dim(g)\, T^2 \qquad [1 \cdot 5]$$

Hence the dimensions of g must be

$$\dim(g) = \frac{L}{T^2} \qquad [1 \cdot 6]$$

Units of measure for this derived quantity g might be feet per second squared, meters per second squared, or miles per hour squared; even kilometers per century squared would be dimensionally consistent.

The requirement for dimensional consistency holds no matter how complex and involved the formula, and regardless of whether it includes constants which also have dimensions. A more general formula for the distance traveled by a falling body is

$$y = y_0 + v_0 t + \tfrac{1}{2}gt^2 \qquad [1 \cdot 7]$$

and every term on the right-hand side has the dimension of length. The constant y_0 represents an initial distance from the origin; the constant v_0 represents an initial velocity with the dimensions L/T, so that $\dim(vt) = (L/T)T = L$; and g has the dimensions L/T^2, so that $\dim(\frac{1}{2}gt^2) = (L/T^2)T^2 = L$. If the second term in this formula had been v_0t^2, we would know that something was wrong with our derivation, because $(L/T)T^2 = LT \neq L$.

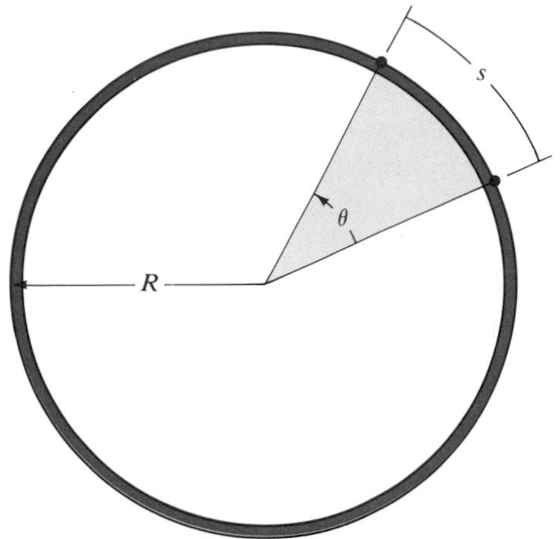

figure 1 • 3 *Radian measure of angular distance;* $\theta = s/R$

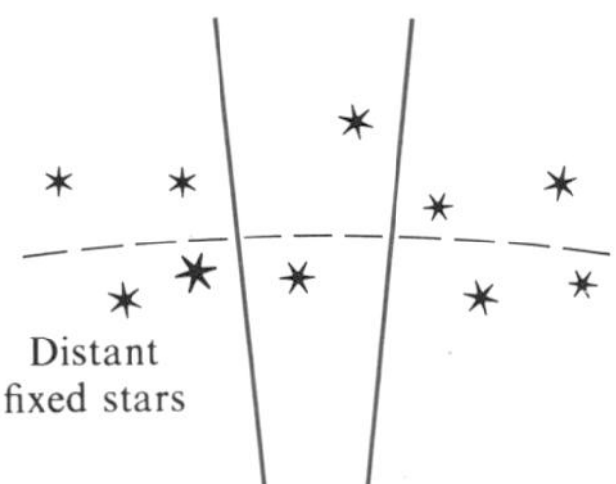

Many quantities used in physics are "dimensionless," which means that they are defined in such a way that their dimensions cancel out. These quantities are usually ratios. One example is the weight-to-drag ratio of a space vehicle reentering the atmosphere, which is critical in determining its trajectory. This is the ratio of the gravitational attraction of the earth to the resistive frictional force of the atmosphere on the vehicle. Since both quantities are forces, their dimensions cancel out, leaving the ratio as a pure number. We shall often make use of dimensionless quantities when we employ computer-based methods to analyze the motions of bodies.

One of the most widely used dimensionless quantities is *angular distance*. We can measure the angle θ in Fig. 1 . 3 either in fixed units (degrees) or by taking θ as the ratio of the two lengths s and R. Since the length of the circular arc s varies directly with the length of the radius R for any angle θ, we can define the derived quantity, angular distance θ, as

$$\theta = \frac{s}{R} \qquad [1 \cdot 8]$$

However, it is commonly assumed that a curved line can be measured in the same units of length as a straight line, and that both have the dimension L. Hence the dimensions of θ cancel,

$$\dim(\theta) = \frac{L}{L} = 1 \qquad [1 \cdot 9]$$

and the derived unit of angular distance, $[\theta] = 1$ radian, is the unit of a dimensionless quantity. The names *radian* and *degree* are not really units, but descriptions of the manner in which angular distance is measured.

One application of radian measure is the use of *parallax,* the apparent change in the position of an object resulting from a change in the position of the observer. Knowing the heliocentric parallax of the nearer stars as the

figure 1 • 4 *Measurement by parallactic displacement*

earth rotates in its orbit enables us to determine their distance from the sun by triangulation. For the near star shown in Fig. 1 . 4, this distance is

$$y = \frac{1\ \text{AU}}{p\ \text{radians}} \qquad [1 \cdot 10]$$

where p is the parallax angle and 1 astronomical unit (AU) is the distance of the earth from the sun, 150 million kilometers.

1 • 4 units of measure

It is not surprising that the earliest units of measure related to the human body or to local artifacts. With the growth of trade and commerce, local standards such as the foot, the fathom, and the furlong became more precisely defined, but it was not until the seventeenth century that the need for more universal standards was recognized. Even the yard, which had been the basic unit of length in English-speaking countries for centuries, was not standardized on an international basis until 1854, when it was defined as the distance between two lines crossing two gold studs set in a certain bar of platinum kept in London.

The present International System of Units, or *Système Internationale d'Unités* (SI), grew out of the original formulation of the metric system, perhaps the most enduring accomplishment of the French Revolution of 1789. Although the metric system may not appear to be a lively topic of study, it would be difficult to convey the passionate intensity that accompanied its development. The meter alone might be the subject of a romance. The political debate which led to the first formal study of the meter was launched in 1790 in the French National Assembly by the great Talleyrand, who evidently did not apply to the sciences his well-known dictum that "language was given to man to conceal his thoughts." The order authorizing official measurement of the meter was signed by the ill-fated King Louis XVI from his prison cell in Paris, six months before his decapitation. The measurement itself was performed by two engineers, Jean Delabre and Pierre Méchain, who were to measure the meridional distance from Barcelona, Spain, to Dunkirk, France. However, the rigors of primitive field conditions and wars, both foreign and domestic, delayed completion of this task until 1799, when the French National Assembly adopted the meter with the motto: "For all people, for all time."

In 1875 an international treaty was signed by seventeen countries, providing for an International Bureau of Weights and Measures to be situated on international territory in Sèvres, near Paris. This bureau, currently maintained by assessed contributions from forty-four signatory governments, serves as the repository for international prototypes of the various primary and secondary standards; it also coordinates national and international measurement techniques and conducts research in such areas of measurement as thermometry, electrical measurements, and photometry.*

The meter had originally been defined by the French Academy of Science as 1/10,000,000 of a meridional quadrant of the earth at sea level, and

*This joint effort illustrates a favorite aphorism of political science: governments have no trouble agreeing on matters they consider of no political importance.

a platinum bar representing this length was used as the standard during most of the nineteenth century. Later it was replaced by an accurate copy, the *international prototypal meter,* made of platinum-iridium, and the meter was redefined as the distance, under specified temperature and pressure conditions, between two lines on this bar. Although the prototypal meter was defined without reference to any other standard, the search continued for some standard that would be universal in nature, and not based on an artifact. One possibility was the wavelength of a particular kind of light, described by A. A. Michelson and E. W. Morley in 1887. Michelson was finally able to determine experimentally the number of wavelengths of the red light in the optical spectrum of cadmium equivalent to the international prototypal meter, and in 1960 the meter was redefined as 1,650,763.73 times the wavelength in vacuum of the orange-red line emitted by krypton-86.

In the meantime, effort had continued on a worldwide basis to identify fundamental physical quantities from which all other quantities and standards could be derived, and also to define standards for them which could be based on universal natural phenomena. In 1971 the Fourteenth General Conference on Weights and Measures adopted the present International System, a highly refined version of the metric system, consisting of the six fundamental dimensions and basic units listed in Table 1 . 1. Since that time a seventh basic unit has been added—the mole, which is a pure number defined to be the number of molecules in exactly 12 g of the isotope carbon-12.

The 1971 Conference also established an extended system of prefixes for derived units which indicate that the basic unit is to be multiplied by some power of 10. These are listed in Table 1 . 2. The prefixes for positive powers of 10 are taken from Greek; those for negative powers of 10 are from Latin, with the exception of *femto-* and *atto-,* which are from Danish. Thus we can express various fractions and multiples of the basic units:

1 micrometer (μm) $= 1 \times 10^{-6}$ m (sometimes called a *micron*)
1 nanosecond (ns) $= 1 \times 10^{-9}$ s
1 kilometer (km) $= 1000$ m
1 gigameter (Gm) $= 1 \times 10^{9}$ m

table 1 • 1 *Fundamental SI units*

physical quantity	*unit*	*abbreviation*
length	meter	m
mass	kilogram	kg
time	second	s
electric current	ampere	A
temperature	kelvin	K
luminous intensity	candela	cd
amount of substance	mole	mol*

*In this text we will never abbrebiate *mole* so that the abbreviation *mol* will stand for the word *molecule.*

table 1 • 2 *SI prefixes and their abbreviations*

prefix	*abbreviation*	*factor*	*prefix*	*abbreviation*	*factor*
deka-	da	10^{1}	deci-	d	10^{-1}
hecto-	h	10^{2}	*centi-	c	10^{-2}
*kilo-	k	10^{3}	*milli-	m	10^{-3}
*mega-	M	10^{6}	*micro-	μ	10^{-6}
*giga-	G	10^{9}	*nano-	n	10^{-9}
tera-	T	10^{12}	*pico-	p	10^{-12}
peta-	P	10^{15}	femto-	f	10^{-15}
exa-	E	10^{18}	atto-	a	10^{-18}

*The asterisks denote the most commonly used prefixes in physics.

Although use of the metric system was legalized in the United States as long ago as 1866, the British system of units remained the official standard in all English-speaking countries until recently. In 1979 the United Kingdom completed conversion to SI units after a decade of gradual implementation, and the same conversion is now under way in the United States. It has proceeded slowly, however, not only because of a natural resistance to change, but also because of the accumulation of a vast industrial base measured in British units.

Since 1971, the official conversion factor for the basic unit of length has been 1 standard yard = 3600/3937 meter, or 0.9144 meter for industrial purposes. Beyond this, however, the various derived units you encounter will depend on both the nature of the problem and historical accident in different fields. In addition to the common conversion factors listed in Table 1 . 3, an intriguing variety of units are in local use all over the world, from the Chinese *ch'ih* (32 cm) and the Egyptian *feddan* (4.193 m^2) to the Nicaraguan *vara* (0.84 m) and the Russian *verst* (1.067 km). There are also units peculiar to certain trades which are unlikely to change for some time; the pica (0.422 cm) is the standard unit of measure for typesetters in all countries, and the nautical mile (1.852 km) is used in both sea and air navigation.

To convert a quantity q from one unit of measure $[Q]$ to another $[Q]^*$, given a conversion factor c such that

$$[\mathrm{Q}] = c[\mathrm{Q}]^* \qquad [1 \cdot 11]$$

we simply multiply the original measurement by a factor of unity in the form

$$1 = \frac{c[\mathrm{Q}]^*}{[\mathrm{Q}]} \qquad [1 \cdot 12]$$

Hence for a given numerical value $\{q\}$ expressed in units $[Q]$, we obtain the equivalent value $\{q\}^*$ in units $[Q]^*$ as

$$q = \{q\}\,[Q] \times \frac{c[Q]^*}{[Q]} = c\{q\}\,[Q]^* = \{q\}^*\,[Q]^* \qquad [1 \cdot 13]$$

table 1•3 *Geometric conversion factors*

equal lengths: 1 m = 100 cm, 1000 m = 1 km

meters (m)	*feet (ft)*	*inches (in)*	*miles (mi)*
1	3.281	39.37	6.214×10^{-4}
0.3048	1	12	1.894×10^{-4}
0.0254	0.0833	1	1.578×10^{-5}
1,609	5,280	63,360	1

equal areas: 1 m^2 = 10^4 cm^2

square meters (m^2)	*square feet* (ft^2)	*square inches* (in^2)
1	10.76	1,550
0.0929	1	144
6.452×10^{-4}	6.944×10^{-3}	1

equal volumes: 1 m^3 = 10^3 *liters* = 10^6 cm^3

cubic meters (m^3)	*cubic feet* (ft^3)	*cubic inches* (in^3)
1	35.31	61,024
0.0283	1	1,728
1.639×10^{-5}	5.787×10^{-4}	1

equal angular distances

radians (rad)	*degrees (°)*	*minutes (′)*	*seconds (″)*	*revolutions (rev)*
1	57.2958*	3438.	2.063×10^5	0.1592
0.0174533*	1	60	3600	2.778×10^{-3}
2.909×10^{-4}	0.01667	1	60	4.630×10^{-5}
4.848×10^{-6}	2.778×10^{-4}	0.01667	1	7.716×10^{-7}
6.28319*	360	21,600	1.296×10^6	1

*Mathematically important constants are given to six significant digits.

and the numerical value of q in the new units is simply

$$\{q\}^* = c\{q\} \qquad [1 \cdot 14]$$

All we are doing, of course, is cancelling dimensions in the form of units. For example, to express the quantity 7 ft^2 in square inches, we would write

$$q = 7\ \text{ft}^2 \times \frac{144\ \text{in}^2}{1\ \text{ft}^2} = 7 \times 144\ \text{in}^2 = 1008\ \text{in}^2$$

Although this procedure may seem unnecessarily cumbersome, it is easy to convert in the wrong direction when you simply multiply numerical values. As a precautionary measure, always carry the units explicitly for each term and make sure that they cancel properly, leaving only the units you want.

1 . 5 preliminary ideas about motion and particles

When we say that a car has traveled a mile, we usually mean that it has moved this distance relative to some point fixed to the surface of the earth. Relative to the sun, the car may be thousands of miles from where it started; relative to its passengers, it has not moved at all. In other words, the position or the motion of an object must refer to its position or its change of position with respect to another object, or to some set of lines or surfaces on the other object, considered as fixed. The way we describe the position or motion of a given object will depend on our choice of this frame of reference; a wise choice often greatly simplifies the problem of description. In all subsequent discussions it will be understood, unless otherwise indicated, that our frame of reference is fixed to the surface of the earth at the locality of interest. We shall also assume that the distances of interest are so small relative to the radius of the earth that we can treat the earth's surface as a plane, rather than as a portion of a sphere.

In the up-and-down motion of an elevator car, or in the motion of a railway car on a straight track, all points in the body of the car move along parallel lines. The same kind of motion is observed in a piece of straight wire that is held in a vertical position, but moved from one place to another in space. Any object which moves in such a way that every point in the body moves the same distance and in the same direction as every other point in the body is said to undergo a *linear motion*, or a *translation* (see Fig. 1 . 5).

In contrast, if a body moves in such a way that, with respect to a fixed axis in the frame of reference, the shortest distance from each point in the body to that axis remains the same, then the motion is called a *rotation*, and the axis is the *axis of rotation*. The axis of rotation need not be a coordinate axis of the frame of reference, although this is helpful in many problems. Examples of this type of motion are the rotation of an automobile wheel about its axle, which is fixed in the frame of reference (the automobile), and the rotation of a door about the imaginary axis through its hinges. The axis

figure 1 • 5 *Translational and rotational motion*

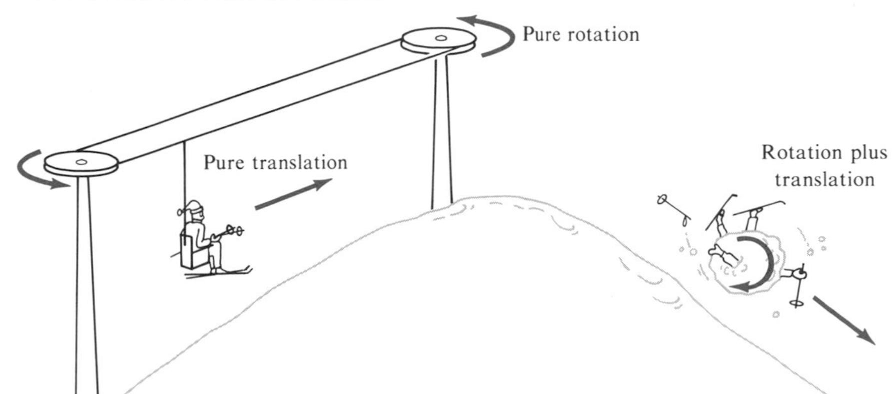

of rotation may be real (the axle) or imaginary (the line through the hinges); it may run through the body (the wheel) or lie wholly outside it (the door). During the rotation all points in the object describe circular arcs about the axis, and all these arcs sweep out equal angles, called the *angular displacement* of the body.

When a solid object or other rigid body undergoes pure translation, every point in the body retains its direction relative to the frame of reference. This means that the motion of the body will be completely described just as soon as the motion of any one of its points is given. In describing this motion, therefore, we can simply treat the body as if it were a single particle. The term *particle* is used here in its technical sense to refer to a moving point, considered as defining the position from time to time of a minute portion of a body. For emphasis, the term *point-particle* is sometimes used. There are circumstances under which any object, whether it is rigid or not, can be treated as a particle. For example, the planets and the sun of our solar system are always very far from one another in comparison with their own size; hence for some purposes their sizes can be neglected, and these bodies can be treated simply as physical particles.

One familiar way of describing the location of a particle near the earth's surface is to give its latitude, its longitude, and its altitude from sea level. These three data elements locate the particle in three-dimensional space relative to a frame of reference fixed to the earth and consisting of two coordinate axes (the meridian through Greenwich, England, and the earth's equator) and a coordinate surface, the mean level of the oceans. Since the particle may change its position with time, the description is made complete if the time is also specified. Four data elements are needed, then, to locate a particle in ordinary space and time: three of these locate the particle in three-dimensional space relative to some frame of reference; and the fourth gives the time of location with respect to some zero or initial time.

In classical physics time is assumed to be an independent parameter; that is, the operations for determining time are independent of those for determining position, and the value resulting from a time measurement will not depend on where the measurement is made. In classical physics it is thus possible to separate the considerations of spatial position from those of time. As we shall see in Chapter 19, however, there are circumstances under which the assumption that time is an independent parameter leads to contradictions between theory and observation. In these cases, positions must be established in space-time frames of reference, rather than in independent spatial and temporal frames.

In many three-dimensional problems it is convenient to orient a rectangular, or *cartesian, coordinate system* so that the position of the particle under discussion is described in two-dimensional space. The projection of the particle's position P along the horizontal axis is its *abscissa,* or x coordinate, and the projection along the vertical axis is its *ordinate,* or y coordinate. The lengths of these projections are the cartesian coordinates (x,y) which locate the particle in space (see Fig. 1 . 6).

A two-dimensional *plane polar coordinate system* may also be convenient, where the two coordinates are r, the radial distance from O to P, and θ, the angle of r relative to the x axis. These polar coordinates are related to the

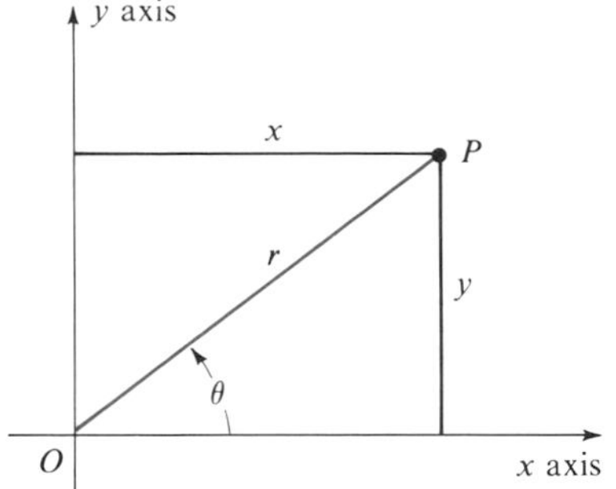

figure 1 • 6 *Two-dimensional cartesian and polar coordinate systems*

cartesian coordinates x and y by Pythagoras' theorem and the definitions of trigonometry:

$$x = r\cos\theta \qquad r = \sqrt{x^2 + y^2}$$

$$y = r\sin\theta \qquad \theta = \tan^{-1}\frac{y}{x} \tag{1 . 15}$$

Thus the position of a point P in the plane is completely described either by (x,y) or by (r,θ).

The corresponding three-dimensional cartesian system adds a third axis, the z axis, perpendicular to the xy plane and drawn from the common origin O. The position P of a particle in three-dimensional space can thus be described by a triplet of three independent values (x,y,z), as shown in Fig. 1 . 7*a*. The directions shown are taken as positive; they would be negative if the coordinates extended in the opposite directions through the origin O. Note that the coordinate axes are so designated that, if viewed from a point on the positive z axis, a counterclockwise rotation would be needed to carry the positive x axis by the shortest way around to the direction of the positive y axis. This kind of rectangular coordinate system, which is the one now generally used in physics, is called a ***right-hand*** system. It can be represented by letting the thumb of your right hand denote the z axis, your forefinger the x axis, and your middle finger the y axis, as shown in Fig. 1 . 7*b*. Evidently

figure 1 • 7 *Three-dimensional cartesian coordinate system*

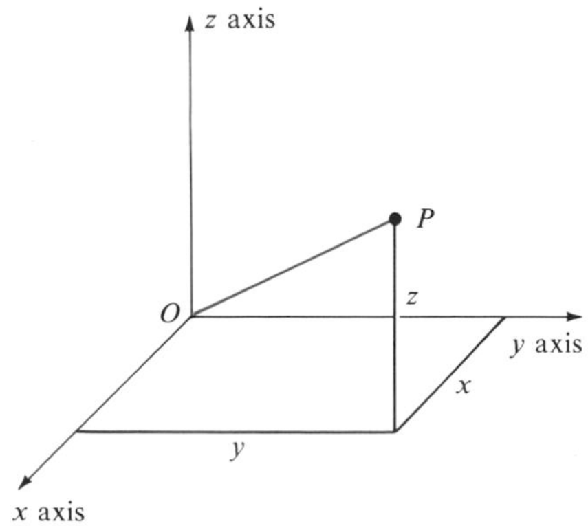

(a) Location of a point P in the xy plane

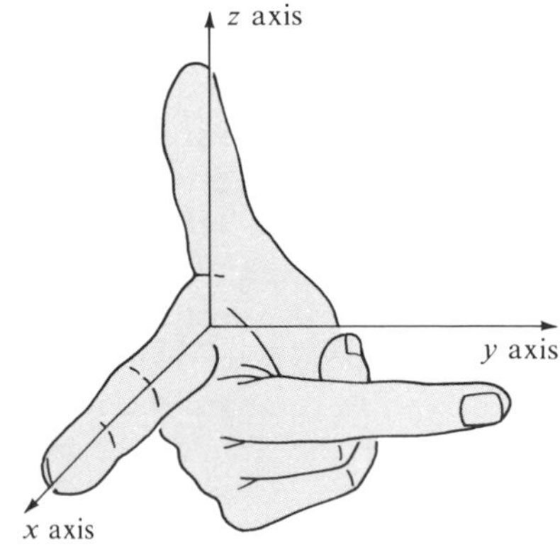

(b) The right-hand coordinate system

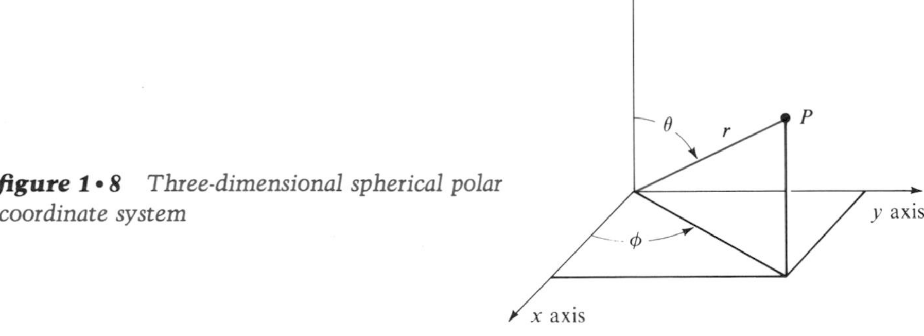

figure 1 • 8 *Three-dimensional spherical polar coordinate system*

the left hand must represent another system of rectangular coordinates which is the mirror image of the right-hand system, and thus can never be made to coincide with it.

Another way of describing the position of a particle, which is often simpler than using rectangular coordinates, is to locate it at the terminal point of a straight line directed radially from the origin. Thus in Fig. 1 . 8 the location of P is completely and uniquely determined by the directed straight line r, which represents the radial distance from O to P in the direction specified by the angles θ and ϕ. Three independent coordinates are needed for this purpose: one length, r, and two direction angles, θ and ϕ. This three-dimensional system is called the *spherical polar coordinate system.* The coordinates (r,θ,ϕ) are related to the cartesian coordinates (x,y,z) as follows:

$$r = \sqrt{x^2 + y^2 + z^2} \qquad \phi = \tan^{-1} \frac{y}{x}$$

$$\theta = \tan^{-1} \frac{\sqrt{x^2 + y^2}}{z} \tag{1 · 16}$$

1 • 6 a note on problem solving

A full understanding of the definitions and principles of physics can be acquired only by using them in the solution of many problems. The ability to solve problems is not only a dependable test of your mastery of the science, but also an index of the growth in your ability to analyze situations and use this ability as a tool in future thinking.

The first step in solving a problem is to represent it graphically, setting down the known and implied relationships. It is also advisable to form the habit of working a problem through in terms of these relationships before you do any computations. The results should be reduced to the simplest terms before you substitute the numerical values and do the actual calculations. This procedure will often save you many tedious computations and enable you to compare the results of alternative methods of solution.

No solution should be considered complete unless the appropriate units of measure are included and all numerical values are expressed to the number of significant digits warranted by the given data. In this text all numerical data given in the problems may be assumed correct to any desired number of significant figures. For example, if a length is given as 4.5 m, you can assume that, to four significant figures, it is 4.500 m. If no decimal point is given, it is assumed to follow the last digit.

In working a problem, especially one that is very involved, you may find that you have made a slight error; factors of 2 and minus signs are often the culprits. Before you do the problem over, work backward from your last step. A fresh viewpoint often helps to uncover careless errors. As a last resort, try working backward from the answer. Except for some of the computer exercises and a few lengthy developments, if you find yourself bogged down in endless calculations, you have probably made a basic error somewhere. In this case try rereading the text, and if the solution still escapes you, note your questions or difficulties for later discussion and go on to something else. A certain amount of frustration is part of the learning process and often helps to fix a concept in memory, but obsession with an intractable problem can be a waste of time put to better use.

If you do arrive at a correct answer, but are not quite sure how you reached it, the solution is worth careful study to be sure you understand it. Wrong answers are plentiful, but correct ones are rarely coincidental. Always remember that *the objective of problem solving is not numbers or formulas, but insight.* For this reason the answers to nearly all the problems are given at the end of each chapter. Although you may not be expected to work every problem, it is advisable that you read through those which are not assigned and scan the answers for whatever insight they give. Where you are expected to solve a problem, it is your own responsibility to use the answers with the utmost integrity.

PROBLEMS

1•1 physical geometry

1•1 According to an early Babylonian formula, translated here into modern notation, the area of any simple quadrilateral having consecutive sides a, b, c, d is given by $\frac{1}{4}(a + c)(b + d)$. Does this formula have any validity?

1•2 What value of π is implied in II Chronicles 4 : 2 and I Kings 7 : 23, which date from about the tenth century B.C.? The quotation is: "Also he made the molten sea of ten cubits from brim to brim, round in compass, and the height thereof was five cubits; and a line of thirty cubits did compass it round about."

1•3 The Rhind papyrus (ca. 1700 B.C.) indicates that the Egyptians used $(\frac{16}{9})^2$ as the value of π. Show that this exceeds the actual value by less than 1 percent.

1•4 Repeated measurements of a sheet of metal in the form of a simple quadrilateral show that the average lengths of the consecutive sides are 121.0, 10.2, 120.3, and 10.0 cm. Compute the area of the equivalent "perfect" rectangle, expressing it to the allowable number of significant figures.

1•2 physical quantities

1•5 If you count a given number of objects—say, pieces of blackboard chalk—a great many times, you will always come out with the same number, barring plain blunders in counting. But even if you exercise the greatest care in measuring the length of a piece of wire to the nearest millimeter, there will be small differences in the values obtained from successive measurements. Does this mean that counting is a more "perfect" operation than measurement? What if you measure the wire to the nearest centimeter?

1•6 State Pythagoras' theorem of plane geometry in operational terms.

1•7 A certain object is weighed on a scale a number of times in an experiment, and its weight is determined to be 150 ± 2 lb. (*a*) If you weighed the same object in 100 separate repetitions of the experiment, how often would you expect its weight to fall between 148 and 152 pounds? (*b*) If you were to perform the experiment 16 times and take the *average* of these measurements as your best estimate of the true weight, what would be the uncertainty in this average value?

1•8 You are told that a new instrument will measure a certain quantity q to within 5 percent uncertainty. (*a*) If you perform such a measurement 1000 times, how often would you expect the measured value of q to exceed $1.05q_0$, where q_0 is the true value of q? (*b*) How often would you expect the measured value of q to be less than $0.95q_0$?

1•3 dimensions

1•9 The orbit of the earth about the sun is approximately 300,000,000 km in diameter. Of the stars that have been measured, the two that are nearest to our sun are the faint star Proxima Centauri, with a heliocentric parallax of 0.783″ (arc-second), and the bright star Alpha Centauri, with a heliocentric parallax of 0.756″. (*a*) Compute their distances in meters and in light years. (*b*) A star at a distance of about 2×10^{18} m has the smallest parallactic orbit that can be measured by the geometric method. Show that the heliocentric parallax for this distance is about 0.015″. (*c*) Show that 0.01″ is equivalent to the angle subtended by the diameter of a U.S. penny at a distance of approximately 250 mi.

1•10 For practical reasons the older sexagesimal system, 1 full circle = 360°, is generally used to express angular distance. It is interesting to note that in this system both the angle of rotation θ and the radial length r are assumed to be fundamental quantities. In that case, if *arc length* s is to have the appropriate dimensions, we must redefine it as $s = 2\pi r\theta/360°$. (*a*) Had we simply defined arc length as $s = r\theta$ in the sexagesimal system, what would be the appropriate unit of measure? (*b*) How many such units are in the circumference of a circle of radius 1 m?

1•11 One of the fundamental laws of physics states that the force exerted on a body equals its mass times its acceleration. (*a*) What are the fundamental dimensions of a force? (*b*) Express the mks unit of force, called a newton (N), in terms of the fundamental units of length, mass, and time.

1•12 An important quantity in fluid mechanics is viscosity η, the "stickiness" of a fluid. Viscosity is defined by the relation

$$\frac{\text{force}}{\text{area}} = \eta \frac{\text{speed}}{\text{length}}$$

(see Prob. 1 . 11 for the dimensions of force).

(*a*) What are the dimensions of viscosity? (*b*) Express the mks unit of viscosity, the dekapoise, in terms of the primary mks units.

1 • 4 units

1 • 13 If a particular piece of square floor tiling of area [*A*] is chosen as a fundamental unit of measure, what is the length of one side of a 400-unit square? How could you measure this length?

1 • 14 An example of a useful "mixed unit" is the secondary, or derived, unit of volume called the acre-foot, used in irrigation engineering and defined as the volume of water covering 1 acre to a depth of 1 ft. Given that 1 acre = 43,560 ft^2 and 1 U.S. gal = 231 in^3, show that, to three significant figures,

$$\begin{aligned} 1\ \text{acre}\cdot\text{ft} &= 12\ \text{acre}\cdot\text{in} \\ &= 4.36 \times 10^4\ \text{ft}^3 = 7.53 \times 10^7\ \text{in}^3 \\ &= 3.26 \times 10^5\ \text{gal} \end{aligned}$$

1 • 15 The speed of light is 3×10^8 m/s, and a light year is the *distance* that light travels in one year. (*a*) What is the length of a light year in kilometers? (*b*) What is its length in earth radii (R_e = 6400 km)? (*c*) What is its length in astronomical units, the average distance between the earth and the sun (1 AU = 150,000,000 km)?

1 • 5 motion and particles

1 • 16 If the polar coordinates of *P* in Fig. 1 . 5 are 50.6 cm and 60°30′, find its rectangular coordinates and express them to the correct number of significant figures.

1 • 17 If the rectangular coordinates of *P* in Fig. 1 . 5 are (7,5), what are its polar coordinates?

1 • 18 The rectangular coordinates of *P* in Fig. 1 . 6 are $(x,y,z) = (1,3,2)$. (*a*) What are the values of r, θ, and ϕ? (*b*) What are their values if the vertical distance z of *P* is doubled?

1 • 19 Determine the cartesian coordinates (x,y,z) of *P* in Fig. 1 . 6 as functions of the spherical polar coordinates (r,θ,ϕ).

answers

1 • 2 $\pi = 3$

1 • 4 1220 cm^2

1 • 7 (*a*) 68 times; (*b*) ±0.5 lb (68 percent of the time our weight-averaging process would yield a value which is within 0.5 lb of the true weight)

1 • 8 (*a*) 160 times; (*b*) 160 times

1 • 10 (*a*) meter-degree; (*b*) 360 m·deg of arc length

1 • 11 (*a*) ML/T^2; (*b*) 1 N = 1 kg·m/s^2

1 • 12 (*a*) M/LT; (*b*) 1 dekapoise = 1 kg/m·s

1 • 13 $20\sqrt{[A]}$

1 • 15 (*a*) 9.46×10^{12} km; (*b*) $1478 \times 10^6\ R_e$; (*c*) 63,072 AU

1 • 16 x = 24.9 cm, y = 44.0 cm

1 • 17 r = 8.602 cm, θ = 35°32′

1 • 18 (*a*) r = 3.742, θ = 57.69°, θ = 71.56°; (*b*) r = 5.099 cm, θ = 38.33°

1 • 19 $x = r \sin\theta \cos\phi$, $y = r \sin\theta \cos\phi$, $z = r\cos\theta$

CHAPTER TWO

vectors

Ignorant people, like Faraday, naturally think in vectors. They may know nothing of their formal manipulation, but if they think about vectors, they think of them as vectors, that is, directed magnitudes. No ignorant man could or would think about the three components of a vector separately, and disconnected from one another. That is a device of learned mathematicians, to enable them to evade vectors. The device is often useful, especially for calculating purposes, but for general purposes of reasoning the manipulation of the scalar components instead of the vector itself is entirely wrong.

Oliver Heaviside, Electromagnetic Theory, *vol. I, 1925*

Modern vector analysis was created independently and nearly simultaneously by two physicists, Josiah W. Gibbs in the United States and Oliver Heaviside in England, at the close of the nineteenth century. Gibbs first published his work at Yale University in 1881 as a pamphlet, of which 130 copies were privately printed and circulated. Heaviside gradually introduced his work paper by paper in journal articles. The great beauty and usefulness of vector analysis in physics will become increasingly apparent as we use it over and over again to describe and solve physical problems.

In this chapter we shall examine the formal system of mathematics known as vector algebra and see how it relates to physical reality and to geometric description.

2 · 1 vector algebra

Those quantities which have only magnitude, and require no specification of direction, are termed *scalars*. Quantities such as time, distance, and temperature, for example, can be completely described in terms of their numerical values. A *vector* is a quantity that has both magnitude and direction. The velocity of a northeast wind blowing at 17 mi/h would be represented by a vector, as would a linear displacement, a force, and any other directed physical quantity. Note that a vector has no specific position in space; the velocity vector of a 17 mi/h northeast wind has the same magnitude and direction no matter where the wind is blowing.

If we are to deal with vector quantities in the abstract, however, we must construct an appropriate language, define unambiguous symbols to express this language, and discover its "grammar"—the rules which ensure that it correctly represents the reality we seek to portray. This language is known as *vector algebra.* We usually deal with vector algebra in two or three dimensions, analogous to the spatial dimensions, but it can be extended in an abstract mathematical sense to any number of dimensions.

A vector quantity is customarily denoted by boldface type (**A**), although you will often see it in written form with an arrow over or under it or indicated by a straight or a wavy underscore. Scalar quantities are simply written as ordinary letter symbols. Since it is often convenient to be able to discuss the magnitude and direction of a vector separately, we also define $|\mathbf{A}|$, or simply A, to be the *absolute value,* or *magnitude,* of **A**. Its direction is then denoted independently by the *unit vector* $\hat{\mathbf{A}}$, defined as a vector which has a magnitude of 1 and points in the direction of **A**. This enables us to write **A** in terms of its length and direction as

$$\mathbf{A} = A\hat{\mathbf{A}} \qquad [2 \cdot 1]$$

As an example of a vector quantity, suppose a particle moves from point P_0 to point P_1 in Fig. 2 . 1 by any one of the curved paths shown. Regardless of the scalar distance s represented by a particular path, the *change of position* of the particle from P_0 to P_1 is its *displacement,* represented by the vector **A**. This change from one position to another is completely described

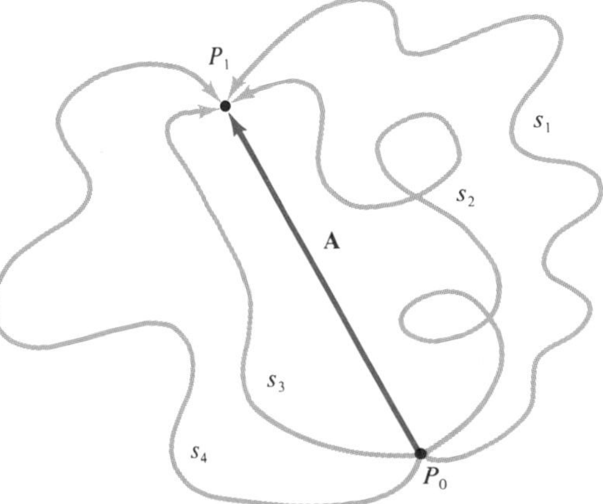

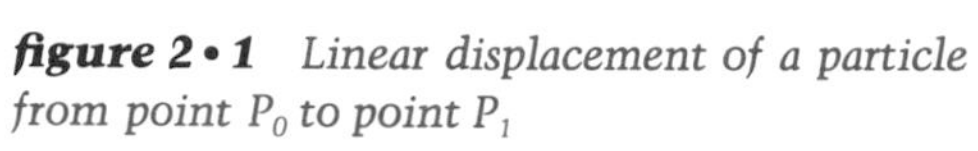
figure 2 • 1 *Linear displacement of a particle from point P_0 to point P_1*

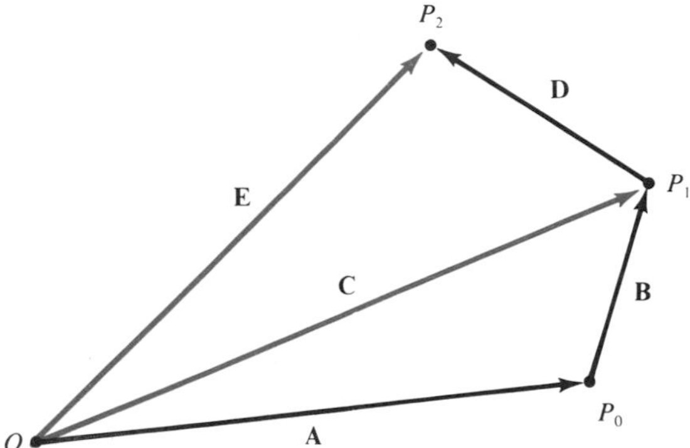

figure 2 • 2 *Vector addition;* **C** = **A** + **B** *and* **E** = **C** + **D**

by specifying the magnitude of the change—which is simply the straight-line distance from P_0 to P_1—and the direction of the change—the direction from P_0 to P_1. To say that a particle has moved 1 m from a certain point in space merely locates it somewhere on the surface of a sphere of radius 1 m; adding the direction of this motion completely describes the change.

Note that the particle in Fig. 2 . 1 might traverse paths of different lengths in moving from P_0 to P_1; hence its travel distance s might vary, depending on the path taken. In all cases, however, its linear displacement is the same. The displacement vector **A** represents the *net effect* of this motion, which was to transport the particle from P_0 to P_1. Although information on the scalar distance traveled or the length of a curved path or trajectory is useful for certain calculations, the general problem in mechanics involves finding changes of position as a function of time. Displacement is therefore of more interest than travel distance in most cases.

Suppose, in Fig. 2 . 2, that the vector **A** locates a particle at point P_0 with respect to some origin O. If the particle is then displaced from P_0 to P_1, its net displacement from the origin O must be the same as if the particle had moved directly from O to P_1. (Why?) Thus we can write the *vector sum,* or *resultant,* of vectors **A** and **B** as equal in magnitude and direction to the net displacement **C**:

$$\mathbf{C} = \mathbf{A} + \mathbf{B} \qquad [2 \cdot 2]$$

This implies that the separate sums of the magnitudes and directions are also equal:

$$C = |\mathbf{A} + \mathbf{B}| \qquad \hat{\mathbf{C}} = \widehat{\mathbf{A} + \mathbf{B}} \qquad [2 \cdot 3]$$

If the particle is now displaced along **D** to point P_2, then its net displacement from O is

$$\mathbf{E} = \mathbf{C} + \mathbf{D} = \mathbf{A} + \mathbf{B} + \mathbf{D} \qquad [2 \cdot 4]$$

In fact, any number of vectors may be added graphically in the same way, and if their net effect is to move the particle to P_2, then the resultant of all these separate vector additions is the vector **E**. This graphical method of vector addition, shown in Fig. 2 . 3, is known as the *polygon,* or *parallelogram method.* It was used by Galileo for adding velocities as early as the sixteenth century and was later used by Newton for adding displacements. To add vectors graphically, we merely place the tail of each successive vector

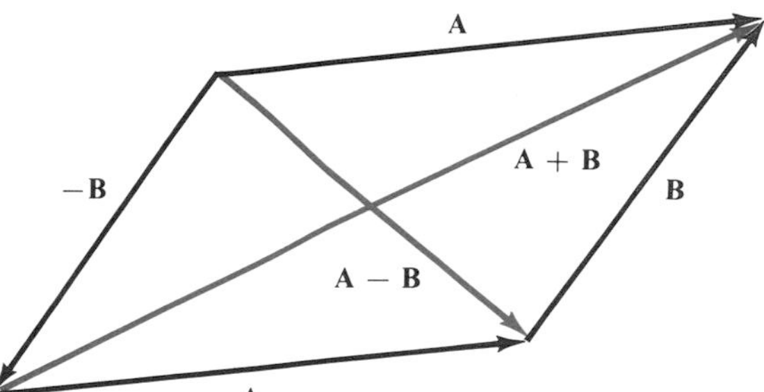

figure 2 • 3 *Parallelogram law of vector addition and subtraction*

at the point of the preceding vector, with the directions and relative magnitudes of each vector unchanged. The definition of a vector does not specify its point of application, so that we can think of it just as an arrow in space, not necessarily tied to any particular origin or coordinate system. For a physical quantity to be represented by a vector, any two of its values must add vectorially in this manner.

It should be apparent from Fig. 2 . 4*a* that the order in which vectors are added does not matter; comparison of the bottom part of the polygon with the top part shows that

$$\mathbf{A} + \mathbf{B} + \mathbf{C} + \mathbf{D} = \mathbf{E} = \mathbf{D} + \mathbf{C} + \mathbf{B} + \mathbf{A} \qquad [2 \cdot 5]$$

Both halves of the polygon are equal to **E**, and hence to each other. Vector addition is thus *commutative*. In Fig. 2 . 4*b* we see that

$$\begin{aligned} \mathbf{A} + \mathbf{B} + \mathbf{C} + \mathbf{D} &= (\mathbf{A} + \mathbf{B}) + \mathbf{C} + \mathbf{D} \\ &= (\mathbf{A} + \mathbf{B} + \mathbf{C}) + \mathbf{D} = \mathbf{E} \qquad [2 \cdot 6] \end{aligned}$$

The grouping of vectors in a sum does not affect the resultant, so that vector addition is also *associative*.

In ordinary scalar algebra we can define an *identity element* with respect to a given operation which leaves the number operated on unchanged. Thus the identity element under addition is zero. The *inverse* of a quantity under addition is a second quantity which, when added to the first, gives zero. The inverse of x under addition is $-x$, and $x + (-x) = 0$. Similarly, it is possible to define an identity element for vector addition; this is the

figure 2 • 4 *Commutative and associative laws of vector addition*

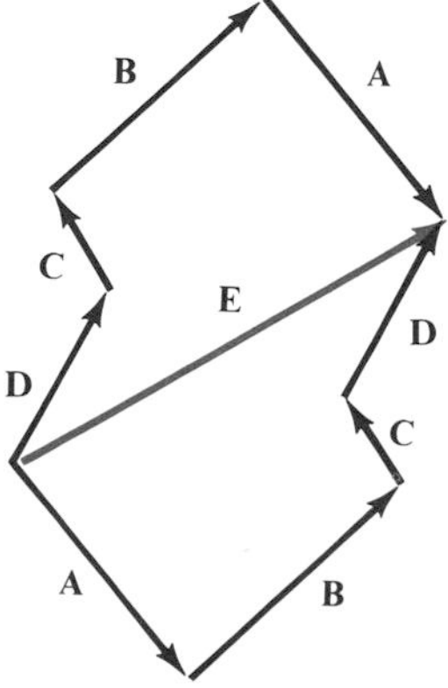

(a) Commutative law

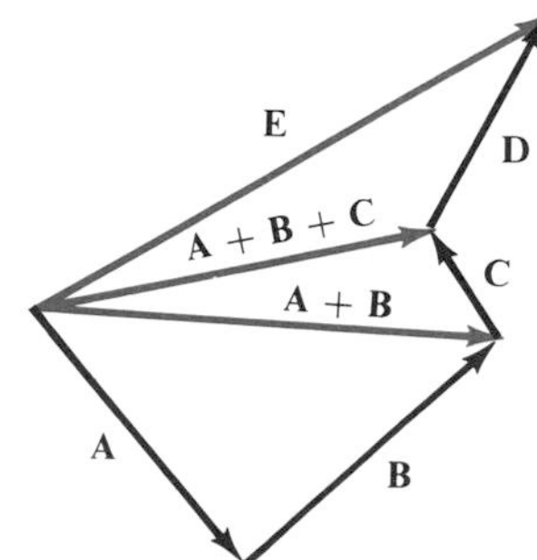

(b) Associative law

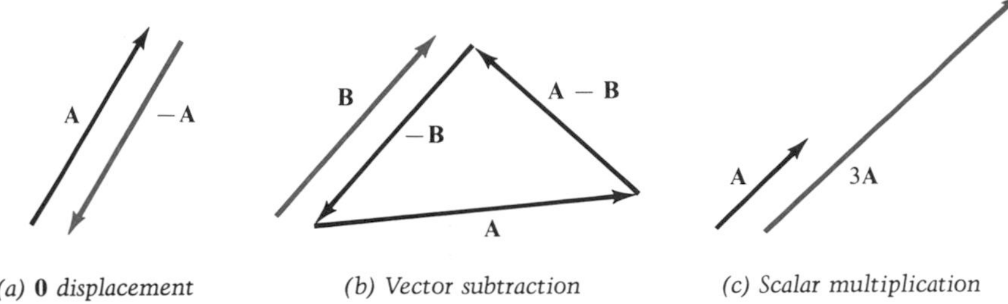

(a) **0** *displacement* *(b) Vector subtraction* *(c) Scalar multiplication*

figure 2 • 5 *Vector subtraction and multiplication by a scalar*

null vector **0**, which has zero length and indeterminate direction. Thus the inverse **A**′ of a vector **A** must satisfy

$$\mathbf{A} + \mathbf{A}' = \mathbf{0} \qquad \mathbf{A}' = -\mathbf{A} \tag{2 . 7}$$

where −**A** is a vector which is identical to **A** in magnitude, but exactly opposite in direction. The parallelogram law then leads us back to the starting point, with a net displacement of **0** (see Fig. 2 . 5). This means that we can define *vector subtraction* as the addition of the negative of a vector:

$$\mathbf{A} + (-\mathbf{B}) = \mathbf{A} - \mathbf{B} = \text{difference vector} \tag{2 . 8}$$

Multiplication of a vector **A** by a positive scalar c multiplies only the magnitude of **A**, leaving its direction unchanged (see Fig. 2 . 5c). For consistency with ordinary algebra, $-\mathbf{A} = (-1)\mathbf{A}$; hence multiplication by a *negative* scalar reverses the direction of **A**:

$$|c\mathbf{A}| = c|\mathbf{A}| \qquad (-c)\mathbf{A} = c(-\mathbf{A}) \tag{2 . 9}$$

It is also apparent that scalar multiplication must be *distributive:*

$$c(\mathbf{A} + \mathbf{B}) = c\mathbf{A} + c\mathbf{B} \qquad (b + c)\mathbf{A} = b\mathbf{A} + c\mathbf{A} \tag{2 . 10}$$

A further implication of scalar multiplication is that we can express a unit vector $\hat{\mathbf{A}}$ in terms of the vector **A** and its magnitude A. Since $\mathbf{A} = A\hat{\mathbf{A}}$ by definition of $\hat{\mathbf{A}}$, scalar multiplication of both sides of this equation by $1/A$ yields

$$\frac{1}{A}\mathbf{A} = \frac{A}{A}\hat{\mathbf{A}} = \hat{\mathbf{A}} \qquad \hat{\mathbf{A}} = \frac{1}{A}\mathbf{A} \tag{2 . 11}$$

The fundamental properties of vectors may be summarized as follows:

1 A vector **A** has both magnitude and direction.
2 Two vectors add to form a third vector, $\mathbf{A} + \mathbf{B} = \mathbf{C}$, according to the parallelogram rule.
3 Vector addition is commutative, $\mathbf{A} + \mathbf{B} = \mathbf{B} + \mathbf{A}$.
4 Vector addition is associative, $(\mathbf{A} + \mathbf{B}) + \mathbf{C} = \mathbf{A} + (\mathbf{B} + \mathbf{C})$.
5 The inverse of a vector **A** under addition is a vector of the same magnitude, but opposite direction, expressed as −**A**.
6 Multiplication of a vector **A** by a positive scalar c multiplies the magnitude of **A** by c; if c is negative, the direction of **A** is reversed.
7 Scalar multiplication is distributive, $c(\mathbf{A} + \mathbf{B}) = c\mathbf{A} + c\mathbf{B}$ and $(b + c)\mathbf{A} = b\mathbf{A} + c\mathbf{A}$.

example 2 • 1 Express the displacement from point P_0 to point P_1 as a difference vector.

solution The change in position from P_0 to P_1 is represented by the displacement vector $\Delta\mathbf{r}$, where the symbol Δ denotes a change in the quantity $\mathbf{r}$. From the parallelogram rule, $\mathbf{r}_1 = \mathbf{r}_0 + \Delta\mathbf{r}$, and adding the negative quantity $-\mathbf{r}_0$ to both sides of the equation yields the difference vector

$$\mathbf{r}_1 - \mathbf{r}_0 = \Delta\mathbf{r}$$

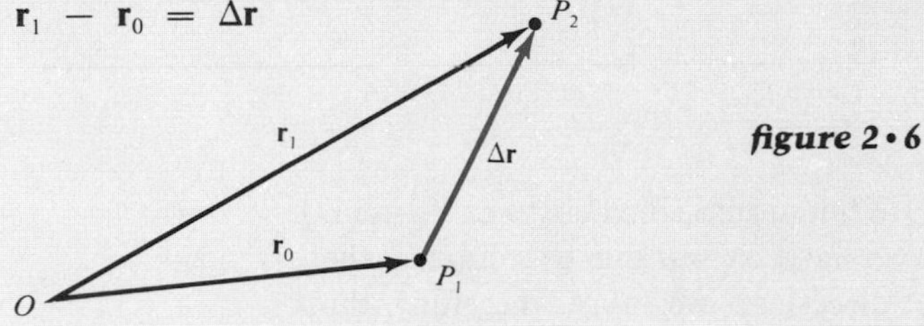

figure 2 • 6

2 • 2 *components of a vector*

Although we have defined the fundamental properties of vectors, we still need some means of performing the algebraic manipulations that make vector algebra a vital theoretical tool. Recall from our discussion in Chapter 1 that to locate a point in three-dimensional space, we must specify its position in terms of three independent coordinates. Thus in Fig. 1 . 7 the position of point P with respect to the origin O was given by its (x,y,z) coordinates. Now consider the radial position vector $\mathbf{r}$ in Fig. 2 . 7; the coordinates x, y, and z are just the projections of $\mathbf{r}$ on the x, y, and z axes, respectively. Since a vector has no definite position in space, we can choose the origin of our coordinate system wherever we like, as long as it remains fixed throughout a given problem.

Where the position of a point P is specified by the cartesian coordinates x, y, and z, we can represent its position vector $\mathbf{r}$ by the vector equation*

$$\mathbf{r} = x\mathbf{i} + y\mathbf{j} + z\mathbf{k} \qquad [2 \cdot 12]$$

where $\mathbf{i}$, $\mathbf{j}$, and $\mathbf{k}$ are unit vectors in the positive x, y, and z directions, as

*Another useful vector representation is $\mathbf{r} = (\mathbf{x},\mathbf{y},\mathbf{z})$.

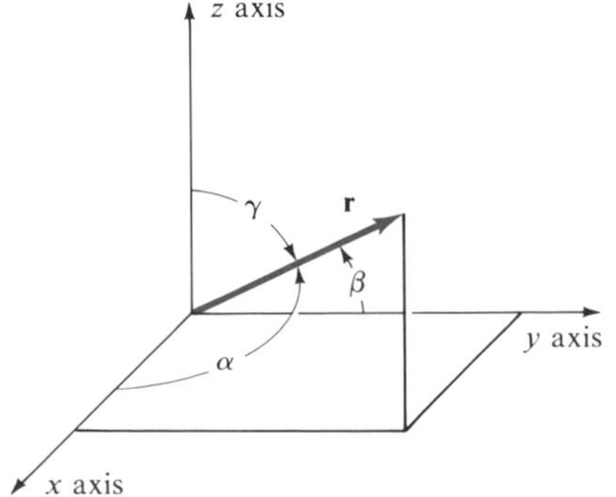

figure 2 • 7 *Direction angles in a cartesian coordinate system*

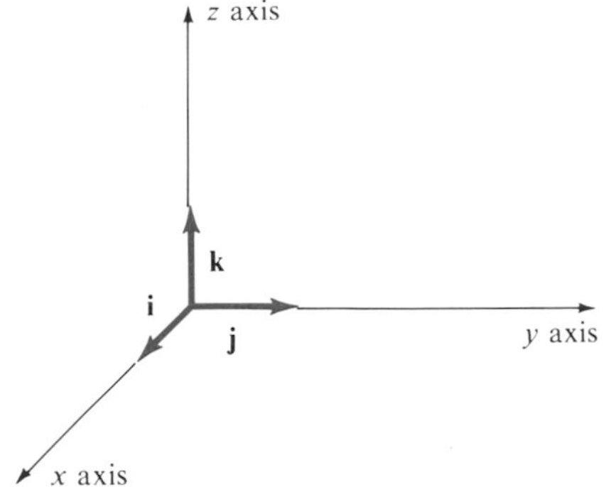

figure 2 • 8 *Unit vectors in a cartesian coordinate system*

shown in Fig. 2 . 8. (Out of deference to long-established custom, we use **i**, **j**, and **k** instead of $\hat{\mathbf{x}}$, $\hat{\mathbf{y}}$, and $\hat{\mathbf{z}}$.) Once we have chosen our coordinates, these unit vectors do not change their direction; we may therefore think of them as a "bookkeeping" device to remind us that x, y, and z are *independent components* of any vector quantity. This viewpoint is important, for it implies that Eq. [2 . 12] is really three independent equations, each of which represents the projection of the vector **r** onto the corresponding coordinate axis.

The scalar components of **r** are related to the *direction angles* α, β, and γ of the cartesian coordinate system by the law of cosines (see Appendix B):

$$x = r\cos\alpha \qquad y = r\cos\beta \qquad z = r\cos\gamma \qquad [2 \cdot 13]$$

The direction angles α, β, and γ are not all independent, however, a point to which we shall return shortly.

We can now add two or more vectors analytically, since, by the associative law, the components of each vector can be grouped according to direction. That is, given two vectors

$$\mathbf{A} = A_x\mathbf{i} + A_y\mathbf{j} + A_z\mathbf{k} \qquad \mathbf{B} = B_x\mathbf{i} + B_y\mathbf{j} + B_z\mathbf{k}$$

we can write their vector sum as

$$\mathbf{C} = \mathbf{A} + \mathbf{B} = (A_x + B_x)\,\mathbf{i} + (A_y + B_y)\,\mathbf{j} + (A_z + B_z)\,\mathbf{k} \qquad [2 \cdot 14]$$

This implies that the scalar components of **C** are related to the scalar components of **A** and **B** by three independent *scalar* equations:

$$C_x = A_x + B_x \qquad C_y = A_y + B_y \qquad C_z = A_z + B_z \qquad [2 \cdot 15]$$

For the sum of any number of vectors, then, we can find the components of the resultant vector simply by adding up the x, y, and z components separately. In Fig. 2 . 9, for example, the sum of the x components, $A_x + B_x = C_x$, gives us the magnitude of **C** in the x direction, and $A_y + B_y = C_y$ gives us its magnitude in the y direction. Multiplication of each scalar component by the appropriate unit vector then gives us the components of **C** in vector form:

$$\mathbf{C} = C_x\mathbf{i} + C_y\mathbf{j} \qquad [2 \cdot 16]$$

The same result is illustrated for three dimensions in Fig. 2 . 10. We can see

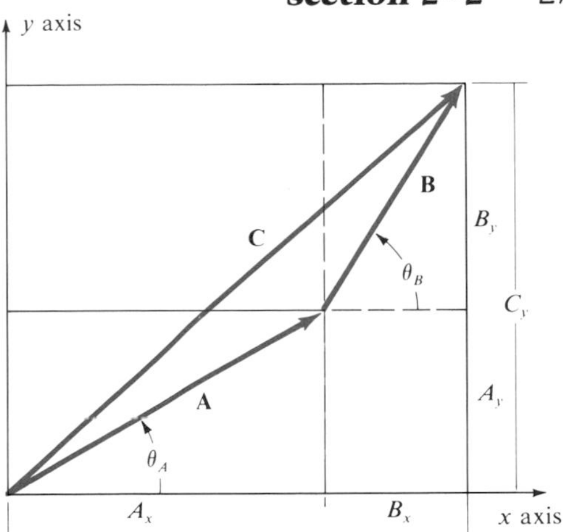

figure 2 • 9 *Vector addition by components in two dimensions*

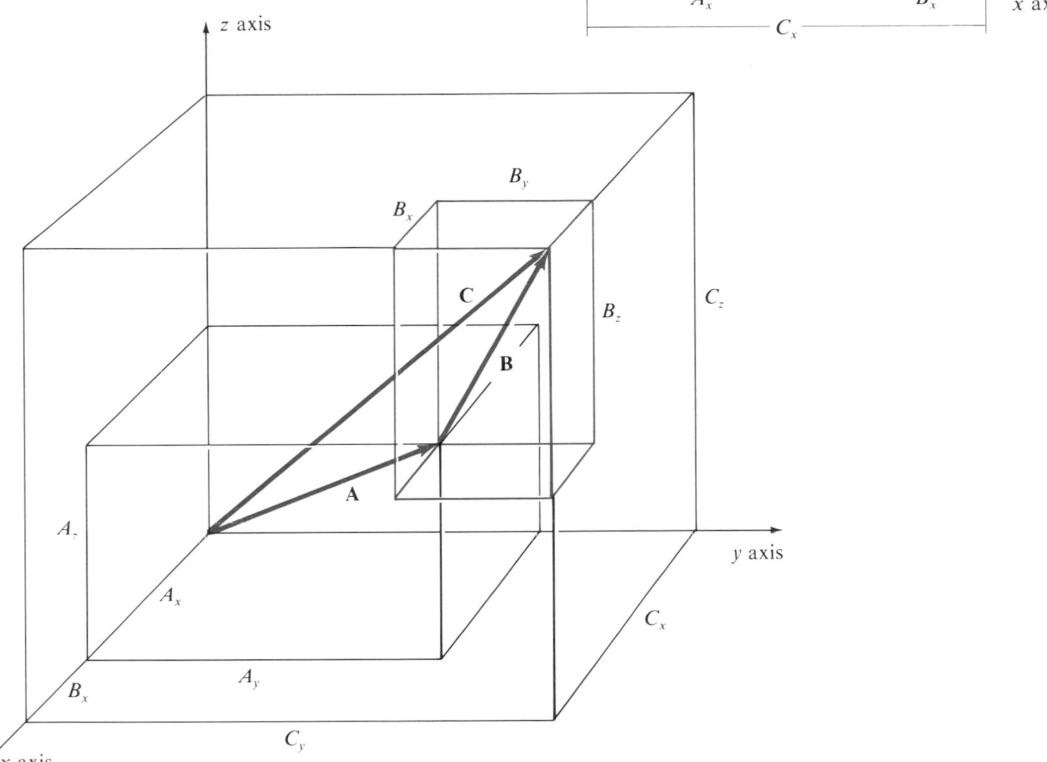

figure 2 • 10 *Vector addition by components in three dimensions*

now that the notation $\mathbf{A}' = -\mathbf{A} = -A_x\mathbf{i} - A_y\mathbf{j} - A_z\mathbf{k}$ was indeed appropriate for the inverse of **A** such that $\mathbf{A} + \mathbf{A}' = \mathbf{0}$.

Since multiplication of a vector by a positive scalar does not change the direction of the vector, it must preserve the *ratios* of its components. This means that each component must be multiplied by the same scalar, so that, given a vector

$$\mathbf{C} = C_x\mathbf{i} + C_y\mathbf{j} + C_z\mathbf{k}$$

scalar multiplication is distributive:

$$b\mathbf{C} = bC_x\mathbf{i} + bC_y\mathbf{j} + bC_z\mathbf{k} \qquad [2 \cdot 17]$$

example 2 • 2 If vectors **A** and **B** in Fig. 2 . 9 have the magnitudes $A = 4$ units and $B = 3$ units and they form angles $\theta_A = 30°$ and $\theta_B = 60°$, respectively, with the x axis, find their vector sum **C**.

solution We must first find the scalar components of **A** and **B**:

$$A_x = A\cos\theta_A = 4 \times 0.866 = 3.464$$
$$A_y = A\sin\theta_A = 4 \times 0.500 = 2.000$$

$$B_x = B\cos\theta_B = 3 \times 0.500 = 1.500$$
$$B_y = B\sin\theta_B = 3 \times 0.866 = 2.598$$

Adding the x and y components separately then gives us the scalar components of $\mathbf{C} = \mathbf{A} + \mathbf{B}$,

$$C_x = A_x + B_x = 4.964 \qquad C_y = A_y + B_y = 4.598$$

and the sum of these components, multiplied by their unit vectors, is

$$\mathbf{C} = C_x\mathbf{i} + C_y\mathbf{j} = 4.964\mathbf{i} + 4.598\mathbf{j}$$

example 2 • 3 Given the three vectors

$$\mathbf{A} = 2\mathbf{i} - 7\mathbf{k} \qquad \mathbf{B} = -\mathbf{i} + 3\mathbf{j} + 5\mathbf{k} \qquad \mathbf{C} = \mathbf{i} - \mathbf{j} - \mathbf{k}$$

find $\mathbf{D} = \mathbf{A} + 2\mathbf{B} - 3\mathbf{C}$.

solution Summing the scalar components separately, we obtain the components of **D** as

$$D_x = A_x + 2B_x - 3C_x \qquad D_y = A_y + 2B_y - 3C_y$$
$$D_z = A_z + 2B_z - 3C_z$$

and their vector sum is

$$\mathbf{D} = -3\mathbf{i} + 9\mathbf{j} + 6\mathbf{k} = -3(\mathbf{i} - 3\mathbf{j} - 2\mathbf{k})$$

The magnitude of a vector, often loosely termed its "length," is one of the three independent quantities needed to specify the vector in spherical polar coordinates. In cartesian coordinates, we must use Pythagoras' theorem to express the length in terms of its components. Thus $A_\perp$ in Fig. 2 . 11, the projection of **A** perpendicular to the z axis, is given by

$$A_\perp^2 = A_x^2 + A_y^2 \qquad [2 \cdot 18]$$

However, $A_\perp$, A_z, and $|\mathbf{A}| = A$ are also sides of a right triangle; therefore we obtain

$$A^2 = A_\perp^2 + A_z^2 = A_x^2 + A_y^2 + A_z^2$$
$$A = \sqrt{A_x^2 + A_y^2 + A_z^2} \qquad [2 \cdot 19]$$

From Eq. [2 . 11], the unit vector in the direction of **A** is then

$$\hat{\mathbf{A}} = \frac{1}{A}\mathbf{A} = \frac{A_x\mathbf{i} + A_y\mathbf{j} + A_z\mathbf{k}}{\sqrt{A_x^2 + A_y^2 + A_z^2}} \qquad [2 \cdot 20]$$

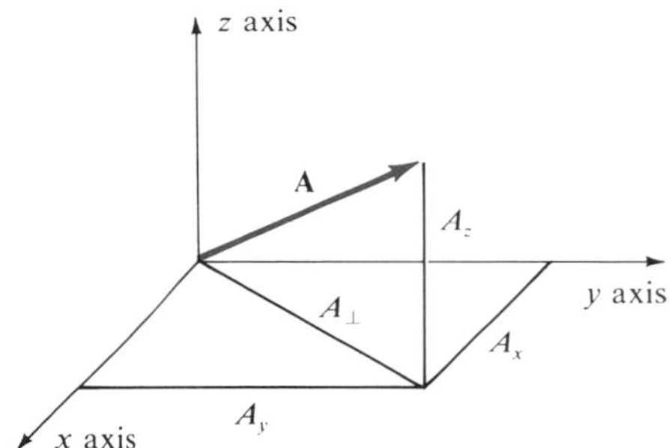

figure 2 • 11 *The magnitude of a vector* **A**; $A = (A_\perp^2 + A_z^2)^{1/2} = (A_x^2 + A_y^2 + A_z^2)^{1/2}$

As we saw in Sec. 2 . 2, a vector may be described either in terms of its rectangular components or in terms of its length and direction angles. If the two formulations are to be equivalent, there must be precise mathematical relations which enable us to go from one representation to the other. Given the components A_x, A_y, and A_z of a vector **A**, we can use Eq. [2 . 19] to find its magnitude, and Eq. [2 . 13] then yields the direction cosines:

$$\cos\alpha = \frac{A_x}{A} \qquad \cos\beta = \frac{A_y}{A} \qquad \cos\gamma = \frac{A_z}{A} \qquad [2 \cdot 21]$$

Conversely, given the magnitude and any two direction angles—say, α and β—we can find the two components A_x and A_y and then use Eq. [2 . 19] to find the third component, A_z:

$$A_z = \pm\sqrt{A^2 - A_x^2 - A_y^2} \qquad [2 \cdot 22]$$

(note the ambiguity in the sign of A_z). Although the three rectangular components are independent, the direction cosines are not independent:

$$\cos^2\alpha + \cos^2\beta + \cos^2\gamma = \frac{A_x^2 + A_y^2 + A_z^2}{A^2} \qquad [2 \cdot 23]$$

At least one length is always needed to specify a vector.

example 2 • 4 Find the magnitude and direction of the vector **C** in Example 2 . 2.

solution The magnitude of the vector sum $\mathbf{C} = 4.964\mathbf{i} + 4.598\mathbf{j}$ is

$$C = \sqrt{4.964^2 + 4.598^2} = 6.766$$

and the direction angle of **C** relative to the x axis is

$$\alpha = \cos^{-1}\frac{C_x}{C} = \cos^{-1}\frac{4.964}{6.766} = 42.8°$$

If we express the squared magnitude of **C** in general terms, we find

$$C^2 = C_x^2 + C_y^2 = (A\cos\theta_A + B\cos\theta_B)^2 + (A\sin\theta_A + B\sin\theta_B)^2$$
$$= A^2 + B^2 + 2AB\cos(\theta_A - \theta_B)$$

Does this look familiar? Yes and no; see Prob. 2 . 23 at the end of the chapter.

example 2 • 5 A fighter pilot scrambles to intercept. Within a few minutes he is flying at an altitude of 3 mi, a distance of 6 mi south and 8 mi west of his base. What are his direction cosines and angles?

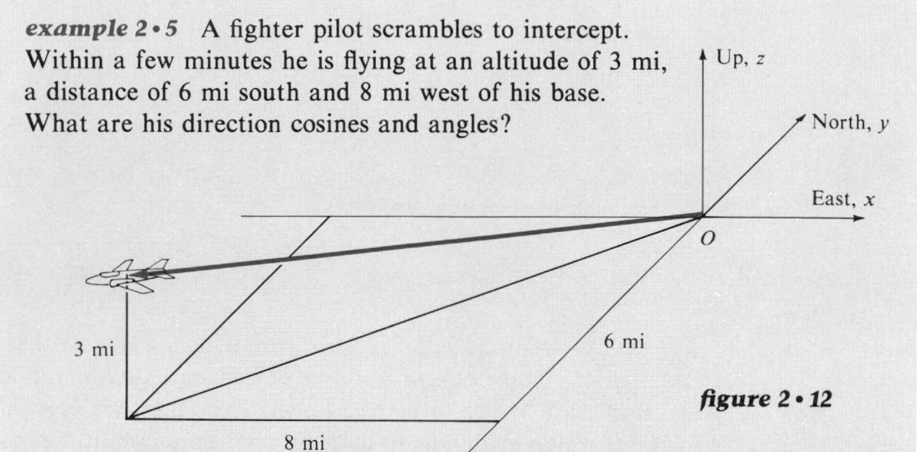

figure 2 • 12

solution We assume that east, north, and up are positive directions. The magnitude of his net displacement **A** from the base is

$$A = \sqrt{64 + 36 + 9} = 10.44 \text{ mi}$$

Relative to the x axis,

$$\cos \alpha = \frac{-8}{10.44} = -0.7662 \qquad \alpha = 140.0°$$

and relative to the y and z axes,

$$\cos \beta = \frac{-6}{10.44} = -0.5747 \qquad \beta = 125.1°$$

$$\cos \gamma = \frac{3}{10.44} = 0.2874 \qquad \gamma = 73.3°$$

When a vector is specified in spherical polar coordinates, its length is the magnitude r, and its direction is described by angles θ and ϕ. In plane polar coordinates (two dimensions), only one angle is needed, and in this case θ is the angular distance of $\mathbf{r}$ from the positive x axis. In spherical coordinates, however, θ is the angular distance of $\mathbf{r}$ from the z axis, and ϕ is the angle from the x axis to a projection of $\mathbf{r}$ perpendicular to the z axis (compare Figs. 1 . 8 and 2 . 7). Thus in two dimensions the coordinate angle θ corresponds to the direction angle α, but in three dimensions θ corresponds to γ.

2 . 3 vector multiplication: scalar product

We can refer any two vectors in space to the same origin. These two vectors then determine a unique plane in space, and the smallest angle between them is the angle θ through which one vector must be rotated in the plane to coincide with the direction of the other vector (see Fig. 2 . 13). The *scalar,* or *dot, product* of two vectors is defined by the relation

$$\mathbf{A} \cdot \mathbf{B} = AB \cos \theta \qquad [2 \cdot 24]$$

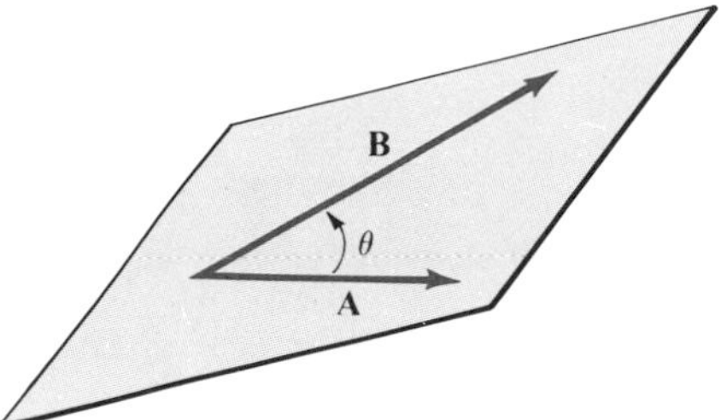

figure 2 • 13 *The angle θ between two vectors* **A** *and* **B**

The dot product is a scalar quantity equal to the product of the magnitudes of the vectors and the cosine of the angle between them. Since the value of the cosine is the same regardless of the order of the two vectors, the dot product follows the commutative law:

$$\mathbf{A} \cdot \mathbf{B} = \mathbf{B} \cdot \mathbf{A} \qquad [2 \cdot 25]$$

For the special case when $\mathbf{A} = \mathbf{B}$, then $\theta = 0$, and we have a useful result:

$$\mathbf{A} \cdot \mathbf{A} = A^2 \qquad [2 \cdot 26]$$

and if the direction of **A** is **i**, **j**, or **k**, then

$$\mathbf{i} \cdot \mathbf{i} = 1 \qquad \mathbf{j} \cdot \mathbf{j} = 1 \qquad \mathbf{k} \cdot \mathbf{k} = 1 \qquad [2 \cdot 27]$$

Another consequence of this definition is that the dot product of two perpendicular vectors is zero (why?):

$$\mathbf{i} \cdot \mathbf{j} = 0 \qquad \mathbf{i} \cdot \mathbf{k} = 0 \qquad \mathbf{j} \cdot \mathbf{k} = 0 \qquad [2 \cdot 28]$$

It can also be shown that the distributive law holds for dot products:

$$\mathbf{A} \cdot (\mathbf{B} + \mathbf{C}) = \mathbf{A} \cdot \mathbf{B} + \mathbf{A} \cdot \mathbf{C} \qquad [2 \cdot 29]$$

Thus we can think of the dot product as the scalar product of the magnitude of **A** with the projection $B_\parallel$ of **B** along **A** (see Fig. 2 . 14):

$$\mathbf{A} \cdot \mathbf{B} = AB_\parallel = A(B \cos\theta) \qquad [2 \cdot 30]$$

A useful relationship can be derived by using the commutative and distributive laws to expand the dot product. From the unit-vector relations in Eqs. [2 . 27] and [2 . 28],

$$\mathbf{A} \cdot \mathbf{B} = (A_x\mathbf{i} + A_y\mathbf{j} + A_z\mathbf{k}) \cdot (B_x\mathbf{i} + B_y\mathbf{j} + B_z\mathbf{k})$$

which yields

$$\mathbf{A} \cdot \mathbf{B} = A_xB_x + A_yB_y + A_zB_z \qquad [2 \cdot 31]$$

This formula enables us to determine the cosine of the angle between two

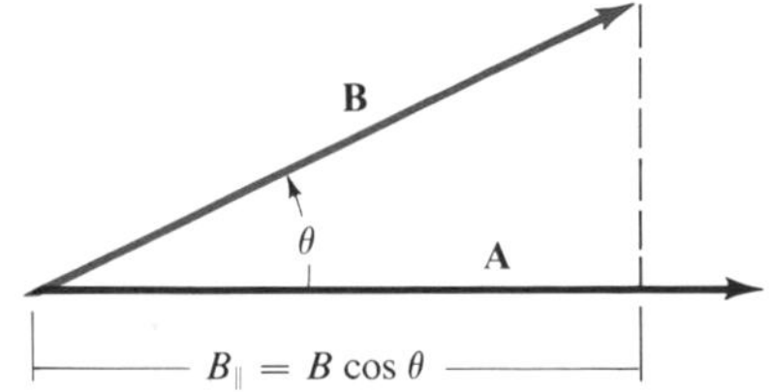

figure 2 • 14 *The scalar, or dot, product of* **A** *and* **B**; $\mathbf{A} \cdot \mathbf{B} = AB_\parallel = AB\cos\theta$

vectors in space if we know their components with respect to any coordinate system.

example 2 • 6 Find the cosine of the angle between the two vectors

$$\mathbf{A} = 3\mathbf{i} - \mathbf{j} - 2\mathbf{k} \qquad \mathbf{B} = -\mathbf{i} + 2\mathbf{j} + 7\mathbf{k}$$

solution From the definition in Eq. [2 . 24],

$$\cos\theta = \frac{\mathbf{A}\cdot\mathbf{B}}{AB}$$

where, from Eq. [2 . 31],

$$\mathbf{A}\cdot\mathbf{B} = -3 - 2 - 14 = -19$$

and the magnitudes of **A** and **B** are

$$A = \sqrt{9 + 1 + 4} = 3.742 \qquad B = \sqrt{1 + 4 + 49} = 7.348$$

Therefore

$$\cos\theta = \frac{-19}{3.742 \times 7.348} = -0.6910$$

and the *smallest* angle between **A** and **B** is $\theta = 133.7°$ ($\cos 226.3°$ is also equal to -0.6910).

example 2 • 7 For the vectors **A** and **B** in Example 2 . 6, find the vector component of **A** parallel to **B** and express it in rectangular components.

solution To find the vector component $\mathbf{A}_\parallel$ of **A** in the **B** direction, we find its scalar projection $A_\parallel = A\cos\theta$ along **B** and then multiply it by the unit vector in the **B** direction:

$$\mathbf{A}_\parallel = (A\cos\theta)\hat{\mathbf{B}}$$

Since $\hat{\mathbf{B}} = (1/B)\mathbf{B}$, we can write

$$\mathbf{A}_\parallel = \frac{(A\cos\theta)B}{B^2}\mathbf{B} = \frac{\mathbf{A}\cdot\mathbf{B}}{B^2}\mathbf{B}$$

and substituting the values from Example 2 . 6, we obtain

$$\mathbf{A}_\parallel = \frac{-19}{54}(-\mathbf{i} + 2\mathbf{j} + 7\mathbf{k}) = 0.35\mathbf{i} - 0.70\mathbf{j} - 2.46\mathbf{k}$$

2 • 4 vector multiplication: vector product

The *vector,* or *cross, product* has a more complicated definition than the scalar product, but one which is extremely useful, for it grew out of the necessities of physical description. We define the cross product of two vectors **A** and **B** to be a third vector,

$$\blacktriangleright\ \mathbf{C} = \mathbf{A} \times \mathbf{B} \qquad [2 \cdot 32]$$

having the magnitude

$$C = |\mathbf{A} \times \mathbf{B}| = AB\sin\theta \qquad [2 \cdot 33]$$

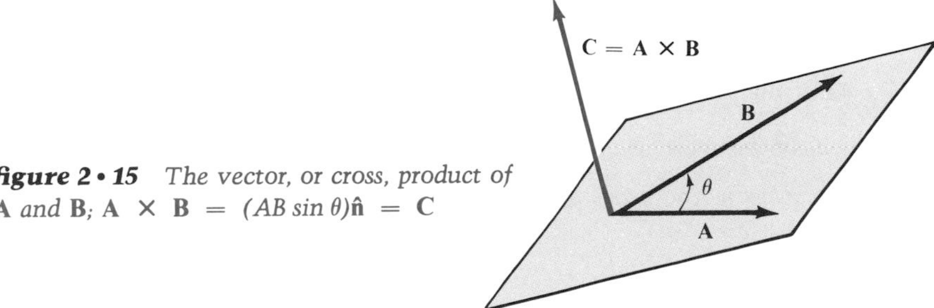

figure 2 • 15 *The vector, or cross, product of* **A** *and* **B**; $\mathbf{A} \times \mathbf{B} = (AB \sin\theta)\hat{\mathbf{n}} = \mathbf{C}$

and a direction *normal* (perpendicular) to the plane determined by **A** and **B** (see Fig. 2 . 15). However, this leaves us with an ambiguity: is the normal direction to be considered up from this plane or down from it? To overcome this difficulty, we need a new concept—the *sense* of a rotation.

We define a rotation in the *xy* plane as positive if its direction is such that it would move a right-hand screw (the ordinary kind) in the direction of the positive *z* axis. In Fig. 2 . 16 rotation of the *x* axis onto the *y* axis is positive; rotation of the *y* axis onto the *x* axis is negative, for it would send a right-hand screw down the negative *z* axis. If the curled fingers of your right fist represent the direction of a positive rotation in the *xy* plane, then your thumb, fully extended, will point in the direction of the positive *z* axis. Thus the vector **C** in Fig. 2 . 15 is defined to be in the direction in which the screw will travel when it is rotated from **A** to **B** through the *smallest* angle θ.

As a consequence of these definitions, the vector product is *anticommutative* (why?):

$$\mathbf{A} \times \mathbf{B} = -\mathbf{B} \times \mathbf{A} \tag{2 . 34}$$

This means that the vector product of two parallel vectors is zero,

$$\mathbf{A} \times \mathbf{A} = \mathbf{0} \tag{2 . 35}$$

figure 2 • 16 *The right-hand-screw rule*

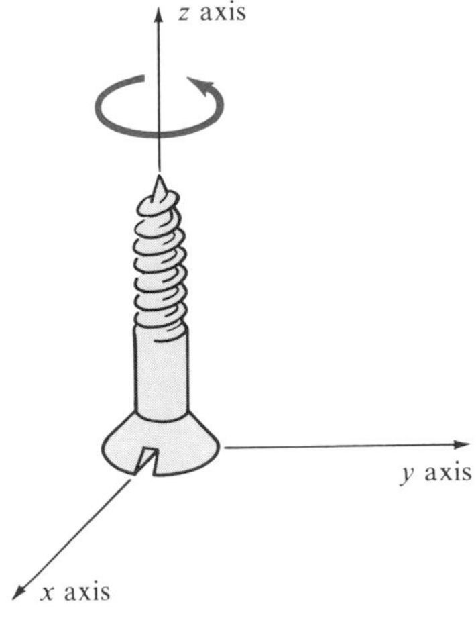

(a) Positive rotation in xy plane

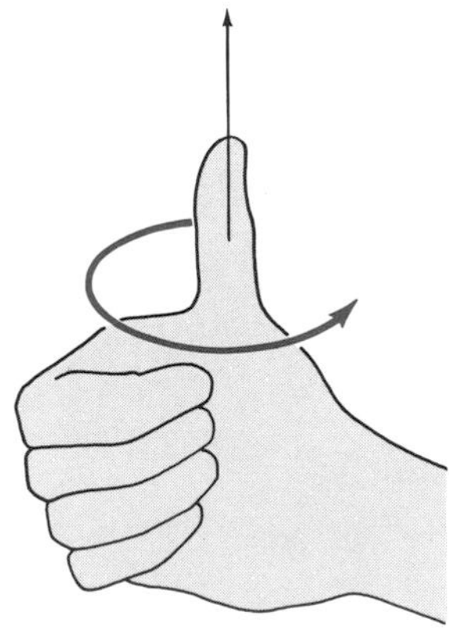

(b) First right-hand rule

and therefore

$$\mathbf{i} \times \mathbf{i} = 0 \qquad \mathbf{j} \times \mathbf{j} = 0 \qquad \mathbf{k} \times \mathbf{k} = 0 \qquad [2 \cdot 36]$$

Scalar multiplication by a positive scalar leaves the direction unchanged,

$$c\mathbf{A} \times \mathbf{B} = c(\mathbf{A} \times \mathbf{B}) = \mathbf{A} \times c\mathbf{B} \qquad [2 \cdot 37]$$

and, by definition, the cross products of the unit vectors **i**, **j**, and **k** obey the following relations:

$$\begin{aligned} \mathbf{i} \times \mathbf{j} &= \mathbf{k} & \mathbf{j} \times \mathbf{k} &= \mathbf{i} & \mathbf{k} \times \mathbf{i} &= \mathbf{j} \\ \mathbf{j} \times \mathbf{i} &= -\mathbf{k} & \mathbf{k} \times \mathbf{j} &= -\mathbf{i} & \mathbf{i} \times \mathbf{k} &= -\mathbf{j} \end{aligned} \qquad [2 \cdot 38]$$

It can also be shown that vector cross products follow the distributive law:

$$\mathbf{A} \times (\mathbf{B} + \mathbf{C}) = \mathbf{A} \times \mathbf{B} + \mathbf{A} \times \mathbf{C} \qquad [2 \cdot 39]$$

Thus we can use the relations among the unit vectors to evaluate a vector product in terms of its components:

$$\begin{aligned} \mathbf{A} \times \mathbf{B} &= (A_x\mathbf{i} + A_y\mathbf{j} + A_z\mathbf{k}) \times (B_x\mathbf{i} + B_y\mathbf{j} + B_z\mathbf{k}) \\ &= (A_yB_z - A_zB_y)\,\mathbf{i} - (A_xB_z - A_zB_x)\,\mathbf{j} \\ &\quad + (A_xB_y - A_yB_x)\,\mathbf{k} \end{aligned} \qquad [2 \cdot 40]$$

The minus sign preceding the **j** component emphasizes its formal similarity to a determinant, a useful mnemonic in this case:

$$\mathbf{A} \times \mathbf{B} = \begin{vmatrix} \mathbf{i} & \mathbf{j} & \mathbf{k} \\ A_x & A_y & A_z \\ B_x & B_y & B_z \end{vmatrix} \qquad [2 \cdot 41]$$

example 2 • 8 Use the vector cross product to find the sine of the angle between vectors **A** and **B** in Example 2 . 6.

solution From Eq. [2 . 33], we can express $\sin\theta$ as

$$\sin\theta = \frac{|\mathbf{A} \times \mathbf{B}|}{AB}$$

Then, from the values for **A** and **B** in Example 2 . 6, the components of the cross product **A** × **B** are

$$\begin{aligned} (A_yB_z - A_zB_y)\,\mathbf{i} &= [(-1 \times 7) - (-2 \times 2)]\,\mathbf{i} = -3\mathbf{i} \\ -(A_xB_z - A_zB_x)\,\mathbf{j} &= -[(3 \times 7) - (-2 \times -1)]\,\mathbf{j} = -19\mathbf{j} \\ (A_xB_y - A_yB_x)\,\mathbf{k} &= [(3 \times 2) - (-1 \times -1)]\mathbf{k} = 5\mathbf{k} \end{aligned}$$

or

$$\begin{aligned} \mathbf{A} \times \mathbf{B} &= -3\mathbf{i} - 19\mathbf{j} + 5\mathbf{k} \\ |\mathbf{A} \times \mathbf{B}| &= \sqrt{395} \end{aligned}$$

Therefore

$$\sin\theta = \frac{\sqrt{395}}{3.742 \times 7.348} = 0.7228$$

and the angle θ is 133.7°, as before. A quick sketch or comparison with the dot product will show that θ is not 46.3°.

In conclusion, let us consider some of the properties of a quantity defined as the *scalar triple product,* or *box product*. Note that the volume of a rectangular parallelepiped determined by three mutually perpendicular vectors **A**, **B**, and **C** is given by $V = ABC$ (see Fig. 2 . 17*a*). If we rotate **A** in the *xy* plane, so that it is no longer perpendicular to **B**, then the area of the base is $AB \sin\theta$. If we next tilt **C** out of the perpendicular by an angle ϕ, the volume V depends on its component perpendicular to the base; hence $V = ABC \sin\theta \cos\phi$. However, the direction normal to the bases is given by $\mathbf{A} \times \mathbf{B}$. Therefore ϕ is equal to the angular distance between $\pm\mathbf{A} \times \mathbf{B}$ and **C**, with the sign of **A** depending on the sense of rotation of **A** into **B**. Thus we can form the box product as the *scalar* product of $\mathbf{A} \times \mathbf{B}$ with vector **C**:

$$V = \pm\mathbf{A} \times \mathbf{B} \cdot \mathbf{C} > 0 \qquad [2 \cdot 42]$$

(No parentheses are necessary, since we cannot form the vector cross product of a vector and a scalar quantity.)

figure 2 • 17 *The scalar box product* $V = \mathbf{A} \times \mathbf{B} \cdot \mathbf{C}$

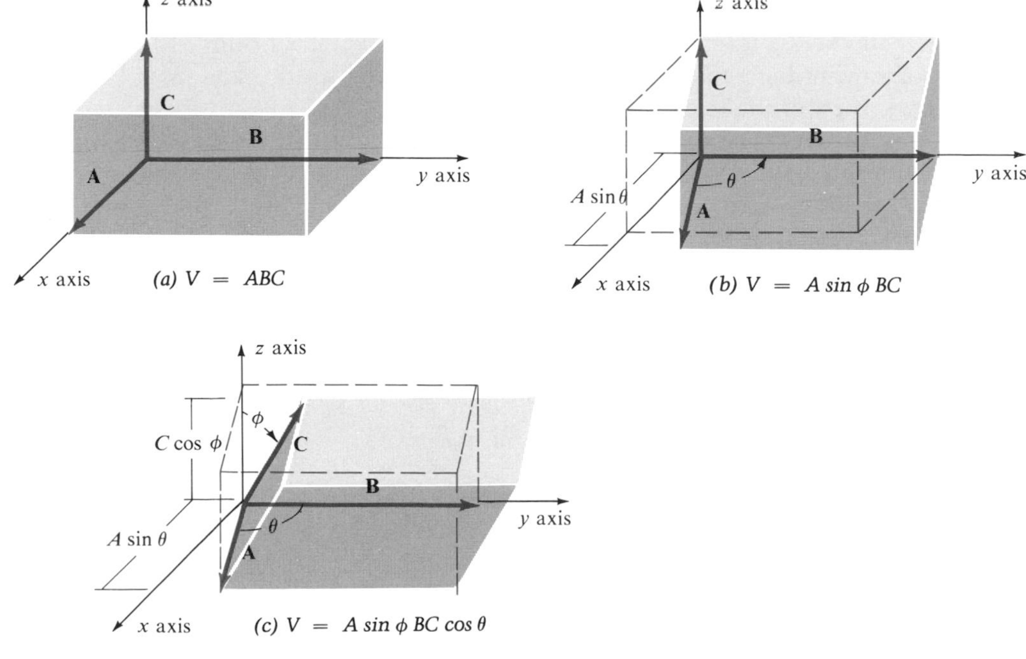

In the example illustrated, the rotations are positive, and we could equally well have chosen the *yz* plane as the base and found

$$V = \mathbf{A} \cdot \mathbf{B} \times \mathbf{C} = \mathbf{A} \times \mathbf{B} \cdot \mathbf{C}$$

However, reversing the senses of the rotations would have required minus signs:

$$V = -\mathbf{B} \times \mathbf{A} \cdot \mathbf{C} = -\mathbf{A} \times \mathbf{C} \cdot \mathbf{B} = \mathbf{C} \times \mathbf{A} \cdot \mathbf{B} \qquad [2 \cdot 43]$$

Note that we can interchange the dot and the cross in the box product, but if we interchange vectors we must observe that dot products commute and cross products anticommute. These geometric arguments determine the rules for manipulation of the box product and prove especially valuable in later chapters.

Another handy mnemonic device is

$$V = \mathbf{A} \cdot \mathbf{B} \times \mathbf{C} = \begin{vmatrix} A_x & A_y & A_z \\ B_x & B_y & B_z \\ C_x & C_y & C_z \end{vmatrix} \qquad [2 \cdot 44]$$

2 · 5 *vectors: the real thing*

Having sampled the pleasures of vector analysis, we would do well to recall Heaviside's admonition in the opening quotation: *vectors are real.* Their reality can be demonstrated in the laboratory. In thinking of vectors it is physically more meaningful to imagine a weathervane—palpable, pointed, massive, with a definite length and direction—than to picture an idealized arrow drawn on a blackboard. Physical quantities which have both magnitude and direction and combine with other such quantities to satisfy the parallelogram rule of vector addition are vector quantities. These include linear displacement, force, velocity, momentum, and gravitational, electrical, and magnetic fields, to name but a few. They must be treated differently from scalar quantities such as mass, time, distance, speed, energy, and temperature. In view of the importance of vectors in physics, it is not surprising that vector analysis should initially have been developed by physicists concerned with the description of physical reality. No matter how elegant and useful the concept of components may be, we must ultimately express our conclusions in terms of reality—that is, the vectors themselves.

A number of useful rules of vector analysis may be found in Appendix E. All of them may be derived from the principles set forth in this chapter.

PROBLEMS

2 • 1 *vector algebra*

2 • 1 Why can we not consider a scalar to be a vector of indeterminate direction?

2 • 2 Using graphical methods, find the displacement of a particle that has moved from the position (100 cm, 60°) to the position (50 cm, 30°) in the same plane.

2 • 3 If a tourist starts out in the morning and drives 15 mi west, 20 mi north, and 25 mi southeast, what is his net displacement?

2 • 4 An object undergoes four successive displacements in the xy plane. The magnitudes of the displacements are 4, 9, 3, and 2 km, and their direction angles relative to the x axis are 25°, 320°, 180°, and 60°, respectively. (*a*) Find the vector sum of these displacements by plotting them on a convenient scale in the order given. (*b*) Plot them in some other order and verify the validity of both the commutative and associative laws for vector addition.

2 • 5 Two sides of a triangle are formed by vectors **A** and **B**. Determine the displacement from O to the midpoint of the third side. (*Note:* No coordinate system is necessary.)

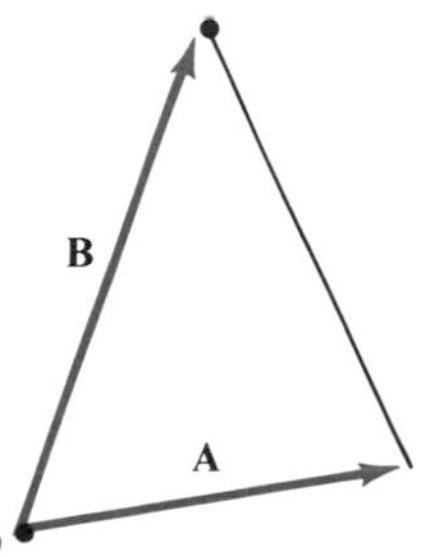

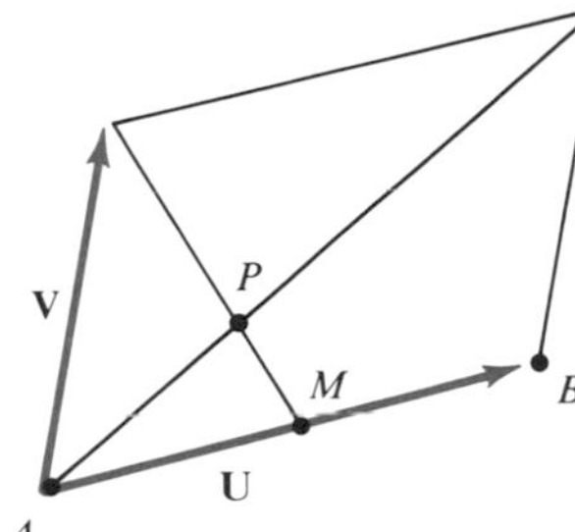

2 • 6 If point M is at the midpoint of side AB of the parallelogram, what is the displacement of point P from A, expressed in terms of vectors **U** and **V**?

2 • 7 If vectors are drawn from the vertices of a triangle to the midpoints of the opposite sides, show that their vector sum is zero. (It may help to do this graphically first.)

2 • 2 *components of a vector*

2 • 8 Derive the vector equation of a straight line passing through two points $A = (x_1,y_1)$ and $B = (x_2,y_2)$ by expressing the position vector **C** of an arbitrary point $P = (x,y)$ on the line $\overline{AB}$ in terms of vectors **A** and **B**. [*Hint:* Let f be the fractional distance of point P along the line $\overline{AB}$.] Show that, in terms of x and y components, this vector equation reduces to the conventional algebraic expression

$$y = y_1 + \frac{y_2 - y_1}{x_2 - x_1}(x - x_1)$$

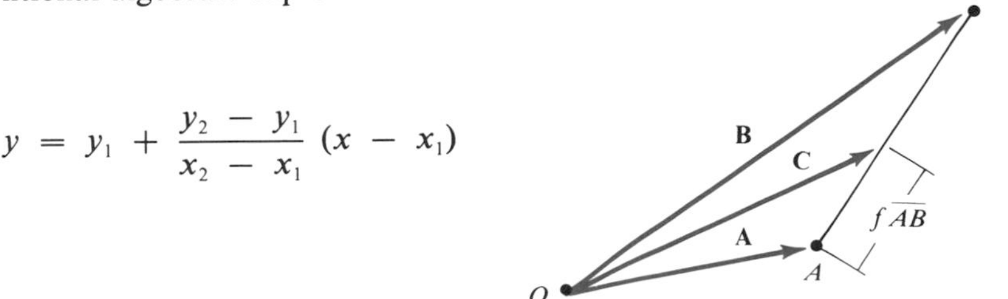

2 • 9 Two vectors, each of magnitude 30 cm, include an angle of 60° between them. Find the magnitude of their vector sum (*a*) graphically and (*b*) algebraically.

2 • 10 Consider five unit vectors pointing away from the origin of coordinates at 45° intervals ranging from $\theta = 0°$ to $\theta = 180°$. Find their sum (*a*) graphically and (*b*) algebraically.

2 • 11 Two vectors with the same origin are $\mathbf{A} = 3\mathbf{i}$ and $\mathbf{B} = 4\mathbf{j}$. Find the vector to the midpoint of a line joining the ends of **A** and **B**.

2 • 12 A small airplane, during a part of its flight that begins at a point O in space, flies 40 mi due south, then turns left 90° and immediately goes into a climb 6 mi long at angle of 10° with the horizontal, and finally turns left and flies horizontally 32 mi due north. Taking the initial point O as the origin of coordinates and the axes x and z as the directions east and south, respectively, show that the final position of the airplane relative to O is approximately 10 mi at direction angles of 54°, 84°, and 37°.

2 • 13 At a certain moment an aircraft is observed by radar to be 80 mi away at 60°10′ east of north and 83°40′ from the local vertical. Find the northward, eastward, and vertical components of its positions.

2 • 14 If two vectors are given as

$$\mathbf{A} = \mathbf{i} + 7\mathbf{j} - 3\mathbf{k} \qquad \mathbf{B} = -3\mathbf{i} + \mathbf{j} + 4\mathbf{k},$$

find the length and the direction angles of the vector $\mathbf{C} = \frac{1}{2}(\mathbf{A} + \mathbf{B})$.

2 • 15 Three vectors are given as

$$\mathbf{A} = \mathbf{i} + 7\mathbf{j} - 3\mathbf{k}$$
$$\mathbf{B} = -3\mathbf{i} + \mathbf{j} + 4\mathbf{k}$$
$$\mathbf{C} = 2\mathbf{i} + 2\mathbf{j} + \mathbf{k}$$

(*a*) Find the vector

$$\mathbf{D} = \tfrac{1}{2}(\mathbf{A} + \mathbf{B} + \mathbf{C}) - (\mathbf{B} + \mathbf{C}) + 3\mathbf{B} - 2\mathbf{A}$$

in component notation. (*b*) What is the magnitude of **D**?

2 • 16 Show that the following vectors **A**, **B**, and **C** are the sides of a triangle:

$$\mathbf{A} = 2\mathbf{i} - 4\mathbf{j} + 3\mathbf{k}$$
$$\mathbf{B} = \mathbf{i} + 2\mathbf{j} - \mathbf{k}$$
$$\mathbf{C} = -3\mathbf{i} + 2\mathbf{j} - 2\mathbf{k}$$

2 • 17 Relate the direction angles α, β, and γ to the spherical polar coordinates (r,θ,ϕ). Compare Prob. 1 . 19 in Chapter 1 and see Figs. 1 . 7 and 2 . 7.

2 • 18 Find the unit vectors $\hat{\mathbf{A}}$ and $\hat{\mathbf{B}}$ of the two vectors **A** and **B** in Prob. 2 . 14.

2 • 3 *the scalar product*

2 • 19 Which of the following vectors are mutually perpendicular? Each triplet of numbers gives the components of the vector.

$$\mathbf{A} = (2,1,1) \qquad \mathbf{B} = (0,0,2) \qquad \mathbf{C} = (1,-2,0)$$
$$\mathbf{D} = (1,1,-3) \qquad \mathbf{E} = (9,5,3)$$

2·20 Given the five vectors in Prob. 2.19, find the vector components of **B**, **C**, **D**, and **E** parallel to **A**.

2·21 Find the cosine of the angle θ between $\mathbf{A} = 3\mathbf{i} + \mathbf{j} + 2\mathbf{k}$ and $\mathbf{B} = -3\mathbf{i} + \mathbf{j} + 2\mathbf{k}$. Draw a figure showing this angle.

2·22 Show that if one of the vectors of a dot product $\mathbf{A} \cdot \mathbf{B}$ is multiplied by a scalar c, the angle between **A** and **B** is unchanged.

2·23 If $\mathbf{C} = \mathbf{A} + \mathbf{B}$, show that the law of cosines holds in the following form:

$$C^2 = A^2 + B^2 + 2AB\cos\theta$$

where θ is the angle between **A** and **B**.

2·24 A rectangular solid of cross section l^2 is cut at an angle θ to the vertical. What is the area of the surface thus created?

2·25 What is the general form of any vector perpendicular to the following two vectors?

$$\mathbf{A} = 3\mathbf{i} - \mathbf{j} + 5\mathbf{k} \qquad \mathbf{B} = 2\mathbf{i} + \mathbf{j} - \mathbf{k}$$

2·4 *the vector product*

2·26 Does the sun appear to circle the earth clockwise or counterclockwise (*a*) as seen by an observer in the continental United States? (*b*) As seen by an observer in Russia? (*c*) As seen by an observer in Argentina?

2·27 To which of the five vectors of Prob. 2.19 is the vector $\mathbf{F} = (-2,-2,6)$ parallel? Can you find a relationship among the components of the parallel vectors?

2·28 The three points (1,2,2), (2,0,−2) and (3,1,1) determine a plane. Find the unit vector $\hat{\mathbf{n}}$ perpendicular to this plane. Is it unique?

2·29 Three vectors have the following components:

$$\mathbf{A} = \mathbf{i} - 2\mathbf{j} - 4\mathbf{k}$$
$$\mathbf{B} = 4\mathbf{i} - 2\mathbf{j} - 2\mathbf{k}$$
$$\mathbf{C} = -\mathbf{i} - \mathbf{j} - 3\mathbf{k}$$

(*a*) Do all three vectors lie in the same plane? (*b*) If $a\mathbf{A} + b\mathbf{B} + c\mathbf{C} = \mathbf{0}$, find b/a and c/a.

2·30 Given two vectors $\mathbf{A} = (2,1,1)$ and $\mathbf{B} = (1,-1,-1)$, (*a*) find the sine of the angle between them. (*b*) What is the area of the triangle determined by these two vectors and the line joining their endpoints?

2·31 One line in space passes through points (0,2,4) and (−1,−2,2) and another passes through points (2,1,1) and (1,0,1). (*a*) Find the direction perpendicular to *both* lines. (*b*) Construct two different vectors which begin on one of these lines and end on the other and find their projections in the mutually perpendicular direction. (*c*) Explain the significance of the result in part (*b*). (*d*) What would be the solution in part (*b*) if the original two lines had intersected at some point?

2 • 32 Perform the indicated cross product:

$$\mathbf{D} = (\mathbf{A} + \mathbf{B}) \times (\mathbf{A} - \mathbf{B})$$

How would you state this result in geometric terms?

2 • 33 Derive the most general formula possible (*a*) for a vector $\mathbf{V}_{\parallel}$ which is parallel to $\mathbf{A} = (1,2,3)$. (*b*) For a vector $\mathbf{V}_{\perp}$ which is perpendicular to **A**. (*c*) For a vector **V** which lies in the plane determined by the answers to parts (*a*) and (*b*).

2 • 34 Find the volume of the parallelepiped determined by

$$\mathbf{A} = (2,1,1) \qquad \mathbf{B} = (1,-1,-1) \qquad \mathbf{C} = (-2,2,-3)$$

2 • 35 (*a*) Show how the determinant representation of the box product in Eq. [2 . 44] can be used to derive the identities of Eq. [2 . 43]. (*b*) Use the box product to prove that the x component of

$$\mathbf{A} \times (\mathbf{B} + \mathbf{C}) = \mathbf{A} \times \mathbf{B} + \mathbf{A} \times \mathbf{C}$$

holds true. (*c*) Prove that Eq. [2 . 40] is true in general.

2 • 36 A vector **V** and a unit vector $\hat{\mathbf{n}}$ are neither parallel nor perpendicular to each other. (*a*) Determine $\mathbf{V}_{\parallel}$, the vector component of **V** parallel to $\hat{\mathbf{n}}$. (*b*) Determine $\mathbf{V}_{\perp}$, the vector component of **V** perpendicular to $\hat{\mathbf{n}}$. Express your answers in terms of **V** and $\hat{\mathbf{n}}$.

answers

2•2 $(-6.7\mathbf{i} - 61.6\mathbf{j})$ cm or (62.0 cm, 263.8°)
2•3 (3.52 mi, $\theta = 40°56'$)
2•4 (*a*) (8.8 km, 345°)
2•5 $\frac{1}{2}(\mathbf{A} + \mathbf{B})$
2•6 $\frac{1}{3}(\mathbf{U} + \mathbf{V})$
2•9 52 cm
2•10 $(1 + \sqrt{2})\mathbf{j}$
2•11 $1.5\mathbf{i} + 2\mathbf{j}$
2•13 39.56 mi N, 68.97 mi E, 8.825 mi up
2•14 $C = 4.153$, $\alpha = 103.9°$, $\beta = 15.62°$, $\gamma = 83.09°$
2•15 (*a*) $\mathbf{D} = -10\mathbf{i} - 9\mathbf{j} + 14\mathbf{k}$; (*b*) $D = 19.42$
2•16 $\mathbf{A} + \mathbf{B} + \mathbf{C} = \mathbf{0}$
2•17 $\cos\alpha = \sin\theta\cos\phi$, $\cos\beta = \sin\theta\sin\phi$, $\gamma = \theta$
2•18 $\hat{\mathbf{A}} = (0.130, 0.911, -0.391)$, $\hat{\mathbf{B}} = (-0.588, 0.196, 0.785)$
2•19 $0 = \mathbf{A}\cdot\mathbf{C} = \mathbf{B}\cdot\mathbf{C} = \mathbf{A}\cdot\mathbf{D} = \mathbf{D}\cdot\mathbf{E}$
2•20 $\mathbf{B}_{\parallel} = 0.816$, $\mathbf{C}_{\parallel} = 0$, $\mathbf{D}_{\parallel} = 0$, $\mathbf{E}_{\parallel} = 6.53$
2•21 $\cos\theta = 0.2857$
2•24 Area $= l^2 \sec\theta$
2•25 $-4c\mathbf{i} + 13c\mathbf{j} + 5c\mathbf{k}$, where c is an arbitrary constant

2•26 (*a*) Clockwise; (*b*) clockwise; (*c*) counterclockwise
2•27 $\mathbf{F} = -2\mathbf{D}$
2•28 $\hat{\mathbf{n}} = \pm(0.254\mathbf{i} + 0.889\mathbf{j} - 0.381\mathbf{k})$
2•29 (*a*) Yes; (*b*) $b/a = -\frac{1}{2}$, $c/a = -1$
2•30 (*a*) $\sin\theta = 0.8199$; (*b*) area $= 1.871$
2•31 (*a*) Take the cross product of (1,4,2) and (1,1,0) to find (−2,2,−3); (*b*) 0.728; (*c*) the minimum distance between the two lines; (*d*) 0
2•32 (*a*) $\mathbf{D} = 2\mathbf{B} \times \mathbf{A}$
2•33 (*a*) $\mathbf{V}_\parallel = c\mathbf{A}$; (*b*) $\mathbf{V}_\perp = (a, b, -a - 2b)$; (*c*) $\mathbf{V} = e\mathbf{V}_\parallel + f\mathbf{V}_\perp$, where a, b, c, e, and f are arbitrary constants
2•34 Volume $= 15$
2•36 (*a*) $\mathbf{V}_\parallel = (\mathbf{V} \cdot \hat{\mathbf{n}})\hat{\mathbf{n}}$; (*b*) $\mathbf{V}_\perp = \hat{\mathbf{n}} \times (\mathbf{V} \times \hat{\mathbf{n}})$

CHAPTER THREE

particle kinematics: I

There is, in nature, perhaps nothing older than motion, concerning which the books written by philosophers are neither few nor small; nevertheless I have discovered by experiment some properties of it which are worth knowing and which have not hitherto been either observed or demonstrated. Some superficial observations have been made, as, for instance, that the free motion of a heavy falling body is continuously accelerated; but to just what extent this acceleration occurs has not yet been announced; for so far as I know, no one has yet pointed out that the distances traversed, during equal intervals of time, by a body falling from rest, stand to one another in the same ratio as the odd numbers beginning with unity.

Galileo Galilei, Two New Sciences,
Third Day, 1638

As we saw in Chapter 1, there were three discernible stages in the development of mechanics—geometry, kinematics, and dynamics. The first, physical geometry, concerned the measurement of space and the description of spatial relationships in terms of coordinate systems. With the preceding chapter on vectors and their algebra as preparation, let us now examine, in this chapter and the next, the second stage in the development of mechanics, *kinematics.*

Kinematics differs from physical geometry in that it includes the concept of motion taking place in *time.* The concern of kinematics is purely motion, not the nature of the moving bodies or the causes of their movements. Of the various kinematic concepts, speed is much the oldest. Acceleration, the most important concept of kinematics, was defined at least as early as the fourteenth century, but it was not clearly treated until late in the sixteenth century, when Galileo attacked the problem of a body falling freely toward the earth. It was Galileo who first emphasized the necessity of investigating the "how" of phenomena, rather than attempting to cope first with the question of "why." For this reason he has been called the founder of modern physics by philosopher Ernst Mach.

In this chapter we shall consider *particle kinematics,* the motion of objects which are considered to be infinitesimal in extent—that is, "point" masses. The device of approximating a real body by assuming all its mass to be concentrated at one point is a useful way of simplifying problems in which the size of the body is assumed to be of little or no importance.

3·1 *time*

Time, the dimension of the physical universe which orders the sequence of events at a given place, is measured in terms of some selected phenomenon which recurs at regular intervals. Since antiquity people have looked to the movements of celestial bodies for their standard of time. Thus the *apparent solar day* is the interval of time between successive passages of the sun across any given meridian of longitude on the earth's surface.

With the development of the clock, it became evident that the solar day does not represent a constant time interval throughout the year, because the axis of the earth's revolution is tilted at an angle of 23.5° to the plane of its orbit about the sun. This observation led to the adoption of the *mean solar day,* which is the average of all apparent solar days in a mean solar year of 365.2422 days, with the *mean solar second* defined as 1/86,400 of a mean solar day. In 1956 the International Committee of Weights and Measures redefined the mean solar second as 1/31,556,925.9747 of the earth's tropical year at the fundamental epoch of January 0, 1900. This hypothetical year is one that the earth would have defined if it had continued its apparent instantaneous orbital rate, corrected for polar and orbital eccentricities. At first glance such a ponderous astronomical definition for so commonplace a notion may seem excessive. However, it illustrates the central importance of time in physics and the degree of precision needed in its measurement.

Although this definition, based on astronomical measurements of the orbital motions of the earth, moon, and planets was vastly superior to any local standard, such as a pendulum, there are indications that even the earth's rate of rotation is changing with time. The braking effect of tidal friction is increasing the day by an amount estimated as 1 second every 120,000 years. There are also indications of fluctuations in the rate of rotation due, at least in part, to shifts of matter within or on the earth. Consequently, in 1967 the standard second was redefined as 9,192,631,770 oscillations of the atom of cesium-133. The National Bureau of Standards in Washington, D.C., maintains a cesium-beam clock as its atomic-frequency standard and broadcasts carrier and modulation frequencies through its radio stations to permit frequency comparison by its users. The broadcast frequencies of these stations are offset from atomic frequency by a precisely known amount, so that they correspond as closely as possible to the mean solar second corrected for polar motion and for seasonal variations in the earth's rotation.

Time is also measured by controlled alternating electric current, the vibration frequency of a tuning fork or a quartz crystal, and periods of radioactive decay. However, there is no unit of absolute time, and from an operational standpoint, we must define *time* T as "what the clock reads." In general, the more stable the cyclic system, the more accurate the clock.

3 • 2 speed

Speed is most commonly defined as a derived quantity, the ratio of distance traveled to time elapsed. As we saw in Sec. 2 . 1, the travel distance s along a particular path is a scalar quantity; it has the dimension of length, but this length is independent of any direction. Since both length and time are considered to be positive scalars, speed is also a positive scalar; it is simply the *time rate of motion,* irrespective of direction. Since the corresponding *directed* motion is velocity, speed is denoted by the symbol V and can be defined in terms of its dimensions as

$$V = \frac{L}{T} \qquad [3 \cdot 1]$$

A given speed v would thus be expressed in units of length per unit time, as shown in Table 3 . 1.

In physics, however, it is important to distinguish between average and instantaneous speed. For example, when a racing car on the Indianapolis speedway is clocked at a speed of 200 mi/h, we understand that this is an average speed $\bar{v}$, which represents the distance around the track divided by the elapsed time for that lap. If the car were clocked at regular time intervals as it moved around the track, the average speed during each of these intervals would not necessarily be the same as the overall average for the lap. Thus, at various instants of time t, the car must have been traveling at instantaneous speeds v which are not directly measurable. We can, however, define an instantaneous speed v.

Suppose a particle is moving along the curved path in Fig. 3 . 1, and at a time t_0 it is observed at some reference point at a distance s_0 from its origin; then, at a later time t_1, it passes another point a distance s_1 from its origin. In the time from t_0 to t_1 the particle has traveled a distance Δs which is the difference between the two distances s_1 and s_0. Its *average speed* $\bar{v}$ during this interval is defined as

$$\bar{v} = \frac{s_1 - s_0}{t_1 - t_0} = \frac{\Delta s}{\Delta t} \qquad [3 \cdot 2]$$

where Δs is the change in distance along the path corresponding to the change Δt in time. In this case Δs refers specifically to a change in s with

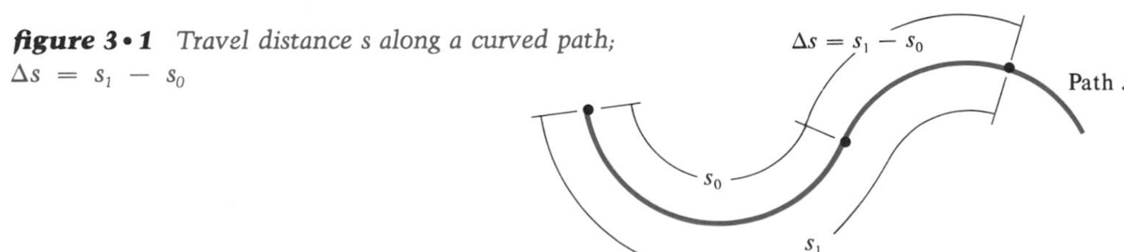

figure 3 • 1 *Travel distance s along a curved path;* $\Delta s = s_1 - s_0$

table 3 • 1 *Speed conversion factors*

meters/second (m/s)	feet/second (ft/s)	kilometers/hour (km/h)	miles/hour (mi/h)
1	3.281	3.600	2.237
0.3048	1	1.097	0.6818
0.2778	0.9113	1	0.6214
0.4470	1.467	1.609	1

Note: See Table 1 . 3 for length conversion factors.

time; that is, the distances s_0 and s_1 imply "initial s" and "later s." In general, however, the symbol Δ (delta) simply denotes an incremental change in a physical quantity, sometimes with no explicit mention of another variable.

If the speed of a moving particle varies significantly along the path, we can obtain a more precise profile of its motion by subdividing the intervals Δs into smaller intervals and taking the average speed $\bar{v}$ over each one. Clearly, if we choose a small enough interval, there will be no significant difference between the instantaneous speed v at the start of this interval and at the end of it (if we can imagine moving points, we might as well endow them with speedometers). This means that we can consider $\bar{v}$ approximately equal to v over this very small interval, and as the interval becomes infinitesimal, this approximation approaches an exact equality. This process of *taking the limit* is, in fact, the central idea of calculus.

Imagine that we take the average speed over successively smaller time intervals, so that t_1 approaches t_0 as Δt approaches zero. By this process we would arrive at the value of the instantaneous speed v at time t_0. This *limiting process* is expressed mathematically as

$$\lim_{t_1 \to t_0} \frac{s_1 - s_0}{t_1 - t_0} = \lim_{\Delta t \to 0} \frac{\Delta s}{\Delta t} = \left(\frac{ds}{dt}\right)_{t = t_0} = v(t_0) \qquad [3 \cdot 3]$$

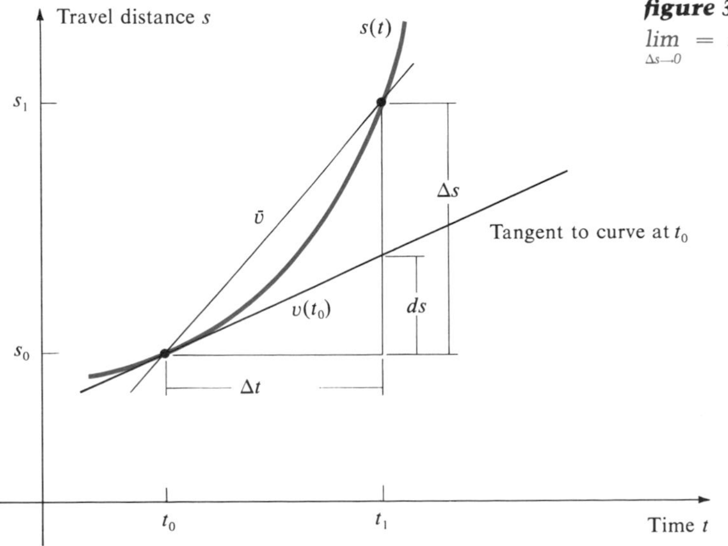

figure 3 • 2 *Instantaneous speed v at t_0; $\lim_{\Delta s \to 0} = \Delta s/\Delta t = ds/dt = v(t_0)$*

where dt is an infinitesimal time interval which vanishes to a point at t_0, and ds is the corresponding infinitesimal change in distance. As illustrated in Fig. 3 . 2, Δs is the increase in the distance ordinate from s_0 to s_1 during time Δt; hence the slope of a straight line passing through the corresponding points on the curve $s(t)$ is the average speed $\bar{v}$ during this interval. As t_1 approaches t_0, s_1 changes accordingly, and the slope $\bar{v}$ approaches the value of the slope of a line tangent to the curve at t_0. The slope of this tangent line, ds/dt, is defined to be the instantaneous speed v at t_0.

Strictly speaking, $\Delta s/\Delta t$ is an approximation to ds/dt which becomes increasingly better as Δt approaches zero. However, just as Δ enables us to express an increment of change without explicit mention of a second variable, the notation ds, dt, dx, etc., is useful to express the concept of *vanishingly small* increments. These terms are known as *differentials*, and ratios such as ds/dt are an economical means of indicating that $v = ds/dt$ is not a single measurement, but the end result, or limit, of a sequence of measurements or computations. In practice, we rarely compute a sequence of smaller and smaller values of $\bar{v}$, but instead use the rules of calculus to derive the desired limiting value. The limiting process expressed in Eq. [3 . 3] is known as "differentiating s with respect to t," or "taking the derivative of $s(t)$." The instantaneous speed $v = ds/dt$ is then known as the *first derivative of travel distance with respect to time* and is itself generally a function of time; that is, $v = v(t)$.

In the simplest case, that of uniform motion at constant speed, the average and instantaneous speed are the same for all intervals $\Delta s/\Delta t$. Thus at any time t the instantaneous speed will be

$$v(t) = \bar{v} = \frac{s - s_0}{t - t_0} = \text{constant} \qquad [3 \cdot 4]$$

and given a known reference distance s_0 at time t_0, the travel distance s is simply a product of the speed and the elapsed time:

$$s(t) = s_0 + v(t - t_0) \qquad [3 \cdot 5]$$

It is often convenient in such problems to choose the starting time as $t_0 = 0$, so that the elapsed time $t - t_0$ becomes just t. If we also take the origin as the starting point, then $s_0 = 0$ and Eq. [3 . 5] reduces to the form

$$s(t) = vt \qquad [3 \cdot 6]$$

example 3•1 A particle moves along a curved path in such a way that its distance from the origin is in constant ratio to the square of its travel time. (*a*) How does the speed of this particle vary with time? (*b*) If the constant ratio of distance to squared travel time is $k = 1.4$ ft/s², find the distance of the particle along the path and its speed 5 s after it starts to move.

solution (*a*) If we set $s_0 = 0$, the travel distance at any time t is $s(t) = kt^2$, and between time t and $t + \Delta t$ the particle will travel

$$\begin{aligned}\Delta s &= s(t + \Delta t) - s(t) \\ &= k(t + \Delta t)^2 - kt^2 = 2kt\,\Delta t + k\,(\Delta t)^2\end{aligned}$$

Then, from Eq. [3 . 3], the speed at time t is the instantaneous speed in the limit as Δt approaches zero:

$$v(t) = \frac{ds}{dt} = \lim_{\Delta t \to 0} \frac{2kt\,\Delta t + k\,(\Delta t)^2}{\Delta t} = 2kt$$

(*b*) At $t = 5$ s (if we set $t_0 = 0$), the distance of the particle from the origin is

$$s(5) = kt^2 = (1.4 \text{ ft/s}^2)(5 \text{ s})^2 = 35 \text{ ft}$$

and its speed is

$$v(5) = 2kt = 2(1.4 \text{ ft/s}^2)(5 \text{ s}) = 14 \text{ ft/s}$$

Note that the dimensions expressed in seconds cancel, just like algebraic terms in ordinary fractions.

example 3 • 2 Find an expression for the speed of a particle that starts from rest at time $t_0 = 0$ and moves in such a way that its distance from the origin is given by $s(t) = kt^3$.

solution In a time interval t to $t + \Delta t$ the particle will travel a distance

$$\Delta s = s(t + \Delta t) - s(t) = k(t + \Delta t)^3 - kt^3$$
$$= 3kt^2\,\Delta t + 3kt\,(\Delta t)^2 + k\,(\Delta t)^3$$

Its speed at any time t is therefore

$$v(t) = \lim_{\Delta t \to 0} [3kt^2 + 3kt\,\Delta t + k\,(\Delta t)^2] = 3kt^2$$

or, in the notation of calculus,

$$v(t) = \frac{ds}{dt} = k\frac{d(t^3)}{dt} = 3kt^2$$

3 • 3 *rectilinear motion: displacement and velocity*

So far we have considered only travel distance and its rate of change with time, without regard for the direction of this motion. To include the concept of direction in our discussion, we must employ vectors. In this section and the next we shall utilize vector algebra to describe the three directed properties of particle motion: displacement, velocity, and acceleration. In the case of rectilinear motion, the particle moves only in a straight line. Hence we can simplify matters by describing these properties in terms of one direction coordinate, chosen for convenience as the x axis.

Since motion takes place over time, the positions of a moving particle can be represented as a function of time. Thus, at any time t, the position of a particle with respect to the origin of the x axis is given by its *position vector,*

$$\mathbf{x}(t) = x(t)\mathbf{i} \qquad [3 \cdot 7]$$

Recall from Sec. 2 . 3 that the unit vector $\mathbf{i}$ is always directed to the right;

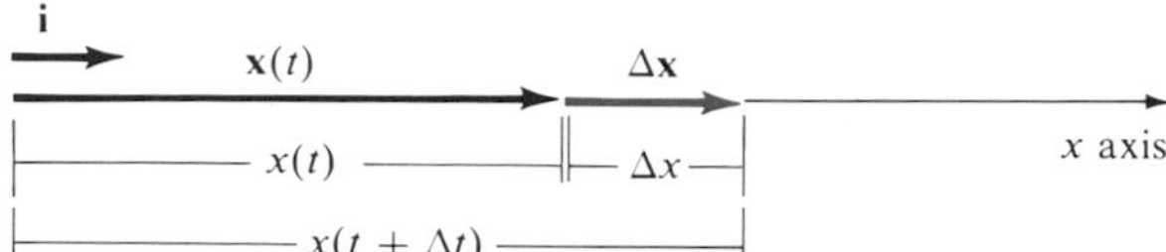

figure 3 • 3 *The position vector* $\mathbf{x}(t)$ *and a displacement vector* $\Delta\mathbf{x}$

when the scalar component $x(t)$ is positive, the position vector $\mathbf{x}(t)$ points to the right, and when $x(t)$ is negative, $\mathbf{x}(t)$ is directed to the left.

The displacement of the particle is the net change in its position vector with time. Thus if a particle moves from one position to another in a time $\Delta t = t_1 - t_0$, its *displacement* $\Delta\mathbf{x}$ is, by definition,

$$\Delta\mathbf{x} = (x_1 - x_0)\,\mathbf{i} \qquad [3 \cdot 8]$$

However, if we express t_1 as $t_1 = t_0 + \Delta t$, we can also define $\Delta\mathbf{x}$ as the difference between the two position vectors $\mathbf{x}(t)$ and $\mathbf{x}(t + \Delta t)$, as shown in Fig. 3 . 3:

$$\Delta\mathbf{x} = \mathbf{x}(t + \Delta t) - \mathbf{x}(t) = [x(t + \Delta t) - x(t)]\,\mathbf{i} = \Delta x\mathbf{i} \qquad [3 \cdot 9]$$

Note that Δx is the magnitude of the displacement vector $\Delta\mathbf{x}$, not the travel distance. Because distance is a positive scalar quantity, Δx and Δs will be the same only for motion in the positive direction along the x axis.

Just as we defined speed as the time rate of change of distance, velocity is the time rate of change of position. The *average velocity* $\bar{\mathbf{v}}$ of a particle is the ratio of its displacement $\Delta\mathbf{x}$ to the time interval Δt in which this motion occurs:

$$\bar{\mathbf{v}} = \frac{\Delta\mathbf{x}}{\Delta t} = \frac{\Delta x}{\Delta t}\mathbf{i} \qquad [3 \cdot 10]$$

Since displacement is a vector and time is a positive scalar, the product of $\Delta\mathbf{x}$ and $1/\Delta t$ is a vector quantity which has the same direction as the displacement. This means that if a particle undergoes a negative displacement, its velocity will also be negative.

As in the case of speed, we derive the *instantaneous velocity* $\mathbf{v}$ by taking the average velocity over increasingly smaller time intervals Δt measured from some time t_0. Then, in the limit as Δt approaches zero,

$$\mathbf{v}(t_0) = \lim_{t_1 \to t_0} \frac{\mathbf{x}_1 - \mathbf{x}_0}{t_1 - t_0} = \lim_{\Delta t \to 0} \frac{\Delta\mathbf{x}}{\Delta t} = \frac{dx}{dt}\mathbf{i} \qquad [3 \cdot 11]$$

Thus instantaneous velocity $\mathbf{v}(t)$ is the *first derivative of position with respect to time.* Its magnitude, $|\mathbf{v}(t)| = v(t)$, is the instantaneous speed at that point.

We could also have found $\mathbf{v}(t)$ from the expanded form of $\Delta\mathbf{x}$ in Eq. [3 . 9]:

$$\blacktriangleright \quad \mathbf{v}(t) = \lim_{\Delta t \to 0} \left[\frac{x(t + \Delta t) - x(t)}{\Delta t}\right]\mathbf{i} = \lim_{\Delta t \to 0} \frac{\Delta x}{\Delta t}\mathbf{i} = \frac{dx}{dt}\mathbf{i} \qquad [3 \cdot 12]$$

In cartesian coordinates, the derivative of a vector can be reduced to individual components, which can be differentiated like any ordinary scalar and then directed by the appropriate unit vector. This property will be particularly useful when we consider vectors in more than one dimension.

The defining formula for the derivative of a mathematical function $f(x)$,

$$\frac{df}{dx} = \lim_{\Delta x \to 0} \frac{f(x + \Delta x) - f(x)}{\Delta x} \qquad [3 \cdot 13]$$

leads to some useful general rules for differentiation (see Appendix F):

$$\frac{d[f(x) + g(x)]}{dx} = \frac{df(x)}{dx} + \frac{dg(x)}{dx} \qquad [3 \cdot 14]$$

$$\frac{d[kf(x)]}{dx} = k\frac{df(x)}{dx} \qquad k = \text{constant} \qquad [3 \cdot 15]$$

$$\frac{d[x^n]}{dx} = nx^{n-1} \qquad n = \text{constant} \neq 0 \qquad [3 \cdot 16]$$

example 3 • 3 The position of a moving particle with respect to the origin is given by

$$\mathbf{x}(t) = (At^2 + Bt + C)\,\mathbf{i}$$

where $A = -2$ ft/s^2, $B = 5$ ft/s, and $C = -2$ ft. (*a*) What is the position of the particle at $t = 2$ s? At $t = 3$ s? (*b*) What is its average velocity $\bar{\mathbf{v}}$ during the interval $2 \leq t \leq 3$ s? Its average speed? (*c*) Using the addition rule of differentiation, Eq. [3 . 14], find the instantaneous velocity of the particle at $t = 2.5$ s. How does this compare with the average velocity found in part (*b*)? (*d*) Calculate the position and velocity of the particle at $t = 0.5$ s. (*e*) At what time is the instantaneous velocity zero, and what does this signify?

solution (*a*) Substituting $t = 2$ in the expression for $\mathbf{x}(t)$, we have

$$\mathbf{x}(2) = (-8 + 10 - 2)\,\mathbf{i}\text{ ft} = 0\mathbf{i}\text{ ft}$$

That is, after 2 s the particle is at the origin. After 3 s its position is

$$\mathbf{x}(3) = (-18 + 15 - 2)\,\mathbf{i}\text{ ft} = -5\mathbf{i}\text{ ft}$$

or 5 ft left of the origin.

(*b*) The average velocity of the particle is, from Eq. [3 . 9],

$$\bar{\mathbf{v}} = \frac{\Delta \mathbf{x}}{\Delta t} = \frac{(-5 - 0)\,\mathbf{i}\text{ ft}}{1\text{ s}} = -5\mathbf{i}\text{ ft/s}$$

and its average speed is $\bar{v} = 5$ ft/s.

(*c*) According to the addition rule, the instantaneous velocity $v(t)$ can be expressed as

$$v(t) = \frac{dx}{dt} = A\frac{d(t^2)}{dt} + B\frac{dt}{dt} + 0 = 2At + B$$

Therefore the instantaneous velocity at $t = 2.5$ s is

$$\mathbf{v}(2.5) = (-10 + 5)\,\mathbf{i}\text{ ft/s} = -5\mathbf{i}\text{ ft/s}$$

Since velocity is a linear function of time in this case, $\mathbf{v}(2.5)$ is equal to the average velocity in the interval $2 \leq t \leq 3$ s.

(*d*) The position of the particle at $t = 0.5$ s is

$$\mathbf{x}(0.5) = (-0.5 + 2.5 - 2)\,\mathbf{i}\text{ ft} = 0\mathbf{i}\text{ ft}$$

and its velocity at this point is

$$\mathbf{v}(0.5) = (-2 + 5)\,\mathbf{i}\text{ ft/s} = 3\mathbf{i}\text{ ft/s}$$

Note that the particle is at the origin, just as it was at $t = 2$ s, but it has a different velocity. There is nothing inconsistent in the particle's being in the same position, but with different velocities, at different times.

(*e*) From part (*c*) we see that the velocity of the particle is

$$\mathbf{v}(t) = (2At + B)\,\mathbf{i} = 0\mathbf{i}\text{ ft/s}$$

at time $t = -B/2A$, or in this case, at time

$$t = \frac{-5\text{ ft/s}}{-4\text{ ft/s}^2} = 1.25\text{ s}$$

(which is dimensionally consistent). This is the point at which the velocity changes from positive to negative; that is, the particle, initially traveling to the right, comes to a stop and reverses direction.

3 • 4 *rectilinear motion: acceleration*

Galileo's clarification of the concept of acceleration and its fundamental role in mechanics was an important step in the development of the science. It made possible the modern conception of mass and force and thus marks the beginning of the science of dynamics. In everyday language the term *acceleration* refers only to increases of speed, but in physical science it is employed in a much wider sense to denote any *change of velocity.* Thus a physicist would say that an automobile is accelerating when it slows down or turns a corner, as well as when it speeds up. The foot throttle of an automobile is an "accelerator," but so, too, are the steering wheel and brakes.

In this section we shall consider acceleration only as it relates to one-dimensional motion, and from a purely mathematical standpoint; in later chapters we shall examine its physical significance. This approach has one distinct advantage: the hard work has already been done. This is because acceleration is defined to be a vector quantity, the *first derivative of velocity with respect to time.* The *average acceleration* $\bar{\mathbf{a}}$ of a particle is simply the change in its velocity vector over a given time interval Δt:

$$\bar{\mathbf{a}} = \frac{\mathbf{v}_1 - \mathbf{v}_0}{t_1 - t_0} \qquad [3 \cdot 17]$$

where $\mathbf{v}_0$ is the instantaneous velocity measured at time t_0 and $\mathbf{v}_1$ is the instantaneous velocity measured at a later time t_1. We can also define t_1 as $t + \Delta t$ and calculate the average acceleration as

$$\bar{\mathbf{a}} = \frac{\mathbf{v}(t + \Delta t) - \mathbf{v}(t)}{\Delta t} = \frac{\Delta\mathbf{v}}{\Delta t} \qquad [3 \cdot 18]$$

We then take the limit of $\Delta\mathbf{v}/\Delta t$ as Δt approaches zero to derive the *instantaneous acceleration* $\mathbf{a}(t)$:

$$\blacktriangleright \quad \mathbf{a}(t) = \lim_{\Delta t \to 0} \frac{\Delta \mathbf{v}}{\Delta t} = \frac{dv}{dt}\mathbf{i} \qquad [3 \cdot 19]$$

Instantaneous acceleration is also called the *second derivative of position with respect to time,* since differentiating $\mathbf{x}$ twice in succession with respect to t yields $\mathbf{a}(t)$:

$$\blacktriangleright \quad \mathbf{a}(t) = \frac{d}{dt}\left(\frac{dx}{dt}\right)\mathbf{i} = \frac{d^2x}{dt^2}\mathbf{i} \qquad [3 \cdot 20]$$

The dimensions of acceleration must therefore be L/T^2—or properly speaking, $(L/T)/T$, since acceleration is a time rate of change in a quantity (velocity) which is itself a time rate of change.

example 3•4 Find the instantaneous acceleration of the particle in Example 3 . 3.

solution From part (*c*) of Example 3 . 3, the instantaneous velocity is

$$\mathbf{v}(t) = (2At + B)\,\mathbf{i}$$

and the change in this velocity as a function of time is

$$\mathbf{a}(t) = \frac{dv}{dt}\mathbf{i} = \left(2A\frac{dt}{dt} + 0\right)\mathbf{i} = 2A\mathbf{i}$$

(*Note:* In differentiating $\mathbf{v}$ to find $\mathbf{a}$, we treat it just like any other function; how it was derived in the first place is immaterial.)

example 3•5 Convert the acceleration rate $a = 1{,}000{,}000\ \text{mi/y}^2$ to more reasonable units.

solution The given acceleration rate must be multiplied by ratios which are equal to 1, since we do not wish to change its magnitude, but only the units in which it is expressed. We can choose conversion factors to suit our needs from Table 1 . 1. Thus, to convert miles per year squared to feet per second squared, we need the equalities 1 mi = 5,280 ft and 1 y = 31,567,000 s. We then combine them in such a way that the units we do not want cancel out:

$$a = 1{,}000{,}000\ \cancel{\text{mi}}\left(\frac{5{,}280\ \text{ft}}{1\ \cancel{\text{mi}}}\right)\left(\frac{1}{\cancel{\text{y}}}\right)^2\left(\frac{1\ \cancel{\text{y}}}{31{,}567{,}000\ \text{s}}\right)^2$$

$$= 5.30 \times 10^{-6}\ \text{ft/s}^2$$

a figure that is much more impressive on an astrophysical scale than on a laboratory scale.

Historically and scientifically, the most important case of one-dimensional motion is the freely falling body, which we shall treat in full in the next chapter. Here the physicist begins with the assumption of *constant*

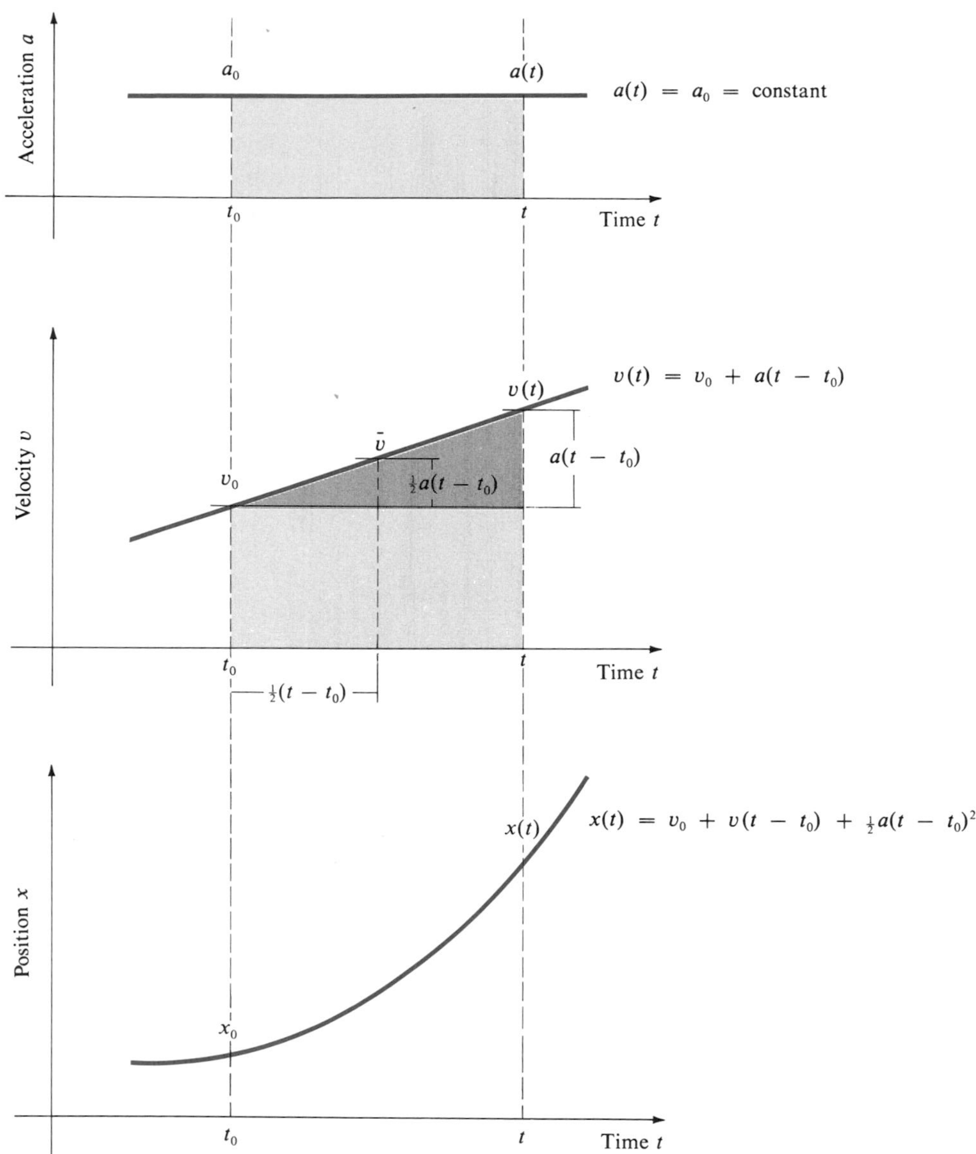

figure 3 • 4 *Graphs of acceleration, velocity, and position for constant acceleration in one-dimensional motion*

acceleration (downward) and uses this to predict the position and velocity of the body as a function of time. The assumption of constant acceleration means that the average and instantaneous accelerations are the same for all intervals of time:

$$\mathbf{a} = \bar{\mathbf{a}} = \frac{\mathbf{v} - \mathbf{v}_0}{t - t_0} = \text{constant} \qquad [3 \cdot 21]$$

Thus, if we know the initial velocity $\mathbf{v}_0$, we can predict the instantaneous velocity at any later time t from the formula

$$\mathbf{v}(t) = \mathbf{v}_0 + \mathbf{a}(t - t_0) \qquad [3 \cdot 22]$$

This relationship is illustrated in Fig. 3 . 4, which shows the graphs of acceleration, velocity, and position plotted against time. Although the values shown are positive for convenience, they could equally well be negative (below the t axis), since we are dealing with vector quantities.

Similarly, if we know the average velocity in an interval $t - t_0$, we can also predict the position of a body at time t. From the definition of average velocity in Eq. [3 . 10], we can write $\Delta\mathbf{x}$ in expanded form as

$$\bar{\mathbf{v}} = \frac{\Delta\mathbf{x}}{\Delta t} = \frac{\mathbf{x}(t) - \mathbf{x}_0}{t - t_0}$$

where $\mathbf{x}_0$ is the initial position at time t_0. Then any position $\mathbf{x}(t)$ is given by

$$\mathbf{x}(t) = \mathbf{x}_0 + \bar{\mathbf{v}}(t - t_0) \qquad [3 \cdot 23]$$

However, it is apparent from Fig. 3 . 4 that when velocity is a linear function of time, the value of the average velocity can also be expressed as

$$\bar{\mathbf{v}} - \mathbf{v}_0 = \tfrac{1}{2}[\mathbf{v}(t) - \mathbf{v}_0]$$

or, from Eq. [3 . 22],

$$\bar{\mathbf{v}} = \mathbf{v}_0 + \tfrac{1}{2}\mathbf{a}(t - t_0) \qquad [3 \cdot 24]$$

If we now substitute this expression into Eq. [3 . 23], we obtain an equation for the position of a uniformly accelerating body:

$$\blacktriangleright \quad \mathbf{x}(t) = \mathbf{x}_0 + \mathbf{v}_0(t - t_0) + \tfrac{1}{2}\mathbf{a}(t - t_0)^2 \qquad [3 \cdot 25]$$

Thus we have a general formula for position as a function of time for motion in one dimension, given a known initial position $\mathbf{x}_0$, a known initial velocity $\mathbf{v}_0$, and constant acceleration. If we also choose thc initial time as $t_0 = 0$, then we can write $t - t_0$ as t and express Eq. [3 . 25] in the simpler form

$$\mathbf{x}(t) = \mathbf{x}_0 + \mathbf{v}_0 t + \tfrac{1}{2}\mathbf{a}t^2 \qquad [3 \cdot 26]$$

example 3 • 6 Show that the area under the velocity curve in Fig. 3 . 4 is equal to the magnitude of the displacement $\Delta\mathbf{x}$.

solution The area between the velocity curve and the time axis from t_0 to t consists of a rectangle of area $v_0 \times (t - t_0)$ and a triangle of area $\frac{1}{2}(t - t_0) \times a(t - t_0)$, or $\frac{1}{2}a(t - t_0)^2$. The sum of these two areas is equal to the linear distance $x(t) - x_0$, which represents the net change on the position ordinate from x_0 to $x(t)$:

$$x(t) - x_0 = v_0(t - t_0) + \tfrac{1}{2}a(t - t_0)^2$$

This is just the value of $\mathbf{x}(t)$ given by Eq. [3 . 25].

Instead of substituting for the average-velocity term in Eq. [3 . 23], we could have eliminated time as a variable by writing Eq. [3 . 22] in scalar form as

$$t - t_0 = \frac{v - v_0}{a}$$

With this expression for time, Eq. [3 . 25] yields

$$x = x_0 + v_0 \frac{v - v_0}{a} + a \frac{(v - v_0)^2}{2a}$$

which we can solve for v to obtain

$$v^2 = v_0^2 + 2a(x - x_0) \qquad [3 \cdot 27]$$

We now have a useful equation for finding the speed of a particle in terms of its displacement.

Since all rectilinear motion can be described in terms of a single spatial coordinate, the vectors we have discussed thus far have only x components; that is, $\mathbf{a} = a_x\mathbf{i}$ and $\mathbf{v} = v_x\mathbf{i}$. Thus we could have described the acceleration and velocity vectors in terms of their single scalar component and treated them as ordinary scalar variables. Motion along a single axis has only two possible directions, and we can account for the direction of the vector simply by designating its value as positive or negative. However, a clear grasp of vector concepts in one dimension will prove advantageous when we extend these concepts to two- and three-dimensional situations in the next chapter.

example 3•7 The speed of a westbound train is uniformly reduced from 45 mi/h to 35 mi/h in 5.5 s. (*a*) What is its rate of acceleration? (*b*) Plot the graph of velocity against time and describe the position graph during the 5.5-s time interval.

solution (*a*) Take west as the positive direction of the velocity; then at t_0 the velocity is $v_0 = 45$ mi/h. Since the time rate of change of velocity is negative, the acceleration is eastward:

$$a(t) = \frac{35 - 45 \text{ mi/h}}{5.5 \text{ s}} = -1.8 \text{ mi/h} \cdot \text{s}$$

or, converting by the factor 1 mi/h = 1.47 ft/s,

$$a(t) = \frac{-1.8 \times 1.47 \text{ ft/s}}{1 \text{ s}} = -2.6 \text{ ft/s}^2$$

(*b*) The velocity graph is a straight line sloping downward, as shown.

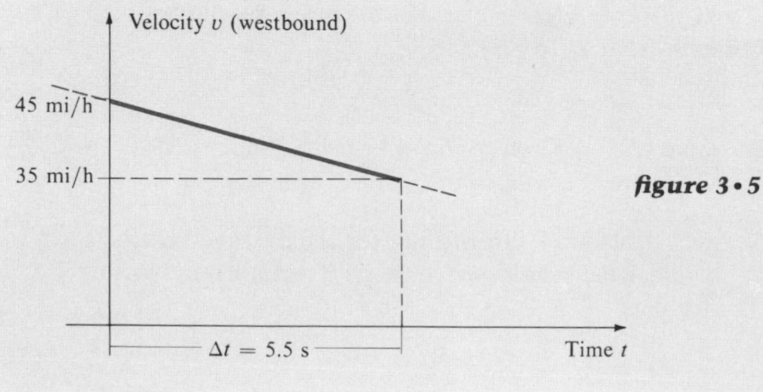

figure 3•5

According to Eq. [3 . 26], the position graph would be a downward-curving parabolic arc represented by

$$x(t) = x_0 + 66t - 1.3t^2$$

where x_0 gives the initial position at $t_0 = 0$. (Remember that west is the positive direction in this case.)

example 3•8 A high-speed train makes a nonstop run on a straight 250-mi track from Washington, D.C., to New York City in 2.5 h. It accelerates uniformly over the first half of the trip and decelerates uniformly at the same rate over the second half of the trip. (*a*) Find its rate of acceleration and its maximum speed. (*b*) Write the pair of equations describing its position at any time during the trip.

solution (*a*) From Eq. [3 . 26], the position of the train halfway through its trip, at $t = 1.25$ h, is

$$x(t) = \tfrac{1}{2}at^2 = \tfrac{1}{2}a(1.25\text{ h})^2 = 125\text{ mi}$$

Hence during the first half of the trip its acceleration is $a = 160\text{ mi/h}^2$ and during the second half of the trip it is $-a$. With the initial conditions all set to zero, its maximum speed is, from Eq. [3 . 27],

$$v^2 = 2ax \qquad v = \sqrt{2 \times 160 \times 125} = 200\text{ mi/h}$$

(*b*) During the first half of the trip we set the initial conditions to zero to predict the distance x_1 from Washington. Thus for any time during the interval $0 \le t \le 1.25$ h the position of the train is given by Eq. [3 . 25] as

$$x_1(t) = \tfrac{1}{2}at^2$$

For the second half of the trip the initial conditions are $x_0 = 125$ mi and $v_0 = 200$ mi/h. Hence for any time in the interval $1.25 \le t \le 2.50$ h, the distance x_2 from Washington is given by Eq. [3 . 25]:

$$x_2(t) = 125\text{ mi} + (200\text{ mi/h})\,(t - 1.25\text{ h}) - \tfrac{1}{2}a(t - 1.25\text{ h})^2$$

example 3•9 Two particles, P_1 and P_2, move along the same straight line. P_1 has an initial velocity of 15 cm/s east and a constant acceleration of 6 cm/s² east. P_1 starts out from the origin at $t_0 = 0$, and P_2 starts out 3 s later from a point 200 cm east of the origin, with an initial velocity of 7 cm/s east and a constant acceleration of 8 cm/s² west. (*a*) When and

figure 3•6

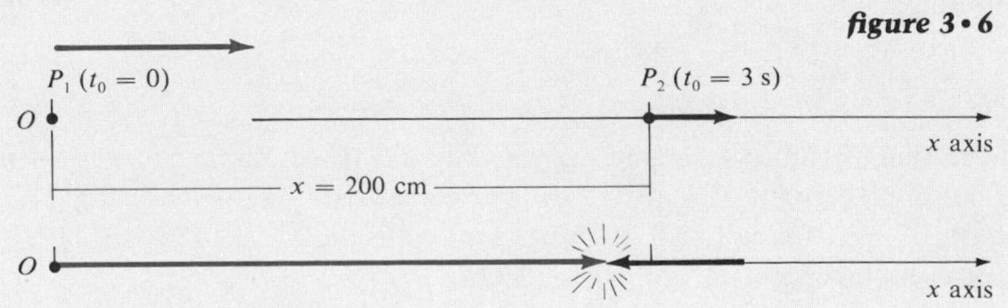

where will the two particles collide? (*b*) What will their velocities be at the time of collision? (*c*) Compute the total distance that particle P_2 will travel and its net displacement at the time of collision.

solution The first step is to sketch the probable situation, as shown. Since both particles must be at the same point with respect to the origin when they collide, we can use Eq. [3 . 25] to locate each particle at the time of collision, $t = t_c$, and then equate these expressions to find t_c. The position of P_1 as a function of time is given by

$$x_1(t) = 0 + (15\text{ cm/s})t + \tfrac{1}{2}(6\text{ cm/s}^2)t^2$$

and for P_2 we obtain

$$x_2(t) = 200\text{ cm} + (7\text{ cm/s})(\text{t} - 3\text{ s}) - \tfrac{1}{2}(8\text{ cm/s}^2)(\text{t} - 3\text{ s})^2$$

Now, setting $t = t_c$ and equating the two positions, $x_1(t_c) = x_2(t_c)$, we have a quadratic equation that we can solve for t_c to obtain $t_c = 5.8$ s (the other root, $t_c = -3.5$ s, has no physical meaning). Substitution of this value into either of the above equations then gives the point of collision with respect to the origin:

$$x_1 = x_2 = 188\text{ cm}$$

(*b*) From Eq. [3 . 22], the velocity of P_1 at time t_c is

$$v_1(t_c) = 15\text{ cm/s} + 6\text{ cm/s}^2 \times 5.8\text{ s} = 49.8\text{ cm/s}$$

and the velocity of P_2 is

$$v_2(t_c) = 7\text{ cm/s} - 8\text{ cm/s}^2 \times (5.8 - 3\text{ s}) = -15.4\text{ cm/s}$$

In other words, P_2 first slowed down to a velocity of zero, and then sped up in the westward direction.

(*c*) To find the total distance traveled by P_2, we must determine its turning point. If we apply Eq. [3 . 27] in the form

$$x - x_0 = \frac{v^2 - v_0^2}{2a}$$

we find the travel distance (in a straight line) from P_2's starting point x_0 to its turning point x' to be

$$x' - x_0 = \frac{0\text{ cm/s} - 7^2\text{ cm/s}}{2(-8\text{ cm/s}^2)} = \frac{7^2}{16}\text{ cm}$$

The distance traveled from the turning point to the collision point is then computed as

$$x' - x_c = \frac{15.4^2}{16}\text{ cm}$$

and the sum of the two distances gives the total travel distance as 16.9 cm. The net displacement of P_2 is the difference between its own starting position, $x_0 = 200$ cm, and its position at collision, $x_c = 188$ cm, or a negative displacement of $\Delta x = -12$ cm.

3 • 5 *integral calculus*

The *integral calculus,* the inverse of the differential calculus we have employed thus far, provides the most powerful and economical means of finding velocities and positions as a function of time. In this section we shall see how the integral calculus can be used to derive the formulas we have just discussed and to find the position of a particle at any time, given its initial velocity and a known rate of acceleration.

As we saw in Example 3 . 6, the area below the velocity curve from a time t_0 to t is equal to the value of the linear displacement during this time interval:

$$A = v_0(t - t_0) + \tfrac{1}{2}a(t - t_0)^2 = x(t) - x_0$$

We shall see that this is true in all cases, not just for constant acceleration.

If we partition the total time from t_0 to t into small segments Δt, as shown in Fig. 3 . 7, then each vertical bar under the velocity curve has an area $\Delta A = v\Delta t$, where v is the value of the instantaneous velocity at the beginning of each interval. For small enough increments of time, $v\ \Delta t$ is a good approximation to the displacement Δx over that particular interval. Hence the sum of all the areas ΔA is approximately equal to

$$A = x(t) - x_0$$

which is the algebraic sum of all the individual displacements Δx, both positive and negative, during the time $t - t_0$. The value obtained in this way includes all but the shaded areas between the velocity curve and the top of each rectangle. If we now take the limit of $v\ \Delta t$ as Δt approaches zero, we obtain

$$\lim_{\Delta t \to 0} v\,\Delta t = v\,dt = dx \qquad [3 \cdot 28]$$

or $v = dx/dt$, which is consistent with our earlier definition of instantaneous velocity. Although the differential notation represents a *limiting process,* one useful feature of differentials is that they can be treated algebraically like ordinary scalars.

It is intuitively obvious that as we take successively smaller intervals of Δt, the shaded areas above the rectangles will vanish, and also that as Δt

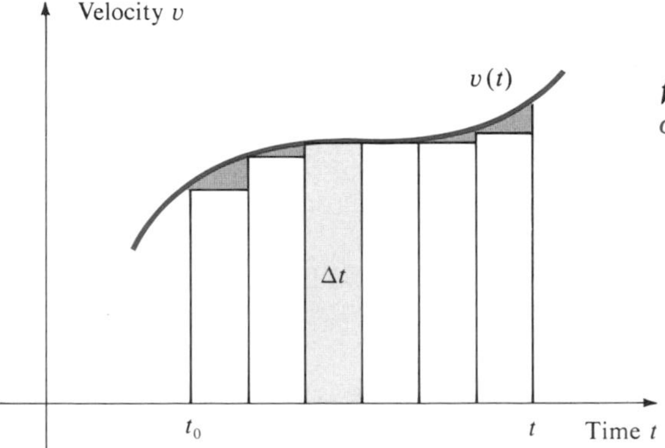

figure 3 • 7 *Integration of the velocity curve over the interval t_0 to t_1*

approaches zero, Δx will approach zero. Therefore

$$\lim_{\Delta t \to 0} \begin{bmatrix} \text{sum of all } v\,\Delta t \\ \text{from } t_0 \text{ to } t \end{bmatrix} = \lim_{\Delta x \to 0} \begin{bmatrix} \text{sum of all } \Delta x \\ \text{from } x_0 \text{ to } x(t) \end{bmatrix}$$

$$= x(t) - x_0 \qquad [3 \cdot 29]$$

or, expressed in the notation of integral calculus,

$$\blacktriangleright \quad \int_{t_0}^{t} v\,dt = \int_{x_0}^{x(t)} dx = x(t) - x_0 \qquad [3 \cdot 30]$$

where the integral sign $\int$ is just an elongated S from *sum*.

If the velocity curve falls below the time axis, then $v(t)$ has a negative value, which implies a negative displacement Δx (motion toward the left on the x axis). For consistency, therefore, we must consider areas ΔA as negative when they are below the horizontal axis of a velocity graph.

The process of integration is the inverse of the process of differentiation. Hence for every rule of differentiation there must be a corresponding rule of integration. The rules corresponding to Eqs. [3 . 14], [3 . 15], and [3 . 16] are as follows:

$$\blacktriangleright \quad \int_{x_1}^{x_2} [f(x) + g(x)]\,dx = \int_{x_1}^{x_2} f(x)\,dx + \int_{x_1}^{x_2} g(x)\,dx \qquad [3 \cdot 31]$$

$$\blacktriangleright \quad \int_{x_1}^{x_2} kf(x)\,dx = k\int_{x_1}^{x_2} f(x)\,dx \qquad k = \text{constant} \qquad [3 \cdot 32]$$

$$\blacktriangleright \quad \int_{x_1}^{x_2} x^{n-1}\,dx = \left.\frac{x^n}{n}\right|_{x_1}^{x_2} = \frac{1}{n}[x_2^n - x_1^n] \qquad n \neq 0 \qquad [3 \cdot 33]$$

Let us now see how use of the integral calculus simplifies our work in deriving the equations of motion for the case of constant acceleration. The *differential equation of motion* for constant acceleration in one dimension is

$$a = \frac{dv}{dt} \qquad \text{or} \qquad dv = a\,dt \qquad [3 \cdot 34]$$

If we simply integrate both sides of the second equation over corresponding intervals $v(t) - v_0$ and $t - t_0$, we obtain

$$v(t) - v_0 = \int_{v_0}^{v(t)} dv = \int_{t_0}^{t} a(t)\,dt$$

$$= a\int_{t_0}^{t} dt = a(t - t_0) \qquad [3 \cdot 35]$$

which is the equivalent of Eq. [3 . 21]. Although we are considering the special case of constant acceleration, the first two integral equations hold even when a is a function of time.

To find the equation for $x(t)$, we first write the velocity in differential form as

$$v = \frac{dx}{dt} \qquad \text{or} \qquad dx = v\,dt = v_0\,dt + a(t - t_0)\,dt \qquad [3 \cdot 36]$$

Since v is a function of time, we can again integrate between the corresponding limits $x(t) - x_0$ and $t - t_0$ to obtain

$$x(t) - x_0 = \int_{x_0}^{x(t)} dx = \int_{t_0}^{t} v(t)\,dt$$

$$= v_0(t - t_0) + \tfrac{1}{2}a(t - t_0)^2 \qquad [3 \cdot 37]$$

which is the equivalent of Eq. [3 . 25]. The x integral is just the linear distancc $x(t) - x_0$, and the t integral is the entire area under the velocity curve, as derived in Example 3 . 6.

By utilizing the algebraic properties of differentials, we can also derive an expression for acceleration as a function of displacement. From the definitions of acceleration in Eq. [3 . 19] and [3 . 20], we obtain

$$a = \frac{dv}{dt} = \frac{dv}{dt}\frac{dx}{dx} = \frac{dx}{dt}\frac{dv}{dx} = v\frac{dv}{dx}$$

or

$$v\,dv = a\,dx \qquad [3 \cdot 38]$$

Integrating both sides of thc last equation then gives

$$\int_{v_0}^{v} v\,dv = \tfrac{1}{2}(v - v_0)^2 = a\int_{x_0}^{x} dx = a(x - x_0) \qquad [3 \cdot 39]$$

which is the equivalent of Eq. [3 . 27]. In this case the v integral is simply the area under a trapezoid varying in height from v_0 to v.

example 3 • 10 A body starting from rest at the origin of coordinates moves along thc x axis with an acceleration given (in mks units) by

$$a = 6t - 3t^2$$

At what time t will it return to the origin and what will its velocity be at that time?

solution Since $t_0 = 0$ and $v_0 = 0$, Eq. [3 . 35] yields

$$v(t) = \int_0^t a(t)\,dt = \int_0^t (6t - 3t^2)\,dt = 3t^2 - t^3$$

Substituting this value of $v(t)$ into Eq. [3 . 37], with $x_0 = 0$, we obtain the integral

$$x(t) = \int_0^t v(t)\,dt = \int_0^t (3t^2 - t^3)\,dt = t^3 - \tfrac{1}{4}t^4$$

Therefore $x(t) = x_0 = 0$ when $t = 4$ s, and the velocity of the body at that time is

$$v(4) = 3 \times 4^2 - 4^3 = -16 \text{ m/s}$$

*3 . 6 numerical analysis of motion

Up to this point we have considered motion in very abstract terms. We have reduced reality to a consideration of points, lines, and calculus, disregarding all the ragged edges, the errors, and the possible discrepancies and inadequacies in our observations and measurements. In short, we have deleted all consideration of those very things which make physics a vital and challenging discipline. But the real world is rarely this tractable, and neat, closed formulas are difficult to find. In actual research the physicist is often called on to interpret a body of data for which there may be no satisfactory theory. When this happens the aid of a computer may be necessary to acquire data, to organize them, and even to develop theories to explain them. In classroom problems as well, numerical calculations are often necessary to verify or illustrate the principles involved. Let us therefore consider here how velocities and accelerations might be derived from position-time data in the *general* rectilinear case, $\mathbf{a} \neq$ constant. We shall construct a step-by-step procedure, or *algorithm,* for performing the calculations. This algorithm may be used for hand calculation or translated into a computer language such as FORTRAN, ALGOL, or BASIC (see Appendix I) for use on an electronic computer. The translated algorithm is known as a *program.*

The data in the second column of Table 3 . 2 are the observed positions of a particle as a function of time. The other columns give the values of the average velocity and the average acceleration determined from the position data over each 1-s time interval. These average values are used as *approximations* to the actual instantaneous values of velocity and acceleration at the midpoints of their respective time intervals. While such approximations must have some error, we shall assume the error to be negligible for our purposes.

table 3 • 2 *Numerical analysis of motion*

t, s	*x, cm*	*v, cm/s*	*a, cm/s²*
0	1.00000		
		−.03407	
1	.96593		−.06583
		−.09990	
2	.86603		−.05902
		−.15892	
3	.70711		−.04819
		−.20711	
4	.50000		−.03407
		−.24118	
5	.25882		−.01764
		−.25882	
6	.00000		.00000
		−.25882	
7	−.25882		.01764
		−.24118	
8	−.50000		.03407
		−.20711	
9	−.70711		.04819
		−.15892	
10	−.86603		.05902
		−.09990	
11	−.96593		.06583
		−.03407	
12	−1.00000		

The average values for velocity and acceleration were computed according to the defining formulas given in Eqs. [3 . 10] and [3 . 17]. This computational procedure may be diagrammed as a *flowchart* showing each of the steps actually performed. Imagine that you have been provided with a careful, precise, but simple-minded assistant who can do only exactly as told. The flowchart represents the set of instructions you must provide your helper and is always the first step in coaxing a computer into doing your bidding.

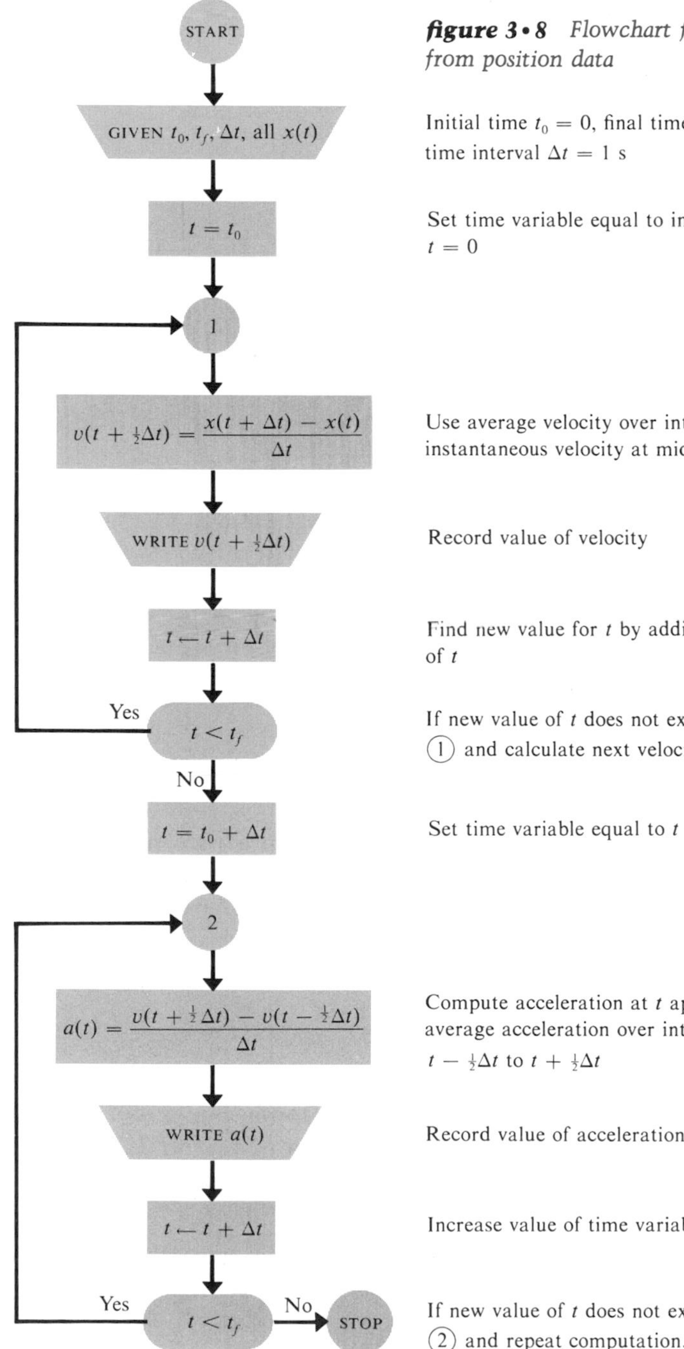

figure 3•8 *Flowchart for numerical computation of velocity from position data*

Initial time $t_0 = 0$, final time $t_f = 12$ s, time interval $\Delta t = 1$ s

Set time variable equal to initial value, $t = 0$

Use average velocity over interval to approximate instantaneous velocity at midpoint of interval

Record value of velocity

Find new value for t by adding Δt to old value of t

If new value of t does not exceed t_f, return to (1) and calculate next velocity value

Set time variable equal to $t = 1$ s

Compute acceleration at t approximated by average acceleration over interval $t - \frac{1}{2}\Delta t$ to $t + \frac{1}{2}\Delta t$

Record value of acceleration

Increase value of time variable by $\Delta t = 1$ s

If new value of t does not exceed t_f, return to (2) and repeat computation. When t exceeds t_f, stop.

The flowchart symbols are fairly standard conventions. A *trapezoid* represents input or output information and a *lozenge* is used for conditional or "if . . . then" statements whose outcome determines which path of the flowchart the subsequent computation follows. The *circles* denote reference points within the program or the end of the program, and *rectangles* are used for nearly everything else. Some additional flowcharting conventions are described in Appendix H. The flowchart in Fig. 3 . 8 shows how you might go about computing the velocity and acceleration from position data. The algorithm could be made more efficient for use on a computer by combining the computations for v and a, but this is a bit tricky. Can you see why?

The motion described by the data in Table 3 . 2 is an example of *simple harmonic motion,* in which the particle oscillates back and forth along the x axis like a pendulum. The "data" for x were actually taken from the formula $x(t) = \cos(\pi t/12)$ for simplicity, although any arbitrary collection of laboratory data would have served equally well.

In this table we should find our worst approximation errors to be 0.29 percent for $v(t)$ and 0.59 percent for $a(t)$. Had we taken the interval $\Delta t =$ 2 s between observations, the maximum errors would have been 1.15 percent and 2.33 percent, respectively. Hence the approximation used here is called a *first-order method,* because it is correct to first order in Δt; that is, the errors are proportional to the square of the interval size, $(\Delta t)^2$.

PROBLEMS

Note: Problems denoted by an asterisk are especially applicable to machine computation.

3 • 2–3 • 3 speed and velocity

3 • 1 Show that the result of Example 3 . 2 is the same if the particle starts at some point $s_0 \neq 0$. What if it starts at some time $t_0 \neq 0$?

3 • 2 If the length of a railroad rail is 33 ft, show that the speed of a train, in miles per hour, is numerically equal to three-eighths the number of rails passed over per minute.

3 • 3 A particle starts from a point O and moves in such a way that its travel distance s is given by $s = (20\ \text{cm/s}^2)t^2$. (*a*) Assuming the numerical coefficient 20 to be accurate to four significant figures, compute the average speed $\bar{v}$ of the particle over each of the following time intervals:

$$2.0 \le t \le 2.1\ \text{s} \qquad 2.00 \le t \le 2.01\ \text{s} \qquad 2.000 \le t \le 2.001\ \text{s}$$

(*b*) What is the instantaneous speed v at $t = 2.0$ s? (*c*) What general conclusion can you draw from this problem?

3 • 4 A skier drives 100 mi from Pasadena, California, to Big Bear Lake in 2 h. The return trip two days later takes the usual 6 h. (*a*) What is his average speed going? (*b*) His average speed returning? (*c*) His average speed for the entire trip? (*d*) His average velocity for the entire trip?

3 • 5 A particle moves in a straight line to the left with a constant speed v_1 for a time t_1, and then it moves with a speed v_2 for a time t_2. What is its average speed over the entire travel time?

3 • 6 The position of a moving particle with respect to the origin is given by

$$\mathbf{x}(t) = \left(3t^4 + 2t - \frac{10}{t + 3}\right)\mathbf{i}$$

(*a*) What is the instantaneous velocity of the particle at $t = 0$? (*b*) At $t = 2$ s? (*c*) At $t = 7$ s?

3 • 7 The positions of two particles, P_1 and P_2, are given by

$$\mathbf{x}_1(t) = (5 + 3t + 2t^2)\,\mathbf{i} \qquad \mathbf{x}_2(t) = (1 - t + 5t^2)\,\mathbf{i}$$

(*a*) At what time t_c do the two particles collide? (*b*) What is the difference in their velocities at this instant?

3 • 8 One particle moves along the x axis at a speed of $v_x = -v_0 \cos 2t$, while another particle travels on the y axis at a speed of $v_y = v_0 \sin 2t$. What is their mutual speed of separation at any given instant as a function of time?

3 • 9 The position of a particle moving along the y axis is given (in centimeters) by

$$y = t^2 - t$$

(*a*) What is its average speed over the interval $0 \leq t \leq 1$ s? (*b*) Over the interval $0 \leq t \leq 2$ s? (*c*) What is its average velocity over some interval $0 \leq t \leq T$ s?

3 • 4 acceleration

3 • 10 Compute the average acceleration of the particle in Prob. 3 . 6 over each of the three 2-s time intervals from $t = 0$ to $t = 6$ s.

3 • 11 Convert the acceleration rate $a = 1{,}000{,}000$ mi/y^2 into centimeters per second squared.

3 • 12 A car traveling at a velocity of 33 mi/h westward suddenly starts to lose speed at a constant rate, and 3 s later its velocity is 15 mi/h westward. (*a*) How long after the speed starts to drop will it take the car to come to a stop? (*b*) What is its total stopping distance? (*c*) What would the stopping time and distance be with the same acceleration, but with an initial speed of 66 mi/h?

3 • 13 A bullet acquires a speed of 185 m/s while traversing a revolver barrel 20.5 cm long. What is its average acceleration (*a*) in meters per second per second? (*b*) In meters per minute per second? (*c*) In meters per second per minute? (*d*) In feet per minute per minute?

3 • 14 An object is dropped from a height of 30 m and falls with a constant acceleration of 9.8 m/s^2 under the influence of gravity. (*a*) How long does it take the object to reach the ground? (*b*) What is its speed on impact?

3 • 15 At a certain point on a straight road the speedometer of an automobile registered 12 mi/h, and at a point 650 ft farther along the road it registered 45 mi/h. (*a*) Assuming that the acceleration was constant, compute the travel time for the 650-ft run and the rate of acceleration. (*b*) Using a precision stopwatch, the driver had found that his actual travel time was

18 s. To test his speedometer, he drove the car over a 1-mi speedometer test section on this road and found that, with the speedometer kept at a constant reading of 45 mi/h, his travel time for the mile was 86 s. Compute the actual initial speed and acceleration of the car in the 650-ft run.

3 • 16 A freight train has a maximum speed of 50 km/h. At constant acceleration, it can attain this speed in 5 min. It also decelerates at the same rate. Compared with a nonstop run of 70 km, how much time is added to the trip if the train makes a 1-min stop midway in the journey?

3 • 17 A car passes an intersection at a speed of 72 km/h and continues on at this rate. Five seconds later a traffic officer waiting at the intersection starts after the car with a constant acceleration of 2 m/s^2. (*a*) When and where will the traffic officer overtake the car? (*b*) How fast will the traffic officer be driving at that point?

3 • 18 A particle is moving southward with a constant acceleration of 5 cm/s^2 north. (*a*) Compute the time required for the particle to move a distance of 48 cm if its velocity after 24 cm is 17 cm/s south. (*b*) For what additional time and through what additional distance will the particle move before coming to rest?

3 • 19 The graph below shows the positions of a moving particle as a function of time. (*a*) Sketch the velocity curve for the same time intervals (see Fig. 3 . 4). (*b*) Sketch the corresponding graph of acceleration. (*c*) Describe the motion of the particle in words.

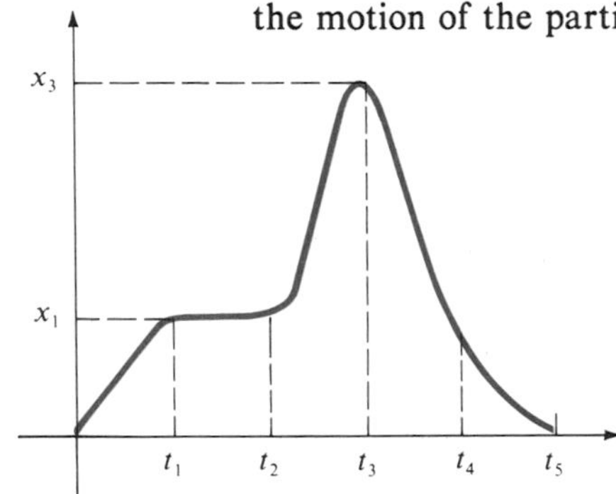

3 • 20 A particle moving on the x axis has the following positions as a function of time:

t, s	0	1	2	3	4
x, m	3	0	−1	0	3

(*a*) Find the average velocity in each 1-s time interval. (*b*) Find the average acceleration in each time interval. (*c*) Write an analytic expression for $\mathbf{x}(t)$ which fits the data in the table.

3 • 21 A particle moving along the x axis leaves the origin at time $t_0 = 0$ and returns to it at a later time T. (*a*) If the rate of acceleration is constant during this entire motion, what is the equation for $x(t)$? (*b*) What is the value of $x(t)$ at time $2T$?

3 • 22 A hare and a tortoise are having a footrace over a 440-m course. The armor-plated competitor takes off in a cloud of dust and gravel; his furry adversary nonchalantly departs from the starting line 20 min later. The hare, steadily accelerating at a constant rate over the entire distance, just loses to the tortoise, who, maintaining a steady pace over the 440 m, wins by a nose in 25 min. (*a*) What was the speed of the tortoise? (*b*) What was the acceleration of the hare? (*c*) What was the final speed of the hare across the finish line?

3 • 23 A car is braked while traveling at a speed of 30 mi/h. (*a*) What is its acceleration if it stops in 10 s? (*b*) If it stops in 44 ft?

3 • 24 An object moves from point A to point B with constant acceleration. Show that $v = \bar{v}$ at the point midway in *time* between A and B, whereas $v^2 > \bar{v}^2$ at the point midway in *distance* between A and B.

3 • 25 A body starts from rest and moves along a straight line with a constant acceleration of 2.5 $\mathrm{cm/s^2}$ for 1 min. It moves with constant velocity for another minute and then slows uniformly to a halt over a period of 1.5 min. (*a*) How far did the body move? (*b*) What was its average acceleration over the entire motion? (*c*) What was its average speed?

3 • 5 *integral calculus*

3 • 26 Use geometry to find the area under the curve given by $y = x$ over the region $x_1 \leq x \leq x_2$ of the x axis and compare your results with the left-hand integral of Eq. [3 . 39]. Can you suggest why the x variable, which appears as a differential dx in the integral $\int x\,dx$, is sometimes called a "dummy variable?"

3 • 27 Use integral calculus to find (*a*) the area under the parabola $y = 2(2x - 3)^2$ from $1 \leq x \leq 2$. (*b*) The area under the parabola $y = 2(2x - 5)^2$ from $2 \leq x \leq 3$. (*c*) The area under the parabola $y = x^2 - 3$ from $0 \leq x \leq 3$.

3 • 28 The graph below describes the acceleration of a particle as it moves along the x axis. (*a*) Sketch the corresponding graph of the particle's velocity. (*b*) Sketch the graph of the particle's position as a function of time. (*c*) Describe the motion in words. Did you have to make any assumptions in sketching the velocity and position curves?

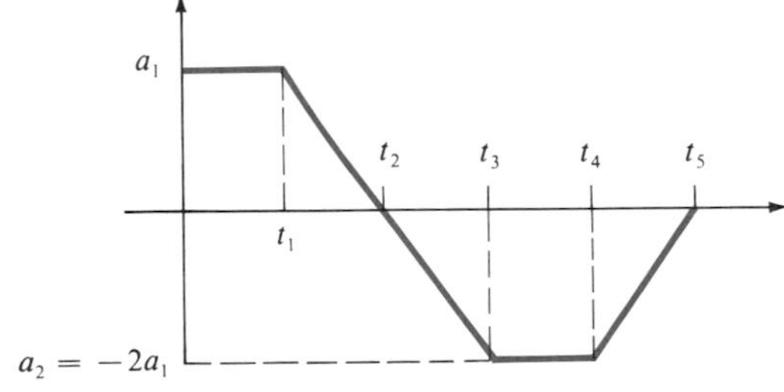

3 • 29 Assume that the equation of motion for the particle in Prob. 3 . 28 is $dv/dt = a$, where $a_1 = 2\ \mathrm{m/s^2}$ and t has the values

$$t_1 = 2\ \mathrm{s} \qquad t_3 = 5\ \mathrm{s} \qquad t_4 = 6\ \mathrm{s}$$

If the particle starts from rest, what is its speed at time t (*a*) for the interval $0 \leq t \leq 2$ s? (*b*) For the interval $2 \leq t \leq 5$ s? (*c*) For the interval $5 \leq t \leq 6$ s?

3 • 30 If the initial position of the particle in Prob. 3 . 29 is $x_0 = 0$, use your solutions for $v(t)$ to find its position $x(t)$ for each of the given time intervals.

3 • 31 A particle starts from rest at $x_0 = 2$ and moves in such a way that its rate of acceleration is given by

$$a(t) = \frac{d^2x}{dt^2} = \frac{dv}{dt} = 4t - 4t^3 \qquad t \geq 0$$

(*a*) Find the velocity of the particle as a function of time. (*b*) Find its position as a function of time. (*Hint:* Use the algebraic properties of differentials.)

3 • 32 A particle moves along the x axis in such a way that its acceleration rate is related to its position $x(t)$ by a *linear restoring force,*

$$\frac{d^2x}{dt^2} = \frac{dv}{dt} = -\alpha^2 x$$

where α is a positive constant. Assume that the particle leaves the origin with an initial velocity $v_0 = V$. Find the relationship between v and x and describe the motion of the particle. (*Hint:* Use the algebraic properties of differentials.)

3 • 33 Using the general formulas for integrals, find $\int_1^3 P(u)\,du$, where P is a function of u defined as $P(u) = 2u^4 - 3/u^2$.

3 • 6 numerical analysis

***3 • 34** A particle is observed at regular time intervals and found at the following positions:

t, s	0.0	0.5	1.0	1.5	2.0	2.5	3.0	3.5	4.0
x, m	0.0	0.61	0.18	2.13	3.63	6.05	10.02	16.54	27.29

Compute the average speeds and accelerations of the particle (*a*) over 1-s time intervals. (*b*) Over 0.5-s time intervals. (*c*) What is the largest percentage difference between corresponding values for (*a*) and (*b*)? How do you interpret this?

***3 • 35** Construct the flowchart for an algorithm to compute and print the positions of a falling body at intervals of time Δt, if the body starts from rest at a height h and has an acceleration rate g. Assume that the computer can be instructed to print out letter symbols, as well as numerical values of specified quantities. Instruct your computer to print out a table of values with appropriate column headings. Designate letter symbols in your flowchart by quotation marks—"HEIGHT," "TIME," etc.—and use parentheses to indicate any lines to be left blank in the output; for example, "(SKIP 3 LINES)."

***3 • 36** As a numerical exercise, evaluate $\Delta y = \Delta(x^3)$ over small intervals Δx to confirm that $\Delta y/\Delta x$ is approximately equal to $dy/dx = 3x^2$, as predicted by Eq. [3 . 16]. Then compute the percentage of error in your approximation.

answers

3•3 (*a*) 82.00 cm/s; (*b*) 80.20 cm/s; (*c*) 80.02 cm/s

3•4 (*a*) 50 mi/h; (*b*) 16.7 mi/h; (*c*) 25 mi/h; (*d*) 0

3•5 $\bar{v} = (v_1 t_1 + v_2 t_2)/(t_1 + t_2)$

3•6 (*a*) 3.1111; (*b*) 98.400; (*c*) 4118.1

3•7 (*a*) $t_c = 2$; (*b*) 8

3•8 v_0

3•9 (*a*) $\bar{v} = 0.5$ cm/s; (*b*) $\bar{v} = 1.25$ cm/s; (*c*) $\bar{\mathbf{v}} = (T - 1)\mathbf{j}$ cm/s

3•10 47.64, 335.9, 912.0

3•11 1.62×10^{-4} cm/s^2

3•12 (*a*) $t = 5.5$ s; (*b*) 133 ft; (*c*) 110 s, 532 ft

3•13 (*a*) 8.35×10^4 m/s^2; (*b*) 5.01×10^6 m/min·s; (*c*) 5.01×10^6 m/s·min; (*d*) 9.86×10^8 ft/min^2

3•14 (*a*) 2.47 s; (*b*) 24.2 m/s

3•15 (*a*) 15.6 s, 3.11 ft/s^2; (*b*) 7.37 mi/h, 2.81 ft/s

3•16 6 min

3•17 (*a*) 24.14 s after he starts, 582.8 m from intersection; (*b*) $v = 173.8$ km/h

3•18 (*a*) 3.2 s; (*b*) 1.4 s later and 4.9 cm farther

3•20 $\bar{v} = 3, -1, 1, 3$ m/s; $\bar{a} = +2$ m/s^2

3•21 (*a*) $x(t) = \frac{1}{2}at(t - T)$; (*b*) $x(t) = aT^2$

3•22 (*a*) 29.3 cm/s; (*b*) 0.978 cm/s^2; (*c*) 293 cm/s

3•23 (*a*) -4.4 ft/s^2; (*b*) -22 ft/s^2

3•25 (*a*) 202.5 m; (*b*) 0; (*c*) 96.4 cm/s

3•26 $A = \frac{1}{2}(x_2^2 - x_1^2)$

3•27 $A = \frac{2}{3}$; (*b*) $A = \frac{2}{3}$; (*c*) 0

3•29 (*a*) $v = 2t$; (*b*) $v = 6t - t^2 - 4$; (*c*) $v = 21 - 4t$

3•30 (*a*) $x = t^2$; (*b*) $x = 3t^2 - t^3/3 - 4t + \frac{8}{3}$; (*c*) $x = 21t - 2t^2 - 39$

3•31 (*a*) $v = 2t^2 - t^4$; (*b*) $x = 2 + 2t^3/3 - t^5/5$

3•32 $v^2 = V^2 - \alpha x^2 > 0$; the motion repeats itself over the interval $-V/\alpha \leq x \leq V/\alpha$

CHAPTER FOUR

particle kinematics: II

It has been observed that missiles and projectiles describe a curved path of some sort; however no one has pointed out the fact that this path is a parabola. But this and other facts, not few in number or less worth knowing, I have succeeded in proving; and what I consider more important, there have been opened up to this vast and most excellent science, of which my work is merely the beginning, ways and means by which other minds more acute than mine will explore its remote corners.

Galileo Galilei, Two New Sciences, *Third Day, 1638*

In this chapter we shall extend the ideas of one-dimensional particle kinematics to three-dimensional space. After generalizing the vector concepts developed in the last chapter, we shall utilize vector algebra to treat the classical problem of freely falling bodies—bodies moving only under the influence of gravity. Finally, we shall employ the same techniques to describe the motion of a particle as viewed by a moving observer, a form of description known as *Galilean relativity.*

4 • 1 motion in three dimensions: velocity

In Sec. 3 . 3 we defined the displacement of a particle moving in a straight line as the net change in its position vector during a time interval Δt; we then defined velocity as the time rate of this change. Let us now see how these concepts apply to motion in more than one spatial dimension. Suppose a particle is moving along the curved path in Fig. 4 . 1, and at some time t_0 its position with respect to an observer at O is specified by the radial position vector $\mathbf{r}_0$. When the particle is next observed at time t_1, it is at a point specified by the position vector $\mathbf{r}_1$. The difference between these two position vectors is its *linear displacement* $\Delta\mathbf{r}$, represented by the difference vector

$$\Delta\mathbf{r} = \mathbf{r}_1 - \mathbf{r}_0 \qquad [4 \cdot 1]$$

The displacement $\Delta\mathbf{r} = \mathbf{r}_1 - \mathbf{r}_0$ is, by definition, the net change of position that takes place in the time interval $\Delta t = t_1 - t_0$, and the ratio of these two quantities is the average velocity $\bar{\mathbf{v}}$ during this interval:

$$\bar{\mathbf{v}} = \frac{\mathbf{r}_1 - \mathbf{r}_0}{t_1 - t_0} = \frac{\Delta\mathbf{r}}{\Delta t} \qquad [4 \cdot 2]$$

The average velocity is therefore a vector pointing in the direction of the linear displacement $\Delta\mathbf{r}$.

In the case of motion in one direction along a straight path, the travel distance and the displacement coincide; hence the magnitude of the average velocity is simply the average speed. Without this restriction, however, we must assume that the displacement of a particle and its path of motion will not necessarily have the same length. For this reason, we must define an instantaneous velocity $\mathbf{v}$, which enables us to find the precise position of the particle as a function of time. As before, we take the average velocity over vanishingly small increments $\Delta\mathbf{r}/\Delta t$, and in the limit as Δt approaches zero, the average velocity approaches its instantaneous value at a specific time t:

$$\mathbf{v}(t) = \lim_{\Delta t \to 0} \frac{\Delta\mathbf{r}}{\Delta t} = \frac{d\mathbf{r}}{dt} \qquad [4 \cdot 3]$$

In other words, as t_1 approaches t_0, the position vector $\mathbf{r}_1$ approaches $\mathbf{r}_0$, and the direction of the displacement $\Delta\mathbf{r}$ approaches that of a line *tangent* to the curve at that point. Since the velocity vector has the same direction as the displacement vector, the instantaneous-velocity vector also coincides with the tangent at any position on the path (see Fig. 4 . 2). Thus instantaneous velocity is the *first derivative of position with respect to time,* and its magnitude at each *point* on the path is equal to the instantaneous speed at that point.

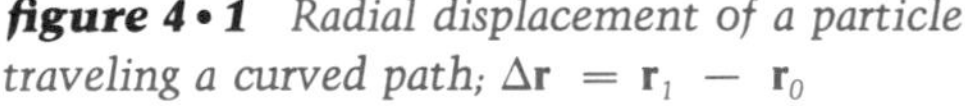

figure 4 • 1 *Radial displacement of a particle traveling a curved path;* $\Delta\mathbf{r} = \mathbf{r}_1 - \mathbf{r}_0$

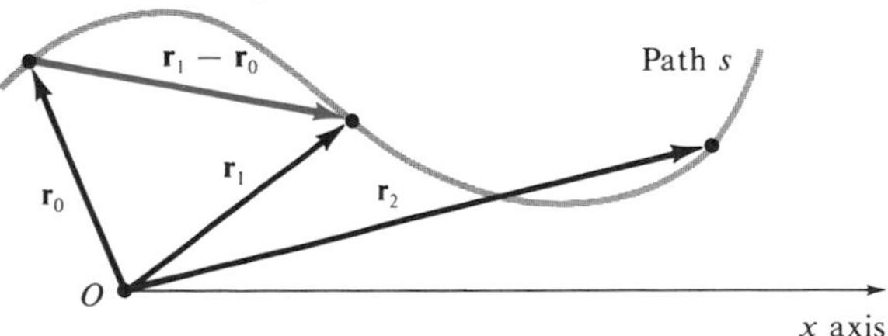

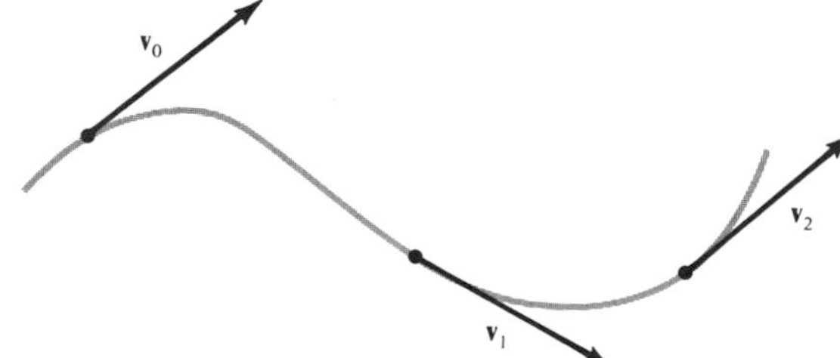

figure 4 • 2 *Instantaneous-velocity vectors of a particle traveling a curved path*

As we saw in Chapter 2, we can locate a vector in three-dimensional space by specifying its components in cartesian coordinates. For example, a velocity vector **v** can be considered the sum, or resultant, of three independent vectors:

$$\mathbf{v} = v_x\mathbf{i} + v_y\mathbf{j} + v_z\mathbf{k} \qquad [4 \cdot 4]$$

From the rules of vector algebra in Sec. 2 . 2, if we represent a position vector $\mathbf{r}(t)$ as the sum of its direction components,

$$\mathbf{r}(t) = x(t)\mathbf{i} + y(t)\mathbf{j} + z(t)\mathbf{k}$$

we can derive each scalar component of the corresponding velocity $\mathbf{v}(t)$ independently from the scalar components of $\mathbf{r}(t)$:

$$v_x(t) = \frac{dx}{dt} \qquad v_y = \frac{dy}{dt} \qquad v_z = \frac{dz}{dt} \qquad [4 \cdot 5]$$

These scalar quantities can then be multiplied by the appropriate unit vectors and added, as in Eq. [4 . 4], to obtain the resultant velocity $\mathbf{v}(t)$.

example 4 • 1 A bullet is fired with a muzzle speed of 500 m/s from a gun pointed eastward and elevated 31° from the horizontal. Find the horizontal and vertical components of the muzzle velocity **v** and write an expression for **v**.

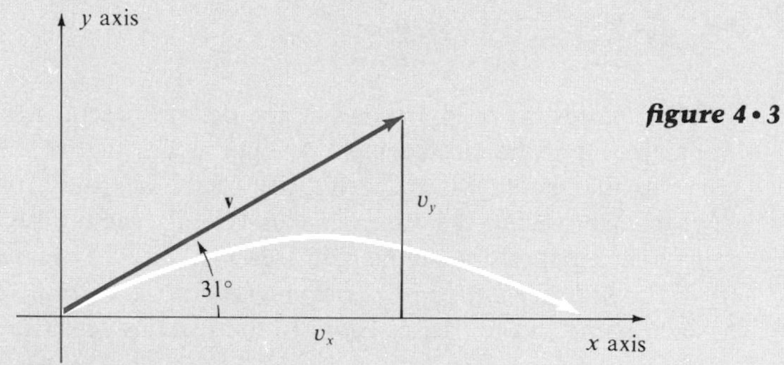

figure 4 • 3

solution If we take the gun muzzle as the origin of coordinates and eastward as the x direction, then the x and y components of the velocity vector are

$$v_x = 500 \cos 31° = 428.6 \text{ m/s} \qquad v_y = 500 \sin 31° = 257.5 \text{ m/s}$$

The total velocity **v** is therefore

$$\mathbf{v} = (v_x\mathbf{i} + v_y\mathbf{j}) \text{ m/s} = (428.6\mathbf{i} + 257.5\mathbf{j}) \text{ m/s}$$

example 4 • 2 The Cascade railroad tunnel in the state of Washington is 8 mi long and runs downhill from northeast to southwest at a 1.5 percent grade. If a train is heading southwest through the tunnel at an average speed of 50 mi/h, describe its average velocity in cartesian coordinates, where the positive x axis is east, the y axis is north, and the z axis is up. (Sketch the coordinate system to orient yourself.)

solution Assume that the train starts at the origin of coordinates and comes out of the tunnel at some point (x,y,z). Since the train is heading southwest, we know that its x and y components are the same and that both are negative. Furthermore, the train travels 8 m down a 1.5 percent grade, and so $z = -0.015 \times 8 \text{ mi} = -0.12$ mi. From Pythagoras' theorem,

$$x^2 + y^2 + z^2 = 64 \text{ mi}^2$$

In this case $z^2 = 0.0144 \text{ mi}^2$ is negligible, and since $x = y$, we can solve $2x^2 = 64 \text{ mi}^2$ to obtain $x = 5.66$ mi and $y = 5.66$ mi. The point at which the train comes out of the tunnel is therefore given by the position vector

$$\mathbf{r}(t) = (-5.66\mathbf{i} - 5.66\mathbf{j} - 0.12\mathbf{k}) \text{ mi}$$

Since the travel time is 8 mi at 50 mi/h, or $t = 0.16$ h, the average velocity is approximately

$$\bar{\mathbf{v}} = \frac{\Delta\mathbf{r}}{\Delta t} = \frac{\mathbf{r}(t) \text{ mi}}{0.16 \text{ h}} = -(35.4\mathbf{i} + 35.4\mathbf{j} + 0.75\mathbf{k}) \text{ mi/h}$$

Using the rules of vector algebra, you should be able to derivc the following useful identities (use cartesian component notation):

$$\frac{d(\mathbf{A} + \mathbf{B})}{dt} = \frac{d\mathbf{A}}{dt} + \frac{d\mathbf{B}}{dt} \qquad [4 \cdot 6]$$

$$\frac{d(f\mathbf{A})}{dt} = f\frac{d\mathbf{A}}{dt} + \frac{df}{dt}\mathbf{A} \qquad [4 \cdot 7]$$

There is also an important theorem of calculus which states that every function $f(x)$ has an *extremum*—a local maximum or a local minimum—at every value of x for which its first derivative vanishes and its second derivative is nonzero. That is, if at $x = x_0$,

$$\frac{df}{dx} = 0 \qquad \frac{d^2f}{dx^2} \neq 0 \qquad \text{then } f(x_0) = \text{max or min} \qquad [4 \cdot 8]$$

We shall assume throughout that $d^2f/dx^2 \neq 0$ unless otherwise stated.

4 • 2 *motion in three dimensions: acceleration*

The acceleration of a particle is defined as the *time rate of change of velocity.* Like velocity, acceleration is the product of a vector and a scalar and is therefore a vector quantity, having both magnitude and direction. However, whereas velocity always has the same direction as the displacement, this is not true of acceleration. Since acceleration represents a *change* in the velocity vector, this change may be in either the magnitude or the direction of the velocity, or both. Thus there are three possible cases of acceleration: a change only in the *speed* of motion, a change only in the *direction* of motion, or a change in both the *speed* and the direction of motion.

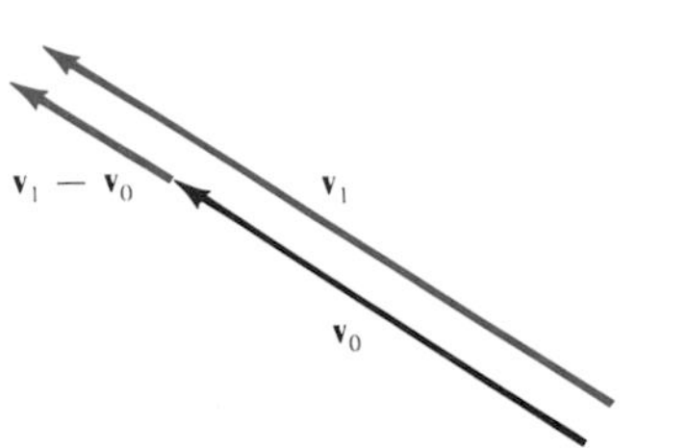

(a) Change in speed, $\hat{\mathbf{r}}$ = *constant*

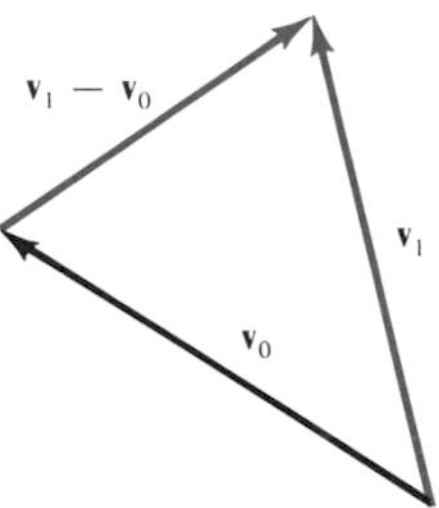

(b) Change in direction, v = *constant*

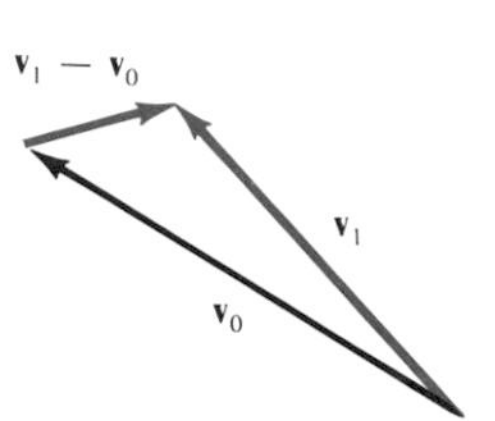

(c) Change in both speed and direction

figure 4 • 4 *Changes of velocity as the vector difference between* $\mathbf{v}_1$ *and* $\mathbf{v}_0$

Case 1 A change in the speed of motion, with no change in direction, represents a change only in the magnitude of the velocity vector, as illustrated in Fig. 4 . 4*a*. Hence the average acceleration, or time rate of change from the initial velocity $\mathbf{v}_0$ to a later velocity $\mathbf{v}_1$, is expressed by the scalar equation

$$\bar{a} = \frac{v_1 - v_0}{t_1 - t_0} = \frac{\Delta v}{\Delta t} \qquad [4 \cdot 9]$$

The instantaneous acceleration at the beginning of this interval is then given by

$$a(t) = \lim_{t_1 \to t_0} \frac{v_1 - v_0}{t_1 - t_0} = \lim_{\Delta t \to 0} \frac{\Delta v}{\Delta t} = \frac{dv}{dt} \qquad [4 \cdot 10]$$

Here the acceleration vector points in the same direction as the velocity, since the change is an increase in speed. Where the acceleration represents a decrease in speed—that is, when the magnitude of $\mathbf{v}_1$ is less than the magnitude of $\mathbf{v}_0$—the acceleration vector will be *antiparallel* to the velocity vector.

Case 2 Acceleration may also change the direction of motion, with no change in speed. In Fig. 4 . 4*b* the velocity vector $\mathbf{v}_1$ has the same magnitude as the initial velocity $\mathbf{v}_0$, and the difference vector $\Delta\mathbf{v} = \mathbf{v}_1 - \mathbf{v}_0$ represents a change only in the direction of the velocity. If we define the average acceleration as $\bar{\mathbf{a}} = \Delta\mathbf{v}/\Delta t$, we obtain an acceleration vector pointing in the direction of $\mathbf{v}_1 - \mathbf{v}_0$. If we then take the limit of this quantity as t_1 approaches t_0, the difference vector becomes increasingly perpendicular to $\mathbf{v}_0$, and as $\Delta\mathbf{v}/\Delta t$ vanishes to zero, we find that the instantaneous acceleration $\mathbf{a}$ is at *right angles* to the instantaneous velocity vector at time t_0:

$$\mathbf{a} = \lim_{t_1 \to t_0} \frac{\Delta\mathbf{v}}{\Delta t} \qquad \mathbf{a} \cdot \mathbf{v} = 0 \qquad [4 \cdot 11]$$

This is the situation when a particle moves uniformly in a circle. Although there is no change in the magnitude of its velocity vector, there is a constant change in its direction; hence at every point along the circumference the acceleration vector has a direction perpendicular to the velocity.

Case 3 The most general case of acceleration is a change in both the speed and the direction of motion, as illustrated in Fig. 4 . 4*c*. In this situation $\mathbf{v}_1$ differs from $\mathbf{v}_0$ in both magnitude and direction, and the difference vector

$\mathbf{v}_1 - \mathbf{v}_0$ represents both changes. A typical example is a car that changes speed as it rounds a curve. Then, for any instantaneous change in velocity, the instantaneous acceleration is defined by the vector equation

$$\mathbf{a}(t) = \lim_{\Delta t \to 0} \frac{\Delta \mathbf{v}}{\Delta t} = \frac{d\mathbf{v}}{dt} \qquad [4 \cdot 12]$$

This constitutes the most general definition of acceleration possible; Eqs. [4 . 10] and [4 . 11] are special cases, both of which can be derived from Eq. [4 . 12].

As we have seen, instantaneous acceleration is the first derivative of velocity with respect to time, and it is also the second derivative of position with respect to time:

$$\blacktriangleright \quad \mathbf{a}(t) = \frac{d\mathbf{v}}{dt} = \frac{d}{dt}\left(\frac{d\mathbf{r}}{dt}\right) = \frac{d^2\mathbf{r}}{dt^2} \qquad [4 \cdot 13]$$

Thus any acceleration vector can be found by differentiating the scalar components of either velocity or position:

$$a_x = \frac{dv_x}{dt} = \frac{d^2x}{dt^2} \qquad a_y = \frac{dv_y}{dt} = \frac{d^2y}{dt^2} \qquad a_z = \frac{dv_z}{dt} = \frac{d^2z}{dt^2} \qquad [4 \cdot 14]$$

example 4•3 If the position of a particle with respect to some origin is given by the position vector

$$\mathbf{r}(t) = t^2\mathbf{i} - 4t\mathbf{j} + (2 + t^3)\,\mathbf{k}$$

find its velocity and acceleration.

solution We can find the velocity components by differentiating each component of $\mathbf{r}$ separately to obtain

$$v_x = \frac{d(t^2)}{dt} = 2t \qquad v_y = \frac{d(-4t)}{dt} = -4 \qquad v_z = \frac{d(2 + t^3)}{dt} = 3t^2$$

The velocity is then the vector sum of these components:

$$\mathbf{v} = 2t\mathbf{i} - 4\mathbf{j} + 3t^2\mathbf{k}$$

To find the acceleration, we take derivatives of the components of $\mathbf{v}$ with respect to time:

$$a_x = \frac{d(2t)}{dt} = 2 \qquad a_y = \frac{d(-4)}{dt} = 0 \qquad a_z = \frac{d(3t^2)}{dt} = 6t$$

Their vector sum is the acceleration $\mathbf{a} = 2\mathbf{i} + 6t\mathbf{k}$, which has no y component and a constant x component.

In the important special case of *constant* acceleration,

$$\mathbf{a} = a_x\mathbf{i} + a_y\mathbf{j} + a_z\mathbf{k} \qquad a_x, a_y, a_z = \text{constant}$$

we can write the three component equations of $\mathbf{a} = d\mathbf{v}/dt$ in differential form as

$$dv_x = a_x dt \qquad dv_y = a_y dt \qquad dv_z = a_z dt$$

and integrate each equation separately over corresponding intervals of $\mathbf{v} - \mathbf{v}_0$ and $t - t_0$. The vector sum of these components then yields the velocity equation

$$\mathbf{v} = \mathbf{v}_0 + \mathbf{a}t \qquad [4 \cdot 15]$$

where $\mathbf{v}_0$ is the initial velocity and $t_0 = 0$. Similarly, if we express velocity as $\mathbf{v} = d\mathbf{r}/dt$ and integrate

$$\frac{d\mathbf{r}}{dt} = \mathbf{v}_0 + \mathbf{a}t \qquad [4 \cdot 16]$$

we obtain three component equations for the position vector which sum to the single vector relation

$$\mathbf{r} = \mathbf{r}_0 + \mathbf{v}_0 t + \tfrac{1}{2}\mathbf{a}t^2 \qquad [4 \cdot 17]$$

where $\mathbf{r}_0$ is the initial position at $t_0 = 0$.

4 · 3 *freely falling bodies and projectiles*

One of the first problems the physicist faces in analyzing a new phenomenon is determining *what* to measure; the next is devising ways to measure it. A good example is the relatively simple and now thoroughly familiar problem of a freely falling body. It is a matter of ordinary observation that a stone, a feather, and a snowflake fall through the air at different rates; however, it is not obvious that these different rates result from differences in air resistance and have nothing to do with the weight of the object. Moreover, objects fall so rapidly that with early measuring techniques, experiments were difficult to devise. Clarification of the fundamental problem and formulation of the laws of falling bodies required a person of Galileo's insight and ingenuity.

Galileo's pioneer study of motion was published in 1638. His primary concern was the flight of projectiles, a problem of as much practical and political importance then as now. He reasoned that such flight might be regarded as consisting of two separate motions—one the projectile's motion in a horizontal direction and the other its initial rise and subsequent fall back to earth. To analyze the vertical component of this motion, Galileo first sought ways to describe the fall of a body. He had no way to produce a vacuum, but by experimenting with various objects in a series of fluids of different densities, he observed that as the density of the fluid decreased, all objects fell more nearly at the same rate. From this he concluded that in a medium totally devoid of resistance, all bodies, regardless of their weights, would attain the same speed when dropped from the same height.

In the seventeenth century there was no way to measure speeds of fall, and the most accurate instrument available for measuring small increments of time was the water clock, invented by the Babylonians or Egyptians some 3000 years earlier. Galileo therefore did what every theoretical physicist has done in a similar situation; he tried to deduce some other relation that could be tested instead. After experimenting with several hypotheses, he eventually arrived at the postulate that the speed acquired by any freely falling body will be proportional to the time of fall. As a second postulate he proposed a relation equivalent to $\bar{v} = \frac{1}{2}(v_0 + v)$ and deduced that the vertical dis-

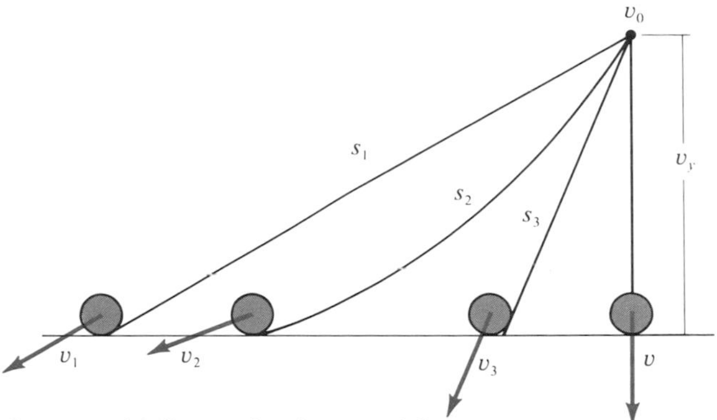

figure 4 • 5 *Travel distance s versus distance of fall y. In the absence of friction, the final speed v will be the same in all cases.*

tance traveled by such a body should therefore be proportional to t^2. Now, by measuring total distances and times of fall, he could test his theory.

From a combination of observations and idealized "thought experiments" on the motions of pendulum bobs and balls rolling down smooth inclined planes, Galileo concluded that, in the absence of friction, the speed acquired by a falling object would depend only on the height of a straight path, not on its length (Fig. 4 . 5). Thus an object sliding down a smooth inclined plane should follow the same rules as an object falling vertically, and the results obtained for longer intervals on inclined planes should be transferable to a freely falling body. The validity of his reasoning was confirmed by experiments which showed that the distance traversed varied as t^2.

Since Galileo's time, many different methods have been devised for studying falling bodies. The general findings show that, as long as air resistance is negligible, a body near the earth falls with a constant acceleration of approximately 32 ft/s^2, or 9.8 m/s^2. The value of this *acceleration due to gravity* is universally denoted by the symbol g.

Galileo's general problem was the kinematics of *ballistic projectiles*—projectiles that are not self-propelled, but are shot or hurled into space with some initial velocity. His study of freely falling bodies and objects sliding on inclined surfaces convinced him that the vertical and horizontal motions of the projectile were independent and could be analyzed separately. His next step was to assume that these motions can occur simultaneously "without altering, disturbing, or hindering each other." This assumption that the component motions are *superposable* is by no means self-evident; however, it has been amply verified by experiment and led ultimately to the concept of independent components of a vector.

To see how this assumption applies to the analysis of projectile motion, suppose a projectile is fired with an initial velocity $\mathbf{v}_0$, as shown in Fig. 4 . 6. Since we are dealing with only two direction components, we can represent this situation easily enough in two-dimensional space, where the x axis denotes horizontal distance and the y axis denotes vertical distance. The acceleration due to gravity is downward, or

$$\mathbf{a} = \frac{d\mathbf{v}}{dt} = -g\mathbf{j} \qquad [4 \cdot 18]$$

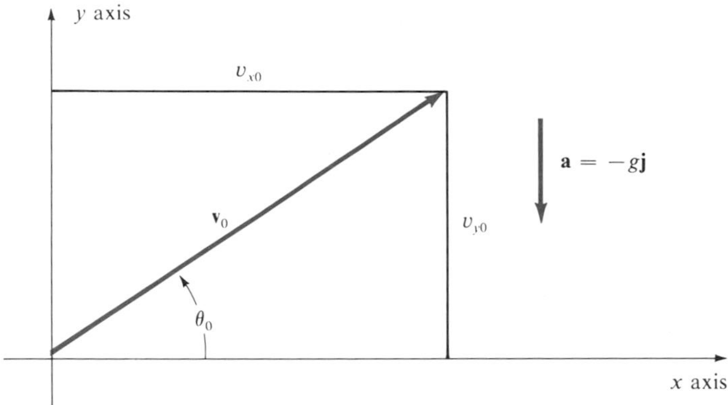

figure 4 • 6 *A ballistic projectile with an initial velocity* $\mathbf{v}_0$ *and a constant acceleration* $\mathbf{a} = -g\mathbf{j}$

Hence the scalar components of the acceleration vector are

$$a_x = \frac{dv_x}{dt} = 0 \qquad a_y = \frac{dv_y}{dt} = -g \tag{4 . 19}$$

If we now integrate each of these equations separately, with $t_0 = 0$, we obtain the two velocity components

$$v_x = v_{x0} \qquad v_y = v_{y0} - gt \tag{4 . 20}$$

where v_{x0} is the horizontal component of the initial velocity $\mathbf{v}_0$ and v_{y0} is the vertical component of $\mathbf{v}_0$. We can also locate the projectile at any time during its flight by finding the components of its position vector $\mathbf{r}(t)$. From Eq. [4 . 16], the position coordinates at any time t are

$$x(t) = x_0 + v_{x0}t \qquad y(t) = y_0 + v_{y0}t - \tfrac{1}{2}gt^2 \tag{4 . 21}$$

where x_0 and y_0 are the coordinates of the initial position at time $t_0 = 0$. These equations are perfectly general and describe, with appropriate values for the constants, the motion of *any* ballistic projectile, neglecting air resistance and assuming a "flat earth" (that the range of the projectile is negligibly small compared to the radius of the earth, $R_e = 4000$ mi).

If we are not interested in exact time dependence of the projectile's position, but in the geometric representation of its path, or *trajectory,* then we can eliminate time as a variable and express one position coordinate as a function of the other. From the first of Eqs. [4 . 21],

$$t = \frac{x - x_0}{v_{x0}} \tag{4 . 22}$$

Substituting this expression for t in the equation for $y(t)$ then gives us

$$y(x) = y_0 + v_{y0}\frac{x - x_0}{v_{x0}} - \tfrac{1}{2}g\frac{(x - x_0)^2}{v_{x0}^2} \tag{4 . 23}$$

This is the equation of a parabola, regardless of the values of the constants, since it can be written in the form $y = A + Bx + Cx^2$.

At the apex of the parabola, the highest point on the trajectory, the vertical component of the velocity is $y_0 = 0$, and from Eq. [4 . 20], the time t_h at which this occurs is

$$t_h = \frac{v_{y0}}{g} \qquad [4 \cdot 24]$$

When we substitute this expression for t in Eqs. [4 . 21], we have the coordinates of the apex:

$$x_h = x_0 + v_{x0}\frac{v_{y0}}{g} \qquad y_h = y_0 + \frac{v_{y0}^2}{2g} \qquad [4 \cdot 25]$$

Before employing these equations, however, it is important to make sure that t_h is not less than the time t_0 at which the motion began or greater than the time t_f at which it ends. If it is, then the mathematical point (x_h, y_h) has no physical significance. This is the sort of hazard against which physicists must continually be on guard: is our mathematics, however elegant, physically meaningful?

The velocity of the projectile at any time during its flight can be found from the instantaneous velocity $\mathbf{v}(t) = v_x\mathbf{i} + v_y\mathbf{j}$, which has the magnitude

$$v = \sqrt{v_x^2 + v_y^2} \qquad [4 \cdot 26]$$

and a direction tangent to the trajectory at that instant,

$$\tan \alpha = \frac{v_y}{v_x} \qquad [4 \cdot 27]$$

as illustrated in Fig. 4 . 7. (Note that the length of the velocity vector is not directly related to the spatial coordinates on the graph of the trajectory.)

If the initial velocity is given as a speed v_0 and an angle of elevation θ_0, its components are easily found by substituting the values

$$v_{x0} = v_0 \cos \theta_0 \qquad v_{y0} = v_0 \sin \theta_0 \qquad [4 \cdot 28]$$

figure 4 • 7 *The trajectory of a ballistic projectile*

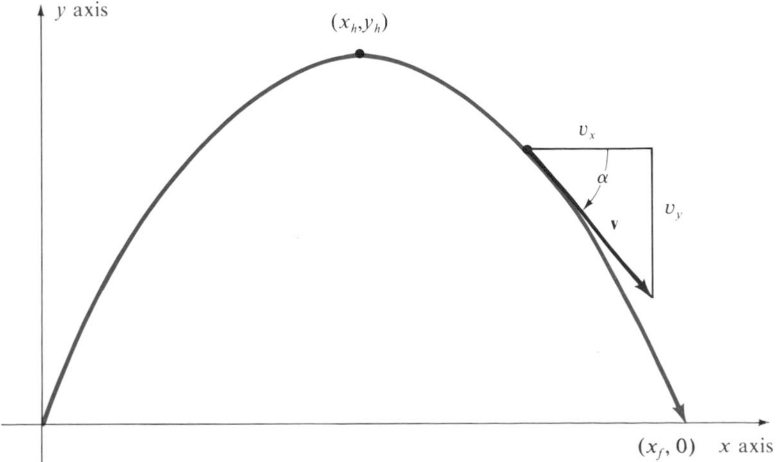

example 4 • 4 A ball is thrown straight up into the air from ground level at a speed of 30 m/s. (*a*) Give its height as a function of time. (*b*) What total time t_f will the ball be in the air before it hits the ground? (*c*) What is the maximum height it will reach?

solution (*a*) From Eq. [4 . 21], the height of the ball at any time t will be

$$y(t) = 30t - 4.9t^2 \text{ m}$$

(*b*) Since $y = 0$ when the ball hits the ground, we have

$$0 = 30t_f - 4.9t_f^2$$

which has two solutions:

$$t_f = 0 \qquad t_f = 6.1 \text{ s}$$

The total time of flight must be greater than $t_0 = 0$, so $t_f = 6.1$ s. We could also have obtained this value by symmetry, since the total time of flight is twice the time of the vertical ascent, $t_f = 2t_h$. (*c*) The maximum height, from Eq. [4 . 23], is

$$y_h = \frac{v_{y0}^2}{2g} = \frac{(30 \text{ m/s})^2}{19.6 \text{ m/s}^2} = 45.9 \text{ m}$$

example 4 • 5 The general case of projectile motion above a flat earth is described in Fig. 4 . 7. (*a*) Write expressions for the total flight time and the horizontal range of a projectile. (*b*) What angle of elevation θ_0 from the earth will result in the maximum horizontal range for a given initial velocity $\mathbf{v}_0$?

solution (*a*) At the time the projectile lands, its vertical position component is $y(t_f) = 0$. Hence if we substitute $v_{y0} = v_0 \sin\theta_0$ in Eq. [4 . 21], we have

$$y(t_f) = 0 = t_f(v_0 \sin\theta_0 - \tfrac{1}{2}gt_f)$$

and since the total flight time is not zero,

$$t_f = \frac{2v_0}{g}\sin\theta_0$$

Substituting for t_f in the horizontal component then gives the horizontal range as

$$x(t_f) = x_f = t_f v_0 \cos\theta_0 = \frac{v_0^2}{g}\sin 2\theta_0$$

(*b*) The sine function takes on a maximum value of unity when its argument is 90°, and setting $2\theta_0 = 90°$ in the equation above yields a maximum horizontal range of

$$x_f = \frac{v_0^2}{g}$$

for an initial elevation of $\theta_0 = 45°$.

4 • 4 Galilean relativity

The way we describe the motion of a particular object depends on the frame of reference we choose for our coordinate system, which may itself be in motion. From the physicist's point of view, the best frame of reference is the one that most simplifies the calculations; however, there is no one "correct" reference frame in an absolute sense. One is just as correct as another, provided the equations of motion are consistent with the frame of reference chosen. Thus a motorboat in a river will have one velocity with respect to the moving water and another velocity with respect to the river banks. The question then arises concerning the relationship between two observations of the same object by observers whose reference frames are in relative motion to each other.

The case in which one reference frame is moving with a *constant velocity* **V** with respect to the other has special significance. In Fig. 4 . 8 the reference frames S and S' are represented by two coordinate systems—xyz, perhaps fixed to the earth's surface, and $x'y'z'$, perhaps fixed to an object or vehicle moving with a constant velocity relative to reference frame S. For simplicity, the corresponding axes of the two systems are parallel, so that the x and x' axes coincide. This means that the whole system S' will be seen by an observer in S as moving in the positive x direction with a constant velocity **V**. Note that the motion of system S' relative to S is one of pure translation; that is, every point on the coordinate axes of S' moves with the same velocity **V**. Thus, if we take $t_0 = 0$ as the instant when the two origins, O and O', coincide, then at any time t the coordinates of O' will be measured by an observer at O as $x = Vt$, $y = 0$, and $z = 0$.

Suppose the observers at O and O' are operating with synchronized clocks, and they measure the position of a moving particle P at the same instant t (see Fig. 4 . 9). The observer at O will locate the particle in terms of a position vector $\mathbf{r}(t)$, with rectangular components (x,y,z), while the observer at O' will describe the position of the particle in terms of a vector $\mathbf{r}'(t)$, with components (x',y',z'). These two position vectors are related by the transformation equation

$$\mathbf{r} = \mathbf{r}' + \mathbf{V}t \qquad [4 \cdot 29]$$

figure 4 • 8 *Two reference frames S and S′; at any time after $t_0 = 0$, the distance of O′ from O is $\Delta x = Vt$*

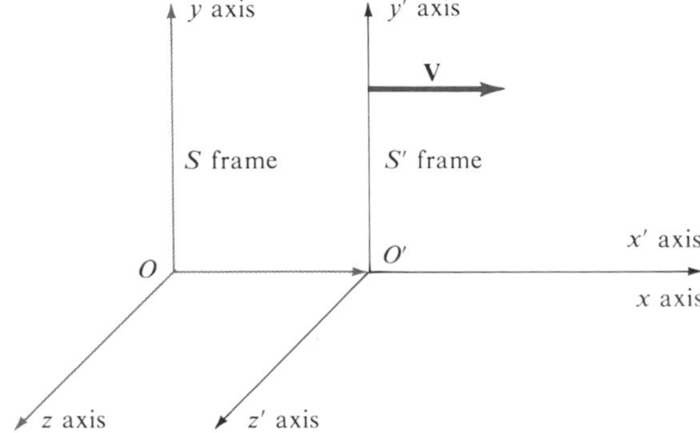

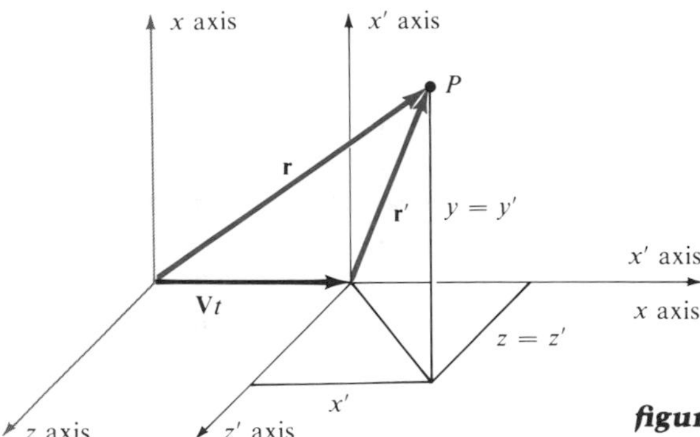

figure 4 • 9 *The position vectors* **r** *and* **r**′ *of a particle P relative to O and O′;* **r**′ = **r** − **V**t

and the rectangular coordinates of the particle in the two systems are related by the three scalar transformation equations

$$x = x' + Vt \qquad y = y' \qquad z = z' \tag{4 . 30}$$

At some later time, $t + \Delta t$, the particle will have moved to a new position, given by a position vector $\mathbf{r} + \Delta\mathbf{r}$ relative to O and a second position vector $\mathbf{r}' + \Delta\mathbf{r}'$ relative to O'. However, O' has also moved in the time interval Δt, so that $\mathbf{r}' + \Delta\mathbf{r}'$ is measured from a new origin relative to O. The position vector $\mathbf{r}' + \Delta\mathbf{r}'$ will therefore differ from the actual displacement of the particle by an amount equal to the displacement of O', so that the relation in Eq. [4 . 29] applies to the later time as well:

$$\mathbf{r} + \Delta\mathbf{r} = \mathbf{r}' + \Delta\mathbf{r}' + \mathbf{V}(t + \Delta t) \tag{4 . 31}$$

The difference between Eqs. [4 . 31] and [4 . 29] then gives two displacement vectors related by

$$\begin{aligned} &\Delta\mathbf{r} = \Delta\mathbf{r}' + \mathbf{V}\,\Delta t \\ &\Delta x = \Delta x' + V\,\Delta t \qquad \Delta y = \Delta y' \qquad \Delta z = \Delta z' \end{aligned} \tag{4 . 32}$$

To obtain the transformation equation relating the two simultaneously observed velocities of the particle, we divide Eq. [4 . 32] by the time interval Δt and take the limit as Δt approaches zero. Then, at any time t, the velocity of the particle will be measured as

$$\mathbf{v} = \mathbf{v}' + \mathbf{V} \tag{4 . 33}$$

where $\mathbf{v}' = \mathbf{v} - \mathbf{V}$ is the particle's velocity relative to the moving system at that instant.

The particle may also undergo an acceleration relative to either system. Suppose $\mathbf{v}$ and $\mathbf{v}'$ represent the velocities measured in each system at time t, and at a later time $t + \Delta t$ the particle has a velocity $\mathbf{v} + \Delta\mathbf{v}$ relative to O and $\mathbf{v}' + \Delta\mathbf{v}'$ relative to O'. Then, from Eq. [4 . 33], this new velocity is

$$\mathbf{v} + \Delta\mathbf{v} = \mathbf{v}' + \Delta\mathbf{v}' + \mathbf{V} \tag{4 . 34}$$

However, when we subtract Eq. [4 . 33] from Eq. [4 . 30] to obtain the *change* in velocity over the time interval Δt, we have

$$\frac{\Delta\mathbf{v}}{\Delta t} = \frac{\Delta\mathbf{v}'}{\Delta t} \tag{4 . 35}$$

Thus, in the limit as Δt approaches zero, *the acceleration is the same,* whether it is measured by a stationary observer at O or the observer moving with constant velocity at O':

$$\mathbf{a} = \mathbf{a}'$$
$$a_x = a'_x \qquad a_y = a'_y \qquad a_z = a'_z \qquad [4 \cdot 36]$$

It is important to realize that these transformation equations are based on an implicit assumption that time is an absolute constant—that is, $t = t'$. This assumption was stated explicitly by Newton, who posited the existence of an "absolute, true and mathematical time [which] from its own nature flows equably without relation to anything external." However, as we shall see when we discuss Einstein's theory of special relativity, when the speed of one system relative to another approaches the speed of light (3×10^8 m/s), there is an appreciable difference between the time measurements in the two systems.* Thus, according to the Galilean principle of relativity:

> If the measurement of time is unaffected by motion, the acceleration of a particle at any given instant is the same, both in magnitude and in direction, with respect to all reference frames moving relative to one another with constant velocity.

This principle is illustrated in Fig. 4 . 10, which shows the velocities of a particle on its trajectory as measured by observers in S and S'. Although the initial velocities $\mathbf{v}$ and $\mathbf{v}'$ and the new velocities $\mathbf{v} + \Delta\mathbf{v}$ and $\mathbf{v}' + \Delta\mathbf{v}'$ differ in the two systems, the *net change* in velocity is the same: $\Delta\mathbf{v} = \Delta\mathbf{v}'$. And if

*A second, more subtle assumption is that length is the same when measured in either system, regardless of the state of motion of the object. This assumption also breaks down near the speed of light.

figure 4 • 10 *The velocity vectors of a moving particle measured at times t_1 and t_2. During the time Δt the origin O' has moved a distance equal to $\mathbf{V}\,\Delta t$; hence the relative velocity at both times differ by a constant amount $\mathbf{V}$, and the same change in velocity, $\Delta\mathbf{v}' = \Delta\mathbf{v}$, is observed from both reference frames*

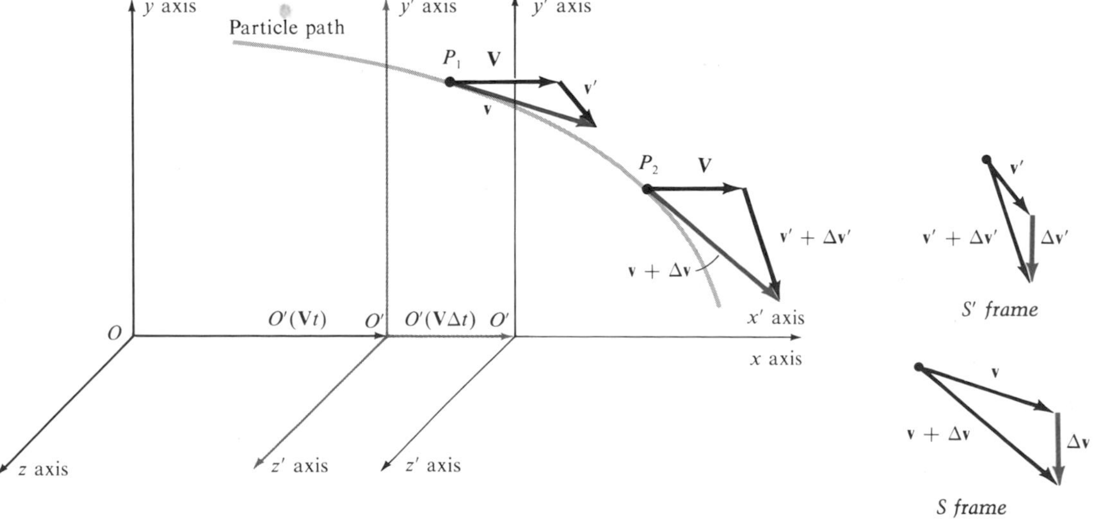

this change corresponds to the same time interval Δt, the observed acceleration is the same in both systems.

Equations [4 . 29], [4 . 34], and [4 . 36] are known as the *Galilean transformation equations*. Although we have derived them here in terms of two reference systems whose coordinate axes are parallel, these equations would have exactly the same form for any fixed orientation of one set of spatial coordinates with respect to the other, and for a constant velocity $\mathbf{V}$ in any direction. In their vector form, the equations are completely general.

example 4 • 6 An airplane is observed from the control tower to be flying northeast at a ground speed of 3 km/min. If the wind is blowing steadily from the south at 52 km/h, what is the airplane's velocity relative to the air?

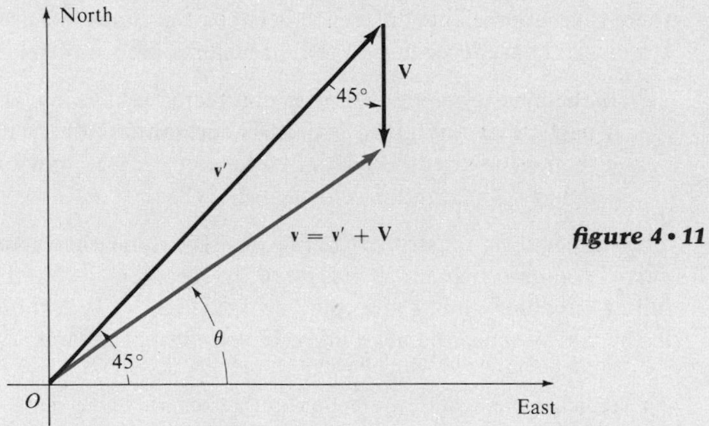

figure 4 • 11

solution If we take the ground as the moving reference system S', then the tower observer at O' is moving at a constant velocity of $\mathbf{V} = 52$ km/h south relative to the air. Since he sees the airplane moving at a velocity of $\mathbf{v}' = 3$ km/min $= 180$ km/h northeast, the sum of these two velocities is the airplane's velocity relative to the air:

$$\mathbf{v} = \mathbf{v}' + \mathbf{V} = 180 \text{ km/h} (\cos 45° \mathbf{i} + \sin 45° \mathbf{j}) - 52\mathbf{j} \text{ km/h}$$

From the law of cosines, the air speed is

$$v = (180^2 + 52^2 - 2 \times 180 \times 52 \cos 45°)^{1/2} = 148 \text{ km/h}$$

Since the eastward components of $\mathbf{v}$ and $\mathbf{v}'$ are equal, the direction of this velocity is found by setting $v_x = v'_x$:

$$148 \cos \theta = 180 \cos 45°$$

or $\theta = 31.5°$ north of east. Note that we could also have found the direction angle θ by applying the law of sines (see Appendix E).

Recall that Galileo's initial interest was in projectile motion. Suppose a projectile is launched straight upward from the bed of a moving truck, as illustrated in Fig. 4 . 12. The truck is moving with a constant horizontal velocity $\mathbf{V}$; hence it constitutes a moving reference system S', with coordinate axes x', y', and z'. Time is measured from the moment of launching, $t_0 = 0$,

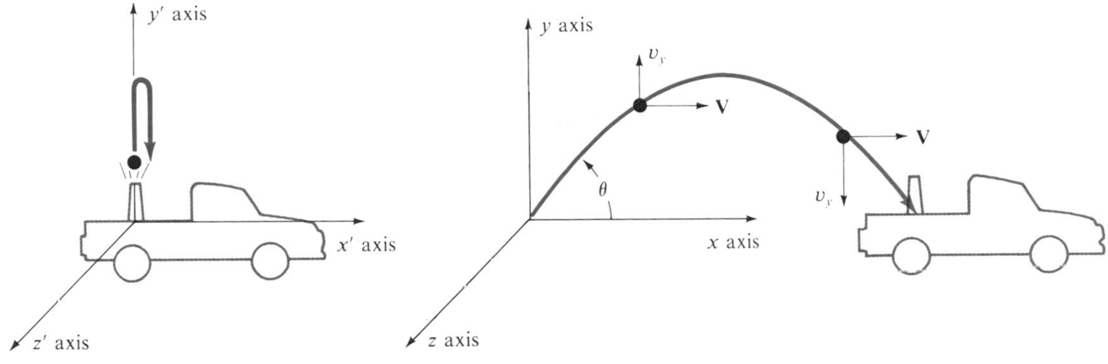

(a) Trajectory relative to S' frame

(b) Trajectory relative to S frame

figure 4 • 12 *Flight of projectile launched from a system S' moving with a constant horizontal velocity* **V**

when the projectile rises with an initial velocity $\mathbf{v}_0$. Since the horizontal velocity component at the time of launching will be the same as the truck's velocity **V**, an observer at O' on the moving truck will see only the vertical component v'_{y0}; the horizontal component v_{x0} will be seen as $v'_{x0} = 0$. Thus, at any time t during its flight, the projectile will be directly over the truck, and relative to the truck, its trajectory will be a straight line up and down. Its position coordinates (in two dimensions) will be

$$x' = 0 \qquad y' = v'_{y0}t - \tfrac{1}{2}gt^2$$

its velocity components will be

$$v'_x = 0 \qquad v'_y = v'_{y0} - gt$$

and its acceleration components will be

$$a'_x = 0 \qquad a'_y = -g$$

If the same projectile is observed from a stationary reference frame S with its origin of coordinates at the launch site ($O = O'$ at $t_0 = 0$), then the positions, velocities, and accelerations relative to O will be as given by the Galilean transformation equations. In this case the initial velocity $\mathbf{v}_0$ as seen from the ground will have the magnitude and direction

$$v_0 = \sqrt{v_{y0}^2 + V^2} \qquad \theta_0 = \tan^{-1}\frac{v_{y0}}{V}$$

The path of the projectile will be a parabola, as shown in Fig. 4 . 12*b*.

These observations bring us to an important general conclusion. Suppose an experimenter in a completely enclosed laboratory sets an object inside the laboratory in motion and measures successive positions, velocities, and accelerations. These kinematic data, and the trajectory they describe, will be the same with respect to the observed initial conditions whether the experimenter's reference system is stationary or is moving with any constant velocity relative to some fixed, or inertial, reference frame. The experimenter can detect jolts, rolls, or other *changes* in the velocity of the closed system, but not the velocity of the system itself. This general conclusion, together with the Galilean transformation equations, is referred to as the *Galilean relativity principle of kinematics.*

PROBLEMS

4 • 1–4 • 2 velocity and acceleration

4 • 1 In an hour you walk 1 km west and then 2 km north. (*a*) What is your average speed? (*b*) Express your average velocity in vector form, with **i** as the unit vector pointing east and **j** as the unit vector pointing north.

4 • 2 After reaching your destination in Prob. 4 . 1, you spend the next hour walking 4 km east. Write your average acceleration $\bar{\mathbf{a}}$ as the difference between the two 1-h velocity vectors.

4 • 3 A helicopter flies 6 km due south in 0.5 h, 10 km east in 0.75 h, and 2 km straight up in 0.1 h. (*a*) What is its average velocity? (Let south be the positive x direction.) (*b*) What is its average speed?

4 • 4 A sperm whale is swimming south and dives to a depth of 600 m for cuttlefish, descending at an angle of 40° from the vertical. It ascends immediately at the same angle, still heading south, and arrives at the surface 20 min after the dive began. However, the whale has also drifted 1500 m eastward with the current during that time. (*a*) What is the whale's average velocity? (*b*) Its average speed?

4 • 5 A school of tuna is traveling northeast off the Peruvian coast at a speed of 5 knots. A fishing trawler capable of a top speed of 20 knots is located 100 nautical mi due north of the school. (*a*) What course should the trawler set to intercept the school as quickly as possible? (*b*) How long will it take? (1 knot = 1 nautical mi/h = 6080 ft/h.) (*Hint:* Set up an equation between two position vectors.)

4 • 6 It is a sunny autumn day on a beach in southern California, and you are running barefoot along the water's edge, just one step ahead of a wave coming in at an angle of 60° to the shore. If the wave is moving shoreward at a speed of 5 mi/h, how fast are you running? Draw a vector diagram.

4 • 7 The position of a moving particle is given by

$$\mathbf{r}(t) = 6t\mathbf{i} - 3t^2\mathbf{j} + t^{3/2}\mathbf{k}$$

Find (*a*) its velocity and (*b*) its acceleration.

4 • 8 A particle is located by the position vector

$$\mathbf{r}'(t) = -3\mathbf{i} + (2t - 3t^2)\,\mathbf{j} + t^{3/2}\mathbf{k} \qquad -\infty < t < \infty$$

(*a*) At what time t will it be closest to the particle in Prob. 4 . 7? (*b*) What is the distance of closest approach? (*c*) What is the relative velocity $\mathbf{v} - \mathbf{v}'$? Convince yourself, by imagining one particle to be stationary, that the relative velocity need not be zero at the distance of closest approach.

4 • 9 A person standing on a pier 10 m above the water hauls in the bow line of a rowboat at the rate of 80 cm/s. (*a*) How fast is the boat moving when it is 40 m from the pier? (*b*) When it is 6 m from the pier? (*c*) Derive a general expression for the speed v' of the boat in terms of the speed v with which the line is hauled in, the height y of the pier, and the distance x of the boat from the pier.

4 • 10 The acceleration of a moving body is

$$\mathbf{a}(t) = (3\mathbf{i} + 2\mathbf{j} + \mathbf{k})\ \text{cm/s}^2$$

(*a*) If the body starts from rest, what is its velocity after 3 s? (*b*) What is its position after 10 s? (*c*) What was its average speed during the first 10 s?

4 • 11 The circles on the graph represent the positions of a body determined by photographing its motion in a dark room illuminated by a stroboscopic light flashing at 0.1-s intervals. Reproduce this figure on a large piece of graph paper and draw vectors representing the average velocity and average acceleration of the body, with the velocity vectors that connect successive points drawn to scale. Indicate the scales for the magnitudes of velocity and acceleration by suitable lengths drawn below the graph, with the magnitude and units noted on each scale.

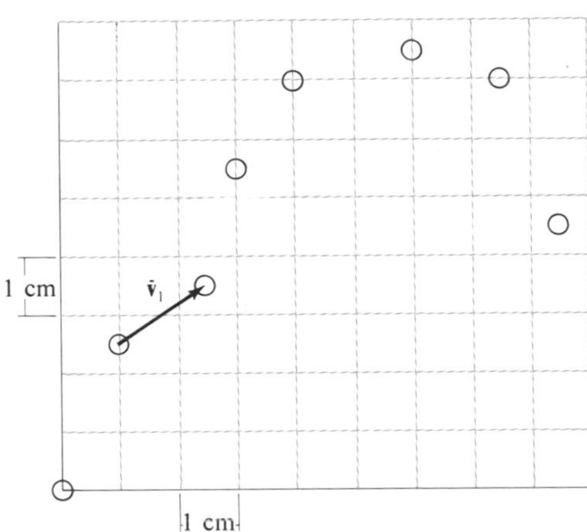

4 • 12 A given function $f(x)$ is defined as

$$f(x) = x^4 - 4x^3 + 4x^2 - 1$$

(*a*) Find its extrema and determine whether they are maxima or minima. (*b*) Determine whether the second derivative, d^2f/dx^2, is positive or negative at each extremum and explain the significance of your result.

4 • 13 A particle is located at a position given by

$$\mathbf{r}(t) = (3 - 2t + t^2)\,\mathbf{i} + (7 - 4t + t^2)\,\mathbf{j}$$

Find the values of t and $\mathbf{r}$ for which (*a*) the x component is a minimum. (*b*) The y component is a minimum. (*c*) The speed of the particle is a minimum.

4 • 3 *freely falling bodies and projectiles*

4 • 14 If the mean solar second were redefined as 1/28,800 of the mean solar day, how would this affect the numerical value of g?

4 • 15 A ball is thrown straight up from the ground with an initial speed of 96 ft/s in a locality where $g = 32\ \text{ft/s}^2$. (*a*) Show that its heights and speeds relative to the ground are as follows:

t, s	1	3	5	6
y, ft	80	144	80	0
v, ft/s	64	0	−64	−96

(*b*) At what times will the ball be at $y = 128$ ft? (*c*) Find the acceleration when the ball is at its maximum height, and therefore momentarily at rest. (*d*) How much does the speed change during successive 1-s intervals?

4 • 16 Show that the distances traversed during successive equal intervals of time by a body falling from rest have the same ratio to each other as the odd numbers beginning with 1.

4 • 17 A stone is dropped from a height of 30 m at the same instant that another stone is hurled upward from the ground. (*a*) If they meet at a height of 15 m, what was the initial speed of the second stone? (*b*) What is the velocity of each stone when they meet?

4 • 18 An arrow is shot with an initial speed of 40 m/s from the top of the Eiffel Tower, which is 335 m high. How much time will elapse before the arrow reaches the ground (*a*) if it is shot vertically upward? (*b*) If it is shot vertically downward? (*c*) If it is shot horizontally? (*d*) In each case, what will its velocity be as it hits the ground?

4 • 19 Complete the square in the quadratic formula for $y(x)$ in Eq. [4 . 23] and use the result to prove that $y - y_0$ has its maximum value when $y = y_h$ and $x = x_h$, as given in Eqs. [4 . 25].

4 • 20 Find the apex coordinates (x_h, y_h) given in Eqs. [4 . 25] by setting $(dy/dx)_{x = x_h} = 0$ in the formula for $y(x)$.

4 • 21 A stone thrown straight up from the ground returns to its starting point in 7 s. (*a*) What was its initial speed? (*b*) What total distance does it travel during the fourth second? (*c*) What is its net displacement during the fourth second?

4 • 22 A juggler keeps five balls continuously in the air, throwing each one to a height of 10 ft. (*a*) What is the time between successive throws? (*b*) What are the heights of the other balls at the moment when one ball reaches his hand?

4 • 23 A bullet is fired with a muzzle speed of 200 m/s at an angle of 30° from the horizontal. (*a*) How high will it rise? (*b*) At what distance from the gun will the bullet strike the earth? (*c*) If the gun had been on top of a tower 335 m high, how far from the base of the tower (neglecting air resistance) would the bullet have struck the ground?

4 • 24 A cannon fires a ball with a speed of 400 m/s at an elevation of 47° and hits a buzzard 400 m from the ground. (*a*) What is the time required for the ball to reach the buzzard? (*b*) What is the greatest *horizontal* distance that the buzzard can be from the gun?

4 • 25 A projectile is launched horizontally from a height y above the ground. (*a*) Show that the total time required for it to reach the ground is $t = \sqrt{2y/g}$. (*b*) Show that the horizontal range is $x = v_0 \sqrt{2y/g}$.

(*c*) Show that the velocity at any time t during the flight is of magnitude $v = \sqrt{v_0^2 + g^2t^2}$ and direction $\theta = \tan^{-1}(gt/v_0)$, where the angle θ is measured clockwise from the horizontal.

4 • 26 For a gun with a muzzle speed of 300 m/s, (*a*) what two firing angles would be satisfactory to hit a target at 5000 m on level ground? (*b*) What is the time of flight in each case? (*c*) What are the maximum heights of the respective trajectories?

4 • 27 A projectile is fircd with a muzzle speed v_0 at an angle θ_0 above the horizontal, and it lands on a slope that is at a constant angle ϕ.

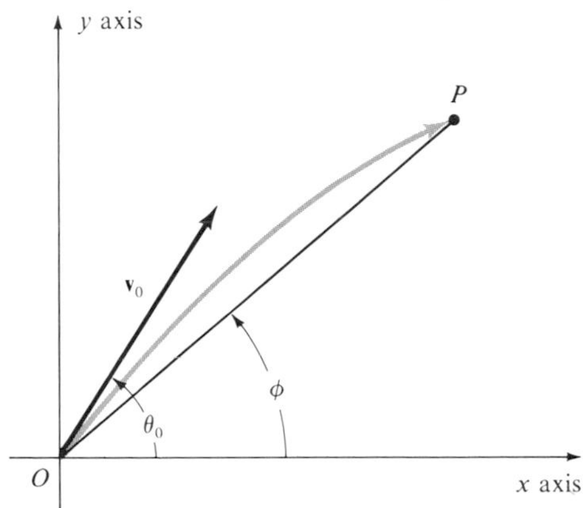

(*a*) Show that its range OP is given by

$$r = \frac{2y_0^2 \sin(\theta_0 - \phi)\cos\theta_0}{g\cos^2\phi}$$

and that

$$2\sin(\theta_0 - \phi)\cos\theta_0 = \sin(2\theta_0 - \phi) - \sin\phi$$

(*b*) Show that the range OP is maximum when $\theta_0 = \frac{1}{2}\phi + 45°$ and then has the value

$$r = \frac{v_0^2}{g + g\sin\phi}$$

4 • 28 When a certain self-propelled rocket reaches a point O' above the ground, its fuel is exhausted and it becomes a ballistic missile with an initial velocity $\mathbf{v}_0$, directed at angle θ_0.

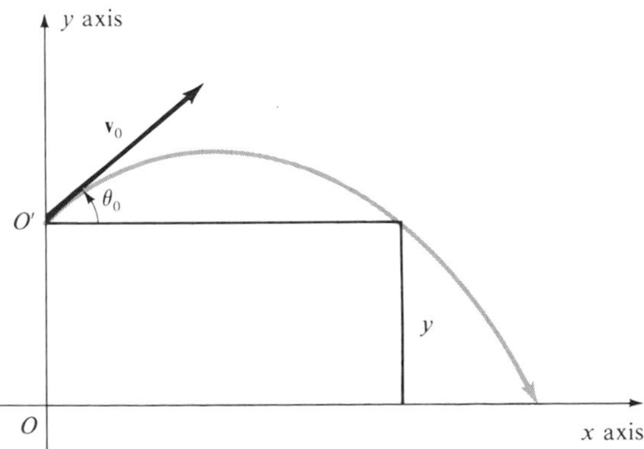

Show that its horizontal range, measured from point O, is

$$x = \frac{v_0 \cos\theta_0}{g}\left(v_0 \sin\theta_0 + \sqrt{v_0^2 \sin^2\theta_0 + 2gy}\right)$$

4 • 29 A shell is fired with an initial speed of 9100 m/min at an elevation of 47°. If the gun is on a cliff 150 m above the sea, (*a*) when will the shell strike the water? (*b*) What is the horizontal distance to the point of impact? (*c*) What is the greatest altitude the shell will reach?

4 • 4 *Galilean relativity*

4 • 30 What is the relative speed of two automobiles, each traveling at a speed of 120 km/h, (*a*) when they collide head-on? (*b*) When they collide at right angles?

4 • 31 A steamship has a velocity of 15 knots northward, and the smoke from its funnels stretches 35° south of west. (*a*) If the wind is blowing from due east, what is its speed according to the weather report? (*b*) What is its speed relative to the ship?

4 • 32 Suppose a train is moving at 108 km/h northeast and a passenger throws a ball down the train corridor in the opposite direction, at a speed of 10 m/s. What is the velocity of the ball relative to an observer on the ground?

4 • 33 A hydrofoil is traveling 30° west of north at a constant speed of 30 m/s relative to the water. The water is flowing at a uniform velocity of 5 m/s south. If an object slides across the deck in a direction 120° west of north, at a speed of 6 m/s relative to the deck, what is its velocity relative to the river banks?

4 • 34 Two motorboats which operate at the same speed v in still water are in a straight channel in which the water is flowing with a uniform velocity **V**. The boats start simultaneously from the same point. Boat A makes the round trip to a buoy located at a distance l directly upstream in a time t_A, while boat B makes a round trip to a buoy located at the same distance directly across the channel in a time t_B. Neglect any time lost in reversing course at each buoy. (*a*) If t is the time needed for either boat to travel the distance $2l$ on a still lake, show that

$$\frac{t_A}{t_B} = \frac{t_B}{t} = \frac{1}{\sqrt{1 - \mathrm{V}^2/\mathrm{v}^2}}$$

(*b*) Show that

$$\frac{t_A}{t} = \frac{1}{1 - \mathrm{V}^2/\mathrm{v}^2}$$

(*c*) Show that the ratio of the speeds of A and B on their outward trips to the buoy is

$$\frac{v_A}{v_B} = \sqrt{\frac{v - V}{v + V}}$$

and the ratio on their return trips is

$$\frac{v_A}{v_B} = \sqrt{\frac{v + V}{v - V}}$$

4•35 An object is projected straight upward with a speed of 34 m/s from a pickup truck that is moving at a constant velocity of 72 km/h eastward on level ground. (*a*) Where will the object land and where will the truck be at that time relative to the projection point? (*b*) What is the initial velocity of the object as observed from the ground? (*c*) What is its velocity at the moment when the truck is 100 m east of the projection point?

4•36 Two cars, C_1 and C_2, start from the same point with the same initial speed of 8 mi/h. C_1 moves due west with a constant acceleration of 3 ft/s^2, while C_2 moves due south with a constant acceleration of 5 ft/s^2. (*a*) What is the velocity of C_1 relative to C_2 after 9 s? (*b*) What is the acceleration of C_1 relative to C_2 at that time? (*c*) The acceleration of C_2 relative to C_1? (*d*) If the velocity of C_2 were constant, what would be the acceleration of C_1 relative to C_2?

4•37 Two particles are seen by a stationary observer to be moving with velocities $\mathbf{v}_1 = 2\mathbf{i}$ m/s and $\mathbf{v}_2 = 2\mathbf{j}$ m/s—that is, at equal speeds, but at right angles to each other. What are their velocities and the angle between them, as seen by another observer moving with a constant velocity of $\mathbf{V} = 1\mathbf{i}$ m/s?

4•38 A jogger traveling at 4 km/h east finds that the wind seems to blow directly from the north. If she doubles her speed, the wind then appears to be coming from the northeast. (*a*) What is the wind velocity relative to the ground? (*b*) From what direction would the wind seem to be blowing if the jogger were moving due west with a speed of 8 km/h?

4•39 The captain of a ship S_1, which is traveling south at a constant engine speed of $15\sqrt{2}$ knots, spots a ship S_2 off his starboard (right) bow. S_2 is heading southeast at an engine speed of 15 knots and has an acceleration $\mathbf{a}$. (*a*) What is the velocity of S_2 relative to S_1 at this instant? (*b*) What is the acceleration of S_2 relative to S_1? (*c*) What is the velocity of S_2 relative to S_1 with respect to the ocean surface? (*d*) Which of the foregoing values would change if S_1 were also accelerating?

4•40 Gordon Goodheart is flying his Sopwith Camel at a speed of 100 mi/h, overtaking the Red Baron, whose Fokker Triplane is cruising along at 90 mi/h and 1000 ft below, on exactly the same course. Attaching a surrender demand to a rock, Gordon hurls it into the Baron's cockpit. (*a*) With what velocity (neglecting air resistance) must he hurl it relative to the Camel when the Baron is directly below him? (*b*) When the Baron is still 1000 ft ahead of him?

answers

4•1 (*a*) 3 km/h; (*b*) $(-\mathbf{i} + 2\mathbf{j})$ km/h
4•2 $\bar{\mathbf{a}} = (5\mathbf{i} - 2\mathbf{j})$ km/h^2
4•3 (*a*) $\bar{\mathbf{v}} = (4.44\mathbf{i} + 7.41\mathbf{j} + 1.48\mathbf{k})$ km/h; (*b*) $v = 3.33$ km/h
4•4 (*a*) $(4290\mathbf{i} + 4500\mathbf{j})$ m/h; (*b*) 6217 m/h, pretty leisurely
4•5 (*a*) 10.08° east of south; (*b*) 4:18:23
4•6 10 mi/h
4•7 (*a*) $\mathbf{v}(t) = 6\mathbf{i} - 6t\mathbf{j} + \frac{1}{2}\sqrt{t}\,\mathbf{k}$; (*b*) $\mathbf{a}(t) = -6\mathbf{j} + \frac{1}{4}\sqrt{t}\,\mathbf{k}$
4•8 (*a*) $t = -0.45$; (*b*) 0.95; (*c*) $6\mathbf{i} - 2\mathbf{j}$
4•9 (*a*) 0.825 m/s; (*b*) 1.56 m/s; (*c*) $v' = v\sqrt{y^2 + x^2}/x$
4•10 (*a*) $\mathbf{v} = \mathbf{a}(3\text{ s})$; (*b*) $\mathbf{r} = \mathbf{a}(50\text{ s}^2)$; (*c*) $\bar{v} = 18.71$ cm/s
4•12 (*a*) $x = 0$ min, 1 max, 2 min; (*b*) $f''(0) > 0, f''(1) < 0, f''(2) > 0$
4•13 (*a*) $t = 1, \mathbf{r} = 2\mathbf{i} + 4\mathbf{j}$; (*b*) $t = 2$, $\mathbf{r} = 3\mathbf{i} + 3\mathbf{j}$; (*c*) $t = 1.5, \mathbf{r} = 2.25\mathbf{i} + 3.25\mathbf{j}$
4•14 g would be 9 times as large
4•15 (*b*) $t = 2$ s, $t = 4$ s; (*c*) $\mathbf{a} = -32$ ft/s^2 throughout the motion; (*d*) -32 ft/s, since the motion is one-dimensional
4•17 (*a*) $v_0 = 17.5$ m/s; (*b*) $\mathbf{v} = -17.15\mathbf{j}$ m/s and **0**
4•18 (*a*) $t = 13.30$ s; (*b*) $t = 5.14$ s; (*c*) $t = 8.27$ s; (*d*) $\mathbf{v} = -90.4\mathbf{j}$ m/s, $\mathbf{v} = -90.4\mathbf{j}$ m/s, $v = 90.4$ m/s at 26°16′ with the vertical
4•21 (*a*) $v_0 = 34.3$ m/s; (*b*) $\Delta s = 2.45$ m; (*c*) 0
4•22 (*a*) 0.32 s; (*b*) 6.4, 9.6, 9.6, and 6.4 ft
4•23 (*a*) $y_h = 510$ m; (*b*) $x = 3535$ m; (*c*) $x = 4042$ m
4•24 (*a*) $t = 56.8$ s; (*b*) $x = 15{,}503$ m
4•26 (*a*) $\theta_0 = 16.5°$, $\theta_0 = 73.5°$; (*b*) $t = 17.4$ s, $t = 58.7$ s; (*c*) $y_h = 370$ m, $y_h = 4222$ m
4•29 (*a*) $t = 23.9$ s; (*b*) $x = 2474$ m; (*c*) 778 m above the sea
4•30 (*a*) $v = 240$ km/h; (*b*) $v' = 170$ km/h
4•31 (*a*) 21.4 knots; (*b*) 26.2 knots
4•32 $\mathbf{v} = 72$ km/h northeast
4•33 $\mathbf{v} = -20.2\mathbf{i} + 18.0\mathbf{j}$ m/s
4•35 (*a*) In the truck, 139 m east of the starting point; (*b*) $\mathbf{v}_0 = 39.4$ m/s, $\theta_0 = 59°32'$; (*c*) $\mathbf{v} = 25$ m/s, $\theta = -36°52'$
4•36 (*a*) $\mathbf{v} = 68.7$ ft/s, 34°19′ west of north; (*b*) $\mathbf{a} = 5.8$ ft/s^2, 31° west of north; (*c*) equal and opposite to (*b*); (*d*) $\mathbf{a} = 3.0$ ft/s^2 west

4 • 37 (*a*) $\mathbf{v}'_1 = 1\mathbf{i}$ m/s, $\mathbf{v}'_2 = (-\mathbf{i} + 2\mathbf{j})$ m/s, $\theta' = 116°34'$
4 • 38 (*a*) $\mathbf{v} = 4 \times \sqrt{2}$ km/h southeast; (*b*) 18°26′ north of west
4 • 39 (*a*) $\mathbf{v}'_2 =$ 15 knots northeast; (*b*) **a**, southeast; (*c*) the same as (*a*); (*d*) part (*b*)
4 • 40 (*a*) $\mathbf{v}_0 =$ 10 mi/h backward; (*b*) $\mathbf{v}_0 =$ 76.5 mi/h forward

CHAPTER FIVE

mass, momentum, and force

Definition I *The quantity of matter is a measure of the same, arising from its density and bulk conjointly. . . . It is this quantity that I mean hereafter everywhere under the name of body or mass. And the same is known by the weight of each body; for it is proportional to the weight.*

Definition II *The quantity of motion [momentum] is the measure of the same, arising from the velocity and mass conjointly. . . .*

Definition IV *An impressed force is an action exerted upon a body, in order to change its state, either of rest, or of uniform motion in a right line.*

Law I *Every body continues in its state of rest or of uniform motion in a right line, unless it is compelled to change that state by forces impressed upon it.*

Law II *The change of motion is proportional to the motive force impressed; and is made in the direction of the right line in which that force is impressed.*

Isaac Newton, Principia, *1686,*
translated by Andrew Motte, 1729

Our discussions of kinematics have led to descriptions of the motions of bodies in terms of such concepts as time, velocity, and acceleration. We also found that the observer's description of displacement and velocity depends on his frame of reference and its own motion. In this chapter we take up the study of *dynamics,* adding the concepts of mass, momentum, and force to those of kinematics. We shall also see how an observer's frame of reference affects the measurement of forces.

The central concept of Newtonian dynamics is force. But how do we define a force? What is it? Long before Galileo's time, the motion of a falling body had been attributed to the "force of gravity." At that time, however, it was thought that the speed of the body was proportional to the force acting on it. When Galileo showed decisively that bodies near the earth fall with constant *acceleration,* not constant speed, he opened the way for an entirely new idea: that the force produced the acceleration. This pointed to change of motion, rather than motion itself, as the criterion for the existence of a force. Thus the ancient view that the absence of force meant the absence of motion gave way to the modern viewpoint: without a force there can be no *change* of motion. Out of this recognition grew the first great law of mechanics, the law of inertia.

5 • 1 the law of inertia

If force causes acceleration, rather than motion, why is it that some kind of engine or other active agent is always needed to keep a body in steady motion? Galileo conjectured that this is because a body left to itself is slowed by friction and other retarding forces. Although such hindering influences are always present, they can be diminished. For example, it is harder to drag a stone over a rough floor than over smooth ice. If the stone is given a push along the floor, it is soon stopped by friction, but on the ice it keeps moving for a considerable time. One begins to speculate on what would happen—or rather, what would not happen—if the ice were perfectly smooth, flat, and of infinite extent and there were no resistance due to the air.

It was by similar reasoning that Galileo had arrived at his principle of the constancy of the horizontal component of projectile motion. The wider significance of this principle was first recognized by René Descartes, and through Descartes' work, by Isaac Newton. Newton restated this principle as his first law of motion, now known as the law of inertia:

> Every body continues in a state of rest or of motion in a straight line with constant speed, except when it is compelled by force to change that state.

Since no real body is ever entirely free of the influence of its surroundings, this principle, like any other physical axiom, is necessarily an idealization, abstracted from observation and experiment. However, the innumerable deductions which have since been made from it are in accord with experience.

Newton's first law describes the motion of an *isolated body,* one which is unaffected by its surroundings. Given such an ideal body, the law of inertia states that its velocity will remain constant. Any reference frame in which Newton's first law of motion is valid is called an *inertial frame* or *inertial system.* Galileo used a reference frame fixed to the earth's surface in applying his limited principle of inertia to projectile motions. However, he was aware that the principle holds only approximately in such an earthbound system; he cited examples to show that, because of the earth's axial rotation and its orbital motion about the sun, the velocity of an isolated particle should be assumed to be constant with respect to a reference system centered at the sun. Later Newton used the sun-centered (heliocentric) coordinate system as an inertial frame when he applied his principles of dynamics, which include the first law of motion, to the orbits of planets.

The sun-centered inertial frame is adequate for motions in our solar system, but is there perhaps a more general reference frame in which the first law holds more exactly, or even absolutely? The significance of this question led Newton to postulate an ultimate reference frame that is at rest in "absolute space" and in which the first law is completely valid. Einstein later used the experimental results of Michelson and Morley and their predecessors to prove that there is no absolute reference frame in the physical universe. However, the system in which Newton's first law holds most accurately consists of a rigid coordinate system moving with any constant velocity relative to the so-called "fixed stars." This reference frame is called the *primary inertial frame.*

While a truly inertial reference frame is a convenient fiction, in practice, reference frames that are not fixed in any absolute sense may be regarded as inertial systems for a given set of problems. Similarly, a reference frame whose velocity is not constant relative to the primary inertial system is often treated as an inertial system if the velocity changes are either negligible or easily corrected. For example, an earth-centered (geocentric) set of nonrotating axes is sometimes used for analysis of the motions of satellites orbiting the earth, since for most purposes the relatively small accelerations of the earth's center in its orbital motion about the sun can be ignored. On a smaller scale, measurements of objects near or on the earth are usually made relative to coordinate systems fixed to the earth's surface. In this case errors due to rotation of the coordinate systems are significant only when the object moves a great distance or for a considerable length of time.

According to Newton's first law, every body continues in its state of rest or uniform motion in the absence of an applied force. This resistance to a change in the state of motion is what we mean by the *inertia* of a body. In order to compare inertias quantitatively, Newton introduced the concept of *mass.* Through a series of experiments with various materials, he found that in any given locality the mass, or amount of inertia, of a body is proportional to its weight, the attractive force exerted on it by the earth. Moreover, he found the ratio of mass to weight to be independent of the chemical composition of the bodies. This furnished him with a practical method for measuring the mass of one body in terms of the mass of another body: the two masses could be compared simply by putting the bodies on an ordinary balance scale and comparing their weights. Even today the beam balance provides one of the most convenient and accurate methods for comparing masses.

The unit of mass in the metric system is the *international prototype kilogram* (kg), a cylinder of platinum-iridium kept at the International Bureau of Weights and Measures. It is the only SI unit still defined by an artifact. Originally intended to be, and very nearly equal to, the mass of 1000 cm^3 of water at 4°C (39.2°F), the kilogram and its decimal subdivisions are the most common units of mass in the world today. For commercial

table 5 • 1 *Mass conversion factors*

equal masses: 1 kg = 1000 g = 10^6 mg = 10^9 μg

kilograms (kg)	*pound-masses (lbm)*	*slugs*
1	2.205	0.0685
0.4536	1	0.03108
14.58	32.17	1

and some engineering purposes, masses are expressed in terms of a 1-lb weight, or *pound-mass* (lbm),* defined by the relation

$$1 \text{ lbm} = 0.4535924277 \text{ kg} \qquad 1 \text{ kg} = 2.205 \text{ lbm}$$

Several of the engineering sciences employ a British unit of mass called the *slug*, which is equivalent to 32.1734 lbm. In other words, the slug has the numerical value of g_s, the international standard acceleration due to gravity:

$$g_s = 9.80665 \text{ m/s}^2 = 32.1734 \text{ ft/s}^2$$

Conversion factors for the mass units are listed in Table 5 . 1.

5 • 2 *inertial mass*

Newton's first law implies that an object will change its velocity only if it interacts in some way with its surroundings. Thus it is natural to wonder if there is some relationship between the velocity changes that occur when two given bodies interact, say, in a collision. In 1668 John Wallis, Christopher Wren, and Christian Huygens made a systematic study of this and related questions at the request of the newly established Royal Society of London. From their work and his own, Newton concluded that when two isolated bodies collide, the velocity change $\Delta\mathbf{v}_1$ of one body will have a constant ratio to the velocity change $\Delta\mathbf{v}_2$ of the other body along a line that passes through their centers (see Fig. 5 . 1). This observation was found to hold for other

*Usually (incorrectly) referred to as a "pound."

figure 5 • 1 *Two pendulum bobs immediately before and after collision. The velocities may differ in any collision, but the ratio $\Delta v_1/\Delta v_2$ is the same for all collisions of these two bodies.*

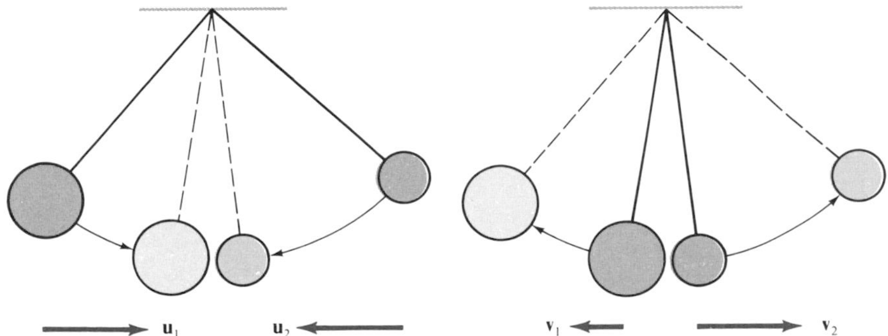

kinds of interactions as well, such as that between two bodies connected by a spring of negligible mass.*

The resulting *interaction postulate* is really the essence of Newtonian, or classical, dynamics:

> When any two given particles interact in isolation over any time interval, the ratio of their velocity changes will have the same numerical value.

In scalar terms,

$$\frac{\Delta v_1}{\Delta v_2} = -k_{12} \qquad [5 \cdot 1]$$

where k_{12} is a constant for a given *pair* of particles 1 and 2. Since Δv_1 and Δv_2 are always opposite in direction, the minus sign in Eq. [5 . 1] makes k_{12} a positive scalar.

If we assign an arbitrary mass m_1 to one particle, we can then define a mass m_2 for the other particle in terms of this constant. The velocity changes are opposite in direction; therefore the masses are related inversely,

$$k_{12} = \frac{m_2}{m_1} \qquad [5 \cdot 2]$$

and by substitution we have

$$\frac{\Delta v_1}{\Delta v_2} = \frac{-m_2}{m_1} \qquad [5 \cdot 3]$$

If the assigned quantity m_1 is exactly the mass of the first particle, then the defined quantity m_2 will be the mass of the second particle, and Eq. [5 . 3] may be read:

> During any interaction between two isolated particles, any changes in velocity that occur are opposite in direction to each other and inversely proportional in magnitude to the respective masses of the particles.

It is important to be aware of the distinction between *inertial mass* and *gravitational mass,* the property of matter accounting for gravitational attraction of bodies to the earth. As an illustration of inertial mass, suppose the two cars in Fig. 5 .2 are connected by a string which is held taut by a compressed spring. The cars have been placed on a carefully leveled track with so little friction that it will not influence their motions appreciably. When the string between the two cars is burned, the small compressed spring pushes the cars apart before it falls to the track. During this time the velocities of the cars change from zero to some values v_1 and v_2 which can be measured. Thus if the mass m_1 of one car is known (or assigned), then the mass m_2 of the other car can be found from Eq. [5 . 3]. No weighing process is involved, and so gravitational effects are irrelevant.

When various particles $P_1, P_2, P_3, \ldots$ interact, one at a time, with some arbitrarily selected reference particle, the values of $m_1, m_2, m_3, \ldots$ determined from their velocity changes will be the same as when any two of

*It is natural to think of particles as having mathematical centers, but as we shall see in Chapter 10, this ratio also applies to the centers of mass of extended bodies.

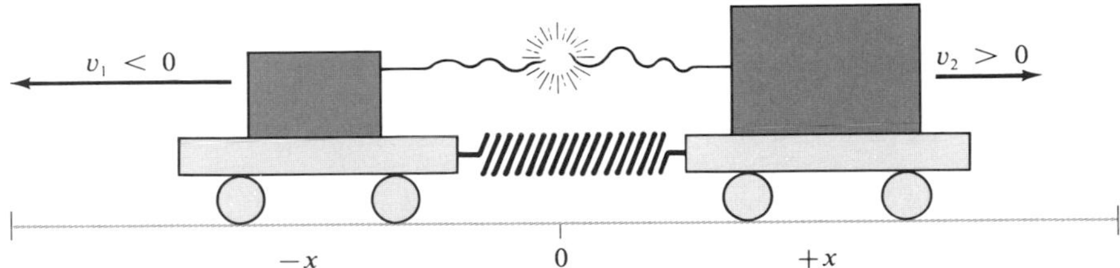

figure 5 • 2 *A direct comparison of the inertial masses of two bodies*

the particles, say P_2 and P_3, interact with each other. This follows from Eq. [5 . 3], since any two relations of the form m_1/m_2 = constant and m_2/m_3 = constant must imply $(m_1/m_2)(m_2/m_3)$ = constant. Moreover, if two or more particles or bodies with masses $m_1, m_2, \ldots$ are in some way bound together to form a single body, the total mass M of this body will be equal to $m_1 + m_2 + \cdots$, the arithmetic sum of the individual masses. This fact is not necessarily self-evident, but it has been determined by experiment. It can also be shown to follow from the dynamical definition of mass, although the proof is somewhat lengthy.

The foregoing conclusion represents a special case of a general principle known as the *conservation of mass*. According to this extremely useful principle, the total mass of any isolated system remains constant, whether the system consists of a single particle or of many bodies, and independently of any changes, either physical or chemical, that may take place within the system. For all "ordinary" physical and chemical processes—those processes not dependent on subatomic interactions—we can safely assume that the classic law of conservation of mass applies. However, one of the most unexpected (and thoroughly confirmed) results of Einstein's special theory of relativity is that this law does not hold under certain conditions.

example 5 • 1 After the spring has ceased to act on the cars in Fig. 5 . 2, they move with practically constant velocities $\mathbf{v}_1$ and $\mathbf{v}_2$ until they reach the ends of the track. Suppose the two cars are initially placed at some point on the track from which they will reach the opposite ends simultaneously. If their final displacements from this origin are $\mathbf{x}_1$ and $\mathbf{x}_2$, respectively, show that in the same time interval Δt each car will have traveled a distance Δx inversely proportional to its mass.

solution Since the motion is in one dimension and the cars move in opposite directions, their positions with respect to the origin can be stated in scalar form as

$$x_1 = -v_1 t \qquad x_2 = v_2 t$$

Both cars reach these positions at the same time t; therefore

$$\frac{x_1}{x_2} = -\frac{v_1 t}{v_2 t} = -\frac{\Delta v_1}{\Delta v_2} = \frac{m_2}{m_1}$$

Note that this answer in no way depends on the compression or strength of the spring.

5 • 3 conservation of linear momentum

The *momentum* **p** of a particle is defined as the product of its mass m and its velocity **v**:

$$\mathbf{p} = m\mathbf{v} \tag{5·4}$$

Momentum is a vector quantity, for it is the product of a scalar and a vector. We can now restate Newton's first law in terms of the concept of momentum:

> The momentum of an isolated particle or body is conserved; that is, it does not change in either magnitude or direction with the passage of time.

Since the mass of a particle or body may ordinarily be regarded as constant, a change in its momentum must reflect a change in its velocity; that is,

$$\Delta\mathbf{p} = m\,\Delta\mathbf{v}$$

Of course, the momentum of a body kept at constant velocity will also change if matter is added to or subtracted from the body. To find the time rate of change of momentum, we use the algebraic rules of differentials to obtain the component form

$$\begin{aligned}\frac{d\mathbf{p}}{dt} &= \frac{d}{dt}(mv_x)\mathbf{i} + \frac{d}{dt}(mv_y)\mathbf{j} + \frac{d}{dt}(mv_z)\mathbf{k} \\ &= \left(\frac{dm}{dt}v_x + m\frac{dv_x}{dt}\right)\mathbf{i} + \left(\frac{dm}{dt}v_y + m\frac{dv_y}{dt}\right)\mathbf{j} \\ &\qquad + \left(\frac{dm}{dt}v_z + m\frac{dv_z}{dt}\right)\mathbf{k}\end{aligned} \tag{5·5}$$

The instantaneous change of momentum is then given by the single vector expression

$$\frac{d\mathbf{p}}{dt} = \frac{dm}{dt}\mathbf{v} + m\mathbf{a} \tag{5·6}$$

In the case of a particle or body of constant mass, $dm/dt = 0$.

example 5 • 2 Suppose ice forms on the wings of an airplane in flight, increasing its mass at an average rate of 7.0 kg/s while the plane maintains a constant velocity of 80 km/h east. What is the change in its momentum?

solution Since there is no acceleration, $\mathbf{a} = \mathbf{0}$; hence the instantaneous change of momentum is

$$\frac{d\mathbf{p}}{dt} = \frac{dm}{dt}\mathbf{v} = 1.56 \text{ kg} \cdot \text{km/s}^2 \text{ east}$$

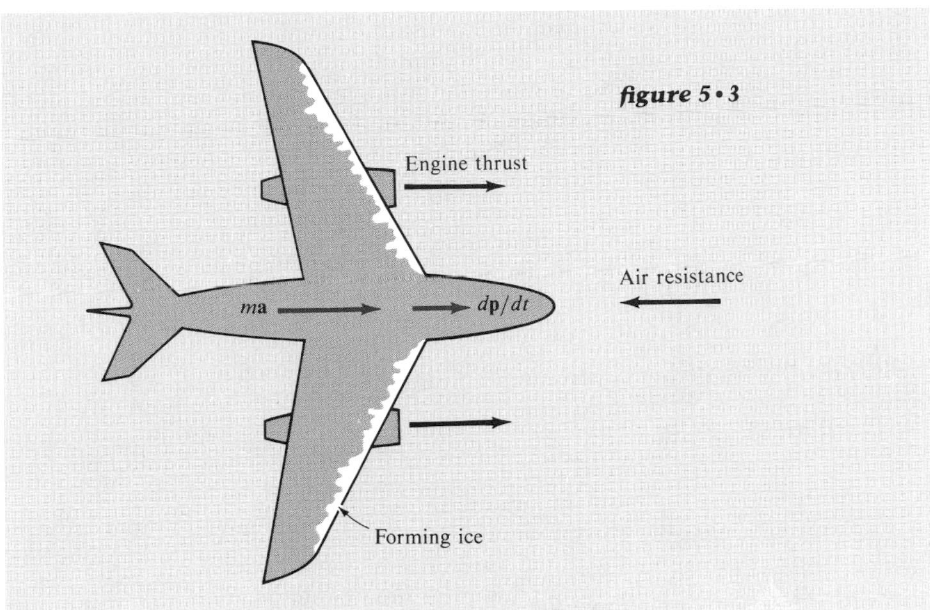

figure 5 • 3

Now let us consider a pair of interacting particles, each moving with constant velocity. The result of their interaction is described by the interaction postulate, Eq. [5 . 3], which may also be written as

$$m_1\,\Delta\mathbf{v}_1 = -m_2\,\Delta\mathbf{v}_2$$

$$\Delta\mathbf{p}_1 = -\Delta\mathbf{p}_2 \qquad [5 \cdot 7]$$

Thus the interaction postulate may also be stated in terms of Eq. [5 . 7]:

> In any interaction between an isolated pair of particles, the change in momentum of one particle is always equal in magnitude, but opposite in direction, to the change in momentum of the other particle.

Since we have referred to $\mathbf{v}_1$ and $\mathbf{v}_2$ as the velocities of the two particles after an interaction, let us denote their respective velocities before this interaction by $\mathbf{u}_1$ and $\mathbf{u}_2$. Then, from Eq. [5 . 7],

$$m_1(\mathbf{v}_1 - \mathbf{u}_1) = -m_2(\mathbf{v}_2 - \mathbf{u}_2)$$

(note that here the subscripts refer to the particles, not to time). We can write this equality as

$$\blacktriangleright \quad m_1\mathbf{u}_1 + m_2\mathbf{u}_2 = m_1\mathbf{v}_1 + m_2\mathbf{v}_2 \qquad [5 \cdot 8]$$

where the left-hand member of the equation is the vector sum of the two momenta before the interaction, and the right-hand member is the vector sum of the two momenta after the interaction. This equation asserts that these two sums are equal; the *total* momentum of the pair of particles does not change, however much the momentum of either particle changes:

> The total momentum of an isolated pair of particles is conserved.

This is the principle of the *conservation of momentum*.

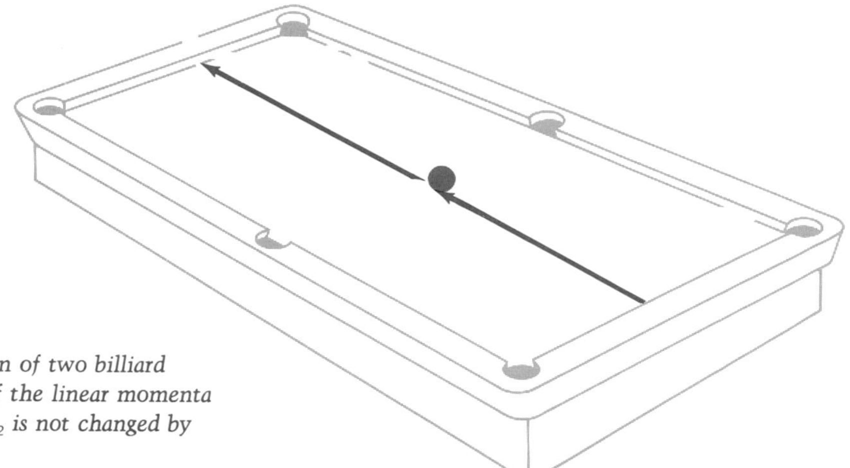

figure 5 • 4 *A collision of two billiard balls. The vector sum of the linear momenta of two bodies m_1 and m_2 is not changed by their collision.*

To visualize this principle, consider the collision of two smooth billiard balls on a horizontal table, as shown in Fig. 5 . 4. If their total momentum just before impact is represented by the vector **p**, then their total momentum just after impact is represented by the vector **p**′, which has the same magnitude and direction as **p**. It is assumed that external forces, such as friction and gravity, do not appreciably alter the momenta of the balls during the fraction of a second they are in contact. Experience shows this assumption to be valid for practically any collision of particles, provided the duration of impact is very short.

Although the isolated system we have dealt with thus far consists of only two particles, conservation of momentum holds for any isolated system of n particles which are interacting simultaneously. It has been verified experimentally that Eq. [5 . 7] is applicable independently of pairs of particles, and that, during any given time interval, the net change in momentum of any one particle is equal to the vector sum of the separate momentum changes it undergoes due to its interactions with all the other particles of the system. That is, if $\Delta\mathbf{p}_1$ represents the total change in momentum of particle 1 during a given time interval, then

$$\Delta\mathbf{p}_1 = \Delta\mathbf{p}_{12} + \Delta\mathbf{p}_{13} + \Delta\mathbf{p}_{14} + \cdots + \Delta\mathbf{p}_{1n}$$

where the terms in the right-hand member represent the momentum changes resulting from the interactions of particle 1 with particles 2, 3, 4, . . . , n.

Similar equations apply for all the other particles of the system during the same time interval:

$$\Delta\mathbf{p}_2 = \Delta\mathbf{p}_{21} + \Delta\mathbf{p}_{23} + \Delta\mathbf{p}_{24} + \cdots + \Delta\mathbf{p}_{2n}$$

$$\Delta\mathbf{p}_3 = \Delta\mathbf{p}_{31} + \Delta\mathbf{p}_{32} + \Delta\mathbf{p}_{34} + \cdots + \Delta\mathbf{p}_{3n}$$

and from Eq. [5 . 7],

$$\Delta\mathbf{p}_{12} = -\Delta\mathbf{p}_{21} \qquad \Delta\mathbf{p}_{13} = -\Delta\mathbf{p}_{31} \qquad \Delta\mathbf{p}_{1n} = -\Delta\mathbf{p}_{n1}$$

Thus if $\Delta\mathbf{p}$ is the total change of momentum of the entire system during a given time interval, then each $\Delta\mathbf{p}_{ij}$ cancels each $\Delta\mathbf{p}_{ji}$, and we have

$$\Delta\mathbf{p} = \Delta\mathbf{p}_1 + \Delta\mathbf{p}_2 + \Delta\mathbf{p}_3 + \cdots + \Delta\mathbf{p}_n = \mathbf{0} \qquad [5 \cdot 9]$$

In other words, if one particle of an isolated system gains momentum as a result of internal interactions, others in the system will simultaneously lose an equal amount; the net momentum **p** does not change with time:

$$\mathbf{p} = \mathbf{p}_1 + \mathbf{p}_2 + \mathbf{p}_3 + \cdots + \mathbf{p}_n = \sum_{i=1}^{n} \mathbf{p}_i = \text{constant} \qquad [5 \cdot 10]$$

The total momentum **p** of an isolated system of interacting particles is conserved.

The principle of conservation of momentum is a useful aid in solving problems concerning isolated systems because it asserts, as does any conservation principle, that there is a certain quantity which remains constant, regardless of variations in other properties of the system. Moreover, we can apply this principle to any isolated system without having to make detailed analyses of its *internal* interactions. This is a great advantage, for the details of most such interactions are extremely complicated and, in many instances, not yet completely understood.

example 5 • 3 A car with a mass of 1700 lbm was traveling east at a speed of $u_1 = 25$ mi/h when it was struck by a larger car entering traffic from the southwest at a speed u_2. The two cars hooked together and skidded off in a direction 18° north of east (see Fig. 5 . 5*a*). If the mass of the larger car was 3270 lbm, how fast was it moving at the time of impact? What was the combined speed V of the two cars after collision?

figure 5 • 5

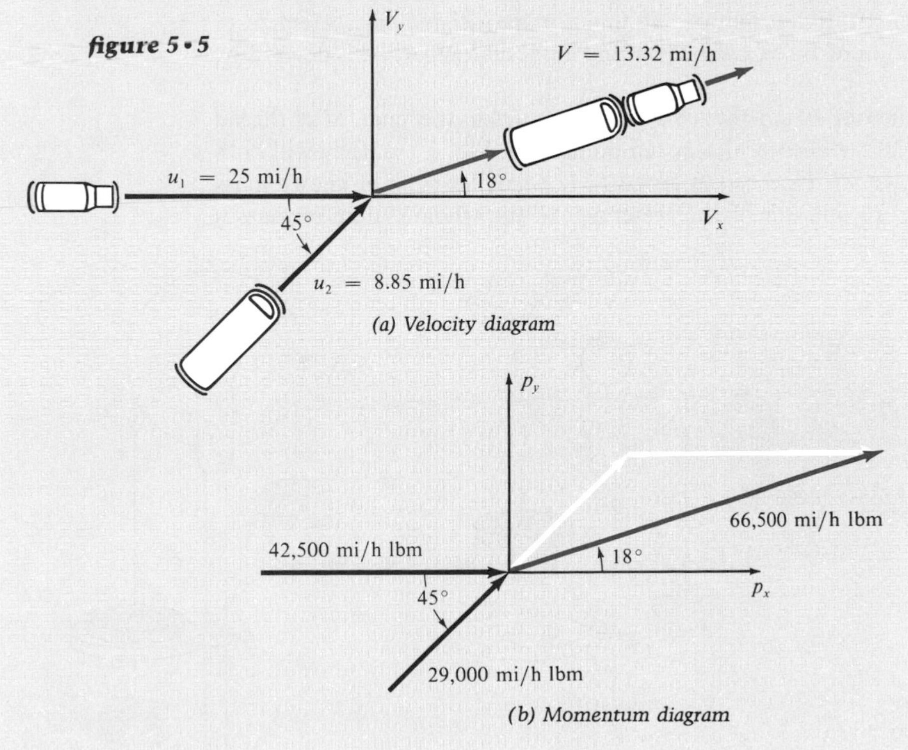

(a) Velocity diagram

(b) Momentum diagram

solution We can consider the total momentum of the two-car system in component form, with east and north as the positive x and y directions, respectively. Then the x component is, from Eq. [5 . 8],

$$p_x \text{ (initial)} = p_x \text{ (final)}$$

$$1700 \times 25 + 3270 \times u_2 \cos 45^\circ = (1700 + 3270)\, V \cos 18^\circ$$

and since the smaller car is traveling due east, the y component is

$$p_y \text{ (initial)} = p_y \text{ (final)}$$

$$3270 u_2 \sin 45^\circ = (1700 + 3270)\, V \sin 18^\circ$$

Solving simultaneously for u_2 and V gives $u_2 = 8.85$ mi/h and $V = 13.32$ mi/h. Note in Fig. 5 . 5*b* that the total final momentum can be found by the parallelogram rule of vector addition.

5 • 4 *force*

It is a human tendency to view phenomena in terms of cause and effect, and in the field of mechanics causes have been termed *forces*. But what is a force? In the first instance man must have found perceptible reasons for motion in the exertions of his own muscles. So deep-rooted is this notion, and so fundamental is the idea of force, that we must be wary of definitions such as Newton's Definition IV in the opening quotation. This concept of force clearly reflects its anthropomorphic origins and provides neither an operational nor a constitutive definition. To find a more satisfactory statement of our intuitive notion of force, let us examine an actual experiment devised by George Atwood in 1784.

If two bodies of equal mass m_1 are hung from the ends of a thread passing over a light, delicately balanced pulley (see Fig. 5 . 6), they will both move up and down with a constant speed V. If a smaller rider of known mass m_2 is now added to one side of the system, then the whole system of masses

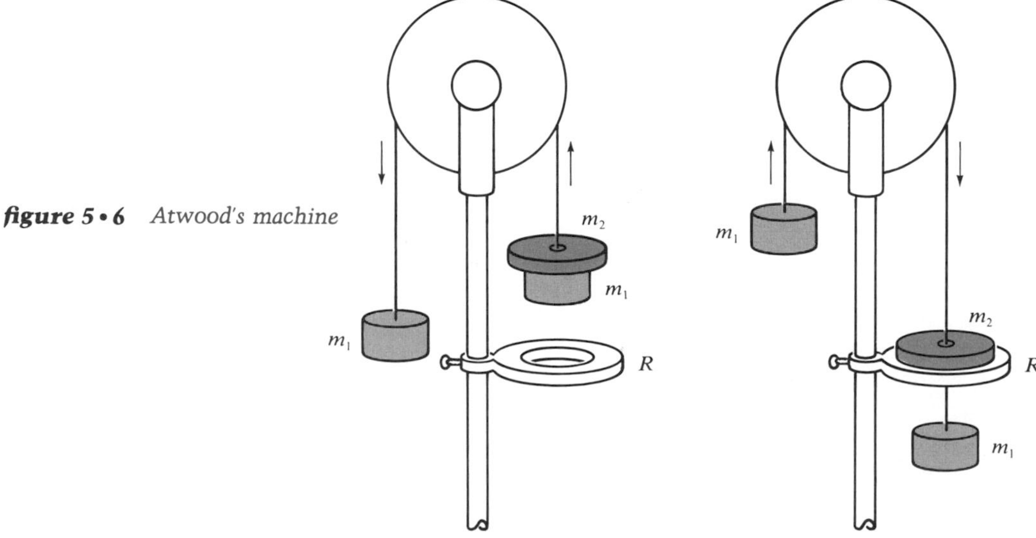

figure 5 • 6 *Atwood's machine*

will be given an acceleration due to the weight of the rider. The total mass M undergoing the acceleration is $M = 2m_1 + m_2$, if we neglect the mass of the rotating pulley. At R there is a horizontal ring through which the attached mass m_1 can pass, but which is too small for the rider m_2; so the rider is caught by the ring and can be removed from the system. Once the rider is removed, the system will now move with a new constant speed V. If we change the length of time the rider acts on the system by changing the position of the ring, we find that the difference between the initial and final speeds is proportional to the amount of time the rider was part of the system:

$$\Delta V \sim \Delta t \qquad [5 \cdot 11]$$

Suppose we increase the total mass of the system by adding two more bodies of mass m_1, one on each side, so that now $M = 4m_1 + m_2$. If we compare the results with those of the previous experiment, we find that when a given rider acts for the same time Δt on each system, the resulting change in speed is inversely proportional to the total mass M set in motion, or

$$\Delta V \sim \frac{1}{M} \qquad [5 \cdot 12]$$

Combining these two experimental relations into a single expression, we obtain

$$\Delta V \sim \frac{\Delta t}{M} \quad \text{or} \quad \frac{M\,\Delta \mathbf{V}}{\Delta t} = \frac{\Delta \mathbf{p}}{\Delta t} = \text{constant} \qquad [5 \cdot 13]$$

Thus, within the limits of accuracy of the measurements, for a given accelerating rider m_2, the change in momentum per unit time remains constant.

If the rider were sliced into three thin wafers, with two placed on one side of the pulley and one on the other, the total mass of the system would be unchanged (see Fig. 5 . 7). However, the difference in weight between the two sides would be one-third of the original. Now suppose that one of these riders is made small enough to pass through the ring along with mass m_1. In

figure 5 • 7 *Experiments with Atwood's machine*

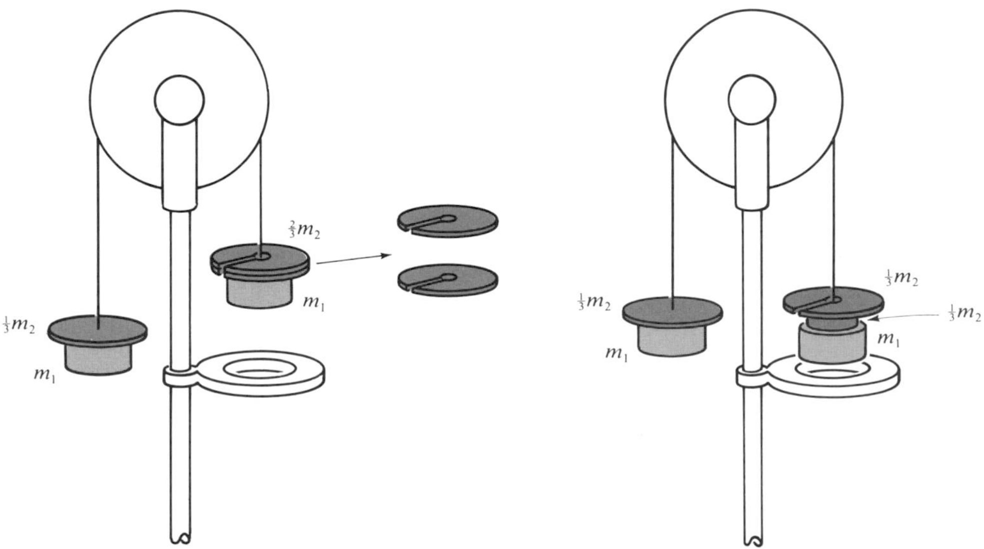

this case the speed V generated by the action of the three riders in time Δt will be found to be *one-third* of the speed generated by the action of the original rider m_2. It is apparent, therefore, that the change in momentum per unit time is proportional to the force applied by the rider, or

$$\mathbf{F} \sim \frac{\Delta \mathbf{p}}{\Delta t} \tag{5 · 14}$$

In modern terms, Newton's second law of motion can be stated as follows:

> The force acting on a body is proportional to and in the same direction as the time rate of change of its momentum.

Hence force must be a vector quantity:

$$\mathbf{F} \sim \lim_{\Delta t \to 0} \frac{\Delta \mathbf{p}}{\Delta t} = \frac{d\mathbf{p}}{dt}$$

or

$$\mathbf{F} = k\frac{d\mathbf{p}}{dt} = k\frac{d(m\mathbf{v})}{dt} \tag{5 · 15}$$

where k is a constant of proportionality whose value depends on the units used in measuring force, momentum, and time.

In the usual case, in which the mass of the particle or body does not change appreciably with time, Eq. [5 . 15] can also be expressed as

$$\mathbf{F} = km\frac{d\mathbf{v}}{dt} = km\mathbf{a} \tag{5 · 16}$$

where $\mathbf{a}$ is the instantaneous acceleration of the particle. In this situation a force $\mathbf{F}$ is the same in any reference frame, since acceleration is invariant under a Galilean transformation (see Sec. 4 . 5) and m and k are both constants. Where the mass of a body does change with time, Eq. [5 . 16] does not hold, and the force must instead be obtained from Eq. [5 . 6]:

$$\mathbf{F} = k\left(\frac{dm}{dt}\mathbf{v} + m\mathbf{a}\right) \tag{5 · 17}$$

In this general case, the applied force and the acceleration of the body are not necessarily in the same direction.

We can now see the distinction between mass and weight. A body of mass m, falling freely under the attraction of gravity, has a constant acceleration rate $a = g$. Therefore, according to Eq. [5 . 16], it must be experiencing a gravitational force $\mathbf{F}_g$ equal in magnitude to kmg. This force is what we term the *weight* of the object. While the mass of a given object remains constant, its weight may vary slightly, depending on the local value of the acceleration due to gravity.

example 5 • 4 Suppose the engines of the airplane in Example 5 . 2 are already operating at capacity when ice begins to form on the wings. At this point the available thrust (force) of the engines is just sufficient to maintain constant speed, so that the net force on the airplane is zero. If the airplane has a mass of 50,000 kg and the ice forms at a constant rate, what is the acceleration?

solution Since the net force on the airplane is zero, there is no change in its momentum:

$$\frac{d\mathbf{p}}{dt} = \frac{dm}{dt}\mathbf{v} + m\mathbf{a} = \mathbf{0}$$

Therefore the velocity must decrease with increasing mass, and the airplane has a negative acceleration of

$$\mathbf{a} = \frac{-1560 \text{ kg} \cdot \text{m/s}^2}{50{,}000 \text{ kg}} = -0.0312 \text{ m/s}^2$$

5 • 5 *units of force*

In mechanics it is usually sufficient to choose units for three fundamental dimensions, called *primary units*. The units for all other mechanical quantities are then expressed in terms of these primary units and are derived, or *secondary, units*. In the physical sciences, electrical engineering, and much of mechanical engineering, the primary dimensions are usually chosen as length, mass, and time. In those branches of engineering in which forces play a more important role than masses, a length-force-time system proves to be convenient. For some purposes a system based on four primary dimensions is advantageous; one example is the length-mass-force-time system.

If all our units of measurement were defined independently of $\mathbf{F} = k\, d\mathbf{p}/dt$, we would not be free to assign to the constant k any value we please. If we do want to choose an arbitrary value, the simplest one possible is $k = 1$, so that Eq. [5 . 15], for instance, becomes

$$\mathbf{F} = m\mathbf{a} \qquad [5 \cdot 18]$$

However, this can be done only if we use the equation $\mathbf{F} = m\mathbf{a}$ to define a new secondary unit for one of the three quantities it contains. If force is chosen for this purpose, the new unit is called an *absolute,* or *dynamical, unit of force* and is defined to be that force which is required to impart to a unit of mass one unit of linear acceleration. In more general terms, it is that force which is required to change the linear momentum of a body at the rate of one momentum unit per unit of time.

If the kilogram is taken as the unit of mass and acceleration is expressed in meters per second per second, the new unit of force in the meter-kilogram-second (mks) system is, by definition, 1 kg · m/s^2, called the *newton* (N). In the centimeter-gram-second (cgs) system the unit of force is 1 g · cm/s^2, called the *dyne*. Matters would be much simpler if we had to contend only with these metric units of length, mass, and time. Unfortunately, however, much of the English-speaking world still clings to the British system of primary units, the foot-pound-second (fps) system. What makes this system particularly confusing is that the unit of force, the pound or pound-force (lbf), is taken as primary, and it is the unit of *mass* that is secondary. This unit, chosen in such a way that $k = 1$ when the acceleration is 1 ft/s^2, has been named the *slug*. It represents the mass of a body that undergoes an acceleration of 1 ft/s^2 when acted upon by a force of 1 lbf; in other words,

$$1 \text{ lbf} = 1 \text{ slug} \cdot \text{ft/s}^2$$

To understand these distinctions it is necessary to separate the two concepts of force and mass. This is difficult to do, partly because we have been confusing them all our lives, but also because the gravitational force, or weight, of an object is proportional to its mass. Imagine, for instance, that you are walking down a street in Istanbul, Turkey, and you stop at a fruit stand to buy a kilo of Marmora peaches. You are buying a definite mass (1 kg) of peaches. This mass weighs

$$F_g = mg = (1 \text{ kg})(9.8 \text{ m/s}^2) = 9.8 \text{ N}$$

Of course, the vendor knows nothing about newtons; he merely reads "1 kg" from his scale, which is calibrated in units of *kilogram-force* (kgf), the weight of a 1-kg mass on the earth's surface. Thus 1 kgf $= \{g'\}$ N, where the numerical value $\{g'\} = 9.80665$ is the magnitude of the acceleration due to gravity, expressed in mks units.*

In contrast, if you are walking down a street in Atlanta, Georgia, and you stop to buy a pound of Georgia peaches, you buy them by weight. Their actual mass is a rather awkward

$$m = \frac{F_g}{g} = \frac{1 \text{ lbf}}{32.1734 \text{ ft/s}^2} = 0.0310816 \text{ slugs}$$

However, we customarily think of this simply as a "mass whose weight is one pound-force (lbf)"—that is, as a pound-mass (lbm). Thus the weight of 1 slug is

$$F_g = \{g''\} \text{ lbf} = 32.1734 \text{ lbf} \qquad 1 \text{ lbm} = 1/\{g''\} \text{ slugs}$$

where $\{g''\} = 32.1734$ is the acceleration due to gravity expressed in fps units. Measurement will show that a mass of 453.6 g weighs 1 lbf, which implies the conversion relationships shown in Table 5 . 2. Table 5 . 3 summarizes the dimensional relations in the force equation $\mathbf{F} = km\mathbf{a}$. After Chapter 7 we shall use the metric system almost exclusively, but it is important to have a clear grasp of the distinction between primary and secondary units in these systems.

*See the discussion of Fig. 1 . 1 in Sec. 1 . 2.

table 5 • 2 *Force conversion factors*

equal forces: 1 N $=$ 1 *kg m/s*^{2} $=$ 10^3 *g* 100 *cm/s*^{2} $=$ 10^5 *dynes*

newtons (N)	*kilogram-forces (kgf)*	*poundforces (lbf)*
1	0.1020	0.2248
9.807	1	2.205
4.448	0.4536	1

table 5 • 3 *Force units in different mechanical systems*

unit system	*unit of mass*	*unit of force*
absolute		
mks	kilogram	newton (N) $=$ kg $\cdot$ m/s^2
cgs	gram	dyne $=$ g $\cdot$ cm/s^2
fps	slug	poundforce (lbf) $=$ g $\cdot$ cm/s^2
*gravitational**		
mks	kilogram	kilogram-force (kgf) $= 1/\{g'\}$ kg $\cdot$ m/s^2
cgs	gram	gram-force (gf) $= 1/1000\{g'\}$ g $\cdot$ cm/s^2
fps	pound-mass	poundforce (lbf) $= 1/\{g''\}$ lbm $\cdot$ ft/s^2

*$g' = 9.80665$ m/s^2, $g'' = 32.1734$ ft/s^2

example 5 • 5 A body of mass 5 lbm is accelerating at a rate of 2 ft/s^2. (*a*) What must be the force acting on it, in pounds? (*b*) If we doubled the mass and tripled the force, what would the acceleration be? (*c*) What would this new mass be in fps units? (*d*) If the force is again increased, and during the next 10 s this new mass accelerates from rest to a speed of 50 ft/s, what is the new force? (*e*) Convert the force, mass, and acceleration of the body in part (*d*) to mks units. How many kilogram-forces does the body experience?

solution (*a*) The amount of force acting on the original body is

$$F = ma = (5\text{ lbm})\left(\frac{1\text{ slug}}{\{g''\}\text{ lbm}}\right)(2\text{ ft/s}^2) = 0.3108\text{ lbf}$$

(*b*) Since the force is now $3F$ and the mass is $2m$, the new acceleration is

$$a = \frac{F}{m} = \tfrac{3}{2}(2\text{ ft/s}^2) = 3\text{ ft/s}^2$$

(*c*) In fps units the doubled mass is

$$10\text{ lbm} = (10\text{ lbm})(1\text{ slug}/\{g''\}\text{ lbm}) = 0.3108\text{ slugs}$$

The consistency can be checked by noting that

$$a = \frac{F}{m} = \frac{0.9324\text{ lbf}}{0.3108\text{ slugs}} = 3\text{ ft/s}^2$$

(*d*) Since $v = at$, a uniform acceleration to 50 ft/s in 10 s means $a = 5$ ft/s^2; hence the force is

$$F = ma = (10\text{ lbm})(5\text{ ft/s}^2) = 1.554\text{ lbf}$$

(*e*) In mks units, by simple substitution from Table 5 . 2,

$$F = 1.554 \text{ lbf} = (1.554 \text{ lbf})(4.448 \text{ N/lbf}) = 6.912 \text{ N}$$

$$m = 10 \text{ lbm} = (10 \text{ lbm})(0.4536 \text{ kg/lbf}) = 4.536 \text{ kg}$$

$$a = 5 \text{ ft/s}^2 = (5 \text{ ft/s}^2)(0.3048 \text{ m/ft}) = 1.524 \text{ m/s}^2$$

and the number of kilogram-forces is

$$6.912 \text{ N} = (6.912 \text{ N})(0.1020 \text{ kgf/N}) = 0.7050 \text{ kgf}$$

5 · 6 Newtonian mechanics and noninertial frames of reference

One major problem in physics is translating what we observe into objective statements. In general, we try to work in frames of reference that do not require correction of the results for the motion of the observer. However, this is not always possible. For example, if we are considering falling bodies (such as reentry vehicles) which travel distances comparable to the earth's radius, it becomes necessary to consider the effects of the earth's rotation. Furthermore, Newton's laws pertain to isolated particles or systems, and any principle or law deduced from them pertains to such particles or systems only as viewed from an inertial reference frame.

Although inertial frames and isolated particles or systems are never completely realizable in practice, their conceptualization represented an enormous step in the development of the theory of dynamics. This becomes evident when we see that all the postulates, principles, and laws of Newtonian mechanics have the same form (are *invariant*) when they are expressed relative to inertial frames of reference; that is, any physical principle which holds true in one inertial frame may be stated in the same form in all inertial frames.

For example, the position and velocity of a moving particle differs with respect to any two inertial frames S and S', but changes in velocity are invariant:

$$\Delta \mathbf{v} = \Delta \mathbf{v}' \qquad [5 \cdot 19]$$

In Newtonian mechanics the mass of an object is independent of its motion, so that the interaction postulate, Eq. [5 . 3], is also invariant:

$$\frac{m_2}{m_1} = \frac{-\Delta \mathbf{v}_1}{\Delta \mathbf{v}_2} = \frac{-\Delta \mathbf{v}_1'}{\Delta \mathbf{v}_2'} \qquad [5 \cdot 20]$$

Similarly, the equations expressing the conservation of momentum are invariant, since $\Delta \mathbf{p} = m\,\Delta \mathbf{v}$.

Because time is invariant in Newtonian mechanics, the *form* of Newton's second law is also invariant in any inertial system. This is an important point, because it allows us to extrapolate physical laws verified in our own system to all other inertial systems. These laws should hold on the farthest stars.

In the case of transformations from inertial frames of reference to noninertial ones—those whose velocity may be changing in magnitude or in direction (rotating) or both—it turns out that the only Newtonian quantities that remain unchanged in value are time, the lengths of rigid rods, and the masses of isolated particles or bodies. It follows that the Newtonian postulates, principles, and laws are *not* invariant when they are transformed to noninertial reference frames. Thus an observer in a noninertial frame of reference can easily be deceived about the nature of the physical world.

Imagine that you are in a train accelerating uniformly eastward through a long, dark tunnel (we shall neglect the earth's rotation and assume that the tunnel is in an inertial frame). You cannot see the sides of the tunnel or the headlights of your own train shining on the track, but you can see the lights of an approaching train, which is traveling at a constant velocity westward. Under these circumstances the approaching train will appear to be *accelerating* westward. Now consider the conclusions you might draw as a physicist if you mistakenly assume that you are situated in an inertial frame—that is, that your own train is moving with constant velocity.

First, you will assume that the approaching train is subject to a net force which accelerates it westward. You can feel the solid objects around you, and since you are actually accelerating eastward, the back of your seat will be pushing against you. However, if you believe you are moving at constant velocity, you must also assume that some mysterious force is pushing you westward against your seat. You might hear your engine straining to accelerate, but it would labor just as much if you were going uphill at constant velocity—or moving against a westward force. Thus you might conclude that in your tunnel universe there is a mysterious force that tends to accelerate all objects westward. Needless to say, if you ever tried to make use of this force in a truly inertial frame—like throwing the dishwater into a real west wind—the results could prove disappointing.

In physics such fictitious forces are termed *inertial reaction forces*. Although they feel quite real to the accelerating observer, they actually arise from his own inertial resistance to being accelerated and are not independent

figure 5 • 8 *Inertial reaction forces*

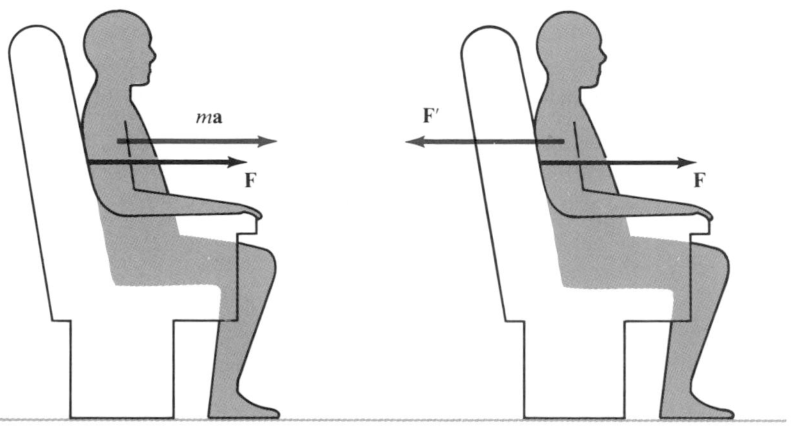

(a) Inertial system

(b) Noninertial system

forces, like gravitational attraction. With respect to an inertial reference frame, such as the tunnel, your seat is exerting an eastward force $\mathbf{F} = m\mathbf{a}$ on you, which causes you to accelerate along with the train. Your resistance to this acceleration causes you to exert an opposing inertial reaction force, $\mathbf{F}' = -m\mathbf{a}$, on the seat. However, if you believe that your accelerating train system is moving with constant velocity, you will have to conclude that the force $\mathbf{F}$ you feel is acting against an opposing accelerating force $\mathbf{F}''$ pushing you westward. In your own perception, both of these forces will be acting on you to keep you in equilibrium:

$$\mathbf{F} + \mathbf{F}'' = \mathbf{0} \qquad \mathbf{F}'' = -\mathbf{F} = -m\mathbf{a} = \mathbf{F}' \tag{5 . 21}$$

The fictitious force $\mathbf{F}''$ you perceive, of course, would be measured by a stationary observer as the real inertial reaction force $\mathbf{F}'$ you are exerting on the seat. Equation [5 . 21] is just another way of stating Newton's second law of motion. It is known as *D'Alembert's principle* and states that the second law may be viewed as a balance between real and fictitious inertial reaction forces.

PROBLEMS

5 • 1–5 • 2 inertial mass

5 • 1 The *density* ρ of an object is defined as the ratio of its mass to its volume. Careful measurements show that a certain cube of metal is 2.2 cm on a side and has a mass of 96.24 g. (*a*) Find its density. (*b*) What is this density in pound-masses per cubic foot? (*c*) In slugs per cubic foot?

5 • 2 If 1 g of water occupies a volume of 1 cm^3, what mass, in pound-masses, occupies a cubic foot?

5 • 3 In the experiment shown in Fig. 5 . 2, it is found that a certain pair of cars must be placed at the center of the track in order to reach the track ends simultaneously when they are sprung apart. A replica of the prototype kilogram is added to one of the cars, and it is then found that this car and the empty car must move −27 and 72 cm, respectively, from the origin to reach the track ends simultaneously. (*a*) What is the mass of each car? (*b*) What is the ratio of their velocity changes as a result of their interaction?

5 • 4 A man of mass 70 kg and a woman of mass 50 kg are ice skating together at a speed of 10 mi/h. If the man gives the woman a push forward so that she attains a speed of 13 mi/h, what is his final speed?

5 • 5 In an experiment with the reaction cars of Fig. 5 . 2, does the amount that the spring is compressed affect (*a*) the distances the two cars move? (*b*) The ratio of Δx_1 to Δx_2? (*c*) What if a stronger spring were used?

5 • 3 conservation of momentum

5 • 6 A 10-ton truck moving with a velocity of 14 ft/s northward is stopped by buffers in 0.3 s. Compute the initial momentum of the truck, its average acceleration, and the average time rate of change of its momentum.

5 • 7 A train carrying empty hopper cars is moving with a constant velocity of 100 km/h at 10° north of east and acquires additional mass at the rate of

10 kg/s from a vertically falling rain. Find the time rate of change of its momentum.

5•8 Suppose the train in Prob. 5 . 7 has been accelerating at a rate of 0.2 m/s^2, and at the moment it reaches a speed of 100 km/h it has a mass of 50,000 kg. What is the time rate of change of its momentum?

5•9 A 4.2-lbm stone is thrown from a 450-ft tower. (*a*) What is the change in its momentum between the time of projection and the time it hits the ground if its initial velocity is zero? (*b*) If the velocity of projection is 40 ft/s vertically downward? (*c*) If the velocity of projection is 40 ft/s vertically upward?

5•10 What would be the change in momentum of the stone in Prob. 5 . 9 if its velocity of projection were 40 ft/s in a horizontal direction?

5•11 A 4-kg projectile is fired horizontally with a speed of 350 m/s from a gun of mass 3000 kg. What is the initial recoil speed of the gun?

5•12 A 4-g putty ball moving southward at a speed of 2000 cm/s strikes and sticks to a 100 g billiard ball rolling eastward at a speed of 200 cm/s. What is the combined velocity of the balls immediately after impact?

5•13 A 10-g bullet is shot horizontally with a speed of 2400 m/min into the center of a 500-g wooden ball which is rolling along a horizontal plane at a speed of 400 cm/s. The path of the bullet is at a 45° angle to the path of the ball before impact. (*a*) If the bullet remains in the ball, how much is the direction of the ball changed by the impact? (*b*) What is the speed of the ball after impact?

5•14 Two toy railroad cars on the same track have masses of 300 g and 100 g, respectively. The lighter car is initially at rest, and the heavier car moves toward it at a speed of 3 m/s. A 10-g ball is fired from the heavier car into the lighter car, causing it to move at a speed of 1 m/s. What is the speed of the heavier car now?

5•15 Explain the statement that the principle of conservation of momentum is valid for a single isolated particle.

5•16 Show that under a Galilean transformation between systems S and S', when $\mathbf{p}_1' = \mathbf{p}_1 + \mathbf{k}$, then $\mathbf{p}_2' = \mathbf{p}_2 + \mathbf{k}$, where $\mathbf{k}$ is a constant momentum vector. Then show that $\Delta\mathbf{p}' = \Delta\mathbf{p}$—that is, that changes of momentum are invariant.

5•4–5•5 force

5•17 Why must a graph of position versus time, $x(t)$, for a particle of mass m be single-valued and continuous and have a continuous first derivative?

5•18 What acceleration is imparted to an object of 5 lbm by a force of 1 lbf?

5•19 A friend pushes steadily forward on your car, which has a mass of 1100 kg and a dead battery. After 5 s the car is moving at a speed of 45 cm/s. Neglecting friction, compute the amount of force your friend has exerted.

5 • 20 A 90-kg projectile strikes an embankment at a speed of 400 m/s and penetrates 4 m. What is the average resisting force, in kilogram-forces, that the embankment offered to its motion?

5 • 21 A particle with a mass of 5 g is moving in a straight line with an instantaneous velocity given by

$$\mathbf{v} = (12\mathbf{i}\ \text{ft/s}) - (6\mathbf{i}\ \text{ft/s}^2)t$$

(*a*) Write the expression for the accelerating force as a function of time. (*b*) Find the magnitude of the velocity, the acceleration, and the accelerating force at times $t = 0$, $t = 1$ s, $t = 2$ s, and $t = 3$ s.

5 • 22 A 12-g particle is moving in a straight line with a momentum of 480 g · m/s, and a force of 2000 dynes is applied to the particle in opposition to this motion. (*a*) How long will it take to bring the particle to rest? (*b*) How long would it take an 18-g particle moving with the same speed to come to rest?

5 • 23 A force of 1 megadyne (10^6 dynes) acts for 1 min on a stationary object of mass 1 metric ton (1000 kg). (*a*) What speed does the object acquire in this time? (*b*) How far does it move? (*c*) What is the momentum it acquires? (*d*) The rate of change of its momentum?

5 • 24 A 75-kg parachutist, wearing a parachute of negligible mass, jumps from a balloon, and after a free fall of 100 m, the parachute opens. In the next 3 s the parachutist is slowed down to a velocity of 5 m/s downward. Assuming that the acceleration was constant, how much force did the parachute exert on the parachutist during the 3-s interval?

5 • 25 A force of $\mathbf{F} = (2\mathbf{i} + 3\mathbf{j})$ dynes acts on a particle of mass m, which starts from the rest at the origin of the coordinates. After 10 s the position of the particle is given by the coordinates (300 cm, 450 cm). What is its mass?

5 • 26 Careful measurements show that the mass of a certain metal specimen is 48.121 g. What is its weight, in gram-forces, (*a*) at a place where local gravity is $g = 980.572$ cm/s^2? (*b*) At a place where local gravity is equal to standard gravity, $g_s = 980.665$ cm/s^2.

5 • 27 The ratio of local gravity g to standard gravity g_s usually differs from unity by less than 0.25 percent. (*a*) When may the mass of a body and its weight in the corresponding gravitational units be considered numerically equal? (*b*) When may the local weight be taken as equal to the standard weight?

5 • 28 In 1919 France legalized the *meter–metric-ton–second* system of units, the metric ton being a mass equivalent to 1000 kg. (*a*) Show that in this system the unit of force, which is called the *athene,* is 1 metric ton · m/s^2, or 1000 N. (*b*) Show that the density of water at 4°C (1 g/cm^3) is 1 metric ton/m^3. (*c*) Make a listing for this system similar to those in Table 5 . 3.

5 • 29 On the European continent use is sometimes made of a unit of mass called the *metric slug,* defined as the mass of a body that would be accelerated 1 m/s^2 by an applied force of 1 kgf. (*a*) What is the equivalent of this unit in kilograms? (*b*) Make a listing similar to those in Table 5 . 3 for the system that includes this unit.

5•6 noninertial frames of reference

5•30 How much force is exerted on the floor of an elevator by a passenger weighing 150 lbf (*a*) if the elevator is ascending with an acceleration of 3.22 ft/s^2? (*b*) If it is descending with an acceleration of 3.22 ft/s^2? (*c*) If it is moving with constant speed? (*d*) If the elevator cable breaks (free fall)?

5•31 Describe and interpret the forces you would actually *feel* if you were blindfolded and (*a*) standing on a high platform. (*b*) Falling freely through the atmosphere. (*c*) Sitting on the floor of a rotating platform, such as a carousel, at a distance from its center.

5•32 Suppose a friend of yours is standing on a bathroom scale inside an elevator which is ascending with an acceleration of $a_y = 5\text{ m/s}^2$. (*a*) If the scale reads 200 lbf, what is your friend's actual weight? (*b*) If the reading suddenly drops to 100 lbf, what is the new acceleration of the elevator? (*c*) If the elevator cable snaps, what is the reading on the scale?

5•33 An astronaut accidentally "drops" a pencil inside a space capsule and watches it rise slowly to the top of the compartment, rotating clockwise in an expanding spiral. It takes the pencil 10 s to complete a loop of the spiral and 4 s to rise 3 m to the top of the compartment. Describe the motion of the ship with respect to an inertial observer. Would you advise the astronaut to take a coffee break?

answers

5•1 (*a*) $\rho = 9.038\text{ g/cm}^3$; (*b*) $\rho = 564.3\text{ lbm/ft}^3$; (*c*) $\rho = 17.55\text{ slugs/ft}^3$
5•2 62.43 lbm
5•3 (*a*) $m_1 = m_2 = 600$ g; (*b*) $-\Delta v_1/\Delta v_2 = \frac{3}{8}$
5•4 $v = 7.86$ mi/h
5•5 (*a*) Yes; (*b*) no; (*c*) travel time might be affected, but not $\Delta x_1/\Delta x_2$
5•6 $\mathbf{p}_0 = 2.8 \times 10^5\text{ lbm}\cdot\text{ft/s}^2$ north, $\mathbf{a} = 4.67\text{ ft/s}^2$ south, $\Delta\mathbf{p}/\Delta t = 2.90 \times 10^4\text{ slug}\cdot\text{ft/s}^2$ south
5•7 $d\mathbf{p}/dt = 277.8\text{ kg}\cdot\text{m/s}^2$ 10° north of east
5•8 $d\mathbf{p}/dt = 10{,}278\text{ kg}\cdot\text{m/s}^2$ 10° north of east
5•9 (*a*) $\Delta\mathbf{p} = -715\mathbf{j}\text{ lbm}\cdot\text{ft/s}^2$; (*b*) $\Delta\mathbf{p} = -566\mathbf{j}\text{ lbm}\cdot\text{ft/s}^2$; (*c*) $\Delta\mathbf{p} = -902\mathbf{j}\text{ lbm}\cdot\text{ft/s}^2$
5•10 Same as Prob. 5 . 9(*a*)
5•11 $v = 0.47$ m/s
5•12 $\mathbf{V} = 2.07$ m/s 21.8° south of east
5•13 (*a*) $\Delta\theta = -37.9°$; (*b*) $v = 451$ cm/s
5•14 $v_1 = 2.72$ m/s
5•15 $d\mathbf{p}/dt = \mathbf{0}$, $\mathbf{p}$ = constant
5•18 $a = 6.43\text{ ft/s}^2$
5•19 $F = 99$ N or 10.1 kgf
5•20 $\bar{F} = 1.84 \times 10^6$ kgf
5•21 (*a*) $F = -30\text{ g}\cdot\text{ft/s}^2$; (*b*) $v = 12, 6, 0, -6$ ft/s, $a = -6\text{ ft/s}^2$, $F = -30\text{ g-ft/s}^2$
5•22 (*a*) $\Delta t = 24$ s; (*b*) $\Delta t = 36$ s
5•23 (*a*) $v = 60$ cm/s; (*b*) $\Delta x = 1800$ cm; (*c*) $\Delta p = 6 \times 10^7\text{ g}\cdot\text{cm/s}^2$; (*d*) $dp/dt = 10^6\text{ g}\cdot\text{cm/s}^2$
5•24 $F = 1226$ N or 125 kgf
5•25 $m = \frac{1}{3}$ g
5•26 (*a*) $F_g = 48.116$ gf; (*b*) $F_g = 48.121$ gf
5•27 When accuracy is limited to three significant figures
5•29 (*a*) 9.8 kg
5•30 (*a*) $F = 165$ lbf; (*b*) $F = 135$ lbf; (*c*) $F = 150$ lbf; (*d*) zero
5•31 (*a*) Upward reaction force of platform; (*b*) no force, "weightlessness"; (*c*) reaction force of platform and "centrifugal" force outward from center of platform
5•32 (*a*) 132.4 lbf; (*b*) $\mathbf{a} = 240\text{ m/s}^2$ downward; (*c*) zero
5•33 Counterclockwise rotation at one revolution every 10 s (approx.) and downward acceleration at 1.5 m/s^2

CHAPTER SIX

dynamics of particles

Corollary I *A body, acted on by two forces simultaneously, will describe the diagonal of a parallelogram in the same time as it would describe the sides by those forces separately.*

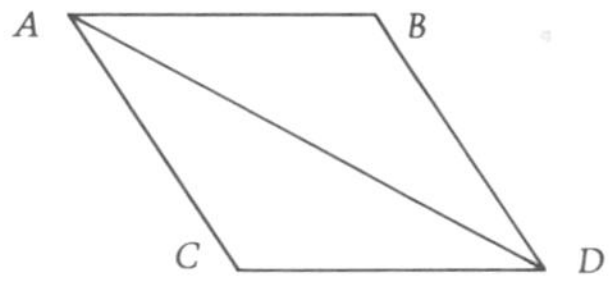

If a body in a given time, by the force M impressed apart in the place A, should with an uniform motion be carried from A to B, and by the force N impressed apart in the same place, should be carried from A to C, let the parallelogram ABCD be completed, and, by both forces acting together, it will in the same time be carried in the diagonal from A to D. For since the force N acts in the direction of the line AC, parallel to BD, this force (by the Second Law) will not at all alter the velocity generated by the other force M, by which the body is carried towards the line BD. The body therefore will arrive at the line BD in the same time, whether the force N be impressed or not; and therefore at the end of that time it will be found somewhere in the line BD. By the same argument, at the end of the same time it will be found somewhere in the line CD. Therefore it will be found in the point D, where both lines meet. But it will move in a right [straight] line from A to D, by the First Law.

Isaac Newton, Principia, *1686,*
translated by Andrew Motte, 1729

In this chapter we shall consider the behavior of particles under the action of external forces. As used here, the term *particle* encompasses not just single bodies, but any *system* of bodies which can be isolated from the rest of the universe except for a small number of forces acting on it. Any such system can be treated mathematically as a point mass, with all the applied forces considered to be concentrated at its geometric center. The reason for this will become clear when we discuss centers of mass in detail in Chapter 10; for the time being, we shall rely on intuition in discussing such particles.

6 • 1 the vector nature of forces

In 1586 Simon Stevin (Stevinus), in his study of the equilibrium of bodies on an inclined plane, derived the method for determining the effect of several forces acting at a point. He did not, however, expressly formulate the method for adding these forces. This was done by Newton, who also demonstrated its general application to dynamics, in his Corollary I to the laws of motion.

The second law of motion implies that each force acting on a particle produces its own effect, independently of the action of any other force, and regardless of whether the particle is at rest or in motion. The independence of forces is not a self-evident physical phenomenon, but it is easily established by experiment. One common experimental device is the *force table,* shown in Fig. 6 . 1. Weights are suspended through pulleys, which can be moved freely around the circumference of the table. A pin through a ring at the center of the table holds the cords in position; when the three forces are perfectly balanced, the ring will remain stationary when the pin is removed. It can be verified graphically that the forces exerted by the cords form a triangle—that is, their vector sum is zero.

It follows from this experimental principle that several individual forces

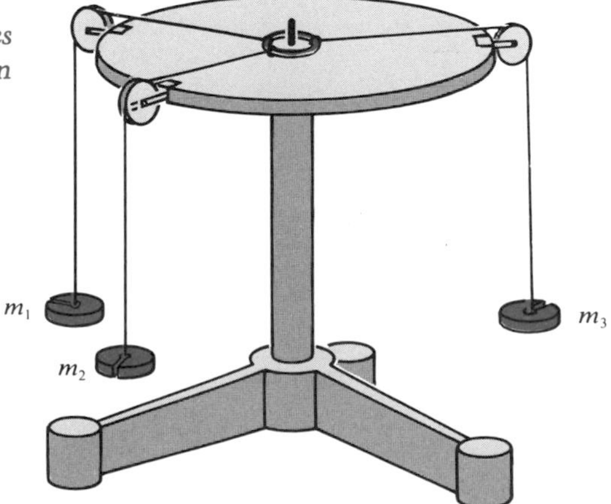

figure 6 • 1 *The force table, a device for applying at a common point several known forces that make known angles with one another. When a test for equilibrium is to be made, the pin holding the cords in position at the center is removed.*

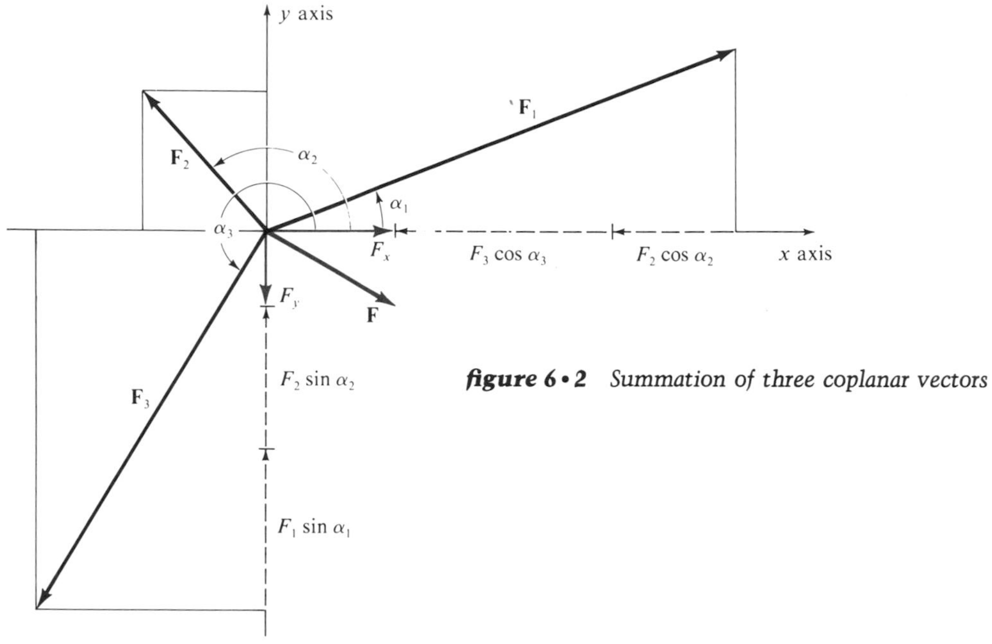

figure 6 • 2 *Summation of three coplanar vectors*

$\mathbf{F}_i$ acting simultaneously on a body impart one resultant acceleration to the body which is the vector sum of the individual accelerations $\mathbf{a}_i$ produced by the separate forces. Since each force is proportional to and in the same direction as the acceleration it produces, it follows that the vector sum $\mathbf{F}$ of the separate forces $\mathbf{F}_i$ is proportional to and in the same direction as the vector sum $\mathbf{a}$ of the separate accelerations $\mathbf{a}_i$. Thus the vector sum, or *resultant,* of any number of forces is that single force which would produce the same acceleration that is produced by the joint action of the several forces. In the notation of vector algebra:

$$\mathbf{F}_i = m\mathbf{a}_i$$

where $i = 1, 2, \ldots, n$, and *

$$\mathbf{F} = \mathbf{F}_1 + \mathbf{F}_2 + \cdots + \mathbf{F}_n = m(\mathbf{a}_1 + \mathbf{a}_2 + \cdots + \mathbf{a}_n)$$

$$= \sum_{i=1}^{n} \mathbf{F}_i = m \sum_{i=1}^{n} \mathbf{a}_i = m\mathbf{a} \qquad [6 \cdot 1]$$

The forces acting on a particle are added just like any other vector quantities: either graphically by the parallelogram or polygon rule or by resolution into cartesian components and straightforward addition of the components. Thus for the sum $\mathbf{F}$ of three forces lying in a single plane (see Fig. 6 . 2),

$$\mathbf{F} = \mathbf{F}_1 + \mathbf{F}_2 + \mathbf{F}_3$$

and the cartesian components of this vector sum are

$$F_x = \sum_{i=1}^{3} F_i \cos \alpha_i \qquad F_y = \sum_{i=1}^{3} F_i \sin \alpha_i \qquad [6 \cdot 2]$$

*For a brief description of summation notation, see Appendix D.

When the forces to be added do not all lie in the same plane, it is generally necessary to resolve each individual force into its direction components. If the direction angles of each force $\mathbf{F}_i$ relative to the x, y, and z axes are, respectively, α_i, β_i, γ_i, then

$$F_x = \sum_{i=1}^{n} F_i \cos\alpha_i \qquad F_y = \sum_{i=1}^{n} F_i \cos\beta_i \qquad F_z = \sum_{i=1}^{n} F_i \cos\gamma_i \qquad [6 \cdot 3]$$

The magnitude of the resultant force is

$$F = \sqrt{F_x^2 + F_y^2 + F_z^2} \qquad [6 \cdot 4]$$

and its direction angles α, β, and γ are given by

$$\cos\alpha = \frac{F_x}{F} \qquad \cos\beta = \frac{F_y}{F} \qquad \cos\gamma = \frac{F_z}{F} \qquad [6 \cdot 5]$$

example 6 • 1 A force of 75 N has direction angles $\alpha = 35°$ and $\beta = 70°$. Find (*a*) its cartesian components and (*b*) the direction angle γ which it makes with the z axis.

solution (*a*) The x and y components of the force can be found from

$$F_x = F\cos\alpha = 75\cos 35° = 61.44 \text{ N}$$

$$F_y = F\cos\beta = 75\cos 70° = 25.65 \text{ N}$$

Since $F_z^2 = F^2 - F_x^2 - F_y^2$, we have two possible values for F_z:

$$F_z = \pm\sqrt{5625 - 3774.9 - 657.9} = \pm 34.53 \text{ N}$$

(*b*) The direction angle with the z axis may also have two values. From Eq. [6 . 5],

$$\cos\gamma = \frac{F_z}{F} = \pm 0.4604$$

Hence $\gamma = 62.59°$ or $117.41°$.

example 6 • 2 A particle having a mass of 3 g is acted on by three forces, measured in dynes:

$$\mathbf{F}_1 = 5\mathbf{i} - 3\mathbf{k} \qquad \mathbf{F}_2 = 6\mathbf{j} \qquad \mathbf{F}_3 = \mathbf{i} - 3\mathbf{j} + 6\mathbf{k}$$

Find the resultant acceleration **a** of the particle and the magnitude of the resultant force.

solution Adding the three forces vectorially gives the resultant force:

$$\mathbf{F} = \mathbf{F}_1 + \mathbf{F}_2 + \mathbf{F}_3 = (6\mathbf{i} + 3\mathbf{j} + 3\mathbf{k}) \text{ dynes} = m\mathbf{a}$$

The resultant acceleration is $\mathbf{a} = \mathbf{F}/m$, or

$$\mathbf{a} = (2\mathbf{i} + \mathbf{j} + \mathbf{k}) \text{ cm/s}^2$$

and the magnitude of the force is $F = 3\sqrt{6}$ dynes.

6 · 2 statics of a particle

A particle is said to be in *equilibrium* when the vector sum of the forces acting on it is zero. Thus equilibrium does not necessarily mean a state of rest; rest is merely one case of constant momentum. When a particle is acted on by two forces that have the same magnitude, but opposite directions, the particle must be in equilibrium, for the sum of the forces is zero, and these two forces just balance each other. Conversely, if a particle is in equilibrium under the action of two forces, these forces must necessarily be equal in magnitude and opposite in direction. In general, when any number of forces acting on a particle can be represented by directed lines that form a closed polygon, the particle is in equilibrium, for the vector sum of the forces is zero. If the directions of *three* forces producing equilibrium are known, their relative magnitudes can be found by constructing a triangle with its sides in the directions of the forces (see Fig. 6 . 3).

In a rectangular coordinate system the condition for equilibrium under the action of any number of forces $\mathbf{F}_i$ is

$$\sum_{i=1}^{n} \mathbf{F}_i = \mathbf{F} = \mathbf{0} \qquad [6 \cdot 6]$$

which is equivalent to the three scalar equations

$$F_x = \sum_{i=1}^{n} F_{ix} = 0 \qquad F_y = \sum_{i=1}^{n} F_{iy} = 0 \qquad F_z = \sum_{i=1}^{n} F_{iz} = 0 \qquad [6 \cdot 7]$$

This relation must hold for each component, or the resultant force $\mathbf{F}$ cannot vanish. If all the forces acting on the particle lie in the xy plane, then the condition $F_z = 0$ is trivially true.

A *nonrigid connector,* such as a rope or a wire, can exert a force only in a direction parallel to its length. Such a force, or *tension* $\mathbf{T}$, acts as a pull on whatever is attached to either end of the rope and is transmitted equally along its entire length. Thus if a weight is hung from a rope, as in Fig. 6 . 4, the rope will exert a tension on the weight which is equal and opposite to the gravitational force $\mathbf{F}_g$ exerted by the weight. Similarly, it will exert a tension on the hook which exactly balances the opposite pull of the hook. Both tensions, of course, are merely the transmitted forces exerted by the weight and the hook. However, they are exerted equally on each segment of the rope, so that the forces acting on any one segment must be equal to the total tension $\mathbf{T}$ in either direction and must counterbalance each other if the

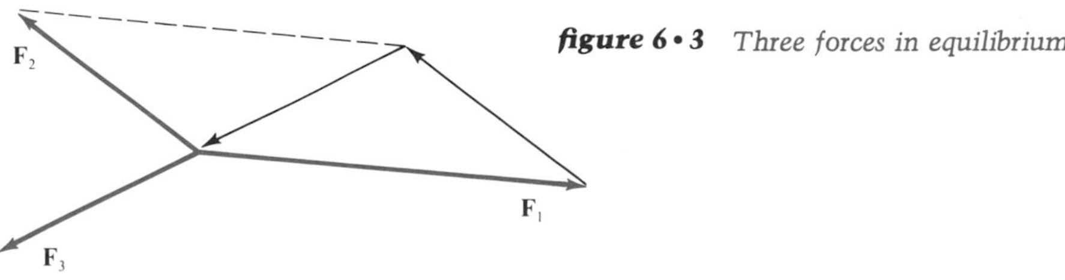

figure 6 • 3 *Three forces in equilibrium*

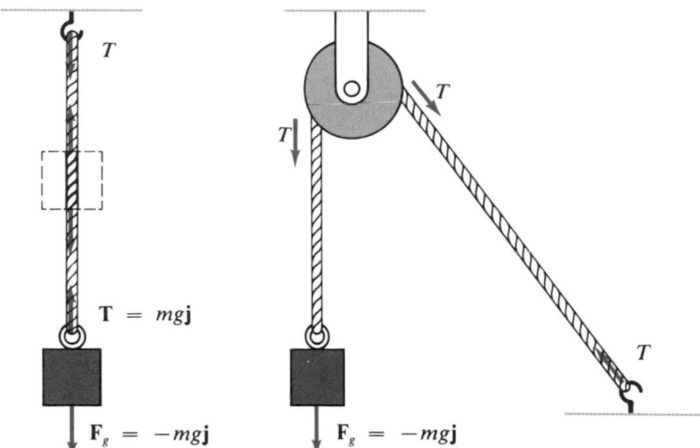

figure 6 • 4 *A rope under tension*

system is to remain in equilibrium. If one point of attachment is a pulley, the direction of the tension is changed, but not its magnitude.

In contrast, a *rigid connector,* such as a rod, can exert either a pull (tension) or a push (compression) on something attached to its ends. In this case the resultant force will not necessarily be parallel to the length of the rod; its direction will depend on the nature of the attachment—a hinge, a bolt, and so on. In this chapter all structural elements, such as ropes, rods, and pulleys, will be considered weightless. We shall assume that connectors have constant length, whether they are rigid or nonrigid, and that all pulleys are completely frictionless (smooth).

example 6 • 3 A rock weighing 450 lbf is fastened at point O to a wire which is strung between two posts as shown. Find the magnitudes of the tensions in parts OA and OB of the wire.

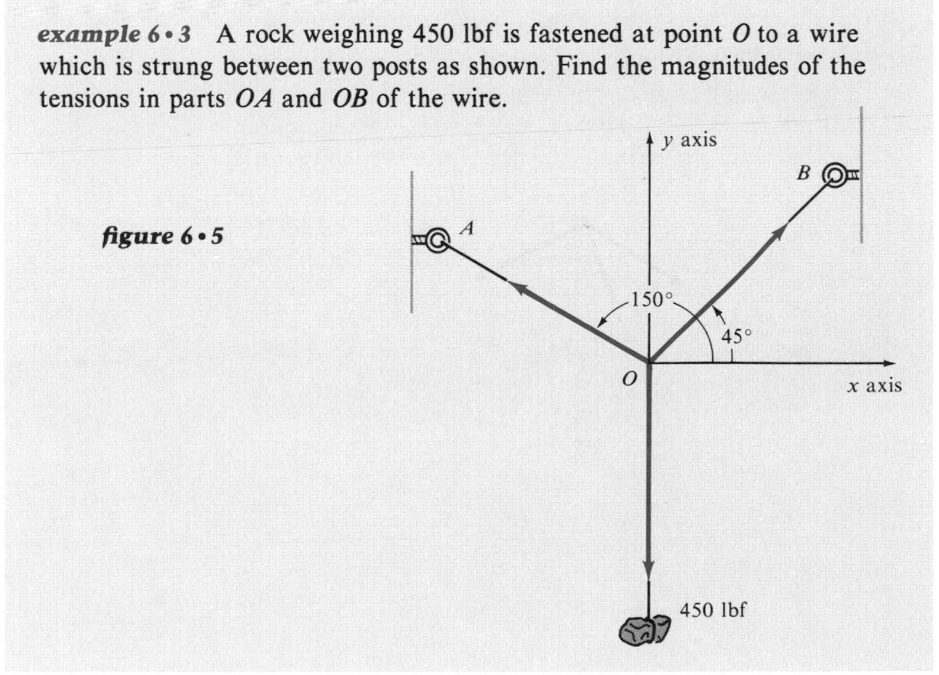

figure 6 • 5

solution Take O as the origin of coordinates and let $\mathbf{T}_1$ and $\mathbf{T}_2$ denote the tensions between the rock and each post. The forces involved in the problem are then $\mathbf{T}_1$, $\mathbf{T}_2$, and $\mathbf{F}_g$, the weight of the rock. All three of these forces act on the same particle O of the wire, which is, by hypothesis, in equilibrium. Hence the resultant force on this particle O is

$$\mathbf{F} = (\mathbf{T}_1 + \mathbf{T}_2 - 450\mathbf{j})\ \text{lbf} = \mathbf{0}$$

In component form this vector sum decomposes to the two scalar equations

$$F_x = T_1 \cos 45° + T_2 \cos 150° = 0$$

$$F_y = T_1 \sin 45° + T_2 \sin 150° = 450\ \text{lbf}$$

and solving for the two unknown quantities T_1 and T_2, we obtain

$$T_1 = 400\ \text{lbf} \qquad T_2 = 330\ \text{lbf}$$

Although the algebraic sum of these two tensions is considerably greater than 450 lbf, it is only their y components that support the rock; their x components counterbalance each other.

example 6 • 4 The maypole at the annual physics picnic is 3 m high and is held in place by three guy wires attached to the top of the pole, symmetrically spaced and each 5 m long. The tension in one wire is 150 N and the tension in the other two is 100 N each. Find the resultant force exerted on the pole by the three guy wires.

solution The triangles OAP, OBP, and OCP are all 3-4-5 right triangles with a horizontal leg of 4 m. We can therefore construct unit vectors $\hat{\mathbf{t}}_1$, $\hat{\mathbf{t}}_2$, and $\hat{\mathbf{t}}_4$ in the direction of each of the three tensions $\mathbf{T}_1$, $\mathbf{T}_2$, and $\mathbf{T}_3$. Since the tension in the first wire must point from P to A, it is parallel to the displacement vector

$$\mathbf{r}_A - \mathbf{r}_P = 4\mathbf{i} - 3\mathbf{k}$$

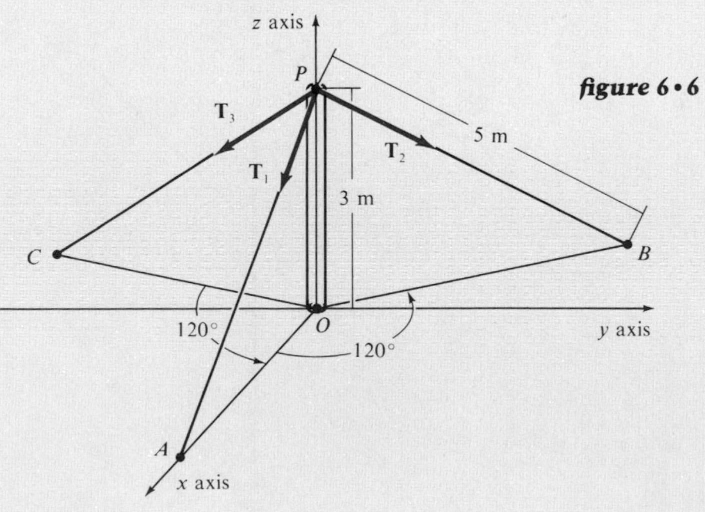

figure 6 • 6

The unit vector $\hat{\mathbf{t}}_1$ is therefore

$$\hat{\mathbf{t}}_1 = \tfrac{4}{5}\mathbf{i} - \tfrac{3}{5}\mathbf{k}$$

Similarly, the tension in the second wire must be parallel to the vector from P to B:

$$\mathbf{r}_B - \mathbf{r}_P = 4(-\cos 60°\ \mathbf{i} + \sin 60°\ \mathbf{j}) - 3\mathbf{k}$$

Since the length of this vector is 5 m, the unit vector is

$$\hat{\mathbf{t}}_2 = \tfrac{4}{5}(-\cos 60°\ \mathbf{i} + \sin 60°\ \mathbf{j}) - \tfrac{3}{5}\mathbf{k}$$

The unit vector in the direction of the third wire is

$$\hat{\mathbf{t}}_3 = \tfrac{1}{5}(-4 \cos 60°\ \mathbf{i} - 4 \sin 60°\ \mathbf{j} - 3\mathbf{k})$$

With the directions of the three tensions, we can now find the resultant force:

$$\begin{aligned}\mathbf{F} &= \mathbf{T}_1 + \mathbf{T}_2 + \mathbf{T}_3 \\ &= \frac{150\ \text{N}}{5}(4\mathbf{i} - 3\mathbf{k}) + \frac{100\ \text{N}}{5}(-4 \cos 60°\ \mathbf{i} + 4 \sin 60°\ \mathbf{j} - 3\mathbf{k}) \\ &\qquad + \frac{100\ \text{N}}{5}(-4 \cos 60°\ \mathbf{i} - 4 \sin 60°\ \mathbf{j} - 3\mathbf{k})\end{aligned}$$

or

$$\mathbf{F} = (40\mathbf{i} - 210\mathbf{k})\ \text{N}$$

example 6 • 5 If the system of pulleys shown is in static equilibrium, what must be the relation between masses m_1 and m_2?

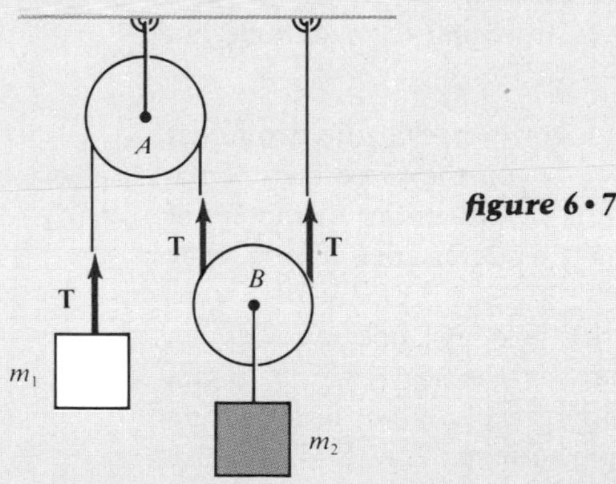

figure 6 • 7

solution The tension in the rope must be sufficient to maintain mass m_1 in equilibrium; therefore $T = m_1 g$. This tension is transmitted throughout the length of the rope, so that there is a net upward force of $2T$ on the movable pulley B. If this pulley is not accelerating, then

$$m_2 g = 2T = 2m_1 g$$

Thus $m_2 = 2m_1$.

6 • 3 Newton's third law of motion

In any interaction between two particles, the resulting changes in their momentum obey the relation

$$\Delta \mathbf{p}_1 = -\Delta \mathbf{p}_2$$

Thus the rate of such changes during any time interval Δt is given by

$$\frac{\Delta \mathbf{p}_1}{\Delta t} = -\frac{\Delta \mathbf{p}_2}{\Delta t} \qquad [6 \cdot 7]$$

Newton made the assumption that this relation holds for every instant of the interaction, so that at any given time t the momentum changes of the two particles are related by

$$\frac{d\mathbf{p}_1}{dt} = -\frac{d\mathbf{p}_2}{dt} \quad \text{or} \quad \mathbf{F}_1 = -\mathbf{F}_2 \qquad [6 \cdot 8]$$

The fact that both momenta change means that each of the particles is exerting a force on the other, and since the changes are equal and opposite, the two forces are equal in magnitude and opposite in direction. If the forces acting *on* each body are related by $\mathbf{F}_1 = -\mathbf{F}_2$, then the forces that are exerted *by* each body must be the corresponding reaction forces $\mathbf{F}_1'$ and $\mathbf{F}_2'$, where

$$\mathbf{F}_1' = \mathbf{F}_2 \qquad \mathbf{F}_2' = \mathbf{F}_1 \qquad [6 \cdot 9]$$

Equation [6 . 8] is the mathematical statement of Newton's third law of motion, the law of *action and reaction:**

> For every force acting on a body, there exists a corresponding force exerted by the body, and these two forces are equal in magnitude but opposite in direction.

All forces, in other words, exist in pairs. Note, however, that the two forces of a pair of acting and reaction forces do not act on the same body. Each body is *subjected* to one of the forces and *exerts* the other. For two forces of equal magnitude and opposite direction to have a vector sum of zero, they must both act on the *same* body.

There are many familiar illustrations of the third law: a person sitting in a chair cannot lift himself by pulling upward on the seat (why?); a gun recoils when it is fired; it is impossible to jump gracefully out of a drifting rowboat because it moves away from you as you leap. The third law also holds when the interacting bodies are not in direct contact. Whenever one body attracts or repels another from a distance, the other body attracts or repels the first body with a force that is equal in magnitude and opposite in direction.

*Note that we have derived the third law from the interaction postulate, Eq. [5 . 1]. Newton in fact began with the second and third laws as basic postulates and used these to derive the interaction postulate. The disadvantage of this order of development, as Ernst Mach has pointed out, is that it does not provide an explicit definition of mass.

example 6 • 6 A ball is suspended above the earth from a string. Identify the forces indicated and state their relationships.

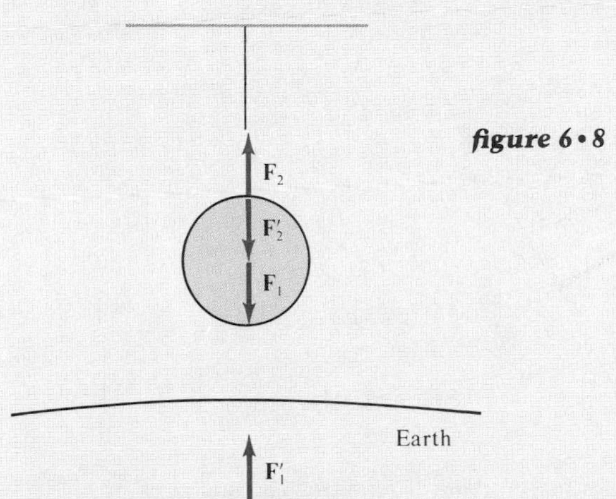

figure 6 • 8

solution Forces $\mathbf{F}_1$ and $\mathbf{F}_1'$ are the mutual gravitational pulls of the earth and the ball on each other; hence, from the third law, $\mathbf{F}_1 = -\mathbf{F}_1'$. Similarly, forces $\mathbf{F}_2$ and $\mathbf{F}_2'$ are the mutual pulls of the string and ball on each other, and so $\mathbf{F}_2 = -\mathbf{F}_2'$. (Note that $\mathbf{F}_2$ is what we have previously called tension **T**.) Since the ball has no acceleration relative to the earth, according to the second law, $\mathbf{F}_1 = -\mathbf{F}_2$, and thus $\mathbf{F}_1' = -\mathbf{F}_2'$.

There is sometimes a tendency to view reaction forces as a convenient fiction, or at least as something not quite as real as a force applied to a body by an external agent. This is a mistake: reaction forces are real. When you walk, it is the reaction force exerted by the earth on your feet that moves you along; when you steer your car, it is the reaction force exerted by the road surface on the sides of your wheels that provides the turning force. The basic requirement in dealing with reaction forces is to be able to use your physical intuition to tell you where such forces are likely to be acting, and if possible, in what directions. For example, a perfectly smooth surface exerts a force, but only in a direction *normal,* or perpendicular, to the surface. This type of reaction force is customarily denoted by **N**, to indicate its direction with respect to the surface. A rough surface may exert an additional reaction force which is tangent to its average surface; these forces are termed *frictional,* or resistive, forces because they tend to resist motion. We shall discuss frictional forces in detail in the next section.

There are a few practical rules for treating reaction forces. When you are in doubt about the direction, or even the existence, of a force at a given point, assume that it exists and see if you can set up enough equations to solve for it. If you assume the force to be pointing in a certain direction and set up a consistent set of equations for its components, the correct solutions will give the proper force, even if it actually points the opposite way (has a negative value) for some components. It is better to assume too many unknowns than too few; the superfluous ones can always be discarded later.

example 6•7 A smooth cylinder is at rest as shown. What are the magnitudes of the reaction forces acting on it?

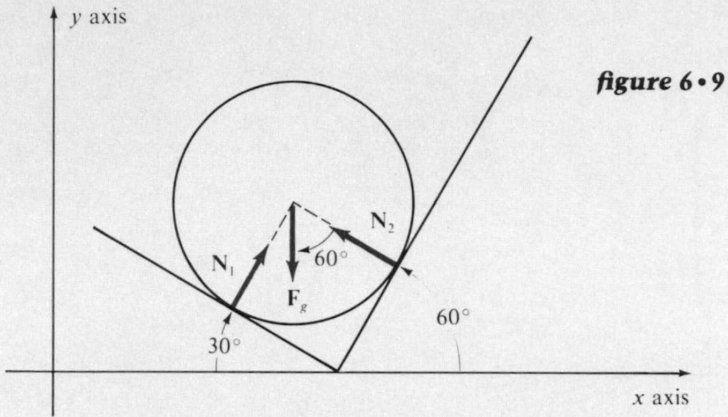

figure 6•9

solution The reaction forces are normal to each of the surfaces on which the cylinder is resting, and since these surfaces are tangential to the cylinder, the force vectors $\mathbf{N}_1$ and $\mathbf{N}_2$ point toward the center of the cylinder. The pull of gravity on the cylinder exerts a force $\mathbf{F}_g = -mg\mathbf{j}$, and since the system is in equilibrium, the total force is

$$\mathbf{F} = \mathbf{N}_1 + \mathbf{N}_2 - mg\mathbf{j} = \mathbf{0}$$

In component form this vector sum is equivalent to

$$F_x = N_1 \cos 60° - N_2 \cos 30° = 0$$

$$F_y = N_1 \sin 60° + N_2 \sin 30° = mg$$

and the magnitudes of the two reaction forces are therefore

$$N_1 = mg \sin 60° = \tfrac{1}{2}\sqrt{3}\, mg \qquad N_2 = mg \cos 60° = \tfrac{1}{2}mg$$

We could have seen this immediately if, instead of using the usual xy coordinates, we had decomposed the acting force $\mathbf{F}_g$ into components perpendicular to the two surfaces, which form a right angle. Then each component of $\mathbf{F}_g$ would be balanced by its corresponding reaction force.

6•4 *frictional forces*

There are several kinds of frictional forces, but they all resist motion, and the mathematical techniques used to handle them are often similar. In this text we shall consider four general types of friction: sliding, static, rolling, and viscous friction. The first three refer to friction between dry solid bodies, and the last is the resistance which fluids (gases or liquids) offer to bodies passing through them. We shall discuss viscous resistance in Chapter 15.

When solid surfaces are in contact, the main causes of friction are the intermolecular forces that tend to hold together portions of surfaces which are extremely close to each other. These forces are generally lumped under the term *adhesion*. Laboratory studies have shown that most frictional forces

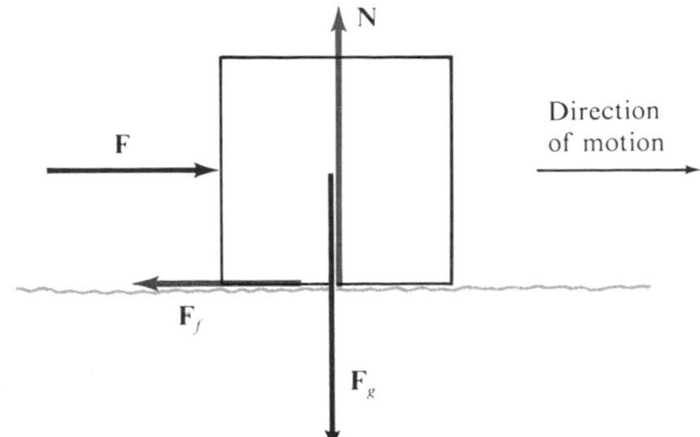

figure 6 • 10 *Motion of a block along a rough plane*

are due to adhesion, with only about 5 percent due to "roughness," or the interlocking of minute irregularities on surfaces.* Frictionless surfaces are referred to as "smooth." Since the surface characteristics which are the main causes of friction are so nonuniform and variable for most solid materials, the theory of frictional effects has been slow to develop, and empirical laws based on experiments still contain most of our knowledge of frictional forces.

Consider the block in Fig. 6 . 10, which is being accelerated by a force **F** applied parallel to the surface. Its motion is opposed by a frictional force $\mathbf{F}_f$. Thus the resulting acceleration has the magnitude

$$a = \frac{F - F_f}{m} \qquad [6 \cdot 10]$$

If the applied force **F** causes the block to move with constant speed, then evidently **F** and $\mathbf{F}_f$ must be exactly equal in magnitude and opposite in direction. We therefore have a means of determining the frictional force between any two solid surfaces experimentally.

The magnitude of $\mathbf{F}_f$ has been found to differ for different substances and to vary with the condition of the contacting surfaces. For a given pair of surfaces, however, the parallel frictional force F_f is nearly proportional to the normal reaction force N pressing the two surfaces together:

$$F_f = \mu_k N \qquad [6 \cdot 11]$$

where μ_k is a constant called the *coefficient of kinetic friction*. Within wide limits, μ_k has been found to be independent of the amount of surface area in contact, and for moderate pressures it is nearly independent of the speed. (As we shall see later, neither of these statements holds for viscous friction.)

Suppose the block in Fig. 6 . 10 is initially at rest, and the force **F** applied to it is at first too small to start it moving. Since the block is in equilibrium, the frictional force $\mathbf{F}_f$ and the applied force **F** must be equal in magnitude and opposite in direction. As **F** is increased, the reaction force $\mathbf{F}_f$ also increases until it reaches a maximum *static value*

$$F_f = \mu_s N$$

where μ_s is the *coefficient of static friction* (generally, $\mu_s > \mu_k$). Once the applied force **F** exceeds this maximum value of $\mathbf{F}_f$, the block will start to

*For a guide to the literature on friction, see Ernest Rabinowicz, Resource letter F-1 on friction, *American Journal of Physics*, vol. 31, pp. 897–900, 1963.

move. At very low speeds the frictional force will not differ appreciably from its static value $\mu_s N$, but as the speed increases, F_f decreases until it reaches its *kinetic value* $\mu_k N$, and thereafter it remains at this level.

When one solid *rolls* on another without slipping, as when a ball rolls along a horizontal surface, there is no sliding friction, and yet a force is required to keep the rolling body moving with constant speed. This is because of slight deformations which always occur where both surfaces are in contact, so that the rolling body, in effect, is always rolling uphill. This effect is actually visible in the case of a heavy wheel rolling on a soft surface. If the moving object is mounted on wheels, of course, part of the resistance is also due to friction in the bearings. When a lubricant is used in the wheel bearings, the laws of friction between solids no longer hold, for the friction then depends in a complicated way on both the load and the speed. Despite these complications, frictional effects in lubricating fluids, unlike those for solids, are susceptible to theoretical treatment.

example 6•8 A block is at rest on a rough inclined plane which is slowly being tilted upward. At what minimum angle θ will the block begin to slide down the plane?

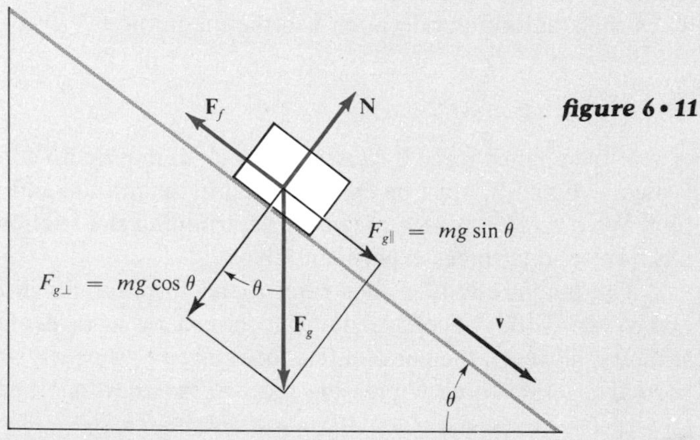

figure 6•11

solution The stationary block is subjected to a downward force $\mathbf{F}_g = -mg\mathbf{j}$, and the normal reaction force of the surface opposes the component of $\mathbf{F}_g$ perpendicular to the plane,

$$F_{g\perp} = mg\cos\theta = N$$

The frictional force $\mathbf{F}_f$ opposes $F_{g\parallel}$, the component of $\mathbf{F}_g$ parallel to the plane; hence $F_f = mg\sin\theta$. Until the block begins to slide, these forces will counterbalance each other, $F_f = -F_{g\parallel}$, and their magnitude will be

$$mg\sin\theta = F_f \leq \mu_s N \qquad \text{or} \qquad \tan\theta \leq \mu_s$$

As we steadily increase the angle θ, the block will just begin to slide when $\tan\theta = \mu_s$. It will then continue to slide even if we lower the plane again until $\tan\theta = \mu_k < \mu_s$. (Why?)

6 • 5 dynamics of a particle

The central problem in dynamics is to predict the motion of a given object from a knowledge of its mass and the forces acting on it. This is achieved by setting up the second law in the form of one or more *differential equations*—equations which may include first and second time derivatives of a quantity (such as velocity, acceleration, position), as well as the quantity itself. These equations are then solved, or *integrated,* to yield the quantity as a function of time. This is exactly the process we carried out in Sec. 4 . 4 to determine the parabolic trajectory of a ballistic missile. In that case the second law, in the form $m\mathbf{a} = -mg\mathbf{j}$, gave us the *equations of motion,* which reduce to

$$a_x = 0 \qquad a_y = -g$$

which we solved to find the position,

$$\mathbf{r}(t) = x(t)\mathbf{i} + y(t)\mathbf{j}$$

Most equations of motion encountered in real situations are quite complicated and can be solved only approximately by the use of computers. For systems of particles it may be necessary to solve many interdependent equations of motion simultaneously. However, even without solving the equations of motion completely, it is often possible to derive quite a bit of useful information from them if they are properly formulated.

Whether you are going to solve a problem analytically or numerically, it is wise to set up the solution in a methodical and unambiguous way. Even if the right answer immediately comes to mind, as it so often does in elementary mechanics, you should still try to systematize the process of deduction. The following steps are suggested:

1 Choose the body whose motion you wish to determine, isolate it from all other bodies, and sketch it.
2 Represent the effects of all other bodies on the body of interest by means of vector forces. Sketch each force at its point of application to make a "free-body" diagram. If you are uncertain of the direction of a force or whether a force exists, include it in your sketch as an unknown quantity. Omit only those components which you are certain are zero or which cancel in pairs, such as internal forces between parts of the body.
3 Choose a coordinate system which simplifies the problem as much as possible. Sketch it on the free-body diagram.
4 Write down the equations of motion from Newton's second law and solve them either graphically, analytically, or numerically.
5 The problem does not end when you get the right answer. Ask yourself if your answer seems reasonable. Try to understand the physical meaning of your answer and the significance of the various quantities. Are the dimensions of the answer correct? If there are variable quantities in your answer, test it by substituting some values for which the answer should be obvious.

Not every step will be applicable to every problem, but this general procedure, in conjunction with the suggestions in Sec. 1 . 5, should help you to systematize your work. However, there is no substitute for understanding the basic principles of physics.

example 6•9 A sled of 400 lbm carrying a 600-lbm load is being pulled along a level road by a tractor. The coupling between the sled and the tractor is at an angle of 23° and exerts a horizontal pull of 110 lbf. The frictional resistance of the road is 70 lbf. (*a*) What is the acceleration of the sled, and how long will it take to reach a speed of 12 mi/h, starting from rest? (*b*) What is the coefficient of kinetic friction μ_k between the sled and the road?

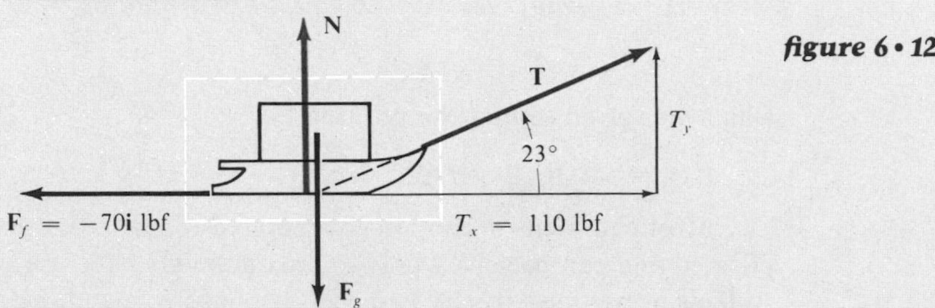

figure 6•12

solution We are not concerned with the forces on the tractor, so we can isolate the sled and its load, with a total mass of 1000 lbm, as shown in Fig. 6 . 12. The vector sum of all the forces crossing this area then gives us the equation of motion for the sled,

$$\mathbf{F} = \mathbf{T} + \mathbf{N} + \mathbf{F}_f + \mathbf{F}_g = m\mathbf{a}$$

which we can express in terms of rectangular components as

$$\mathbf{F} = (110\mathbf{i}\ \text{lbf} + T_y\mathbf{j}) + N\mathbf{j} - 70\mathbf{i}\ \text{lbf} - mg\mathbf{j} = ma\mathbf{i}$$

Since the horizontal and vertical components are independent, they can be treated separately.

(*a*) The horizontal component of **F** consists only of

$$F_x = 110\ \text{lbf} - 70\ \text{lbf} = (1000\ \text{lbm})a$$

and since 1 lbf = 32.2 lbm · ft/s², the acceleration is

$$a = 40\ \text{lbf}/1000\ \text{lbm} = 1.29\ \text{ft/s}^2$$

Therefore, from $v = at$ (and 30 mi/h = 44 ft/s), the sled will be moving at 12 mi/h at the time

$$t = \frac{12\ \text{mi/h}}{1.29\ \text{ft/s}^2} = 13.6\ \text{s}$$

(*b*) To find the coefficient of kinetic friction, we must first find the normal reaction force **N**. Since the direction of motion is horizontal, we know that the vertical component of the resultant force does not contribute to the acceleration:

$$F_y = T_y + N - F_g = 0$$

The x and y components of **T** are related by

$$\frac{T_y}{T_x} = \tan 23° = 0.4245$$

Hence the normal force **N** has a magnitude of

$$N = F_g - T_y = 1000 \text{ lbf} - 46.7 \text{ lbf} = 953.3 \text{ lbf}$$

and the coefficient of kinetic friction is

$$\mu_k = \frac{F_f}{N} = \frac{70}{953.3} = 0.0734$$

example 6 • 10 Three masses of 10 kg each are connected by two ropes, one of which passes over a pulley. The first two masses are placed on a smooth horizontal surface, and the third mass hangs free. (*a*) What is the acceleration of the entire mass system? (*b*) Find the internal forces in the system—the tensions $\mathbf{T}_1$ and $\mathbf{T}_2$ exerted by the connecting ropes.

solution (*a*) The only accelerating force acting on the entire system is the pull of gravity on the unsupported 10-kg mass, $\mathbf{F}_g = -98\mathbf{j}$ N. The weight of each of the other two masses is exactly balanced by the reaction force exerted by the surface. However, the gravitational force of 98 N must accelerate all three masses at once. Thus this force is acting on a total mass of $3m = 30$ kg, and since

$$F_g = 3ma = 98 \text{ N}$$

the acceleration is

$$a = \frac{F}{3m} = \frac{98 \text{ N}}{30 \text{ kg}}$$
$$= 3.27 \text{ m/s}^2$$

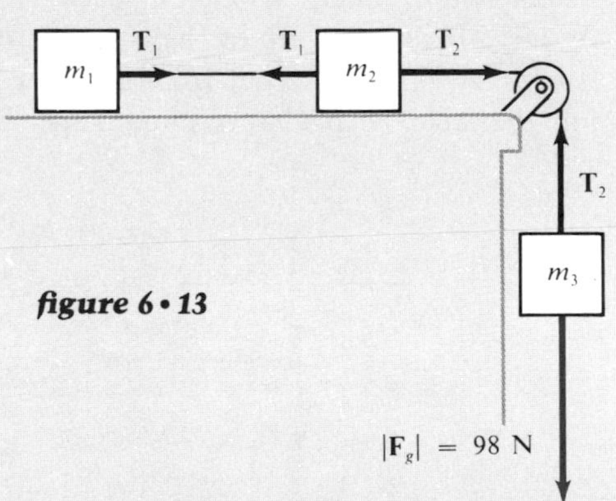

figure 6 • 13

(*b*) Although the reasoning in part (*a*) is correct, much is left to physical intuition, in view of the fact that the masses are not all accelerating in the same direction. When intuition fails, or when we need to understand the forces *within* a system, it becomes necessary to break the system down in more detail. This has been done in Fig. 6 . 14 to obtain three equations of motion, one for each mass, which we can state in scalar form as

$$T_1 = ma_x \qquad T_2 - T_1 = ma_x \qquad F_g - T_2 = -ma_y$$

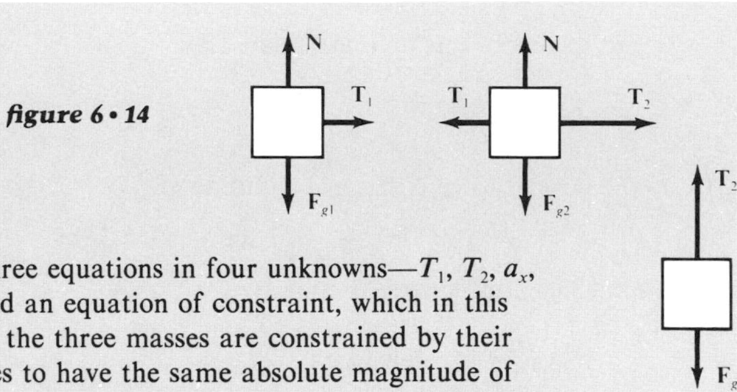

figure 6 • 14

To solve three equations in four unknowns—T_1, T_2, a_x, and a_y—we need an equation of constraint, which in this case states that the three masses are constrained by their connecting ropes to have the same absolute magnitude of acceleration, regardless of the direction of motion. That is, $a_x = -a_y = a$. Thus, if we add the three equations of motion, we get the same result as in part (*a*): $F_g = 3ma = 98$ N. Then, solving the first two equations for T_1 and T_2, we obtain

$$T_1 = ma = \frac{F_g}{3} = 32.7 \text{ N} \qquad T_2 = 2ma = \frac{2F_g}{3} = 65.4 \text{ N}$$

Excessive reliance on physical intuition might have led you to guess that $T_1 = T_2$; analysis shows that this is not so.

example 6 • 11 A dish of mass m is placed in the middle of a table which is covered just to its edges by a tablecloth. If the tablecloth is pulled with a slight horizontal force **F**, the surface friction between the dish and the tablecloth will cause the dish to move with the tablecloth. (*a*) Given a coefficient of kinetic friction μ_k between the dish and the tablecloth, what is the differential form of the equation of motion for the dish? (*b*) In order to pull the tablecloth out from under the dish without having the dish fall off the table, with what minimum speed must you move the tablecloth?

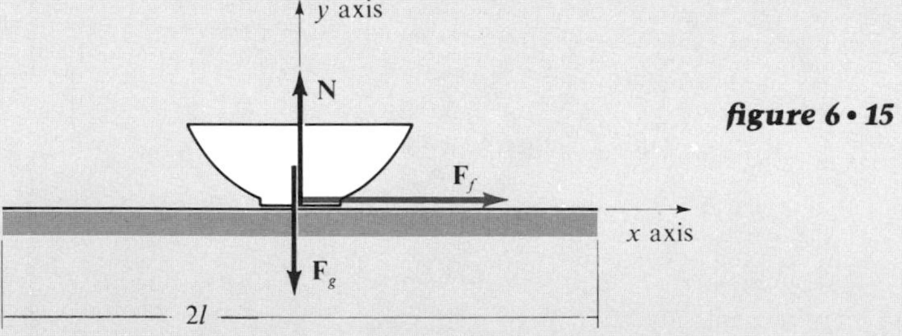

figure 6 • 15

solution (*a*) Since the motion of the dish is due to frictional force, the accelerating force has the magnitude

$$F_f = \mu_k N = \mu_k mg = ma$$

Hence we can state the equation of motion as

$$a_x = \frac{d^2x}{dt^2} = \mu_k g$$

(*b*) The tablecloth must travel twice the distance of the dish in the same time t. Therefore, if the dish is to stay on the table, it must move no farther than

$$x(t) = \tfrac{1}{2}at^2 = \tfrac{1}{2}\mu_k g t^2 \leq l$$

Hence the operation must be executed by the time

$$t \leq \sqrt{\frac{2l}{\mu_k g}}$$

and the tablecloth must be pulled smoothly at a speed of

$$v = \frac{2l}{t} \geq \sqrt{2l\mu_k g}$$

Note that the speed of the tablecloth is independent of the mass of the dish. (You can confirm these equations empirically while the rest of your family is not at home.)

*6 • 6 numerical integration of equations of motion

Thus far we have dealt only with forces that give rise to constant acceleration. However, if we are willing to tolerate small errors in our results, we can find approximate solutions to more difficult problems by using the computer to perform a *numerical integration* of the differential equations of motion. Problems that involve a linear force called the *Hooke's law force* are soluble exactly. The required mathematics for an analytic solution are beyond the scope of this chapter, but we can study the behavior of this phenomenon by means of the computer.

To carry out numerical integration of the equations of motion, we must return to our original definitions of velocity and acceleration:

$$\mathbf{v} = \lim_{\Delta t \to 0} \frac{\Delta \mathbf{x}}{\Delta t} \qquad \mathbf{a} = \lim_{\Delta t \to 0} \frac{\Delta \mathbf{v}}{\Delta t} \qquad [6 \cdot 12]$$

Our numerical approximation rests on the assumption that we can take an interval Δt small enough so that the instantaneous values of $\mathbf{v}$ and $\mathbf{a}$ are approximately equal to the average values over this interval. Thus we ignore the limit symbols in Eqs. [6 . 12] and multiply both sides of each equation by Δt to obtain the approximations

$$\Delta \mathbf{x} \doteq \mathbf{v}\,\Delta t \qquad \Delta \mathbf{v} \doteq \mathbf{a}\,\Delta t \qquad [6 \cdot 13]$$

Suppose now that we can partition the entire interval of interest, $0 \leq t \leq T$, into a number of smaller intervals Δt. Thus we have n instants of time t_i where $i = 0, 1, 2, \ldots, n$, at which to compute the position $\mathbf{x}$ and the velocity $\mathbf{v}$. We assume that the acceleration is given by $\mathbf{a} = \mathbf{F}/m$, according to Newton's second law, so we can write the approximate velocity change $\Delta \mathbf{v}$ as

$$\Delta \mathbf{v} \doteq \frac{\mathbf{F}}{m}\Delta t \qquad [6 \cdot 14]$$

Equations [6 . 13] and [6 . 14] imply that if we know the position and velocity of the particle at a time t_i, then we can calculate its approximate position and velocity at the next instant of time, $t_{i+1} = t_i + \Delta t$. If we substitute $\mathbf{x}_{i+1} - \mathbf{x}_i$ for $\Delta\mathbf{x}$ and $\mathbf{v}_{i+1} - \mathbf{v}_i$ for $\Delta\mathbf{v}$ in Eq. [6 . 13], we obtain approximate equations for the position and velocity at any time t_{i+1}:

$$\mathbf{x}_{i+1} = \mathbf{x}_i + \mathbf{v}_i \Delta t \qquad [6 \cdot 15]$$

$$\mathbf{v}_{i+1} = \mathbf{v}_i + \left(\frac{\mathbf{F}}{m}\right)_i \Delta t \qquad [6 \cdot 16]$$

expressed as equalities, since we shall now use the equations in this form. For simplicity, the values of $\mathbf{v}$ and $\mathbf{F}/m$ are assumed to be the instantaneous values at time t_i, and Eqs. [6 . 15] and [6 . 16] give us a prescription for computing the (approximate) values of position and velocity at the next instant t_{i+1}. This method of numerical integration, known as *Euler's method,* is the least accurate procedure, but the simplest. More accurate methods consist, in essence, of evaluating "average" values of $\mathbf{v}$ and $\mathbf{F}/m$ over the interval $t_i \leq t \leq t_{i+1}$ before applying Eqs. [6 . 15] and [6 . 16].

Let us apply Euler's method to the case of a linear force discovered empirically by Robert Hooke in 1660. Hooke found that when a spring is placed on a smooth horizontal surface with one end attached to a wall (see Fig. 6 . 16), the force required to stretch or compress the spring varies directly with the displacement from its original unstretched length. Thus this force is given by *Hooke's law* as

$$\mathbf{F} = -k\mathbf{x} \qquad [6 \cdot 17]$$

where k is the constant of proportionality between the force exerted by the spring and its stretched or compressed position. This constant, often called the *spring constant* or *restoring-force constant,* is a direct measure of the stiffness of the spring. The force described by Eq. [6 . 17] is known as a *linear restoring force;* it always acts in opposition to the deformation, tending to restore the spring to its unstretched length. As we shall see later, this type of force is highly significant in all areas of physics.

The one-dimensional differential equation of motion for a mass m attached to the end of a massless spring is

$$m\frac{d^2x}{dt^2} = -kx \qquad [6 \cdot 18]$$

figure 6 • 16 *Hooke's law for a linear restoring force*

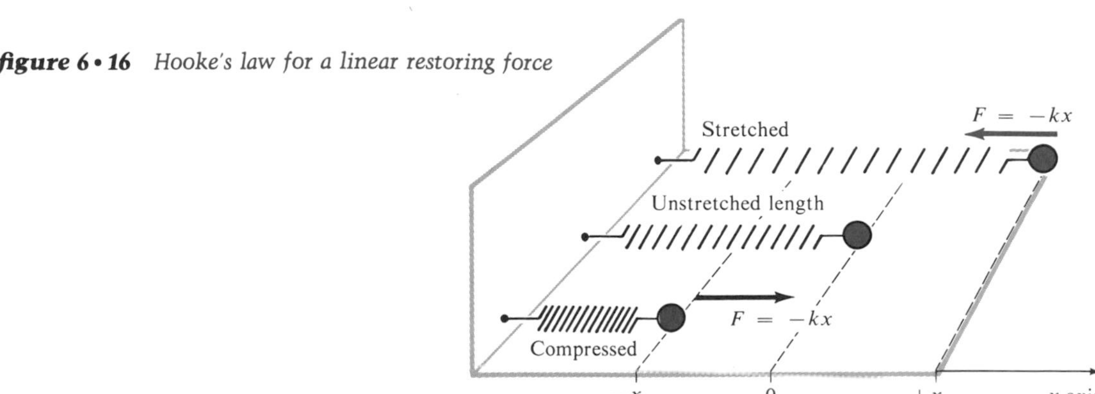

The constant k therefore has the dimensions

$$\dim(k) = \frac{F}{L} = \frac{M}{T^2} \qquad [6 \cdot 19]$$

To avoid having to recompute the solution to Eq. [6 . 18] for every combination of k/m, it is customary to "scale" the equation to dimensionless quantities by substituting new variables for x and t; in this way, one solution can serve for all cases.* Careful scaling can often provide useful insights as well as economy of effort.

Let us begin by defining a dimensionless length variable $\lambda = x/L$ and a dimensionless time variable $\tau = t/T$. We can treat each differential operator d as an algebraic factor, so that $dx = L\,d\lambda$ and $dt^2 = T^2\,d\tau^2$. Substitution of these values in Eq. [6 . 18] then gives us

$$m\frac{L\,d^2\lambda}{T^2\,d\tau^2} = -kL\lambda \qquad [6 \cdot 20]$$

From Eq. [6 . 19], we can also define $T^2 = m/k$ (verify that m/k has the dimension of time). This substitution yields

$$m\frac{L\,d^2\lambda}{(m/k)\,d\tau^2} = -kL\lambda$$

which reduces by cancellation to the dimensionless acceleration equation

$$\frac{d^2\lambda}{d\tau^2} = -\lambda \qquad [6 \cdot 22]$$

If we now define a dimensionless velocity $\nu = d\lambda/d\tau$, we can also write the acceleration as

$$\frac{d^2\lambda}{d\tau^2} = \frac{d\nu}{d\tau} = -\lambda$$

and Eqs. [6 . 15] and [6 . 16] take the dimensionless form†

$$\lambda_{i+1} = \lambda_i + \nu_i\,\Delta\tau \qquad [6 \cdot 23]$$

$$\nu_{i+1} = \nu_i - \lambda_i\,\Delta\tau \qquad [6 \cdot 24]$$

These two equations give us the prescription for the numerical solution of Eq. [6 . 22]. Such expressions are called *recursion formulas;* recursive methods are at the heart of computer usage.

Unlike the numerical differentiation we performed in Sec. 3 . 6, in this case each value of λ and ν we compute depends indirectly on all the preceding values. Thus small errors in the computed values of Eqs. [6 . 23] and [6 . 24] for each time τ_{i+1} will tend to accumulate and result in progressively larger errors in succeeding values. One type of error, called a *truncation error,* arises from the use of approximations such as Eq. [6 . 13] in place of

*Most computers will handle an extremely wide range of numerical values (10^{-38} to 10^{33} or greater) by means of exponential notation, so that scaling is not always strictly necessary for simple problems.

†In computer work another symbol is often substituted for the subscript i, since it sometimes looks like a 1 on the printout. However, the more familiar mathematical form has been retained here for simplicity.

the exact equalities in Eqs. [6 . 12]. It is therefore necessary to take fairly small intervals $\Delta\tau$ to minimize the error for each interval. A simple way to determine whether $\Delta\tau$ has been chosen small enough is to recompute the approximate solution for successively smaller values of $\Delta\tau$ until the percentage of difference between successive solutions becomes tolerable.

A second type of error is *roundoff error,* which arises from the fact that the computer can handle only about ten decimal digits, or fewer, depending on the type of machine. Succeeding digits in a number are either dropped or rounded off, again depending on the type of machine. The result is a gradual accumulation of random roundoff errors; the more operations the machine performs on a particular quantity, the greater the error. For this reason we cannot solve the problem of accuracy by taking $\Delta\tau$ as small as possible,

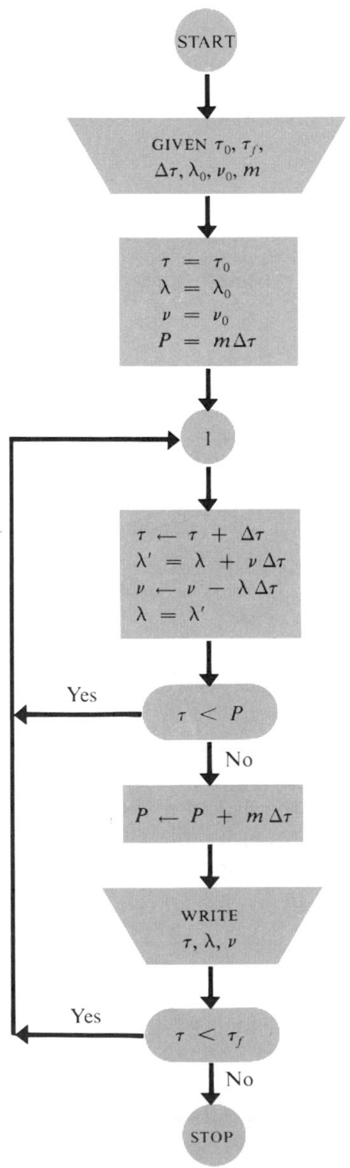

figure 6 • 17 *Flowchart for numerical integration of a Hooke's law force; data are given in Table 6 . 1.*

Initial time $\tau_0 = 0.0$, final time $\tau_f = 6.4$; time interval $\Delta\tau = 0.01$; $\lambda_0 = 1.0$, $\nu_0 = 0.0$, $m = 20$. (You may also want to have these values printed out.)

Set time, position, and velocity variables equal to initial values. P is print time at which current values will be printed out.

Reference point in program

Find new values of λ and ν at next instant of time. New value of λ is temporarily stored as λ' until v is calculated.

Is it time to print yet?

Compute next print time P.

Print out current values of all variables.

If τ does not exceed final time, return to (1) and repeat computation; otherwise stop.

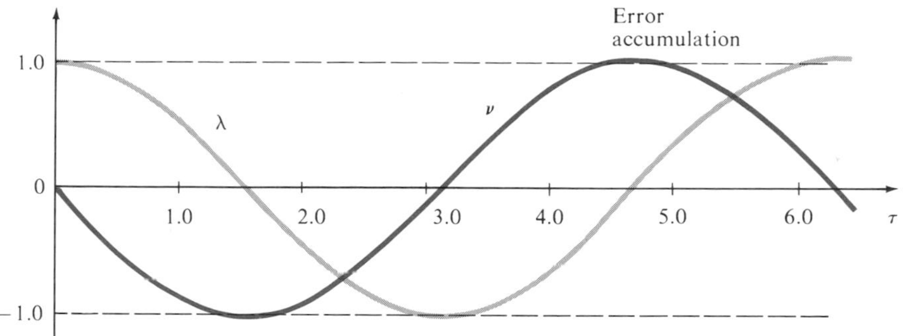

figure 6 • 18 *Graphs of dimensionless position λ and velocity ν as functions of time τ for numerical integration of the second law in the case of a Hooke's law force. Initial conditions* $\lambda_0 = 1.0$ *and* $\nu_0 = 0.0$.

because so many numerical operations would be needed to integrate over a reasonable length of time that the increased accuracy would be offset by cumulative roundoff errors.

The flowchart for the numerical integration of Eq. [6 . 22] is shown in Fig. 6 . 17. One more practical detail must be mentioned here. Since we will be taking $\Delta\tau$ fairly small, we do not wish to write out the result for every single value of τ_i. Therefore a test is included (the first lozenge in Fig. 6 . 17), so that results are printed out only when i is a multiple of some value m.

Table 6 . 1 gives the results of a computation for the case of initial values $\lambda_0 = 1$ and $\nu_0 = 0$. You can easily verify the first few values with a hand calculator. To illustrate the accumulation of truncation errors, the numerical result is compared with the correct analytic solution, $\lambda = \cos\tau$. Even without the correct solution, it is apparent from the graph of the numerical result, shown in Fig. 6 . 18, that by interpolated time $\tau = 6.283$ (actually 2π) the system has returned to its initial conditions (neglecting error) of $\lambda = 1$ and $\nu = 0$. However, since the equation of motion remains the same, the motion can only repeat itself. Thus we have a *periodic motion,* one that is repeated at regular intervals. This particular type of periodic motion is known as *simple harmonic oscillation.* As we shall see in Chapter 16, when we study such motions in detail, regardless of the initial conditions, the *period,* the time of one complete cycle, is given by $2\pi T = 2\pi(m/k)^{1/2}$.

Scaling gives us immediate insight into the physical problem, since it indicates that the period of a harmonic oscillation must be independent of the length scale L and of the order of the time scale T.

example 6 • 12 A spring with a restoring-force constant $k = F/x =$ 2.3 N/m has its right end attached to a wall; attached to its other end is a mass $m = 9.2$ kg. The spring-mass system is stretched to the left a distance of 10 cm and then released. If the position of the mass at equilibrium is $x = 0$, what are its position and velocity at $t = 4$ s, according to the scaled results in Table 6 . 1?

table 6 • 1 *Numerical integration of Hooke's law force*

τ	λ	ν	$\cos \tau$	% error
0.0	1.0000	0.0000	1.0000	0.0
0.2	0.9810	−0.1989	0.9801	0.1
0.4	0.9229	−0.3902	0.9211	0.2
0.6	0.8278	−0.5663	0.8253	0.3
0.8	0.6995	−0.7202	0.6967	0.4
1.0	0.5430	−0.8457	0.5403	0.5
1.2	0.3646	−0.9376	0.3623	0.6
1.4	0.1712	−0.9924	0.1700	0.7
1.6	−0.0294	−1.0076	−0.0292	0.6
1.8	−0.2292	−0.9827	−0.2272	0.9
2.0	−0.4203	−0.9185	−0.4161	1.0
2.2	−0.5950	−0.8175	−0.5885	1.1
2.4	−0.7462	−0.6837	−0.7374	1.2
2.6	−0.8681	−0.5223	−0.8569	1.3
2.8	−0.9555	−0.3398	−0.9422	1.4
3.0	−1.0049	−0.1434	−0.9900	1.5
3.2	−1.0144	0.0592	−0.9983	1.6
3.4	−0.9834	0.2598	−0.9668	1.7
3.6	−0.9131	0.4504	−0.8968	1.8
3.8	−0.8062	0.6235	−0.7910	1.9
4.0	−0.6670	0.7720	−0.6536	2.0
4.2	−0.5008	0.8900	−0.4903	2.1
4.4	−0.3143	0.9727	−0.3073	2.3
4.6	−0.1149	1.0168	−0.1122	2.4
4.8	0.0895	1.0204	0.0875	2.3
5.0	0.2907	0.9832	0.2837	2.5
5.2	0.4807	0.9068	0.4685	2.6
5.4	0.6519	0.7940	0.6347	2.7
5.6	0.7975	0.6493	0.7756	2.8
5.8	0.9115	0.4784	0.8855	2.9
6.0	0.9894	0.2881	0.9602	3.0
6.2	1.0279	0.0859	0.9965	3.1
6.4	1.0255	−0.1201	0.9932	3.3

solution The spring has been stretched a distance $x = L = -0.1$ m, and its time scale is given by

$$T^2 = \frac{m}{k} = 4 \text{ kg} \cdot \text{m/N}$$

or $T = 2$ s. Then, from Table 6 . 1, its position at $t = 4$ s is

$$x(2T) = 2L\lambda(2) = +0.042 \text{ m}$$

or 4.2 cm to the right of the origin. Since the scaled velocity is $\nu = Tv/L$, the velocity of the mass at this time is

$$v(2T) = \frac{Lv(2)}{T} = \frac{(-0.1 \text{ m})(-0.92 \text{ kg})}{2 \text{ s}} = 0.046 \text{ m/s}$$

That is, the mass is moving to the right.

PROBLEMS

6 • 1 *vector addition of forces*

6 • 1 A particle is acted on by the following three forces, expressed in dynes:

$$\mathbf{F}_1 = 3\mathbf{i} - 8\mathbf{j} + 2.4\mathbf{k}$$

$$\mathbf{F}_2 = 1.3\mathbf{i} + 7\mathbf{j} + 5.1\mathbf{k}$$

$$\mathbf{F}_3 = -2\mathbf{i} - \mathbf{j} - 3.3\mathbf{k}$$

(*a*) Find the resultant force **F** in component form. (*b*) What are its magnitude and direction angles?

6 • 2 Three forces with magnitudes of 2, 4, and 6 N, respectively, act on a particle at the corner of a cube and are directed as shown. Determine the resultant force on the particle and give its direction angles.

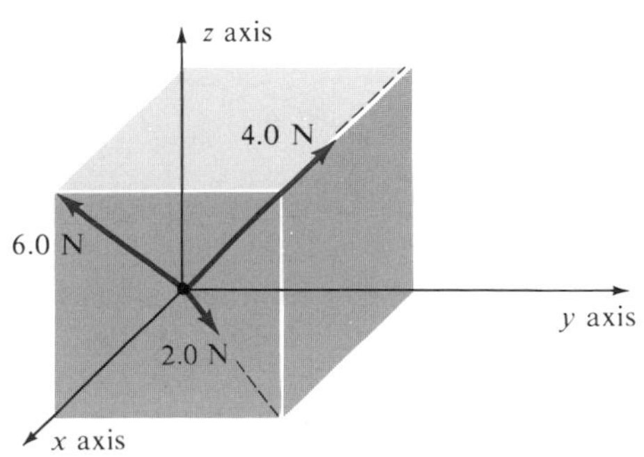

6 • 3 A particle having a mass of 300 mg is acted on by forces of magnitudes 1, 2, 3, 4, 5, and 6 dynes, all in the same plane. The angle between the first force and the x axis, and between each force and the next, is 30°. (*a*) What is the resultant force on the particle? (*b*) Find the acceleration of the particle.

6 • 4 Three forces are acting on a particle at point O, as shown. The constant of proportionality between force and length on the diagram is 1 N/cm. What is the net force vector acting on the particle?

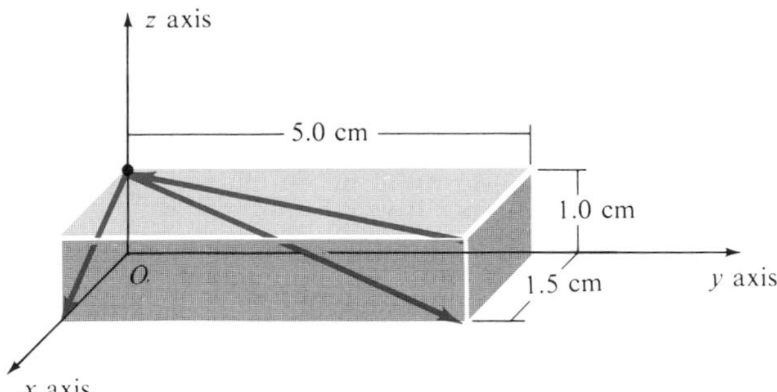

6 • 2 particle statics

6 • 5 Three forces, of magnitudes 10, 25, and 30 dynes, are in equilibrium on the force table shown in Fig. 6 . 1. (*a*) Find the angles between the forces algebraically. (*b*) Determine these angles graphically. (*Hint:* Put three lengths of appropriate sizes together to form a triangle.)

6 • 6 An acrobat standing in the middle of a 60-ft tightrope exerts a force of 150 lbf. If his weight depresses the middle of the rope 5 ft below the fixed ends, what is the tension in the rope?

6 • 7 An object of mass m hangs from one end of a uniform rope of length l. (*a*) Neglecting the mass of the rope, show that the tension in the rope must be the same at every point along its length. (*b*) If the rope has a mass of m', what is the tension in the rope as a function of the vertical distance y measured from the lower end of the rope?

6 • 8 A bridge span is 5 m high and 20 m long. Find the vertical force and the horizontal thrust on each of the piers P in terms of the weight suspended from the center of the span.

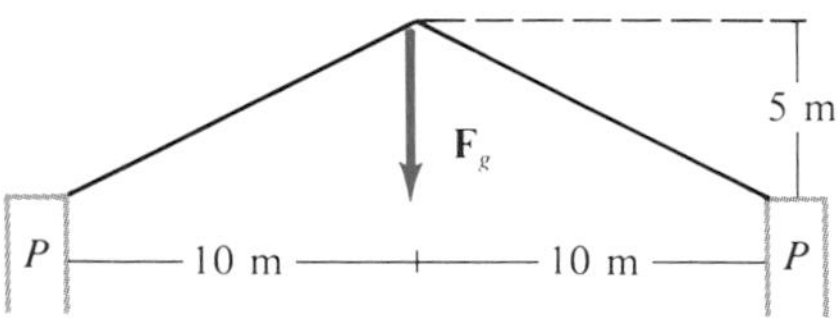

6 • 9 A 100-g object, suspended 48 cm below a horizontal beam, is supported by two strings, one attached to each end of the beam. One string is 60 cm long and has twice the tension of the other string. (*a*) How long is the other string? (*b*) What are the tensions in each string?

6 • 10 A 20-kg block is placed on a smooth inclined plane 10 m long and 6 m high. (*a*) How much force parallel to the plane is required to keep the block from sliding, and what is the force with which the block presses against the plane? (*b*) How much horizontal force would hold the block, and what would the force against the plane be in this case?

6 • 11 A weight of 17.3 N is supported by a pulley which is free to move along the string fastened at points *A* and *B*. Angle *BAC* is 90°, and *BC* is twice the length of *AC*. (*a*) Find the inclinations of the two parts of the string relative to the vertical. (*b*) What is the amount of tension in the string?

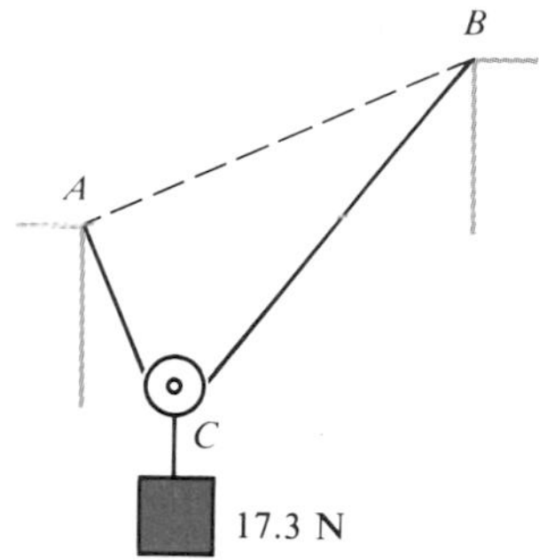

6 • 12 A 1000-lbf cargo container is suspended as shown for loading. (*a*) Find the amount of tension in the cable *OB*. (*b*) How much force does the boom *OA* exert on the post?

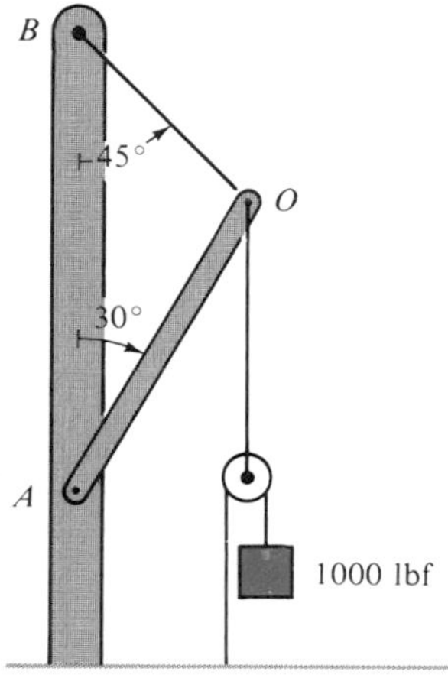

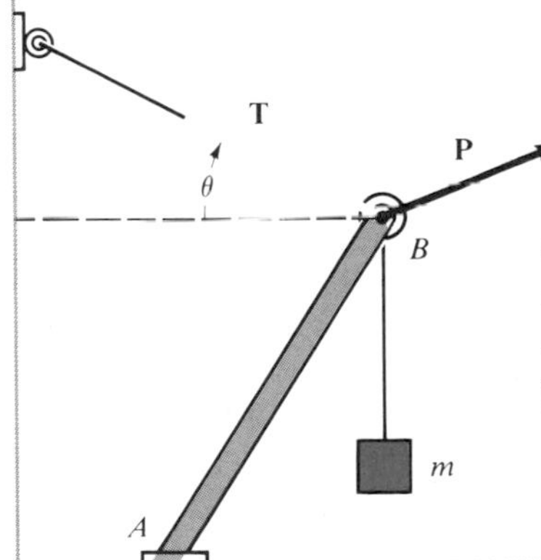

6 • 13 Find the scalar components of the force exerted by the rod *AB* on the pulley at *B*. Assume that all the forces on the pulley pass through its center, *B*.

6 • 14 A mass *m* is suspended from a rope which is passed through a pulley supported by a flexible wire. The end of the rope is held horizontal as shown. (*a*) Find the amount of tension in the rope and the net force **F** it exerts on the pulley. (*b*) What is wrong with the figure?

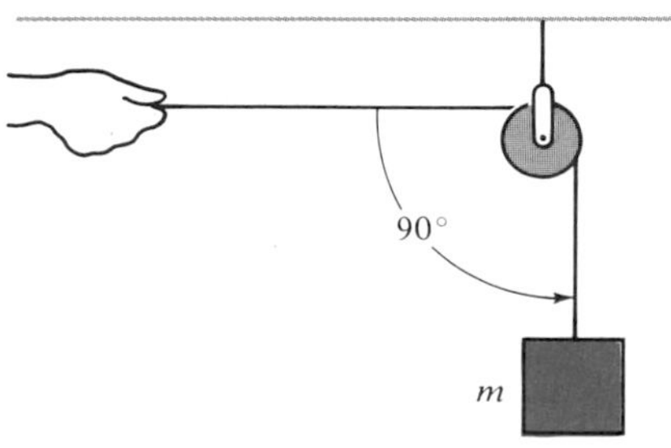

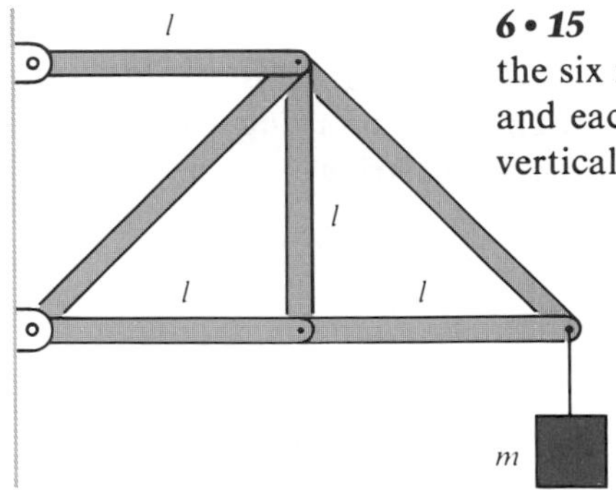

6 • 15 Find the magnitudes of the forces acting in each of the six massless rigid bars if they are joined at smooth pivots and each short bar is of length l. What would happen if the vertical bar were removed?

6 • 3–6 • 5 friction and particle dynamics

6 • 16 A block with a mass of 75 kg is drawn steadily up a 25° rough incline by means of a rope. The coefficient of kinetic friction between the block and the incline is $\mu_k = 0.2$. (*a*) What is the tension in the rope if it is held parallel to the incline? (*b*) If it is held horizontal to the incline? (*c*) What would the answers to these questions be if the coefficient of friction were zero?

6 • 17 A block that has a weight of 60 N is held against a vertical wall by a horizontal force of 15 N. The coefficient of kinetic friction between the block and the wall is $\mu_k = 0.15$. (*a*) How much vertical force is needed to pull the block up the wall with constant speed? (*b*) To lower it with constant speed? (*c*) If a horizontal force of 270 N is needed to keep the block from falling when it is at rest, what is the coefficient of static friction?

6 • 18 A particle of mass m is moving in a straight line. Its position as a function of time is given by the equation $x(t) = A \sin Bt$, where A and B are constant quantities. Write expressions for the momentum of the particle and the accelerating force as functions of the time.

6 • 19 A block having a mass of 5 kg is projected up a 30° incline with an initial speed of 3 m/s. The coefficient of kinetic friction is $\mu_k = 0.3$. (*a*) How far up the incline will the block slide before it comes to rest? (*b*) Will it remain at rest or start to slide back down? Why?

6 • 20 A certain object slides with constant speed down an inclined plane when the plane is inclined at a given angle θ. Show that for an inclination $\theta' > \theta$ the block will accelerate at a rate

$$a = g\frac{\sin(\theta' - \theta)}{\cos\theta}$$

6 • 21 An inclined plane is set up over a given horizontal distance x, as shown. (*a*) If there is no friction, at what angle of inclination θ will an object slide down the plane in the shortest time? (*b*) What must the value of θ be if the coefficient of kinetic friction is $\mu_k = 0.2$?

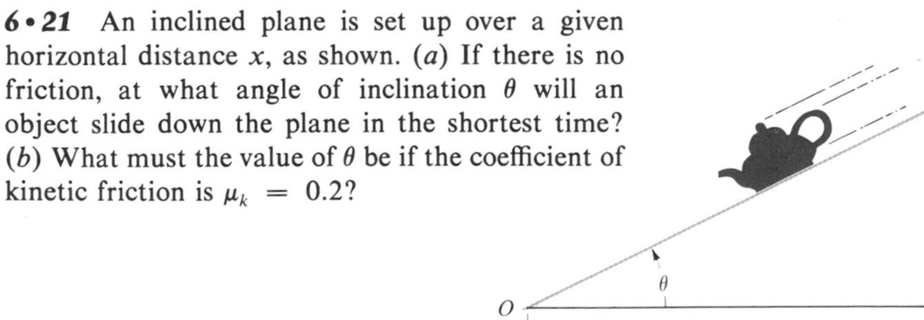

6 • 22 In an experiment with Atwood's machine (see Fig. 5 . 6), two masses of 200 g each are suspended on either side of a pulley, and a rider of mass 100 g is added to one side. (*a*) What is the resulting acceleration? (*b*) How far does each mass move from rest in 1.5 s? (*c*) What is the tension in the cord while the masses are moving freely?

6 • 23 What is the least acceleration with which a 75-kg person can slide down a rope that will sustain a tension of only 490 N? What will the person's speed be after sliding 20 m?

6 • 24 Three automobiles, each weighing 3500 lbf, are tied together bumper to bumper, and the engine of the first car is used to accelerate all three from rest to a speed of 20 mi/h. (*a*) If this is accomplished in a distance of 200 ft, what is the average force that has been applied? (*b*) Neglecting friction, find the tensions in the towlines connecting the cars.

6 • 25 A 500-g car is connected to two weights by a cable-and-pulley system as shown. (*a*) Assuming that the track is frictionless, find the magnitude of the acceleration. (*b*) What is the tension in each cable?

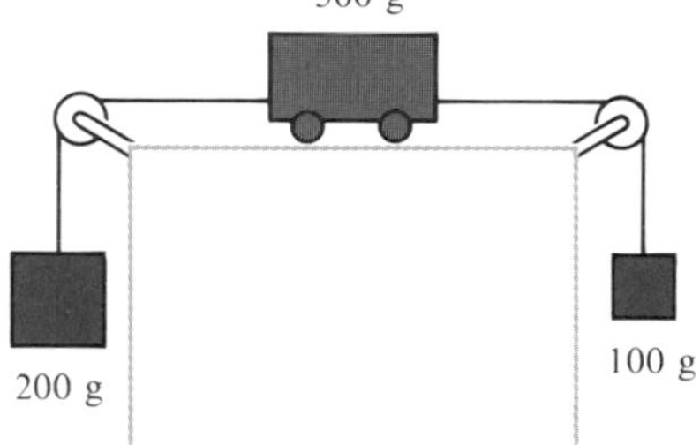

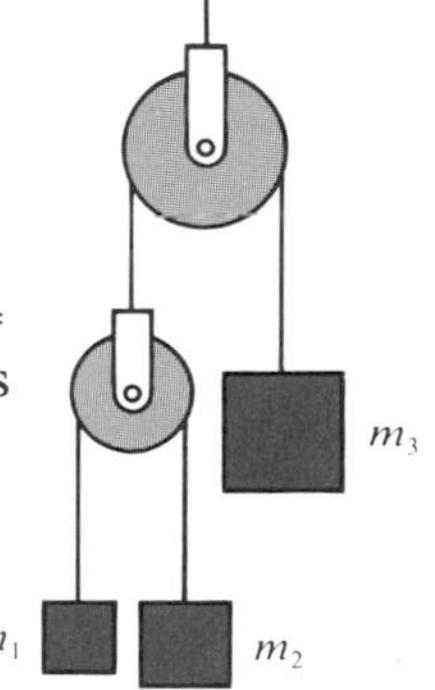

6 • 26 In the pulley system shown, $m_1 = 100$ g, $m_2 = 200$ g, and $m_3 = 300$ g. Neglecting friction and the masses of the pulleys and cords, what is the acceleration **a** of m_3?

6 • 27 Neglecting friction, find the magnitude of the acceleration and the tension in the cord (*a*) when $\theta_1 = 30°$ and $\theta_2 = 90°$. (*b*) When $\theta_1 = 30°$ and $\theta_2 = 25°$.

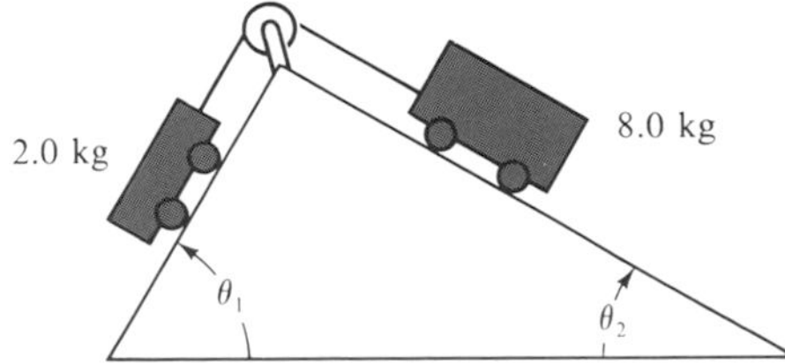

6 • 28 A monkey climbs up a rope of negligible mass that passes over a light, frictionless pulley and has a counterweight on the other end which is the same mass as the monkey. Show that if the monkey and the counterweight are initially at rest, they will both ascend at the same rate.

6 • 29 Given a vertical circle (a wheel) of radius R, show that the time of descent to the end of any smooth chord which starts from the top of the circle is the same.

6 • 30 A load of 1500 N is being pulled up a 5 percent grade (percentage of horizontal distance) at a constant speed. (*a*) If friction is neglected, how much force is needed? (*b*) What total amount of force would be needed to increase the speed uniformly from 8 to 24 km/h in 12 s? (*c*) If the force of friction is 10 N per hundredweight of load, how much force is needed to pull the load at a constant speed?

6 • 31 Two balls of mass m are connected by a string, and one ball is attached by a separate string to a support as shown. (*a*) If a slotted board is placed over the lower ball and slowly pressed down, which string will break? Why? (*b*) If the board is rapped sharply with a hammer, the lower string will break. Why?

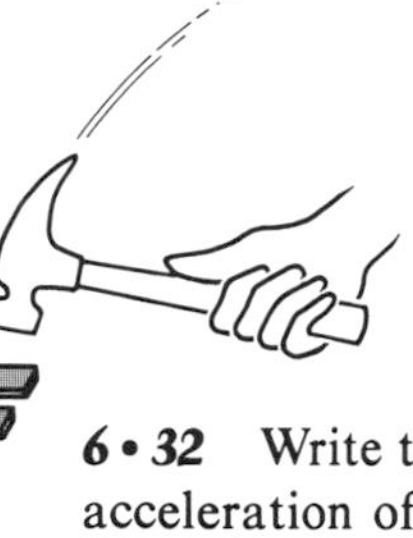

6 • 32 Write the differential equations of motion in component form for the acceleration of a particle of mass m sliding on a horizontal surface with a coefficient of kinetic friction μ_k. (*Hint:* The frictional force must always be opposed to the direction of motion.) Use only first and second time derivatives of x and y in your equations, not **a** or **v**.

6 • 33 The position of a particle of mass m is given by

$$\mathbf{r}(t) = \cos t\,\mathbf{i} + \sin t\,\mathbf{j}$$

(*a*) Describe the motion of the particle. (*b*) What is the force on it as a function of time?

6 • 6 *numerical integration*

6 • 34 If $y = \frac{1}{2}x^2$, (*a*) what is the finite change Δy due to a finite change Δx? (*b*) What is the size of the error in applying a formula such as Eq. [6 . 12] to the integration of $dy/dx = x$ over a *single* interval Δx?

***6 • 35** Integrate the equation $dy/dx = -2xy^2$ numerically over the interval $0 \leq x \leq 2$, for $y = 1$ at $x = 0$, (*a*) in steps of $\Delta x = 0.5$. (*b*) In steps of $\Delta x = 0.25$. (*c*) Compare the computed values of y for $x = 2$ with the exact solution, which is given by $y = 1/(1 + x^2)$, expressing the error as a percentage of the exact value. (*Hint:* This is easily done on a hand calculator.)

***6 • 36** Replace the variables x, y, and t by the dimensionless forms $\xi = x/L$, $\eta = y/L$, and $\tau = t/T$, and then scale the following equations to eliminate the parameters a and b:

$$\frac{dx}{dt} = \frac{ay}{x} \qquad \frac{dy}{dt} = bx^2 + by^2$$

***6 • 37** Compute and graph $x(t)$, given the initial condition $x_0 = 0$ and $v_0 = 0$ and the equation of motion

$$\frac{d^2x}{dt^2} = A - Bv$$

where A and B are positive constants. Stop the computation when v becomes approximately constant, as indicated by your graph. (*a*) What is the final value of v? (*b*) At what time does v become equal to half its final value? (*c*) At what time does v become equal to three-fourths its final value?

***6 • 38** Double the interval size and repeat the computation of Table 6 . 1. Compare the errors. What percentage of error do you find at $\tau = 6.4$?

answers

6 • 1 (*a*) $\mathbf{F} = 2.3\mathbf{i} - 2\mathbf{j} + 4.2\mathbf{k}$; (*b*) $F = 5.189$ dynes, $\alpha = 63.7°$, $\beta = 112.7°$, $\gamma = 36.0°$
6 • 2 $F = 10$ N, $\alpha = 55.6°$, $\beta = 64.9°$, $\gamma = 45°$
6 • 3 (*a*) $\mathbf{F} = (15.3 \text{ dynes}, 133°4')$; (*b*) $\mathbf{a} = (51 \text{ cm/s}^2, 133°4')$
6 • 4 $\mathbf{F} = (1.5\mathbf{i} - 2\mathbf{k})$ N
6 • 5 (*a*) 69.5°, 161.8°, 128.7°
6 • 6 $T = 450$ lbf
6 • 7 (*b*) $T(y) = (m + m'y/l)g$
6 • 8 $F_y = \frac{1}{2}F_g$, $F_x = F_g$
6 • 9 (*a*) 50.3 cm; (*b*) $T_1 = 0.3626$ N, $T_2 = 0.7242$ N
6 • 10 (*a*) $F_\parallel = 12$ kgf, $F_\perp = 16$ kgf; (*b*) $F_x = 15$ kgf, $F_\perp = 25$ kgf
6 • 11 (*a*) 30°, 30°; (*b*) $T = 10$ N
6 • 12 (*a*) $T = 1035$ lbf; (*b*) $F = 1464$ lbf
6 • 13 $F_x = mg\cos\theta$, $F_y = mg(1 - \sin\theta)$
6 • 14 (*a*) $T = mg$, $\mathbf{F} = (\sqrt{2}\,mg, 45°)$; (*b*) wire holding pulley should be at 45° angle
6 • 15 $F = mg$ in two lower bars, $2mg$ in upper bar, and $\sqrt{2}\,mg$ in the two diagonal bars; nothing
6 • 16 (*a*) $T = 45.3$ kgf; (*b*) $T = 55.1$ kgf; (*c*) 31.7 kgf, 35.0 kgf
6 • 17 (*a*) $F_y = 62.25$ N; (*b*) $F_y = 57.75$ N; (*c*) $\mu_s = 0.22$
6 • 18 $p = BAm\cos Bt$, $F = -B^2Am\sin Bt$
6 • 19 (*a*) 60.4 cm; (*b*) it will slide, unless $\mu_s > 0.57$
6 • 21 (*a*) $\theta = 45°$; (*b*) $\theta = 51°$
6 • 22 (*a*) $a = 1.96$ m/s²; (*b*) $\Delta y = 220.5$ cm; (*c*) $T = 240$ gf
6 • 23 $a = 3.27$ m/s², $v = 11.4$ m/s
6 • 24 (*a*) $F = 702$ lbf; (*b*) $T_1 = 234$ lbf, $T_2 = 468$ lbf
6 • 25 (*a*) $a = g/8$; (*b*) $T_1 = 112.5$ gf, $T_2 = 175.0$ gf
6 • 26 $\mathbf{a} = -57.7\mathbf{j}$ cm/s²
6 • 27 (*a*) $a = 0.7g$, $T = 23.5$ N; (*b*) $a = 233$ cm/s², $T = 14.5$ N
6 • 29 $t = (4R/g)^{1/2}$
6 • 30 (*a*) $F = 75$ N; (*b*) $F = 131.7$ N; (*c*) $F = 281.7$ N
6 • 31 (*a*) Upper string, because tension is greater; (*b*) inertial reaction force of upper ball increases resistance to rapid acceleration
6 • 32

$$\sqrt{\left(\frac{dx}{dt}\right)^2 + \left(\frac{dt}{dt}\right)^2}\left(\frac{d^2y}{dt^2}\mathbf{i} + \frac{d^2y}{dt^2}\mathbf{j}\right) + \mu_k g\left(\frac{dx}{dt}\mathbf{i} + \frac{dy}{dt}\mathbf{j}\right) = \mathbf{0}$$

6 • 33 (*a*) Counterclockwise, one revolution every 2 s; (*b*) $\mathbf{F}(t) = -m\mathbf{r}$
6 • 34 (*a*) $\Delta y = x\,\Delta x + (\Delta x)^2$; (*b*) error $= \Delta y - x\,\Delta x = \frac{1}{2}(\Delta x)^2$
6 • 35 (*a*) $y(2) \doteq 0.156$, 22 percent error; (*b*) $y(2) \doteq 0.182$, 9 percent error
6 • 36 $T = 1/(ab)^{1/2}$, $L = (a/b)^{1/2}$
6 • 37 (*a*) $v = A/B$; (*b*) exact solution is $x = At/B - (A/B^2)(1 - e^{-Bt})$, $v = (A/B)(1 - e^{-Bt})$, scale with $T = 1/B$, $L = A/B^2$, hence $t = 0.693/B$; (*c*) $t = 1.386/B$

CHAPTER SEVEN

work, power, and energy

If the activity of an agent be measured by its amount and its velocity conjointly; and if, similarly, the counteractivity of the resistance be measured by the velocities of its several parts and their several amounts conjointly, whether these arise from friction, cohesion, weight, or acceleration—activity and counteractivity, in all combinations of machines will be equal and opposite.

Isaac Newton, Scholium of the Third Law, Principia, *1680, translated by Wm. Thompson (Lord Kelvin) and Peter Taite*

The basis of modern civilization lies in the use of machinery, and there are no notions more intimately associated with a machine than those of work, power, and energy. Energy is not a concept that can easily be abstracted from everyday experience, for it is not readily apparent, like force or velocity. It was first used in physics almost as a "bookkeeping" device. Not until the middle of the nineteenth century did the concept of energy attain the importance it has in science today. This was brought about by formulation of the physical law of conservation of energy, a comprehensive principle growing out of the work of several great physicists and chemists, and perhaps the single most important generalization in the history of the sciences. In this chapter we shall develop the definitions of work, power, and energy needed to understand this great principle.

7 • 1 work

When we speak of "doing work," we imply a process that takes place over time and involves some movement. Thus, if it were possible to define a physical quantity that represented work, you might expect it to have certain properties. For example, if you carry a mass m up a flight of stairs, let us say you have done a certain amount of work W. If you carry a mass $2m$ up the same flight of stairs, you have done work $2W$. Similarly, if you had carried m up the stairs twice, or up two flights of stairs, you would also have done work $2W$. Such simple observations suggest that work is a physical quantity which is proportional to force and distance, but can be added up like a scalar.* Of

*It will be helpful to review Sec. 2 . 3 on scalar products.

course, it is harder to run up the stairs than to walk up them, but that additional strain is related to the *rate* at which the work is being done, dW/dt, or the *power* you are putting out. Work itself is defined only in terms of force and distance.

When a constant force F_x moves a body through a displacement which is in the *same direction* as the force, then we define the amount of work done by this force as

$$\Delta W = F_x \Delta x \qquad [7 \cdot 1]$$

Suppose the block in Fig. 7 . 1 is being pulled along a horizontal surface by a force $F_x\mathbf{i}$ parallel to the surface; this force does work $F_x \Delta x$ on the block. Now suppose an independent vertical force F_y is applied which is less than F_g, the weight of the block. Since this vertical force will not cause any vertical motion, it is doing no work. The vector sum of these two applied forces is

$$\mathbf{F} = F_x\mathbf{i} + F_y\mathbf{j}$$

However, the displacement of the block is only in the x direction, $\Delta\mathbf{r} = \Delta x\mathbf{i}$. Therefore the work done by the applied force is the *scalar dot product* of force and the displacement:

$$\Delta W = \mathbf{F} \cdot \Delta\mathbf{r} = F_x \Delta x + F_y \times 0 = F_x \Delta x \qquad [7 \cdot 2]$$

Next, imagine that the horizontal force $F_x\mathbf{i}$ vanishes, leaving only the vertical force $F_y\mathbf{j}$. Since this remaining force has no effect on the motion of

figure 7 • 1 *The work done in moving a mass by an amount* $\Delta\mathbf{x} = \Delta x\mathbf{i}$

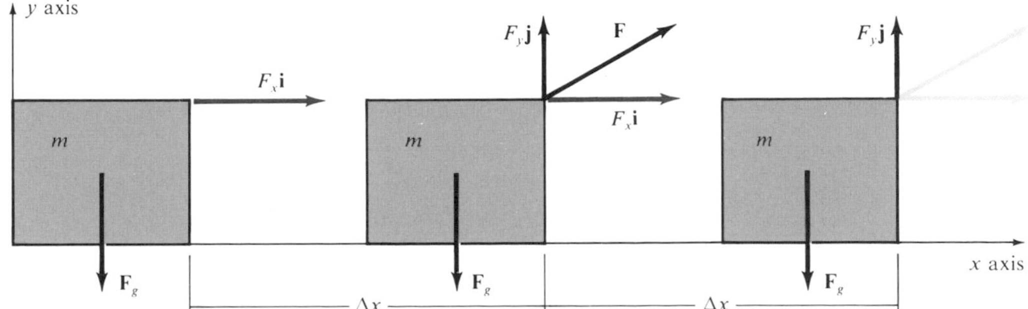

the block in the x direction, in the absence of an applied force $F_x\mathbf{i}$, no work is being done. Even if the block now slides an additional distance Δx, there is no force moving it in this direction, and the scalar product ΔW is therefore zero:

$$\Delta W = \mathbf{F} \cdot \Delta\mathbf{r} = 0 \times \Delta x + F_y \times 0 = 0 \qquad [7 \cdot 3]$$

If the vertical force in Fig. 7 . 1 were greater than the weight of the block, then a net vertical force of magnitude $F_y - F_g$ would raise the block a distance Δy, doing work against gravity:

$$\Delta W = (F_y - F_g)\,\Delta y \qquad [7 \cdot 4]$$

In this case, although the work done by the applied force $\mathbf{F}$ alone is $\Delta W = \mathbf{F} \cdot \Delta\mathbf{r}$, the total work done on the body will be the algebraic sum of the work done by all the forces acting on it:

$$\begin{aligned} \Delta W_{\text{net}} &= F_x\Delta x + F_y\Delta y - F_g\Delta y \\ &= \mathbf{F} \cdot \Delta\mathbf{r} - \mathbf{F}_g \cdot \Delta\mathbf{r} \end{aligned} \qquad [7 \cdot 5]$$

Note that in all three cases the work ΔW done by a constant force $\mathbf{F}$ on a body moving through a displacement $\Delta\mathbf{r}$ can be represented by a scalar product:

$$\Delta W = \mathbf{F} \cdot \Delta\mathbf{r} \qquad [7 \cdot 6]$$

If you push a heavy trunk across a floor, you do work. If you push it back into place, your total work is not zero, but twice as much; hence work cannot be a vector quantity. Moreover, the definition of work does not require that a force $\mathbf{F}$ necessarily cause a displacement, or even that it point in the direction of motion. A force may act on a moving body without doing any work on it, as we saw in Fig. 7 . 1, or it can do *negative* work. In Eq. [7 . 5], for example, the gravitational force $\mathbf{F}_g$ makes a negative contribution in the vertical displacement Δy of the block.

example 7 • 1 Find the work done by the force of gravity on a body of mass m as it slides down a smooth inclined plane.

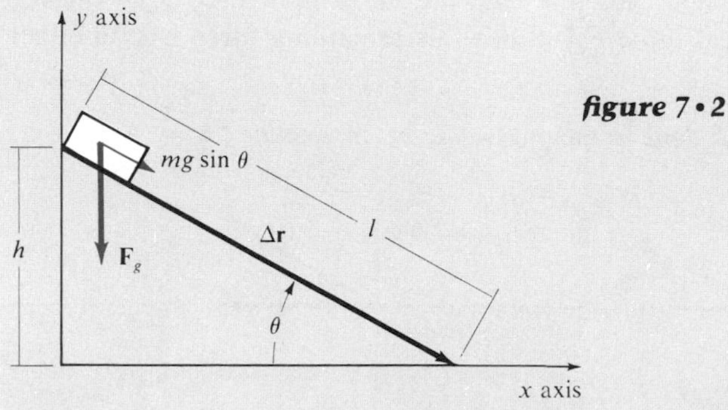

figure 7 • 2

solution The total displacement down the length of the incline is

$$\Delta\mathbf{r} = l\,(\cos\theta\,\mathbf{i} - \sin\theta\,\mathbf{j})$$

However, the work done depends only on the displacement in the direction of the acting force, $\mathbf{F}_g = -mg\mathbf{j}$:

$$\Delta W = \mathbf{F}_g \cdot \Delta\mathbf{r} = -mg\mathbf{j} \cdot \Delta\mathbf{r}$$
$$= mgl \sin\theta = mgh$$

In other words, the work ΔW done by gravity depends only on the height of the inclined plane, and not on its length. Note that we could have obtained the same result by considering the force component

$$F_{\parallel} = mg \sin\theta$$

parallel to the displacement $\Delta\mathbf{r}$.

Figure 7 . 3 shows a particle traversing a curved path under the influence of a constant force in the x direction. Since the motion of the particle is not one-dimensional, there must be other forces acting on it as well. Here, however, we are interested only in the work done by the horizontal force $\mathbf{F} = F_x\mathbf{i}$ as the particle moves from x_1 to x_4. Adding together the work done over the separate segments of the path,

$$\Delta W = F_x[(x_2 - x_1) + (x_3 - x_2) + (x_2 - x_3) + (x_3 - x_2) + (x_4 - x_3)]$$

we obtain

$$\Delta W = F_x(x_4 - x_1) \qquad [7 \cdot 7]$$

Although the path is curved, the work done by a constant force depends only on the *net displacement in the direction of the force.* Note that any retracing of the path, such as that between x_2 and x_3, simply cancels out. This effect is important, because we live in a gravitational force field which is, for most purposes, constant in magnitude and direction. Thus, no matter how a body moves through a net vertical distance Δy, the gravitational force $\mathbf{F}_g$ must always do net work equal to $mg\,\Delta y$ if the body is falling and work equal to $-mg\,\Delta y$ if the body is rising. For a horizontal displacement where $\Delta y = 0$, gravity does no work.

figure 7 • 3 *Constant force, curvilinear path. Arrows indicate linear displacements Δx.*

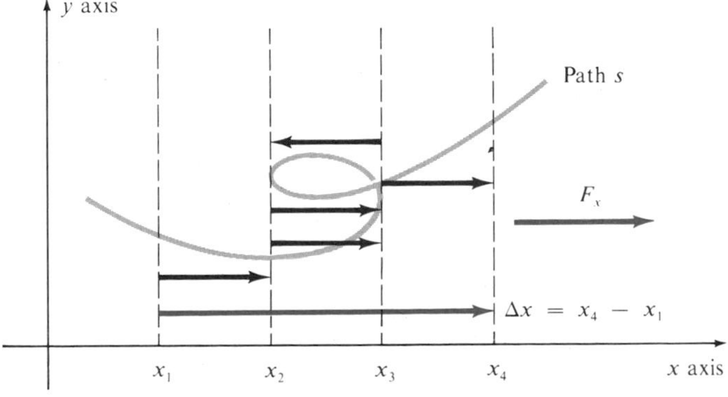

***example* 7 • 2** A steel ball bearing of mass m is placed on the lip of a hemispherical porcelain bowl of radius R and then released. (*a*) How much work is done by gravity as the ball bearing falls to the bottom of the bowl? (*b*) What is the net work done by the time the ball bearing rolls up to the lip of the bowl on the opposite side?

solution (*a*) Since the bowl is a hemisphere, its radius R is equal to the vertical distance through which the ball bearing is displaced. Thus the work done by gravity is simply

$$\Delta W = mg\,\Delta y = mgR$$

(*b*) At the time the ball bearing reaches the opposite lip of the bowl its net vertical displacement is zero; hence the net work done by gravity is also zero:

$$\Delta W = mg \times 0 = 0$$

To measure work we use a derived unit consisting of the product of any unit of force and any unit of length. In the mks system the unit of work is the *newton-meter* (N · m), or *joule* (J), which represents the work done by a constant force of 1 N in moving any object a distance of 1 m in the direction of the force. In the cgs system the unit of work is the *dyne-centimeter,* or *erg,* defined as 1 erg $= 10^{-7}$ J. In the fps system the unit of work is the foot-pound (ft · lbf or ft · lb).

7 • 2 work done by a variable force

Since forces may vary in magnitude and direction along a particle's path, we must generalize the concepts discussed thus far to include such cases. We can begin by thinking of any curved path as a succession of straight-line segments so short that the force acting along any one segment may be regarded as constant in direction and magnitude. We can then apply Eq. [7 . 1] to each of these segments and add the resulting increments of work to obtain the total work done in moving the particle along the path.

Suppose we partition the curved path in Fig. 7 . 4 into successively smaller segments Δs. As Δs becomes vanishingly small,

$$\lim_{\Delta s \to 0} \Delta s = ds$$

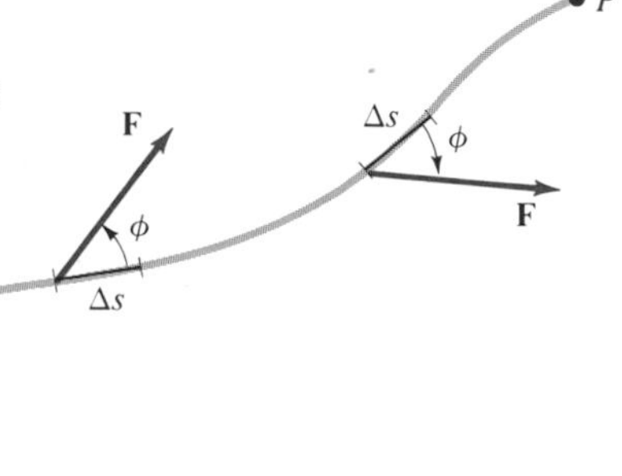

***figure* 7 • 4** *Work done by a variable force, showing vectors tangent to path segments Δs*

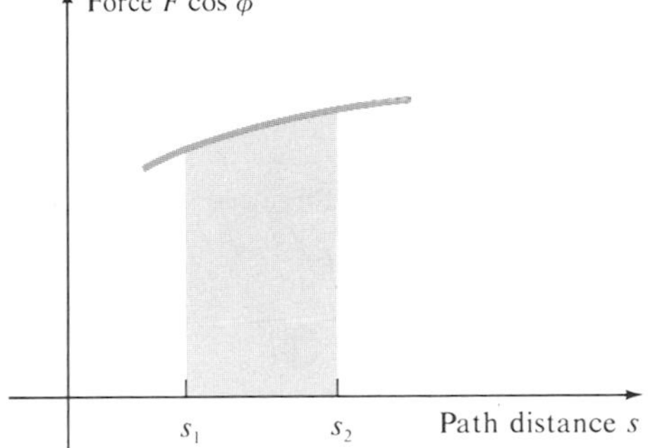

figure 7 • 5 *Work done over the path distance* $s_2 - s_1$

the force component $F \cos \phi$ parallel to the direction of motion at that point will do work $dW = F \cos \phi \, ds$. The total work W done on the particle over the whole path length s is then exactly the sum of the increments of work dW_i done over each infinitesimal segment ds_i:

$$W = \lim_{n \to \infty} \sum_{i=1}^{n} F_i \cos \phi_i \, ds_i \qquad [7 \cdot 8]$$

where ϕ may be constantly changing along the path. This equation simply defines the integral

$$W = \int_0^s F \cos \phi \, ds \qquad [7 \cdot 9]$$

In vector notation Eq. [7 . 9] would be expressed as the integral of the scalar product of force and displacement $d\mathbf{r}$ between the endpoints of the path:

$$\blacktriangleright \quad W = \int_0^W dW = \int_O^P \mathbf{F} \cdot d\mathbf{r} \qquad [7 \cdot 10]$$

This equation may be viewed as a prescription for finding the total work done by summing all the work differentials $dW = \mathbf{F} \cdot d\mathbf{r}$ along the path between the endpoints O and P.

Although the integrals of Eqs. [7 . 9] and [7 . 10] appear disarmingly simple, care must be taken to account for both the magnitude of $\mathbf{F}$ and its direction relative to the path. Since the path may be curved, the applied force will not necessarily be the same everywhere along the path; hence the work done by this force may depend on the path itself, and not just on the linear distance between its endpoints. This is especially true of work done against friction. The magnitude of a frictional force may vary along the path, and its direction is tangential to the path at every point.

If we know $F \cos \phi$ as a function of the distance s along the path, then we can use Eqs. [7 . 9] and [7 . 10] to calculate W. In Fig. 7 . 5, for example, the area under the curve is just the work represented by the integral

$$W = \int_{s_1}^{s_2} F \cos \phi \, ds$$

Thus the work done by $F \cos \phi$ in moving a particle through the distance $\Delta s = s_2 - s_1$ can be calculated by measuring this area. Since any path in

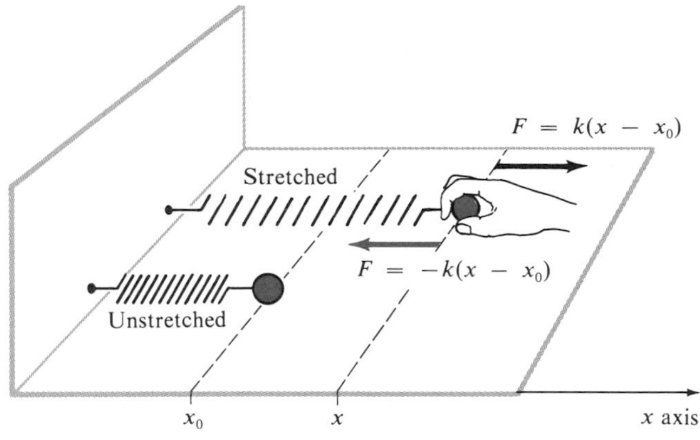

figure 7 • 6 *A Hooke's law linear restoring force*

space can, in principle, be represented by a one-dimensional curve, the graphical approach is quite general. However, it is sometimes difficult to determine $F \cos \phi$ at every point along the path in order to perform the integration.

As an example of a variable force, consider the relatively simple case of motion along a straight line, assumed to be the x axis, under the influence of a force $F_x = F(x)$. That is, F is a function of x which can take on either positive or negative values as it acts on an object moving from x_0 to some position x. Thus the work done is simply

$$W = \int_{x_0}^{x} F(x)\,dx \qquad [7 \cdot 11]$$

Suppose an unstretched spring lies at a position x_0, as shown in Fig. 7 . 6. When it is stretched to some position x, it will exert a linear restoring force given by *Hooke's law* as

$$F = -k(x - x_0) \qquad [7 \cdot 12]$$

where $x - x_0$ is the net displacement and k is the restoring-force constant of the spring. As we saw in Sec. 6 . 6, such a force always acts to restore the spring to its original unstretched length.* The amount of force applied to the spring to stretch it must therefore be $F = k(x - x_0)$, and the work done *on the spring* by this applied force is

$$W(x) = \int_{x_0}^{x} k(x - x_0)\,dx$$

$$= \tfrac{1}{2}k(x - x_0)^2 \qquad [7 \cdot 13]$$

Note that the work done in stretching or compressing a spring is always positive but has a minimum value of zero when the spring is left in its undisturbed state.

*Hooke's law is one of the most fundamental relations in physics. It gives rise to periodic motions or vibrations known as *simple harmonic oscillations*. This type of motion was analyzed numerically in Sec. 6 . 6.

***example* 7 • 3** A spring has been stretched upward to some position y from its horizontal position x_0. How much work was done on the spring?

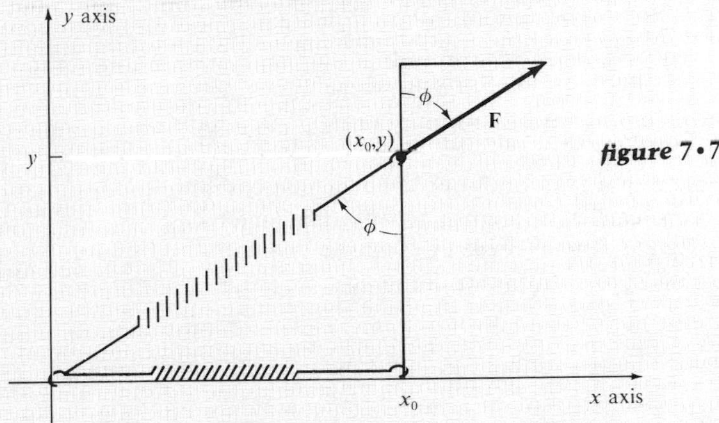

***figure* 7 • 7**

solution Work can be done on the spring only by the force component in a direction parallel to the spring. Hence we can express this work entirely as a function of the extension of the spring, with an expression identical to Eq. [7 . 13], but with Δx replaced by the net extension, regardless of direction. We immediately obtain

$$W = \tfrac{1}{2}k\left(\sqrt{x_0^2 + y^2} - x_0\right)^2$$

Had we simply followed the prescription of Eq. [7 . 10], we would have obtained exactly the same result.

***example* 7 • 4** A mass m slides down a smooth quadrant of a circle of radius R as shown. Use Eq. [7 . 9] to find the work done on the mass by gravity.

***figure* 7 • 8**

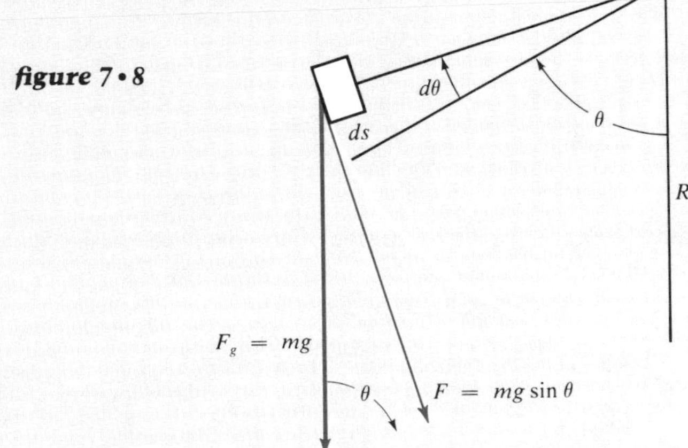

solution If we use the angle θ to describe the motion, then a small displacement of the particle in the direction of decreasing θ is given by $ds = -R\,d\theta$, where $R = \Delta y$ and $d\theta > 0$. The accelerating force is

$\mathbf{F}_g = -mg\mathbf{j}$ and its component in the direction of motion is $F = mg \sin \theta$. Hence Eq. [7 . 9] gives the work done by gravity as

$$W = \int_{\pi/2}^{0} mg \sin \theta \, (-R \, d\theta) = \int_{0}^{\pi/2} mgR \sin \theta \, d\theta = mgR$$

Note that reversing the limits on the first integral cancels the minus sign. This result agrees with the simpler approach used in Example 7 . 2.

example 7 • 5 The inhabitants of the land of Pypyk live in a vast basin of radius R at the rim, created eons earlier by a collision with the planetoid Zdrafstvwy. Local physicists have determined the existence of two gravitational forces in the basin. Expressed in two dimensions, one is a vertical force,

$$\mathbf{F}_1 = -mg\left(1 - \frac{2y}{R}\right)\mathbf{j}$$

and the other is a horizontal force radiating outward from the y axis:

$$\mathbf{F}_2 = \frac{mg}{R} x\mathbf{i}$$

After a state ceremony at the center of the basin, an intrepid band of explorers of total mass m ascends the side of the basin, traveling straight up to the rim along a smooth curve given by $y = 3x^2/4R$. (*a*) Find the work done by the explorers against the vertical force $\mathbf{F}_1$. (*b*) Find the work done against force $\mathbf{F}_2$. (*c*) What is the work done against the total gravitational force?

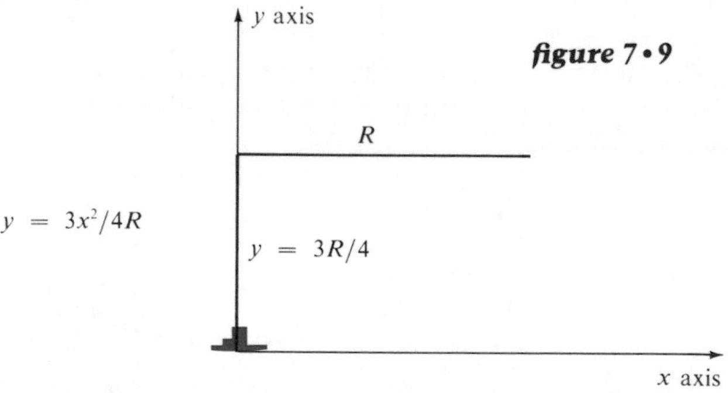

figure 7 • 9

solution The net outward distance from the center of the basin to the rim is $\Delta x = R$, so the height of the ascent is $\Delta y = 3R/4$. The work done against the vertical force $\mathbf{F}_1$ is therefore

$$W_1 = \int_0^{3R/4} mg\left(1 - \frac{2y}{R}\right) dy = mg\left(\frac{3R}{4} - \frac{9R^2}{16R}\right) = \frac{3mgR}{16}$$

(*b*) The outward force $\mathbf{F}_2$ aids the explorers, so that

$$W_2 = \int_0^R \frac{-mg}{R} x \, dx = -\frac{mgR}{2}$$

(*c*) The total work done against gravity is

$$W = W_1 + W_2 = -\frac{5mgR}{16}$$

After the explorers pass a certain point, gravity will start pulling them away from the center, resulting in negative work.

7 • 3 power

The growing use of machines in the latter part of the eighteenth century brought a need for ways of describing the *rapidity* with which they could do work. Thus James Watt, wishing to compare the practical value of his steam engine with that of a horse, made some actual experiments and arrived at the estimation of 550 ft · lbf/s for the average *power,* or work output per second, of a horse; he called this unit the *horsepower* (hp). In ordinary language the word *power* has a variety of meanings, but as a physical term it means the *time rate at which work is done.* Thus average power is a derived physical quantity, defined as the ratio of work to time:

$$\overline{P} = \frac{W}{t} \qquad [7 \cdot 14]$$

Obviously power is a scalar quantity.

In mks units power is measured in *joules per second* (J/s), usually called *watts* (W). Other power units in use are ergs per second, and foot-pounds per second (ft · lbf/s). For practical situations in which large amounts of power are involved, the power units of most convenient size are the *kilowatt* (1 kW = 10^3 W), the *megawatt* (1 MW = 10^6 W), and the *horsepower* (1 hp = 550 ft · lbf/s = 745.70 W). Since the product of a unit of power and a unit of time is work, mixed units such as the kilowatt-hour, 3.6 × 10^6 J, are often used in engineering practice. Table 7 . 1 gives conversion factors between various units of work. For convenience, the units

table 7 • 1 *Work conversion factors*

equal powers: 1 J/s = 1 W = 10^7 ergs/s, 1 hp = 550 ft · lbf/s

joules (J)	calories (cal)	kilowatt-hours (kW · h)	foot-pounds (ft · lbf)	British thermal units (Btu)	horsepower-hours (hp · h)
1	0.2389	2.778×10^{-7}	0.7376	9.481×10^{-4}	3.725×10^{-7}
4.186	1	1.163×10^{-6}	3.087	3.968×10^{-3}	1.559×10^{-6}
3.600×10^{6}	8.601×10^{5}	1	2.655×10^{6}	3413	1.341
1.356	0.3239	3.766×10^{-7}	1	1.285×10^{-3}	5.051×10^{-7}
1055	252.0	2.930×10^{-4}	777.9	1	3.929×10^{-4}
2.685×10^{6}	6.414×10^{5}	0.7457	1.980×10^{6}	2545	1

of heat, the calorie and the British thermal unit (Btu), are also included; as we shall see later, heat and work are interchangeable and can therefore be measured in the same units. To convert units of work to units of power, simply divide by the appropriate time unit.

When a constant force **F** is acting on a moving body, and the point to which the force is applied moves through a displacement $\Delta\mathbf{r}$, the work done by the force is $\Delta W = \mathbf{F}\cdot\Delta\mathbf{r}$. If this work is accomplished in a time Δt, the *average power* expended during that interval is

$$\overline{P} = \frac{\Delta W}{\Delta t} = \mathbf{F}\cdot\frac{\Delta\mathbf{r}}{\Delta t} \qquad [7\cdot 15]$$

The *instantaneous power* is therefore

$$\blacktriangleright\; P = \lim_{\Delta t\to 0}\mathbf{F}\cdot\frac{\Delta\mathbf{r}}{\Delta t} = \mathbf{F}\cdot\lim_{\Delta t\to 0}\frac{\Delta\mathbf{r}}{\Delta t} = \mathbf{F}\cdot\frac{d\mathbf{r}}{dt} = \mathbf{F}\cdot\mathbf{v} \qquad [7\cdot 16]$$

where **v** is the velocity of the moving point. Thus, if the force is constant in magnitude and direction, the power expended in moving a body is directly proportional to the speed of the body. In fact, Eq. [7 . 16] holds for *any* force, whether it is constant or variable.

***example* 7 • 6** Find the power expended by the earth's gravity on an object of mass m sliding down the inclined plane in Example 7 . 1. Assume that the object starts from a position of rest.

solution The acceleration of the body down the incline is given by the gravitational component $g\sin\phi$; hence the position of the body along the plane at any time t is $r(t) = \frac{1}{2}(g\sin\phi)\,t^2$. Since the acting force $F = mg\sin\phi$ is constant, the power, or time rate of work, is

$$P = \frac{dW}{dt} = \frac{d[\frac{1}{2}m\,(g\sin\phi)^2t^2]}{dt} = m(g\sin\phi)^2t$$

or, in terms of the height h and length l of the inclined plane,

$$P = m\left(g\frac{h}{l}\right)^2 t$$

That is, as the body moves faster, the power expended is not constant, as you might think, but increases accordingly.

7 • 4 kinetic energy

Although we customarily speak of many forms of energy—food energy, fuels, explosives, moving bodies, electricity, nuclear power—there are only two basic forms: *kinetic energy* and *potential energy*. The first is the energy of motion, which we shall discuss here; the second is *stored* energy, as in a stretched or compressed spring, and will be discussed in the next chapter. When we come to the study of atomic physics and light, we shall see that all forms of energy can be reduced to kinetic and/or potential energy. Like work, energy is a scalar quantity, and the units of energy are the same as those of work. Although we have discussed work first, energy is the more

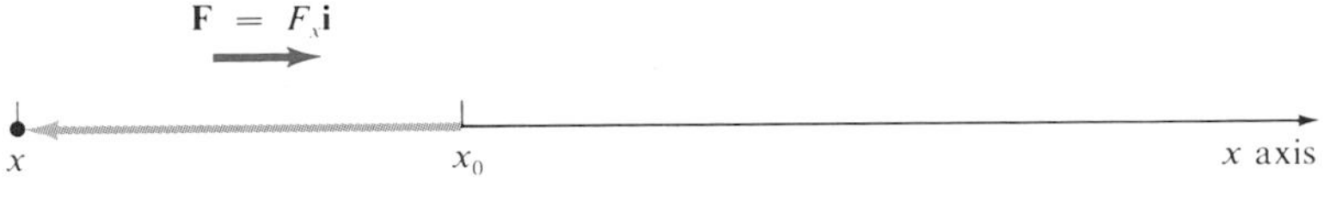

figure 7 • 10 *Kinetic energy*

basic concept; in a sense, work is the transfer of energy from one system to another by *directed motion.**

Since work is measurable, we can operationally define any body or system with the capacity to do work as one that possesses energy. Clearly, a moving body must possess energy, since it can move against friction, compress a spring on impact, fly upward against gravity, and so on. This capacity of a body to do work by virtue of its motion is called *kinetic energy.* In the preceding sections we have been intuitively aware that a force which is doing positive work on a body must be increasing the speed of that body; conversely, a force doing negative work is decreasing its speed.

Consider the one-dimensional motion shown in Fig. 7 . 10, where a particle initially moving to the left is continuously acted upon by a force $\mathbf{F} = F_x\mathbf{i}$ directed to the right. As the particle moves to the left from the origin, the force does negative work, slowing down the particle until it reverses its direction. As the particle then moves back toward the origin, the force does positive work because its direction now parallels the particle's direction of motion. Since the work done is a scalar quantity, we would like to be able to relate it to some scalar measure of the particle's motion. The *speed* of the particle is such a measure, and it seems logical to look for some relationship between changes in speed and the work done on the particle. Let us derive such a relationship with the aid of the second law, and in the process, precisely define the kinetic energy of the particle.

In the case of constant acceleration illustrated in Fig. 7 . 10, the work done in a net displacement from $\mathbf{x}_0$ to $\mathbf{x}$ is

$$\Delta W = \mathbf{F} \cdot (\mathbf{x} - \mathbf{x}_0) = F(x - x_0) = ma(x - x_0)$$

Substituting this expression into Eq. [3 . 27] from Sec. 3 . 4, we obtain

$$v^2 = v_0^2 + 2a(x - x_0) = v_0^2 + \frac{2W}{m}$$

or

$$W = \tfrac{1}{2}mv^2 - \tfrac{1}{2}mv_0^2 \qquad [7 \cdot 17]$$

Thus we have a measure for work done on the particle expressed solely in terms of a change in the quantity $\frac{1}{2}mv^2$. This quantity is defined as the *kinetic energy K* of the particle:

$$K = \tfrac{1}{2}mv^2 \qquad [7 \cdot 18]$$

Kinetic energy is a general property of particle motion; it is literally the *energy of motion* and has the dimensions of work.

*Heat, as we shall see in Chapter 21, is the transfer of energy from one system to another by *random motion.* Heat and work are interchangeable.

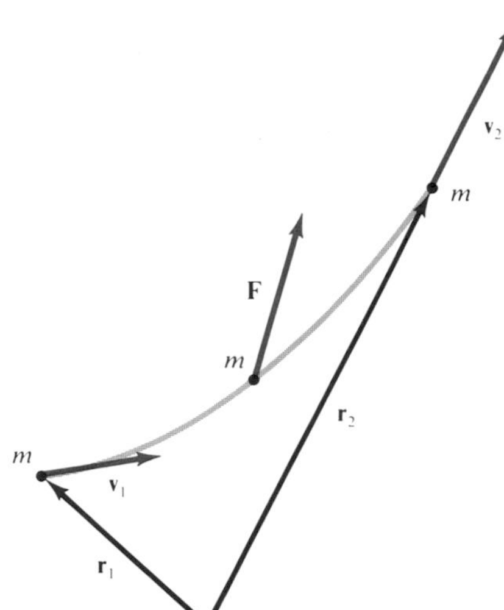

figure 7 • 11 *A particle moving under a general force* F

Suppose a particle of mass m is moving under the influence of a net force $\mathbf{F}$ as shown in Fig. 7 . 11. According to the second law, we can state this force as

$$\mathbf{F} = m\frac{d\mathbf{v}}{dt}$$

If we substitute this expression for force in Eq. [7 . 16], we obtain an expression for power, the rate at which work is done on the particle by the force:

$$P = \mathbf{F} \cdot \mathbf{v} = m\frac{d\mathbf{v}}{dt} \cdot \mathbf{v} = \frac{dW}{dt} \qquad [7 \cdot 19]$$

Then, making use of the derivative of a scalar product, we have

$$m\mathbf{v} \cdot \frac{d\mathbf{v}}{dt} = \frac{m}{2}\frac{d}{dt}(\mathbf{v} \cdot \mathbf{v}) = \frac{m}{2}\frac{d(v^2)}{dt} = \frac{dW}{dt} \qquad [7 \cdot 20]$$

We can now integrate the instantaneous values on both sides of the last equation over entire interval corresponding to $\Delta\mathbf{r} = \mathbf{r}_2 - \mathbf{r}_1$:

$$\tfrac{1}{2}m\int_{v_1}^{v_2} d(v^2) = \tfrac{1}{2}mv_2^2 - \tfrac{1}{2}mv_1^2$$

$$= \Delta K = \int_{r_1}^{r_2} dW = \Delta W_{12} \qquad [7 \cdot 21]$$

Thus we see that the change ΔK in the particle's kinetic energy as it moves from $\mathbf{r}_1$ to $\mathbf{r}_2$ always equals the net work done *on* the particle by the applied force $\mathbf{F}$, whether $\mathbf{F}$ is constant or variable, one force or the resultant of several forces, or in three dimensions or fewer. This is known as the *work–kinetic-energy theorem.*

Note in Eq. [7 . 21] that the work done on the particle during a displacement from $\mathbf{r}_1$ to $\mathbf{r}_2$ is represented by the symbol ΔW_{12}, and not as $W_2 - W_1$. This is because, in general, we cannot define a work function which is uniquely dependent on position; instead, ΔW_{12} may depend on the path between these two endpoints. However, we shall find a way to circumvent this difficulty in the important special case of "conservative" systems, discussed in the next chapter.

In Sec. 7 . 1 we saw that the work done by a constant force over any path, no matter how circuitous, is just the force multiplied by the net displacement in the direction of the force. Thus the work done by gravity on a body moving down an inclined plane was equal to $mg\,\Delta y$. In other words, when a body starting from rest slides down a smooth incline, its speed when it reaches the bottom can be calculated from $\frac{1}{2}mv^2 = mg\,\Delta y$, regardless of the angle of the incline. The change in kinetic energy for a body moving from *any* height y_1 to another height y_2 under the influence of constant gravity is, from Eq. [7 . 21],

$$\Delta K = \tfrac{1}{2}mv_2^2 - \tfrac{1}{2}mv_1^2 = mg(y_1 - y_2) \qquad [7 \cdot 22]$$

***example* 7 • 7** Use the results of Sec. 4 . 3 to find the change in the kinetic energy of a ballistic projectile as a function of its height, where the components of its velocity are $v_x = v_{x0}$ and $v_y = v_{y0} - gt$.

solution First we must express the kinetic energy of the projectile as a function of time:

$$K = \tfrac{1}{2}mv^2 = \tfrac{1}{2}m(v_x^2 + v_y^2)$$
$$= \tfrac{1}{2}m(v_{x0}^2 + v_{y0}^2 - 2gtv_{y0} + g^2t^2)$$

From Eq. [4 . 21] we see that

$$-2gtv_{y0} + g^2t^2 = -2g(y - y_0)$$

and substituting this value into the above expression for K, we find the *change* in K to be

$$\Delta K = \tfrac{1}{2}mv^2 - \tfrac{1}{2}mv_0 = -mg(y - y_0)$$

Some advantages of the energy method are immediately apparent. When the projectile falls back to its original height y_0, the net change in its kinetic energy is zero, and it is again moving at its initial speed v_0, even though its *direction* has changed. Thus we can easily find the maximum height the projectile could attain if it were launched vertically by setting $v = 0$ at that height and then solving the kinetic-energy equation for $y = \frac{1}{2}v_0^2/g + y_0$.

7 • 5 kinetic energy and momentum

Kinetic energy was first used in its present sense of $K = \frac{1}{2}mv^2$ by Lord Kelvin in 1884. Still earlier, in 1695, Leibnitz had referred to mv^2 as the "living force"—*vis viva*. For over half a century a controversy had raged between the followers of Descartes and the followers of Leibnitz over whether the ability of a moving body to overcome opposing forces and produce changes in other bodies was proportional to the velocity or to the square of the velocity. It turns out that both viewpoints are correct. Every moving body has both momentum and kinetic energy, and we can determine force in terms of either quantity. Which one is emphasized in a given case depends on the problem at hand.

For example, consider a bullet that enters a fixed wooden block at an impact velocity **v** and penetrates the block through a displacement $\Delta\mathbf{r}$. The bullet experiences an average force $\overline{\mathbf{F}}$ resisting its motion. From the work–kinetic-energy theorem, Eq. [7 . 21], this average force is

$$\Delta W = \int_{x_0}^{x} \mathbf{F} \cdot d\mathbf{r} = \overline{F}\,\Delta r = \tfrac{1}{2}mv^2 \qquad \text{or} \qquad \overline{F} = -\frac{\Delta K}{\Delta r} \qquad [7 \cdot 23]$$

If the penetration takes place in a time $\Delta t = t - t_0$, the second law yields

$$\int_{t_0}^{t} \mathbf{F}\,dt = \overline{\mathbf{F}}\,\Delta t = -m\mathbf{v} \qquad \text{or} \qquad \overline{\mathbf{F}} = \frac{\Delta\mathbf{p}}{\Delta t} = -\frac{m\mathbf{v}}{\Delta t} \qquad [7 \cdot 24]$$

The time required for the bullet to come to rest depends on its momentum, and the distance it penetrates depends on its kinetic energy. The force is expressible in terms of either. The quantity

$$\Delta\mathbf{p} = \overline{\mathbf{F}}\,\Delta t = \int_{t_0}^{t} \mathbf{F}\,dt \qquad [7 \cdot 25]$$

is called the *impulse* of the force. The total change in momentum $\Delta\mathbf{p}$ equals the impulse, whereas the change in kinetic energy ΔK depends on the work done by the force.

The momentum and the kinetic energy of a particle of mass m at any given instant are related by the classical *momentum-energy equation:*

$$K = \frac{p^2}{2m} = \frac{1}{2m}\mathbf{p} \cdot \mathbf{p} \qquad [7 \cdot 26]$$

This is an extremely useful equation, since it is sometimes possible to measure the kinetic energy of an object without knowing its momentum, and Eq. [7 . 26] then provides a way to evaluate the magnitude of the momentum. Differentiating both sides of Eq. [7 . 26] gives us another expression for power:

$$\frac{dK}{dt} = \frac{1}{m}\mathbf{p} \cdot \frac{d\mathbf{p}}{dt} = \mathbf{v} \cdot \mathbf{F} = P$$

Since $P = dW/dt$, this relation simply becomes a restatement of the work–kinetic-energy theorem in differential form:

$$\frac{dK}{dt} = \frac{dW}{dt} \qquad [7 \cdot 27]$$

***example** 7 • 8* An object of mass m is dropped from a height h above a massive spring and is brought to rest as it compresses the spring below y_0 by a slight amount $\delta y \ll h$. How long does this compression process take?

solution To find the compression time $t = p/\overline{F}$, we note that the force of gravity has done work approximately equal to

$$\Delta W = mg(h + \delta y) \doteq mgh$$

Hence the object has lost kinetic energy $K \doteq mgh$ when it is brought to rest by the spring, and the average force during the brief period of compression is

$$\overline{F} = \frac{K}{\delta y} = \frac{mgh}{\delta y}$$

The object therefore strikes the spring with a momentum

$$p = \sqrt{2mK} = m\sqrt{2gh}$$

and from Eq. [7 . 24], the duration of compression is

$$t = \frac{p}{\overline{F}} = \delta y \sqrt{\frac{2}{gh}}$$

*7 • 6 numerical integration and quadrature

In Eq. [7 . 6] we have our first physical quantity expressly defined in terms of a definite integral between two points. In general, the procedure for calculating work done is to set up an expression for the integrand $F \cos \phi$ in terms of the variable of integration by careful analysis of the relationships between path, force, and the spatial variables x, y, z. These relationships can become quite complicated, and neat formulas often prove impossible to find or prohibitively difficult to derive. Sometimes only a table of experimental data for the force is available, which cannot be represented algebraically. In such cases we must resort to computer-oriented methods of numerical analysis.

The evaluation of definite integrals $\int_{x_1}^{x_2} f(x)\,dx$ is also called *quadrature,* a term that refers to the fact that the definite integral can be represented as an area under a curve. We shall employ this term as distinct from *integration,* which refers to the solution of differential equations, such as the second law. (You may also encounter the term *quadrature* used to mean the evaluation of indefinite integrals $I = \int^{x} f(x)\,dx$, in which the endpoint of the interval is variable.)

It is possible to find the area A under the curve of a function $f(x)$ which is continuous over $a \leq x \leq b$ by either of two basic approaches. One method is to approximate the *integrand* $f(x)$ by some other function—say, a polynomial

$$f(x) \doteq \sum a_i x^i \qquad i = 0, 1, 2, \ldots, n$$

which is easy to integrate. The other method is to approximate the *integral* itself by a "weighted" sum of the ordinate values $f_i = f(x_i)$,

$$\blacktriangleright \quad \int_a^b f(x)\,dx = A \doteq \sum_{i=0}^{n} c_i f(x_i) \qquad [7 \cdot 28]$$

where the interval from a to b has been divided into $n + 1$ points x_i, with $x_0 = a$ and $x_n = b$ (see Fig. 7 . 12a). The coefficients c_i are the weights of the ordinates $f(x_i)$.

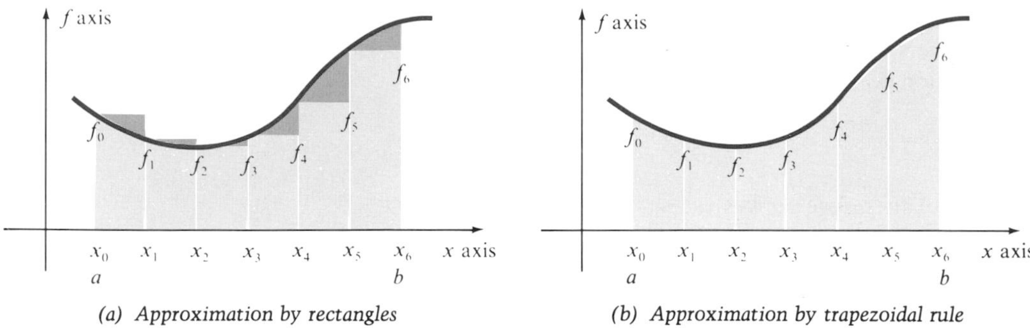

(a) Approximation by rectangles

(b) Approximation by trapezoidal rule

figure 7 • 12 *Quadrature. The darker areas show the errors in determining area A.*

For our first method of quadrature we shall derive a form of Eq. [7 . 28] known as the *trapezoidal rule*. The method is simple and still affords a high degree of accuracy. The form of Eq. [7 . 28] is particularly useful in working with experimental data, such as a table of $f(x)$ for x at selected points, and also closely resembles the mathematical definition of the integral,

$$A = \lim_{\Delta x \to 0} \sum_{i=0}^{n-1} f(x_i)\,\Delta x \qquad [7 \cdot 29]$$

where $n\,\Delta x = b - a$. If we remove the limit symbol, the equality becomes an approximation, with $c_i = \Delta x$ for every value of c (except that $c_n = 0$; that would be one rectangle too many).

The error in using this method is indicated by the dark shaded areas in Fig. 7 . 12*a*. However, instead of taking the height of each rectangle as the value of $f(x)$ at the *start* of each interval Δx, we can set it equal to an "average" value of f over each interval:

$$\bar{f}(x_i) = \tfrac{1}{2}[f(x_i) + f(x_{i+1})] \qquad [7 \cdot 30]$$

Then Eq. [7 . 29] yields the approximation

$$A \doteq \sum_{i=0}^{n-1} \tfrac{1}{2}[f(x_i) + f(x_{i+1})]\,\Delta x \qquad [7 \cdot 31]$$

Each term within the summation is then the area of a trapezoid with "bases" $f(x_i)$ and $f(x_{i+1})$ and "height" Δx (Fig. 7 . 12*b*). If you write out a few terms in succession, you will see that they can be combined as

$$A \doteq \tfrac{1}{2}\Delta x\,(f_0 + 2f_1 + \cdots + 2f_{n-1} + f_n) \qquad [7 \cdot 32]$$

It can be shown that for small enough values of Δx the error in computing A this way is proportional to $(\Delta x)^2$. Hence the trapezoidal rule is known as a *first-order method,* since errors are of second order in powers of Δx.

Figure 7 . 13 gives the general flowchart for trapezoidal-rule quadrature of $A = \int_a^b f(x)\,dx$. Since many different forms of input and output are possible with computers, the general terms INPUT and OUTPUT are used in place of READ (which usually implies a card reader or magnetic tape) and WRITE (which implies a line printer).* Note that in the case of subscripted

*Both operations may also take place on a single teletype.

values such as f_i, the value of the index is used to determine when to leave the loop which goes from point 1 to the branch point. Since we are not testing x against x_n, but comparing i to n, the sign of Δx does not matter, and the algorithm is quite general. To take advantage of the computer's ability to perform simple instructions repeatedly with speed and accuracy, Eq. [7 . 31] has been used in the heart of the loop. Thus it is not necessary to stipulate special treatment for the coefficients of f_0 and f_n. However, for a computation on a hand calculator, Eq. [7 . 32] would be more convenient.*

Consider a body such that a force $F = k(x)x$ is required to produce an extension to x, where

$$k(x) = 2x^2 - 3x + 4$$

(in cgs units). If you did not know this formula and had only the experimental data given in Table 7 . 2, application of the trapezoidal rule would give the third column of figures as output for the total work done *on* the body. You can verify by exact integration that the absolute error is never greater than 0.0012 ergs. In this particular case, the positive and negative errors exactly balance at $x = 1.0$ cm (merely a coincidence).

*A calculator cannot store instructions for branching decisions.

figure 7 • 13 *Flowchart for trapezoidal rule quadrature of* $\int_a^b F(x)\,dx$

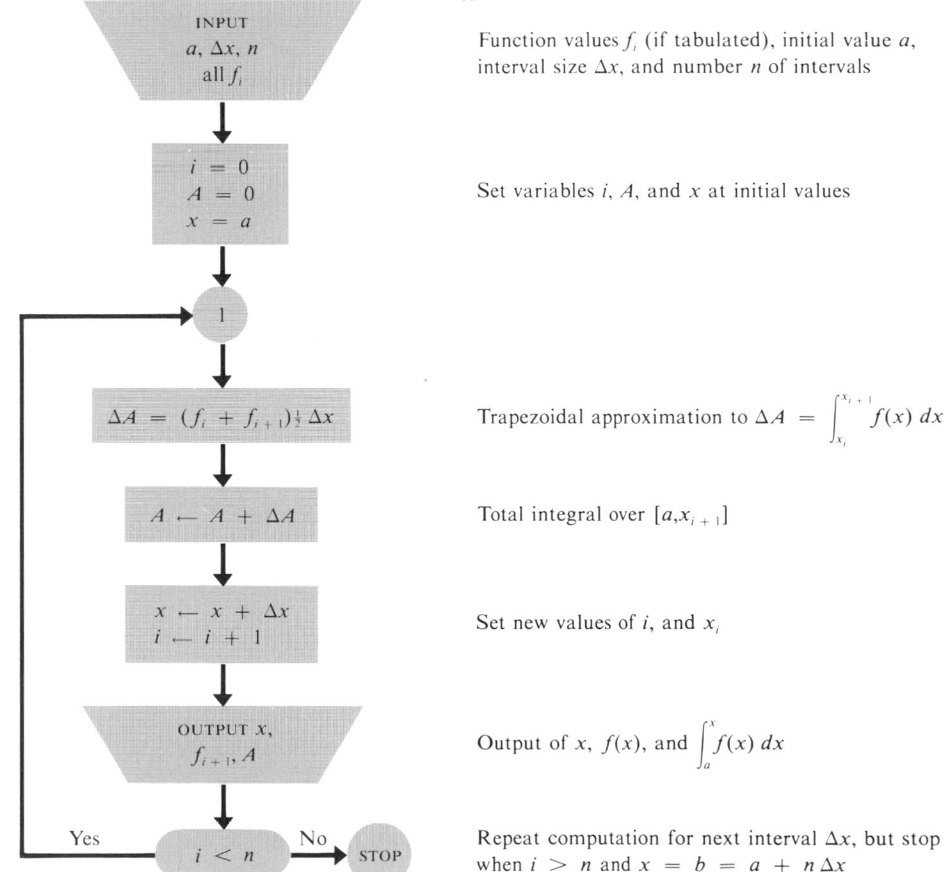

table 7 • 2 *Hypothetical experimental data, force versus extension*

x, cm	*F, dynes*	*total work W, ergs*
0.0	0.000	0.0000
0.1	0.372	0.0186
0.2	0.696	0.0720
0.3	0.984	0.1560
0.4	1.248	0.2676
0.5	1.500	0.4050
0.6	1.752	0.5676
0.7	2.016	0.7560
0.8	2.304	0.9720
0.9	2.628	1.2186
1.0	3.000	1.5000

As a second example, consider a particle sliding along a smooth straight track with speed v. The only force acting on it is air resistance, which, at low speeds, is shown experimentally to be $F = -Cv$, (Stokes' law), where C is a positive constant. We want to know how long it will take for the particle to lose half its initial kinetic energy K_0.

The motion is one dimensional, and so the equation of motion is

$$m\frac{dv}{dt} = -Cv \qquad [7 \cdot 33]$$

Although we could integrate this equation numerically and find the time when $v/v_0 = 1/\sqrt{2}$, we can convert it directly into an equation in $K = \frac{1}{2}mv^2$ by multiplying both sides by v:

$$mv\frac{dv}{dt} = \frac{dK}{dt} = -Cv^2 = -\frac{2C}{m}K \qquad [7 \cdot 34]$$

We can scale this equation by substituting $\tau = 2Ct/m$ as the unit of time, and if we assume that K is measured in units of K_0, we can substitute $\kappa = K/K_0$ and $\kappa_0 = 1$. This gives us the dimensionless form

$$\frac{d\kappa}{d\tau} = -\kappa \qquad [7 \cdot 35]$$

and integrating numerically by the recursion formula, Eq. [6 . 13], we obtain the approximation

$$\kappa_{i+1} - \kappa_i = \Delta\kappa \doteq -\kappa_i\,\Delta\tau \qquad [7 \cdot 36]$$

The flowchart of this computation is shown in Fig. 7 . 14.

If we take $\Delta\tau = 0.1$ as our first choice of the dimensionless time interval, we compute $\kappa(0.6) = 0.5314$ and $\kappa(0.7) = 0.4783$. Then, by interpolation, $\kappa(0.649) = 0.5$; hence the (scaled) time required for the particle to lose half its initial kinetic energy is $\tau_{1/2} \doteq 0.649$. If we reduce the step size

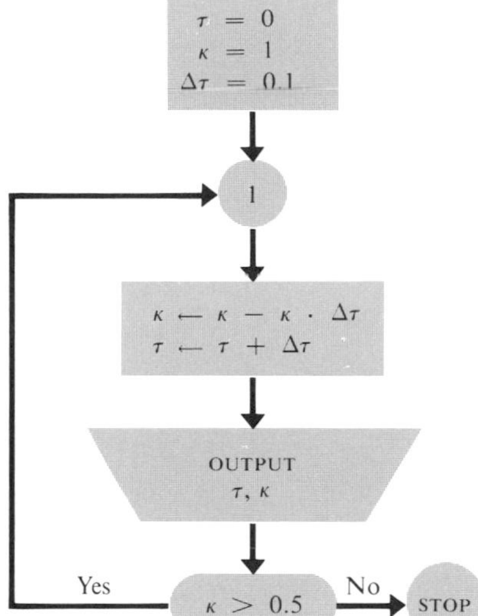

figure 7 • 14 *Flowchart for* $d\kappa/d\tau = -\kappa$

to $\Delta\tau = 0.05$, then we find $\kappa(0.676) = 0.5$, a difference of 4.2 percent in $\tau_{1/2} = 0.676$. We can sharpen the approximation by reducing the step size to $\Delta\tau = 0.025$, and from Table 7 . 3 we see that $\tau_{1/2} = 0.685$, a change of 1.3 percent, which is tolerable. This method of numerical integration, known as *Euler's method,* is a first-order method in which errors decrease as $(\Delta\tau)^2$. We might have guessed this, since the percent changes in $\tau_{1/2}$ were in the ratio $4.2/1.3 \doteq 4$ when we halved the interval size.

Since we approximate $d\kappa/d\tau$ by the value of κ at the beginning of each interval, and κ is monotonically decreasing, we expect that Euler's method will show the kinetic energy dissipating somewhat faster than it actually does. In fact, as we decrease $\Delta\tau$, the value $\tau_{1/2}$ steadily increases. Hence we would guess that the *true* value is $\tau_{1/2} > 0.685$, within an error of around 1 percent.

We could stop here, but the trapezoidal rule allows us to improve our approximation further. Since $d\kappa/d\tau$ represents a ratio of two differentials, we can invert it:

$$\frac{d\tau}{d\kappa} = \frac{-1}{\kappa} \qquad [7 \cdot 37]$$

This form can then be integrated directly by quadrature,

$$\int_0^\tau d\tau = \tau = -\int_1^\kappa \frac{d\kappa}{\kappa} = \int_\kappa^1 \frac{d\kappa}{\kappa} \qquad [7 \cdot 38]$$

and evaluated for any desired κ by means of the trapezoidal rule. If we take $\kappa = 0.5$ and the interval size as $\Delta\kappa = 0.1$, it is very easy to find $\tau_{1/2} = 0.695$. Furthermore, on a graph of $f(\kappa) = 1/\kappa$ versus κ the curve bends away from the κ axis, so that the trapezoids approximating the area under the curve always lie above it; hence the trapezoidal rule must give a value *larger* than the actual area under the curve.

We know, therefore, that the true value of $\tau_{1/2}$ must lie within the range

table 7 • 3 *Numerical integration of $d\kappa/d\tau = -\kappa$, $\Delta\tau = 0.025$*

τ	κ	τ	κ
0.000	1.000	0.400	0.6670
0.025	0.9750	0.425	0.6502
0.050	0.9506	0.450	0.6340
0.075	0.9269	0.475	0.6181
0.100	0.9037	0.500	0.6027
0.125	0.8811	0.525	0.5876
0.150	0.8591	0.550	0.5729
0.175	0.8376	0.575	0.5586
0.200	0.8167	0.600	0.5446
0.225	0.7963	0.625	0.5310
0.250	0.7763	0.650	0.5178
0.275	0.7569	0.675	0.5048
0.300	0.7380	0.700	0.4922
0.325	0.7195		
0.350	0.7016		
0.375	0.6840		

$0.685 < \tau_{1/2} < 0.695$, which is very helpful, indeed, since this means we cannot be in error by more than 0.73 percent if we simply take the average of the extreme values. However, we should give more weight to the upper value, since with Euler's method each value of κ depends on preceding values of κ, so that truncation errors will accumulate. This does not happen with quadrature, and since this procedure entails far fewer arithmetic operations, there is also less chance for roundoff error to creep in.

The loss in kinetic energy shown here is an example of a well-known type of mathematical behavior known as *exponential decay,* and the true value of $\tau_{1/2}$ is known to be $\tau_{1/2} = 0.693$. Although both methods described here are first-order, with the errors per step proportional to the square of the step size, the method of quadrature has clear advantages. This is generally the case, although numerical integration may give more information in a more usable form.

The trapezoidal rule can also be applied directly in the integration of the equation of motion as in Sec. 6 . 6 to find the work done by a force as the body moves over a previously unknown path. As the velocity and position of the body are computed for successive times, we can also compute the increment of the work,

$$\begin{aligned} \Delta W &= \overline{F}\,\Delta x \\ &= \tfrac{1}{2}[F(t) + F(t + \Delta t)]\,[x(t + \Delta t) - x(t)] \end{aligned} \qquad [7 \cdot 39]$$

and add it onto the previous value of ΔW.

PROBLEMS

7 • 1 *work*

7 • 1 A force of 500 N is acting on a certain body. What is the work done by the force if the body moves a distance of 60 m (*a*) in the direction of the force? (*b*) At a 60° angle to the force? (*c*) At right angles to the force? (*d*) In a direction opposite that of the force?

7 • 2 A force of $\mathbf{F} = (2\mathbf{i} - \mathbf{j} + 3\mathbf{k})$ dynes is required to move an object from one corner of a cube at the origin to the opposite corner (l, l, l). (*a*) If the cube has a side length of l cm, how much work is done by this force? (*b*) If a second force $\mathbf{F}' = -4\mathbf{k}$ dynes is added to the first, what is the work done by the resultant force as the object moves from the origin to the opposite corner?

7 • 3 A Mongol pony standing 5 ft high at the shoulders pulls a loaded sled across the tundra of northern Siberia against an average frictional force of 300 lbf. If the pony weighs 400 lbf and the sled is always 7 ft behind him, how much work does he do on the sled in pulling it 6 versts?

7 • 4 A spiral stairway of average radius 1 m winds up to the top of a lighthouse 30 m from the ground. How much work does the 150-lbm caretaker do against gravity in climbing to the top?

7 • 5 A force of 1 N is applied to a 50-g particle which is initially at rest on a smooth horizontal surface. If the force is tangential to the surface, how much work does it do on the particle in 10 s?

7 • 6 Suppose the applied force in Prob. 7 . 5 makes an angle of 30° with the normal to the horizontal surface. (*a*) How much work is done on the particle in 10 s? (*b*) How much work is done on the particle if the horizontal surface is not smooth, but has a coefficient of kinetic friction $\mu_k = 0.25$? (*c*) How much work is done in overcoming friction?

7 • 2 *work by a variable force*

7 • 7 A famous baseball pitcher has had his fast ball clocked at 90 mi/h. If his hand moves through a net horizontal displacement of about 9 ft during the throw, and the ball weighs 0.3 lbf, how much work does he do on the ball?

7 • 8 A sorcerer's apprentice must haul a leaky gallon bucket of water daily from the well in the courtyard up to the top of a tower 50 m high. The rate of leakage is such that the apprentice usually arrives at the tower with only half a bucketful left. Assuming that the apprentice's rate of climb and the rate of water loss are both constant and that the bucket itself is weightless, how much work does this task entail? Express your answer in joules (1 gal of water = 3785 g).

7 • 9 The position of a particle in a plane is given by $\mathbf{r} = 2t\mathbf{i} - t^3\mathbf{j}$, and the force exerted on the particle at some position (x, y) is $\mathbf{F} = x^2\mathbf{i} - xy\mathbf{j}$. How much work, expressed in cgs units, is done on the particle during the interval $1\text{ s} \leq t \leq 3\text{ s}$?

7 • ***10*** A cylindrical tank of height h is half full of water, all of which is to be pumped up over the side of the tank. If the total mass of water is m, how much work must be done by the pump?

7 • ***11*** An auto accelerates from rest at a constant rate a. Because of fluctuations in the wind, the force exerted by air resistance varies with the speed v of the car and with time according to $F_f = -v(A + B \sin t)$. What is the work done by the car against friction in a time $\Delta t = \pi/2$? (*Hint*: See Appendix F and evaluate the work integral with all quantities expressed in terms of t.)

7 • ***12*** A spring of negligible mass with restoring-force constant k hangs from the cabin of an airplane and suspends a mass m. When the airplane moves in a horizontal line for t seconds with a uniform acceleration of $a = \frac{1}{2}g$, the inertial reaction of the mass causes the spring to stretch. (*a*) How much work does the airplane do on the spring-mass system? (*b*) How much work does it do on the spring? Neglect oscillations of the spring.

7 • ***13*** Rework Example 7 . 3, using the integral of Eq. [7 . 10] in the form

$$W = \int_{y_0}^{y} F_y \, dy$$

and show that the results are the same.

7 • ***14*** In Example 7 . 5, a variable force defined by

$$\mathbf{F} = \frac{mgx}{R}\mathbf{i} + mg\left(\frac{2y}{R} - 1\right)\mathbf{j}$$

is acting along a curved path $y = 3x^2/4R$ in the xy plane. Suppose a particle of mass m is moving outward from the origin along this path. (*a*) For what value of x will the net work vanish? (*b*) For what value of x will the net force begin pulling the mass away from the center? (*Hint*: Look for x such that the net work begins decreasing with increasing x. Can you see why this should be so?)

7 • 3–7 • 5 *power, kinetic energy, and momentum*

7 • ***15*** What would be the values of the standard units of length, mass, and time in a system in which the foot-pound was the unit of work, the horsepower the unit of power, and the acceleration due to gravity (32 ft/s^2) the unit of acceleration?

7 • ***16*** Find the horsepower of a steam pump that lifts 10^5 liters of water an hour from a well 30 m deep.

7 • ***17*** The average rate of flow over Niagara Falls is 270,000 ft^3/s and the height of the falls is 160 ft. (*a*) If all the energy of the falling water were utilized, what power could be developed? (*b*) If all the energy from the falls were converted into electrical energy and sold at 7 cents a kilowatt-hour, how much money would be realized in a day? Assume accuracy to two significant figures.

7•18 The firing mechanism on a pinball machine consists of a plunger with a spring wrapped around it, which propels a small steel ball of mass 60 g lying against the front end of the plunger. A player pulls the plunger back 6 cm by applying a force of 900,000 dynes and then releases it suddenly to propel the ball into the machine. (*a*) What work does the player do in compressing the spring, which lies almost horizontal? (*b*) Neglecting the mass of the spring and plunger, with what speed does the ball leave the front end of the plunger?

7•19 At what rate must energy be expended to raise 1000 kg/min of water to a height of 22 m if the water is discharged from the top of the pipe at a speed of 4.0 m/s?

7•20 During a tornado a 25-g chunk of metal strikes a wooden post at a speed of 120 m/s and penetrates 10 cm into the post. At what speed would this object have to be traveling to penetrate 18 cm into the post?

7•21 If an object starting from rest slides down a smooth inclined plane of height h, find the power P expended by gravity as a function of the object's vertical position y relative to the bottom of the incline.

7•22 Using the cartesian components of $\mathbf{v}$, prove that

$$\frac{d}{dt}(\mathbf{v} \cdot \mathbf{v}) = 2\mathbf{v} \cdot \frac{d\mathbf{v}}{dt} = 2v\frac{dv}{dt}$$

7•23 A train with a mass of 2.0×10^5 kg is traveling at a uniform speed of 65 km/h. (*a*) What is its kinetic energy in ergs? (*b*) In joules? (*c*) In kilowatt-hours? (*d*) In meter-kilogram forces?

7•24 A bullet of mass 12 g is fired with a muzzle speed of 30 km/min. The gun has a smooth bore 750 mm long and 9 mm in internal diameter. (*a*) What is the energy of the bullet in joules? (*b*) Neglecting friction, find the average pressure (force per unit area) inside the barrel during firing.

7•25 A 25-g bullet is moving with a speed of 500 m/s and a freight train of mass 10^6 kg is moving in the same direction with a speed of 1 cm/s. (*a*) Compare their momenta and kinetic energies. (*b*) If a constant retarding force of 100 N is applied to each object, what will be the elapsed times and distances traversed before they are brought to rest?

7•26 How far will an automobile traveling at 65 mi/h skid if the driver brakes suddenly and the wheels lock? The coefficient of kinetic friction of the tires on the road is $\mu_k = 0.75$.

7•27 A constant horizontal force of magnitude F acts on an object of mass m which is initially at rest on a smooth horizontal surface. Find the instantaneous power P expended by the force at any time t.

7•28 If the clutch is disengaged on a certain automobile of weight $F_g =$ 9800 N when it is moving with a speed of 48 km/h on a level road, the automobile will coast 0.8 km before stopping. (*a*) Assuming that the net frictional force is independent of speed, compute its average magnitude. (*b*) What horsepower must be expended to keep this car moving on a level road at a speed of 48 km/h? (*c*) Considering the nature of the frictional forces involved, do you think it likely that these forces are independent of speed?

7 • 29 The top of an incline is 1 m higher than the bottom. If a 100-g body sliding down this incline acquires a speed of 180 m/min, how much work has been done against friction?

7 • 30 Rewrite Eq. [7 . 22], the work–kinetic-energy theorem as it applies to a body under the influence of gravity, so that the subscript 1 appears only on the left side of the equation and the subscript 2 only on the right. How do you interpret the quantity E in the equation $\frac{1}{2}mv^2 + mgy = E$ for a falling body? E is known as the total mechanical energy of the body.

7 • 31 Write the work–kinetic-energy theorem as it applies to a mass m attached to a spring of restoring-force constant k resting on a smooth horizontal table. (*Hint*: See Eq. [7 . 9].) Compare your result with the equation for E in Prob. 7 . 30 and see if you can write a similar equation for the spring-mass system.

7 • 6 *numerical integration and quadrature*

***7 • 32** The speed of a particle in the x direction is given by

$$v = \frac{t^3}{\sqrt{1 + 3t^2}} \text{ cm/s}$$

Find the distance traveled during the time interval $0 \leq t \leq 5$ s to within 1 percent.

***7 • 33** Use the trapezoidal rule to integrate $f(x) = x^3$ for the interval $1 \leq x \leq 2$, by dividing the interval into two parts. Compare the accuracy with the correct value given by the exact integral.

***7 • 34** The restoring force exerted on a particle of mass m is given by $F = -kx^3$, and its initial position and speed are $x = x_0$ and $v_0 = 0$. (*a*) Set up the equations of motion and (*b*) scale them to dimensionless variables. (*c*) Construct a flowchart for numerical integration of the motion from $t = 0$ to $t = 13T$, where $T = \sqrt{m/kx_0^2}$, simultaneously computing the work done on the particle. (*d*) Compute the motion of the particle. (*e*) Graph the results and discuss any significant differences between this situation and that of the harmonic oscillation discussed in Sec. 6 . 6.

***7 • 35** The rate of loss of fluid through a hole in the bottom of a bucket is proportional to the square root of the amount of fluid in the bucket. This can be expressed as a differential equation:

$$\frac{dm}{dt} = -\beta \sqrt{m}$$

where β is a positive constant. (*a*) If the sorcerer's apprentice in Prob. 7 . 8 takes 30 min to carry his water bucket up to the top of the tower, what is the value of β? (This can also be solved exactly as a check on accuracy.) (*b*) How much work does the apprentice do in carrying the bucket to the top of the tower? (*Hint*: First write the differential equation for dW/dt.)

***7 • 36** Suppose rate of loss of fluid in Prob. 7 . 35 is proportional to the mass of the fluid in the bucket:

$$\frac{dm}{dt} = -\alpha m$$

where α is the constant of proportionality. (*a*) What is the value of α? (*b*) How much work does the apprentice do in carrying the bucket to the top of the tower?

***7 • 37** Pancratos, the world's strongest trained flea, pulls a 1-mg solid steel block across a steel tabletop. His harness is attached to a silk thread which makes an angle of 30° with the horizontal. The coefficient of kinetic friction is given by $\mu_k = 0.5\,(1 - \sqrt{v}/10)$, where v is fleaspeed, measured in millimeters per second. If Pancratos' acceleration is a steady 2 mm/s, how much work does he do against friction over a 1-m path, starting from rest? Apply the trapezoidal rule and then compare your results with an exact analytic solution.

answers

7 • 1 (*a*) $\Delta W = 30{,}000$ J; (*b*) $\Delta W = 15{,}000$ J; (*c*) $\Delta W = 0$; (*d*) $\Delta W = -30{,}000$ J
7 • 2 (*a*) $\Delta W = 4l$ ergs; (*b*) zero; resultant is perpendicular to diagonal
7 • 3 $\Delta W = 1800$ lbf · versts or 6.3×10^6 lbf · ft
7 • 4 $\Delta W = 20{,}000$ J
7 • 5 $\Delta W = 1$ kJ
7 • 6 (*a*) $\Delta W = 250$ J; (*b*) $\Delta W = 143$ J; (*c*) $\Delta W = 46$ J
7 • 7 $\Delta W = 81.24$ lbf · ft
7 • 8 $\Delta W = \frac{3}{4}mgh = 1391$ J
7 • 9 $\Delta W = -1804$ ergs
7 • 10 $\Delta W = \frac{3}{4}mgh$
7 • 11 $\Delta W = \frac{1}{24}a^2A\pi^3 + a^2B(\pi - 2)$
7 • 12 (*a*) $\Delta W = \frac{1}{4}mg^2t^2$; (*b*) $\Delta W = \frac{1}{8}m^2g^2/k$
7 • 14 (*a*) $x = \frac{2}{3}R$; (*b*) $x = \sqrt{2}\,R/3$
7 • 15 $[L] = 32/550^2$ ft, $[M] = 550^2/32$ lbm, $[T] = 1/550$ s
7 • 16 $P = 11$ hp
7 • 17 (*a*) 4.9×10^6 hp or 3.7×10^6 kW; (*b*) $6,140,000
7 • 18 (*a*) $\Delta W = 2.7 \times 10^6$ ergs; (*b*) $v = 300$ cm/s
7 • 19 $P = 3.7$ kW
7 • 20 $v = 161$ m/s
7 • 21 $P = \sqrt{2g(h - y)}\, g \sin\phi$
7 • 23 (*a*) $K = 3.3 \times 10^{14}$ ergs; (*b*) $K = 3.3 \times 10^7$ J; (*c*) $K = 9.2$ kW · h; (*d*) $K = 3.3 \times 10^6$ m · kgf
7 • 24 (*a*) $K = 1500$ J; (*b*) $3.14 \times 10^7\ \mathrm{N/m^2}$
7 • 25 (*a*) $p_1 = 1.25 \times 10^6$ g · cm/s, $K_1 = 3.1 \times 10^{10}$ ergs, $p_2 = 10^9$ g · cm/s, $K_2 = 5 \times 10^8$ ergs; (*b*) $\Delta t_1 = 0.125$ s, $\Delta x_1 = 31$ m, $\Delta t_2 = 100$ s, $\Delta x_2 = 0.5$ m
7 • 26 $\Delta x = 58$ m
7 • 27 $\overline{P(t)} = F^2t/m$
7 • 28 (*a*) $\overline{F_f} = 110$ N; (*b*) $P = 2.0$ hp
7 • 29 $\Delta W = 0.53$ J
7 • 30 $E =$ constant
7 • 31 $\frac{1}{2}mv^2 + k(x - x_0)^2 = E$
7 • 32 $\Delta x = 23.6442$ cm
7 • 33 $f(x) = \int_1^2 x^3\,dx = 3.75$
7 • 34 (*a*) $md^2x/dt^2 = -kx^3$; (*b*) $d^2\lambda/d\tau^2 = -\lambda^3$, where $x = x_0\lambda$, $t = T\tau$, $T^2 = mx_0^2/k$; (*d*) periodic motion, with period $7.4T$ and amplitude x_0
7 • 35 (*a*) $\beta = 1.20\ \mathrm{g^{1/2}/min}$; (*b*) $W = 1.365 \times 10^{10}$ ergs
7 • 36 $\alpha = 0.0231$/min, $W = mgh/1.386$
7 • 37 $W(t) = 0.098\,(\frac{1}{2}t^2 - \sqrt{2}\,t^{5/2}/25)$ erg, $W(31.6\text{ s}) = 17.825$ ergs

CHAPTER EIGHT

potential energy and conservative systems

We must now raise the question whether it is not true that, in all the interactions of bodies, the total quantity of living absolute force is conserved. And, first, let it be remarked that if this living force could ever be augmented, then ... the effect could reproduce its cause and something more, which would lead to the absurdity of perpetual mechanical motion. ... But if the force could be diminished, then it would perish at last entirely ... which is without doubt contrary to the order of things.

Secondly, this conclusion is confirmed by experiment also, and we shall find always that if bodies should convert their horizontal into ascending motions, they could always raise on the whole the same weight to the same height before or after an impact, supposing that no living force has been absorbed in the impact ... and ignoring any absorption of living force by friction with the medium or other circumstances.

Gottfried Wilhelm Leibnitz, "Essays on dynamics,"
New Essays Concerning Human Understanding,
1700–1705

A *conservative system* is one in which the work done by the system forces, such as restoring forces or gravity, is completely independent of the path of motion from one position to another. That is, there are no frictional or dissipative forces involved which can cause a loss of kinetic energy. The only forces acting in such a system are *conservative forces.*

In a conservative system the work done on a body depends only on the *endpoints* of the path through which it moves from one position to another. It follows, therefore, that this work can be expressed as the difference in values of some function which is evaluated at these endpoints. The existence, construction, and use of such functions, known as *potential-energy functions,* is the central mathematical feature of the analysis of conservative systems and a topic of considerable importance in physics and mathematics alike. Furthermore, the insights gained from the study of conservative systems can be applied to systems in which frictional forces are negligible or very small. In such cases, formulas derived from conservative systems may become the mathematical basis for approximate solutions to the behavior of nonconservative systems. Potential energy is not merely a convenience, but a quantity that is quite real to the physicist.

The reason path-independent forces are termed "conservative" is related to the work–kinetic-energy theorem. Imagine that the endpoints of the work integral of a conservative force approach each other:

$$\Delta W = \int_{\mathbf{r}}^{\mathbf{r}'} \mathbf{F} \cdot d\mathbf{r} \qquad \mathbf{r}' \longrightarrow \mathbf{r}$$

Since ΔW depends only on the two endpoints of the integral, $\mathbf{r}$ and $\mathbf{r}'$, then in the limit as $\mathbf{r}'$ approaches $\mathbf{r}$, the path becomes a closed curve, and the integral must vanish:

$$\Delta W = \oint \mathrm{F} \cdot d\mathbf{r} = 0$$

where the symbol $\oint$ denotes integration over a closed path. Thus, in accordance with the work–kinetic-energy theorem, the kinetic energy of the system does not change over this path—that is, the kinetic energy (Leibnitz, "living force") is conserved.

We shall restrict our discussion here to one-dimensional conservative systems—those which can be described completely in terms of a single spatial variable, whether it is x, y, or some generalized variable s. However, we shall touch on the application of the principle of conservation of energy to three-dimensional situations and to thermodynamic processes in general, those processes which involve either the transfer or the generation of heat energy.

8 • 1 potential energy in one dimension

One simple example of a conservative system is a body under the influence of gravity. Consider a body of mass m resting on the ground, as shown in Fig. 8 . 1*a*. The gravitational force $\mathbf{F}_g = -mg\mathbf{j}$ is opposed by an equal and opposite reaction force $\mathbf{N}$ exerted on the body by the ground; hence the system is in equilibrium at $y = 0$. Suppose we now apply a gradually increasing external force $\mathbf{F}_{\text{ext}}$ to lift the body. As we increase $\mathbf{F}_{\text{ext}}$, the reaction force $\mathbf{N}$ decreases accordingly; then, when $\mathbf{F}_{\text{ext}}$ becomes slightly larger than the downward force $\mathbf{F}_g$, the body begins to rise. Finally, when the accelerating force $\mathbf{F}_{\text{ext}}$ has lifted the body to some height y, we reduce the applied force to an amount slightly smaller than $\mathbf{F}_g$ to bring the body to a stop, after which we keep $\mathbf{F}_{\text{ext}}$ exactly equal in magnitude to $\mathbf{F}_g$.

Any kinetic energy imparted to the body by the accelerating force $\mathbf{F}_{\text{ext}}$ is reduced to zero when the body comes to rest. For the operation as a whole, then, no net speed has been imparted to the body, and since the only resisting force in the system has been gravity (in the absence of air resistance), the work done on the body is equal to the work done against gravity, $\Delta W = mg\,\Delta y$. If we now remove the external force $\mathbf{F}_{\text{ext}}$, the body falls (Fig. 8 . 1*c*). Its speed just before it hits the ground is given by $v = \sqrt{2gy}$, and its kinetic energy at that time is $K = \frac{1}{2}mv^2 = mgy$. This energy is exactly equal to the work that was done by the external force in raising the body and can be used to do work on another body. The fact that a body at rest has the potential to do work by virtue of its position indicates that energy can be *stored*. Thus we say that the work ΔW done by the applied force $\mathbf{F}_{\text{ext}}$ is stored as *potential energy*, denoted by the symbol U.

figure 8 • 1 *Gravitational potential energy*

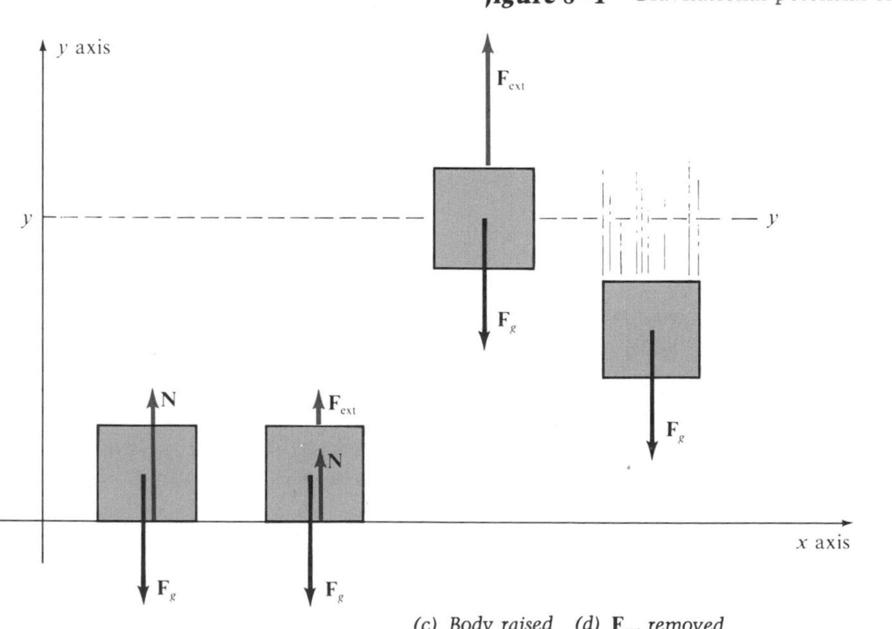

(a) Body at rest *(b) $\mathbf{F}_{ext}$ increased* *(c) Body raised to height h* *(d) $\mathbf{F}_{ext}$ removed and body released*

Mathematically, we have not accomplished anything new, but we have introduced a new idea—that a system may possess energy which is potential rather than kinetic, and that changes in this energy are measured by the amount of work the system force *could* perform if the system were free to move. Furthermore, we have an operational definition for determining the change in the potential energy of a conservative system as its parts move from one position to another. The change in potential energy is just the amount of work an external force would have to do to move the parts of a system against the resistance of its internal forces—in this case the force of gravity—without changing the net kinetic energy of the parts.

Now, with the aid of this definition, let us derive a mathematical expression for changes in the potential energy of a particle moving in one dimension in a conservative system. Under these circumstances we can expect the potential energy U of the particle to be a function of its position. From the work–kinetic-energy theorem, Eq. [7 . 21], the net work done in moving the particle from x_1 to x_2 is, in scalar form,

$$K_2 - K_1 = \Delta W_{12} = \int_{x_1}^{x_2} F_x \, dx \qquad [8 \cdot 1]$$

where $F_x\mathbf{i}$ is the force that the system exerts on the particle. However, the work this system force can do in moving the particle from x_1 to x_2 is just the amount of work an applied external force $F_{ext}\mathbf{i}$ must have done in moving the particle from x_2 to x_1:

$$\int_{x_1}^{x_2} F_x \, dx = \int_{x_2}^{x_1} F_{ext} \, dx = U_1 - U_2 \qquad [8 \cdot 2]$$

Therefore the change in potential energy resulting from the action of the system force can be defined as

$$U_2 - U_1 = -\int_{x_1}^{x_2} F_x \, dx \qquad [8 \cdot 3]$$

This mathematical definition of the change in potential energy of a conservative system arises from, and is equivalent to, our physical definition. Potential energy is any capacity for doing work that is put into the system by a change in the positions of its parts against the forces that hold them together. Figure 8 . 2 illustrates the sign convention for potential energy, where the system force $F_x\mathbf{i}$ is the force exerted by a compressed spring. Note that the force required to compress the spring must be supplied by some external agent.

If we substitute Eq. [8 . 3] into Eq. [8 . 1], we have

$$K_2 - K_1 = \Delta K = \Delta W = U_1 - U_2 = -\Delta U \qquad [8 \cdot 4]$$

or

$$\Delta K + \Delta U = 0 \qquad [8 \cdot 5]$$

so that only *differences* in potential energy have any physical significance. This equation implies that the potential energy of a system in a particular configuration may be computed with respect to some other arbitrarily selected configuration of the system. That is, we are free to select any position

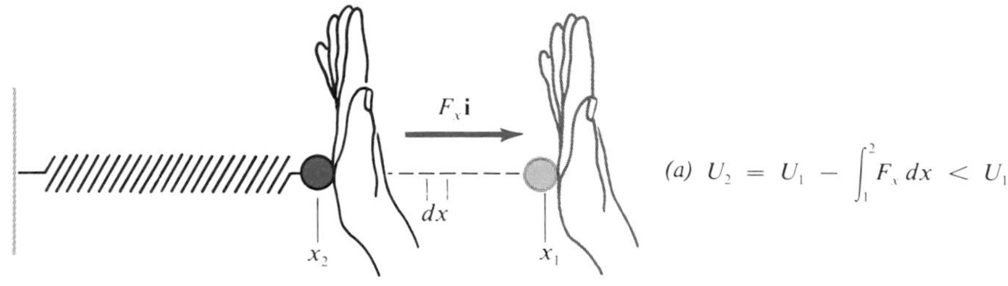

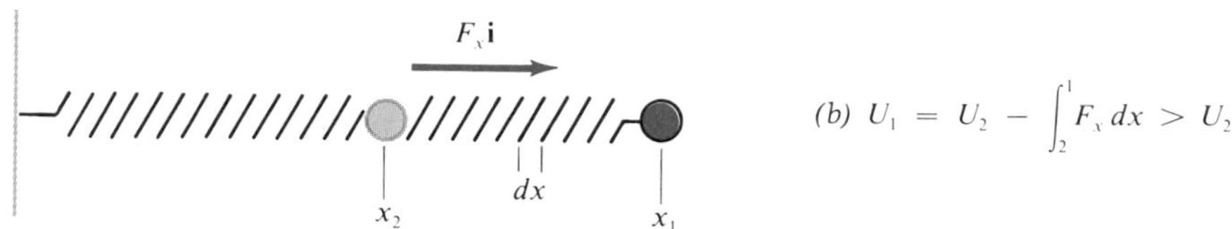

figure 8 • 2 *Changes in potential energy in a spring-mass system under compression by an external force and free expansion.* $\mathbf{F}_x$ *is the force exerted by the spring.*

$x = x_0$ and any potential-energy value $U = U_0$ as reference values, so that we can define the potential energy of the system as a function of relative position:

$$U(x) = U_0 - \int_{x_0}^{x} F_x\,dx \qquad [8 \cdot 6]$$

Note that if $U(x)$ is known, we can also determine the system force as a function of position:

$$F_x(x) = -\frac{dU(x)}{dx} \qquad [8 \cdot 7]$$

By analogy with the gravitational case, where y_0 is usually chosen at ground level, we can call x_0 the "floor" of the potential and U_0 its "ground" value.

Let us now use Eq. [8 . 6] to compute the gravitational potential energy U_g of the body in Fig. 8 . 1. In this case the spatial variable is y and the system force in the y direction is $\mathbf{F}_g = -mg\mathbf{j}$. The gravitational potential energy as a function of vertical position is therefore

$$U_g(y) = U_0 + mg(y - y_0) \qquad [8 \cdot 8]$$

If we take ground level as the floor of the potential, $y_0 = 0$, and set the potential-energy value at zero at this level, $U_0 = 0$, then Eq. [8 . 8] is simply $U_g(y) = mgy$. Correspondingly, we can determine the kinetic energy at any point y from Eq. [8 . 4] if we know the height from which the body was released from rest. For example, if the body was released from height y_1, then $K(y_1) = 0$ and at any $y < y_1$

$$K(y) - K(y_1) = U_g(y_1) - U_g(y)$$

or

$$K(y) - 0 = mgy_1 - mgy \qquad [8 \cdot 9]$$

We could equally well have chosen the height from which the body was released as the floor of the potential, $y_1 = y_0$, with $U_0 = 0$ at this height. In that case, Eqs. [8 . 8] and [8 . 9] would become

$$U_g(y) = 0 + mg(y - y_1)$$

$$K(y) - 0 = 0 - U_g(y) = mgy_1 - mgy$$

Although this equation for $K(y)$ is identical to Eq. [8 . 9], it is somewhat harder on the intuition, since at ground level, $y = 0$, the potential would be $U_g = -mgy_1$, or $U_g < 0$.

In the case of a horizontal spring-mass system with a Hooke's law restoring force,

$$F_x = -k(x - x_0)$$

the system is said to possess *elastic potential energy,* or a potential energy U_s due to extension of the spring. The work done against the molecular forces within the spring can be regained by allowing the spring to return to its original undisturbed length. If we choose this length as the reference distance x_0, then $U_0 = U_s(x_0) = 0$, and the potential energy of the spring as a function of any position x relative to x_0 is

$$\blacktriangleright \quad U_s(x) = \int_{x_0}^{x} k(x - x_0)\, dx = \tfrac{1}{2}k(x - x_0)^2 \qquad [8 \cdot 10]$$

As we saw in Eq. [7 . 13], this is just the work required to stretch or compress the spring.

example 8 • 1 When an archer draws his bow, the force he applies is approximately a linear function of the length of draw. If we neglect air resistance and assume that all the potential energy of the drawn bow is converted into kinetic energy of the arrow, what is the maximum range the archer can attain with an arrow of mass 0.25 kg drawn back 0.9 m by a force of 450 N?

solution By analogy with a spring, the restoring-force constant of the bow must be the ratio of the applied force to the length of draw, or $k = F_x/(x - x_0) = 500\ \text{N/m}$. The potential energy of the drawn bow is therefore

$$U_s = \tfrac{1}{2}k\,(\Delta x)^2 = (250\ \text{N/m})(0.81\ \text{m}^2) = 202.5\ \text{J}$$

The kinetic energy imparted to the arrow when the bowstring is released is also 202.5 J; thus the initial speed of the arrow is given by

$$v_0^2 = \frac{2 \times 202.5\ \text{J}}{0.25\ \text{kg}} = 1620\ \text{m}^2/\text{s}^2$$

In Chapter 4 (Example 4 . 5) we found that a launch angle of 45° results in the maximum range for any projectile, which is given by

$$\frac{v_0^2}{g} = \frac{1620\ \text{m}^2/\text{s}^2}{9.8\ \text{m/s}^2} = 165.3\ \text{m}$$

This explains the outcome of the Battle of Agincourt.

8 · 2 conservation of energy

Now that we have discussed the conversion of potential energy to kinetic energy, and vice versa, we are ready to define the *total energy* of a system. If we rearrange Eq. [8 . 4], we obtain

$$K_1 + U_1 = K_2 + U_2 \tag{8 · 11}$$

This equation was derived generally for any two endpoints of a particle's path. It must therefore follow that, in the absence of work done on the system by an external force, the sum of the kinetic and potential energies of a conservative system, called its *total mechanical energy E,* remains constant:

$$K + U = E \tag{8 · 12}$$

Although the value of E will depend on our choice of the potential floor and ground value, once this choice is made, E does not change, provided there are no frictional forces that would dissipate kinetic energy.

If we had used Eq. [8 . 12] instead of Eqs. [8 . 5] and [8 . 6] to solve the gravitational example in the preceding section, all the results would have been the same. Here again, no new mathematical insights have been introduced, but simply by rearranging a formula, we have obtained a new physical concept, the principle of *conservation of energy:*

> The total mechanical energy of any isolated system remains constant (is conserved), provided there are no dissipative (frictional) forces within the system.

Note, however, that an external force may be applied to the system to change its kinetic energy or its potential energy or both; this will change the total energy of the system by an amount equal to the work done on the system. This was the case when we applied an external force to lift the body in Fig. 8 . 1, changing the total energy of the gravitational system by an amount $\Delta W = mg\,\Delta y$, the work done on the system by the applied force.

The law of conservation of energy was definitively established through the work of J. R. Mayer, James P. Joule, Hermann von Helmholtz, Rudolf Clausius, and William Thomson (Lord Kelvin) around the middle of the nineteenth century. After Joule's experiments, it became generally recognized as one of the most fundamental and fruitful of all our physical laws. Many diverse physical phenomena can be accounted for in energy terms. Electrically charged bodies may be assigned a potential energy defined by Eq. [8 . 6], and application of the same equation to the magnetic forces of flowing electrical currents defines magnetic potential energies. The heat and motion developed in chemical reactions can be ascribed to the release of chemical potential energies locked into a substance by intermolecular forces on an atomic scale. In fact, by measuring the energy changes in such reactions, we can deduce the nature of the forces. In every case it has proved possible to define potential energies such that the total energy of an isolated system is conserved, or else that losses in this total energy can be clearly identified with an equivalent amount of dissipation as heat.

One major advantage of the energy principle is that it enables us to eliminate time as an explicit variable. Moreover, since Eq. [8 . 12] is a scalar equation, we can avoid dealing with vectors. Although energy equations do not yield as much information as we could get from a complete integration of the vector equation of motion, $\mathbf{F} = m\mathbf{a}$, this second-order differential equation is often difficult to integrate, except by numerical methods. However, Eq. [8 . 6] does provide us with an analytic expression for potential energy as a function of position, rather than time. Thus we can discuss a complicated system in terms of the net changes in its energies, without having to concern ourselves with the minute details of its motion. For example, in the gravitational case, where $\Delta y = y - y_0$ is the height above the ground, we can write the general form $K + U = E$ as

$$K + U_g = \tfrac{1}{2}mv^2 + mg(y - y_0) = E \qquad [8 \cdot 13]$$

And in the case of a spring-mass system, with a force constant k and an extension $\Delta x = x - x_0$, we can write the total energy as

$$K + U_s = \tfrac{1}{2}mv^2 + \tfrac{1}{2}k(x - x_0)^2 = E \qquad [8 \cdot 14]$$

example 8 • 2 Galileo performed an experiment with a pendulum of mass m hung from a smooth nail driven into the wall at point O, as shown in Fig. 8 . 3. When the pendulum was raised to height y_1 and released, its downward swing was arrested by another nail driven into the wall at P. His question was: how high will the pendulum rise after the string is arrested by the nail at P? If this nail had been driven in at P' instead of P, how high would the pendulum rise? What if there were nails at both P and P'? Assume that the string is not extensible, so that the tension in it does no work.

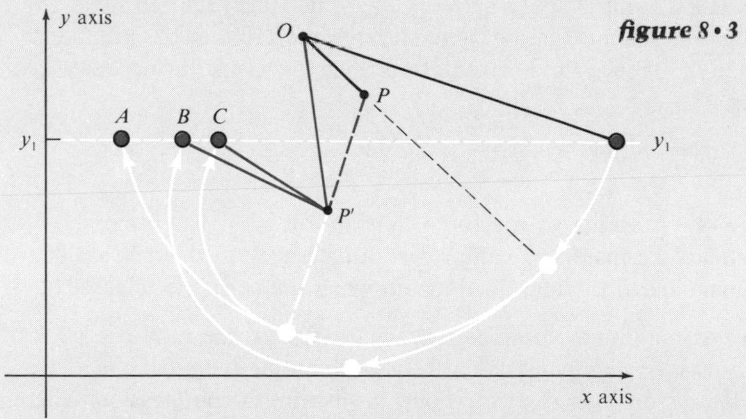

figure 8 • 3

solution Since the potential energy of the raised pendulum is $U_g = mgy$, the total energy of the system is $E = U_g = mgy_1$ in all three cases. Therefore, since $K = E - U_g \geq 0$, the pendulum will always come to rest at height y_1, where $K = 0$ and $E = mgy_1$ (just as it would if there were no nail at P or P'). Points A, B, and C represent the rest positions for the swing with a nail at P, the swing with a nail at P', and the swing with nails at both P and P', respectively. Once the energy principle gives us the rest height, the remainder of the trajectory can be determined geometrically with a ruler and compass; to accomplish as much by integration from the second law would be fairly tedious.

example 8 • 3 An object slides down a smooth curve onto a smooth tabletop as shown, and then slides off the table and falls to the ground. At what horizontal distance Δx from the table will the object strike the ground?

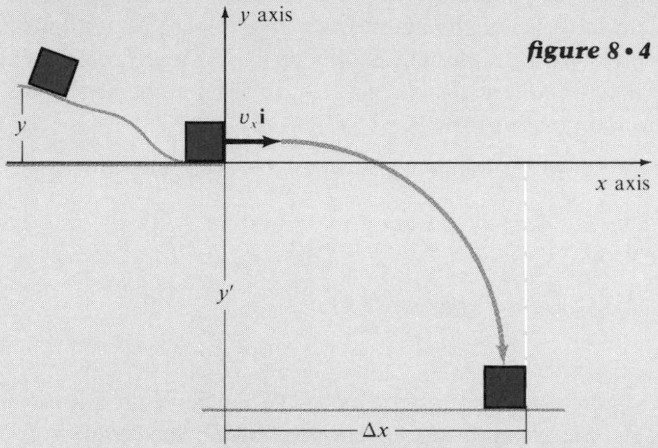

figure 8 • 4

solution First we must find the horizontal speed of the object as it slides off the table. At this point it has acquired a kinetic energy equal to the loss in potential energy as it fell through the vertical distance y to the tabletop:

$$K = \tfrac{1}{2}mv^2 = mgy$$

Since there is no frictional loss to the tabletop, the object therefore leaves the table with an initial speed $v_x = \sqrt{2gy}$. The time required for it to fall through the distance $y' = \frac{1}{2}g(\Delta t)^2$ from the table to the ground is $\Delta t = \sqrt{2y'/g}$. Hence the horizontal distance it travels during this interval is

$$\Delta x = v_x \Delta t = \sqrt{2gy} \times \sqrt{\frac{2y'}{g}} = 2\sqrt{yy'}$$

example 8 • 4 A spring with a force constant of $k = 100$ N/m is attached to a 4-kg mass and compressed a distance of 0.1 m. What is the maximum speed the mass will attain when the spring is released?

solution At maximum compression, $\Delta x = 0.1$ m, the total energy E of the spring-mass system consists entirely of potential energy. Thus, if we evaluate the system energy at the point of maximum compression, where $v = 0$, we can express E as

$$E = \tfrac{1}{2}mv^2 + \tfrac{1}{2}k(\Delta x)^2 = 0 + (50\text{ N/m})(0.1\text{ m})^2 = 0.5\text{ J}$$

As the system moves toward its equilibrium position x_0, all its potential energy is converted to kinetic energy. At x_0 the potential energy reaches zero, and the kinetic energy is therefore at its maximum:

$$K = \tfrac{1}{2}mv^2 = 0.5\text{ J} \qquad \text{or} \qquad v = \sqrt{\frac{1\text{ J}}{4\text{ kg}}} = 0.5\text{ m/s}$$

As the inertia of the mass stretches the spring past its equilibrium position, the restoring force then slows it down and stops it when the spring reaches an extension of 0.1 m. At this point the mass begins to return to its starting position.

8 • 3 energy diagrams

As we saw in Eq. [8 . 10], the potential energy of a spring-mass system with a linear restoring force is a function of position:

$$U_s(x) = \tfrac{1}{2}k(x - x_0)^2$$

If we plot this function on a graph, with y axis representing U_s and the x axis representing the extension of the spring, we obtain a parabolic curve, as shown in Fig. 8 . 5. Such a diagram is of great utility, because it enables us to visualize the motion of a particle "through a potential" which is defined at every point in space (in this case we are considering a one-dimensional space). Just as we can analyze the motion of a body in terms of the forces acting on it, without reference to their source, once the mathematical form of $U(x)$ is specified, we can ignore the source of the potential energy and concentrate on its effects.

The lower portion of the figure shows a spring-mass system compressed from its equilibrium position, which we shall choose here as $x_0 = x_3$, and traces its motion to subsequent positions corresponding to those shown on the potential-energy curve. In this case, for simplicity, let us consider x_1 the starting position at time $t = t_1$:

1 $t = t_1$: The compressed spring is released and begins to move the mass from its position at x_1. Although the mass has an initial velocity $v = 0$, the restoring force of the spring gives it an acceleration to the right. Since the initial kinetic energy is zero, the total system energy equals the potential energy, $E = U_1$.

2 $t = t_2$: By this time the spring has moved the mass to x_2, and the restoring force and acceleration have decreased by an amount corresponding to the increase in speed. Thus the system now has kinetic energy $K_2 = E - U_2$, as indicated by the distance of the energy line from the curve.

3 $t = t_3$: The spring has reached its equilibrium position x_3. At the minimum of the potential-energy curve the derivative of $U(x)$ must be zero, and from Eq. [8 . 7], the total force on the mass is

$$F_x = -\left(\frac{dU}{dx}\right)_{x = x_3} = 0$$

since the mass is located at that instant at the equilibrium point for the undisturbed spring. However, because its kinetic energy is now at a maximum, $K_3 = E$, the mass passes through this point and continues on to the right.

4 $t = t_4$: As the mass continues moving to the right, the spring is now being extended. Since the restoring force of the spring is now opposing the motion of the mass, and therefore decelerating it, the kinetic energy decreases and the potential energy increases accordingly. The total energy E, however, remains constant, $E = U_4 + K_4$.

5 $t = t_5$: When the spring reaches position x_5, the potential energy of the system is once again at a maximum: $U_5 = U_1 = E$. Hence $K_5 = E - U_5 = 0$, and at that instant the motion has stopped. However, the restoring force is now also at a maximum,

$$F_x = -\left(\frac{dU}{dx}\right)_{x=x_5} = -k(x_5 - x_3)$$

and will start the mass moving back to the left.

If the mass had continued moving to the right once it reached x_5, this would imply a potential energy U_5 greater than E, and therefore a kinetic energy $K = E - U < 0$, which is impossible, since $\frac{1}{2}mv^2$ cannot be less

figure 8 • 5 *Comparison of the motion of a spring-mass system and its corresponding potential-energy diagram. The velocity* **v** *and acceleration* **a** *of the system at each position are shown below the energy curve.*

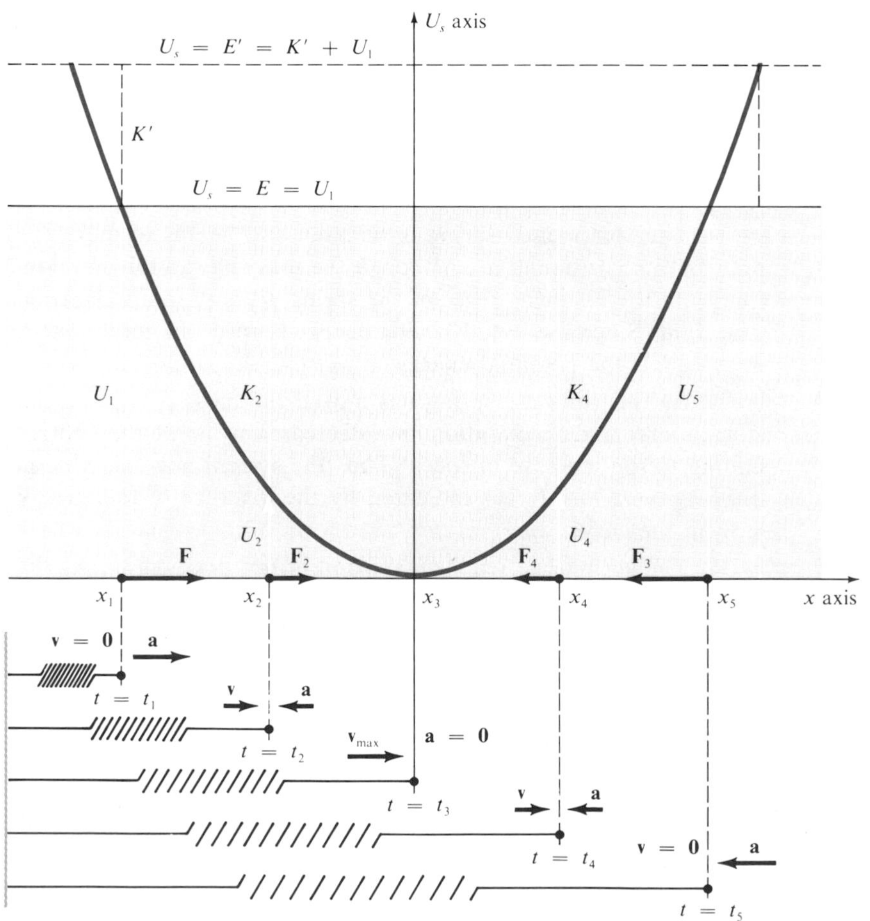

than zero. Thus the system can only reverse its direction at x_5, the *turning point* of the motion. To move the turning point farther right, we would have to raise the total mechanical energy. For example, had we imparted some initial kinetic energy K' to the mass when it was released at time t_1, the total energy of the system would have been $E' = K' + U_1'$, allowing the mass to go as far to the right as x' in Fig. 8 . 5. The total energy provides a "ceiling" on its motion, for it cannot travel into regions where $E < U$.

The parabolic potential-energy curve of this system, characterized by a Hooke's law restoring force, is known as a *harmonic-oscillator potential well.* Under the influence of the restoring force which gives rise to the potential energy, the mass will oscillate periodically back and forth, as we saw in Sec. 6 . 6, where we solved the equation of motion for this system. The harmonic-oscillator potential is of great importance in physics and has many applications.

Since $F = -dU/dx$, the restoring force always points in the same direction on this graph, as if $U(x)$ were actually the profile of a valley, with the mass rolling down one side and up the other. This is a very useful notion, because it means you can use your physical intuition as a guide to the expected motion of the object. In fact, it is customary for physicists to speak of potential-energy valleys, hills, wells, and barriers, by analogy with the earth's topography. But beware: the energy diagram is a one-dimensional plot; the particle actually moves along the x axis, not on the potential-energy curve.

In the hypothetical curve of Fig. 8 . 6 we see the effect of different total energies E on the system motion. This graph might represent the potential energy of a two-atom molecule as function of the separation distance r, between the atoms, which exert attractive and repulsive forces on each other. The potential-energy curve approaches two limiting values:

$$U(r) = \begin{cases} \infty & r \to 0 \\ 0 & r \to R \end{cases}$$

where R is the point of dissociation of the two-atom system. On one hand, as the separation decreases, very strong repulsive forces arise which keep the atoms apart; on the other hand, the attractive forces which hold the molecule together vanish if the separation distance reaches R. At that point the molecule dissociates, or breaks up into two separate atoms. The reference level is chosen to be $U_0 = 0$ at the point of dissociation R; this implies that $U(r) < 0$ over those regions in which the mutual attraction of the particles holds the molecule together.

The total energy of the system may be less than zero, as it is, for example, at $E = E_1 < 0$ in Fig. 8 . 6. Under these circumstances the system can exist only at separation distance r_A, the minimum of potential energy, since it has no kinetic energy. A system at energy level E_2 can exist at separations in region A or in region C, but it cannot go spontaneously from one "well" to the other, because it does not have enough energy to cross the potential-energy barrier in region B, where $U > E_2$. Thus at this energy level we can distinguish two different subspecies of molecule: one with an average size r_A and the other with an average size r_C. At E_3, which is still a negative energy level, we have a molecule whose size fluctuates between turning points r_2 and r_3. The maximum kinetic energy of the system is $K = E_3 - E_1$ at point r_A.

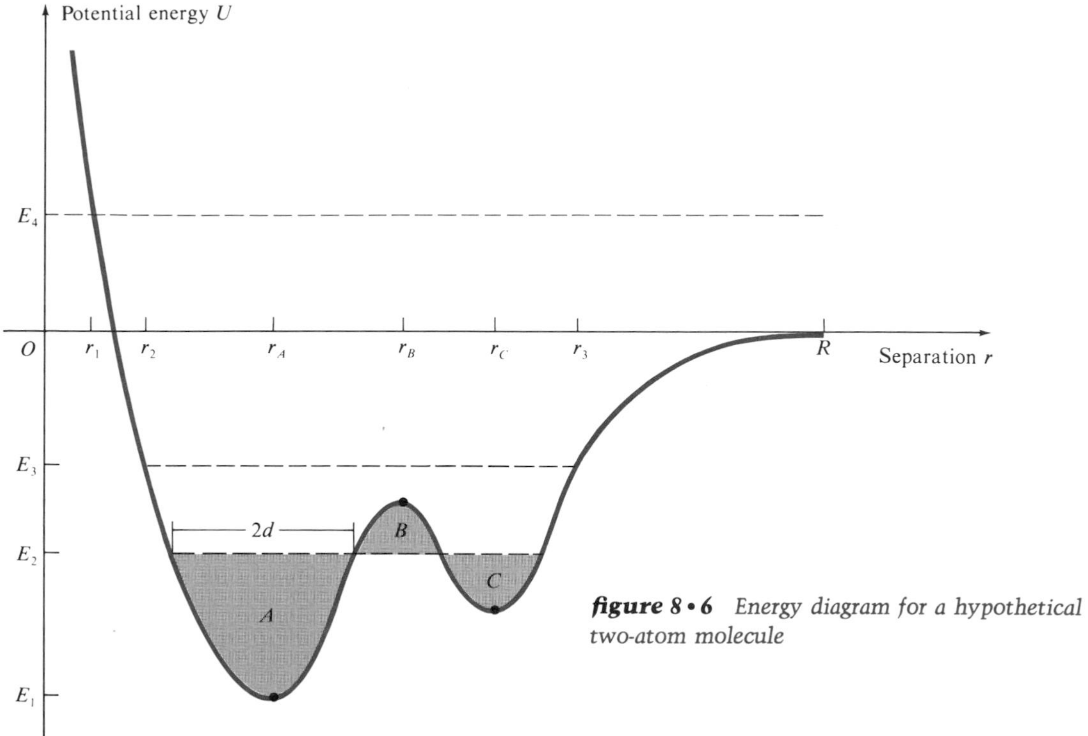

figure 8 • 6 *Energy diagram for a hypothetical two-atom molecule*

A boost to energy level E_4, where $E > 0$, might occur when a stable molecule suddenly acquires additional kinetic energy from some external source, such as a collision with another molecule. Since the system can now separate to distances greater than R, the molecule dissociates. This could also represent the case in which one free atom collides with another, forming a temporary molecule until it reaches r_1, where the repulsive forces reflect it back beyond R.

At the extrema of potential energy, r_A, r_B, and r_C, the system force is $F = -dU/dr = 0$, so that these are equilibrium positions. However, r_A and r_C are positions of *stable* equilibrium, because for small enough changes Δr, restoring forces merely cause the system to oscillate about the equilibrium position. Separation r_B is a point of unstable equilibrium, because at this point a slight change Δr in either direction will cause the system to move into region A or C, like a ball rolling downhill.

While the actual motion of the system may be difficult to describe in detail, its motion in a potential well can often be approximated by fitting a parabola to the shape of U over the restricted region of the well. Imagine that the curve in Fig. 8 . 6 represents the motion of a particle of mass m with total energy E_2, located at approximately r_A relative to some center of force. Since the equilibrium position of the well occurs at r_A, we can approximate the potential energy as

$$U(r) \doteq U_A(r) = \tfrac{1}{2}k(r - r_A)^2 + E_1 \qquad [8 \cdot 15]$$

a typical harmonic-oscillator potential. If the curve is not too asymmetrical about r_A, we can also take the average distance between turning point and equilibrium point to be the half-width d of the well at energy level E_2. Thus we see that

$$E_2 \doteq \tfrac{1}{2}kd^2 + E_1$$

and from the principle of conservation of energy, the kinetic energy at the equilibrium point r_A must be

$$K_A = E_2 - E_1 \doteq \tfrac{1}{2}kd^2$$

The force constant is therefore

$$k \doteq \frac{2K_A}{d^2} \qquad [8 \cdot 16]$$

and we can predict the duration, or *period T*, of one complete oscillation as

$$T = 2\pi\sqrt{\frac{m}{k}} = 2\pi d\sqrt{\frac{m}{2K_A}} \qquad [8 \cdot 17]$$

This is the same formula we derived numerically in Sec. 6 . 6. We shall see an exact derivation later, but meanwhile note that the approximation can be made as precise as we wish if the amplitude of oscillation is small enough.

example 8 • 5 A particle of mass 1 kg is released from rest at $x = 0$. After traveling 1 m in an unknown potential, the particle strikes a fixed target, and its kinetic energy is measured. This experiment is repeated at $x = 2$ m, with the following results:

x, m	0	1	2
K, J	0	12	8

Fit these data to a parabolic potential-energy curve and find the equilibrium position (no accelerating force) of the particle, its maximum speed, its turning point, and its period of oscillation.

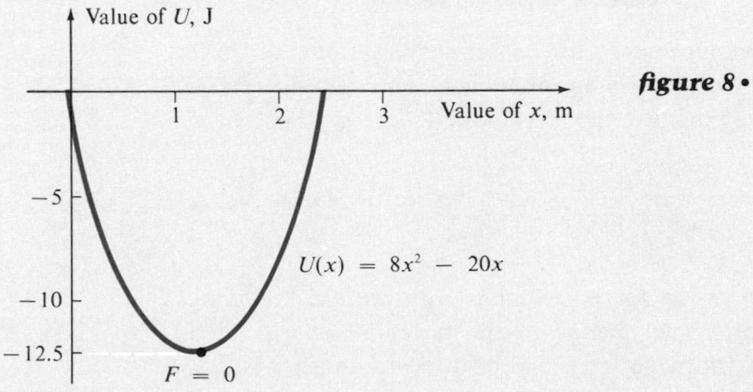

figure 8 • 7

solution The general form of a parabolic curve is given by the quadratic equation

$$U(x) = ax^2 + bx + c$$

We are at liberty to choose our reference value as $U_0 = 0$, so that $c = 0$. Since the particle started from rest, it must have gained its kinetic energy from the change in potential. Thus $K = -U$, and we can find a and b by evaluating $U(x)$ at $x = 1$ m and $x = 2$ m:

$$U_1 = a + b = -12 \text{ J} \qquad U_2 = 4a + 2b = -8 \text{ J}$$

Then, solving for a and b, we find

$$U(x) = 8x^2 \times 20x \text{ J} \qquad F(x) = -\frac{dU}{dx} = (20 - 16x) \text{ N}$$

The equilibrium position, $F = dU/dx = 0$, is at the point of minimum potential energy, or $x = 1.25$ m, where $U = -12.5$ J.

Since the potential energy is at its minimum at $x = 1.25$ m, the kinetic energy must be at its maximum:

$$K_{\text{max}} = E - U_{\text{min}} = 0 - (-12.5 \text{ J}) = 12.5 \text{ J}$$

The maximum speed of the particle is therefore

$$v_{\text{max}} = \sqrt{2mK_{\text{max}}} = \sqrt{25} \text{ m/s} = 5 \text{ m/s}$$

The first turning point is at x_0. The second turning point must occur where $E = U = 0$, or $x = 2.50$ m. Hence the half-width of the well is

$$d = 2.50 - 1.25 = 1.25 \text{ m}$$

its depth is $K = K_{\text{max}}$, and the oscillation period T as the particle bounces back and forth within this well is, from Eq. [8 . 17]:

$$T = 2\pi d\sqrt{\frac{m}{2K_{\text{max}}}} = 1.57 \text{ s}$$

8 • 4 *potential energy in three dimensions*

In general, we can define potential energy for conservative systems in three dimensions just as we did for the one-dimensional case. The only difference is that in place of Eq. [8 . 1] we must use the work–kinetic-energy theorem in its general vector form:

$$\blacktriangleright \quad K_2 - K_1 = \int_{\mathbf{r}_1}^{\mathbf{r}_2} \mathbf{F} \cdot d\mathbf{r} \qquad K = \tfrac{1}{2}m(v_x^2 + v_y^2 + v_z^2) \qquad [8 \cdot 18]$$

To derive a mathematical expression for potential energy, we use the same operational definition as before: The change in the potential energy of a system when its component parts move from one position to another is just the amount of work an external force would have to do against the system forces when there is no net change in kinetic energy. If we recognize the work integral in Eq. [8 . 18] as the three-dimensional generalization of the integral $\int_{x1}^{x2} F_x \, dx$ in Eq. [8 . 1], then it must represent the decrease in the system's potential energy when the kinetic energy increases from K_1 to K_2.

Suppose we represent the three-dimensional potential energy $U(x,y,z)$ as $U(\mathbf{r})$, where U is a function of position,

$$\mathbf{r} = x\mathbf{i} + y\mathbf{j} + z\mathbf{k}$$

We can then write the equivalent mathematical definition of the change in potential energy for a displacement from $\mathbf{r}_1$ to $\mathbf{r}_2$ as

$$U(\mathbf{r}_2) - U(\mathbf{r}_1) = \int_{\mathbf{r}_1}^{\mathbf{r}_2} \mathbf{F} \cdot d\mathbf{r} \qquad [8 \cdot 19]$$

or for the general case,

$$U(\mathbf{r}) = U(\mathbf{r}_0) - \int_{\mathbf{r}_0}^{\mathbf{r}} \mathbf{F} \cdot d\mathbf{r} \qquad [8 \cdot 20]$$

Note that, as in the one-dimensional case, $U(\mathbf{r})$ is a scalar, and the arbitrary reference value $U_0 = U(\mathbf{r}_0)$ is assigned to a single point in space located by the reference vector

$$\mathbf{r}_0 = x_0\mathbf{i} + y_0\mathbf{j} + z_0\mathbf{k}$$

In order to derive the three-dimensional form of the principle of conservation of energy, we note that Eqs. [8 . 18] and [8 . 19] are equivalent to $\Delta K + \Delta U = 0$. That is, the sum of the kinetic and potential energies at any instant in a conservative system is a constant equal to the total energy E:

$$\begin{aligned} K + U &= \tfrac{1}{2}m(v_x^2 + v_y^2 + v_z^2) + U(\mathbf{r}_0) - \int_{\mathbf{r}_0}^{\mathbf{r}} \mathbf{F} \cdot d\mathbf{r} \\ &= E \end{aligned} \qquad [8 \cdot 21]$$

This equation is the general expression for the total energy of a particle of mass m subject to a conservative force $\mathbf{F}(\mathbf{r})$.

Although the integral of Eq. [8 . 21] is generally difficult to evaluate, we may draw two important conclusions. First, if there is more than one force, then the system potential is the algebraic sum of the potentials due to the separate conservative forces $\mathbf{F}_i$ taken individually. Furthermore, each separate potential energy U_i may have its own separate reference level $\mathbf{r}_{0i}$. From the distributive law for scalar dot products, if $\mathbf{F}$ is the *resultant* of the forces $\mathbf{F}_i$, where $i = 1, 2, \ldots, n$, then

$$\mathbf{F} \cdot d\mathbf{r} = \left(\sum_i F_i\right) \cdot d\mathbf{r} = \sum_i (\mathbf{F}_1 \cdot d\mathbf{r})$$

and we can express the potential energy $U(\mathbf{r})$ as

$$U(\mathbf{r}) = U(\mathbf{r}_0) - \int_{r_0}^{r} \mathbf{F} \cdot d\mathbf{r} = U(\mathbf{r}_0) - \int_{r_0}^{r} \left(\sum_i F_i\right) \cdot d\mathbf{r}$$

If now we take the summation outside of the integral, we may write

$$\begin{aligned} U(\mathbf{r}) &= U(\mathbf{r}_0) - \sum_i \left(\int_{\mathbf{r}_0}^{\mathbf{r}} \mathbf{F}_i \cdot d\mathbf{r}\right) \\ &= U(\mathbf{r}_0) + \sum_i [U_i(\mathbf{r}) - U_i(\mathbf{r}_{0i})] \end{aligned} \qquad [8 \cdot 22]$$

Since our choice of $U(\mathbf{r}_0)$ is entirely arbitrary, we may just set it equal to the sum of the reference values of the individual potentials:

$$U(\mathbf{r}_0) = \sum_i U_i(\mathbf{r}_{0i})$$

Hence the term in brackets in Eq. [8 . 22] vanishes,

$$U(\mathbf{r}) = \sum_i U_i(\mathbf{r}) \qquad [8 \cdot 23]$$

and the potential-energy functions may simply be added as algebraic scalars to determine the potential energy for the entire system. We shall see the advantages of this in the next section.

The second conclusion we can draw from Eq. [8 . 21] is that if we know the system potential energy, we can derive the forces within the system. The work done—the loss in potential energy—in moving a particle through a displacement $d\mathbf{r} = dx\,\mathbf{i} + dy\,\mathbf{j} + dz\,\mathbf{k}$ is given by the scalar product

$$-dU = F_x\,dx + F_y\,dy + F_z\,dz \qquad [8 \cdot 24]$$

In the special case that there is no change in the y or z coordinates, then

$$-dU = F_x\,dx \qquad \text{or} \qquad F_x = -\left(\frac{dU}{dx}\right)_{y,\,z\,=\,\text{const}} \qquad [8 \cdot 25]$$

That is, to obtain the net force in the x direction we differentiate $U(x,y,z)$ with respect to x only, as if y and z were just constant quantities. This type of differentiation is known as *partial differentiation* and is denoted by the special symbol ∂:

$$F_x = -\frac{\partial U}{\partial x} \qquad F_y = -\frac{\partial U}{\partial y} \qquad F_z = -\frac{\partial U}{\partial z} \qquad [8 \cdot 26]$$

If the total potential of a system is known in the appropriate three-dimensional coordinates, partial differentiation with respect to each of the three coordinates in turn gives the *vector* components of the resultant system force.

example 8 • 6 A mass m hangs from a spring which is displaced to one side of the vertical, as shown in Fig. 8 . 8. Find the total potential energy $U(\mathbf{r})$ of the system and the components of the resultant force on the mass.

solution If we take y as positive in the downward direction, the gravitational potential energy of the hanging mass is $U_g = -mgy$, where y is the distance below the ceiling. However, the spring potential in two dimensions is

$$U_s(\mathbf{r}) = \tfrac{1}{2}k(|\mathbf{r}| - r_0)^2 = \tfrac{1}{2}k(r - r_0)^2$$

where $r = \sqrt{x^2 + y^2}$ and r_0 is the length of the unstretched spring relative to the origin. The total potential energy of the system is therefore

$$U(x,y) = U_s + U_g = \tfrac{1}{2}k\left(x^2 + y^2 + r_0^2 - 2r_0\sqrt{x^2 + y^2}\right) - mgy$$

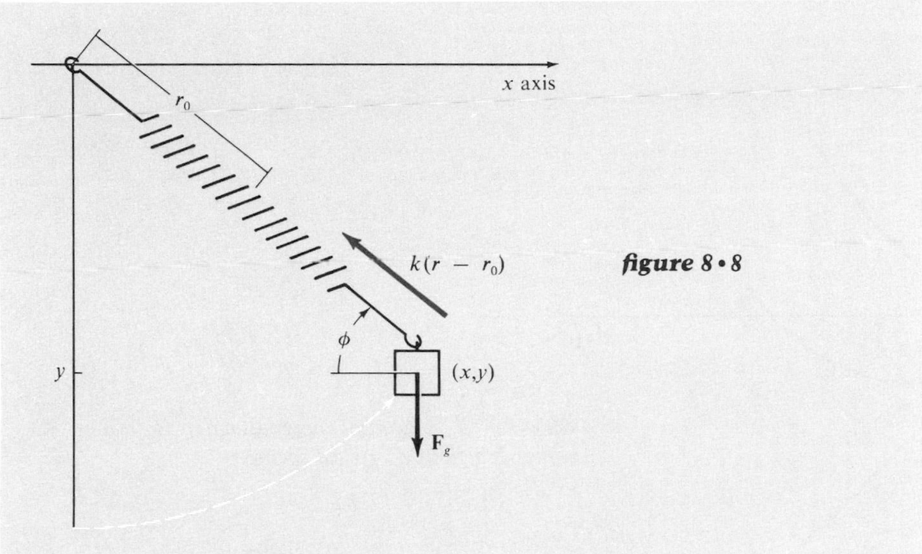

figure 8 • 8

We can now find the components of the resultant force by partial differentiation:

$$F_x = -\frac{\partial U}{\partial x} = -k\frac{x}{r}(r - r_0) = -k\cos\phi\,(r - r_0)$$

$$F_y = -\frac{\partial U}{\partial y} = -k\frac{y}{r}(r - r_0) + mg = -k\sin\phi\,(r - r_0) + mg$$

It is easy to verify that these are the components of the forces shown in the figure by a straightforward decomposition of the forces due to gravity and the extension of the spring.

8 • 5 some applications of the energy principle

In the preceding section we saw that individual potentials due to different conservative forces can be added algebraically to find the total system potential. Let us analyze such a case in detail. Consider a mass m hanging vertically from a spring which is attached to the ceiling, with y taken as positive in the downward direction (see Fig. 8 . 9). If the unstretched length of the spring is y_0, then the potential energy of the system when it has been stretched to some lower point y is $U(y) = U_s(y) + U_g(y)$, the sum of the elastic and gravitational potential energies.

Suppose, for mathematical convenience, that we choose the reference levels of both potentials to be at the ceiling and we arbitrarily take U_g and U_s to be zero there as well. Then these potentials take the form

$$U_g(y) = -mgy \qquad U_s(y) = -\tfrac{1}{2}ky_0^2 + \tfrac{1}{2}k(y - y_0)^2 \qquad [8 \cdot 27]$$

Hence the total potential energy is

$$U(y) = \tfrac{1}{2}ky^2 - kyy_0 - mgy \qquad [8 \cdot 28]$$

The graph of potential energy versus position y is a parabola, since U is

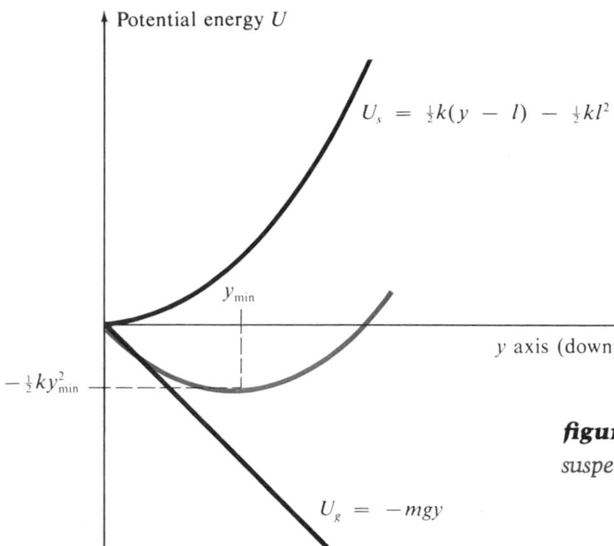

figure 8 • 9 *Potential-energy diagram for a mass suspended from a vertical spring*

quadratic in y. This is shown by the middle curve in Fig. 8 . 9, which has its minimum at y_{min}, the equilibrium position of the spring-mass system under the action of gravity. This point is computed from the total system force at equilibrium:

$$-F_y = \frac{dU}{dy} = ky_{min} - ky_0 - mg = 0$$

$$k(y_{min} - y_0) = mg \qquad [8 \cdot 29]$$

Since the total potential is quadratic, we can expect that if the mass is raised and then released, it will oscillate on the end of the spring about the equilibrium position $y_{min} = y_0 + mg/k$. If we substitute $y_{min} - mg/k$ for y_0 in Eq. [8 . 28], we obtain

$$U(y) = \tfrac{1}{2}k(y - y_{min})^2 - \tfrac{1}{2}ky^2_{min}$$

The total system force is therefore

$$F_y(y) = -\frac{dU}{dy} = -k(y - y_{min}) \qquad [8 \cdot 30]$$

a linear restoring force, and the oscillation has the same period we found for a horizontal spring-mass system, $T = 2\pi\sqrt{m/k}$. Surprisingly enough, the only effect of gravity is to shift the equilibrium position of the spring-mass system.

When the analytic form of a one-dimensional potential-energy function is known, we can integrate directly to find the relationship between position and time. If we write $K + U = E$ as

$$\tfrac{1}{2}mv^2 + U(x) = E \qquad [8 \cdot 31]$$

where E can ordinarily be evaluated from the initial conditions of a problem, we obtain

$$v = \frac{dx}{dt} = \pm\sqrt{\frac{2}{m}[E - U(x)]}$$

or

$$\blacktriangleright \quad \pm\int_{x_0}^{x} \frac{dx}{\sqrt{(2/m)[E - U(x)]}} = \int_{t_0}^{t} dt = t - t_0 > 0 \qquad [8 \cdot 32]$$

The sign of the left-hand member depends on the direction of motion at time t_0; the energy equation involves the square of the velocity and is therefore indeterminate with regard to sign. Although this function may be difficult, or sometimes impossible, to integrate analytically, we can always tabulate values of $x(t)$ by means of the trapezoidal rule (see Sec. 7 . 6). However, it is necessary to test whether $E - U > 0$, so that we do not attempt to carry the integration past a turning point at $E = U$. At this point the sign of the integral changes, and special techniques are necessary to avoid errors, since the integrand goes to infinity, while the integral remains finite.

If $U(x)$ is known, we can also derive $F_x(x)$ by differentiating Eq. [8 . 7] to obtain

$$F_x(x) = -\frac{dU(x)}{dx} \qquad [8 \cdot 33]$$

Since we derived $U(x)$ from $F_x(x)$ in the first place, this may not seem very useful. However, experimental data are often in the form of direct measurements of the energies of the system under study, and if we can determine $\Delta U = \Delta E - \Delta K$, we can assign some arbitrary reference level and tabulate values of $U(x)$. This enables us to determine values of F_x by the numerical-differentiation method discussed in Sec. 3 . 6, using the approximation

$$F_x(x) \doteq -\frac{\Delta U}{\Delta x} \qquad [8 \cdot 34]$$

A somewhat better approximation is obtained by letting $-\Delta U/\Delta x$ be the force at the midpoint of the interval (see Fig. 3 . 8):

$$F_x(x + \tfrac{1}{2}\Delta x) \doteq -\frac{U(x + \Delta x) - U(x)}{\Delta x} \qquad [8 \cdot 35]$$

It is often easier to derive the forces in complicated systems from potential-energy differences than to perform a detailed dynamic analysis.

example 8 • 8 An object is stretched along its x dimension by an external force, and measurements of the energy changes show that the following work is required for each 2 cm of additional stretch:

x, cm	0	2	4	6	8	10
W, ergs	0	9	42	91	159	245

Calculate approximate values of the *internal* restoring force as a function of x.

solution Taking the force at the midpoint of each 2-cm interval, we have

$$F \doteq -\frac{\Delta U}{\Delta x} = -\frac{\Delta W}{\Delta x}$$

since the work done in stretching increases the potential energy of the body. Therefore

x, cm	1	3	5	7	9
$-F$, dynes	4.5	16.5	24.5	34.0	43.0

8 · 6 heat and the conservation of energy

When an object is pulled across a rough horizontal surface at constant velocity, work is being done. However, there is no corresponding increase in kinetic energy because a frictional force is resisting the motion. Instead, heat is generated at the surface. The "disappearance" of mechanical energy is always accompanied by the production of heat. Even before Newton, Francis Bacon had suggested in 1620 that heat is simply a mode of motion, a view also held by later physicists such as Robert Boyle and Robert Hooke. Toward the end of the eighteenth century, however, the doctrine of *caloric theory,* that heat is an indestructible fluid, came into favor and led to the idea that the energy expended against friction simply disappears.

The person who revived the notion that heat is *not* a fluid substance was the versatile Count Rumford (Benjamin Thompson), an American Tory, who after having fled Massachusetts during the Revolution enlisted in the service of the Elector of Bavaria and was made head of the state arsenal in Munich.* From experiments on the production of heat in the boring of cannon, which he reported to the Royal Society of London in 1798, he was led to conclude that "anything which any isolated body, or system of bodies, can continue to furnish without limitation, cannot possibly be a material substance." Heat must be motion, Rumford decided, and this view received complete quantitative confirmation when James Prescott Joule, in a series of famous experiments extending from 1840 to 1878, demonstrated the equivalence of heat and work by showing that for every definite amount of work done against friction there always appears a definite quantity of heat (see Table 7 . 1). We shall discuss heat at length in Chapter 21.

Not only can we produce heat from work, but the reverse is also true, as in the case of a steam engine. If the two are interconvertible, then it must follow that heat is also a form of energy *in the process of transfer.* The distinction is that heat may be transferred from one body to another without any physical displacement of either the body or its boundaries. Heat absorbed by a body goes into increasing the internal kinetic and potential energies of the particles constituting the body. If we recognize the existence of

*Kaleidoscopic might be a better descriptive term for Count Rumford, a great physicist whose career also encompassed espionage, intrigue, command of the Imperial Army of Bavaria, and the invention of coffeepots, vegetable stews, kitchen stoves, and the first successful antipoverty program. Sanborn Brown's very readable *Count Rumford: Physicist Extraordinary* (Doubleday Anchor Books, 1962) should whet your appetite to learn more about this most unusual man.

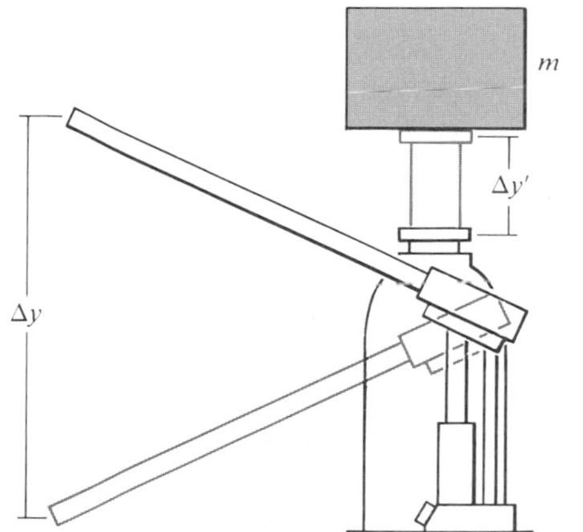

figure 8 • 10 *Mechanical advantage* $\Delta y/\Delta y'$

such a *total internal energy* $\mathcal{U}$, as distinguished from any kinetic or potential energy possessed by the system as a whole, then we can extend the principle of conservation of energy to thermal as well as mechanical phenomena.* Indeed, all such processes can be grouped under one law: the *first law of thermodynamics:*

> The change in the internal energy of a system in some physical process equals the heat absorbed by the system minus the work done by the system on its external surroundings.

This law, expressed in mathematical terms, is simply

$$\Delta\mathcal{U} = Q - W \qquad [8 \cdot 36]$$

Where Q is the heat *absorbed by* the system and W is the work *done by* the system on its external surroundings. The first law of thermodynamics generalizes the energy principle to *all* energy phenomena, whether the system is conservative ($Q = 0$) or nonconservative ($Q \neq 0$). Thus any of the terms in Eq. [8 . 36] may be zero, positive, or negative, depending on the circumstances. The importance of this generalization is that once we recognize heat as a form of energy, interconvertible with all other forms, we can postulate that *energy is neither created nor destroyed.*†

One concept easily derived from Eq. [8 . 36] is that of *mechanical advantage*. Imagine that the "black box" in Fig. 8 . 10 contains some mysterious device such that when the handle is pushed down through a distance Δy by a force **F**, the platform above the box is raised a distance $\Delta y'$ into the air. If there are no heat losses in the device (check its temperature), then the work put in at the handle must equal the work done by the platform in lifting the object placed on it:

$$W_{\text{in}} = F\,\Delta y = F'\,\Delta y' = W_{\text{out}} \qquad [8 \cdot 37]$$

*Although the total internal energy is conventionally represented by the symbol U, in this book U is used for potential energy. Therefore, to avoid confusion, total internal energy is represented by $\mathcal{U}$.

†This is strictly true if we include mass as a form of energy (see Chapter 19).

This enables us to evaluate the mechanical advantage of a machine in terms of the ratio of distances:

$$\frac{F'}{F} = \frac{\Delta y}{\Delta y'} \qquad [8 \cdot 38]$$

If $\Delta y = 2$ ft and $\Delta y' = 1$ in, the mechanical advantage is 24/1, which means that an applied force of $F = 50$ lbf can raise a 1200-lbm object.

A *machine* can be defined as a device that transmits mechanical energy from one point to another. It differs from an *engine,* in that an engine converts energy, usually from nonmechanical into mechanical form. In a machine an operating force $\mathbf{F}_{in}$ does work on the machine at one point, and the machine does work against a load, or output force $\mathbf{F}_{out}$, at another point. These two forces usually differ from each other in both magnitude and direction. In all machines the output of work W_{out} is inevitably less than the input of work W_{in}, simply because there is always some friction which dissipates part of the input as heat. The *mechanical efficiency* η of a machine is thus the ratio of its work output to the work input,

$$\eta = \frac{W_{out}}{W_{in}} \qquad [8 \cdot 39]$$

which is always less than unity. (What bearing does this have on the idea of a perpetual-motion machine?)

PROBLEMS

8 • 1 potential energy

8 • 1 The restoring force of a nonlinear spring (one that does not follow Hooke's law) is given by

$$F(x) = -k\,\Delta x\,[1 - a\,(\Delta x)^2]$$

where $\Delta x = x - x_0$ is the linear extension of the spring. Find the potential energy of the spring as a function of Δx.

8 • 2 It can be shown theoretically and experimentally that it takes work to form a liquid surface; therefore liquid surfaces represent potential energy. In the case of mercury at ordinary temperatures and pressures, this potential energy is 520 ergs/cm^2. If a hemispherical drop of mercury of radius 0.25 cm is resting on a smooth horizontal surface, how much work must be done in cutting it into eight equal but smaller hemispherical drops?

8 • 3 A piledriver with a 1000-kg driving head drops 1.2 m onto a pile and drives it 8 cm into the earth. What weight object would it take to drive the pile that deep if the object were merely resting on top of the pile? (Neglect the mass of the pile and assume that the force necessary to move the pile is constant while the pile is being driven.)

8 • 4 An automobile traveling at 45 mi/h runs into a telephone pole, and a 150-lbf passenger in the back seat is driven 5 ft by the impact. (*a*) What is the applied force per square inch if the passenger is stopped by a knob of 1 in diameter projecting from the dashboard? (*b*) By a safety harness 2 ft long and 3 in wide? Express your answer in atmospheres (1 atm = 14.7 lbf/in^2, the force per unit area exerted by the earth's atmosphere at sea level). Which mode of deceleration would you recommend?

8 • 5 The *Yukawa potential* can be used to approximate the force between two particles of the atomic nucleus in terms of their separation distance r:

$$U(r) = -U_0\left(\frac{r_0}{r}\right)\exp\left(-\frac{r}{r_0}\right)$$

(a) Draw a graph of this potential as a function of r/r_0. (b) What is the force $F(r)$ between the particles?

8 • 6 A particle of mass m on the x axis is attracted toward the origin by a force $\mathbf{F}(x) = -m/x^2\mathbf{i}$. Construct a potential-energy function $U(x)$ for this particle. Try to choose a reference point for the potential which will simplify its form as much as possible.

8 • 2 *conservation of energy*

8 • 7 A physics student fashions a primitive ballistic device by stretching a rubber band between her thumb and forefinger, a distance of 10 cm. A 10-g wad of paper is placed in the middle of the band and drawn backward a distance of 5 cm by an applied force of 5 N. Neglecting air resistance, what is the speed of the wad when it passes its original position after the rubber band is released? Assume a linear restoring force.

8 • 8 A particle moving along the x axis is subject to a conservative system force corresponding to the potential-energy function

$$U(x) = a + bx^2 - cx^4$$

Determine the coefficients a, b, and c if it is known (1) that the potential vanishes at the origin, (2) that $x = 2$ m is an equilibrium position, and (3) that a 5-kg particle with speed of 2 m/s at the origin comes to rest at $x =$ 1 m. Include appropriate dimensions with your answer.

8 • 9 A ball of mass 1 g (whose dimensions may be neglected) slides down a frictionless trough as shown. Starting from rest, the ball drops from a height of 200 cm and leaves the trough at angle θ with the horizontal. At the highest point in its free trajectory the ball strikes a spring mounted on the wall and compresses the spring 2 cm. The spring-force constant is $k = 49{,}000$ dynes/cm. (a) At what height is the spring mounted? (b) What is the angle θ?

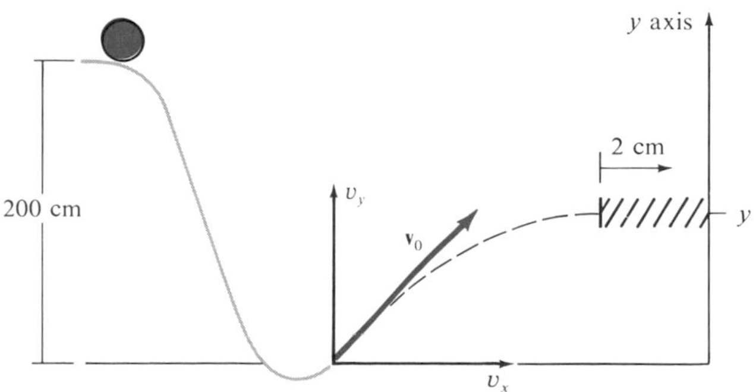

8 • 10 A 20-g bullet is fired from a rifle with a muzzle speed of 250 m/s, and at a height of 2680 m the bullet is moving at an angle of 45° with the horizontal. (*a*) What is its speed at this point? (*b*) What is its potential energy? (*c*) What is its kinetic energy? (*d*) What maximum height will the bullet attain and how fast will it be moving then?

8 • 11 What average force must be applied to a 1-kg mass to raise it 5 m above the earth and give it an upward speed at that point of 10 m/s?

***8 • 12** Using the trapezoidal rule, tabulate $U(x)$ over the interval $0 \leq x \leq 10$, given that the system force acts according to the law

$$F(x) = x^2 \cos^3 \tfrac{1}{2}\pi x$$

Compute $U(x)$ to within 5 percent maximum error, as determined by comparing errors for various interval sizes. Let $U_0 = 0$.

8 • 13 If one of the nails in Example 8 . 2 were set so low that the pendulum could not reach its maximum height of swing, what would happen to its motion?

8 • 3 *energy diagrams*

8 • 14 Suppose the particle in Example 8 . 5 starts its motion at $x_0 = 0$ with a kinetic energy of $K = 37.5$ J. Redraw Fig. 8 . 7 to show its total energy and turning point under these circumstances.

8 • 15 A one-dimensional potential-energy function $U(x)$ is given by the formula

$$U(x) = U_0 \sin^2 x > 0$$

for any x in the interval $-\pi \leq x \leq +\pi$, and $U(x) = 0$ for all other values of x. Two particles are traveling through this potential-energy barrier, coming from $x = -\infty$. Both particles are of mass m, but one has an initial kinetic energy equal to $\frac{1}{2}U_0$, and the other has an initial kinetic energy equal to $2U_0$.

(*a*) Draw the graph of $U(x)$, indicating the total energy of the incoming particles on it. (*b*) Describe the behavior of the slower particle: what is the minimum value of its kinetic energy? Where does it occur? What does it signify? (*c*) What is the force on the slower particle when it reaches the point of minimum kinetic energy? Indicate the force by drawing a tangent to the curve, and explain its significance. (*d*) What is the minimum value of the kinetic energy of the faster particle and at what point does it occur? (*e*) What is the maximum kinetic energy of the faster particle and where does it occur? (*f*) What is the force on the faster particle at the point in the field where the slower particle has its minimum kinetic energy? How does this force compare with the force on the slower particle?

8 • 16 A one-dimensional potential-energy function is given by

$$U(x) = -U_0 \cos^2 x \qquad U_0 > 0$$

for any x in the interval $-\frac{1}{2}\pi \leq x \leq +\frac{1}{2}\pi$ and is zero everywhere else. A particle with total mechanical energy $E = \frac{1}{2}U_0$ at $x = \pi$ approaches the origin from the right. (*a*) Describe the motion of the particle and find its kinetic energy at the origin. (*b*) Describe its subsequent motion if U_0 joules of kinetic energy is suddenly lost by the particle when it arrives at the origin.

8 • 17 The collision of two real, solid spheres of radius R can be described in terms of a large repulsive force of magnitude F acting over a very short distance $\Delta x \ll R$. Draw the potential-energy diagram for the spheres.

8 • 18 The force acting on a particle can be described approximately as follows:

$$F(x) = \begin{cases} F_0 & \text{for } -2l \leq x \leq -l \\ -F_0 & \text{for } -l \leq x \leq l \\ F_0 & \text{for } l \leq x \leq 2l \end{cases}$$

and $F(x) = 0$ everywhere else. Graph (approximately) the potential energy of the particle.

8 • 19 The potential energy between two neighboring molecules is given by the *Lennard-Jones potential,*

$$U(r) = -\frac{A}{r^6} + \frac{B}{r^{12}}$$

where r is the separation distance between the molecules. (*a*) Express the force between them as a function of r. (*b*) What is the equilibrium position of the two molecules? (*c*) What dissociation energy would be necessary to pull them apart from their equilibrium position?

8 • 20 Imagine that particles of mass m are strung out on the x axis, each one subject to the system potential-energy function, which is

$$U(x) = U_0 \cos \frac{\pi x}{l}$$

for all values of x. Each minimum of U is a well containing one particle of potential energy $-\frac{1}{2}U_0$. (*a*) What is the kinetic energy of each particle at the equilibrium position? (*b*) What is the average separation between particles? (*c*) How much energy would a particle have to gain to break free of the restoring forces which keep it in its own particular well?

8 • 4–8 • 5 addition of potentials and derivation of forces

8 • 21 What is the vector force on a particle when its potential energy is given (*a*) by $U = xyz$? (*b*) By $U = x^2 + 2xy + z^2$? (*c*) By $U = x \sin z + \cos y$?

8 • 22 Find the conservative force which gives rise to the potential-energy function

$$U(\mathbf{r}) = 3x^2y + \frac{zy}{x} - y^2$$

8 • 23 The potential energy of a particle is given by $U = -a/r$, where a is a position constant and r is the length of the radius vector in two dimensions. Derive the x and y components of force on the particle.

8 • 24 A mass m at the end of a freely pivoting spring moves around the perimeter of a wheel of radius R. If the unstretched length is l_0 as shown, derive the potential energy as a function of the angle θ.

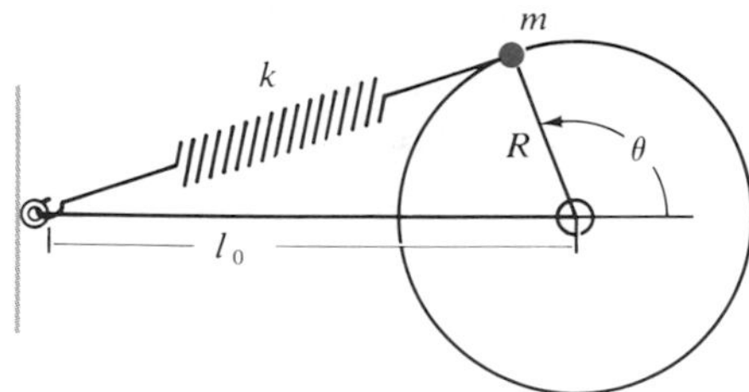

8 • 25 A mass m on top of a freely pivoting pole of length l_0 is connected to the ground by two springs of equal unstretched length l_0 and force constant k. Since the mass moves back and forth in a circular arc, we can express its path length s in terms of the angular distance θ, $s = r\theta$. (*a*) Find the formula for the potential energy U of the system as a function of θ. (*b*) Draw a graph of $U(\theta)$ and compute values in some appropriately normalized coordinate system. (*c*) Find the frequency of small-amplitude oscillations about $\theta = 0$ rad.

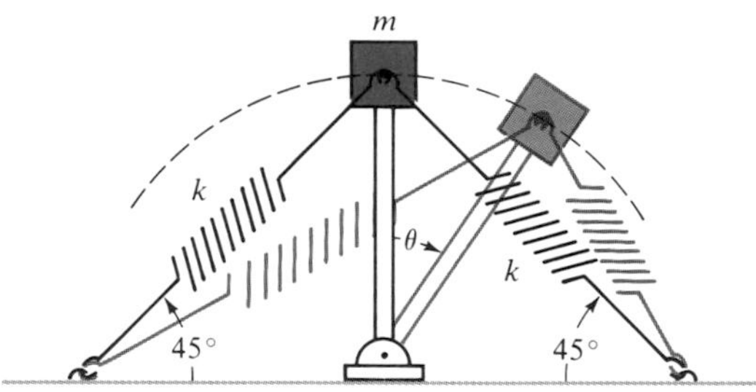

8 • 5 *applications*

8 • 26 An acrobat of mass m tumbles off a platform at height y onto a trampoline, which is supported at a height y_0 above the ground. If the trampoline springs have an effective restoring-force constant k in the vertical direction, what is the lowest point the acrobat will reach before bouncing back into the air?

8 • 27 A body moves through two potentials,

$$U_1(x) = \sin^2 x \qquad U_2(x) = \cos^2 x$$

(*a*) What is its total potential energy? (*b*) Explain your answer in terms of the forces on the body.

8 • 28 A ball of mass m is thrown upward with an initial kinetic energy K_0, and it reaches a height at which its gravitational potential is 0.9 K_0 before it drops back to the ground. What is the average force on the ball due to air resistance?

8 • 29 A mass m lying on a smooth tabletop is fastened to two springs of equal unstretched length x_0 and force constant k. (*a*) Derive the potential energy of the system. (*b*) Find the period of oscillation. (*c*) If the mass is released from a position $x = \frac{3}{2}x_0$, what will be its greatest speed and at what point in its motion will it attain this speed?

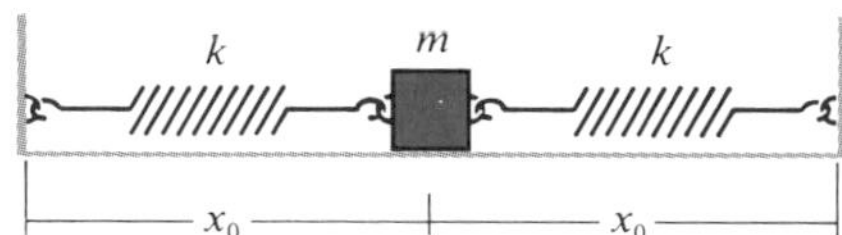

***8 • 30** A mass m moves through two potentials,

$$U_1(x) = U_0 \frac{\sin \pi x}{\pi x} \qquad U_2(x) = U_0 x^2$$

(*a*) Graph both functions and their sum over the range $-2 \leq x \leq +2$. (*b*) Determine the minimum value of the total potential and the points of stable equilibrium, to within $\Delta x = \pm 0.01$ error. (*c*) Determine the maximum values of the total potential and the points of unstable equilibrium, to within $\Delta x = \pm 0.01$ error. (*d*) Determine the approximate frequency of oscillation about the points of stable equilibrium for small displacements.

8 • 31 A 500-g object suspended from a spring on the moon stretches the spring 40 cm at equilibrium and oscillates with a period of 3.1 s. What is the acceleration due to the moon's gravity?

8 • 32 The potential energy of a particle of mass m is given by $U(x) = cx(x - x_0)$, where c is a constant. If the particle is initially at rest at $x_0 > 0$, find its speed v as a function of position x. (*b*) How long does it take the particle to reach the origin from the starting point. (*Hint:* To visualize the motion, graph this situation.)

8 • 6 heat and the conservation of energy

8 • 33 Explain why the efficiency of the block-and-tackle system shown is likely to be higher for moderate loads than for very small or very large loads. Are there any simple machines for which it is desirable that the mechanical efficiency be small?

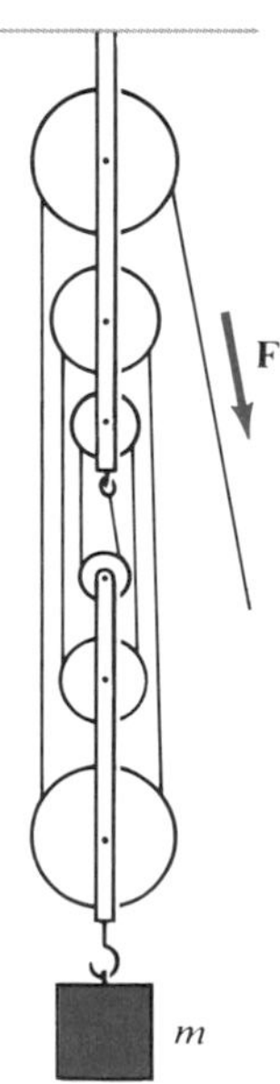

8 • 34 Three concepts other than efficiency are often useful for describing the performance of a machine: *actual mechanical advantage,* defined as the ratio of the output force to the input force, F_{out}/F_{in}, *ideal mechanical advantage,* the force-multiplying capability that the machine would have if its mechanical efficiency were unity, and the *speed ratio,* v_{out}/v_{in}, of those parts of the machine to which the output and input forces are applied. Show that for any machine (*a*) the efficiency is equal to the ratio of the actual mechanical advantage to the ideal mechanical advantage, (*b*) the speed ratio is inversely proportional to the ideal mechanical advantage, and (*c*) the efficiency is equal to the product of the actual mechanical advantage and speed ratio.

8 • 35 With the pulley system shown in Prob. 8 . 33 an applied force of 100 N is needed to hoist a safe weighing 2500 N. Compute the actual mechanical advantage, the ideal mechanical advantage, the speed ratio, and the mechanical efficiency as defined in Prob. 8 . 34.

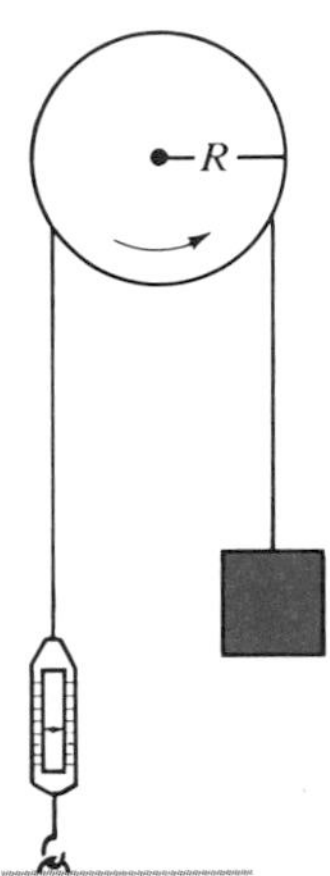

8 • 36 The power delivered by the shaft of an engine or motor is known as the *braking horsepower* (check your automobile registration). It is measured by applying known tangential forces on opposite sides of the shaft as shown, while the shaft turns at a constant rate. This apparatus is known as a *Prony brake*. If the shaft has a cross section of radius R and is turning at ν revolutions per second, the mass on the free end is represented by m, and the spring scale shows a reading F, what is the formula for braking horsepower?

8 • 37 Consider the piledriver described in Prob. 8 . 3. (*a*) Neglecting any heat loss to the earth and air, what is the increase in the internal energies of the piledriver and the pile? (*b*) If the piledriver rebounds 0.8 m into the air after hitting the pile, what is the total increase in the internal energies of the piledriver and the pile?

8 • 38 A vertical cylinder containing a gas is sealed by a movable piston which weighs 200 N and is under an atmospheric force of 100 N. When the gas is heated from below, it expands, forcing the piston 20 cm upward. (*a*) If the internal energy of the gas increases by 90 J during this process, how much heat is absorbed by the gas? (*b*) If the same process is repeated, this time keeping the piston fixed, what is the change in the internal energy of the gas? Assume that no heat is lost to the environment.

answers

8 • 1 $U_s(x) = \frac{1}{2}k(\Delta x)^2[1 - \frac{1}{2}a(\Delta x)^2]$
8 • 2 $W = 204$ ergs
8 • 3 $F_g = 147{,}000$ N or 15 metric tons
8 • 4 (*a*) $F = 176$ atm; (*b*) $F = 1.92$ atm
8 • 5 (*b*) $F = -(U_0/r_0)[(r_0^2/r^2) + (r_0/r)]e^{-r/r_0}$
8 • 6 $U = m/x$, $U(\infty) = U_0 = 0$
8 • 7 $v = 5$ m/s
8 • 8 $a = 0$, $b = \frac{80}{7}$ J/m^2, $c = \frac{10}{7}$ J/m^4
8 • 9 (*a*) $y = 100$ cm; (*b*) $\theta = 45°$
8 • 10 (*a*) $v = 100$ m/s; (*b*) $U = 525$ J; (*c*) $K = 100$ J; (*d*) $y_{max} = 3189$ m, $v_{min} = 70.7$ m/s
8 • 11 $\overline{F} = 19.8$ N
8 • 12 $U(x) = -(2/\pi^2)^3[(\frac{3}{4}y^2 - \frac{3}{2})\sin y + (\frac{1}{12}y^2 - \frac{1}{54})\sin 3y + \frac{3}{2}y\cos y + \frac{1}{18}y\cos 3y]$, where $y = \frac{1}{2}\pi x$
8 • 13 It would wind around the nail, rotating faster as its radius of rotation decreased.
8 • 14 $E = 37.5$ J, $x_{max} = 3.75$ m
8 • 15 (*b*) $K = 0$ at $x = -\frac{3}{4}\pi$, particle turns around; (*c*) $F_x = -\{U_0\}$ units of force;
(*d*) U_0 at $x = \pm\frac{1}{2}\pi$;
(*e*) $2U_0$ at $x = 0$ and $x > \pi$;
(*f*) $F_x = -\{U_0\}$ units of force

8 • 16 (*b*) $\frac{3}{2}U_0$; (*c*) it oscillates between $x = \pm\frac{1}{4}\pi$ with total energy $E = -\frac{1}{2}U_0$
8 • 19 (*a*) $F(r) = -6A/r^7 + 12B/r^{13}$; (*b*) $r = (2B/A)^{1/6}$; (*c*) $\Delta E = A^2/4B$
8 • 20 $K = \frac{1}{2}U_0$; (*b*) $2l$; (*c*) $\Delta E = \frac{3}{2}U_0$
8 • 21 (*a*) $\mathbf{F} = -zy\mathbf{i} - xz\mathbf{j} - xy\mathbf{k}$; (*b*) $\mathbf{F} = -(2x + 2y)\mathbf{i} - 2x\mathbf{j} - 2z\mathbf{k}$; (*c*) $\mathbf{F} = -\sin z\,\mathbf{i} + \sin y\,\mathbf{j} - x\cos z\,\mathbf{k}$
8 • 22 $\mathbf{F} = y[(z/x^2) - 6x]\mathbf{i} + [2y - 3x^2 - (z/x)]\mathbf{j} - (y/x)\mathbf{k}$
8 • 23 $\mathbf{F} = ar^{-3}(x\mathbf{i} + y\mathbf{j})$
8 • 24 $U = U_0 + kl_0 \times [R\cos\theta - \sqrt{R^2 + l_0^2 + 2l_0\cos\theta}] + mgR\sin\theta$, U_0 arbitrary
8 • 25 (*a*) $U(\theta) = k\,l_0^2\,[3 - \sqrt{2 + 2\sin\theta} - \sqrt{2 - 2\sin\theta}]$; (*c*) $2^{-1/4}\sqrt{k/m}$ rad/s
8 • 26 $y_{\min} = y_0 - mg/k \times [1 + \sqrt{1 + (2k/mg)(y - y_0)}]$
8 • 27 (*a*) $U = 1$; (*b*) the net force is zero
8 • 28 $\overline{F} = mg/9$
8 • 29 (*a*) $U = kx^2$; (*b*) $T = 2\pi\sqrt{m/2k}$; (*c*) $v_{\max} = x_0\sqrt{k/2m}$ at equilibrium
8 • 30 (*b*) $x = \pm 0.6843$; (*c*) $x = 0$; (*d*) $v = 0.23\sqrt{U_0/\mathrm{m}}$
8 • 31 $a = 164.3$ cm/s^2
8 • 32 (*a*) $v(x) = (2c/m)^{1/2}(xx_0 - x^2)^{1/2}$; (*b*) $\Delta t = \pi(m/2c)^{1/2}$
8 • 35 (*a*) $F_{\text{out}}/F_{\text{in}} = 5$; (*b*) ideal $= 6$; (*c*) $v_{\text{out}}/v_{\text{in}} = \frac{1}{6}$; (*d*) $\eta = \frac{5}{6}$
8 • 36 $P = 2\pi Rv(mg - F)$
8 • 37 (*a*) $\Delta\mathcal{U} = 12{,}550$ J; (*b*) $\Delta\mathcal{U} = 3{,}920$ J
8 • 38 (*a*) $Q = 150$ J; (*b*) $\Delta\mathcal{U} = 150$ J

CHAPTER NINE

momentum

Thus if a sphaerical body A with two parts of velocity is triple of a sphaerical body B which follows in the same right line with ten parts of velocity; the motion [momentum] of A will be to that of B as 6 to 10. Suppose then their motions to be of 6 parts and of 10 parts, and the sum will be 16 parts. Therefore upon the meeting of the bodies, if A acquire 3, 4, or 5 parts of motion, B will lose as many; and therefore after reflexion A will proceed with 9, 10, or 11 parts, and B with 7, 6, or 5 parts, the sum remaining always of 16 parts as before.

Isaac Newton, Explanation of Corollary III to Laws of Motion, Principia, *1686, translated by Andrew Motte, 1729*

In the preceding chapters we have analyzed the motion of particles in terms of the forces between them, the forces acting on them, and their resultant accelerations, velocities, and displacements as functions of time. In establishing the second law of motion, however, we used Newton's "quantity of motion," or momentum, to obtain $\mathbf{F} = d\mathbf{p}/dt$. Although displacement and its time derivatives are easier to visualize, the idea of momentum is more important in physics. Numerous situations which resist detailed analysis of time-dependent behavior can be profitably discussed in terms of net momentum changes in the system and its parts. These include *macroscopic,* large-scale, impact problems in which the surfaces of two bodies—like billiard balls—actually strike each other, as well as *microscopic* collisions in chemical and

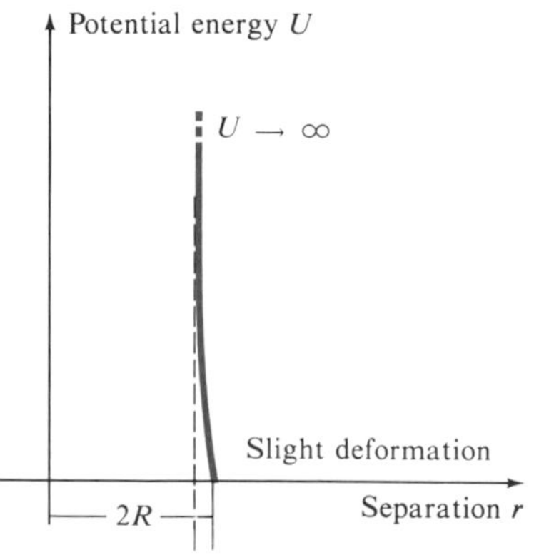

figure 9 • 1 *Potential-energy diagram for the force between two rigid spheres of radius R*

nuclear physics, where the relatively long-range forces between atoms, molecules, and subatomic particles cause them to deflect each other from their paths without direct physical contact, as we customarily think of it. Actually, "physical contact" is any interaction involving changes in momentum.

Both types of problems are just special cases of collision phenomena. In the macroscopic case the repulsive forces between objects increase very rapidly for relatively small deformations of their surfaces (no object can be perfectly rigid). Thus an energy diagram of potential energy versus the separation between the centers of two billiard balls of radius R would look like a nearly vertical line rising up from $U_0 = 0$ to $U \to \infty$ at separation distance equal to $2R$ (Fig. 9 . 1).

By focusing our attention on the net changes in momentum of a system and its parts, we can also determine the reaction forces necessary to make a fluid flow in a curved path—for example, the number of firefighters needed to hold a fire hose—as well as the reaction forces necessary for jet propulsion. The key ideas here will be momentum, impulse, the conservation of momentum, and the second law of motion.

9·1 *impulse*

The derivative of a vector is the vector sum of the derivatives of its separate components; hence the second law can be written as three separate differential equations relating the components of the net force on a particle and its momentum:

$$\frac{dp_x}{dt} = F_x \qquad \frac{dp_y}{dt} = F_y \qquad \frac{dp_z}{dt} = F_z \qquad [9 \cdot 1]$$

For a particle moving from position $\mathbf{r}_1 = (x_1, y_1, z_1)$ to $\mathbf{r}_2 = (x_2, y_2, z_2)$ during the time interval $t_2 - t_1$, we may, in principle, integrate these equations,

$$\begin{aligned} \int_{\mathbf{r}_1}^{\mathbf{r}_2} dp_x &= p_{x2} - p_{x1} = \int_{t_1}^{t_2} F_x\, dt \\ \int_{\mathbf{r}_1}^{\mathbf{r}_2} dp_y &= p_{y2} - p_{y1} = \int_{t_1}^{t_2} F_y\, dt \qquad [9 \cdot 2] \\ \int_{\mathbf{r}_1}^{\mathbf{r}_2} dp_z &= p_{z2} - p_{z1} = \int_{t_1}^{t_2} F_z\, dt \end{aligned}$$

and summarize the results as a single vector expression:

$$\int_{\mathbf{r}_1}^{\mathbf{r}_2} d\mathbf{p} = \mathbf{p}_2 - \mathbf{p}_1 = \int_{t_1}^{t_2} \mathbf{F}\, dt = \mathbf{P} \qquad [9 \cdot 3]$$

The integral of the net force over time is known as the *impulse* **P**, and it is equal to $\Delta\mathbf{p}$, the net change in momentum which takes place in the time interval. Both impulse and change in momentum are vectors, and both have the same units and dimensions.

Equation [9 . 3] would not be of much use for the type of dynamics problem which we have been considering, since the forces are either constant or given in terms of position, so that we would have to solve the problem first, before being able to evaluate the impulse. However, where the data of a physical problem specify the net change in momentum, then the notion of impulse is exceedingly useful in determining the average effective force $\overline{\mathbf{F}}$ acting during the time interval:

$$\overline{\mathbf{F}}(t_2 - t_1) = \int_{t_1}^{t_2} \mathbf{F}\, dt \qquad [9 \cdot 4]$$

or

$$\overline{\mathbf{F}} = \frac{\Delta\mathbf{p}}{\Delta t} \qquad [9 \cdot 5]$$

As the name *impulse* implies, this approach is particularly relevant to collision forces which act only for brief periods of "collision time," during which, however, the forces may be quite large. In a collision between two particles, where external forces are negligibly small during the collision, the third law shows that the impulses on the two bodies must be equal and opposite. This also follows directly from the principle of conservation of momentum, which we derived in Sec. 5 . 3 from the interaction postulate.

As an example of the utility of this approach, consider the case of *specular reflection* (from the Latin *speculum,* mirror) of particles from a stationary wall, as shown in Fig. 9 . 2. If a body of mass m approaches a rigid

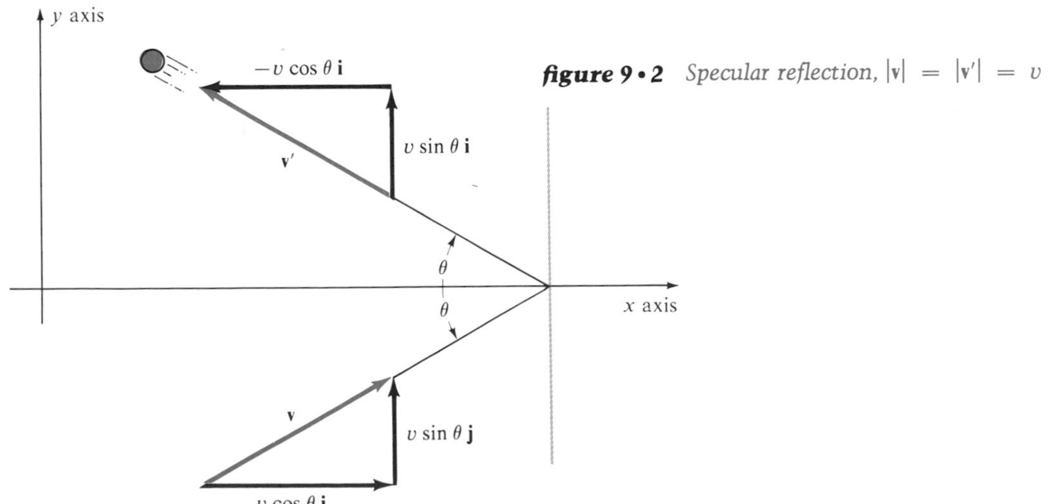

figure 9 • 2 *Specular reflection,* $|\mathbf{v}| = |\mathbf{v}'| = v$

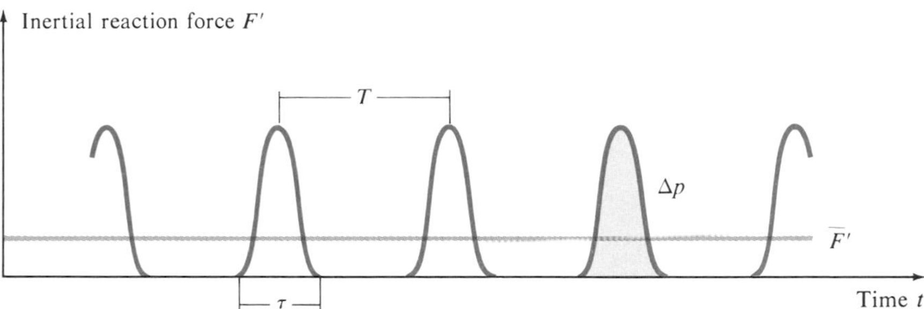

figure 9 • 3 *Force-time graph of particle impulses*

wall with velocity $\mathbf{v}$ at an angle θ, and then bounces off again at the same angle with its speed unchanged, the net change in its momentum is

$$\begin{aligned}\Delta\mathbf{p} &= \mathbf{p}' - \mathbf{p} \\ &= mv(-\cos\theta\,\mathbf{i} + \sin\theta\,\mathbf{j}) - mv(\cos\theta\,\mathbf{i} + \sin\theta\,\mathbf{j}) \\ &= -2mv\cos\theta\,\mathbf{i} = -2mv_x\,\mathbf{i}\end{aligned} \qquad [9 \cdot 6]$$

where v_x is the x component of the velocity. Note that only the component of momentum perpendicular to the surface has changed; the tangential component remains the same. If the brief period of actual collision is defined as $\Delta t = \tau$, then the impulse on the body is

$$\mathbf{F} = \frac{\Delta\mathbf{p}}{\Delta t} = -\frac{2mv_x}{\tau}\,\mathbf{i} \qquad [9 \cdot 7]$$

Thus the inertial reaction force which the particle exerts *on the wall* is $\mathbf{F}' = -\mathbf{F} = (2mv_x/\tau)\mathbf{i}$, which acts only during the collision time τ and is normal to the surface.

Now imagine that this body is one of a stream of identical particles striking the wall of their container at the same angle and at regular intervals. The force-time graph of the inertial reactions at the wall is shown in Fig. 9 . 3. The area under a single pulse is $|\Delta\mathbf{p}| = (2mv_x/\tau)$, as shown. If we approximate a succession of n impulses by a straight line subtending the same area over the t axis, its height is just the average reaction force $\bar{\mathbf{F}}'$. This is the force which, to the observer, seems to be acting over the *entire* time nT without any perceptible change, if we assume individual impulses to be small and indistinguishable. Therefore

$$nT\bar{\mathbf{F}}' = n\,\Delta\mathbf{p} \qquad \bar{F}' = \frac{2mu_x}{T} \qquad [9 \cdot 8]$$

In fact, since all reaction forces are normal to the surface for specular reflection, the particles could be striking the wall at any angle, with any velocity, and not necessarily at regular intervals. To find the average force on the wall, we would just sum all the impulses due to the separate collisions and divide by the entire period of time over which they occur. This will be of great importance in Chapter 22, when we derive an expression for the pressure of a gas on the walls of its container.

example 9 • 1 Just in the nick of time, Superman interposes his manly bosom between a band of villains and their intended victim, so that their machine-gun bullets bounce specularly (and harmlessly) off his chest. If the scoundrels are firing 3-g bullets at the rate of five per second with a speed of 500 m/s directly at our hero, what average force does he feel?

solution From Eq. [9 . 8], the average reaction force is the momentum change per bullet multiplied by the number of bullets per second, or

$$\overline{F} = 2(0.003 \text{ kg})(500 \text{ m/s})(5/\text{s}) = 15 \text{ N}$$

If the bullets had merely flattened against his chest instead of bouncing off, Superman would feel an average force of only 7.5 N. It's not as hard as it looks—if you assume that no bullet will exceed the average force.

example 9 • 2 A fire hose is held so that water coming horizontally out of the fire hydrant is directed upward at an angle of 60°, as shown. If the water emerges at a speed of 100 ft/s and at a rate of 50 lbm/s, neglecting friction, what force must the firefighters apply to hold the hose in this position?

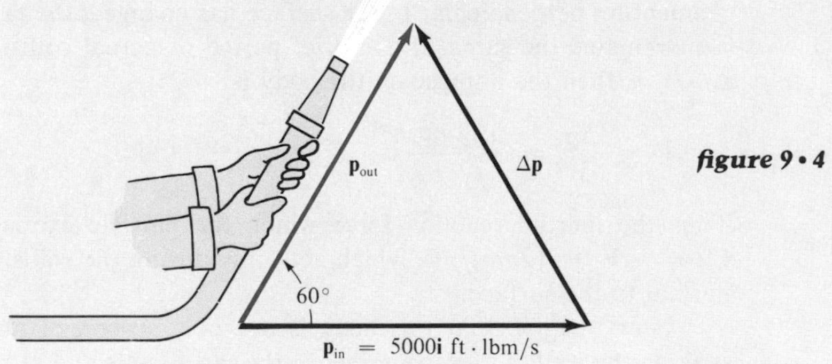

figure 9 • 4

solution During each second, 50 lbm of water in the hose system is changing direction, although its rate of flow remains constant:

$$\mathbf{p}_{in} = (50 \text{ lbm} \times 100 \text{ ft/s})\mathbf{i}$$

$$\mathbf{p}_{out} = (50 \text{ lbm} \times 100 \text{ ft/s})(\cos 60° \, \mathbf{i} + \sin 60° \, \mathbf{j})$$

The net rate of change of momentum in the time $\Delta t = 1$ s is therefore

$$\frac{\Delta\mathbf{p}}{\Delta t} = (5000 \text{ lbm} \cdot \text{ft/s})(-0.5\mathbf{i} + 0.866\mathbf{j})$$

Since 1 lbm · ft/s^2 = 1/32.2 lbf, the force the firefighters must exert to accomplish this change is

$$\mathbf{F} = \frac{\Delta\mathbf{p}}{\Delta t} = (-77.6\mathbf{i} + 134.5\mathbf{j}) \text{ lbf}$$

(Data courtesy of Baltimore Fire Department.)

9 . 2 elastic collisions

We saw in Sec. 5 . 3 that the linear momentum of any isolated pair of interacting particles is always conserved. Hence for all two-body collisions,

$$m_1\mathbf{v}_1 + m_2\mathbf{v}_1 = m_1\mathbf{u}_1 + m_2\mathbf{u}_2 \qquad [9 \cdot 9]$$

where $\mathbf{u}$ is the initial velocity and $\mathbf{v}$ is the velocity after the interaction (note that in this case the subscripts refer to the two bodies). Thus Eq. [9 . 9] involves four velocities. If only two of these are known in advance, as is usually the case, this equation alone does not afford a complete solution. If all the kinetic energy of the colliding particles is conserved, we can express this information as a second, independent equation involving the unknown velocities. In most collisions, however, some undeterminable amount of the mechanical energy is converted into heat, which rules out the possibility of an energy equation that involves the conservation of kinetic energy alone. Later on we shall see how to circumvent this difficulty by means of an empirical rule that provides the necessary second equation.

A binary collision, a collision of two particles, is said to be direct, or *collinear,* if the velocities both before and after the impact are along the same straight line. An example would be two homogeneous, smooth spheres moving along the line passing through their centers (see Fig. 9 . 5) and not appreciably affected by outside forces during the very brief time they are in contact. For all cases of direct impact, the momentum equation reduces to the one-dimensional scalar form

$$m_1u_1 + m_2u_2 = m_1v_1 + m_2v_2 \qquad [9 \cdot 10]$$

where the approach velocity u and the separation velocity v may have either positive or negative values along the common line of motion. An impact is said to be *perfectly elastic* if the total kinetic energy of the colliding particles is conserved; that is, if

$$\tfrac{1}{2}m_1u_1^2 + \tfrac{1}{2}m_2u_2^2 = \tfrac{1}{2}m_1v_1^2 + \tfrac{1}{2}m_2v_2^2 \qquad [9 \cdot 11]$$

Perfectly elastic collisions are known to occur under certain circumstances between atoms and between subatomic particles. For macroscopic bodies the collisions are seldom, if ever, perfectly elastic, although the loss of energy is very small for materials such as glass and ivory.

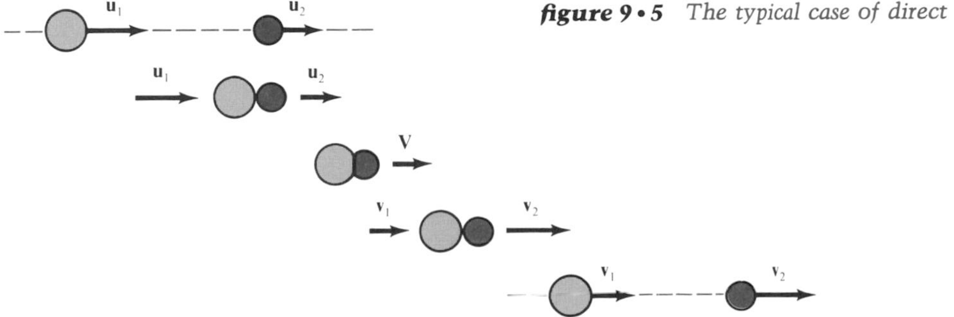

figure 9 • 5 *The typical case of direct impact*

If one body has a smooth surface and is so massive that its recoil after collision can be neglected, then a perfectly elastic collision necessarily implies specular reflection of the smaller impinging body, since the surface can exert force only in a normal direction, while the total kinetic energy of the smaller body must remain constant. Under these circumstances,

$$u_y = v_y \qquad \tfrac{1}{2}mu_x^2 = \tfrac{1}{2}mv_x^2 \qquad v_x = -u_x \tag{9 . 12}$$

This is the reason we assumed specular reflection in calculating the average force on the wall of a rigid container in Fig. 9 . 3. Suppose we rearrange Eqs. [9 . 10] and [9 . 11] into the forms

$$m_1(u_1 - v_1) = -m_2(u_2 - v_2)$$

$$m_1(u_1^2 - v_1^2) = -m_2(u_2^2 - v_2^2)$$

If we then divide the second of these equations by the first, we obtain

$$u_1 + v_1 = u_2 + v_2$$

or

$$v_1 - v_2 = -(u_1 - u_2) \tag{9 . 13}$$

In other words,

> The relative velocity of two bodies after a direct, perfectly elastic collision is equal in magnitude but opposite in direction to the relative velocity of the two bodies before impact.

Equation [9 . 13] is independent of the masses of the bodies; moreover, the pair of simultaneous equations [9 . 10] and [9 . 11] has now been reduced to the pair of simultaneous *linear* equations [9 . 10] and [9 . 13].

example 9 • 3 Show that when two identical particles collide elastically, they simply exchange velocities on impact, and the result is the same as if one had passed through the other without in any way influencing it.

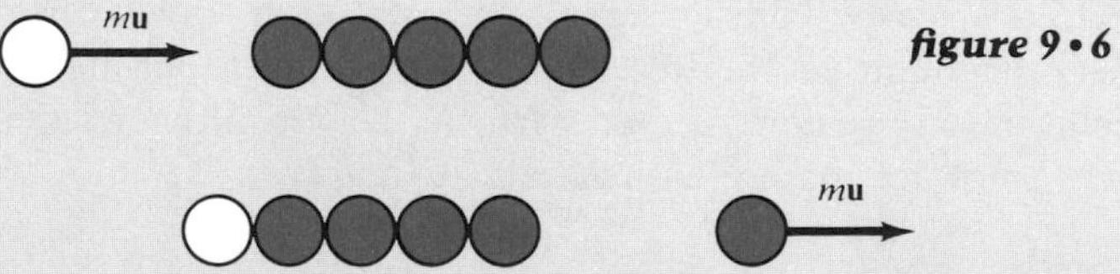

figure 9 • 6

solution Equations [9 . 10] and [9 . 13] reduce to the forms

$$u_1 + u_2 = v_1 + v_2 \qquad u_2 - u_1 = v_1 - v_2$$

and solving for the velocities after impact, we obtain $v_1 = u_2$ and $v_2 = u_1$. If one of the masses were initially stationary, for example, as a result of the collision it would acquire all the momentum and kinetic energy of the impinging mass. The classic example is what happens when a billiard ball is rolled into the end of a line of billiard balls; the only effect is to drive off the last ball at the opposite end of the line. If you experiment further, you will see that two balls rolled into the line will drive off the two end balls, three balls will drive off three balls, and so on.

example 9 • 4 If the machine gun in Example 9 . 1 has a mass of 15 kg, what force is needed to hold it in place as it fires?

solution The recoil momentum of the gun must have the same magnitude as the momentum of the bullets it discharges, regardless of the mass of the gun. Although the gun jerks back and forth in the gunner's hands, the average force he exerts in firing five 3-g bullets per second with a muzzle velocity of 500 m/s is

$$\overline{F}' = (0.003 \text{ kg})(500 \text{ m/s})(5/\text{s}) = 7.5 \text{ N}$$

example 9 • 5 Derive Eq. [9 . 10] by considering the impulses as they would be felt by the colliding bodies during the actual collision.

solution Assuming that all vectors lie on the same axis, from Eq. [9 . 3] we have

$$(m_1 v_1 - m_1 u_1)\mathbf{i} = \mathbf{P}_1 \qquad (m_2 v_2 - m_2 u_2)\mathbf{i} = \mathbf{P}_2$$

Since the forces felt by the two colliding bodies must be equal and opposite, so are their time integrals. Therefore

$$\mathbf{P}_2 = -\mathbf{P}_1 \qquad \text{or} \qquad m_1 v_1 - m_1 u_1 = m_2 u_2 - m_2 v_2$$

and rearranging these terms gives Eq. [9 . 10].

9 • 3 *inelastic collisions*

If the two colliding bodies are of soft clay or some similar pliable material, they are deformed by the impact, but show little or no tendency to recover their form and thrust each other apart again. Instead, they adhere together after impact and move forward as a single body. Such a collision is said to be *perfectly inelastic*. In this special case the velocity of each body after impact is the same as their common velocity V,

$$\blacktriangleright \quad v_1 = v_2 = V \qquad [9 \cdot 14]$$

and this equation replaces Eq. [9 . 13] as the second independent equation needed to determine the velocities. Substitution into Eq. [9 . 10] gives

$$m_1 u_1 + m_2 u_2 = (m_1 + m_2)V \qquad [9 \cdot 15]$$

from which the speed after impact may be calculated by solving for V.

example 9 • 6 A 50-g ball of putty is thrown at a speed of $u_1 = 40$ cm/s and makes direct and perfectly inelastic impact with a stationary 500-g billiard ball. Calculate their common speed V after impact and the kinetic energies before and after impact.

solution Since the billiard ball is initially at rest, $u_2 = 0$, Eq. [9 . 15] gives

$$m_1 u_1 = (m_1 + m_2)V$$

$$(50g)(40 \text{ cm/s}) = (550 \text{ g})V$$

so that $V = 3.64$ cm/s. The total kinetic energy before impact is 40,000 ergs, but after impact it is only

$$\tfrac{1}{2}(m_1 + m_2)V^2 = 3640 \text{ ergs}$$

or a net loss of 36,360 ergs. Even though momentum is conserved, 90 percent of the kinetic energy is converted to heat, which means that the temperature of the colliding bodies must rise.

The *ballistic pendulum,* invented by Benjamin Robins in 1742, is a classic example of perfectly inelastic impact (see Fig. 9 . 7). A bullet of mass m_1 and unknown speed u is fired horizontally into a wooden pendulum bob of much larger mass m_2, which is initially at rest. Since no appreciable external forces are brought into play during the very brief interaction, the total momentum is conserved. If the frictional forces in the pendulum supports are negligibly small, all the kinetic energy of the bob immediately after impact is converted to gravitational potential energy when the bob swings up to height y. Thus

$$\tfrac{1}{2}(m_1 + m_2)V^2 = (m_1 + m_2)gy$$

$$V = \sqrt{2gy}$$

To express y in terms of the horizontal position x of the bob and the length l of the pendulum, we use Pythagoras' theorem:

$$l^2 = x^2 + (l - y)^2 = x^2 + l^2 - 2yl + y^2$$

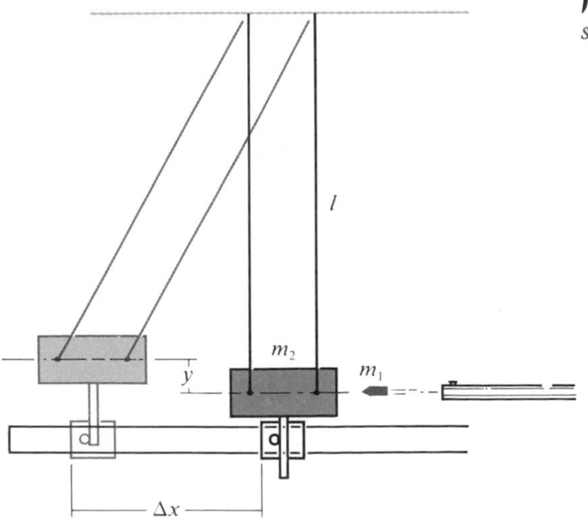

figure 9 • 7 *Ballistic pendulum for measuring the speed of a rifle bullet*

However, y is so small in comparison to x and l that we can ignore y^2 and reduce this expression to $2ly \doteq x^2$. Therefore

$$V \doteq \frac{\sqrt{2gx}}{\sqrt{2l}} = x\sqrt{\frac{g}{l}}$$

By conservation of momentum, $m_1 u = (m_1 + m_2)V$; therefore the speed u of the bullet is given by

$$u = \frac{(m_1 + m_2)}{m_1} V \doteq \frac{(m_1 + m_2)x}{m_1}\sqrt{\frac{g}{l}} \qquad [9 \cdot 16]$$

As Fig. 9 . 7 shows, these are all readily measured quantities.

In general, the change in kinetic energy in a perfectly inelastic collision is negative:

$$\Delta K = \tfrac{1}{2}(m_1 + m_2)V^2 - \tfrac{1}{2}m_1u_1^2 - \tfrac{1}{2}m_2u_2^2 \qquad [9 \cdot 17]$$

If we substitute for V from Eq. [9 . 15] and collect terms over the lowest common denominator, $2(m_1 + m_2)$, we obtain the energy lost in the collision:

$$\Delta K = -\frac{m_1m_2}{2(m_1 + m_2)}(u_1 - u_2)^2 \leq 0 \qquad [9 \cdot 18]$$

Most impacts are neither perfectly elastic nor perfectly inelastic, and in such cases another independent equation is needed in addition to Eq. [9 . 10] to solve for the two unknown velocities. In principle, an energy equation of the form

$$(\tfrac{1}{2}m_1v_1^2 + \tfrac{1}{2}m_2v_2^2) - (\tfrac{1}{2}m_1u_1^2 + \tfrac{1}{2}m_2u_2^2) = \Delta K \qquad [9 \cdot 19]$$

would be adequate if the change in kinetic energy were known. The quantity ΔK is usually negative; that is, kinetic energy is converted into such forms as heat and noise. However, we can also broaden our concept of collision to include the kind of "collision" that results from firing a gun or from the breaking up of a radioactive nucleus into two or more smaller particles. Under these circumstances some form of *internal* energy is converted into kinetic energy, and ΔK in Eq. [9 . 19] is a positive quantity.

example 9 • 7 What is the value of ΔK in the "collision" in which a gun of mass 3000 kg fires a projectile of mass 4 kg with a muzzle speed of 350 m/s relative to the earth?

solution From Newton's third law, the recoil speed of the gun is

$$\frac{(4\ \text{kg})(350\ \text{m/s})}{3000\ \text{kg}} = 0.47\ \text{m/s}$$

Since the initial speeds of gun and missile were zero, Eq. [9 . 19] reduces to

$$\Delta K = (\tfrac{1}{2}m_1v_1^2 + \tfrac{1}{2}m_2v_2^2) = 2.45 \times 10^5\ \text{J}$$

Usually the mechanisms that produce or consume kinetic energy are so complex that we cannot evaluate ΔK independently, but must calculate it from Eq. [9 . 19] after all the velocities are known. For this reason Eq. [9 . 19] is not usable for determining the velocities themselves. Newton succeeded in obtaining a second equation for inelastic impact that is usable, but this required the definition of a new constant for pairs of colliding bodies. He found that where the impact is not so violent as to produce permanent deformation of either body, the approach and separation speeds of the bodies are related by their *coefficient of restitution* e:

$$v_1 - v_2 = -e(u_1 - u_2) \qquad [9 \cdot 20]$$

The constant e is a measure of the resilience, or *elasticity,* of bodies and is used only in dealing with collision phenomena. Like the coefficient of friction, the value of e depends on the characteristics of both bodies involved in the interaction.

Equation [9 . 20] is known as *Newton's rule:*

> The relative separation velocity of two bodies after direct impact is equal in magnitude (but opposite in direction) to their relative approach velocity multiplied by their coefficient of restitution.

Note that this rule applies only to *direct* impact—that is, for the condition that $\mathbf{u}_1$, $\mathbf{u}_2$, $\mathbf{v}_1$, and $\mathbf{v}_2$ all have a common line of motion. In general, it must be modified for oblique impact.

In Eq. [9 . 20] we have a second relation between v_1 and v_2, and the solution of any problem of direct impact may be obtained by solving Eqs. [9 . 10] and [9 . 20] as simultaneous equations. This leads to the following scalar expressions for the final velocities:

$$v_1 = \frac{m_1u_1 + m_2u_2 - em_2(u_1 - u_2)}{m_1 + m_2}$$
$$v_2 = \frac{m_1u_1 + m_2u_2 + em_1(u_1 - u_2)}{m_1 + m_2} \qquad [9 \cdot 21]$$

example 9 • 8 A body moving with a velocity of

$$\mathbf{u} = (10\mathbf{i} - 10\mathbf{j} + 14.14\mathbf{k})\ \text{m/s}$$

collides with an identical stationary body. If the impact is direct and the coefficient of restitution for the two bodies is $e = 0.5$, find their velocities after collision.

solution The direction of motion is defined by the unit vector $\hat{\mathbf{u}}$ in the direction of the initial velocity of the moving body:

$$\hat{\mathbf{u}} = \mathbf{u}/u = 0.5\mathbf{i} - 0.5\mathbf{j} + 0.71\mathbf{k}$$

All the velocities in this one-dimensional problem lie in this direction. Since $u^2 = 10^2 + 10^2 + 14.14^2 = 400\ \text{m}^2/\text{s}^2$, the speed of approach is $u = 20$ m/s. Substituting $u_1 = u$ and $u_2 = 0$ in Eqs. [9 . 21] gives the speeds after impact as

$$v_1 = \tfrac{1}{4}u = 5\ \text{m/s} \qquad v_2 = \tfrac{3}{4}u = 15\ \text{m/s}$$

and in three-dimensional coordinates the two velocities after impact are

$$\mathbf{v}_1 = v_1\hat{\mathbf{u}} = (2.5\mathbf{i} - 2.5\mathbf{j} + 3.55\mathbf{k})\ \text{m/s}$$

$$\mathbf{v}_2 = v_2\hat{\mathbf{u}} = (7.5\mathbf{i} - 7.5\mathbf{j} + 10.65\mathbf{k})\ \text{m/s}$$

oblique impact

When the centers of two homogeneous spherical bodies are not moving along the same line before impact, the impact is described as *oblique*. In this case the approach velocity of each body may be resolved into two components, one along the line through the center of the two bodies at the moment of impact and the other in a direction perpendicular to that line. If the spheres are perfectly smooth, only the first of these components will be affected by the impact, since the reaction forces affecting the motion act along the line of centers. The change in this velocity component may be calculated just as for the case of direct impact. (If the spheres are *not* frictionless, there will be tangential forces between the bodies which give rise to rotations, and the problem becomes much more complex.)

Figure 9 . 8 shows the unit vectors $\hat{\mathbf{n}}$ and $\hat{\mathbf{t}}$ normal and tangential to the surfaces of two spheres colliding at an angle θ to their line of centers. The impulses $\mathbf{P}$ and $-\mathbf{P}$ act along the line of centers to change the normal components of the *momentum* of the colliding bodies. The larger sphere recoils along the line of centers with a speed v_2' which is less than the normal component v_n for the smaller sphere. The smaller sphere is therefore deflected from its original path through the *scattering angle* ϕ.

figure 9 • 8 *Elastic collision; oblique impact with a sphere initially at rest*

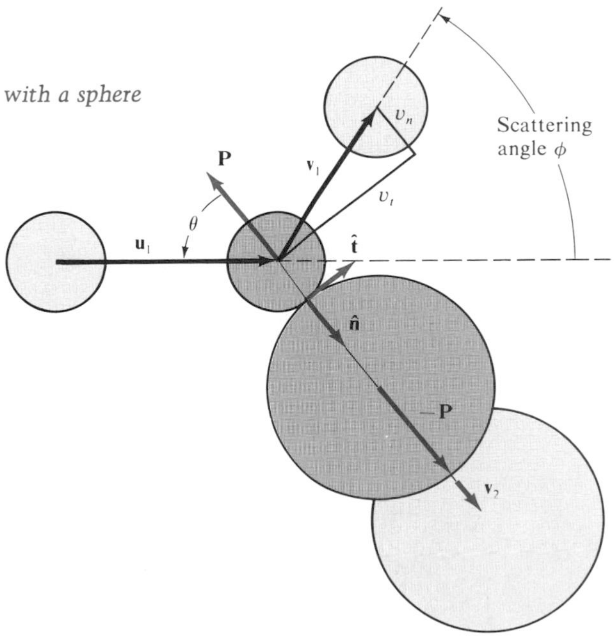

Since the angle of deflection is a function of the physical parameters u, θ, and e which describe the interaction between the two colliding bodies, it must be possible, in principle, to reason backward from a knowledge of ϕ to determine the nature of the collision interaction. Imagine that the collision takes place in such a small region that it cannot be observed directly, but we can observe the angle of deflection ϕ and the speed v of the particle deflected. If we were to observe the collision in Fig. 9 . 8 many times, with values of θ chosen at random (this is the actual case in the laboratory, where many particles may be undergoing binary interactions), the resulting collection of ϕ values would be found distinctively characteristic of a hard-sphere collision and no other. Similarly, by trying many hypothetical models of the interaction forces until we find one that fits the data, we can determine the nature of the interacting particles from the results of their collision. Such experiments, called *scattering experiments,* are very important in atomic and nuclear physics and enable physicists to determine the nature of the forces acting between elementary particles, and consequently the structure of the nuclei and atoms themselves.

example 9 • 9 A smooth ball with a mass of 100 g is rolling on a table with a velocity $\mathbf{u}_1 = 100$ cm/s east. It strikes a second smooth ball with a mass of 200 g which is moving with a velocity $\mathbf{u}_2 = 100$ cm/s northeast. If the line of centers is in the east-west direction at the moment of impact and the coefficient of restitution is $e = 0.75$, what are the velocities of the two balls after impact?

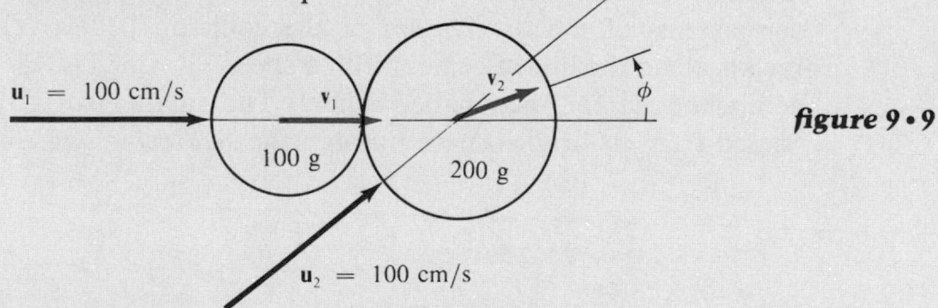

figure 9 • 9

solution Assume that the line of centers for the two balls is the x axis. Let u_{x1}, u_{y1} and u_{x2}, u_{y2} represent the x and y components of the velocities before impact and v_{x1}, v_{y1} and v_{x2}, v_{y2} the corresponding components after impact. Then

$$u_{x1} = 100 \text{ cm/s} \qquad u_{y1} = 0 \qquad u_{x2} = 70.7 \text{ cm/s} \qquad u_{y2} = 70.7 \text{ cm/s}$$

The x components of the velocities after impact can be computed from Eqs. [9 . 10] and [9 . 20],

$$m_1 u_{x1} + m_2 u_{x2} = m_1 v_{x1} + m_2 v_{x2}$$

$$v_{x1} - v_{x2} = -0.75(100 - 70.7) \text{ cm/s}$$

and simultaneous solution of these two equations yields

$$v_{x1} = 65.8 \text{ cm/s} \qquad v_{x2} = 87.8 \text{ cm/s}$$

If the balls are assumed to be perfectly smooth, the y components of the velocities will be unchanged by the impact, since they are tangential to the surfaces at the point of impact:

$$v_{y1} = 0 \qquad v_{y2} = 70.7 \text{ cm/s}$$

The velocity of the first ball after impact must therefore be $\mathbf{v}_1 =$ 65.8 cm/s east and that of the second ball is

$$\mathbf{v}_2 = (87.8\mathbf{i} + 70.7\mathbf{j}) \text{ cm/s} \qquad v_2 = 112.7 \text{ cm/s}$$

The second ball is moving as shown in Fig. 9 . 9, at an angle

$$\phi = \tan^{-1}\frac{70.7}{87.8} = 38°51' \text{ north of east}$$

While the application of Newton's rule seems very straightforward in the case of spherical bodies, it is not so obvious for the more general case of smooth but irregular bodies. However, Newton's rule can be applied to this case as well. Although the proof is complicated, the general principle can be stated as follows:

> When any two smooth bodies collide, their velocity components tangential to the surfaces at the point of contact are unchanged, but their relative velocities of approach and separation normal to the surface at the point of contact are related by Newton's rule.

9 • 5 kinetic-energy loss

Generally, a direct impact involves a loss of kinetic energy, $\Delta K < 0$. To determine ΔK in terms of initial velocities and the coefficient of restitution e, we can begin by writing

$$\Delta K = \tfrac{1}{2}m_1v_1^2 + \tfrac{1}{2}m_2v_2^2 - \tfrac{1}{2}m_1u_1^2 - \tfrac{1}{2}m_2u_2^2 \qquad [9 \cdot 22]$$

and obtain the binomial expression

$$\Delta K = \tfrac{1}{2}m_1(v_1 - u_1)(v_1 + u_1) + \tfrac{1}{2}m_2(v_2 - u_2)(v_2 + u_2) \qquad [9 \cdot 23]$$

In most problems the final velocities v_1 and v_2 are unknown, so we should like some way to express ΔK without using them. From Eqs. [9 . 21], we can write the impulse of the collision in scalar form as

$$P = m_1(v_1 - u_1) = -m_2(v_2 - u_2)$$
$$= \frac{m_1m_2(1 + e)(u_2 - u_1)}{m_1 + m_2} \qquad [9 \cdot 24]$$

so that Eq. [9 . 23] becomes

$$\Delta K = \frac{m_1m_2}{2(m_1 + m_2)}(1 + e)(u_2 - u_1)(v_1 - v_2 + u_1 - u_2) \qquad [9 \cdot 25]$$

Then, substituting $v_1 - v_2 = -e(u_1 - u_2)$, we can express ΔK in terms of the conditions before collision as

$$\Delta K = -\frac{m_1 m_2}{2(m_1 + m_2)}(1 - e^2)(u_2 - u_1)^2 \qquad [9 \cdot 26]$$

Compare this equation with Eq. [9 . 18]; here again we see that the sole condition for zero loss of kinetic energy in an impact is that $e = 1$, the case of perfectly elastic impact, and that for any $e < 1$ kinetic energy is decreased by the interaction.

These conclusions hold equally well for oblique impact, where we can also express the kinetic energy in terms of the normal and tangential components of velocity, u_n and u_t:

$$K = \tfrac{1}{2}m(u_n^2 + u_t^2) \qquad [9 \cdot 27]$$

The tangential components of velocity are unchanged, so that the energy loss is due to changes in the normal components u_n; hence in Eq. [9 . 26], we merely substitute u_{1n} and u_{2n} for u_1 and u_2.

example 9 • 10 Find the fractional energy loss $-\Delta K/K$ for the special case where m_2 is initially at rest before m_1 collides with it.

solution Dividing Eq. [9 . 26] by $K = \frac{1}{2}m_1 u_1^2$ gives

$$\frac{-\Delta K}{K} = (1 - e^2)\frac{m_2}{m_1 + m_2}$$

The *fractional* energy loss is independent of the speed of the striking body.

The result of Example 9 . 9 reflects the fact that the energy loss ΔK depends on the relative approach velocity of the colliding bodies. The fractional loss is large or small according as $m_1 \ll m_2$ or $m_1 \gg m_2$. In driving a nail much of this energy loss goes into deforming the nail; since this diminishes the supply of energy available for driving the nail, it is clearly advantageous to use a hammer of large mass. However, if the objective is to shape the object being struck, as in forging, the work of deformation should be made large by using repeated blows with a hammer of very small mass.

example 9 • 11 Suppose a small steel ball is dropped from a height h onto a smooth, horizontal massive steel plate. If the ball rebounds to a height h' above the plate, determine the coefficient of restitution e and the fractional energy loss in terms of h and h'.

solution If $m_2 \gg m_1$, we can neglect the motion of m_2, the steel plate. Then, by definition, e is the negative ratio of the speeds of the ball before and after impact:

$$e = -\frac{v}{u}$$

For the case of uniform acceleration, $u^2 = 2gh$ and $v^2 = 2gh'$, and therefore

$$e = -\sqrt{\frac{h'}{h}}$$

Since the fractional energy loss is independent of the speed of the ball, setting $m_1/m_2 = 0$ in Example 9 . 9, we have simply

$$\frac{-\Delta K}{K} = 1 - e^2 = 1 - \frac{h'}{h}$$

9 • 6 motion with changing mass

If a body at rest can spontaneously eject a fraction of its mass due to some internal source of energy, the remainder of the mass will recoil with a momentum opposite and equal to that of the ejected fraction. If this process can be sustained for a time, as in the case of a rocket, the remaining portion of the original system will appear to an observer at rest to be accelerating. This is expressed by the most general form of the second law

$$\frac{d\mathbf{p}}{dt} = m\frac{d\mathbf{v}}{dt} + \mathbf{v}\frac{dm}{dt} = \mathbf{F} \qquad [9 \cdot 28]$$

which enables us to determine the motion of a particle whose mass is changing as it moves.

It is reassuring to see a neat, precise, and familiar equation. However, when we actually apply it to a problem such as an accelerating rocket, troublesome questions arise concerning the meaning of m, $\mathbf{v}$, and $\mathbf{F}$. It is natural to think of m as the instantaneous mass of the rocket and its fuel, but where should the exhaust velocity of the burnt fuel appear? It is clearly not the same thing as $\mathbf{v}$, the velocity of the rocket itself. Also, if we ignore air resistance, there is no external applied force on the system; so should not $\mathbf{F} = \mathbf{0}$? But in that case, according to Eq. [9 . 28], a rocket starting from rest would never acquire any velocity! Equally puzzling is the question of simultaneously formulating the motion of the entire system—rocket plus fuel plus exhaust gases—assuming that the momentum of the entire system is conserved.

In fact, we can approach the problem either way—by considering either the entire system or the forces on its parts—provided we define the system carefully. Let us first consider the consequences of conservation of momentum for the entire system in one dimension, assuming that m and v refer to the rocket and the fuel it contains and that the exhaust gases have a constant velocity V relative to the rocket. (We would therefore expect $V < 0$, although in the case of retrorockets $V > 0$.) The actual exhaust velocity relative to a stationary observer is then $v + V$. Since an amount of gas dm' exhausts with a velocity $v + V$ in a time dt, the rocket's mass m decreases by an amount $dm = -dm'$ (see Fig. 9 . 10). Therefore the principle of conservation of momentum gives us

$$mv = (m + dm)(v + dv) + dm'(v + V) \qquad [9 \cdot 29]$$

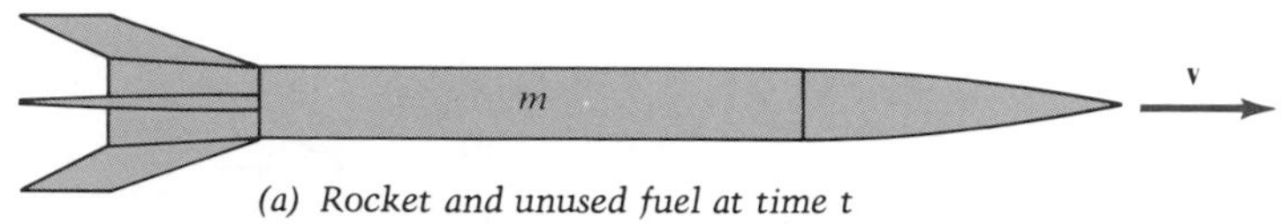

(a) Rocket and unused fuel at time t

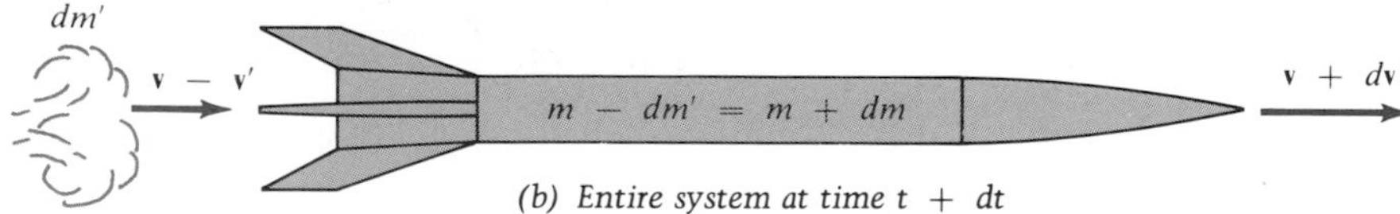

(b) Entire system at time t + dt

***figure* 9 • 10** *Conservation of momentum applied to the rocket problem*

or, neglecting the second-order terms $dm\,dv$ in the right-hand member,

$$\begin{aligned} 0 &\doteq m\,dv + v\,dm - v\,dm - V\,dm \\ &= m\,dv - V\,dm \end{aligned} \qquad [9.30]$$

Dividing by dt then gives

$$m\frac{dv}{dt} = V\frac{dm}{dt} \qquad [9.31]$$

an equation which seems to bear only passing resemblance to the expression in Eq. [9 . 28].

Alternatively, we can use the second approach and apply Eq. [9 . 28] to the forces on the system parts. The difficulty is in recognizing that at any given instant this equation refers only to the rocket plus the unused fuel. Thus, if we analyze the motion of the rocket alone, then in addition to any external forces on the rocket system (zero in this case), **F** must include a reaction force due to the exhaust gas. In scalar form, the force F' on the exhaust gas $dm' = -dm$ during the time dt is

$$\frac{dp'}{dt} = (v + V)\frac{-dm}{dt} = F' \qquad [9.32]$$

However, from the third law, $F' = -F$, and substituting for F' in Eq. [9 . 28] yields

$$m\frac{dv}{dt} + v\frac{dm}{dt} = (v + V)\frac{dm}{dt}$$

or

$$m\frac{dv}{dt} = V\frac{dm}{dt}$$

in agreement with Eq. [9 . 31], which we obtained for the system as a whole. The initial force, $F = V\,dm/dt > 0$ starting from rest at $v_0 = 0$, is known as the *thrust* of the rocket.

Given the exhaust velocity V and the initial mass m_0 and initial velocity v_0 of the rocket system, we can solve Eq. [9 . 31] by analytic methods or by the numerical methods of Sec. 7 . 6. Elimination of dt and substitution of the scaled variables $\kappa = m/m_0$ and $\nu = v/V$ gives us a form similar to Eq. [7 . 33]:

$$d\nu = \frac{d\kappa}{\kappa} < 0 \qquad [9.33]$$

table 9 • 1 *Rocket motion as computed from the trapezoidal rule*

κ	$\nu_0 - \nu$	κ	$\nu_0 - \nu$
1.0	0.0000	0.5	0.6907
0.9	0.1055	0.4	0.9157
0.8	0.2186	0.3	1.2074
0.7	0.3526	0.2	1.6240
0.6	0.5074	0.1	2.3740

Thus for $\nu_0 = v_0/V$ and $\kappa_0 = 1$, we have

$$\nu - \nu_0 = \int_1^\kappa \frac{d\kappa}{\kappa} < 0 \qquad [9 \cdot 34]$$

and we can tabulate the quantity $\nu - \nu_0$ against κ (which decreases from unity) by using the trapezoidal rule, as in Table 9 . 1. For any given time t we compute $v = V\nu$ from the corresponding value of $\kappa = m(t)/m_0$, where $m(t)$ is a known quantity.

The integral of Eq. [9 . 34], which we have already encountered in Sec. 7 . 6, is a very important one in the calculus, especially in the theory of differential equations. It defines a number which is the logarithm of κ to the base e, where $e = 2.71828 \ldots$:

$$\int_1^\kappa \frac{d\kappa}{\kappa} = \log_e \kappa \qquad \frac{d \log_e \kappa}{d\kappa} = \frac{1}{\kappa} \qquad [9 \cdot 35]$$

This number, usually written as $\ln \kappa$, is known as the *natural logarithm* of κ.* It has all the properties of a bona fide logarithm: for example, $\ln ab = \ln a + \ln b$ and $\ln 1 = 0$. Thus, in the language of the calculus, we may write the relationship expressed by Eq. [9 . 34] in the following form:

$$\nu - \nu_0 = \ln \kappa \qquad \kappa = \exp(\nu - \nu_0) \qquad [9 \cdot 36]$$

The derivative of κ also has an interesting property:

$$\frac{d\kappa}{d\nu} = \frac{1}{d\nu/d\kappa} = \kappa \quad \text{or} \quad \frac{de^\nu}{d\nu} = e^\nu \qquad [9 \cdot 37]$$

Therefore, if we assume that the mass of the rocket is a linear function of time, $m = m_0 - At$, where $A > 0$ is a constant rate of fuel consumption, we can solve Eq. [9 . 34] analytically:

$$v = v_0 + V \ln \frac{m}{m_0} = v_0 + V \ln \frac{m_0 - At}{m_0} \geq v_0 \qquad [9 \cdot 38]$$

(recall that $V < 0$).

*Because of the unusual property of its derivative, this function is tabulated in mathematics and physics handbooks, and a function key for computing it is included on scientific hand calculators.

example 9 • 12 The Saturn V launch vehicle used for the United States Apollo program, which landed a man on the moon on July 20, 1969, was the largest space vehicle yet developed. Its first-stage rockets had a thrust of 34 MN (meganewtons) and burned fuel at the rate of 13.8 metric tons per second for a total of 150 s before burnout and separation of the first stage. The initial mass of the vehicle was 2850 metric tons. Neglecting air resistance and gravity, predict the velocity of the vehicle at first-stage burnout.

figure 9 • 11

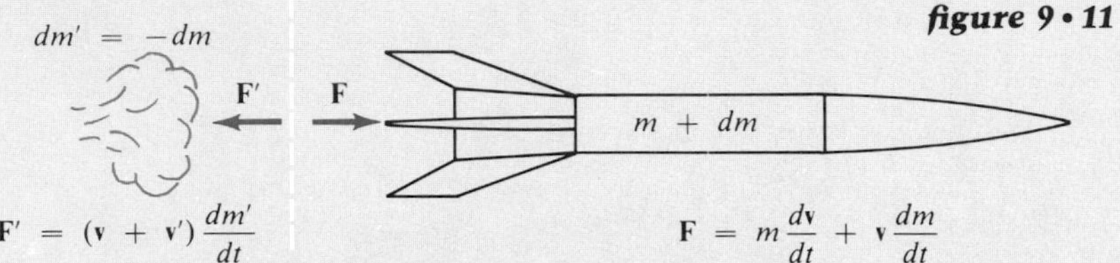

solution Since the rate of fuel consumption is $A = 13{,}840$ kg/s, the exhaust velocity at first-stage burnout is

$$AV = -34{,}000{,}000\text{ N} \qquad \text{or} \qquad V = -2464\text{ m/s}$$

and the mass of the vehicle at this time is

$$m = 2850 - (150 \times 13.84) = 774\text{ metric tons}$$

Therefore the velocity of the vehicle at burnout is

$$v = V \ln \frac{m}{m_0} = -2464 \ln \frac{774}{2850} = 3212\text{ m/s}$$

The actual velocity of the Saturn V at separation was only 2400 m/s, which is accounted for by gravity and air resistance.

If you have ever seen a rocket launching, you know how slowly the rocket seems to lift off its pad, apparently accelerating more and more rapidly as it goes. This behavior is explained by observing that

$$a = \frac{dv}{dt} = \frac{V}{m}\frac{dm}{dt}$$

so that, other things being equal, the acceleration of the rocket increases as its mass m decreases. Thus at launching almost all the energy is imparted to the kinetic energy of the exhaust gases. As the rocket accelerates, the exhaust gases leave it more and more slowly *relative to the earth,* hence more and more energy, as seen by a terrestrial observer, goes into the motion of the rocket, until its fuel is exhausted and $dm/dt = 0$.

example 9 • 13 A jet plane of mass m scoops up still air at a constant rate of A kg/s and ejects it at the same rate with an exhaust speed V. What is the equation of motion for the airplane and its maximum attainable speed?

solution Consider the airplane, incoming air, and exhaust as a single system, with momentum conserved over time dt. Just prior to dt, the incoming air is at rest, while the air about to be ejected is moving with the plane at velocity v. At the end of dt the plane has reached speed $v + dv$, the exhaust is moving with speed $V + v$, and the fresh air taken in is moving with speed v. Equating the momenta and substituting $dm = A\,dt$ for air passing through the plane, we have

$$mv + Av\,dt = m(v + dv) + Av\,dt + (V + v)A\,dt$$

$$m\frac{dv}{dt} = -A(V + v)$$

and when $v = v_{max}$, then $dv/dt = 0$. Therefore

$$v_{max} = -V > 0$$

and the maximum speed is attained when the exhaust gases are actually stationary relative to an observer on the ground!

PROBLEMS

9 • 1 *impulse*

9 • 1 A 300-g baseball approaches a bat with a velocity of 50 m/s and leaves with an oppositely directed velocity of 100 m/s. What was the average force of the hit if the impact lasted 0.02 s?

9 • 2 A force $\mathbf{F} = t\mathbf{i} + t^2\mathbf{j} + t^3\mathbf{k}$ acts on a body for $0 \le t \le 5$ s. Find the impulse it gives the body.

9 • 3 A 15,000-ton battle cruiser at rest fires a salvo of six 8-in shells at a 45° angle with a muzzle velocity of 1000 ft/s. If each shell weighs 100 lbf, how far has the cruiser recoiled by the time all six shells strike the target? Neglect all frictional or inertial effects of the water.

9 • 4 A 1-kg pail will hold 10 kg of water and is filled from a faucet in 12 s. At the instant the pail is half full, the scales read 6.5 kg. What is the velocity of the flowing water at that instant? (Assume there is no splashing.)

9 • 5 The ball in a pinball machine is shot off, hits the lights and bumpers, drops to rest, and is pushed back into the firing position. What total impulse has it received?

9•6 A 6-kg sphere slides along a smooth track which consists of two straight sections joined at an angle of 30° by a circular arc of radius 3 m. Describe and evaluate the impulse received by the sphere if its speed is 9 m/s. What is the average impulsive force $\overline{\mathbf{F}}$ acting on the sphere? (Ignore gravity.)

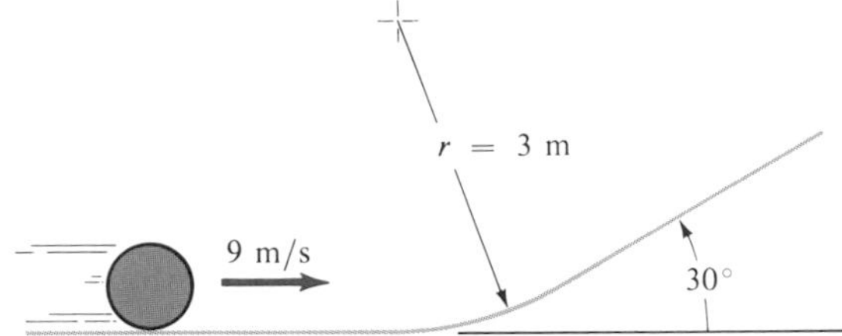

9•7 A defensive linebacker tackles a 200-lb fullback who runs into him head-on at a speed of 10 yards/s. (*a*) If he stops the fullback cold in 0.5 s, what impulse does he impart to the fullback? (*b*) What average force does the linebacker *feel*?

9•8 In Example 9 . 1 we found that if 3-g bullets fired with a speed of 500 m/s flattened against Superman's chest, the average force of their impact was 7.5 N. However, this is an average over time at the rate of five impacts per second. Suppose each bullet is 2 cm long and flattens to a negligibly thin disk on impact. (*a*) How long does each impact take? (*b*) What is the impulse of each impact? (*c*) What is the average force during each impact?

9•2 *elastic collision*

9•9 A particle moving at a speed u collides elastically with another particle which is stationary. The impact propels the second particle in the same direction with a speed of $\frac{1}{3}u$. What is the ratio of the masses of the two particles?

9•10 A mass m_1 traveling at a velocity u_1 collides with a mass $m_2 = 2m_1$ traveling in the same direction at a velocity of $u_2 = \frac{1}{2}u_1$. What are the velocities of the two masses after collision?

9•11 A comet of mass m enters the solar system from the outer galaxy with a speed u relative to the sun and is deflected through an angle of 56° by the gravitational attraction of the sun. What is the change in momentum of the solar system relative to the fixed stars due to this "collision"?

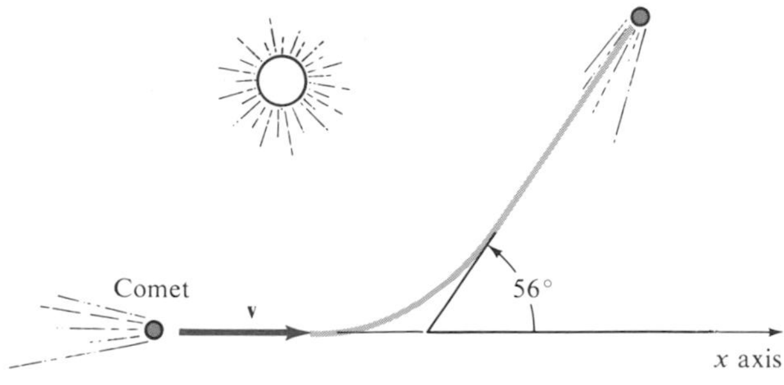

9•12 Two perfectly elastic balls having masses in the ratio of 3 to 1 are suspended by threads so as to swing with pendular motions in a plane and make direct impacts. (*a*) Prove that if the initial speeds of the balls are equal, on the first impact the more massive ball will come to rest, whereas the less massive one will rebound with twice its initial speed; then on the second impact the two balls will rebound with equal speeds, establishing a cycle of impacts that consists of two different types of motion. (*b*) Show that a different cycle of impacts occurs if the less massive ball is initially at rest and the more massive one swings down upon it.

9 • 13 Particles from the sun are specularly reflected from a comet-shaped magnetic field, the *magnetosphere,* which encloses and protects the earth. The protective field has a circular cross section of radius $10R_e$ (R_e = 6400 km) and is struck by about 10^{12} particles per square meter of area every second. Each particle has mass $m = 1.66 \times 10^{-27}$ kg and strikes at a speed of u = 300 km/s radially outward from the sun. (*a*) If the earth's mass is $M_e = 6 \times 10^{24}$ kg, what is its approximate momentum, in kilogram-kilometers per second? Assume a circular orbit of 150,000,000 km radius about the sun. (*b*) It can be shown that the average momentum change is as if each particle were simply absorbed by the earth. What total outward impulse does the earth receive from the solar particles yearly? (*c*) What conclusions do you draw from the answers to (*a*) and (*b*)?

9 • 14 A particle of mass m_1 makes a completely elastic head-on impact with a particle of mass m_2 that is initially at rest. (*a*) Prove that the fractional loss in kinetic energy for particle m_1 is

$$\frac{u_1^2 - v_1^2}{u_1^2} = 4\frac{m_1}{m_2(1 + m_1/m_2)^2}$$

(*b*) What is the fractional loss in kinetic energy of a neutron when it makes completely elastic impact with an initially stationary hydrogen nucleus for which $m_1/m_2 = 1.0$? (*c*) With a boron nucleus, for which $m_1/m_2 = 0.092$? (*d*) With a lead nucleus, for which $m_1/m_2 = 0.0049$?

9 • 15 Fred, a knight by trade, is out questing for dragons. Complete with arms, armor, horse, and lunchbox, Fred weighs 2000 lb. Espying his quarry, he charges it furiously at a speed of 5 mi/h. If the dragon is observed to recoil from the encounter at 1.0 mi/h, what is its mass?

9 • 16 A 75-kg fisherman dives off the stern of his 50-kg rowboat. The horizontal component of his motion is 1 m/s relative to the boat. Find the horizontal velocities of fisherman and boat relative to an observer on the dock (*a*) if the boat is initially at rest and (*b*) if the boat is initially moving forward at a speed of 2 m/s. Ignore energy losses to the water.

9 • 3 *inelastic collisions*

9 • 17 A 4-kg projectile is shot with a muzzle velocity of 350 m/s from a 3000-kg gun. What is the initial recoil speed of the gun?

9 • 18 Two identical cars of cops and robbers are conducting a running chase. Both cars are traveling at 30 m/s, and the occupants of each car fire five shots a second, all of which are embedded in the opposing vehicles. In order not to fall behind, how much more power must the police car expend? The muzzle velocity of each bullet is 1 m/s and its mass is 7 g.

9•19 A 50-g sphere traveling at a speed of 600 cm/s makes direct impact with a 100-g sphere traveling in the same direction at a speed of 350 cm/s. (*a*) What are the respective speeds of the spheres after impact if $e = 1$? (*b*) If $e = 0.9$?

9•20 A 20-g rifle bullet is fired into a 4-kg block of wood which is suspended by a cord 1 m long. If the block is moved through an angle of 20° by the impact of the bullet, what was the speed of the bullet?

9•21 A 500-g can rests on top of a pole 30 m tall. A boy standing 20 m from the base of the pole shoots at the can with a 10-g bullet, which strikes it at a speed of 450 m/s. Assuming that the bullet lodges in the can, (*a*) how far does the can rise above the pole? (*b*) How far from the base of the pole does the can strike the ground?

9•22 A train robber tosses a 50-lbm mail sack to a confederate waiting beside the track while the train is moving at 20 ft/s. The robber is somewhat startled to see the confederate receding at a speed of 15 ft/s after the catch. What is the confederate's mass? (Remember that the train is a moving coordinate system.)

9•23 A ball is dropped to the floor from a height h. If the coefficient of restitution is e, write expressions for the total time it will take the ball to stop bouncing and the total scalar distance it will travel in this time. (*Hint:* Apply the formula for a geometric series.)

9•24 One of the heaviest rainfalls on record anywhere in the United States was that reported for a storm occurring in the mountains of California, when 1.02 in of rain fell in 1 min. Suppose the speed of the raindrops was 300 ft/s, that they struck a 60 × 150-ft horizontal roof at an angle of 75° with the roof, and that the coefficient of restitution was $e = 0.2$. Find the total force exerted on the roof, assuming that each drop struck only once during the minute of measured rainfall.

9•25 In the impoverished mountain kingdom of Pontevedro, the Royal Artillery has developed a technique to conserve gunpowder: its cannon is fired horizontally from a mountaintop, with the muzzle velocity controlled so that the cannonball will strike the target on the plain below on the second bounce. The two Royal Engineers agree that the value of e in this instance is 0.6. (*a*) If the mountain is 1 km high, what muzzle speed is required to hit a target at a range of 9 km? (*b*) If the initial energy of the cannonball is proportional to the mass of gunpowder used, what is the percentage of gunpowder saved by the bounce method?

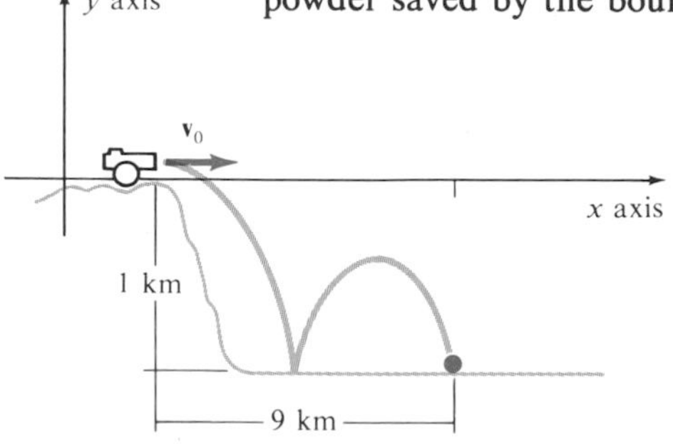

9 • 4 *oblique impact*

9 • 26 A fire engine throws water at the rate of 16 liters/s (1 liter = 1 kg) from a hose furnished with a nozzle 3 cm in inside diameter. (*a*) If inelastic impact is assumed, what force is experienced by a wall against which the jet is directed horizontally and at short range? (*b*) If each particle of water rebounded elastically, what would the force be? (*c*) If the coefficient of restitution is $e = 0.5$, what normal force would the wall experience if the jet were directed at short range and at an angle of 30° with the wall?

9 • 27 A smooth ball moving at a speed of 6 m/s collides with a stationary smooth ball of twice its mass. If its approach velocity makes an angle of 30° with the line of centers at the moment of impact, and if $e = 0.5$, what is the velocity of the smaller ball after impact?

9 • 28 Two cars of mass 2000 and 1000 kg, respectively, collide inelastically. Their velocity after impact is 4 m/s, at an angle of 30° with the initial direction of the 2000-kg car. The initial speed of the heavier car was 10 m/s. Find the initial velocity of the lighter car.

9 • 29 A ball strikes a smooth plane at an angle ϕ_1 with the normal to the plane and rebounds at an angle ϕ_2. Write the expression for the coefficient of restitution in terms of these angles.

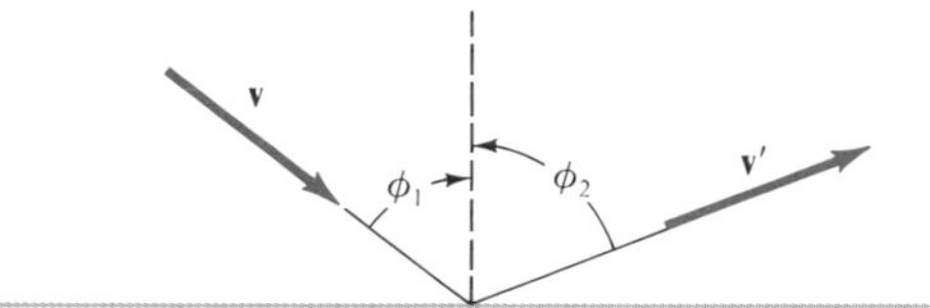

9 • 5 *kinetic-energy loss*

9 • 30 Suppose that one particle is initially at rest and that a second particle makes a perfectly inelastic impact with it. (*a*) What is the relation between their masses when the fractional part of the kinetic energy transformed into heat is one-half? (*b*) When it is one-fourth? (*c*) When it is three-fourths?

9 • 31 If a 50-g bullet is fired into a 125-g block, what is the fractional loss in kinetic energy?

9 • 32 A 200-g billiard ball slides on a smooth floor with a speed of 1 m/s and strikes a smooth wall at a 45° angle. (*a*) If $e = 0.5$, find the direction of motion after impact. (*b*) Find the loss in kinetic energy.

9 • 33 A 5-kg object moving at a speed of 120 cm/s collides directly with a 20-kg object which is at rest. The smaller object is observed to rebound at a speed of 60 cm/s. (*a*) What is the kinetic-energy loss from the impact? (*b*) What is the coefficient of restitution?

9 • 6 *motion with changing mass*

9 • 34 A rocket losing mass at a constant rate k is subject to a constant external force of magnitude F in addition to the reaction force of the exhaust gases. The initial mass of the rocket and fuel is m_0 and the configuration of the exhaust is changed in such a way that $V = -v$ at every instant

($V \neq$ constant). (*a*) Write the equation of motion, assuming that the rocket travels on a smooth, horizontal track. (*b*) Integrate the equation of motion by quadrature to obtain $v(t)$.

9 • 35 A rocket of mass m_r is loaded with fuel of mass m_f and has a constant exhaust speed v' relative to the rocket. (*a*) Develop an expression for the final speed of the rocket at burnout if it starts from rest. (*b*) If $v' = 10{,}000$ km/h of kerosene-lox (liquid oxygen) propellant and the initial mass is 90 percent fuel, find the burnout speed.

***9 • 36** A 40-metric-ton missile is subjected to a thrust of $F = 850{,}000$ N as it burns 17 metric tons of fuel uniformly in 68 s. (*a*) Assuming a flat earth and constant g, write the two-dimensional equations of motion in terms of the velocity components of the missile. (*b*) If the missile takes off vertically, what is its velocity at burnout? (*c*) What is its altitude?

9 • 37 A rocket is fired from a tower at a height h above the ground. Its mass at any time t after firing is

$$m(t) = m_0 \exp\left(-\frac{at}{V}\right)$$

where V is the constant speed of the exhaust gases relative to the rocket and a is a constant with the dimensions of acceleration. Assume that the rocket's direction is at a constant angle to the horizontal as shown. (*a*) Show that the rocket travels in a straight line under the influence of its thrust and gravity and find the equation of its trajectory. (*b*) Is this result consistent with the prediction that the rocket may hit the ground (assume a flat earth and constant g)? If so, at what distance from the launch tower will it hit?

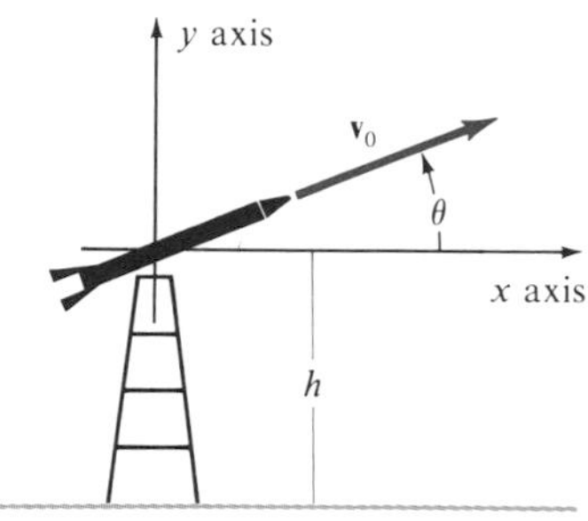

9 • 38 The actual velocity of the Saturn V space vehicle after burnout and separation of the first-stage rockets was stated in Example 9 . 11 to be 2400 m/s. The configuration of the rest of the vehicle is as follows:

Stage 2	43 metric tons plus 427 tons of fuel, thrust 4.5 MN
Stage 3	15 metric tons plus 105 tons of fuel, thrust 0.9 MN
Payload	47 metric tons

(*a*) If the second stage burns for 400 s, exhausting all its fuel, what is the speed of the vehicle at burnout? (*b*) How long would the third stage have to burn for the payload to attain an escape velocity of 11,200 m/s? Neglect the effects of gravity and air resistance.

answers

9•1 $\overline{F} = 22.5$ N
9•2 $\mathbf{P} = 12.50\mathbf{i} + 41.67\mathbf{j} + 156.25\mathbf{k}$
9•3 $\Delta x = 7.5$ in
9•4 $\mathbf{v} = -5.88\mathbf{j}$ m/s
9•5 $\mathbf{P} = \mathbf{0}$
9•6 $\overline{\mathbf{F}} = (-41.45\mathbf{i} + 154.70\mathbf{j})$ N
9•7 (*a*) $\overline{\mathbf{P}} = 186.5$ lbf · s;
(*b*) $\overline{F} = 375$ lbf
9•8 (*a*) $\tau = 40\ \mu$s; (*b*) $P = 1.5$ N · s;
(*c*) $\overline{F} = 37{,}500$ N
9•9 $m_1/m_2 = \frac{1}{5}$
9•10 $v_1 = \frac{1}{3}u_1, v_2 = \frac{5}{6}u_1$
9•11 $\Delta\mathbf{p} = mv(0.441\mathbf{i} - 0.829\mathbf{j})$
9•13 (*a*) $p_e = 18 \times 10^{25}$ kg · km/s;
(*b*) $P = 2 \times 10^{12}$ kg · km/s
9•14 (*b*) $\Delta K/K = 1.0$; (*c*) $\Delta K/K = 0.31$;
(*d*) $\Delta K/K = 0.019$
9•15 $m = 16{,}000$ lbm
9•16 (*a*) $v_1 = 0.4$ m/s, $v_2 = -0.6$ m/s;
(*b*) $v_1 = 2.4$ m/s, $v_2 = 1.4$ m/s
9•17 $v = 47$ cm/s
9•18 $\Delta P = 630$ W
9•19 (*a*) $v_1 = 267$ cm/s, $v_2 = 517$ cm/s;
(*b*) $v_1 = 283$ cm/s, $v_2 = 508$ cm/s
9•20 $u_1 = 219$ m/s
9•21 (*a*) $\Delta y = 2.75$ m; (*b*) $\Delta x = 16.3$ m
9•22 $m = 150$ lbm
9•23 $\Delta t = \sqrt{\frac{2y}{g}}\frac{1+e}{1-e}, \quad s = y\frac{1+e^2}{1-e^2}$
9•24 $F = 4.45$ tons of force
9•25 (*a*) $v_0 = 286$ m/s; (*b*) 79 percent
9•26 (*a*) $F_x = 37.0$ kgf; (*b*) $F_x = 74.0$ kgf;
(*c*) $F_x = 27.7$ kgf
9•27 $\mathbf{v} = 3$ m/s normal to line of centers
9•28 $\mathbf{v}_2 = 11.3$ m/s at 148°
9•29 $e = \tan\phi_1/\tan\phi_2$
9•30 (*a*) $m_1 = m_2$; (*b*) $m_1 = 3m_2$;
(*c*) $m_2 = 3m_1$
9•31 $-\Delta K/K = 0.714$
9•32 (*a*) 26°34′; (*b*) $\Delta K = 3.75 \times 10^{-2}$ J
9•33 (*a*) $\Delta K = -0.675$ J; (*b*) $e = 0.875$
9•34 (*a*) $F = (m_0 - kt)(dv/dt) - kv$;
(*b*) $v(t) = Ft/m_0$
9•35 (*a*) $v = V\ln(1 + m_f/m_r)$, from
Eq. [9 . 38]; (*b*) $v = 14{,}500$ m/h
9•36 (*a*) $a_x = (F/m)(v_x/v)$,
$a_y = (F/m)(v_y/v) - g$,
$m = 40{,}000\ \text{kg} - 250t\ \text{kg/s}, v^2 = v_x^2 + v_y^2$;
(*b*) $v = 1215$ m/s; (*c*) $y = 35{,}443$ m
9•37 (*a*) $y = y_0 + Ax$,
$A = \tan\theta - (g/a)\sec\theta$;
(*b*) $\Delta x = -y_0/A$ for $A < 0$
9•38 (*a*) $v = 7078$ m/s; (*b*) $\Delta t = 379$ s
(actual values were 6593 m/s and 460 s)

CHAPTER TEN

rigid bodies and center of mass

Theorem I *The state of motion or of rest of the center of gravity of several bodies is not altered in any way by the mutual interactions of these bodies, provided that the system is completely free; that is, if it be not constrained to move about some fixed point. . . .*

Theorem III *If arbitrarily many bodies are tied together in any way, and one or several of these bodies forced to move in the same or parallel planes, then I assert that the center of mass moves parallel to these planes in the same way.*

Jean Le Ronde d'Alembert, Traité de Dynamique, *1743*

So far we have treated all bodies as if they were particles, with their entire mass concentrated at a mathematical point, the *geometric center,* and with all the applied forces acting on them through that point. In this chapter we shall consider the more realistic situation of *extended bodies,* those whose physical dimensions and mass distribution cannot be neglected in an analysis of their motion. Real bodies are always slightly deformable under the impact of a force, but to simplify our discussion, we shall assume that they are perfectly *rigid*—that is, that the distance between any pair of points in the body remains the same at all times. As we shall see, the motion of any body or system of bodies can be described in terms of a fictitious point particle, the mass of which is equal to the total mass of the body or system and the location of which is called the *center of mass.* The center of mass does not necessarily correspond to the geometric center, but depends on the extent and mass distribution of the particular system.

As with particles, it is the center of mass that is acted upon by the net external force on the system. The total effect of an applied force depends, of course, on its point of application. In general, a force will not only cause the center of mass to accelerate, but also cause the body or system to rotate about an axis through its center of mass. We shall put off discussion of such rotations until the next chapter and concentrate in this chapter on the concept of the center of mass.

10 • 1 *center of mass*

When you balance an object on end with one finger, it is as if all the mass of the object were concentrated at some point along the vertical line above your finger. In other words, the center of mass is located somewhere on that vertical line. If the object starts to tip, you can move your finger to keep it directly beneath this point so that the object remains balanced. An experimental way to locate the center of mass would be to turn the object lengthwise and balance it again. The vertical line along which the center of mass lies must intersect the previous vertical line at the exact location of the center of mass (see Fig. 10 . 1). It is important to recognize that the center of mass is a fictitious particle; that is, it is a mathematical abstraction, rather than a physical entity. The center of mass of a billiard ball is no more a real mass element than the center of mass of a rubber tire, which lies in the empty space at the geometric center of the tire.

In previous chapters we have ignored the spatial extent of all bodies and have treated them simply as particles. Any physical body, however, may be thought of as a system of particles whose individual masses sum to equal the total mass of the body. Our intuition correctly tells us that the location of the center of mass depends on the mass *distribution* of the material that makes up the body. Consider the two uniform slabs in Fig. 10 . 2, one of lead and the other of wood. Since the slabs are uniform, the center of mass of each slab will be at its geometric center. However, if we now make up one body

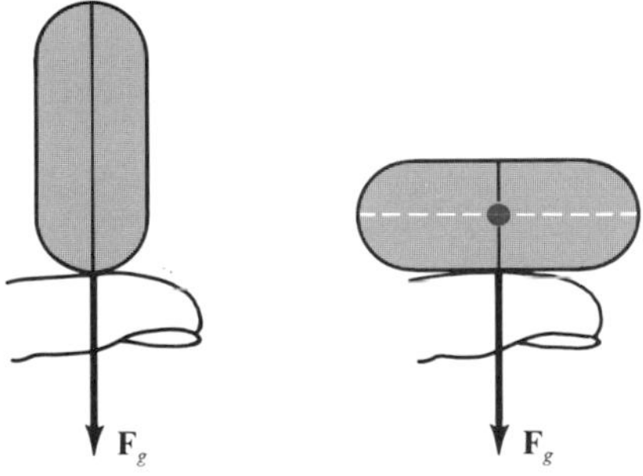

figure 10 • 1 *Center of mass*

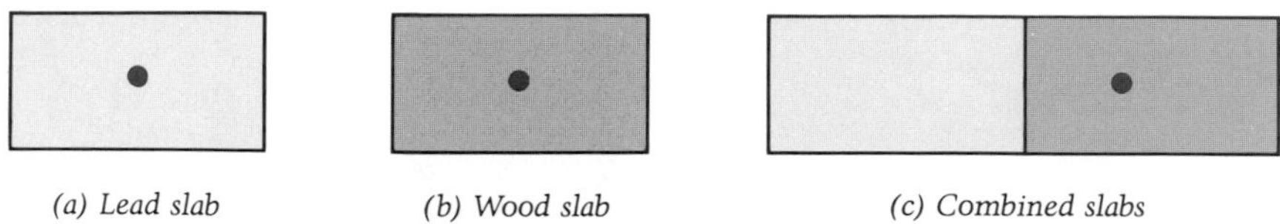

figure 10 • 2 *Center of mass of a nonhomogeneous body*

from these two pieces, the center of mass of the combination will not be at the wood-lead interface; it will be located at some point very close to the center of mass of the lead slab. With this example in mind, we can define the location of the center of mass as a kind of mass-weighted average of the positions of the individual particles that make up the system. If each individual particle has a mass m_i and a position vector $\mathbf{r}_i$ with respect to some coordinate system, then the position vector $\mathbf{r}_c$ of the center of mass is given by

$$\mathbf{r}_c = \frac{\sum_{i=1}^{n} m_i \mathbf{r}_i}{M} \qquad [10 \cdot 1]$$

where M is the total mass of the system,

$$M = \sum_{i=1}^{n} m_i$$

Here we are assuming that the system is made up of a finite number, $i = 1, 2, \ldots, n$, of discrete particles, as shown in Fig. 10 . 3.

Differentiating Eq. [10 . 1] with respect to time yields an expression for the velocity $\mathbf{v}_c$ of the center of mass:

$$\mathbf{v}_c = \frac{d\mathbf{r}_c}{dt} = \frac{\sum_{i=1}^{n} m_i (d\mathbf{r}_i/dt)}{M} = \frac{\sum_{i=1}^{n} m_i \mathbf{v}_i}{M} \qquad [10 \cdot 2]$$

The center of mass must then have a momentum given by

$$M\mathbf{v}_c = \sum_{i=1}^{n} m_i \mathbf{v}_i \qquad [10 \cdot 3]$$

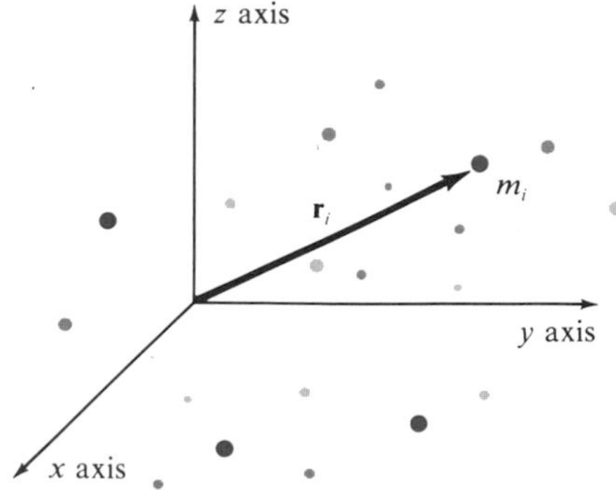

figure 10 • 3 *A particle of mass m_i in a system of total mass $M = \sum_{i=1}^{n} m_i$*

As expected, this equation simply states that the momentum of the center of mass is equal to the total momentum of all the particles in the system.

Now let us consider the total force acting on the ith particle. This particle is subject both to external forces and to internal forces from all the other $n - 1$ particles that make up the system. If we denote the internal force of the jth particle on the ith particle as $\mathbf{F}_{ij,\,\text{int}}$, then from Newton's law of action and reaction,

$$\mathbf{F}_{ij,\,\text{int}} = -\mathbf{F}_{ji,\,\text{int}}$$

and we can express the total force on the ith particle as

$$\mathbf{F}_{i,\,\text{tot}} = m_i\mathbf{a}_i = \mathbf{F}_{i,\,\text{ext}} + \sum_{j=1}^{n}\mathbf{F}_{ij,\,\text{int}} \qquad i \neq j \qquad [10 \cdot 4]$$

In order to find the total force on the system, we must sum up all the forces acting on its particles:

$$\begin{aligned}\mathbf{F}_{\text{tot}} &= \sum_{i=1}^{n}\mathbf{F}_{i,\,\text{tot}} \\ &= \sum_{i=1}^{n}\mathbf{F}_{i,\,\text{ext}} + \sum_{i=1}^{n}\sum_{j=1}^{n}\mathbf{F}_{ij,\,\text{int}} \qquad j \neq i \qquad [10 \cdot 5]\end{aligned}$$

The second summation in this equation is zero, since the internal forces cancel in pairs, and so we finally have

$$\mathbf{F}_{\text{tot}} = \sum_{i=1}^{n}\mathbf{F}_{i,\,\text{ext}} \qquad [10 \cdot 6]$$

We also know from Eq. [10 . 3] that one more differentiation yields

$$M\frac{d\mathbf{v}_c}{dt} = \sum_{i=1}^{n}m_i\frac{d\mathbf{v}_i}{dt} = \sum_{i=1}^{n}m_i\mathbf{a}_i = M\mathbf{a}_c \qquad [10 \cdot 7]$$

Putting these last two equations together implies that

$$\mathbf{F}_{\text{tot}} = M\mathbf{a}_c \qquad [10 \cdot 8]$$

In other words, the center of mass moves precisely as if all the mass of the system and all the external forces on it were concentrated at that point, regardless of the system or the actual point at which the external forces are applied. This is a surprisingly simple and useful result, as we shall see.

The coordinates of the center of mass are given by the three scalar equations equivalent to Eq. [10 . 1]:

$$x_c = \frac{\sum_{i=1}^{n}m_ix_i}{M} \qquad y_c = \frac{\sum_{i=1}^{n}m_iy_i}{M} \qquad z_c = \frac{\sum_{i=1}^{n}m_iz_i}{M} \qquad [10 \cdot 9]$$

Note that since we have derived the center of mass in terms of a system of particles, the definition in Eqs. [10 . 1] and [10 . 9] is valid for any system, including nonrigid (deformable) bodies.

example 10 • 1 Particles of mass m, $2m$, and $3m$ are located at the vertices of an equilateral triangle with sides of length l. Find the center of mass.

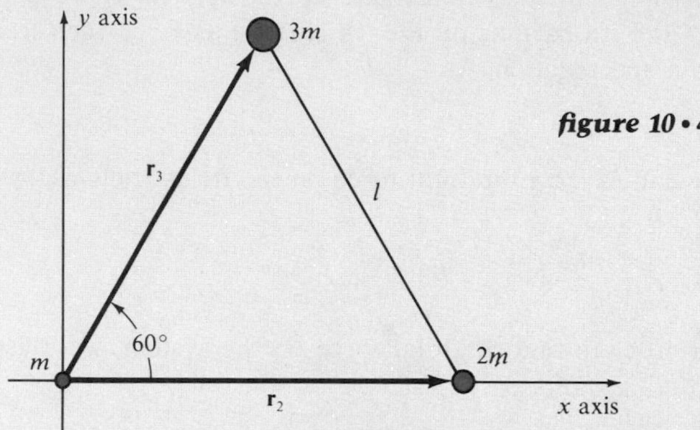

figure 10 • 4

solution If we choose the origin of coordinates at one vertex, the position vectors of the three particles are

$$\mathbf{r}_1 = \mathbf{0} \qquad \mathbf{r}_2 = l\mathbf{i} \qquad \mathbf{r}_3 = \tfrac{1}{2}l\mathbf{i} + \tfrac{1}{2}\sqrt{3}l\mathbf{j}$$

The position vector of the center of mass is therefore

$$\mathbf{r}_c = \frac{(2ml + \frac{3}{2}ml)\mathbf{i} + (\frac{3}{2}m\sqrt{3}l)\mathbf{j}}{m + 2m + 3m}$$
$$= 0.583l\mathbf{i} + 0.433l\mathbf{j}$$

example 10 • 2 A solid ball of mass $m_b = 6$ kg is dropped to the earth from a height $h = 100$ m. (*a*) Given that the mass of the earth is $M_e = 6 \times 10^{24}$ kg and its radius is $R_e = 6400$ km, how far is the center of mass from the earth's center? (*b*) If we ignore any external forces on the earth-ball system, how far does the earth move as the ball falls to the ground?

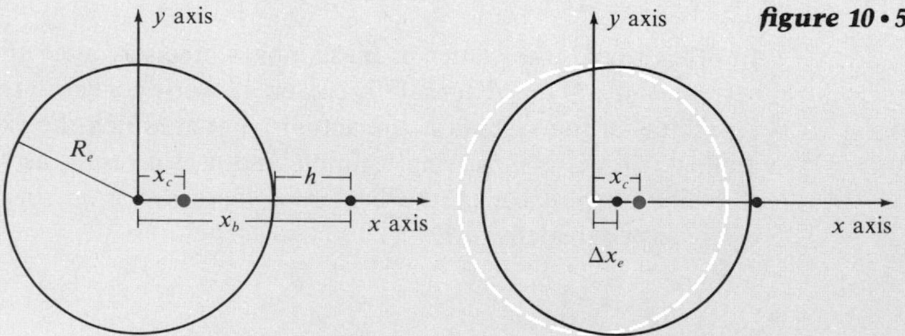

figure 10 • 5

solution (*a*) Assume the origin of coordinates to be at the initial position of the earth's center, as shown. The position of the center of the earth is then $x_e = 0$ and the position of the ball is

$$x_b = R_e + h$$

The center of mass of the earth-ball system is given by

$$x_c = \frac{M_e x_e + m_b x_b}{M_e + m_b}$$

However, since $m_b \ll M_e$, the mass of the ball is a negligible part of the total mass of the system. Therefore,

$$x_c \doteq \frac{M_e \times 0 + m_b x_b}{M_e} = \frac{m_b x_b}{M_e}$$

The distance h from the earth is also negligible in relation to R_e, so that $x_b \doteq R_e$. The approximate location of the center of mass is therefore

$$x_c \doteq \frac{m_b R_e}{M_e} = \frac{6\text{ kg} \times 6400\text{ km}}{6 \times 10^{24}\text{ kg}} = 6.4 \times 10^{-18}\text{ m}$$

(*b*) Since we are assuming that the earth-ball system is isolated from external forces, $\mathbf{F}_{\text{ext}} = \mathbf{0}$ and therefore $\mathbf{a}_c = \mathbf{0}$. This means that $\mathbf{v}_c$ is constant, and since the center of mass started out at rest, it must remain at rest: $\mathbf{v}_c = \mathbf{0}$. Therefore x_c *cannot change*. Instead, as the falling ball moves to the left, it will pull the earth a very slight distance Δx_e to the right. The final position of the center of mass is

$$x_c = \frac{M_e \Delta x_e + m_b R_e}{M_e + m_b} \doteq \frac{M_e \Delta x_e + m_b R_e}{M_e}$$

and the distance the earth moves is

$$\Delta x_e = \frac{M_e x_c - m_b R_e}{M_e} = \frac{m_b (x_b - R_e)}{M_e} = \frac{6\text{ kg} \times 100\text{ m}}{6 \times 10^{24}\text{ kg}} = 10^{-22}\text{ m}$$

10 • 2 center-of-mass coordinates

As we have just seen, if the net external force on a system is zero, the velocity of its center of mass must be constant. For example, if a rocket explodes while traveling through gravity-free interstellar space, the fragments will explode outward in all directions because of the internal forces from the explosion. However, the center of mass of the rocket system will continue along the same straight line and at the same speed as if no explosion had occurred. Until the first fragment hits some external object, so that an external force begins to act on the system, the velocity of the center of mass cannot change. The internal interactions may change the momenta of the individual fragments, but the velocity of the center of mass can be changed only by an applied external force.

Consider a system of particles that is subject to no net external force, so that $\mathbf{v}_c$ is constant. If we were to attach a coordinate system to the center of mass, it would be an inertial reference frame; moreover, it would have the unique property that the total momentum of the particles of the system would be zero. Such a frame is called a *barycentric* coordinate system, more commonly known as *center-of-mass coordinates*.

In discussing Galilean transformations in Chapter 4, we referred to S' as a reference frame moving with some constant velocity relative to a frame S. This latter frame can usually be thought of as a system of *laboratory coordinates,* the stationary reference frame of an observer in the laboratory. In this case the velocity of S' relative to S will be just $\mathbf{v}_c$. Hence the velocity of an individual particle of the moving system would be measured by an observer in this system as $\mathbf{v}_i'$, whereas an observer in the stationary laboratory system would see the same velocity as $\mathbf{v}_i = \mathbf{v}_i' + \mathbf{v}_c$. From this transformation equation for velocity, we can compute the momentum of the system of particles as seen by an observer moving with the center of mass:

$$\mathbf{p}'_{\text{tot}} = \sum_i m_i \mathbf{v}_i' = \sum_i m_i (\mathbf{v}_i - \mathbf{v}_c) = \sum_i m_i \mathbf{v}_i - \sum_i m_i \mathbf{v}_c \qquad [10 \cdot 10]$$

Since $\mathbf{v}_c$ is a constant, however, the last summation becomes

$$\sum_i m_i \mathbf{v}_c = \mathbf{v}_c \sum_i m_i = M\mathbf{v}_c$$

and from Eq. [10 . 3] we know that $M\mathbf{v}_c = \Sigma_i\, m_i \mathbf{v}_i$. Therefore the total momentum of the system in center-of-mass coordinates is

$$\blacktriangleright \quad \mathbf{p}'_{\text{tot}} = \sum_i m_i \mathbf{v}_i - \sum_i m_i \mathbf{v}_c = \mathbf{0} \qquad [10 \cdot 11]$$

Moreover, after a purely internal interaction, such as an explosion or a collision between two particles of the system, the total momentum of the system remains zero in center-of-mass coordinates. If there is no net external force on the system, $\mathbf{v}_c$ remains constant and the system remains an inertial reference frame. We must therefore conclude that in a two-particle collision the individual momenta after impact are equal in magnitude and opposite in direction:

$$m_1 \mathbf{v}_1' + m_2 \mathbf{v}_2' = \mathbf{0} \qquad [10 \cdot 12]$$

To solve for the final momenta in terms of the initial momenta, we need one more equation, usually obtained from energy considerations.

Consider the kinetic energy of a two-particle system as measured in both laboratory and center-of-mass coordinates:

$$K = \tfrac{1}{2} m_1 v_1^2 + \tfrac{1}{2} m_2 v_2^2 \qquad K' = \tfrac{1}{2} m_1 v_1'^2 + \tfrac{1}{2} m_2 v_2'^2 \qquad [10 \cdot 13]$$

To relate these two values of kinetic energy, we must substitute $\mathbf{v}_1 = \mathbf{v}_1' + \mathbf{v}_c$ and $\mathbf{v}_2 = \mathbf{v}_2' + \mathbf{v}_c$ into the first equation. Recall from Chapter 2 that the general definition of v^2 is the scalar product

$$v^2 = \mathbf{v} \cdot \mathbf{v}$$

With this in mind we can write the equation for K as

$$\begin{aligned} K &= \tfrac{1}{2} m_1 (\mathbf{v}_1' + \mathbf{v}_c) \cdot (\mathbf{v}_1' + \mathbf{v}_c) + \tfrac{1}{2} m_2 (\mathbf{v}_2' + \mathbf{v}_c) \cdot (\mathbf{v}_2' + \mathbf{v}_c) \\ &= \tfrac{1}{2} m_1 v_1'^2 + \tfrac{1}{2} m_2 v_2'^2 + m_1 (\mathbf{v}_1' \cdot \mathbf{v}_c) + m_2 (\mathbf{v}_2' \cdot \mathbf{v}_c) \\ &\qquad + \tfrac{1}{2}(m_1 + m_2)\, v_c^2 \end{aligned}$$

The third and fourth terms in the second equality can be rewritten in the form $(m_1\mathbf{v}_1' + m_2\mathbf{v}_2') \cdot \mathbf{v}_c$, and we know from Eq. [10 . 12] that this expression is equal to zero. Therefore our final transformation equation is simply

$$K = K' + \tfrac{1}{2}Mv_c^2 \qquad [10 \cdot 14]$$

In other words, the kinetic energy of the system as seen by the stationary observer is equal to the kinetic energy as measured in center-of-mass coordinates *plus* the kinetic energy associated with a single particle of mass M moving with velocity $\mathbf{v}_c$. This means that if K is conserved in an elastic collision, then so is K'. We can generalize Eq. [10 . 14] to a system of any number n of discrete particles in the usual manner:

$$K = \tfrac{1}{2}\left(\sum_i m_i v_i'^2\right) + \tfrac{1}{2}\left(\sum_i m_i\right)v_c^2$$

$$= \tfrac{1}{2}\left(\sum_i m_i v_i'^2\right) + \tfrac{1}{2}Mv_c^2 \qquad [10 \cdot 15]$$

example 10 • 3 A particle of mass m and approach velocity $\mathbf{u}_1 = u\mathbf{i}$ collides with a stationary particle of mass $2m$. The collision is elastic, and the x components of the velocity after impact are equal:

$$v_{x1} = v_{x2} = v_x$$

Draw the momentum vector as seen in the center-of-mass system and then find the velocity vectors $\mathbf{v}_1$ and $\mathbf{v}_2$ as seen by a stationary observer.

solution The first step is to find the velocity of the center of mass. Since the second particle is stationary, $\mathbf{u}_2 = \mathbf{0}$ and

$$\mathbf{v}_c = \frac{mu\mathbf{i} + 2m \times \mathbf{0}}{3m} = \tfrac{1}{3}u\mathbf{i}$$

In center-of-mass coordinates, the approach velocity of particle m is

$$\mathbf{u}_1' = \mathbf{u}_1 - \mathbf{v}_c = \tfrac{2}{3}u\mathbf{i}$$

For particle $2m$,

$$\mathbf{u}_2' = \mathbf{u}_2 - \mathbf{v}_c = 0 - \tfrac{1}{3}u\mathbf{i}$$

$$= -\tfrac{1}{3}u\mathbf{i}$$

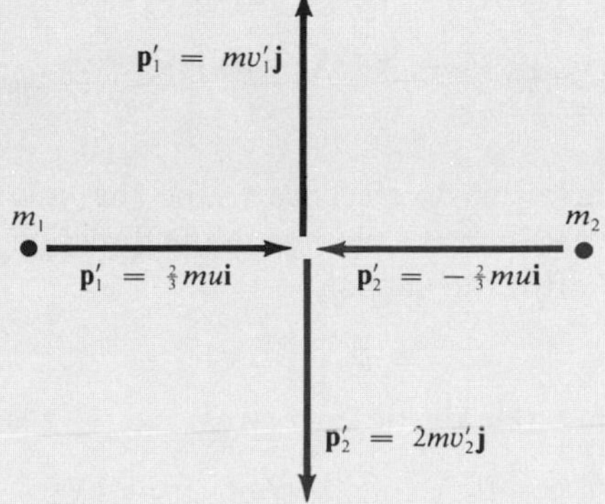

figure 10 • 6

and the two momenta before collision are therefore

$$m_1\mathbf{u}_1' = \tfrac{2}{3}mu\mathbf{i} \qquad m_2\mathbf{u}_2' = 2m(-\tfrac{1}{3}u\mathbf{i}) = -\tfrac{2}{3}mu\mathbf{i}$$

In other words, in center-of-mass coordinates *both* particles are moving horizontally before collision and their momenta are equal and opposite, as shown.

The two velocities after the collision have the same horizontal component $v_x = v_c$, and

$$v_x' = v_x - v_c = 0$$

Neither particle has a horizontal velocity component v_x' in the center-of-mass system. Hence both must now be moving in the y' direction, and since their two momenta must sum to zero, they must be moving in opposite directions:

$$m\mathbf{v}_1' = -2m\mathbf{v}_2'$$

or

$$mv_1'\mathbf{j} = -2mv_2'\mathbf{j} \qquad v_2' < 0$$

To relate the final momenta to the initial momenta, we can express the initial kinetic energy as

$$K' = K - \tfrac{1}{2}Mv_c^2 = \tfrac{1}{2}mu^2 - \tfrac{1}{2}(3m)(\tfrac{1}{3}u)^2$$
$$= \tfrac{1}{3}mu^2$$

and the final kinetic energy as simply

$$K' = \tfrac{1}{2}mv_1'^2 + \tfrac{1}{2}(2m)v_2'^2$$

If we now solve these two equations, substituting $v_1' = -2v_2'$, we obtain the final velocity vectors

$$\mathbf{v}_1' = \tfrac{2}{3}u\mathbf{j} \qquad \mathbf{v}_2' = -\tfrac{1}{3}u\mathbf{j}$$

The direction of $\mathbf{v}_c$, however, has not changed; hence, when we transform back to laboratory coordinates, the velocities as seen by a stationary observer are

$$\mathbf{v}_1 = \mathbf{v}_1' + \mathbf{v}_c = \tfrac{2}{3}u\mathbf{j} + \tfrac{1}{3}u\mathbf{i}$$
$$\mathbf{v}_2 = \mathbf{v}_2' + \mathbf{v}_c = -\tfrac{1}{3}u\mathbf{j} + \tfrac{1}{3}u\mathbf{i}$$

In the center-of-mass reference frame the only effect of a completely elastic two-body collision is a change in the direction of the initial velocities. Both before and after the impact,

$$\mathbf{p}_1' + \mathbf{p}_2' = \mathbf{0} \qquad p_1'^2 = p_2'^2$$

and we can express the kinetic energy as

$$K' = \frac{1}{2}\left(\frac{p_1'^2}{m_1} + \frac{p_2'^2}{m_2}\right) = \frac{1}{2}p_1'^2\left(\frac{m_1 + m_2}{m_1m_2}\right) = \frac{1}{2}\frac{p_1'^2}{\mu} \qquad [10 \cdot 16]$$

The constant

$$\mu = \frac{m_1 m_2}{m_1 + m_2}$$

is known as the *reduced mass* and is useful in more advanced work in physics. Since the *expression* for K' is unchanged in a collision, and the *value* of K' is unchanged when the collision is elastic, it follows that

$$p_1' = |\mathbf{p}_1'| \qquad p_2' = |\mathbf{p}_2'|$$

is also unchanged by the collision. Therefore, as we saw in the preceding example,

$$m_1^2 u_1'^2 = m_2^2 u_2'^2 = m_1^2 v_1'^2 = m_2^2 v_2'^2$$

and the only change the elastic collision produces in center-of-mass coordinates is a change in the direction of the particle velocities.

10 • 3 center of mass of a continuous solid

The concept of *density* is important in many branches of science and has particular significance in the treatment of solid bodies. The *average density* $\bar{\rho}$ of a body is defined as the mass per unit volume of the body:

$$\bar{\rho} = \frac{m}{V} \qquad [10 \cdot 17]$$

Note that here V stands for volume, not velocity. Obviously the mass need not be distributed uniformly throughout the body. The earth's atmosphere, for example, "thins out" with increasing altitude. We therefore need a definition of the *local density* ρ at any given point of the body:

$$\blacktriangleright \quad \rho = \lim_{\Delta V \to 0} \frac{\Delta m}{\Delta V} = \frac{dm}{dV} \qquad [10 \cdot 18]$$

where Δm is the amount of mass included in the volume ΔV which contains the point in question. In general, ρ is a function of position in the body; in the case of nonrigid bodies, it may also be a function of time. Throughout this text we shall assume that ρ is a continuous function of position, without any abrupt changes, since even for gases under most conditions, volumes ΔV as small as 10^{-15} cm^3 contain thousands of particles.

Although the mass of a body is not affected by changes in temperature and pressure, this is not true of volume, especially in the case of gases. The density values listed in physical tables should therefore include temperature-pressure specifications (see Table 10 . 1). In the case of solids, the density may also depend on the manner in which samples have been prepared or treated.

Sometimes it is more convenient to determine the ratio of the mass of some volume of a substance at a given temperature and pressure to the mass of an equal volume of another substance taken as a standard. This ratio, which is a pure number, is commonly called the *specific gravity* of the substance relative to the standard. For solids and liquids, the standard usually chosen is pure air-free water at 3.98°C and 1 atm of pressure. For gases the

table 10 • 1 *Densities of various substances*

equal densities: 1000 g/cm^3 = 62.43 lbm/ft^3 = 1.940 $slugs/ft^3$

substance	*density, g/cm³*		*substance*	*density, g/cm³*	
solids at 20°C, 1 atm					
aluminum, hard-drawn	2.70		iron, cast, gray	7.03–7.13	
balsa wood, seasoned	0.11–0.14		iron, wrought	7.80–7.90	
brick	1.4–2.2		lead, compressed	11.35	
copper, hard-drawn	8.89		lithium	0.53	
cork	0.22–0.26		oak, seasoned	0.60–0.90	
glass, common crown	2.4–2.8		pine, white, seasoned	0.35–0.50	
glass, flint	2.9–5.9		platinum	21.37	
gold, cast	19.3		steel, 99 Fe, 1 C	7.83	
hydrogen (−260°C)	0.763		tungsten	19.3	
ice (0°C)	0.917		uranium	18.90	
liquids at 0°C and 20°C, 1 atm					
alcohol, ethyl	0.807	0.789	kerosene	—	0.82
carbon tetrachloride	1.633	1.595	mercury	13.5951	13.546
gasoline	—	0.66–0.69	water, air-free	0.9998	0.9982
hydrogen (−252°C)	0.070	—	water, sea	—	1.03
gases at 0°C, 1 atm					
air	1.293×10^{-3}		hydrogen	0.0899×10^{-3}	
carbon dioxide	1.977×10^{-3}		nitrogen	1.251×10^{-3}	
helium	0.178×10^{-3}		oxygen	1.429×10^{-3}	

standard is often hydrogen, oxygen, or air at a specified temperature and pressure. The use of water as a standard is especially advantageous in the metric system. Since the density of water is 1.00 g/cm^3 over a wide range of temperatures, the density and specific gravity of the substance being measured are approximately numerically equal in cgs units.

Let us now see how the concept of density relates to the calculation of the center of mass of a continuous body. In Sec. 10 . 1 we defined the center of mass in terms of a finite number of discrete particles whose individual masses m_i summed to the total mass of the system:

$$M = \sum_{i=1}^{n} m_i \qquad i = 1, 2, \ldots, n$$

However, we may also regard a continuous body as being made up of individual volume elements ΔV_i, each of which contains a mass element Δm_i.

Then, in the limit as ΔV approaches zero, the mass elements become infinitesimal and their number approaches infinity and from Eq. [10 . 18],

$$\lim_{n\to\infty} \Delta m_i = dm_i = \rho(x_i, y_i, z_i)\, dV$$

where $\mathbf{r}_i = (x_i, y_i, z_i)$ is the location of dm_i.

To generalize Eq. [10 . 1] to continuous bodies, therefore, we must replace the summation over the range of i by a continuous integral of mass elements dm over the whole body. The individual position vectors $\mathbf{r}_i$ also take on continuous values over the whole body. The equation for the position vector of the center of mass then becomes

$$\mathbf{r}_c = \lim_{n\to\infty} \frac{\sum_{i=1}^{n} \mathbf{r}_i\, dm_i}{M} = \frac{\int_{\text{body}} \mathbf{r}\, dm}{M} = \frac{\int_{\text{body}} \rho \mathbf{r}\, dV}{M} \qquad [10 \cdot 19]$$

In component form the coordinates of the center of mass would be

$$x_c = \frac{\int_{\text{body}} \rho x\, dV}{M} \qquad y_c = \frac{\int_{\text{body}} \rho y\, dV}{M} \qquad z_c = \frac{\int_{\text{body}} \rho z\, dV}{M} \qquad [10 \cdot 20]$$

A note of caution is in order regarding Eqs. [10 . 19] and [10 . 20]: Although they are complete and correct, the integrals can be quite formidable to evaluate. In its most general form, the volume element is

$$dV = dx\, dy\, dz$$

an infinitesimal parallelepiped whose sides are dx, dy, and dz. The resulting integration actually necessitates performing three different integrations (three-dimensional multiple integration), the limits of which may be interdependent. In this chapter we shall consider only cases that can be reduced to integrals over a single variable of integration.

example 10 • 4 A straight rod of length l has a density which varies as the nth power of the distance x from one end, $\rho = kx^n$. If the cross-sectional area of the rod is A, find the point x_c which represents the center of mass.

solution The mass element $dm = \rho\, dV$ at a distance x from one end occupies a volume $dV = A\, dx$. Therefore

$$dm = Akx^n\, dx$$

and

$$x_c = \frac{A\int_0^l kx^{n+1}\, dx}{A\int_0^l kx^n\, dx} = \frac{n+1}{n+2}\, l$$

If $n = 0$, so that the rod has uniform density, then $x_c = \frac{1}{2}l$, as expected.

example 10 • 5 Find the center of mass of the very thin homogeneous cylindrical arc in Fig. 10 . 7.

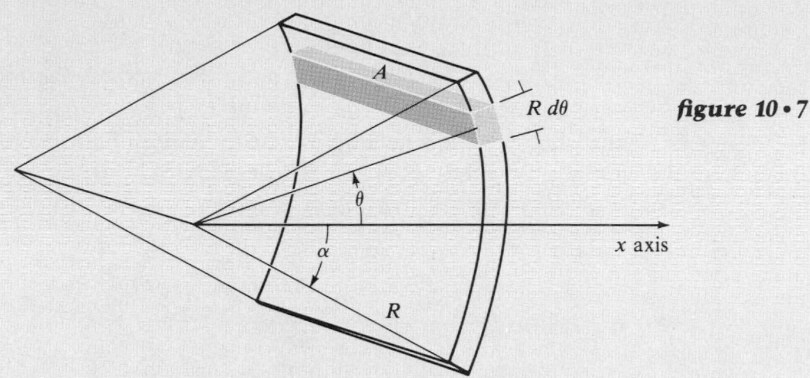

figure 10 • 7

solution Let A be the uniform cross-sectional area perpendicular to the page, so that the volume element dV is a very thin rod of area A and thickness $R\,d\theta$, located a distance R from the origin. The volume element

$$dV = AR\,d\theta$$

therefore contains a mass element

$$dm = \rho\,dV = \rho AR\,d\theta$$

For the coordinate system chosen, $y_c = 0$, since the distribution of mass is the same above and below the x axis. Therefore, from Eq. [10 . 20],

$$x_c = \frac{\int \rho x\,dV}{M} = \frac{\int \rho x\,dV}{\int \rho\,dV} = \frac{\int x\,dV}{\int dV}$$

The x coordinate of each segment of arc is $x = R\cos\theta$, so that the location of the center of mass is

$$x_c = \frac{\int_{-\alpha}^{+\alpha} xAR\,d\theta}{\int_{-\alpha}^{+\alpha} AR\,d\theta} = \frac{R\int_{-\alpha}^{+\alpha} \cos\theta\,d\theta}{2\alpha}$$

$$= \frac{R}{2\alpha}[\sin\alpha - \sin(-\alpha)] = \frac{R\sin\alpha}{\alpha}$$

For a full circle $\alpha = \pi$, and the center of mass would be located at $x_c = 0$, the center of the circle. For a semicircle $\alpha = \frac{1}{2}\pi$ and

$$x_c = \frac{2R}{\pi} = 0.6366R$$

For a very small arc, $\alpha \to 0$, and $x_c \to R$, as expected.

10 • 4 symmetry

In finding centers of mass, and in many other problems in physics, our work is greatly simplified by elements of symmetry. *Symmetry* refers to the correspondence of physical or geometric properties on opposite sides of a plane, line, or point. In most situations the symmetries will be immediately apparent; in others the mathematical forms of the functions of interest provide clues. A simple rule of thumb, however, is to ask yourself the following question: If I change my position and the coordinate system which I use to describe the physical situation in such a way as to invert (reverse) one or more of the coordinates, does the situation appear any different? If not, it is symmetrical with respect to the appropriate plane, line, or point, depending on whether one, two, or three coordinates were inverted.

As an example, consider a homogeneous cylindrical solid whose axis lies along the z axis of coordinates (Fig. 10 . 8). The cylinder will look exactly the same when you stand on your head as it does when you stand on your feet, although you have inverted both the x and y axes by turning your frame of reference upside-down (see Fig. 10 . 9). The cylinder is therefore symmetrical with respect to the z axis. Alternatively, you might simply have rotated the cylinder around the z axis by 180° and noted that it still looks exactly the same.

figure 10 • 8 *Symmetries of a right circular cylinder*

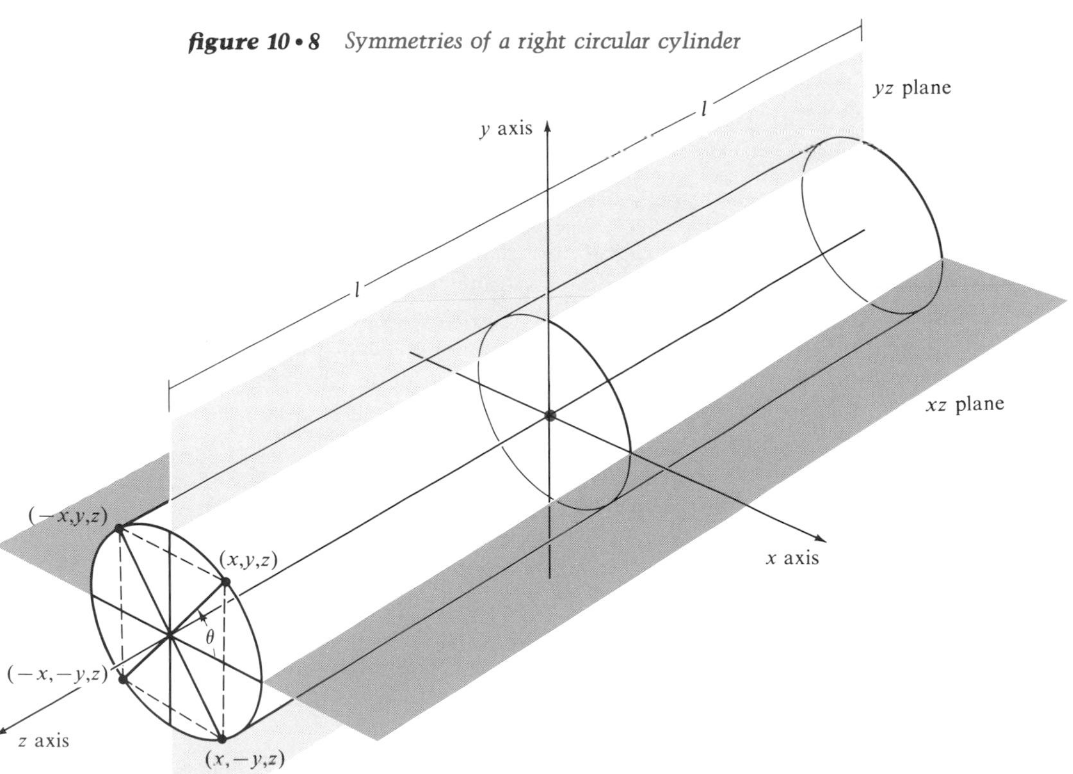

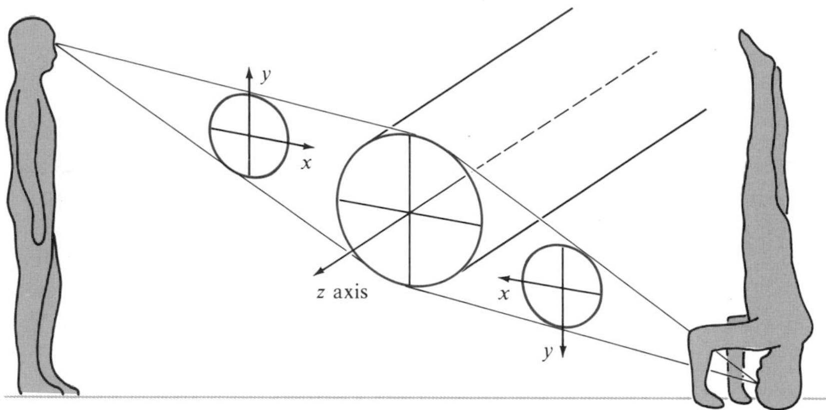

figure 10 • 9 *Heuristic proof of symmetry with respect to the z axis*

Using the cylinder as an example, we can formalize our conception of its symmetries in mathematical terms. Since the density of a uniform cylinder is the same whether x is positive or negative,

$$\rho(x,y,z) = \rho(-x,y,z) \qquad [10 \cdot 21]$$

the density is an *even function of* x; or equivalently, we can say that the cylinder is *symmetrical with respect to the yz* plane. It also has symmetry with respect to the xz plane,

$$\rho(x,y,z) = \rho(x,-y,z) \qquad [10 \cdot 22]$$

and therefore, in general,

$$\rho(x,y,z) = \rho(-x,y,z) = \rho(-x,-y,z) \qquad [10 \cdot 23]$$

The cylinder is symmetrical with respect to *reflection through the z axis,* which is the intersection of the two planes of symmetry.

A special case of this kind of symmetry is symmetry with respect to rotations about the z axis, in which the density function is independent of the angle of rotation $\theta = \tan^{-1}(y/x)$ and depends only on the distance $\sqrt{x^2 + y^2}$ from the axis of rotation, z. A rigid body that displays such *axial symmetry* is called a *body of revolution.*

If the plane which bisects the length of the cylinder happens to be the xy plane, as in Fig. 10 . 8, then we can also say that the cylinder is symmetrical with respect to the origin, since

$$\rho(x,y,z) = \rho(x,y,-z) = \rho(-x,-y,-z) \qquad [10 \cdot 24]$$

That is, the mass element $dm = \rho\, dV$ located by a position vector $\mathbf{r} = x\mathbf{i} + y\mathbf{j} + z\mathbf{k}$ is equivalent to the element located by the opposite vector $-\mathbf{r}$, which exactly cancels its contribution to the position vector $\mathbf{r}_c$ of the center of mass.

Let us see how we can make use of symmetry in evaluating Eqs. [10 . 20], the scalar components of $\mathbf{r}_c$. Suppose the distribution of mass is symmetrical with respect to the x axis. In that case, for every contribution

$\rho(x,y,z)x\,dV$ to the integral for x_c there is, by symmetry, an equal but opposite contribution which exactly cancels it,

$$-\rho(x,y,z)x\,dV = \rho(-x,y,z)(-x)\,dV$$

and the result $x_c = 0$ is immediately achieved by symmetry arguments (these are usually intuitively obvious). Similarly, if there is symmetry with respect to the y axis, then the contributions to the integral for y_c will also cancel, giving us $y_c = 0$. If both conditions of symmetry apply, then the center of mass must lie on the z axis, which is the intersection of the two planes of symmetry. In the very special case that the distribution of mass is also symmetrical with respect to the origin, then $\mathbf{r}_c = \mathbf{0}$, since every contribution to the integral for $\mathbf{r}_c$ is cancelled by an equal but opposite element:

$$-\rho(x,y,z)\mathbf{r}\,dV = \rho(-x,-y,-z)(-\mathbf{r})\,dV$$

It should be clear by now that, given some point for which the density is a symmetrical function about that point, as in Eq. [10 . 24], then that point is the center of mass. Moreover, if the body is of *uniform density,* then the center of mass coincides with the geometric center of symmetry. This concept also simplifies finding the center of mass of a body which is not itself symmetrical, but is composed of symmetrical pieces, as in Fig. 10 . 2. We can locate the center of mass of each piece separately and then add the position vectors for the pieces to find the center of mass of the entire body.

example 10 • 6 Find the center of mass of a homogeneous isosceles triangle of uniform density ρ and thickness τ, with base angle α and height h.

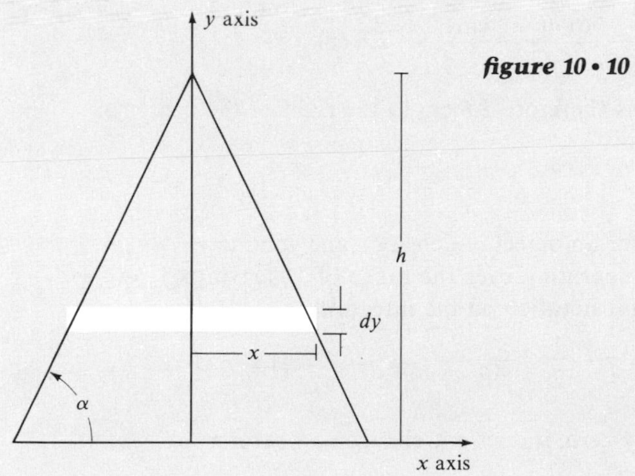

figure 10 • 10

solution Common sense based on symmetry conditions tells us that the center of mass will be on the vertical axis of the triangle at some height y_c above its base. If we subdivide the triangle into horizontal strips of width $2x$ and height dy, we can write the mass of the triangle as

$$M = \rho\tau h^2 \cot\alpha$$

The center of mass is then located at

$$y_c = \frac{\rho\tau}{M}\int_0^h 2x\,dy = \frac{2\rho\tau\cot\alpha}{M}\int_0^h (h - y)y\,dy = \tfrac{1}{3}h$$

example 10 • 7 A homogeneous disk D, of diameter 16 cm, contains a circular hole D' which is 12 cm in diameter and is tangent to the circumference of the disk as shown. If the surface density, or mass per unit area, is σ, locate the center of mass.

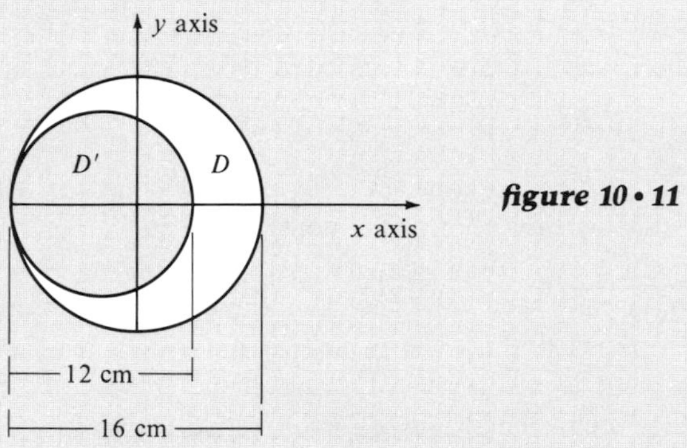

figure 10 • 11

solution We can first treat the body as if it had no cavity and find its center of mass, in this case $x_c = 0$. We may then consider as a "negative mass" the imaginary body which would just fill the cavity D' and has its center at $x_c = -2$ cm. If we now apply Eq. [10 . 9] to the two centers of mass as if they were particles, we obtain

$$x_c = \frac{64\pi\sigma \times 0 - 36\pi\sigma(-2\text{ cm})}{64\pi\sigma - 36\pi\sigma} = 2.6\text{ cm}$$

The center of mass is therefore 2.6 cm to the right of the center of disk D.

In the language of the calculus, for an object of density ρ and volume V containing a hole V', we find $\mathbf{r}_c$ by integrating over the range of V, leaving out the region V', as indicated by limit notation on the integral:

$$M\mathbf{r}_c = \int_{V-V'} \rho\mathbf{r}\,dV = \int_{V-V'} \rho\mathbf{r}\,dV + \int_{V'} (\rho - \rho)\mathbf{r}\,dV \qquad [10 \cdot 25]$$

Since the second integral is identically zero, it does not change $\mathbf{r}_c$. Therefore

$$M\mathbf{r}_c = \int_{V-V'} \rho\mathbf{r}\,dV + \int_{V'} \rho\mathbf{r}\,dV - \int_{V'} \rho\mathbf{r}\,dV \qquad [10 \cdot 26]$$

If we now combine the positive integral over the hole V' with the first integral, we obtain

$$M\mathbf{r}_c = \int_{V} \rho\mathbf{r}\,dV - \int_{V'} \rho\mathbf{r}\,dV \qquad [10 \cdot 27]$$

This means that we can separate the calculation into two relatively easy parts such that

$$M\mathbf{r}_c = M_V \mathbf{r}_{cV} - M_{V'} \mathbf{r}_{cV'} \qquad [10 \cdot 28]$$

This formula can be applied to any object of volume V containing a hole V'. We just think of the hole as a "negative mass" superimposed over an equal "positive mass" which it cancels exactly.

*10 • 5 numerical computation of the center of mass

In the case of a homogeneous rigid planar body we can construct an algorithm for numerical computation of the center of mass. This algorithm illustrates some useful computational concepts, such as *subprograms, dimensioned* (subscripted) *variables, table look-up* and *interpolation,* and *nested loops.* With these procedures we can perform a two-dimensional integration simply by counting squares in order to evaluate x_c and y_c.

Assume that we know the size and shape of a planar object in Fig. 10 . 12. First we enclose it in a rectangle in the first quadrant of the xy plane. Two sides of the rectangle are formed by the x and y axes, and the other two sides are given by x_m and y_m, the maximum x and y values of any point on the boundary of the body. It is assumed that the upper boundary values y_2 and the lower boundary values y_1 are known, either as functions of x or as values tabulated for selected x, between which we can interpolate to find the boundary value for any given x.

The next step is to subdivide the rectangle in smaller rectangles of area ΔA. If we divide the x axis into n_x intervals of size $\Delta x = X/n_x$ and the y axis into n_y intervals of size $\Delta y = Y/n_y$, then for each value of $i = 1, 2, \ldots, n_x$ and $j = 1, 2, \ldots, n_y$ we can determine the coordinates of the center of the rectangle $\Delta A_{ij} = \Delta x\,\Delta y$, which is the intersection of the two intervals Δx and Δy:

$$x_i = (i - \tfrac{1}{2})\,\Delta x \qquad y_j = (j - \tfrac{1}{2})\,\Delta y \qquad [10 \cdot 29]$$

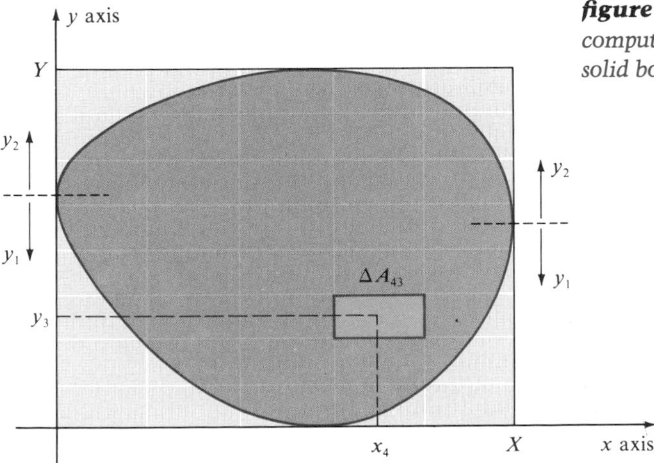

figure 10 • 12 *Partition for numerical computation of the center of mass of a planar solid body;* $n_x = 5, n_y = 8$

Then each area ΔA is uniquely specified by the values of two indices i and j, as illustrated for ΔA_{43}. These indices provide a more natural way of counting our steps than keeping track of the values of x and y.

To compute the coordinates (x_c, y_c) of the center of mass, we note that the density and thickness of the planar body are assumed to be constant. Therefore we can approximate Eqs. [10 . 20] numerically in terms of the areas ΔA_{ij}:

$$x_c = \frac{\int \rho x \, dV}{\int \rho \, dV} \doteq \frac{\sum_{i,j} x_i \, \Delta A_{ij}}{\sum_{i,j} \Delta A_{ij}} \qquad y_c = \frac{\int \rho y \, dV}{\int \rho \, dV} \doteq \frac{\sum_{i,j} y_j \, \Delta A_{ij}}{\sum_{i,j} \Delta A_{ij}} \qquad [10 \cdot 30]$$

where the summation notation means to sum over all possible values of i and j, for a total of $n_x n_y$ terms. The assignment statements for the recursive evaluation of $x_i \, \Delta A_{ij}$ and $y_j \, \Delta A_{ij}$ are

$$\begin{aligned} x_m &\leftarrow x_m + x_i \, \Delta A_{ij} \qquad y_m \leftarrow y_m + y_j \, \Delta A_{ij} \\ A &\leftarrow A + \Delta A_{ij} \end{aligned} \qquad [10 \cdot 31]$$

and at the end of this computation we obtain

$$x_c = \frac{x_m}{A} \qquad y_c = \frac{y_m}{A} \qquad [10 \cdot 32]$$

For each value of i we determine x_i and the corresponding boundary values $y_1(x_i)$ and $y_2(x_i)$. Then, counting through the values of j, we compute x_m and y_m for each ΔA_{ij} such that $y_1(x_i) \leq y_j \leq y_2(x_i)$, ignoring those for which y_j falls outside the boundaries of the object. Remember that the computer's forte is rapid iteration of recursive formulas. Thus, although the continuous testing of y_j required by our algorithm would be wasteful in terms of human effort, it is quite appropriate to computer usage. As we take finer and finer subdivisions of the rectangle, our approximation should begin to approach the correct answer with an error inversely proportional to the number of subdivisions. We assume that $y_1(x)$ and $y_2(x)$ are single-valued functions.

Since $\Sigma_{i,j}$ entails a *double* summation, we must "nest" one sum, or recursive loop, within another. To clarify this nesting process we need an additional flowchart symbol which indicates that the computations within the loop are reiterated for all values of i from 1 to m in increments of unity.* The program continues on past the loop only when i becomes greater than or equal to m. Then, as shown in Fig. 10 . 13, for each value of i we sum over all the possible values of j; that is, the j loop is nested within the i loop.

The last new flowchart symbol we shall need in this text is the *hexagon,* which represents a subprogram, or subroutine, referred to from the main program (see Fig. 10 . 14). The variable x above the line represents input to

*Although earlier computer languages, such as FORTRAN II, allowed for loop structures only with increments of unity, present versions of BASIC and FORTRAN will accept both fractional and negative increments.

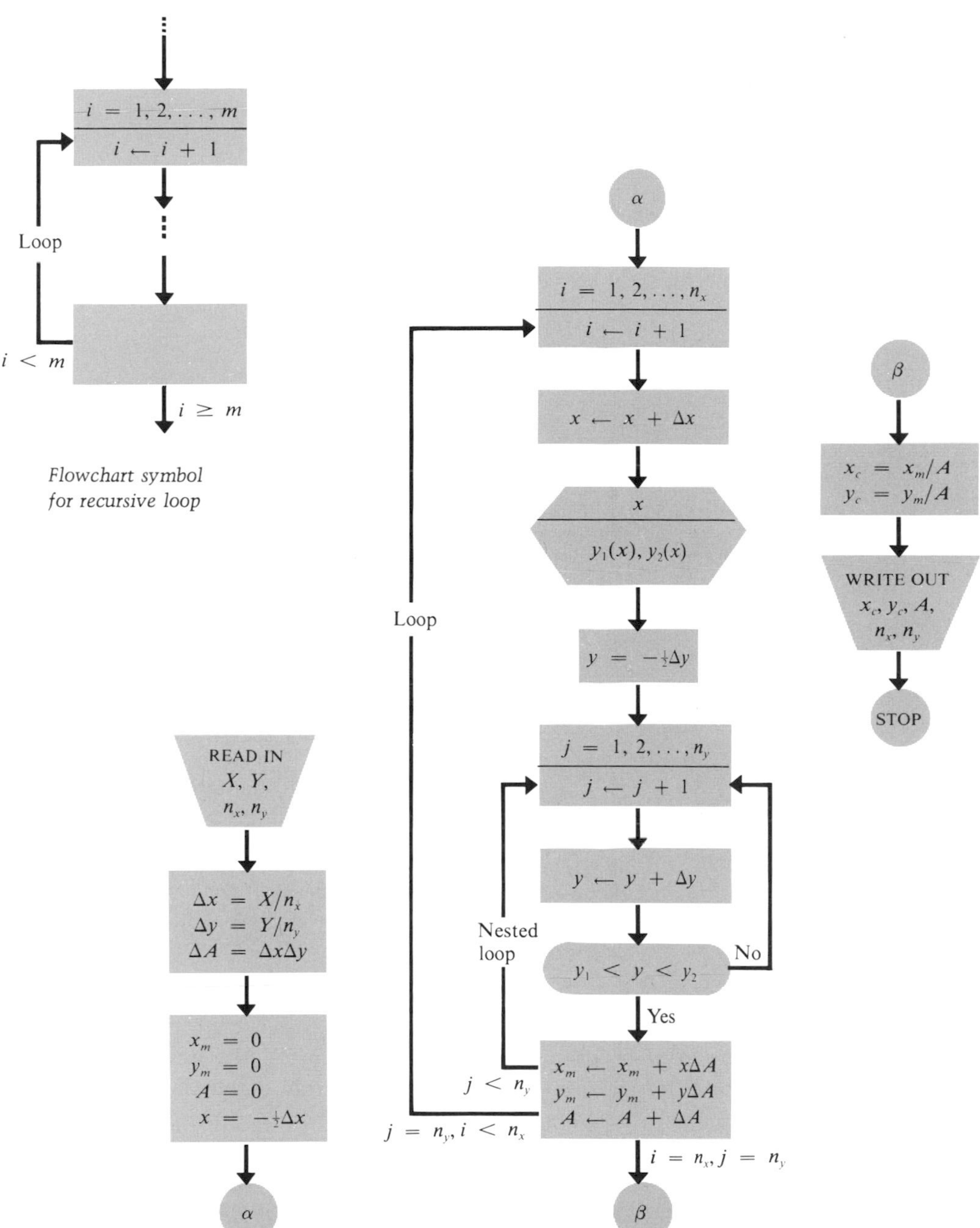

figure 10 • 13 *Flowchart for numerical computation of center of mass of a planar body. The symbol for a recursive loop is shown above, and the subprogram (hexagon) for evaluation of* $y_1(x)$ *and* $y_2(x)$ *is shown in Fig. 10 . 14.*

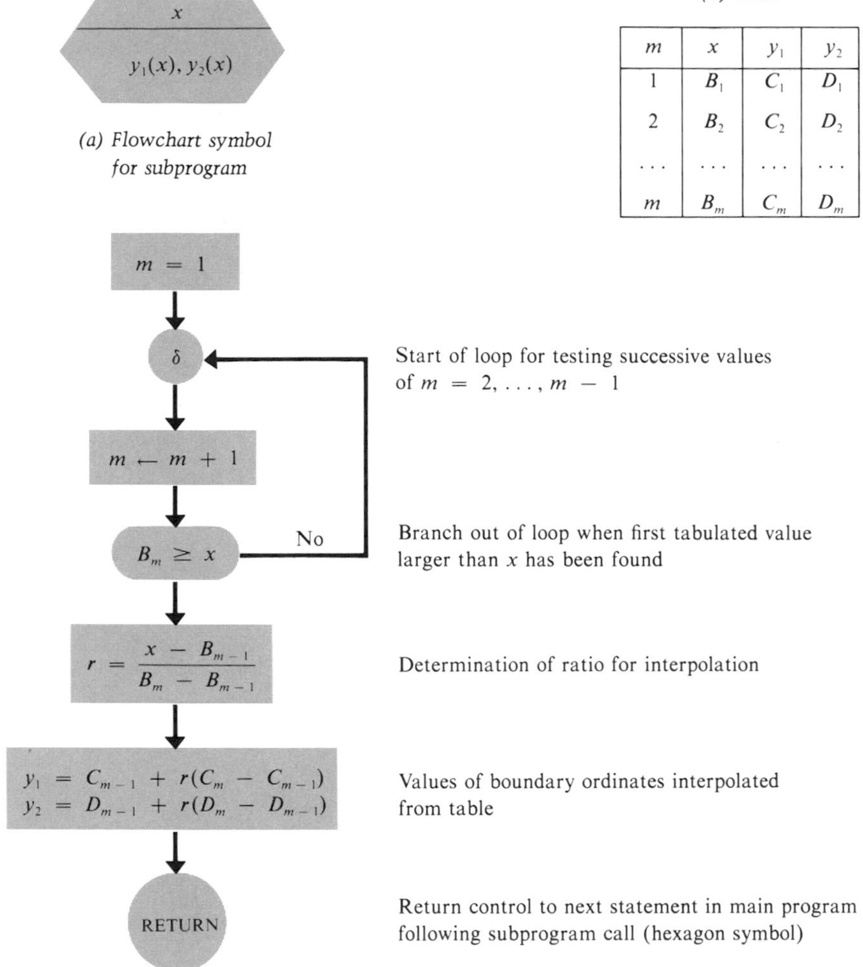

figure 10 • 14 *Subprogram for the special case of tabulated boundary ordinates*

the subprogram from the main program, and the variables below the line represent output from the subprogram back to the main program. After the subprogram has been executed, the computer returns to the next statement in the main program, bringing the new information with it. All modern computer languages make provision for subprograms; they greatly simplify the construction of complicated algorithms, and separate general-purpose subprograms may be kept on file for use as required.

The evaluation of $y_1(x)$ and $y_2(x)$ is simple and straightforward if the boundaries are known functions of x. In the case of tabulated values, shown in the subprogram flowchart of Fig. 10 . 14, for each value of x_i we must test the tabulated values B_m until we find the interval in the table (which may even be subdivided into irregular intervals) which contains that x_i. Then a linear interpolation on the tabulated values C_m of $y_1(x)$ and D_m of $y_2(x)$ yields approximate values of $y_1(x_i)$ and $y_2(x_i)$. To do this we compute

$$r = \frac{x_i - B_{m-1}}{B_m - B_{m-1}}$$

table 10 • 2 *Numerical computation of the center of mass of a quadrant of the unit circle*

Analytic value: $x_c = 0.4244 = y_c$; $A = 0.7854$

$n_x = n_y$	A	x_c	y_c
1	1.0000	0.5000	0.5000
2	0.7500	0.4167	0.4167
4	0.8125	0.4327	0.4327
8	0.8125	0.4327	0.4327
16	0.7930	0.4266	0.4266
32	0.7881	0.4251	0.4251
64	0.7869	0.4248	0.4248

the fraction of the interval $B_{m-1} < x \leq B_m$ which is included in the interval B_{m-1} to x_i. We then add this fraction r of the corresponding tabulated change in y to the value of y for $x_i = B_{m-1}$. For example, as illustrated in the subprogram,

$$y_1(x_i) = C_{m-1} + r(C_m - C_{m-1}) = rC_m + (1 - r)C_{m-1}$$

The quantities B_m, C_m, and D_m for values 1, 2, . . . , m are called *dimensioned*, or *subscripted, variables*. The computer stores the value of B_1, for example, in a memory cell called B, and when the value B_m is called up, the computer goes to B and counts $m - 1$ cells to the memory cell in which the value of B_m is stored. In a computation of this sort dimensioned variables are a great convenience in programming; otherwise we should need a separate name and a separate algorithm for each variable. With dimensioned variables we perform an iterative computation on the *index* of the variable and use the same command repetitively. Table 10 . 2 shows the results of such a computation for the center of mass of the quadrant of a circle, with different choices of interval size. In this case the boundary values are

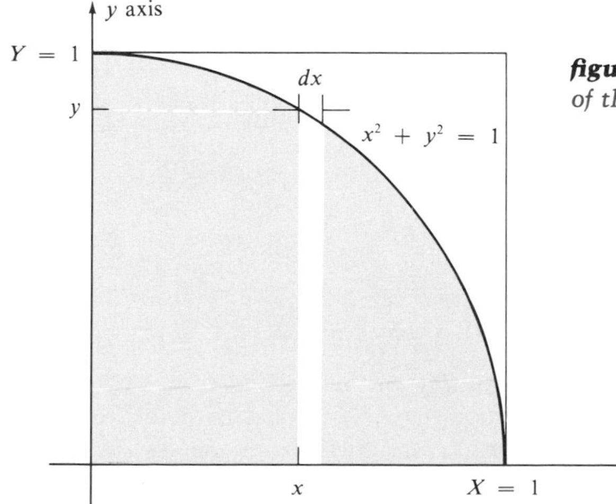

figure 10 • 15 *Computation of the center of mass of the quadrant of a circle*

$X = Y = 1$, and we take $\Delta x = \Delta y$ (see Fig. 10 . 15). The actual value of x_c, computed from Eq. [10 . 20], is

$$x_c = \frac{\int_0^X xy\,dx}{\frac{1}{4}\pi X^2} = \frac{4X}{3\pi} = 0.4244 \qquad [10 \cdot 33]$$

The integral is most easily evaluated by substituting $x = X\cos\theta$ and $y = X\sin\theta$ for $0 \leq \theta \leq \frac{1}{2}\pi$. Since the quadrant is symmetrical with respect to the axis for which $y = x$ (the 45° line), it follows that $y_c = x_c = 0.4244$. This basic technique can readily be extended to three dimensions to find (x_c, y_c, z_c).

PROBLEMS

10 • 1 center of mass

10 • 1 A cube has one corner at the origin of coordinates and the diagonally opposite corner located at (l,l,l). (*a*) If a mass m is placed at each of these two corners, where is the center of mass? (*b*) If these masses are removed and identical masses are placed at every *other* corner of the cube, where is the center of mass?

10 • 2 A mass m_1 is located at (x_1,y_1,z_1) and another mass m_2 is located at (x_2,y_2,z_2). (*a*) Find the distance r_0 between m_1 and m_2. (*b*) The distance r_1 between m_1 and the center of mass. (*c*) The distance r_2 between m_2 and the center of mass. Express your answers in terms of r_0 and masses m_1 and m_2 and state their meaning in words.

10 • 3 A water molecule is in the shape of an isosceles triangle, with an oxygen atom of mass 16 μ (atomic mass units) located at the vertex, and a hydrogen atom of mass 1 μ located at each angle of the base. The height of the triangle is 18.8 Å (angstroms) and the vertex forms an angle of 105°. (*a*) Locate the center of mass of the molecule. (*b*) Locate the center of mass of a molecule of *heavy water,* in which the two hydrogen atoms are replaced by deuterium atoms (an isotope of hydrogen) of mass 2 μ. Express your answer in angstroms. ($1\ \mu = 1.66 \times 10^{-27}$ kg and $1\ \text{Å} = 10^{-10}$ m.)

10 • 4 Two particles of masses $m_1 = 1$ kg and $m_2 = 3$ kg move through space with position vectors given by

$$\mathbf{r}_1 = 3\mathbf{i} + t\mathbf{j} - \sqrt{t}\,\mathbf{k}$$

$$\mathbf{r}_2 = \sin t^2\mathbf{i} + t\mathbf{k}$$

(*a*) Find the center of mass. (*b*) What is its acceleration? (*c*) Find its acceleration as seen by an observer moving with a constant velocity $\mathbf{v} = \mathbf{j} + 3\mathbf{k}$.

10 • 5 Two objects, with masses m and $2m$, are suspended from opposite ends of a rope passed over a light, frictionless pulley. What is the acceleration of the center of mass of the system? (*Note:* First define the system.)

10 • 6 Two identical squares of mass m and side l are located side by side at the origin, and one square is being pulled to the right with a constant force $\mathbf{F}$. (a) Find the momentum of the center of mass of the system. (b) What is the position of each square's center in center-of-mass coordinates? (c) What is the momentum of each square in center-of-mass coordinates? (*Hint:* Treat each square as a particle with its mass at its geometric center.)

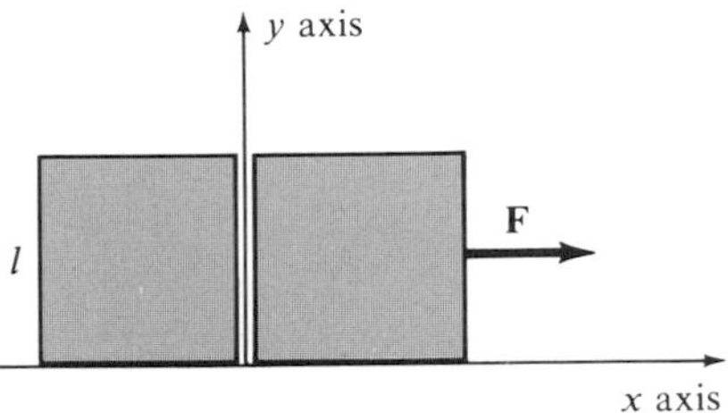

10 • 7 Two identical particles of mass m make a direct collision, and according to an observer moving with the center of mass, they lose half their energy on impact. What is the fractional kinetic energy loss as seen by an observer moving with the initial velocity of one of the particles?

10 • 2 center-of-mass coordinates

10 • 8 A 100-g mass m_1 with a velocity $\mathbf{u}_1 = 40\mathbf{i}$ cm/s strikes a 500-g mass m_2 that is initially at rest. The two objects stick together and after the impact are observed to move with velocity $\mathbf{V}$. (a) What is their combined velocity $\mathbf{V}$ after collision? (b) What is the speed v_c of the center of mass? (c) Compute the initial velocities $\mathbf{u}_1'$ and $\mathbf{u}_2'$ in center-of-mass coordinates. (d) Compute the initial total momentum of the two-particle system in center-of-mass coordinates. (e) Compute the initial kinetic energy of the system in both stationary and center-of-mass coordinates. (f) Compute the final kinetic energy in both coordinate systems. (g) The result in part (f) indicates why the center-of-mass system is particularly handy in dealing with inelastic collisions; explain this.

10 • 9 Two particles of masses m_1 and m_2 collide elastically with *speeds* u_1' and u_2' in the center-of-mass system. (a) Find their recoil speeds v_1' and v_2'. (b) If the particles recoil at some angle θ' to their initial directions as shown, find their recoil speeds v_1' and v_2'. (c) A stationary observer sees the two-particle system as moving with some constant speed V parallel to the initial direction of m_1; find the relationship between v_1' and V. (d) What is the recoil angle θ of particle m_1 as seen by this observer?

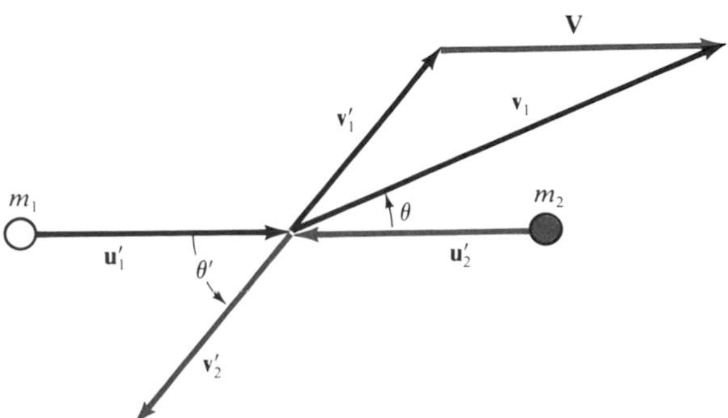

10 • 3–10 • 4 center of mass of a solid and symmetry

10 • 10 Find the center of mass of a right circular cylinder of height h with a density that decreases uniformly from ρ at the bottom to $\frac{1}{2}\rho$ at the top.

10 • 11 Two homogeneous rods of equal density ρ are joined together as shown. One rod is of length l and diameter d and the other rod is of length $2l$ and diameter $\frac{1}{2}d$. (*a*) How far from the left end of the first rod is the center of mass of the combination? (*b*) If a hole of diameter $\frac{1}{4}d$ is drilled through both pieces along their axis, how much will the center of mass be displaced?

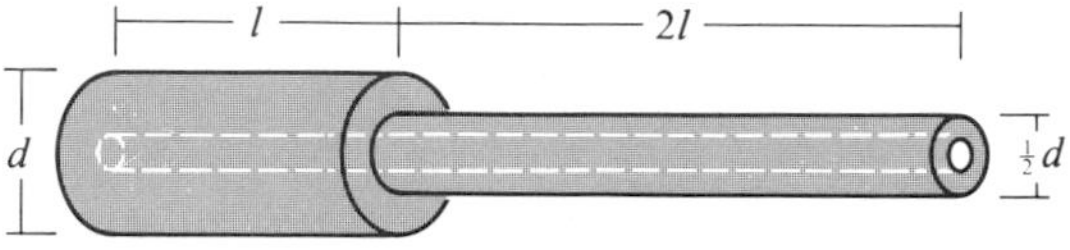

10 • 12 Suppose the disk in Example 10 . 7 is 1.06 cm thick and weighs 0.8136 kg. (*a*) What is its surface density σ? (*b*) What is its volume density ρ? (*c*) What is its specific gravity relative to mercury at 20°C, for which $\rho = 13.546 \text{ g/cm}^3$?

10 • 13 A thin iron rod 63 cm long is bent into an L shape. If the two segments are 36 and 27 cm long, respectively, find the coordinates of the center of mass relative to the vertex.

10 • 14 Find the center of mass of a right circular cone of height h, with a base of diameter d.

10 • 15 The equation for a cardioid, expressed in polar coordinates, is $r = 1 + \cos\theta$. Locate its center of mass, (r_c, θ_c).

10 • 16 A 1-ft cube has a cylindrical cavity 10 in across and 8 in deep cut centrally in the top. Show that if this cavity is half filled with a liquid one-fourth as dense as the material composing the vessel, the center of mass will be 7 in below the top of the vessel.

10 • 17 Show that the center of mass of a triangular plate lies at the intersection of the lines joining the three vertices to the midpoints of the opposite sides.

10 • 18 Find the coordinates of the center of mass of a sector of a circle of radius r which subtends an angle of 2α. Let the x axis bisect the sector. (*Hint:* Use the results of Example 10 . 5.)

10 • 19 Locate the center of gravity of a homogeneous plane sheet of metal shaped as shown. Note the hole centered in the lower portion.

6 m
10 m
4 m
4 m
2 m

10 • 20 Four thin disks of uniform density and masses of 1, 2, 3, and 4 kg, respectively, are arranged so that their centers form a square of area $A = 4\ \text{m}^2$. (*a*) Find the center of mass of this system. (*b*) Find the center of mass relative to a coordinate system with origin O' at the center of the square. (*c*) If a sphere of mass 6 kg is supported with its center 2 m above the center of the square, where is the center of mass of the five bodies?

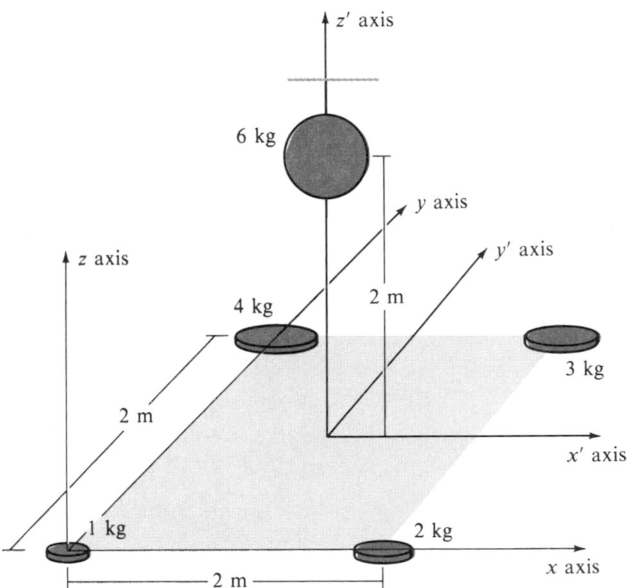

10 • 21 Consider the three two-dimensional curves given by

$$(1)\ \frac{x^2}{a^2} + \frac{y^2}{b^2} = 1$$

$$(2)\ x^2 + y^2 = y$$

$$(3)\ (x^2 + y^2)^{3/2} = 2xy$$

(*a*) Which are symmetrical with respect to the y axis? (*b*) Which are symmetrical with respect to the x axis? (*c*) Which are symmetrical with respect to the origin?

10 • 5 numerical computation of center of mass

10 • 22 By appropriate scaling of lengths, use Eq. [10 . 33] to locate the center of mass of the upper right quadrant of an ellipse

$$\frac{x^2}{a^2} + \frac{y^2}{b^2} = 1$$

****10 • 23*** Find the center of mass of a planar body shaped like the curve $y = \cos^2 \pi x$ cm over the interval $0 \leq x \leq 0.5$ cm.

****10 • 24*** Assume that a piece is cut from the planar body of Prob. 10 . 23. The cutout portion is the upper right quadrant of a circle of radius $R = 0.25$ cm about the origin. Find the center of mass of the remainder of the body.

10 • 25 Assume that the quadrant in Fig. 10 . 13 represents a body of homogeneous density which is symmetrical about the xy plane but tapers linearly with the radius from a thickness of two units at the center to a thickness of one unit at the bounding arc. Find its center of mass.

10 • 26 If the outer boundary of the quadrant in Fig. 10 . 13 is replaced by six equal straight-line segments, where will the center of mass of this body be located?

answers

10 • 1 (*a*) $\mathbf{r}_c = \frac{1}{2}l(\mathbf{i} + \mathbf{j} + \mathbf{k})$
(*b*) same as (*a*)
10 • 2 (*a*) $r_0 = [(x_2 - x_1)^2 + (y_2 - y_1)^2 + (z_2 - z_1)^2]^{1/2}$;
(*b*) $r_1 = m_2 r_0/(m_1 + m_2)$;
(*c*) $r_2 = m_1 r_0/(m_1 + m_2)$
10 • 3 (*a*) $y_c = 16.7$ Å; (*b*) $y_c = 15.0$ Å, x_c directly below the vertex
10 • 4 (*a*) $\mathbf{r}_c = \frac{1}{4}[(3 + 3\sin t^2)\mathbf{i} + t\mathbf{j} + (3t - \sqrt{t})\mathbf{k}]$;
(*b*) $\mathbf{a}_c = \frac{3}{2}(\cos t^2 - 2t^2 \sin t^2)\mathbf{i} + \frac{1}{16}t\mathbf{k}$;
(*c*) same as (*b*)
10 • 5 $\mathbf{a}_c = -\frac{1}{9}g\mathbf{j}$
10 • 6 (*a*) $\mathbf{p}_c = \mathbf{F}t$;
(*b*) $r_1' = \frac{1}{2}l + (F/4m)t^2$, $r_2' = r_1'$, but on the opposite side; $\mathbf{p}' = \pm\frac{1}{2}\mathbf{F}t$
10 • 7 $\Delta K/K = 25$ percent
10 • 8 (*a*) $\mathbf{V} = \frac{40}{6}\mathbf{i}$ cm/s;
(*b*) $v_c = \frac{40}{6}$ cm/s;
(*c*) $\mathbf{u}_1' = \frac{200}{6}\mathbf{i}$ cm/s, $\mathbf{u}_2' = -\frac{40}{6}\mathbf{i}$ cm/s;
(*d*) $\mathbf{p}_1' + \mathbf{p}_2' = \mathbf{0}$;
(*e*) $K = 80{,}000$ ergs, $K' = 66{,}667$ ergs;
(*f*) $K = 13{,}333$ ergs, $K' = 0$
10 • 9 (*a*) $v_1' = u_1'$, $v_2' = u_2'$; (*b*) same as (*a*);
(*c*) $v_1' = m_2 V/m_1$;
(*d*) $\tan\theta = \sin\theta' (\cos\theta' + m_1/m_2)^{-1}$
10 • 10 $y_c = \frac{4}{9}h$
10 • 11 (*a*) $x_c = l$; (*b*) $\Delta x_c = -\frac{1}{14}l$
10 • 12 (*a*) $\sigma = 9.25$ g/cm^2;
(*b*) $\rho = 8.73$ g/cm^3; (*c*) 0.644
10 • 13 $x_c = 5.8$ cm, $y_c = 10$ cm (for shorter arm as x axis)
10 • 14 $y_c = \frac{1}{4}h$ on the axis of the cone
10 • 15 $r_c = 0.833$, $\theta_c = 0$
10 • 18 $x_c = (2r/3\alpha)\sin\alpha$, $y_c = 0$
10 • 19 $\mathbf{r}_c = 0.983$ m above center of hole
10 • 20 $\mathbf{r}_c = (1\mathbf{i} + 1.4\mathbf{j})$ m; (*b*) $\mathbf{r}_c' = 0.4\mathbf{j}$ m;
(*c*) $\mathbf{r}_c' = (0.25\mathbf{j} + 0.75\mathbf{k})$ m
10 • 21 (*a*) Curves 1 and 2; (*b*) curve 1;
(*c*) curves 1 and 3

10 • 22 $x_c = 0.4244a$, $y_c = 0.4244b$
10 • 23 $x_c = 0.1487$ cm, $y_c = 0.3750$ cm
10 • 24 $x_c = 0.1591$ cm, $y_c = 0.4408$ cm
10 • 25 $x_c = y_c = 0.398$
10 • 26 $x_c = y_c = 0.4256$

CHAPTER ELEVEN

rotations and torques

I postulate the following:

I Equal weights at equal distances are in equilibrium, and equal weights at unequal distances are not in equilibrium but incline towards the weight which is at the greater distance.

II If, when weights at certain distances are in equilibrium, something be added to one of the weights, they are not in equilibrium but incline towards that weight to which the addition was made.

Archimedes, On the equilibrium of planes or the centres of gravity of planes, ca. 287–212 B.C.

In the preceding chapter we described the *translational motion* of extended bodies in terms of their centers of mass. In this chapter and the next we shall develop a description of *rotational motions* of particles and extended bodies. A rotating body may be regarded as a system of particles to which Newton's laws of motion apply. However, we shall find it more efficient to define *rotational analogs* of the mechanical quantities we have been using so far, in which angles take the place of distances. Although the point of application of an external force has no effect on the motion of the center of mass, it does affect the rotational motion of the body. We must therefore incorporate the concept of distance into our treatment of forces. This leads to a new physical quantity, the *moment of force,* or *torque,* which we shall find essential in describing the forces on an extended body in static equilibrium.

We may define *equilibrium* as a situation in which the total kinetic energy of a system is constant. In the case of *static equilibrium,* that constant is zero. In this situation the observer must be stationary with respect to the center of mass of the system, so that he can distinguish between cases in which there is no rotational motion and those in which there may be rotation even though the center of mass is at rest or moving with constant velocity.

Although the science of dynamics in its broadest sense goes back only to Galileo, the scientific treatment of *statics,* the branch of physics dealing with static equilibrium, began with the ancient Greeks. Archimedes, the greatest mathematician, physicist, and engineer of antiquity, developed a mathematical theory of levers and of the center of mass which remained substantially unaltered for 1800 years. In the sixteenth century Stevin again studied the lever and attacked such problems as the inclined plane. The revived theory of statics was later absorbed as a special case into the more fundamental and powerful dynamics of Galileo and Newton.

11 • 1 rotational kinematics

Imagine that a peanut is mounted on a stiff wire along the z axis, as shown in Fig. 11 . 1a. The axis of symmetry of the peanut between its endpoints A and B is neither perpendicular nor parallel to the wire. If we now cause the peanut to rotate about the wire, its opposite ends describe circles about the z axis. Suppose there is a light shining along the z axis, so that the peanut casts a shadow on the xy plane. We can see that the circles of rotation of the

figure 11 • 1 *Rotation of an asymmetrically mounted rigid body about a fixed axis in space*

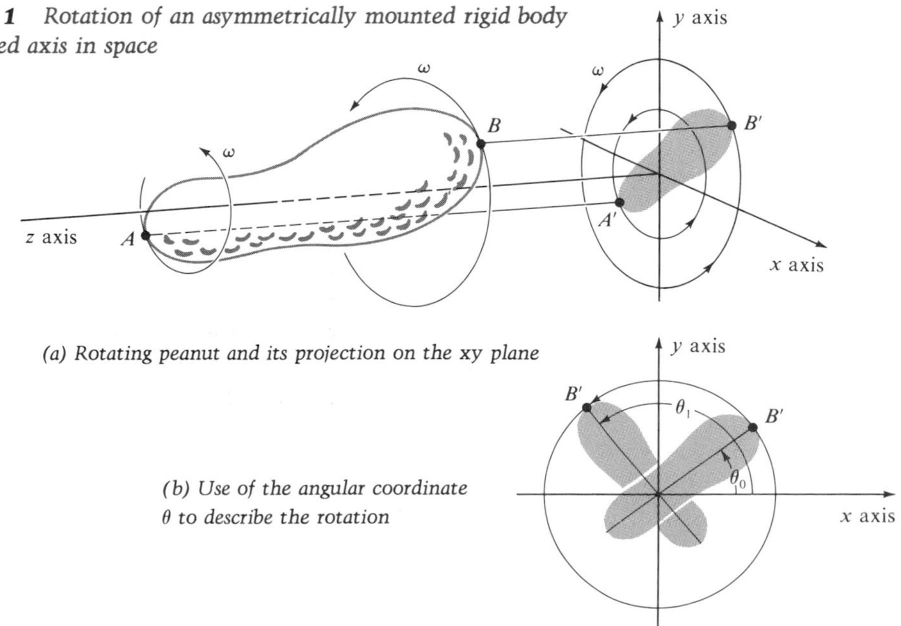

(a) Rotating peanut and its projection on the xy plane

(b) Use of the angular coordinate θ to describe the rotation

endpoints A and B are concentric about the z axis and the circles of rotation of their shadows A' and B' in the xy plane are concentric about the origin. In fact, every point of the peanut rotates about the z axis, forming concentric circles whose centers lie on the z axis.

Now, instead of using our intuitive understanding of rotation to determine its mathematical properties, let us retrace our reasoning in order to arrive at a mathematical definition of the rotation of a rigid body about a fixed axis:*

> If a rigid body moves such that every particle in the body describes an arc of a circle and all such circles are concentric about a fixed axis, then the body is said to be rotating about a fixed axis, which is the axis of rotation.

Note from the figure that points on the axis of rotation may or may not lie within the body. For example, when a tire rotates, its axis of rotation is the axle, which is not part of the tire, whereas when a top spins very rapidly in place, its axis of rotation lies along the axis of symmetry of the top.[†]

When a rigid body rotates about a fixed axis, we do not need a separate set of coordinates for each point, since, by definition, the distance between any two points is constant. In fact, we do not even need three coordinates to describe its motion. All we need is one coordinate—an angle in a plane perpendicular to the axis of rotation. Figure 11 . 1*b* shows the motion of the shadow of the peanut in the xy plane and the angular coordinate θ which describes that motion. For example, if θ_0 is the angular position of the shadow B' point B at time t_0 and θ_1 is its position at time t_1, then in the time interval $\Delta t = t_1 - t_0$ the shadow—and, of course, the peanut—has undergone an angular displacement $\Delta\theta = \theta_1 - \theta_0$. Its *average angular velocity* during this interval is

$$\bar{\omega} = \frac{\Delta\theta}{\Delta t} \qquad [11 \cdot 1]$$

and in the limit as Δt approaches zero, we obtain the *instantaneous angular velocity* ω:

$$\omega = \lim_{\Delta t \to 0} \frac{\Delta\theta}{\Delta t} = \frac{d\theta}{dt} \qquad [11 \cdot 2]$$

This is precisely analogous to velocity v in one dimension. As in the case of linear speed, instantaneous angular speed $|\omega|$ is the magnitude of angular velocity, since the direction of rotation may be either clockwise or counterclockwise. For motion in only one dimension we can represent all angular quantities in scalar form, with the direction of motion specified by sign. If the rotation is counterclockwise, ω is positive; if it is clockwise, ω is negative.

*This definition does not allow for the case where every point in the body also has the same translational component of velocity, as with a spinning projectile, but it is correct as far as it goes. We will discuss combined rotations and translations at the end of the next chapter.

[†]When the top starts to wobble, or *nutate,* matters become more complicated. Whole books have been written on this subject.

If the speed of rotation is uniform, then ω is constant. For convenience, we measure angular speed ω in radians per second (rad/s). The number of revolutions per second, or *frequency of rotation* is denoted by ν,

$$\nu = \frac{\omega}{2\pi} \qquad [11 \cdot 3]$$

with a unit of 1 hertz (Hz) equal to 1 cycle per second (cps).

If θ is expressed in radians, then segments of arc length are given by $\Delta s = r\,\Delta\theta$ and the following calculus formulas apply:

$$\frac{d(\sin\theta)}{d\theta} = \cos\theta \qquad \frac{d(\cos\theta)}{d\theta} = -\sin\theta \qquad [11 \cdot 4]$$

Also, for $\theta \ll 1$, $\sin\theta \doteq \theta$ and $\cos\theta \doteq 1 - \frac{1}{2}\theta^2$.

When a rigid body rotates about a fixed axis with varying speed, its *angular acceleration* α is given by

$$\blacktriangleright \quad \alpha = \lim_{t_1 \to t_0} \frac{\omega_1 - \omega_0}{t_1 - t_0} \lim_{\Delta t \to 0} \frac{\Delta\omega}{\Delta t} = \frac{d\omega}{dt} = \frac{d^2\theta}{dt^2} \qquad [11 \cdot 5]$$

If ω is measured in radians per second, α will be expressed in radians per second squared.

If the angular *velocity* is either constant or a linear function of time, then the angular acceleration is obviously constant. Therefore, just as in the case of constant linear acceleration, if we integrate $\alpha = d\omega/dt$ directly,

$$\int_{\omega_0}^{\omega} d\omega = \int_{t_0}^{t} \alpha\, dt$$

we obtain an expression for the angular *velocity* at any time t:

$$\blacktriangleright \quad \omega = \omega_0 + \alpha(t - t_0) \qquad [11 \cdot 6]$$

Similarly, setting $\omega = d\theta/dt$ in Eq. [11 . 6] and integrating,

$$\int_{\theta_0}^{\theta} d\theta = \omega_0 \int_{t_0}^{t} dt + \alpha \int_{t_0}^{t} (t - t_0)\, dt$$

gives us the angular position at any time t:

$$\blacktriangleright \quad \theta = \theta_0 + \omega_0(t - t_0) + \tfrac{1}{2}\alpha(t - t_0)^2 \qquad [11 \cdot 7]$$

a form exactly analogous to Eq. [3 . 21] for uniformly accelerated translational motion. Use of the chain rule

$$\alpha = \frac{d\omega}{dt} = \frac{d\omega}{d\theta}\frac{d\theta}{dt} = \omega \frac{d\omega}{d\theta}$$

yields

$$\int_{\omega_0}^{\omega} \omega\, d\omega = \int_{\theta_0}^{\theta} \alpha\, d\theta$$

or

$$\blacktriangleright \quad \omega^2 = \omega_0^2 + 2\alpha(\theta - \theta_0) \qquad [11 \cdot 8]$$

analogous to Eq. [3 . 33]. We could also have obtained this result from Eq. [11 . 5] by writing Eq. [11 . 6] as

$$t - t_0 = \frac{\omega - \omega_0}{\alpha}$$

and substituting for $t - t_0$ in Eq. [11 . 7].

example 11 • 1 When the switch of a certain motor is opened, its frequency of rotation ν is observed to change uniformly from 33 Hz to 3 Hz in 15 s. Find (*a*) the angular acceleration of the motor and (*b*) its angular displacement $\Delta\theta$ during the 15-s interval.

solution (*a*) From Eq. [11 . 5] and the fact that $\omega = 2\pi\nu$, the angular acceleration is

$$\alpha = 2\pi \frac{3 - 33}{15} = -4\pi \text{ rad/s}^2$$

(*b*) From Eq. [11 . 7], the angular position at $t = 15$ s is

$$\theta = \omega_0 t + \tfrac{1}{2}\alpha t^2 = 66\pi \times 15 - \tfrac{1}{2} \times 4\pi \times 225 = 540\pi \text{ rad}$$

or, from Eq. [11 . 8],

$$\theta - \theta_0 = \frac{\omega^2 - \omega_0^2}{2\alpha} = (2\pi)^2 \frac{9 - 1089}{-8\pi} = 540\pi \text{ rad}$$

11 • 2 uniform circular motion

An important special case of rotational motion is that of motion around a circle at constant speed. As the particle moves, its velocity vector is at all times tangent to the circle. Although the magnitude of the velocity does not change, its direction does; hence the particle must be continuously accelerating. We wish to determine that acceleration and to be able to express the position, velocity, and acceleration of the particle in terms of its angular velocity. The angular displacement of the particle in a time interval $t - t_0$ is given by

$$\theta - \theta_0 = \omega(t - t_0) \qquad [11 \cdot 9]$$

where θ_0 is the initial angular position and $\omega = d\theta/dt$. For positive ω the rotation is counterclockwise in the direction of increasing θ. Every time θ increases by 2π, the particle has made one complete rotation about the circle. The time required for this is called the *period of rotation:**

$$T = \frac{2\pi}{\omega} \qquad [11 \cdot 10]$$

*Strictly speaking, conventional formulas such as Eq. [11 . 10] should refer to the absolute value $|\omega|$, since in one-dimensional problems ω may be negative but T is always taken to be a positive quantity.

Conversely, if T is the time required for one rotation, then its inverse must be the number of rotations per unit time, or the frequency ν:

$$\nu = \frac{1}{T} = \frac{\omega}{2\pi} \qquad [11 \cdot 11]$$

If the radius of the circle is given, the position $\mathbf{r}$ of the particle can be specified as a function of its angular distance θ (in radians) from the x axis:

$$\mathbf{r}(\theta) = R(\cos\theta\,\mathbf{i} + \sin\theta\,\mathbf{j}) = R\hat{\mathbf{r}} \qquad [11 \cdot 12]$$

where R is the *radius of rotation* and $\mathbf{r}$ is the position vector of the particle relative to the origin of coordinates (see Fig. 11 . 2).

When the particle moves through an angular distance $\Delta\theta$, it traverses an arc of length $\Delta s = R\,\Delta\theta$. Therefore its constant tangential speed v is

$$v = \lim_{\Delta t \to 0} \frac{\Delta s}{\Delta t} = R\left|\frac{d\theta}{dt}\right| = R|\omega| \qquad [11 \cdot 13]$$

To determine the velocity vector $\mathbf{v}$, we can apply the chain rule of calculus to differentiate the position vector:

$$\mathbf{v} = \frac{d\mathbf{r}}{dt} = \frac{d\mathbf{r}}{d\theta}\frac{d\theta}{dt} \qquad [11 \cdot 14]$$

Then, since $d(\cos\theta)/d\theta = -\sin\theta$ and $d(\sin\theta)/d\theta = \cos\theta$,

$$\mathbf{v} = R\omega(-\sin\theta\,\mathbf{i} + \cos\theta\,\mathbf{j}) = R\omega\hat{\boldsymbol{\theta}} \qquad [11 \cdot 15]$$

where the unit vector $\hat{\boldsymbol{\theta}}$ always points in the direction of increasing θ. The unit vectors

$$\hat{\mathbf{r}} = \cos\theta\,\mathbf{i} + \sin\theta\,\mathbf{j} \qquad \hat{\boldsymbol{\theta}} = -\sin\theta\,\mathbf{i} + \cos\theta\,\mathbf{j} \qquad [11 \cdot 16]$$

are always mutually perpendicular, so that the velocity is always tangential to the circle, $\mathbf{v} \cdot \mathbf{r} = 0$.

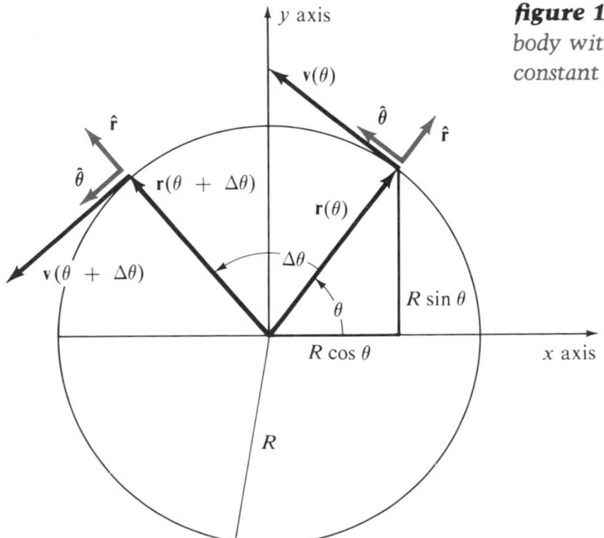

figure 11 • 2 *Uniform circular motion of a body with constant tangential speed v and constant angular velocity $\omega = d\theta/dt$*

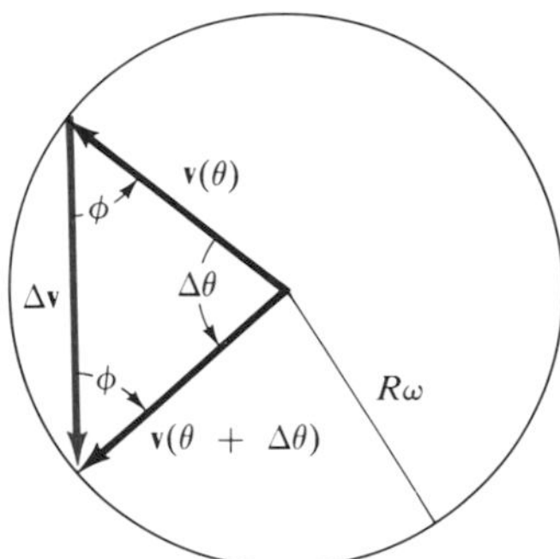

figure 11 • 3 *Reference circle for velocity in the case of uniform circular motion*

In a similar fashion we can find the acceleration **a** of the particle, since the vector **v** *is also rotating.* When the two velocity vectors $\mathbf{v}(\theta)$ and $\mathbf{v}(\theta + \Delta\theta)$ in Fig. 11 . 2 are placed at the same origin for comparison as in Fig. 11 . 3, it is apparent that we can also view them as two radii of a second reference circle of "radius" $R\omega$, since both vectors are of the same length. Simple geometry requires that $\Delta\theta$ also be the angle between them. Hence

$$\Delta v \doteq v\,\Delta\theta \qquad [11 \cdot 17]$$

and the magnitude of the acceleration must be

► $$a \lim_{\Delta t \to 0} \frac{\Delta v}{\Delta t} = v\omega = R\omega^2 = \frac{v^2}{R} \qquad [11 \cdot 18]$$

As $\Delta\theta$ approaches zero, the angles ϕ in Fig. 11 . 3 approach right angles, so that the instantaneous acceleration must be perpendicular to the tangential velocity—that is, directed *back toward the center.* This *centripetal acceleration vector* can be obtained by straightforward differentiation of Eq. [11 . 15]:

► $$\mathbf{a} = \frac{d\mathbf{v}}{dt} = R\omega^2(-\cos\theta\,\mathbf{i} - \sin\theta\,\mathbf{j}) = -R\omega^2\hat{\mathbf{r}} \qquad [11 \cdot 19]$$

In nature, uniform circular motion will not occur of itself, but must be produced by some *centripetal force* $\mathbf{F}_c$, which always points toward the center of rotation. The force may be independent, like gravity, or it may be a reaction force or a resistive force; whatever its origin, it must satisfy

► $$F_c = ma = \frac{mv^2}{R} = m\omega^2 R \qquad [11 \cdot 20]$$

It is often difficult to visualize centripetal force as pulling inward, toward the center. After all, if you twirl a mass around on a string, you can feel the string trying to pull out of your fingers. However, the force you feel in this case is the equal and opposite *reaction force* to the centripetal force that you are actually exerting on the mass through the medium of the string. If you suddenly release the string, the mass will fly off in a direction tangent to the point on its circle at which it was released, and not radially outward, as you might expect.* Similarly, when you drive around a curve you feel an outward "centrifugal" force. However, since you are accelerating, your

*If David had not known this, instead of slaying Goliath, he probably would have demolished the Israelite standing on his left.

frame of reference is noninertial, and what you feel is your own inertial reaction to a change in direction (see Sec. 5 . 6). The actual force acting on you is the centripetal one causing your moving frame of reference to rotate about some center.

example 11 • 2 A high-speed ultracentrifuge of radius $R = 10$ cm used in a chemistry laboratory has a frequency of $\nu = 1000\ \mathrm{s^{-1}}$. (*a*) What is the centripetal force on a 0.01-g particle being rotated in the ultracentrifuge? (*b*) If the particle is suspended in a liquid solution inside the ultracentrifuge, what is the effective force causing it to settle out of solution? Express the force in units of $g = 980\ \mathrm{cm/s^2}$.

solution (*a*) At this frequency the centripetal acceleration of the particle is given by $a = R\omega^2$. Hence the force acting on it must be

$$F_c = ma = mR\omega^2 = mR(2\pi\nu)^2 = 3.95 \times 10^6 \text{ dynes}$$

(*b*) In units of particle weight, the force is $F/mg = 4.03 \times 10^5$, or 400,000 times the pull of gravity. The inertial reaction to this rotation causes the particles to move into the outer end of their container so that they can be separated from the liquid.

When a driver turns the front wheels of an automobile in rounding a curve, static friction arises as a reaction force opposing the original direction of motion, and it is this reaction force that causes the car to turn. The inertial reaction of the driver, $F_c' = mv^2/R$, is felt as a force pushing him sideways relative to the turn, and if the road is slippery, the car and driver may slide off altogether. Actually, the car is merely continuing on a straight path, and it is the road that is turning away. The solution to this hazard is to bank the road at an angle θ_b such that the resultant of the gravitational and inertial centrifugal reaction forces is directed normal to the roadbed (see Fig. 11 . 4*a*). This requires that

$$\tan\theta_b = \frac{F_c}{mg} = \frac{v^2}{Rg} \qquad [11 \cdot 21]$$

figure 11 • 4 *Views of banked curve*

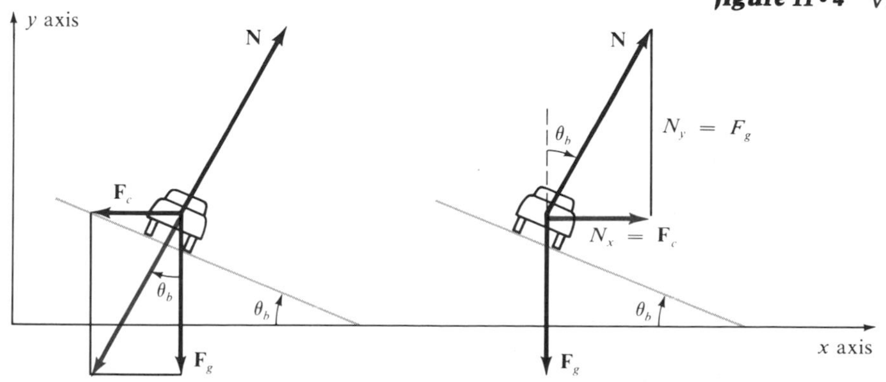

(*a*) *Noninertial view*

(*b*) *Inertial view*

Note that in the noninertial system of the automobile, this is essentially a static problem.

Since it does not matter which reference frame we use as long as the physical reasoning is correct, we should obtain the same result if we approach the problem from the reference frame of a stationary observer (Fig. 11 . 4*b*). From this viewpoint, the normal force supplied by the road surface must be sufficient both to support the vehicle against gravity and to turn it in a circle, so that it is unnecessary to rely on friction to make the turn. The result is the same as in Eq. [11 . 21].

example 11 • 3 A truck full of sausages, each of mass m, rounds a curve of radius R at a speed v. If the sausages are hanging by strings from the roof of the truck, what angle ϕ do they make with the vertical as the truck drives around the curve?

solution The tension **T** in the rope must balance the weight of the sausage *and* provide centripetal acceleration. Components of **T** are

$$T_x = \frac{mv^2}{R} \qquad T_y = mg$$

and from Eq. [11 . 21],

$$\tan\phi = \frac{T_x}{T_y} = \frac{v^2}{Rg}$$

Therefore

$$\phi = \tan^{-1}\frac{v^2}{Rg}$$

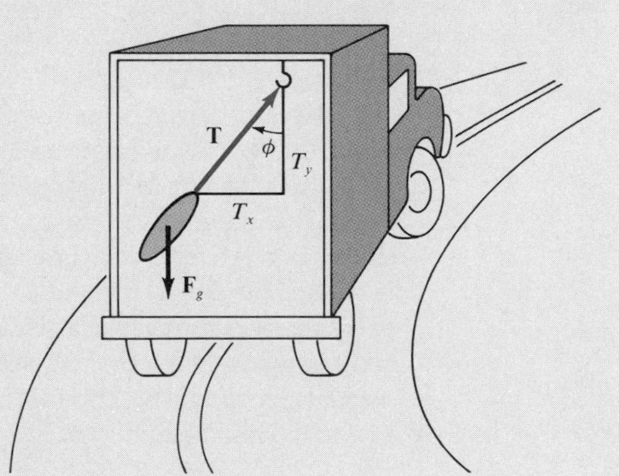

figure 11 • 5

example 11 • 4 A small metal object of mass m starts from rest at the top of a smooth incline of height h. When it reaches the bottom of the incline, it slides around the inside of a smooth hoop of radius R and is then released. (*a*) What is its speed when it reaches the top of the hoop? (*b*) What is the normal force exerted on the object at that point? (*c*) What is the minimum value of y such that it will not leave the track?

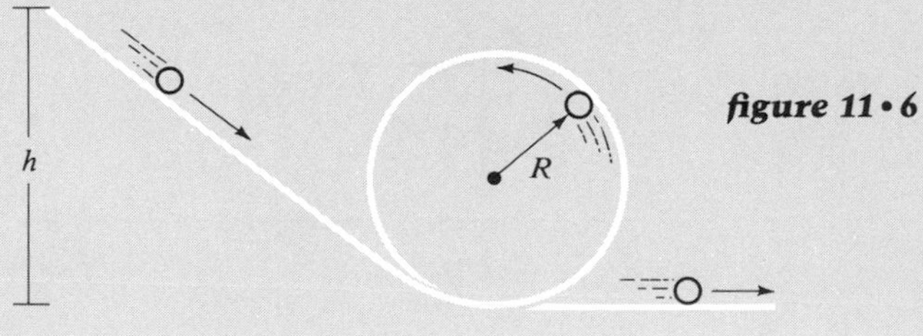

figure 11 • 6

solution (*a*) At the bottom of the incline the kinetic energy of the object is $K = \frac{1}{2}mv^2 = mgh$. Therefore it enters the hoop with a speed $v = \sqrt{2gh}$. Since there is no energy loss as a result of friction, at the top of the hoop the speed of the object is

$$K = \tfrac{1}{2}mv^2 = mg(h - 2R) \qquad \text{or} \qquad v = \sqrt{2g(h - 2R)}$$

(*b*) The centripetal force on the object is provided by the normal force N and the force of gravity $F_g = mg$:

$$N + F_g = \frac{mv^2}{R} \qquad \text{or} \qquad N = \frac{2mgh}{R} - 5mg$$

(*c*) The object just barely keeps to the track when N vanishes, or $h_{\min} = \frac{5}{2}R$, half a radius higher than the hoop.

11 • 3 *rotational vector quantities*

Let us now take a more formal approach to the use of vectors to represent angular velocity and acceleration. The two specifications that describe an angular displacement $\Delta\theta$ are (1) the angle θ through which a body has rotated and (2) the direction of the axis of rotation. These two independent specifications can be represented graphically by a directed line of length numerically equal to the number of radians in θ, laid off *along the axis of rotation* in accordance with the right-hand screw rule (see Sec. 2 . 4).

Both angular velocity and angular acceleration may be represented by vectors. Angular displacement $\Delta\theta$ is not a vector, however, since the addition of two successive finite angular displacements about different axes is not commutative. To see this, hold a book in any fixed position and picture a set of rectangular coordinate axes drawn in it. Rotate the book 90° in a clockwise sense about the x axis and then, from this position, rotate it 90° counterclockwise about the y axis (Fig. 11 . 7*a*). Note the final position of

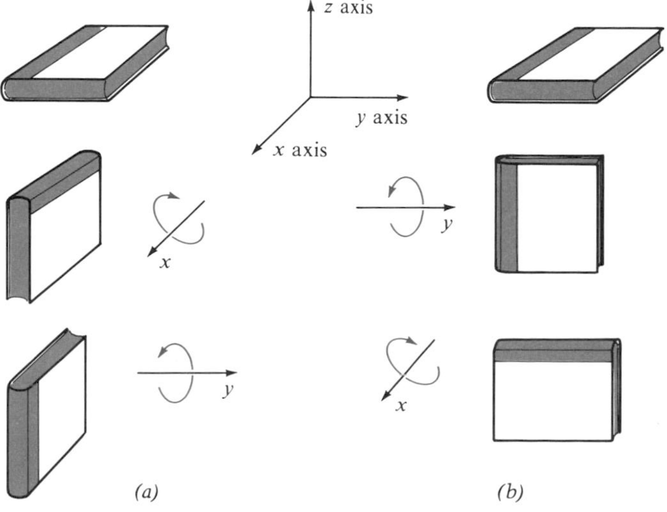

figure 11 • 7 *Rotations of a book through 90°, (a) first clockwise about the x axis, then counterclockwise about the y axis, and (b) the same rotations performed in reverse order*

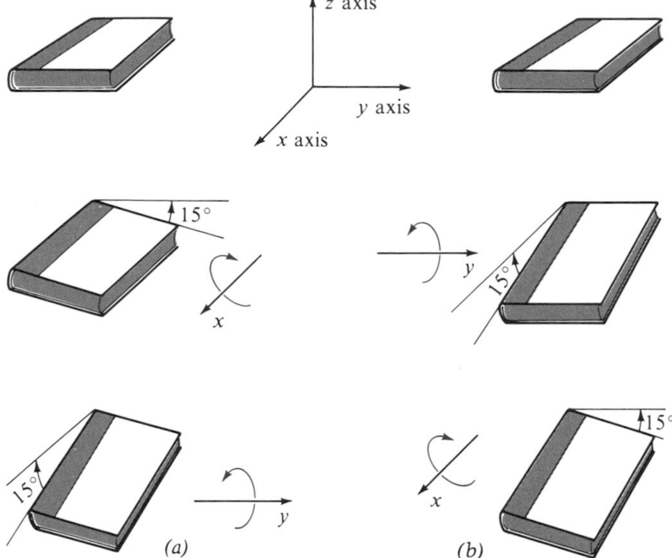

figure 11 • 8 *Rotations of a book through 15°, (a) first clockwise about the x axis, then counterclockwise about the y axis, and (b) the same rotations performed in reverse order*

the book. Now reverse the order of these two rotations, starting with the book in the same initial position as before. The resulting final position in Fig. 11 . 7*b* is quite different. However, if you repeat the experiment, this time with the two displacements made very small, as in Fig. 11 . 8, you will find that the sum of two successive angular displacements is nearly independent of the order in which the quantities are added. Infinitesimal angular displacements do commute, and for this reason an *infinitesimal displacement* $d\theta$ may be regarded as a vector quantity.

Figure 11 . 9 shows a particle located by $\mathbf{r}$ and rotating counterclockwise, with radius of rotation R, about an axis that is normal to the plane of rotation and passes through O. The change $d\mathbf{r}$ in $\mathbf{r}$ due to an infinitesimal angular displacement $d\theta$ about the axis is perpendicular to the plane formed by the directions $\hat{\mathbf{n}}$ and $\hat{\mathbf{r}}$ and thus points in the direction given by $\hat{\mathbf{n}} \times \hat{\mathbf{r}}$ (see Fig. 11 . 10). If this is not intuitively obvious, just remember that the plane of rotation, which contains $d\mathbf{r}$, must be perpendicular to $\hat{\mathbf{n}}$, and also that the *length* of the position vector is unchanged:

$$(\mathbf{r} + d\mathbf{r}) \cdot (\mathbf{r} + d\mathbf{r}) = \mathbf{r} \cdot \mathbf{r} \qquad [11 \cdot 22]$$

If we ignore $d\mathbf{r} \cdot d\mathbf{r}$ as a second-order infinitesimal, then

$$\mathbf{r} \cdot d\mathbf{r} = 0 \qquad [11 \cdot 23]$$

which is equivalent to saying that $d\mathbf{r}$ is perpendicular to $\mathbf{r}$, and hence to the plane containing $\hat{\mathbf{r}}$ and $\hat{\mathbf{n}}$. Therefore $d\mathbf{r}$ is parallel to the direction of $\hat{\mathbf{n}} \times \hat{\mathbf{r}}$.

Since the arc $ds = R\, d\theta$ which subtends $d\theta$ approaches dr in length, and $R = r \sin\phi = |\hat{\mathbf{n}} \times \mathbf{r}|$, we can represent the displacement vector by

$$d\mathbf{r} = ds\, \widehat{\mathbf{n} \times \mathbf{r}} = \frac{R\, d\theta}{r \sin\phi} \hat{\mathbf{n}} \times \mathbf{r} \qquad [11 \cdot 24]$$

$$d\mathbf{r} = (d\theta)\, \hat{\mathbf{n}} \times \mathbf{r} = (d\theta\, \hat{\mathbf{n}}) \times \mathbf{r} = d\boldsymbol{\theta} \times \mathbf{r} \qquad [11 \cdot 25]$$

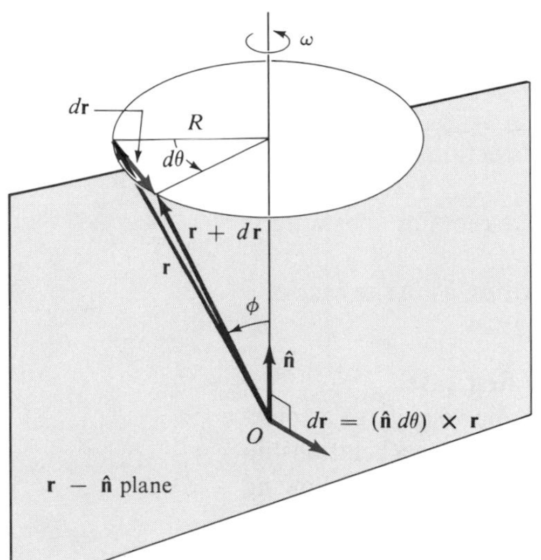

figure 11 • 9 *Relationship between angular and linear displacement*

Thus we can define the *vector infinitesimal* angular displacement

$$d\boldsymbol{\theta} = d\theta\, \hat{\mathbf{n}} \qquad [11 \cdot 26]$$

as equal in magnitude to $d\theta$ and pointing in the direction $\hat{\mathbf{n}}$ of the axis of rotation. The resultant infinitesimal linear displacement is given by Eq. [11 . 25]. Equation [11 . 26] is also consistent with clockwise rotation, $d\theta < 0$.

The construction of the angular velocity vector $\boldsymbol{\omega}$ is immediate: multiplying Eqs. [11 . 25] and [11 . 26] by the scalar $1/dt$, we obtain

$$\blacktriangleright \quad \boldsymbol{\omega} = \frac{d\boldsymbol{\theta}}{dt} = \frac{d\theta}{dt}\hat{\mathbf{n}} \qquad [11 \cdot 27]$$

and the instantaneous tangential velocity of the rotating particle is

$$\blacktriangleright \quad \frac{d\mathbf{r}}{dt} = \mathbf{v} = \boldsymbol{\omega} \times \mathbf{r} \qquad [11 \cdot 28]$$

We may also define the *vector angular acceleration* $\boldsymbol{\alpha}$ as

$$\blacktriangleright \quad \boldsymbol{\alpha} = \frac{d\boldsymbol{\omega}}{dt} = \frac{d\omega}{dt}\hat{\mathbf{n}} + \omega\frac{d\hat{\mathbf{n}}}{dt} \qquad [11 \cdot 29]$$

figure 11 • 10 *Second right-hand rule for vector cross products* **A** × **B**

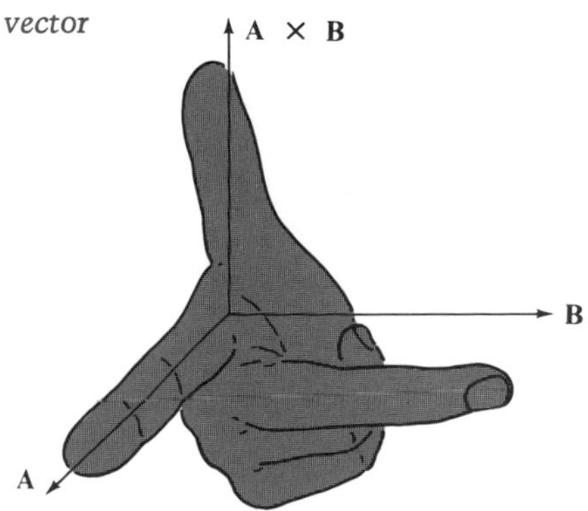

which is the rotational analog of linear acceleration. From the preceding vector equation we can see that there are three separate cases of angular acceleration:

1 A change $\Delta\omega$ in the angular speed while the direction $\hat{\mathbf{n}}$ of the axis of rotation remains fixed
2 A change in the direction $\hat{\mathbf{n}}$ of the axis while ω remains constant, as for a rapidly spinning top
3 A simultaneous change in both of these quantities, as in the case of a slowly spinning top which is wobbling to a halt

In this text we shall be primarily concerned with the first case.

Equation [11 . 28] gives the relationship between the velocity of a rotating particle and its angular velocity. It is also possible to derive a relationship between angular and linear acceleration. First let us state the following relationship:

$$\frac{d}{dt}(\mathbf{A} \times \mathbf{B}) = \frac{d\mathbf{A}}{dt} \times \mathbf{B} + \mathbf{A} \times \frac{d\mathbf{B}}{dt} \qquad [11 \cdot 30]$$

The instantaneous acceleration of a rotating particle is given by

$$\mathbf{a} = \frac{d\mathbf{v}}{dt} = \frac{d\boldsymbol{\omega}}{dt} \times \mathbf{r} + \boldsymbol{\omega} \times \frac{d\mathbf{r}}{dt}$$

$$= \boldsymbol{\alpha} \times \mathbf{r} + \boldsymbol{\omega} \times \mathbf{v} \qquad [11 \cdot 31]$$

and substituting from Eq. [11 . 28], we obtain

$$\blacktriangleright \quad \mathbf{a} = \boldsymbol{\alpha} \times \mathbf{r} + \boldsymbol{\omega} \times (\boldsymbol{\omega} \times \mathbf{r}) \qquad [11 \cdot 32]$$

The parentheses define the sequence of vector products necessary to evaluate the vector triple product; they are essential. Equation [11 . 32] is the general formula for the acceleration of a particle which is rotating with instantaneous angular velocity $\boldsymbol{\omega} = \omega\hat{\mathbf{n}}$ about some axis $\hat{\mathbf{n}}$, not necessarily fixed, which passes through the origin.

In the common special case of rotation about a *fixed* axis, $\boldsymbol{\omega} = \omega\hat{\mathbf{n}}$ and $\boldsymbol{\alpha} = \alpha\hat{\mathbf{n}}$, and the acceleration vector can be decomposed into tangential and radial components:

$$\mathbf{a} = \mathbf{a}_t + \mathbf{a}_r = \alpha\hat{\mathbf{n}} \times \mathbf{r} + \omega^2\hat{\mathbf{n}} \times (\hat{\mathbf{n}} \times \mathbf{r}) \qquad [11 \cdot 33]$$

From Fig. 11 . 9 it is apparent that $\hat{\mathbf{n}} \times \mathbf{r}$ is parallel to $d\mathbf{r}$, which lies along the tangent to the circle at $\mathbf{r}$. Now imagine taking the cross product $\hat{\mathbf{n}} \times d\mathbf{r}$ which is parallel to $\hat{\mathbf{n}} \times (\hat{\mathbf{n}} \times \mathbf{r})$; it points directly back along the radius of rotation R. Thus we have the following relationships between linear and angular quantities:

$$v = v_t = R\omega$$

$$\alpha = \frac{d\omega}{dt} \qquad a_t = R\alpha \qquad a_r = -R\omega^2 \qquad [11 \cdot 34]$$

In the special case of uniform circular motion, $\alpha = 0$ and the above relations agree with those of the previous section.

example 11 • 5 A motorcycle daredevil rides around the inside of a vertical circular track (loop-the-loop) of radius $R = 10$ m. The speed of the bike is $v = 20$ m/s and the outer radius of the tires is $R' = 30$ cm. (*a*) What is the angular velocity of the cyclist? (*b*) Assuming no slipping, what is the angular velocity of each wheel?

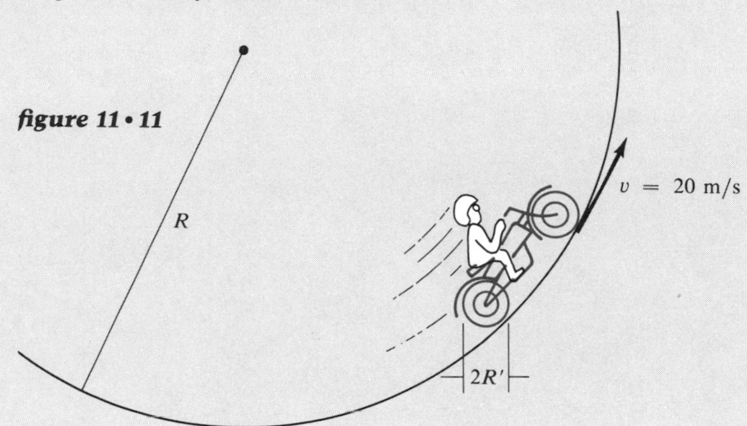

figure 11 • 11

solution (*a*) It takes the cyclist a time $T = 2\pi R/v$ to make one loop, so his angular velocity is

$$\omega_{\text{cy}} = \frac{2\pi}{T} = \frac{2\pi}{2\pi R/v} = \frac{v}{R} = 2 \text{ rad/s, directed out of the page}$$

(*b*) If the motorcycle does not slip as it goes around the loop, then during the time T of one revolution, the outer surface of each tire covers a distance $s = 2\pi R$. However, since the circumference of each tire is $2\pi R'$, that means that each tire must make $s/2\pi R' = R/R'$ revolutions. Hence the frequency of these revolutions is

$$\nu = \frac{R/R'}{T} = \frac{R}{TR'} = \frac{v}{2\pi R'}$$

The *total* angular velocity of each wheel is therefore

$$\omega_{\text{wh}} = 2\pi\nu - \omega_{\text{cy}} = \frac{v}{R'} - \frac{v}{R}$$

$$= 66.7 - 2 = 64.7 \text{ rad/s, directed into the page}$$

since the wheels rotate about about their axles in the opposite sense to the rotation of the entire motorcycle around the track.

example 11 • 6 A person opens a door by turning the doorknob in the direction shown in Fig. 11 . 12. The angular speed of the door is $\omega_d = 0.8$ rad/s, while the doorknob turns through 30° in one second. (*a*) What is the angular velocity of the doorknob at the instant the door begins to open? (*b*) At some time t after the door begins to open? (*c*) What is the angular acceleration of the doorknob?

solution (*a*) If we take the z direction vertically upward, then the angular velocity of the door is

$$\boldsymbol{\omega}_d = 0.8\mathbf{k} \text{ rad/s}$$

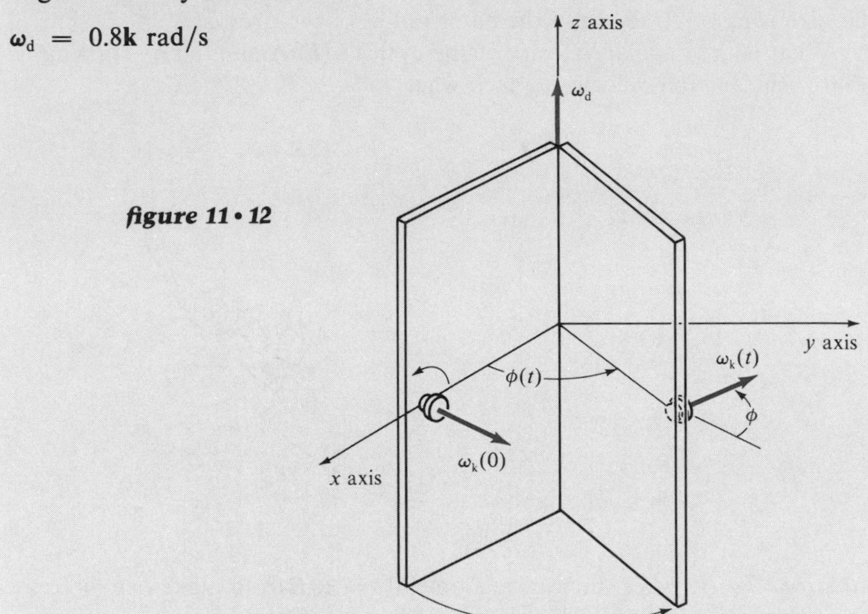

figure 11 • 12

At the instant the door begins to open, the axis of rotation of the knob is along the y axis. Therefore it has a component ω_k of angular velocity

$$\boldsymbol{\omega}_k = \tfrac{30}{180}\pi\mathbf{j} = 0.524\mathbf{j} \text{ rad/s}$$

and the total vector angular velocity of the doorknob at time $t = 0$ is

$$\boldsymbol{\omega} = \boldsymbol{\omega}_d + \boldsymbol{\omega}_k = (0.524\mathbf{j} + 0.8\mathbf{k}) \text{ rad/s}$$

(*b*) At some later time the axis of symmetry of the doorknob has rotated with the door through an angle ϕ. However, since $\phi = \omega_d t$, we have, taking the components of $\boldsymbol{\omega}_k$ in the $\mathbf{i}$ and $\mathbf{j}$ directions:

$$\begin{aligned}\boldsymbol{\omega}_k &= (0.524 \text{ rad/s})(-\sin\phi\,\mathbf{i} + \cos\phi\,\mathbf{j})\\ &= (0.524 \text{ rad/s})(-\sin 0.8t\,\mathbf{i} + \cos 0.8t\,\mathbf{j})\end{aligned}$$

and the total angular velocity of the doorknob is now a function of time:

$$\begin{aligned}\boldsymbol{\omega} &= \boldsymbol{\omega}(t) = \boldsymbol{\omega}_d + \boldsymbol{\omega}_k(t)\\ &= (0.8\mathbf{k} - 0.524\sin 0.8t\,\mathbf{i} + 0.524\cos 0.8t\,\mathbf{j}) \text{ rad/s}\end{aligned}$$

(*c*) To obtain the angular acceleration, we simply differentiate

$$\boldsymbol{\alpha} = \frac{d\boldsymbol{\omega}}{dt} = \frac{d\boldsymbol{\omega}_k}{dt} = (-0.419 \text{ rad/s}^2)(\cos 0.8t\,\mathbf{i} + \sin 0.8t\,\mathbf{j})$$

Note that $\boldsymbol{\alpha}$ does not have a constant direction, so that the doorknob does not have a fixed axis of rotation when the door is swinging open. As we can see, the angular acceleration $\boldsymbol{\alpha}$ is itself a vector which is itself rotating as the door opens.

11 • 4 torque

If the resultant of several forces acting on a stationary body is zero, the center of mass of the body does not move. However, if the separate counter-balancing forces are applied at different points on the body and the lines of force do not intersect at a single point, then the body may rotate about some axis that passes through the stationary center of mass. For example, consider the rigidly mounted wheel and axle in Fig. 11 . 13. Assume that the circumference of the wheel is smooth and there is no friction in the axle bearings. A weight $F_1 = mg$ is attached to the cord wound around the axle, and a second cord is wound around the rim of the wheel and subjected to an applied force F_2.

If the weight F_1 is raised at a constant speed through some distance h_1 by a steady rotation through an angle θ, the work done on the weight is $F_1 h_1 = F_1 r_1 \theta$. The applied force F_2 moves in the same time through some distance h_2, and the work it does on the weight is $F_2 h_2 = F_2 r_2 \theta$. No other work is done, for there is neither acceleration nor friction. Therefore $F_1 h_1 = F_2 h_2$, or

$$F_1 R_1 = F_2 R_2 \qquad [11 \cdot 35]$$

That is, these two forces will have equal turning effects about a given axis if the *products* of force times the distance from the axis to the point of application are equal. The product FR is known as the *moment* of force *with respect to the axis,* or the *torque* τ. Torque bears the same relation to rotation that force does to translational motion; it is a measure of the effectiveness of a force in producing rotation about a given axis.

The *line of action* of a force is defined to be an infinite geometric line in space which is parallel to the force and passes through its point of application. In Fig. 11 . 13 the lines of action are tangent to the wheel and the axle, respectively. The *lever arm* l of a force that produces a rotation is defined to be the *perpendicular* distance from the axis of rotation to the line of action of the force. In Fig. 11 . 13 we have a kind of continuous lever, with arms $l_1 = r_1$ and $l_2 = r_2$. The *torque* is generally defined to be the product of the force and its lever arm with respect to a given axis. Note that we have here, for the first time, explicitly involved the frame of reference in our

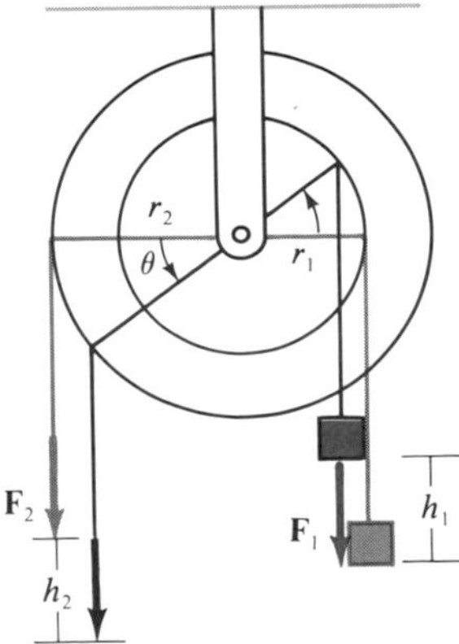

figure 11 • 13 *Equal moments of force,* $F_1 r_1 = F_2 r_2$

definition of a physical concept. That is, the moment of a force is defined with respect to some axis chosen by the observer. As we shall see, the description of torques and rotational motion depends much more intimately on the observer's choice of coordinates than does the description of translational motion.

Figure 11 . 14 shows five coplanar forces acting on a rigid rod pivoted to rotate about point A in a plane parallel to the paper. The lever arms of forces $\mathbf{F}_1$, $\mathbf{F}_2$, $\mathbf{F}_3$, $\mathbf{F}_4$, and $\mathbf{F}_5$ are, respectively l_1, l_2, r_3, zero, and zero, with the lines of action indicated by dashes. Suppose the rod is rotating with a constant angular velocity $\omega = d\phi/dt$ counterclockwise about an axis passing through A. The forces $\mathbf{F}_1$ and $\mathbf{F}_3$ aid the motion, $\mathbf{F}_2$ opposes it, and the other forces have no effect on the rotation, since their lines of action pass through A.

If ω is constant, then the speed $v = r\omega$ of each particle of the rigid body remains constant, although the direction of motion of each particle is continuously changing. Thus the total kinetic energy of the body is constant, and the forces do no work. If the rod is rotating through an infinitesimal angle $d\phi$, then the work done is

$$dW = (F_1 \sin \theta_1)r_1\, d\phi - (F_2 \sin \theta_2)r_2\, d\phi + F_3 r_3\, d\phi = 0 \qquad [11 \cdot 36]$$

The force $\mathbf{F}_4$ does not appear in the work equation, since it is only forces perpendicular to the length of the rod that do any work, and $\mathbf{F}_5$ does not appear because point A does not move during the rotation.

The geometry of Fig 11 . 14 shows that the lever arms are related to the distance r from the axis to the point of application by

$$l_1 = r_1 \sin \theta_1 \qquad l_2 = r_2 \sin \theta_2 \qquad l_3 = r_3 \qquad l_4 = 0 = l_5$$

so that we may rewrite Eq. [11 . 36] as

$$F_1 l_1 - F_2 l_2 + F_3 l_3 + F_4 l_4 + F_5 l_5 = 0$$

If, as with rotation, we assign a sense to each torque which agrees with the sense of the rotation it would tend to produce, then the torque is positive if it tends to produce counterclockwise rotation and negative if it tends to

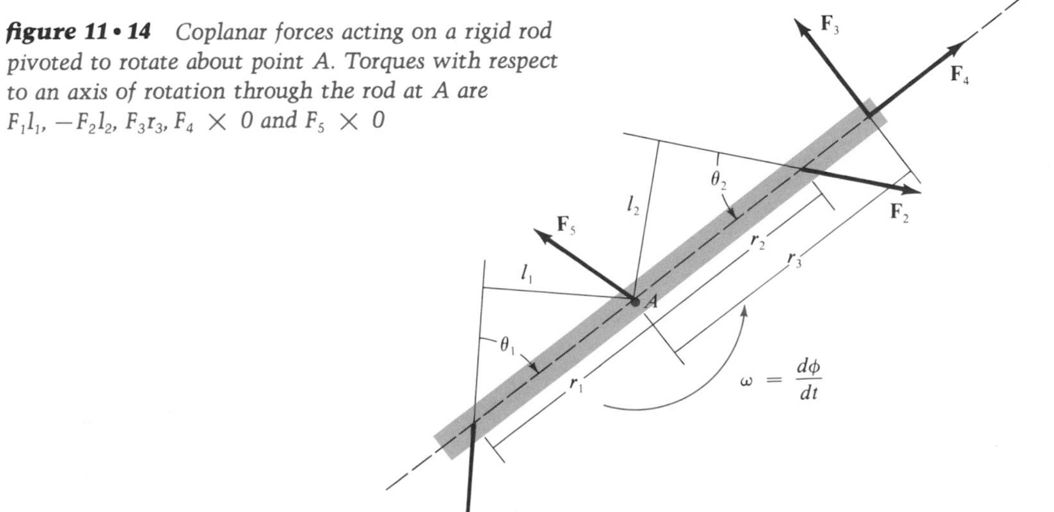

figure 11 • 14 *Coplanar forces acting on a rigid rod pivoted to rotate about point A. Torques with respect to an axis of rotation through the rod at A are* $F_1 l_1$, $-F_2 l_2$, $F_3 r_3$, $F_4 \times 0$ *and* $F_5 \times 0$

produce clockwise rotation. If we then set $\tau_1 = F_1 l_1$, $\tau_2 = -F_2 l_2$, and so on, this equation becomes

$$\sum_i \tau_i = 0 \qquad [11 \cdot 37]$$

which is the equilibrium condition for steady rotation, of which the static situation $\omega = 0$ is a special case.

example 11 • 7 Suppose a homogeneous, rigid, symmetrical Y-shaped body with arms of equal length l is in equilibrium under the action of four forces, one at the center of mass and the others applied at the ends of the three arms. (*a*) If $\mathbf{F}_1 = -10\mathbf{j}$ N and $\mathbf{F}_2 = 10\mathbf{i}$ N, what is the minimum force $\mathbf{F}_3$ required to maintain rotational equilibrium about the center of mass? (*b*) What force $\mathbf{F}_0$ must be applied at the center to maintain translational equilibrium as well?

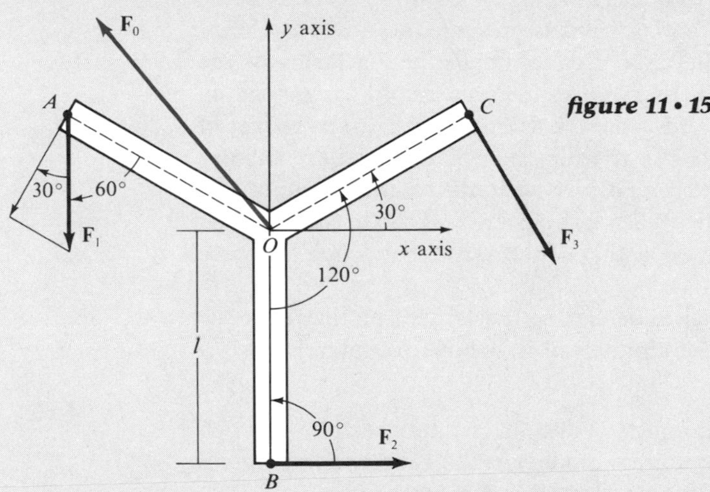

figure 11 • 15

solution (*a*) Since any component of $\mathbf{F}_3$ parallel to the arm length l will have no effect on the rotational equilibrium, $\mathbf{F}_3$ will be a minimum when it is directed perpendicular to the arm. Therefore the condition for rotational equilibrium is, from Eq. [11 . 37],

$$\sum_i \tau = 0 = F_1 l \sin 60^\circ + F_2 l - F_3 l$$

$$F_3 = 8.66 + 10 = 18.66 \text{ N}$$

(*b*) The condition for translational equilibrium is found by substituting $\mathbf{F}_3 = 18.66 (\cos 60^\circ\, \mathbf{i} - \sin 60^\circ\, \mathbf{j})$:

$$\sum_i \mathbf{F}_i = \mathbf{0} = \mathbf{F}_0 + \mathbf{F}_1 + \mathbf{F}_2 + \mathbf{F}_3$$

$$= \mathbf{F}_0 + (F_2 + F_3 \cos 60^\circ)\mathbf{i} - (F_1 + F_3 \sin 60^\circ)\mathbf{j}$$

Therefore

$$\mathbf{F}_0 = (-19.33\mathbf{i} + 26.17\mathbf{j})\ \text{N}$$

which has no effect on the rotational equilibrium about O, since it passes through O; that is, its lever arm is zero.

11 . 5 torque as a vector

The most general definition of torque is as a vector cross product:

$$\boldsymbol{\tau} = \mathbf{r} \times \mathbf{F} \qquad [11 \cdot 38]$$

This equation is a good illustration of the beauty and economy of vector notation, because it is packed with all the properties we have so far ascribed to torques. First, the presence of $\mathbf{r}$, the *position vector of the point at which the force is applied,* shows that torque is always defined with respect to some center about which the force $\mathbf{F}$ acting alone is presumed to produce a rotation. Second, the presence of the cross product implies the inclusion of the sine of the angle between $\mathbf{r}$ and $\mathbf{F}$, which gives the fraction of $\mathbf{r}$ corresponding to the lever arm of the force. Third, from the definition of a cross product in Sec. 2 . 4, this angle is measured in the direction of the smallest angular displacement from $\mathbf{r}$ to $\mathbf{F}$, so that it is positive when the rotation produced by $\mathbf{F}$ is counterclockwise and negative when it is clockwise. Fourth, the direction of the vector $\boldsymbol{\tau}$ in space is the same as that of the axis about which the force produces rotation.

Now let us use Eq. [11 . 38] to determine the torques produced by the forces in Fig. 11 . 14, taking z as the axis of a positive (counterclockwise) rotation about point A:

1 Turning from the direction of $\mathbf{r}_1$ into $\mathbf{F}_1$ counterclockwise through the angle θ_1 gives a vector cross product in the z direction, in agreement with the sense of the rotation produced by the force:

$$\tau_1 = r_1 \sin\theta_1\, F_1\mathbf{k}$$

2 Turning from $\mathbf{r}_2$ into $\mathbf{F}_2$ clockwise through the negative angle $-\theta_2$ gives a vector cross product antiparallel to the axis of rotation, corresponding to a negative sense of rotation:

$$\tau_2 = -r_2 \sin\theta_2\, F_2\mathbf{k}$$

Note that if we had simply rotated counterclockwise from $\mathbf{r}_2$ to $\mathbf{F}_2$ through $2\pi - \theta_2$ rad, we would have had

$$\tau_2 = r_2 \sin(2\pi - \theta_2)\, F_2\mathbf{k} = -r_2 \sin\theta_2\, F_2\mathbf{k}$$

The sign of the cross product will be consistent even if we rotate the first vector of the cross product counterclockwise into the second in order to determine the direction of the cross-product vector.

3 We describe the angle from $\mathbf{r}_3$ to $\mathbf{F}_3$ by turning clockwise through $90°$, so that the torque is simply of magnitude r_3F_3 parallel to the

axis of rotation, in agreement with the sense of the rotation produced by F_3:

$$\boldsymbol{\tau}_3 = r_3 F_3 \mathbf{k}$$

4 Since the angle between the position and force vectors is zero, the cross product vanishes and the torque is zero:

$$\boldsymbol{\tau}_4 = \mathbf{0}$$

5 The force F_5 passes through the origin at A; hence $r_5 = 0$ and there is no torque:

$$\boldsymbol{\tau}_5 = \mathbf{0}$$

In every instance the definition of Eq. [11 . 38] is consistent with our concept of torque as derived from energy considerations in the case of motion in a plane.

Although we have discussed Eq. [11 . 38] in terms of a planar example, it applies equally well in three-dimensional space. Consider the force **F** applied at point P in Fig. 11 . 16*a*; we wish to determine its torque about some axis passing through point O. We know from solid geometry that when two straight lines intersect, they determine a unique plane in space, and the unit vector $\hat{\mathbf{n}}$ normal to that plane defines a unique direction in space. Figure 11 . 16*b* shows the plane determined by **F** and the position vector **r** which locates point P. In Fig. 11 . 16*c*, **F** is decomposed into components parallel and perpendicular to **r**. The $F_\parallel$ component produces no rotation; the $F_\perp$ component, of magnitude $F \sin\theta$, produces rotation about the $\hat{\mathbf{n}}$ vector as shown. Therefore, as in Eq. [11 . 38],

$$\boldsymbol{\tau} = rF\sin\theta\,\hat{\mathbf{n}} = \mathbf{r} \times \mathbf{F} \qquad [11 \cdot 39]$$

figure 11 • 16 *Determination of the torque* $\boldsymbol{\tau}$ *produced by a force* **F**. *The force component* $F_\perp$ *gives rise to the torque and rotation about the* $\hat{\mathbf{n}}$ *vector normal to the plane.*

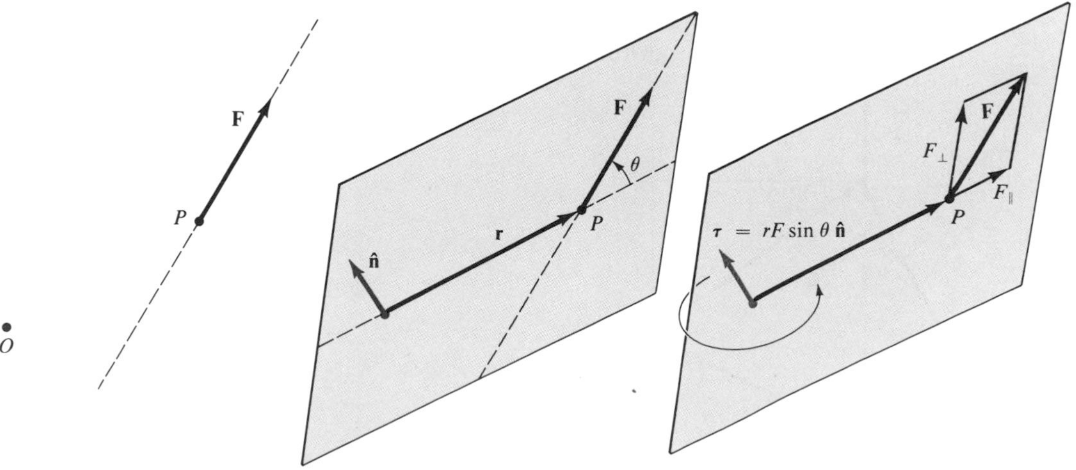

(a) Force **F** *applied at point P* *(b) Determination of the plane of* **F** *(c) Decomposition of* **F**

In general, if the vectors **r** and **F** are described in cartesian coordinates, then the torque $\boldsymbol{\tau}$ will also have x, y, and z components. We could have derived these components separately by decomposing **F** and considering the effect of F_y and F_z in producing rotations about the x axis, the effect of F_x and F_z in producing rotations about the y axis, and the effect of F_x and F_y in producing rotations about the z axis. The total torque is the vector sum of its components:

$$\begin{aligned}\boldsymbol{\tau} &= \mathbf{r} \times \mathbf{F} \\ &= (yF_z - zF_y)\mathbf{i} + (zF_x - xF_z)\mathbf{j} + (xF_y - yF_x)\mathbf{k} \qquad [11 \cdot 40]\end{aligned}$$

The resultant torque on a body due to several forces acting at different points is the *vector sum* of the individual torques. When the applied forces do not produce rotation about any one of the three coordinate axes, then there is complete rotational equilibrium; the body is either perfectly stationary or rotating with constant angular velocity. Equation [11 . 37] for rotational equilibrium becomes a vector equation,

$$\sum \boldsymbol{\tau} = \mathbf{0} \qquad [11 \cdot 41]$$

which is the equivalent of three scalar equations, one for each independent component of the torque:

$$\sum_i \tau_{xi} = 0 \qquad \sum_i \tau_{yi} = 0 \qquad \sum_i \tau_{zi} = 0 \qquad [11 \cdot 42]$$

Now let us look at an interesting special case of torques. When you wind your watch, your thumb and forefinger exert equal and opposite forces on the crown of the watch. Such a pair of parallel forces which do not act along the same line is called a *couple*. The two forces constituting a couple add up to zero, and so have no tendency to produce translational motion; however, they do produce rotation. For the couple shown in Fig. 11 . 17, the net torque

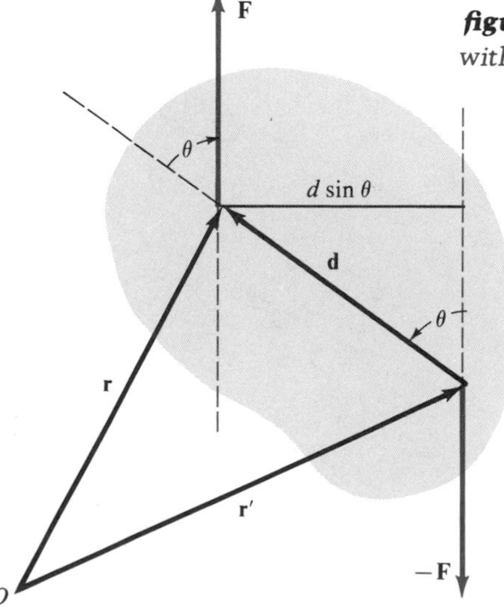

figure 11 • 17 *A couple of forces acting on a rigid body with the torque $\tau_c = (d \sin \theta)F$ directed into the paper*

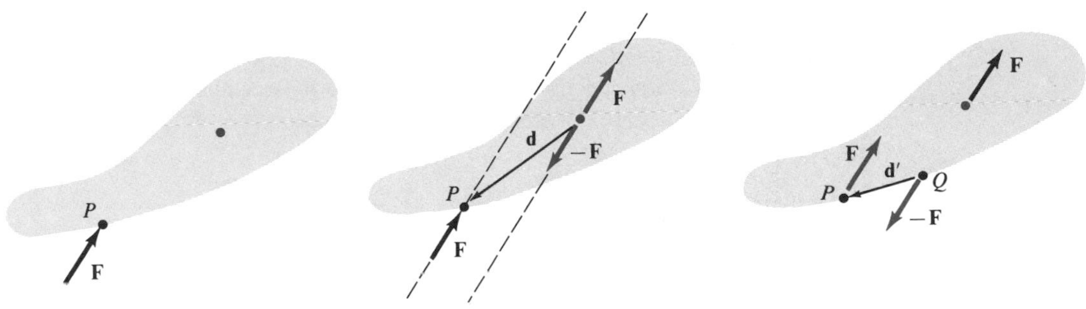

(a) Force **F** applied at point P

(b) Three equivalent forces

(c) A force at the center of mass and a couple at points P and Q

figure 11 • 18 *Decomposition of a force* **F** *into three equivalent forces*

about any arbitrary point O is

$$\begin{aligned}\boldsymbol{\tau}_c &= \mathbf{r} \times \mathbf{F} + \mathbf{r}' \times (-\mathbf{F}) \\ &= \mathbf{r} \times \mathbf{F} - \mathbf{r}' \times \mathbf{F} \\ &= (\mathbf{r} - \mathbf{r}') \times \mathbf{F} = \mathbf{d} \times \mathbf{F} \qquad [11 . 43]\end{aligned}$$

Hence the torque is always perpendicular to the plane containing the couple, since $\boldsymbol{\tau}_c$ is independent of the choice of coordinates, and the torque of any given couple has the same magnitude and direction for all parallel axes of rotation.

It can be seen from Eq. [11 . 43] that the rotational effects of a couple are not changed:

1 If the couple is moved to any position in space without altering the magnitudes or directions of **d** and **F**
2 If the values of $d \sin \theta$ and F vary inversely, although the direction of **F** and the plane of the couple do not change
3 If the couple rotates in its plane (as when you turn a nut or a bolt with your fingers)

That is, the rotational effects are unchanged as long as the orientation of the *plane of the couple* remains the same. A couple is a nonlocalized, or free, vector quantity. It produces rotation of an isolated body about some axis *through the center of mass,* since the net external force is zero, so that the center of mass is not accelerated.

Consider the simple case where a single force **F** is acting on a point P of a rigid body (Fig. 11 . 18*a*), and we add a net force $\mathbf{F} - \mathbf{F} = \mathbf{0}$ acting on the center of mass of the body (Fig. 11 . 18*b*). The effect of the three forces is equivalent to the translational motion produced by **F** alone acting on the center of mass and the couple of torque $\boldsymbol{\tau}_c = \mathbf{d} \times \mathbf{F}$. We can then, if desired, move the force couple to points P and Q, as shown in Fig. 11 . 18*c*, since

$$\mathbf{d}' \times \mathbf{F} = \mathbf{d} \times \mathbf{F}$$

We are thus able conceptually to separate the translational and rotational effects about the center of mass due to the original force $\mathbf{F}_1$.

example 11 • 8 A vertical force **F** is acting at A on the homogeneous rectangular solid shown. Replace this force by a force whose line of action passes through the center of mass and a couple acting horizontally at A and B.

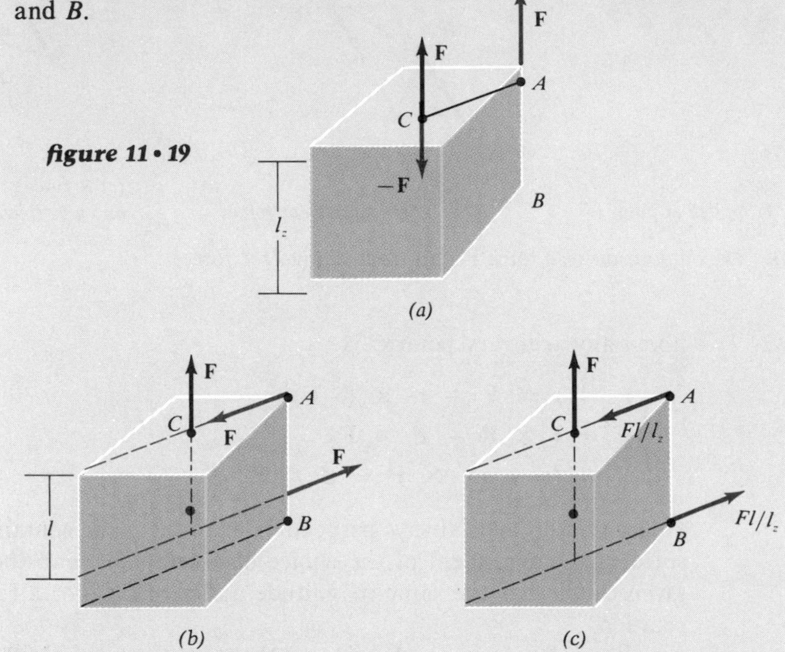

figure 11 • 19

solution The first step is to replace the given vertical force **F** with an equal force at the center C of the upper face of the solid, so that, by symmetry, its line of action passes through the center of mass. The couple, of torque Fl, acts as shown at points C and A. Next the couple is rotated in its plane, the diagonal plane of the solid, and displaced to the line AB (Fig. 11 . 19*b*). The magnitudes of the forces are then changed to Fl/l_z (Figure 11 . 19*c*), so that the torque remains equal to $l_z(Fl/l_z) = Fl$.

The foregoing technique may readily be extended to cover the general case of any number of forces acting on a rigid body. Each of these forces may be replaced by a single force together with a couple. Summing the force and torque vectors separately, we find that:

> Any system of forces acting on a rigid body is always reducible to a single force passing through an arbitrarily chosen point of the body and a single couple.

Furthermore, we can resolve the force and the couple to a single resultant force along some unique line of action, such that the total effect of the force, both *translational and rotational,* is the same as that of the entire system of forces. This can, of course, all be done analytically; however, the method of couples lends itself to graphical solutions which are very useful in engineering.

example 11 • 9 Four forces of magnitudes $F_A = 28.3$ dynes, $F_B = 60$ dynes, $F_C = 20$ dynes, and $F_D = 50$ dynes act on a solid plate as shown. Find the resultant force on the plate and determine its line of action by analytic means.

figure 11 • 20

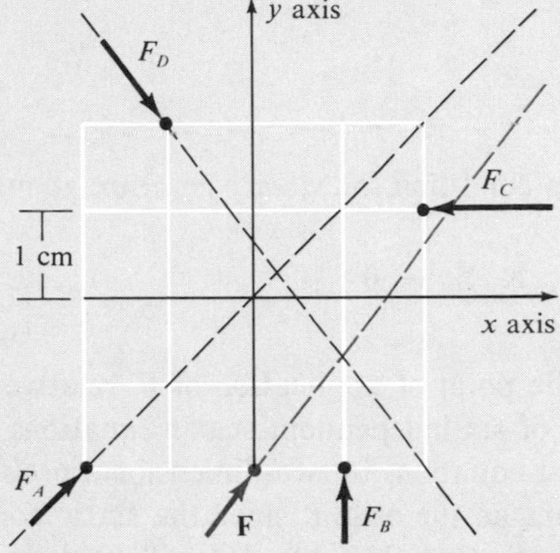

solution The *xy* grid can be used to determine the scalar components of the forces. The resultant force is then

$$\mathbf{F} = \mathbf{F}_A + \mathbf{F}_B + \mathbf{F}_C + \mathbf{F}_D$$
$$= (20 - 20 + 30)\mathbf{i} + (20 + 60 - 40)\mathbf{j}$$
$$= (30\mathbf{i} + 40\mathbf{j}) \text{ dynes}$$

The resultant torque about the center of mass is the sum of the individual torques, or

$$\tau = \tau_A + \tau_B + \tau_C + \tau_D$$
$$= \mathbf{0} \times \mathbf{F}_A + 60\mathbf{k} + 20\mathbf{k} + (-\mathbf{i} + 2\mathbf{j}) \times (30\mathbf{i} - 40\mathbf{j})$$
$$= 60\mathbf{k} \text{ dyne} \cdot \text{cm}$$

To provide this torque with force **F**, the point of application $\mathbf{r} = x\mathbf{i} + y\mathbf{j}$ must be chosen such that $\mathbf{r} \times \mathbf{F} = 60\mathbf{k}$ dyne · cm; that is, we have

$$xF_y - yF_x = 60 \text{ dyne} \cdot \text{cm}$$

$$40x - 30y = 60 \text{ cm}$$

However, this is just the equation of the line of action of the force. Therefore, if the force is to be applied at some point on the lower edge of the plate, where $y = -2$ cm, then

$$x = \frac{60 + 30y}{40} = 0$$

and it must be placed directly beneath the center of the plate, as shown. Actually, the force can be applied at any point along its line of action to obtain the desired torque.

11 . 6 equilibrium

In discussing the motion of a rigid body in Secs. 6 . 2 and 10 . 1, we saw that the equilibrium condition for *uniform translational motion* of the center of mass is

$$\sum_i \mathbf{F}_i = \mathbf{0} \qquad [11 \cdot 44]$$

and the equilibrium condition for *steady rotation* about some axis is

$$\sum_i \boldsymbol{\tau}_i = \sum_i \mathbf{r}_i \times \mathbf{F}_i = \mathbf{0} \qquad [11 \cdot 45]$$

where $\mathbf{r}_i$ specifies the point of application of $\mathbf{F}_i$ relative to a selected origin. This makes a total of six independent scalar equations in three dimensions, or three independent equations in two-dimensional problems. In this case we can choose any point as the origin, since the static condition of no motion obtains in any case, so that Eq. [11 . 45] will hold. In solving problems in static equilibrium it is often advantageous to take torques about some axis such that one or more forces pass through that axis, and are thereby eliminated from the resulting form of Eq. [11 . 45].

Equation [11 . 45] holds even if the center of mass of the rotating body is moving with some constant translational velocity. Consider what happens if we refer the torques to an inertial frame of reference with origin at $\mathbf{R} = \mathbf{V}t$. Note that since forces are unchanged in a Galilean transformation, the condition for uniform translational motion of the center of mass, $\Sigma\mathbf{F} = \mathbf{0}$, still applies.

$$\mathbf{r}' = \mathbf{r} - \mathbf{V}t = \mathbf{r} - \mathbf{R}$$

the net torque relative to the moving coordinate system is

$$\begin{aligned} \sum_i \boldsymbol{\tau}_i' &= \sum_i \mathbf{r}_i' \times \mathbf{F}_i = \sum_i (\mathbf{r}_i - \mathbf{R}) \times \mathbf{F}_i \\ &= \sum_i \mathbf{r}_i \times \mathbf{F}_i - \mathbf{R} \times \sum_i \mathbf{F}_i \\ &= \sum_i \mathbf{r}_i \times \mathbf{F}_i = \sum_i \boldsymbol{\tau}_i = \mathbf{0} \end{aligned} \qquad [11 \cdot 46]$$

That is, whether the motion is described in terms of an inertial system or whether the body is moving with constant velocity past a stationary observer and is not rotating, the sum of the torques is zero.

It is fairly straightforward to list general rules for employing Eqs. [11 . 44] and [11 . 45] to solve problems in equilibrium:

1 Select a body, or part of one, for which all external forces are known or can be calculated.
2 Make a free-body diagram showing all forces acting on that body.
3 Write out the equilibrium equations in terms of these forces and their torques about the chosen axis.

4 Compare the number of unknowns with the number of independent equilibrium equations. If there are as many equations as unknowns, proceed to solve; otherwise isolate another part of the system and repeat the foregoing steps.

In this section we shall work some examples of problems in static equilibrium which demonstrate the application of these principles. The problems are given in order of increasing complexity, with symmetry arguments used freely to simplify wherever possible.

example 11•10 What are the forces on a pair of pliers with identical arms if they are gripped with an average force **A** as shown?

solution The reaction forces **B** at the mouth of the pliers are equal and opposite. If we focus attention on one arm, the equilibrium conditions are

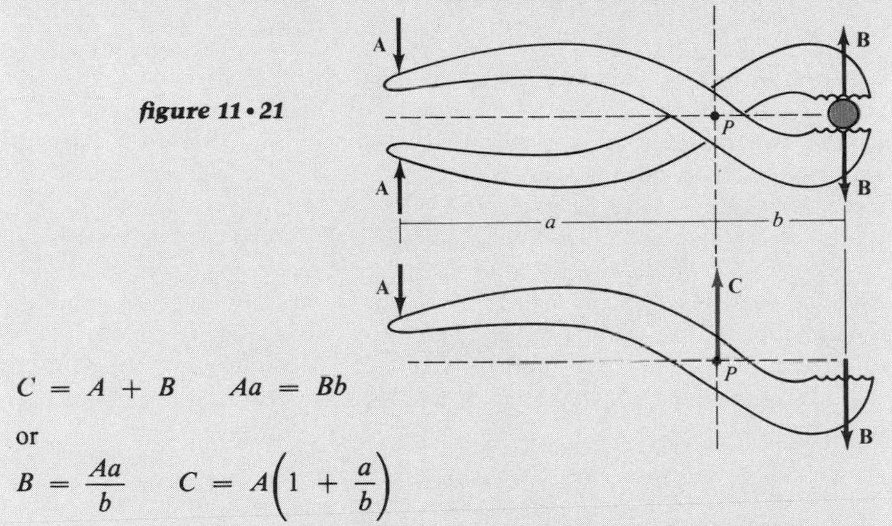

figure 11•21

$$C = A + B \qquad Aa = Bb$$

or

$$B = \frac{Aa}{b} \qquad C = A\left(1 + \frac{a}{b}\right)$$

C is an internal force, of interest only to the manufacturer, since it is counterbalanced by an equal but opposite force from the other arm.

The next example, the "leaning ladder," is a classic, because it affords one of the simplest, yet general, illustrations of a two-dimensional problem. It also demonstrates how a static approach can answer what is essentially a dynamic question: under what conditions will the static situation become a dynamic one?

example 11•11 A uniform ladder of length l and weight $\mathbf{F}_1$ stands on a horizontal floor and leans against a vertical wall, the coefficients of static friction at the floor and wall being μ_s and μ_s', respectively. The ladder makes an angle θ with the floor. If a man of weight $\mathbf{F}_2$ starts up the ladder, how far can he climb before the ladder starts to slip?

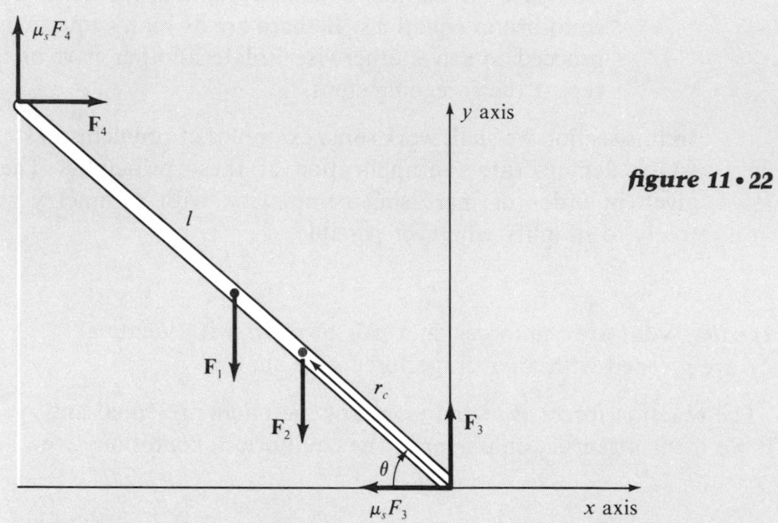

figure 11 • 22

solution The free-body diagram of the isolated ladder shows $\mathbf{r}_c$, the position of the center of mass of the climber and the forces acting on the ladder. The choice of origin is suggested by the fact that two of the forces act through this point; the axes are chosen to make the determination of force components as simple as possible. The floor and the wall are represented by the normal reaction forces, $\mathbf{F}_3$ and $\mathbf{F}_4$, and the frictional forces, $\mu_s F_3$ and $\mu_s' F_4$, at each end of the ladder. Since we are concerned with the point at which the ladder just begins to slip, this fact determines the directions of the maximum frictional forces, which oppose the onset of motion.

Under the conditions for equilibrium, the components of the resultant force are

$$F_x = -\mu_s F_3 + F_4 = 0 \qquad F_y = F_3 - F_1 - F_2 + \mu_s' F_4 = 0$$

and in three-dimensional problems,

$$F_z = \tau_x = \tau_y = 0$$

with the torque about the z axis given by

$$\tau_z = \tfrac{1}{2} F_1 l \cos\theta + F_2 r_c \cos\theta - F_4 l \sin\theta - \mu_s' F_4 l \cos\theta = 0$$

Eliminating the unknown forces F_3 and F_4 and solving for r_c, we obtain

$$r_c = \frac{l}{F_2}\left[\frac{\mu_s(F_1 + F_2)}{1 + \mu_s \mu_s'}(\tan\theta + \mu_s') - \tfrac{1}{2}F_1\right]$$

(What is the effect on r_c of placing the ladder closer to the wall?) By setting $F_2 = 0$ in the equilibrium equations above and solving for θ, we could also show that the ladder falls under its own weight when

$$\tan\theta \leq \frac{1 - \mu_s \mu_s'}{2\mu_s}$$

More often than not, it is necessary to "dismantle" a structure conceptually, piece by piece, in order to solve a problem, as in the next example. This example also indicates the misconceptions to which naive intuition is prone.

example 11 • 12 A rigid horizontal bar of uniform cross section is fixed to a wall at point A. At point C it is attached by a smooth pin to a rigid conical bar that tapers uniformly from B to D. The tapered bar is fixed to the ground at B, and at point D it supports a pulley and rope which sustains a weight F_g. The mass of the horizontal bar is m_1 and the mass of the tapered bar is m_2. Ignoring the mass of the rope and pulley, compute the force the wall exerts on the structure at A and the force the ground exerts on it at B. The distances l are as shown.

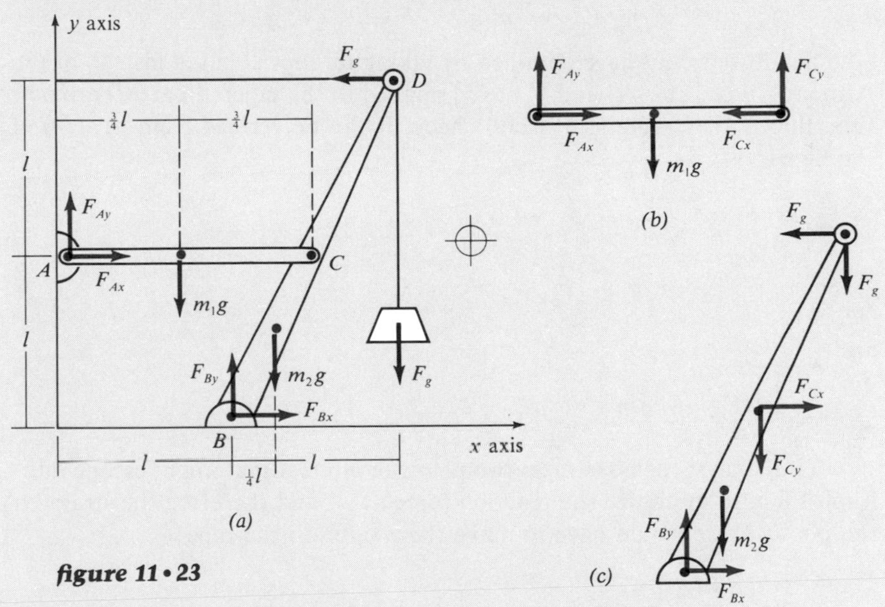

figure 11 • 23

solution The center of mass of the horizontal bar is its geometric center. Let us consider the forces on the entire structure $ABCD$ and take the torques about B. In component form, the force equation is

$$\sum F_x = F_{Ax} + F_{Bx} - F_g = 0$$

$$\sum F_y = F_{Ay} + F_{By} - m_1 g - m_2 g - F_g = 0$$

The equation for the torque about B is

$$\sum \tau_B = \tfrac{1}{4} m_1 g l + 2 l F_g - l F_{Ax} - l F_{Ay} - \tfrac{1}{4} m_2 g l - l F_g = 0$$

which is a system of three equations in four unknowns, F_{Ax}, F_{Ay}, F_{Bx}, and F_{By}.

Since we do not yet have enough equations to solve the system, let us equate the torques about C on the horizontal bar to zero:

$$\sum \tau_C = \tfrac{3}{4} m_1 g l - \tfrac{3}{2} F_{Ay} l = 0$$

which gives a fourth independent condition. We can solve, without going any further, to find

$$F_{Ax} = F_g - \tfrac{1}{4}(m_1 + m_2)g \qquad F_{Bx} = \tfrac{1}{4}(m_1 + m_2)g$$

$$F_{Ay} = \tfrac{1}{2} m_1 g \qquad F_{By} = F_g + \tfrac{1}{2} m_1 g + m_2 g$$

The internal forces F_{Cx} and F_{Cy} can be found immediately from the components of the forces on the horizontal bar:

$$F_{Ax} = F_{Cx} \qquad F_{Ay} + F_{Cy} = m_1 g \qquad F_{Cy} = \tfrac{1}{2} m_1 g$$

(the result we would have obtained by taking torques about A instead of C). Although Eqs. [11 . 44] and [11 . 45] applied to the tapered bar BD are now superfluous, they serve as a useful check. It can be verified from part (c) of the figure that

$$\sum F_x = F_{Bx} + F_{Cx} - F_g = 0$$

$$\sum F_y = F_{By} - m_2 g - F_{Cy} - F_g = 0$$

and

$$\sum \tau_C = l(F_{Bx} + \tfrac{1}{4} m_2 g + F_g - \tfrac{1}{2} F_{By} - \tfrac{1}{2} F_g) = 0$$

This analysis demonstrates two points of interest that might escape intuition. First, to minimize the reaction force at A, and therefore the strain on the pin at C, we would have to make the weight on the rope

$$F_g = \tfrac{1}{4}(m_1 + m_2)g$$

This would make

$$F_{Ax} = F_{Cx} = 0$$

with F_{Ay} and F_{Cy} constant. Second, the geometry is such that as F_g increases, the horizontal bar tends to press harder against the wall instead of pulling on it as one might expect.

Since the gravitational force acting at the center of mass of a solid object is equal to its weight, this point is often termed the *center of gravity* of the object. In the first mathematical treatment of the lever (see the opening quotation), Archimedes defined this as the point such that the body will balance in all positions when supported at this point. Since it applies only to bodies subject to an approximately uniform gravitational force, we shall employ the more general center-of-mass concept, although it is not uncommon to hear scientists and engineers speak of the center of gravity.

example 11 • 13 A 100-kg chest is pushed across a floor with a coefficient of friction $\mu_k = 0.25$. (*a*) What force is exerted if the chest moves with constant speed? (*b*) If the chest moves with an acceleration of 0.98 m/s²? (*c*) What force would be needed to tip the chest?

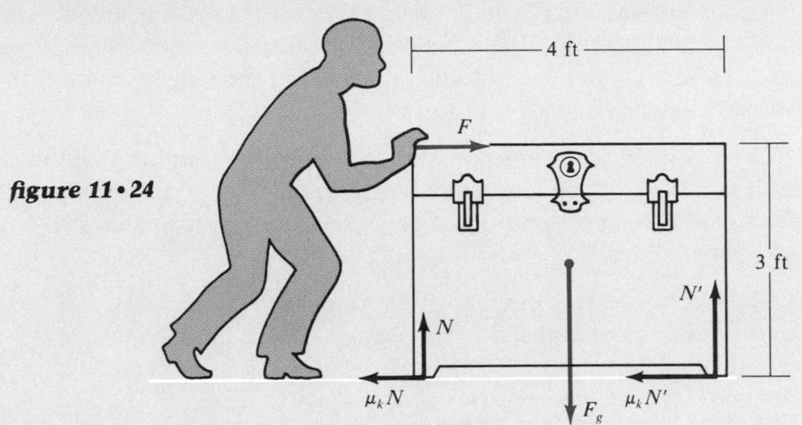

figure 11 • 24

solution (*a*) First we must set up general equations for uniform acceleration and no rotation about the forward edge of the chest:

$$\sum F_x = F - \mu_k(N + N') = ma \qquad \sum F_y = N + N' - mg = 0$$

and

$$\sum \tau_A = -3F + 2mg - 4N = 0$$

We can solve the force equations for

$$F = m(a + \mu_k g)$$

For the case of constant speed, $a = 0$ and

$$F = 25\text{ kg} \times g = 245\text{ N}$$

(*b*) For $a = 0.98\text{ m/s}^2$, the same equations yield

$$F = 100\text{ kg} \times 0.35g = 343\text{ N}$$

(*c*) When the chest just begins to tip it must start to rotate clockwise; hence $N = 0$. Substituting this in the third equation yields

$$F = \tfrac{2}{3}mg = 653\text{ N}$$

$$N' = mg = 980\text{ N}$$

$$a = \frac{F}{m} - \mu_k g = (\tfrac{2}{3} - \mu_k)g = 4.08\text{ m/s}^2$$

Note that raising or lowering the center of mass has no effect on the stability of the chest when the force is exerted as shown.

PROBLEMS

11 • 1 *rotational kinematics*

11 • 1 The flywheel of an engine runs with a constant angular speed of 150 rpm. When the engine is shut off, the friction of the bearings and of the air brings the wheel to rest in 2.2 h. (*a*) What is the average angular acceleration of the wheel? (*b*) How many revolutions will it make before coming to rest? (*c*) What is the linear acceleration, along the tangent to its path, of a flywheel particle 50 cm from the axis of rotation?

11 • 2 (*a*) What is the angular speed of the earth's rotation about its north-south axis? (*b*) Taking leap years into account, what is the average angular speed of the earth's revolution about the sun in radians per year? (*c*) In radians per second?

11 • 3 The drive shaft of an engine rotates with constant angular acceleration within its circular mounting. At time $t = 0$ a mark on the shaft is lined up with an identical mark on the mounting. As the shaft rotates, the marks are lined up at consecutive times $t = 4$, 5.66, 6.93, and 8.05 s. (*a*) What are the angular displacements corresponding to these times? (*b*) What is the angular acceleration of the shaft? (*c*) What are the angular speeds corresponding to these times?

11 • 4 A windmill rotating in a variable breeze has an angular displacement in degrees given by

$$\theta = 3t^4 + 2t - \frac{10}{t + 3}$$

(*a*) What is its instantaneous angular velocity at $t = 0$? (*b*) At $t = 2$ s? (*c*) At $t = 7$ s?

11 • 5 The initial angular displacement of a particle is zero, its initial angular velocity is 3 rad/s, and its angular acceleration is given by $\alpha(t) = \sin^2 t$, where t is a number of radians equal to the time in seconds. Use quadrature to find (*a*) the angular velocity as a function of time. (*b*) The angular displacement as a function of time.

11 • 2 *uniform circular motion*

11 • 6 A disk of radius $R = 3$ m is mounted on a shaft 2 cm in diameter. A point on the circumference of the shaft has a tangential speed of $v = 0.7$ m/s. (*a*) What is the angular speed ω in radians per second? (*b*) What is the tangential speed of a point on the circumference of the disk? (*c*) What is the acceleration of a point on the circumference of the disk?

11 • 7 The centripetal acceleration of a body in uniform circular motion was first derived by Christian Huygens and published in 1673. His method was to consider the moving particle to be in a state of free fall and to ask himself what would have to be the acceleration due to a hypothetical gravitational attraction that would ensure that a body projected tangentially to the earth dropped just enough to keep it moving in a circle at the same altitude, neither leaving nor striking the earth. Use his method to derive Eq. [11 . 18]. (*Hint:* Consider a path much smaller than the radius of the earth.)

11 • 8 Satellites such as Syncom and Early Bird are in synchronous orbits with the earth. If such a satellite is moving eastward over the equator at an altitude of 36,000 km, it appears to stay over one position on the earth. (*a*) What is the frequency of its orbit? (*b*) What is its radial acceleration?

11 • 9 A pilot flies an airplane above the earth at a ground speed of 1 km/s and at a latitude of 45°. (*a*) If there were a scale in the plane, what would be the fractional change in the pilot's weight on the scale when the plane changed direction from westward to eastward? (*b*) From northward to southward? (*c*) Could the pilot use the scale as a speedometer? Explain.

11 • 10 For an object traveling at a constant speed v, a smooth curve of radius R should be banked at an angle $\theta_b = \tan^{-1}(v^2/Rg)$. However, if friction is taken into account, there is a range of bank angles θ at which the frictional force will keep the object from sliding as it rounds the curve. Determine this range for a given coefficient of friction μ_k and angle θ_b. (*Hint:* First solve for the actual normal and frictional forces and then relate them.)

11 • 11 A cylinder of radius R contains a ball of mass m suspended from a string of length $l > R$. The string is attached to a rod fixed along the axis of the cylinder as shown. With what frequency must the cylinder be rotating before the ball will touch it?

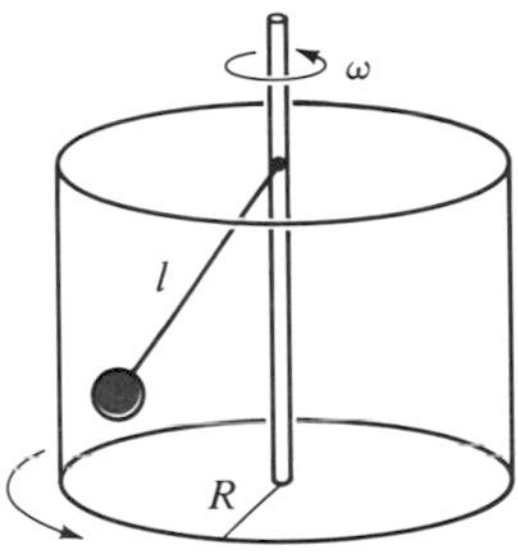

11 • 12 A ball of mass m traveling at speed v enters a horizontal semicircular track of radius R. (*a*) What is its average time rate of change of momentum during its trip along the track? (*b*) Compare this to the inertial reaction force which the ball exerts on the track when it is halfway around and explain any discrepancy between the two forces.

11 • 13 Assume that the sphere in Example 11 . 4 starts at the minimum height for a complete rotation. (*a*) What normal reaction force does the track exert on the sphere when the sphere reaches the lowest point of the circle? (*b*) When the sphere is halfway up the circle?

11 • 14 A particle located by $\mathbf{r} = \mathbf{i} + \mathbf{j} + \mathbf{k}$ and moving with angular speed ω rotates clockwise about an axis that passes through the origin in the direction given by $\mathbf{i} - \mathbf{j} + \mathbf{k}$. (*a*) What is the radius of rotation? (*b*) What is the angular velocity $\boldsymbol{\omega}$? (*c*) What is the velocity $\mathbf{v}$ of the particle? (*d*) Draw a careful sketch of the situation, including $\boldsymbol{\omega}$ and $\mathbf{v}$ as a rough check on the correctness of your answers.

11 • 15 At any given moment the velocity and acceleration vectors of a particle may be represented in terms of a unit vector $\hat{\mathbf{t}}$ tangent to its path and another unit vector $\hat{\mathbf{n}}$ which is normal to its path and points instantaneously toward the center of a circle, an arc of which instantaneously coincides with the path element ds. Suppose a particle of mass 5 g is moving in a curved path, and its total acceleration at a given moment is $\boldsymbol{\alpha} = (3\hat{\mathbf{t}} + 4\hat{\mathbf{n}})$ cm/s^2. Find (*a*) the tangential acceleration, (*b*) the centripetal acceleration, (*c*) the magnitude of the total acceleration, (*d*) the angle ϕ which the total acceleration makes with the tangent to the curve, (*e*) the tangential component of the accelerating force, (*f*) the centripetal component of the accelerating force, and (*g*) the total accelerating force.

11 • 16 A horizontal turntable has an angular acceleration of $\alpha = 3$ rad/s^2. At the instant when the angular speed is 2.4 rad/s, a particle of mass 1.8 kg will rest without slipping on the turntable, provided that it is situated no more than 50 cm from the vertical axis of rotation. (*a*) What is the magnitude of the frictional force $\mathbf{F}_f$? (*b*) Find the coefficient of static friction μ_s between the object and turntable.

11 • 3 rotational vector quantities

11 • 17 To complete the proof that an infinitesimal angular displacement can be represented by a vector, it is necessary to show that the effect of two successive displacements $d\boldsymbol{\theta}_1$ and $d\boldsymbol{\theta}_2$ is the same as the effect of their vector sum $d\boldsymbol{\theta} = d\boldsymbol{\theta}_1 + d\boldsymbol{\theta}_2$. Show that the linear displacement $d\mathbf{r}$ of the position vector $\mathbf{r}$ due to two successive rotations $d\boldsymbol{\theta}_1 + d\boldsymbol{\theta}_2$ about axes through the origin of coordinates is the same as the linear displacement due to a rotation which is equal to the vector sum of the first two rotations. (*Hint:* See Eq. [11 . 25].)

11 • 18 Prove that for any two vectors $\mathbf{A}(t)$ and $\mathbf{B}(t)$ whose components are differentiable functions of time,

$$\frac{d(\mathbf{A} \times \mathbf{B})}{dt} = \frac{d\mathbf{A}}{dt} \times \mathbf{B} + \mathbf{A} \times \frac{d\mathbf{B}}{dt}$$

(*Hint:* Show that the relationship holds for the x component of the derivative vector.)

11 • 19 A uniform circular disk of radius R and mass m is rigidly mounted on one end of a thin, massless shaft of length l as shown. The shaft is normal to the disk at its center C. The disk rolls in the xy plane, being fixed at the origin O by a smooth universal joint. Assume that the disk rolls without slipping, rotating about OC with angular frequency ω, so that during one revolution it covers an arc of length $2\pi R$ over the plane.

(*a*) What is the angular velocity $\boldsymbol{\Omega}$ of point C about the z axis? (*b*) Assuming that OC is in the xz plane at $t = 0$, what is the angular velocity $\boldsymbol{\omega}$ of a point on the disk due to its rotation around C? (*c*) What is the total angular velocity of a point on the disk? (*d*) What is the angular acceleration of a point on the disk?

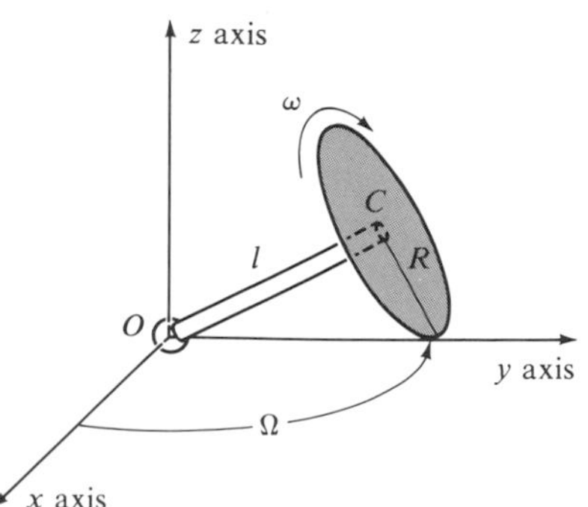

11•20 The mechanical linkage shown rotates in the plane about two points: arm *AB* rotates counterclockwise about point *A* with an angular speed $\omega = 4$ rad/s and arm *CD* rotates clockwise about point *D* with the same speed. The link arm *BC* also performs a rotational motion with constant angular speed. Find this speed. (*Hint:* If we consider each arm as a vector, their vector sum from *A* to *D* must be constant in time.)

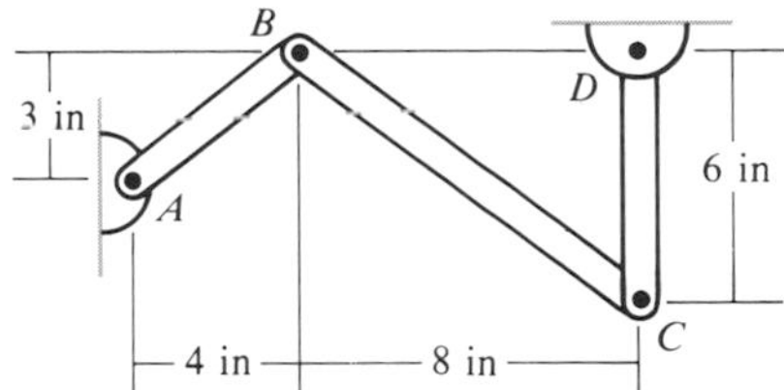

11•4–11•5 torque

11•21 The line of action of a 1000-dyne force lies in the *xz* plane and cuts the *z* axis at a point 6 cm from the origin. (*a*) What is the torque about the *y* axis if the angle between the direction of the force and the *z* axis is 60°? (*b*) If the angle is 180°? (*c*) If the angle is 330°?

11•22 A force $\mathbf{F} = 3\mathbf{i} + 2\mathbf{j} + \mathbf{k}$, measured in mks units, is applied to a body at a point given by $\mathbf{r} = \mathbf{i} - \mathbf{j} + \mathbf{k}$. (*a*) Compute the torque. (*b*) If another force is applied at $\mathbf{r}' = \mathbf{i} + \mathbf{j} + \mathbf{k}$ so that the total torque vanishes, what is this second force?

11•23 A torque vector $\boldsymbol{\tau}$ is the sum of 3 units in the direction of $\mathbf{n} = \mathbf{i} + \mathbf{j} + \mathbf{k}$ and 1 unit in the direction of $\mathbf{m} = \mathbf{i} - \mathbf{j} + \mathbf{k}$. It also has a component equal to -2 units perpendicular to the plane containing **n** and **m** (the positive sense rotating from **n** to **m**). (*a*) Find $\boldsymbol{\tau}$. (*b*) If the force producing this torque is exerted at the point $(9,4,-1)$, find the minimum force required.

11•24 Show that the torques relative to points *A* at $\mathbf{r}_A$ and *B* at $\mathbf{r}_B$ are related by $\boldsymbol{\tau}_A = \boldsymbol{\tau}_B + (\mathbf{r}_B - \mathbf{r}_A) \times \mathbf{F}$, where **F** is the force applied at some arbitrary point at **r**. Generalize this relation to several forces $\mathbf{F}_i$ applied at points $\mathbf{r}_i$.

11•25 A planar body is subjected to a 5-N horizontal force through *A*, a 5-N diagonal force, an 8-N horizontal force, and a 4-N vertical force acting as shown. (*a*) Find the magnitude and direction of the resultant torque about an axis perpendicular to the plane and passing through point *A*. (*b*) About an axis passing through point *B*. (*c*) About an axis passing through point *C*.

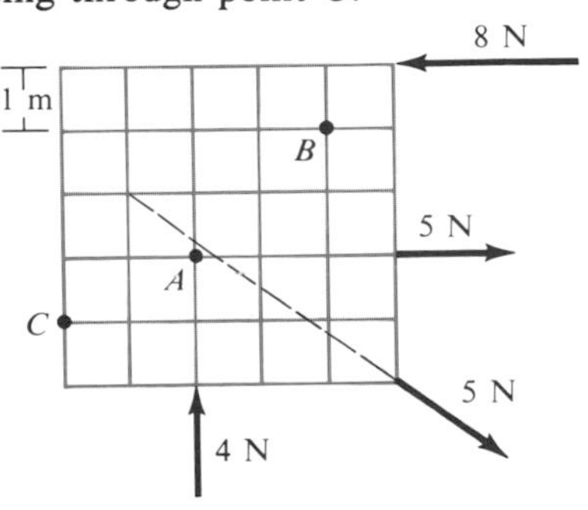

11 • 26 For the planar body in Prob. 11 . 25, (*a*) determine the equivalent force and its line of action. (*b*) Replace this force by a force F_A acting at A (the center of mass) and a couple acting between points B and C and find the (minimum) forces F_B and F_C.

11 • 27 A planar body is acted on by four forces, $F_1 = 12$ lbf, $F_2 = 6$ lbf, $F_3 = 10$ lbf, and $F_4 = 8$ lbf, as shown. (*a*) Find the magnitude and direction of the resultant torque about a perpendicular axis passing through point A. (*b*) About an axis passing through point B.

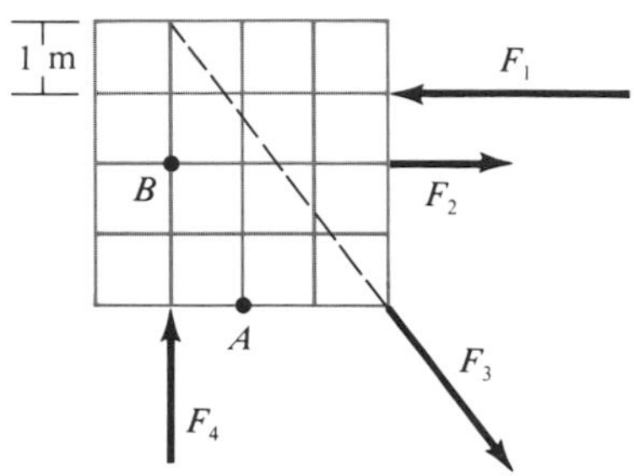

11 • 28 Since a force may act at any point along its line of action without changing the net force or torque on a body, a system of nonconcurrent forces can be resolved into a single equivalent force by adding successive vectors at the points of intersection of their lines of action (each succeeding vector is added to the *resultant* of previous additions). Use this graphic technique to demonstrate that the square body of Prob. 11 . 27 is indeed in equilibrium.

11 • 29 Two parallel forces of magnitude $F_1 = 20$ N and $F_2 = 15$ N act on a rigid bar as shown. (*a*) Calculate the magnitude, direction, and point of application of the third force required to balance the bar. (*b*) What single force is equivalent to $\mathbf{F}_1$ and $\mathbf{F}_2$? (*c*) What single force would balance the two forces if $\mathbf{F}_1$ and $\mathbf{F}_2$ were equal in magnitude?

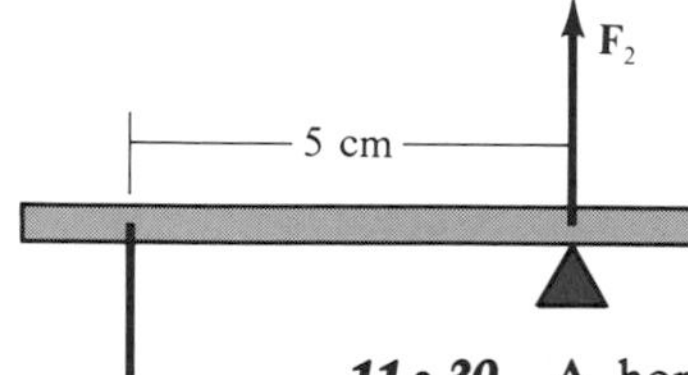

11 • 30 A horizontal bar of negligible weight is acted on by two forces, $F_1 = 800$ kgf and $F_2 = 500$ kgf, directed as shown. Find the magnitude, direction, and point of application of the single additional force that will prevent the bar from moving.

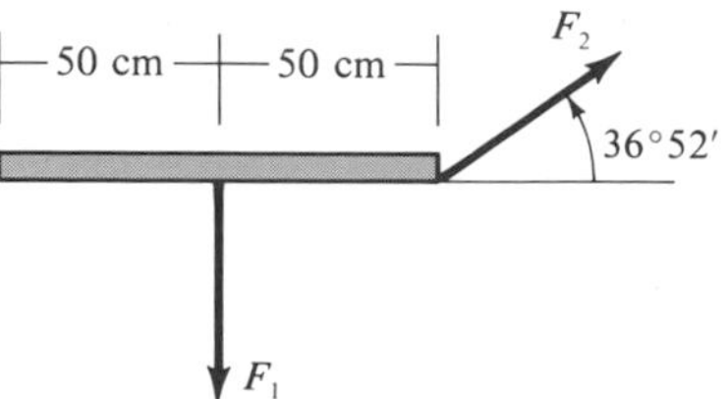

11 • 6 *equilibrium*

11 • 31 Two people carry between them a load of 500 N supported on a uniform pole of length l and weight 100 N. Where must the load be placed so that one person will be carrying twice as much of the total weight as the other?

11 • 32 Suppose you want to overturn a hollow cubical block which has a 152-cm edge and a mass of 100 kg. (*a*) Where, in what direction, and with what force must you push in order to do this most easily? (*b*) Once the block has been set in motion, will more or less force be required? Why? (*c*) If the coefficient of friction is 0.4, will the block slide or tip when you push horizontally? (*d*) When you push at the best angle for overturning the block?

11 • 33 A circular ring of weight 5 N rests horizontally on three points of support 120° apart as shown. What is the smallest downward force, applied to the ring in a direction perpendicular to its plane, that will cause it to leave one of the points of support?

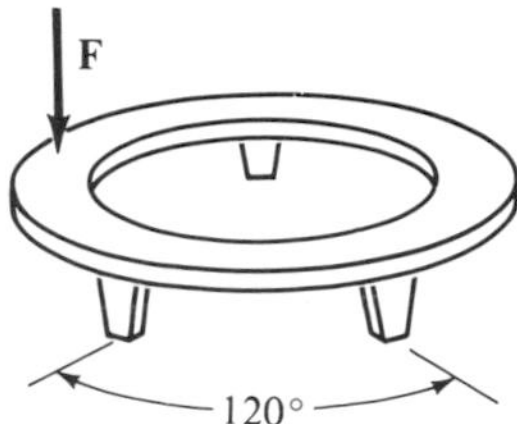

11 • 34 A uniform square board weighing 25 N rests on a block at A and is kept from rotating by a horizontal force at B. Find the force at B and the vertical and horizontal forces on the block at A.

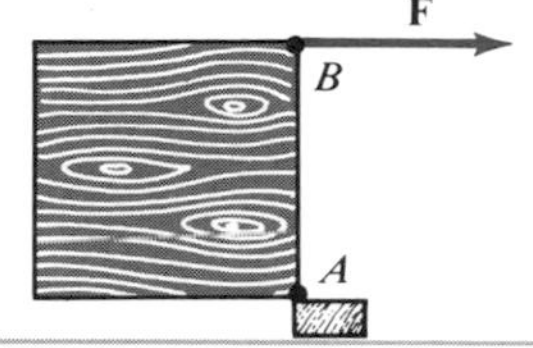

11 • 35 A 60-kg hiker walks at a pace of 2 m/s across a 30-kg plank that is 10 m long. (*a*) What is the force on support B as a function of time? (*b*) If the maximum force that B can withstand is 490 N, when and where will the hiker fall into the stream? Assume that the hiker's weight always acts in a vertical line through the center of mass.

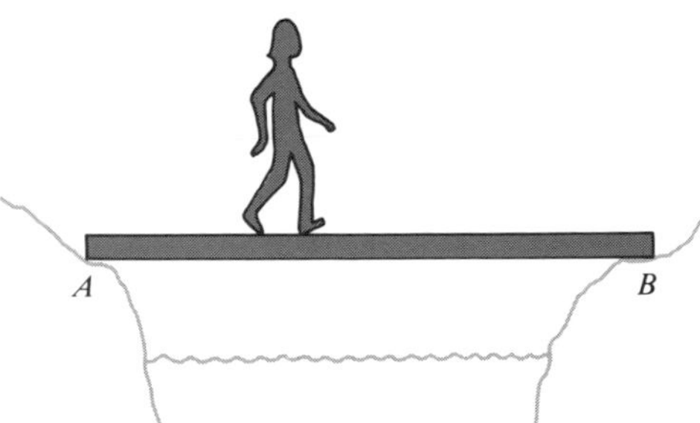

11 • 36 A square wire frame of side length l is hung on a rough peg with a coefficient of static friction μ_s. How close to the corner is the peg if the square is just about to slip?

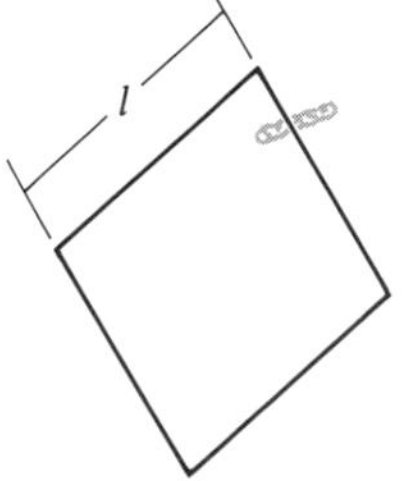

11 • 37 Two smooth spheres of radius r and weight F are placed inside a hollow cylinder of radius R which is open at both ends and rests on end on a horizontal plane. Show that to avoid being overturned the cylinder must weigh at least $2F(1 - r/R)$, where $R > r > \frac{1}{2}R$.

11 • 38 A door 3 ft wide and 7 ft high weighs 150 lbf. If the hinges are 10 in from the ends and the weight is carried entirely by the upper hinge, what is the total force on each hinge?

11 • 39 The axle of a wheel 50 cm in radius carries a load of 500 N as shown. What horizontal force must be applied to the axle to raise the wheel over an obstacle 12 cm high?

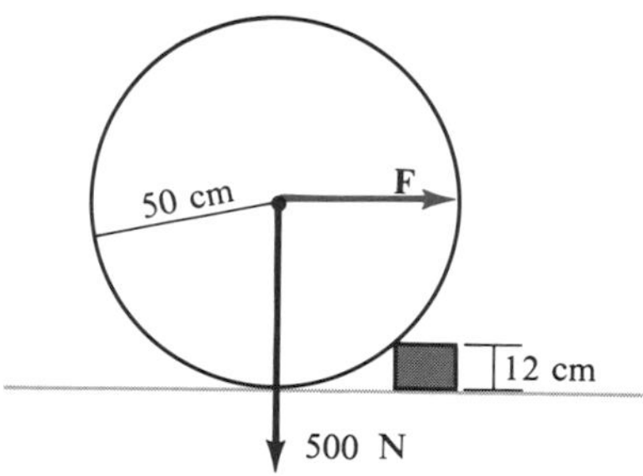

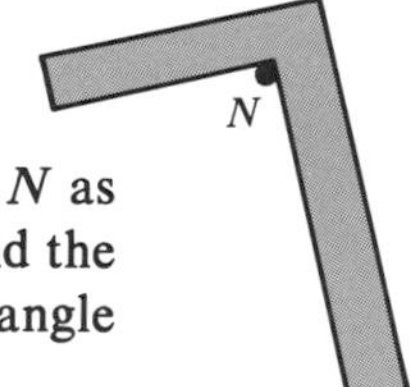

11 • 40 A carpenter's square is supported on a nail at N as shown. The short blade is 45 cm long and 3.5 cm wide, and the long blade is 58 cm long and 5 cm wide. Calculate the angle that the larger blade makes with the vertical.

11 • 41 A metal workbench is dragged steadily along the floor by a force **F** directed as shown. The center of mass of the bench is at C and the coefficient of kinetic friction between the bench legs and the floor is $\mu_k = 0.1$. If the weight of the bench is 100 N, find the magnitude of **F** and the downward push, F_A and F_B, of each leg on the floor.

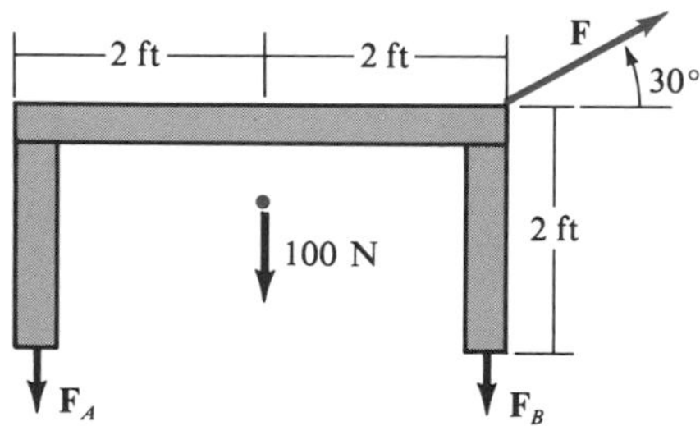

11 • 42 Two uniform planks of the same length l and weight **F** are joined at one end by a hinge and stand with their free ends on a smooth floor. They are prevented from slipping by a rope which is tied to each plank at the same distance y above the floor. Find the tension in the rope and the force at the hinge.

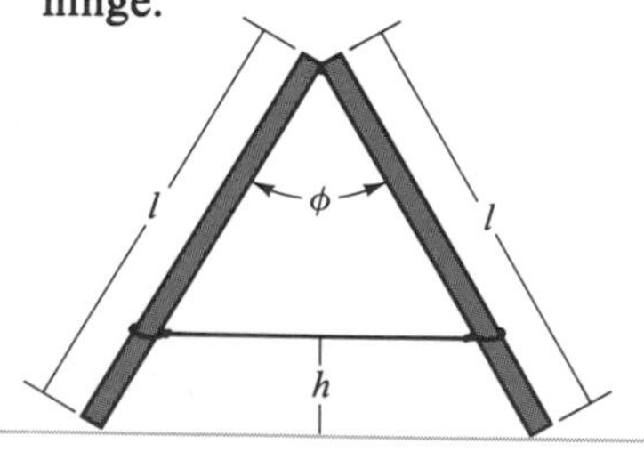

11•43 A car weighs 10,000 N and the distance between its right and left wheels is 3 m. If the largest coefficient of friction encountered between the wheels and the road is $\mu_k = 1.5$, how close to the ground must the center of gravity be to ensure that any centrifugal force generated in a turn will not overturn the car, but will only make it skid?

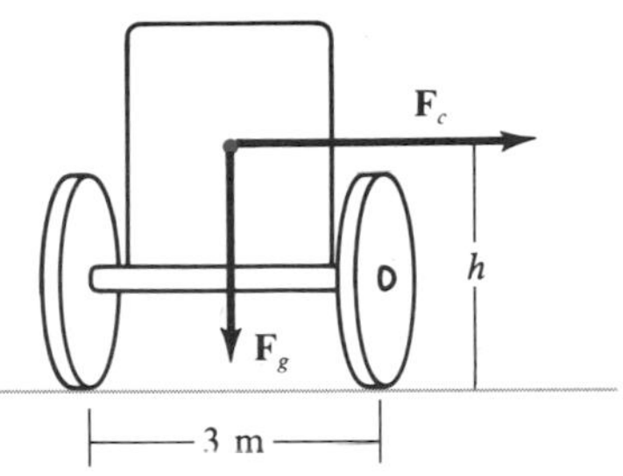

11•44 The bus in which you are standing is traveling at a speed of 50 km/h when the driver suddenly applies the brakes. The bus decelerates uniformly in a stopping distance of 15 m. At what angle from the vertical must you lean to keep from falling over? (*Hint:* You are in a noninertial frame.)

answers

11•1 (*a*) $\bar{\alpha} = -0.002\ \text{rad/s}^2$; (*b*) 9900 revolutions; (*c*) $a = -1.0\ \text{mm/s}^2$
11•2 (*a*) $\omega = 7.27 \times 10^{-5}$ rad/s; (*b*) $\bar{\omega} = 2\pi$ rad/year; (*c*) $\bar{\omega} = 2.0 \times 10^{-7}$ rad/s
11•3 (*a*) $\theta = 2\pi, 4\pi, 6\pi, 8\pi$ rad; (*b*) $\alpha = \frac{1}{4}\pi$ rad/s²; (*c*) $\omega = \pi, \sqrt{2}\,\pi, \sqrt{3}\,\pi, 2\pi$ rad/s
11•4 (*a*) $\omega = 3.11$ deg/s; (*b*) $\omega = 98.4$ deg/s; (*c*) $\omega = 4118.1$ deg/s
11•5 (*a*) $\omega = 3 + \frac{1}{2}t - \frac{1}{4}\sin 2t$; (*b*) $\theta = 3t + \frac{1}{4}t^2 + \frac{1}{8}\cos 2t$
11•6 (*a*) $\omega = 35$ rad/s; (*b*) $v = 105$ m/s; (*c*) $a = 3675\ \text{m/s}^2$
11•8 (*a*) $\nu = 1.157 \times 10^{-5}$ Hz; (*b*) $a_r = -22.42\ \text{cm/s}^2$
11•9 (*a*) -2.1 percent; (*b*) no change; (*c*) $v = \sqrt{R_e g - R_e W/m}$, where W is the scale reading and v is the speed relative to the fixed stars
11•10 $|\theta - \theta_b| \le \tan^{-1}\mu$
11•11 ω, where $\omega^2 = g/\sqrt{l^2 - R^2}$
11•12 (*a*) $\Delta mv/\Delta t = 2mv^2/\pi R$; (*b*) $F' = mv^2/R$
11•13 (*a*) $F = 6mg$; (*b*) $F = 3mg$
11•14 (*a*) $R = 1.633$; (*b*) $\boldsymbol{\omega} = -(\omega/\sqrt{3})(\mathbf{i} - \mathbf{j} + \mathbf{k})$; (*c*) $\mathbf{v} = 1.155\omega(\mathbf{i} - \mathbf{k})$
11•15 (*a*) $a_t = 3\ \text{cm/s}^2$; (b) $a_r = -4\ \text{cm/s}^2$; (*c*) $a = 5\ \text{cm/s}^2$; (*d*) $\phi = 53.1°$; (*e*) $F_t = 15$ dynes; (*f*) $F_r = -20$ dynes; (*g*) $F = 25$ dynes
11•16 (*a*) $F_f = 7.9$ N; (*b*) $\mu_s = 0.45$
11•19 (*a*) $\boldsymbol{\Omega} = (R\omega/l^2)\sqrt{R^2 + l^2}\,\mathbf{k}$; (*b*) $\boldsymbol{\omega} = -\omega(\cos\theta\cos\Omega t\,\mathbf{i} + \cos\theta\sin\Omega t\,\mathbf{j} + \sin\theta\,\mathbf{k})$, where $\tan\theta = R/l$; (*c*) $\boldsymbol{\omega}_{\text{tot}} = \boldsymbol{\Omega} + \boldsymbol{\omega}$; (*d*) $\boldsymbol{\alpha} = \omega\Omega(\cos\theta\sin\Omega t\,\mathbf{i} - \cos\theta\cos\Omega t\,\mathbf{j})$
11•20 $\omega = 2$ rad/s, clockwise
11•21 (*a*) $\tau = 5196$ dyne · cm; (*b*) $\tau = 0$; (*c*) $\tau = -3000$ dyne · cm
11•22 (*a*) $\boldsymbol{\tau} = -3\mathbf{i} + 2\mathbf{j} + 5\mathbf{k}$; (*b*) any vector of form $\mathbf{F} = A\mathbf{i} + (\frac{5}{3} - A)\mathbf{j} + (A - \frac{2}{3})\mathbf{k}$, where A is an arbitrary constant
11•23 (*a*) $\boldsymbol{\tau} = 2\mathbf{j} - 8\mathbf{k}$; (*b*) $\mathbf{F} = \frac{1}{49}(-17\mathbf{i} + 36\mathbf{j} - 9\mathbf{k})$, $F = 0.833$
11•25 (*a*) $\tau = 23$ N · m; (*b*) $\tau = 23$ N · m; (*c*) $\tau = 24$ N · m
11•26 (*a*) $\mathbf{F} = \mathbf{i} + \mathbf{j}$, $y = x - 23$; (*b*) $\mathbf{F}_B = \frac{23}{25}(-3\mathbf{i} + 4\mathbf{j}) = -\mathbf{F}_C$
11•27 (*a*) Zero; (*b*) zero
11•29 (*a*) 5.0 N upward, 20 cm to the left of $\mathbf{F}_2$; (*b*) 5.0 N downward, 20 cm to the left of $\mathbf{F}_2$; (*c*) no single force can balance them
11•30 $F = 640$ kgf at an angle of 129°, 20 cm from left end
11•31 A distance of $0.3l$ from one end
11•32 (*a*) At the upper edge of the block, at 45°, with $F = 35.35$ kgf; (*b*) less force; (*c*) slide; (*d*) tip
11•33 $F = 5$ N
11•34 $F_B = 12.5$ N, $F_{Ax} = 25.0$ N, $F_{Ay} = 12.5$ N
11•35 (*a*) $F_B = 12t + 15$ kgf; (*b*) $t = 2.92$ s, $x = 5.83$ m from A
11•36 A distance of $\frac{1}{2}l(1 - \mu_s)$ from the corner
11•38 $F_{\text{top}} = 155.8$ lbf, $F_{\text{bot}} = 42.2$ lbf
11•39 $F = 428$ N
11•40 $\theta = 16°$
11•41 $F = 11$ N, $F_A = 49$ N, $F_B = 45$ N
11•42 $T = \frac{1}{2}Fl\sin\frac{1}{2}\phi/(l\cos\frac{1}{2}\phi - h) = -F_{\text{hinge}}$
11•43 $h \le 1$ m
11•44 $\theta = 33.27°$ backward

CHAPTER TWELVE

rotational dynamics

To those who study the progress of exact science, the common spinning top is a symbol of the labours and the perplexities of men who had successfully threaded the mazes of the planetary motions. The mathematicians of the last age, searching through nature for problems worthy of their analysis, found in this toy of their youth, ample occupation for their highest mathematical powers.

No illustration of astronomical precession can be devised more perfect than that presented by a properly balanced top, but yet the motion of rotation has intricacies far exceeding those of the theory of precession.

James Clerk Maxwell, On a dynamical top, Transactions of the Royal Society of Edinburgh, *vol. 21, part 4, 1857*

The general problem of rotations may be approached by regarding a rotating body as a system of particles and applying Newton's laws of motion to them. Since the linear speeds of the particles will vary with their distances from the axis of rotation, this procedure would be laborious and complicated. However, the fact that, in equilibrium, torques bear the same relationship to rotation that forces bear to translation suggests the possibility of defining still other rotational analogs of translational quantities. In the case of rigid bodies, it turns out that these new quantities can be defined so that they reduce the equations of motion to forms which not only are relatively simple, but are exact analogs of the familiar equations for translation. Although no new principles are involved, this process greatly simplifies the solution of problems involving rotation. In this chapter we shall develop rotational analogs for mass, momentum, and Newton's second law and apply them to problems of considerable interest in physics.

12 · 1 moment of inertia

Any moving body may be considered as a system of individual particles, each with kinetic energy $\frac{1}{2}mv^2$. The kinetic energy of the body itself is equal to the sum of the kinetic energies of its particles:

$$K = \tfrac{1}{2}(m_1v_1^2 + m_2v_2^2 + \cdots + m_nv_n^2) \qquad [12 \cdot 1]$$

If the motion of the body is pure translation, so that all particles have the same linear speed v, then Eq. [12 . 1] reduces to $K = \frac{1}{2}Mv^2$, where M is the total mass of the body. If the motion is pure rotation, as in the case of the rotating peanut in Fig. 12 . 1, the linear speeds v_i of the particles differ according to the distance R_i of the particle from the axis of rotation. However, the angular speeds ω are the same, provided the body is rigid, and since $v_i = R_i\omega$ for pure rotation (Eq. [11 . 15]), we can express the kinetic energy of a rotating body as

$$K = \tfrac{1}{2}\sum_i m_i R_i^2 \omega^2 \qquad [12 \cdot 2]$$

The summation term is known as the *moment of inertia* of the body about the axis of rotation, denoted by the symbol I:

$$I = \sum_i m_i R_i^2 \qquad [12 \cdot 3]$$

The equation for the kinetic energy of a rotating rigid body then reduces to the form

$$K = \tfrac{1}{2} I \omega^2 \qquad [12 \cdot 4]$$

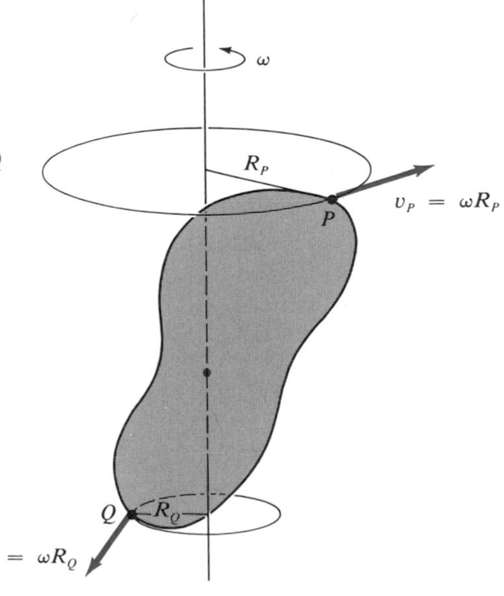

figure 12 • 1 *Speeds of particles at points P and Q in a rotating body*

Note the similarity of this expression to that for kinetic energy of translation, $K = \frac{1}{2}Mv^2$; the quantity I corresponds to mass M and the angular speed ω corresponds to linear speed v. We shall encounter the moment of inertia I in all problems that involve rotation. The concept was familiar to Huygens through his study of the physical pendulum, but was first named by Leonhard Euler. It is evident from the definition in Eq. [12 . 3] that I may be found by imagining the rotating body divided into minute particles, multiplying the mass of each particle by the square of its distance from the axis, and then adding these products. It is important to note that R_i does not refer to the position vector of m_i; it is the *perpendicular* distance from m_i to the axis of rotation. Hence I depends on the position as well as the direction of the axis.

Consider a rigid body mounted on a fixed horizontal axis, with a moment of inertia I about that axis, and acted on by a constant torque of magnitude $\tau = F_g R$. While the body is turning through an angle θ, the mass m, in giving up its gravitational potential energy $mg\,\Delta y = F_g\,\Delta y$, does an equal amount of work on the rotating body:

$$W = F_g\,\Delta y = F_g R\theta = \tau\theta \qquad [12 \cdot 5]$$

If frictional forces are negligible, the work done will equal the increase in the kinetic energy, or

$$\tau\theta = \tfrac{1}{2}(I\omega^2 - I_0\omega_0^2) \qquad [12 \cdot 6]$$

In Eq. [12 . 5] and Fig. 12 . 2 the direction of the torque is always parallel to the axis of rotation. However, if the torque vector $\boldsymbol{\tau}$ makes an angle ϕ with the fixed axis at a given instant, then it is the torque component parallel to the axis that is associated with the work done in an infinitesimal angular displacement $d\boldsymbol{\theta}$. That is, dW is given by the scalar product

$$dW = \tau \cos\phi\, d\theta = \boldsymbol{\tau} \cdot d\boldsymbol{\theta} \qquad [12 \cdot 7]$$

The torque parallel to the axis of rotation arises from forces perpendicular to that axis, as in Fig. 12 . 2. Any torque components normal to the fixed axis would arise from the components of force parallel to the axis. These do no

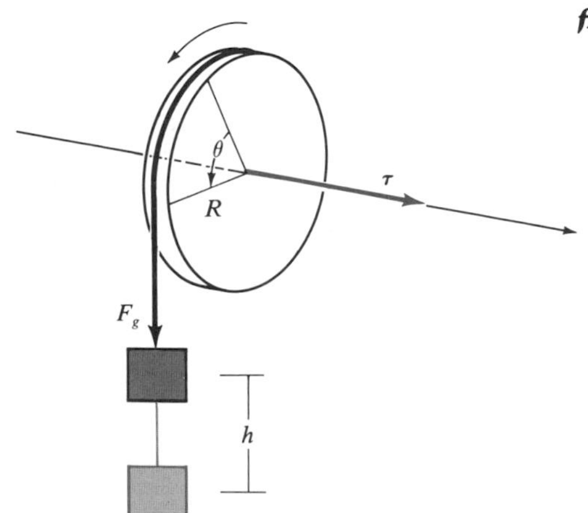

figure 12 • 2 *Work done by a constant torque*

work, since they are counterbalanced by reaction forces from the bearings and other elements which hold the rotating body in place.

Given a torque τ that is constant in magnitude and direction relative to the fixed axis of rotation, Eq. [12 . 7] shows that the power expended by the agent maintaining the torque is given by the scalar product

$$P = \frac{dW}{dt} = \boldsymbol{\tau} \cdot \frac{d\boldsymbol{\theta}}{dt} = \boldsymbol{\tau} \cdot \boldsymbol{\omega} \qquad [12 \cdot 8]$$

where ω is the instantaneous angular velocity. This equation is the rotational analog of $P = \mathbf{F} \cdot \mathbf{v}$, the more general power equation. It is good to keep in mind that *all* motions can be treated by Newton's laws. The rotational analogs as derived here are restrictive, since they generally assume a fixed axis of rotation. Given that condition, however, they afford considerable economies of thought.

example 12 • 1 A uniform rod of length l and mass M rotates about an axis perpendicular to it with an angular speed ω. (*a*) Find the kinetic energy of the rotating rod if the axis of rotation passes through its center of mass. (*b*) Find the kinetic energy if the axis passes through one end of the rod, as shown in Fig. 12 . 3*b*.

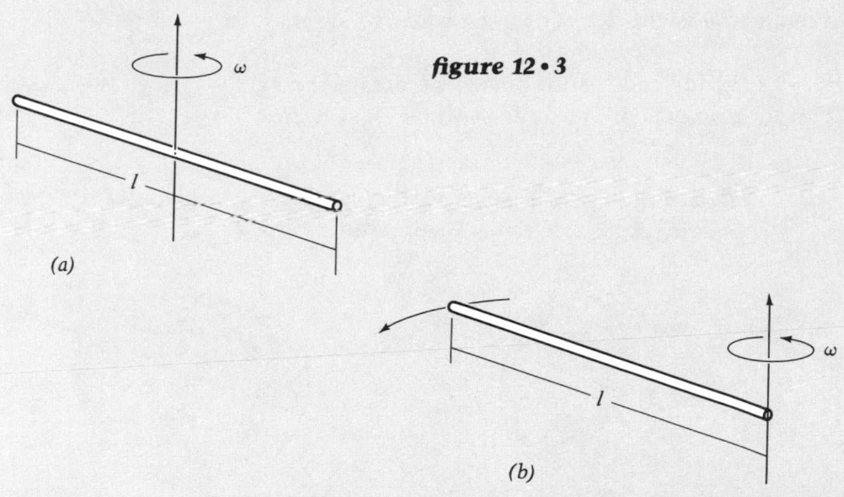

figure 12 • 3

solution (*a*) At some instant of time let the positive x axis lie along the rod, with its origin at the center. An element of rod at position x has mass $dm = (M/l)\,dx$ and a tangential velocity of $v = \omega x$. The kinetic energy of this element is therefore

$$dK = \tfrac{1}{2} dm\, v^2 = \frac{M}{2l} (\omega x)^2\, dx$$

and the kinetic energy of the rod is

$$K = \frac{M\omega^2}{2l} \int_{-l/2}^{l/2} x^2\, dx = \frac{Ml^2}{24} \omega^2$$

(*b*) For the axis passing through one end of the rod, we just repeat the argument and integral of part (*a*) with altered limits, in effect moving the origin from the center to one end:

$$K = \frac{M\omega^2}{2l}\int_0^l x^2\,dx = \frac{Ml^2}{6}\omega^2$$

Just as the inertia or mass of a body is a measure of the resistance it offers to linear acceleration, the moment of inertia I is a measure of its resistance to angular acceleration. It is evident from Eq. [12 . 3], however, that I is not proportional to mass alone; moment of inertia is a function both of the total mass and of the *distribution* of mass—that is, of the distances of the mass elements dm from the axis of rotation. In the case of continuous bodies, it is generally necessary to calculate I with the aid of the integral calculus. For this reason, the summation for I in Eq. [12 . 3] becomes an integral as the masses m_i become infinitesimal mass elements dm:

$$\lim_{m_i \to dm} I = \int_0^M R^2\,dm = \int_V R^2 \rho\,dV \qquad [12 \cdot 9]$$

where M is the total mass of the body, R is the perpendicular distance of dm from the axis of rotation, and ρ is the density of the body, which may be a function of position.

By employing Eq. [12 . 9] for continuous bodies we obtain the expressions listed in Fig. 12 . 4 for moments of inertia for various rigid bodies. The

figure 12 • 4 *Moments of inertia I_c of various homogeneous rigid bodies rotating about axes through their centers of mass*

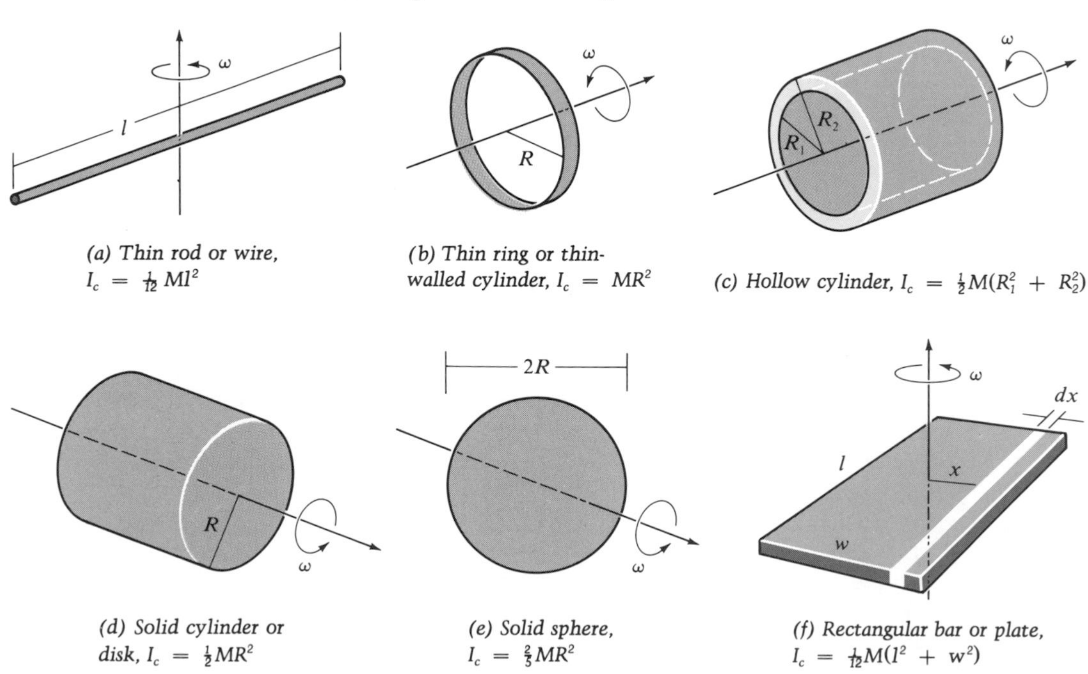

(a) Thin rod or wire, $I_c = \frac{1}{12}Ml^2$

(b) Thin ring or thin-walled cylinder, $I_c = MR^2$

(c) Hollow cylinder, $I_c = \frac{1}{2}M(R_1^2 + R_2^2)$

(d) Solid cylinder or disk, $I_c = \frac{1}{2}MR^2$

(e) Solid sphere, $I_c = \frac{2}{5}MR^2$

(f) Rectangular bar or plate, $I_c = \frac{1}{12}M(l^2 + w^2)$

formula for the moment of inertia about the center of mass of a thin rod has already been derived, in essence, in Example 12 . 1. The formula for a thin ring or cylinder rotating about its longitudinal axis (Fig. 12 . 4*b*) is obvious; derivation of the formula for a hollow cylinder (Fig. 12 . 4*c*) is slightly more complicated. If ρ is the density of the material, and we regard the cylinder as made up of a series of concentric cylinders with radii R in the range $R_1 \leq R \leq R_2$, each of thickness dR and length l, then we can write the corresponding mass element as

$$dm = \rho(2\pi Rl\, dR)$$

and from Eq. [12 . 9],

$$I_c = \int_0^M R^2\, dm = 2\pi l\rho \int_{R_1}^{R_2} R^3\, dR = \tfrac{1}{2}\pi l\rho(R_2^4 - R_1^4) \qquad [12 \cdot 10]$$

Since $M = \pi l\rho(R_2^2 - R_1^2)$ and $R_2^4 - R_1^4 = (R_2^2 - R_1^2)(R_2^2 + R_1^2)$,

$$I_c = \tfrac{1}{2}M(R_1^2 + R_2^2) \qquad [12 \cdot 11]$$

The solid cylinder or disk (Fig. 12 . 4*d*) is just a special case of this equation for $R_1 = 0$. The case of a solid sphere is left as an exercise (see Prob. 12 . 7).

Suppose the element of mass dm in Fig. 12 . 5 is rotating about an axis that passes through point A, perpendicular to the plane of the figure. We can then regard dm as at a distance R from this axis and a distance R_c from a parallel axis passing through the center of mass at the origin. If D is the distance between the two axes and R_c makes an angle θ with D, then by the law of cosines,

$$R^2 = R_c^2 + D^2 - 2DR_c \cos\theta \qquad [12 \cdot 12]$$

Substituting this expression in Eq. [12 . 9], we have

$$I_A = \int_0^M R^2\, dm$$

$$= \int_0^M R_c^2\, dm + MD^2 - 2D \int_0^M R_c \cos\theta\, dm \qquad [12 \cdot 13]$$

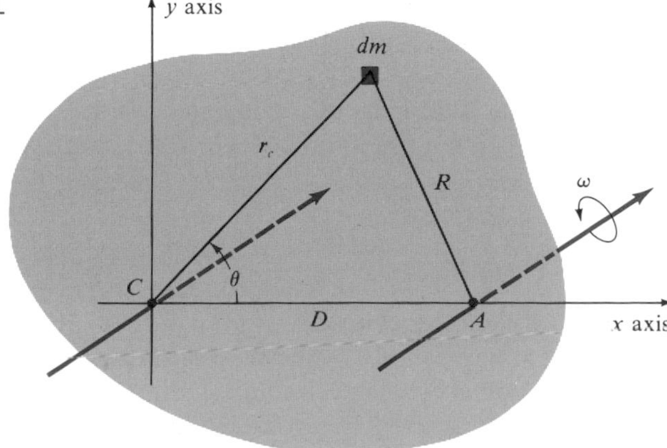

figure 12 • 5 *Proof of the Lagrange parallel-axis theorem*

The last integral is zero, since $R_c \cos\theta = x$ and, by definition, when the origin of coordinates is at the center of mass,

$$\int_0^M x\,dm = 0$$

The remaining integral is simply the moment of inertia about the center of mass,

$$\int_0^M R_c^2\,dm = I_c$$

and we have derived the *Lagrange parallel-axis theorem:*

$$I_A = I_c + MD^2 \qquad [12 \cdot 14]$$

The problem of deriving the formula for any rigid body is now simplified to finding moments of inertia about fixed axes passing in different directions through the center of mass. It is easy to verify that Eq. [12 . 14] agrees with the result of Example 12 . 1 for a rod pivoting about one end. Note that the moment of inertia of any body rotating about an axis through the center of mass is smaller than the moment about any other *parallel* axis.

To return to Fig. 12 . 4*f*, we can compute the moment of inertia of a uniform rectangular plate by subdividing it into infinitesimal bars of width dx. According to the parallel-axis theorem, a bar of length l, width dx, thickness t, and uniform density ρ, located at a distance x from its axis of rotation, should have a moment of inertia dI about this axis

$$dI = dI_c + dm\,x^2$$

where $dm = lt\rho\,dx$. Therefore

$$dI = \left(\frac{l^2}{12} + x^2\right) lt\rho\,dx \qquad [12 \cdot 15]$$

Integrating over the entire plate from $-\frac{1}{2}w \leq x \leq \frac{1}{2}w$, we then have

$$I = lt\rho\left(\frac{wl^2}{12} + \frac{w^3}{12}\right) = \frac{M}{12}(l^2 + w^2) \qquad [12 \cdot 16]$$

the moment of inertia with respect to the center of mass of the plate. In the case of irregular planar bodies which rotate about some axis, the numerical method of Sec. 10 . 5 may readily be adapted to furnish an approximate solution for I.

If M is the total mass of a body and I is its moment of inertia about a given axis, it is often useful in engineering applications to define a *radius of gyration*

$$k = \sqrt{\frac{I}{M}} \qquad [12 \cdot 17]$$

For some rotational computations, we can then idealize the body as a particle rotating at a distance k from the axis, since $I = Mk^2$. For example, for a solid homogeneous cylinder of radius R rotating about its axis, $k = 0.707R$. The rotational kinetic energy of a body can be expressed as

$$K = \tfrac{1}{2}Mk^2\omega^2 \qquad [12 \cdot 18]$$

example 12 • 2 Derive the expression for the moment of inertia of a thin homogeneous disk of mass M and radius R, with its vertical diameter as an axis of rotation.

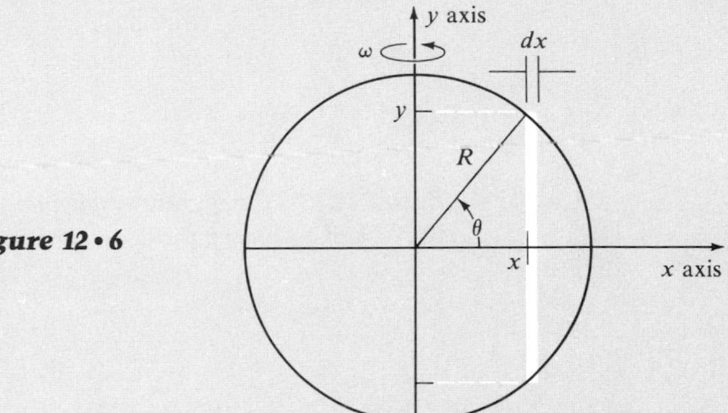

figure 12 • 6

solution The quadrature of Eq. [12 . 9] is performed by dividing the disk into vertical strips dx of height $2y$ and distance x from the axis. Since the surface density of the disk is $\sigma = M/\pi R^2$, the moment of inertia of the strip dx is

$$dI = 2\sigma y \, dx \, x^2$$

Letting $x = R\cos\theta$ and $y = R\sin\theta$, we can integrate to find I_c:

$$I_c = 2\int_0^{\pi/2} 2\sigma R^4 \sin^2\theta \cos^2\theta \, d\theta$$

Using the identities $\sin 2\theta = 2(\sin\theta\cos\theta)$ and $d(2\theta) = 2d\theta$, we can then evaluate this integral from the formula in Appendix F:

$$I_c = \tfrac{1}{2}\int_0^{\pi} \frac{MR^2}{\pi} \sin^2 2\theta \, d(2\theta) = \tfrac{1}{4}MR^2$$

Note that this is only half the value of I_c in Fig. 12 . 4*d*.

12 • 2 *angular momentum*

Angular momentum, the rotational analog of linear momentum, is an important quantity in physics. We define the *angular momentum* $\mathbf{L}$ of a rotating particle as the cross product of its position vector $\mathbf{r}$ and its linear momentum $\mathbf{p} = m\mathbf{v}$:

$$\blacktriangleright \quad \mathbf{L} = \mathbf{r} \times \mathbf{p} = \mathbf{r} \times m\mathbf{v} \qquad [12 \cdot 19]$$

The time-dependent behavior of angular momentum is closely related to the torque defined with respect to the same origin of coordinates. This is apparent if we substitute the force equation $\mathbf{F} = d\mathbf{p}/dt$ from Newton's second law and express the torque as

$$\boldsymbol{\tau} = \mathbf{r} \times \mathbf{F} = \mathbf{r} \times \frac{d(m\mathbf{v})}{dt} \qquad [12 \cdot 20]$$

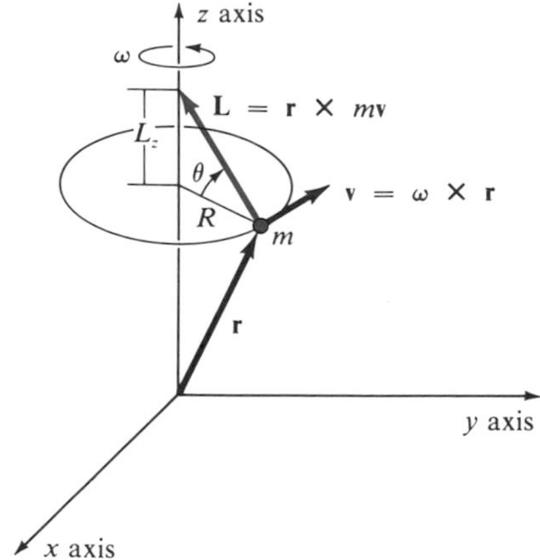

figure 12 • 7 *Component of angular momentum along the axis of a rotating particle,* $L_z = mR^2\omega$

We can use the vector identity from Appendix F to write the instantaneous angular momentum as

$$\frac{d\mathbf{L}}{dt} = \frac{d(m\mathbf{r} \times \mathbf{v})}{dt} = \frac{d\mathbf{r}}{dt} \times m\mathbf{v} + \mathbf{r} \times \frac{d(m\mathbf{v})}{dt}$$

$$= \mathbf{v} \times m\mathbf{v} + \mathbf{r} \times \mathbf{F} \qquad [12 \cdot 21]$$

However, since the cross product of two parallel vectors is zero,

$$\mathbf{v} \times m\mathbf{v} = \mathbf{0}$$

and the last equality in Eq. [12 . 21] reduces to $\mathbf{r} \times \mathbf{F}$. Substitution of this result into Eq. [12 . 20] then yields the rotational analog of the force equation $\mathbf{F} = d\mathbf{p}/dt$:

$$\tau = \frac{d\mathbf{L}}{dt} \qquad [12 \cdot 22]$$

Therefore, when there is no torque ($\tau = 0$), the angular momentum **L** must be constant over time.

For purely rotational motion of a particle m at a distance R from a fixed axis of rotation (see Fig. 12 . 7), the angular-momentum component along that axis is

$$L_z = L \sin\theta = mvr \sin\theta$$

For pure rotation, however, $r \sin\theta = R$ and $v = R\omega$, so that

$$L_z = mR^2\omega$$

For a rigid body, then, in which all particles m_i rotate with the same angular velocity $\boldsymbol{\omega}$ about an axis of rotation that is fixed with respect to the body, the angular momentum of the body along the axis of rotation is

$$\mathbf{L} = \sum_i m_i R_i^2 \omega = I\omega \qquad [12 \cdot 23]$$

where the moment of inertia I is computed with respect to the axis of rotation.

For rotational motion about a fixed axis, Eqs. [12 . 22] and [12 . 23] combine to yield the corresponding equation for torque:

$$\tau = \frac{d(I\omega)}{dt} \qquad [12 \cdot 24]$$

In the case of *constant* I, the torque can also be expressed as

$$\tau = I\alpha \qquad [12 \cdot 25]$$

the rotational analog of $\mathbf{F} = m\mathbf{a}$. These equations all assert one fact:

> The time rate of change of the angular momentum of a body is proportional to and in the same direction as the impressed torque.

example 12 • 3 A flywheel mounted on a shaft is set in rotation by a driving weight of mass m. The wheel and shaft have a total moment of inertia I about the axle, and τ_f is the magnitude of the resistive torque due to the frictional forces at the axle bearing. (*a*) What is the magnitude of the angular acceleration of the wheel? (*b*) What is the angular speed when the driving weight has descended a distance h from rest?

solution (*a*) If T is the tension in the string attached to mass m and a is its acceleration, then $ma = mg - T$, or $T = m(g - a)$, which results in a torque of magnitude $\tau = TR$ applied to the shaft. From Eq. [12 . 25],

$$TR - \tau_f = mR(g - a) - \tau_f = I\alpha$$

Since $a = R\alpha$, we have

$$I\alpha + mR^2\alpha = mgR - \tau_f$$

and the angular acceleration is

$$\alpha = \frac{mgR - \tau_f}{I + mR^2}$$

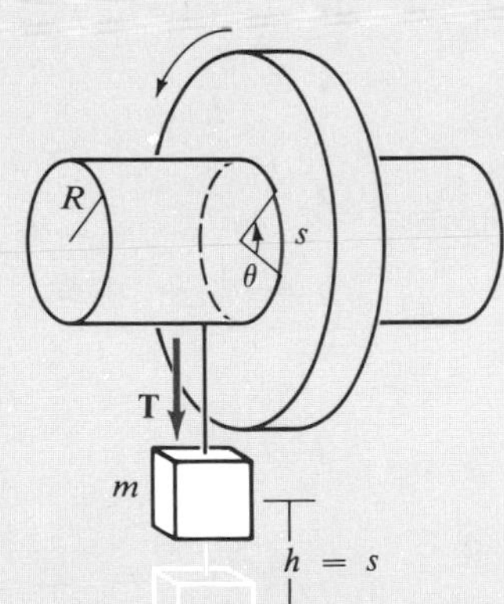

figure 12 • 8

(*b*) In this case $\alpha = d\omega/dt$ is constant, and since

$$\omega_0 = 0 = \theta_0 \qquad h = s = R\theta$$

we finally obtain $\omega^2 = 2\alpha\theta$, or an angular speed

$$\omega = \left(2\frac{mgR - \tau_f}{I + mR^2}\frac{s}{R}\right)^{1/2}$$

A second, and more direct, way of getting this result is to equate the initial potential energy of the system, $U_0 = mgh$, to the sum of the final kinetic energy and the work done against friction, $\tau_f\theta$:

$$mgh = \tfrac{1}{2}mv^2 + \tfrac{1}{2}I\omega^2 + \tau_f\theta$$

where $v = R\omega$ and $\theta = s/R$. This can immediately be solved for the angular speed ω.

For some purposes, the concept of *angular impulse* $\mathcal{P}$ is useful. This is the change in angular momentum, defined as the integral of the torque over the time during which it acts:

$$\mathcal{P} = \int_0^t \tau\, dt = \mathbf{L} - \mathbf{L}_0 \qquad [12 \cdot 26]$$

That is, the change in angular momentum of the body is equal to the angular impulse associated with the applied torque. Evidently angular impulse is analogous to the linear impulse **P** defined earlier in Eq. [9 . 3]. The various rotational and linear analogs we have discussed thus far are summarized in Table 12 . 1.

table 12 • 1 *Rotational analogs of linear mechanical quantities and expressions*

quantity	*linear*	*rotational*
displacement	$d\mathbf{r}$	$d\boldsymbol{\theta}$
velocity	$\mathbf{v} = \dfrac{d\mathbf{r}}{dt}$	$\omega = \dfrac{d\theta}{dt}$
acceleration	$\mathbf{a} = \dfrac{d\mathbf{v}}{dt}$	$\alpha = \dfrac{d\omega}{dt}$
constant acceleration	$v = v_0 + at$, etc.	$\omega = \omega_0 + \alpha t$, etc.
inertia	m	$I = \Sigma\, mr^2$
momentum	$\mathbf{p} = m\mathbf{v}$	$\mathbf{L} = I\omega = \sum_{i=1}^{n} \mathbf{r} \times \mathbf{p}$
impulse	$\mathbf{P} = \mathbf{F}t$	$\mathcal{P} = \tau t$
Newton's second law	$\mathbf{F} = \dfrac{d\mathbf{p}}{dt} = m\mathbf{a}$	$\tau = \dfrac{d\mathbf{L}}{dt} = I\alpha$
element of work	$dW = \mathbf{F} \cdot d\mathbf{r}$	$dW = \tau \cdot d\boldsymbol{\theta}$
power	$P = \dfrac{dW}{dt} = \mathbf{F} \cdot \mathbf{v}$	$P = \dfrac{dW}{dt} = \tau \cdot \omega$
kinetic energy	$K = \tfrac{1}{2}mv^2 = \dfrac{p^2}{2m}$	$K = \tfrac{1}{2}I\omega^2 = \dfrac{L^2}{2I}$

Pure rotational interrelations; R = perpendicular distance to axis of rotation:

$s = R\theta \qquad v = R\omega \qquad a = R\alpha \qquad \omega = 2\pi\nu$

If a body is rigid and of constant mass, the moment of inertia of the body about any given axis is constant. In this case, if $\boldsymbol{\tau} = \mathbf{0}$, Eq. [12 . 25] reduces to $\boldsymbol{\alpha} = d\boldsymbol{\omega}/dt = \mathbf{0}$, or $\boldsymbol{\omega} =$ constant:

> For a rigid body, the angular velocity as well as the angular momentum remains constant if the vector sum of the externally applied torques is zero.

If a body is not rigid, its moment of inertia about a given axis may change during rotation because of a redistribution of mass about the axis. Suppose the moment of inertia about a fixed axis changes from its initial value I_0 to some value I_1 in the time $t_1 - t_0$. Then, if $\boldsymbol{\tau} = \mathbf{0}$, there is no change in angular momentum $L = I\omega$; hence there must be a change in angular velocity such that the angular momentum is conserved:

$$I_1\omega_1 = I_0\omega_0 = L \qquad [12 \cdot 27]$$

> For a nonrigid body upon which no externally applied torques are acting, the angular momentum **L** remains constant, but the change in angular velocity is inversely proportional to the change in moment of inertia.

Note that in this case, although there are no external torques acting, the kinetic energy will change with I, since

$$K = \frac{I\omega^2}{2} = \frac{L^2}{2I} \qquad [12 \cdot 28]$$

Thus if I decreases and L remains constant, then K must increase, as does $\omega = L/I$. When an acrobat pulls his knees toward his chest in order to turn more rapidly in the air, he is using his muscles to supply the required increase in his kinetic energy. That is, in order to change I, some *force* must be exerted. However, since this force is directed inward *toward* the axis of rotation, it changes the kinetic energy without supplying additional torque, since the lever arm about the axis is zero. Therefore L remains constant.

example 12 • 4 A ballerina performs a pirouette, beginning with arms extended and spinning with an angular speed ω_0 (neglect friction). (*a*) If, by folding her arms, she decreases her moment of inertia about her vertical axis of rotation by one-ninth, what is her new speed ω_1? (*b*) How much energy does she expend in this process?

solution (*a*) There are no torques, and since angular momentum must be conserved,

$$I_1\omega_1 = I_0\omega_0 \qquad \omega_1 = \frac{I_0}{I_1}\omega_0 = \frac{I_0}{\frac{8}{9}I_0}\omega_0 = \tfrac{9}{8}\omega_0$$

(*b*) From Eq. [12 . 27], when L remains constant,

$$\frac{K_1}{K_0} = \frac{I_0}{I_1} = \tfrac{9}{8}$$

hence $\Delta K = \frac{1}{8}K_0$.

As we have seen, for a particle or a rigid body, angular momentum is conserved in the absence of any external torques. In fact, we can state a more general conclusion:

> The total angular momentum of a system of bodies with respect to any fixed axis remains unchanged so long as the total external torque on the system with respect to that axis is zero.

This principle of the conservation of angular momentum, like its linear analog, is one of the most important generalizations of mechanics. It has applications in all of physical science, notably in celestial mechanics and in atomic and nuclear physics.

Since, by definition, angular momentum is a vector, the angular momentum of a system of particles is the vector sum of the individual angular momenta with respect to a given origin:

$$\mathbf{L} = \sum_i \mathbf{L}_i = \sum_i m_i \mathbf{r}_i \times \mathbf{v}_i \qquad [12 \cdot 29]$$

If the total external torque on the system is zero, then the total angular momentum of the system will remain unchanged, although individual angular momenta may change due to internal interactions. Furthermore, the angular momentum of a rigid body or a system of particles can be divided into two parts—the angular momentum of the center of mass,

$$\mathbf{L}_c = M\mathbf{r}_c \times \mathbf{v}_c$$

and the angular momentum $\mathbf{L}'$ of all the particles *with respect to the center of mass.* If we transform each position vector $\mathbf{r}_i$ to center-of-mass coordinates,

$$\mathbf{r}'_i = \mathbf{r}_i - \mathbf{r}_c$$

then from the distributive law for the vector product, Eq. [12 . 29] becomes

$$\begin{aligned} \mathbf{L} &= \sum_i m_i(\mathbf{r}_c + \mathbf{r}'_i) \times (\mathbf{v}_c + \mathbf{v}'_i) \\ &= \sum_i m_i \mathbf{r}_c \times \mathbf{v}_c + \left(\sum_i m_i \mathbf{r}'_i\right) \times \mathbf{v}_c \\ &\qquad + \mathbf{r}_c \times \left(\sum_i m_i \mathbf{v}'_i\right) + \sum_i m\mathbf{r}'_i \times \mathbf{v}'_i \end{aligned} \qquad [12 \cdot 30]$$

Therefore, since the terms in $\Sigma m\mathbf{r}'$ and $\Sigma m\mathbf{v}'$ are identically zero from the definition of center of mass (Sec. 10 . 1),

$$\mathbf{L} = M\mathbf{r}_c \times \mathbf{v}_c + \sum_i m_i \mathbf{r}'_i \times \mathbf{v}'_i = \mathbf{L}_c + \mathbf{L}' \qquad [12 \cdot 31]$$

In the special case in which all the parts of a rigid body are moving with the same velocity as the center of mass, there is no rotation about the center of mass, and $\mathbf{L}' = \mathbf{0}$. Hence the total angular momentum $\mathbf{L}$ with respect to some arbitrary point in space would be just the angular momentum $\mathbf{L}_c$ of the center of mass with respect to that point.

example 12 • 5 A physics student is observed to run down the sidewalk at a constant velocity **v** with arms outstretched perpendicular to his velocity vector. The distance from fingertip to fingertip is $2l_0$. As the student speeds past a street sign, he grasps the post with his left hand, lifts his feet from the ground, and swings around the signpost. (*a*) If his mass is M, what is the magnitude of his angular momentum relative to the signpost as he runs down the sidewalk? (*b*) If the reaction force of the signpost not only causes him to revolve, but also provides an impulsive force which slows his forward motion, so that his kinetic energy is reduced to four-fifths of its original value, what is his moment of inertia with respect to the signpost?

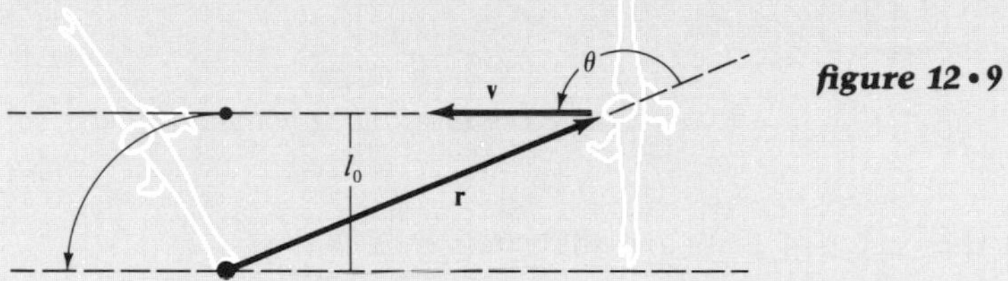

figure 12 • 9

solution (*a*) The student approaches the signpost with an angular momentum of $\mathbf{L} = M\mathbf{r} \times \mathbf{v}$, or

$$L = M(r \sin \theta)v = Ml_0v$$

independently of θ. It may surprise you to think of a purely translational motion as having angular momentum with respect to some point. However, an observer at a point off the path would have to keep rotating his line of sight in order to follow the motion. In atomic and nuclear physics, l_0 is a very important quantity known as the *impact parameter.*

(*b*) Since the lines of action of the centripetal and impulsive reaction forces must pass through the signpost, it can exert no torque on the student relative to an origin at the post. His angular momentum is therefore constant, and his rotational kinetic energy is

$$K = \frac{L^2}{2I} = \tfrac{4}{5}(\tfrac{1}{2}Mv^2)$$

his moment of inertia must be

$$I = \frac{5L^2}{4Mv^2} = \tfrac{5}{4}Ml_0^2$$

12 • 3 *combined translations and rotations*

If a body is not mounted on a fixed axis, but is free to move in space, the applied forces will generally produce translational motion of the center of mass, as well as rotational motion about some axis through the center of mass. The independence of these two motions was first demonstrated in 1834 by Louis Poinsot. He showed that in considering the motion of any rigid body, we may discuss the translation as if the body were a single particle

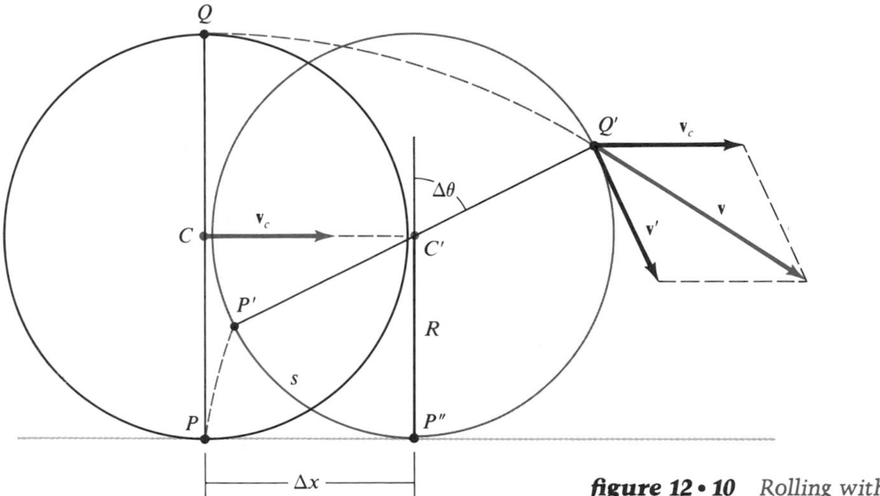

figure 12 • 10 *Rolling without slipping*

with its mass concentrated at the center of mass and the rotation as if the center of mass were fixed. In describing translation, we treat the applied force as acting on the center of mass,

$$\mathbf{F} = M\mathbf{a}_c$$

and in describing rotation we utilize the definitions of torque and angular momentum in Eqs. [12 . 22] and [12 . 23]. In addition, we may use the energy relation of Eq. [10 . 15] to write

$$K = \tfrac{1}{2}I_c\omega^2 + \tfrac{1}{2}Mv_c^2 \qquad [12 \cdot 32]$$

That is, the total kinetic energy of a moving body is the sum of its kinetic energy of rotation *about* the center of mass and its kinetic energy of translation associated with the motion *of* the center of mass.

One of the simplest forms of combined translation and rotation is *rolling without slipping,* defined as motion such that the point or line on the body which is in contact with the surface is instantaneously at rest, although the center of mass of the body continues to move uniformly parallel to the surface. Consider a cylinder that is rolling at a constant speed $v = v_c$ across a flat surface, as in Fig. 12 . 10. The horizontal distance traveled by the center of the cylinder in moving from C to C' must be equal to the arc length $s = R\,\Delta\theta$ through which a contact point turns in moving from P to P'. Therefore the translational motion in this special (but very practical) case is related to the rotation about the center of mass by the equation

$$v_c = R\frac{\Delta\theta}{\Delta t} = R\omega \qquad [12 \cdot 33]$$

and the kinetic energy of the cylinder is

$$K = \tfrac{1}{2}I_c\omega^2 + \tfrac{1}{2}MR^2\omega^2 = \tfrac{1}{2}(I_c + MR^2)\omega^2 \qquad [12 \cdot 34]$$

However, from the parallel-axis theorem, the quantity in parentheses is equal to the moment of inertia I_P about an instantaneous axis of rotation through P. Hence the motion can also be described in terms of a purely rotational motion about a moving axis which translates from P to P'' with speed v_c, even though the material point in contact with the surface is instantaneously at rest.

The velocity $\mathbf{v}$ of a point on the surface of the cylinder is the vector sum of $\mathbf{v}_c = v_c\mathbf{i}$ and a tangential velocity $\mathbf{v}'$ in center-of-mass coordinates. Thus the initial velocity of point Q, at the top of the cylinder, is

$$\mathbf{v}_c + R\omega\mathbf{i} = 2\mathbf{v}_c$$

while the initial velocity of P is

$$\mathbf{v}_c - R\omega\mathbf{i} = \mathbf{0}$$

The velocity vector $\mathbf{v}$ in Fig. 12 . 10 could also have been found by constructing a vector perpendicular to a line passing through P'' and Q' of length $\overline{P''Q'}\ \omega$, or

$$v = 2R\omega \cos(\tfrac{1}{2}\Delta\theta) \qquad [12 \cdot 35]$$

which is easily derived from the principles of plane geometry.

example 12 • 6 A homogeneous solid cylinder of radius R is rolling without slipping down a plane inclined at an angle θ, as shown. (*a*) What is the translational acceleration of the cylinder? (*b*) What minimum frictional force F_f is necessary between the cylinder and the plane to ensure that there is no slipping?

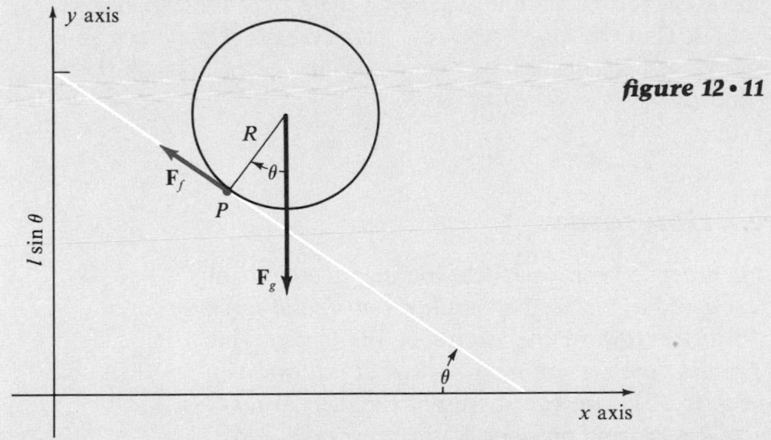

figure 12 • 11

solution (*a*) Consider the cylinder to be rotating about its geometric axis. Its potential energy when it is at the top of the incline is $U_g = Mgl \sin\theta$, where l is the length of the incline. Since $I_c = \frac{1}{2}MR^2$ and $\omega = v/R$,

$$K = \tfrac{1}{2}Mv^2 + \tfrac{1}{2}I_c\omega^2 = \tfrac{3}{4}Mv^2 = Mgl\sin\theta$$

and therefore $v^2 = \frac{4}{3}gl\sin\theta$. The acceleration is constant, so that

$$v^2 = 2al = \tfrac{4}{3}gl\sin\theta$$

Hence $a = \frac{2}{3}g\sin\theta$.

Alternatively consider rotation about the instantaneous axis through P. The weight $F_g = Mg$ of the cylinder acts vertically through the center of mass to yield a torque $\tau = MgR \sin\theta$ about P. Since the moment of inertia about P is $I_P = \frac{3}{2}MR^2$, the angular acceleration is

$$\alpha = \frac{\tau}{I_P} = \frac{MgR\sin\theta}{\frac{3}{2}MR^2} = \frac{2g\sin\theta}{3R}$$

and the linear acceleration is then

$$a = \alpha R = \tfrac{2}{3}g\sin\theta$$

(*b*) The only force whose line of action does not pass through the center of the cylinder is the frictional force F_f. Therefore this force must supply the torque necessary to cause the angular acceleration α:

$$F_f R = I_c\alpha = \tfrac{1}{2}MR^2\frac{2g\sin\theta}{3R}$$

or a minimum force of

$$F_f = \tfrac{1}{3}Mg\sin\theta$$

Alternatively, we could have found F_f by applying the second law of motion to components tangential to the surface of the incline:

$$F_f = Mg\sin\theta - Ma$$

Note that the accelerations do not depend on the size of the cylinder, although F_f does. Had the object been a sphere instead of a cylinder, we should only have had to substitute $I_c = \frac{2}{5}MR^2$ and carry through the analysis as above.

12 · 4 rotational reaction forces

In Sec. 11 . 3 we found that in order to keep a particle rotating in a circle of radius R with uniform angular speed ω, we need an applied *centripetal* force of magnitude $F = mR\omega^2$, directed toward the center of the circle. This force provides the continual radial acceleration $a = -R\omega^2$ of uniform circular motion. According to Eq. [5 . 21], we can represent the inertial reaction by a fictitious force which is equal and opposite to the centripetal force. Therefore, expressed in plane polar coordinates, the general equation of motion along the radial coordinate r is

$$m\frac{d^2r}{dt^2} = F_r + mr\omega^2 \qquad [12 \cdot 36]$$

where F_r is the radial component of the external force acting on the particle. In the special case that $F_r = -mr\omega^2$, we have uniform circular motion for which $r = R$ is the radius of rotation. Equation [12 . 36] can also be derived by a straightforward but lengthy consideration of the derivative $d(r\hat{\mathbf{r}})/dt$, taking into account the fact that the unit vector $\hat{\mathbf{r}}$ is changing direction with time—as is the unit vector $\hat{\boldsymbol{\theta}}$ in the direction of increasing θ.

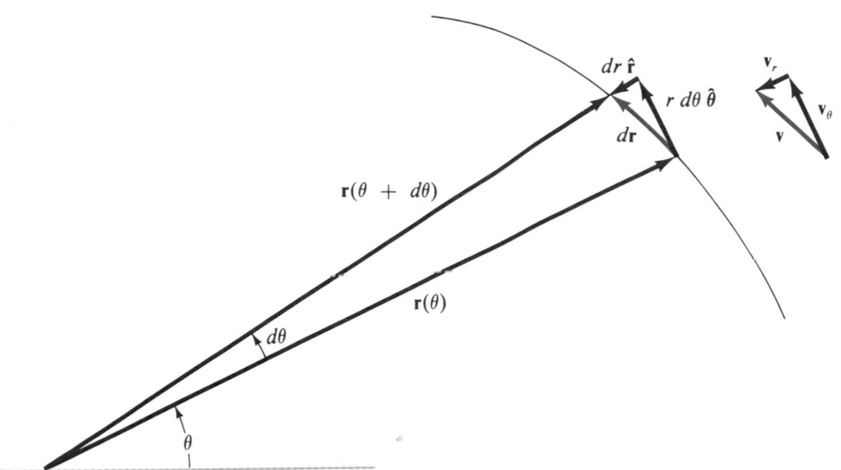

figure 12 • 12 *Differential displacements and corresponding velocity vectors in polar coordinates*

We can generalize this equation to the radial movement of *any* particle as described in plane polar coordinates (r,θ). Suppose the position of the particle can be represented by $\mathbf{r}(\theta)$, as in Fig. 12 . 12. If $ds = |d\mathbf{r}|$ is the infinitesimal distance the particle travels in time dt, then in polar coordinates Pythagoras' theorem gives

$$(ds)^2 = (dr)^2 + r^2(d\theta)^2$$

and the speed of the particle is found by dividing each term by $(dt)^2$:

$$v^2 = \left(\frac{ds}{dt}\right)^2 = \left(\frac{dr}{dt}\right)^2 + r^2\left(\frac{d\theta}{dt}\right)^2 \qquad [12 \cdot 37]$$

We can identify two components of velocity $\mathbf{v}$, a radial component v_r and a tangential component v_θ:

$$v_r = \frac{dr}{dt} \qquad v_\theta = r\frac{d\theta}{dt} = r\omega \qquad [12 \cdot 38]$$

For any particle motion, then, we can also express Eq. [12 . 36] as

$$m\frac{d^2r}{dt^2} = F_r + mr\left(\frac{d\theta}{dt}\right)^2 \qquad [12 \cdot 39]$$

Note that this form of the equation of motion holds in any inertial reference frame.

If the displacements of the particle occur between points on a trajectory given by $r(\theta)$, we can use the algebraic properties of differentials to substitute

$$\frac{dr}{dt} = \frac{dr}{d\theta}\frac{d\theta}{dt} = \frac{dr}{d\theta}\omega \qquad [12 \cdot 40]$$

so that the velocity components in Eqs. [12 . 38] become

$$v_r = \frac{dr}{d\theta}\omega \qquad v_\theta = r\omega \qquad [12 \cdot 41]$$

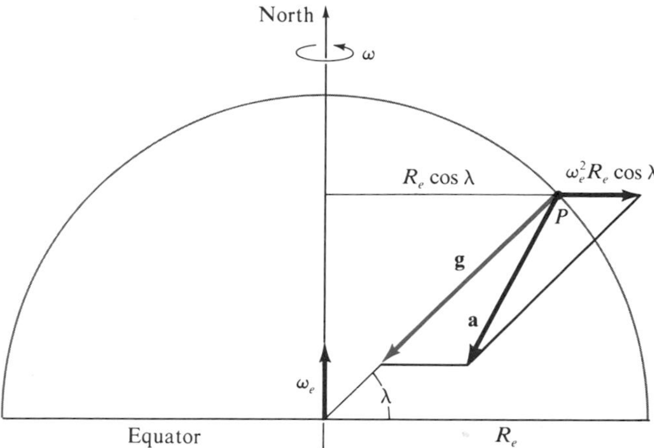

figure 12 • 13 *Acceleration of a particle at P on the earth's surface*

The kinetic energy of the moving particle may then be expressed as the sum of the squared velocities in polar coordinates:

$$K = \tfrac{1}{2}m\omega^2\left[\left(\frac{dr}{d\theta}\right)^2 + r^2\right] \qquad [12 \cdot 42]$$

We shall find this expression very useful in the next chapter.

As an example of the effect of the centrifugal reaction force in a rotating system, consider the acceleration due to gravity at some northern latitude λ on the earth's surface, as shown in Fig. 12 . 13. Since the earth is rotating with angular velocity $\omega = \omega_e \mathbf{j}$, an observer of mass m at this latitude will experience a fictitious force $F' = m\omega_e^2 R_e \cos\lambda$, (which seems quite real to him) directed away from the earth's axis. If we assume that the acceleration due to gravity is $-g(\cos\lambda\,\mathbf{i} + \sin\lambda\,\mathbf{j})$ at the earth's surface, then the actual acceleration experienced by the observer is

$$\mathbf{a} = (\omega_e^2 R_e - g)\mathbf{i}\cos\lambda - g\mathbf{j}\sin\lambda \qquad [12 \cdot 43]$$

The effective acceleration of gravity directed toward the earth's center is

$$g_{\text{eff}} = g - \omega_e^2 R_e \cos^2\lambda \qquad [12 \cdot 44]$$

This means that the effective pull of gravity is greatest and directly inward only at the poles. Since a liquid surface at rest will be normal to the actual force on it, this explains why, as the earth cooled and solidified, it assumed the shape of an oblate spheroid, slightly flattened at the poles, instead of a perfect sphere. The radius of the earth is 6357 km at the poles and 6378 km at the equator.

There is one more fictitious force which is encountered in a *noninertial* coordinate system rotating with constant angular velocity, such as our planet. This is known as the *Coriolis force,* discovered by Gustave-Gaspard Coriolis in 1835, and it depends on v_r as well as v_θ. We shall consider only a heuristic derivation of the Coriolis force, since the mathematical derivation is beyond the scope of this text. Imagine a projectile initially located at the North Pole, so that it has no tangential velocity. As the projectile moves horizontally away from the pole with speed v_r, the earth turns underneath it, as shown in Fig. 12 . 14. To an observer rotating with the earth, the projectile will appear to be deflected through some angle $\Delta\theta = \omega_e \Delta t$ while it moves outward by an amount $v_r \Delta t$. The approximate distance Δs through which it is deflected is

$$\Delta s = (v_r \Delta t)(\omega_e \Delta t) = v_r \omega_e (\Delta t)^2 = \tfrac{1}{2} a (\Delta t)^2 \qquad [12 \cdot 45]$$

Therefore the Coriolis force is given by

$$F_c = ma = 2m\omega_e v_r$$

or in vector form,

$$\mathbf{F}_c = -2m\boldsymbol{\omega}_e \times \mathbf{v} \qquad [12 \cdot 46]$$

This last equation is the general form of the Coriolis force.

Now consider a situation in which the observer is rotating with angular speed ω as he observes a particle which is actually stationary. Imagine an observer seated on a rotating stool. As he spins around, the particle appears to him to be rotating with angular speed ω in the sense opposite his own sense of rotation, as shown in Fig. 12 . 15. He will feel a centrifugal *outward* force in his system, of magnitude

$$F = mr\omega^2$$

figure 12 • 14 *Flight of a projectile fired from the North Pole as seen by an observer directly above the pole and rotating with the earth at angular frequency* ω_e

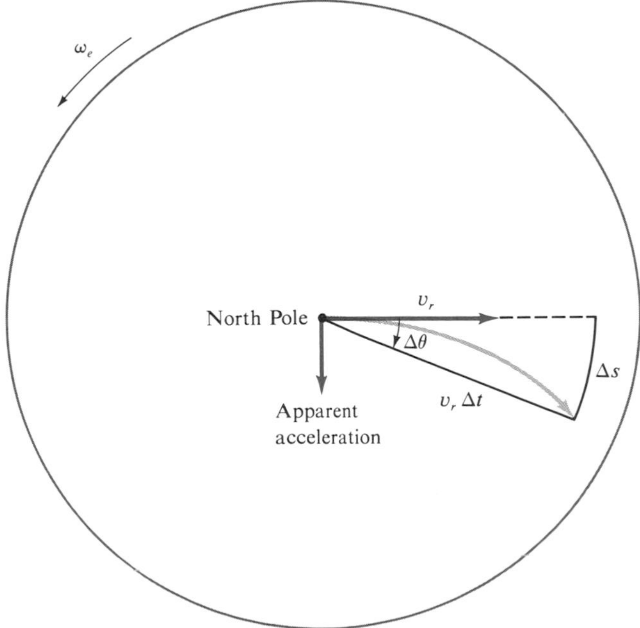

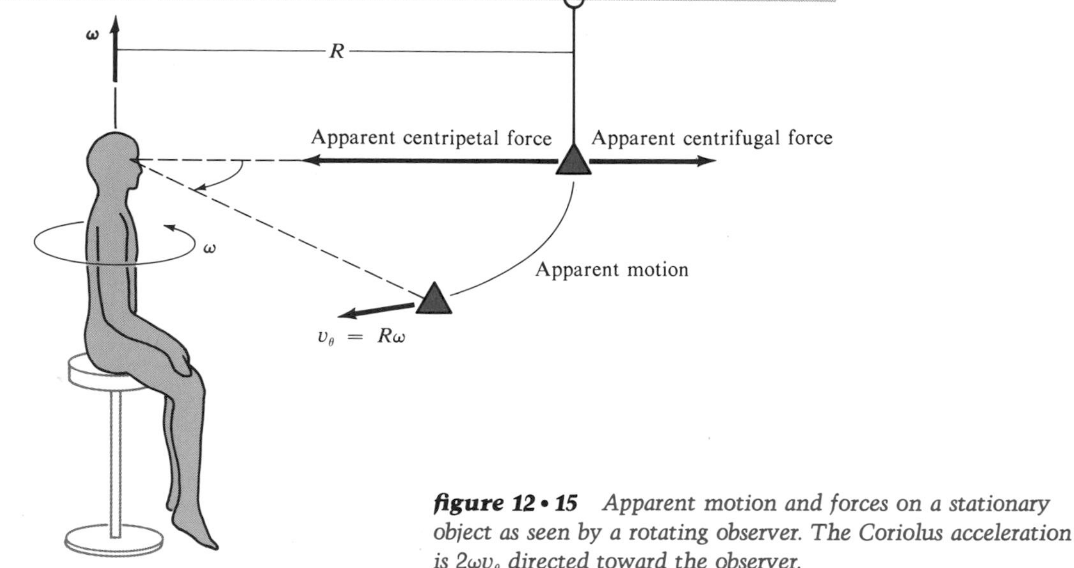

figure 12 • 15 *Apparent motion and forces on a stationary object as seen by a rotating observer. The Coriolus acceleration is $2\omega v_\theta$ directed toward the observer.*

which acts on any body displaced from the origin. However, he also knows that for uniform circular motion there must be a *net inward* force of the same magnitude. Therefore his observation of the particle from a rotating system is in fact an experimental demonstration of the existence of a fictitious force given by

$$F_r = -2mr\omega^2 = +2m\omega v_\theta$$

or in vector notation,

$$\mathbf{F} = -2m\boldsymbol{\omega} \times \mathbf{v} \qquad [12 \cdot 47]$$

Therefore the general expression for the coriolis force is stated as

$$\mathbf{F}_c = -2m\boldsymbol{\omega} \times \mathbf{v} \qquad [12 \cdot 48]$$

as seen by an observer in a coordinate system rotating with constant angular velocity $\boldsymbol{\omega}$. Note that $\mathbf{v}$ is the velocity vector as observed from the *rotating system.*

The Coriolis force always acts in a direction normal to the velocity vector. For this reason, in the Northern Hemisphere it tends to turn a moving body to the right, and in the Southern Hemisphere it tends to turn a moving body to the left. This force has a significant effect on large-scale free or fluid motions at the earth's surface, such as long-range projectiles, wind patterns, and ocean currents. It provides the mechanism which triggers the formation of hurricanes and typhoons. It also explains why wind patterns on weather maps generally circulate around regions of high or low pressure in the atmosphere, rather than blowing directly from high-pressure to low-pressure regions, as one might expect. In fact, if you tilt your head while you are sitting on a rapidly spinning stool, you will experience a sensation of falling owing to your Coriolis acceleration with respect to the stool (seatbelts are recommended).

example 12 • 6 A projectile is fired horizontally outward from a cannon at the North Pole. What is the approximate angular deflection of the projectile relative to the earth after an elapsed time of 2 min?

solution The angular deflection is just that due to the rotation of the earth:

$$\Delta\theta = \omega\,\Delta t = \frac{360^\circ}{24\text{ h}}\,2\text{ min} = 0.5^\circ$$

example 12 • 7 What is the total force on an object traveling southward with a speed v at latitude λ as seen in the Northern Hemisphere?

solution The Coriolis force will point directly *westward* with magnitude

$$F_w = 2m\omega_e v \sin\lambda$$

Remember that to a rotating observer the object's velocity points southward. The centrifugal force depends on the position of the body as seen in the rotating system, so it has a *southward* component of

$$F_s = m\omega_e^2 R_e \cos\lambda \sin\lambda$$

The *downward* component of the force is given by Eq. [12 . 44]:

$$F_d = m(g - \omega_e^2 R_e \cos^2\lambda)$$

12 . 5 *gyroscopic phenomena*

The gencral case of rotation of a free body is much more involved than the foregoing considerations might indicate. Although the motion does consist of a simultaneous translation of the center of mass and a rotation about an axis through this point, the direction of the axis through the center of mass may change with time. As a result, both the moment of inertia and the angular velocity will change with time. This means that the angular momentum **L** may not always have the same direction as $\boldsymbol{\omega}$, and Eq. [12 . 26] will not give the angular momentum unless I is interpreted as the moment of inertia about some instantaneous axis of rotation. Furthermore, owing to its inertia, every particle in the body tends to get as far as possible from the axis of rotation, and so to increase the moment of inertia about that axis. The body will therefore not spin stably until the moment of inertia is as large as possible.

The general case of rotation of a free body is very complex, but let us consider one especially important type of rotational motion in three dimensions. This is the motion of a rigid symmetrical body that is spinning rapidly about its axis of symmetry, with only one point of this axis kept fixed. The prime example is the motion of an ordinary top. Our treatment, although incomplete, will provide a first approach to an understanding of such important phenomena as the astronomical precession of the equinoxes, the behavior of electrified subatomic particles spinning in a magnetic field, and the operation of the complicated gyroscopic instruments used for guidance of ships, aircraft, and rockets.

The simple model in Fig. 12 . 16 depicts a bicycle wheel supported by a universal joint at end O of its axle x, a distance l from its hub at O'. The torque due to gravity is

$$\tau = l\mathbf{i} \times (-Mg\mathbf{k}) = Mgl\mathbf{j} \qquad [12 \cdot 49]$$

about the horizontal y axis. If the wheel is spinning on its axle x, however, it does not fall. Instead, it revolves, or *precesses,* about the vertical z axis, its center O' rotating slowly in the *horizontal xy* plane. This is because the change $\Delta\mathbf{L}$ in angular momentum must be along the y axis, parallel to the torque which generates it.

To understand this apparently extraordinary result, suppose Fig. 12 . 17 is the face of the wheel as viewed from the positive end of the y axis, with the wheel spinning about this axis with constant angular speed ω_s and supported by a second universal joint at O'. If the second joint were suddenly removed, the torque due to the wheel's weight would cause it to fall, and in so doing, to rotate about the axis MN in such a way that the upper quadrants, a and b, would tilt toward you and quadrants c and d would tilt away from you.

Consider now the instantaneous effect of the *inertia* of a particle in each of the four quadrants separately in resisting this motion about the MN axis. Since the wheel is simultaneously spinning on its axle x, a particle in quadrant a is being carried farther away from the MN axis, and hence is being made to move faster in whatever motion there is about this axis. Because of inertia, it resists this increase in speed, and so opposes the motion about MN. This inertial resistance may be represented as a force acting on the wheel in a direction perpendicular to the plane of the paper and *away from* the observer. At the same time, the spin is carrying a particle in quadrant d closer to the MN axis, and so its speed at right angles to the plane of the paper

figure 12 • 16 *Theory of the simple top*

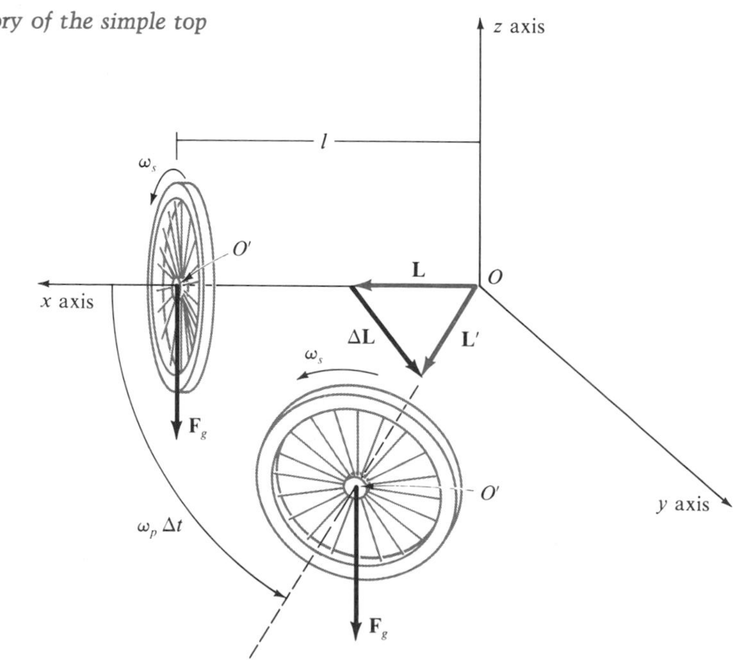

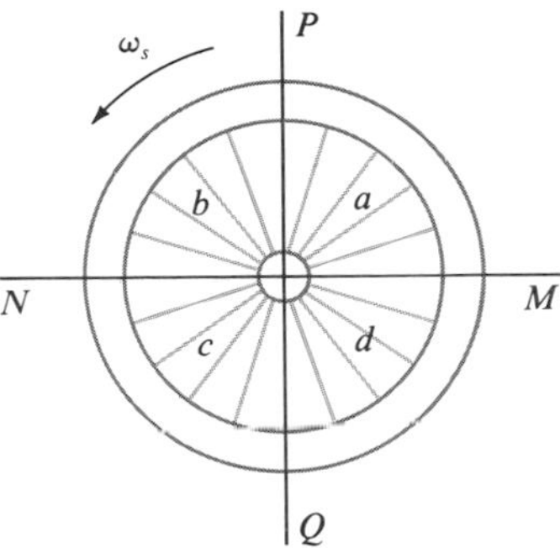

figure 12 • 17 *The wheel of Fig. 12 . 16 as seen from the positive end of the spin axis*

must diminish. Hence this particle, by virtue of its inertia, also reacts on the wheel in a direction *away from* the observer. However, the reactions of particles in quadrants *b* and *c* are *toward* the observer, and the net effect of these four reactions is to rotate the wheel about the axis *PQ* instead of about *MN*; the wheel therefore precesses as shown in Fig. 12 . 16. The strength of the inertial reactions depends on the rate at which the spin of the wheel causes the gravitationally induced velocities to change. Hence the rate of precession should be proportional to ω_s, the spin angular velocity of the wheel on its axis.

Assume that ω_s is very large, as is usually true in practice. The total angular momentum L will then be directed very nearly along the spin axis x,

$$L \doteq I\omega_s$$

where I is the moment of inertia about the spin axis. The vector sum is $\mathbf{L}' = \mathbf{L} + \Delta\mathbf{L}$, but $L' \doteq L$, so that the angular-momentum vector may be said to revolve approximately in a circle about the z axis:

$$\Delta L = L\Omega\,\Delta t \qquad [12 \cdot 50]$$

where Ω is the angular speed of precession. Therefore the torque about the z axis is

$$\tau = \frac{dL}{dt} = L\Omega = I\omega_s\Omega \qquad [12 \cdot 51]$$

and the angular speed of precession is

$$\Omega = \frac{Mgl}{I\omega_s} \qquad [12 \cdot 52]$$

The condition on this theory is that $\Omega \ll \omega_s$ or $Mgl \ll I\omega_s^2$. Note that the last inequality has the dimensions of energy.

In the case of a spinning top, the torque vector $\boldsymbol{\tau}$ is at every instant perpendicular to the plane of spin velocity $\boldsymbol{\omega}_s$ and precession $\boldsymbol{\Omega}$, and the direction of $\mathbf{L}$ is always in the direction of a right-handed rotation from $\boldsymbol{\Omega}$ to $\boldsymbol{\omega}_s$ (Fig. 12 . 18*a*). The vector equation of motion for the top therefore relates its angular momentum to the applied torque:

$$\frac{d\mathbf{L}}{dt} = \boldsymbol{\tau} = \boldsymbol{\Omega} \times I\boldsymbol{\omega}_s = \boldsymbol{\Omega} \times \mathbf{L} \qquad [12 \cdot 53]$$

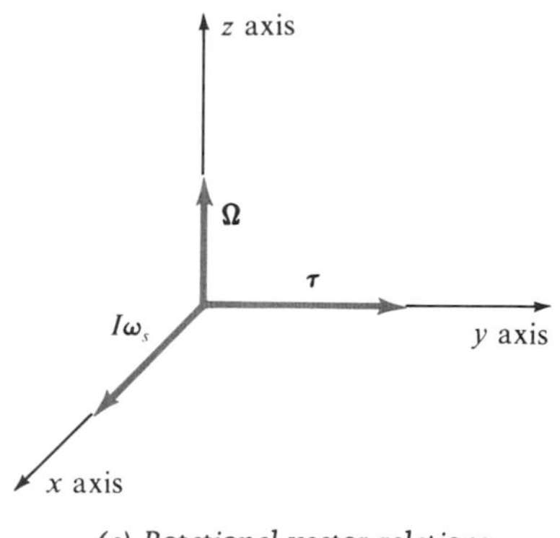

(a) Rotational vector relations

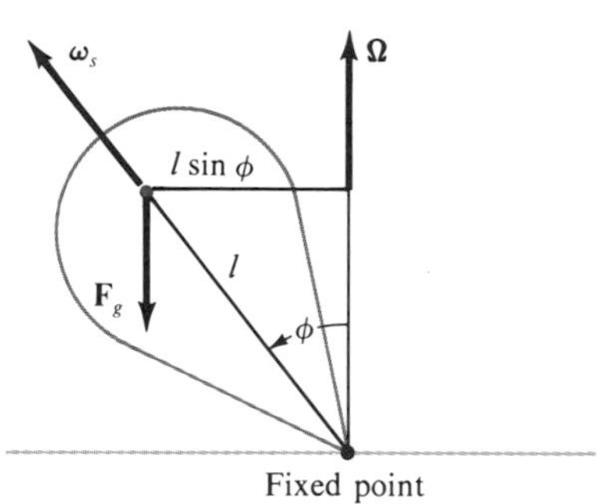

figure 12 • 18 A spinning top

(b) Nutation angle ϕ

This is precisely the equation for a vector **L** rotating with angular velocity $\mathbf{\Omega}$. If the spinning body has $\boldsymbol{\omega}_s$ inclined at some other angle ϕ to the precession axis, as in Fig. 12 . 18*b*, then $\tau = Mgl \sin \phi$, and in time Δt,

$$\Delta L = I\omega_s \Omega \sin \phi \, \Delta t = \tau \, \Delta t = Mgl \sin \phi \, \Delta t \qquad [12 \cdot 54]$$

which reduces to Eq. [12 . 52], regardless of the value of ϕ. Thus the vector relation in Eq. [12 . 53] also holds for any angle ϕ.

Suppose that the bicycle wheel in Fig. 12 . 16 is initially supported at O' as it spins about its axis. When this support is withdrawn, Ω does not immediately reach the steady-state value Mgl/L. Instead, the axle tilts downward under gravity and at the same time begins to move sideways about z with a gradually increasing rate of precession. The fixed end of the axle bears down more and more on the support at O until the support exerts an upward reaction force on the axle which becomes greater than Mg. Then the center of mass of the wheel rises and the axle tilts upward. However, this reduces the rate of precession and the axle again begins to fall. This rising and falling motion continues to take place between certain limiting values of ϕ, known as the *nutation* (nodding) *angle*. This periodic motion continues during precession until frictional forces at the support reduce the spin velocity $\boldsymbol{\omega}_s$ and finally allow the body to fall.

Another example is the *precession of the equinoxes*, the time at which the sun crosses the equator (about March 21 for the vernal equinox, September 23 for the autumnal equinox). Hipparchus, the greatest astronomer of antiquity, observed in the second century B.C. that the position of the sun relative to the fixed stars at the time of equinox was changing at the rate of about 1.5° per century. This change occurs because the earth's spin axis, the line between the North and South Poles, makes an angle of 23.5° with the normal to the plane of the *ecliptic*, the earth's yearly path around the sun. Since the earth is not perfectly spherical, there is a small net torque on it due to the attraction of the sun and moon. This causes our axis to precess with a period of 26,000 years.

An important practical application of these principles is the *gyroscope*, a term often loosely applied to any symmetrical body with one point of its

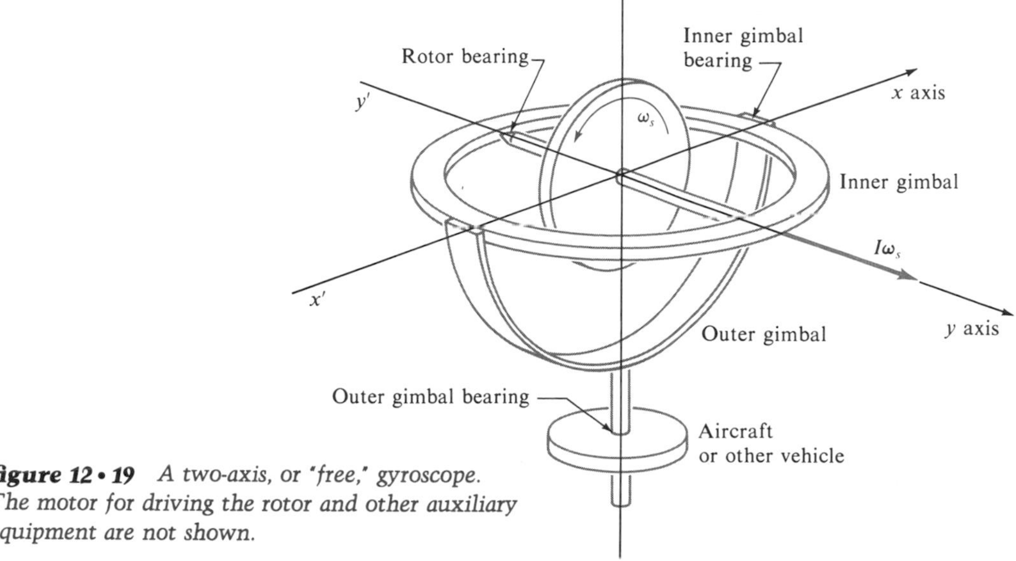

figure 12 • 19 *A two-axis, or 'free,' gyroscope. The motor for driving the rotor and other auxiliary equipment are not shown.*

axis of symmetry kept fixed. We shall restrict the term to cases where this point coincides with the center of mass. The gyroscope shown in Fig. 12 . 19 has a spinning rotor mounted on gimbals which are pivoted inside one another. The several axes of rotation intersect at the center of mass of each rotating part, about which there is consequently no net torque due to gravity. Therefore the spin angular momentum always points in the same direction relative to the fixed stars, and the y axis is oriented accordingly, even though the vehicle containing the gyroscope may change direction. This makes it possible to use the gyroscope as a steering mechanism in rockets, satellites, and space vehicles. The stabilizing effect of spin is also utilized in bicycle riding, football throwing (American style), and rifled guns (spiral bores).

PROBLEMS

12 • 1 moment of inertia

12 • 1 The base of a large centrifuge weighs 20 metric tons and has a diameter of 9 m. If its mass is considered uniformly distributed over its surface and we neglect friction, what angular speed will a 375-W engine impart to it in 2 min?

12 • 2 A hydrogen molecule consists of two hydrogen atoms, each of mass 1.7×10^{-24} g, with a separation distance between their centers of mass kept at 10^{-8} cm by interatomic forces. Nearly all the mass of each atom is concentrated in its nucleus, which has a diameter of approximately 1.5×10^{-13} cm. (*a*) Compute the moment of inertia of a single hydrogen nucleus about any axis through its center. (*b*) The moment of inertia of the pair of nuclei about the line connecting them. (*c*) The moment of inertia of the pair of nuclei about an axis at right angles to the line connecting them and through the center of mass of the pair.

12 • 3 Using the "negative-mass" technique of Example 10 . 7, find the moment of inertia I_c of a thick ring or annulus of mass M with inner and outer radii R_1 and R_2 about its diameter.

12 • 4 A closed cubical box of volume V is made of thin sheet metal of constant surface density σ. Find the moment of inertia of the box about an axis passing through its center of mass and perpendicular to two opposite faces.

12 • 5 (*a*) How could you distinguish a gold sphere from a silver sphere if they had the same radius, the same weight, and were painted the same color? (*b*) If a hard-boiled egg were in the same basket with some uncooked eggs, how would you go about picking it out? (See S. P. Thompson, *Life of William Thomson, Baron Kelvin of Largs,* vol. 2, p. 740.)

12 • 6 If the thickness of a solid disk of mass M and radius R varies directly with the distance from its axis of rotation, what is the expression for the moment of inertia about the axis?

12 • 7 A thin spherical shell has a radius r, thickness dr, and density ρ. (*a*) Show that the moment of inertia about its center of mass is given by

$$dI_c = \tfrac{8}{3}\pi r^4 \rho \, dr$$

(*Hint:* Consider the shell as made up of thin rings about the diameter of rotation.) (*b*) Use this result to show that the moment of inertia of a sphere of mass M and radius R about an axis through its center is

$$I_c = \tfrac{2}{5}MR^2$$

12 • 8 Assume that the earth is a perfect sphere of radius $R_e = 6370$ km and mass $M_e = 6 \times 10^{24}$ kg. (*a*) Use the result of Prob. 12 . 7 to find the kinetic energy K of the rotation of the earth on its axis. (*b*) The earth is actually slightly flattened at the poles and bulges at the equator. What is the percentage of change in K if we adjust for this by slicing off the North and South Poles of our perfect sphere, so that the polar radius becomes 15 km less, and then transfer the same amount of mass to the equator?

12 • 9 A body of mass M starts from rest at the rim of a hemispherical bowl, and its center of mass moves in an arc of radius R as it slides down the side. (*a*) If the coefficient of kinetic friction is μ_k, find the differential equation for angular displacement $d\theta$. (*b*) Assuming that μ_k is small enough so that M reaches the bottom of the bowl, how much work is done against friction? (*c*) What is the angular speed of the particle at the bottom of the bowl? (*d*) Find the maximum angle θ' which the particle will attain before it begins to slide back.

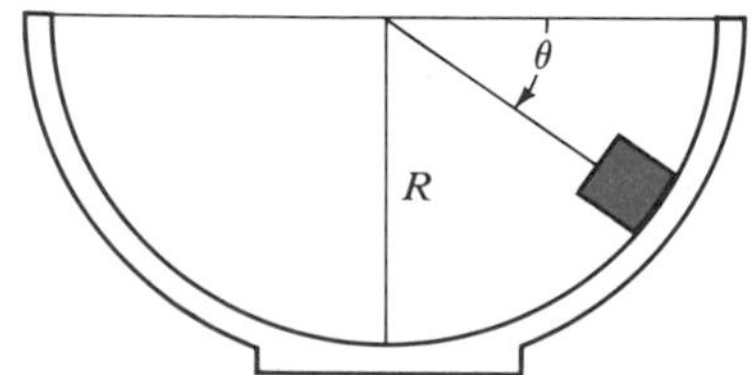

12 • 10 Use the box-product identities $\mathbf{A} \cdot \mathbf{B} \times \mathbf{C} = \mathbf{A} \times \mathbf{B} \cdot \mathbf{C}$ to show that for the case of an infinitesimal rotation $d\boldsymbol{\theta}$ displacing a particle at $\mathbf{r}$ by an amount $d\mathbf{r}$, the work done by a force $\mathbf{F}$ on the particle can be expressed as $dW = \boldsymbol{\tau} \cdot d\boldsymbol{\theta}$.

12 • 11 A thin homogeneous rod of length l is standing on end on a perfectly smooth floor. If it is allowed to fall, with what angular speed ω will it strike the ground?

12 • 2 *angular momentum*

12 • 12 A constant pull of $F = 1960$ N applied tangentially to the rim of a wheel of radius $R = 100$ cm changes the rotation frequency ν of the wheel from 2 to 4 Hz in 30 s. (*a*) What is the moment of inertia I_c of the wheel about its axle? (*b*) What is the change ΔL in the angular momentum during the 30 s? (*c*) Through what angle θ does the wheel turn during this time? (*d*) How much energy is expended in producing this increase in angular momentum?

12 • 13 A solid copper cylinder of radius $R = 2$ cm and length $l = 30$ cm rotates about a longitudinal axis through the center of mass. (*a*) Find the moment of inertia about the axis. (*b*) If friction is negligible, what angular acceleration α is produced about this axis by a 2-kg driving weight suspended from a cord wound around the cylinder? (*c*) What is the angular speed after the weight has acted for 3 s? (*d*) If the driving weight were removed and a force of 2 kgf were applied to the cord, what would be the resulting angular acceleration?

12 • 14 The pulley of a certain Atwood's machine consists of a uniform disk of mass $m_1 = 150$ g. The masses on the ends of the cord are $m_2 = 200$ g and $m_3 = 250$ g. (*a*) Considering bearing friction and the mass of the cord as negligible, find the linear acceleration a of the moving system. (*b*) What is the equivalent mass of the pulley—that is, the mass which, if placed at the circumference of the pulley, would offer the same resistance to acceleration as the system shown?

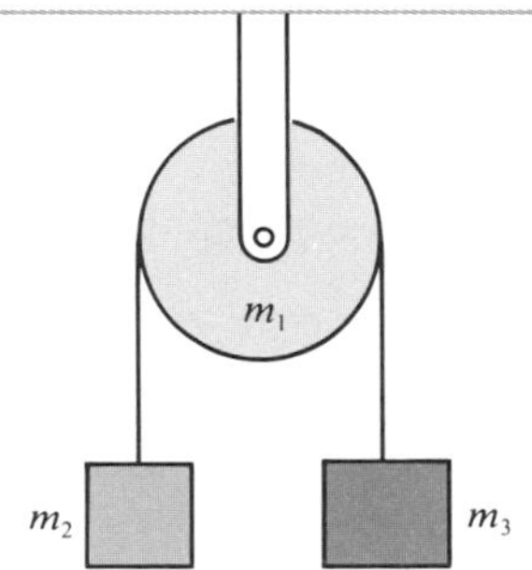

12 • 15 A homogeneous solid cylinder has a perfectly flexible but inextensible cord wound around it. One end of the cord is fastened to a fixed point and the cylinder is allowed to fall. Write the expression for the angular acceleration.

12 • 16 A putty ball of mass m is thrown with speed v normal to a long thin bar of mass M and length $2l$. The ball makes a completely inelastic impact, causing it to rotate about an axis through its center. What is the angular speed ω imparted to the bar?

12 • 17 A uniform plank which is slightly bowed falls from an upright position, pivoting around its bottom end so that only its ends strike the ground. What is the ratio of the impulses delivered to the ground by each end as the plank falls?

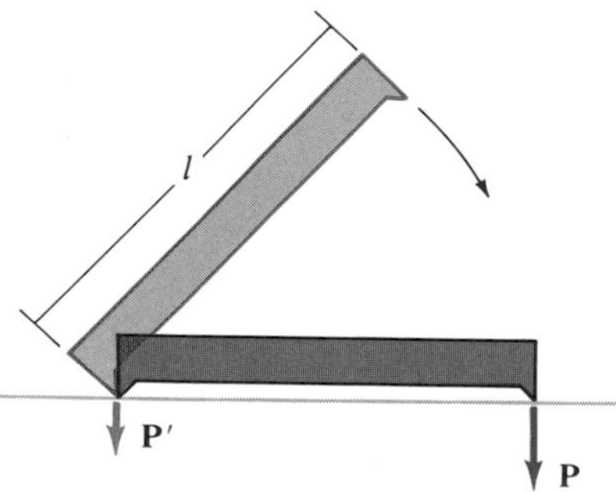

12 • 18 A uniform thin bar of mass M and length l, which is lying on a smooth table, is given an impulse $\mathbf{P}$ tangent to the table and perpendicular to the bar at one end. How far will the center of mass of the bar slide before it completes one revolution?

12 • 19 As the earth cooled it steadily shrank. What was the period of a rotation at a time when the density of the earth was 1.1 g/cm^3? Assume that the mass was the same as now and homogeneously distributed. The present density of the earth is 5.522 g/cm^3.

12 • 20 An object is dropped from a high tower in the Northern Hemisphere. Because of the rotational motion of the earth, it does not fall toward the exact center of the earth. In which direction does it drift as it falls? Explain your reasoning.

12 • 21 A 5-g bullet is moving with a speed of 100 m/sec toward a projection on a wheel mounted on a fixed axis. The bullet makes a completely inelastic impact with the projection. If the wheel has a moment of inertia $I_c = 2 \times 10^{-2}$ kg · m^2 and a radius $R = 20$ cm, what is its angular speed after the impact?

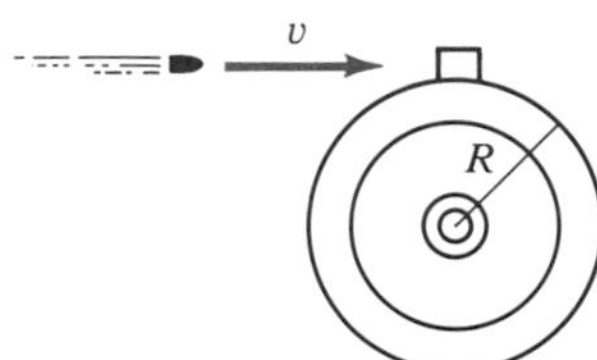

12 • 22 A uniform hollow tube of mass M and length l rotates freely about its center in a horizontal plane with angular velocity ω_0. There is a sphere of mass m at the center of the tube. If the sphere is displaced slightly from the center, it will leave the tube. Find the angular velocity of the tube after the sphere has been ejected. Assume that the sphere is small enough that its mass can be considered to be concentrated at a point.

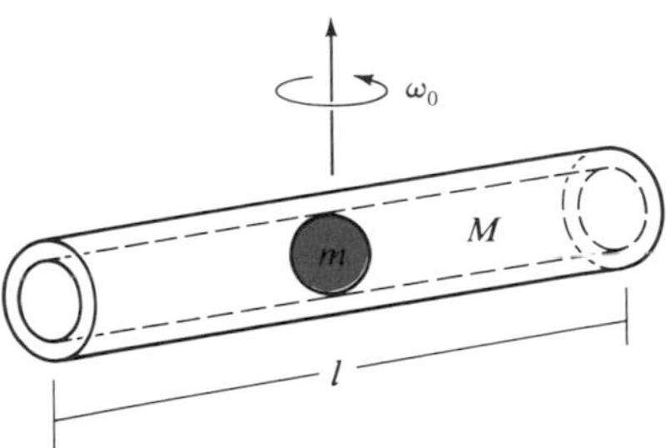

12 • 23 It is believed that the planets may have been formed as a result of highly unstable motions in the sun and that ejection of the planets carried away enough angular momentum to relieve the instability. Using the data of Table 13 . 1 compute the ratio of the angular momentum of the earth to that of the sun, given that the sun makes a complete revolution about its north-south axis (approximately perpendicular to the plane of the earth's orbit) every two weeks.

12 • 3 *combined translation and rotation*

12 • 24 A bicycle with wheels 70 cm in diameter is ridden at a speed of 20 km/h. (*a*) Find the angular speed ω_s of the wheel about its axle. (*b*) Find the linear speed of a particle at its highest point with respect to the axle. (*c*) Find the linear speed of the same particle with respect to the ground. (*d*) Find the linear speed with respect to the ground of a particle in contact with the ground. (*e*) Find the linear speed with respect to the ground of any particle on the rim, expressed in terms of the angle θ which a line drawn from the axle to the point makes with the vertical. (*f*) Find the angular speed ω of *any* particle on the rim with respect to an axis passing through the point on the rim that is in contact with the ground.

12 • 25 What fraction of the total kinetic energy is translational energy in the case of (*a*) a rolling hoop? (*b*) A rolling solid cylinder? (*c*) A rolling solid sphere?

12 • 26 A sphere of mass M and radius R rolls along the ground with an angular speed ω. (*a*) Find its total kinetic energy by adding the kinetic energy of translation to that of rotation about the axis of the sphere. (*b*) Find its total kinetic energy by considering the motion of the sphere as a pure rotation about a point in contact with the ground.

12 • 27 What is the ratio of the linear speed acquired by a body sliding down a smooth inclined plane to that acquired in rolling without slipping down the same incline, ifthe body is (*a*) a hoop? (*b*) A solid cylinder? (*c*) A solid sphere?

12 • 28 A hoop and a solid disk start from rest and roll without slipping down an incline. (*a*) Which one will reach the bottom first? (*b*) What is the ratio of their speeds at the bottom? (*c*) When frictional effects are negligibly small, why can a heavy man on a bicycle always coast with greater acceleration on a hill than a light man on the same bicycle? (*d*) If heavier tires were used on a bicycle, how would the coasting acceleration be affected? (*e*) What if the weight of the frame were increased?

12 • 29 Find the tension in the cord in Prob. 12 . 15.

12 • 30 Two solid wheels, each 50 cm in diameter and weighing 10 kgf are attached by an axle 20 cm in diameter and weighing 20 kgf. A string wrapped around the axle is pulled on with a force of 22 N in a horizontal direction, as shown. Find the linear acceleration of the center of mass of the system as it rolls without slipping.

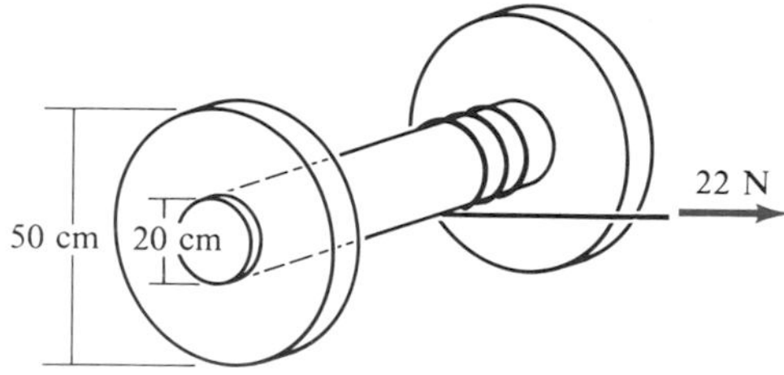

12 • 31 A small cylinder of mass m and radius r rolls without slipping from A to B on a block of wood (cross-sectional view shown). The block of wood has a mass M and is hollowed out to form a quadrant with a cylindrical surface of radius R. The block of wood rests on a frictionless surface. If both m and M are at rest when the cylinder is at A, find their linear speeds v and V when the small cylinder m reaches point B.

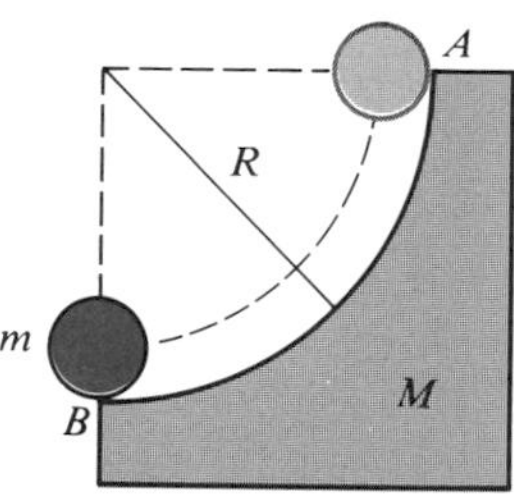

12 • 32 A board of mass m_1 rests on two solid cylinders of equal mass $m_2 = m_3$ and diameter d. Starting from rest in the position shown, what is the speed v of the board when the mass m_4 has fallen through a distance h? Assume that the cylinders roll without slipping on *both* the table and the board, and that the pulley is massless and frictionless.

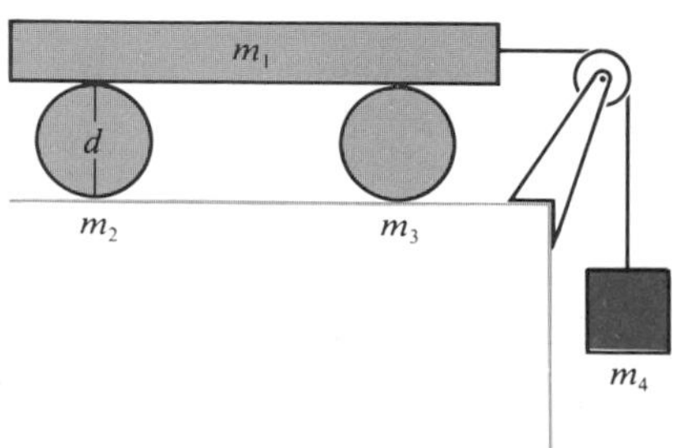

12 • 33 Write the parametric equations which describe the motion, as seen by a stationary observer, of a particle on the rim of a wheel of radius R which is rolling without slipping at constant velocity $\mathbf{v}_c$ to the right. Assume the particle to be initially located at the origin of coordinates, directly below the center of the wheel. Sketch the curve defined by the equations.

12 • 4 *rotational reaction forces*

12 • 34 What is the percentage of difference between the real acceleration due to gravity and the apparent acceleration due to gravity as seen by an observer on the earth's surface halfway between the North Pole and the equator?

12 • 35 The unit vectors in plane polar coordinates may be expressed as

$$\hat{\mathbf{r}} = \cos\theta\,\mathbf{i} + \sin\theta\,\mathbf{j}$$
$$\hat{\boldsymbol{\theta}} = -\sin\theta\,\mathbf{i} + \cos\theta\,\mathbf{j}$$

For $\omega = d\theta/dt$ (*a*) find $d\hat{\mathbf{r}}/dt$. (*b*) Find $d\hat{\boldsymbol{\theta}}/dt$. (*c*) Find $d\mathbf{r}/dt$. (If you can do this, you can probably also derive Eqs. [12 . 38]).

12 • 36 An object is dropped from a tower of height y located at the equator. During the object's fall its velocity vector is nearly vertically downward, but it drifts a distance eastward due to Coriolis forces. Find the amount of drift s in terms of y, ω_e, and g.

12 • 37 A person is standing at the center of a turntable that is 6 m in diameter and is rotating with a frequency of $\nu = 0.2$ Hz. He fires a bullet horizontally with a muzzle velocity of 300 m/s directly at the center of a target mounted at the rim of the turntable. (*a*) What is the Coriolis acceleration of the bullet relative to the turntable? (*b*) Find the displacement of the point of impact from the center of the target.

12 • 38 A pilot of weight F_g on the ground flies west at night over the equator at a speed of 1000 km/h. (*a*) If the airplane is kept pointed at a distant star above the horizon, what would the pilot weigh on a scale inside the plane? (*b*) Express the Coriolis force as a fraction of this weight. (*c*) What is the pilot's weight on the scale if the plane is kept pointed at a distant light on the earth's equator? (*d*) What is the Coriolis force in this case?

12 • 5 *gyroscopic phenomena*

12 • 39 A bicycle wheel 82 cm in diameter has a heavy wire wrapped around the rim so that the resulting mass of the system, $M = 7.3$ kg, may be regarded as located entirely at the rim. One end of the axle is suspended in a loop of cord placed 15 cm from the center of mass of the system. Using the rectangular coordinate system of Fig. 12 . 16, show that when the spin frequency of the wheel is $\boldsymbol{\nu}_s = -4\mathbf{i}$ Hz, the precessional frequency will be $\boldsymbol{\nu} = -0.554\mathbf{k}$ Hz.

12 • 40 Suppose you support the bicycle wheel in Prob. 12 . 39 in a horizontal position by holding the two ends of the axle in your hands. The axle projects 15.2 cm on each side of the wheel. While the wheel is spinning at a frequency of 4 Hz, you rotate the axle by hand in a horizontal plane about its center. Compute the magnitude and direction of the force that you must exert with each hand to produce a precession rate 0.1 Hz about the center.

12 • 41 A solid disk of mass 1.2 kg and diameter 10 cm is mounted on one end of an axle of negligible mass which is pivoted at a point 6 cm from the center of the disk. From the other end of the axle, at a distance of 10 cm from the pivot, is suspended an object of mass 960 g. If the rate of spin of the disk is $\nu = 11.9$ Hz, what is the rate of precession?

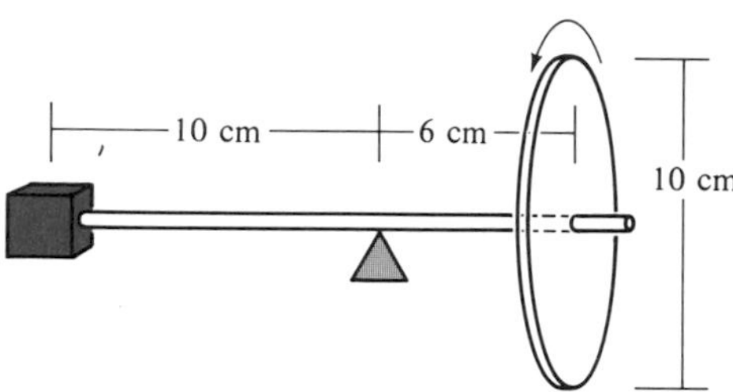

12 • 42 (*a*) If an electric fan is spinning clockwise and at the same time is rotating slowly counterclockwise about a vertical axis, as viewed from above, what is the direction of the torque exerted on the shaft of the motor? (*b*) If a device includes a rotating part with large angular momentum and it is fastened rigidly to an airplane, why must the shaft bearings be designed to withstand unusually large loads?

answers

12 • 1 $\omega = 0.667$ rad/s
12 • 2 (*a*) $I = 1.53 \times 10^{-50}$ g · cm^2;
(*b*) $I = 3.06 \times 10^{-50}$ g · cm^2;
(*c*) $I = 8.5 \times 10^{-41}$ g · cm^2
12 • 3 $I_c = \frac{1}{2}M(R_1^2 + R_2^2)$
12 • 4 $I_c = \frac{5}{3}\sigma V^{4/3}$
12 • 6 $I = \frac{3}{5}MR^2$
12 • 8 (*a*) $K_e = 2.58 \times 10^{29}$ J; (*b*) 0.88 percent
12 • 9 (*a*) $d^2\theta/dt^2 = (g/R)(\cos\theta - \mu_k \sin\theta)$;
(*b*) $W = \mu_k mgR$;
(*c*) $\omega = \sqrt{2g(1 - \mu_k)/R}$;
(*d*) $\theta' = 2 \cot^{-1} \mu_k$
12 • 11 $\omega = \sqrt{3g/l}$
12 • 12 (*a*) $I_c = 4679$ kg · m^2;
(*b*) $\Delta L = 58{,}800$ kg · m^2/s; (*c*) $\theta = 180\pi$ rad;
(*d*) $W = 1.11 \times 10^6$ J
12 • 13 (*a*) $I_c = 6700$ g · cm^2;
(*b*) $\alpha = 270$ rad/s^2; (*c*) $\omega = 800$ rad/s;
(*d*) $\alpha = 580$ rad/s^2
12 • 14 (*a*) 93.3 cm/s^2; (*b*) $m = 75.0$ g
12 • 15 $\alpha = \frac{2}{3}g/R$
12 • 16 $\omega = mvl/(I_c + ml^2)$
12 • 17 $P'/P = \frac{1}{2}$

12•18 $\Delta x = \pi l/3$
12•19 $T = 2.93$ days
12•20 Eastward
12•21 $\omega = 4.95$ rad/s
12•22 $\omega = M\omega_0/(M + 3m)$
12•23 $L_e/L_s = 0.0132$
12•24 (*a*) $\omega_s = 16$ rad/s; (*b*) $v' = 20$ km/h; (*c*) $v = 40$ km/h; (*d*) $v = 0$; (*e*) $v = 40 \cos \frac{1}{2}\theta$ km/h; (*f*) $\omega = 16$ rad/s
12•25 (*a*) $K_{tr}/K = \frac{1}{2}$; (*b*) $K_{tr}/K = \frac{2}{3}$; (*c*) $K_{tr}/K = \frac{5}{7}$
12•26 $K = \frac{7}{10}MR^2\omega^2$ (by either method)
12•27 (*a*) $v_s/v_r = \sqrt{2}$; (*b*) $v_s/v_r = \sqrt{\frac{3}{2}}$; (*c*) $v_s/v_r = \sqrt{\frac{7}{5}}$
12•28 (*a*) Hoop; (*b*) $v_h/v_d = \frac{1}{2}\sqrt{3}$
12•29 $T = \frac{1}{3}mg$
12•30 $a = 25.6$ cm/s^2
12•31 $v = \sqrt{2MgR/(m + M)}$, $V = mv/M$
12•32 $v = \sqrt{4m_4gh/(2m_1 + m_2 + 2m_4)}$
12•33 $\omega = v_c/R$, $x(t) = v_ct - R\sin \omega t$, $y(t) = R - R\cos \omega t$
12•34 0.17 percent
12•35 (*a*) $d\hat{\mathbf{r}}/dt = \omega\hat{\boldsymbol{\theta}}$; (*b*) $d\hat{\boldsymbol{\theta}}/dt = -\omega\hat{\mathbf{r}}$; (*c*) $d\mathbf{r}/dt = (dr/dt)\hat{\mathbf{r}} + r\omega\hat{\boldsymbol{\theta}}$
12•36 $s = \omega_e\sqrt{\frac{8}{9}y^3/g}$
12•37 (*a*) $a = 754$ m/s^2; (*b*) $\Delta s = 3.77$ cm
12•38 (*a*) $W = 1.00346\ F_g$; (*b*) $F_c = 0.00414\ F_g$; (*c*) $W = 1.00290\ F_g$; (*d*) same as part (*b*) since it depends only on the terrestrial observer, not on the pilot
12•40 A couple of ± 64.6 N applied at each end of the axle
12•41 $\Omega = \frac{1}{3}$Hz

CHAPTER THIRTEEN

gravity and central forces

Hitherto we have explained the phenomena of the heavens and of our sea by the power of gravity, but have not yet assigned the cause of this power. This is certain, that it must proceed from a cause that penetrates to the very center of the sun and planets, without suffering the least diminution of its force; that operates not according to the quantity of the surfaces of the particles upon which it acts (as mechanical causes used to do), but according to the quantity of solid matter which they contain, and propagates its virtue on all sides to immense distances, decreasing always as the inverse square of the distances.

Isaac Newton, Principia, *1686,*
translated by Andrew Motte, 1729

In this chapter we consider the general problem of central forces. A *central force* is a force on a body which always points directly toward or away from a fixed point, the *center of force,* in a given frame of reference. A *conservative* central force is a function only of the distance from the center of force. For example, in a reference frame with the center of the earth as the origin of coordinates, we can consider weight as a conservative central force, directed toward the center of mass of the earth, at approximately its geometric center. Clearly, in central-force problems the choice of the center of force as the origin of coordinates is not arbitrary. One implication, as we shall see, is that angular momentum is constant during the motion of a body acted on by the central force. Thus this description provides us with another invariant of motion besides energy.

Central-force problems are quite important in physics and encompass a range of phenomena, from nuclear physics to astrophysics. The gravitational attraction between two masses can be treated as a particular case of a central force, provided the origin is appropriately chosen. Gravity is of such fundamental significance that it warrants a chapter in itself. It was Newton's formulation of the law of gravitational attraction between the sun and the planets, and his consequent prediction of elliptic planetary orbits, which led to the development of the physical sciences as we know them today. This monumental achievement had a profound influence on the philosophers of the humanistic Enlightenment in the eighteenth century, who in turn laid the ideological foundations for both the virtues and vices of modern Western civilization.

13 • 1 Kepler's laws

Newton formulated his laws of motion in the course of his investigation of the motions of the heavenly bodies. More than a hundred years earlier, the Polish astronomer Nicolaus Copernicus (Mikolaj Kopernik) had published (posthumously) a *heliocentric,* or sun-centered, theory of the solar system. The Ptolemaic view, which was generally accepted at that time, was *geocentric;* it postulated that the sun and planets moved in orbits about the earth. Copernicus suggested that the planets, including the earth, orbit the sun. This simple change in the frame of reference reduced the description of the complex motions of the planets to comparative order and simplicity.

Copernican doctrine had a profound influence on man's view of the universe and his place in it, but it was by no means a perfect system, as Copernicus himself realized. Among its hostile critics was Tycho Brahe, a Danish astronomer who proposed a theory of his own. In order to test this theory, Brahe spent twenty-five years making extraordinarily accurate measurements of planetary positions—all without benefit of the telescope, which had not yet been invented. In the hands of the German mathematician Johannes Kepler, these data became the foundation of the Copernican doctrine.

Copernicus, in the absence of data to the contrary, had retained the ancient notion that the planets move in circles with constant speeds. Kepler found that such an assumption led to a difference of as much as 8 minutes of arc in the computed and observed positions of Mars. "Out of these eight minutes," he said "we will construct a new theory that will explain the motions of all the planets." It took two decades of the most patient study of

Tycho Brahe's data, but Kepler finally arrived at the three empirical laws that bear his name:

1 The orbit of each planet is an ellipse with the sun at one focus.
2 The speed of a planet in its orbit varies in such a manner that the radius vector joining the planet with the sun sweeps over equal areas in equal times.
3 The squares of the periods of revolution of the several planets are proportional to the cubes of the semimajor axes of their elliptic orbits.

The first two laws, which Kepler published in 1609, were based on his study of Brahe's data for the planet Mars. Later Kepler extended these laws explicitly, although without adequate proof at the time, to the motions of the other planets known in his day (see Fig. 13 . 1), the motion of the moon about the earth, and the motions of the four satellites of Jupiter, which Galileo had discovered in 1610.

Kepler's first law specifies the shape of the orbit in which a planet moves. However, it gives no information about variations in the planet's speed along its path. The shaded triangles of Fig. 13 . 2 illustrate the *law of equal areas,* which supplies this information. In an infinitesimal time dt the position vector $\mathbf{r}$ connecting the sun and planet sweeps out an area of magnitude $dA = \frac{1}{2}r^2\,d\theta$. Therefore, for any given planet,

$$\frac{dA}{dt} = \tfrac{1}{2}r^2\omega = k_p = \text{constant} \qquad [13 \cdot 1]$$

figure 13 • 1 *Relative sizes of the orbits of the six planets known to Kepler*

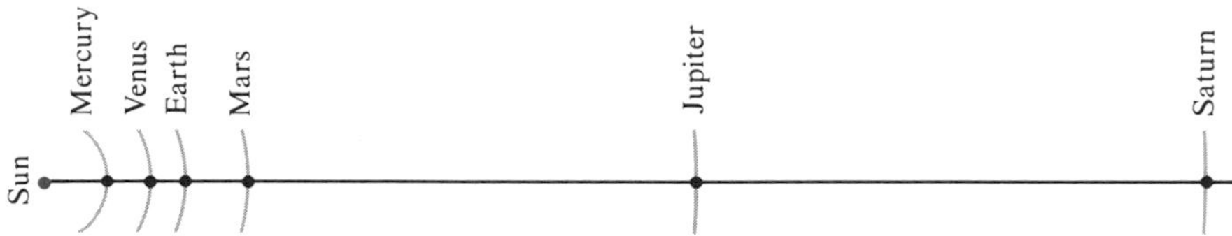

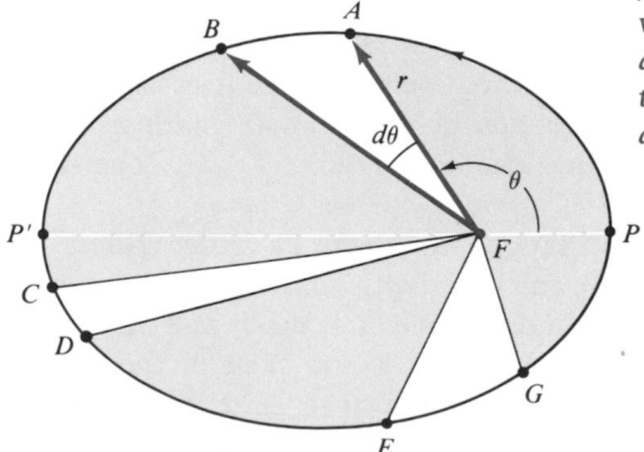

figure 13 • 2 *Kepler's second law. The radius vector* $\mathbf{r}$ *which joins the sun at F to the planet sweeps out equal areas FAB, FCD, and FEG in equal times. The height of triangle FAB (with base r) is approximately r dθ and its area is* $dA = \frac{1}{2}r^2\,d\theta$.

table 13 • 1 *Solar system data*

body	*semimajor axis, AU*	*eccentricity*	*period of rotation* *seconds*	*period of rotation* *tropical years*	*mass* *kilograms*	*mass* *earth masses*
Mercury	0.3871	0.206	7.60×10^6	0.241	3.18×10^{23}	0.532
Venus	0.7233	0.007	1.94×10^7	0.615	4.88×10^{24}	0.816
Earth	1.0000	0.017	3.16×10^7	1.000	5.98×10^{24}	1.000
Mars	1.524	0.093	5.94×10^7	1.881	6.42×10^{23}	0.107
Jupiter	5.203	0.048	3.74×10^8	11.86	1.90×10^{27}	318.
Saturn	9.539	0.056	9.30×10^8	29.46	5.68×10^{26}	95.0
Uranus	19.18	0.047	2.65×10^9	84.01	8.68×10^{25}	14.5
Neptune	30.06	0.009	5.20×10^9	164.8	1.03×10^{26}	17.2
Pluto	39.44	0.250	7.84×10^9	247.7	1.08×10^{24}	0.181
Moon	0.002570	0.055	2.36×10^6	0.0748	7.35×10^{22}	0.0123
Sun	—	—	—	—	1.99×10^{30}	332,776

Earth radius = 6370 km (avg), 6378 km (equatorial), 6357 (polar)
Lunar radius = 1738 km
Solar radius = 695,950 km
1 astronomical unit (AU) = 1.496×10^8 km
1 tropical year = 31,556,926 s

although r and ω may vary along the path. The value of the planetary constant k_p is different for each planet and may be calculated from the data given in Table 13 . 1.

Kepler's third law, known as the ***harmonic law,*** was not published until a decade later. This law completed the kinematic picture by showing the relation between the periods and sizes of the orbits of the different planets about the sun or of the different satellites about their parent planet. This relation, stated algebraically, is

$$\frac{T^2}{a^3} = k_s = \text{constant} \qquad [13 \cdot 2]$$

where T is the period of revolution of a given body about its parent body, a is the semimajor axis of the body's elliptic orbit, and k_s is the ***system constant*** that characterizes a particular system of revolving bodies. Thus k_s has one value for all the planets orbiting about the sun, another value for the satellite system orbiting Jupiter, and so on.

Using Kepler's three laws, we can immediately derive the law of gravity (see the opening quotation), provided we make the simplifying assumption that the earth's orbit about the sun is a *circle* of radius R (this is true within 2 percent error), which is a special case of an ellipse. Then we can let M_e be the mass of earth and M the mass of the sun. Kepler's second law implies that $\omega = 2k_p/R^2$ is constant. Hence, to maintain this orbit, the centripetal force on the earth due to the sun's gravitational attraction must be

$$F_c = M_e\omega^2 R = \frac{4\pi^2 M_e R}{T^2} \qquad [13 \cdot 3]$$

Since the semimajor axis of a circle is its radius, from Kepler's third law, $T^2 = k_s R^3$. Substitution into Eq. [13 . 3] then yields

$$F_c = \frac{4\pi^2}{k_s}\frac{M_e}{R^2} \sim \frac{M_e}{R^2} \qquad [13 \cdot 4]$$

That is, the gravitational force on the earth is proportional to its mass and the inverse square of its distance from the attracting center of force. By the same arguments, the reaction force which the earth exerts on the sun must be similar, but proportional instead to the mass M of the sun:

$$F_c' \sim \frac{M}{R^2} \qquad [13 \cdot 5]$$

We can combine Eqs. [13 . 4] and [13 . 5] as a single *law of gravity,* stated in vector form as

$$\mathbf{F}_g = -G\frac{mM}{R^2}\hat{\mathbf{r}} \qquad [13 \cdot 6]$$

where G is a positive constant of proportionality and $\hat{\mathbf{r}}$ is the unit vector in the direction from the attracting center M to the mass m.

Although Eq. [13 . 6] is derived here under the special assumption of a circular orbit, we shall see in the next section that it is rigorously true for elliptic orbits and that G is a universal constant, called the *gravitational constant.* Newton further postulated that the gravitational force between *any* two bodies is attractive and depends only on their masses and separation distance, as in Eq. [13 . 6].

example 13 • 1 Given that the earth has a radius R_e and a mass M_e, (*a*) find the relation between G and k_s. (*b*) The relation between G and g. (*c*) Compute the value of G from each of these formulas.

solution (*a*) A comparison of Eqs. [13 . 4] and [13 . 6] gives us the relation

$$G = \frac{4\pi^2}{k_s M}$$

where M is the mass of the sun. (*b*) At the earth's surface the gravitational force on any small body m is

$$F_g = mg = G\frac{mM_e}{R_e^2}$$

The gravitational acceleration g is therefore

$$g = G\frac{M_e}{R_e^2}$$

(*c*) From Table 13 . 1, $k_s = 1\ \text{y}^2/1\ \text{AU}^3$ and $M = 1.99 \times 10^{30}$ kg. Hence from the expression for G in part (*a*),

$$G = \frac{4\pi^2}{k_s M} = \frac{39.5\,(1.5 \times 10^{11}\ \text{m})^3}{(3.16 \times 10^7\ \text{s})^2(1.99 \times 10^{30}\ \text{kg})}$$
$$= 6.71 \times 10^{-11}\ \text{m}^3/\text{kg}\cdot\text{s}^2$$

and from the relation in part (*b*),

$$G = \frac{gR_e^2}{M_e} = \frac{(9.80\ \text{m/s}^2)(6.37 \times 10^6\ \text{m})^2}{5.98 \times 10^{24}\ \text{kg}}$$
$$= 6.65 \times 10^{-11}\ \text{m}^3/\text{kg}\cdot\text{s}^2$$

The difference of 0.9 percent between the two answers is due to roundoff of the various constants. The true value of G is $6.67 \times 10^{-11}\ \text{m}^3/\text{kg}\cdot\text{s}^2$.

13 • 2 dynamic consequences of Kepler's laws

Under the special assumption of circular orbits, Kepler's empirical laws lead directly to the law of gravity for the attraction between two masses. Furthermore, this assumption implies a central force on the orbiting body which is directed toward the center of the circle, thus producing uniform rotation (see Sec. 11 . 2). However, to develop the law of gravity in all its generality, we need to show that for an *elliptical* planetary orbit, Eq. [13 . 6] still holds true. We shall assume that the planets are at such great distances from the sun that they may be treated as particles, and that the sun is so large that it may be considered stationary relative to an inertial frame of reference. Then we can show that Kepler's laws imply that gravity is a central force. On this basis we can use the gravitational potential energy U_g of the orbiting planet to show that the gravitational attraction between sun and planet is inversely proportional to the square of their separation distance r. Let us neglect interplanetary forces, so that we have a two-particle problem.

For any central force—one which is always directed toward or away from a fixed point—we can choose that point as the origin of coordinates and represent the central force on a body as $\mathbf{F} = F\hat{\mathbf{r}}$, where $\hat{\mathbf{r}}$ is the direction of the radius vector $\mathbf{r}$ from the force center to the orbiting body (see Fig.

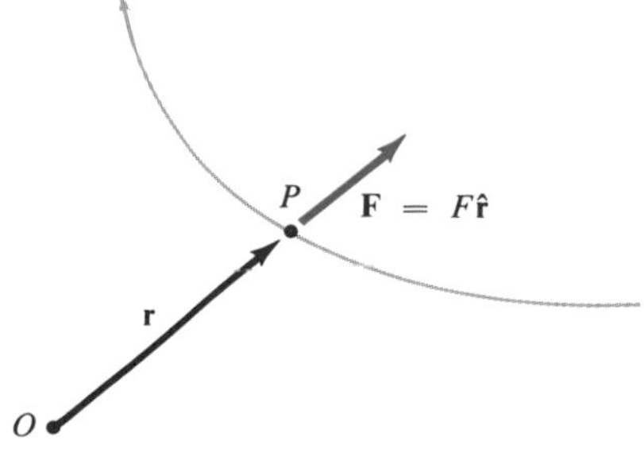

figure 13 • 3 *A particle at P passing a center of force at O which exerts a repulsive force* $\mathbf{F} = F\hat{\mathbf{r}}$. *Since* $\mathbf{F}$ *and* $\mathbf{r}$ *are parallel, there is no torque about O.*

13 . 3). Hence there is no torque exerted on the body by the central force:

$$\boldsymbol{\tau} = \mathbf{r} \times \mathbf{F} = \frac{F}{r}\mathbf{r} \times \mathbf{r} = \mathbf{0} \qquad [13 \cdot 7]$$

Since torque represents a change in angular momentum, $\boldsymbol{\tau} = d\mathbf{L}/dt$, this means that the *angular momentum* **L** *of the body remains constant.* The orbiting body must therefore be moving in a constant plane in space which is normal to its angular momentum.

The angular momentum of an orbiting body of mass m is

$$L = |m\mathbf{r} \times \mathbf{v}| = mrv_\theta = mr^2\omega \qquad [13 \cdot 8]$$

where $v_\theta = r\omega$ is the velocity component perpendicular to **r**, as we saw in Fig. 12 . 12. Comparison with Kepler's second law shows that

$$\frac{L}{m} = 2k_p = \text{constant} \qquad [13 \cdot 9]$$

In effect, Kepler's second law states an empirical result which is equivalent to saying that the angular momentum of a planet is conserved. Therefore *the gravitational force acting on it must be a central force* attracting it to the sun taken as the origin. A noncentral force would give rise to a torque, $\mathbf{r} \times \mathbf{F} \neq \mathbf{0}$, and thus to a nonconstant angular momentum.

Let us now use Kepler's laws to derive the law of gravity. The following five steps will show:

1 That the gravitational force can be represented as a function of separation distance, $F = F(r)$
2 The statement of the equations of motion in plane polar coordinates for a body subject to a central force
3 The representation of an ellipse in plane polar coordinates
4 That for an elliptic orbit the gravitational force must necessarily be

$$\mathbf{F} = -\frac{GmM}{r^2}\hat{\mathbf{r}}$$

5 That the equations for the orbit of a planet, $r = r(\theta)$, can be stated in terms of its angular momentum, total energy, and other physical parameters

Step 1 The stability of planetary orbits allows us to assume that the gravitational force is conservative, and hence derivable from a potential-energy function U_g (see Sec. 8 . 4). Since there are no unique directions in this problem, we can argue from symmetry that U_g is a function of r—that is, that the potential energy does not depend on angles, but only on the distance between the two bodies. Therefore $F = -dU_g/dr$ must also be a function of distance alone, even though we do not yet know its precise form.

Step 2 In Sec. 12 . 4 we argued that the equation of motion for the radial plane polar coordinate may be obtained by moving an inertial-reaction term to the right side of the force equation, as if it were centrifugal force:

$$m\frac{d^2r}{dt^2} = F + mr\omega^2 \qquad [13 \cdot 10]$$

where F is the force in the radial direction *away from* the origin. The equation for the angular coordinate θ is given by the constancy of angular momentum:

$$\frac{d\theta}{dt} = \omega = \frac{L}{mr^2} \qquad [13 \cdot 11]$$

Since we wish to eliminate any explicit time dependence in order to find $F(r)$, we must replace the time derivatives of r and θ with spatial derivatives. To do so, we can use the simple algebraic properties of differentials. Thus for the velocity components

$$v_r = \frac{dr}{dt} = \frac{d\theta}{dt}\frac{dr}{d\theta} = \omega\frac{dr}{d\theta} \qquad v_\theta = r\omega \qquad [13 \cdot 12]$$

we can substitute to obtain the radial acceleration:

$$\frac{d^2r}{dt^2} = \frac{dv_r}{dt} = \frac{d\theta}{dt}\frac{dv_r}{d\theta} = \omega\frac{d}{d\theta}\left(\omega\frac{dr}{d\theta}\right) \qquad [13 \cdot 13]$$

From the constancy of angular momentum, we know that $\omega = L/mr^2$ at any point on the orbit, so that we can rewrite Eq. [13 . 10] as

$$m\frac{d^2r}{dt} - mr\omega^2 = \frac{L}{r^2}\frac{d}{d\theta}\left(\frac{L}{mr^2}\frac{dr}{d\theta}\right) - \frac{L^2}{mr^3} = F \qquad [13 \cdot 14]$$

We have now replaced two time-dependent equations of motion, Eqs. [13 . 10] and [13 . 11], by a single time-independent equation for the orbit $r = r(\theta)$. Solely for convenience, let us substitute

$$\frac{d(1/r)}{d\theta} = -r^{-2}\frac{dr}{d\theta}$$

and rewrite Eq. [13 . 14] as

$$\frac{d^2(1/r)}{d\theta^2} + \frac{1}{r} = -\frac{mr^2}{L^2}F \qquad [13 \cdot 15]$$

In principle, if we knew $F(r)$, we could derive the orbit $r(\theta)$. Our problem, however, is to derive $F(r)$, and to do so we must make use of Kepler's first law: the orbit is an ellipse for a particle under the influence of a gravitational force.

Step 3 The ellipse belongs to a family of curves known as *conic sections,* which also include hyperbolas and parabolas. Their cartesian equations should be familiar from analytic geometry. In polar coordinates, their general equation is

$$r = \frac{l}{1 + e\cos\theta} \qquad [13 \cdot 16]$$

where the *scale* l and the *eccentricity* e are constants. The origin is at one focus of the conic and does not coincide with the geometric center, as it does in cartesian coordinates. The polar equations for conics, shown in Table 13 . 2, are most easily derived from their geometric definitions, or else by substituting

$$x = c + r\cos\theta \qquad y = r\sin\theta$$

table 13 • 2 *Conic sections in polar form, $r = l/(1 + e\cos\theta)$, origin at focus*

conic section	cartesian form, $a > b > 0$	eccentricity e	$l = l(e)$	graph
ellipse	$\frac{x^2}{a^2} + \frac{y^2}{b^2} = 1$	$c^2 = a^2e^2$ $= a^2 - b^2$, $0 \leq e < 1$	$a(1 - e^2) > 0$	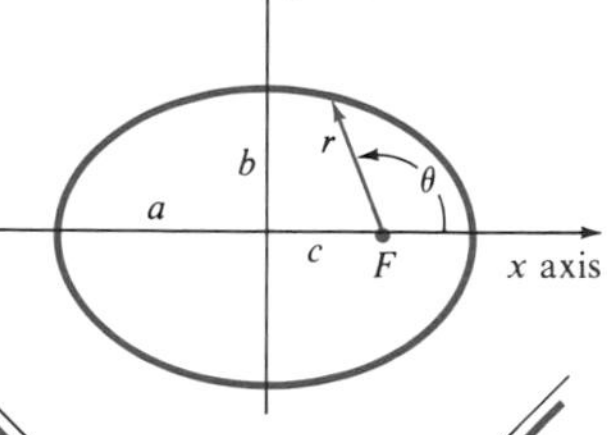
hyperbola*	$\frac{x^2}{a^2} + \frac{y^2}{b^2} = 1$	$c^2 = a^2e^2$ $= a^2 + b^2$, $1 < e$	$a(1 - e^2) < 0$	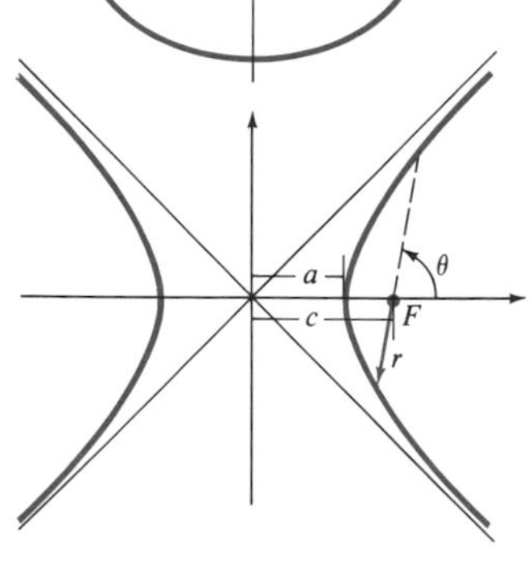
parabola	$y^2 = 4ax$	$c = a$, $e = 1$	$-2a$	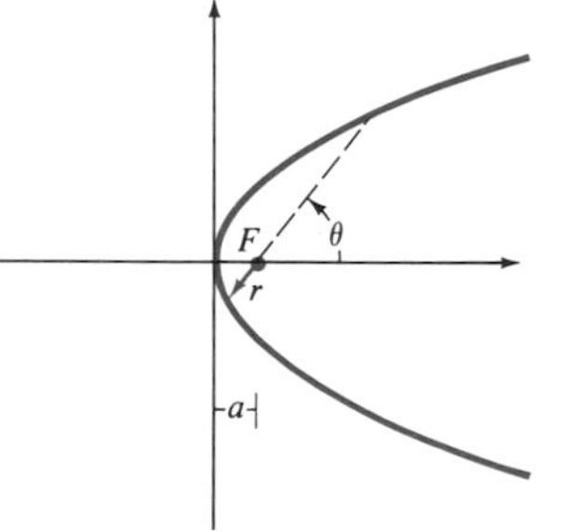

*Negative values of r are to be reflected through the focus.

in the standard cartesian forms. Changing the sign of e has the same effect as adding π to θ and does not change the final value. (Try sketching the ellipse for $e = 0.5$ by plotting r from the focus for $\theta = 0, \frac{1}{4}\pi, \frac{1}{2}\pi, \frac{3}{2}\pi$.)

Step 4 Assuming the orbit to be an ellipse, we can multiply the orbit Equation [13 . 15] by the scale l and we can evaluate the second derivative of $l/r = 1 + e\cos\theta$:

$$\frac{d^2(l/r)}{d\theta^2} + \frac{l}{r} = -e\cos\theta + 1 + e\cos\theta$$

$$= 1 = -l\frac{mr^2}{L^2}F \qquad [13 \cdot 17]$$

We have now proved, from Kepler's first two laws, that the gravitational force is proportional to the inverse square of the separation between the attracting bodies:

$$\blacktriangleright \quad \mathbf{F} = -\frac{L^2}{mlr^2}\hat{\mathbf{r}} = -\frac{GmM}{r^2}\hat{\mathbf{r}} \qquad [13 \cdot 18]$$

where the solar-mass factor M can be deduced by the same action-reaction argument used in Sec. 13 . 1. Therefore the scale of the orbit is given by

$$l = \frac{L^2}{Gm^2M} \qquad [13 \cdot 19]$$

This still does not prove that G is constant throughout the entire solar system. Kepler's third empirical law is crucial to demonstrate the constancy of G, as we shall see shortly. For the moment, let us assume it as given.

Step 5 To complete our derivation we need only express the eccentricity e in terms of the physical parameters of our two-body gravitational system. We cannot use the equation of motion for this, since the eccentricity canceled out of Eq. [13 . 17]. However, we can use the fact that the force is conservative and derivable from a potential, as defined in Eq. [8 . 6]:

$$U_g(r) = -\int_{\infty}^{r} F(r)\,dr = -\frac{GmM}{r} \qquad [13 \cdot 20]$$

where we have taken the reference level of U_g to be at $r = \infty$ and its reference value as $U_g(\infty) = 0$. We can now use the expression for kinetic energy in polar coordinates, Eq. [12 . 42], to express the conservation of energy of the orbiting mass in terms of the geometric characteristics of its elliptic orbit. The total energy E of the orbiting planet is

$$E = K + U_g = \tfrac{1}{2}m\omega^2\left[\left(\frac{dr}{d\theta}\right)^2 + r^2\right] - \frac{GmM}{r} \qquad [13 \cdot 21]$$

Substituting $\omega = L/mr^2$ and $r = l(1 + e\cos\theta)^{-1}$ into this equation, we find that terms in $1/r$ cancel when the eccentricity is

$$\blacktriangleright \quad e = \sqrt{1 + \frac{2EL^2}{G^2m^3M^2}} \qquad [13 \cdot 22]$$

Therefore the orbit of a planet of mass m, angular momentum L, and total energy E can be stated in polar coordinates as

$$\blacktriangleright \quad r = \frac{L^2/Gm^2M}{1 + \sqrt{1 + (2EL^2/G^2m^3M^2)}\cos\theta} \qquad [13 \cdot 23]$$

where the sun M at one focus of the ellipse is taken as the origin of coordinates.

Comparison of this result with Table 13 . 2 shows that the solution of the problem is not necessarily an ellipse. It may be any conic section, depending on the value of the eccentricity e, which depends in turn on the total energy E:

> Any attractive inverse-square central force gives rise to an orbit that is elliptic, hyperbolic, or parabolic, according as the total mechanical energy E is, respectively, less than, greater than, or equal to zero. A repulsive central force can give rise only to hyperbolic orbits.

The quantity $-E$ is sometimes referred to as the *binding energy* of the system, especially in atomic and nuclear physics. This is the amount of work that would have to be done on the system to remove a body from an elliptic orbit to a state of rest an infinite distance away.

If E is less than zero, we can solve the energy equation for the *apsides,* the minimum distance r_p (the *perigee*) and the maximum distance r_a (the *apogee*) from the focus. At these distances the ellipse must have turning points, at which $dr/d\theta = 0$. Therefore Eq. [13 . 21] yields a quadratic,

$$Er^2 + GmMr - \frac{L^2}{2m} = 0 \qquad [13 \cdot 24]$$

with solutions

$$\left.\begin{matrix} r_a \\ r_p \end{matrix}\right\} = -\frac{GmM}{2E}\left(1 \pm \sqrt{1 + \frac{2EL^2}{G^2m^3M^2}}\right) \qquad [13 \cdot 25]$$

example 13 • 2 Derive Kepler's third law from Newton's theory of the ellipse.

solution The area of an ellipse is given by $A = \pi ab = \pi a^2\sqrt{1 - e^2}$. However, this must equal the period T, multiplied by the constant rate of change, $dA/dt = L/2m$:

$$A^2 = \pi^2 a^4(1 - e^2) = \pi^2 a^3[a(1 - e^2)] = \frac{L^2T^2}{4m^2}$$

Since $a(1 - e^2) = l = L^2/Gm^2M$, according to Table 13 . 2, we substitute in the above equation to find

$$\frac{T^2}{a^3} = \frac{4\pi^2}{GM} = k_s$$

which is Kepler's third law, with k_s constant for a given stationary attracting center. Thus G is constant throughout the solar system, and so far as we know, throughout all of physics.

13 • 3 general law of gravity

Newton's success in explaining Kepler's empirical laws led him to the important generalization that the attractions of the heavenly bodies for one another are merely special cases of gravitational force which exists between *all* particles of matter:

> For any two particles of mass m and M separated by a distance r, there is a gravitational force of attraction, $F = GmM/r^2$, directed along the line joining the particles.

Note that this general statement of the law of gravity must refer to particles, rather than to extended bodies, due to the uncertainty of the definition of the "distance between" such bodies. Table 13 . 3 summarizes the Newtonian theory of gravity.

The gravitational constant G is one of the *universal constants* of physics, because its value depends only on the units used, and not on locality,

table 13 • 3 *Summary of central forces and gravity*

central forces	
$\mathbf{F} = F\hat{\mathbf{r}}$	$\hat{\mathbf{r}}$ is the unit radial vector in plane polar coordinates
$\boldsymbol{\tau} = \mathbf{r} \times \mathbf{F} = 0$	a central force exerts no torque about the origin
$\mathbf{L} = mr^2\omega\mathbf{k},\ \omega = \dfrac{d\theta}{dt}$	constant angular momentum vector normal to plane of orbit
$\dfrac{d^2(1/r)}{d\theta^2} + \dfrac{1}{r} = -\dfrac{mr^2}{L^2}F$	equation of a particle orbit $r(\theta)$ under a central force F
gravity	
$\mathbf{F} = -\dfrac{GmM\hat{\mathbf{r}}}{r^2}$	force of gravity on m due to M, $\hat{\mathbf{r}}$ directed from M to m
$U_g = -\dfrac{GmM}{r}$	gravitational potential energy, $\lim_{r\to\infty} U_g = 0$
$E = K + U_g$	constant total energy E
orbits	
$r = \dfrac{l}{1 + e\cos\theta}$	equation of a conic section, one focus at the origin
$l = \dfrac{L^2}{Gm^2M}$	scale of orbit
$e = \left(1 + \dfrac{2EL^2}{G^2m^3M^2}\right)^{1/2}$	eccentricity of orbit: $E < 0$, orbit is an ellipse $E > 0$, orbit is a hyperbola $E = 0$, orbit is a parabola
$r_p = \dfrac{l}{1 + e}$	minimum radius of orbit of conic section
$r_a = \dfrac{l}{1 - e}$	maximum radius of elliptic orbit

time, temperature, or any particular properties of matter. Therefore the gravitational force between any two particles is not affected by any motions of the particles or by the presence of other matter. The mutual attraction of the earth and the sun apparently is not changed in the slightest by the passage of the moon between them. In these ways gravitational phenomena differ from all other interactions between matter.

The first accurate determination of the value of G was made by Henry Cavendish in 1798 by means of a *torsion balance*. Determinations have also been made with sensitive beam balances (see Fig. 13 . 4). Subsequent refinements have led to the present accepted value:

$$G = (6.672 \pm 0.004) \times 10^{-11}\ \mathrm{N \cdot m^2/kg^2} \qquad [13 \cdot 26]$$

This is the force, expressed in newtons, with which two homogeneous spheres, each of mass 1 kg, attract each other when their centers are 1 m apart.

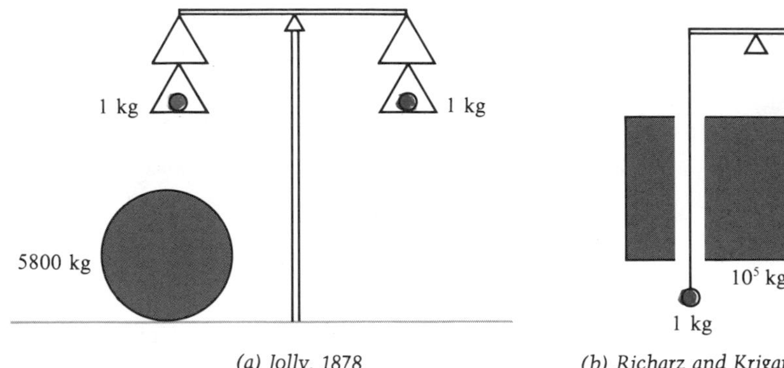

(a) Jolly, 1878

(b) Richarz and Krigar-Menzel, 1896

figure 13 • 4 *Simplified versions of apparatus used for determining the constant of gravitation*

Emphasis on the universality of Newton's law of gravity does not mean, of course, that the law must necessarily apply outside the ranges for which it has been tested experimentally. In the stellar regions, however, nearby binary (double) star systems have long been known to exhibit internal motions that are explainable in the same terms as are the motions of bodies in our solar system. Within the solar system, the excellence of Newton's law was strikingly confirmed when the existence of Neptune (1846) and of Pluto (1930) was predicted before these planets were ever seen; their positions were calculated from the effects they produced on the orbits of the planets nearest to them.

According to Newton's law, the gravitational attraction between any two bodies is entirely independent of their physical and chemical natures. This means, for example, that the earth exerts the same pull on all bodies of mass 1 kg placed at a given distance from its center, whether the bodies consist of lead, water, ice, air, wood, or other matter; and the force on 2 kg of any material is just twice as much. The mass of a body as determined by weighing on a beam balance is sometimes referred to as the *gravitational mass,* in order to distinguish it from the *inertial mass* of the body, which appears in Newton's second law, $\mathbf{F} = m\mathbf{a}$. The former is a measure of the ability of the body to attract other bodies by gravitation, whereas the latter is a measure of its inertia. Support for the assertion that the gravitational and inertial masses of any given body are equal was furnished in 1894 by the experiments of Roland von Eötvös, who showed gravitational and inertial mass to be identical to within one part in 10^8. More refined measurements made by Robert Dicke in the early 1960s established equivalence to within three parts in 10^{11}. For our purposes we shall consider them identical, although their equivalence is still the subject of scientific research.

In the late eighteenth and early nineteenth centuries there was further theoretical development of Newton's work, notably by Joseph Lagrange and Pierre Simon de Laplace, which had far-reaching effects. One outcome was the powerful *potential theory,* which is not restricted to problems in gravitation, but has important applications in many areas of physics and pure mathematics. Another was a gradual shift of attention to the space surrounding bodies, the starting point of *field theory,* which plays a basic role in physics.

Thus far we have interpreted gravitational forces as *action at a distance,* in which there is no contact between the interacting bodies. In field theory, however, each body is regarded as "conditioning" the region of space surrounding it in such a way that another body will experience gravitational force if placed anywhere in this region. The region is called a *field of force,* and any body placed in it is most usefully regarded as interacting with the field itself, rather than with the body or bodies that set up the field. The measure of the field generated by a point mass M is called its *gravitational field strength* **f**, defined by the equation

$$\mathbf{f} = \frac{\mathbf{F}}{m} \qquad f = \frac{GM}{r^2} \qquad [13 \cdot 27]$$

Operationally, the field strength is the *gravitational force per unit mass* acting on a test particle m at each point of the field. If **f** is known at a particular field point, the gravitational force **F** on a particle of *any* mass m placed at that point can be calculated. Since $\mathbf{F} = m\mathbf{a}$, we see that **f** is the *acceleration* that the particle would have at the field point in question if gravitation were the only force acting.

13 • 4 extended bodies, rings, and shells

Although the law of gravity is stated in terms of particles, we usually have to deal with extended bodies in which the matter is continuously distributed. The gravitational forces that such bodies can exert are not obvious from intuition. However, bodies that are rigid, uniform in density, and regular in shape can be dealt with by certain methods that have fairly wide applicability in physics. These methods involve the assumption that a continuously distributed body can be divided into infinitesimal elements dm whose fields are independent of each other and can be superposed by vector addition.

In the case of a thin narrow ring of radius R and mass M, our problem is to find the gravitational force exerted by the ring on a particle of mass m placed at any point P on the x axis of the ring (see Fig. 13 . 5). If we imagine the ring M divided into infinitesimal parts of equal length and equal mass

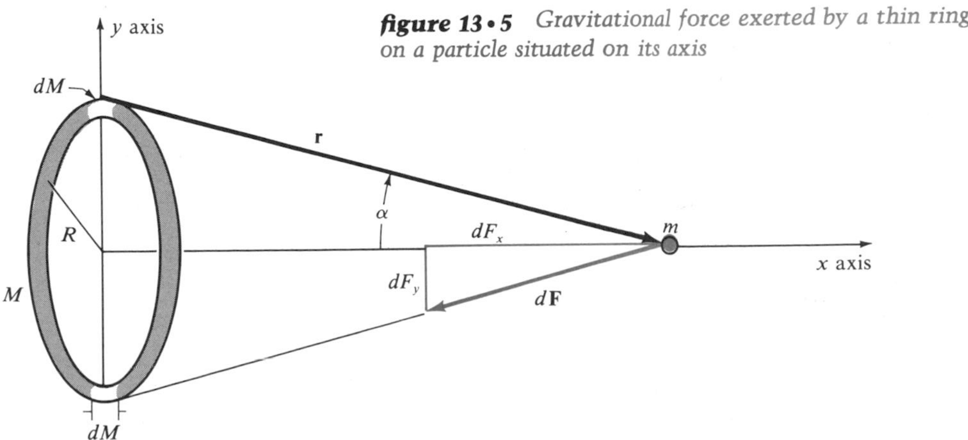

figure 13 • 5 *Gravitational force exerted by a thin ring on a particle situated on its axis*

dM, each element dM will exert a force on the particle m given by

$$d\mathbf{F} = -\frac{Gm\,dM}{r^2}\hat{\mathbf{r}} \qquad [13 \cdot 28]$$

The forces due to all the mass elements dM can readily be added vectorially by resolving them into components parallel to and normal to the x axis. Consideration of the symmetry about the axis shows that the normal components will cancel in pairs, leaving only the components dF_x to be added. Since

$$dF_x = dF\cos\alpha = -\frac{Gm\,dM}{r^2}\cos\alpha \qquad [13 \cdot 29]$$

the total gravitational force exerted by the ring is

$$F_x = -\frac{Gm}{r^2}\cos\alpha\int_0^M dM = -\frac{GmM}{r^2}\cos\alpha$$

$$= -\frac{GmM}{(x^2 + R^2)^{3/2}}x \qquad [13 \cdot 30]$$

in the direction toward the origin. Note that the ring does *not* attract as if its mass were all concentrated at the center of mass.

The same result can be derived from the potential energy U_g. For a particle m in the gravitational field of an extended body, we can consider U_g to be the algebraic sum of the contributions due to the interaction of m with each mass element dM of the body. Therefore $dU_g = -Gm\,dM/r$, and the total potential energy of the body is

$$U_g = -\frac{Gm}{r}\int_0^M dM = -\frac{GmM}{r} = -\frac{GmM}{\sqrt{x^2 + R^2}} \qquad [13 \cdot 31]$$

From Eq. [8 . 34], the force in the x direction at P is

$$F_x = -\frac{\partial U_g}{\partial x} = -\frac{GmMx}{(x^2 + R^2)^{3/2}} \qquad [13 \cdot 32]$$

in agreement with Eq. [13 . 30]. Note that for $x \gg R$, F_x approaches $-GmM/x^2$, and the ring may be treated as a point mass.

Let us now use this result to calculate the gravitational attraction to the thin-walled spherical shell in Fig. 13 . 6. Imagine that a narrow ring of mass dM is cut from this shell. When a particle of mass m is placed at point P, its gravitational potential energy due to the ring dM is, from Eq. [13 . 31], $U_g = -Gm\,dM/r$. If R is the radius of the shell and σ is its surface density, then the area dA of the ring is $2\pi R\sin\theta\,R\,d\theta$ and its mass is

$$dM = \sigma(2\pi R\sin\theta\,R\,d\theta) = 2\pi R^2\sigma\sin\theta\,d\theta \qquad [13 \cdot 33]$$

For triangle OPQ the law of cosines gives

$$r^2 = x^2 + R^2 - 2xR\cos\theta \qquad [13 \cdot 34]$$

If we now integrate over the shell, we have

$$\int_0^{U_g} dU_g = -Gm2\pi R^2\sigma\int_0^{\pi}\frac{\sin\theta\,d\theta}{\sqrt{x^2 + R^2 - 2xR\cos\theta}}$$

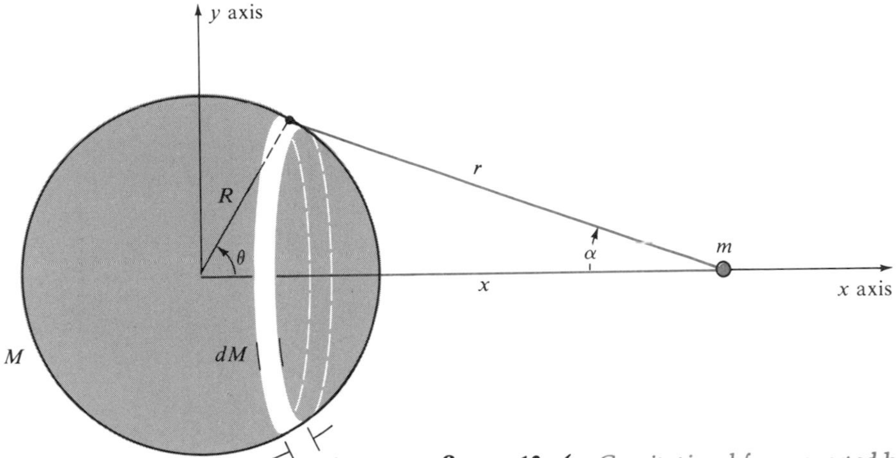

figure 13 • 6 *Gravitational force exerted by a thin-walled hollow sphere on a particle situated outside the sphere*

or a total potential energy

$$U_g = -Gm\frac{2\pi R\sigma}{x}\left(\sqrt{x^2 + R^2 + 2xR} - \sqrt{x^2 + R^2 - 2xR}\right) \tag{13 . 35}$$

In evaluating the formulas found in integral tables it is important to remember that the radical sign, by convention, denotes a positive quantity; hence Eq. [13 . 35] has two different forms. Since

$$\sqrt{x^2 + R^2 - 2xR} = \begin{cases} x - R & \text{for } x \geq R \\ R - x & \text{for } x \leq R \end{cases} \tag{13 . 36}$$

the corresponding potential-energy formulas are

$$U_g = \begin{cases} -\dfrac{Gm4\pi R^2\sigma}{x} = -\dfrac{GmM}{x} & \text{for } x \geq R \\ -\dfrac{GmM}{R} & \text{for } x \leq R \end{cases} \tag{13 . 37}$$

These somewhat surprising results indicate that a uniform spherical shell does attract external matter as if all its mass M were concentrated at its center. Once the shell is penetrated, however, the potential energy U_g is constant and is equal to its value at the surface. Therefore

$$F_x = -\frac{\partial U_g}{\partial x} = 0$$

everywhere *inside* the shell, and a particle of mass m inside the shell would experience no force at all! This result is peculiar to forces that follow the inverse-square law.

The gravitational attraction and potential energy of a mass in the field of a spherical shell are shown in Fig. 13 . 7 as a function of the distance r from the center (the direction is irrelevant). The sudden sharp rise in the

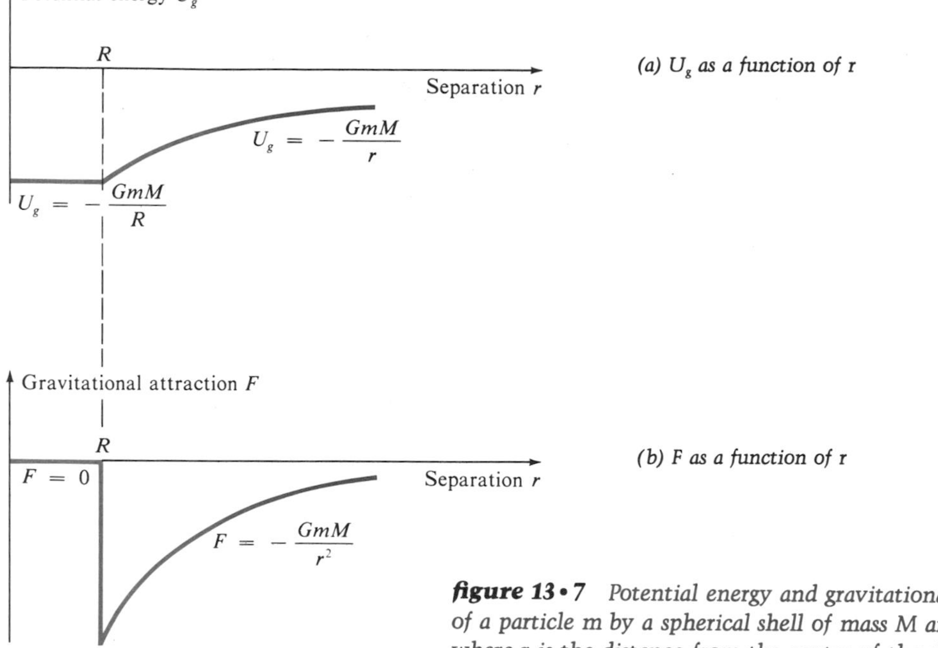

(a) U_g as a function of r

(b) F as a function of r

figure 13 • 7 *Potential energy and gravitational attraction of a particle m by a spherical shell of mass M and radius R, where r is the distance from the center of the shell*

force at $r = R$ is to be viewed as a slope which approaches the vertical as the shell becomes infinitely thin.

We could have derived the foregoing results, with significantly greater effort, by considering the vector sum of the gravitational forces $d\mathbf{F}$ due to each ring dM comprising the shell. The mathematics, although more tedious, is essentially the same.

13 . 5 gravitational field of the earth

Any spherical body that is uniform in density, or that consists of uniform concentric layers, can be imagined as subdivided into a continuous series of thin concentric shells. Such a sphere attracts *outside* matter as if all its mass were concentrated at its center. Let us now consider what happens to a particle that is embedded in a solid sphere.

Figure 13 . 8 shows a particle of mass m embedded in a solid sphere of mass M and uniform density ρ. The gravitational force on m due to the outer

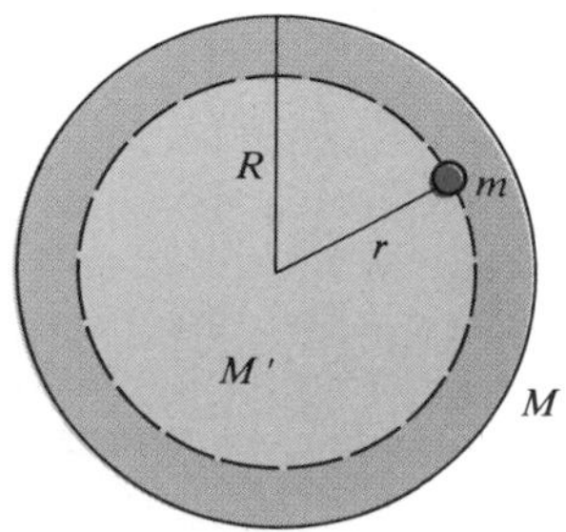

figure 13 • 8 *A particle embedded in a solid sphere*

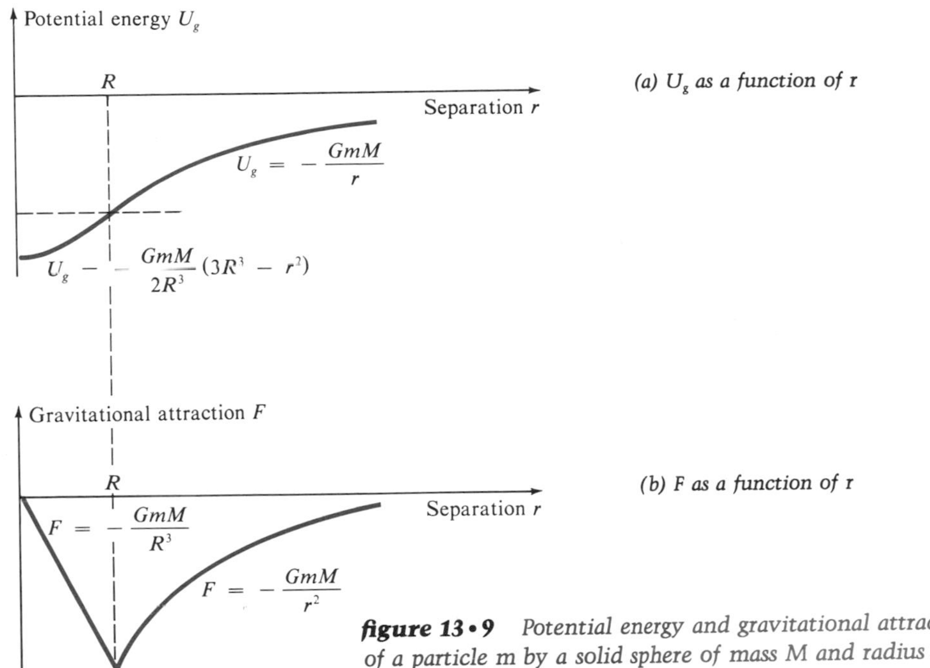

figure 13 • 9 *Potential energy and gravitational attraction of a particle m by a solid sphere of mass M and radius r*

shell, of radius R, is zero, as we saw in the preceding section. The force due to the "inner sphere" of radius r and mass M' is given by

$$F = -Gm \int_0^{M'} \frac{dM}{r^2} = -\frac{Gm\rho}{r^2}\frac{4\pi r^3}{3} \qquad [13 \cdot 37]$$

since $M' = \rho(4\pi r^3/3)$. Similarly, for the entire sphere, $M = \rho(4\pi R^3/3)$. Therefore, as shown in Fig. 13 . 9, the force on a particle within the sphere, at a distance $r \leq R$ from the center, is

$$F = -\frac{GmM}{R^3}r \qquad r \leq R \qquad [13 \cdot 38]$$

For a particle at any point outside the sphere, $r > R$, the attracting mass M of the sphere is a constant and the integral of Eq. [13 . 37] is simply

$$F = -\frac{GmM}{r^2} \qquad r > R \qquad [13 \cdot 39]$$

By similar reasoning, it is easy to show that gravitational potential outside the sphere is

$$U_g = -\frac{GmM}{r} \qquad r > R \qquad [13 \cdot 40]$$

However, the computation of U_g at an internal point is a bit more subtle. Although each outer concentric shell contributes a constant amount to the potential energy, the amount varies with the radius of the shell. Therefore, for a particle at any internal point,

$$U_g(r) = -\frac{Gm}{r}\int_0^{M'} dM - Gm\int_{M'}^{M} \frac{dM}{x} \qquad [13 \cdot 41]$$

where x is the variable of integration. Substitution of $dM = 4\pi\rho x^2\,dx$ in the integrals then yields

$$U_g = -\frac{Gm}{r}\frac{4\pi r^3\rho}{3} - Gm\int_r^R 4\pi\rho x\,dx$$

$$= -\frac{GmM}{2R^3}(3R^2 - r^2) \qquad r \leq R \qquad [13 \cdot 42]$$

It is easy to verify that $F = -dU_g/dr$.

example 13 • 3 Find the radius R of the orbit of a synchronous communications satellite, which appears to hang in a stationary positon above the earth's surface.

solution We can consider the earth's mass M_e to be concentrated at the center of the earth and apply Kepler's third law. From Eq. [13 . 2],

$$R^3 = \frac{GM_eT^2}{4\pi^2} = (6.673 \times 10^{-11}\,\text{N}\cdot\text{m/kg}^2)(6 \times 10^{24}\,\text{kg})\left(\frac{1\text{ day}}{2\pi\text{ rad}}\right)^2$$

$$= 75.7 \times 10^{21}\,\text{m}^3$$

or

$$R = 4.23 \times 10^7\,\text{m} = 6.65R_e$$

which must be the distance of the satellite from the attracting center, the geocenter. This value can be checked by comparing it with the distance of the moon from the earth, which is about $60R_e$. The period of the moon is 27.3 days, and

$$\frac{GM_e}{4\pi^2} = \frac{60^3}{27.3^2} = 290 \doteq 6.65^3 = 294$$

in units of R_e^3 per day squared (the discrepancy is due to roundoff).

example 13 • 4 A straight evacuated tunnel is bored through the earth and passes at some minimum distance $r_{\min}$ from the geocenter as shown.

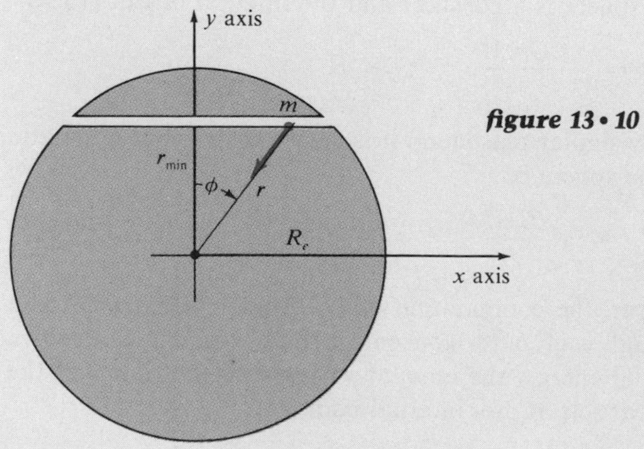

figure 13 • 10

(*a*) Describe the motion, neglecting friction, of an object of mass m dropped into the tunnel. (*b*) Find its maximum speed if the minimum distance is $r_{\min} = \frac{3}{5}R_e$.

solution (*a*) Since the potential energy of the object must be the same at each end of the tunnel, the motion must have turning points at these positions. The movement of the object must therefore be some kind of oscillation between these two points. From Eq. [13 . 38],

$$F = \frac{GmM_e}{R^3}r = \frac{mgr}{R_e}$$

and the force component moving the object is

$$F_x = -\frac{mgr\sin\phi}{R_e} = -\frac{mgx}{R_e}$$

where $g = GM_e/R_e^2$ is acceleration at the earth's surface. Hence the equation of motion of the object is

$$F_x = m\frac{d^2x}{dt^2} \qquad \text{or} \qquad \frac{d^2x}{dt^2} + \frac{gx}{R_e} = 0$$

since it can move only in the tunnel. From Sec. 6 . 6, we can recognize this as the equation for harmonic oscillation, with $\omega^2 = g/R_e$ and an oscillation period given by

$$T = 2\pi\sqrt{\frac{R_e}{g}} \doteq 1600\pi \text{ s} = 84 \text{ min}$$

(If you did not read Sec. 6 . 6, accept this on faith until Sec. 16 . 1.) Note that the oscillation period is independent of $r_{\min}$; it is also a lot faster than air travel.

(*b*) From the law of conservation of energy,

$$\tfrac{1}{2}mv^2 + U_g(r) = E = U_g(R_e)$$

so that maximum speed occurs just as the object passes $r_{\min}$, where

$$U_g(r_{\min}) = U_g(R_e) - \frac{\frac{1}{2}mg}{R_e}(R_e^2 - r_{\min}^2)$$

Therefore

$$U_g(R_e) - U_g(r_{\min}) = \tfrac{1}{2}mv_{\max}^2 = \frac{\frac{1}{2}mg}{R_e}(R_e^2 - r_{\min}^2)$$

and the maximum speed is

$$v_{\max} = \sqrt{\frac{g(R_e^2 - r_{\min}^2)}{R_e}}$$

For $r_{\min} = \frac{3}{5}R_e$, the maximum speed of the object is

$$v_{\max} = \tfrac{4}{5}\sqrt{gR_e} \doteq 6300 \text{ m/s}$$

nearly nineteen times the speed of sound.

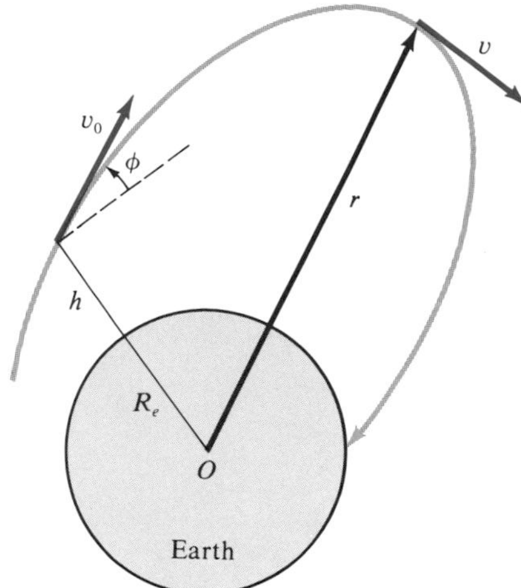

figure 13 • 11 *A projectile launched at an altitude h above a homogeneous and motionless spherical earth*

In analyzing the motion of long-range projectiles, Galileo's flat-earth model is not valid (see Fig. 13 . 11). For this type of analysis we must use Newton's central-force treatment and the fact that the orbits must be conic sections. Assume that the earth is a homogeneous sphere of radius R_e and mass M_e and is motionless with respect to an inertial frame of reference. Let us also assume that the atmosphere offers no appreciable resistance to the projectile. The projectile path must be part of a conic section, and since

$$E = \tfrac{1}{2}mv_0^2 - \frac{GmM_e}{R_e + h}$$

where h is the distance above the earth's surface, this path depends on the initial speed v_0 as follows:

1 The path is a *parabola* ($E = 0, e = 1$) with focus at the geocenter O if

$$v_0 = \sqrt{\frac{2GM_e}{R_e + h}}$$

If $h \ll R_e$, then $v_0 \doteq 11.2$ km/s (see Fig. 13 . 12). With this initial speed and any projection angle ϕ, the projectile will escape from the earth's gravitational field. Since this is the smallest value of v_0 for which escape is possible, it is commonly called the *escape velocity*.

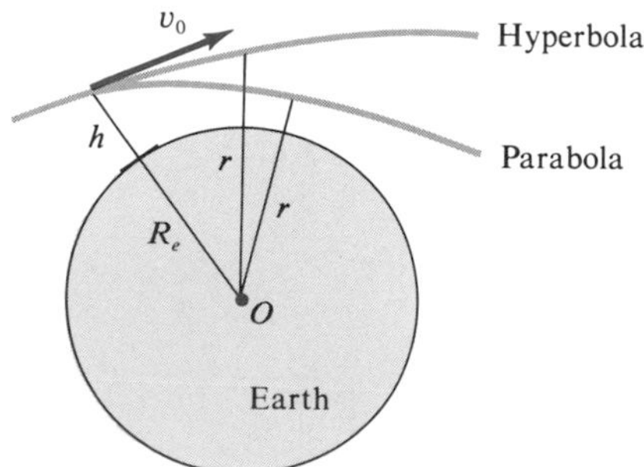

figure 13 • 12 *Parabolic and hyperbolic projectile paths. For $v_0 = 11.2$ km/s and any value of θ, the path is a parabola; for $v_0 > 11.2$ km/s, the path is a hyperbola.*

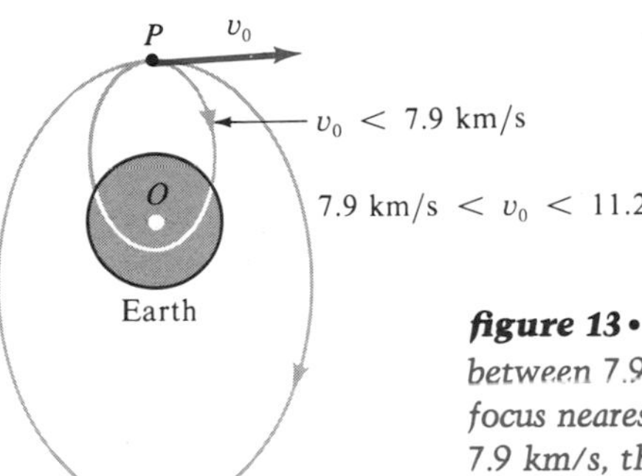

figure 13 • 13 Elliptic paths of a projectile. If v_0 is between 7.9 and 11.2 km/s, the geocenter O is at the focus nearest the point of projection P; if $v_0 <$ 7.9 km/s, the geocenter is at the farther focus of the ellipse.

2 The path is a *hyperbola* ($E > 0, e > 1$) with a focus at the earth's center if

$$v_0 > \sqrt{\frac{2GM_e}{R_e + h}} \doteq 11.2 \text{ km/s}$$

3 The path is an *ellipse* ($E < 0, e < 1$) if $v_0 < 11.2$ km/s and if $h \ll R_e$ (see Fig. 13 . 13). When $\phi = 0$ and

$$v_0 = \sqrt{\frac{GM_e}{R_e + h}} \doteq 7.9 \text{ km/s}$$

the path will be a *circle* ($e = 0$) surrounding the earth. For all other combinations of ϕ and $v_0 \leq 11.2$ km/s the path will be elliptical, but not circular. At low speeds the elliptic trajectory is approximated by the parabolic trajectory of Sec. 4 . 4.

Kepler's three laws obviously are applicable to all elliptic paths, including those of the moon and artificial satellites. For very small values of v_0 the Galilean parabolic paths discussed in Chapter 4 closely approximate the actual elliptic paths.

*13 • 6 numerical orbit computations

In Sec. 6 . 6 we considered a first-order method for numerical integration in which the errors per step Δx are proportional to $(\Delta x)^2$. This method is known generally as *Euler's method* and can, if $f(x,y)$ is a smooth continuous bounded function of x and y, be used to obtain an approximate solution to

$$\frac{dy}{dx} = f(x,y) \qquad [13 \cdot 43]$$

by means of the recursion formula

$$y_{n+1} = y_n + f(x_n,y_n)\,\Delta x \qquad [13 \cdot 44]$$

where $x_{n+1} = x_n + \Delta x$ and $y_n \doteq y(x_n)$. We also saw that second-order differential equations of the form

$$\frac{d^2y}{dx^2} = g\left(x,y,\frac{dy}{dx}\right) \qquad [13 \cdot 45]$$

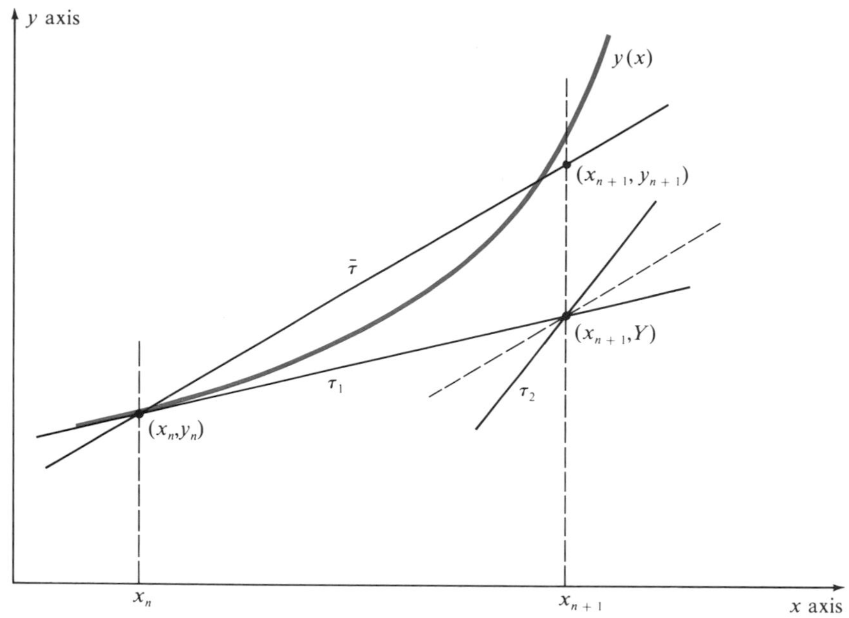

figure 13 • 14 *Second-order improved Euler method for orbit calculations. The average slope over the interval is that of lines τ_3 and $\bar{\tau}$.*

can be treated by considering the function y and its derivative $v = dy/dx$ as *two separate functions* related as

$$\frac{dy}{dx} = v \qquad \frac{dv}{dx} = g(x,y,v) \tag{13 . 46}$$

In this section we shall examine the *improved Euler method,* a second-order method in which the errors per step are proportional to $(\Delta x)^3$. This method is a member of a class of higher-order methods known as Runge-Kutta methods. A fourth-order Runge-Kutta method enjoys widespread use in physics and mathematics and can be found in the program library of every well-equipped computing center.

In Eq. [13 . 44], y_{n+1} is found from the derivative evaluated at the beginning of the nth interval. In the improved method we shall use an *average* value of the derivative $f(x,y)$ over the nth interval (Fig. 13 . 14).

First draw the tangent to the curve through (x_n,y_n) with slope $f(x_n,y_n)$. This line, τ_1, determines Y, the first-order approximation to $y(x_{n+1})$

$$Y = y_n + f(x_n,y_n)\,\Delta x \tag{13 . 47}$$

Now, if we pass a line τ_2 with slope $f(x_{n+1},Y)$ through this point, it should be very nearly parallel to the actual tangent to the curve at $f(x_{n+1}, y(x_{n+1}))$. Thus the average slope over the interval is

$$\bar{f} = \tfrac{1}{2}[f(x_n,y_n) + f(x_{n+1},Y)] \tag{13 . 48}$$

illustrated by the dashed line τ_3. We draw the line $\bar{\tau}$ parallel to τ_3 through (x_n,y_n) and use it to determine

$$y_{n+1} = y_n + \bar{f}\Delta x \doteq y(x_{n+1}) \tag{13 . 49}$$

This formula resembles the trapezoidal rule,

$$y_{n+1} - y_n = \int_{x_n}^{x_{n+1}} \frac{dy}{dx}\,dx \doteq \tfrac{1}{2}\Delta x\,[f(x_n,y_n) + f(x_{n+1},y_{n+1})]$$

except that we use Y instead of y_{n+1} in the last term of Eq. [13 . 48].

We begin the integration of Eq. [13 . 46] with the approximations

$$Y = y_n + v_n\,\Delta x \qquad V = v_n + g(x_n,y_n,v_n)\,\Delta x \qquad [13 \cdot 50]$$

Then we find the average slopes,

$$\bar{v} = \tfrac{1}{2}(v_n + V) \qquad \bar{g} = \tfrac{1}{2}[g(x_n,y_n,v_n) + g(x_{n+1},Y,V)] \qquad [13 \cdot 51]$$

and compute

$$y_{n+1} = y_n + \bar{v}\,\Delta x \qquad v_{n+1} = v_n + \bar{g}\,\Delta x \qquad [13 \cdot 52]$$

figure 13 • 15 *Improved Euler method for solution of $d^2y/dx^2 = g(x, y, dy/dx)$ with increasing x, $x_0 \leq x \leq x_f$*

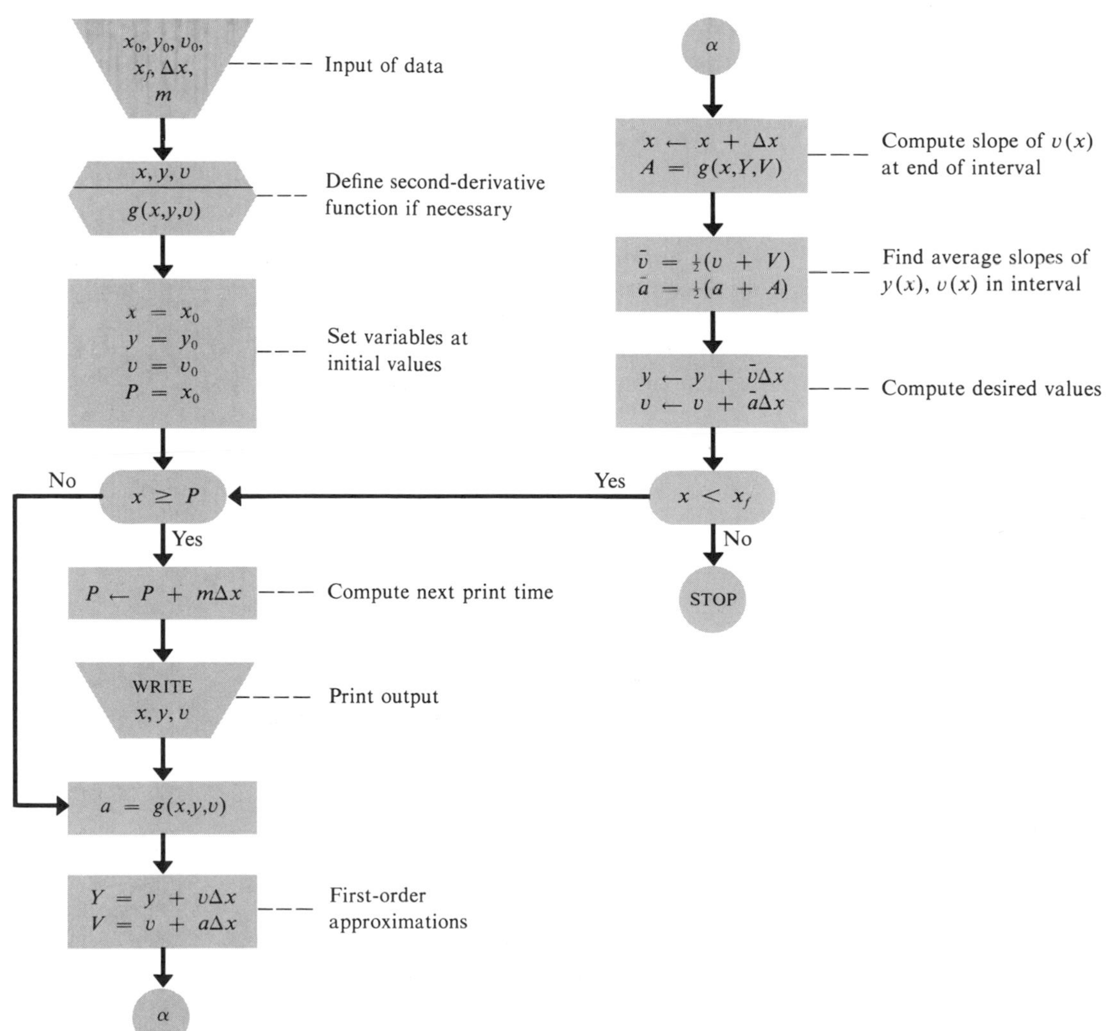

table 13 • 4 *Numerical solution to* $d^2y/d\theta^2 + y = 1$, $y_0 = 1.25$, $v_0 = 0$ *by the improved Euler method*

analytic solution: $y = 1 + 0.25 \cos\theta$

θ, °	y	v	$1 + 0.25 \cos\theta$	error, %
0	1.25000	0.00000	1.25000	—
30	1.21625	−0.12561	1.21651	−0.021
60	1.12394	−0.21731	1.12500	−0.094
90	0.99802	−0.25025	1.00000	−0.198
120	0.87256	−0.21548	0.87500	−0.280
150	0.78149	−0.12235	0.78349	−0.256
180	0.74951	0.00396	0.75000	−0.065
210	0.78531	0.12928	0.78349	0.232
240	0.87925	0.21970	0.87500	0.486
270	1.00594	0.25071	1.00000	0.594
300	1.13111	0.21388	1.12500	0.543
330	1.22088	0.11913	1.21651	0.360
360	1.25092	−0.00793	1.25000	0.074

In the flowchart of Fig. 13 . 15, as usual, requirements on the computer's memory are minimized by using as few variable names as possible.

We can now apply this method to solve the orbital equation for a particle of mass m moving under a central force $F(r)$. If we substitute $u = 1/r$ in the orbital Equation [13 . 15] for $r(\theta)$, we obtain $F(r) = F(1/u)$ and

$$\frac{d^2u}{d\theta^2} = -\frac{m}{u^2L^2}F\left(\frac{1}{u}\right) - u = \frac{Gm^2M}{L^2} - u \qquad [13 \cdot 53]$$

Scaling this equation by substituting $y = ul$ gives

$$\frac{d^2y}{d\theta^2} = 1 - y \qquad [13 \cdot 54]$$

Note that $F(r)$ may be any central force, not just the force of gravity.

As an example, let us choose the initial conditions as $y_0 = 1.25$ and $(dy/d\theta)_0 = 0 = (dr/d\theta)_0$, corresponding to the conditions $l = 1.25r_0$ and $L^2 = 1.25Gm^2Mr_0$. From Eq. [13 . 24], this implies a total energy of

$$E = \frac{L^2}{2mr_0^2} - \frac{GmM}{r_0} = -\frac{3GmM}{8r_0}$$

Using the improved Euler method with $g = 1 - y$ and $\Delta\theta = 10° = 0.17453$ rad (printing every third line), we obtain the results given in Table 13 . 4. The correct analytic solution is an ellipse with eccentricity $e = 0.25$ and $y = l/r = 1 + 0.25 \cos\theta$. The results in the table show that, even for such relatively large intervals, the maximum error is less than 0.6 percent. A first-order computation for the same interval size displays a maximum error of 14 percent. Even with $\Delta\theta = 45°$, the second-order method still yields accuracy better than 12 percent, so that, if necessary, we could sketch a rough approximate solution from computations on a calculator.

PROBLEMS

13 • 1 *Kepler's laws*

13 • 1 The mean distance between the centers of the earth and the moon is 3.84×10^8 m, and the moon's period is 27.3 days. (*a*) Find the Kepler constant k_s for the earth and its satellites. (*b*) What is the mean distance of an artificial satellite from the center of the earth if the observed period of the satellite is 100 min? (*c*) If the period is one day?

13 • 2 A satellite circles the earth in a very low orbit, $R \doteq R_e$. Express its period of rotation in terms of the density ρ_e of the earth.

13 • 3 Jupiter, the largest of the planets, has twelve satellites. The largest of these, Ganymede (of course), was discovered by Galileo in 1610 (it is large enough to be seen with good binoculars). It is located 15 Jovian radii R_J from the center of the planet and has a period of 620,000 s. Find the average density and radius of the planet.

13 • 4 A particle of mass m rotates in a circle of radius R under the influence of a central force F. If the angular momentum L of the particle is found to have the same value for every R, how does F depend on R?

13 • 5 Compare the earth's acceleration toward the sun during a solar eclipse with that during a lunar eclipse.

13 • 6 A binary star consists of two stars revolving about their common center of mass under their mutual gravitational attraction. Often such systems appear as a single star because of their great distance from the sun. Suppose an astronomer observes such a system for which the period of rotation is T and the speeds of the two stars are v_1 and v_2, respectively. Suppose also that the orbits are circular. What is the separation of the two stars and what are their individual masses?

13 • 7 The two shaded areas in the figure are equal in size and occur at the turning points of an elliptic orbit. (*a*) If the orbiting particle "falls" toward the focus through distances s_a and s_p at the turning points for *very small* angles δ_a and δ_p, show that the forces on the particle at the turning points are in the ratio $F_a/F_p = s_a/s_p$. Newton showed that in this case

$$\frac{s_a}{s_p} = \frac{r_a^2\delta_a^2}{r_p^2\delta_p^2}$$

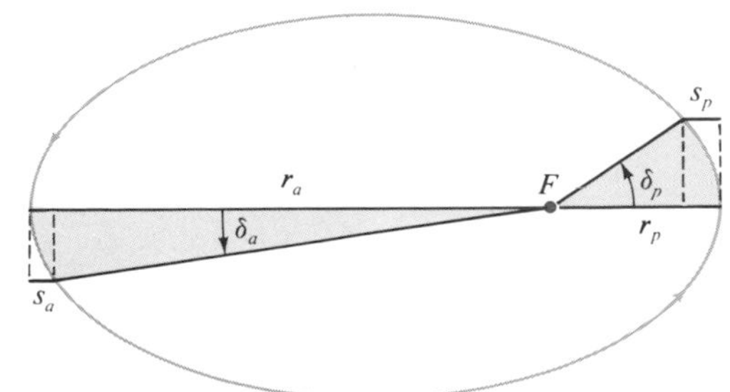

(*b*) Use the relationship in part (*a*) to show that

$$\frac{F_a}{F_p} = \frac{r_p^2}{r_a^2}$$

13 • 2 dynamic consequences of Kepler's laws

13 • 8 Verify the geometric relationships in the third and fourth columns of Table 13 . 2.

13 • 9 The orbit of the moon may be represented by an ellipse that is continuously changing in position and form, has the earth at one of its foci, and has an average eccentricity $e = 0.055$. (*a*) Why should we expect such changes to occur? (*b*) Show that the moon usually is more than one-tenth farther from the earth at r_a than it is at r_p. (*c*) Although the sidereal period of the moon (relative to the fixed stars) is 27.3 days, the synodic period (as seen from earth) is 29.5 days (one month). Why?

13 • 10 A body is moving with period T in an elliptic orbit of semimajor axis a. (*a*) Show that its acceleration toward the central attracting body is $4\pi^2 a^3/T^2 r^2$. (*b*) What is the maximum value of the acceleration of the earth toward the sun?

13 • 11 Show the derivation of Eq. [13 . 22] given in the text, going through the steps in explicit detail, to find the expression for eccentricity.

13 • 12 (*a*) Express the scale l and eccentricity e in terms of the minimum and maximum radii r_p and r_a of the elliptic orbit. (*b*) Now express U_g in terms of k_p, r_a, and r_p.

13 • 13 A long pendulum of length l is suspended from a ceiling with the bob almost touching the floor. If it is pulled away through a small angle ϕ and given a push which is horizontal, but directed normal to the initial horizontal position vector, as shown, the shadow of the bob will traverse an ellipse whose center is *directly below* the point of suspension. Find the dependence of the central force on the distance from this center as seen in the horizontal plane of the floor and show that the trajectory must indeed be an ellipse, ignoring the slight vertical component of the motion.

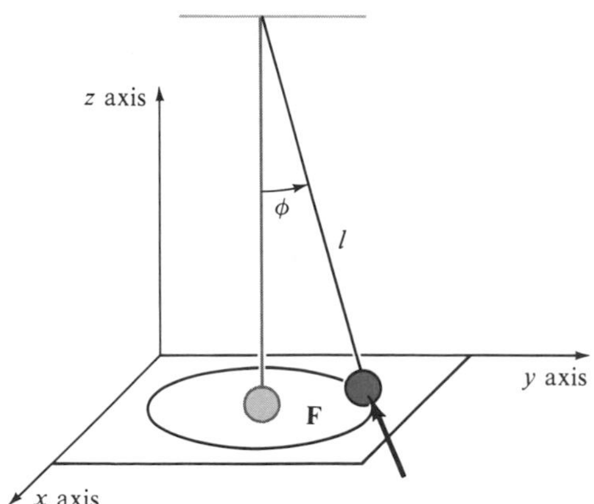

13 • 14 In plane polar coordinates, kinetic energy is $K = \frac{1}{2}m(v_r^2 + v_\theta^2)$. (*a*) What is the relationship between the kinetic energy and the force for a particle under the influence of a central force? (*b*) Use the algebra of differentials to find the relationship between dv_r/dr and d^2r/dt^2. (*c*) Derive Eq. [13 . 10] for a particle subject to a central force by performing the differentiation indicated in the answer to part (*a*) above.

13•15 Two particles of masses $m_1 = 10$ g and $m_2 = 40$ g are located in a rectangular coordinate system with its origin at the center of mass of the two particles. The location of particle m_1 is (4 cm,0,0). (*a*) Find the rectangular components of the total gravitational force exerted by the two particles on a 1-g particle placed at the point (−5 cm,0,0). (*b*) At the point (0,0,6 cm). (*c*) At what point on the *x* axis is the force on a 1-g particle zero? (*d*) What is the total potential energy of the *system* in part (*b*)?

13•16 (*a*) Show that the total energy of an orbiting mass under the influence of gravity can be given in terms of the minimum and maximum radii as

$$E = -\frac{GmM}{r_p + r_a}$$

(*b*) Also show that the kinetic energies at these points are in the ratio $K_a/K_p = r_p^2/r_a^2$. (*c*) How is E related to the semimajor axis of the ellipse?

13•17 If the largest and smallest speeds of a planet in its elliptic orbit about the sun are v_1 and v_2, what is the eccentricity of the ellipse?

13•18 Explain how the question of whether a newly discovered comet is a permanent or temporary member of the solar system can be settled by measuring its speed v at some known distance r from the sun.

13•19 A body of mass m is in an elliptic orbit about a mass $M \gg m$. Measurements at its minimum radius r_p show that its kinetic energy at that point is three times the magnitude of the total energy. Express the following quantities in terms of r_p: (*a*) The total energy E. (*b*) The angular momentum L. (*c*) The scale l and the eccentricity e of the orbit.

13•20 A rocket of mass m approaches a planet of mass M in a hyperbolic orbit with energy E and angular momentum L. At the minimum radius r_p, the retrorockets fire briefly to slow down the rocket enough to inject it into a circular orbit of radius r_p around the planet. Find the new values L', E', and K' of the rocket's angular momentum, energy, and kinetic energy, all in terms of E and L.

13•4 *extended bodies*

13•21 The endpoints of a thin rod of length l and uniform linear density ρ_l have the rectangular coordinates (0,0,0) and (l,0,0), respectively. (*a*) Prove that the gravitational field strength produced by the rod at any fieldpoint (x,0,0) not in the rod is given by

$$f_x = -\frac{Gl\rho_l}{x(x - l)}$$

where $f_y = 0$ and $f_z = 0$. (*b*) Show that the rod attracts gravitationally as if all its mass were concentrated at the distance $\sqrt{x(x - l)}$ from the fieldpoint P, which is the geometric mean of the distances of the ends of the rod from P.

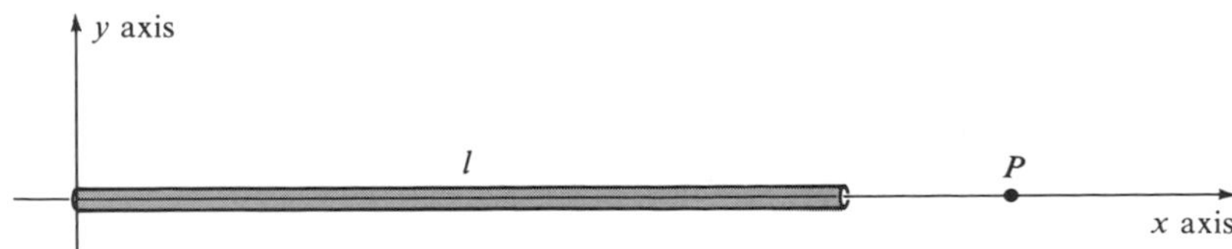

13 • 22 For the gravitational field due to the thin rod of Prob. 13 . 21, prove that the field strength at a point P on the perpendicular bisector of the rod and at a distance y from the rod's center is given by

$$f_x = 0 = f_z \qquad f_y = -\frac{2Gl\rho_l}{y\sqrt{y^2 + l^2}}$$

Show that this field strength is the same as if all the mass of the rod were concentrated at a point whose distance from P is the geometric mean of the nearest and farthest points of the rod from P. Also show that this "equivalent particle" is not in the rod, but is beyond it as viewed from P.

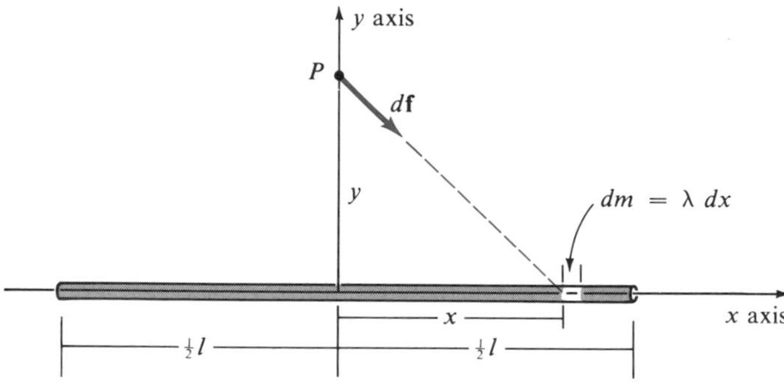

13 • 23 For the gravitational field produced by a thin ring of radius R, show that the field strength is maximum at the point $x = R/\sqrt{2}$.

13 • 24 For the rod of Prob. 13 . 21 find the potential distribution along the x axis and then deduce from this an expression for the field strength at any fieldpoint $(x,0,0)$.

13 • 25 Find the gravitational force per unit mass on a particle on the axis of a disk of radius R and mass M as a function of distance r from the center of the disk.

***13 • 26** A 1-kg point mass is situated 1 m above the center of a plate 2 m square and of mass 4 kg. Establish a mesh of points (cf. Fig. 10 . 9), as fine as you like, on the plate. Give each point a mass equal to that of the small square of which it is the center, and then compute the total gravitational force on the point mass by adding the contribution from each mesh point considered as a point mass. Will your answer be an upper or a lower bound to the gravitational attraction? (If you cannot answer this question theoretically, try smaller and smaller meshes to see how the force converges.)

13 • 5 *the earth's field*

13 • 27 Show that the period of oscillation in the tunnel of Example 13 . 4 is the same as the period of the satellite in Prob. 13 . 2.

13 • 28 Find the gravitational potential energy of a 500-kg object when it is 1000 km above the surface of the earth and the zero level of potential energy is taken (*a*) at infinity. (*b*) At the earth's surface.

13 • 29 An artificial satellite of mass 30 kg is moving about the earth in an orbit of $r_p = 6800$ km and $r_a = 9000$ km. Using the results of Prob. 13 . 12, find its kinetic energy at each of these points.

13 • 30 In *The System of the World,* Newton wrote: "... the attractions of homogeneous spheres near their surfaces are as their diameters. Hence a sphere of one foot in diameter, and of a like nature to the earth, would attract a small body placed near the surface with a force 20,000,000 times less than the earth would if placed near its surface.... If two such spheres were distant but by $\frac{1}{4}$ of an inch, they would not, even in space void of resistance, come together by the force of their mutual attraction in less than a month's time...." Verify the accuracy of this statement.

13 • 31 Assume the earth to be a homogeneous sphere of mass M_e and radius R_e and let the reference level for the gravitational potential energy U_g of objects in the earth's field be taken at the earth's surface. (*a*) Show that at distance $R_e + h$ only slightly greater than R_e, the potential energy of a particle of mass m is $U_g \doteq GmMh/R_e^2$. (*b*) Use this result to show that $U_g \doteq mgh$.

13 • 32 (*a*) If a projectile is fired at an altitude of 200 km and at any angle of projection θ_0, what minimum initial speed must it have in order to escape from the earth's gravitational field? (*b*) Prove that if a long-range projectile is traveling in a parabolic path, its kinetic and potential energies will be of equal magnitude at every point of the path. (*c*) If an object escapes from the earth along a parabolic path, what is its speed when it is 10^7 km away from the earth's center?

13 • 33 A projectile is fired from the surface of the earth with an initial speed v_0 in a direction making an angle θ_0 with the horizontal and is observed to travel an elliptical path. (*a*) Show that the semimajor and semiminor axes of the ellipse are given by

$$a = \frac{GM_e}{2GM_e/R_e - v_0^2} \qquad b = \frac{v_0 R_e \cos\theta_0}{\sqrt{2GM_e/R_e - v_0^2}}$$

where M_e and R_e are the mass and radius of the earth. (*b*) Show that the orbit will be circular if $v_0 = \sqrt{GM_e/R_e}$ and $\theta_0 = 0$.

13 • 34 An object is dropped from a captive balloon over the equator at an altitude of 120 m. Since its angular momentum is conserved as it falls with acceleration g (neglecting air resistance), it appears to drift *eastward*. Show that the eastward drift is $s(t) = \frac{1}{3}\omega g t^3$, where $\omega = 7.3 \times 10^{-5}$ rad/s is the earth's frequency of rotation and t is measured from the moment the object is released. Compute the total eastward drift during the fall.

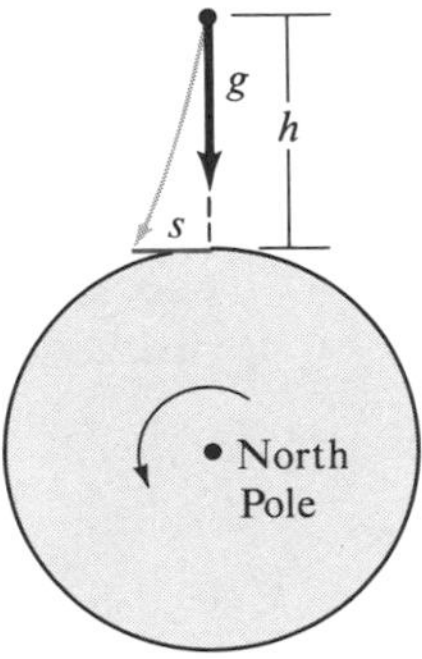

13 • 6 numerical computations

***13 • 35** A particle of mass m moves in a circular orbit of radius R about the origin of coordinates under the influence of gravitational attraction of a fixed mass M. Suddenly, when the body is moving through the point $(x,y) = (R,0)$ a "wind" starts blowing against it in the x direction, exerting a constant force $F_x = 0.03(GmM/R^2)$ on the particle. (*a*) Compute the subsequent motion of the particle for four orbits, using cartesian coordinates. (*b*) Show that the motion tends toward an ellipse with its long axis in the y direction, *normal* to the wind. (*Hint:* Use loops in your algorithm, as in Fig. 10. 13, and take smaller intervals as $r_{\min} \to 0$, or your solution will diverge.)

***13 • 36** Examine the orbits for the inverse-cube law of force, $F = -k/r^3$, for a range of positive, zero, and negative energies, starting from a turning point of $v_0 = 0$, as in the example following Eq. [13 . 54]. (*Hint:* Express the energy in units of K_0.)

13 • 37 Newton pointed out that any slight deviation from an inverse-square force would cause the nearly circular orbits of the planets to rotate very slowly in their planes about their focus (the sun). Let us assume that the force law is of the form

$$F(r) = -\frac{Gm}{r^2}\frac{1 + C}{r}$$

where C is a constant. Thus Eq. [13 . 54] becomes

$$\frac{d^2u}{d\theta^2} + u = \frac{Gm^2M}{L^2}(1 + Cu)$$

Explain Newton's statement by showing that with the appropriate scaling of *both* u and θ, the above equation for $u(\theta)$ reduces to the form of that for an ellipse, Eq. [13 . 54]. Show that for $C/l \ll 1$ and $l = L^2/Gm^2M$, the ellipse rotates at the rate of $\pi C/l$ rad per orbit.

***13 • 38** Compute several orbits for the force law of Prob. 13 . 37 for the case $C/l = 0.1$ and $l = 1.25r_0$. Sketch your results and compare them with Table 13 . 4.

***13 • 39** With the origin on the (assumed) stationary mass M, describe the motion of its satellite m, using Newton's second law to give two acceleration equations, one for d^2x/dt^2 and one for d^2y/dt^2. Let x be in the direction $\theta = 0$, and compute $x(t)$ and $y(t)$ for the initial conditions of the orbit in Table 13 . 4:

$$r_0 = x_0 = \tfrac{4}{5}l \qquad y_0 = 0 \qquad v_{x0} = 0 \qquad v_{y0} = \frac{L}{mx_0}$$

You will need a total of *four* first-order equations to describe the motion. As a check on consistency, recompute E and L from the numerical data as you proceed.

answers

13 • 1 (*a*) $k_s = 1.32 \times 10^{-23}$ day^2/m^3; (*b*) $R = 7156$ km; (*c*) $R = 42{,}356$ km

13 • 2 $T = \sqrt{3\pi/G\rho_e}$

13 • 3 $\rho_J = 1.24$ g/cm^3, $R_J = 76{,}800$ km

13 • 4 $F = -L^2/mR^3$

13 • 5 1.1 percent greater during solar eclipse

13 • 6 $R = \frac{1}{2}T(v_1 + v_2)/\pi$, $m_1 = \frac{1}{2}(v_1 + v_2)^2 v_2 T/\pi G$, $m_2 = m_1 v_1/v_2$

13 • 10 (*b*) $a_{\max} = 5.72$ mm/s^2

13 • 12 (*a*) $l = 2r_a r_p/(r_a + r_p)$, $e = (r_a - r_p)/(r_a + r_p)$; $U_g = -[2mk_p^2(r_a + r_p)/r_a r_p]/r$

13 • 13 $F_r = -mgr/l$

13 • 14 (*a*) $F = dK/dr$; (*b*) $d^2r/dt^2 = v_r dv_r/dr$; (*c*) Note that $v_\theta dv_\theta/dr = -L^2/m^2r^3$

13 • 15 In cgs units: (*a*) $\mathbf{F} = 2.62G\mathbf{i}$; (*b*) $\mathbf{F} = -0.071G\mathbf{i} - 1.266G\mathbf{k}$; (*c*) $x = 2.33$; (*d*) $U_g = -81.27G$

13•16 (*c*) $E = -GmM/2a$
13•17 $e = (v_1 - v_2)/(v_1 + v_2)$
13•19 (*a*) $E = -\frac{1}{4}GMm/r_p$;
(*b*) $L = \sqrt{1.5GMm^2r_p}$; (*c*) $l = 1.5r_p$, $e = 0.5$
13•20 $L' = L/\sqrt{1 + e}$, $E' = -\frac{1}{2}m^3M^2G^2/L'^2$,
$K' = -\frac{1}{2}E'$, where $e =$
$\sqrt{(1 + 2EL^2)/G^2m^3M^2}$
13•24 $U_g = -G\rho_l \ln(1 - l/x)$
13•25 $F = (2MG/R^2)\left(1 - r/\sqrt{r^2 + R^2}\right)$
13•26 $F = \frac{2}{3}\pi G$ downward
13•28 (*a*) $U_g = -2.72 \times 10^{10}$ J;
(*b*) $U_g = 4.3 \times 10^9$ J
13•29 $K_p = 1.00 \times 10^9$ J,
$K_a = 5.70 \times 10^8$ J
13•30 $g_e/g_{sph} \doteq 42 \times 10^6$, $\Delta t \doteq 475$ s
13•32 (*a*) $v_0 = 11.0$ km/s;
(*c*) $v = 10.38$ km/s
13•34 $s = 2.9$ cm (in agreement with Prob. 12.35)
13•36 $r \to \infty$ for $E > 0$; $r = r_0$ for $E = 0$;
$r \to 0$ for $E < 0$
13•37 $r = (l - C)/(1 + e\cos\phi)$,
$\phi = \theta\sqrt{1 - C/l}$, $e = (l - r_0 - C)/r_0$
13•38 Some sample points:

θ	0°	95°	189°	360°	378°
r	1	1.125	1.286	1.006	1.000

CHAPTER FOURTEEN

elasticity

About two years [ago] I printed this theory in an anagram at the end of my book of the descriptions of helioscopes, viz. . . . the power of any spring is in the same proportion with the tension thereof; that is, if one power stretch or bend it one space, two will bend it two, and three will bend three, and so forward.

Robert Hooke, De Potentia Restitutiva, *1678*

In the preceding chapters we have, for the most part, treated all bodies as if they remained perfectly rigid when acted on by forces. This simplification has been exceedingly useful in our initial approach to the study of mechanics. However, every real body undergoes some deformation—a change in size or shape—under the impact of a force or torque or when it is heated or cooled. A body that tends to return to its original size and shape when the deforming forces or torques are removed is termed *elastic.* It is *completely elastic* if the magnitudes of the internal restoring forces and torques depend only on the extent and character of the deformation, and not on the rate of return to its original form. For example, a wire stretched by a number of weights would show complete elasticity if the successive removal of each weight caused it to resume exactly the length it had before that weight was added, no matter how long the weights had been in place. Some materials that are not completely elastic, such as glass, show a marked elastic aftereffect, or *elastic lag,* a time delay in returning to their original size or shape.

If sufficient force or torque is applied to a solid body, it will acquire a permanent set and will never completely return to its original configuration. For materials such as putty or clay, the force needed to produce a permanent set is so small that the material is regarded as *inelastic.*

The theory of elasticity is based primarily on generalizations of *Hooke's law* to a number of different types of deformations. In this chapter we shall consider three basic types of deformation: (1) linear stretching or compression, (2) changes in volume, and (3) twisting, or *torsion.* We will see how, within certain limits, the deformation is related to the applied force causing it and, in the closing section, relate this behavior to the structure of crystalline solids, those materials whose molecules form a regular, ordered structure. Crystalline solids are of great interest because of their many uses in engineering, and developments in atomic theory have given rise to a better understanding of crystal elasticity.

We shall limit our considerations here to the elastic properties of bodies in both *static* and *thermal equilibrium*—that is, bodies held at rest under the action of applied forces and torques and kept at a constant temperature during and after deformation. We shall also confine ourselves to materials that are *homogeneous* and *isotropic*—materials that have the same elastic properties at every point and for all directions of applied force and deformation.

14 • 1 Hooke's law

Suppose a horizontal metal rod of length l_0 is clamped at one end and is then stretched by a horizontal force applied at the other end. By progressively increasing the force F applied to the rod and observing the corresponding change in its length, we obtain data for a curve like that in Fig. 14 . 1. If the rod is maintained at constant temperature and the load is not too large, this curve will be practically a straight line. Moreover, for the small loads that correspond to the straight-line portion of the curve, gradual removal of the load is accompanied by a shortening of the rod to its original length. For small loads, then, the amount of elongation Δl is proportional to the deforming force F, or

$$F = k_s \Delta l \qquad [14 \cdot 1]$$

where k_s is the restoring-force constant, also called the *stiffness coefficient*

figure 14 • 1 *Force-displacement curve for a mild steel wire of length 2 m and diameter 1 mm, with stiffness coefficient k_s = 75,000 N/m*

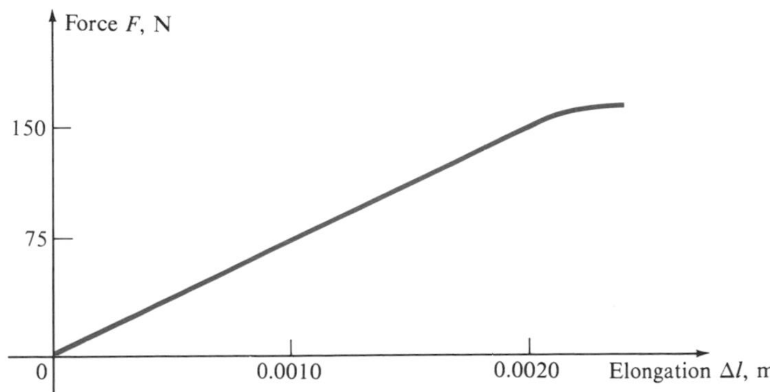

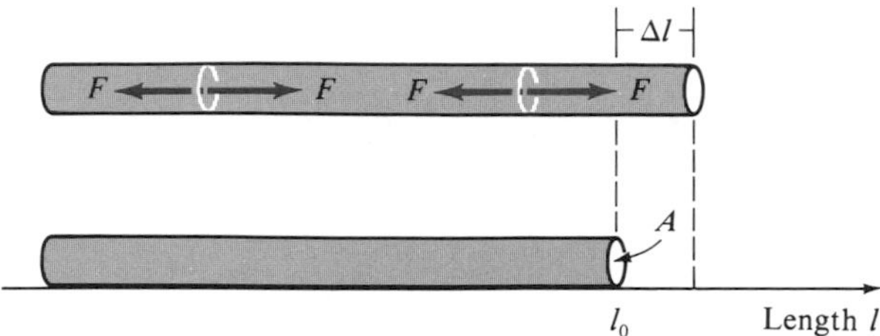

figure 14 • 2 *A tensile force F applied to a uniform rod of length l_0, causing a longitudinal strain $\Delta l/l_0$ and a longitudinal stress F/A*

of the rod. This equation expresses the famous law of proportionality of force and elastic deformation that Robert Hooke discovered empirically in 1660.

Since stiffness coefficients have been found to depend on the geometry of the particular body involved, it would be useful to have elastic constants that characterize all bodies made of a given material. Such constants can be determined if we introduce two new physical quantities, *stress* and *strain.* When an elastic body is subjected to external forces, internal restoring forces arise between all adjacent parts of the body, in addition to those which may have existed before applying the forces. If we imagine a plane drawn through any point in the body, the part of the body on one side of the plane acts with a certain force on the part on the other side, which reacts with an equal but opposite force (Fig. 14 . 2). A body under the action of these restoring forces is said to be in a state of stress, and we define the *stress* as F/A, the force per unit area of the imaginary plane across which the force is transmitted. In the case of the rod we have a *longitudinal,* or *tensile, stress.*

The term *strain* pertains to any change occurring in the relative positions of the parts of an elastic body. A rod that has been stretched or compressed in the direction of its length has undergone longitudinal, or tensile, strain. The greater the original length l_0 of the rod, the greater will be the change of length Δl produced by a given deforming force. Therefore we define *longitudinal strain* as $\Delta l/l_0$, the fractional change in length, or change in length per unit length, independent of the size of the particular rod.

If we now express Hooke's law in terms of stress and strain, we find that

$$\frac{F}{A} = k_s \frac{l_0}{A}\frac{\Delta l}{l_0} = Y\frac{\Delta l}{l_0} \qquad [14 \cdot 2]$$

where Y is a constant of elasticity known as *Young's modulus.* If stretching experiments are conducted with rods of different sizes, but composed of the same elastic material, the data will result in a stress-strain diagram like that in Fig. 14 . 3. If these tests are carried out under constant temperature conditions, in all cases the first part of the curve will be a straight line, with an identical slope Y for each rod. Within this region, Y is a constant of the material, and Eq. [14 . 2] can be interpreted as the general form of Hooke's law for all longitudinal deformations:

Stress is proportional to strain.

In fact, this principle is generally valid for other types of deformation as well. Table 14 . 1 gives some selected values of Y for various materials.

table 14 • 1 *Elastic constants of various metals at room temperature (All values are approximate, since these constants depend critically on the previous histories and exact compositions of the test samples.)*

material	*Young's modulus Y, 10^{10} Pa*	*elastic limit for tension, 10^{10} Pa*	*bulk modulus B, 10^{10} Pa*	*shear modulus S, 10^{10} Pa*
aluminum, rolled	7.0	0.013	7.5	2.4
brass, cold-rolled	9.0	0.038	11.0	3.5
copper, rolled	12.0–13.0	0.015	14.0	4.2
iron, cast	8.4–9.8	0.0035–0.0040	9.6	5.3
magnesium	4.2	—	3.4	1.7
silver, hard-drawn	7.7	—	—	2.0
steel, mild	21.0	0.017–0.021	16.0	8.0
steel, tempered spring	20.0	0.076–0.12	17.0	7.6
tungsten, hard-drawn	35.0	—	37.0	15.0

Note: 1 Pa = 1 N/m^2.

As indicated in Fig. 14 . 3, the *proportional limit* for any given material is the maximum stress for which the stress is proportional to the strain; that is, it is the stress at which the stress-strain curve first deviates from a straight line. The *elastic limit* of the material is the largest stress it can sustain without acquiring a permanent set. If the applied load does not exceed the elastic limit, the stress-strain curve retraces itself as the load is reduced until the length of the test sample is the same as it was before the load was applied. If the stress does exceed the elastic limit, subsequent removal of the load may leave a permanent set, as illustrated in Fig. 14 . 4. Here, instead of returning to its original length, the material returns to a new length, l_0'. If the load on the sample is again increased, now from l_0', the stress-strain curve has

figure 14 • 3 *Typical stress-strain curve for a metal in tension*

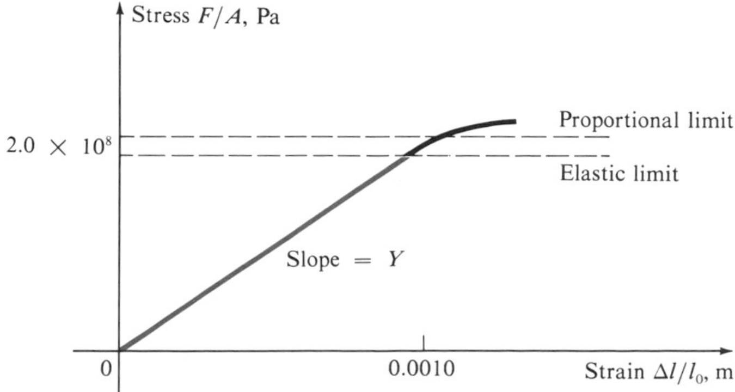

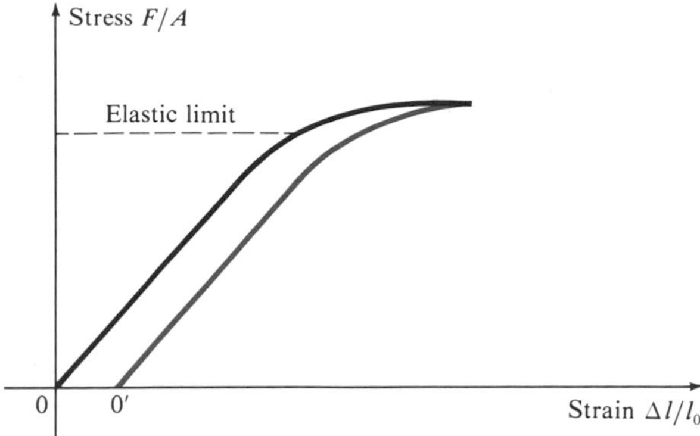

figure 14 • 4 *An example of elastic hysteresis*

a different slope from the original one and the elastic limit is different. This phenomenon is termed *elastic hysteresis.*

We may state Hooke's law in its broadest terms as

$$\text{Stress} = \text{modulus} \times \text{strain} \qquad [14 \cdot 3]$$

where the modulus—in this case, Young's modulus—is constant for a given material, rather than merely for a particular body. In this form Eq. [14 . 3] applies to many kinds of elastic changes. There are as many kinds of moduli as there are kinds of strains. Although the moduli must be determined from measurements made on particular test samples, they really are properties of the material of which the sample is made, and not of its size and shape. They do depend, however, on the exact composition and *history* of the material, on its temperature, and on the ambient pressure to which it is exposed. In the case of solids, for example, the values of the moduli diminish as the temperature rises, so that a spring balance stretches a little farther under the same force when it is hot than when it is cold. The effect of increased ambient pressure on a solid is usually to increase the values of the moduli.

14 • 2 volume elasticity

A strain that consists of a change in size of the volume of a body without change in its shape is called a *volume strain,* or *dilation* (see Fig. 14 . 5). It is measured by $\Delta V/V_0$, the fractional change in the volume of the body. In elastic bodies such a strain is caused by a stress *F/A*, which, for homogeneous and isotropic materials, will have the same magnitude regardless of

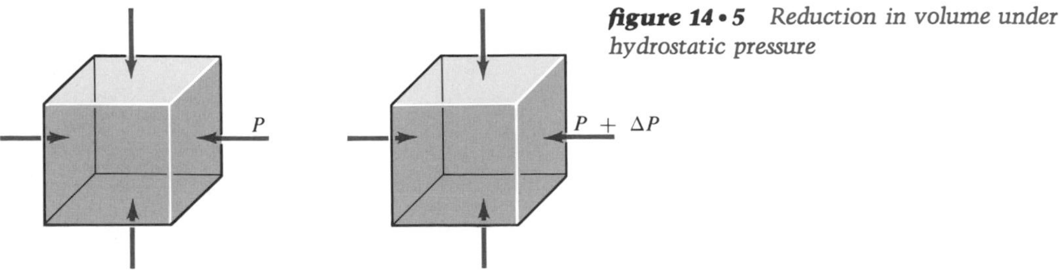

figure 14 • 5 *Reduction in volume under hydrostatic pressure*

table 14 • 2 *Pressure conversion factors*

equal pressures: 1 *Pa* = 1 N/m^2 = 10 *dynes/cm*^{2}; 1 *mm Hg* = 0.535 *in* H_2O

Pa	*atm*	*mm Hg*	*lbf/in*^{2}
1	9.869×10^{-6}	7.501×10^{-3}	1.450×10^{-4}
1.013×10^5	1	760.0	14.70
133.3	1.316×10^{-3}	1	0.01934
6895	0.06804	51.71	1

the orientation of the surface across which it is transmitted. Such a stress is termed a *hydrostatic pressure,* because it is the only kind of stress that can exist in fluids at rest. In fact, the simplest way to apply a uniform pressure to a solid body is to submerge it in a confined liquid and apply a force to the liquid.

The term *pressure* as used here refers to a force applied normal to a planar surface of area A, as shown in Fig. 14 . 5. The average pressure $\overline{P}$ over a surface area A is therefore

$$\overline{P} = \frac{F_\perp}{A} \qquad [14 \cdot 4]$$

The standard unit of measure for pressure is the *pascal* (Pa), which is equal to 1 N/m^2. In the British system, pressure is most often expressed in pounds per square inch (lbf/in^2, or more commonly, psi). Other units in common use are the *bar,* equal to 10^5 Pa, and the standard atmosphere (atm), defined as the pressure at the base of a column of mercury (Hg) 760 mm in height, at 0°C, subject to the standard acceleration due to gravity. Since the density of mercury under these conditions is 13.60 g/cm^3, 1 atm = 1.013×10^6 dynes/cm^2 = 1.013 bars. The relationships of various units of pressure are summarized in Table 14 . 2. The unit 1 mm Hg = 1 torr, although anachronistic, is still in wide use and testifies to the great influence which Torricelli's invention of the mercury barometer had on the subsequent developments of concepts of pressure.

In many cases the pressure is not uniform over a surface area—as, for example, with stresses in a nonhomogeneous body. We therefore need the more precise concept of local pressure, or *pressure at a point:*

$$\blacktriangleright \quad P = \lim_{\Delta A \to 0} \frac{\Delta F}{\Delta A} = \frac{dF}{dA} \qquad [14 \cdot 5]$$

where dF is the normal force on a vanishingly small portion dA of the surface.

All materials, whether solid, liquid, or gas, possess elasticity of volume, and experiments show that the general form of Hooke's law in Eq. [14 . 3] also applies to volume strain. If an increase in stress ΔP causes a strain $\Delta V/V_0$, then

$$\blacktriangleright \quad \Delta P = -B\frac{\Delta V}{V_0} \qquad [14 \cdot 6]$$

table 14 • 3 *Compressibilities of some common liquids*

liquid	*compressibility κ at 20°C and 1 atm,* m^2/N
benzene	8.9×10^{-10}
ethyl alcohol	9.0×10^{-10}
glycerin	2.2×10^{-10}
mercury	0.38×10^{-10}
water	5.0×10^{-10}
sea water	6.3×10^{-10}

where B is a constant of proportionality called the *volume,* or *bulk, modulus.* Since an increase in pressure produces a reduction in volume, the minus sign makes B a positive constant. The harder it is to compress a substance, the larger the value of its bulk modulus (see Table 14 . 1).

As a rule, liquids offer less resistance to compression than solids, but even this resistance is so large that for many purposes liquids may be treated as incompressible. For all *fluids*—a term which refers to substances that flow and includes gases as well as liquids—it is customary to tabulate the *compressibility* κ:

$$\kappa = \frac{1}{B} = -\frac{\Delta V/V_0}{\Delta P} \qquad [14 \cdot 7]$$

These values refer to the relative ease which various substances can be compressed (see Table 14 . 3). However, for gases under ordinary conditions, this factor is approximately $\kappa \doteq 1/P$.

Although the bulk modulus B is approximately constant for liquids and solids, it can be quite variable for a gas. According to *Boyle's law,* formulated in 1662:

> If the temperature of a given mass of gas remains unchanged, then the product of the pressure and volume of the gas is constant.

The constant itself is proportional to the mass of gas used, for if the pressure and temperature are kept the same, twice the original volume would be needed for twice the mass of gas. We may therefore express Boyle's law as

$$PV = MC \qquad [14 \cdot 8]$$

where P is the pressure and V is the volume of a given mass M of gas. The constant C depends on the kind of gas and the temperature.

If we define the *specific volume* v as the volume per unit mass,

$$v = \frac{1}{\rho} \qquad [14 \cdot 9]$$

then we can rewrite Eq. [14 . 8] as

$$\frac{P}{\rho} = Pv = C \qquad [14 \cdot 10]$$

We can now restate Boyle's law in terms of the density and specific volume of the gas as follows:

> Under constant temperature conditions the pressure on a gas is proportional to its density (or inversely proportional to its specific volume) and is independent of its mass.

For gases such as hydrogen, helium, and air at ordinary temperatures, the departures from Boyle's law are less than 1 percent up to 10 atm of pressure. We shall often find it convenient to speak of an *ideal gas* (sometimes called a *perfect gas*), which is an imaginary gas of point masses whose behavior under constant temperature conditions is described by Boyle's law. To find the bulk modulus of such a gas, we can write Eq. [14 . 6] in differential form as

$$dP = -B\frac{dV}{V} \qquad [14 \cdot 11]$$

and from Eq. [14 . 8] we have

$$dP = -MC\frac{dV}{V^2} = -P\frac{dV}{V} \qquad [14 \cdot 12]$$

Hence the bulk modulus of an ideal gas is the pressure itself,

$$B = P \qquad [14 \cdot 13]$$

This accords with the observation that tires or balloons become harder to deform as the pressure inside them is increased.

example 14 • 1 What is the pressure exerted at its base by a column of material of density ρ, height h, and cross-sectional area A?

solution The force on the base is due to the weight of material in the column,

$$F_g = mg = \rho(Ah)g$$

Hence the pressure on the base is

$$P = \frac{mg}{A} = \rho gh$$

If the material is a fluid, such as a gas or a liquid in a container, then at any depth h it also exerts the *same* pressure in all directions, and in particular against the sides of its container.

example 14 • 2 To study the effects of pressures greater than atmospheric pressure on a gas confined to a volume V, Boyle used a J tube, with the pressure measured in centimeters of mercury (cm Hg). If P_0 is the atmospheric pressure in standard units, what is the pressure P in the bulb of the J tube? What is the constant MC? What is the atmospheric pressure P_0' expressed in centimeters of mercury?

solution The pressure in the bulb is the sum of the atmospheric pressure P_0 and the pressure $P = \rho g h$ of the column of mercury:

$$P = P_0 + \rho g h = \frac{MC}{V}$$

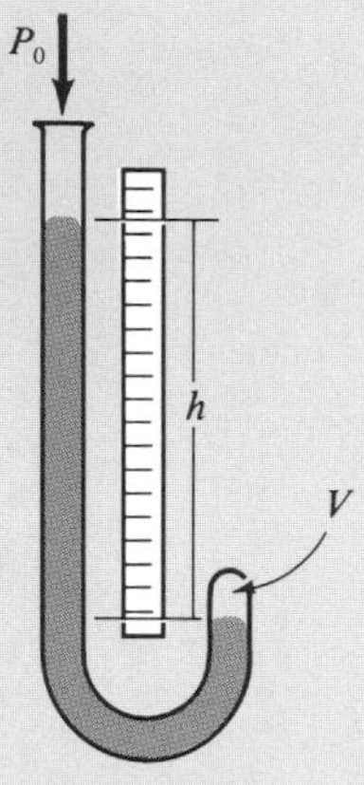

figure 14 • 6

Therefore

$$MC = (P_0 + \rho g h)V = \text{constant}$$

If we express ρg in cgs units, we can consider the pressure of a column of mercury 1 cm high to be ρg. Then

$$\left(\frac{P_0}{\rho g} + h\right)V = (P_0' + h)V = \frac{MC}{\rho g}$$

and the atmospheric pressure P_0' is

$$P_0' = \frac{P_0}{\rho g} = \frac{76 \text{ cm Hg} \times P_0}{1.013 \times 10^6 \text{ dynes/cm}^2} = 7.5 \times 10^{-5}\{P_0\} \text{ cm Hg}$$

where $\{P_0\}$ is the measure of P_0 in pascals.

14 • 3 elasticity of shape

A strain that consists of a change only in the shape of a body, without any change of volume, is called a *shear strain.* Since our everyday ideas of change of shape are essentially qualitative, we must first agree on a way of measuring the amount of shear existing at any point in a distorted body. Suppose the shape *abcd* in Fig. 14 . 7 represents a small rectangular element of a solid body. If the base *ab* is firmly fixed, a force *F* applied tangentially to the upper face will strain the element into the shape *abc'd'*, roughly in the same way a closed book can be pushed out of square by applying a tangential force to the top cover. Such a deformation, in which the adjacent layers of a body merely slip past one another, is obviously a shear, since only the configuration has been changed; the volume remains the same. As the measure of this shear strain, we shall take the ratio $\overline{dd'}/\overline{da}$, or $\tan \phi$. When the shear is small, as is usually the case, this ratio is approximately equal to the *shear angle* ϕ, expressed in radians.

If the shear tends to disappear when the forces that produce it cease to act, the material possesses *shear elasticity,* or *rigidity.* Imagine a horizontal plane drawn at any height in the element *abc'd'* in Fig. 14 . 7. This plane must have the same area *A* as the upper surface, and the portion of the element above this plane will exert the same tangential force *F* on the portion

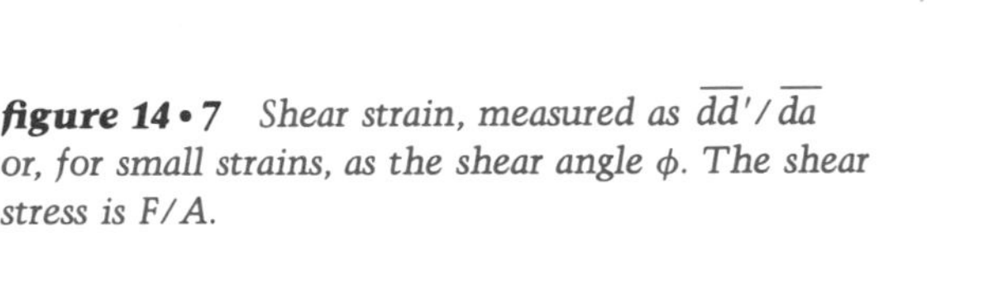

figure 14 • 7 *Shear strain, measured as $\overline{dd'}/\overline{da}$ or, for small strains, as the shear angle ϕ. The shear stress is F/A.*

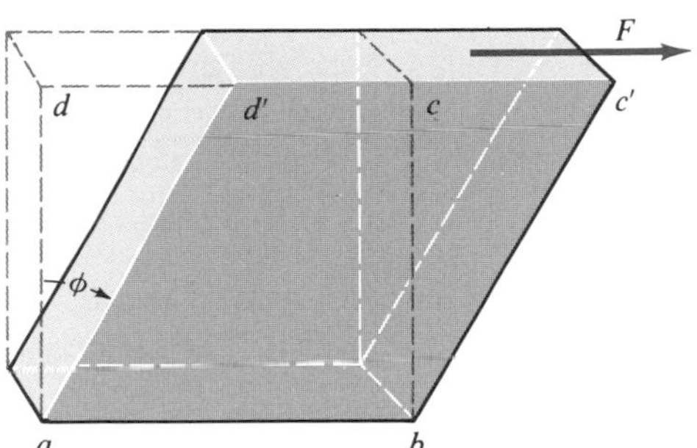

below it as is applied to the upper surface. The tangential force per unit area of any such horizontal surface is therefore $F_\parallel/A$, the ***shear stress.*** If we now apply the general Hooke's law to the material, then the shear stress must be proportional to the shear strain, or

$$\frac{F_\parallel}{A} = S\phi \qquad [14 \cdot 14]$$

where S is the ***shear modulus*** of the material (Table 14 . 1).

It is this elasticity of shape, or the ability to "sustain" a shear, which distinguishes solids from fluids, which flow freely. A body is called a *solid* if it possesses elasticity of both size and shape; it is called a *fluid* if it exhibits elasticity of size, but not of shape. Although a material is classed as a fluid if it does not permanently resist change of shape, this yielding to shear forces is found to occur at different *rates* in different fluids. Consequently, there is an intermediate class of materials, such as glass, pitch, and wax, that will act as solids when stressed only briefly, but gradually yield and flow, or *creep,* when stressed for a considerable period of time. A "true solid" will show no tendency to creep as long as it is not stressed beyond its proportional limit.

As an example, Fig. 14 . 8 illustrates the torsion of a thin-walled tube of mean radius R subjected to the net tangential shear force F of a couple. When the tube is twisted about the x axis by holding one end fixed and applying a torque $\tau = RF$ to the other end, a pure shear results, with point P displaced to P'. Imagine that the whole length of the tube is divided into thin rings of thickness ΔR and equal width Δx, as shown in Fig. 14 . 8*b*. If a

figure 14 • 8 *Torsion of a thin-walled tube, an example of pure shearing action*

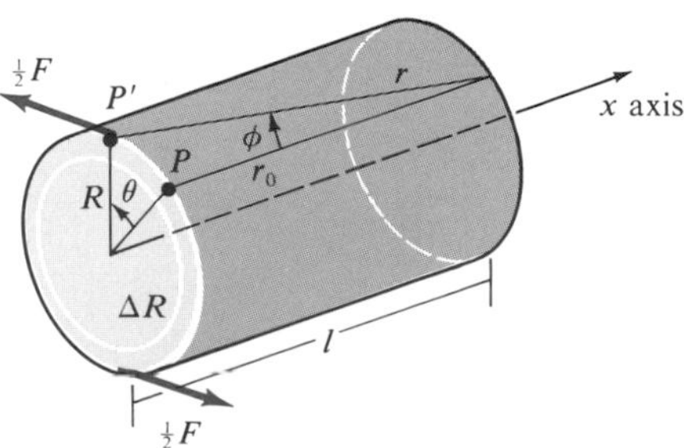

(a) Displacement of a point P to P^- by an applied torque $\tau = RF$

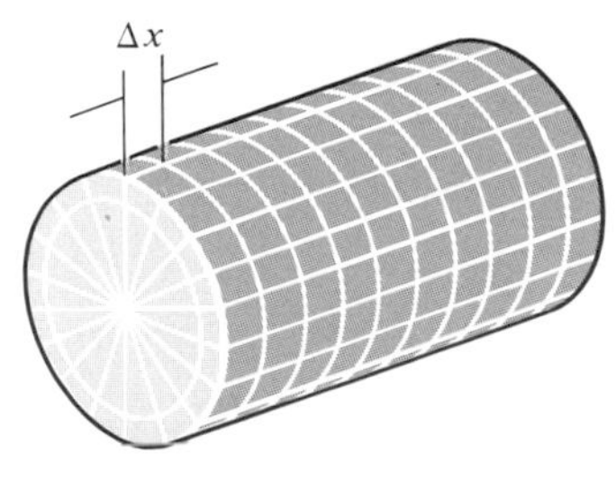

(b) Division of tube into equal volume elements

series of imaginary radial planes containing the x axis is drawn before the tube is twisted, they will divide each ring into very small, nearly rectangular volume elements. After the torsion, each of these elements will have a shear strain like that of the rectilinear element in Fig. 14 . 7.

In this case it is more difficult to measure ϕ directly than to measure θ, the angle through which the end of the tube is twisted. Point P is displaced through an arc of length $s = R\theta = l\phi$, where R is the mean radius of the tube and l is its length. The surface area to which $F_{\parallel}$ is applied is the end of the tube, $A = 2\pi R\,\Delta R$. Therefore Eq. [14 . 14] becomes

$$\frac{F}{A} = \frac{F}{2\pi R\,\Delta R} = S\phi = S\frac{R\theta}{l} \qquad [14 \cdot 15]$$

and the shear modulus S is given by

$$S = \frac{RFl}{2\pi R^3\,\Delta R\,\theta} = \frac{\tau l}{2\pi R^3\,\Delta R\,\theta} \qquad [14 \cdot 16]$$

Not only does this equation give an easily measured expression for S, but it will also allow us to derive the relationship between the total torque and the angular displacement θ of a solid cylinder, which is the prototype of the *torsion balance,* an extremely important tool in the development of physics.

Imagine a solid cylinder of radius R whose base is shown in Fig. 14 . 9 and which is twisted about its longitudinal axis. We may assume it to be made up of a very large number of infinitesimally thin tubes of radius r and thickness dr. The torque $d\tau$ needed to twist each tube through an angle θ is, from Eq. [14 . 16],

$$d\tau = 2\pi S\frac{\theta}{l}r^3\,dr \qquad [14 \cdot 17]$$

and integrating over all tubes $0 \leq r \leq R$, we find the applied torque needed to twist the solid cylinder through a displacement θ:

$$\tau = \frac{2\pi SR^4\,\theta}{4l} = \tau_0\,\theta \qquad [14 \cdot 18]$$

Note that τ is directly proportional to θ through the *torsion constant* τ_0:

$$\tau_0 = \frac{\pi SR^4}{2l} \qquad [14 \cdot 19]$$

This is the rotational analog of the spring, where the applied force is counterbalanced by a linear restoring force $F = -k_s\,\Delta x$, and provides us with a method for relating S to the frequency of torsional oscillations of a cylinder

figure 14 • 9 *Cross section of a solid cylinder*

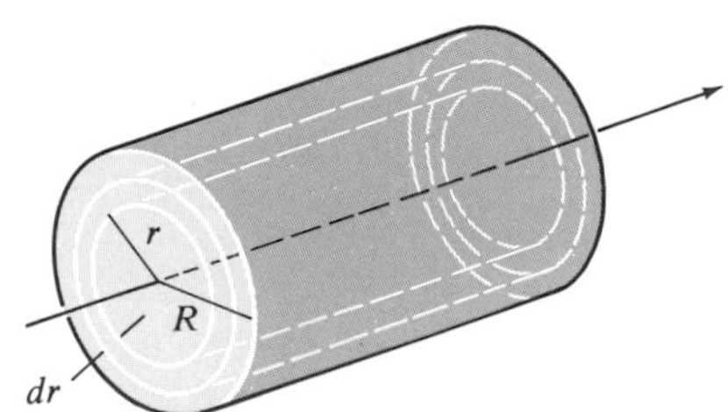

when it is twisted and released. Like the linear oscillations in a spring, these oscillations serve as a very sensitive balance for measuring torques, since the deforming couple can be applied at any distance R' from the central axis to produce a torque $\tau = R'F$. Hence, with a long lever arm, very small forces can be measured in terms of the angular displacement θ.

example 14 • 3 A solid shaft 2 m long, made of steel with shear modulus $S = 8 \times 10^{10}$ Pa, is to transmit 8 kW of power at an angular speed $\omega = 22$ rad/s. If the shaft is not to twist more than $\theta = 0.02$ rad during operation, what should its minimum radius be?

solution From Eq. [14 . 18],

$$R^4 = \frac{2l\tau}{S\pi\theta}$$

and since the power is given by $P = \tau\omega$,

$$R^4 = \frac{2lP}{S\pi\theta\omega} = \frac{2(2\text{ m} \times 8\text{ kW})}{(8 \times 10^{10}\text{ Pa})\pi(0.02\text{ rad})(22\text{ rad/s})}$$
$$= 29 \times 10^{-8}\text{ m}^4$$

The minimum radius is therefore $R = 0.023$ m, provided that this does not exceed the shear proportional limit for the steel.

14 • 4 *relationships of the elastic constants*

Of all possible elastic changes in an isotropic material, the only two that are independent are *volume strain* and *shear strain.* Other changes involve both of these. For instance, the stretching of a rod involves a change in shape as well as in volume. Instead of increasing or decreasing in all dimensions, the rod increases in length and decreases in cross-sectional area. This lateral, or *transverse*, contraction is obvious with a stretched rubber band. The change in the longitudinal direction is describable in terms of Young's modulus alone, but to take the transverse effect into account as well, we need two elastic constants. When a tensile or compressive force is applied to a rod, the longitudinal strain $\Delta l/l_0$ is accompanied by a transverse strain $\Delta d/d_0$, where d_0 is the initial diameter (Fig. 14 . 10). Experiment shows that, for an isotropic material strained within its proportional limit, the transverse strain is proportional to the longitudinal strain, or

$$\frac{\Delta d}{d_0} = -\sigma\frac{\Delta l}{l_0} \qquad [14 \cdot 20]$$

The proportionality constant σ is called the *Poisson ratio,* first noted by Poisson in 1828. Since an increase in length always produces a reduction in diameter, the minus sign in Eq. [14 . 20] makes σ a positive number. For most metals the value of σ lies between 0.25 and 0.45. For fluids it is, of course, meaningless.

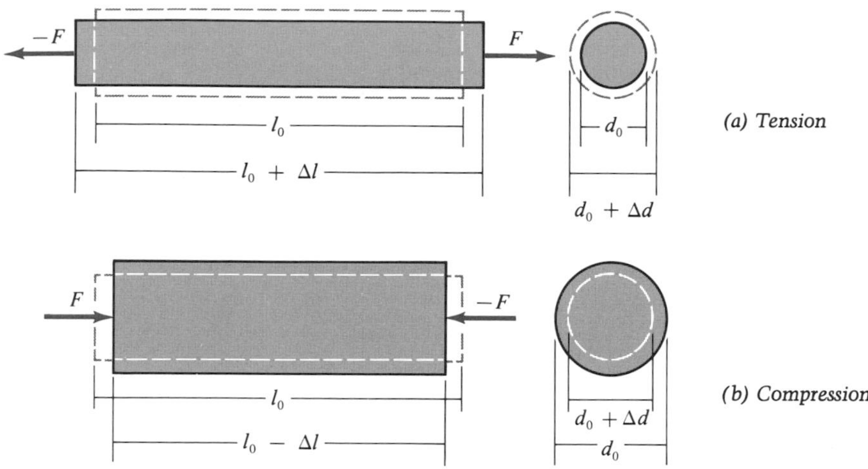

figure 14 • 10 *Transverse strain as a result of longitudinal strain for a rod under tension and under compression*

When a rod or bar is subjected to a longitudinal stress, it undergoes a fractional change in volume $\Delta V/V_0$. Consider the special case of a rectangular bar of length l_0, width d_0, and height h_0. When it is subjected to a stress over its cross-sectional area A, the fractional change in volume is

$$\frac{\Delta V}{V_0} = \frac{(l_0 + \Delta l)(d_0 + \Delta d)(h_0 + \Delta h)}{l_0 d_0 h_0} - 1$$

Since its changes in *shape* are related by

$$\frac{\Delta d}{d_0} = \frac{\Delta h}{h_0} = -\sigma \frac{\Delta l}{l_0}$$

if we neglect squares and cubes of the small quantity $\Delta l/l_0$, we have

$$\frac{\Delta V}{V_0} = \left(1 + \frac{\Delta l}{l_0}\right)\left(1 - \sigma \frac{\Delta l}{l_0}\right)^2 - 1 \doteq (1 - 2\sigma)\frac{\Delta l}{l_0} \qquad [14 \cdot 21]$$

However, for longitudinal stress, $F/A = Y \Delta l/l_0$, so that

$$\frac{\Delta V}{V_0} = \frac{1 - 2\sigma}{Y}\frac{F}{A} \qquad [14 \cdot 22]$$

To express the bulk modulus B for a solid material in terms of the Poisson ratio σ and Young's modulus Y, let a cubical block with faces of area A be subjected to a change ΔP in hydrostatic pressure, as in Fig. 14 . 5. The compressive forces $F = A\,\Delta P$ on each pair of opposite faces produce a fractional change in volume given by Eq. [14 . 22]. And since there are *three* pairs of faces, the total volume strain is three times that due to compression in one dimension:

$$\frac{\Delta V}{V_0} = -3\frac{1 - 2\sigma}{Y}\Delta P \qquad [14 \cdot 23]$$

From Eq. [14 . 11], $\Delta P = -B\,\Delta V/V_0$, so that the bulk modulus B is given by

$$B = \frac{Y}{3(1 - 2\sigma)} \qquad [14 \cdot 24]$$

This relation holds for an isotropic solid of any form whatever, since a body can always be viewed as made up of infinitesimal cubes. Clearly $\sigma < \frac{1}{2}$. A similar expression exists for the shear modulus:

$$S = \frac{Y}{2(1 + \sigma)} \qquad [14 \cdot 25]$$

Its derivation, which is somewhat lengthy, also involves consideration of a cubical block, but with the simultaneous application of tensile forces to one pair of opposite faces and compressive forces to another pair.

Of the various elastic constants, Y and S are the easiest to determine. B and σ can then be found from the above equations. Note that the elastic constants of metals as given in Table 14 . 1 do not agree precisely with these equations. This is because metals are not isotropic, owing to their crystalline structure. They do agree, however, with the less stringent conditions that

$$\tfrac{1}{3}Y < S < \tfrac{1}{2}Y \quad \text{and} \quad \sigma \doteq \tfrac{1}{3},\ B \doteq Y \qquad [14 \cdot 26]$$

which may serve as a useful rule of thumb. The exact theory of such substances is more complex and may involve as many as nine elastic constants and the interrelations between them.

14 • 5 *atomic-molecular model of elasticity*

Our treatment of elastic properties has so far related entirely to bodies large enough to be assumed continuous in structure. Moreover, the laws we have been dealing with provide quantitative descriptions of macroscopic events, but we have not considered the underlying explanations of these events. In this section we shall see how deductive reasoning based on the atomic-molecular model of matter can lead to an expression for Young's modulus. That is, we shall now relate the microscopic structure of matter to its macroscopic elastic properties.

Physicists frequently explain phenomena by formulating a conceptual model which displays properties analogous to those to be explained. It is important to be aware of the use of conceptual models and to guard against mistaking the model for the reality it represents. No model, however venerable, can ever be immune from scrutiny. At best, it shows only that some mechanism can be imagined which accounts for all known observations of a phenomenon.

Solids are generally classified as *amorphous* or *crystalline*. Some amorphous solids, such as glass, consist of collections of molecules in completely random arrangements, like a supercooled fluid, which is too cold to flow but has not frozen into a true solid structure. Other amorphous substances, such as long-chain molecules (polymers) of rubber, glue, or other organic materials, show ordered structure over regions which contain many molecules, but

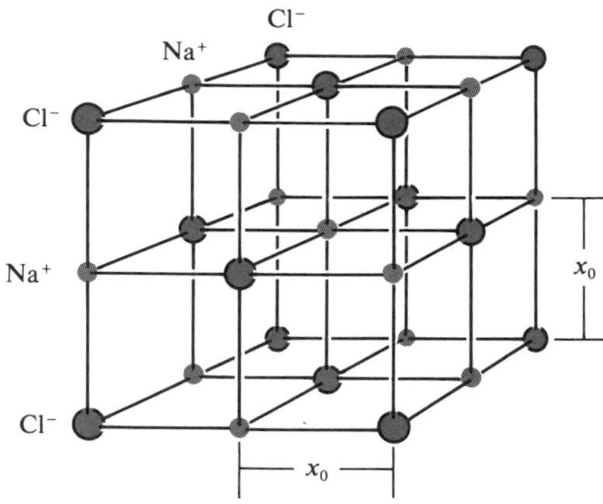

figure 14 • 11 *Simple cubic crystal lattice of common salt, NaCl*

are so small that the substance appears to consist of disordered aggregates of small, highly ordered regions.

Crystalline solids are highly ordered symmetric structures of molecules which may be classified, first, according to the nature and strength of the forces, or *bonds*, that bind the molecules together, and second, according to the types of patterned structures that occur. The study of crystals and their properties is highly complex and forms the bulk of the branch of physics known as *solid-state physics*. Since we are interested only in an explanation of the basic elastic properties of structural materials, let us concentrate on the simplest possible crystal, the simple cubic lattice, of which common table salt (NaCl) is a good example (Fig. 14 . 11).

The atoms within a crystal retain their average positions under some set of complicated interatomic forces. If the situation is stable, however, we can be sure that the potential energy of each atom at its equilibrium distance from all other atoms is a minimum (see Sec. 8 . 3). A typical atom or molecule in the crystal possesses no "knowledge" of the total extent of the crystal; it responds only to the potential well, which is the resultant of the individual forces on it due to all the other particles in the crystal.

Figure 14 . 12 illustrates this situation for motion in the x direction: the particle has some thermal kinetic energy K_0, so that it oscillates about its equilibrium position $x = 0$. If the crystal were heated sufficiently, the kinetic energy would rise to K_1 and the particle would be free; that is, the

figure 14 • 12 *Potential well for motion in the x direction for a particle in a crystal lattice*

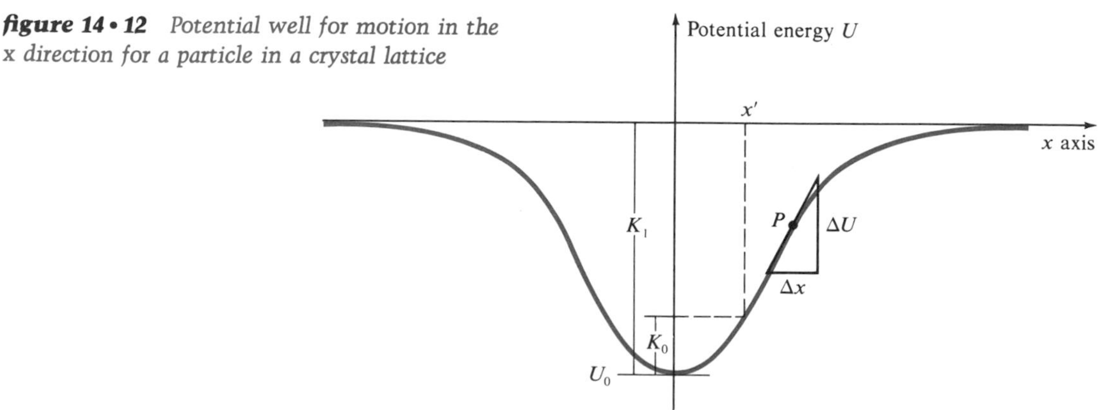

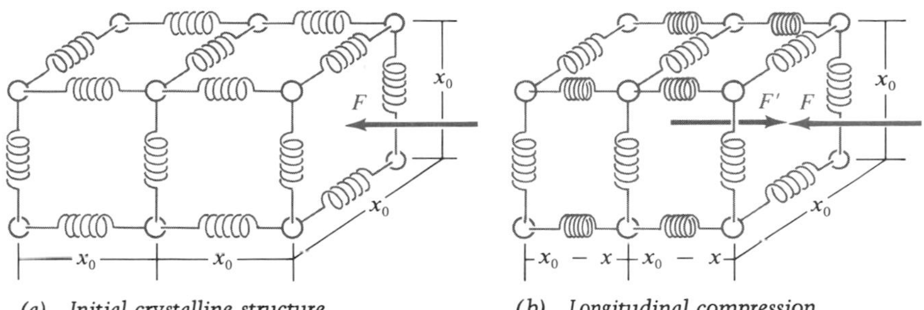

(a) Initial crystalline structure

(b) Longitudinal compression

figure 14 • 13 *Bedspring model of simple longitudinal deformation of a cubic crystal along its x axis*

crystal would melt or vaporize. Or one might pull the particle out of its potential well by brute force—that is, rupture the crystal. This could be done by applying a force f_{max} to each particle in a cross-sectional plane equal to the maximum slope of the potential-energy curve which occurs at the inflection point P.

For forces much smaller than f_{max}, we can approximate the potential well near equilibrium by a parabola, as in Eq. [8 . 15]

$$U(x) \doteq U_0 + \tfrac{1}{2}kx^2 \qquad [14 \cdot 27]$$

so that the Hooke's law restoring force which keeps the particle in its equilibrium position is $f = -kx$. This law of force holds out to approximately $x = \pm x'$, which may be recognized as the proportional limit of the material. Beyond this point the force remains approximately constant at f_{max} until the elastic limit is exceeded and the molecule is steadily drawn out of the crystal. The internal forces then go to zero as the molecule separates from its neighbors.

The Hooke's law force for small oscillations about an equilibrium point gives rise to the "bedspring" model of the crystal shown in Fig. 14 . 13. In this model the solid behaves like an array of particles connected by springs, which represent the internal forces that hold the crystal together. Now let us see how Young's modulus Y can be related to the structure of such a crystal and the strength of its intermolecular forces.

Suppose the crystalline structure in Fig. 14 . 13 is part of the cross-sectional area A of a metal bar of length l_0. Each atom in the structure is separated by a distance x_0, and the number of atoms per unit volume is $n = 1/x_0^3$. Then the number of atoms in a single linear chain is $n_l = l_0/x_0$ and the number of chains over area A is $n_{ch} = A/x_0^2$. The average applied force per chain is therefore

$$\bar{f} = \frac{F}{n_{ch}} = kx \qquad [14 \cdot 28]$$

transmitted equally by action and reaction to each atom in the chain (ignoring any effects of lateral forces). The change in length of each chain is then

$$\Delta l = n_l x = \frac{l_0}{x_0}\frac{\bar{f}}{k} = \frac{F}{A}\frac{l_0 x_0}{k} \qquad [14 \cdot 29]$$

If we rewrite this as

$$\frac{F}{A} = \frac{k}{x_0}\frac{\Delta l}{l_0} \qquad [14 \cdot 30]$$

we can immediately recognize Young's modulus for this crystal as

$$Y = \frac{k}{x_0}$$

We now have a relationship between a measurable macroscopic property of the material and its internal structure and intermolecular forces (at least, for elastic deformations). If we know the density of the material and the mass of its atoms (its atomic weight), we can compute the initial separation distance—typically in the range of 1 to 10 Å—and so compute k. There are also other ways of measuring k, and if the results do not agree, we adjust the model until it works.

PROBLEMS

14 • 1 Hooke's law

14 • 1 A wire 80 cm long and 0.3 cm in diameter is stretched 0.3 mm by a force of 20 N. If another wire of the same material, temperature, and previous history is 180 cm long and 8 mm in diameter, what force is required to stretch it to a length of 180.1 cm?

14 • 2 A thin wire of length l_0, Young's modulus Y, and cross-sectional area A has a heavy mass m attached at one end. If the mass is swung in a *horizontal* circle of radius R with angular speed ω, what is the strain in the wire. (Assume the mass of the wire is negligible.)

14 • 3 A copper wire 31 cm long and 0.5 mm in diameter is joined to a drawn brass wire 108 cm long and 1 mm in diameter. If a certain stretching force produces an elongation of 0.5 mm in the whole wire and $Y = 12 \times 10^{10}$ Pa, what is the elongation of each part?

14 • 4 A mild steel wire 4 m long and 1 mm in diameter is passed over a light pulley and weights of 30 and 40 kg, respectively, are attached to its ends. The weights are supported so that the system is in static equilibrium. When the support is removed, how much does the wire change in length?

14 • 5 The elasticity of a rubber cord of length l_0 is such that a tensile force F applied to each end produces a longitudinal strain of unit amount. Two weights, each of magnitude F_g, are fastened to the cord, one at an end and one in the middle, and the cord with the weights is then lifted from the ground by its free end. What is the least amount of work that will lift both weights from the ground?

14 • 6 A wire consists of a mild steel core 1.3 cm in diameter, to which is fused an outer shell of copper, $Y = 12 \times 10^{10}$ Pa, 0.26 cm thick. A tensile force of 9000 N is applied uniformly to each end of the wire. If the resulting strain is the same in both the steel and the copper, what is the force on the steel core?

14•7 A loaded elevator with a total mass of 2000 kg is supported by a cable 3.5 cm^2 in cross section. The cable material has an elastic limit of 2.5×10^8 Pa, and for this material $Y = 2 \times 10^{10}$ Pa. It is specified that the stress in the cable shall never exceed 0.3 of the elastic limit. (*a*) Find the stress in the cable when the elevator is at rest. (*b*) What is the largest allowable upward acceleration? (*c*) The shortest allowable stopping distance when the velocity of the elevator is 15 m/s downward?

14•8 A heavy cable of initial length l_0 and cross-sectional area A is of uniform density ρ and has a Young's modulus Y. The cable hangs vertically and supports a load F_g at its lower end. The tensile force at any given point in the cable evidently is the sum of the load F_g and the weight of the part of the cable that is below this point. Assuming that the *average* tensile force in the cable acts on the entire length l_0, find the resulting elongation.

14•9 If the stress in a rod does not exceed the elastic limit, we can assign an *elastic energy density,* or elastic potential energy per unit volume, to the deformation. Show that this energy is

$$U_s = \tfrac{1}{2}(\text{stress} \times \text{strain})$$

14•10 Rework Prob. 14 . 7, this time taking into account the weight of the cable when it is at or near its maximum length of 150 m. The density of the cable material is 7.8×10^3 kg/m^3. If the maximum load is exceeded, will the cable break near its top or near the elevator?

14•11 The elastic energy density acquired by a material that has been strained to its elastic limit is called the *modulus of resilience* of the material. (*a*) Prove that, for a longitudinal deformation, this modulus is directly proportional to the square of the elastic limit and inversely proportional to Young's modulus. (*b*) Show that it is represented on a stress-diagram by the triangle included between the stress-strain curve, the axis of strain, and the ordinate representing the elastic limit. (*c*) If Young's modulus for a certain kind of steel is 2×10^{11} Pa and the elastic limit is 4.3×10^8 Pa, what is the modulus of resilience for this steel?

14•2 volume elasticity

14•12 At ocean depths of about 10 km, the pressure is approximately 1 kilobar. (*a*) If a piece of mild steel sinks to this depth, how much is its density changed? (*b*) What is the density of sea water at this depth if its density at the surface is 1.04 g/cm^3?

14•13 A steel storage tank with a capacity of 60 liters contains oxygen under a gage pressure of 140 Pa. What volume will the oxygen occupy if it is allowed to expand at constant temperature until its gage pressure is zero? (*Gage pressure* is the difference between the actual pressure in a container and that of the outside atmosphere.)

14•14 A glass bulb of known volume V is weighed when filled with dry air at atmospheric pressure P_0 and is weighed again after enough air is pumped out to reduce the pressure in it an observed amount ΔP. If Δm is the mass of air removed, as revealed by the difference between the two weighings, show that the density of air at atmospheric pressure is given by $\rho = P_0 \, \Delta m / V \, \Delta P$.

14 • 15 The space above the mercury in a certain barometer tube contains some air. When the volume of this space is 10 cm^3, the barometer indicates a pressure of 70 cm Hg. When the space is reduced to a volume of 5 cm^3 by pushing the barometer tube down into the cistern of mercury, the pressure reading is 69.5 cm Hg. What would the original barometer reading have been if the space above the mercury had not contained air?

14 • 16 A glass tube 60 cm long is open at one end, and the inside of the tube is coated with a soluble pigment. After a sea sounding, in which the tube is lowered open end down, the pigment was found to be dissolved to within 5 cm of the top. If the average density of sea water is 1.03×10^3 kg/m^3, how deep is the water in the location sounded?

14 • 17 (*a*) Prove that when a body is subjected to a hydrostatic pressure that does not exceed the volume elastic limit for the material, the body acquires an elastic energy density equal to one-half the product of the bulk modulus B and the square of the volume strain. (*b*) Show that if the energy density is plotted as a function of the volume strain, the resulting curve will be a parabola with vertex at the origin and focus at $(0,\frac{1}{2}\kappa)$, where κ is the compressibility.

14 • 18 A solid of mass m, density ρ, and compressibility κ has its pressure raised under constant temperature from a pressure P_1 to a new pressure P_2. Find the work done on the solid.

14 • 3 shape elasticity

14 • 19 A cube of gelatin 30 cm on a side has one face held while a tangential force of 1 N is applied to the opposite face. The surface to which this force is applied is displaced 1 cm. (*a*) What is the shear stress? (*b*) What is the shear strain? (*c*) What is the shear modulus S?

14 • 20 Prove that when an elastic body is subjected to a pure shear stress that does not exceed the elastic limit of the material for shear, the elastic energy density of the body is equal to one-half the product of the shear stress and the shear strain.

14 • 21 A rod that is 100 cm long and 1 cm in diameter is gripped at one end and twisted through an angle of 1°. If the same force is applied to the circumference of a rod of the same material that is 80 cm long and 2 cm in diameter, what is the resulting twist?

14 • 22 The steel propeller shaft of a ship is designed to be 30 ft long and to be driven by a 1200-hp engine. What is the minimum diameter of the shaft if the twist is not to exceed 1° when the speed of the shaft is 200 rpm?

14 • 23 A certain wire 50.1 cm long and 2.732 mm in diameter has a torsion constant $\tau_0 = 0.721$ in mks units. Find the torque required to twist a wire of the same material, but 4 cm long and 1.000 mm in diameter, through an angle of 90.0°.

14 • 24 Torques of the same magnitude are applied to a solid glass rod 100 cm long and 2 cm in diameter and to a hollow glass tube having the same length and mean diameter and a wall thickness of 0.1 cm. Compare the twists produced.

14 • 25 The torsion balance below consists of a 40-cm crossbar with lead balls 2 cm in diameter at each end. The bar is suspended by 100 cm of silver wire which has a diameter of 0.5 mm. When two larger lead balls, each 30 cm in diameter ($\rho = 11.4 \text{ g/cm}^3$), are brought very close to the small balls from opposite sides, their gravitational attractions tend to turn the bar in the same direction. How much is the silver wire twisted?

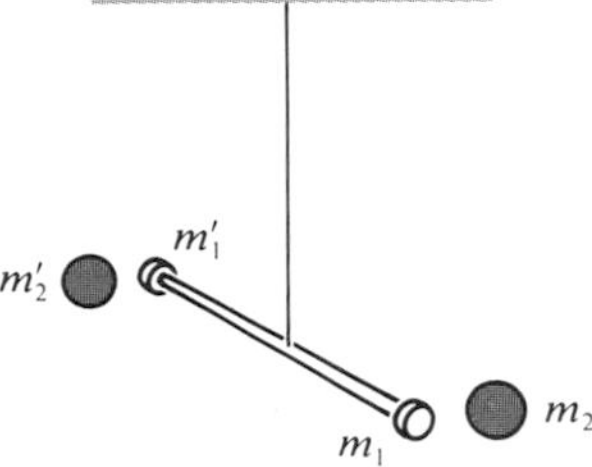

14 • 26 Experiment shows that the shear modulus of hard-drawn silver at a centigrade temperature of τ is $S_\tau = S_{15}[1 - 4.8 \times 10^{-4}(\tau - 15°\text{C})]$, where S_{15} is the shear modulus at 15°C. In Prob. 14 . 25, what would be the angle of twist if the temperature of the wire were 30°C?

14 • 27 (*a*) Develop an expression for the torsion constant of a hollow cylinder as a function of its inner radius R_0, outer radius R_1, length l, and shear modulus S. (*b*) What would be the radius of a solid cylinder of the same length and material and having the same torsion constant? (*c*) What would be the saving in mass if the hollow cylinder rather than the solid one were used as a driveshaft?

14 • 4 relationships of elastic constants

14 • 28 A tensile force of 2800 N is applied to each end of a horizontal bar 1.5 m long, 1.6 cm wide, and 1 cm high. Young's modulus and the Poisson ratio for the material in the bar are $Y = 2 \times 10^6$ Pa and $\sigma = 0.3$. (*a*) Find the transverse strain in the bar. (*b*) What are the fractional changes in width and height? (*c*) What is the increase in volume? (*d*) What is the potential energy acquired by the bar?

14 • 29 (*a*) Show that the Poisson ratio is given by

$$\sigma = \frac{3B - 2S}{2(3B + S)}$$

(*b*) Show how it follows from this equation that the Poisson ratio must lie between -1 and $\frac{1}{2}$. (*c*) Experiments show that bars or rods subjected to tensile forces (positive values of F) always increase in volume, and when they are subjected to compressive forces (F negative), they always decrease in volume. Does this lend support to the assertion that there is no material for which $\sigma \geq \frac{1}{2}$?

14 • 30 A materials handbook lists these data for rolled sheet aluminum:

Young's modulus, 7×10^{10} Pa
Elastic limit for tension, 7.2×10^7 Pa
Poisson ratio, 0.33
Ultimate tensile stress, 14×10^7 Pa
Allowable tensile stress, 0.4 of ultimate tensile stress

The allowable tensile stress is the maximum stress considered to be safe when this material is used in structures subject to constant, known tensile loads. A strip of this aluminum 76 cm long, 2.5 cm wide, and 0.8 mm thick is gradually stretched until the tensile stress in it reaches the allowable limit. Compute (*a*) the change in length, (*b*) the change in volume, (*c*) the work done, and (*d*) the gain in elastic energy density.

14 • 5 atomic molecular model

14 • 31 The potential well for an atom in a cubic lattice with atomic separation x_0 is given by

$$U(x) = \frac{k}{d^2 + x^2} \qquad k < 0,\ d > 0$$

(*a*) Find the ultimate tensile stress the crystal can sustain. (*b*) Determine the exact stress-strain relationship for longitudinal deformations. (*c*) Find the value of Y for $x \ll d$ such that Hooke's law holds approximately (below the proportional limit). (*d*) For what value of x/d is the proportional relationship in error by 20 percent?

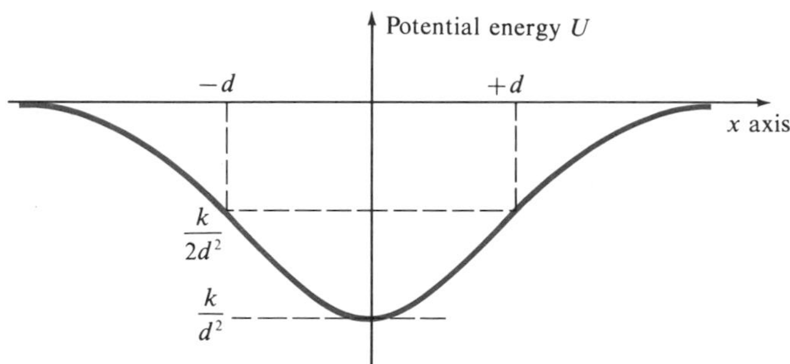

14 • 32 Copper has a density of 8.9 g/cm^3 and each atom weighs 1.05×10^{-22} g. (*a*) What is the separation distance between atoms in a cubic copper crystal? (*b*) What is the oscillation frequency of an atom undergoing simple harmonic motion in its potential well?

answers

14 • 1 $F = 211$ N
14 • 2 $\Delta l/l = m\omega^2 R/AY$
14 • 3 $\Delta l = 0.27$ mm for brass, $\Delta l = 0.23$ mm for copper
14 • 4 $\Delta l = 1.0$ mm
14 • 5 $W = \frac{7}{4} l_0 F^2/F_g$
14 • 6 $F = 5812$ N
14 • 7 (*a*) $F/A = 5.6 \times 10^7$ Pa; (*b*) $a = 3.33$ m/s^2; (*c*) $\Delta y = 33.8$ m
14 • 8 $\Delta l = (l_0/Y)(F_g/A + \frac{1}{2}\rho g l_0)$
14 • 10 (*a*) $F/A = 6.75 \times 10^7$ Pa; (*b*) $a = 1.32$ m/s^2; (*c*) $\Delta y = 85.3$ m
14 • 11 (*c*) Modulus of resilience $=$ 462,250 J/m^3

14 • 12 (*a*) 0.062 percent; (*b*) $\rho = 1.105\ \mathrm{g/cm^3}$
14 • 13 $V = 889$ liters
14 • 15 $P = 705$ mm Hg
14 • 16 Depth $= 110$ m
14 • 18 $W = \frac{1}{2}(\kappa m/\rho)(P_2^2 - P_1^2)$
14 • 19 (*a*) $F/A = 11.1$ Pa;
(*b*) $\phi = 0.0333$ rad; (*c*) $S = 333$ Pa
14 • 21 $\theta = 0.1°$
14 • 22 $2R = 9.1$ inches
14 • 23 $\tau = 0.255\ \mathrm{N \cdot m}$
14 • 24 $\theta_{\mathrm{tube}} = 2.5\theta_{\mathrm{rod}}$
14 • 25 $\theta = 0.00422°$
14 • 26 $\theta = 0.00424°$
14 • 27 (*a*) $\tau_0 = (\frac{1}{2}\pi S/l)(R_1^4 - R_0^4)$;
(*b*) $R = (R_1^4 - R_0^4)^{1/4}$;
(*c*) saving $=$
$100[1 - \sqrt{(R_1^2 - R_0^2)/(R_1^2 + R_0^2)}]$ percent
14 • 28 (*a*) $\Delta d/d_0 = -2.625 \times 10^{-4}$;
(*b*) $\Delta d = -4.2 \times 10^{-4}$ cm,
$\Delta h = -2.625 \times 10^{-4}$ cm;
(*c*) $\Delta V = 8.4 \times 10^{-4}\ \mathrm{cm^3}$; (*d*) $\Delta U_s = 1.84$ J
14 • 30 (*a*) $\Delta l = 0.608$ mm;
(*b*) $\Delta V = 0.0041\ \mathrm{cm^3}$; (*c*) $W = 0.341$ J;
(*d*) $\Delta U_s = 22{,}400\ \mathrm{J/m^3}$
14 • 31 (*a*) $F/A = 0.65\ k/d^3x_0^2$;
(*b*) $F/A = 2kx/x_0^2(d^2 + x^2)^2$,
where $x = x_0\,\Delta l/l$;
(*c*) $Y = 2k/x_0 d^4$; (*d*) $x = 0.344\ d$
14 • 32 (*a*) $x_0 = 2.276 \times 10^{-8}$ cm;
(*b*) $\nu = 8.1 \times 10^{10}$ Hz

CHAPTER FIFTEEN

mechanics of fluids

If the tube AB *of some convenient capacity, which is not definitely fixed, is assumed to be always full of water up to the level* A, *and if it is pierced at* B *with a small orifice, we assume that the water which issues at* B *will have the same speed that any*

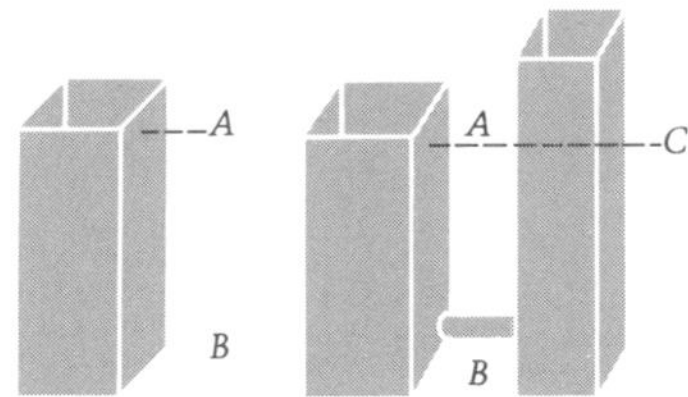

heavy body would have if it were to fall from A *to* B. . . . *[If] another tube is connected with the opening* B *and accurately fitted to it, the water flowing from* B *into the tube* C *has such a force that it lifts itself to the level* A. *Thus it seems very probable that even when the water issues freely from* B, *it will have a force that will carry it up to the horizontal line drawn through* A; *or what is the same thing, that it will have the same speed as that of any body or of one drop of liquid, falling freely from* A *to* B. . . . *These things being supposed, we shall demonstrate certain theorems about liquids issuing from orifices that seem in a wonderful way to fit in with the theory of projectiles.*

Evangelista Torricelli, Trattato del moto dei gravi, *1641*

Fluids, substances which cannot permanently sustain a shearing stress, can be divided broadly into two categories: liquids, which are essentially incompressible (have constant density), and gases, which are compressible. The *statics of fluids* is a branch of fluid mechanics which deals with fluids at rest and with bodies at rest in these fluids. The term *statics* here is more restricted than in the mechanics of solids, where it may also pertain to bodies in uniform motion without acceleration.

The *dynamics of fluids* is concerned with fluids in motion and with the forces acting on bodies moving through fluids. This field is subdivided into *hydrodynamics* and *gas dynamics,* in which liquids and gases, respectively, are the moving fluids. Fluid dynamics is one of the most vigorous fields of research at present. It encompasses, among other things, the motions of the atmosphere, the oceans, and the fluids within the human body. It is still the spiritual home of some of the most sophisticated areas of applied mathematics, and it has given great impetus to the development of techniques for high-speed computation. In these days of rapid motion below, on, and above the seas, fluid dynamics has acquired major importance in many areas. Even in space, where there may be only 1 to 10 electrically charged particles *(ions)* per cubic centimeter, very weak magnetic fields existing on a vast scale can provide the "glue" which allows the assemblage of particles to behave as a *plasma,* a fluid of charged particles. Many of the concepts of conventional fluid dynamics have proved applicable to plasmas, sometimes called "the fourth state of matter," even on a scale ten times the radius of the earth!

What characterizes a fluid as a distinct system, rather than an assemblage of independent particles, is its ability to transmit stresses. In an incompressible liquid, this is accomplished by the short-range intermolecular forces, which, as Pascal showed in 1653, transmit pressures very effectively in all directions. In a compressible gas, interactions among parts of the fluid are accomplished by means of relatively rare collisions between moving particles. In a plasma, magnetic fields and relatively weak long-range electrical interactions among charged particles provide the basis for the fluid properties of the plasma as a whole. *Viscosity* is the stickiness of a fluid, the property which enables it to transmit stresses in a direction perpendicular to the direction of the applied stress.

In this chapter we shall begin by considering the laws of pressure as they apply to fluids at rest, particularly the buoyant force these pressures exert on a body immersed in the fluid (Archimedes' principle). We shall then develop those laws which apply to the steady flow of an incompressible *nonviscous* liquid. This simplification allows us to omit time as a significant variable, treat density as constant, and ignore viscosity altogether. Afterwards we shall consider viscous effects in fluids and the problems of fluid flow.

15 · 1 *fluid pressure*

Fluids, unlike solids, cannot permanently sustain a shearing stress. Therefore the force exerted by a fluid at rest can have no component parallel to the surface on which it acts:

> The direction of the force exerted by a fluid at rest upon any element of surface is perpendicular to the surface.

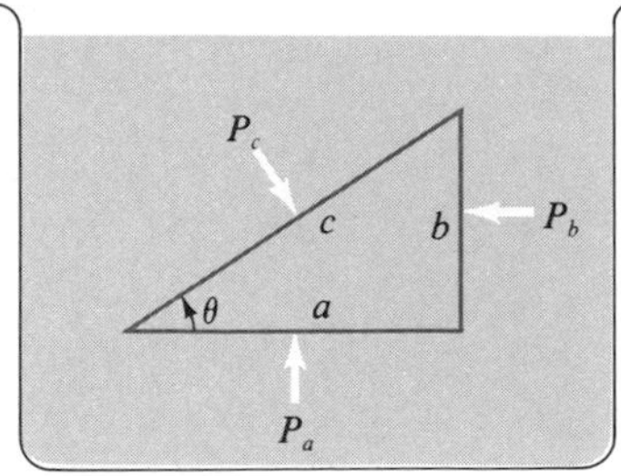

figure 15 • 1 *Isotropy of hydrostatic pressure for static equilibrium* $P_a = P_b = P_c$

If there were a parallel component of force, the fluid would start to flow and we would no longer be dealing with static conditions. The pressure, or force per unit area, in a fluid at rest is a *hydrostatic pressure:*

At a given point in a fluid at rest, the pressure has the same magnitude regardless of the orientation of the surface on which it acts.

This principle can easily be proved by applying the conditions for static equilibrium to any very small portion of the fluid. Consider the small right-triangular prism of unit thickness shown in Fig. 15 . 1. Since the element of fluid is at rest, the forces on the different surfaces of the triangle must cancel each other:

$$P_c c \cos\theta = P_a a \qquad P_c c \sin\theta = P_b b \tag{15 . 1}$$

Therefore the magnitudes of the pressures are

$$P_a = P_c \frac{c}{a} \cos\theta = P_c \qquad P_b = P_c \frac{c}{b} \sin\theta = P_c$$

so that the pressure is the same in all directions, $P_a = P_b = P_c$.

In discussing the elastic properties of fluids in Sec. 14 . 2, for the most part we ignored the effects of gravitational forces on bodies. Suppose, however, that gravity cannot be ignored, as in the situation in Fig. 15 . 2. The fluid in the thin hollow cylinder on the left is subjected to horizontal pressures P_1 and P_2 at its two ends. Since the forces on the curved side surfaces of the cylinder have no components in the direction of the length of the cylinder, the conditions for equilibrium give $P_1 = P_2$:

The pressure in a fluid at rest under gravity is the same at all points in the same horizontal plane.

The cylinder on the right is in equilibrium in a vertical position. For

figure 15 • 2 *Pressure in a fluid at rest under gravity*

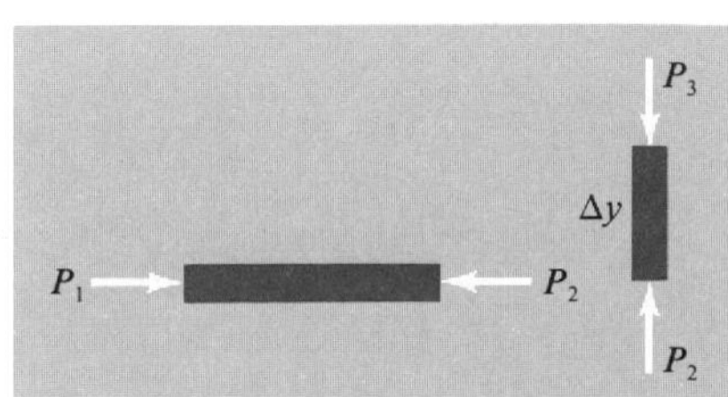

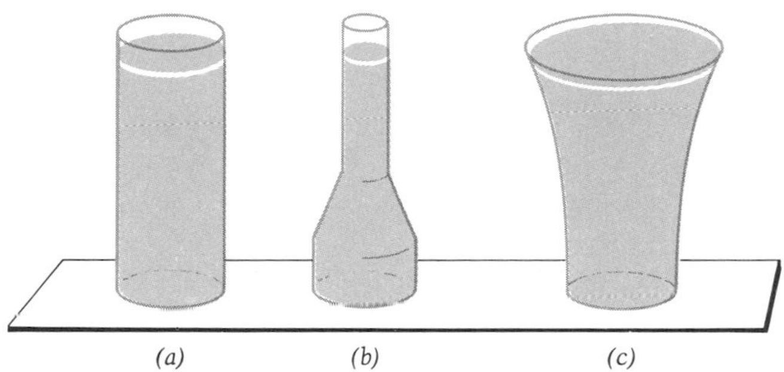

figure 15 • 3 *The hydrostatic paradox. In vessels of equal base area A, liquid of the same depth exerts the same force* $A\rho gh$*, although this force may be (a) equal to, (b) greater than, or (c) less than the weight of the liquid present.*

simplicity, let us assume that the fluid in the cylinder is of average density ρ and that the height Δy of the cylinder is small enough that the gravitational acceleration g does not vary with height. If A is the surface area of the lower and upper faces, then the upward force $F_y = (P_3 - P_2)A$ must equal the weight $F_g = \rho g A\ \Delta y$ of the fluid in the cylinder, or, cancelling A, we can write

$$P_3 - P_2 = \rho g\ \Delta y \qquad [15 \cdot 2]$$

In other words:

> The difference in pressure between two points at different levels in a fluid at rest under gravity is equal to the weight of a column of the fluid of unit cross-sectional area reaching vertically from one level to the other.

This principle holds for liquids over fairly large vertical distances Δy, since a liquid is so nearly incompressible as to have almost the same density at any depth. Even at ocean depths the density change is never so great that changes in g need to be taken into account. The fact that ρ and g can be treated as constants under ordinary conditions leads to the so-called "hydrostatic paradox" shown in Fig. 15 . 3. Regardless of the shape of the container, and therefore of the amount of liquid it contains, in vessels of the same base area A a liquid of the same depth exerts the same gravitational force $\rho g A\ \Delta y$. If a fluid completely fills a *closed* container and the pressure at any point in the fluid is increased, there will be a corresponding increase of pressure at every other point (Pascal's law).

The principle for finding the *buoyant force* that a fluid exerts on a body at rest was first stated in the third century by Archimedes:

> The buoyant force on a body immersed in a fluid is equal in magnitude and opposite in direction to the weight of the fluid displaced by the body.

In 1586 Stevin demonstrated this theoretically by means of his "*solidification principle.*" Imagine that in a stationary fluid of density ρ_f we can isolate some volume V of the fluid by an impermeable massless bounding surface of

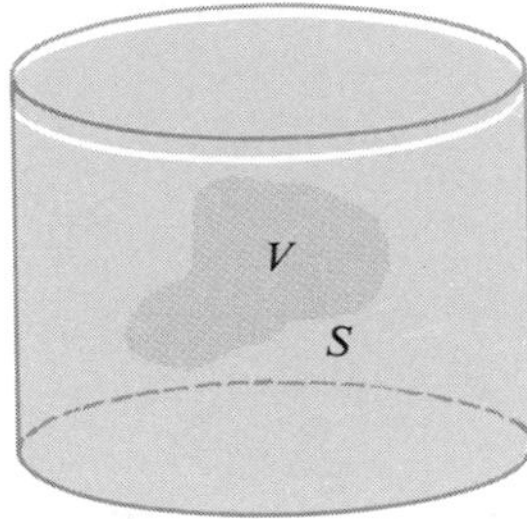

figure 15 • 4 *Stevin's proof of Archimedes' principle*

area S (see Fig. 15 . 4). Since the mass m_f of the fluid within S is in equilibrium, its weight $m_f g = \rho_f V g$ must be balanced by the forces exerted on it by the surrounding fluid. However, these forces depend only on the conditions existing *outside* of S; hence *any* object with the same surface area S must be buoyed up by forces whose resultant is equal to the weight of the displaced fluid. Moreover, since the displaced fluid was in equilibrium, the buoyant force must pass upward through its center of mass. The same geometric point in an immersed body is called the *center of buoyancy* of the body. It does *not* necessarily coincide with the body's center of mass.

To see why this distinction is important, consider the situation illustrated in Fig. 15 . 5. Unless the center of buoyancy B and the center of mass C are in the same vertical line, the buoyant and gravitional forces will exert a torque on the body. The subsequent stability or instability of the body will depend on whether the torque tends to restore it to the equilibrium position, or to rotate it into a new position, as in Fig. 15 . 5*c*.

Archimedes' principle furnishes a convenient and accurate method for determining the *densities* of irregular solids and liquids. Suppose a solid object of mass m, volume V, and density ρ requires a weight of mass m_a to balance it on an equal-arm balance in air (the air also exerts buoyant forces, but these are negligible). When the object is suspended by a light thread and immersed in a liquid of density ρ_f, a weight of mass m_l is needed to balance it. Since the object is buoyed up by the fluid and with a force $F_y = \rho_f V g$, we have

$$mg = m_l g + \rho_f V g \qquad \text{or} \qquad m = m_l + \rho_f V$$

figure 15 • 5 *In a surface vessel the center of mass C is typically higher than the center of buoyancy B. When the vessel rolls, the center of buoyancy shifts to some point B′ that is not on the vertical line through C. The point at which a vertical line through B′ intersects the line BC is called the metacenter.*

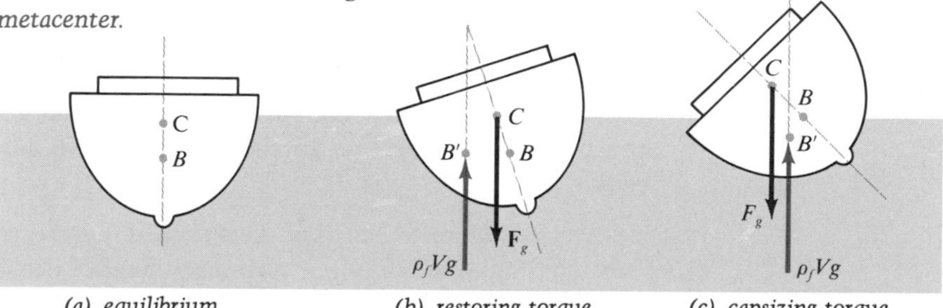

If we substitute $V = m/\rho$ and solve for the density ρ of the object, then, since $m = m_a$, we obtain the density in terms of measurable quantities:

$$\rho = \frac{m_a \rho_f}{m_a - m_l} = \frac{m_a \rho_f}{\Delta m} \qquad [15 \cdot 3]$$

where $\Delta m\ g$ is the apparent "weight loss" of the object immersed in the liquid.

example 15 • 1 If an object of density ρ and volume V is floating in a liquid of density ρ_f, what fraction V_f/V of the object is submerged?

solution If we equate the weight of the object with the weight of the liquid displaced,

$$\rho V g = \rho_f V_f g$$

then the fractional volume of the object submerged is inversely proportional to the density ratio:

$$\frac{V_f}{V} = \frac{\rho}{\rho_f}$$

example 15 • 2 Show that a body of volume V will "float" at the interface of two fluids if one fluid has a density greater than that of the body and the other fluid has a density less than that of the body.

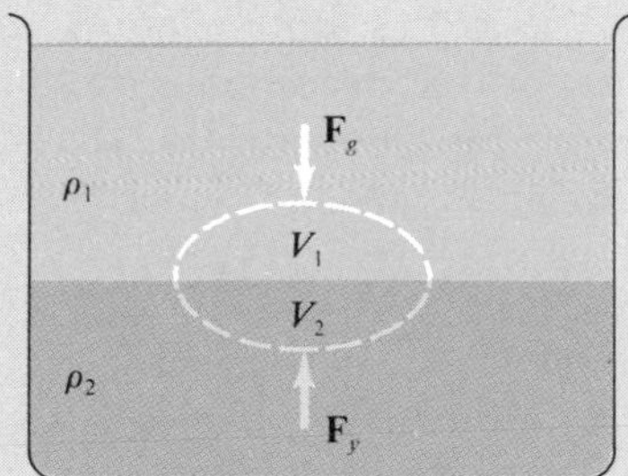

figure 15 • 6

solution Let us assume a static interface between the two fluids of density ρ_1 and ρ_2, as shown. Then the fluid within the bounded volume $V = V_1 + V_2$ has a mass $m = \rho_1 V_1 + \rho_2 V_2$ and is in equilibrium, held up by a buoyant force $F_y = (\rho_1 V_1 + \rho_2 V_2)g$. If we *change* V_1 by an amount ΔV_1, and also change V_2 by an amount $\Delta V_2 = -\Delta V_1$, so that $\Delta V = 0$, then the net upward force on the object is changed correspondingly by an amount

$$\Delta F_y = (\rho_1 - \rho_2)\ \Delta V_1\ g$$

Therefore, if $\rho_1 < \rho_2$, then a decrease in V_1 (sinking of volume V) results in a net increase in the upward force, restoring the equilibrium, and an increase in V_1 (rising of volume V) results in a net downward force. If $\rho_1 > \rho_2$, we have an unstable equilibrium; the slightest disturbance would cause the fluids to invert spontaneously.

If we now apply Stevin's solidification principle, we can see that a solid body of mass m and density ρ, where $\rho_1 < \rho < \rho_2$, will be in equilibrium at the interface. Since

$$V_1 + V_2 = V = \frac{m}{\rho} \qquad m = \rho_1 V_1 + \rho_2 V_2$$

we find

$$V_1 = \frac{m}{\rho}\frac{\rho_2 - \rho}{\rho_2 - \rho_1} \qquad V_2 = \frac{m}{\rho}\frac{\rho - \rho_1}{\rho_2 - \rho_1}$$

When we cannot neglect changes in the density of a fluid with height, or when the height is so great that changes in g must be taken into account, then we must write Eq. [15 . 2] in the form of a differential equation:

$$dP = -\rho g\, dy \qquad [15 \cdot 4]$$

where the minus sign signifies that the pressure decreases as the height increases. For gases the density is usually so small that the pressure of a gas in an ordinary-size container is practically constant throughout. Up to altitudes of several dozen meters Eq. [15 . 2] may suffice, but in the field of meteorology and in work with high-altitude rockets and space vehicles, it is essential to use Eq. [15 . 4].

If we assume that the earth's atmosphere is an ideal gas and that its temperature and gravitational acceleration g do not vary appreciably with height ($y \ll R_e$), then we can integrate Eq. [15 . 4] to find pressure as a function of altitude. Taking P_0 and ρ_0 as the pressure and density at sea level, we have, from Boyle's law (Eq. [14 . 10]),

$$\frac{P}{\rho} = \frac{P_0}{\rho_0} = \text{constant} \qquad [15 \cdot 5]$$

so that

$$dP = -P\frac{\rho_0}{P_0} g\, dy \qquad [15 \cdot 6]$$

Integration over the entire range of pressure,

$$\int_{P_0}^{P} \frac{dP}{P} = \ln\frac{P}{P_0} = -\frac{\rho_0 g y}{P_0}$$

then gives us

$$P = P_0 \exp\left(-\frac{\rho_0 g y}{P_0}\right) = P_0 \exp\left(-\frac{y}{h}\right) \qquad [15 \cdot 7]$$

called the *law of atmospheres*, or the *isothermal barometric equation.*

The quantity h is called the *scale height* of the gas. For dry air at 1 atm and 20°C (room temperature),

$$h = \frac{P_0}{\rho_0 g} \doteq \frac{10^5\ \text{N/m}^2}{1.2\ \text{kg/m}^3 \times 9.8\ \text{m/s}^2} = 8.5\ \text{km}$$

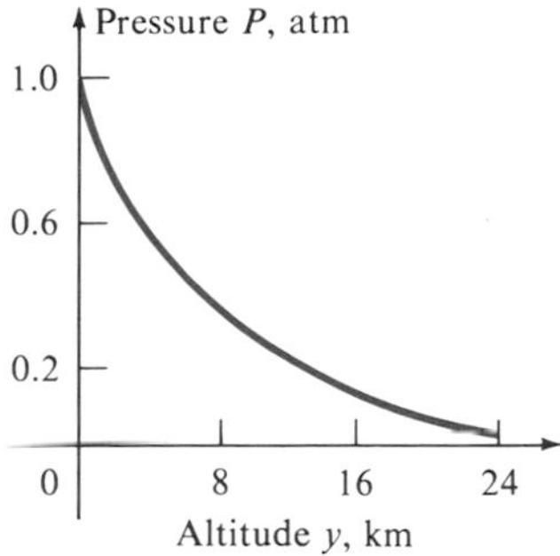

figure 15 • 7 *The law of atmospheres, P atm = exp ($-y$/8.5 km), for temperature of 20°C*

Scale height h does vary slowly with altitude, as do temperature, gravity, and the chemical composition of the atmosphere (see Fig. 15 . 7). Table 15 . 1 shows that at around 100 km the *ionosphere* begins, in which temperatures increase rapidly, diatomic molecules begin to dissociate into individual atoms, and the scale height changes markedly, while locally, $P \sim e^{-y/h}$.

15 • 2 steady flow: conservation of mass

Imagine an observer in a boat moving slowly with constant velocity across the surface of a lake. As the water flows smoothly around the sides of the boat, its speed and appearance relative to the boat do not change. To the moving observer the flow appears steady and unvarying with time. However, an observer standing on the shore would not see the flow of the water as steady; the position of the boat, and therefore the movement of the water at

table 15 • 1 *Atmospheric properties (average)*

altitude h, km	*pressure* P, bars	*density* ρ, kg/m^3	*temperature* τ, °C	*scale height* h_0, km
0	1.01325	1.2250	15	8.43
2	0.7950	1.0066	15	8.06
4	0.6166	0.8195	−11	7.68
8	0.3565	0.5258	−37	6.93
12	0.1940	0.3119	−56	6.37
16	0.1045	0.1665	−56	6.37
20	0.0553	0.0889	−56	6.38
30	0.0119	0.0179	−42	6.83
40	0.0030	0.0040	−12	7.73
60	0.0002	0.0003	−19	7.57
80	1.00×10^{-5}	2.12×10^{-5}	−107	4.97
100	2.14×10^{-7}	3.73×10^{-7}	−74	6.02
150	5.33×10^{-9}	1.76×10^{-9}	+758	32.40
200	1.63×10^{-9}	3.67×10^{-10}	+1131	48.12

various points, would instead appear to be changing with time. In describing the flow of the water it is clearly advantageous to use the boatman's moving frame of reference, since we then eliminate time from our description of the liquid motion. Once we have obtained such a description, we can always relate it afterward to observations made in any other frame of reference. Therefore we shall confine ourselves to flows which are steady in our frame of reference.

It is of interest to note that, historically, there have been two different approaches to fluid dynamics: the Lagrangian and the Eulerian. In the Lagrangian approach a *fluid particle* is considered to move under the influence of gravity, external pressure, viscous stresses, and so on as it passes through the fluid. The term *particle* is used here in its Newtonian sense of a point mass and refers to a volume element dV of fluid containing a mass $dm = \rho dV$. The volume dV is small enough that all forces acting on it can be considered to be applied at the same point, and yet large enough to contain so many millions of molecules that their random velocities average to zero. The motion of the particle is then determined by the solution of the time-dependent equation of motion, $\mathbf{F} = m\mathbf{a}$. The Eulerian approach focuses on a geometric *control volume* in the space through which the fluid is passing, a sort of window through which we may look into the region of flow. We then ask what conditions on the surface and within the interior of the control volume must hold in order to satisfy Newton's second law and the conservation of mass and energy. Both methods yield the same equations, but the Eulerian approach makes it possible to eliminate any explicit dependence on time in steady-flow problems and is therefore more widely used.

By a *steady flow,* to be more precise, we mean one that does not change its properties with time at any given point in the fluid path. Although the velocity of an individual fluid particle may change as it moves along, and its other properties, such as density and pressure, may vary from point to point along the path, there is no variation in a steady flow at any *fixed* point in space. Hence there is no need for the time variable in our equations for steady flow.

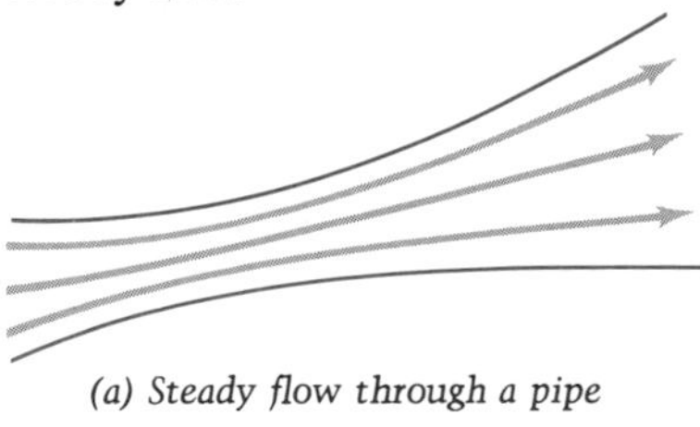

(a) Steady flow through a pipe

figure 15 • 8 *Streamlines of a fluid in steady flow. A stream tube is any portion of the fluid that is bounded by streamlines.*

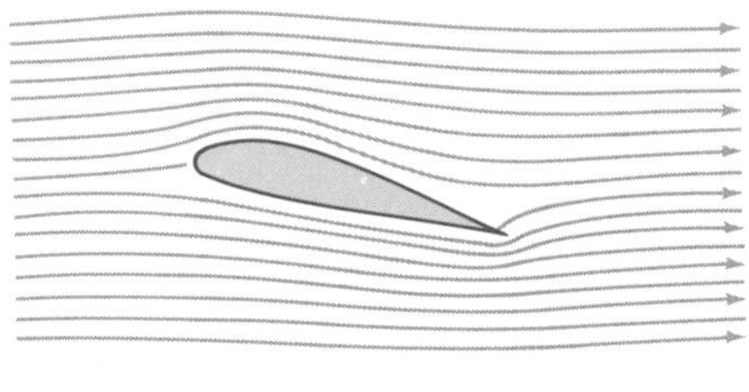

(b) Steady flow about an airfoil

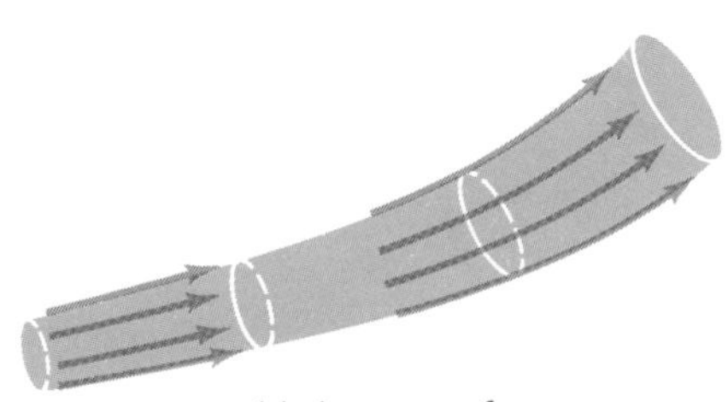

(c) A stream tube

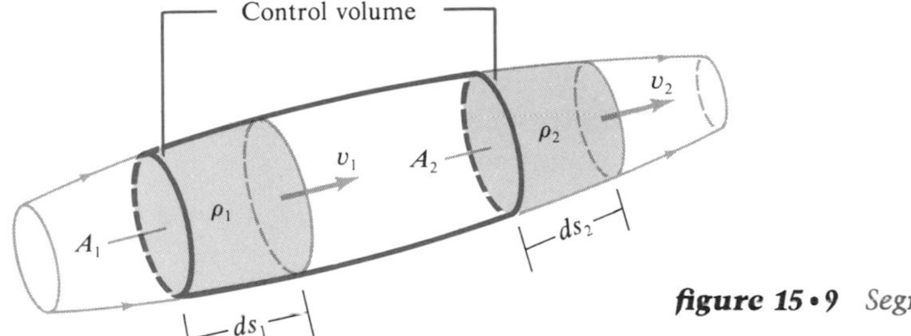

figure 15 • 9 Segment of a stream tube

A *streamline* is an imaginary curve drawn tangent to the velocity of the fluid at every point in its flow (see Fig. 15 . 8). When the flow is steady, the streamlines are identical with the actual paths of the fluid particles and are fixed in position, providing an unchanging pattern characterizing the particular flow under study. A *stream tube* is any portion of the fluid bounded by streamlines. In steady flow, any stream tube evidently has the property that a particle of fluid that starts out in a given tube *remains* in that tube during the whole course of the flow. An ordinary pipe is the largest stream tube for fluid flowing through it.

In the stream tube of Fig. 15 . 9, A_1 and ρ_1 are the cross-sectional area and fluid density at one end of the tube and A_2 and ρ_2 are the area and density farther downstream. Suppose that during the instant dt a mass of fluid enters the tube at A_1, occupying the shaded volume $A_1\,ds_1$, while another mass leaves the tube at A_2, occupying the shaded volume $A_2\,ds_2$. Since there is no creation or destruction of matter within the control volume between A_1 and A_2, these two masses must be identical:

$$\rho_1 A_1\,ds_1 = dm_1 = dm_2 = \rho_2 A_2\,ds_2$$

Dividing both sides by dt gives the *equation of continuity* for steady fluid flow:

$$\rho_1 A_1 v_1 = \rho_2 A_2 v_2 \qquad [15 \cdot 8]$$

where A_1 and A_2 are the areas normal to the flow velocities $\mathbf{v}_1$ and $\mathbf{v}_2$. The quantity $\rho A\mathbf{v}$ is the *flux* and $\rho\mathbf{v}$ is the *flux density*. This equation states that the *mass current* ρAv, which is the mass of fluid flowing past a given point per second, is constant for steady flow throughout a given stream tube. In fact, the equation of continuity and the shape of the fluid boundaries physically determine the shape of a given stream tube and the streamlines bounding it. The streamlines in Fig. 15 . 8 are not just arbitrary in shape.

If the fluid is incompressible, then $\rho_1 = \rho_2$ and Eq. [15 . 8] reduces to

$$A_1 v_1 = A_2 v_2 \qquad [15 \cdot 9]$$

According to the continuity equation for an incompressible fluid in steady flow, Av (the volume of fluid crossing all sections of a stream tube per unit time) is constant; in other words, the speeds in a stream tube vary *inversely* as the cross-sectional areas. It follows that, in the pattern of streamlines for incompressible steady flow, widely spaced lines indicate regions of low speeds, whereas closely spaced ones indicate regions of high speeds.

example 15 • 3 Water shoots upward out of a spout of cross-sectional area A_1 in a drinking fountain. If the spout area is decreased by half to A_2, what is the relative change in the height reached by the stream? Assume that the rate of discharge remains constant.

solution The speed of flow is inversely proportional to the area:

$v_2/v_1 = A_1/A_2 = 2$

Hence the kinetic energy of the fluid particles increases by $(v_2/v_1)^2 = 4$ when the area is halved. The particles should rise four times as high as they did originally.

15 • 3 conservation of energy: the Bernoulli equation

We have obtained a continuity equation for steady flow based on the principle of conservation of mass. Now let us derive a second and independent relation based on the conservation of energy. We shall limit our consideration to the steady flow of a fluid that is both *nonviscous* and *incompressible*, $\rho_1 = \rho_2 = \rho$. In Fig. 15 . 9, when a volume dV of fluid flows in across A_1, an equal volume must flow out across A_2. The fluid behind A_1 acts like a piston, forcing the volume $dV = A_1\ ds_1$ of the fluid across the boundary A_1. If the fluid pressure at A_1 is P_1, then the work done on the fluid by this pressure is

$$dW_1 = F_1\ ds_1 = P_1A_1\ ds_1 = P_1\ dV \qquad [15 \cdot 10]$$

At the same time, the fluid between A_1 and A_2 does work $dW_2 = P_2\ dV$ in forcing fluid out across A_2. The net work done on the fluid in the control volume is therefore $W = (P_1 - P_2)\ dV$. Since the frictional resistance is assumed to be negligible, all this net work must appear in the form of increased kinetic and potential energy of the fluid. If the heights of sections A_1 and A_2 above some arbitrarily chosen reference level are y_1 and y_2, then the increase in the kinetic energy of dV in passing from A_1 to A_2 is

$$dK = \tfrac{1}{2}\rho(v_2^2 - v_1^2)\ dV$$

and the increase in the gravitational potential energy is

$$dU_g = \rho g(y_2 - y_1)\ dV$$

Since work done equals energy gained, $dW = dK + dU_g$, we obtain

$$(P_1 - P_2)\ dV = \tfrac{1}{2}\rho(v_2^2 - v_1^2)\ dV + \rho g(y_2 - y_1)\ dV$$

so that

$$P_1 + \tfrac{1}{2}\rho v_1^2 + \rho g y_1 = P_2 + \tfrac{1}{2}\rho v_2^2 + \rho g y_2 \qquad [15 \cdot 11]$$

This is the famous *Bernoulli equation*, derived by Daniel Bernoulli in 1738. If we divide through by ρ, we can recognize $P/\rho = PV/m$ as the work done *on* a unit mass of fluid in forcing it through a section of tube. Accord-

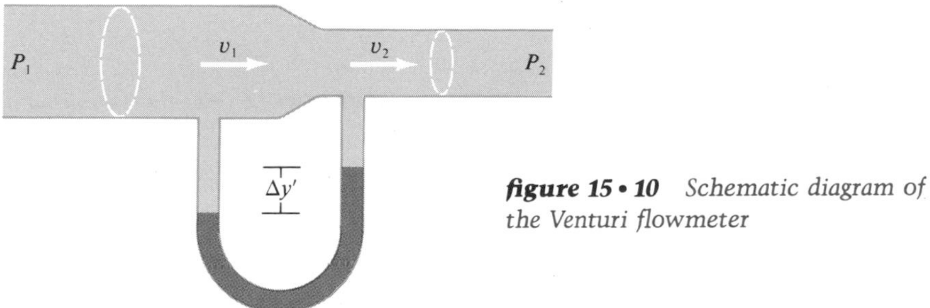

figure 15 • 10 *Schematic diagram of the Venturi flowmeter*

ing to Eq. [15 . 11], the sum of this work, the kinetic energy per unit mass, and the gravitational potential energy per unit mass is constant *throughout* the region of steady flow. In many practical situations Bernoulli's equation is found to hold for the region between any two points in the fluid, whether or not they are in the same stream tube. For an incompressible fluid at rest, Eq. [15 . 11] reduces to the hydrostatic equation [15 . 2].

Let us now consider some applications of Bernoulli's equation. Figure 15 . 10 shows the steady flow of a liquid through a horizontal *Venturi flowmeter.* If the cross section of the tube is small enough that we can ignore any differences in the heights of the liquid at various points, the Bernoulli equation becomes

$$P_1 + \tfrac{1}{2}\rho v_1^2 = P_2 + \tfrac{1}{2}\rho v_2^2 \qquad [15 \cdot 12]$$

When the tube is entirely filled with the liquid, Eq. [15 . 9] shows that the speed v_2 in the constricted region must exceed that in the wider part, hence the pressure P_2 inside the constricted region is *less*. This causes a difference $\Delta y'$ in the levels of fluid in the manometer tube which is directly proportional to the pressure difference $P_1 - P_2$.

The fact that the pressure decreases as the speed increases due to a constriction in the flow is a principle also utilized in aspirators and in various spray attachments for garden hoses (Fig. 15 . 11). Another application is the pivoted "no-draft" windows on automobiles, which, when partially open, constrict the flow of air alongside the automobile and create a pressure difference which results in a mild circulation of air within the vehicle. The next time you are driving with a smoker, open your window a bit and see how the smoke is drawn to the low-pressure opening.

To measure flow speed, consider a blunt-nosed object immersed in a

figure 15 • 11 *In the aspirator, water is forced through the constriction; the side tube is connected to the vessel which is to be evacuated.*

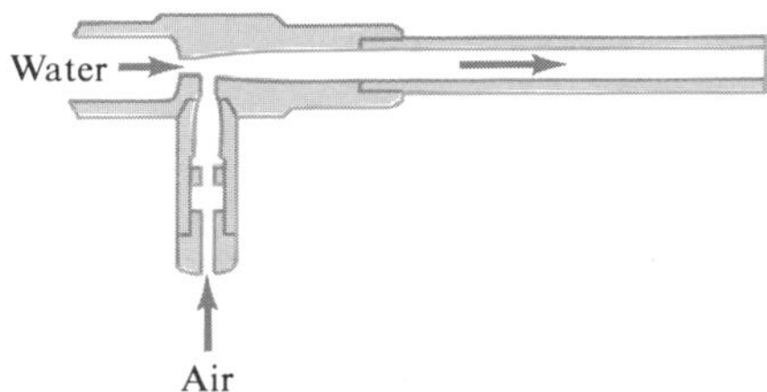

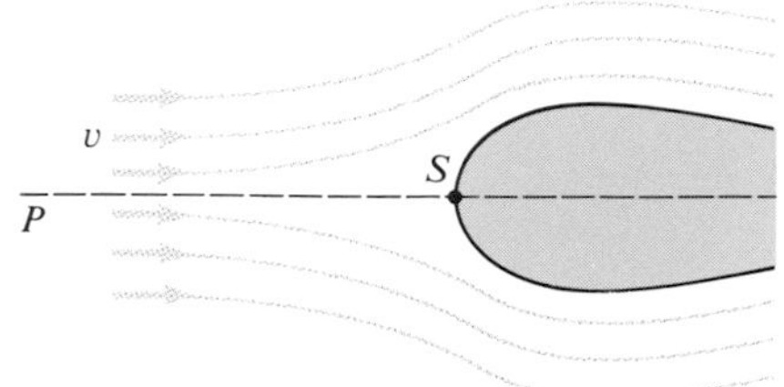

figure 15 • 12 *Steady flow around a blunt-nosed object*

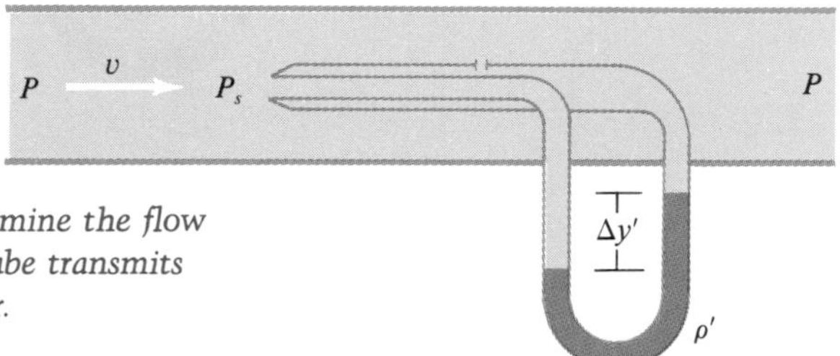

figure 15 • 13 *A pitot static tube, used to determine the flow speed of a fluid. The opening at the top of the tube transmits the pressure P to the right arm of the manometer.*

fluid which, from the reference frame of the object, is in steady flow (see Fig. 15 . 12). If the object's shape is a surface of revolution, then, by symmetry, the velocity at point S must be zero. This point is called the ***stagnation point.*** If P_s is the ***stagnation pressure*** and P is the pressure in the free stream, then

$$P_s = P + \tfrac{1}{2}\rho v^2 \qquad [15 \cdot 13]$$

The difference $P_s - P$ is easily measurable (Fig. 15 . 13) and can be computed from

$$P_s - P = \rho' g \,\Delta y'$$

where ρ' is the density of the liquid in the glass pitot tube. Then the flow speed v of the undisturbed stream can be found from Eq. [15 . 13]. Whenever a fluid divides in steady flow around a blunt object, a stagnation point will appear, where the velocity is zero or very nearly zero. The exact location of this point will depend on the geometry of the object.

Let us now consider another application of Bernoulli's equation. Suppose a liquid is allowed to drain from a small opening in the uncovered tank in Fig. 15 . 14. The upper surface of the liquid is falling with speed v at the moment this surface passes height y. Since the pressure is atmospheric both

figure 15 • 14 *Efflux of liquid from an orifice in an open tank*

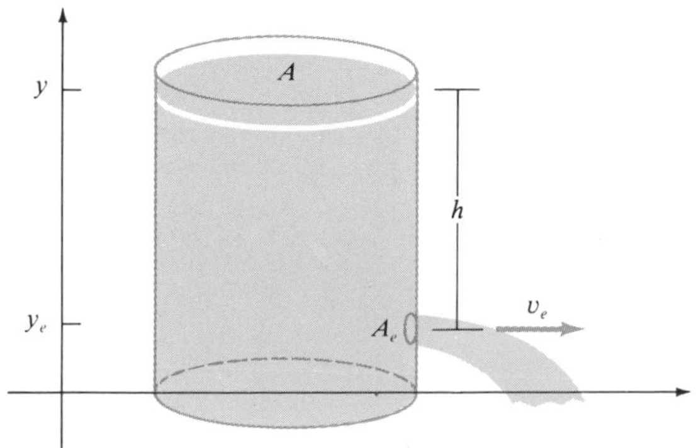

at the top surface and in the free jet emerging from the opening at height y_e, the Bernoulli equation gives the speed of outflow or *efflux* v_e as

$$v_e = \sqrt{v^2 + 2gh} \qquad [15 \cdot 14]$$

where $h = y - y_e$ is the distance from the upper surface to the exit level. If the cross-sectional area A of the tank is very large in comparison to the area A_e of the exit, the rate of fall of the upper surface will be negligible, and Eq. [15 . 14] reduces to

$$v_e = \sqrt{2gh} \qquad [15 \cdot 15]$$

This result, called *Torricelli's theorem,* represents a special case of the Bernoulli equation and was first derived almost a century earlier by Evangelista Torricelli. As Torricelli pointed out (see the opening quotation), the flow speed of liquid from a small opening in a large vessel is equal to the speed gained by a solid body in falling freely from rest through a distance equal to the depth of the orifice below the liquid surface.

example 15 • 4 A large closed chamber is filled with a liquid of density ρ to a height $y = h$ above an opening, as shown. The space above the liquid contains gas under pressure P. (*a*) Show that the speed of efflux is given by

$$v_e = \sqrt{2\left(gh + \frac{P - P_0}{\rho}\right)}$$

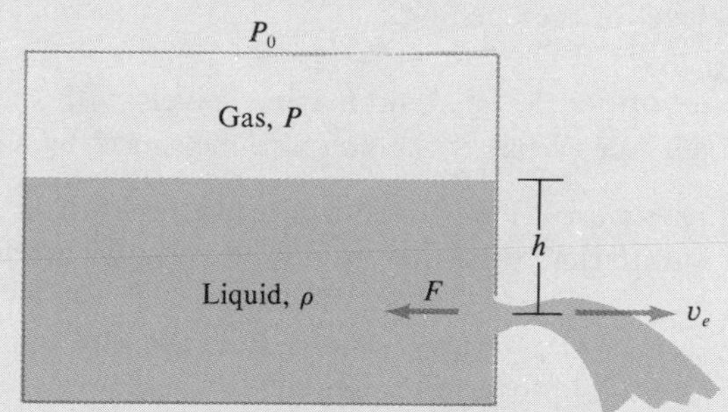

figure 15 • 15

where P_0 is atmospheric pressure. (*b*) If the gas pressure P is so high that the effect of gravity may be neglected, show that the magnitude of the thrust, or reaction force F, exerted by the jet on the chamber is

$$F = 2A_eP_g$$

where $P_g = P - P_0$ is the gage pressure of the gas in the chamber. (*c*) If the chamber were open to the atmosphere and initially filled with liquid to the same height h as before, approximately how long would it take the liquid to drop to the level of the opening?

solution (*a*) If we ignore the motion of the liquid surface, Bernoulli's equation gives

$$\tfrac{1}{2}\rho v_e^2 + P_0 = \rho gh + P$$

and solving for v_e yields the given equation. (*b*) If we let A be the cross-sectional area of chamber and A_e the area of the exit, then a mass of liquid $m = \rho A_e v_e$ flows out through the opening each second; hence there is a net momentum outflow per second of $d(mv)/dt = \rho A_e v_e^2$. From Newton's third law, the magnitude of the reaction force exerted by the jet is

$$F = \frac{d(mv)}{dt} = \rho A_e v_e^2$$

Then, substituting from part (*a*) and ignoring the effect of gravity, we obtain

$$F \doteq 2A_e(P - P_0) = 2A_e P_g$$

(*c*) From the equation of continuity, $A_e v_e = Av$, where $v = -dy/dt$ is the downward speed of the liquid surface. Since $v = v_e A_e/A \ll v_e$, we can use Torricelli's theorem, Eq. [15 . 15], to obtain

$$\frac{dy}{dt} = -\frac{v_e A_e}{A} \doteq -\frac{A_e}{A}\sqrt{2gy}$$

Rearranging, we can then integrate to find the total time of flow:

$$-\frac{A}{A_e\sqrt{2g}}\int_h^0 \frac{dy}{\sqrt{y}} = \frac{A}{A_e\sqrt{2g}}\,2\sqrt{h} = \int_0^t dt$$

or an elapsed time of $t = (A/A_e)\sqrt{2h/g}$ for the liquid to drop to the level of the opening.

example 15 • 5 A submarine travels with speed v at a depth h below the surface. What is the pressure measured by a pitot tube located in its bow?

solution If we transform to the coordinate system moving with the submarine, then the water is moving with velocity v past its bow. The pitot tube measures stagnation pressure, $P_s = P + \frac{1}{2}\rho v^2$. If the flow pressure is $P = P_0 + \rho gh$, where P_0 is the atmospheric pressure at sea level, then the total pressure at depth h is

$$P_s = P_0 + \rho(gh + \tfrac{1}{2}v^2)$$

In laboratory studies we generally observe fluid flow around or through a stationary object. In wind tunnels, for example, a model of the test object is held in place while air is blown past it. Furthermore, when the *Mach number,* the ratio of the fluid velocity to the speed of sound in that fluid, is less than about 0.5, the air can usually be considered as incompressible. It can get out of the object's way before it is "squeezed" appreciably against the object owing to its own inertia. At speeds greater than the speed of sound, the molecules of the fluid medium pile up around the object like snow in front of a shovel, forming a *shock wave.* Sonic booms and blast waves from explosions are shock waves.

15 . 4 viscosity

It was not until the nineteenth century, after the theory for the steady flow of nonviscous fluids had been developed, that fluid mechanics was extended to account for the effects of *viscosity.* Imagine two planes, an infinitesimal distance ds apart, in a viscous fluid that is flowing horizontally in parallel layers *(laminar flow)*, as shown in Fig. 15 . 16. The layer of fluid in the lower plane is moving with speed v, while that in the upper plane is moving with speed $v + dv$. The faster-moving fluid in the upper plane exerts a tangential, or shearing, force of magnitude F on the fluid layer below it, tending to speed it up. At the same time, the upper fluid layer is acted on by a retarding force of equal magnitude. This shearing force F is proportional to the area A of the plane, and therefore F/A clearly is in the nature of a shearing stress (see Sec. 14 . 3).

As a result of laminar flow, the upper layer of fluid also undergoes a shearing strain $d\phi$ in time dt. Experiments show that in a *fluid,* as contrasted with a solid, the shearing stress is not a function of the shearing strain $d\phi$, but of its *time rate of change,* $d\phi/dt$. For gases under all ordinary conditions and for many common liquids, the shearing stress during laminar flow is given by

$$\frac{F}{A} = \eta \frac{d\phi}{dt} \qquad [15 \cdot 16]$$

where η is a constant of proportionality called the *coefficient of viscosity.* Fluids obeying Eq. [15 . 16] are termed *Newtonian,* or *true, fluids.* For non-Newtonian fluids, such as very viscous liquids, gels, and plastic materials, the stress is not simply proportional to the rate of shearing strain. We shall confine ourselves here to Newtonian fluids.

The *difference* in the distance moved in time dt by the particles in the two horizontal planes in Fig. 15 . 16 is $dx = dv\ dt$, so that (in radians)

$$d\phi = \frac{dv\ dt}{dy} \qquad \text{and} \qquad \frac{d\phi}{dt} = \frac{dv}{dy}$$

Therefore the viscous stress is proportional to the *velocity gradient* dv/dy, the change in the speed of flow per unit distance normal to the direction of flow:

$$\blacktriangleright \quad \frac{F}{A} = \eta \frac{dv}{dy} \qquad [15 \cdot 17]$$

figure 15 • 16 *A viscous fluid in laminar flow—that is, flow in which layers, or laminas, of the fluid flow steadily over each other*

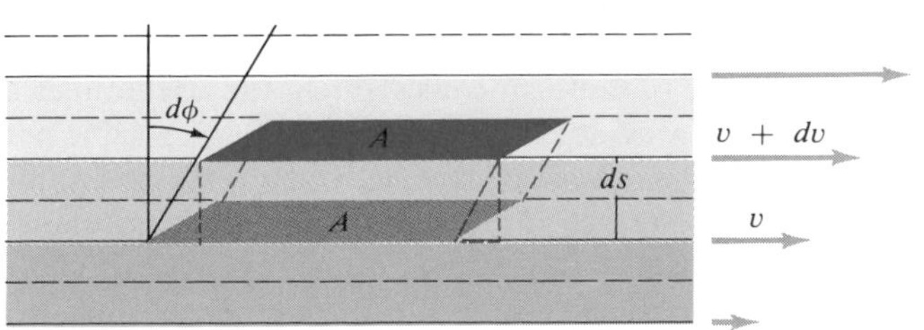

table 15 • 2 *Coefficients of viscosity for common substances*

substance	*coefficient of viscosity η, dekapoises*	*temperature, °C*
alcohol, ethyl	1.20×10^{-3}	20
castor oil	986×10^{-3}	20
glycerin	1490×10^{-3}	20
oil, SAE 20*	*c*280×10^{-3}	40
SAE 30*	*c*400×10^{-3}	40
mercury	1.55×10^{-3}	20
water	1.79×10^{-3}	0
	1.005×10^{-3}	20
	1.000×10^{-3}	20.20
	0.299×10^{-3}	95
air	1.71×10^{-5}	0
	1.84×10^{-5}	20
carbon dioxide	1.48×10^{-5}	20
hydrogen	0.876×10^{-5}	21

equal viscosities: 1.00 *dekapoise* $= 1.00\ N \cdot sec/m^2 = 10.0$ *poises*
$= 2.09 \times 10^{-2}\ lbf \cdot s/ft^2$

*Society of Automotive Engineers (SAE) viscosity numbers are used to describe crankcase lubricating oils.

Equation [15 . 16] is the more general form, however, for there are situations to which Eq. [15 . 17] is not applicable. Units for the coefficient of viscosity η include the *dekapoise* ($1\ N \cdot s/m^2$); the *poise* ($1\ dyne \cdot s/cm^2$), named after the French physiologist J. L. M. Poiseuille (pronounced *pwahz-yu-wee*); and the British engineering unit $1\ lbf \cdot s/ft^2$. Some typical viscosity coefficients are listed in Table 15 . 2.

There is weighty experimental evidence that very little or no slip occurs between a solid surface and the layer of fluid immediately in contact with it. Instead, a thin film of fluid, a *boundary layer* only a few molecules thick, adheres to the solid surface, and the fluid motion takes place relative to this film. Even the degree of roughness of the solid surface has little effect on laminar flow, although the effect of surface roughness becomes significant when the flow becomes turbulent.

When a viscous fluid moves with constant speed through a narrow straight pipe of uniform cross section, the streamlines are parallel to the axis of the pipe, provided a certain critical flow speed is not exceeded. Moreover, the flow is then laminar; that is, if you imagine an infinite number of cylinders to be described in the fluid about the longitudinal axis of the tube, the motion consists in the slipping of one cylinder through another, the way the tubes of a pocket telescope slip through each other.

If the fluid is incompressible, the volume of fluid that crosses any section of the pipe per unit time is given by the *Poiseuille equation*:

$$\frac{V}{t} = \frac{\pi R^4}{8\eta}\frac{\Delta P}{l} \qquad [15 \cdot 18]$$

where R and l are the inner radius and length of the pipe, and ΔP is the difference between the pressure of the fluid as it enters the pipe and its exit pressure. The rate of flow varies directly with the fourth power of the radius R of the pipe and directly with the *pressure gradient* $\Delta P/l$, but inversely with the viscosity coefficient η of the fluid. This provides the basis for capillary-flow methods used to determine η for liquids (see Fig. 15 . 17). Incidentally, the Poiseuille equation does not apply rigorously to blood flow, because blood is a suspension of particles in a fluid, and hence non-Newtonian in the sense of Eq. [15 . 16]. The speed of each slipping cylinder of fluid decreases with the square of its radius, from a maximum value in the center of the pipe to zero at the walls. This is indicated in Fig. 15 . 17 by the parabolic *velocity profile* of the fully developed laminar flow.

The derivation of Eq. [15 . 18] is tedious, but not difficult. For our purposes, however, it will suffice to demonstrate the reasonableness of this equation by means of *dimensional analysis*. This technique was first successfully applied to fluid mechanics in 1892 by Lord Rayleigh and is often used when analytic derivations of complex situations are extremely difficult.* Recall from Sec. 1 . 2 that all the terms of a physical equation must have *dimensional consistency,* even though the units of the various terms are not necessarily the same. For example,

$$V = V_1 + V_2 = 4 \text{ ft}^3 + 20 \text{ cm}^3$$

the unit for each of the terms involved has the dimensional formula $V = L^3$.

*For comprehensive and rigorous treatment of dimensional analysis and its applications, see, for example, H. L. Langhaar, *Dimensional Analysis and Theory of Models* (Wiley, 1951) or the standard treatise of P. W. Bridgman, *Dimensional Analysis* (Yale University Press, rev. ed., 1931).

figure 15 • 17 *One type of capillary-flow viscometer. The manometer is placed downstream where the flow has become fully laminar.*

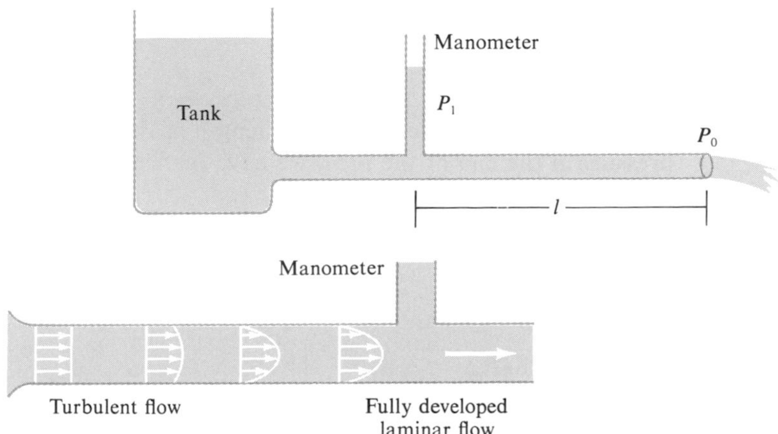

In the case of Poiseuille's equation we can make the reasonable assumption that the *mass* of incompressible fluid crossing a section of pipe per second is directly proportional to its density and to powers, as yet unknown, of the viscosity, the radius of the pipe, and the pressure gradient:

$$\frac{m/\rho}{t} = \frac{V}{t} = k\eta^a R^b \left(\frac{\Delta P}{l}\right)^c \qquad [15 \cdot 19]$$

where k is the unknown dimensionless constant of proportionality. Reducing all quantities to their basic dimensions of mass, length, and time, we have the relationship

$$L^3T^{-1} = (ML^{-1}T^{-1})^a L^b (ML^{-2}T^{-2})^c \qquad [15 \cdot 20]$$

Since the dimensions of corresponding factors in both members of the equation must agree, we have

For L $\quad 3 = -a + b - 2c$
For T $\quad -1 = -a - 2c$
For M $\quad 0 = a + c$

Therefore $a = -1$, $b = 4$, $c = 1$, as in Eq. [15 . 18]. The value of $k = \frac{1}{8}\pi$ must be determined by experiment or analysis, because dimensional analysis cannot yield the values of *numerical factors* in equations. The methods of dimensional analysis can be successfully applied in the construction of an equation only if the basic theory and experience with the physical situation is sufficient to determine the variables that must enter the equation.

example 15 • 6 Find the formula for steady flow of a compressible ideal gas through a pipe corresponding to Poiseuille's equation for a liquid.

solution Consider a very short section dl of pipe, where the density of the gas is ρ and the mass current is

$$\frac{dm}{dt} = \frac{\rho V}{t} = -\rho\,\frac{\pi R^4}{8\eta}\,\frac{dP}{dl} \qquad \left(\frac{dP}{dl} < 0\right)$$

From Boyle's law, $\rho = CP$, where $C =$ constant, and if P_1 is the pressure at the entrance of the pipe and $\rho_1 V_1/t$ the mass current entering the pipe, then conservation of mass requires

$$\frac{CP_1V_1}{t} = -CP\,\frac{\pi R^4}{8\eta}\,\frac{dP}{dl}$$

Integrating this equation along the length of the pipe, where P_2 is the exit pressure at the end of the pipe, we have (cancelling C)

$$\frac{P_1V_1}{t}\int_0^l dl = -\frac{\pi R^4}{8\eta}\int_{P_1}^{P_2} P\,dP$$

or

$$\frac{V_1}{t} = \frac{\pi R^4(P_1^2 - P_2^2)}{16\eta P_1 l}$$

The result derived in Example 15 . 6 provides the basis for capillary-flow methods of determining the viscosity of gases.

The problem of the force exerted by a viscous fluid on an object moving *through* it with constant velocity **v** was investigated theoretically by Stokes in 1845. The general problem is so difficult that Stokes began with the special, although important, case of a *spherical* object completely immersed in a fluid of constant density, behaving as if it were incompressible, and in laminar flow relative to the sphere. The result he obtained, now known as *Stokes' law,* is

$$\mathbf{F}_r = -6\pi\eta R\mathbf{v} \qquad [15 \cdot 21]$$

where $\mathbf{F}_r$ is the retarding force, or *drag,* η is the viscosity coefficient of the fluid, and R and $\mathbf{v}$ are the radius and the constant velocity of the sphere relative to the stagnant fluid. Stokes' law, despite its specialized character, applies in a wide variety of situations: precipitation phenomena in meteorology, the movement of small particles suspended in a fluid, and motions of small particles in gravitational and electric fields. An example of the last application is the classic oil-drop experiment of Robert Andrews Millikan in 1911, which yielded the first accurate determination of the charge on an electron.

The derivation of Stokes' law requires mathematical techniques beyond our present scope. However, since only three variables enter the equation, we can perform a dimensional analysis, assuming that the retarding force is proportional to the viscosity and to the size of the sphere and is opposed to the direction of motion:

$$\mathbf{F}_r = k\eta^a R^b v^c \left(-\frac{\mathbf{v}}{v}\right)$$

or, in dimensional form,

$$MLT^{-2} = (ML^{-1}T^{-1})^a L^b (LT^{-1})^c$$

Because the dimensions of corresponding factors in both members of this relation must agree, we have

$$\begin{array}{lrcl} \text{For } M & 1 & = & a \\ \text{For } L & 1 & = & -a + b + c \\ \text{For } T & -2 & = & -a - c \end{array}$$

Therefore $a = 1$, $b = 1$, and $c = 1$, and the equation for $\mathbf{F}_r$ must be of the form

$$\mathbf{F}_r = -k\eta R\mathbf{v} \qquad [15 \cdot 22]$$

in agreement with Eq. [15 . 21], where $k = 6\pi$.

Consider a sphere of mass m and radius R, falling from rest in a stagnant fluid of density ρ. If Stokes' law applies, we can find the downward speed of the sphere as a function of time. There is a constant *net* downward force

$$F = mg - \tfrac{4}{3}\pi R^3 \rho g$$

due to gravity less buoyancy. The equation of motion can be expressed in terms of the downward speed as

$$m\frac{dv}{dt} = F - 6\pi\eta Rv \qquad [15 \cdot 23]$$

If we rewrite this equation, we can then integrate to obtain

$$m\int_0^v \frac{dv}{F - 6\pi\eta Rv} = \int_0^t dt = t \qquad [15 \cdot 24]$$

From any handbook of integrals we find the integral on the left to be a natural logarithm,

$$-\frac{m}{6\pi\eta R}\ln\left(1 - \frac{6\pi\eta Rv}{F}\right) = t$$

and raising both sides to a power of e and rearranging gives us the downward speed as a function of time:

$$v(t) = \frac{F}{6\pi\eta R}\left[1 - \exp\left(-\frac{6\pi\eta Rt}{m}\right)\right] \qquad [15 \cdot 25]$$

As we could have predicted from Eq. [15 . 23] by setting $dv/dt = 0$, regardless of the time of fall, the sphere will reach a *terminal speed* given by

$$v_T = v(\infty) = \frac{F}{6\pi\eta R} \qquad [15 \cdot 26]$$

This equation provides the basis of various methods for determining η for liquids, especially very viscous liquids. A small sphere is allowed to fall in the liquid, or alternatively, a small bulb or bubble is allowed to rise in it, and the constant terminal speed it reaches in that liquid is measured.

Experiment and theory show that Eq. [15 . 26] is also valid for particles of water, ice, and dust falling through still air of uniform density, provided that the radius of the particle does not exceed about 40 μm (see Fig. 15 . 18). For larger particles the air flow about the sphere is no longer laminar. Water drops of very large radius (above 500 μm) tend to flatten in falling and can no longer be treated as spherical. However, if the droplet size is of the order of molecular size, then it and the air must be treated as molecular, rather than continuous in structure as assumed in the exact theoretical derivation of Stokes' law.

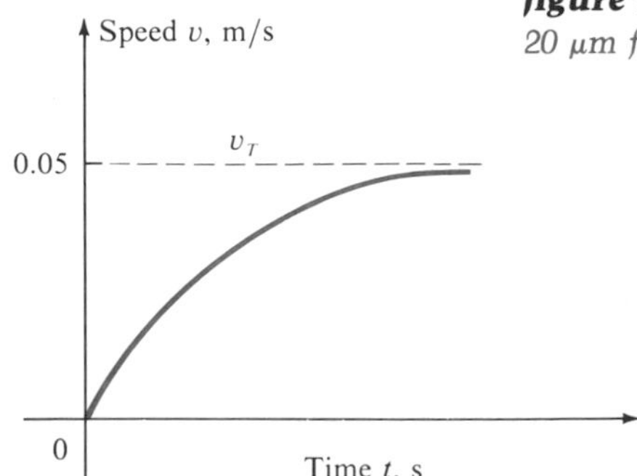

figure 15 • 18 *Speed-time graph for a water droplet of radius 20 μm falling from rest through stagnant air of temperature 20°C*

15 • 5 laminar and turbulent flow

In any fluid, the flow relative to solid boundary surfaces is laminar only up to some critical flow speed. Beyond this speed, which differs for different fluids and different boundary surfaces, the flow becomes *turbulent*. The particles of the fluid begin to fluctuate in a random manner, causing disordered whirling and eddying of the fluid as it proceeds in its forward flow. The rising smoke of a cigarette in the still air of a room exhibits both types of flow—laminar flow near the cigarette and turbulent flow higher up.

In laminar flow the shearing stresses between adjacent layers of the fluid are due both to cohesive forces between the fluid molecules and to momentum exchanges which occur as the molecules, in their random motions, diffuse from faster- to slower-moving layers, and vice versa. In highly turbulent flow, however, the shearing stresses are not primarily molecular in origin; instead they stem from momentum exchanges occurring because whole portions of fluid, each consisting of millions of molecules, move between adjacent parts of the fluid as a result of the turbulent mixing process. In brief, fluid resistance in turbulent flow is predominantly an inertial phenomenon on a large, or macroscopic, scale. In contrast, viscosity in laminar flow is a measure of the tendency for stresses to be transmitted throughout the fluid—that is, for the fluid to behave as a continuous whole, with conditions in one region predictable from a knowledge of conditions in any other. Seen in this light, viscosity, despite the unpleasant connotations of "resistance" and "stickiness" takes on new meaning.

Stokes noted that a flow may change rather suddenly from laminar to turbulent as conditions are changed. The exact nature of these conditions was investigated both experimentally and theoretically by Osborne Reynolds in 1883, first by experiments in which the flow of water at different speeds in glass pipes was observed by injecting dye into the water. As a result of such studies, Reynolds was able to introduce a dimensionless parameter later called the *Reynolds number* N_{Re}:

$$N_{\text{Re}} = \frac{\rho \bar{v} l'}{\eta} \qquad [15 \cdot 27]$$

where ρ and η are the density and the viscosity coefficient of the fluid, v is the flow speed relative to the particular solid body exposed to the fluid, and l' is a length "characteristic" of that solid body. For flow through a straight pipe of circular cross section, l' is the *diameter* $2R$ of the pipe and $\bar{v}$ is the *average speed* of the fluid in the pipe, which can be shown to be equal to one-half the speed at the axis. For flow past a sphere, l' is the diameter $2R$ of the sphere.

The physical significance of the Reynolds number N_{Re} is that it represents the ratio of inertial force F_i to viscous force F_v in the fluid. To see this, let us set

$$N_{\text{Re}} = F_i/F_v$$

and derive Eq. [15 . 27] for the case of a liquid that is flowing with average speed $\bar{v}$ in a pipe of cross-sectional area A and radius R. From Newton's second law, the order of magnitude of the inertial force may be assumed to be given by*

$$F_i = \bar{v}\,\frac{dm}{dt} \approx \bar{v}\,(\rho A\bar{v}) = \rho A\bar{v}^2$$

The order of magnitude of the viscous force is, from Eq. [15 . 16],

$$F_v = A\eta\,\frac{dv}{dr} \approx \frac{A\eta\bar{v}}{2R}$$

Therefore, for pipes,

$$\frac{F_i}{F_v} \approx \frac{\rho A\bar{v}^2}{A\eta\bar{v}/2R} = \frac{2\rho\bar{v}R}{\eta} = \mathrm{N}_{\mathrm{Re}} \qquad [15 \cdot 28]$$

Since inertial force predominates in turbulent flow and viscous force predominates in laminar flow, large values of N_{Re} are associated with turbulence and small values with laminar flow.

The value of N_{Re} at the onset of turbulence is the *critical Reynolds number.* Experience shows that for any straight pipe of circular cross section the flow will be laminar if $N_{\mathrm{Re}} \leq 2000$. Therefore, from Eq. [15 . 27], the critical *average* flow speed $\bar{v}_{cr}$ at and below which the Poiseuille equation holds, is

$$\bar{v}_{cr} = \frac{2000\eta}{2\rho R}$$

For $N_{\mathrm{Re}} > 2000$, the flow may be either laminar or turbulent, depending on such circumstances as the way in which the flow was started in the pipe and the roughness of the pipe. Another way to view turbulence is to say that

*In first encountering a complex problem it is helpful to ascertain the gross effects of the variables in terms of their order-of-magnitude values. In this text, the order of magnitude of some quantity refers in general to the power of 10 nearest to its actual value.

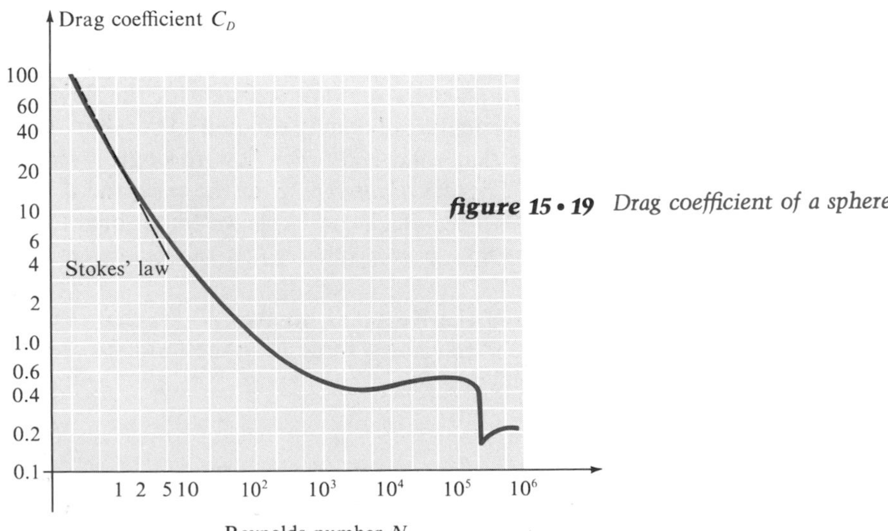

figure 15 • 19 *Drag coefficient of a sphere*

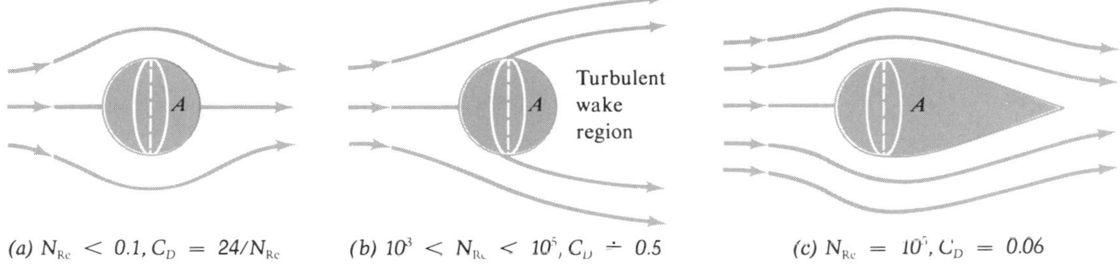

figure 15 • 20 *Examples of the dependence of the drag coefficient C_D on body shape and Reynolds number N_{Re}*

when inertial effects dominate, viscosity does not "have time" to transmit stress throughout the fluid in order to produce a smooth laminar flow around or through an obstacle.

Now imagine that all the fluid in a cylinder of fluid impinging upon a sphere of cross section A loses, on the average, half its forward momentum on striking the sphere. Then $\rho v A$ is the mass per unit time losing speed $\frac{1}{2}v$, and the inertial reaction force on the sphere is proportional to $\frac{1}{2}\rho v^2 A$. In this model the constant of proportionality is known as the *drag coefficient* C_D:

$$\mathbf{F}_r = -\tfrac{1}{2}C_D \rho v A \mathbf{v} \qquad [15 \cdot 29]$$

This formula, first proposed by Newton, is known as *Newton's resistance law*. The drag coefficient is a modern development and depends on a number of variables, but mainly on the Reynolds number. For a sphere it varies only slowly over a fairly large range of values of $N_{Re} > 10$ (see Fig. 15 . 19). In the limit of $N_{Re} \leq 1$, we would expect to recover Stokes' law, which requires $C_D = 24/N_{Re}$. A more general form of the resistance law for a sphere would therefore be

$$\mathbf{F}_r = -6\pi\eta R\mathbf{v}\tfrac{1}{24}C_D N_{Re} \qquad [15 \cdot 30]$$

For values of $N_{Re} > 500{,}000$, turbulence appears, as indicated by an abrupt drop in the drag coefficient. However, with appropriate average values of C_D, Newton's resistance law still applies. Figure 15 . 20 illustrates the dependence of the flow and the drag coefficient on body shape and the Reynolds number.

example 15 • 7 A water droplet of density $\rho_1 = 1000\ \text{kg/m}^3$ is falling with constant velocity through air of uniform density $\rho_2 = 1.2\ \text{kg/m}^3$ and viscosity $\eta = 1.8 \times 10^{-5}$ dekapoise. (*a*) Derive the equation for the largest radius that the droplet can have if Stokes' law is to apply. (*b*) Compute the radius.

solution (*a*) Since buoyancy is negligible, the downward force on the drop is

$$F_g = mg = \tfrac{4}{3}\pi R^3 \rho_1 g$$

Then, from Eqs. [15 . 26] and [15 . 27], the final speed of the droplet is

$$v_T = \frac{\frac{4}{3}\pi R^3 \rho_1 g}{6\pi\eta R} = \frac{\eta N_{\text{Re}}}{2R\rho_2}$$

Stokes' law holds only up to $N_{\text{Re}} \leq 1$, so that, solving for R and cancelling terms, we obtain

$$R = \left(\frac{9N_{\text{Re}}\eta^2}{4\rho_1\rho_2 g}\right)^{1/3} \leq \left(\frac{9\eta^2}{4\rho_1\rho_2 g}\right)^{1/3} = R_{\text{max}}$$

(*b*) Setting $N_{\text{Re}} = 1$, we can compute the largest radius for the droplet as

$$R_{\text{max}} = \left(\frac{9 \times 1 \times (1.8 \times 10^{-5}\ \text{N}\cdot\text{s/m}^2)^2}{4 \times (1.2 \times 1000)\ \text{kg}^2/\text{m}^6 \times 9.8\ \text{m/s}^2}\right)^{1/3} = 40\ \mu\text{m}$$

example 15 • 8 An earth satellite 1 m in diameter is moving at an altitude of 50 mi with a speed of 8 km/s relative to the atmosphere. If the atmosphere is assumed isothermal at 15°C and the scale height is $h = 5$ mi, what are the Reynolds number and drag coefficient of the satellite?

solution The density of the atmosphere at sea level is $\rho = 1.2\ \text{kg/m}^3$, and it varies as $e^{-y/h}$ with altitude. Since the viscosity coefficient is constant in isothermal atmosphere (see Table 15 . 1), the Reynolds number is, from Eq. [15 . 28],

$$N_{\text{Re}} = \frac{(1.2e^{-10}\ \text{kg/m}^3)(8000\ \text{m/s})(1\ \text{m})}{1.8 \times 10^{-5}\ \text{N}\cdot\text{s/m}^2} \doteq 24{,}000$$

and from Fig. 15 . 21, the drag coefficient is approximately $C_D = 0.45$.

PROBLEMS

15 • 1 fluid pressure

15 • 1 A jar 10 cm in diameter and 12 cm tall is half full of mercury and half full of water. Find the magnitude of the force against the side walls.

15 • 2 The areas of the large and small cylinders of a press are A_1 and A_2, respectively. What is its mechanical advantage?

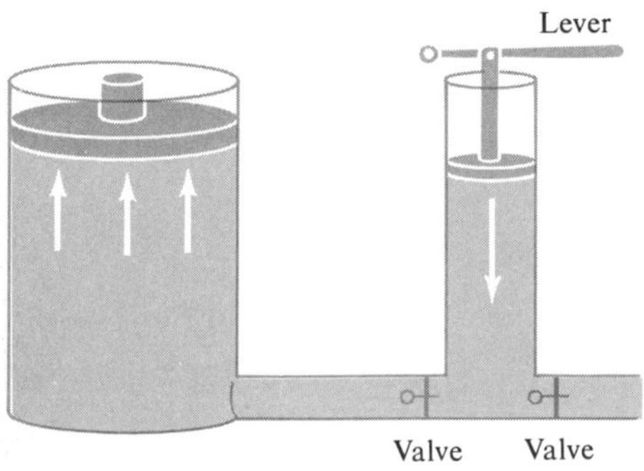

15 • 3 Find the equivalent force on a wall of height y and width x owing to the pressure of water contained by the wall. Also find the point of application of the equivalent force.

15 • 4 A solid homogeneous dam is built so that its cross section is a right-triangular wedge 3 m thick at the base, 4 m tall, and 12 m wide, as shown. The water comes to the top of the dam. (a) What is the total force exerted by the water on the dam? (b) What is the overturning torque, about an axis through O, exerted by the water? (c) Find the depth below the water surface at which a single force equal in magnitude to the force of the water would have to act to produce the overturning torque. (d) Find the least density ρ that the dam material can have if the dam is to resist being toppled by the overturning torque.

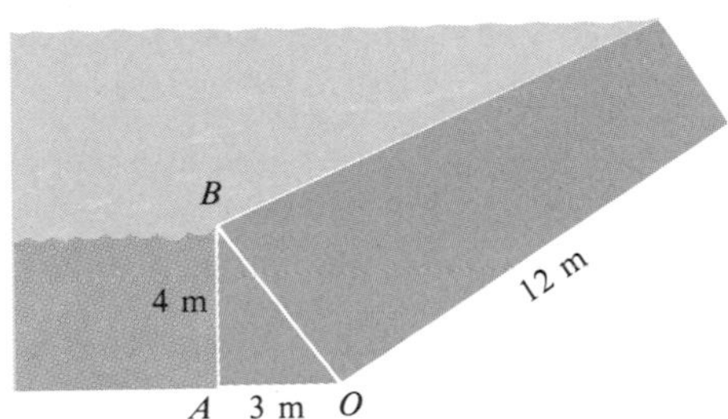

15 • 5 Explain the hydrostatic paradox illustrated in Fig. 15 . 3 in terms of the reaction forces of each of these containers on the fluid.

15 • 6 A cylinder of cork is floating upright in a closed container half full of water. (a) If the air above the water is removed, will the cork rise or fall? (b) Derive an expression for the ratio of the lengths of the unsubmerged portions in the two cases.

15 • 7 It is said that Archimedes discovered his principle while seeking to detect a suspected fraud in the construction of a crown made for Hiero of Syracuse. The crown was thought to have been made from an alloy of gold and silver instead of from pure gold. If it weighed 1000 gf in air and 940 gf in water, how much gold and how much silver did it contain? Assume that the volume of the alloy was the combined volumes of the components.

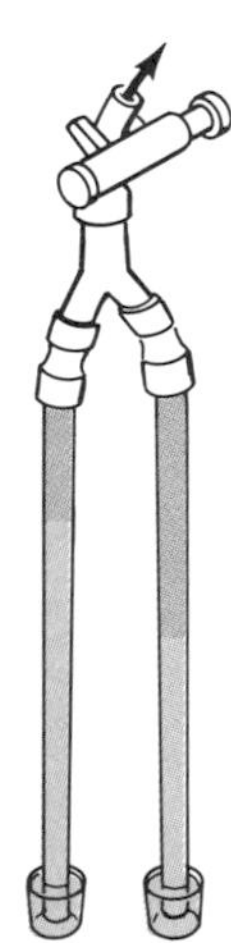

15 • 8 One arm of an inverted Y tube is placed in a liquid of unknown density, while the other arm is placed in water as shown. A portion of the air is pumped from the tube until the unknown liquid stands at a level 33.1 cm above that in its open vessel and the water stands at a level 30 cm above that in the vessel of water. The temperature is 20°C. (a) What is the density of unknown liquid? (b) Is it possible to find the pressure of the air in the tubes above the liquids without knowing the pressure of the outside atmosphere?

15 • 9 Rederive Eq. [15 . 3] including the correction for ρ_a, the density of air. (*Hint:* If your answer is correct, it should reduce to Eq. [15 . 3] when $\rho_a = 0$.)

15 • 10 A 10-g object placed on a 30-g block of waterproofed wood sinks the block to a certain point in water. A 15-g object must be placed on the block to sink it to the same point in a given salt solution. Find the specific gravity of the salt solution.

15 • 11 How much work must be done in order to sink a 50-kg wood cube 1 m^3 in volume to the bottom of a pond of water 2 m deep?

15 • 12 A skindiver finds that in fresh water of density 1.00 g/cm^3 she can neutralize the buoyant force on her by carrying a 2-kg solid object 1 liter in volume. When she is under seawater of density 1.03 g/cm^3, she needs an object 2.5 times this weight and volume to neutralize her buoyancy. (*a*) What is the diver's volume? (*b*) Her average density? (*c*) Her weight?

15 • 13 A brass sphere 1 cm in radius and of density 8.4 g/cm^3 is dropped from the surface of the water into a tank of water 8 m deep. Neglecting friction, find the time required for the weight to reach the bottom.

15 • 14 Considering the earth to be a homogeneous spherical mass of radius R_e and density ρ_e, show that the hydrostatic pressure at a distance r from the center is given by

$$P = \tfrac{2}{3}\pi\rho_e^2 G(R_e^2 - r^2)$$

If the pressure at the center is 1.7×10^{12} $dynes/cm^2$, what is the average density of the earth?

15 • 15 If a liquid is of density ρ_0 at its free surface, the increase in its density with depth y is given by

$$\rho(y) = \frac{\rho_0}{1 - \kappa\rho_0 g y}$$

provided there is no appreciable variation in g and in the temperature and compressibility κ of the liquid with depth. (*a*) Use Eq. [15 . 4] and the fundamental differential equation for a fluid at rest under gravity to prove this relation. (*b*) Show that this expression can be given the approximate form $\rho \doteq \rho_0(1 + cy)$. (*c*) Use this approximate formula to compute the density of seawater at a depth of 1.5 km. (*d*) Derive an approximate expression for the pressure of a compressible liquid at any depth y. State the important assumptions involved and compute the pressure at a depth of 1.5 km in seawater, where

$$\rho_0 = 1030\ kg/m^3$$

15 • 16 If we neglect the effect of the earth's rotation, then for altitudes $y \ll R_e$ the variation in g is given approximately by

$$g(y) \doteq g_0\left(1 - \frac{2y}{R_e}\right)$$

where g_0 is the value of g at the earth's surface, $y_0 = 0$. (*a*) Prove this approximate relation. (*b*) Show that if it is taken into account, the isothermal barometric equation becomes

$$\ln \frac{P(y)}{P_0} = -\frac{\rho_0 g_0 y}{P_0}\left(1 - \frac{y}{R_e}\right)$$

15 • 2–15 • 3 *steady flow*

Unless otherwise specified in a problem, assume the flow to be steady and the effects of fluid compressibility and of all the dissipative frictional forces to be negligible.

15 • 17 A vertical pipe 78 mm in diameter is constricted to 13 mm in diameter. Water is flowing steadily in the pipe at the rate of 1400 cm^3/s, and the pressure at a point 122 cm above the constriction is 6.2×10^6 dynes/cm^2. (*a*) What are the mass flow and flow speed in the pipe? (*b*) What are the flow speed and pressure in the constricted section?

15 • 18 Show that the volume of water flowing per unit time through the horizontal Venturi flowmeter in Fig. 15 . 10 is given by

$$\frac{V}{t} = A_1 A_2 \sqrt{\frac{2g\ \Delta y'\ (\rho'/\rho - 1)}{A_1^2 - A_2^2}}$$

where A_1 and A_2 are the cross-sectional areas of the wide and constricted sections, and $\Delta y'$ is the difference in the heights of the liquid (of density ρ') in the manometer.

15 • 19 Old Faithful geyser in Yellowstone National Park shoots water 124 ft into the air, emerging from the ground with a speed of 89.7 ft/s. What is the gage pressure in the emerging jet?

15 • 20 Suppose the pitot tube in Fig. 15 . 13 is equipped with a differential manometer containing colored water and is inserted in a stream of water. (*a*) Show that the speed of the stream is $v = \sqrt{2g\,\Delta y'}$ where $\Delta y'$ is the difference in heights of the water in the manometer arms. (*b*) If the difference between the stagnation and free-stream pressures in a stream of water is measured as $\Delta y' = 254$ mm, what is the flow speed? (*c*) In 1732 Henri Pitot showed experimentally that if a pitot tube is placed in a liquid flowing with speed v in an *open* channel, the liquid in the tube rises above the free surface a distance $\Delta y' = v^2/2g$. Derive this result theoretically.

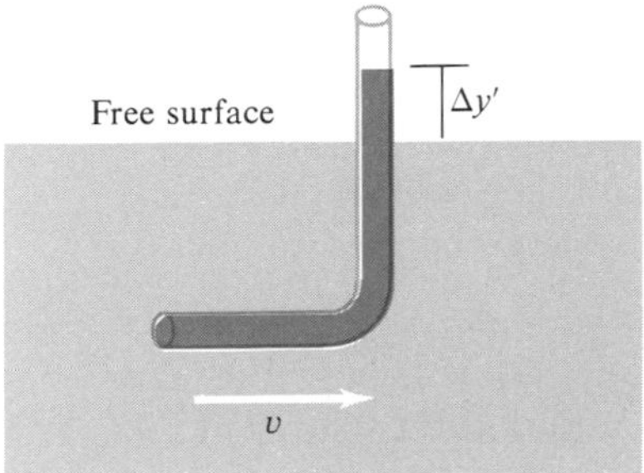

15 • 21 Prove that the speed of efflux of water from an opening in the wall of a reservoir is given by

$$v_e = \sqrt{\frac{2g\,\Delta y'}{1 - (A_2/A_1)^2}}$$

15 • 22 The surface of the water in a large open standpipe is at a height H above the ground. (*a*) If there is a small hole in the pipe at a depth h below the water surface, at what distance Δx from the base of the standpipe will the emerging jet of water strike the ground? (*b*) At what depth from the water surface should the hole be for the emerging jet to reach the ground at the maximum horizontal distance from the base?

15 • 23 A turbine is attached to an opening of cross-sectional area $A_e = 120\ \text{cm}^2$ at the bottom of a reservoir, which is kept filled to a constant height of 20 m. What is the maximum power the water jet can deliver to the turbine?

15 • 24 If the pressure at the top of a siphon filled with liquid becomes zero, the liquid will pull apart and bubbles will form, interrupting the flow. This formation of vapor bubbles in a liquid as a result of a drop in pressure is called *cavitation.* (*a*) Assuming streamline flow and a tube of uniform bore, show that the siphon flow cannot continue if the height of the siphon above the surface of the liquid in the tank exceeds $h_2 = (P_0/\rho g) - h_1$, where P_0 is the atmospheric pressure, ρ is the density of the liquid, and h_1 is the distance of the opening below the liquid level in the tank. (*b*) Find the limiting height h_2 if $h_1 = 3\text{m}$ and $P_0 = 1$ atm. (*c*) If the top of the siphon is actually 1 m above the liquid level in the tank, what is the pressure at that point?

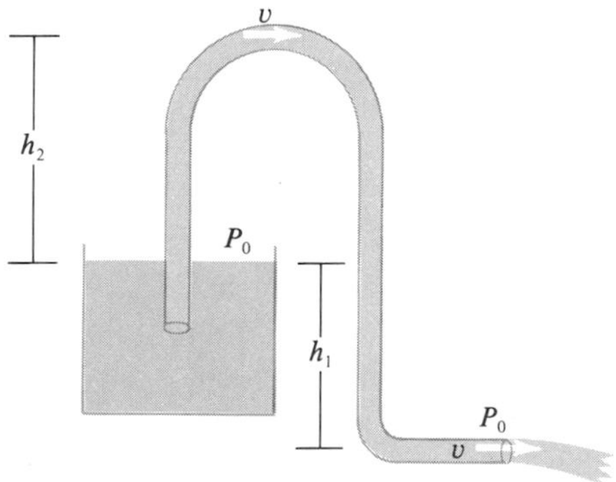

15 • 25 A large closed chamber contains water, and above the water, air at a constant gage pressure of 2 atm. An opening of area $A_e = 10\ \text{cm}^2$ is at a depth of 1 m below the water surface. (*a*) Compute the speed of efflux v_e of the water jet. (*b*) Find the total thrust exerted by the jet on the chamber.

****15 • 26*** A closed vertical cylindrical tank 20 m high and 1 m in radius is half full of water and contains air at 2 atm pressure. How long will it take the tank to empty through a circular hole 1 cm in radius at the bottom of the tank? (*Hint:* $T = \int(dy/v)$; make an analytical approximation over the range $0 \le y \le 0.5$.)

15 • 27 Prove the following theorem, deduced by Torricelli in 1641: For a hole situated in the horizontal bottom of a vessel, the volumes of water escaping in successive equal intervals of time fall off in proportion to successive odd numbers.

15 • 4 viscosity

15 • 28 Consider a very thin cylinder of incompressible fluid in steady laminar flow through a pipe of radius R. Show that if the radius of the cylinder is r, equating the accelerating force on the cylinder to the viscous drag due to the slower-moving liquid outside it yields the differential equation for speed,

$$\frac{dv}{dr} = -\frac{r\,\Delta P}{2\eta l}$$

and find v as a function of the radius of the cylinder.

15 • 29 (*a*) Using the results of Prob. 15 . 28, show that the average flow speed $\bar{v}$ in the pipe is $\bar{v} = \frac{1}{2}v_0$, where v_0 is the flow speed along the central axis of the pipe, $r = 0$. (*b*) Show that the Poiseuille equation can be written in the form $V/t = \frac{1}{2}\pi R^2 v_0$. (*c*) Show that the corresponding equation for flow of a gas through the pipe can be derived by multiplying the Poiseuille equation by the ratio of mean pressure to the entrance pressure.

15 • 30 Using the results of Probs. 15 . 28 and 15 . 29, (*a*) calculate the volume rate of flow. (*b*) Calculate v_0. (*c*) Find $\bar{v}$, the volume rate of flow per unit cross section of pipe, for water of viscosity 10^{-3} dekapoise escaping from a tank filled to an initial height of 40 cm, if the water escapes through a horizontal pipe at the base which is 15 cm long and 0.4 mm in diameter.

15 • 31 At 20°C, a volume V of water takes 25 s to pass through the capillary-flow viscometer in Fig. 15 . 17. (*a*) What time would be needed for passage of the same volume of glycerin of density $\rho = 1.3$ g/cm^3? (*b*) By what factor should the radius of the tube be changed so that the glycerin passes through it in approximately 25 s?

15 • 32 A water droplet with a radius of 40 μm falls through air at 1 atm and 20°C. The density of air under these conditions is 1.2 kg · m^{-3}. (*a*) Compute the final speed v_T of the droplet. (*b*) Experiment shows that the final speed of a water droplet 100 μm in radius is about 0.6 m/s. How does this value compare with that computed from Stokes' law?

15 • 33 Experience indicates that the downward distance y covered in time t by an object projected toward the earth with an initial downward speed v_0 depends only on v_0, g, and t. (*a*) Show by dimensional reasoning that $y = v_0 t\, f(gt/v_0)$, where f is an unknown function of gt/v_0. (*b*) Find f.

15 • 34 (*a*) Scale the differential equation [15 . 23] to dimensionless variables in the form $dw/d\tau = 1 - w$, integrate, and convert back to the form in the text. (*b*) Find the expression for the actual *distance* traveled in a time t if $v(t)$ is given by Eq. [15 . 25].

15 • 35 (*a*) Rework Prob. 15 . 13, this time including the effect of viscosity. (*b*) If the water temperature is 20.2°C, will the sphere approach its terminal speed? (*c*) If the sphere has a radius of 1 mm, how long will it take to reach the bottom of the tank and what will be its final speed?

15 • 36 An air bubble 1 cm in radius is formed at the bottom of a 5-m column of glycerin, of density $\rho = 1.25\ \text{g/cm}^3$ and viscosity $\eta = 1.5\ \text{N}\cdot\text{s/m}^2$. If the density of air at the surface is $1.2\ \text{kg/m}^3$ and the bubble expands as it rises, how long will it take to reach the surface? Use Boyle's law. (*Hint*: How long will it take the bubble to reach terminal speed?)

15 • 5 laminar and turbulent flow

15 • 37 A horizontal pipe 10 cm in diameter is kept a constant temperature of 20°C. What is the maximum average flow speed $\bar{v}$ to which the Poiseuille equation can be applied ($N_{\text{Re}} < 2000$) (*a*) when the fluid is water? (*b*) When the fluid is linseed oil, with a density $\rho = 0.94\ \text{g/cm}^3$ and viscosity coefficient $\eta = 0.44$ poise? (*c*) When the fluid is glycerin, with a density $\rho = 1.3\ \text{g/cm}^3$ and viscosity coefficient $\eta = 14.9$ poise?

15 • 38 The drag coefficient C_D for a body depends not only on its shape, but also on its orientation to the direction of flow. Consider a fluid composed of particles of mass m, initially flowing horizontally with speed v against a square metal plate at some angle θ to its normal. There are n particles per unit volume of fluid, and each reflects specularly from the plate. (*a*) If a single particle collides with the plate in a time Δt, what impulse does it impart to the plate? (*b*) What is the average pressure on the plate? (*c*) What is the drag coefficient for the plate?

15 • 39 A fluid consists of a stream of independent particles of mass m and density $\rho = nm$, all traveling in the same direction with speed v and reflected specularly from a sphere of radius R. (*a*) Find the average change in momentum of a particle reflected from the sphere. (*b*) What is the drag coefficient of the sphere?

15 • 40 A steel sphere of density $\rho = 7.8\ \text{g/cm}^3$ is observed to have a terminal speed of 10 cm/s when it falls from rest through lubricating oil of density $\rho_f = 0.9\ \text{g/cm}^3$ and viscosity coefficient $\eta = 0.357\ \text{N}\cdot\text{s/m}^2$. (*a*) Compute the Reynolds number. (*b*) Compute the drag coefficient. (*c*) Why is the Stokes' law method of determining η especially useful for liquids that are very viscous?

15 • 41 The drag force on a sphere of density ρ moving with constant velocity in a fluid of density ρ_f is given by Eq. [15 . 30] instead of by Stokes' law. (*a*) Show that its terminal speed, when it falls from rest in the fluid, is given by

$$v_T = \frac{2}{9}\frac{R^2 g}{\eta}(\rho - \rho_f)\frac{24}{C_D N_{\text{Re}}}$$

(*b*) Experiment shows that raindrops 2800 μm in diameter, falling in air of temperature 20°C and practically uniform density $\rho = 1.2 \times 10^{-3}\ \text{g/cm}^3$, have a terminal speed of about 7 m/s. What is the resulting value of $\frac{1}{24}C_D N_{\text{Re}}$ for the sphere? (*c*) The value of N_{Re}? (*d*) What would have been the terminal speed if Stokes' law had been used for this case?

15 • 42 In designing the prototype of an aircraft, spacecraft, or vessel, tests are usually carried out with a scale model, which is *geometrically similar* to the prototype. If the model and the prototype are exposed to fluid flows such that the forces on them are everywhere proportional, then they are *dynamically similar*. This is the case when both systems have the same Reynolds number. The *similarity analysis* leading to this principle is, in effect, the scaling of equations of motion so they have the same dimensionless form. (*a*) Water at 40°C, with density $\rho = 0.99$ g/cm^3 and viscosity coefficient $\eta = 0.0065$ poise, flows in a pipe 1 cm in diameter. What must be the flow speed of castor oil, with $\rho = 0.95$ g/cm^3 and $\eta = 2.31$ poise, through a pipe 10 cm in diameter for the two flows to be dynamically similar? (*b*) In wind-tunnel tests of aircraft models, why might it be desirable to pressurize the air in the tunnel?

15 • 43 A skydiver plummeting earthward reaches a terminal speed of 120 mi/h. If he weighs 180 lbf and his effective area is 8 ft^2, what is his drag coefficient? Take the density of air as 1.2 kg/m^3 (where 1 kg/m^3 = 0.0624 lbm/ft^3).

15 • 44 A sphere of radius R and mass m moves along a frictionless horizontal track in air, starting with an initial speed v_0. (*a*) Show that a force obeying Stokes' law leads to a predicted position along the track given by the relation

$$x(t) = \frac{v_0}{\alpha}(1 - e^{-\alpha t})$$

where $\alpha = 6\pi\eta R/m$ is a constant and $x(\infty) = v_0/\alpha$. (*b*) Show that a force obeying Newton's law leads to a predicted position along the track given by the relation

$$x(t) = \frac{2m}{C_D A\rho} \ln\left(\frac{C_D A\rho}{2m} v_0 t + 1\right)$$

where $A = \pi R^2$ and $x(\infty) = \infty$.

***15 • 45** Suppose the sphere of Prob. 15 . 44 begins its motion with a Reynolds number of $N_{\text{Re}} = 10{,}000$. (*a*) Show that Stoke's law predicts a maximum displacement of 833 $(m/\rho A)$, where $A = \pi R^2$. (*b*) Do a numerical computation of the maximum displacement, using $x(t)$ from part (*b*) of Prob. 15 . 44 *until* $t = 18T$, taking the time scale as $T = 2m/\rho A v_0$, and the length scale as $L = v_0 T$. (*c*) *After* $t = 18T$ perform an improved Euler method integration, using table lookup to determine $C_D(N_{\text{Re}})$ from the graph of Fig. 15 . 21 and going over to the Stokes' law formula when $N_{\text{Re}} = 1$. You should find a maximum displacement of about $19.5(m/\rho A)$. Note that according to Newtonian prediction the sphere would *never* stop moving. A naïve approach might then assume an actual maximum displacement between Stokes' prediction of $x(\infty) = 833(m/\rho A)$ and infinity. Can you explain why this is not the case?

***15 • 46** Write the equations of motion for a space vehicle reentering the atmosphere in terms of C_D and density of air as predicted by the law of atmospheres. Assume a flat earth and treat the vehicle as a particle. Set up an algorithm for solving the equations numerically in the form of a flow diagram.

answers

15 • 1 $F = 92$ N
15 • 2 Mechanical advantage $= A_1/A_2$
15 • 3 $F = \frac{1}{2}\rho g x y^2$ at height $y/3$
15 • 4 (*a*) $F = 940{,}800$ N;
(*b*) $\tau = 1{,}254{,}400$ N · m; (*c*) $h = 267$ cm;
(*d*) $\rho_{\min} = 890$ kg/m^3
15 • 5 The sloped container walls exert reaction forces downward (Fig. 15 . 3*b*) or upward (Fig. 15 . 3*c*)
15 • 6 (*a*) fall; (*b*) $h'/h = (\rho_l - \rho_a)/\rho_l$
15 • 7 811 g gold, 189 g silver
15 • 8 (*a*) $\rho = 0.906$ g/cm^3; (*b*) $P = P_0 - \rho g h$
15 • 9 $\rho = (m_a \rho_f - m_l \rho_a)/\Delta m$
15 • 10 Specific gravity $= 1.13$
15 • 11 $W = 13{,}720$ N · m
15 • 12 (*a*) $V = 47.5$ liters;
(*b*) $\bar{\rho} = 0.979$ g/cm^3; (*c*) $F_g = 456$ N
15 • 13 $\Delta t = 1.361$ s
15 • 14 $\bar{\rho} = 5.5$ g/cm^3
15 • 15 (*c*) $\rho = 1040$ kg/m^3;
(*d*) $P = 150.2$ atm
15 • 17 (*a*) $dm/dt = 1400$ g/s, $v = 29.3$ cm/s;
(*b*) $v = 1055$ cm/s, $P = 5.76 \times 10^6$ dynes/cm^2
15 • 19 $P_g = -0.0311$ atm
15 • 20 (*b*) $v = 223$ cm/s
15 • 22 (*a*) $\Delta x = 2\sqrt{h(H - h)}$;
(*b*) depth $= \frac{1}{2}H$
15 • 23 $dW/dt = 46.57$ kW
15 • 24 (*b*) $h_2 = 7.34$ m; (*c*) $P = 0.613$ atm
15 • 25 (*a*) $v_e = 20.61$ m/s; (*b*) $F = 424.8$ N
15 • 26 $\Delta t = 3.1$ h
15 • 28 $v = (\Delta P/4\eta l)(R^2 - r^2)$
15 • 30 (*a*) $dV/dt = 16.42$ mm^3/s;
(*b*) $v_0 = 26.14$ cm/s; (*c*) $\bar{v} = 13.07$ cm/s
15 • 31 (*a*) $\Delta t = 10$ h 21 min; (*b*) $R'/R = 6.2$
15 • 32 (*a*) $v_T = 18.9$ cm/s; (*b*) half as much (because the droplet flattens as it falls)
15 • 33 (*b*) $f = 1 + \frac{1}{2}gt/v_0$
15 • 34 $x = (FT/m)[t - T\exp(-t/T)]$, where $T = m/6\pi\eta R$
15 • 35 (*a*) $\Delta t = 1.363$ s; (*b*) no;
(*c*) $\Delta t = 1.527$ s to reach bottom at speed $v = 0.57 v_T$

15 • 36 $\Delta t = 23.9$ s
15 • 37 (*a*) $\bar{v} = 2$ cm/s; (*b*) $\bar{v} = 94$ cm/s; (*c*) $\bar{v} = 2300$ cm/s
15 • 38 (*a*) $\Delta p = 2mv \cos\theta$; (*b*) $\overline{P} = 2nmv^2 \cos^2\theta$; (*c*) $C_D = 4\cos^2\theta$
15 • 39 (*a*) $\overline{\Delta p} = mv$; (*b*) $C_D = 2$
15 • 40 (*a*) $N_{\text{Re}} = 0.777$; (*b*) $C_D = 30.9$
15 • 41 (*b*) $C_D N_{\text{Re}}/24 = 0.03$; (*c*) $N_{\text{Re}} = 1278$; (*d*) $v_T = 232$ m/s
15 • 42 (*a*) $v = 3.70$ m/s
15 • 43 $C_D = 0.624$
15 • 46 $d^2x/dt^2 = -B\sqrt{u^2 + v^2}\,u$; $d^2y/dt^2 = -g - B\sqrt{u^2 + v^2}\,v$; where $u = dx/dt$, $v = dy/dt$, and $B = [C_D A\rho_s \exp(-y/h_0)]/2\pi$.
Algorithm: solve equations for dy/dt, du/dt, and dv/dt simultaneously and $dx/dt = u$ separately; suggested scaling factors: $L = 1/A\rho_s$, $T = \sqrt{L/g}$

CHAPTER SIXTEEN

oscillations

Given a pendulum composed of any number of weights, if each of the weights be multiplied by the square of its distance from the axis of oscillation and if the sum of these products be divided by that obtained by multiplying the sum of the weights by the distance of the common center of gravity of all the weights from the same axis of oscillation, there results the length of a simple pendulum which is isochronous with the compound pendulum, or in other words the distance between the axis and the center of oscillation of the given compound pendulum.

Christian Huygens, Horologium Oscillatorium,
part IV, proposition V, 1673

Periodic phenomena are those that are repeated at regular intervals of some independent variable—time, space, or some combination of both. We have already seen examples of time-dependent periodicity in the orbital motions of the planets (which, in fact, define time for us) and in the vibratory motion of a spring-mass system. Motion that is periodic in time is called *oscillatory motion.* In the next chapter we shall discuss the *propagation,* or motion in space, of the disturbances caused by these oscillations—the phenomenon known as *wave motion.*

The simplest and most important oscillatory motion is the vibration of a particle about an equilibrium position under the influence of a Hooke's law restoring force which is proportional to the displacement of the particle from its equilibrium position. This type of vibration is called *simple harmonic motion.** As long as the effects of *damping* forces due to friction remain small, all oscillations ordinarily arising from the elasticity of matter are either simple harmonic, or a composite of simple harmonic motions. The swings of a pendulum, the vibrations of a tuning fork or the strings of a musical instrument, or a building swaying in a high wind are all examples of simple harmonic motion, as are many other phenomena in acoustics, optics, and electromagnetism.

The physical essentials of simple harmonic motion are already familiar

*This term was used by Lord Kelvin and P. G. Tait because the simplest musical sounds are produced by such vibrations. See their *Treatise on Natural Philosophy,* part I (Cambridge University Press).

from preceding chapters. It is natural to wonder at all this attention to the simplest type of periodic motion. Many recurring motions are only "quasi-periodic," because of the presence of dissipative forces, while others are not even periodic at all. In 1822, however, Joseph Fourier demonstrated the following theorem:

> Any finite periodic motion can be represented as the summation of a number of suitably chosen simple harmonic motions.

This is the famous *Fourier series* (see Appendix K). We shall also examine another principle, illustrated in Sec. 8 . 3:

> In any physical situation, sufficiently small displacements from a condition of stable equilibrium result in approximately simple harmonic motion about the potential minimum.

An understanding of simple harmonic motion, then, is essential to the comprehension of all periodic phenomena in physics.

16 · 1 simple harmonic motion

One complete cycle of an oscillatory motion takes place in a time interval $\Delta t = T$, known as its *period*. The *frequency* of the motion, ν, is the number of cycles per unit time, $\nu = 1/T$; hence the appropriate unit of frequency is $[\nu] = \mathrm{s}^{-1}$. However, frequency ν is more often expressed in hertz (Hz) or cycles per second (cps), where 1 Hz = 1 cps, to distinguish it from the angular frequency $\omega = 2\pi\nu$, where $[\omega] = \mathrm{rad/s}$. Note that the same symbol is used for the oscillatory frequencies as was used for rotational frequencies in Chapter 11. This is because the two quantities have a close mathematical similarity even though their physical meanings differ. As we shall see, any uniform circular rotation can be decomposed into two component oscillatory motions at right angles to each other.

The first-order numerical integration in Sec. 6 . 6 demonstrated, within the limits of numerical accuracy, that the motion of a mass m attached to a stretched spring of force constant k is periodic, with a period $T = 2\pi\sqrt{m/k}$. The velocity of the spring is also periodic, but the velocity curve appears to be shifted by $\frac{1}{4}T$ with respect to the graph of displacement. Furthermore, in scaling the differential equation of motion we found that the length scale

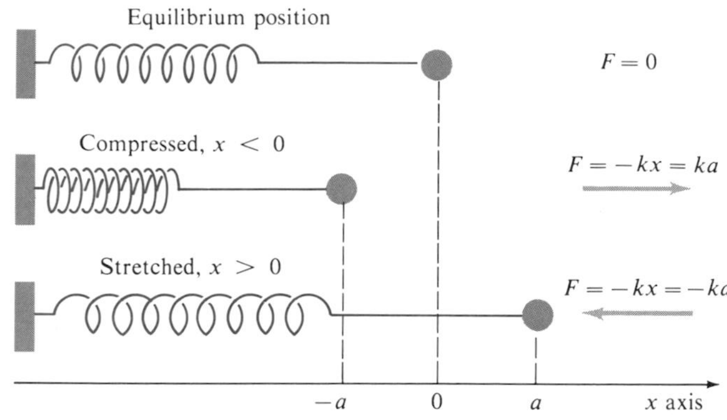

figure 16 • 1 *Restoring force of a spring*

canceled and was therefore arbitrary. This implies that the *amplitude,* or maximum displacement, of the motion is independent of the frequency. Let us now derive these facts analytically.

Consider again the simple spring-mass system in Fig. 16 . 1, subject to a Hooke's law restoring force $F_x = -kx$, where x is the position of the mass m with respect to the equilibrium position $x = 0$. From Newton's second law, the acceleration of the mass is

$$\frac{d^2x}{dt^2} = -\frac{k}{m}x \qquad [16 \cdot 1]$$

which is a second-order linear differential equation in x. However, since this equation is difficult to integrate analytically, let us instead use the energy equation for such a system, Eq. [8 . 14]:

$$\tfrac{1}{2}mv^2 + \tfrac{1}{2}kx^2 = K + U = E \qquad [16 \cdot 2]$$

where E is a constant whose value depends on the initial conditions x_0 and v_0.

No matter what the initial conditions are, we can see from Eq. [16 . 2] that the maximum extension or *amplitude* A of the motion occurs at the turning points, where $v = 0$. Hence the total mechanical energy E can be written as

$$E = \tfrac{1}{2}kA^2 \quad \text{or} \quad A^2 = \frac{2E}{k} \qquad [16 \cdot 3]$$

Substituting $E = \frac{1}{2}kA^2$ in Eq. [16 . 2] and solving for v, we obtain

$$v = \pm\sqrt{\frac{k}{m}}\sqrt{A^2 - x^2} \qquad [16 \cdot 4]$$

It is clear that a real solution requires that $|x| \leq A$, and this implies that

$$|v| \leq A\sqrt{k/m}$$

Therefore the velocity is also periodic and has its extreme values, $v = \pm A\sqrt{k/m}$ at $x = 0$, when the mass is passing through the equilibrium point in one direction or the other.

To obtain an expression for the position x at any given moment, we substitute dx/dt for v in Eq. [16 . 4] and rearrange to obtain

$$\frac{dx}{\sqrt{A^2 - x^2}} = \pm\sqrt{\frac{k}{m}}\,dt \qquad [16 \cdot 5]$$

Integration of both members (omitting the sign) yields

$$\sin^{-1}\frac{x}{A} = \sqrt{\frac{k}{m}}\,t + \phi$$

or

$$x = A\sin\left(\sqrt{\frac{k}{m}}\,t + \phi\right) \qquad [16 \cdot 6]$$

where the angle ϕ is the constant of integration of Eq. [16 . 5]. We can derive the general expressions for the velocity and acceleration of the mass by differentiating with respect to time:

$$v = \frac{dx}{dt} = A\sqrt{\frac{k}{m}}\cos\left(\sqrt{\frac{k}{m}}\,t + \phi\right) \qquad [16 \cdot 7]$$

$$a = \frac{dv}{dt} = -A\frac{k}{m}\sin\left(\sqrt{\frac{k}{m}}\,t + \phi\right) \qquad [16 \cdot 8]$$

It is easy to verify, by substituting these equations into Eq. [16 . 1], that Eq. [16 . 6] is the general solution of the equation of motion for the mass, since A and ϕ may take any values. Whenever you see an equation of the form

$$\frac{d^2x}{dt^2} + \omega^2 x = 0$$

you know that its most general solution is

$$x = A\sin(\omega t + \phi)$$

We shall see later how A and ϕ can be determined from x_0 and v_0.

The argument $\sqrt{k/m}\,t + \phi$ in Eqs. [16 . 6] to [16 . 8] is called the *phase* of the motion, and ϕ is termed the *phase constant,* or *epoch angle;* obviously it is the phase at the instant $t_0 = 0$. Note that $x(t)$ can equally well be represented by a cosine function,

$$x = A\cos\left(\sqrt{\frac{k}{m}}\,t + \phi'\right) \qquad [16 \cdot 9]$$

if the phase constant is taken as $\phi' = \phi - \frac{1}{2}\pi$. The motion described by Eqs. [16 . 6] and [16 . 9] is known as *simple harmonic motion.* Since it is oscillatory, the motion can be represented by either of these two functions, regardless of the sign of v in Eq. [16 . 4]. In fact, uniform motion in a circle with angular velocity ω can be represented by two such oscillations at right angles. Thus, from Eq. [11 . 12],

$$\mathbf{r}(t) = R(\cos\theta\mathbf{i} + \sin\theta\mathbf{j}) = R(\cos\omega t\mathbf{i} + \sin\omega t\mathbf{j})$$

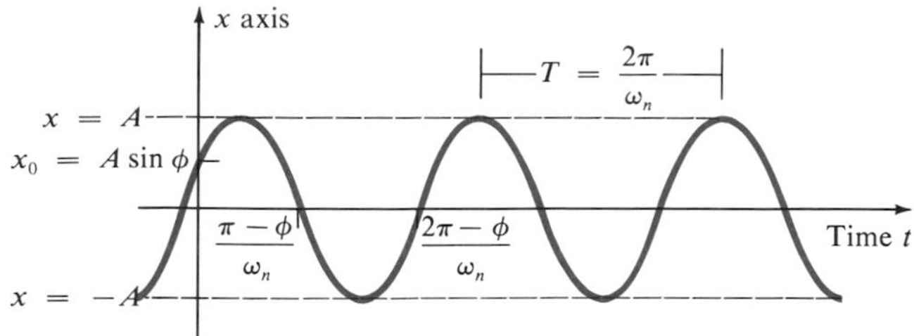

figure 16 • 2 *Displacement-time graph of simple harmonic motion*

where **r** is the position of the moving point and R is the radius of rotation.

The rules of calculus require that the arguments of the trigonometric functions in Eqs. [16 . 6] to [16 . 8] be measured in radians, so the factor $\sqrt{k/m}$ must have the unit radians per second, like an angular speed. Let us therefore define

$$\omega_n = \sqrt{\frac{k}{m}} \qquad [16 \cdot 10]$$

as the *natural angular frequency* of the motion. The quantity ω_n is the most fundamental of the various concepts associated with simple harmonic motion, apart from the idea of periodicity itself.

It is evident that a given value of x occurs periodically; for if t is increased by $T = 2\pi/\omega_n$, then Eq. [16 . 6] becomes

$$x(t + T) = A \sin\left[\omega_n\left(t + \frac{2\pi}{\omega_n}\right) + \phi\right]$$

$$= A \sin\left[\omega_n t + 2\pi + \phi\right] = A \sin\left[\omega_n t + \phi\right] = x(t)$$

Therefore the period T is exactly

$$\blacktriangleright \quad T = \frac{2\pi}{\omega_n} = 2\pi\sqrt{\frac{m}{k}} \qquad [16 \cdot 11]$$

as we found, approximately, in the numerical integration of Sec. 6 . 6. Note that as the inertial mass m is increased, the system becomes more sluggish and T increases; conversely, T decreases for "stiffer" springs (larger k). Note also that T depends only on the parameters of the physical system, not on the initial conditions. No matter how the motion is initiated—by pushing the mass, pulling it, or throwing stones at it—the period is the same.

Figure 16 . 2 is a graph of simple harmonic motion as a function of time. The initial value of x is $x_0 = A \sin \phi$, and the function $x(t)$ oscillates beween $x = +A$ and $x = -A$, crossing the equilibrium position $x = 0$ at times where the argument $\omega_n t + \phi$ is equal to $n\pi$, where n is an integer. As shown on the graph, the period is $T = 2\pi/\omega_n$.

example 16 • 1 Find the frequency of oscillation of a vertical spring-mass system under the influence of gravity.

solution Assume that l is the length of the unstretched spring and that y is the position of the mass under the influence of gravity, where y is measured positively downward from the fixed end. Therefore

$$m\frac{a^2y}{dt^2} = -k(y - l) + mg$$

When the spring is at rest under the influence of gravity, its equilibrium position y_0, at which $d^2y/dt^2 = 0$, is found by equating the weight to the restoring force:

$$mg = k(y_0 - l)$$

Hence the equation of motion can be written as

$$m\frac{d^2y}{dt^2} = -k(y - l) + k(y_0 - l) = -k(y - y_0)$$

Setting $y' = y - y_0$, the displacement from equilibrium under gravity, we obtain

$$m\frac{d^2y}{dt^2} = m\frac{d^2y'}{dt^2} = -ky'$$

This equation, like Eq. [16 . 1], represents simple harmonic motion with a frequency $\omega_n = \sqrt{k/m}$, the same frequency as in the case of a horizontal spring. This result, obtained by considering the force and the law of motion of the system, also agrees with that of Sec. 8 . 5, where we saw that the total gravitational and spring potential could be cast into the form of a quadratic equation, with the mass m performing simple harmonic oscillations within the parabolic potential well.

16 • 2 *energy: initial conditions*

The energies of an oscillating system can also be expressed as functions of time:

► $$U(t) = \tfrac{1}{2}kx^2 = \tfrac{1}{2}kA^2\sin^2(\omega_n t + \phi) \qquad [16 \cdot 12]$$

► $$K(t) = E - U = \tfrac{1}{2}kA^2[1 - \sin^2(\omega_n t + \phi)]$$
$$= \tfrac{1}{2}kA^2\cos^2(\omega_n t + \phi) \qquad [16 \cdot 13]$$

Using the half-angle formulas

$$\cos^2\theta = \tfrac{1}{2}(1 + \cos 2\theta) \qquad \sin^2\theta = \tfrac{1}{2}(1 - \cos 2\theta)$$

(see Appendix E), we can show that the kinetic and potential energies also vary sinusoidally, but with *half* the period of the displacement itself. Substituting for the squared sine and cosine terms in Eqs. [16 . 12] and [16 . 13] gives the energies in terms of a simple cosine function

$$U(t) = \tfrac{1}{4}kA^2[1 - \cos 2(\omega_n t + \phi)] \qquad [16 \cdot 14]$$

$$K(t) = \tfrac{1}{4}kA^2[1 + \cos 2(\omega_n t + \phi)] \qquad [16 \cdot 15]$$

These two formulas are particularly useful in computing the *time-average*

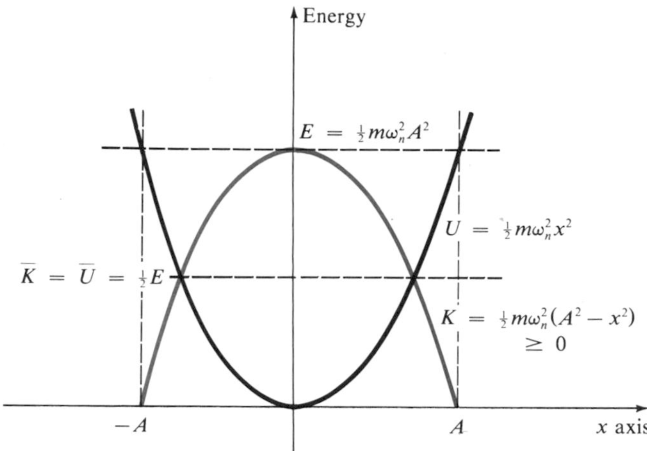

figure 16 • 3 *Energy relations for a simple harmonic oscillator*

values of the energies, $\overline{U}$ and $\overline{K}$, to which we are most directly sensitive in those physical applications in which harmonic time-varying phenomena are encountered. Since natural phenomena such as sound or light often oscillate so rapidly, we see, hear, or measure energy values which are averaged over times that are much larger than the period T.

For periodic motions it is sufficient to determine the average energies over a single period. These can be computed from the formulas

$$\overline{K} = \frac{1}{T}\int_0^T K(t)\,dt \qquad \overline{U} = \frac{1}{T}\int_0^T U(t)\,dt \qquad [16 \cdot 16]$$

The average value of a sine or cosine over any integral number of periods must be zero, since for every positive value there is a corresponding negative value of the function over the range of integration. Hence, substituting Eqs. [16 . 14] and [16 . 15] into the forms in Eqs. [16 . 16] gives no contribution from the cosine term:

$$\overline{K} = \overline{U} = \tfrac{1}{4}kA^2 = \tfrac{1}{2}E \qquad [16 \cdot 17]$$

The total energy of the oscillator is, on the average, half kinetic and half potential, a fact that will be of primary importance when we discuss the specific heats of substances in Chapter 21. If we plot actual (not average) potential and kinetic energy as a function of position, we obtain the parabolic curves of Fig. 16 . 3. In Fig. 16 . 4 the kinematic and energetic relations are plotted for the special case where the phase constant is $\phi = \frac{1}{2}\pi$.

example 16 • 2 Find $\overline{K}$ by direct integration of $\frac{1}{2}m\int_0^T v^2\,dt$.

solution From Eq. [16 . 7], if we set $\omega_n^2 = k/m$, we have $v^2 = \omega_n^2A^2\cos^2(\omega_nt + \phi)$, and since $\int_0^T\cos^2(\omega_nt + \phi)\,dt = \frac{1}{2}T$ (see Appendix F), Eq. [16 . 16] yields

$$\overline{K} = \frac{m}{2T}\int_0^T v^2\,dt = \tfrac{1}{4}m\omega_n^2A^2 = \tfrac{1}{2}E$$

While the frequency of a simple harmonic oscillation is independent of the way in which the motion is initiated, the initial conditions will determine the values of the amplitude and phase constant, and consequently the precise details of displacement, velocity, and acceleration as functions of time. In

figure 16 • 4 *Kinematic and energy relations for a simple harmonic oscillator, $\phi = \pi/2$. Note that the period of U is one-half the period of x. The kinetic energy K is the distance from U(t) up to the line U = E, so the graph of K(t) is identical to that of U(t), but displaced by $\frac{1}{2}T$ along the time axis.*

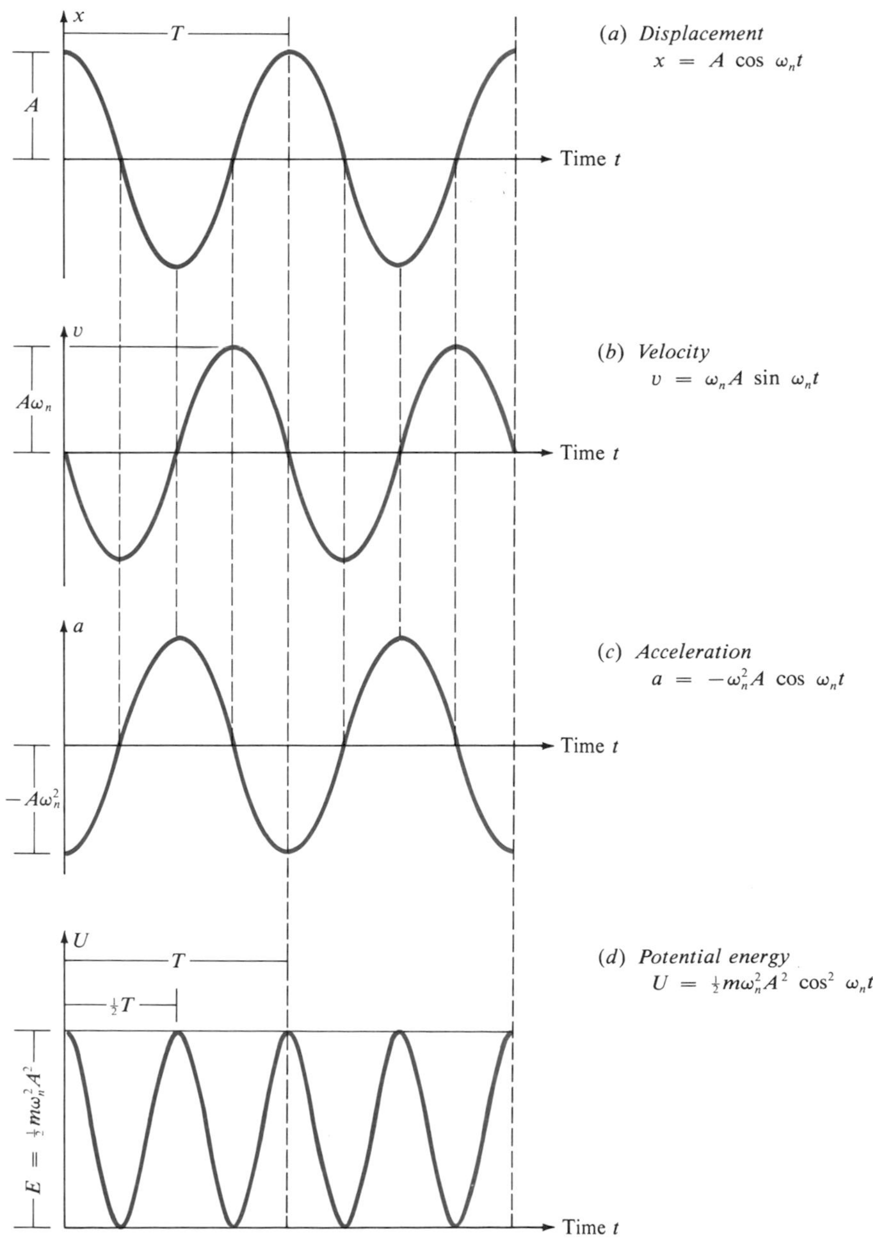

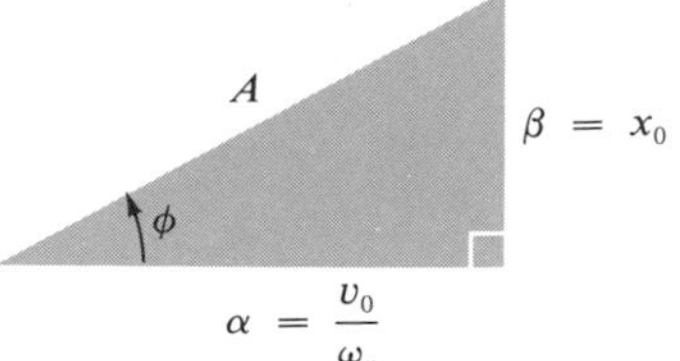

figure 16 • 5 *Mnemonic for relationships between constants of integration for simple harmonic motion:*

$$x = A \sin(\omega_n t + \phi) = \alpha \sin \omega_n t + \beta \cos \omega_n t$$

order to determine the motion completely, we must know *two* independent facts, usually given as the initial position x_0 and the initial velocity v_0. The amplitude can be determined from the energy equation [16 . 2], evaluated at time $t = 0$:

$$\tfrac{1}{2}mv_0^2 + \tfrac{1}{2}kx_0^2 = E = \tfrac{1}{2}m\omega_n^2 A^2$$

or

$$A = \sqrt{x_0^2 + \frac{v_0^2}{\omega_n^2}} \qquad [16 \cdot 18]$$

Then at time $t = 0$ the initial position is

$$x_0 = A \sin \phi$$

and the phase constant is

$$\phi = \sin^{-1} \frac{x_0}{A} = \tan^{-1} \frac{\omega_n x_0}{v_0} = \cos^{-1} \frac{v_0}{\omega_n A} \qquad [16 \cdot 19]$$

where the last two equalities can be derived by substituting Eq. [16 . 18] into the identities for inverse trigonometric functions (see Appendix E). Figure 16 . 5 is a useful mnemonic for the above relationships.

Sometimes it is more useful to represent $x(t)$ as a sum,

$$x(t) = \alpha \sin \omega_n t + \beta \cos \omega_n t \qquad [16 \cdot 20]$$

which can be related to our original formula, Eq. [16 . 6], by expanding the trigonometric function:

$$\begin{aligned} x(t) &= A \sin(\omega_n t + \phi) \\ &= (A \cos \phi) \sin \omega_n t + (A \sin \phi) \cos \omega_n t \end{aligned} \qquad [16 \cdot 21]$$

Then

$$\alpha = A \cos \phi \qquad \beta = A \sin \phi$$

so that

$$A = \sqrt{\alpha^2 + \beta^2} \qquad \phi = \tan^{-1} \frac{\beta}{\alpha} \qquad [16 \cdot 22]$$

From the expressions for ϕ in Eq. [16 . 19] we also see that

$$\alpha = \frac{v_0}{\omega_n} \qquad \beta = x_0 \qquad [16 \cdot 23]$$

which completes the set of mathematical relations most fundamental to the discussion of oscillations. These relations are summarized in Table 16 . 1.

table 16 • 1 *Definitions and equations of simple harmonic motion*

equation of motion	$\frac{d^2x}{dt^2} + \frac{k}{m}x = 0$
period	$T = 2\pi\sqrt{\frac{m}{k}}$
natural frequency	$\nu_n = \frac{1}{T} = \frac{1}{2}\pi\sqrt{\frac{k}{m}}$
natural angular frequency	$\omega_n = 2\pi\nu_n = \sqrt{\frac{k}{m}}$
restoring force	$F = -kx = -m\omega_n^2 x$
displacement	$x = A\sin(\omega_n t + \phi) = \alpha\sin\omega_n t + \beta\cos\omega_n t$
	$\alpha = A\cos\phi,\ \beta = A\sin\phi$
	$A^2 = \alpha^2 + \beta^2,\ \phi = \tan^{-1}\frac{\beta}{\alpha}$
velocity	$v = A\omega_n\cos(\omega_n t + \phi)$
	$= \alpha\omega_n\cos\omega_n t - \beta\omega_n\sin\omega_n t$
acceleration	$a = -A\omega_n^2\sin(\omega_n t + \phi) = -\omega_n^2 x$
speed	$\lvert v\rvert = \omega_n\sqrt{A^2 - x^2}$
potential energy	$U = \frac{1}{2}kx^2 = \frac{1}{2}kA^2\sin^2(\omega_n t + \phi)$
	$= \frac{1}{4}kA^2[1 - \cos 2(\omega_n t + \phi)]$
kinetic energy	$K = \frac{1}{2}kA^2 - U = \frac{1}{4}kA^2[1 + \cos 2(\omega_n t + \phi)]$
total energy	$E = K + U = \frac{1}{2}mv^2 + \frac{1}{2}kx^2 = \frac{1}{2}kA^2$
initial conditions: amplitude	$A = \sqrt{x_0^2 + \left(\frac{v_0}{\omega_n}\right)^2}$
phase constant	$\phi = \sin^{-1}\frac{x_0}{A} = \cos^{-1}\frac{v_0}{\omega_n A}$
	$= \tan^{-1}\frac{\omega_n x_0}{v_0}$
	$\alpha = \frac{v_0}{\omega_n} \quad \beta = x_0$
time-average energies	$\overline{U} = \overline{K} = \frac{1}{2}E$

example 16 • 3 A small hemispherical cup of mass m_1 is attached to the free end of a spring, as shown, and a ball of mass m_2 is pushed snugly into the cup, compressing the spring by an amount Δx from its equilibrium position. The ball is then released. (*a*) How much time elapses before the ball and the cup begin to separate? (*b*) What is the energy of the ball at separation? (*c*) What is the equation for the position of the spring-cup system just after separation?

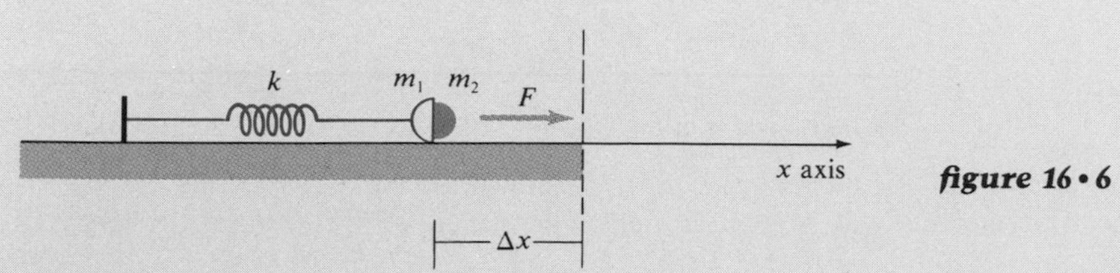

figure 16 • 6

solution (*a*) Once the cup ceases to accelerate the ball, the two objects will separate. This occurs when they pass through the equilibrium position of the spring, one-fourth of a period after release:

$$\tfrac{1}{4}T = \tfrac{1}{2}\pi\sqrt{\frac{m_1 + m_2}{k}}$$

(*b*) The total kinetic energy of the system is at a maximum at this point and equals the total initial potential energy of the compressed spring, $U_0 = \tfrac{1}{2}k\,(\Delta x)^2$:

$$K_1 + K_2 = \tfrac{1}{2}(m_1 + m_2)v^2 = \tfrac{1}{2}k\,(\Delta x)^2$$

Therefore, at separation speed v, the kinetic energy of the ball is

$$K_2 = \frac{m_2}{m_1 + m_2}\tfrac{1}{2}k\,(\Delta x)^2$$

(*c*) If we now take the time of separation as $t_0 = 0$, then $x_0 = 0$ and the speed of the system predicted by the first equation in part (*b*) is

$$v_0 = \Delta x\sqrt{\frac{k}{m_1 + m_2}}$$

Since $\alpha = v_0/\omega_n$ and $\beta = x_0 = 0$, substitution into Eq. [16 . 20] gives the position of the spring-cup system after separation as

$$x(t) = \Delta x\sqrt{\frac{m_1}{m_1 + m_2}}\sin\omega_n t \qquad \text{where} \qquad \omega_n = \sqrt{\frac{k}{m_1}}$$

16 • 3 *pendulums*

So far we have considered simple harmonic motion as it relates to linear displacement. In the case of a pendulum we are concerned with the oscillations of an *angular displacement*. One example is the *torsion pendulum*. As we saw in Sec. 14 . 3, the torque exerted on a twisted solid cylinder of shear modulus S, radius of rotation R, and length l is related to the angular displacement θ of the free end by

$$\tau = \tau_0\theta \qquad [16 \cdot 24]$$

where $\tau_0 = \pi R^4 S/2l$. When the cylinder is released, the restoring forces arising from the elasticity of the material will cause it to oscillate, twisting back and forth. If the cylinder is actually a very thin rod or wire, and we

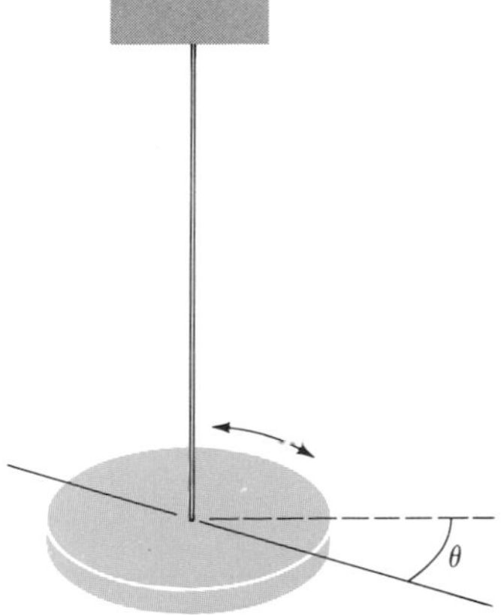

figure 16 • 7 Torsion pendulum

suspend a large disk from its free end, the result is the torsion pendulum in Fig. 16 . 7. The rotational analog of the second law of motion then gives

$$I\frac{d^2\theta}{dt^2} = -\tau_0\theta \qquad [16 \cdot 25]$$

where the moment of inertia of the pendulum with respect to the axis of the wire is

$$I = I_{\text{wire}} + I_{\text{disk}} \doteq I_{\text{disk}}$$

Comparison with Eq. [16 . 1] shows that the oscillations are simple harmonic with natural angular frequency

$$\omega_n = \sqrt{\frac{\tau_0}{I}} \qquad [16 \cdot 26]$$

and the following rotational analogs:

$$\theta \longrightarrow x \qquad \tau \longrightarrow F \qquad \tau_0 \longrightarrow k \qquad I \longrightarrow m \qquad \frac{d\theta}{dt} = \omega \longrightarrow v \qquad [16 \cdot 27]$$

The general solution of Eq. [16 . 25] therefore has the form

$$\theta(t) = \theta_m \sin(\omega_n t + \phi) \qquad [16 \cdot 28]$$

where θ_m is the maximum angular displacement, the *angular amplitude.* Similarly, the total energy is

$$\tfrac{1}{2}I\omega^2 + \tfrac{1}{2}\tau_0\theta^2 = E = \tfrac{1}{2}\tau_0\theta_m^2 \qquad [16 \cdot 29]$$

This analysis holds as long as the elastic limit of the wire for shear is not exceeded, and it affords a useful method for determining the shear modulus of the wire, as well as for measuring small torques.

In the case of an ideal, or *simple, pendulum,* a material particle (the bob) is suspended by an inextensible cord of negligible mass. Although no actual pendulum has such idealized properties, a small heavy ball suspended from a fixed point by a thread can, with little error, be treated as a simple pendulum. The position of such a pendulum is described by its angular distance θ from the vertical, as shown in Fig. 16 . 8. The gravitational torque on the pendulum bob relative to point P is $\tau = mgl \sin\theta$, always tending to restore it to the vertical equilibrium position. Since the moment of inertia of

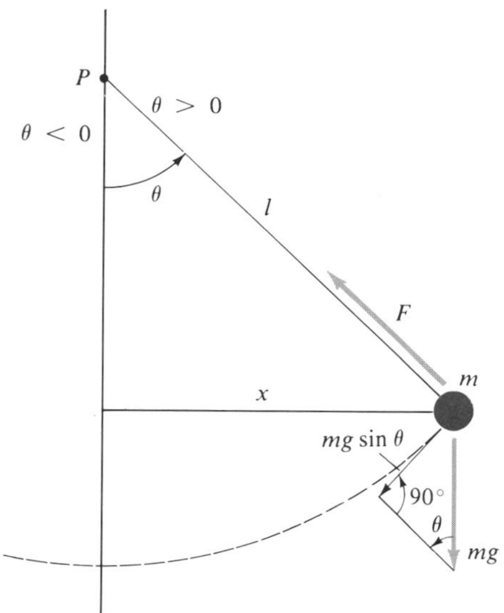

figure 16 • 8 *Simple pendulum under the influence of gravity and suspended from P*

the pendulum is $I = ml^2$, the equation of motion $I\alpha = \tau$ becomes

$$ml^2 \frac{d^2\theta}{dt^2} = -mgl \sin \theta$$

or

$$\frac{d^2\theta}{dt^2} + \frac{g}{l} \sin \theta = 0 \qquad [16 \cdot 30]$$

Although this equation looks deceptively simple (it isn't), it is not the differential equation for simple harmonic motion. In the next section we shall examine the solution of this type of equation by means of numerical integration.

If θ is small, we can use the approximation $\sin \theta \doteq \theta$ (radians), which holds to within 10 percent for $\theta < 45°$ and to within 1 percent for $\theta < 14°$. Substituting θ for $\sin \theta$ in Eq. [16 . 30], we obtain a harmonic oscillator equation:

$$\frac{d^2\theta}{dt^2} + \omega_n^2\theta = 0 \qquad [16 \cdot 31]$$

where $\omega_n^2 = g/l$. Note that the period of the motion is independent of the mass of the bob. In the limit of small oscillations, the general solution $\theta(t)$ has the form of Eq. [16 . 28],

$$\theta = \theta_m \sin (\omega_n t + \phi)$$

example 16 • 4 Show that the assumption that the simple pendulum in Fig. 16 . 8 moves approximately in a horizontal plane is equivalent to the assumption of small angular oscillations.

solution Let x be the horizontal displacement of the bob. The assumption of a purely horizontal displacement implies that the acceleration d^2x/dt^2 is due to the horizontal component of the tension **F** in the string. Since $F \cos\theta = mg$ if vertical motion is neglected, the horizontal component is

$$F \sin\theta = mg \tan\theta = mg\frac{x}{l}$$

where l is the vertical length of the pendulum at rest. Hence the equation of motion is

$$m\frac{d^2x}{dt^2} + m\left(\frac{g}{l}\right)x = 0$$

which gives $\omega_n^2 = g/l$, as before. If for small oscillations, we approximate θ (radians) $\doteq x/l$, we recover Eq. [16 . 30].

A *physical,* or *compound, pendulum* consists of any rigid body suspended from a horizontal axis of support and free to swing about its equilibrium position under the action of its own weight and the reaction of the axis of support. An ordinary clock pendulum is an example. Unlike the case of an ideal pendulum, the mass of a physical pendulum cannot be treated as a particle. The mathematical theory of physical pendulums was first worked out by Huygens and published in his *Horologium Oscillatorium* in 1673. This was the second major problem to be solved in kinetics, the first (that of falling bodies) having been solved by Galileo. It was the first successful attempt to deal with the kinetics of a rigid body; in fact, the concept of the moment of inertia first arose in this connection.

Figure 16 . 9 shows a vertical section taken through a compound physical pendulum, which is suspended from an axis passing through O and has its center of mass at point C. The torque due to gravity may be considered as applied at the center of mass and opposed to the motion of the pendulum,

$$\tau = -mgr_c \sin\theta \qquad [16 \cdot 32]$$

where m is the mass of the entire pendulum and r_c is the distance of the center of mass from O. The general equation of motion for the physical pendulum is therefore

$$\blacktriangleright \quad \frac{d^2\theta}{dt^2} + \frac{mgr_c}{I}\sin\theta = 0 \qquad [16 \cdot 33]$$

where I is the moment of inertia of the body about the axis of suspension.

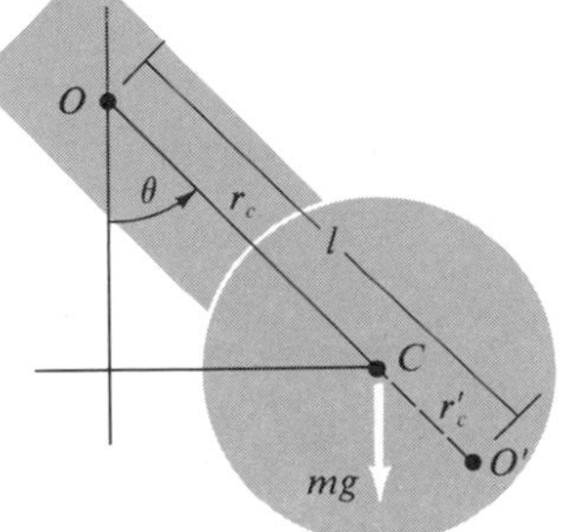

figure 16 • 9 *Physical pendulum*

For small θ this results in simple harmonic oscillation with period

$$T = 2\pi\sqrt{\frac{I}{mgr_c}} \qquad [16 \cdot 34]$$

This corresponds to the oscillation frequency of an ideal pendulum of length l, where

$$l = \frac{I}{mr_c} \qquad [16 \cdot 35]$$

In other words, a physical pendulum suspended from O will oscillate as if it were an ideal pendulum with all its mass concentrated at a point O', located a distance l from O on the extension through C. The point O' is therefore called the *center of oscillation* of the pendulum.

The point of suspension O and the center of oscillation O' are *conjugate points;* if the body were suspended instead at O', it would oscillate as if O were its center of oscillation and would have the same period of oscillation. To see this, note that the period of oscillation about O' is

$$T' = 2\pi\sqrt{\frac{I'}{mgr'_c}} \qquad [16 \cdot 36]$$

where I' is the moment of inertia about a horizontal axis through O' and r'_c is the distance from O' to the center of mass. Then, from Fig. 16 . 9,

$$r'_c = l - r_c = \frac{I - mr_c^2}{mr_c} = \frac{I_c}{mr_c} \qquad [16 \cdot 37]$$

Since the Lagrange parallel-axis theorem (see Sec. 11 . 3) gives

$$I = I_c + mr_c^2 \qquad I' = I_c + mr'^2_c$$

when we substitute for the primed quantities in Eq. [16 . 36], we obtain

$$T' = 2\pi\sqrt{\frac{I_c + m(I_c/mr_c)^2}{mg(I_c/mr_c)}}$$

$$= 2\pi\sqrt{\frac{mr_c^2 + I_c}{mgr_c}} = 2\pi\sqrt{\frac{I}{mgr_c}} = T \qquad [16 \cdot 38]$$

When the body is suspended from either conjugate point, O or O', the period of oscillation T is the same.

Suppose a body is suspended from some point O and is struck sharply at its center of oscillation O', as shown in Fig. 16 . 10. It receives an impulse $\mathbf{P}$, which imparts a velocity $\mathbf{v}_c = \mathbf{P}/m$ to the center of mass. At the same time, this impulse imparts an angular velocity $\omega = Pr'_c/I_c$ to the body with respect to the center of mass, and the body will begin to rotate as it translates. The linear velocity $\mathbf{v}$ of any point of the body is the vector sum of $\mathbf{v}_c$ and the tangential velocity $\mathbf{v}'$ about the center of mass due to the rotation. At O' the velocities $\mathbf{v}'$ and $\mathbf{v}_c$ add, but at O they oppose each other, so that

$$v = v_c - v' = v_c - \omega r_c = \frac{P}{m} - \omega r_c = \frac{P}{m}\left(1 - \frac{mr_c r'_c}{I_c}\right) \qquad [16 \cdot 39]$$

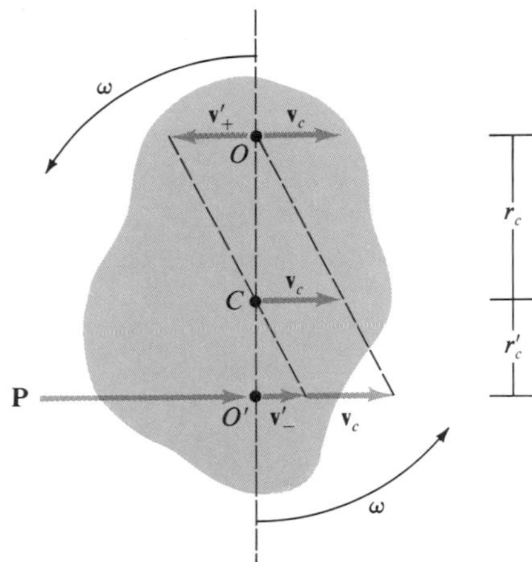

figure 16 • 10 *Motion of a physical pendulum immediately after receiving an impulse* **P** *at O′, the center of oscillation when suspended at O. At O, the speed* $v = v_c - \omega v_c = 0$*; and at O′, speed* $v = v_c + \omega r'_c = (P/m)(1 + r'_c/r_c)$.

However, from Eq. [16 . 37], $r_c r'_c = I_c/m$, so that $v_c = \omega r_c$ and $v = 0$. Consequently, the entire object seems to be rotating about the axis through O. For this reason, the center of oscillation is also called the *center of percussion.* You know that your bat has really connected with a good pitch—has hit squarely at the center of percussion—when the bat handle rests securely in your grip and does not twist or sting on impact, since $v = 0$.

example 16 • 5 What is the smallest possible period of harmonic oscillation about some horizontal axis through the line from O to O' in Fig. 16 . 9? Through what point must this axis pass?

solution From the parallel-axis theorem, if the moment of inertia about an axis through some point a distance r_c from the center of mass is I, then

$$I = I_c + mr_c^2$$

where $I_c = mk_c^2$ is the moment of inertia of the body about its center of mass. Substituting for I and cancelling m yields

$$T = 2\pi\sqrt{\frac{k_c^2 + r_c^2}{gr_c}}$$

where k_c is the *radius of gyration* of the pendulum about an axis through its center of mass and parallel to the axis of suspension.

Now imagine that we extend a line from O through the center of mass of the body and suspend the pendulum from any arbitrary point on that line. The possible periods are given by the above equation for $T(r_c)$, with the minimum value T_{min} found by differentiation:

$$\frac{dT}{dr_c} = \frac{2\pi^2}{gr_c^2T}(2r_c^2 - k_c^2 - r_c^2) = 0 \qquad T = T_{\text{min}}$$

Therefore $r_c = k_c$ and

$$T_{\text{min}} = T(k_c) = 2\pi\sqrt{\frac{2k_c}{g}}$$

This gives us a way to determine k_c, and hence I_c, experimentally, without knowing C or even making any measurements of length. The center of mass is then known to be located a distance $r_c = k_c$ from the point of suspension when T is found to be a minimum.

*16 . 4 physical pendulums: Simpson's rule

In the preceding section we found the exact equation of motion of a physical pendulum to be

$$\frac{d^2\theta}{dt^2} + \omega_n^2 \sin\theta = 0 \qquad [16 \cdot 40]$$

where $\omega_n^2 = mgr_c/I$. However, we could solve this equation only in the approximate form of Eq. [16 . 31] for small θ. In fact, there is no way to solve Eq. [16 . 40] explicitly for $\theta(t)$, although $\theta(t)$ can be tabulated. We cannot even scale θ with respect to an arbitrary amplitude, as we did in the numerical integration of x in Sec. 6 . 6, because even if we substituted a dimensionless variable—say, $\phi = \theta/\theta_m$—there would be no way to factor θ_m out of the equation. We can therefore expect to find that our solutions depend on angular amplitude θ_m, which is unlike the case of simple harmonic oscillation.

We can, of course, scale the time to a dimensionless variable τ such that

$$\tau = t/T' \qquad T' = 2\pi/\omega_n \qquad [16 \cdot 41]$$

where the factor of 2π is retained for comparison with the simple harmonic

table 16 • 2 *Integration of* $d^2\theta/d\tau^2 + 4\pi^2 \sin\theta = 0$, *for* $\theta_0 = \frac{1}{2}\pi$ *and* $\omega_0 = 0$ *by the improved Euler method with step size* $\Delta\tau = 0.05$ *(see Fig. 16 . 11)*

τ	θ	$d\theta/d\tau$	$\frac{1}{2}\pi \cos 2\pi\tau$
0.0	1.57080	0.00000	1.57080
0.1	1.37346	−3.93584	1.27080
0.2	0.79294	−7.44014	0.48540
0.3	−0.06012	−8.87693	−0.48540
0.4	−0.89435	−7.08292	−1.27080
0.5	−1.43270	−3.47093	−1.57080
0.6	−1.58338	0.46492	−1.27080
0.7	−1.33959	4.39456	−0.48540
0.8	−0.71632	7.78988	0.48540
0.9	0.15764	8.90023	1.27080
1.0	0.97654	6.80001	1.57080
1.1	1.47979	3.09673	1.27080
1.2	1.59258	−0.84242	0.48540
1.3	1.31110	−4.76565	−0.48540
1.4	0.65361	−8.06288	−1.27080

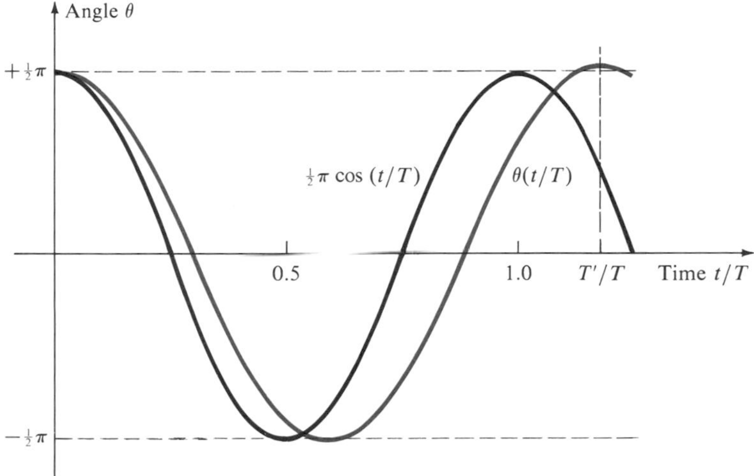

figure 16 • 11 *Comparison of $\theta(t/T)$ (Table 16 . 2) and $\frac{1}{2}\pi \cos (t/T)$. The true shape resembles the simple harmonic approximation, but stretched out along the time axis. The accumulation of small errors in the numerical integration can be judged from the fact that the computed extrema of θ are not quite tangential to $\theta = \pm\frac{1}{2}\pi$.*

oscillator. The resulting differential equation is

$$\frac{d^2\theta}{d\tau^2} + 4\pi^2 \sin\theta = 0 \qquad [16 \cdot 42]$$

To solve this equation for $\theta(\tau)$, given θ_0 and $(d\theta/dt)_{t=0} = \omega_0$, let us employ the *improved Euler method* of Sec. 13 . 6. Note that the time scale requires

$$\left(\frac{d\theta}{d\tau}\right)_{\tau=0} = T'\omega_0 = \frac{2\pi\omega_0}{\omega_n}$$

In this case, the only difference is that

$$g = g\left(\tau, \theta, \frac{d\theta}{d\tau}\right) = -4\pi^2 \sin\theta$$

If you have already constructed the program of Fig. 13 . 15, you will find it necessary to alter only the two or three statements containing $g(x,y,v)$.

Table 16 . 2 lists the results of a computation with $\theta_0 = \frac{1}{2}\pi$, $\omega_0 = 0$, and $\Delta\tau = 0.05$ (printing every second time), in comparison with the harmonic-oscillator approximation, $\theta = \frac{1}{2}\pi \cos 2\pi\tau$. Figure 16 . 11 shows a graphic comparison of the two solutions. It is interesting to note that, although the motion is still periodic, the true period T is approximately $1.17638T'$, determined by interpolating near the first turning point, $\theta \doteq -\frac{1}{2}\pi$ (to within 0.8 percent) to find τ such that $d\theta/d\tau = 0$. It appears, therefore, that increasing the amplitude actually has the effect of increasing the period. We might have suspected something like this had we paused to consider the extreme case in which $\theta_0 = \pi$ and $\omega_0 = 0$, where the pendulum is squarely balanced (although unstable) on the point of suspension—giving, according to this model, an infinite period!

The quantity of particular interest is usually the true period T, which can be calculated from the energy principle without the need to compute all the details of the motion. If we take the gravitational potential to be zero at the height of the point of suspension, then

$$K + U = \tfrac{1}{2}I\left(\frac{d\theta}{dt}\right)^2 - mgr_c \cos\theta = E = -mgr_c \cos\theta_m \qquad [16 \cdot 43]$$

where

$$E = \tfrac{1}{2}I\omega_0^2 - mgr_c \cos\theta_0$$

is a constant determined by the initial conditions and θ_m is the maximum angular displacement. Following the same procedure as in Sec. 16 . 1, we obtain the implicit solution for $\theta(t)$:

$$t = \pm\sqrt{\frac{I}{2mgr_c}}\int_{\theta_0}^{\theta} \frac{d\theta}{\sqrt{\cos\theta - \cos\theta_m}} \qquad [16 \cdot 44]$$

To obtain an expression for T we need only consider the time of a quarter-oscillation, beginning the swing at $\theta_0 = 0$ and integrating to $\theta = \theta_m$. The value of ω_0 is implicit in the value of $\theta_m = \cos^{-1}(-E/mgr_c)$, from conservation of energy. The integral expression for T, assuming $\theta_m < \pi$, is then

$$T = 4\sqrt{\frac{I}{2mgr_c}}\int_{0}^{\theta_m} \frac{d\theta}{\sqrt{\cos\theta - \cos\theta_m}} \qquad [16 \cdot 45]$$

which is an *improper integral*—one for which the integrand is undefined somewhere in the range of the integral. In this case the denominator of the integrand vanishes as θ approaches θ_m. On physical grounds, however, we expect T to be finite, except for $\theta_m = \pi$.

To put this integral into a form which is not improper, it is customary to use the half-angle identities of trigonometry to show that

$$\sin\phi = \frac{\sin\frac{1}{2}\theta}{\sin\frac{1}{2}\theta_m} \qquad \cos\phi = \sqrt{\frac{\cos\theta - \cos\theta_m}{2\sin^2\frac{1}{2}\theta_m}} \qquad [16 \cdot 46]$$

and substitute in Eq. [16 . 45] to obtain

$$T = \frac{2T'}{\pi}\int_0^{\pi/2} \frac{d\phi}{(1 - k^2\sin^2\phi)^{1/2}} \qquad [16 \cdot 47]$$

where $0 < k = \sin\frac{1}{2}\theta_m < 1$. This integral is known as the *complete elliptic integral of the first kind,* and extensive tabulations of it are given in handbooks. As the amplitude becomes small, k approaches zero and T' approaches T, but for any nonzero k, the integrand is always greater than unity, so that $T > T'$.

To compute the integral in Eq. [16 . 47] numerically, we shall employ a new, very accurate method of quadrature known as *Simpson's rule.* Like the trapezoidal rule discussed in Sec. 7 . 6, Simpson's rule takes the form of a weighted average of values of $f(x)$:

$$\int_a^b f(x)\,dx \doteq \sum_{i=0}^{n} c_i f(x_i) \qquad [16 \cdot 48]$$

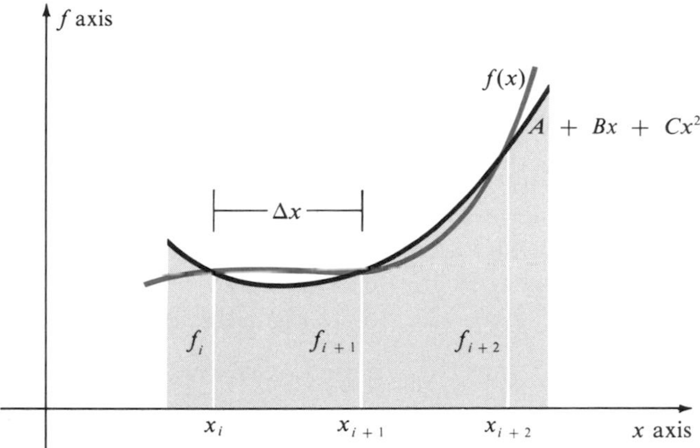

figure 16 • 12 *Approximating $f(x) \doteq A + Bx + Cx^2$ over $x_i \leq x \leq x_{i+2}$, we obtain Simpson's rule for quadrature, Eq. [16 . 50]*

where $n = (b - a)/\Delta x$ and Δx is the interval size. Whereas the trapezoidal rule is equivalent to replacing, or "fitting," the curve $f(x)$ on each interval Δx by a straight line (Fig. 7 . 7*b*), in the case of Simpson's rule we "fit" $f(x)$ by a parabola spanning *two* intervals Δx, as shown in Fig. 16 . 12.

If we approximate

$$f(x) \doteq A + Bx + Cx^2 \qquad [16 \cdot 49]$$

on some pair of intervals $[x_i, x_{i+1}]$ and $[x_{i+1}, x_{i+2}]$, the requirement that the parabola intersect the curve on the interval boundaries enables us to set up three independent equations for the evaluation of A, B, and C in terms of f_i, f_{i+1} and f_{i+2}. If we then approximate the integrand by

$$\begin{aligned}\int_{x_i}^{x_{i+2}} f(x)\,dx &\doteq \int_{x_i}^{x_{i+2}} (A + Bx + Cx^2)\,dx \\ &= \tfrac{1}{3}\Delta x\,[f_i + 4f_{i+1} + f_{i+2}] \qquad [16 \cdot 50]\end{aligned}$$

You can verify Eq. [16 . 50] by substituting for the values of f_i, f_{i+1}, and f_{i+2} from Eq. [16 . 49]. What is particularly delightful about this formula is that even if we had used a cubic approximation to $f(x)$ instead of a quadratic, we would have obtained the same formula.

If we take an *even* number of intervals in Eq. [16 . 48] and integrate in $n/2$ pairs, the formula for Simpson's rule is

$$\begin{aligned}\int_a^b f(x)\,dx &= \tfrac{1}{3}\Delta x\,[(f_0 + 4f_1 + f_2) + (f_2 + 4f_3 + f_4) + \cdots] \\ &= \tfrac{1}{3}\Delta x\,[f_0 + 4f_1 + 2f_2 + 4f_3 + 2f_4 + \cdots \\ &\qquad + 2f_{n-2} + 4f_{n-1} + f_n] \qquad [16 \cdot 51]\end{aligned}$$

Simpson's rule is a third-order method; that is, errors in quadrature are proportional to $(\Delta x)^4$ (the trapezoidal rule is only first-order). The simple formulation of Eq. [16 . 51], which makes it suitable even for hand computation, and its high degree of accuracy have led to the widespread use of this method.

Only minor modifications to the trapezoidal rule flowchart (Fig. 7 . 9) are necessary to generate an algorithm for Simpson's rule. One is shown in Fig. 16 . 13, applied to the computation of T from Eq. [16 . 47]. In this case, since the analytic form of the function is known, it is defined separately as $g(\phi)$. Whenever the computer encounters the designation $g(\phi)$, it returns to the definition, stored in its memory in the form of a subprogram, and generates the desired value of $f = g(\phi)$. The simple algorithm shown here

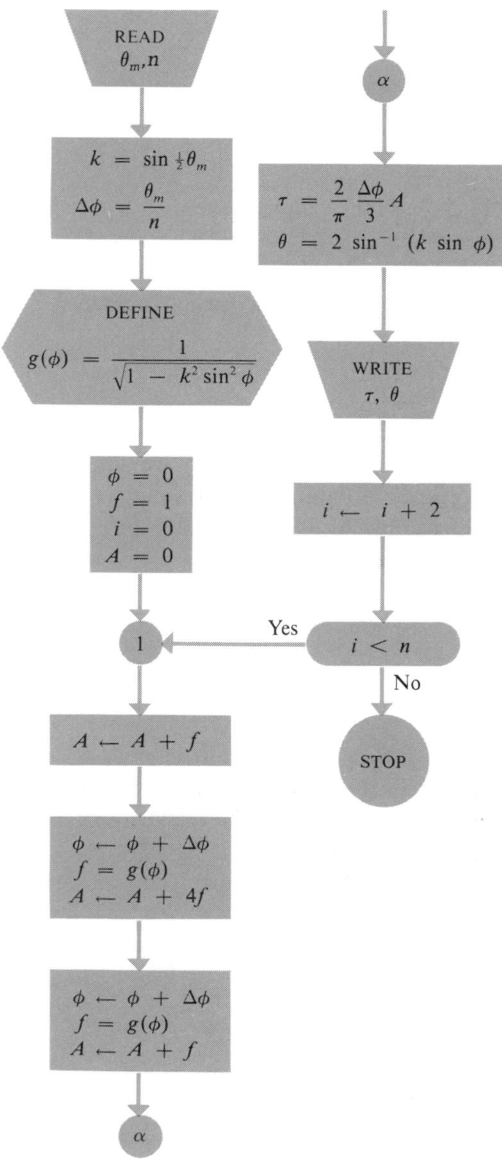

figure 16 • 13 *Algorithm for application of Simpson's rule to the computation of the period of a real physical pendulum with maximum angular displacement* θ_{max}

is efficient and easy to "debug," although it could be more condensed.* It is distinctly different from a hand computation, in which we would first compute each $f_i = g(i\,\Delta\phi)$, then separately add the even-, odd- and end-interval values of f and multiply them by their separate coefficients, 2, 4, or 1.

Although we have used the dimensionless time variable τ and combined all physical data in θ_m, we could have used θ_0, ω_0, I, m, r_c, and g and also performed concurrent computations of K, U, and ω, obtaining values in the appropriate physical units. Most computers will handle a very wide range of numbers to at least six significant digits. However, scaled computations have the advantage of broader applicability, and the results may be saved for future reference. In this case, for $n = 6$ and $\theta_m = \frac{1}{2}\pi$, Simpson's rule gives $T = 1.18034T'$, identical to the handbook value to six significant digits. Recall also that the improved Euler method applied to Eq. [16 . 42] was in error by less than 0.4 percent ($T = 1.17638T'$).

16 • 5 *damped harmonic oscillators*

Up to this point we have disregarded any dissipative, or *damping, forces* that might be present in an oscillating system. Damping forces reduce the total mechnical energy of the system, usually by transforming some of it into heat. A pendulum may continue to swing for a considerable time; however, if no energy is supplied to compensate for air resistance and friction in the pivot, the amplitude of the swings will gradually diminish, since, as we saw in Eq. [16 . 3], the total energy is proportional to the square of the amplitude.

Mathematical analysis of damped oscillatory motion ranges from moderately difficult to impossible, depending on the character of the damping forces involved. These often depend in complicated ways on the speed and shape of the oscillator and the nature of the resisting medium. If we consider only small-amplitude oscillations in which the restoring force obeys Hooke's law, then the damping forces often can be represented by Stokes' law, in which the resistance is proportional to the first power of the speed.

Since the damping force is always opposed to the velocity, we can express it as $\mathbf{F}_v = -\gamma\mathbf{v}$, where γ is a constant called the *damping coefficient*. The equation of motion is, for one-dimensional motion,

$$m\frac{d^2x}{dt^2} = -\gamma\frac{dx}{dt} - kx \qquad [16 \cdot 52]$$

which, after some rearrangement, becomes

$$\blacktriangleright \quad \frac{d^2x}{dt^2} + \mu\frac{dx}{dt} + \omega_n^2 x = 0 \qquad [16 \cdot 53]$$

*The term *debug*—to remove errors—generally refers to errors which may be inherent in the logical structure of the algorithm itself. Etymologically (or entomologically), the term is said to refer to a fly whose untimely demise occurred in an electric relay in an early computer (ca. 1950?) used at the Massachusetts Institute of Technology. However, this account may be apocryphal.

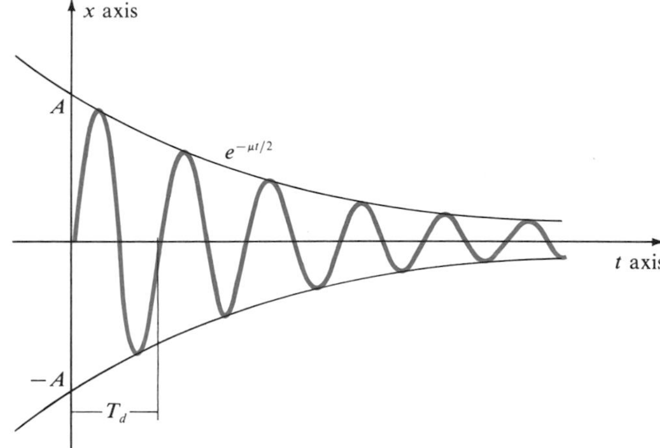

figure 16 • 14 *Underdamped harmonic oscillation, with* $x = A \exp(-\frac{1}{2}\mu t) \sin \omega_d t$; $T_d = 2\pi/\omega_d$; *and* $\mu = \omega_n/10 \doteq \omega_d/10$

where $\mu = \gamma/m$ and $\omega_n^2 = k/m$. A rigorous derivation of $x(t)$ is difficult, so we shall obtain it by reasoning as follows:

1 For $\gamma = 0$ the solution should be that of a simple harmonic oscillator.
2 Stokes' law damping should cause an exponential decay in speed (see Sec. 7 . 7).
3 These two types of behavior should be multiplicative, rather than additive (why?).

Hence, let us assume a solution of the form

$$x(t) = Ae^{-\beta t} \sin(\omega_d t + \phi) \qquad [16 \cdot 54]$$

where A is the undamped amplitude, ω_d is the *damped angular frequency,* and β is a positive constant called the *decay constant.*

To find β and ω_d we can substitute Eq. [16 . 54] into Eq. [16 . 53] and separately set the coefficients of sines and cosines equal to zero to obtain

$$\beta = \tfrac{1}{2}\mu \qquad \omega_d^2 = \beta^2 - \beta\mu + \omega_n^2 = \omega_n^2 - \tfrac{1}{4}\mu^2 < \omega_n^2 \qquad [16 \cdot 55]$$

The solution is therefore

$$x = Ae^{-\mu t/2} \sin(\omega_d t + \phi) \qquad [16 \cdot 56]$$

where $\omega_d = \sqrt{\omega_n^2 - \frac{1}{4}\mu^2}$ and the undamped amplitude A and phase angle ϕ are arbitrary constants which depend on the initial conditions. As shown in Fig. 16 . 14, the motion is periodic, but with an amplitude that becomes smaller and smaller as time progresses, and with a damped frequency ω_d smaller than the undamped natural frequency ω_n. Since the sine function is always less than or equal to unity,

$$-Ae^{-\mu t/2} \leq x \leq Ae^{-\mu t/2} \qquad [16 \cdot 57]$$

The exponential curve is known as the *envelope* of the oscillation. If $\mu^2 \ll \omega_n^2$, then $\omega_d \doteq \omega_n$; however, the effect of damping on the amplitude is quite noticeable, since the ratio of two successive maxima one period apart is

$$\frac{x(t + 2\pi/\omega_d)}{x(t)} = e^{-\mu\pi/\omega_d} = e^{-\delta} \qquad [16 \cdot 58]$$

where δ is the *logarithmic decrement* of the motion. If the resistive effects are negligible or small, then $\delta \ll 1$.

Note that ω_d^2 in Eq. [16 . 55] is a difference of two independent parameters; it need not be positive, but may be zero or negative. We can still obtain meaningful solutions in all cases. If $\omega_d^2 > 0$, the oscillation is known as *underdamped,* since the restoring force is strong enough to keep the oscillations going indefinitely, although the resistance causes them to become smaller and smaller with time. What happens, however, if the resistive force increases to the point where ω_d vanishes? If we expand Eq. [16 . 56] and take the limit of small ω_d, using the approximations $\cos\omega_d \doteq 1$ and $\sin\omega_d \doteq \omega_d$, then

$$\begin{aligned} x &= \lim_{\omega_d \to 0} Ae^{-\beta t}(\sin\omega_d t\cos\phi + \cos\omega_d t\sin\phi) \\ &= Ae^{-\beta t}[(\omega_d\cos\phi)t + \sin\phi] \end{aligned} \qquad [16 \cdot 59]$$

So in the limit as the damped frequency vanishes, we have *critical damping,*

$$x(t) = e^{-\mu t/2}(Bt + C) \qquad [16 \cdot 60]$$

where B and C are two arbitrary constants. You can verify this by substitution into the equation of motion, Eq. [16 . 53], where μ has the critical value $\mu = 2\omega_n$. This case, illustrated in Fig. 16 . 15, is termed critical because it appears to begin as an oscillation, but once the velocity (slope of the curve) becomes large enough, the resistive forces start to dominate the restoring force and quickly damp out the motion.

When the damped frequency becomes imaginary, $\omega_n^2 < \frac{1}{4}\mu^2$, then the resistive forces are completely dominant, and if the body is disturbed, it will return slowly through the resisting medium to its equilibrium position, just like the decaying exponentials that occur in Stokes' law (see Fig. 16 . 16). This is known as *overdamped oscillation.* The general solution to this case is

$$x(t) = e^{-\mu t/2}(De^{\omega' t} + Ee^{-\omega' t}) \qquad [16 \cdot 61]$$

where D and E are arbitrary constants of integration and $\omega'^2 = -\omega_d^2 = \frac{1}{4}\mu^2 - \omega_n^2$.

Sometimes the damped oscillator is also subjected to an external *periodic driving force* which prevents decay of the oscillations, or even acts to increase their amplitude, as in electrically driven clock mechanisms, spring-driven wind-up toys and clocks, the marching of troops on a bridge, or the

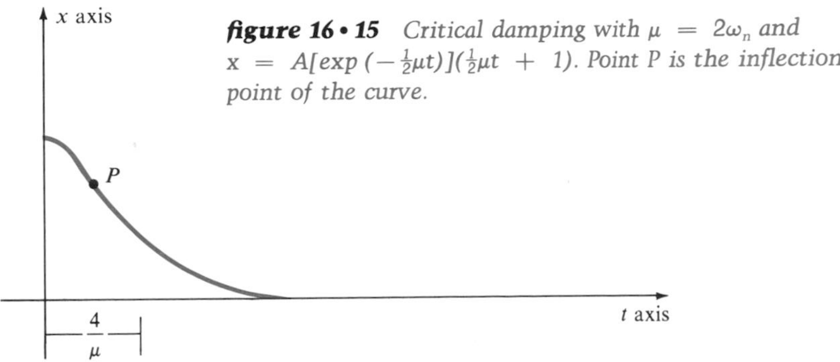

figure 16 • 15 *Critical damping with $\mu = 2\omega_n$ and $x = A[\exp(-\frac{1}{2}\mu t)](\frac{1}{2}\mu t + 1)$. Point P is the inflection point of the curve.*

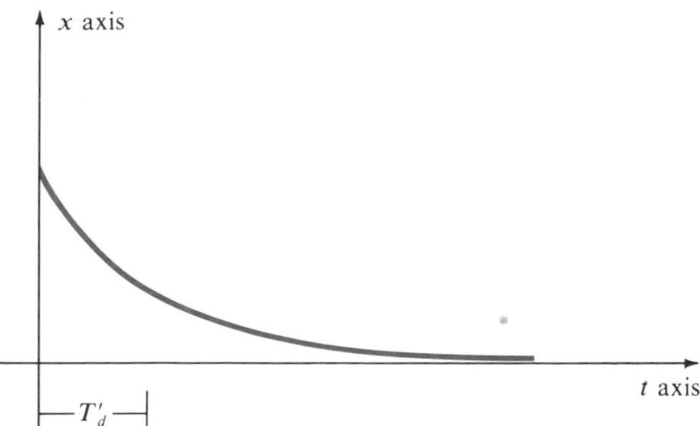

figure 16 • 16 *Overdamped harmonic oscillation, with* $x = A \exp[-(\frac{1}{2}\mu - \omega'_d)t]$; $T'_d = 2\pi/\omega'_d$; *and* $\mu = 4\omega_n$

action in "pumping" a swing. If we represent the force by $F = F_0 \sin \omega_F t$, where ω_F is the frequency of the force, the equation of motion for $f_0 = F_0/m$ becomes

$$\frac{d^2x}{dt^2} + \mu \frac{dx}{dt} + \omega_n^2 x = F_0 \sin \omega_F t \qquad [16 \cdot 62]$$

When this equation is solved, assuming the oscillator to be *underdamped,* the response of the system to the driving force is found to consist of two motions, the *transient response* and the *steady-state response.* The first is a free, underdamped harmonic oscillation of frequency ω_d which dies out with time; the second is a forced oscillation which continues at constant amplitude and at the frequency ω_F of the driving force.

The most general solution of Eq. [16 . 62] is

$$x(t) = \underbrace{Ae^{-\mu t/2} \sin(\omega_d t + \phi)}_{\text{transient}} + \underbrace{A_F \sin(\omega_F t + \phi_F)}_{\text{steady-state}} \qquad [16 \cdot 63]$$

where A_F, the amplitude of the forced steady-state oscillations, is not arbitrary, but can be determined by substitution into Eq. [16 . 62]. Since the transient term does not contribute to A_F we obtain

$$A_F = \frac{f_0}{\sqrt{(\omega_n^2 - \omega_F^2)^2 + \mu^2\omega_F^2}} \qquad [16 \cdot 64]$$

The phase difference, or *phase shift* ϕ_F, between the driving force and the steady-state displacement is

$$-\pi \le \phi_F = \tan^{-1} \frac{\mu\omega_F}{\omega_F^2 - \omega_n^2} \le 0 \qquad [16 \cdot 65]$$

which implies that the displacement will not be exactly synchronized with the force, but "out of step" by a constant angle. Note that A_F and ϕ_F are not arbitrary, but are determined by the physical conditions.

The response of the oscillator is a maximum when the amplitude $A_F(\omega_F)$ is greatest. This phenomenon, known as *resonance,* occurs at frequencies

$$\omega_F = \omega_{\text{res}} = \sqrt{\omega_n^2 - \tfrac{1}{2}\mu^2}$$

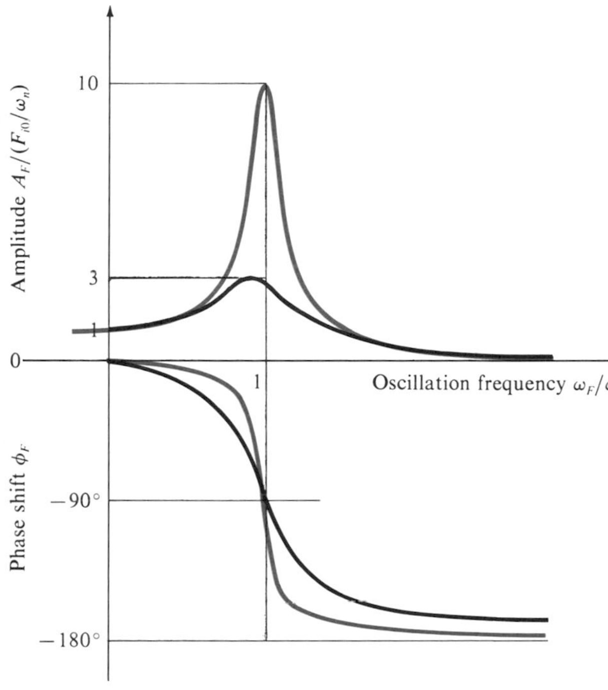

figure 16 • 17 *Amplitude A_F and phase shift ϕ_F of a forced underdamped harmonic oscillator, with $\mu = \omega_n/10$ (colored line) or $\mu = \omega_n/3$ (black line)*

Figure 16 . 17 shows A_F and ϕ_F as functions of ω_F/ω_n for $\mu = \omega_n$ and $\mu = \frac{1}{10}\omega_n$. For extreme values of ω_F the driving force and oscillator velocity are parallel only part of the time, during which the force does positive work. However, as the driving frequency gradually approaches the natural frequency, the phase difference between force and *velocity* rapidly decreases, and the amplitude reaches a maximum value at resonance, $\omega_{res} \doteq \omega_n$.

The actual time-dependent behavior of a forced, damped harmonic oscillator can look quite complex. Figure 16 . 18 shows the response of a damped oscillator to a driving force of frequency $\omega_F = \frac{1}{2}\omega_n$, with the system assumed to be at rest when the force begins to act. Note that once the transient behavior has died out, only the steady-state motion at ω_F remains.

example 16 • 6 Find $x(2T_F)$ where $T_F = 2\pi/\omega_F$, for $\omega_F = \frac{1}{2}\omega_n$, $\mu = \omega_n/\sqrt{5}$, $x_0 = 0$, and $v_0 = 0$.

solution The general solution will include an underdamped transient oscillation and a forced oscillation:

$$x(t) = Ae^{-\mu t/2} \sin(\omega_d t + \phi) + A_F \sin(\omega_F t + \phi_F)$$

where

$$\omega_d = \sqrt{19/20}\,\omega_n \doteq \omega_n$$

$$A_F = \sqrt{80/49}\,\frac{f_0}{\omega_n^2} \doteq \frac{9}{7}\frac{f_0}{\omega_n^2}$$

$$\phi_F = -\tan^{-1} 0.298 = -16.6° \doteq -0.29 \text{ rad}$$

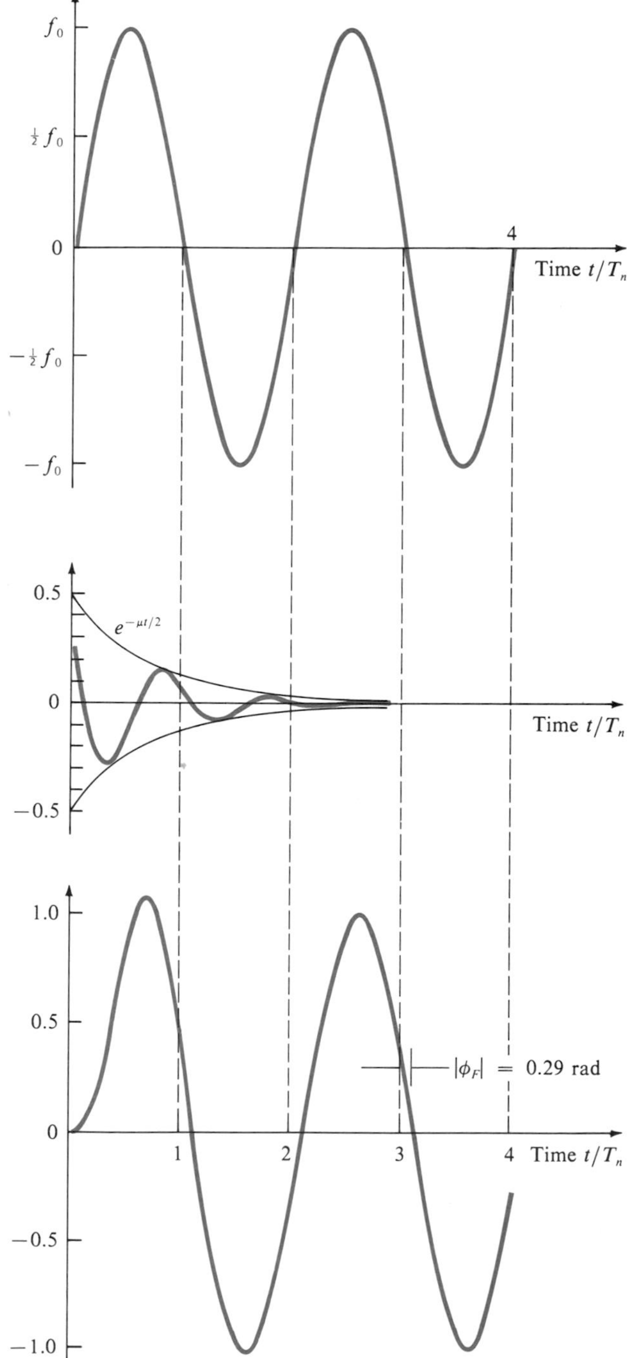

(*a*) *Driving force* $F = f_0 \sin \omega_F t$, *where* $\omega_F = \frac{1}{2}\omega_n$

(*b*) *Transient oscillation* $x = -\frac{1}{2}A_F \exp(-\frac{1}{2}\mu t) \sin(\omega_d t - 0.60)$, *where* $\omega_d \doteq \omega_n$

(*c*) *Combined steady-state and transient oscillation in units of* A_F. *Note initial behavior and slight displacement* ϕ_F *from the sine curve of part* (*a*)

figure 16 • 18 Forced motion of an underdamped harmonic oscillator, with $\mu = \omega_n/\sqrt{5}$

From the initial conditions we have

$$x_0 = 0 = A \sin \phi + A_F \sin \phi_F$$

$$v_0 = 0 = \omega_d A \cos \phi - \tfrac{1}{2}\mu A \sin \phi + \omega_F A_F \cos \phi_F$$

which can be solved for

$$\tan \phi = \frac{\omega_d \tan \phi_F}{\omega_F + \frac{1}{2}\mu \tan \phi_F} \doteq -0.67 \qquad \phi = -33.8^\circ = -0.59 \text{ rad}$$

$$A = -A_F \frac{\sin \phi_F}{\sin \phi} \doteq -\tfrac{1}{2}A_F$$

Hence

$$x(2T_F) = A_F\left[-\tfrac{1}{2}\exp\left(-\frac{4\pi}{\sqrt{5}}\right)\sin \phi + \sin \phi_F\right] \doteq 0.285A_F = 0.364\frac{f_0}{\omega_n^2}$$

Note that the transient component is essentially negligible. In fact, as can be seen from Fig. 16 . 18, it becomes negligible by the time $t = 2T$.

16 • 6 small-amplitude oscillations

The theme of simple harmonic motion has appeared throughout this chapter. One reason, as stated at the beginning of the chapter, is Fourier's theorem that any periodic oscillation can be represented as an appropriate sum of simple harmonic oscillations. The other reason, as we shall now see, is that in any stable physical situation, small-amplitude oscillations will be simple harmonic. Let us examine this second statement by reference to Taylor's theorem:

> Any function $f(x)$ with continuous derivatives can be "expanded in a power series" about any point x_0 in its domain.

Expressed in mathematical terms,

$$f(x) = \sum_{n=0}^{\infty} a_n(x - x_0)^n \qquad [16 \cdot 66]$$

where the constant coefficients a_n can be determined from the derivatives of $f(x)$ evaluated at $x = x_0$:

$$a_n = \frac{1}{n!}\left(\frac{d^n f}{dx^n}\right)_{x = x_0} = \frac{1}{n!}f^{(n)}(x_0) \qquad [16 \cdot 67]$$

(by convention $f^{(1)} = f'$, $f^{(2)} = f''$, and $f^{(3)} = f'''$).

If a function and its derivatives are known at a given point, we can generate an approximate formula for $f(x)$ by setting

$$f(x) \doteq \sum_{n=0}^{N} a_n(x - x_0)^n$$

truncating the series after some Nth term. If $x - x_0$ is small enough, higher powers of $x - x_0$ become progressively more and more negligible. Quite often the sum of the first few terms of the series is sufficiently accurate. Consider a quadratic expansion of a potential-energy function $U(x)$ of a system

$$U(x) \doteq U(x_0) + (x - x_0)U'(x_0) + \tfrac{1}{2}(x - x_0)^2 U''(x_0) \tag{16 . 68}$$

where we have ignored higher-order terms in $(x - x_0)^n$ for $n \geq 3$, on the assumption that we need consider only values of x sufficiently close to x_0 to make the approximation legitimate.

If x_0 is, by definition, the equilibrium coordinate, where all forces balance, the resultant force is

$$F(x_0) = -U'(x_0) = 0$$

Then, as we saw in Sec. 8 . 3, the potential-energy function can always be approximated by a parabola, where $U(x_0) = U_0$ and $U''(x_0) = U_0''$ are constants:

$$U(x) \doteq U_0 + \tfrac{1}{2}U_0''(x - x_0)^2 \tag{16 . 69}$$

Hence, for small oscillations, the net force on the system is a Hooke's law force,

$$F(x) = -\frac{dU}{dx} \doteq -U_0''(x - x_0) \tag{16 . 70}$$

and the system of mass m, if only slightly disturbed, will perform stable simple harmonic oscillations of natural frequency

$$\omega_n = \sqrt{\frac{U_0''}{m}} \qquad U_0'' > 0 \tag{16 . 71}$$

Of course, if $U_0'' < 0$, then $x = x_0$ is the maximum of the potential, ω_n is complex, and we have an *unstable* system. Unstable solutions are meaningless, however, because $x - x_0$ would soon become too large for the approximation of Eq. [16 . 71] to apply. If $U_0'' = 0$, we would have to consider the effect of the next term $\frac{1}{6}U_0'''(x - x_0)^3$ in the Taylor's series expansion of $U(x)$.

Approximating a function—whether it is known or unknown, in algebraic form or merely tabulated—by means of a Taylor's series is a useful technique not only in physics, but in mathematics as well, particularly in the construction of numerical methods.* In all the numerical methods discussed thus far, it is customary to derive the dependence of error on interval size by expanding the function in a Taylor's series and examining the consequences of the method when it is applied to the series term by term. For example, Simpson's rule can be viewed as resulting from the quadrature of a succession of second-order expansions of the integrand around the midpoint of each pair of intervals Δx.

*The section on series in Appendix F includes a discussion of the Taylor's series, and examples of various functions expanded in Taylor's series forms.

example 16 • 7 A hypothetical spring is found to have a restoring force represented by

$$F(x) = \alpha \cot \beta x$$

(*a*) Find its first equilibrium point for $x > 0$ and the frequency of small-amplitude oscillations about this point. (*b*) What is the second-order expansion of the potential $U(x)$ around equilibrium?

solution (*a*) The first equilibrium point occurs at $x_0 = \pi/2\beta$, where $F(x_0) = 0$. Since

$$U''(x) = -F'(x) = \alpha\beta \csc^2 \beta x$$

then $U_0'' = \alpha\beta$, and from Eq. [16 . 71], the frequency of the system for small oscillations about x_0 is

$$\omega_n = \sqrt{\frac{\alpha\beta}{m}}$$

(*b*) If we set the zero of potential at $U_0 = 0$, then the potential energy of the system is given by

$$U(x) \doteq \tfrac{1}{2}\alpha\beta\left(x - \frac{\pi}{2\beta}\right)^2$$

for small displacements from x_0. This formula is fairly good, since the next higher-order term is $U_0''' = 0$.

PROBLEMS

16 • 1–16 • 2 simple harmonic motion

16 • 1 A molecule of an ideal gas is bouncing elastically back and forth with speed v between two fixed points on the opposite walls of a cube of side length l. (*a*) Show that the motion of the molecule is both oscillatory and periodic, but not simple harmonic. (*b*) Derive an expression for the frequency of oscillation. (*c*) Sketch diagrams that show how the displacement and velocity of the molecule vary with time.

16 • 2 A simple harmonic oscillator of mass 5 g has a period of 0.6 s and an amplitude of 18 cm. Find the phase angle, speed, and accelerating force at an instant when the displacement of the oscillator is -9 cm.

16 • 3 A diver of mass m stands on a scale placed on the end of a diving board, which she has previously set in simple harmonic motion, with angular frequency ω and amplitude $A = y_m$. (*a*) What does the scale read? (*b*) Under what conditions is the diver launched from the board?

16 • 4 A 150-g mass on the end of a horizontal spring is displaced 3 cm to the left of the equilibrium position by a force of 60 N. (*a*) Find the natural angular frequency ω_n. (*b*) Find the amplitude of the subsequent motion if the mass is suddenly released. (*c*) What are the position and speed of the mass 10 s after its release?

16 • 5 A wrought-iron shelf is moving horizontally with a simple harmonic motion. A block of soft steel resting on the shelf begins to slide when the frequency is $\nu = 0.4$ Hz and the amplitude is $A = 0.3\ m$. What is the coefficient of static friction for soft steel on iron?

16 • 6 A force of 10 gf is observed to stretch a certain elastic cord by 50 mm. A 15-g object is suspended from one end of the cord and is set in vertical vibration by pulling down on it and then releasing it. How far down should the suspended object be pulled so that when it reaches the highest point of its vibration there is no tension in the cord?

16 • 7 A particle rotates with constant speed in a circle of radius R. (a) Show that its projections on the horizontal and vertical axes (its x and y components) represent simple harmonic motions with phase constants differing by $\frac{1}{2}\pi$. This is known as the *reference circle* for the horizontal oscillation. (b) If the x axis represents the displacement of a simple harmonic oscillator in units of amplitude A and the y axis represents its *velocity* in units of ωA, show that the graph of the motion in the xy plane is a unit circle.

16 • 8 Consider the simple harmonic oscillator in Prob. 16 . 2. (a) Find the total mechanical energy of the oscillator. (b) What is its initial speed v_0 if the initial displacement is 6 cm?

16 • 9 At time $t = 0$ a simple harmonic oscillator with a frequency of 5 rad/s has a displacement of 25 cm and speed of -10 cm/s. (a) Find the amplitude A of the oscillation. (b) What is its phase constant? (c) If there is a 10-g weight on the oscillator, what is its total mechanical energy?

16 • 10 A simple harmonic oscillator of mass 0.8 kg and frequency $10/3\pi$ Hz is set into motion with an initial kinetic energy $K_0 = 0.2$ J and an initial potential energy $U_0 = 0.8$ J. Compute (a) its initial position and (b) its initial speed. (c) What is the amplitude of the oscillation? (d) The maximum speed? (e) Find the displacement at the moment when the kinetic and potential energies are equal.

16 • 11 A crank of length A rotates with constant angular speed ω, causing point P to oscillate between A and $-A$, as shown. (a) Show that the position of P with respect to the midpoint O of its path is given by

$$x = A\cos\omega t - l + l\sqrt{1 - \left(\frac{A}{l}\right)^2 \sin^2\omega t}$$

That is, the motion of the piston is not simple harmonic. (b) Find the displacements of P when $\omega t = 0°$, $\omega t = 90°$, and $\omega t = 180°$. (c) With ω kept constant, is it ever possible for the motion of the piston to be approximately simple harmonic?

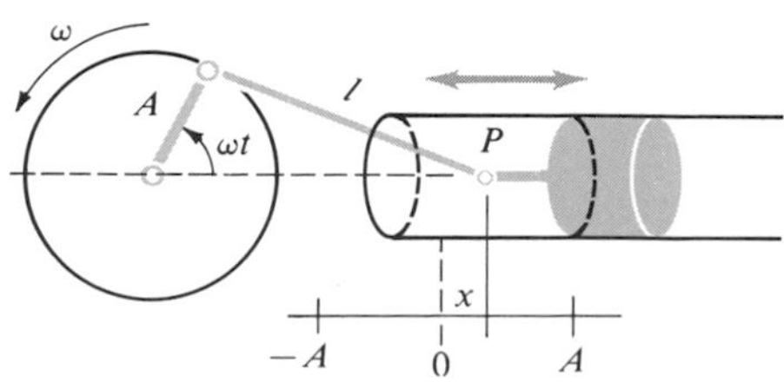

16 • 12 Prove that when averages are taken with respect to displacement over one vibration, instead of time, then $\overline{K} = \frac{1}{3}kA^2$ and $\overline{U} = \frac{1}{6}kA^2$. Explain in physical terms why these values differ from those in Eq. [16 . 17].

16 • 13 A spring of mass m_1 distributed evenly along its length has a mass m_2 suspended from its lower end. If the spring elongates uniformly as the system oscillates, show that when the suspended mass is moving at speed v, the kinetic energy of the system is given by

$$K = \tfrac{1}{2}(m_2 + \tfrac{1}{3}m_1)v^2$$

If the spring-mass system performs simple harmonic oscillations, show that these will have a period

$$T = 2\pi\sqrt{\frac{m_2 + \frac{1}{3}m_1}{k}}$$

16 • 14 The internal motions of a diatomic molecule may be represented by two isolated atoms of masses m_1 and m_2, connected by a light spring of force constant k and unstretched length l. The center of mass is at rest. (*a*) If the atoms oscillate with small amplitudes and displacements x_1 and x_2, from their equilibrium positions, show that their accelerations relative to the center of mass are $a_1 = kx/m_1$ and $a_2 = -kx/m_2$, where $x = x_2 - x_1$. (*b*) Show that the separation between the atoms varies as a simple harmonic oscillation about the mean value l, with frequency $\omega = \sqrt{k/\mu}$, where $\mu = m_1m_2/(m_1 + m_2)$ is the reduced mass of the system. (*c*) Show that the result of part (*b*) is equivalent to an acceleration of m_1 relative to m_2 of kx/μ.

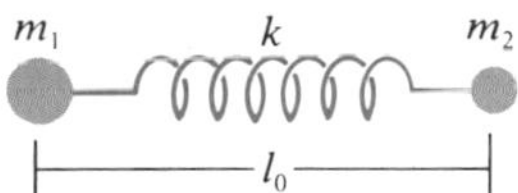

16 • 3 pendulums

16 • 15 The bob of the torsion pendulum in Fig. 16 . 7 is a disk of unknown moment of inertia I. Its period is $T = 3$ s. When a thin ring of mass 3 kg and a radius of 10 cm is placed on the disk, with the suspension wire passing through the exact center of the ring, the new period of oscillation is $T = 4$ s. Find the moment of inertia I.

16 • 16 The period and angular amplitude of the balance wheel of a certain watch are $T = 0.4$ s and $\theta_m = 30°$. (*a*) Find the angular acceleration of the wheel at the instant when its angular displacement is zero. (*b*) If the moment of inertia of the wheel about its axle is $I = 1\ \text{g} \cdot \text{cm}^2$, what is the accelerating torque at the instant when the wheel is at rest?

16 • 17 The period of a certain pendulum was observed to be $T = 1.002$ s at sea level. When the pendulum was carried to the top of a mountain, the period was found to be $T = 1.003$ s. (*a*) How high is the mountain? (*b*) How would a torsion pendulum be affected by altitude?

16 • 18 Prove that if water in a U-shaped tube is displaced, it will oscillate with a period which is the same as that of an ideal pendulum of length equal to one-half the total length of water in the U-tube.

16 • 19 A rocket that delivers a thrust five times its weight is equipped with a vertical pendulum clock. The rocket is fired at time $t = 0$ and rises straight up. After 5 s its fuel is exhausted. What is the time on the pendulum clock when a clock on the ground reads $t = 15$ s?

16 • 20 A thin uniform hoop of diameter d hangs on a nail. It is displaced through a small angle in its own plane and then released. Assuming that the hoop does not slip on the nail, prove that its period of oscillation is the same as that of an ideal pendulum of length d.

16 • 21 (*a*) A thin homogeneous rod of length l oscillates about a horizontal axis passing through one end of it. Find the length of the equivalent ideal pendulum and locate the center of oscillation and center of percussion. (*b*) A solid disk of radius R is oscillating with small amplitude about an axis perpendicular to the plane of the disk and at a distance r from its center. At what distance r' will the frequency be maximum?

16 • 22 A cubical block 20 cm on a side is suspended by two cords 15 cm long, as shown. (*a*) What is the period of oscillation when the motion is parallel to the plane of the figure? (*b*) When the motion is perpendicular to the plane of the figure?

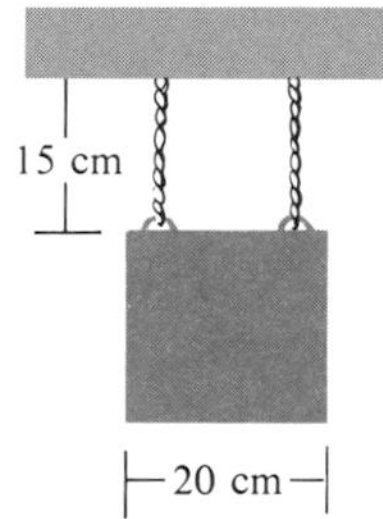

16 • 23 A thin wire is bent in the form of a half-circle of radius R. It is set oscillating in its own plane about an axis perpendicular to its plane and passing through the midpoint of the wire. (*a*) Find the length of the equivalent ideal pendulum. (*b*) Find the radius of gyration k about the axis from which it is swung.

16 • 24 Sketch a graph of the period T in Example 16 . 5 as a function of the distance r_c from the point of suspension to the center of mass, where T is given by

$$T = 2\pi\sqrt{\frac{k_c^2 + r_c^2}{gr_c}}$$

Show that for $T > T_{\text{min}}$ there are four points on an axis passing through O and C in Fig. 16 . 9 which have the same period T, located at distances

$$r_c = r_c'$$

$$= \frac{gT^2}{8\pi^2}\left(1 \pm \sqrt{1 - \frac{64\pi^4 k_c^2}{g^2 T^4}}\right)$$

above and below the center of mass.

16 • 4 physical pendulums: Simpson's rule

16 • 25 Derive Eq. [16 . 47] for T from Eq. [16 . 45] as indicated in text.

16 • 26 Show that if the maximum kinetic energy of a physical pendulum is $K > 2mgr_c$, then the pendulum continues rotating in the direction of increasing θ with period

$$T = 2\sqrt{\frac{I}{2mgr_c}}\int_0^{\pi}\frac{d\theta}{\sqrt{\cos\theta + E/mgr_c}}$$

16 • 27 A real physical pendulum on the end of a weightless rod has a maximum amplitude θ_m. Show that the tension in the rod is given by $F = mg(\cos\theta + \theta_m^2 - \theta^2)$ for small-amplitude oscillations and $F = mg(3\cos\theta - 2\cos\theta_m)$ for larger oscillations. When does the tension become a compression?

***16 • 28** A mass m is hung from the end of a spring of unstretched length $l_0 = mg/k$. (*a*) Taking the point of suspension of the spring as the origin of coordinates, with y positive downward, write the general equations of motion for acceleration in the x and y directions. (*b*) Suppose the stationary spring-mass system is stretched *horizontally* to a point $x = l_0, y = l_0$ and then released. Compute the motion numerically and see if you can detect any periodicity.

***16 • 29** Using the definition $\ln x = \int_1^x (dx/x)$ and Simpson's rule, make a table of the natural logarithm from $x = 1$ to $x = 9$ in intervals of (*a*) $\Delta x = 4$, (*b*) $\Delta x = 2$, and (*c*) $\Delta x = 1$. Compare your results with the tables in your handbook.

16 • 30 Derive Eq. [16 . 50] as indicated in the text by solving for A, B, and C in terms of f_i, f_{i+1}, and f_{i+2}.

16 • 5 damped and forced oscillations

16 • 31 Verify that the general solution of Eq. [16 . 53] is

$$x = Ae^{-\mu t/2}\sin(\omega_d t + \phi)$$

16 • 32 For the critically damped oscillation shown in Fig. 16 . 15, (*a*) find the inflection point P. (*b*) Show that the magnitude of the acceleration is a maximum at $t = 4/\mu$.

16 • 33 Derive Eqs. [16 . 64] and [16 . 65].

16 • 34 Suppose the time for an underdamped harmonic oscillator is reckoned from the instant when its displacement from the equilibrium position is x_0 and its speed is v_0. (*a*) Prove that

$$\phi = \sin^{-1}\frac{x_0}{A} = \cos^{-1}\frac{v_0 + (\gamma/2m)x_0}{A\omega_d}$$

and therefore that

$$A = \sqrt{x_0^2 + \left(\frac{v_0 + (\gamma/2m)x_0}{\omega_d}\right)^2}$$

(*b*) A 50-g harmonic oscillator of force constant $k = 5000$ dynes/cm, and damping coefficient $\gamma = 196$ g/s is set in motion by giving it a positive displacement of 10 cm and then releasing it from rest. (1) Show that the resulting oscillation is underdamped. (2) Compute ω_n, ω_d, A, and ϕ. (3) Write the displacement from equilibrium as a function of time.

16 • 35 Find A_F and ϕ_F (*a*) in the limit $\omega_F \ll \omega_n$. (*b*) In the limit $\omega_F \gg \omega_n$. (*c*) For $\omega_F = \omega_{\text{res}}$.

16 • 36 If $\mu \ll \omega_n$, show that the *average* energy of a damped harmonic oscillator during the interval t to $t + 2\pi/\omega_n$ is approximately

$$\overline{E} = \tfrac{1}{2}m\omega_n^2 A^2 e^{-\mu t} = E_0 e^{-\mu t}$$

16 • 37 If a damped harmonic oscillator is to be maintained in a steady state, then the driving force must supply an average power equal to the rate at which it is being dissipated by friction. (*a*) Use the result of Prob. 16 . 36 to show that the average kinetic energy under these circumstances is $\overline{K} = \frac{1}{4}m(A_F\omega_F)^2$. (*b*) Show that the power supplied by the driving force must be

$$\overline{P} = \overline{Fv} = -\tfrac{1}{2}F_0\omega_F A_F \sin\phi_F = \mu(\tfrac{1}{2}m\omega_F^2 A_F^2)$$

16 • 38 The amplitude of a simple pendulum 1 m long with a 0.6-kg bob is observed to change from 4 cm to 2 cm during nine cycles. (*a*) Find the average power that a driving force would have to supply to maintain the amplitude at 4 cm. (*b*) If a descending weight of 10 kg is used to supply this power, find its time rate of descent.

***16 • 39** In Eq. [16 . 52] assume that the resistive force is Newtonian and proportional to the square of the speed, $(dx/dt)^2$. Examine the resulting oscillations, starting from rest, for $\mu x_0 = 0.1, 1, 10$ and find their periods (if any). (*Hint:* In setting up the equation of motion, be sure the resistive force always opposes the velocity.)

16 • 40 Prove that $\omega_F = \sqrt{\omega_n^2 - \frac{1}{2}\mu^2}$ for forced oscillations gives the maximum response A_F or resonance.

16 • 6 small-amplitude oscillations

***16 • 41** A block is forced to move in a straight horizontal slot under the action of a spring of force constant k. Assume that all frictional forces are negligible and that the elastic limit of the spring is never exceeded. (*a*) Show that the restoring force on the block at any given position $x \leq A \ll l_0$ is $F = -(k/2l_0^2)x^3$, and therefore the block is a simple *nonlinear* oscillator, no matter how small the displacement. (*b*) Show that the total energy of this oscillator at every instant is $E = (k/8l_0^2)A^4$. (*c*) Compute the period of the oscillation for $A/l_0 = 0.1, 0.2, 0.3$.

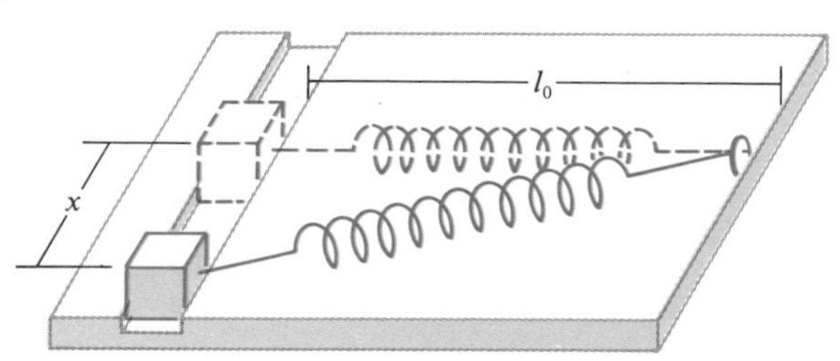

16 • 42 A more exact expression for the period of an ideal pendulum oscillating with angular amplitude θ_m is given by the infinite series

$$T = 2\pi\sqrt{\frac{l}{g}}\left(1 + \tfrac{1}{4}\sin^2 \tfrac{1}{2}\theta_m + \tfrac{9}{64}\sin^4 \tfrac{1}{2}\theta_m + \cdots\right)$$

Derive this formula from Eq. [16 . 47] by expanding the integrand in a Taylor's series about $\phi = 0$ and integrating term by term. Compare the sum of the first three terms with the numerical result $T/T' = 1.18034$ obtained by Simpson's rule for $\theta_m = \frac{1}{2}\pi$.

***16 • 43** The Lennard-Jones potential describing forces between two molecules at a separation distance r is given by

$$U = U_0\left[\left(\frac{r_0}{r}\right)^{12} - 2\left(\frac{r_0}{r}\right)^{6}\right]$$

What is the period of oscillation of a mass m about the equilibrium separation r_0 in the small-amplitude approximation? Compute the *actual* motion, starting from rest, at $0.94r_0$ and compare it to the prediction of simple harmonic motion at the frequency determined above.

***16 • 44** If the potential energy of a spring is

$$U(x) = \tfrac{1}{2}kx^2(1 + cx^2)$$

where c is a constant, we have *anharmonic* oscillations. Compute the motion for $v_0 = 0$ and $c = 0.1/x_0^2, 0.2/x_0^2, 0.3/x_0^2$ and compare with the simple harmonic case. What is the effect of c on the period?

answers

16 • 1 (*b*) $\nu = v/2l$

16 • 2 Phase $= 120°$ or $240°$, $v = 160$ cm/s, $F = 5000$ dynes

16 • 3 $F_g = mg - m\omega^2 y_m \sin \omega t$

16 • 4 (*a*) $\omega_n = 115.47$ rad/s; (*b*) $A = 3$ cm; (*c*) $x = 0.492$ cm left of equilibrium position, $v = -341.66$ cm/s

16 • 5 $\mu_s = 0.20$

16 • 6 $\Delta y = 7.5$ cm

16 • 8 (*a*) $E = 88{,}826$ erg; (*b*) $v_0 = 177.7$ cm/s

16 • 9 (*a*) $A = 25.08$ cm; (*b*) $\phi = 94.6°$; (*c*) $E = 78{,}625$ erg

16 • 10 (*a*) $x_0 = \pm 0.45$ m; (*b*) $v_0 = \pm 1.5$ m/s; (*c*) $A = 0.50$ cm; (*d*) $v_{\max} = \pm 3.3$ m/s; (*e*) $x = 0.35$ m

16 • 11 (*b*) $x = A, \sqrt{l^2 - A^2} - l, -A$; (*c*) yes, for $A \ll l$

16 • 15 $I = 0.0386 \text{ kg} \cdot \text{m}^2$

16 • 16 (*a*) $\alpha = 129 \text{ rad/s}^2$; (*b*) $\tau = 129$ dynes · cm

16 • 17 (*a*) $h = 6.36$ km; (*b*) not at all, except for lower air resistance

16 • 19 $t = 21.2$ s

16 • 21 (*a*) $l_0 = 2l/3$; (*b*) $r' = R/\sqrt{2}$

16 • 22 (*a*) $T = 0.78$ s; (*b*) $T = 1.1$ s

16 • 23 (*a*) $l_0 = 2R$; (*b*) $k = 0.85R$

16 • 32 (*a*) $t_P = 2/\mu$

16 • 34 (*b*)(2) $\omega_n = 10$ rad/s, $\omega_d = 9.81$ rad/s, $A = 10.2$ cm, $\phi = 1.37$ rad; (*b*)(3) $x(t) = (10.2 \text{ cm})[\exp(-1.96\, t)]$ $\{\sin[(9.81 \text{ rad/s})t + 1.37 \text{ rad}]\}$

16 • 35 (*a*) $A_F = f_0/\omega_n^2, \phi_F = -\mu\omega_F/\omega_n^2$; (*b*) $A_F = f_0(\omega_F^4 + \omega_F^2\mu^2)^{-1/2}$, $\phi_F = \tan^{-1}(\mu/\omega_F)$; (*c*) $A_F = (f_0/\mu)(\omega_F^2 + \frac{1}{4}\mu^2)^{-1/2}$, $\phi_F = \tan^{-1}(-2\omega_F/\mu)$

16 • 38 (*a*) $\overline{P} = 0.72$ mW; (*b*) $v_y = 2.645$ cm/h

16 • 39 The first two periods (in units of $1/\omega_n$) are (1) 6.29, (2) 6.58, 6.33, (3) 9.93, 6.34

16 • 41 (*c*) In units of $\sqrt{m/k}$, the periods are 105, 52.5, and 35.0 (proportional to l_0/A)

16 • 43 In units of $r_0(m/12U_0)^{1/2}$, the period is 2.960 (cf 2.565 for small oscillations)

16 • 44 For $cx_0^2 = 0.1$, 0.2, and 0.3, the periods (in units of $2\pi/\omega_n$) are 0.933, 0.894, and 0.832. Small-oscillation approximation: $T = (2\pi/\omega_n)(1 + 6cx_0^2)^{-1/2}$

CHAPTER SEVENTEEN

waves

In the case of sounds, something of the same sort takes place as when a stone is thrown out and falls into a pool or other calm water. The stone first produces a wave with a very small circumference. Then it causes the waves to spread out in ever wider circles until the motion, growing weaker as the waves spread out, finally ceases. The later and larger the wave, the weaker the impulse with which it breaks. Now if there is an object that can block the waves as they grow larger, the motion is at once reversed and forced back, in the same series of waves, to the center from which it originated.

In the same way, then, when air is struck and produces a sound, it impels other air next to it and in a certain way sets a rounded wave of air in motion, and is thus dispersed and strikes simultaneously the hearing of all who are standing around. And the sound is less clear to the one who stands farther away since the wave of impelled air which comes to him is weaker.

Boethius, De institutione musica, *ca. 500 A.D.*

The various ways in which mechanical energy can be transferred from one point to another fall into two general classes: (1) the passage of matter from one point to another and (2) the passage of energy through a continuous material medium containing two points, which leaves the medium essentially unchanged after transfer. This latter process is exemplified by the passage of ocean waves or sound waves. In Example 9.3 we touched on this type of energy transfer when we considered the collision of a hard sphere with a set of identical spheres at rest. There we found that all the energy contained in each sphere was transferred to the succeeding sphere by means of elastic collision, and so passed through the line of spheres like a single pulse of energy passing through a continuous medium. Such a pulse or disturbance is called a *wave.* We shall see that a wave has a *speed of propagation* through the medium which is characteristic both of the nature of the disturbance and the physical properties of the medium.

There are many kinds of waves. We shall be considering only *mechanical waves* propagating through a material medium by virtue of its elasticity. In scientific usage, *elasticity* refers to the ability of a substance to retain its size and shape when stressed and to recover from any deformation without dissipating energy. In this sense a steel ball is more elastic than a rubber one.

The elasticity of a medium, then, is clearly related to its ability to transmit forces and energy. This concept implies that locally the medium returns to its equilibrium condition after the energy associated with a disturbance has propagated to other parts of the medium in the form of a wave. However, owing to its inertia and elasticity, a material medium also has the capacity to store energy, provided the oscillations of the particles of the medium are not too quickly dissipated as heat, turbulence, or other forms of disordered or random motion.

Mechanical waves are of two fundamental types: longitudinal and transverse. In a *longitudinal wave* the oscillating particles of the medium are displaced parallel to the direction of motion (the direction of energy transmission) of the wave. In a *transverse wave* particles of the medium are displaced in a direction perpendicular to the motion of the wave. When a rod is struck on the end with a hammer, a single wave pulse, in the form of a longitudinal compression of the rod, travels along its length. If the rod is struck periodically, a succession of such pulses is sent along it. This constitutes a *wave train.* The idea that sound energy—say, from the vibrating strings of a musical instrument—propagates through the air in a similar manner was probably familiar to Aristotle. It was first clearly discussed, however, by Boethius (see the chapter-opening quotation)*.

The familiar water waves which emanate from the point where a stone has dropped into the water are *transverse* surface waves. Photographs taken at various stages of such disturbances show that the surface of the water under the stone is forced down, leaving a hole. The water forced out of the hole first piles up around its edges and then begins to fall back into the hole and also forward onto the outer undisturbed surface, thus starting a wave that moves out from the hole. Because of the inertia of the water, this action does not cease when the hole has been refilled, but continues until a depression is formed where the water had previously been heaped up. As the liquid piles up on each side of this new depression, the whole process is then repeated, and the disturbance propagates radially outward from its source. For small-amplitude waves, the kind we shall be considering here, the motion of

*Boethius was born in Rome ca. 480 A.D., just a few years after the destruction of the Roman Empire in the West. A great theologian, philosopher, and statesman, his dedicated scholarship was responsible for the preservation of much of ancient Greek thought. He was executed in 524 A.D. for being too smart, among other things.

any individual water molecule is approximately transverse to the wave direction, like the motion of a cork bobbing in place on the water. The molecule moves only up and down, and not in the direction of the wave.

In this chapter we shall discuss the mathematical representation of various types of waves, both longitudinal and transverse, propagating in solid, liquid, or gaseous media. In particular, we shall consider their speeds of propagation and the physical characteristics, such as energy content and pressure variations, associated with them.

17 • 1 mathematical representation of waves

The physical state of a medium is specified by various properties: its geometry, temperature, pressure, density, and so on. Under static conditions, these properties may even show some spatial variation from one point to another in the medium, such as the variation of air pressure with altitude. However, if a physical property at some point in the medium changes suddenly with *time,* then the medium is said to be disturbed and to oscillate. The oscillation may be a single isolated pulse, or it may be many periods in duration; it may be damped, undamped, harmonic, anharmonic (having a nonlinear restoring force), or forced. When the elasticity of the medium causes neighboring particles to display a similar oscillation, a wave is set up, and the oscillation appears to move through the medium with some *speed of propagation* which is characteristic for the medium and the type of disturbance. A single oscillation may set up a pulse, such as a blast wave or a sonic boom; a series of oscillations can set up a wave train, such as a succession of ocean waves.

The movement of such a wave through the medium, in the simplest geometry, may be specified by one spatial variable x, a time variable t, and some dependent variable, preceded by the letter δ, which represents the actual disturbance in the medium as a function of x and t. For example, consider a *transverse* mechanical disturbance, such as a ripple of water on a quiet lake, where $\delta y = f(x, t)$ represents the vertical displacement at x of the surface of the water from its undisturbed position, and x is the direction of propagation of the wave, which we assume here is of infinite extent in the z direction. Then at some given instant of time t_0, $\delta y = f(x, t_0)$ represents the *waveform* or *wave profile,* a cross section of the wave in the xy plane, as shown in Fig. 17 . 1. For simplicity we shall assume here that the waveform

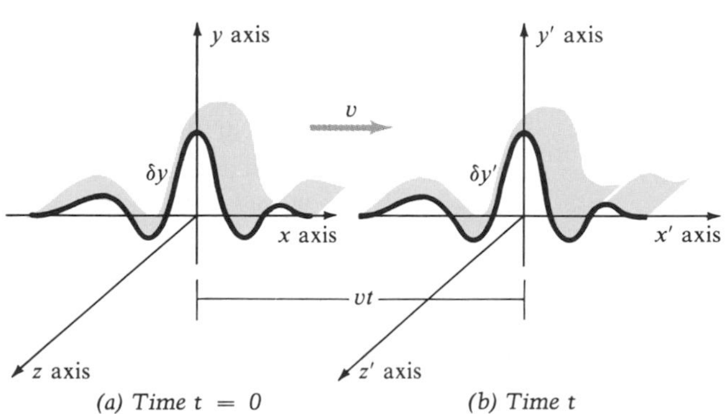

figure 17 • 1 *Transverse water wave of constant profile traveling with speed v*

is unchanged as it moves through the medium. Note that x is independent of t and is merely a descriptive coordinate of position. There is some appropriate value of δy for every value of x, because a wave, unlike a particle, has no definite position in space.

Imagine that at some time $t = 0$ a snapshot of the wave shows its profile to be given by

$$\delta y = f(x, 0) = g(x) \qquad [17 \cdot 1]$$

If the waveform is propagating with some constant speed v to the right, and you are moving along at the *same* speed, the wave in your moving system will always look the same:

$$\delta y = g(x') \qquad x' = x - vt \qquad [17 \cdot 2]$$

(This is the familiar Galilean transformation to a moving coordinate system.) However, if Eq. [17 . 2] is to be consistent with Eq. [17 . 1], then in the stationary system

$$\blacktriangleright \quad \delta y = f(x, t) = g(x - vt) \qquad [17 \cdot 3]$$

Thus the assumption that the waveform propagates with some definite speed implies a constraint on the form of the function $f(x,t)$: it must be a function of $x' = x - vt$, taken as a single argument. Similarly, if the wave is traveling to the left with speed v, then

$$\blacktriangleright \quad \delta y = f(x, t) = g(x + vt) \qquad [17 \cdot 4]$$

These restrictions on the mathematical representations of traveling waves are quite general and apply to any traveling wave that depends on only one spatial dimension. It is not difficult to generalize them to additional dimensions.

The dependence on $x \pm vt$ is a general property of wave propagation, quite unrelated to the waveform itself, which may look like a spike, a chimney, a bell-shaped curve, a sine curve, or any other single-valued function. However, in Chapter 16 we saw that the small oscillations of particles in stable mechanical equilibrium, as in an undisturbed continuous medium, are simple harmonic motions. Furthermore, the entire wave at any given instant is just the composite of the instantaneous displacements of the oscillating particles comprising it. As the most basic type of wave, therefore, let us consider the wave set up by a particle oscillating in place with simple harmonic motion.

We assume the particle at $x = 0$ to have vertical oscillations given by

$$\delta y = f(0, t) = A \sin(\omega t + \phi) \qquad [17 \cdot 5]$$

where A is the amplitude of the oscillation, ω is its angular frequency, and ϕ is its phase angle. For a wave traveling with velocity $\mathbf{v} = v\mathbf{i}$, Eq. [17 . 3] gives

$$\delta y = f\left(0, -\frac{1}{v}(x - vt)\right) = f(x, t) = A \sin\left[\omega\left(t - \frac{x}{v}\right) + \phi\right] \qquad [17 \cdot 6]$$

where the substitution of $t - x/v$ for t is equivalent to our previous substitution of $x - vt$ for x. If the phase angle is initially zero at the origin, then

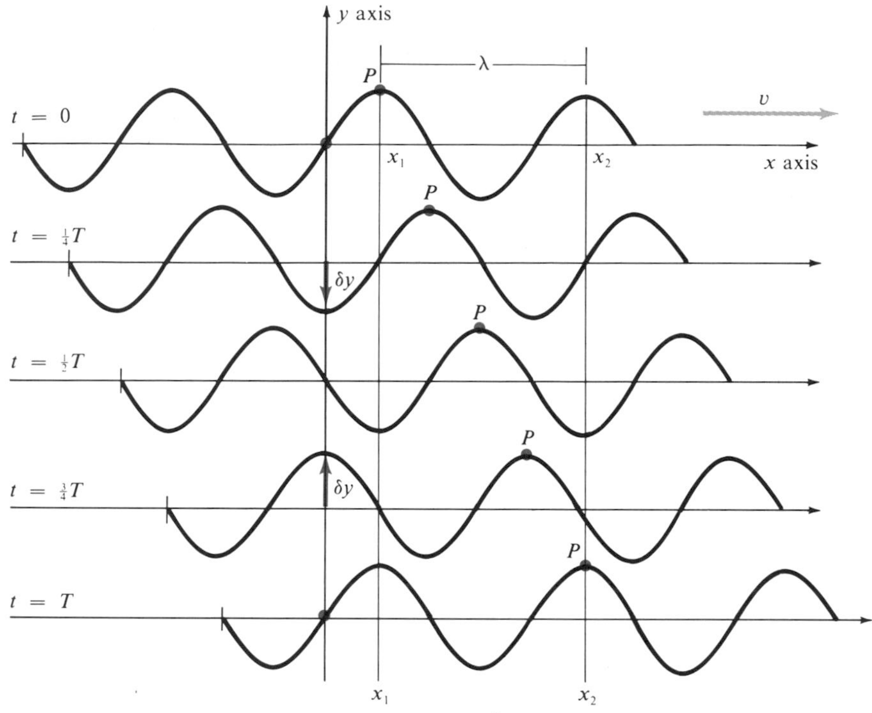

figure 17 • 2 *Traveling wave* $\delta y = A \sin [\omega(x/v - t)]$ *at time intervals* $\Delta t = T/4 = \pi/2\omega$. *The point P on the waveform moves with speed* $v = \lambda/T$.

$$\delta y = A \sin\left(\omega t - \frac{\omega x}{v}\right) \qquad [17 \cdot 7]$$

This is the equation of a simple harmonic, or *sinusoidal, wave* traveling with velocity $\mathbf{v} = v\mathbf{i}$ in an elastic medium.

For some particle located at x_0,

$$\delta y = A \sin\left[\omega t + \left(\phi - \frac{\omega x_0}{v}\right)\right]$$

so the term $\phi - \omega x_0/v$ is a constant phase angle and the particle oscillates with the same period and frequencies as the particle at the origin. Therefore unless otherwise stated, we assume $\phi = 0$. The period of the wave is then just the period of oscillation of any particle of the wave for fixed x:

$$T = \frac{2\pi}{\omega} = \frac{1}{\nu} \qquad [17 \cdot 8]$$

The waveform is also periodic in *space* because of its dependence on x. This period, shown in Fig. 17 . 2, is defined to be the *wavelength* λ. That is, at regular intervals along the x axis particles are in the same phase, and the distance between them is λ. If the first particle is located at x_1 and the second at x_2, then $\lambda = x_2 - x_1$ must correspond to a phase change of 2π, or

$$\omega\left(t - \frac{x_1}{v}\right) - \omega\left(t - \frac{x_2}{v}\right) = \omega\frac{\lambda}{v} = 2\pi$$

Therefore we have the very important relation

$$\blacktriangleright\ \lambda = \frac{2\pi v}{\omega} = vT \qquad \text{or} \qquad \frac{\lambda}{T} = \lambda\nu = v \tag{17.9}$$

which holds for all periodic waveforms $f(x \pm vt)$. We may also express the simple wave in scaled form as

$$\delta y = A \sin\left[2\pi\left(\frac{t}{T} - \frac{x}{\lambda}\right)\right] \tag{17.10}$$

Depending on the phase constant ϕ, a wave can be expressed equally well by a cosine function or by reversing the argument to $x/\lambda - t/T$.

Figure 17 . 2 shows how the traveling wave must appear to a stationary observer after each time interval $\Delta t = \frac{1}{4}T$. If the wave is to look the same after an interval $\Delta t = T$, then point P must have moved a distance $\lambda = x_2 - x_1$ in this time. The speed of the *wave* is therefore $v = \lambda/T$. The actual material particle at x_1 remains at x_1, although its displacement is again a maximum.

example 17 • 1 Neglecting its boundaries, assume that the shape of a lake's surface at time $t = 0$ is given by the function

$$\delta y = f(x, 0) = A \sin\left(\frac{2\pi x}{\lambda}\right)$$

for a transverse wave ($\mathbf{v} = v\mathbf{i}$), where λ is a constant. Find the displacement δy of the surface (*a*) At $x = \frac{3}{4}\lambda$ and $t = 0$. (*b*) At $x = \frac{3}{4}\lambda$ and $t = \frac{1}{2}T$. (c) At $x = 10\lambda$ and $t = \frac{1}{2}T$.

solution (*a*) Evaluating $f(\frac{3}{4}\lambda, 0)$ gives

$$\delta y = A \sin\left(\frac{2\pi}{\lambda}\,\tfrac{3}{4}\lambda\right) = A \sin \tfrac{3}{2}\pi = -A$$

(*b*) Since $v/\lambda = 1/T$, the general equation must be, from Eq. [17 . 3],

$$\delta y = A \sin\left[\frac{2\pi}{\lambda}(x - vt)\right] = A \sin\left[2\pi\left(\frac{x}{\lambda} - \frac{t}{T}\right)\right]$$

Therefore

$$\delta y = A \sin\left[2\pi\left(\tfrac{3}{4} - \tfrac{1}{2}\right)\right] = A \sin \tfrac{1}{2}\pi = A$$

(*c*) Evaluating $f(10\lambda, \frac{1}{2}T)$ gives

$$\delta y = A \sin\left[2\pi\left(10 - \tfrac{1}{2}\right)\right] = A \sin(-\pi) = 0$$

17 • 2 *speed of transverse waves*

Transverse waves can be supported by a solid, since a solid possesses shear elasticity—that is, tangential restoring forces. This is the case with a stretched violin string (essentially one-dimensional) or with an extended solid body. Such waves can also exist at the interface of fluids of different densities, as between air and water. Consider a transverse wave pulse sent

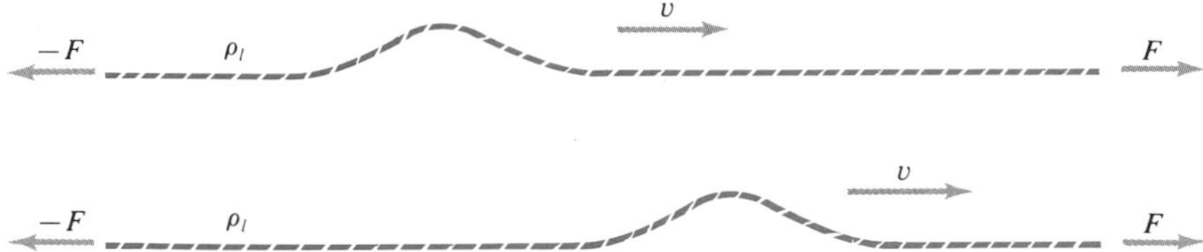

(a) Propagation of a transverse wave down a stretched string

figure 17 • 3 Transverse wave pulse

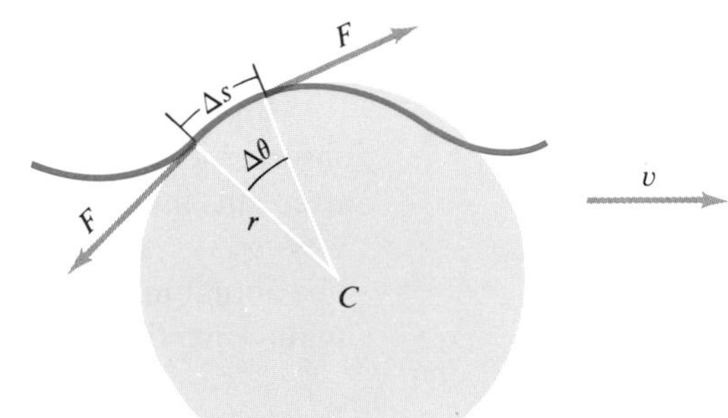

(b) Speed of transverse wave pulse

down a stretched violin string, as in Fig. 17 . 3*a*. Assume that the string is of uniform linear density ρ_l, is perfectly flexible, and is stretched so tightly by a force F that the weight of the wire can be neglected in comparison with F. For small displacements the tension does not change appreciably, and we can deduce the speed of the transverse wave pulse.

The curve in Fig. 17 . 3*b* represents a portion of the stretched string over which the deformation is being propagated, where Δs is a curved element of the string small enough that we can consider it the arc of a circle of radius r. The mass of each element Δs is then $\Delta m = \rho_l \Delta s$. Since the string is flexible, the only force of appreciable magnitude directed toward the center C is the centripetal force F_c due to the tension F acting tangentially at each end of arc Δs. The sum of the centripetal components of these two forces is

$$F_c = 2F \sin\left(\tfrac{1}{2}\Delta\theta\right) \doteq F\,\Delta\theta = F\,\frac{\Delta s}{r} \qquad [17 \cdot 11]$$

since $\Delta\theta = \Delta s/r \ll 1$. The acceleration of Δs is therefore centripetal, $a_c = v^2/r$, and the equation of motion is

$$\Delta m\; a_c = \rho_l\,\Delta s\,\frac{v^2}{r} = F_c = F\,\frac{\Delta s}{r}$$

The speed of a transverse wave on a string is therefore a function only of its tension and linear density:

$$\blacktriangleright \quad v = \sqrt{\frac{F}{\rho_l}} \qquad [17 \cdot 12]$$

If we know the frequency ν of the oscillating source, then we can determine the wavelength from $\lambda = v/\nu$. The validity of this result depends on the fact that the vertical displacements δy of the string are small compared with its overall length, but v is not dependent on the shape of the waveform propagated down the string.

Any type of oscillatory motion may be propagated as a wave, given an appropriate medium. For example, *torsion waves,*

$$\delta\theta = \theta_m \sin\left[\omega\left(t - \frac{x}{v}\right)\right]$$

can be set up in a solid cylinder, such as a heavy rod, by holding one end fixed and applying a sudden twist at the other end. These are purely transverse waves in which the motion is a twisting back and forth in a circular arc about the axis of the rod, as in the case of the torsion pendulum discussed in Sec. 16 . 3. We shall not attempt to prove it, but the speed of such waves is given by the shear modulus S and density ρ of the rod:

$$v = \sqrt{\frac{S}{\rho}} \qquad [17 \cdot 13]$$

This is the speed of a pure transverse wave in any extended homogeneous solid medium.

If such a medium were infinite in extent, then, by symmetry, a transverse wave would spread out radially in all directions, with the components of the wave disturbance perpendicular (transverse) to the radius vector from the source. At large distances from the source, the pulse will be in the form of a spherical surface enclosing the source at its center, the three-dimensional analogy to circular ripples on a pond. This surface, on which all particles are in the same phase of oscillation, is called a *wavefront;* it represents a cross section of the wave profile perpendicular to the direction of propagation (see Fig. 17 . 4). A wavefront is also called a *surface of constant phase,* since it moves along with the wave, and the particles instantaneously within it are always in the same phase of their oscillation. As long as the amplitude of the disturbance is much less than the distance from the source,

figure 17 • 4 *Wavefront of maximum displacement for a transverse plane wave $\delta y = f(x') = f(x - vt)$*

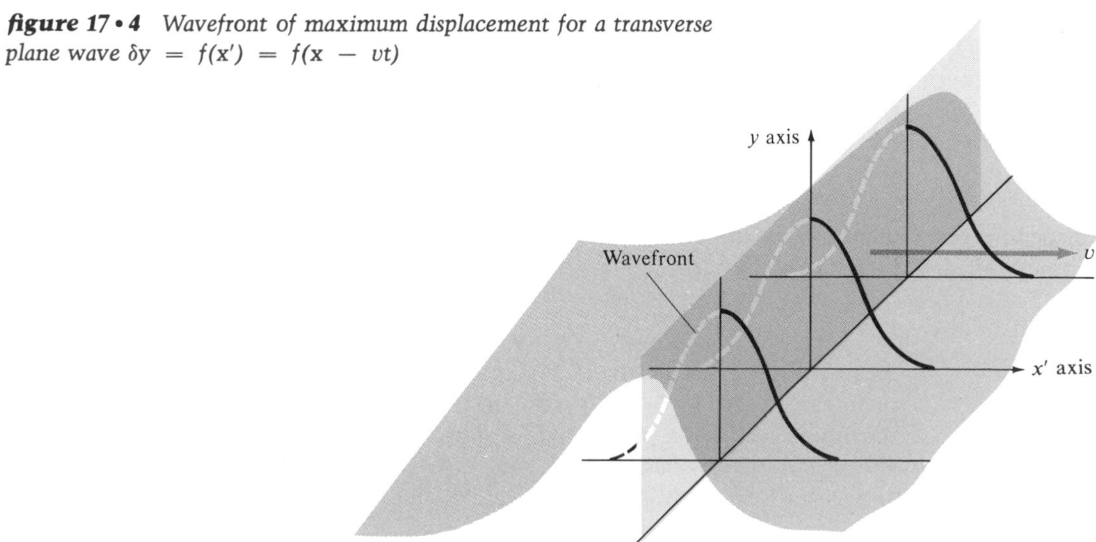

the wavefronts will appear, to a good approximation, plane surfaces.

Although a derivation of the speed of transverse surface waves is beyond our present scope, the results for water waves warrant special mention. The propagation speed for surface waves on water is given by

$$v = \sqrt{\frac{g\lambda}{2\pi}\tanh\frac{2\pi h}{\lambda}} \qquad \tanh x = \frac{e^{x} - e^{-x}}{e^{x} + e^{-x}} \qquad [17 \cdot 14]$$

where h is the depth of the water measured from the surface. This equation has two special cases of interest. For deep water, $h \gg \lambda$,

$$\tanh\frac{2\pi h}{\lambda} \longrightarrow 1 \qquad v = \sqrt{\frac{g\lambda}{2\pi}} \qquad [17 \cdot 15]$$

and for shallow water, $h \ll \lambda$,

$$\tanh\frac{2\pi h}{\lambda} \longrightarrow \frac{2\pi h}{\lambda} \qquad v = \sqrt{gh} \qquad [17 \cdot 16]$$

In deep water the bottom has almost no effect on the wave, but in shallow water it is critical. The crest of a wave coming in to the beach travels with a higher speed than the trough, since its effective depth is significantly greater. Hence the wave steepens and finally "breaks" as it outruns its shallower portion. These are also called *gravity waves,* because it is the force of gravity which governs the phenomenon.

Surface waves on water afford an example of a *dispersive medium.* When wave speed is not a constant, but depends on the wavelength of the disturbance, the medium is said to be dispersive for that type of wave, because waves of different wavelengths starting out together will tend to disperse with time, due to their different velocities. Most wave shapes in nature are not ideal sinusoidal wave trains, but are made up, according to Fourier's theorem, of sinusoidal components of varying wavelength. Consequently, the shape of a wave will change as its different components propagate with varying speeds in a dispersive medium. In all but shallow water, v is proportional to $\sqrt{\lambda}$ and the frequency of an oscillating source of disturbance and the wavelength of the surface waves it produces are related by

$$\nu = \frac{v}{\lambda} = \sqrt{\frac{g}{2\pi\lambda}} \qquad \text{or} \qquad \lambda = \frac{g}{2\pi\nu^2} \qquad [17 \cdot 17]$$

Dispersive phenomena will be important when we discuss the passage of light through glass in later chapters.

example 17 • 2 A weight of 200 N is attached as shown to a steel string of linear density $\rho_l = 0.2$ g/cm. (*a*) If the weight performs one simple harmonic oscillation per second, of amplitude $A = 1$ mm, what are the speed and the wavelength of the wave pulse generated on the string? (*b*) Describe the motion of a particle of the string located 1 m from the end at which the weight is attached. Assume the string to be infinitely long.

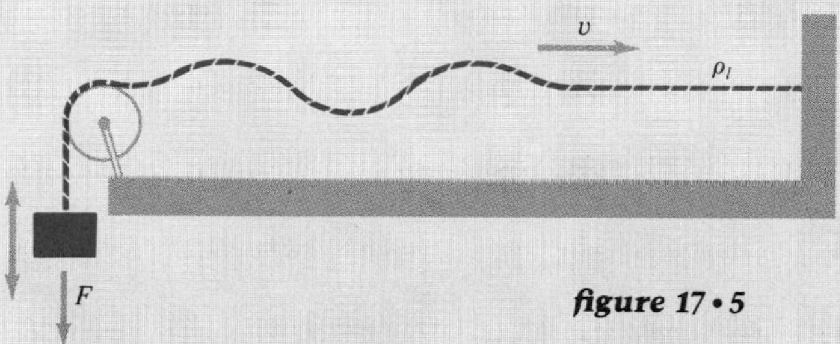

figure 17 • 5

solution (*a*) The speed of propagation is

$$v = \sqrt{\frac{F}{\rho_l}} = \sqrt{\frac{2 \times 10^7 \text{ dynes}}{0.2 \text{ g/cm}}} = 100 \text{ m/s}$$

and the wavelength is therefore

$$\lambda = vT = 100 \text{ m}$$

(*b*) The profile of the pulse is

$$\delta y = \sin\left[2\pi\left(\frac{t}{1 \text{ s}} - \frac{x}{100 \text{ m}}\right)\right] \text{mm}$$

so that for $x = 1$ m,

$$\delta y\,(1 \text{ m}, t) = \begin{cases} \sin\,(2\pi t - 0.01)\text{ mm} & 0.01 \leq t \leq 1.01 \text{ s} \\ 0 & \text{all other times} \end{cases}$$

since the pulse does not reach the point $x = 1$ m until $t = 0.01$ s, and it takes a time $\Delta t = 1$ s to pass the point completely.

17 • 3 *longitudinal compression waves*

The transverse mechanical wave is the simplest waveform to visualize. However, we can apply the same types of mathematical representation to longitudinal waves, in which the disturbance travels in the direction of propagation. For longitudinal waves propagating in the x direction, $\delta x = f(x, t)$ represents the horizontal displacement of a particle originally at x due to the passage of the wave. Like a transverse wave in an extended medium, such waves propagate radially outward from their source. In this case the components of the disturbance are along the radius, rather than perpendicular to it. Nevertheless, at distances from the source which are large compared with the wavelength, the wavefronts appear planar, as in Fig. 17 . 4. That is, the disturbances of particles in a given plane perpendicular to the radius vector from the source will have the same phase, although their motions are longitudinal, rather than transverse.

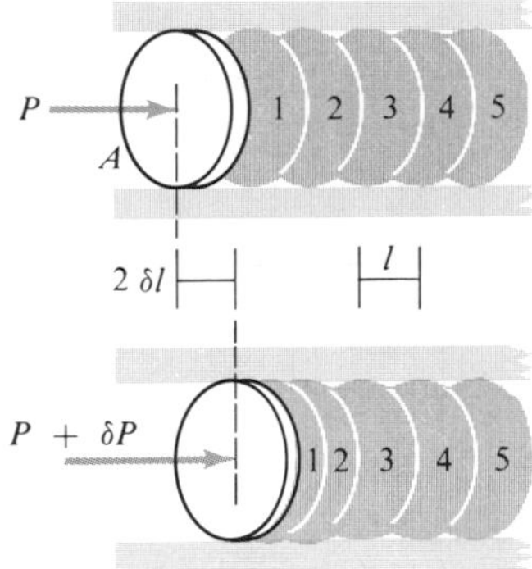

(*a*) *Before application of a pressure pulse* δP

(*b*) *Situation at a time* $t = 2l/v = 2\delta l/v_p$, *after disks 1 and 2 have been compressed to an internal pressure* $P + \delta P$.

figure 17 • 6 *Compression-wave model*

Since the wavefront can be approximated by a plane surface coming directly from the distant source, we can construct a model of the wave in which we imagine a portion of the medium to be contained in an infinitely long, rigid tube of cross-sectional area A, our "control volume." Let us assume that the medium, of density ρ, is at constant temperature and that P is the undisturbed static pressure in the medium, before any compression waves are produced. Imagine that the medium is divided into disks, each of volume $V = Al$, where l is the initial width of each (see Fig. 17 . 6). The tube is closed at one end by a massless, frictionless piston.

Now, suppose we apply a small additional pressure or pulse δP to the piston, moving it to the right and starting a compressional pulse down the tube with some speed v. As soon as the piston starts to move forward, disk 1 is compressed and the pressure inside it rises. When this pressure reaches the value $P + \delta P$, the disk ceases to compress any further and transmits pressure to disk 2. The change δV in volume that disk 1 has experienced is $\delta V = -A\,\delta l$. Disk 2 is in turn compressed by an amount $\delta V = -A\,\delta l$ under the action of $P + \delta P$, transmitting the pressure pulse to disk 3, and so on.

As the pulse moves down the tube, the piston and all the compressed disks will move uniformly in the x direction, with the piston moving an additional amount δl each time another disk is compressed. If, in a time dt, a number dn of disks is compressed, each moving with the speed v_p of the piston to which the force $F = A\,\delta P$ is applied, then Newton's second law of motion gives

$$\frac{d(mv)}{dt} = \rho Al\,\frac{dn}{dt}\,v_p = A\,\delta P \qquad [17 \cdot 18]$$

where each disk has mass $dm = \rho Al$. Now, in order to obtain an expression for the speed of propagation v for the pressure pulse δP being transmitted from disk to disk, we must express v_p and dn/dt in terms of v.

First we note that in a time dt, by definition, the pulse must travel a distance $v\,dt$, traversing dn disks of width l. Thus

$$v\,dt = l\,dn \qquad \text{or} \qquad \frac{dn}{dt} = \frac{v}{l} \qquad [17 \cdot 19]$$

Second, during this same time the piston moves a distance $v_p\;dt$ which must equal the total compression $dn\,\delta l$ of the disks. Therefore, if we substitute dn/dt from Eq. [17 . 19], the piston speed is

$$v_p = \frac{dn}{dt}\delta l = v\frac{\delta l}{l} < v \qquad [17 \cdot 20]$$

Third, we can relate δl to δV by noting that the volume of each disk is $V = Al$, where we have assumed that the cross section A of our control volume is constant. This yields the relation

$$\delta V = -A\,\delta l = -\frac{V}{l}\delta l \quad \text{or} \quad \frac{\delta l}{l} = -\frac{\delta V}{V} \qquad [17 \cdot 21]$$

Finally, we can substitute the preceding equations into Eq. [17 . 18], first cancelling A on both sides, to obtain

$$\rho v^2 = -l\frac{\delta P}{\delta l} = -V\frac{\delta P}{\delta V} \qquad [17 \cdot 22]$$

If we can now relate the ratio $\delta P/\delta V$ to some observable quantity, we can find v.

In Sec. 15 . 2 we found that this ratio is related to B, the bulk modulus of elasticity of the medium, which is easily measured. For small-amplitude oscillations, $|\delta V| \ll V$, which do not significantly alter the properties of the medium,

$$-V\frac{\delta P}{\delta V} \doteq -V\frac{dP}{dV} = B$$

Therefore Eq. [17 . 22] yields

$$\blacktriangleright \quad v = \sqrt{\frac{B}{\rho}} \qquad [17 \cdot 23]$$

for the speed of a compressional wave in the medium. Since this expression depends only on the properties of the medium, we see that a pulse, once started, will travel through the control volume at a speed which has nothing whatever to do with the size and shape of the medium or the motion of the source of disturbance (in this case, the piston).

If the pressure disturbance in Fig. 17 . 6 had been negative, $-\delta P$, the piston would have started back instead of forward, and disk 1 would have expanded until its pressure reached the value $P - \delta P$. This expansion would have been followed by a similar expansion of disk 2, and so on. Thus a pulse of *rarefaction*—a sudden drop in pressure—instead of one of *condensation* (compression) would have traveled down the tube, and by exactly the same reasoning we would have found that the speed of this pulse is also $\sqrt{B/\rho}$.

In a pulse of condensation, the particles of the medium are always moving in the same direction as the pulse, whereas in a pulse of rarefaction, the direction of motion of the particles is always *opposed* to the direction of propagation of the pulse. If a piston is moved alternately forward and backward at regular intervals, a wave train of compression and rarefactions will follow one another down the tube, with the motions of all the particles repeating, in succession, exactly the motion of the piston. Hence the wave is *longitudinal,* with oscillations parallel to the direction of propagation.

example 17 • 3 What is the speed of sound in air under standard conditions, as predicted by Eq. [17 . 21]?

solution In Sec. 14 . 2 we saw that $B = P$ for a perfect gas under constant temperature conditions. Therefore a speed of $v = \sqrt{P/\rho}$ is predicted. Under standard conditions, $P = 1 \text{ atm} \doteq 10^5 \text{ N/m}^2$ and $\rho = 1.29 \text{ kg/m}^3$, hence

$$v \doteq \sqrt{\frac{P}{\rho}} = 278 \text{ m/s}$$

(As we shall soon see, this value is actually about 18 percent too small.)

17 • 4 compression waves in different media

The relationship $v^2 = B/\rho$ holds well for small-amplitude compression waves propagating in an isotropic *liquid* of infinite extent. However, it is not quite accurate in three important cases: in gases, in solids, and in thin solid rods which are "infinite" in extent in only one direction. However, these instances can be treated as already outlined in Sec. 17 . 3 if we are careful to stipulate that

$$v^2 = \frac{B'}{\rho} \qquad B' = -V\left(\frac{\delta P}{\delta V}\right)_{\text{wave}} \qquad [17 \cdot 24]$$

where the process may *not* be isothermal (at constant temperature) and, in the case of solids, may also involve some twisting or lateral expansion of the control-volume boundaries.

When Isaac Newton applied the formula $v = \sqrt{B/\rho}$ to the data at his command, he found the predicted speed of sound in air to be 16 percent less than the best experimental values of his day. More than 120 years after Newton published his calculations, Laplace finally pointed out the reason for this discrepancy. In the passage of sound through a gas the compressions and rarefactions are so rapid that there is not time for heat to flow out of or into the volume of gas under study. Hence—and this is significant in the case of a gas—there are fluctuations in temperature which violate the isothermal assumptions of the preceding derivation. A process in which there is no heat transfer between the system and its surroundings is termed *adiabatic*. If the temperature of a compressed volume of gas increases as a result of the work done on it under adiabatic conditions, a smaller decrease δV_{ad} in volume is required to achieve the same increase δP in pressure. Hence

$$v = \sqrt{-V\frac{\delta P}{\rho\,\delta V_{\text{ad}}}} > \sqrt{-V\frac{\delta P}{\rho\,\delta V}} \qquad [17 \cdot 25]$$

In Chapter 21 we shall see that it is possible to define a bulk modulus B_{ad} for small strains taking place adiabatically. The relation of B_{ad} to the bulk modulus B (see Sec. 14 . 2) is $B_{\text{ad}} = \gamma B$, where, for a perfect gas, γ is a dimensionless constant known as the *ratio of specific heats,* of value between

1 for large molecules and $\frac{5}{3}$ (for monatomic gases), depending on the molecular structure of the gas. The formula

$$v = \sqrt{\frac{\gamma P}{\rho}} \qquad [17 \cdot 26]$$

is so accurate that it is often used to obtain values of γ from measurements of the speed of sound. As we shall see in Chapter 20, for most gases under ordinary conditions, $P/\rho = RT/M_0$, where $R = 8314$ J/K, M_0 is the kilogram-molecular weight of the gas, and T is the absolute temperature in kelvins (K), so the speed of sound is given by

$$v = \sqrt{\frac{\gamma RT}{M_0}} \qquad [17 \cdot 27]$$

If we set $\gamma = 1.4$ for air (diatomic molecules) and use the results of Example 17 . 3, we find that $v = 332$ m/s at 0°C, where $T = 273$ K.

example 17 • 4 What is the rate of change with temperature of the speed of sound in air at 20°C?

solution Since $v = 332$ m/s at 0°C, Eq. [17 . 20] predicts

$$v = (332 \text{ m/s}) \sqrt{\frac{293}{273}} = 343.9 \text{ m/s at } 20°\text{C} = 293 \text{ K}$$

Therefore

$$\frac{dv}{dT} = \frac{v}{2T} = \frac{343.9 \text{ m/s}}{546 \text{ K}} \doteq 0.6 \text{ m/s} \cdot \text{K}$$

Because a disturbance in a solid body is generally transmitted simultaneously by both transverse waves and isothermal longitudinal waves, the boundary of the cylindrical control volume of Fig. 17 . 6 undergoes some distortion. This problem was first treated by Siméon-Denis Poisson in 1829 and Augustin Louis Cauchy in 1830. They found that for a large mass of matter of bulk modulus B and shear modulus S, the speed of the longitudinal wave is

$$v = \sqrt{\frac{B + \frac{4}{3}S}{\rho}} \qquad [17 \cdot 28]$$

In the case of a fluid, which, by definition, has a shear modulus $S = 0$, this reduces to Eq. [17 . 23] for isothermal waves.

Earth tremors, or *seismic waves,* are both transverse and longitudinal. Seismograph records taken at an observing station some distance from the source of an earthquake show three distinct sets of waves: (1) longitudinal waves that come directly through the earth with a speed of about 8 km/s; (2) transverse waves, also direct, with a speed of about 4.5 km/s; and (3) large-amplitude surface waves, known as *Rayleigh waves,* which are analogous to water waves and penetrate to a small depth only. Since the speeds of the various types of waves are known, the difference in their arrival times gives the distance of the disturbance from the station.

example 17 • 5 From the seismographic data and Table 14 . 1, would you say the earth's core is more likely to be composed chiefly of aluminum, copper, iron, magnesium, or tungsten?

solution The ratio of longitudinal and transverse wave speeds gives

$$\frac{B + \frac{4}{3}S}{S} = \left(\frac{8}{4.5}\right)^2 = 3.16$$

and solving for B/S gives $B/S = 1.83$. The known values in Table 14 . 1 show $B/S = 1.81$ for iron, indicating that the earth's core is probably composed mostly of iron.

If a solid medium is confined in a rigid tube, so that there is no possibility of its expanding laterally, the speed of a compression wave depends only on the bulk modulus B and the density ρ of the medium. This is also the case when a disturbance originates in the interior of an elastic medium of great extent in all directions. However, when the wave travels along a thin rod of solid material, slight lateral expansions and contractions occur in that portion of the rod which is undergoing the volume strain. If we imagine a rod divided into disks, as in Fig. 17 . 6, and we apply a small pressure δP at one end, then the longitudinal compression δl of each disk is slightly greater than in the isotropic case, because the rod bulges laterally outward. Hence $-\delta V/V$ does not equal $\delta l/l$. Instead, from the longitudinal stress-strain relation in Sec. 14 . 1, we have $\delta l/l = \delta P/Y$, where Y is Young's modulus. If we express this equality as $l\,\delta P/\delta l = Y$, then Eq. [17 . 22] gives the speed of propagation as

$$v = \sqrt{\frac{Y}{\rho}} \qquad [17 \cdot 29]$$

Such longitudinal waves travel nearly twice as fast as transverse torsion waves on the same rod (compare Eq. [17 . 13]). Table 17 . 1 summarizes the speeds of propagation for various types of transverse and longitudinal waves.

17 • 5 pressure variations in compression waves

In developing the mathematical representation of a traveling wave, particularly one due to a simple harmonic disturbance, we began with the transverse disturbance δy because of its ease of visualization. We next found the speeds of longitudinal disturbances δx in various media, but without discussing the mathematical details of such disturbances and the pressure and density waves that accompany them. Let us now consider the representation of these properties in detail for the case of a simple harmonic longitudinal wave, where each particle is displaced by an amount $\delta x = f(x, t)$ from its equilibrium position x.

table 17 • 1 *Speeds of propagation*

wave	medium	propagation speed
longitudinal	liquid	$\sqrt{\frac{B}{\rho}}$
	perfect gas	$\sqrt{\frac{\gamma P}{\rho}} = \sqrt{\frac{\gamma RT}{M_0}}$
	extended solid	$\sqrt{\frac{B + \frac{4}{3}S}{\rho}}$
	solid rod	$\sqrt{\frac{Y}{\rho}}$
transverse	solid wire or string	$\sqrt{\frac{F}{\rho_l}}$
	extended solid	$\sqrt{\frac{S}{\rho}}$
	surface, deep water	$\sqrt{\frac{g\lambda}{2\pi}}$
	shallow water	$\sqrt{gh}$

Compression waves in a medium excited by the vibrating prongs of a tuning fork form a wave train of alternating regions of condensation and rarefaction (see Fig. 17.7). For example, movement of the prong from A to C causes the condensation region ca, in which particles of the medium are moving to the right; the subsequent movement of the prong from C back to A causes the region of rarefaction Cc, in which particles are moving to the left. Hence each particle oscillates about its equilibrium position. The phase of the particle therefore depends on its equilibrium position x, as well as on the time. If the longitudinal displacement of a particle oscillating about $x = 0$ is given by $\delta x = A \sin \omega t$, then the wave can be expressed as

$$\delta x = A \sin\left[\omega\left(t - \frac{x}{v}\right)\right] \tag{17 · 30}$$

figure 17 • 7 *Wave train maintained by the vibrating prong of a tuning fork*

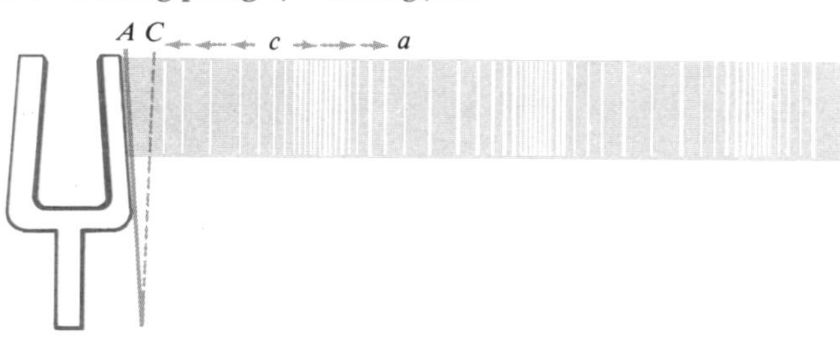

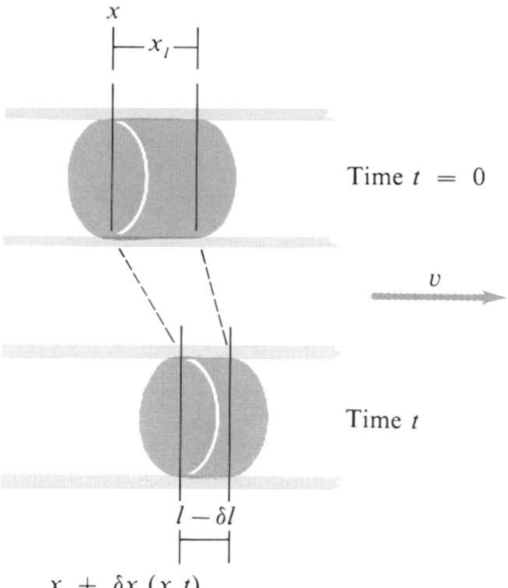

figure 17 • 8 *Change δP in gage pressure with change $\delta V = -A\,\delta l$ in volume of a disk in the compression-wave model: $\delta P = -B\,\delta V/V = -\rho v^2\,\partial(\delta x)/\partial x$*

The actual instantaneous location of the particle is $x + \delta x = x + f(x, t)$; $\nu = \omega/2\pi$ must be the same frequency as that of the tuning fork. As in the transverse case, the distance between two particles on the x axis and in the same phase is the wavelength $\lambda = v/\nu = vT$. Therefore

$$\delta x = A \sin\left[2\pi\left(\frac{t}{T} - \frac{x}{\lambda}\right)\right] \qquad [17 \cdot 31]$$

The relative velocities of different layers of particles are shown by the arrows in Fig. 17 . 7. For small-amplitude oscillations (ordinary sound waves), $A \ll \lambda$.

To see the relationship between displacement and pressure recall that the small but finite pressure variation δP is approximately

$$\delta P \doteq -B\frac{\delta V}{V} = B\frac{\delta l}{l} \qquad [17 \cdot 32]$$

where B is the bulk modulus. As shown in Fig. 17 . 8, l is the initial distance between the particles on the front and rear faces of each disk and δl is the decrease in this distance due to the fact that their displacements are slightly out of phase. If the two particle layers are initially very close together, then the difference in their displacements at some later time t is

$$-\delta l = \Delta(\delta x) = \delta x(x + l, t) - \delta x(x, t) \doteq l\frac{\partial(\delta x)}{\partial x} \qquad [17 \cdot 33]$$

where the partial-derivative symbol ∂ indicates that time is considered constant in the differentiation. This gives us the useful relation

$$\delta P = -B\frac{\delta V}{V} = B\frac{\delta l}{l} = -B\frac{\partial(\delta x)}{\partial x} \qquad [17 \cdot 34]$$

For a simple wave, with $\rho v^2 = B$ and $\delta P = -\rho v^2\,\partial(\delta x)/\partial x$, we obtain

$$\delta P = -\rho v^2\frac{2\pi A}{\lambda}\cos\left[2\pi\left(\frac{t}{T} - \frac{x}{\lambda}\right)\right]$$

$$= \rho v^2\frac{2\pi A}{\lambda}\sin\left[2\pi\left(\frac{t}{T} - \frac{x}{\lambda}\right) - \tfrac{1}{2}\pi\right] \qquad [17 \cdot 35]$$

The phase of the pressure wave lags behind the displacement by $\frac{1}{2}\pi$, and the pileup of particles causes the pressure to increase to its maximum, $\delta P_{max} = \rho v^2(2\pi A/\lambda)$. Equation [17 . 35] also holds for waves in a gas, with the substitutions $\delta V \longrightarrow \delta V_{ad}$ and $B = \gamma B$ in the derivation.

Propagating along with the pressure wave is a wave of density fluctuations $\delta\rho$. Since there is no change in the number of particles contained in a disk of volume Al, when it undergoes a compression (or a rarefaction) δl, its mass remains constant:

$$(\rho + \delta\rho)(V + \delta V) = \rho V \qquad \text{or} \qquad \delta\rho = -\rho\frac{\delta V}{V} \qquad [17 \cdot 36]$$

where we have ignored the second-order term $\delta\rho\ \delta V$. Then, from Eq. [17 . 34],

$$\begin{aligned}\delta\rho &= \frac{\rho}{B}\,\delta P = -\rho\frac{\partial(\delta x)}{\partial x}\\ &= 2\pi\rho\frac{A}{\lambda}\sin\left[2\pi\left(\frac{t}{T} - \frac{x}{\lambda}\right) - \tfrac{1}{2}\pi\right]\end{aligned} \qquad [17 \cdot 37]$$

Note the presence of the term $2\pi A/\lambda$ in this equation. As long as $A \ll \lambda/2\pi$, then $\delta\rho/\rho \ll 1$, which is consistent with the approximation on which we have based our derivation. The wavelength, in a sense, gives the geometric scale of the wave, providing a criterion for what we mean by *small-amplitude waves.* It also gives us some qualitative ideas about how the wave will interact with the matter it encounters as it propagates through the medium. For example, an observer on a boat of length l_0 floating over ocean waves of length $\lambda \ll l_0$ would feel only the average, or equilibrium, height of the water; however, waves of $\lambda \doteq l_0$ will cause very noticeable bobbing of the boat. A *tsunami,* a water wave due to submarine earthquakes or volcanoes, has a wavelength $\lambda \gg l_0$ and may not even be noticeable as it passes under the vessel, although its effect may be catastrophic when it piles up on some distant shore. Similar considerations apply in the interactions of sound waves, and even of electromagnetic waves such as light, radar, and radio waves, with objects. The wavelength λ is crucial in determining the theoretical approach to the interaction of the wave with its medium and other matter.

example 17 • 6 If a change in pressure from 1 atm to 50 atm produces a volume change of 0.00243 cm³ in a gram of water at 10°C, what is the speed of sound in water?

solution Substituting $B = -V\ \delta P/\delta V$ in $v = \sqrt{B/\rho}$, we obtain

$$v = \sqrt{-\frac{V\ \delta P}{\rho\ \delta V}} = \sqrt{\frac{1\text{ cm}^3 \times 49 \times 10^6\text{ dynes/cm}^2}{1(\text{g/cm}^3) \times 0.00243\text{ cm}^3}} = 142{,}000\text{ cm/s}$$

example 17 • 7 If a tuning fork vibrates in water at 1000 Hz with an amplitude of 1 nm $= 10^{-9}$ m, what is the maximum amplitude of the pressure wave?

solution Substitute for the amplitude of δP from Eq. [17 . 35] to obtain

$$\delta P_{max} = 2\pi\rho v^2 \frac{A}{\lambda} = 2\pi\rho v A\nu$$

$$= 6.28(1\ \text{g/cm}^3)(1.42 \times 10^5\ \text{cm/s})(10^{-7}\ \text{cm})(10^3\ \text{Hz})$$

or

$$\delta P_{max} = 89.2\ \text{dynes/cm}^2 \doteq 10^{-4}\ \text{atm}$$

This is a very loud noise. The human ear can detect pressure waves as low as 2×10^{-10} atm at 1000 Hz!

17 · 6 energy, power, and intensity

Any periodic wave possesses, in addition to its speed of propagation, the characteristics of frequency, amplitude, and waveform. Perhaps the most important physical attribute of wave motion, however, is its ability to *transfer energy* from place to place without the transfer of matter. Although the most important quantity in wave motion is the energy transferred per unit time, or the power delivered by the wave, let us first examine the energy "stored" in a wave at any given moment.

Consider a generalized sinusoidal wave whose displacement, whether longitudinal or transverse, we shall simply represent by

$$\delta q(x, t) = A \sin\left[\omega\left(t - \frac{x}{v}\right)\right] \qquad [17 \cdot 38]$$

The same energy relations hold for either type of wave. If a particle of mass m and equilibrium position x is displaced by the wave, then since x is constant for this particle, its instantaneous velocity is $\partial(\delta q)/\partial t$ and its kinetic energy is

$$K = \tfrac{1}{2}m\left[\frac{\partial(\delta q)}{\partial t}\right]^2 = \tfrac{1}{2}m\omega^2 A^2 \cos^2\left[\omega\left(t - \frac{x}{v}\right)\right] \qquad [17 \cdot 39]$$

Similarly, the acceleration of the particle is $\partial^2(\delta q)/\partial t^2$. From Newton's second law, the net disturbing force on the particle due to the wave must be

$$F = m\frac{\partial^2(\delta q)}{\partial t^2} = -m\omega^2 A \sin\left[\omega\left(t - \frac{x}{v}\right)\right] = -m\omega^2\ \delta q \qquad [17 \cdot 40]$$

Hence the increase in elastic potential energy of the particle due to restoring forces in the medium is

$$U_s = -\int_0^\delta F\, d(\delta q) = \tfrac{1}{2}m\omega^2 (\delta q)^2$$

$$= \tfrac{1}{2}m\omega^2 A^2 \sin^2\left[\omega\left(t - \frac{x}{v}\right)\right] \qquad [17 \cdot 41]$$

and the total energy per particle is

table 17 • 2 *Energy stored in a sinusoidal wave*

disturbance $\delta q(x,t)$	*energy stored*	
δy, transverse wave	$\frac{1}{2}m\omega^2A^2$	per particle
δy, transverse wave on string	$\frac{1}{2}\rho_l\omega^2A^2$	per unit length of string
$\delta\theta$, torsional wave on rod	$\frac{1}{2}I_l\omega^2\theta_m^2$	per unit length of rod, I_l
δx, longitudinal wave	$\frac{1}{2}\rho\omega^2A^2$	per unit volume of extended medium

$$K + U_s = E = \tfrac{1}{2}m\omega^2A^2 \qquad [17 \cdot 42]$$

as expected for a harmonic oscillator. Furthermore, if we average K and U_s over one time period T, we find that (see Eq. [16 . 17])

$$\overline{K} = \overline{U_s} = \tfrac{1}{2}E \qquad [17 \cdot 43]$$

Although the energies

$$K = \tfrac{1}{2}m\left[\frac{\partial(\delta q)}{\partial t}\right]^2 \qquad U_s = \tfrac{1}{2}m\omega^2(\delta q)^2 \qquad [17 \cdot 44]$$

reside instantaneously in the particle, once the wave pulse or wave train has passed on, the particle returns to rest. Where has the energy gone? In a nondissipative medium the wave continues to excite particles as it travels on, and we therefore say that the energy *resides in the wave and travels with it*. Actually, in an elastic medium, each particle excites the one next to it, transferring its energy to that particle, which in turn transfers its energy to the next, and so on. The total amount of energy in the medium remains constant, until finally waves break on some distant shore or a sound is heard and energy is thereby removed from the medium.

Equations [17 . 38] to [17 . 44] can be applied to any type of mechanical wave discussed so far (see Table 17 . 2). Consider a general disturbance δq in a cylinder of some medium of cross-sectional area S and length Δl, as shown in Fig. 17 . 9. If the energy in this cylinder travels to the right with speed v, then in a time $\Delta t = \Delta l/v$ all the energy originally in the cylinder

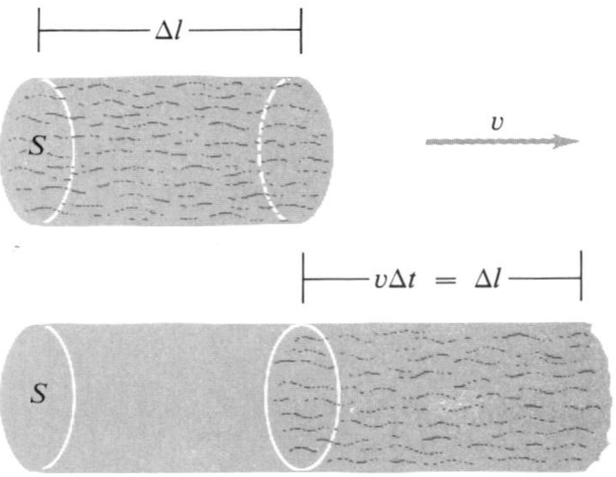

figure 17 • 9 *Flow of wave energy through a medium*

will have crossed its right face. Hence the energy crossing an area S per unit time is given by

$$E_V \frac{S\,\Delta l}{\Delta t} = E_V S v \qquad [17.45]$$

where E_V is the average energy per unit volume, or *energy density* in the medium. Here we have used S to denote area to avoid conflict with amplitude A. We can now define the *intensity* I of a wave as the energy transmitted per second across a unit area perpendicular to the wave direction:

$$I = E_V v = \tfrac{1}{2}\rho\omega^2 A^2 v = \frac{2\pi^2 \rho A^2 v^3}{\lambda^2} \qquad [17.46]$$

where ρ is the density of the medium, ω is the angular frequency of oscillation, and A is the amplitude of the wave. In the case of transverse waves on a wire of linear density ρ_l (density per unit length) it is more useful to specify the total power P_x transmitted by the wire along its length:

$$P_x = \tfrac{1}{2}\rho_l \omega^2 A^2 v \qquad [17.47]$$

Note that since the density and speed are constant for a given medium, the energy transmitted by waves in the medium is proportional to the squares of their amplitudes and frequencies and the inverse squares of their wavelengths.

example 17 • 8 In Example 17 . 7, we considered a sound wave of speed $v = 142{,}000$ cm/s produced in water by a tuning fork oscillating at $\omega/2\pi = 1000$ Hz. (*a*) Find its energy density E_V and the intensity I of the sound wave. (*b*) What is the relation between E_V and the pressure amplitude in the wave?

solution (*a*) From Eq. [17 . 46], $E_V = \tfrac{1}{2}\rho\omega^2 A^2 = \tfrac{1}{2}(\text{g/cm}^3)\,4\pi^2(10^6/\text{s}^2)(10^{-14}\,\text{cm}^2) = 1.97 \times 10^{-7}\,\text{ergs/cm}^3$.

$I = E_V v = 0.00280\,\mu\text{W/cm}^2 \qquad (1\,\mu\text{W} = 10^{-6}\ \text{J/s})$

(*b*) From Eq. [17 . 35], the maximum of the pressure pulse is

$$\delta P_{\text{max}} = \rho v^2 \frac{2\pi A}{\lambda}$$

and since $\omega\lambda = 2\pi v$, substituting for A^2 in E_V gives

$$E_V = \tfrac{1}{2}\rho\omega^2 A^2 = \tfrac{1}{2}\left(\frac{\omega\lambda}{2\pi v^2}\right)^2 \frac{(\delta P_{\text{max}})^2}{\rho} = \frac{\tfrac{1}{2}(\delta P_{\text{max}})^2}{\rho v^2}$$

You can verify this formula by using $\delta P_{\text{max}} = 89.2$ dynes/cm^2 from Example 17 . 7 to recompute E_V and compare it with (*a*) above.

Although we have so far discussed waves traveling in only one direction, Eq. [17 . 46] does enable us to discover an important property of spherical waves, which propagate radially outward and closely approximate normal

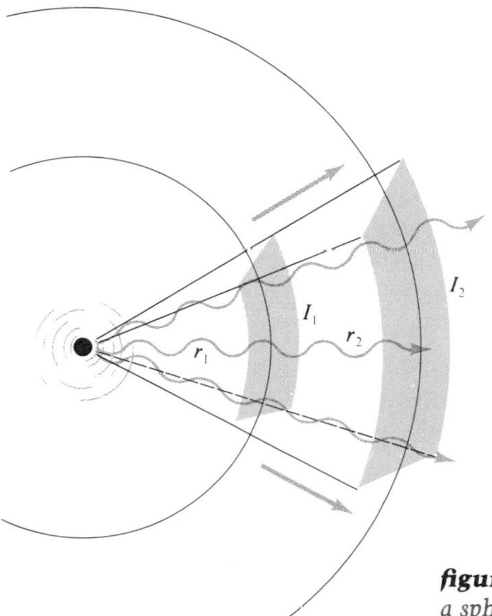

figure 17 • 10 *Propagation and intensity of a spherical wave:* $I_2/I_1 = r_1^2/r_2^2$

sound waves in air. Suppose a spherical wave is traveling out from the small pulsating sphere in Fig. 17 . 10, which transfers average power P_r to the medium. If we assume that no power is dissipated within the medium, then the average power moving across the surface of a sphere of radius r_1 must be equal to the average power moving across the sphere of radius r_2. The average power moving through each sphere is, by definition, the wave intensity at the given radius multiplied by the area of the sphere:

$$P_r = 4\pi r_1^2 I_1 = 4\pi r_2^2 I_2 \qquad [17 \cdot 48]$$

or

$$\frac{I_1}{I_2} = \frac{r_2^2}{r_1^2} \qquad [17 \cdot 49]$$

That is, the intensity of a spherical wave is inversely proportional to the square of the distance from the source (assuming the source is very small when compared to the radius of the wave), whereas the *amplitude* of the wave is inversely proportional to the *distance* from the source.

The *loudness* of a sound is closely related to the intensity of the wave entering the ear. Experiment shows, however, that loudness increases, not directly with intensity, but logarithmically with intensity. On the average, the intensity of a given sound must be increased by about 26 percent before the human ear will perceive a difference in loudness. Moreover, the pitch, or apparent frequency, of a musical note is also somewhat affected by the intensity and form of the incoming wave, while the sensitivity of the ear varies with frequency over the audible range of 20 to 20,000 Hz and 10^{-16} to 10^{-4} W/cm^2. Hence thousands of times as much energy is required to produce an audible sound at 30 Hz as at 2000 Hz.

17 · 7 the Doppler effect

In all of our discussion of wave motion so far we have assumed that the source of the disturbance is at rest in the medium through which the waves are propagated, and also that the frequency and velocity of the waves are measured by an observer at rest in the medium. Let us now consider how wave properties are modified, or *seem* to be modified, when either the source or the detector is moving through the medium. Suppose the source or the detector is moving along the same coordinate as the propagating wave, with a speed much less than the wave speed. The result in this case is that the frequency appears higher when source and detector are approaching and lower when they are separating. This phenomenon is called the *Doppler effect,* after Christian Johann Doppler, who first applied it in 1842 to light waves in analyzing the colors of stars.

To see the effect of this combined motion on the apparent frequency of mechanical waves, let us consider the motion in two steps: first with the source moving and the observer remaining stationary, and then with both the source *and* its waves momentarily frozen while the observer is in motion. It is clear from Fig. 17 . 11 that the motion of the source shortens the wavelength in the direction of motion and lengthens it in the opposite direction. Now consider the situation in Fig. 17 . 12. If the surface wave arriving at E at time $t = \Delta t$ was emitted by a stationary source at A at time $t_0 = 0$, then the observer at D will see waves of the source frequency ν and wavelength $\lambda = v/\nu$. However, suppose the source moves with speed v_s to point B in the same time interval Δt (Fig. 17 . 12*b*). Then the number $\Delta n = \nu \, \Delta t$ of crests emitted must lie in the remaining space from B to E, $\Delta x = (v - v_s) \, \Delta t$. The apparent wavelength λ' is therefore

$$\lambda' = \frac{\Delta x}{\Delta n} = \frac{(v - v_s) \, \Delta t}{\nu \, \Delta t} \qquad [17 \cdot 50]$$

and the apparent frequency ν' as determined by a *stationary* observer is greater than the source frequency ν:

$$\nu' = \frac{v}{\lambda'} = \nu \frac{v}{v - v_s} \qquad [17 \cdot 51]$$

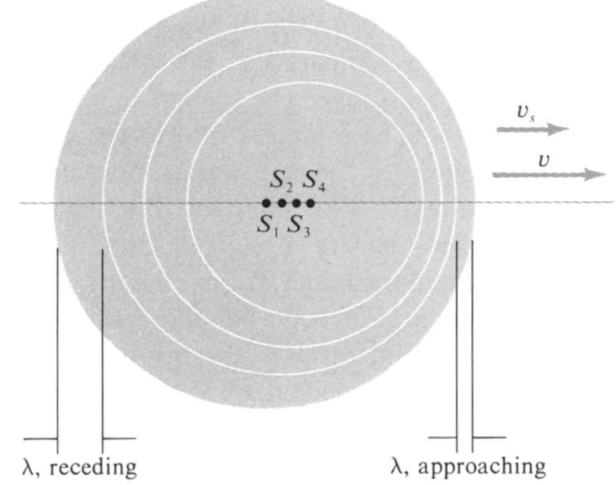

figure 17 • 11 *Doppler effect on wavelength due to a source S moving with speed $v_s < v$. Circles show positions of maximum amplitude emitted when source was at S_1, S_2, S_3, and S_4.*

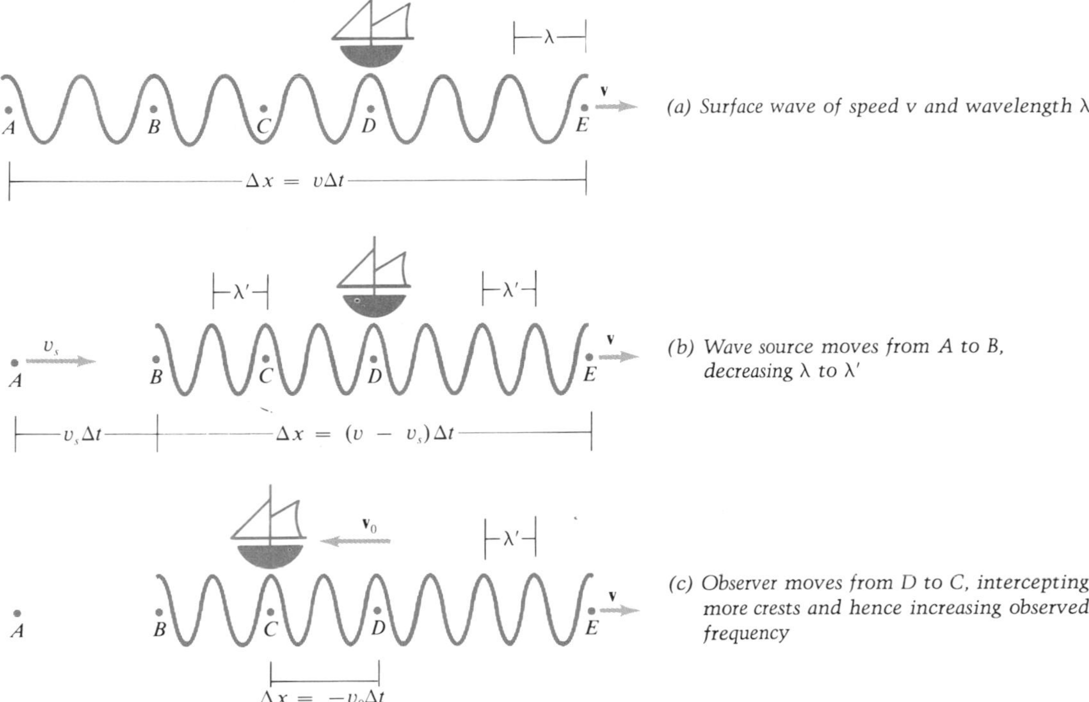

figure 17 • 12 *Doppler effects due to movement of the wave source and movement of the observer, using surface waves on water as an example*

If we now let the observer move *toward* the source with speed $-v_o$, he will pass through $\Delta n' = -v_o \, \Delta t/\lambda'$ waves in time Δt. Therefore, by the time he arrives at point C (Fig. 17 . 12*c*), he will have observed a total of $\Delta n'' = \nu' \, \Delta t + \Delta n'$ waves with an apparent frequency

$$\nu'' = \frac{\Delta n''}{\Delta t} = \nu' - \frac{v_o \nu}{v - v_s} = \nu \frac{v - v_o}{v - v_s} \qquad [17 \cdot 52]$$

If we express this tranformation equation as

$$\frac{\nu''}{\nu} = 1 + \frac{v_s - v_o}{v - v_s}$$

we can see that when the source and detector are approaching, $v_s - v_o > 0$ and the frequency appears higher than the actual source frequency; if they are separating, then $v_s - v_o < 0$ and $\nu'' < \nu$. If the situation is being observed from another system in which the *medium* is moving with velocity v_m, then the wave speed in this system is $v + v_m$ and Eq. [17 . 52] becomes

$$\nu'' = \nu \frac{v + v_m - v_o}{v + v_m - v_s} \qquad [17 \cdot 53]$$

Note that even in this most general case there will not be any change in the observed frequency when the source and observer are at rest relative to each other, $v_o = v_s$. These equations also hold when the v terms are one-dimensional (positive or negative) velocities.

The acoustic Doppler effect is easily observed in the rise in pitch of the whistle of an approaching train. Since light propagates through empty space with properties analogous to those of the mechanical waves we have been discussing, there also exists an *optical* Doppler effect which differs somewhat from the acoustic effect. The speed of approaching missiles can be determined from the Doppler effect on the frequency of reflected radar waves. The light received from the most distant galaxies appears to have a lower frequency than the light of corresponding laboratory samples of incandescent elemental gases. This effect, "the recessional red shift," is cited as evidence that the universe is expanding in all directions at great speed.

example 17•9 The engineer of a high-speed monorail train ($v_s = \frac{1}{5}v$) blows his whistle at frequency ν as he approaches a mountainous curve. What is the frequency of the sound he hears reflected back to him from the mountainside?

solution The frequency "emitted" by the mountainside is that which it appears to receive: from Eq. [17 . 51], $\nu' = \frac{5}{4}\nu$. Since the engineer is approaching the "emitter" with speed $\frac{1}{5}v$, he hears $\nu'' = \frac{5}{4}\nu' = \frac{25}{16}\nu$.

PROBLEMS

17•1 mathematical representation of waves

17•1 When a pebble is dropped into a pond, the ripples travel outward in expanding circles. Assuming the amplitude of the ripples and their speed do not change, how would you express the waveform mathematically?

17•2 Three bathers are floating vertically in a lake, facing the incoming waves. The second bather is 3 m behind the first bather and 3.6 m in front of the third. The second bather always sees the first bather rising or falling relative to herself, whereas the third bather appears always to be at the same height relative to the second. (*a*) What is the *maximum* possible wavelength of the water waves? (*b*) The next largest wavelength possible?

17•3 A cork is bobbing in the water with a maximum vertical speed of 3 cm/s and a maximum acceleration of 2 cm/s^2. (*a*) Find the amplitude and frequency of its motion. (*b*) If the wavelength of the transverse wave is 3 m, what is the speed of propagation of the waves?

17•4 A certain simple wave has an amplitude of 60 cm, a frequency of 0.5 Hz, and a speed of 1.5 m/s. When the displacement δy of one of the particles is a maximum in the negative direction, what is the displacement of a particle at a point 1.2 m forward in the direction of the wave travel?

17•5 In Prob. 17 . 4, what is the *slope* of the waveform at a point 1.2 m forward of the first particle in the direction of the wave?

17 • 6 Express the waveform $f(x,t)$ of a pulse traveling in the $-x$ direction with speed v if it is known that at time $t = 0, f(x,0) = 1$ for $0 \leq x \leq l$ and $f(x,0) = 0$ everywhere else.

17 • 2 *transverse waves*

17 • 7 Derive the equation $v = \sqrt{F/\rho_l}$ for the speed of a wave on a stretched string by dimensional analysis. Include all possible physical parameters which seem relevant, make any assumptions explicit, and give your reasons.

17 • 8 A simple harmonic transverse wave is observed traveling along a string of linear density 0.25 g/cm. After a number of measurements, it is determined that the equation of the traveling disturbance in the wave is given (in cgs units) by

$$\delta y = 0.35 \sin (376.8t + 0.167x)$$

(*a*) Find the speed of propagation of the wave. (*b*) The angular frequency ν. (*c*) The wavelength λ. (*d*) The tension in the string.

17 • 9 Show that any disturbance of the form $\delta q = f(x \pm vt)$ obeys the one-dimensional wave equation

$$\frac{\partial^2(\delta q)}{\partial x^2} = \left(\frac{1}{v}\right)^2 \frac{\partial^2(\delta q)}{\partial t^2}$$

(Recall that the partial-derivative operator $\partial/\partial x$ differentiates with respect to x only, etc.)

17 • 10 Find the relation between the instantaneous speed of an oscillating point on the stretched string and the slope of the string.

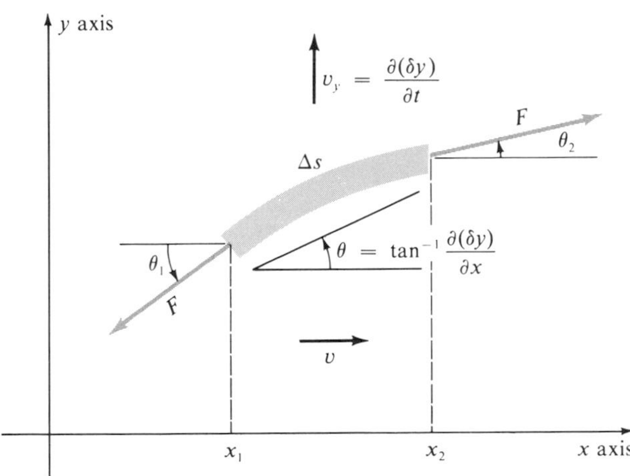

17 • 11 Find the speed of torsional waves on a mild-steel rod. Write the mathematical representation of the simple harmonic torsional waves set up in the rod of Example 14 . 3.

***17 • 12** For what ratio of h/λ does the effect of the bottom on the speed of a gravity water wave become negligible, $dv/dh < 0.1\ v/h$? Justify this criterion. Can you think of some other criterion?

17 • 13 A tsunami is a wave of very long wavelength and may be considered a shallow-water wave. If the average depth of the ocean is 5 km, what is the average speed of the tsunami?

17 • 3–17 • 5 compression waves

17 • 14 The French physicist Émile-Hilaire Amagat found that 1 cm^3 of alcohol at 14°C and 1 atm was decreased in volume by 0.000101 cm^3 for each atmosphere of increase in pressure. If the density of alcohol at 14°C is 0.795 g/cm^3, what is the speed of sound in this medium?

17 • 15 A compression wave in water traveling with speed $v = 1420$ m/s has an amplitude of 0.001 mm and a wavelength of 50 cm. At time $t = 0$ the particle whose undisturbed position is at $x = 0$ is actually located at $x = -0.001$ mm. (*a*) Express the disturbance as a traveling wave. (*b*) What is the exact position at time $t = 0.75$ s of the particle whose equilibrium position is $x = 1$ m? (*c*) The simultaneous position of the particle whose equilibrium position is $x = 1.25$ m?

17 • 16 In echo sounding the ocean depth is measured by noting the time required for a sound wave to leave a ship on the surface, bounce off the ocean floor, and return to the ship. If the density of seawater is 1030 kg/m^3 at the surface and its compressibility is $\kappa = 5 \times 10^{-10}$ m^2/N, compute the time required to measure a depth of 1.5 km in this way, ignoring the change in density with depth.

17 • 17 If the change in the density of seawater with depth y is assumed to be $\rho = \rho_0/(1 - \kappa\rho_0 g y)$, find the approximate error in the depth measurement of Prob. 17 . 16.

17 • 18 Repeat the derivation of v in Sec. 17 . 3 for the case of an expansion wave caused by withdrawing the piston from the tube. Show that the rarefaction pulse propagates to the right with the same speed as the condensation pulse in the original derivation. (*Hint:* Pay careful attention to signs.)

17 • 19 When a certain tuning fork is sounded in air at 0°C, the compression waves produced in the air have a wavelength of 130 cm. What is the period of vibration of the fork?

17 • 20 If the speed of a compression wave in a gas is 320 m/s at 20°C, what will be the speed at 50°C and twice the pressure?

17 • 21 A hollow tube is filled with helium. At one end of the tube is a source which sends sound waves of a frequency ν down the tube, and at the other end is an elastic membrane which transmits the sound into the surrounding air. The pressure and temperature of the gas in the tube are the same as those of the surrounding air. (*a*) What is the ratio of the wavelengths of the sound waves in the helium and in the air, assuming $m_{air}/m_{He} = 7$. (*b*) What is the ratio of the frequencies of the sound waves in the helium and in the air? Assume that helium is a monatomic gas and that air is diatomic.

17 • 22 Show that the ratio of velocities of transverse and longitudinal waves on a string which is stretched by a fractional amount $\Delta l/l$ equals the square root of that fraction.

17 • 23 The Young's modulus of mild steel is $Y = 21 \times 10^{10}$ Pa. (*a*) What is the speed of sound in a mild-steel rod compared to that in a large block of the same material? (*b*) Compared to the speed of sound in air?

17 • 24 Show how, in theory, the ratio of the speed of sound v_b in a block of elastic solid to its speed v_r in a rod of the same material can be computed from the Poisson ratio σ of the material (see Sec. 14 . 4).

17 • 25 Show that the pressure wave in a rod is given by

$$\delta P = -Y \frac{\partial(\delta x)}{\partial x}$$

and the form of Eq. [17 . 35] for δP is unchanged for propagation of a sinusoidal pressure wave in the rod.

17 • 26 Thirteen brass spheres 2 cm in radius are hung in a row from long wires 4 cm apart, so that the spheres are in contact. If a sphere is displaced at one end and then released, how long after impact will it take for the sphere at the other end to rise into the air? Assume that the density of the material is $\rho = 9$ g/cm^3.

17 • 27 The human ear can detect sounds as faint as $\delta P_{max} = 0.0002$ dynes/cm^2 at $\nu = 1000$ Hz. (*a*) What is the corresponding displacement? (*b*) If $\delta P_{max} = 0.002$ dynes/cm^2 is the faintest sound that can be detected at $\nu = 200$ Hz, what is the corresponding displacement?

17 • 28 The Young's modulus of aluminum is $Y = 7 \times 10^{10}$ Pa. What is the amplitude A of a compression wave such that $\delta\rho = 0.01\rho$ is the maximum density change in an aluminum rod at $\nu = 25{,}000$ Hz?

17 • 6 energy, power, and intensity

17 • 29 What is the rate of energy transmission by the transverse wave in Prob. 17 . 8?

17 • 30 An intensity of 1 μW/cm^2 is a very loud, almost painful, noise. (*a*) What is the pressure amplitude of such a sound in air? (*b*) Since the sensation of "loudness" varies approximately as the logarithm of the intensity of the sound, noises are usually rated in *decibels* (db), according to the formula 1 db = 100 log (I/I_0), where $I_0 = 10^{-10}$ μW/cm^2. How many decibels is the noise of part (*a*)? (*c*) If the frequency of the noise is $\nu = 1000$ Hz, what is the amplitude of this sound wave?

17 • 31 Suppose a sound source is a long rod of radius R and the sound travels out along the radii of a cylinder. Use an argument analogous to that for a spherical wave to show that the intensity of a cylindrical wave is inversely proportional to the distance from the source and the amplitude is inversely proportional to $r^{1/2}$.

17 • 32 "Daisycutter" demolition bombs are made to explode just above the ground level in order to flatten wide areas for helicopter landing strips. Such a device releases a total energy E in a hemispherical wave, which we can approximate as a single cycle of a sinusoid of period T and speed V, with compression followed by rarefaction. (*a*) Find the maximum pressure in the blast wave as a function of distance r from the impact point. (*b*) Suppose the bomb contains 900 kg of TNT, an explosive releasing about 3000 J/g with a period of $T_s = 0.1\ \mu s$. If $v = 3000$ m/s, approximately how close can a steel structure be to the blast point without being demolished if the maximum tensile strength of steel is 2×10^8 Pa? Ignore the fact that this is a shock wave, in which density changes significantly, and take the density as $\rho = 1.29\ \text{kg/m}^3$.

17 • 7 the Doppler effect

17 • 33 A locomotive with a whistle is receding from a railway station at a constant speed of 48 km/h. Find the percentage change in pitch of the whistle heard by a person standing at the station.

17 • 34 What is the *fractional* change in wavelength predicted by Eq. [17 . 52]?

17 • 35 The whistle on a certain locomotive has a frequency of 500 Hz. The speed of the train is 64 km/h and the temperature of the air is 20°C. (*a*) What is the frequency of the sound detected by an observer on the train? (*b*) On the track behind the train? (*c*) On the track ahead of the train?

17 • 36 At an outdoor concert a 25-km/h breeze is blowing from the audience toward the musicians. (*a*) If the speed of sound in still air is 348 m/s, what is the frequency of middle C ($\nu = 261.6$ Hz) as heard by a listener seated in the audience? (*b*) What is the change in frequency heard by a latecomer walking down the aisle at 2.7 km/h?

17 • 37 A sound of frequency ν propagates through a medium with wave speed v. (*a*) Find the observed frequency if the source is approaching a stationary observer with constant speed V. (*b*) If the observer is approaching the stationary source with constant speed V. (*c*) Are the two cases equivalent? Explain your answer.

17 • 38 Show that when the source and observer in Fig. 17 . 12 are separating, rather than approaching, and the medium is moving in the direction of the emitter, the formula for ν'' in Eq. [17 . 53] remains valid where the velocities are one-dimensional.

17 • 39 An observer is at rest at the origin of coordinates when an emitter of frequency ν crosses the positive x axis with a constant speed v_s at an angle $\theta < 90°$ to the x direction. (*a*) What frequency does the observer hear? (*b*) What frequency does he hear if the positions and velocities of the source and observer are interchanged?

17 • 40 If the emitter in Fig. 17 . 11 moves with a speed v_s greater than the speed v of propagation in the medium, it will outdistance the wave, leaving in its wake a cone-shaped wavefront. Show that the half-angle of the cone is given by the formula $\sin\theta = v/v_s$.

answers

17•1 $\delta y = A \sin(\omega t - \omega r/v)$
17•2 (*a*) $\lambda = 1.2$ m; (*b*) $\lambda = 0.4$ m
17•3 (*a*) $A = 2.25$ cm, $\omega = 0.667$ rad/s; (*b*) $v = 1/\pi$ m/s
17•4 $\delta y = 48.5$ cm
17•5 slope $= -59°$
17•6 $f(x - vt) = 1$ for $vt \leq x \leq vt + l$, and $f(x - vt) = 0$ everywhere else
17•8 (*a*) $v = -2256$ cm/s; (*b*) $\nu = 60$ Hz; (*c*) $\lambda = 37.6$ cm; (*d*) $F = 12.7$ N
17•10 $d(\delta y)/dt = -v\, d(\delta y)/dx$
17•11 $v = 3190$ m/s
17•12 $h/\lambda \geq 0.358$
17•13 $\bar{v} = 797$ km/h
17•14 $v = 1123$ m/s
17•15 (*a*) $\delta x = -0.001$ mm $\sin[2\pi(2840t - 2x)]$; (*b*) $x = 1$ m; (*c*) $x = 1.125$ m $+ 0.001$ mm
17•16 $\Delta t = 2.153$ s
17•17 2.85 m too deep
17•19 $T = 0.004$ s
17•20 $v = 336$ m/s
17•21 (*a*) $\lambda_{He}/\lambda_{air} = 2.89$; (*b*) $\nu_{He}/\nu_{air} = 1$
17•23 (*a*) $v_r = 5169$ m/s, $v_b = 5825$ m/s; (*b*) $v_{air} = 348$ m/s
17•24 $(v_b/v_r)^2 = 1/(3 - 6\sigma) + 2/(3 + 3\sigma)$
17•26 $\Delta t = 139$ μs
17•27 (*a*) $\delta x = 7.44$ pm; (*b*) $\delta x = 372$ pm
17•29 $P_x = 0.491$ W
17•30 (*a*) $\delta P_{max} - 2.92$ N/m^2; (*b*) loudness $= 400$ db; (*c*) $A = 1.09$ μm
17•32 (*a*) $P_{max} = 5.77 \times 10^9 r^{-1}$ (in mks units); (*b*) 29 m away
17•33 $\Delta\nu/\nu = 3.9$ percent
17•34 $\Delta\lambda/\lambda = -v_s/v$
17•35 (*a*) $\nu = 500$ Hz; (*b*) $\nu = 475$ Hz; (*c*) $\nu = 527$ Hz
17•36 (*a*) $\nu = 261.6$ Hz; (*b*) $\Delta\nu = 0.58$ Hz
17•37 (*a*) $\nu' = \nu v/(v - V)$; (*b*) $\nu' = \nu(v + V)/v$; (*c*) no; the medium introduces asymmetry
17•39 (*a*) $\nu' = \nu v/(v + v_s \cos\theta)$; (*b*) $\nu' = \nu(v - v_s \cos\theta)/v$

CHAPTER EIGHTEEN

interference

To understand the reason for this result, notice that when we pluck the string AB *as a whole, it makes oscillations through all its length; but when we put an obstacle* C *on the first division mark* D *of the string, which I suppose to be divided into 5 equal parts, the total oscillation* AB *is divided first into the 2 oscillations* AD, DB, *and since*

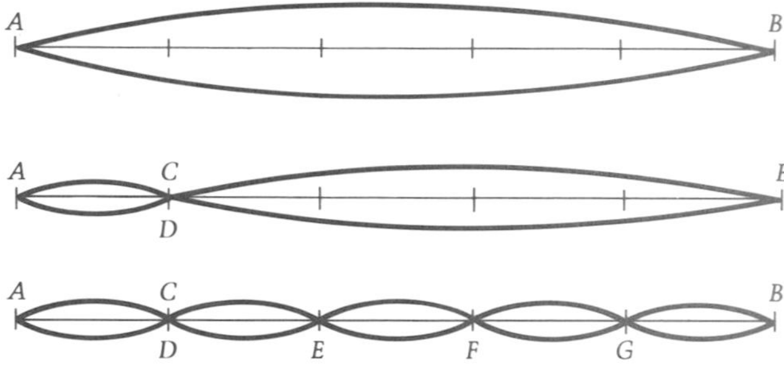

AD *is* $\frac{1}{5}$ *of* AB *or* $\frac{1}{4}$ *of* DB, *it makes its oscillations 5 times as fast as the whole cord* AB, *or 4 times as fast as the other part* DB; *so that the part* AB *carries with it its neighboring part* DB *and compels it to follow its motion. This part* DE *should consequently be equal to it; for a greater part will move more slowly and a smaller part more quickly. Then the part* DE *will compel the next part* EF *to take up the same motion and so on to the last; so that all the parts will make oscillations which will cross at the points of division,* D, E, F, G . . .

I shall call these points A, D, E, F, G, B *the nodes of these oscillations and the middle points of these oscillations will be called the ventres.*

Joseph Sauveur, Système Générale des Intervalles des Sons, et son Applications à tous les Systèmes et à tous les Instruments de Musique, *1701*

In 1822 Joseph Fourier first showed that simple sinusoidal waves may combine to produce waves of any shape whatever, depending on the relationships of their amplitudes and frequencies. Any combination or superposition of two or more waves is known as *interference*. In constructive interference the superposition of two waves produces a greater disturbance than either wave acting alone; in destructive interference the total disturbance is smaller than that produced by either wave acting alone. As we shall see, musical tones are produced by interference, the mechanical and geometric properties of the instrument producing a characteristic disturbance by the superposition of simple waves from some source of energy.

In sinusoidal waves each particle of the transmitting medium oscillates at a single frequency, acted on by an elastic restoring force. Most bodies however, can vibrate at many different frequencies at the same time. If a disturbance consists of simple waves of various frequencies and amplitudes that traverse the same portion of the medium simultaneously, each particle is subjected to as many forces as there are component simple waves, with each independent force producing its own simple harmonic motion. The problem of describing such a compound wave reduces to one of combining the several components by *superposition.* Although any two waves in a medium will superpose, they may not necessarily produce a pattern of variations—one that is regular and predictable. Whether they do or not depends on the relationship between their phases at different points in space. In this chapter we shall consider only *coherent interference,* in which each source has a definite frequency and the difference between their phase constants does not vary with time.

We shall first treat superposition mathematically and discuss its consequences in terms of *standing waves,* waves that oscillate but do not propagate. Such waves are very important in the production of musical tones. We shall also devote considerable attention to *Huygens' principle,* which directly involves the superposition of individual "wavelets" to determine the direction of propagation of a wave. This important principle, as we shall see, enables us to predict *reflection, refraction* (bending of waves), and *diffraction* (bending of waves around obstacles) in terms of interference phenomena. These diverse phenomena, from music to diffraction, are all examples of the interference of coherent waves.

18 · 1 superposition

Imagine a stable, quiescent medium capable of propagating waves from small-amplitude oscillatory disturbances. When a particle in such a medium is displaced by an amount δq in some direction, the other particles in the medium exert a restoring force, tending to move it back to its equilibrium position. We saw in Sec. 16 . 6 that any restoring force can be treated as linear if the displacements are small enough. Hence the equation of motion for a single particle of mass m is of the form

$$\frac{d^2(\delta q)}{dt^2} + \omega_n^2 \, \delta q = 0 \qquad [18 \cdot 1]$$

where ω_n is the natural angular frequency and $k = m\omega_n^2$ is the appropriate restoring-force constant.

We also saw in Sec. 16 . 5 that frictional forces which cause the oscillations to die out with time can be expressed as a damping force, $F_d = -m\mu \, d(\delta q)/dt$. The complete equation of motion of an oscillating particle subject to some external force F_q in the q direction is therefore

$$m\left[\frac{d^2(\delta q)}{dt^2} + \mu \frac{d(\delta q)}{dt} + \omega_n^2 \, \delta q\right] = F_q \qquad [18 \cdot 2]$$

Since the choice of coordinates may be independent of the direction of displacement, we can generalize Eq. [18 . 2] to a vector equation:

$$m\left[\frac{d^2(\delta \mathbf{r})}{dt^2} + \mu \frac{d(\delta \mathbf{r})}{dt} + \omega_n^2 \, \delta \mathbf{r}\right] = \mathbf{F} \qquad [18 \cdot 3]$$

If the applied force is a sinusoidal function of time with driving frequency ω_F, then the long-term steady motion of the particle will be an oscillation at the same frequency, but somewhat out of phase, due to inertial and frictional effects.

There are many situations, especially in acoustics, optics, and electronics, in which a particle is acted on simultaneously by several periodic driving forces, each of which, if acting alone, would cause the particle to oscillate with simple harmonic motion in the direction of that force. Suppose each force $\mathbf{F}_i$ acting alone would produce an acceleration $d^2(\delta \mathbf{r}_i)/dt^2$. Then, from the second law of motion,

$$\mathbf{F}_i = m\left[\frac{d^2(\delta \mathbf{r}_i)}{dt^2} + \mu \frac{d(\delta \mathbf{r}_i)}{dt} + \omega_n^2 \, \delta \mathbf{r}_i\right] \qquad [18 \cdot 4]$$

and the resultant of all the applied forces is

$$\mathbf{F} = \sum_{i=1}^{n} \mathbf{F}_i = \sum_{i=1}^{n} m\left[\frac{d^2(\delta \mathbf{r}_i)}{dt^2} + \mu \frac{d(\delta \mathbf{r}_i)}{dt} + \omega_n^2 \, \delta \mathbf{r}_i\right] \qquad [18 \cdot 5]$$

The derivative is a *linear operator,* so that

$$\sum_{i=1}^{n} \frac{d(\delta \mathbf{r}_i)}{dt} = \frac{d}{dt}\left(\sum_{i=1}^{n} \delta \mathbf{r}_i\right)$$

and we have

$$\mathbf{F} = m\left[\frac{d^2}{dt^2}\left(\sum_{i=1}^{n} \delta \mathbf{r}_i\right) + \mu \frac{d}{dt}\left(\sum_{i=1}^{n} \delta \mathbf{r}_i\right) + \omega_n^2\left(\sum_{i=1}^{n} \delta \mathbf{r}_i\right)\right] \qquad [18 \cdot 6]$$

The terms in parentheses must equal the net displacement $\delta \mathbf{r}$ due to $\mathbf{F}$:

$$\delta \mathbf{r} = \sum_{i=1}^{n} \delta \mathbf{r}_i \qquad [18 \cdot 7]$$

We see, therefore, that the net displacement of the particle is the vector sum of the individual displacements independently derivable from the actions of the individual forces. This is the *principle of superposition* of simple harmonic motions. It holds only for "linear" systems, in which the displacement and its derivatives occur only to the first power in the equations of motion. For this reason we shall be dealing only with small-amplitude oscillations.

A full mathematical description of interference effects would be too difficult at this stage. Instead we shall consider the action of only two forces on a particle, as an illustration of how simple waves may superpose to give some interesting results. Beyond this, we shall make greater use of physical intuition in discussing the acoustics of musical instruments, and use a geometric construction, Huygens' principle, to deal with more complicated interference effects which occur in the propagation of waves. We will use the general symbol δq to indicate either a transverse or a longitudinal wave, since our mathematical representation will be generally valid for either type of wave.

Consider two forces, both acting in the same direction and giving rise to two simple waves. If we characterize each wave by its amplitude A and phase θ, which includes the propagation term $\omega(t \pm x/v)$, we can distinguish three cases of superposition: (1) a difference only in the amplitudes of the two waves, (2) the more complex case of equal amplitudes but unequal phases, and (3) the general case in which both the amplitudes and phases are unequal. In the first case, if $\delta q_1 = A_1 \sin \theta$ and $\delta q_2 = A_2 \sin \theta$, then the resultant disturbance is

$$\delta q = \delta q_1 + \delta q_2 = (A_1 + A_2) \sin \theta \qquad [18 \cdot 8]$$

The resultant wave is also sinusoidal, with the same period and phase as its components; its amplitude is the algebraic sum of the amplitudes of its components (see Fig. 18 . 1).

The second case is of particular interest for sound waves. Given two wave trains of equal amplitudes A_0, but different phases θ_1 and θ_2, their sum is the *product* of two sinusoidal terms. This can be seen from the trigonometric identities

$$\delta q = A_0(\sin \theta_1 + \sin \theta_2) = 2A_0 \cos \tfrac{1}{2}(\theta_1 - \theta_2) \sin \tfrac{1}{2}(\theta_1 + \theta_2) \qquad [18 \cdot 9]$$

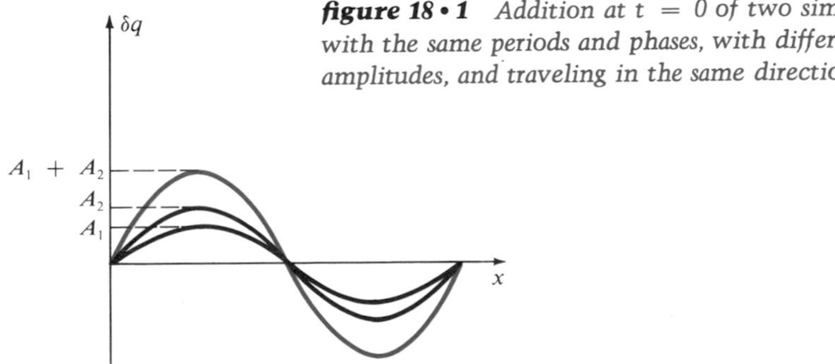

figure 18 • 1 *Addition at t = 0 of two simple waves with the same periods and phases, with different amplitudes, and traveling in the same direction*

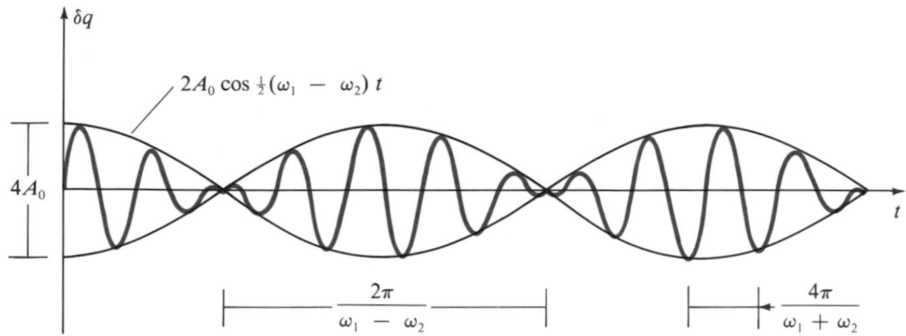

figure 18 • 2 *Superposition of two waves. Black envelope curve depicts the amplitude limits,* $w_2 = \frac{4}{5} w_1$.

Superposing two traveling waves with phases

$$\theta_1 = \omega_1\left(t - \frac{x}{v}\right) \qquad \theta_2 = \omega_2\left(t - \frac{x}{v}\right) + \phi$$

gives a resultant compound wave

$$\delta q = \left\{2A_0 \cos \tfrac{1}{2}\left[(\omega_1 - \omega_2)\left(t - \frac{x}{v}\right) - \phi\right]\right\} \times \sin \tfrac{1}{2}\left[(\omega_1 + \omega_2)\left(t - \frac{x}{v}\right) + \phi\right] \qquad [18 \cdot 10]$$

with an effective frequency equal to the mean of the frequencies of its components. The term in curly brackets, the "amplitude" of the compound wave, oscillates with frequency $|\omega_1 - \omega_2|$ at any given point in space. If $\omega_1 = \omega_2$, then the amplitude of the compound wave is simply $2A_0 \cos \frac{1}{2}\phi$.

An example of the superposition represented by Eq. [18 . 10] is shown in Fig. 18 . 2. Note that there are minima at which the two waves exactly cancel (destructive interference). Moreover, since the expression in curly brackets contains a propagation term, $(t - x/v)$, the entire pattern moves along in its medium with wave speed v. The dashed envelope represents the amplitude of the compound wave, which limits the amplitudes of the more rapid oscillations taking place at the average of the two frequencies.

If the components of a wave are audible sounds, their mean frequency will be too high for the ear to distinguish individual oscillations. However, the variations of the envelope—that is, the cosine term in Eq. [18 . 10]—may be slow enough to be heard. Since the envelope has zero amplitude twice during every period of the cosine term, with a maximum loudness, or beat, between each pair of successive minima, a beat frequency of

$$\nu_b = \frac{|\omega_1 - \omega_2|}{2\pi} \qquad [18 \cdot 11]$$

will be heard. The human ear is sensitive to as many as six or seven beats a second, so that musical instruments can be tuned simply by minimizing the audible beats between a sounded tone and a standard, such as a tuning fork. In the transmission of sound by radio waves, it is common to have a *carrier wave* of average frequency around 10^6 Hz, while the beat frequency of the envelope is of the order of 10^3 Hz, the frequency of the original sound.

If the component waves differ in both phase *and* amplitude, there are several alternative ways to combine them, all involving some trigonometric agility. If we set $\theta_2 = \theta_1 + \Delta$ and expand

$$\sin(\theta_1 + \Delta) = \sin\theta_1 \cos\Delta + \cos\theta_1 \sin\Delta$$

$$\begin{aligned}\delta q &= \delta q_1 + \delta q_2 = A_1 \sin\theta_1 + A_2 \sin(\theta_1 + \Delta) \\ &= (A_1 + A_2 \cos\Delta)\sin\theta_1 + A_2 \sin\Delta \cos\theta_1\end{aligned} \qquad [18 \cdot 12]$$

This can be recombined into a single wave

$$\delta q = A \sin(\theta_1 + \phi) \qquad [18 \cdot 13]$$

where we can use the identity

$$A \sin(\theta_1 + \phi) = A\cos\phi \sin\theta_1 + A \sin\phi \cos\theta_1$$

to show that

$$\begin{aligned}A &= \sqrt{A_1^2 + A_2^2 + 2A_1A_2\cos\Delta} \\ \tan\phi &= \frac{A_2 \sin\Delta}{A_1 + A_2\cos\Delta}\end{aligned} \qquad [18 \cdot 14]$$

If desired, the resultant wave may be added to a third wave, and so on. In general, this superposition is not simple, since both the resultant amplitude and the phase may be functions of time and position.

example 18 • 1 Superpose two waves of amplitudes A_1 and $A_2 = 1.5A_1$ and phases θ_1 and $\theta_2 = \theta_1 - \frac{1}{4}\pi$.

solution From Eqs. [18 . 13] and [18 . 14], we have the resultant amplitude

$$A = A_1\sqrt{1 + 2.25 + 3\cos\tfrac{1}{4}\pi} = 2.32A_1$$

and phase

$$\tan\phi = \frac{1.5 \sin\frac{1}{4}\pi}{1 + 1.5\cos\frac{1}{4}\pi} = 0.515 \qquad \text{or} \qquad \phi = 27.2°$$

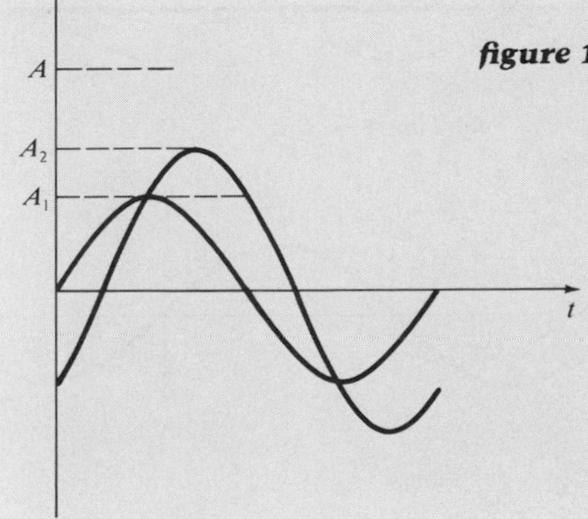

figure 18 • 3

18 • 2 standing waves and Lissajous figures

One application of the superposition of two traveling waves is the production of stationary, or *standing, waves.* A standing wave is the result of the action of two waves of the same period and amplitude that are propagating in *opposite* directions. Consider the shape of a vibrating string supporting a transverse standing wave δy. When the two traveling waves are exactly out of phase (Fig. 18 . 4*a*), they cancel each other everywhere; after a time $\Delta t = \frac{1}{4}T$ each wave has progressed $\Delta x = v\,\Delta t = \lambda/4$, and the string undergoes its maximum displacement (Fig. 18 . 4*b*). Note that the maxima and minima of displacement are at the *antinodes A*, and there is no motion at the *nodes N*. As illustrated in Fig. 18 . 5, when one section between two consecutive nodes of the string moves up, the adjacent sections are moving down, and vice-versa.

Similar considerations hold for a longitudinal standing wave $\delta q = \delta x$, shown in Fig. 18 . 6. In this case the motions of the particles are such that they will accumulate at node N_1, causing a local maximum in the pressure fluctuation δP. At the next node N_2, the particles are moving away in both directions, so that δP has its extreme negative value there. However, in the vicinity of the antinodes the particles are moving in the same direction with approximately the same speed, so that their relative positions are about the same as when they are in equilibrium. Hence the density and pressure fluctuations vanish, and $\delta P = 0$ at such points. In other words, the *pressure* nodes and antinodes are just the reverse of the *displacement* nodes and antinodes. This inverse correspondence between the pressure and the displacement will be quite useful when we consider the characteristic tones of wind instruments.

To represent a standing wave mathematically, we use the trigonometric identity of Eq. [18 . 10], so that the sum of two identical waves traveling in

figure 18 • 4 *Two oppositely directed traveling waves δq_1 and δq_2 with velocities $+v$ and $-v$, respectively. The colored curve is the resultant of the two waves. Nodes are at N, antinodes at A.*

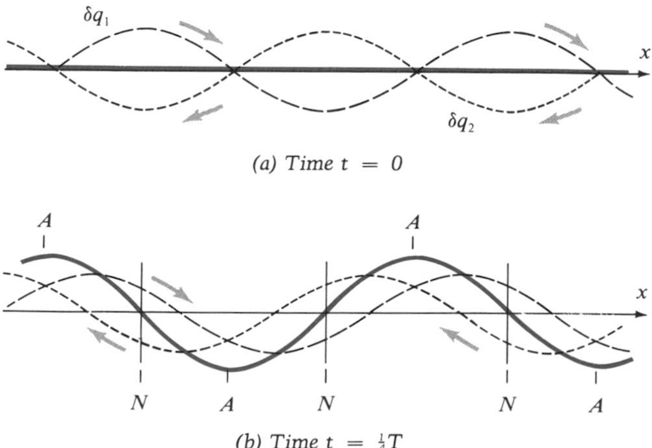

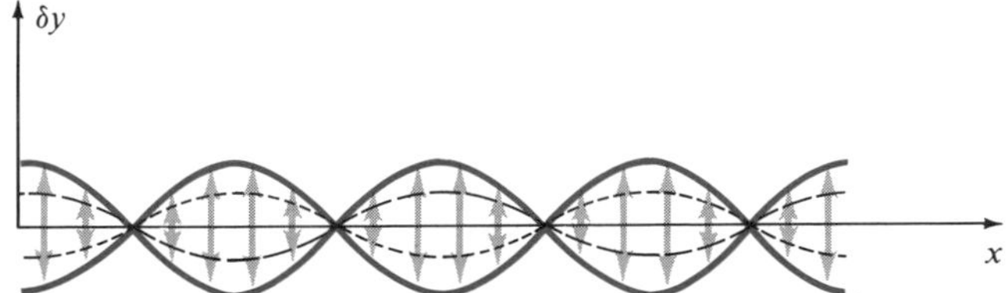

figure 18 • 5 *Transverse standing wave in a stretched string. Double-headed arrows indicate amplitudes of oscillation of individual particles; dashed lines show different positions of the string.*

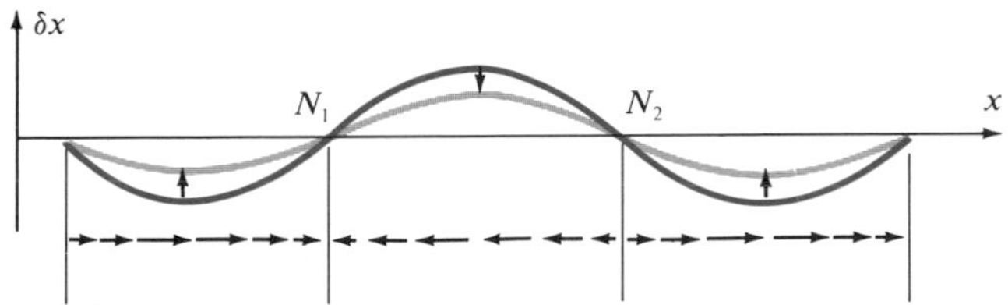

figure 18 • 6 *Displacements of the particles in a stationary longitudinal wave. The particles have just begun to return from their positions of maximum displacement, from the light colored line to the dark colored line. At the node N_1 a condensation is forming, and at N_2 a rarefaction.*

opposite directions is

$$\delta q = A \sin\left[\omega\left(t + \frac{x}{v}\right)\right] + A \sin\left[\omega\left(t - \frac{x}{v}\right)\right]$$

$$= 2A \cos\frac{\omega x}{v} \sin \omega t \qquad [18 \cdot 15]$$

Therefore

$$\delta q = 2A \cos\frac{2\pi x}{\lambda} \sin\frac{2\pi t}{T} \qquad [18 \cdot 16]$$

We can look at this wave in two ways. We may think of it as a trigonometric curve in space,

$$\delta q = \left[2A \sin\frac{2\pi t}{T}\right] \cos\frac{2\pi x}{\lambda}$$

whose amplitude (in brackets) changes with time like a simple harmonic function, or we may consider it a collection of oscillating particles,

$$\delta q = \left[2A \cos\frac{2\pi x}{\lambda}\right] \sin\frac{2\pi t}{T}$$

whose amplitude varies with position x. Certain particles, those located at nodes $x = \pm\frac{1}{4}\lambda, \frac{3}{4}\lambda, \frac{5}{4}\lambda, \ldots$, will always be at rest (note that their separation is $\frac{1}{2}\lambda$. Others located at antinodes, or *loops*, $x = \pm 0, \frac{1}{2}\lambda, \lambda, \frac{3}{2}\lambda, \ldots$, will periodically experience a maximum displacement $2A$. There are also times $t = 0, \frac{1}{2}T, T, \frac{3}{2}T, \ldots$ when $\delta q = 0$ everywhere; that is, the two waves exactly cancel each other. Note that there is no phase difference between particle oscillations at different positions; the time-dependent factor is always the same, sin $(2\pi t/T)$. Hence their oscillations are synchronous and there is *no net wave propagation* at all. This is why such a compound "wave" is called a stationary or standing wave.

Because a standing wave does not propagate, no net energy is transferred; that is, the net effect of the two simple waves is to exactly cancel each other's transfer of energy. Energy is *stored* in the medium, however. For a particle oscillating at x with amplitude $2A\cos(2\pi x/\lambda)$, the total energy is

$$E = \tfrac{1}{2}m\omega^2\left(2A\cos\frac{2\pi x}{\lambda}\right)^2 \qquad [18 \cdot 17]$$

and is continually changing its form between kinetic and potential energy of the particle. Therefore the energy contained in a cylinder of cross-sectional area S and length λ along the x direction is

$$E_\lambda = \tfrac{1}{2}\rho\omega^2 S\int_0^\lambda 4A^2\cos^2\frac{2\pi x}{\lambda}\,dx = \rho\omega^2 A^2 S\lambda \qquad [18 \cdot 18]$$

If the wavelength is short compared to other dimensions of interest, then we may take the energy density, or average energy per unit volume, to be

$$E_V = \frac{E_\lambda}{S\lambda} = \rho\omega^2 A^2 \qquad [18 \cdot 19]$$

exactly *twice* what it would be for a single simple wave (see Sec. 17 . 6). Musical instruments typically owe their production of sound to the generation of standing waves by means of *reflection* of traveling waves at fixed points (stringed instruments) or rigid surfaces. We shall discuss this later.

example 18 • 2 Two waves of amplitudes A_1 and A_2 and equal frequency are propagating in opposite directions. An observer moving along the x axis finds that the amplitude of the compound wave is never more than 4 units and never less than 2. If $A_1 > A_2$, find their values.

solution As in the case of a standing wave,

$$\Delta = \theta_2 - \theta_1 = 2\frac{\omega x}{v}$$

However, the amplitude is, from Eq. [18 . 14],

$$A = \sqrt{A_1^2 + A_2^2 + 2A_1A_2\cos\frac{4\pi x}{\lambda}}$$

which takes on a maximum value $A = A_1 + A_2$ when $4\pi x/\lambda = 0, 2\pi, 4\pi, \ldots$, and minimum values $A = A_1 - A_2$ when $4\pi x/\lambda = \pi, 3\pi, 5\pi, \ldots$. Since these values are known, we can solve for

$$A_1 = 3 \qquad A_2 = 1$$

The ratio $(A_1 + A_2)/(A_1 - A_2)$ is known as the *standing-wave ratio.*

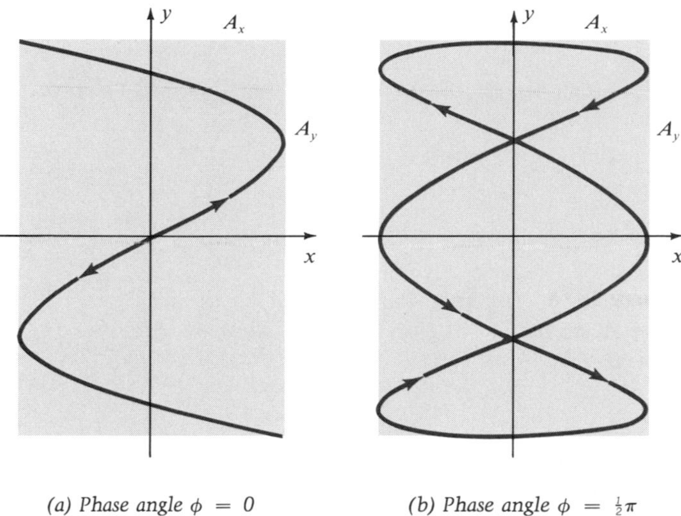

(a) Phase angle $\phi = 0$ (b) Phase angle $\phi = \frac{1}{2}\pi$

figure 18 • 7 *Lissajous figures for* $x = A_x \sin 3\omega t$ *and* $y = A_y \sin(\omega t + \phi)$

In cases where waves may be passing through a three-dimensional medium in several different directions, the vector forces of Eq. [18 . 4] can cause oscillations in mutually perpendicular directions. The resultant path of the oscillating particle as the waves pass is known as a *Lissajous figure*. For example, consider a particle oscillating simultaneously in the x and y directions:

$$x = A_x \sin \omega_x t \qquad y = A_y \sin(\omega_y t + \phi) \tag{18 · 20}$$

where δ is omitted for simplicity. By eliminating time from these two equations, we can obtain the path of the Lissajous figure in the form $f(x, y) = \text{constant}$. It turns out that as long as $\omega_y/\omega_x = m/n$, the ratio of two integers, the resultant motion will be periodic, with the particle continuously retracing a closed path, as shown in Fig. 18 . 7. If the ratio of frequencies is not a rational number, the Lissajous curve never closes and may be extremely complicated.

We shall limit ourselves here to only a few special cases for which $\omega_x = \omega_y = \omega$. We can solve Eq. [18 . 20] for

$$\sin \omega t = \frac{x}{A_x} \qquad \cos \omega t = \frac{(y/A_y) - (x \cos \phi / A_x)}{\sin \phi} \tag{18 · 21}$$

Squaring and adding both equations yields

$$\frac{x^2}{A_x^2} + \frac{y^2}{A_y^2} - \frac{2xy}{A_x A_y} \cos \phi = \sin^2 \phi \tag{18 · 22}$$

This is the general equation of an ellipse whose axes are at some angle to the coordinate axes, and which is contained within a rectangle with horizontal and vertical sides of length $2A_x$ and $2A_y$, respectively (see Fig. 18 . 8). The eccentricity and inclination of the ellipse depends on ϕ and the ratio of amplitudes A_y/A_x.

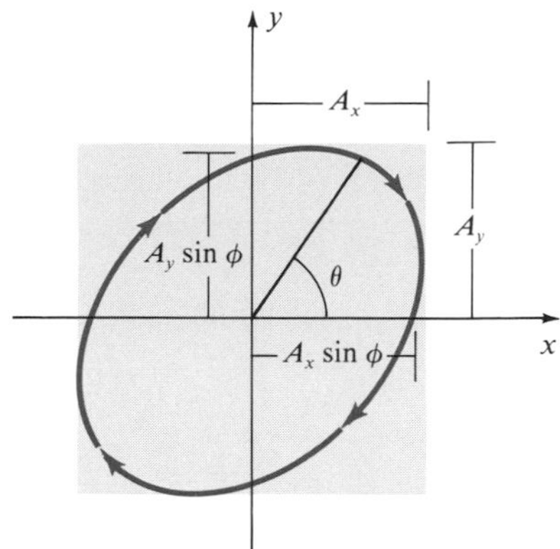

figure 18 • 8 *Superposition of* $x = A_x \sin \omega t$ *and* $y = A_y \sin (\omega t + \phi)$ *gives an ellipse;* $d\theta/dt < 0$ *for* $\phi > 0$

If $\phi = 0$ or π, then Eq. [18 . 22] reduces to $y = \pm A_y x/A_x$; the particle moves back and forth along a straight line passing through the origin, as shown in Fig. 18 . 9*a*. If $\phi = \pm \frac{1}{2}\pi$ rad, then the particle moves around an ellipse with axes of length $2A_x$ and $2A_y$ along the respective coordinate axes (Fig. 18 . 9*b*):

$$\frac{x^2}{A_x^2} + \frac{y^2}{A_y^2} = 1 \qquad [18 \cdot 23]$$

For all other values of ϕ the axes are at angles α and β with the x axis, where setting $x = r \cos \theta$, $y = r \sin \theta$ and $dr/d\theta = 0$ yields

$$\alpha = \tfrac{1}{2} \tan^{-1} \left[\frac{2A_x A_y \cos \phi}{A_x^2 - A_y^2} \right]$$

$$\beta = \alpha + \tfrac{1}{2}\pi \qquad [18 \cdot 24]$$

If the frequencies of the two component motions differ slightly, then one component will gradually gain in phase, and the ellipse will slowly rotate within its confining rectangle, assuming the complete cycle of forms from $\phi = 0$ to $\phi = 2\pi$. This is a beautiful effect to observe on a cathode-ray tube and is the basis of many special visual effects seen in television commercials. It also provides a sensitive technique for tuning electronic circuits.

figure 18 • 9 *Lissajous figures for* $\omega_y = \omega_x$

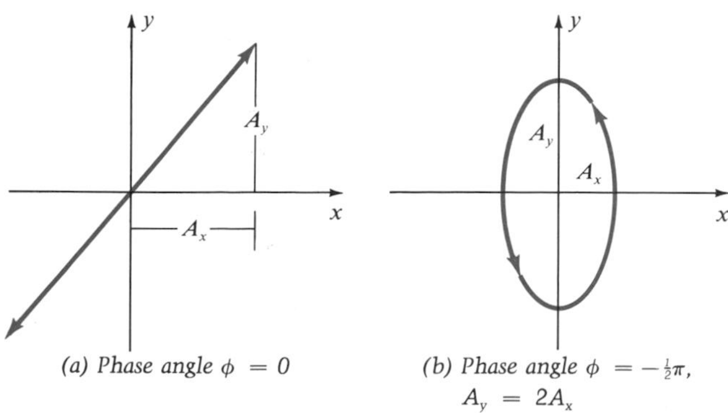

(a) Phase angle $\phi = 0$

(b) Phase angle $\phi = -\frac{1}{2}\pi$, $A_y = 2A_x$

18 . 3 reflection of longitudinal waves

In the last chapter we discussed the transmission of a longitudinal compression wave as a pulse moving down a rigid tube of some material medium. As long as the pulse travels through a medium of uniform density, the situation is directly analogous to the transmission of kinetic energy in a sequence of perfectly elastic collisions. Each "disk," or layer of particles, gives up all its motion to the next layer, just as a moving ball striking a stationary ball of the same mass gives up all its motion to the stationary ball and itself comes to rest. However, if the moving pulse strikes a denser medium than the one in which it is traveling, the situation is analogous to the impact of a ball with one of larger mass. Instead of coming to rest after impact, particles of the less dense medium transmit part of their energy on through the layers of the denser medium and then recoil back toward the source. This backward motion is transferred from layer to layer, causing a reflected pressure pulse of condensation to travel back through the rarer medium, as shown in Fig. 18 . 10*a*. This reflected pressure wave has a smaller amplitude, but the same shape as the incident pressure wave.

Since reflection reverses the direction of displacement in the rarer medium, δx undergoes a change of phase of π rad. If a wave strikes a perfectly rigid wall, it is as if it had encountered a medium of infinite density. All the wave energy is reflected, but the net particle velocity must be zero at the wall; thus a standing compression wave is set up with a node at the wall.

Now suppose the medium on the other side of the interface is less dense than the one in which the pulse is initially traveling. Then the case is analogous to the impact of large ball on a smaller one. When the pulse reaches the interface it continues to move forward, but more slowly, and so produces a diminution of pressure at the interface. The excess of pressure in the layer on the left then drives particles toward the right, and a pressure pulse of *rarefaction* is reflected back toward the source. This reflected pressure wave, as shown in Fig. 18 . 10*b*, has a shape which is the inverse of the shape of the incident wave. However, the displacements δx do not undergo any phase change upon reflection from a rarer medium. If the rarer medium were simply empty space, there would be nothing to hinder the motion of the

figure 18 • 10 *Reflection of a longitudinal pressure pulse at a density boundary, and its mechanical analog showing particle velocities v_x*

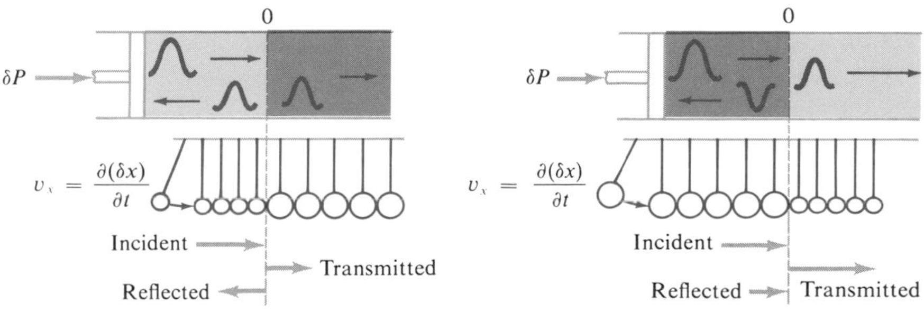

(*a*) *Reflection from a denser medium*

(*b*) *Reflection from a rarer medium*

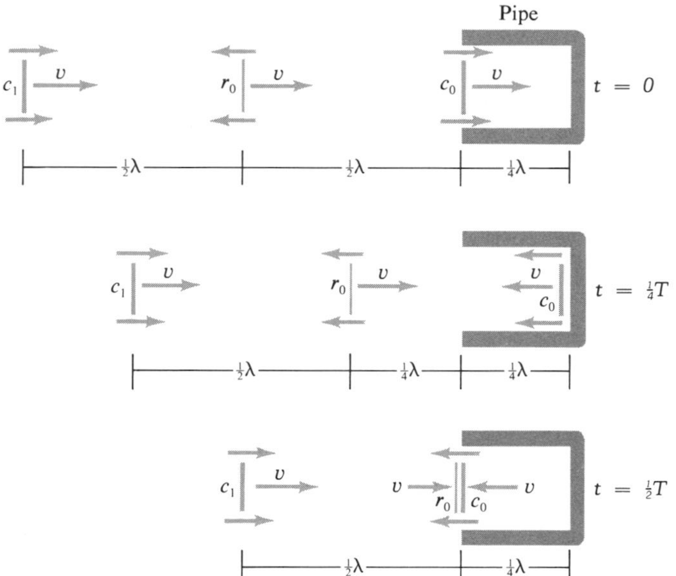

figure 18 • 11 *Wave train entering a pipe that is closed at one end. Pressure maxima (condensations) are at c and c_0; the pressure minimum (rarefaction) between them is at r_0. The* **v** *vectors show the direction of propagation for each pulse; the other vectors indicate the directions of particle motions at the pulse.*

particles; hence their displacement would be a maximum and their density would be unchanged as they came spilling out of the interface. Therefore $\delta P = 0$ at this point, and again we see the general rule:

> A displacement node coincides with a pressure antinode, and a displacement antinode coincides with a pressure node.

Figure 18 . 11 shows a *train* of waves of wavelength λ approaching a pipe which is closed at the far end. In the condensation pulses of the advancing wave train the particles are all moving in the direction of propagation of the wave ($\rightarrow$), but in the rarefaction pulses they are moving in the opposite direction ($\leftarrow$). If the length of the pipe is exactly $l = \frac{1}{4}\lambda$, the condensation pulse c_0 ($\rightarrow$) will move into the pipe at time $t = 0$ and be reflected at the closed end at time $t = \frac{1}{4}T$, with the particles now moving from right to left. This reflected condensation pulse will reach the open end at $t = \frac{1}{2}T$, the same instant when the rarefaction pulse r_0 ($\leftarrow$), which also consists of particles moving from right to left, reaches the open end.

When an acoustic wave c_0 ($\leftarrow$) traveling back out of the pipe reaches the open end, it experiences the same sort of reflection as if it had passed from a denser to a rarer medium, since the wave can then expand laterally, allowing the particles to move leftward more readily. Therefore a rarefaction r ($\leftarrow$) starts back into the pipe which unites with the rarefaction r_0 ($\leftarrow$) which is just entering it, resulting in a rarefaction pulse of increased amplitude. This pulse is reflected at the closed end as a rarefaction ($\rightarrow$) and again at the open end as a condensation ($\rightarrow$), exactly in time to unite with the next condensation c_1 ($\rightarrow$) as it enters the pipe. By this process of continuous union of incident and reflected pulses, the amplitude of vibration in the pipe is increased to a value many times larger than that of the original wave, limited only by the loss of energy transmitted to the outside air. Hence the pipe resonates and, in effect, becomes a source of sound.

The length $l_1 = \frac{1}{4}\lambda$ is the shortest possible resonant length for a given wavelength, but if the pipe is continually lengthened, other resonant lengths are obtained. It is clear that the next possible resonant length is one that permits c_0 to return exactly in time to unite with r_1. Since r_1 is a distance λ behind r_0, the second resonant closed-pipe length must be a half-wavelength greater than the first, or $l_2 = \frac{3}{4}\lambda$. Similarly, it is possible to obtain resonance at lengths of $l_3 = \frac{5}{4}\lambda$, $l_4 = \frac{7}{4}\lambda$, and so on. That is, for a given wavelength λ, the resonant lengths of a closed pipe are

$$l_n = \frac{2n - 1}{4}\lambda \qquad n = 1, 2, 3, \ldots \qquad [18 \cdot 25]$$

Let us examine the particle motions in detail for the second resonant length, $l_2 = \frac{3}{4}\lambda$. In Fig. 18 . 12 the condensation c_0 enters the pipe at time $t = 0$. At time $t = \frac{3}{4}T$ it is reflected from the closed end, and at $t = \frac{5}{4}T$ it collides with condensation c_1 at N, a distance $\frac{1}{2}\lambda$ from the closed end of the pipe. Since particles in the two condensation pulses are moving with equal speeds, but in opposite directions, the "collision" of c_0 and c_1 is entirely analogous to the collision of two oppositely moving identical elastic balls, the effect being the same as if each ball "passed through" the other. The net displacement at N is zero as the condensation pulses pass through each other. Actually, c_1 returns to the left after the collision and unites with r_1, while c_0 is forced back toward the closed end of the pipe again.

One half-period after the collision of condensations c_1 and c_0 at N, the

figure 18 • 12 *Wave train entering a closed pipe of length $\frac{3}{4}\lambda$. Pressure maxima (condensations) are at c_0 and c_1; pressure minima (rarefactions) are at r_0 and r_1. The **v** vectors show the direction of propagation for each pulse; the other vectors show the directions of particle motions at the pulse; N is a displacement node.*

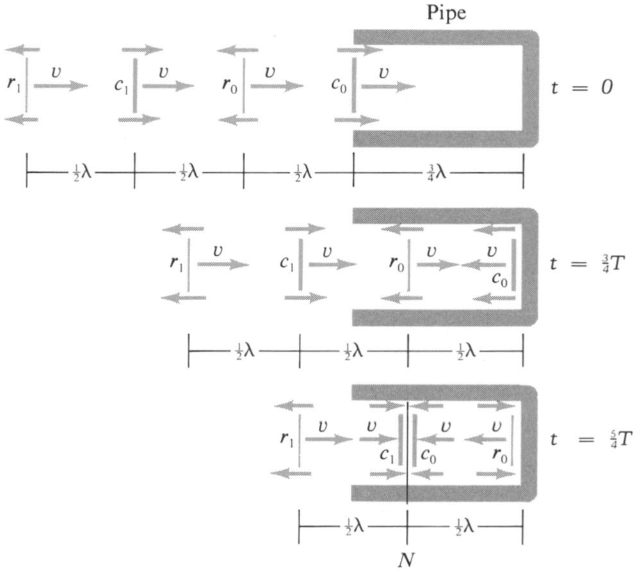

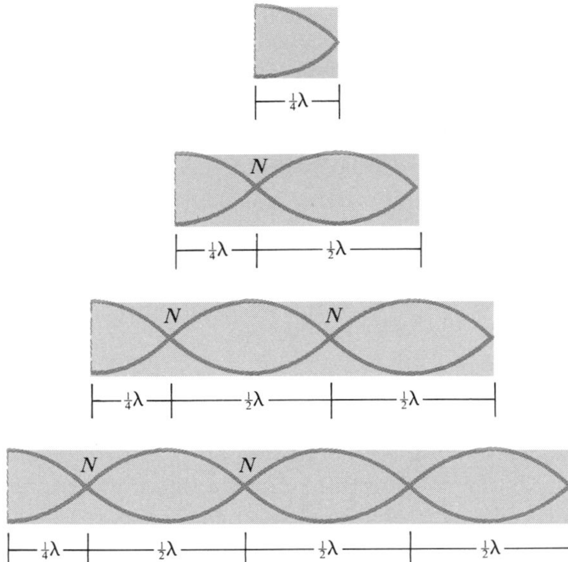

figure 18 • 13 *Stationary waves in closed pipes of resonant lengths $\frac{1}{4}\lambda$, $\frac{3}{4}\lambda$, $\frac{5}{4}\lambda$, and $\frac{7}{4}\lambda$. The curves are proportional to δx.*

rarefactions r_1 and r_0 will collide at this point. Hence the particles near N are first pushed together by equal but opposing forces and then pulled apart by equal but opposing forces. The result is that they do not move at all, so the point N is a node. The closed end of the pipe must also be a node, while the open end is a region of maximum disturbance, an antinode. The points of greatest disturbance, which are midway between nodes, are also antinodes. Therefore, we obtain the standing-wave patterns illustrated in Fig. 18 . 13. These patterns can be represented mathematically for a given pipe of length l by the expressions

$$\delta x = 2A \cos \frac{2\pi x}{\lambda_n} \sin \omega t \qquad \lambda_n = \frac{4l}{2n - 1} \qquad [18 \cdot 26]$$

where the open end is at $x = 0$. Incoming wave trains capable of resonating with a given closed pipe will therefore have frequencies in the ratios 1:3:5: . . . .

If a pipe is *open* instead of closed at the far end, a single condensation (⟶) entering the mouth is partially reflected from the far end as a rarefaction (⟶), which will in turn be reflected again from the mouth of the pipe as a condensation (⟶). After traveling the length of the tube twice, the pulse will again emerge at the far end as a condensation. The wavelength of the train of waves into which the pipe has transformed the single pulse is $\lambda = 2l$. This tone is an octave higher (twice the frequency) than the fundamental of a closed pipe of the same length.

Another way to characterize the foregoing phenomenon is to say that there must be a displacement antinode (or a pressure node) at *each* end of the open pipe with one or more nodes in between. Hence, for a given wavelength λ, open pipes of length

$$l_n = \tfrac{1}{2}n\lambda \qquad n = 1, 2, 3, \ldots$$

will resonate (see Fig. 18 . 14). An open pipe of given length l can support

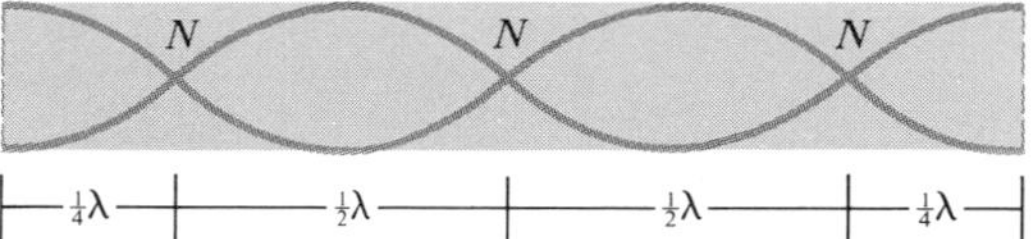

figure 18 • 14 *Stationary waves in an open pipe,* $l = \frac{3}{2}\lambda$

standing waves of the form

$$\delta x = 2A \cos \frac{2\pi x}{\lambda_n} \sin \omega t \qquad \lambda_n = \frac{2l}{n} \qquad [18 \cdot 27]$$

where the open ends are at $x = 0$ and $x = l$. Incoming wave trains capable of resonating with a given open pipe will have frequencies in the ratios 1:2:3 . . . . Wind instruments are open-ended pipes, usually played by changing their effective length by opening and closing holes in their sides.

example 18 • 3 A famous singer hums an aria as he relaxes in his completely tiled shower room. He shuts off the water and turns toward the open entrance directly behind him, humming as he goes a very pure note. At a distance of 1 m from the shower-room wall, his sensitive ear detects an increase in the intensity. What is the frequency of the note he is humming?*

solution The shower room, with its perfectly reflecting tile walls, acts as a closed pipe. The singer first will notice an increase in intensity at the first resonant length for that frequency,

$$l_1 = \tfrac{1}{4}\lambda = \frac{v}{4\nu} = 1 \text{ m}$$

Therefore

$$\nu = \frac{v}{4 \text{ m}} \doteq 85 \text{ Hz}$$

18 • 4 *reflection of transverse waves*

Considerations of reflection apply to transverse waves as well as to longitudinal compression waves. When a transverse pulse is sent along a string, either fastened at one end or attached to a heavier string, a crest in δy is reflected from the boundary as a trough (see Fig. 18 . 15). Just as in the compression wave reflected from a denser medium, the displacements of the particles are reversed in sign by the reflection. If the end of the string is free or is attached to a lighter string, a crest sent along the string is reflected as a crest without change of phase. If the string is fixed at one end, the reflected waveform may be reversed, but conservation of energy requires that it have the same shape

*Lord Rayleigh once observed that "some of the natural notes of the air contained within a room may generally be detected on singing the scale. Probably, it is somewhat in this way that blind people are able to estimate the size of rooms." There are now a number of sophisticated devices for the blind based on this principle.

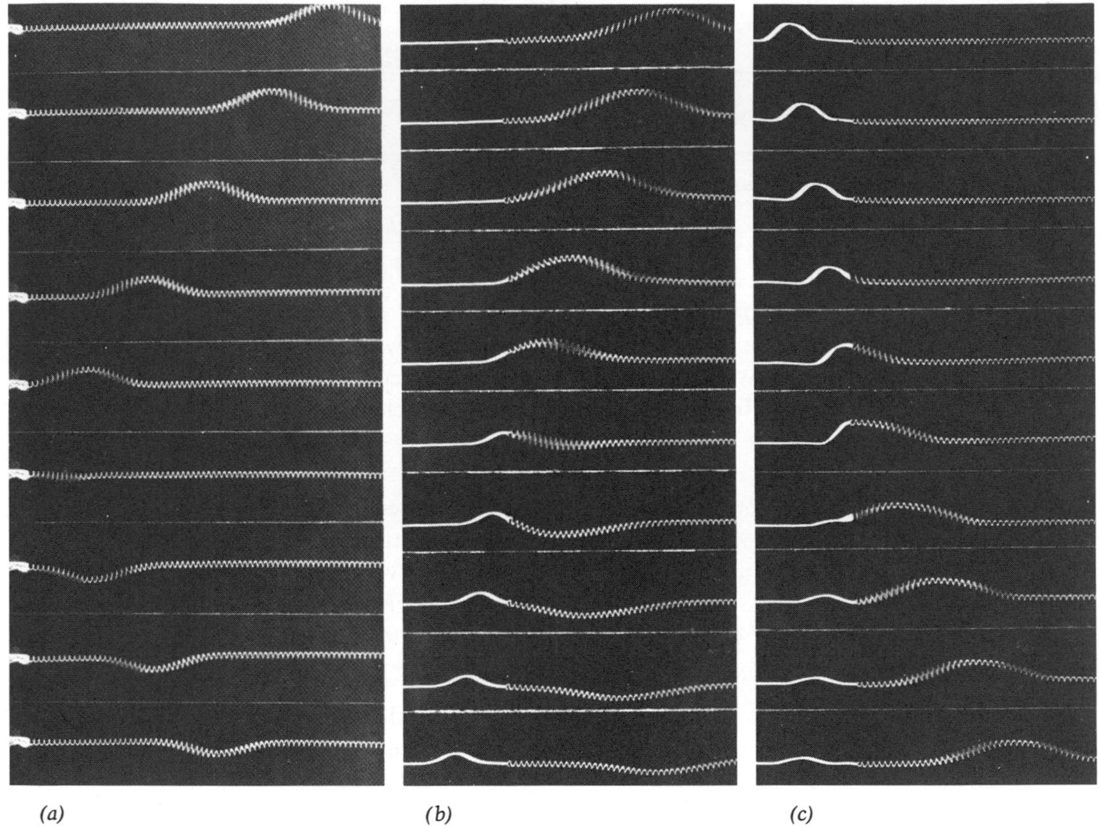

figure 18 • 15 *Reflection of a pulse from a boundary on a stretched string. (a) Reflection from a fixed end; the reflected pulse is upside down. (b) Partial reflection and partial transmission as a pulse passes from a light string (right) to a heavy string (left); the reflected pulse is upside down. (c) Partial reflection and partial transmission as a pulse passes from a heavy string to a light string; the reflected pulse is right side up.*

as the incident wave. If the string is free, no energy can be lost by transmission, and the reflected waveform, although identical in shape to the incident waveform, is not reversed in sign. A standing wave may therefore be set up on the string by the reflection of a simple wave train. If the string is fixed at both ends, then the standing wave must have nodes at both ends.

In Sec. 18 . 2 we saw that nodes must be separated by an integral number of half-wavelengths. Hence only those standing waves are possible for which, for a string of length l and uniform density ρ_l,

$$l = \tfrac{1}{2}n\lambda_n \qquad \text{or} \qquad \lambda_n = \frac{2l}{n} \tag{18 · 28}$$

If, as in Fig. 18 . 16 the string supports a heavy weight mg, then the allowed frequencies are

$$\nu_n = \frac{v}{\lambda_n} = \frac{n}{2l}\sqrt{\frac{mg}{\rho_l}} \tag{18 · 29}$$

If we attempted to set up oscillations of any other frequencies, their successive reflections would interfere destructively and ultimately cancel each other. Another way to view this is to say that a *given* wavelength can be sustained only by fixed strings of length $l_n = \frac{1}{2}n\lambda$.

The lowest frequency ν_1 of a string is its *fundamental,* or *first partial;* the other frequencies are *overtones,* or *upper partials.* When the overtones are integral multiples of the fundamental, they are known as *harmonics* of the fundamental, which is sometimes called the first harmonic. An excited string will usually support a number of standing waves of different frequencies simultaneously; it is the combination of tones which gives each instrument its unique quality, or timbre. One can easily distinguish a concert A (440 Hz) played on a violin from that played on a clarinet or an oboe (as in Prokofiev's symphonic suite, *Peter and the Wolf*). If we vary the mode of excitation or the geometry of a vibrating system, then we vary the allowed frequencies or even select certain frequencies and eliminate others. For example, a violin string will oscillate with most of its energy in the fundamental. However, if the violinist places a finger very lightly over the center of the string, this interferes with the fundamental standing wave, and the string oscillates as shown in Fig. 18 . 16*b* (see the opening quotation from Sauveur). Just as with a musical instrument, it is the shape and weight of the cavities and bones of your head that determines the overtones in your voice and their relative energies, and hence the unique sound of your voice.

A rod surrounded by air is in every respect analogous to an open pipe, for the reflections at the ends are the same as those which occur when a wave passes from a denser to a rarer medium. Therefore a rod will respond to a train of waves if it is of length

$$l_n = \tfrac{1}{2}n\lambda \qquad [18 \cdot 30]$$

figure 18 • 16 *Four modes of vibration of a stretched string*

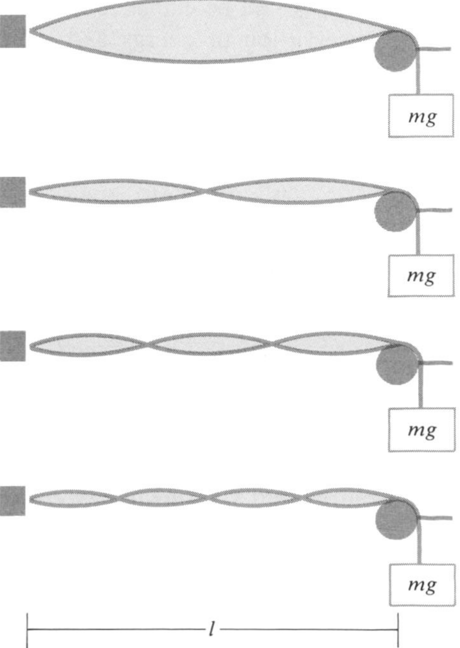

figure 18 • 17 *Transverse waves in rods*

where λ is the wavelength of the train *in* the rod. Transverse waves travel in rods as the result of the shear elasticity of the material. A bar clamped at one end gives off its fundamental frequency when vibrating as shown in Fig. 18 . 17*a*. If it is struck sharply near the free end, it gives off its first overtones, as shown in Fig. 18 . 17*b*. Unlike the case of compression waves, for transverse waves the relations between the frequencies of a fundamental and its overtones are not simple. If the rod is supported at two points, it will vibrate as shown in Fig. 18 . 17*c*, yielding its fundamental; it yields its first overtone when supported as in Fig. 18 . 17*d*.

Vibrating plates and membranes are also capable of oscillating only in certain fixed modes. These are much more complex than those of the vibrating string, since the geometry of these instruments is two-dimensional. When a drumhead is struck, circular waves spread outward from the point of disturbance like the ripples on a pond and are reflected back from the rigid boundaries. The resulting standing waves display nodal *lines,* both straight and circular, which separate depressed (trough) areas from elevated (crest) areas. Figure 18 . 18 illustrates some of these modes. The frequencies of oscillation do not form a simple series, as in the case of a stretched string; however, they can be calculated mathematically.

Conservation of energy and momentum have been mentioned in our discussions of reflection. However, the latter principle cannot be used to derive the amplitudes of reflected and transmitted waves, because we do not know how to estimate the density or the volume of particles which are effectively interacting at the interface. Instead, we simply assume that the *net* particle disturbance is the same on both sides of the interface. That is, for a wave δq normally incident on a boundary (approaching normal to the boundary), there is a reflected wave $\delta q_r = -r\ \delta q$ and a transmitted wave $\delta q_\tau = \tau\ \delta q$ related by

$$\delta q + \delta q_r = \delta q_\tau \qquad \text{or} \qquad 1 - r = \tau \tag{18 · 31}$$

where r and τ are positive numbers less than unity and the minus sign accounts for the change of phase on reflection.

Conservation of energy applies in general to all normally incident waves and requires that the *incident intensity* be equal to the sum of the reflected and transmitted intensities. If the densities of the two mediums in Fig. 18 . 10 are ρ and ρ' and the wave speeds in these two mediums are v and v', respectively, then Eq. [17 . 46] for wave intensity, $I = \frac{1}{2}\rho\omega^2 A^2 v$, yields

$$\rho v = \rho r^2 v + \rho' \tau^2 v' \tag{18 · 32}$$

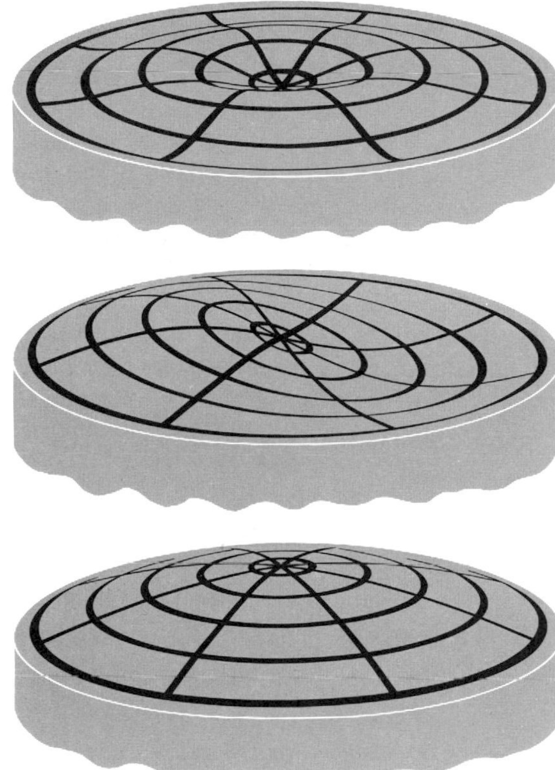

figure 18 • 18 *Instantaneous positions of a vibrating drum head*

and since $1 - r = \tau$,

$$r = \frac{\rho' v' - \rho v}{\rho' v' + \rho v} \qquad \tau = \frac{2\rho v}{\rho' v' + \rho v} \tag{18 . 33}$$

Note that reflection is total for ρ' approaching infinity.

example 18 • 4 Compute the ratio of transmitted energy E_τ to incident energy and the ratio of reflected energy E_r to incident energy for normally incident waves at the interface between two monatomic gases.

solution For perfect gases in equilibrium the pressure is equal across the interface; hence $v = \sqrt{\gamma P/\rho}$ and $\rho'/\rho = v^2/v'^2$. Therefore

$$\frac{E_\tau}{E} = \frac{\rho' \tau^2 v'}{\rho v} = \frac{v \tau^2}{v'} = \frac{4vv'}{(v + v')^2}$$

and

$$\frac{E_r}{E} = r^2 = \frac{(v - v')^2}{(v + v')^2}$$

It is easy to verify that these two equations sum to unity, as required by Eq. [18 . 32].

18 • 5 Huygens' principle: reflection and refraction

There are three fundamental ways in which the direction of a propagating wave in a three-dimensional medium can change: *reflection, refraction,* and *diffraction.* When a wave arrives at the boundary between two different mediums, generally part of it is reflected back into the original medium and part is transmitted into the new medium. It can be shown experimentally that at any angle of incidence other than normal the propagation speed of the reflected wave is unchanged, but its velocity component *normal* to the boundary is reversed in direction. In addition, the direction of the transmitted wave is not the same as that of the incident wave. As Fig. 18 . 19 shows, the lines, or *rays,* that indicate the direction of propagation of the wave trains are bent, or *refracted,* at the reflecting surface.

> In any isotropic medium, the rays are always at right angles to the wavefront.

Huygens' principle, enunciated in 1690, affords a means of locating the reflected and refracted wavefronts at any time, if we know the propagation speed of the incident wave and its position at some previous time:

> "In considering the propagation of waves, we must remember that each particle of the medium through which the wave spreads does not com-

figure 18 • 19 *Shallow water waves generated in a ripple tank whose depth is altered by the presence of a glass plate. Arrows show the directions of the incident, refracted, and reflected waves at the plate boundary.*

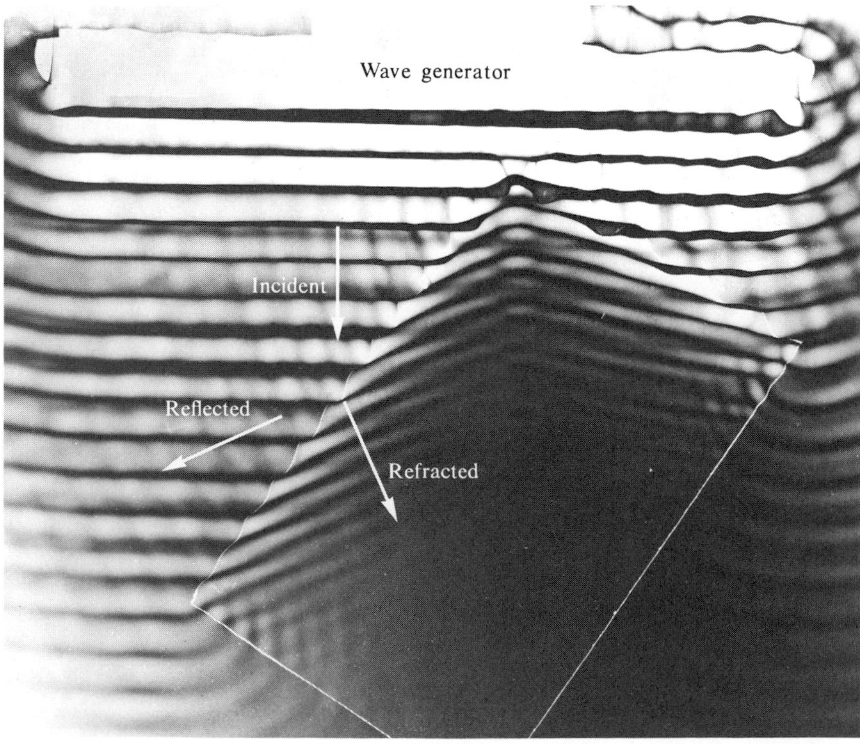

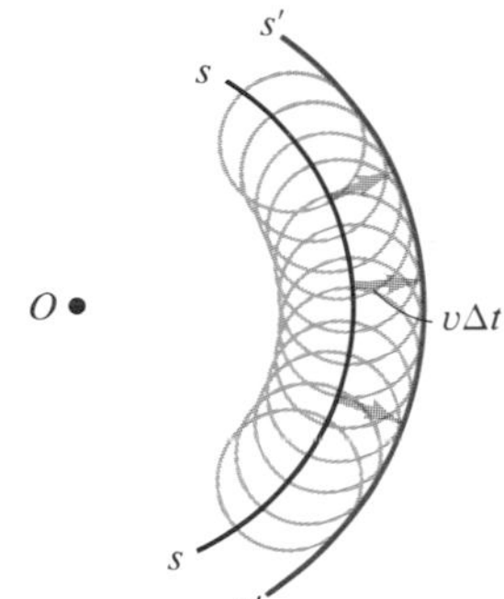

figure 18 • 20 *Huygens' principle applied to the propagation of a spherical wave. Arrows are rays; curves ss and s's' are wavefronts. The new wavefront s's' is found by drawing circles (spheres) of radius v Δt about each point of the previous wavefront ss. The envelope of these secondary waves is the new wavefront s's'.*

municate its motion only to that neighbor which lies in the same straight line drawn from the source of the disturbance but shares it also with all the particles which touch it and resist its motion. Each particle is thus to be considered as the center of a wave."

In other words, any particle in the wavefront of a disturbance may be viewed as the point source of a secondary spherical wave propagating outward from it in all directions. If arc *ss* in Fig. 18 . 20 is the instantaneous position of a wavefront emanating from a disturbance at *O*, then to find the new wavefront after an interval of time Δt, we draw spheres of radius $v\ \Delta t$ about *each point* of *ss*, where v is the wave speed. The sum, or envelope, of these secondary waves is the new wavefront, *s's'*. This is known as *Huygens' construction.*

It would appear from Huygens' construction that a disturbance should be propagated back to its origin as well as forward. In 1826 Fresnel pointed out that this did not happen because of destructive interference effects; later Kirchhoff showed by a mathematical analysis that the secondary waves from the individual sources really do destroy one another by mutual interference except at the surface *s's'*. The wave therefore propagates only in the direction away from the origin.

Huygens' principle may also be employed to predict the direction and curvature of a wavefront completely reflected from a plane surface. The reflection of a spherical wave is shown in Fig. 18 . 21. In this case *sss* is the

figure 18 • 21 *Applying Huygens' principle to trace the path of a wave that strikes a smooth plane surface*

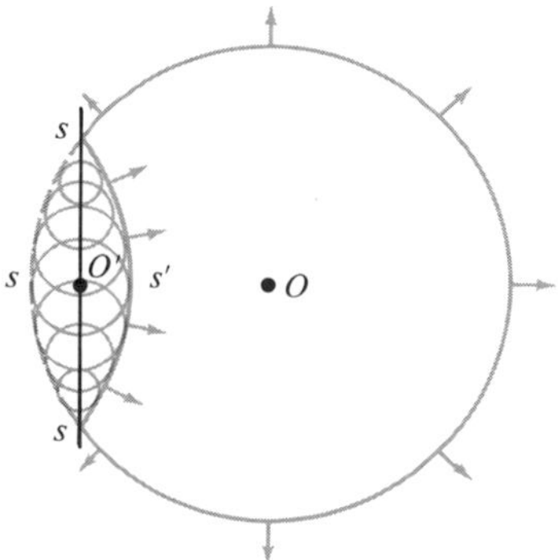

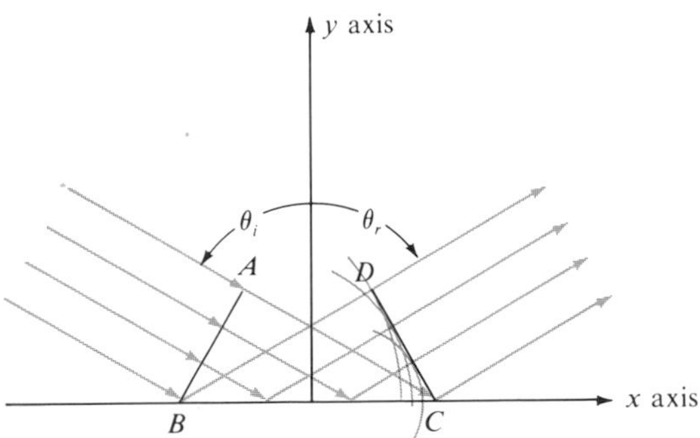

figure 18 • 22 *Huygens' construction of an incident wavefront AB on a reflecting plane surface BC and its reflected wavefront DC;* $\theta_i = \theta_r$

position the wave advancing from O *would* have reached at a certain instant if it had not encountered the plane surface shown. However, when the wave reached O', which is the nearest point on the reflecting surface to O, this point became, according to Huygens' principle, the origin of a secondary wave. At immediately succeeding intervals, the successive points above and below O' were also reached by the incident wavefront and, in turn, became sources of secondary waves. These secondary waves emitted by each successive point on the reflecting surface have for their envelope the spherical surface $ss's$. Notice that the reflected wave appears to come from some point behind the reflecting surface; in the present case of a plane reflector, this point is as far behind the reflector as the source O in front of it.

Huygens' principle can be used to derive the direction of reflected rays. In Fig. 18 . 22, θ_i is the angle of incidence which a ray makes with the normal to a reflecting surface and θ_r is the angle of reflection. The secondary wavelets from the surface BC then form the wavefront DC, and from plane geometry,

$$\angle ABC = \theta_i \qquad \angle BCD = \theta_r \qquad [18 \cdot 34]$$

Since the speed of the reflected wave is the same as that of the incident wave,

$$\overline{AC} = \overline{BD} \quad \text{and} \quad \Delta ABC \cong \Delta BCD \text{ (congruence)} \qquad [18 \cdot 35]$$

because both are right triangles (the rays must be perpendicular to their respective wavefronts AB and DC). Therefore we have the first of two fundamental laws of geometric optics:

The angle of incidence equals the angle of reflection.

$$\theta_i = \theta_r \qquad [18 \cdot 36]$$

When any wave passes from one medium into another in which its speed is different, its direction of propagation changes. Such a change in the direction of the transmitted portion of a wave is called *refraction*. For example, if different layers of the atmosphere are at different temperatures, a sound wave passing through them will tend to travel in a curved path, rather than a straight line. When sound waves in one gas are made to pass through a lens-shaped bag containing another gas of different density, the results are similar to those observed when light passes through a glass lens. Refraction experiments of this kind have also been made with acoustic "lenses" made of pitch, with rubber vessels containing water, and even with an acoustic lens which consisted of a large balloon containing carbon dioxide.

Figure 18 . 23 illustrates Huygens' construction of a refracted wave. When the incident wavefront reaches the surface of the refracting medium at B, the wavelet radiating from B travels more slowly than the wavelet from the front at A. Wavelets from successive portions of the wavefront entering the refracting medium have shorter radii than those traveling in the same time in the original medium. Hence the new wavefront forms at DC at a smaller angle θ' to the normal to the surface. In a given time Δt, the wavelet from A travels to C, while the wavelet from B travels to D:

$$\Delta t = \frac{\overline{AC}}{v} = \frac{\overline{BD}}{v'} \qquad [18 \cdot 37]$$

where v is the original speed above and v' is the speed below the surface through BC. Since ABC and BCD are both right triangles,

$$\theta = \angle ABC = \sin^{-1}\frac{\overline{AC}}{\overline{BC}} \qquad \theta' = \angle BCD = \sin^{-1}\frac{\overline{BD}}{\overline{BC}} \qquad [18 \cdot 38]$$

Therefore

$$\sin\theta = \frac{v\,\Delta t}{\overline{BC}} \qquad \sin\theta' = \frac{v'\,\Delta t}{\overline{BC}} \qquad [18 \cdot 39]$$

and we have *Snell's law of refraction:*

$$\frac{\sin\theta}{\sin\theta'} = \frac{v}{v'} \qquad [18 \cdot 40]$$

Note that in the slower medium the rays turn *toward* the normal. Had the wave been passing from a slower to a faster medium, they would have turned *away* from the normal upon refraction.

This behavior is easily demonstrated for shallow water waves in a "ripple tank." If plane water waves are generated by a vibrating metal plate in a shallow tank, their speed depends on the depth of the water as $v = \sqrt{gh}$ (see Eq. [17 . 34]). If a glass plate of thickness Δh is placed on the bottom of

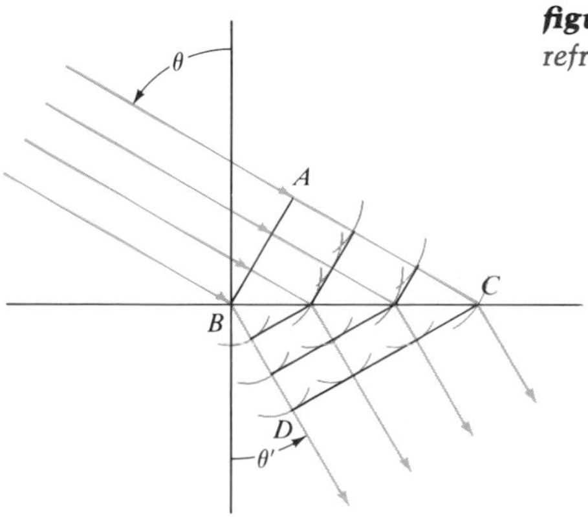

figure 18 • 23 *Huygens' construction of a refracted wavefront:* $\overline{AC}/\overline{BD} = v/v' = \sin\theta/\sin\theta'$

the tank with its upstream edge corresponding to $\overline{BC}$ in Fig. 18 . 23, then the advancing wavefronts will be refracted because of their change in speed as they pass over the glass plate. In that case

$$\frac{\sin\theta}{\sin\theta'} = \sqrt{\frac{h}{h - \Delta h}} \qquad [18 \cdot 41]$$

Note that the wavelength of the refracted wave is shorter than that of the incident wave. Since the frequency of the wave remains constant, we have the relation

$$\frac{\lambda}{\lambda'} = \frac{v}{v'} \qquad [18 \cdot 42]$$

If we apply Snell's law to the case of a wave passing from a slow medium into a fast one, what happens when $\sin\theta' > v'/v$? Equation [18 . 37] predicts $\sin\theta > 1$, an impossibility. However, the theory does not predict an impossibility; rather, it predicts that such refraction is impossible. In fact, when

$$\theta' = \sin^{-1}\frac{v'}{v} \qquad \theta = \tfrac{1}{2}\pi \qquad [18 \cdot 43]$$

we have "total internal reflection"; the wave is not transmitted, but is reflected at the boundary back into the slower medium. This phenomenon is particularly striking in the case of light, which is "piped" around corners by glass fibers or lucite "light pipes."

It is important to bear in mind that Huygens' principle and the laws of reflection and refraction hold for *every wave* we have discussed or will discuss in this text. They are also obeyed by the transverse surface wave which propagates outward in two dimensions as circles, as when a pebble is dropped into still water.

example 18 • 5 Two rays of a circular surface wave form a small angle $\Delta\theta$. What is the angle between them after they have propagated into a second medium whose normal makes an angle with the original direction of propagation?

solution If one ray is incident at angle θ, then the other must be incident at angle $\theta + \Delta\theta$. If we differentiate

$$\sin\theta' = \frac{v'}{v}\sin\theta$$

we obtain

$$\Delta\theta' \cos\theta' \doteq \frac{v'}{v}\cos\theta\,\Delta\theta \qquad \text{or} \qquad \Delta\theta' \doteq \Delta\theta\sqrt{\frac{1 - \sin^2\theta}{(v/v')^2 - \sin^2\theta}}$$

If $v' < v$, then $\Delta\theta' < \Delta\theta$, and the source of the wave, as determined by an observer in the slower medium, unaware of the refraction, would appear to be farther away than it actually is.

18 • 6 diffraction

When waves pass around an obstacle or through an aperture in it, they tend to curl around the edges so that the "shadow" of the obstacle on the downstream side is not sharply defined. This modification of wave behavior is called *diffraction*. Diffracted sound waves can be heard around corners, and water waves entering a harbor spread into the region behind a breakwater (see Fig. 18 . 24). The wavelength λ gives the scale of the wave. If the obstacle or aperture had a length $l \doteq \lambda$, the amount of diffraction is large enough that the notion of a "shadow" becomes meaningless. For example, for sound at 264 Hz (the frequency of middle C) the wavelength $\lambda = 1.3$ m is comparable to room dimensions, so that a voice around the corner of a building in a large open space, although fainter, is still audible. If $l \gg \lambda$, diffraction is negligible and the shadow of the obstacle is relatively sharp; with the appropriate transmitting and receiving apparatus, such waves can be used to determine the geometry of the obstacle.

example 18 • 6 The great pyramid of Khufu at Gizeh, Egypt, has an altitude of approximately 150 m. What range of sound frequencies would cast a fairly sharp acoustic shadow?

solution The sharpest acoustic shadow would be cast by those frequencies for which $\lambda = v/\nu \ll 150$ m; that is,

$$\nu \gg \frac{340 \text{ m/s}}{150 \text{ m}} = 2.3 \text{ Hz}$$

Any audible sound will do (we have chosen approximately 100°F as the temperature). To detect a large tree about 1 m in diameter, frequencies much higher than 340 Hz would be needed.

The classic diffraction experiment is Young's two-slit "interference" experiment, in which a light wave passes simultaneously through two very narrow slits (of width much less than λ), separated by a distance d, as shown in Fig. 18 . 25. The pattern of displacement is formed on the other side by the Huygens wavelets spreading out from each slit. At some angle θ the wavelets that form the wavefront at AB are out of phase by an amount

$$\frac{\Delta s}{\lambda} = \frac{d}{\lambda} \sin \theta \text{ cycles} \qquad [18 \cdot 44]$$

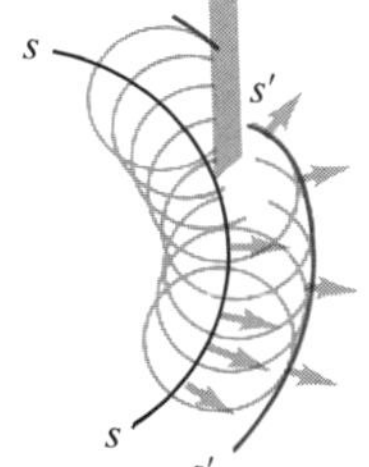

figure 18 • 24 *Huygens' construction of a diffracted wavefront, showing how the wavefront s's' curls in behind the obstacle*

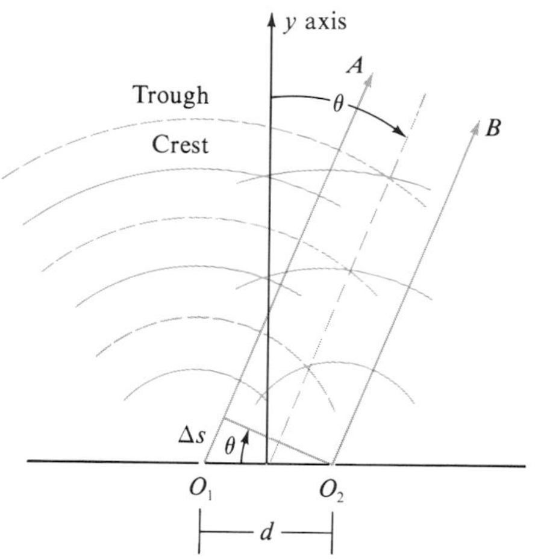

figure 18 • 25 *(a) Photograph of interference pattern formed in a water tank by superposition of circular waves from two point sources. (b) Young's interference experiment using plane water waves falling upon two slits separated by a distance d. Nodal lines occur at angles* $\theta = \sin^{-1}(n\lambda/d)$ *for n = 1, 2, ...*

The displacement will be a maximum for angles θ_m such that the phase difference is some integral number of cycles,

$$\sin \theta_m = m \frac{\lambda}{d} \qquad m = 0, 1, 2, \ldots \qquad [18 \cdot 45]$$

and a minimum for angles θ_n such that the phase difference is an odd number of half-cycles,

$$\sin \theta_n = \frac{2n - 1}{2} \frac{\lambda}{d} \qquad n = 1, 2, 3, \ldots \qquad [18 \cdot 46]$$

where θ_n gives the direction of the *nodal lines*. Note that as the distance d between the slits decreases, the value of θ increases. The integer m or n is the

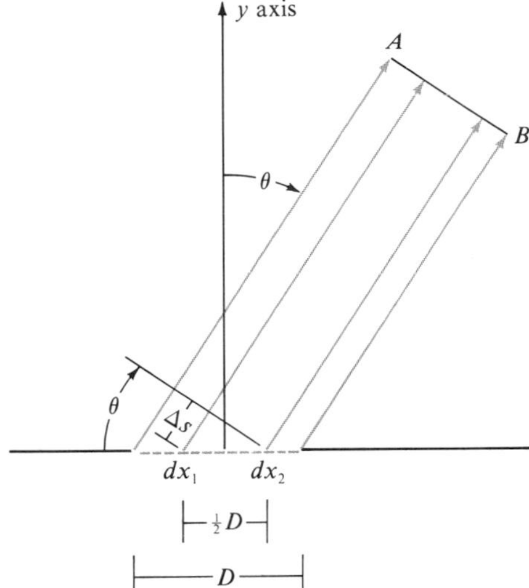

figure 18 • 26 *Single-slit diffraction*

order of a given extremum, with the number of possible orders limited by the fact that $\theta < \frac{1}{2}\pi$.

The same effect is reproducible with two independent coherent waves in phase at O_1 and O_2. It is therefore usually classified as an interference effect, implying two independent wave trains, rather than as a diffraction effect, in which parts of the same wavefront interfere with each other as they propagate. The distinction is largely academic, since Huygens' principle treats each point along a wavefront as the source of an independent wavelet in any event. A less ambiguous example of diffraction is the passage of a wave through a *single* slit of width D, which we can consider as a continuous sequence of "slitlets" of width dx. We divide the slit into left and right halves, as shown in Fig. 18 . 26, and group the slitlets in pairs dx_1, dx_2 separated by $\frac{1}{2}D$.

For wavelets propagating at an angle θ, there is a difference in path length of

$$\Delta s = \tfrac{1}{2}D \sin\theta \qquad [18 \cdot 47]$$

and if $\Delta s = \frac{1}{2}\lambda$, the paired wavelets arrive out of phase and exactly cancel. This means that there is a nodal line along the direction of θ such that

$$\sin\theta = \frac{\lambda}{D} \qquad [18 \cdot 48]$$

with the disturbance from each element in the upper half of the slit canceling that from its corresponding element in the lower half.

Clearly, as θ increases, $\Delta s/\lambda$ goes through a succession of values such that wavefronts at different angles have alternating maxima and minima. It seems obvious that if $\Delta s = \lambda$, we should obtain a maximum of intensity in the direction $\theta = \sin^{-1}(2\lambda/D)$. However, this is *not* the case, because at such a large angle the slitlets within a half-slit can interfere destructively with each other. That is, if the upper half-slit is considered to be a single slit of width $\frac{1}{2}D$, it will have a nodal line in the direction

$$\theta = \sin^{-1}\frac{\lambda}{\frac{1}{2}D} = \sin^{-1}\frac{2\lambda}{D}$$

This type of reasoning leads to the general formula for the directions of nodal lines:

$$\sin\theta_n = \frac{n\lambda}{D} \qquad [18 \cdot 49]$$

If D decreases, the angular spread of a given order of diffraction increases, whereas for $D \gg \lambda$ the angular spread is quite small and many orders are present. Although we shall not prove it here, the intensity of the diffracted wave between successive single-slit minima drops off as $1/n^2$ for higher and higher orders, so that the shadow of such a slit where $D \gg \lambda$ would appear quite sharp.

Although Eq. [18 . 49] resembles [18 . 45], the former is for *minima* and latter is for *maxima,* and a conscious effort is required to keep the two distinct in your mind. The maxima of the single-slit diffraction pattern occur approximately, but not quite, halfway between the minima, in contrast to the regularity of the two-slit interference pattern. In these derivations it has been tacitly assumed that the observer is located at a distance much greater than D, so that the distance between two portions of a diffracted wavefront is negligible.

example 18 • 7 If the first-order minimum for deep-water waves diffracted by a single slit 1 cm wide occurs at $\theta = \frac{1}{6}\pi$, what is the frequency of the waves?

solution From Eq. [18 . 48], the wavelength is $\lambda = \frac{1}{2}D = 0.5$ cm, and from Eq. [17 . 15], the wave speed is $v = \sqrt{g\lambda/2\pi}$. Therefore the frequency is

$$\nu = \frac{v}{\lambda} = \sqrt{\frac{g}{2\pi\lambda}} = 17.7 \text{ Hz}$$

PROBLEMS

18 • 1 superposition

18 • 1 A particle at the origin of coordinates is acted on by two forces, one in the x direction and the other in the y direction. The first force, acting alone, would cause the particle to oscillate with a displacement $\delta x = r\cos\omega t$, and the second force, acting alone, would cause it to oscillate with a displacement $\delta y = r\sin\omega t$. What is the net motion of the particle under the influence of both forces?

18 • 2 A particle is oscillating along the x axis under the influence of two forces, both acting along the x axis. Its net motion can be represented as

$$\delta x = 3 \sin \omega_1 t \sin \omega_2 t$$

where $\omega_1 = 2$ rad/s and $\omega_2 = 100$ rad/s. (*a*) What are the amplitudes of the component oscillations? (*b*) What are their frequencies? (*c*) What is the initial phase difference between the oscillations?

18 • 3 A wave of amplitude 1 cm and phase $\omega t - x/\lambda$ is superposed on a wave of amplitude 2 cm and phase $\omega t - x/\lambda - \frac{1}{3}\pi$. (*a*) What are the amplitude and phase angle of the compound traveling wave? (*b*) If a third wave of amplitude 3 cm and phase $\omega t - x/\lambda + \frac{1}{3}\pi$ is superposed on the first two, what are the amplitude and phase angle of the resultant compound?

18 • 4 Two identical sound sources are located on the x axis at (0, 0) and (D, 0). They emit sounds in phase at the same wavelength λ. (*a*) At what points on the positive y axis do the minima of intensity occur? (*b*) For what range of separation values D will no minima occur?

18 • 5 In Prob. 18 . 4, what is the maximum number n of intensity minima on the y axis for a given value of D?

18 • 6 If two musicians are playing slightly out of tune, so that 4 beats per second occur at $\nu = 256$ Hz, how fast must you travel from one musician toward the other in order to hear no noticeable beats? Assume standard temperature and pressure conditions.

18 • 7 Derive Eqs. [18 . 13] and [18 . 14] in detail and show that they are consistent with the results in the case of equal amplitudes but unequal phases.

18 • 8 Use the reference circle for simple harmonic motion (see Prob. 16 . 7) and the accompanying figure to derive Eqs. [18 . 13] and [18 . 14] from geometric considerations.

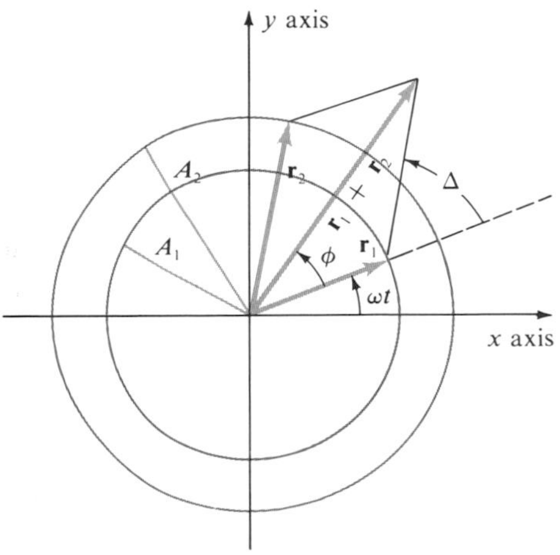

18 • 9 Three vibrators are located at corners of a square with sides of length $l = 3\lambda$. Each vibrator produces water waves of the same amplitude, angular frequency, and phase. What is the total disturbance at the fourth corner of the square?

18 • 10 Two trains are approaching you from the same direction, both traveling at a speed of 129 km/h. Their whistles produce 10 beats per second (sound speed, 1223 km/h), and after the first train passes the beats disappear. (*a*) What are the frequencies of the whistles? (*b*) How many beats per second will you hear after both trains have passed?

****18 • 11*** Draw a flowchart for a computer program that will compute the amplitude and phase of a compound wave at any time t and position x, given the speeds of propagation, amplitudes, frequencies, and phases of n component waves at some specified time t_0 and position x_0. Check the results of your program against the answers to some of the preceding problems. (*Hint:* Mathematical sophistication is not required; just give the computer adequate instructions to perform the drudgery.)

18 • 2 *standing waves and Lissajous figures*

18 • 12 If Eq. [18 . 16] represents a longitudinal standing wave δx, what is the formula for the correspoding pressure wave δP? Find the positions of the pressure nodes and antinodes and compare them to the positions of the displacement nodes and antinodes.

18 • 13 Derive expressions for the instantaneous kinetic and potential energies of a particle in a standing wave.

18 • 14 An acoustic standing wave in air ($\rho = 1.225$ kg/m^3) is observed to have nodes 34 cm apart, and all displacements vanish every millisecond. The energy density of the standing wave is $E_v = 10^{-8}$ J/m^3. (*a*) Find the amplitudes of the component traveling waves. (*b*) Find the speed of sound in air.

18 • 15 The sum of two oppositely directed waves of the same frequency and initial phase at $x = 0$ has a maximum amplitude A_{max} and a standing-wave ratio S. (*a*) Find the amplitudes A_1 and A_2 of the two component waves, where $A_1 > A_2$. (*b*) If the phase constant is found to be greater than 45° at a distance of $\frac{1}{8}\lambda$ to the right of the origin, what is the direction in which the wave of larger amplitude is propagating?

18 • 16 Two identical sound sources are located 9 m apart. Walking along the line from one to the other, an observer notices that the sound is inaudible at two points along the path. What are the wavelength and frequency of the sounds?

18 • 17 The motion of the particle in a Lissajous figure ellipse, $\omega_x = \omega_y = \omega$, can be either clockwise or counterclockwise, depending on which component leads in phase. (*a*) If the angular position of the particle is $\theta(t) = \tan^{-1}(y/x)$, show that

$$\frac{d\theta}{dt} = -\frac{\omega A_x A_y \sin\phi}{x^2 + y^2}$$

so that the *sign* of $d\theta/dt$ depends entirely on the value of ϕ. (*b*) By examining the values of dy/dx and dx/dy, prove that the ellipse is contained within a rectangle of width $2A_x$ and height $2A_y$.

18 • 18 An oscilloscope is adjusted so that the x and y components of the Lissajous curve on the screen have equal amplitudes, frequencies in the ratio $\omega_y/\omega_x = 2$, and the same initial phase ($\phi_0 = 0$). (*a*) Show that $y/x = 2 \cos \omega_x t$ and $\cos \omega_x t = \sqrt{A^2 - x^2}/A$, and therefore $y^2 = 4x^2 (A^2 - x^2)/A^2$. (*b*) Find x when $y = 0$ and when $y = \pm A$. (*c*) Make a rough sketch of the resulting Lissajous figure.

18 • 3–18 • 4 *reflection of waves*

18 • 19 A wave is incident upon one of the legs of an isosceles right triangle, (45°–45°–90°), the hypotenuse being open or transparent to incident waves. Show that no matter what the angle of incidence, the wave ultimately returns parallel to the incident wave if it is reflected successively at both legs of the triangle. (*Hint:* Begin with a sketch.)

18 • 20 A loudspeaker emitting sound at a frequency ν_0 moves away from you with speed v_s toward a large wall. (*a*) If the speed of sound is $v \gg v_s$, will you detect any beats, and if so, how many? (*b*) What if the loudspeaker turns around and starts coming toward you?

18 • 21 New York and London are about 5660 km apart by the great circle route over the earth's surface. Shortwave broadcasts can be transmitted from one city to the other by successively "bouncing" electromagnetic radio waves off the surface of the earth and a layer of charged particles, the ionosphere, which is like a spherical shell about the earth at an altitude of 250 km. (*a*) What angle must the transmitted wave make with the horizon in order to be reflected to earth precisely 5660 km away? (*b*) How many times will the wave bounce off the ionosphere? Take the earth's radius to be 6400 km.

18 • 22 Show that whether a sound originates in air and passes into water or vice-versa, the ratio of transmitted to incident energy has the same value and $E_\tau/E < 0.001$ under standard conditions.

18 • 23 A transverse wave passes down a uniform metal wire under tension. If, at some point, the radius of the wire suddenly increases by a factor of 2, what are the ratios r and τ of the reflected and transmitted wave amplitudes to that of the original wave? Show that Eqs. [18 . 33] are consistent with reflection and transmission of a transverse wave at a fixed or free end of a string as explained in the text.

18 • 24 What are the allowed wavelengths in a pipe that is closed at *both* ends? Explain your reasoning.

18 • 25 How many beats per second are produced by two organ pipes of length l, one closed at both ends and the other closed at one end, if the beats occur between the third harmonics of both pipes?

18 • 26 How much must an organ pipe whose fundamental tone is 273 Hz at 0°C be heated to change the frequency of the tone by 8 Hz?

18 • 27 A whistle ordinarily blown with air is blown with hydrogen of density 0.0692 relative to air at the same temperature and pressure. What change does this produce in the sound?

18 • 28 A brass rod 200 cm long, when stroked longitudinally, is in tune with a 25-cm length of a given wire. A steel rod 300 cm long, when stroked longitudinally, is in tune with a 26-cm length of the same wire. What are the relative speeds of sound in steel and brass.

18 • 29 Notes of frequencies 225 Hz and 336 Hz are sounded simultaneously. Only the even overtones are present in the first note, and both odd and even overtones are present in the second note. How many beats per second occur, and to what overtones are they due?

18 • 30 A mild steel string 0.1 mm in radius is stretched over stops 30 cm apart by a force of 100 kgf. (*a*) If the specific gravity of the string is 8.0, what is the frequency of its fundamental? (*b*) If the tension is increased so that the total length of the string increases by 0.1 percent, what is the relative change in the fundamental frequency? Can we neglect the change in linear density?

18 • 5 Huygens' principle

18 • 31 Repeat Huygens' construction for the passage of a wave from a denser to a rarer medium and show that Snell's law still holds.

18 • 32 When shallow-water waves are incident upon deeper water at an angle of 60°, what minimum percentage change in depth will be needed to cause total reflection?

18 • 33 A wave passes through three successive mediums at speeds v, v', and v'' and angles of incidence θ, θ', and θ''. How are θ and θ'' related?

18 • 34 A brass cube ($\rho = 9\ \text{g/cm}^3$), of side length $l = 50$ cm, rests on a similar cube of aluminum ($\rho = 2.7\ \text{g/cm}^3$). If the brass cube is struck with a hammer on the center of its upper face, what is the position of the center of the spherical compression wave propagating through the aluminum block, as viewed near the vertical axis?

18 • 6 diffraction

18 • 35 Sound of frequency $\nu = 302$ Hz in air under standard conditions passes through two narrow slits 5 m apart. Compute the angular positions of the highest-order maximum and the highest-order minimum in the resulting diffraction pattern.

18 • 36 In Example 18 . 7, how far apart would *two* such slits have to be located to produce a fourth-order *maximum* at $\theta = 30°$? What would be the effect of superimposing the two-slit and single-slit diffraction patterns at this angle?

18 • 37 By adding together the sinusoidal waveforms of amplitude A, frequency ω, and speed v propagating outward as circles from the two slits of Fig. 18 . 25, show that the time-average intensity of the diffraction part should be proportional to $\cos^2(\frac{1}{2}kd\sin\theta)$, where $k = 2\pi/\lambda$.

18 • 38 The barrier in Fig. 18 . 26 may be thought of as a "mask" or "filter" which prevents secondary wavelets from propagating forward from the incoming wavefront, except for those portions "striking" the slits. We can therefore view the diffraction pattern as the *difference* between the original wavefront and the wavefront that would have been produced by the secondary wavelets in the absence of the barrier. Use this principle to describe the diffraction effects to be expected if the two-slit barrier is replaced by two thin posts, masking only the portions of the wavefront that had previously passed through the slits.

answers

18 • 1 Uniform rotation in a circle of radius r, with angular velocity ω

18 • 2 (*a*) $A = 1.5$ for each; (*b*) $\omega = 98$ rad/s, $\omega' = 102$ rad/s; (*c*) $\Delta = \pm\pi$

18 • 3 (*a*) $A = \sqrt{7}$ cm, $\theta = -40.9°$; (*b*) $A = \sqrt{13}$ cm, $\theta = 13.9°$

18 • 4 (*a*) $y_n = (D^2 - \frac{1}{4}n^2\lambda^2)/n\lambda$, for odd n; (*b*) $D < \frac{1}{2}\lambda$

18 • 5 largest integer $n < \frac{1}{2} + D/\lambda$

18 • 6 $v = 2.6$ m/s

18 • 9 $\delta y = 2.277A \sin(\omega t - 26.0°)$

18 • 10 (*a*) $\nu_1 = 50$ Hz, $\nu_2 = 41$ Hz; (*b*) 8 beats/s

18 • 12 $\delta P = (4\pi\rho v^2 A/\lambda) \sin(2\pi x/\lambda) \sin(2\pi t/T)$

18 • 13 $K = E \cos^2 \omega t$, $U = E \sin^2 \omega t$

18 • 14 (*a*) $A = 28.8$ nm; (*b*) $v = 340$ cm/s

18 • 15 (*a*) $A_1 = \frac{1}{2}A_{max}(1 + 1/S)$, $A_2 = \frac{1}{2}A_{max}(1 - 1/S)$; (*b*) to the left

18 • 16 $\lambda = 6$ m, $\nu = 56.7$ Hz

18 • 18 (*b*) $x = 0, \pm A$ when $y = 0$, and $x = \pm A/\sqrt{2}$ when $y = \pm A$

18 • 20 (*a*) $\nu_b \doteq 2\nu_0 v_s/v$; (*b*) same as part (*a*)

18 • 21 (*a*) angle $= 3.55°$; (*b*) twice

18 • 23 $r = 1/3$, $\tau = 2/3$

18 • 24 $\lambda_n = 2l/n$, for $n = 1, 2, 3, \ldots$

18 • 25 $v/4l$ beats/s

18 • 26 16.2°C

18 • 27 $\nu_H = 3.8\nu_{air}$

18 • 28 $v_s/v_b = 1.44$

18 • 29 6 beats/s, $\nu_1 = 6 \times 225 = 1350$ Hz, $\nu_2 = 4 \times 336 = 1344$ Hz

18 • 30 (*a*) $\nu_1 = 3291$ Hz, (*b*) $\Delta\nu_1/\nu_1 =$ 0.29 percent, neglecting ρ_l causes an error of 15% in this result

18 • 32 $\Delta h/h = 3.3$ percent

18 • 33 $v \sin\theta'' = v'' \sin\theta$

18 • 34 33 cm above center of upper face of aluminum block

18 • 35 $m = 4$, $\theta = 61.6°$; $n = 5$, $\theta = 81.7°$

18 • 36 4 cm apart, but a minimum still exists at 30° because of the single-slit effect

18 • 38 Maxima and minima of the diffraction pattern are exactly interchanged in comparison with those for two-slit diffraction

CHAPTER NINETEEN

relativity

The views of space and time which I wish to lay before you have sprung from the soil by experimental physics, and therein lies their strength. They are radical. Henceforth space by itself, and time by itself, are doomed to fade away into mere shadows, and only a kind of union of the two will preserve an independent reality.

H. Minkowski, Space and Time, *1908*

The most beautiful fate of a physical theory is to point the way to the establishment of a more inclusive theory, in which it lives on as a limiting case.

Albert Einstein, 1916

During the last two decades of the nineteenth century and the first few years of the twentieth, several sets of phenomena came under investigation that could not be satisfactorily interpreted in terms of existing physical theories. James Clerk Maxwell predicted in 1865, and Heinrich Hertz demonstrated in 1887, that light is an electromagnetic wave. Maxwell and most other nineteenth-century scientists believed that light must propagate as a transverse wave through some medium, which they called the *ether.* The notion of a single, all-pervasive medium permeating the universe may have originated with René Descartes, who suggested in 1638 that light was a pressure transmitted from the visible object to the observer's eyes by means of the ether. In fact, although light has no mass, it does possess a property resembling a classical momentum.

Despite some successes of the theory, the "imperceptible and invisible ether" proved elusive and evaded detection. Various attempts to demonstrate its existence came to naught, and in 1887 the Americans Albert Michelson and Edward Morley also failed to find it in an experiment designed to measure changes in the speed of light due to the earth's supposed motion through the ether. In the process, however, they discovered that the speed of light was not measurably different in any particular direction, *regardless of the earth's motion through space.* The search for the ether never yielded anything but negative results, and theoretical attempts to explain Michelson's results by assuming that the earth dragged some ether around with it in its motions proved similarly unsuccessful.

It was not until 1905 that Albert Einstein published his *special theory of relativity,* which showed that these contradictions could be resolved by introducing two basic postulates into all physical theories and carefully analyzing the operational definitions corresponding to the theoretical concepts of time and length. Like Galilean, or Newtonian, relativity, Einstein's theory explains the effect of the uniform motion of an inertial frame of reference (a nonaccelerating system in which the law of inertia holds) on the measurement of physical phenomena by an observer situated in that system.*

In discussing relativity we must distinguish carefully between a phenomenon as seen from a system in which it is at rest and the phenomenon as it appears to an observer moving relative to it. By *appearance* we do not mean an optical illusion, but rather the physical properties of the phenomenon as determined by *any* experiment or test whereby we interact with it. Einstein's special theory of relativity enables us to delineate and understand the effect of an observer's uniform, rectilinear motion on his perception—in the broadest possible sense—of physical reality.

In this chapter we shall derive some of the more elementary, but rather startling, consequences of Einstein's theory, including the reformulation of our concepts of mass, length, time, and momentum. The key to this derivation is the *Lorentz transformation,* which relates temporal and spatial coordinates of one inertial system to those of another moving uniformly relative to it. This transformation and its direct consequences—time dilation, length contraction, and the relativity of simultaneous events—are analogous to our discussion of classical kinematics in Chapters 3 and 4. We shall also discuss the dynamic consequences of the Lorentz transformations (analogous to Chapters 5, 6, and 7) for our definitions of mass, momentum, force, and energy. This brings us to Einstein's famous formula for *mass-energy equivalence,* $E = mc^2$, where c is the speed of light. As we shall see, however, all the kinematic and dynamic concepts of Newtonian mechanics represent valid approximations to relativistic quantities in the limit of speeds

*In 1916 Einstein extended the theory of relativity to include noninertial systems. The great success of the *general theory of relativity,* as it is called, has been to provide a theory of gravitation. Although now generally accepted, relativity theory was condemned in the 1930s by Hitler's Germany as "Jewish physics" and by Stalin's Russia as "bourgeois idealism." However, nature, it would appear, is no respecter of ideologies.

much less than that of light. Thus Newtonian mechanics is, in fact, a special case of relativistic mechanics.

The theory of relativity, by its very nature, cannot be proved absolutely by any single experiment. However, no other reasonable alternative has been able to account for the contradictions in Newtonian mechanics, although many have been attempted. Part of the importance of relativity is as an ultimate test of the validity of any proposed theory. Relativity asks such questions as: Is this result invalid in any other inertial system? Does the formula look fundamentally different in any other inertial system? Does it contradict the constancy of the speed of light in free space in any other inertial system? Does it imply any way of determining a unique inertial system in the universe? If the answer to any of these questions is yes, then the theory is relativistically incorrect or, at best, only an approximation to the correct theory.

The theory of relativity has had a powerful influence on twentieth-century thinking, comparable to the impact of Newtonian mechanics on the eighteenth-century Enlightenment. By demonstrating that the universe is not what we perceive it to be, and that no concept, however "intuitively obvious," is immune to scrutiny, the theory of relativity opened wide the doors to bold new concepts in science and philosophy. At the same time, it has reminded us of the conditional nature of everything we know and given us a fresh perspective on the fragility of man's tenure in a vast universe in which no system is absolute.

19 • 1 the Michelson-Morley experiment

Light is a disturbance consisting of interrelated fluctuations in electric and magnetic fields in space. Light propagates like a wave, and in many respects behaves like a wave, exhibiting the full range of interference phenomena: superposition, reflection, refraction, and diffraction. Yet in some situations it also behaves like a particle, exhibiting momentum, energy, and a high degree of localization in space. Until the nineteenth century, Newton's *corpuscular theory* of light assumed it to consist of many small, rapidly moving particles, primarily because light did not seem to exhibit diffraction, but traveled in straight lines. Newtonian mechanics was quite capable of explaining the phenomena of reflection and refraction on the basis of elastic collisions or normal forces at the boundary between two different media. However, in 1802 Thomas Young's two-slit interference experiment definitely established that light was a wave, but with such small wavelengths (in the range of 0.4 to 0.7 μm) that diffraction effects had not previously been measurable. In 1819, following an earlier suggestion by Young, Augustin Fresnel proved that light consists of transverse oscillations—but oscillations in *what?*

The question of exactly what medium was oscillating turned out to be the crucial issue which led to the theory of relativity. Scientists raised in a Newtonian tradition believed that there must be some medium to sustain these waves, the *ether,* which permeated all of space and the objects within it. Many and varied were the attempts to demonstrate the presence of the ether until Michelson and Morley, in 1887, showed conclusively that its existence could not be demonstrated.

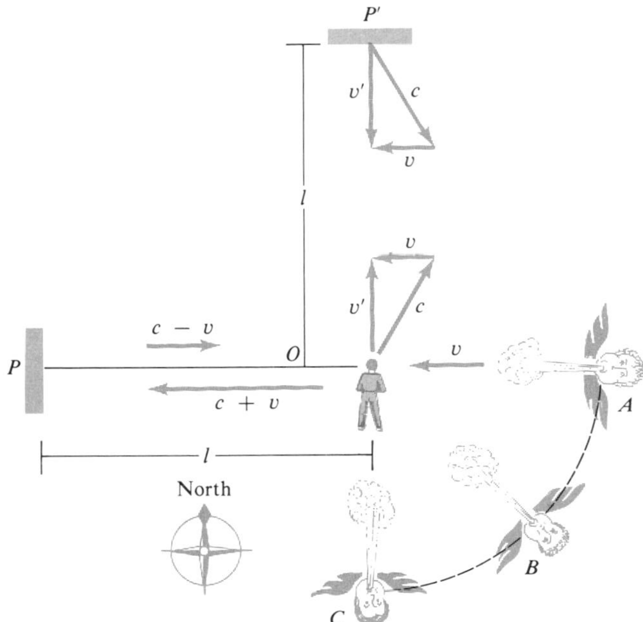

figure 19 • 1 *Propagation of sound from a source O to reflectors at P and P′ and back to O when the wind is blowing from the east (A). Phase difference between returning waves at O is Δθ. When wind is from the southeast (B), the phase difference vanishes. When wind is from the south (C), the phase difference is −Δθ.*

To illustrate the Michelson-Morley experiment, let us construct an analogous experiment with sound waves in air. Imagine yourself situated at a sound source in Fig. 19 . 1 which emits sound waves of a given frequency equally in all directions. These waves bounce off reflectors at P and P' and return to O. In still air symmetry requires that the reflected waves returning along paths $\overline{OP}$ and $\overline{OP'}$ of equal length l will arrive at O in phase and interfere constructively to give a maximum energy density four times greater than the original wave.

But what if a wind of velocity $\mathbf{v}$ is blowing directly toward you from A? The disturbance, once created, moves with propagation speed c *relative to the medium*. Thus a sound wave travels from O to P with speed $c + v$, but it returns with speed $c - v$. A sound wave emitted at time $t = 0$ will therefore travel to P and back in an elapsed time

$$t = l\left(\frac{1}{c + v} + \frac{1}{c - v}\right) = \frac{2l/c}{1 - v^2/c^2} \qquad [19 \cdot 1]$$

Things are a bit more complicated for the wave reflected from P'. To begin with, sound emitted in the direction of P' will drift westward with the windspeed v as it propagates through the moving medium. The portion of the spherical wavefront that reaches P' is the part emitted toward the northeast, so that its velocity has an eastward component which exactly cancels the windspeed. Hence the effective speed from O to P' is

$$v' = \sqrt{c^2 - v^2} \qquad [19 \cdot 2]$$

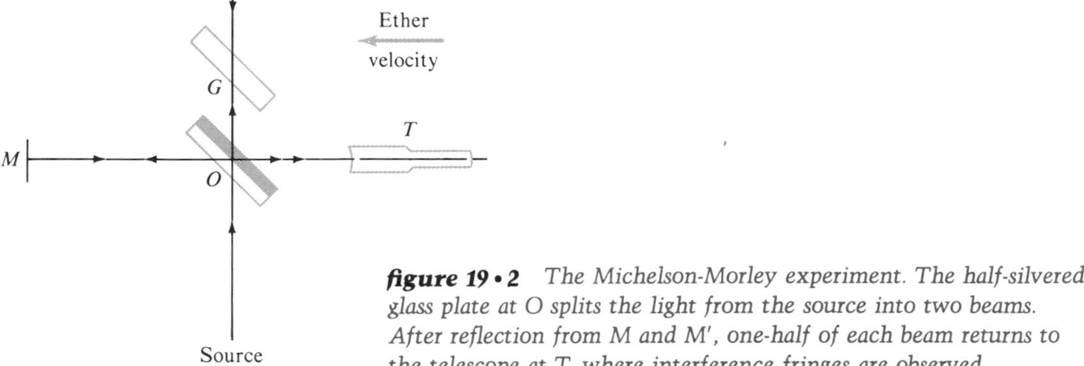

figure 19 • 2 *The Michelson-Morley experiment. The half-silvered glass plate at O splits the light from the source into two beams. After reflection from M and M′, one-half of each beam returns to the telescope at T, where interference fringes are observed.*

But what of the wave reflected back from P'? If it has approached P' along the line from O to P', will it not reflect directly back along the same line and thus be swept westward with the wind, missing O completely?

To answer this question imagine yourself moving with the medium—in this case, the wind, moving at velocity $\mathbf{v}$. *Relative to the medium,* the wave is moving at the angle at which it was originally emitted. In the Huygens construction used to derive the law of reflection, we assumed the medium to be at rest. The fact that the smooth surface of the reflector may be in motion with respect to the medium has no effect on the derivation. Hence the returning wave from P' will be reflected at the angle indicated by the vector c at P'. Since the wave also drifts westward with windspeed v, it returns to O as if it were obeying the law of reflection for normal incidence in the observer's system. The speed of approach is v', so that the time of travel from O to P' and back is

$$t' = \frac{2l}{v'} = \frac{2l/c}{\sqrt{1 - v^2/c^2}} \qquad [19 \cdot 3]$$

When the two reflected waves arrive back at O, the difference in their travel times, $t - t'$, will cause a phase difference

$$\Delta\theta = \omega(t - t') = \frac{4\pi l}{\lambda}\left(\frac{1}{1 - v^2/c^2} - \frac{1}{\sqrt{1 - v^2/c^2}}\right) \qquad [19 \cdot 4]$$

If $\Delta\theta = \pi$, the two waves will interfere destructively and no sound will be heard. In that case, however, look what happens when the wind shifts. When the wind is from the southeast at B, symmetry requires that there be no phase difference between the reflected waves returning to O, and the sound is again at a maximum. When the wind is due south at C, the phase difference is exactly reversed, $\Delta\theta = -\pi$, and there is again destructive interference. Since the length l is at our disposal, this implies that, for the case of sound, the presence of the medium can be detected from interference effects *if it is moving relative to the observer.* All we have to do is rotate the entire experiment through 90° and observe any changes in the intensity of the reflected waves at O. Conversely, the absence of any effect can be taken to mean either that the medium is always stationary with respect to the observer (we only need two observers who are moving relative to each other to demonstrate the absurdity of this notion) or else that the velocity is independent of the velocity of the medium. This means that the medium cannot be anything analogous to water or air—and hence that the concept of a medium may not be useful.

The apparatus used by Michelson to attempt to detect the ether, known as a *Michelson interferometer,* is shown schematically in Fig. 19 . 2. Light from a source strikes at O a "half-silvered" mirror—one with a thin coating of silver that causes it to reflect half the incident light to the mirror at M and to transmit the other half to the mirror at M'. Reflected light then returns to O, where half the beam from each mirror is reflected into the telescope at T. With the glass plate of the same thickness and composition at O and G, both beams undergo the same conditions during their travels and should arrive at the telescope in phase—*unless* their speeds relative to an all-pervading ether are different.

In such an experiment it is practically impossible to achieve the required accuracy. For example, the mirrors cannot be exactly the same distance apart to within a wavelength of light (about 0.55 μm), nor can they be at angles of exactly 90°. What Michelson actually saw was a pattern of light and dark regions of constructive and destructive interference called *fringes,* shown in Fig. 19 . 3. He reasoned that if the entire apparatus were rotated relative to the earth's motion, in an ether which he presumed to be at rest in the solar system, the positions of the fringes would gradually change, as if they were actually moving across his field of vision. The number n of dark fringes moving past a given point would then correspond to the number of increments of phase difference $\Delta\theta = 2\pi$ between the reflected beams. Since the earth's orbital speed about the sun is about 30 km/s, the ratio of this speed to the speed of light is $v/c = 10^{-4}$, so that we can apply the approximation $\sqrt{1 - v^2/c^2} \doteq 1 - \frac{1}{2}v^2/c^2$ to evaluate n:

$$n = \frac{2\,\Delta\theta}{2\pi} \doteq \frac{4l}{\lambda}\left(1 + \frac{v^2}{c^2} - 1 - \frac{\frac{1}{2}v^2}{c^2}\right) = \frac{2l}{\lambda}\frac{v^2}{c^2} \qquad [19 \cdot 5]$$

For Michelson's experiment $l = 11$ m, so that the presence of the ether should yield $n = 0.4$. Michelson's interferometer was capable of detecting shifts one-third as large. However, in what was probably the most famous and important negative result ever obtained, nothing was found to indicate the presence of the ether. In 1930 Joos repeated the Michelson-Morley experiment with an accuracy 200 times as great and still detected no ether. What the experiment did show, however, was that there was no detectable difference in the speed of light due to the *motion of the observer.* This discovery led directly to one of the fundamental postulates of relativity:

The speed of light is the same in every inertial system.

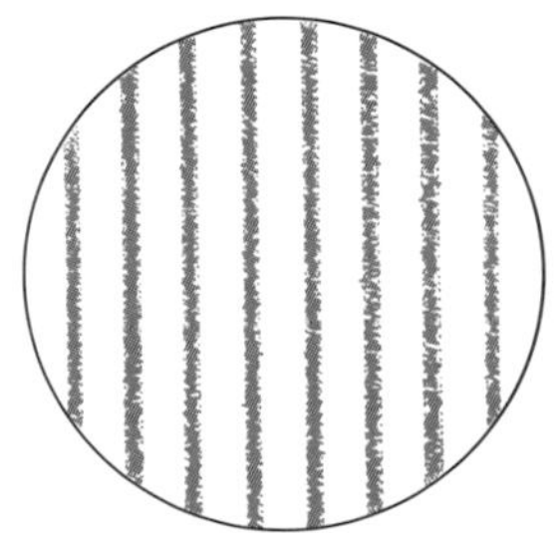

figure 19 • 3 *Interference pattern of fringes in the Michelson-Morley experiment*

19 . 2 the postulates of special relativity

The speed of light c plays a fundamental role in the operational definitions of kinematic quantities. For example, two observers stationed at different points in space must have some way of synchronizing their clocks. This can best be accomplished by exchanging light pulses, and it becomes crucial to know the velocity of these pulses.

It had long been known that the speed of light is large, but finite. At present, the best value of the speed of light, or of any other electromagnetic wave, traveling *in vacuo*, or free space, is

$$c = (2.99792458 \pm 0.00000001) \times 10^8 \text{ m/s}$$

or about 186,300 mi/s. This is one of the most important constants of physics, for no particle has ever been observed to travel faster.

Throughout the development of classical mechanics, it had proved satisfactory to assume that the speed of light was practically infinite. However, problems began to arise in the interpretation of experiments which involved bodies moving with velocities approaching the speed of light, as well as experiments dealing with the propagation of electromagnetic radiation. To illustrate one of the puzzles encountered, imagine that a source of light is fixed at the origin O of an inertial (nonaccelerating) reference frame S, and that another inertial frame S' is moving past the light source with a constant velocity $\mathbf{V}$ (see Fig. 19 . 4). Suppose that at time $t = 0$, when the origins O and O' coincide, the source emits a flash of light. Observers in the S frame will find that the light moves with the same constant speed c in all directions, spreading out from O in a sphere with an increasing radius

$$r = \sqrt{x^2 + y^2 + z^2} = ct$$

According to the Galilean transformation discussed in Sec. 4 . 5,

$$x' = x - Vt \qquad y' = y \qquad z' = z \qquad t' = t \tag{19 . 6}$$

the observer in the S frame would predict that, to an observer moving with the S' frame, the speed of light would appear to be $c - V$ at P_1, $c + V$ at P_2, and $\sqrt{c^2 - V^2}$ at P_3. The S observer would therefore conclude that the

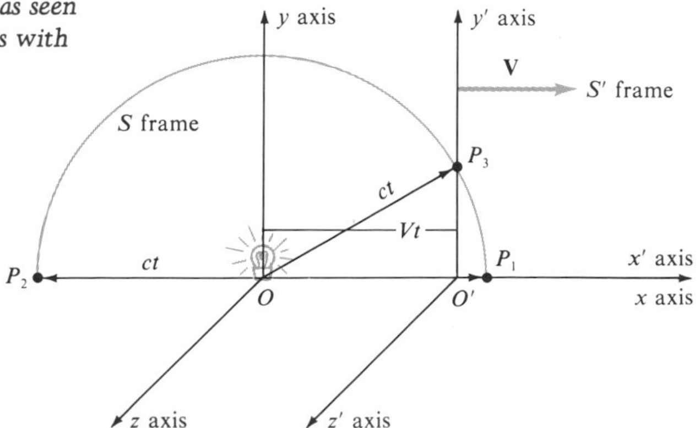

figure 19 . 4 *Propagation of a light pulse as seen in frame of reference S. The S′ frame moves with speed V relative to the S frame.*

speed of light is not an invariant quantity. However, all experimental evidence shows that c is the same whether it is measured at P_1, P_2, or P_3 and no matter what inertial reference system is used. On this basis there seems to be a contradiction between the prediction of theory and the experimental results.

To explain such results, various investigators proposed hypotheses that accounted for a particular phenomenon, rather than for a group of related phenomena. It was Albert Einstein who resolved the paradox by daring to question the validity of the Newtonian theory, sanctified by two centuries of impressive successes in describing the physical universe. Einstein's proposed solution, in the form of a new theory of relativity, was the simplest and most comprehensive of any that had been advanced—and also the most revolutionary.

Einstein based his new conception of relativity on two main postulates. The first of these postulates is known as the *principle of relativity:*

> The basic laws of all physical phenomena, when properly formulated, are the same in all inertial frames of reference. As a consequence of this postulate, an observer who confines his observations to any single inertial frame cannot detect the motion of that frame. That is, no unique absolute or preferred frame of reference exists.
>
> The basic physical laws must be in a form such that they predict the same constant value c for the speed of light at every point in space in all inertial frames of reference, regardless of whether the light source is stationary or moving and regardless of the direction in which the light is traveling.

The theory based on these postulates is called the *special* theory of relativity because, like Newtonian relativity, it is restricted to inertial reference frames and to sets of rigid coordinate axes moving relative to one another with constant linear velocity. Today the word *relativity,* when unqualified, is commonly used in the physical sciences to refer to Einsteinian, not Newtonian, relativity.

The first postulate concerns the invariance of *form* of physical definitions and laws when they are transformed from one inertial system to another. For example, consider a rigid rod of length l_0, measured while it is at rest in the S system. If the endpoints of this rod are located at (x_1,y_1,z_1) and (x_2,y_2,z_2), as shown in Fig. 19 . 5, then its length is defined as

$$l = l_0 = \sqrt{(x_2 - x_1)^2 + (y_2 - y_1)^2 + (z_2 - z_1)^2} \qquad [19 \cdot 7]$$

The value of l_0 is called the *proper length* of the rod, measured when it is at rest in the reference frame with respect to which the measurement is made. If the endpoints of this rod are measured *simultaneously* by an observer in the moving S' frame, he will find its length to be

$$l' = \sqrt{(x_2' - x_1')^2 + (y_2' - y_1')^2 + (z_2' - z_1')^2} \qquad [19 \cdot 8]$$

Note that, except for the primes, this equation is identical in form to Eq. [19 . 7]. That is, the *formula* for length is invariant, or the same in all rectangular inertial coordinate systems.

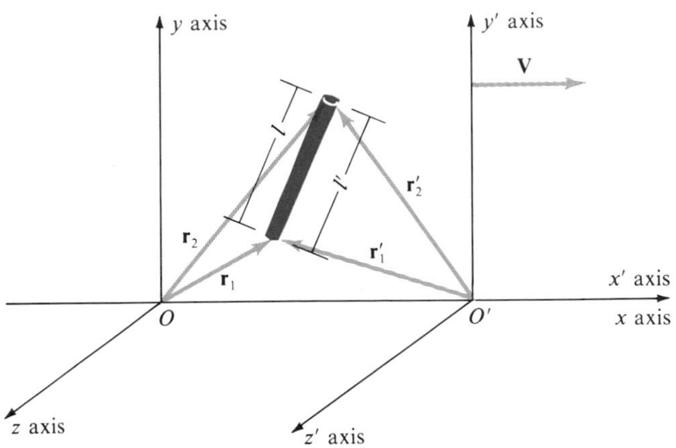

figure 19 • 5 *Galilean transform of proper length* l_0

The numerical value resulting from Eq. [19 . 8] will not necessarily be the same as l_0; this depends on the nature of the transformation between the S and S' systems. For example, if we used the Galilean transformations of Eq. [19 . 6], we would obtain

$$x_2' - x_1' = (x_2 - Vt) - (x_1 - Vt) = x_2 - x_1 \qquad [19 \cdot 9]$$

and since $y' = y$ and $z' = z$,

$$l' = l = l_0 \qquad [19 \cdot 10]$$

This is what intuition, based on the sum of our personal experiences, would lead us to expect. So deeply ingrained is this prejudice in favor of classical mechanics, that it is often difficult to accept the relativistically correct *Lorentz transformations,* which show that l' is not equal to l_0, although the disagreement is negligible at speeds much less than c. However, the formula *defining* length is relativistically invariant and does not change from one rectangular coordinate system to another. As we shall see, the laws of physics must also be relativistically invariant in all inertial systems.

19 . 3 the Lorentz transformation

The Dutch physicist and Nobel laureate (1902) Hendrik Antoon Lorentz made numerous contributions to the theory of electromagnetism and its interactions with matter. He was the first to introduce notions of the relativity of time. On the basis of the Michelson-Morley experiment, he also predicted changes in length in the direction of motion shortly after and independently of the Irish physicist George F. FitzGerald, who had arrived at the same conclusion. Lorentz then pursued the logical consequences of these ideas, deriving a set of mathematical relations which were consistent not only with the Michelson-Morley experiment, but also with the accepted formulation of the electromagnetic theory of light. These relations, the *Lorentz transformations,* are the mathematical basis of the theory of special relativity, and they can be derived directly from Einstein's postulates.

Imagine an experiment in which two meter sticks are placed on end at the origins O and O' of two inertial reference frames, as shown in Fig. 19 . 6.

Let us assume that the proper length l_0 of each meter stick, measured at rest in its own system, has the same numerical value. When the moving S' frame on the left passes the S frame, observers at O and O' will see points O' and P' cross the y axis simultaneously, so that they can both measure l and l' at the same instant. In this case, therefore, we can disregard the possibility of different ways of measuring time in the two systems.

Let us assume that the S frame is stationary as usual. Suppose that, owing to the motion of S', the length l' is greater than l. Since both observers can compare the positions of the endpoints P and P' at the instant of cross-over, this fact would be apparent from both reference frames. However, according to the principle of relativity, this situation is entirely equivalent to S' at rest being approached by the S frame at speed V—in which case l would be the greater length. This dichotomy can be resolved only if $l = l'$, and we must conclude that lengths measured *perpendicular* to the relative motion are unaffected by it. Hence, if the motion is in the x direction, the y and z coordinates must transform as

$$y' = y \qquad z' = z \tag{19.11}$$

Since we know that Galilcan transformations hold at ordinary speeds (by experiment, of course), we assume that x values transform to the S' frame as

$$x' = \gamma(x - Vt) \tag{19.12}$$

where γ is some unknown factor which must approach unity as V/c approaches zero. However, as we saw in Fig. 19 . 4, the Galilean transformation conflicts with Einstein's second postulate, so we shall be cautious and take time as measured in the S' frame to be

$$t' = \gamma_1(t - \alpha x) \tag{19.13}$$

where linearity of the transformation is assumed only for simplicity. The constants γ_1 and α remain to be evaluated, but we also expect that γ_1 will

figure 19 • 6 *Two meter sticks $\overline{OP}$ and $\overline{O'P'}$ have their lengths l and l' compared simultaneously by observers in both the S and S' frames of reference. The observers measure values $l = l' = l_0$, because S' approaching S with speed V is equivalent to S approaching S' with speed V.*

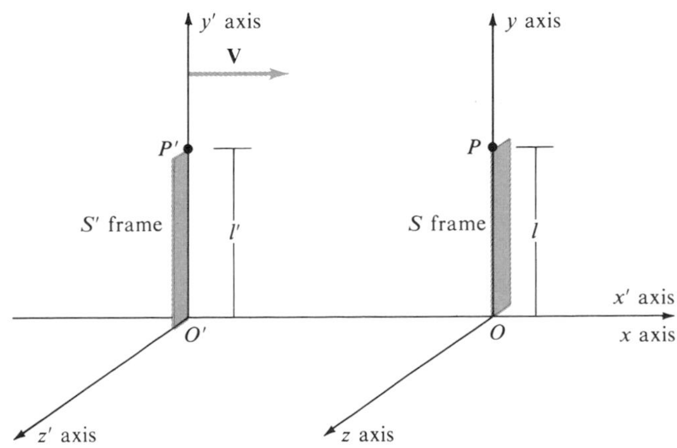

approach unity and α will approach zero as V/c approaches zero, in agreement with the Galilean transformation.

Now let us return to the experiment in Fig. 19 . 4 and interpret it, not by applying the Galilean transformation, but by assuming the validity of Einstein's two relativity postulates. The observer in the S frame finds that the light pulse emitted at O when $t = 0$ spreads in a sphere of increasing radius $r = ct$. The equation of the wavefront is therefore

$$r^2 = x^2 + y^2 + z^2 = c^2t^2 \qquad [19 \cdot 14]$$

The relativity postulates assert that an equation of this same form, and with the light speed equal to c, will be obtained by the observer in the S' frame. But then, instead of a single time scale for both inertial frames, different time coordinates t and t' must exist. That is, for the S' frame,

$$r'^2 = x'^2 + y'^2 + z'^2 = c^2t'^2 \qquad [19 \cdot 15]$$

Each observer (one in S and one in S') will see the same light pulse spreading out in a sphere with its center fixed at *his own* origin of coordinates, since the pulse was emitted when O and O' coincided. However, the same light signal cannot be in two places in the universe at once, so that r cannot equal r'; hence t is not equal to t', as we already suspected.

If we substitute the transformations of Eqs. [19 . 11] to [19 . 13] into Eq. [19 . 15], we obtain

$$r'^2 = \gamma^2(x - Vt)^2 + y^2 + z^2 = c^2\gamma_1^2(t - \alpha x)^2 \qquad [19 \cdot 16]$$

which must reduce to Eq. [19 . 14] for the transformation to be valid, since the speed of light must be the same in S and S' coordinates.

We can now multiply both sides of Eq. [19 . 16] and collect terms:

$$(\gamma^2 - c^2\gamma_1^2\alpha^2)x^2 - 2(\gamma^2 V - c^2\gamma_1^2\alpha)xt + y^2 + z^2 = \left(\gamma_1^2 - \frac{\gamma^2V^2}{c^2}\right)c^2t^2 \qquad [19 \cdot 17]$$

If this is to agree with Eq. [19 . 14], then we must have

$$\gamma^2 - c^2\gamma_1^2\alpha^2 = 1 \qquad \gamma^2V - c^2\gamma_1^2\alpha = 0 \qquad \gamma_1^2 - \gamma^2\frac{V^2}{c^2} = 1 \qquad [19 \cdot 18]$$

and solving these equations yields

$$\gamma = \gamma_1 = \frac{1}{\sqrt{1 - V^2/c}} \qquad \alpha = \frac{V}{c^2} \qquad [19 \cdot 19]$$

The equations for the *Lorentz transformation* are therefore

$$x' = \frac{x - Vt}{\sqrt{1 - \beta^2}} \qquad y' = y \qquad z' = z$$
$$t' = \frac{t - Vx/c^2}{\sqrt{1 - \beta^2}} \qquad [19 \cdot 20]$$

where, for convenience, we have set $\beta = V/c$. Note that in the nonrelativistic limit, for speeds $V \ll c$, β approaches zero, so that γ aproaches unity and α vanishes to zero, as expected. In fact, even for $\beta = \frac{1}{2}$, γ is only 1.155, while for $\beta = \frac{1}{4}$, $\gamma = 1.033$.

These transformations were proposed by Lorentz in 1904 as an *a priori* postulate. A year later Einstein derived them from his relativity postulates. Minkowski referred to the fact that space and time are so interrelated that ct must be treated as a "length," a fourth geometric dimension (see the opening quotation). This four-dimensional space is known as *Minkowski space,* or more simply, *space-time.* A point in space-time is an *event*—something happening at a particular time at a particular place—denoted by $(x,y,z;\, ct)$. The *space-time distance* Δs between two events, one at $(x_1,y_1,z_1;\, ct_1)$ and the other at $(x_2,y_2,z_2;ct_2)$, is defined as

$$(\Delta s)^2 = (c\,\Delta t)^2 - (\Delta x)^2 - (\Delta y)^2 - (\Delta z)^2 \qquad [19 \cdot 21]$$

where $\Delta t = t_2 - t_1$, $\Delta x = x_2 - x_1$, and so on.

example 19 • 1 Show that the space-time distance between two events remains invariant in value when the events are transformed to a moving reference frame S'.

solution From the Lorentz transformation Equations [19 . 20], we have

$$\Delta y' = \Delta y \qquad \Delta z' = \Delta z \qquad \Delta x' = \gamma(\Delta x - V\,\Delta t)$$

$$\Delta t' = \gamma\left(\Delta t - \frac{V\,\Delta x}{c^2}\right)$$

Therefore an observer in the S' frame would measure the distance in Eq. [19 . 22] as

$$\begin{aligned}
(\Delta s')^2 &= (c\,\Delta t')^2 - (\Delta x')^2 - (\Delta y')^2 - (\Delta z')^2 \\
&= \gamma^2\left[c^2\left(\Delta t - \frac{V\,\Delta x}{c^2}\right)^2 - (\Delta x - V\,\Delta t)^2\right] - (\Delta y)^2 - (\Delta z)^2 \\
&= c^2\gamma^2\left(1 - \frac{V^2}{c^2}\right)(\Delta t)^2 - \gamma^2\left(1 - \frac{V^2}{c^2}\right)(\Delta x)^2 \\
&\qquad - (\Delta y)^2 - (\Delta z)^2 \\
&= (c\,\Delta t)^2 - (\Delta x)^2 - (\Delta y)^2 - (\Delta z)^2 \\
&= (\Delta s)^2
\end{aligned}$$

Note that unlike the usual definition of length, in which squared intervals simply add, the minus signs distinguish the special character of the time coordinate from that of the spatial coordinates, so that Δs is invariant under Lorentz transformations.

19 . 4 *time dilation*

As we saw in Sec. 19 . 1, the elapsed time between two given events will be different when it is measured in different inertial frames. Suppose a clock is located at O' in a moving system, and at the instant O' coincides with the origin O of a stationary system, an observer moving with the clock reads the time as $t' = t_0'$. Then at some later time t' the clock has moved a distance $\Delta x = V\,\Delta t$ relative to O; however, its position relative to O' is still $x' = 0$.

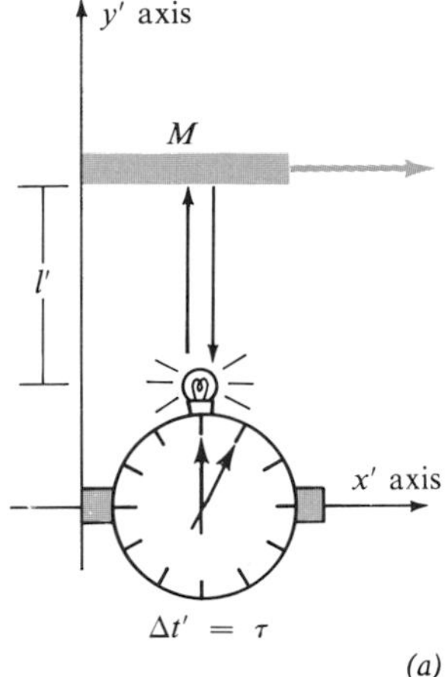

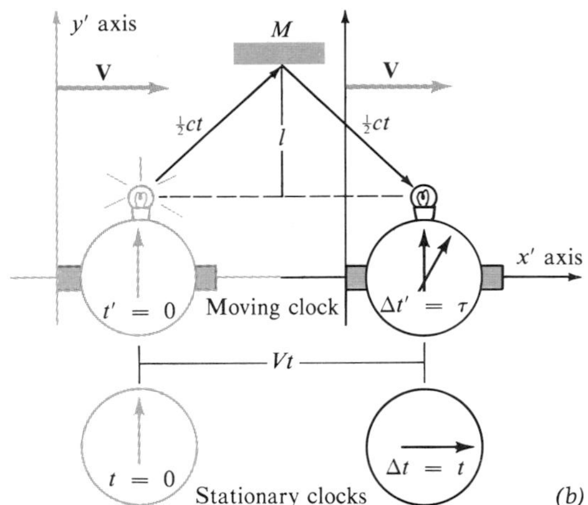

figure 19 • 7 *Direct reflection of a light beam from a moving clock (a) as seen in the moving system and (b) as seen in the stationary system*

The Lorentz transformation yields an elapsed time in the moving system given by

$$\Delta t' = \gamma\left(\Delta t - \frac{V^2\,\Delta t}{c^2}\right) = \frac{\Delta t}{\gamma} \qquad [19 \cdot 22]$$

so that the time intervals $\Delta t = t - t_0$ and $\Delta t' = t' - t_0'$ are related by

$$\Delta t = \gamma\,\Delta t' > \Delta t' \qquad [19 \cdot 23]$$

This effect is known as *time dilation* and we can now refer to γ as the *system dilation factor.*

The fact that the time interval Δt recorded on the stationary clock is longer by a factor of γ than the interval $\Delta t'$ recorded on the moving clock implies that the moving clock is "losing time," or ticking more slowly. This result is consistent with the Lorentz transformation, but at variance with our intuition. Is it really reasonable? Since the clock at O' is in motion, let us compare it with two clocks in the stationary frame. Consider a clock–light source combination in a moving frame S', as shown in Fig. 19 . 7*a*. A light signal emitted in the y' direction strikes a mirror M' moving with the system and returns to the clock at time $t' = 2l'/c$. In the stationary system in Fig. 19 . 7*b* we have two clocks, positioned a distance $\Delta x = V\,\Delta t$ apart. One records the time $t = t' = 0$ that the light signal leaves O' and the other records the time t that it returns. According to Pythagoras' theorem,

$$\tfrac{1}{4}c^2t^2 = l^2 + \tfrac{1}{4}V^2t^2 \qquad [19 \cdot 24]$$

and since $l = l' = \frac{1}{2}ct'$, we can substitute $\frac{1}{4}c^2t'^2$ into this equation to obtain Eq. [19 . 23].

In this case the time t' is the *proper time,* since it is measured between two events occurring at the same position in the S' frame. The time t is not proper, since it is measured by two different clocks in *two different positions* and is therefore affected by the speed of light. Since the proper measurement of a quantity is unique, let us denote the proper time in the inertial system moving with the clock as τ.

If the observer in S' watches a clock at rest in S, whose proper time is $t = \tau$, he will also observe a time dilation $\Delta t' = \gamma\,\Delta\tau$, since the S frame is moving with velocity $-V\mathbf{i}$ relative to S'. This effect is not a paradoxical case of one clock mysteriously registering two different times; the clock itself is always consistent. The effect is due entirely to the measurement process, as in Fig. 19 . 7*b*, and the fact that the speed of light is constant in all inertial frames. Time dilation is therefore a reciprocal phenomenon. The time interval $\Delta\tau$ between two events occurring at any fixed point in a proper reference frame is always less than the elapsed time measured in any other inertial reference frame.

example 19 • 2 Two identical clocks have been adjusted in a laboratory to tick once a second. If one of the clocks is moving with a speed $v' = 0.8c$ and the other is stationary, what is the time between ticks of the moving clock as measured by the stationary clock?

solution The time between two successive ticks of the moving clock is

$$\Delta t = \gamma\,\Delta\tau = \frac{1.0\text{ s}}{\sqrt{1 - 0.8^2}} = 1.7\text{ s}$$

so it appears to have slowed down.

The frequent reference to observers must not be taken to mean that time dilation is more apparent than real. There is nothing illusory about time dilation; its "observers" may be photographic film or other recording devices, and the effect has been amply verified, mainly in experiments with high-speed atomic and nuclear particles.

In one experiment carried out with nuclear particles, measurements showed that the velocity of the particles relative to the laboratory was constant at $v = 2.40 \times 10^8$ m/s, or $v = 0.8c$. The particles were observed to break up (decay) spontaneously into particles of other kinds and were found to have an average life in their original form of $\Delta t = 4.20 \times 10^{-8}$ s. If each particle has its own internal timing mechanism, or proper "clock," which determines its rate of decay, then the measured time must have been the *dilated* proper-time interval, $\Delta t = \gamma\,\Delta\tau$. Thus the mean life of the particles if measured when they are at rest in the laboratory frame should be given by

$$\Delta\tau = \frac{\Delta t}{\gamma} = (4.2 \times 10^{-8}\text{ s})\sqrt{1 - 0.64} = 2.52 \times 10^{-8}\text{ s}$$

Such measurements of particles at rest have verified this predicted time dilation.

The phenomenon of time dilation has given rise to the famous "twin paradox," in which biological aging provides the clock mechanism in both S and S' systems. Suppose you had a twin brother who left earth when you were both 15 years old, and he traveled to the star Alpha Centauri and back in a spaceship that traveled with a speed $v = 0.8c$. Since the distance to Alpha Centauri is about 4.4 light years, you will be 11 years older when the spaceship returns. Your twin, however, who has been moving relative to the earth, will be $\Delta t = 11\sqrt{1 - 0.80^2} = 6.6$ years older. Confusion arises when you begin to look at the problem from your twin's point of view. According to the special theory of relativity, the earth has receded from the origin of *his* proper frame, which was fixed in the spaceship. You left him and returned; consequently, he should have aged 11 years and you only 6.6.

This paradox can be resolved if we recognize that the postulates of special relativity apply only to *inertial* frames of reference, those moving with constant velocity with respect to one another. Since your twin went away and returned, he must have accelerated during part of the trip. This acceleration is detectable and serves as a way of distinguishing between the frames of reference. The age measurements are not comparable, since his frame is *noninertial.* With a treatment of the problem that takes into account Einstein's general theory of relativity, both you and your brother would agree that he does return nearly 4.4 years younger than you.*

*See T. M. Helliwell, *Introduction to Special Relativity* (Allyn and Bacon, 1966), pp. 165–171, for a simple first-order approximation to the solution of the twin problem.

example 19 • 3 Two systems S and S' coincide at time $t = t' = 0$, and at that instant a source at O' emits a train of light waves with proper frequency ν'. What is the value of ν as seen by an observer located a distance x from the origin of the S system?

solution If we consider the emitter as a clock that is ticking ν' times a second, then to an observer watching it pass by, it seems to slow down to ν'/γ ticks a second. Note that this is not contrary to Eq. [19 . 23], since frequency refers to the number of ticks per unit of the *observer's* time. At the instant our observer at x receives the light that was emitted when the source passed his own origin O, he records a time $t = x/c$. However, since he also knows that at that time the source at O' is located at $x_0 = Vt$, he assumes that there are a total of $(\nu'/\gamma)t$ pulses in an interval of length $x - x_0 = (c - V)t$. Their apparent wavelength is therefore

$$\lambda = \frac{(c - V)t}{(\nu'/\gamma)t} = \frac{c}{\nu'}\frac{1 - \beta}{\sqrt{1 - \beta^2}} = \lambda'\sqrt{\frac{1 - \beta}{1 + \beta}}$$

where λ' is the proper wavelength as seen in the S' frame. Here we have the *relativistic Doppler shift* in the frequency of a wave $\lambda'\nu' = \lambda\nu = c$:

$$\frac{\lambda'}{\lambda} = \frac{\nu}{\nu'} = \sqrt{\frac{1 + \beta}{1 - \beta}} = \sqrt{\frac{c + V}{c - V}}$$

If the source O' is receding, we just replace V by $-V$ in this equation. In the classical limit, β goes to zero and the equation becomes

$$\frac{\nu}{\nu'} = \frac{\sqrt{1 - \beta^2}}{1 - \beta} \to \frac{1}{1 - \beta} = \frac{c}{c - V}$$

in agreement with Eq. [17 . 51].

We can reconcile the twin paradox by using the optical Doppler effect and the special theory of relativity if we choose to ignore the traveler twin's periods of acceleration as negligibly small. (We cannot prove this without general relativity, however.) Assume that there is a clock in each system emitting pulses at the rate ν of an average heartbeat, with P pulses equal to one biological year. As the traveler twin recedes from earth, during the time $\Delta t'$ on his clock the traveler receives P_1 pulses,

$$P_1 = \nu \, \Delta t' \sqrt{\frac{1 - \beta}{1 + \beta}}$$

and as he returns he receives P_2 pulses,

$$P_2 = \nu \, \Delta t' \sqrt{\frac{1 + \beta}{1 - \beta}}$$

He therefore sees the earth twin as having aged, in years,

$$\begin{aligned} \frac{P_1 + P_2}{P} &= \frac{\nu \, \Delta t'}{P} \left(\sqrt{\frac{1 - \beta}{1 + \beta}} + \sqrt{\frac{1 + \beta}{1 - \beta}} \right) \\ &= \frac{\nu \, \Delta t'}{P} \left(\frac{1 - \beta}{\sqrt{1 - \beta^2}} + \frac{1 + \beta}{\sqrt{1 - \beta^2}} \right) \\ &= \frac{2\nu \, \Delta t'}{P \sqrt{1 - \beta^2}} \end{aligned}$$

while he has aged only $2\nu \, \Delta t'/P$ years on his clock:

$$\frac{\text{Traveler twin's age}}{\text{Earth twin's age}} = \sqrt{1 - \beta^2}$$

We know that the earth twin has aged by 11 years in his inertial frame; therefore the traveler twin has aged by only 6.6 years in his noninertial frame.

19 • 5 the FitzGerald-Lorentz contraction and simultaneity

FitzGerald and Lorentz had sought independently to account for the puzzling results of the Michelson-Morley experiment by proposing an astonishing hypothesis: Any body in motion with a constant speed relative to the stationary ether will appear shorter in the direction of its motion than when it is at rest in the ether. Lorentz gave a partially satisfactory justification of this contraction hypothesis in terms of the electromagnetic theory of the structure of matter. According to Einstein's relativity postulates, however, this contraction is a property of space-time, rather than of the structure of matter, and applies to motion relative to any inertial reference frame.

Consider a moving rod of proper length $l' = l_0$, equipped with a light source at one end and a mirror at the other, as shown in Fig. 19 . 8*a*. An observer moving with the rod, and therefore measuring events in the S' frame, would see the light leave the source O' at time $t'_0 = 0$, strike the mirror, and return to the source in $\Delta t' = \Delta\tau = 2l_0/c$, the elapsed proper

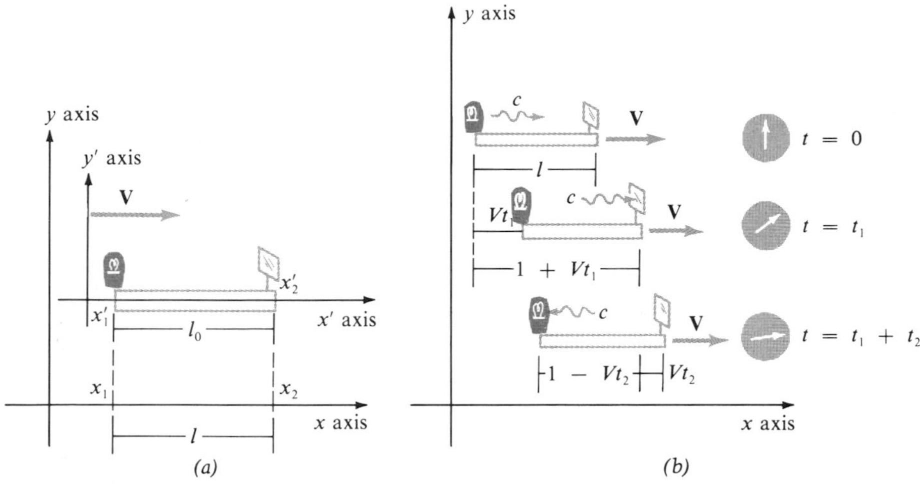

figure 19 • 8 *Moving rod with a light source at one end and a mirror at the other*

time. A stationary observer in the S frame, however, would see the situation in Fig. 19 . 8*b*. He would see the light leave the source at time $t_0 = 0$, but to reach the moving mirror at time t_1, it must travel a distance

$$s_1 = l + Vt_1 = ct_1 \tag{19.25}$$

By time t_2 the source is closer to the point from which the light was reflected, so that the reflected light has to return a distance of only

$$s_2 = l - Vt_2 = ct_2 \tag{19.26}$$

The total travel time would therefore be measured by the observer in the S frame as

$$\Delta t = t_1 + t_2 = l\left(\frac{1}{c - V} + \frac{1}{c + V}\right) = 2\gamma^2 \frac{l}{c} \tag{19.27}$$

According to Eq. [19 . 23], however, the time dilation in the S frame is

$$\Delta t = \gamma\,\Delta\tau = \gamma\frac{2l_0}{c}$$

and substituting this value into Eq. [19 . 27], we find that

$$l = \frac{l_0}{\gamma} < l_0 \tag{19.28}$$

This is the *FitzGerald-Lorentz contraction:*

> A moving body appears to contract in the direction of motion by a factor $1/\gamma$ relative to its proper length.

For instance, if the moving rod in Fig. 19 . 8 is a meter stick with a known proper length of 1 m and if its speed relative to the S frame is $V = 0.8c$, then an observer in the S frame will measure its length as

$$l = \sqrt{1 - 0.8^2} = 0.6 \text{ m}$$

Like time dilation, this effect is reciprocal and does not depend on the sign of V. If the same rod were lying at rest on the x axis in the S system, a moving observer in the S' system would measure its length as $l' = l_0/\gamma$. We refer to γ as the *system* dilation factor.

These relativistic contractions are too small to be detected by direct measurements in the laboratory. Experience shows, however, that it is the contracted length in Eq. [19 . 28] that must be used when the length of a rapidly moving body has to be taken into account in describing its physical effects on its surroundings. In this sense, the contraction in the direction of motion is quite real to the stationary observer. The difference is that the motion, unlike a structural change, has no effect on the *proper* length of the body.

To measure the length l of a rod that is moving relative to the S frame, the positions of its endpoints must be determined *simultaneously* by observers in the S frame. Their clocks must be synchronized to show the same time at every instant, and both observations must be made at the same time t on the synchronized clocks. Synchronization is not as easy as it sounds. An observer at x_2 cannot just set his clock to match that of an observer at x_1; owing to the finite speed of the light by which he reads the clock located at x_1, his own clock will actually be behind. Nor can he carry his clock to x_1, synchronize it, and return to x_2, since, as we saw in the preceding section, nonuniform motion affects the measurement of time.

> We define two events as *simultaneous* if an observer stationed midway between them receives light signals from both events at the same instant.

Simultancity is also relative, however, and two clocks synchronized in their proper system will appear unsynchronized to an observer in another reference frame. To illustrate, suppose the moving rod in Fig. 19 . 9 has a proper length $l' = l_0$ and the clocks on each end of the rod are synchronized in the S' frame. To locate the ends of the moving rod at some time t, an observer stationed midway between x_1 and x_2 in the S frame must send out a light pulse at time $t = 0$, when the midpoint M' of the rod is still a distance Vt to the left (see Fig. 19 . 9*a*). To reach both clocks simultaneously at time t, the pulse travels a distance $\frac{1}{2}l$ in a time interval $\Delta t = \frac{1}{2}l/c$. However, when these light pulses are reflected back to the observer at the source, he will find that they have illuminated the clocks in the S' frame at two different times t'_1 and t'_2 (see Fig. 19 . 9*b*).

The reason for the discrepancy is that in order to strike both clocks at the same time t in the S frame, the light pulses must leave at the time point P' is passing the source at O. Although P' is not the midpoint of the rod, the light must also travel down the rod with speed c in the S' system, and since $l'_1 > l'_2$, it arrives *later* at the first clock than at the second one. To compute the difference between t'_1 and t'_2, consider the situation as seen by an observer moving in the S' frame with the rod (see Fig. 19 . 9*c*). Since the distances l_1 and l_2 in the S frame are $l_1 = \frac{1}{2}l + Vt$ and $l_2 = \frac{1}{2}l - Vt$, in the S' frame, where the lengths are proper, the light must travel distances

$$l'_1 = \gamma(\tfrac{1}{2}l + Vt) \qquad l'_2 = \gamma(\tfrac{1}{2}l - Vt) \qquad [19 \cdot 29]$$

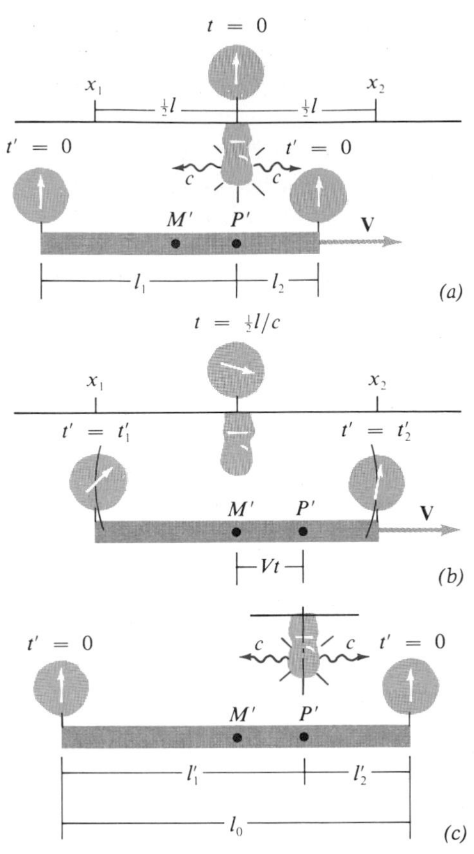

figure 19 • 9 *The Fitzgerald-Lorentz contraction, as viewed in the S frame*

Therefore, when the clocks are illuminated by the light from *O*, their readings will differ by an amount

$$t'_1 - t'_2 = \frac{l'_1}{c} - \frac{l'_2}{c} = \gamma \frac{\frac{1}{2}l + Vt - \frac{1}{2}l + Vt}{c}$$

$$= 2\gamma \frac{Vt}{c} \qquad [19 \cdot 30]$$

An observer in the *S* frame, of course, will see both clocks lit up at $t = \frac{1}{2}l/c$, so that the moving clocks appear to be out of synchronization by

$$t'_1 - t'_2 = \gamma \frac{Vl}{c^2} = \frac{Vl_0}{c^2} \qquad [19 \cdot 31]$$

in agreement with Eq. [19 . 29]. To a stationary observer, the approaching clock appears to be ahead of the departing clock by an amount proportional to their separation distance and the relative velocity of the two inertial frames. (Note that direction is significant in this case.) However, the two moving clocks are synchronized in their own *proper* system.

example 19 • 4 Show that the contraction of length, Eq. [19 . 28], and the apparent lack of synchronization of moving clocks, Eq. [19 . 31], can both be derived from the Lorentz transformation equations.

solution In Fig. 19 . 9, we assumed that the ends of the rod of length $l = x_2 - x_1$ were measured simultaneously at time t in the S system. Transforming to the proper system S' according to Eqs. [19 . 20], we can write the length of the rod as

$$l_0 = x_2' - x_1' = \gamma(x_2 - x_1) = \gamma l$$

where $\gamma = 1/\sqrt{1 - \beta^2}$. Then, from the transformation equation for t',

$$t_1' - t_2' = \gamma\left(t - \frac{Vx_1}{c^2}\right) - \gamma\left(t - \frac{Vx_2}{c^2}\right) = \gamma\frac{V(x_2 - x_1)}{c^2}$$

$$= \gamma\frac{Vl}{c^2} = \frac{Vl_0}{c^2}$$

The Lorentz transformation equations are obviously far more economical of thought and effort than the physical type of derivation we have employed. The transformation equation for $t_1' - t_2'$ also describes the situation in which both clocks are located at x_1 and x_2, respectively, and synchronized to time *t in the S frame.* In that case, two events simultaneous in the S frame, but separated in space, will not appear to be simultaneous in the S' frame.

19 • 6 relativistic transformation of velocities

Effects such as length contraction and time dilation are hard for us to accept, since all our physical intuitions have been educated through experiences at nonrelativistic speeds. Moreover, they give rise to the paradoxical results that a moving clock will appear to stop altogether when it moves with the speed of light, while a length will contract to zero and vanish. Rather than assume that relativity also has its paradoxes, we can reverse the argument and state flatly that no physical object can ever travel at the speed of light. But if this is so, then we cannot accelerate objects indefinitely merely by continuing to apply force to them. This implies that the classical laws of motion and the concepts of mass, momentum, and energy will have to be revised—all because the speed of light is invariant in all inertial systems. As we shall see, this necessary recasting of the fundamental laws of mechanics leads to some important new ideas using the fairly simple kinematic concept of velocity transformations.

If a moving object is being observed from two systems, a stationary S frame and an S' frame moving with constant velocity $\mathbf{V}$, then the velocity of the object will have different measured values in the two frames: it will be $\mathbf{v}$ in the S frame and $\mathbf{v}'$ in the S' frame. The classical Galilean transformation relates these two velocities by the formula $\mathbf{v} = \mathbf{v}' + \mathbf{V}$. However, at relativistic speeds the Lorentz transformation predicts a different form of velocity transformation. Since the Lorentz formulas deal only with time and space, we must return to the original definition of velocity, which is invariant in any system.

Let us use Eqs. [19 . 20] to transform from the moving S' system back to the S system, which we can then say is moving with a constant velocity

$-V\mathbf{i}$ in the x' direction. From the transformation equations for position and time, we have

$$x = \frac{x' + Vt'}{\sqrt{1 - \beta^2}} \qquad t = \frac{t' + (V/c^2)x'}{\sqrt{1 - \beta^2}} \qquad [19 \cdot 32]$$

Therefore, for changes Δx with respect to time,

$$\frac{\Delta x}{\Delta t} = \frac{\Delta x' + V\Delta t'}{\Delta t' + (V/c^2)\,\Delta x'} = \frac{(\Delta x'/\Delta t') + V}{1 + (V/c^2)(\Delta x'/\Delta t')} \qquad [19 \cdot 33]$$

Then, from our original definition of velocity,

$$\lim_{\Delta t \to 0} \frac{\Delta x}{\Delta t} = \frac{dx}{dt} = v \qquad \lim_{\Delta t' \to 0} \frac{\Delta x'}{\Delta t'} = \frac{dx'}{dt'} = v'$$

and the velocity transformation of Eq. [19 . 33] becomes

$$v = \frac{v' + V}{1 + v'V/c^2} \qquad [19 \cdot 34]$$

where the instantaneous velocities $\mathbf{v}$ and $\mathbf{v}'$ are assumed to have the same direction as the system velocity $\mathbf{V}$. This value of v is *less* than we would have predicted from the Galilean formula $v = v' + V$. However, note that the Galilean result is indeed approached as V/c approaches zero in the nonrelativistic limit.

To illustrate, suppose the train in Fig. 19 . 10 is traveling at a speed measured in the S' frame as $dx'/dt' = v'$. A stationary observer in the S frame will measure a speed v as given by Eq. [19 . 34]. The two observations are completely independent, but the relation between them shows that in the S frame $v \leq c$. Even if the train were traveling at the speed of light in the S' frame, with $V = c$ we would still find that $v = c$. Equation [19 . 34] provides additional support for the important assertion that a material particle or body can never attain a speed equal to or greater than the speed of light *in vacuo*.

example 19 • 5 A particle A is moving in the laboratory with a constant speed $v = 0.9c$, and a particle B moving the same direction passes A with a speed $v' = 0.9c$ relative to A. What is the speed of B relative to the laboratory?

solution The speed of B relative to the laboratory is *not* $1.8c$; from Eq. [19 . 34], it is

$$v = \frac{0.9c + 0.9c}{1 + (0.9c)(0.9c)/c^2} = 0.994c$$

If a particle is moving in an *arbitrary* direction relative to the direction of a moving reference frame S', its velocity $\mathbf{v}'$ as measured from S' has rectangular components

$$v_x' = \frac{dx'}{dt'} \qquad v_y' = \frac{dy'}{dt'} \qquad v_z' = \frac{dz'}{dt}$$

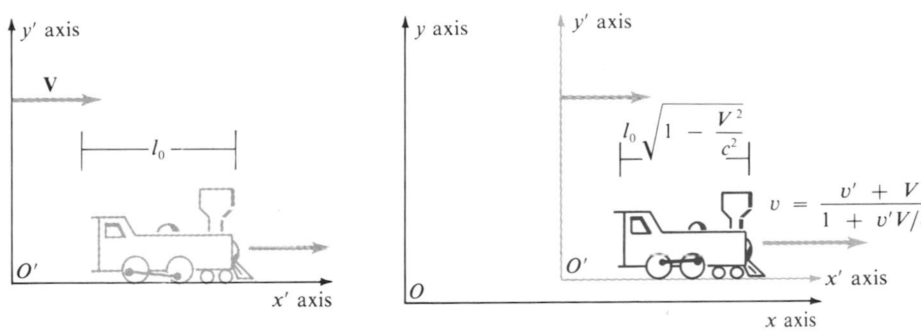

(a) Velocity v'_x in S' frame (b) Velocity v_x in S frame

figure 19 • 10 *Transformation of velocities*

In the S frame the corresponding velocity is $\mathbf{v}$, with components

$$v_x = \frac{dx}{dt} \qquad v_y = \frac{dy}{dt} \qquad v_z = \frac{dz}{dt}$$

If we take the system velocity $\mathbf{V}$ as $V\mathbf{i}$, then we already have the relation between v_x and v'_x in Eq. [19 . 34]:

$$v_x = \frac{v'_x + V}{1 + v'_x V/c^2} \qquad [19 \cdot 35]$$

To find the y and z components, note that from the Lorentz Equations [19 . 20], $dy = dy'$ and $dz = dz'$. Therefore, since

$$v_y = \frac{dy}{dt} = \frac{dy'}{\gamma(dt' + V\,dx'/c^2)}$$

and the same considerations hold for dz/dt,

$$v_y = \frac{v'_y}{\gamma(1 + v'_x V/c^2)} \qquad v_z = \frac{v'_z}{\gamma(1 + v'_x V/c^2)} \qquad [19 \cdot 36]$$

To find the inverse velocity transformations and express the components of $\mathbf{v}'$ in terms $\mathbf{v}$ and $\mathbf{V}$, we simply interchange the primed and unprimed quantities in Eqs. [19 . 35] and [19 . 36] and replace V by $-V$.

19 • 7 relativistic momentum and mass

We have considered the kinematic aspects of the Lorentz transformation. Now let us examine its implications for the classical principle of conservation of momentum. In a simple elastic collision of a particle of mass m with a stationary particle of mass $2m$, if the approach speed of the moving particle is $u_1 = 3V$, then according to classical mechanics, the two particles will rebound with separation speeds $v_1 = -V$ and $v_2 = 2V$. Now let us take the S' frame as the center of mass of the two-particle system, moving with velocity $v_c = V$ relative to a stationary observer. Figure 19 . 11b shows the same collision in center-of-mass coordinates, where the smaller particle has an approach velocity $u'_1 = 2V$ and the larger particle has an approach velocity $u'_2 = -V$. The velocities after collision are $v'_1 = -2V$ and $v'_2 = V$.

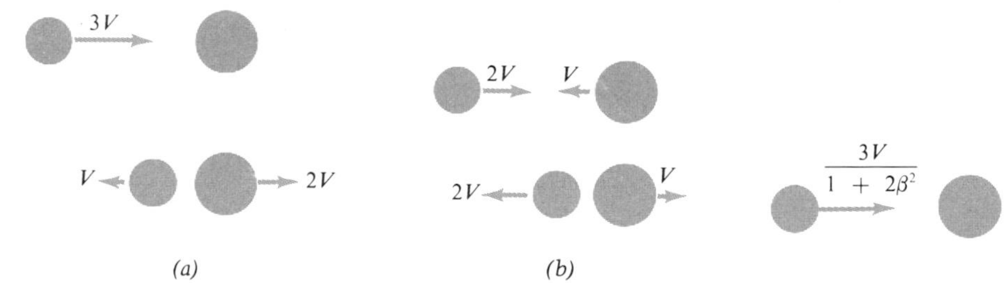

figure 19 • 11 *Relativistic failure of the classical principle of conservation of momentum. (a) Classical binary collision in S system. (b) Same collision in the classical center-of-mass system S′. (c) Relativistic transformation of S′ velocities back to S system.*

As we have already seen, these velocities do not transform back to the S frame according to the simple Galilean transformation $v = v' + V$. Instead, from the relativistic velocity transformation of Eq. [19 . 34], the approach and separation speeds of the smaller particle transform to the S frame as

$$u_1 = \frac{u_1' + v_c}{1 + u_1' v_c/c^2} = \frac{3V}{1 + 2\beta^2} \qquad \beta = \frac{V}{c} \qquad [19 \cdot 37]$$
$$v_1 = \frac{v_1' + v_c}{1 + v_1' v_c/c^2} = -\frac{V}{1 - 2\beta^2}$$

Before we go any further, note that the denominator of v_1 vanishes for $\beta^2 = \frac{1}{2}$, a clear indication that our assumptions for the momentum values in the S' frame are relativistically in error to begin with. For the time being, however, let us focus on the case where $\beta^2 < \frac{1}{2}$ and look at the conservation of momentum under transformation to S coordinates.

The relativistically transformed velocities of the larger particle, which is initially stationary in the S frame, are

$$u_2 = \frac{u_2' + v_c}{1 + u_2' v_c/c^2} = 0 \qquad v_2 = \frac{v_2' + v_c}{1 + v_2' v_c/c^2} = \frac{2V}{1 + \beta^2} \qquad [19 \cdot 38]$$

According to classical mechanics, the total initial momentum of the two-particle system,

$$p_i = mu_1 = \frac{3mV}{1 + 2\beta^2} \qquad [19 \cdot 39]$$

must be equal to the total final momentum after collision,

$$p_f = mv_1 + 2mv_2 = -\frac{mV}{1 - 2\beta^2} + \frac{4mV}{1 + \beta^2} \qquad [19 \cdot 40]$$

It is easy to show that these two momenta are in fact approximately equal for very small values of β, and easier still to compute some representative momenta on a programmable calculator:

β	0.1	0.2	0.3	0.4	0.5	0.6
p_i	2.9412	2.7778	2.5423	2.2727	2.0000	1.7442
p_f	2.9400	2.7592	2.4502	1.9777	1.2000	−0.6302

While the disagreement is only 3.6 percent for $\beta < 0.3$, it becomes increasingly bad for relativistic speeds. On this basis, we must conclude that the classical formula for conservation of momentum is not invariant under transformation to other reference frames and holds only as an approximation for speeds much less than the speed of light.

To preserve the validity of the conservation principle for relativistic processes, we must redefine momentum so that the principle does remain invariant under Lorentz transformations. Let us therefore define the momentum **p** of a particle moving with velocity **v** as

$$\mathbf{p} = \frac{m_0\mathbf{v}}{\sqrt{1 - v^2/c^2}} \qquad [19 \cdot 41]$$

where v is the speed of the particle and m_0 is its proper mass, or *rest mass,* measured when the particle is at rest in the reference frame in which it is being observed. At particle speeds much less than c, this equation reduces to the classical formula $\mathbf{p} = m_0\mathbf{v}$. By definition, the instantaneous velocity $\mathbf{v} = d\mathbf{r}/dt$ is the rate of change of position as measured in the observer's time interval dt. According to the formula for time dilation, however, the new factor in Eq. [19 . 41] is a *particle* dilation factor:

$$\gamma_v = \frac{1}{\sqrt{1 - v^2/c^2}} = \frac{dt}{d\tau} \qquad [19 \cdot 42]$$

where $dt/d\tau$ is the ratio between the observer's time and the *proper time* in the reference frame moving with the particle at instantaneous speed v. Therefore we can rewrite our definition of relativistic momentum as

$$\mathbf{p} = m_0\gamma_v\frac{d\mathbf{r}}{dt} = m_0\frac{d\mathbf{r}}{d\tau} \qquad [19 \cdot 43]$$

This is an important relation, because $d\tau$ is unique to a given particle at a given instant of its motion and is invariant. Note that, unlike γ, which is a function of the constant relative velocity V of two inertial reference frames, γ_v is a variable that depends on the instantaneous speed v of the object viewed in the S frame. This behavior is shown in Fig. 19 . 12.

To see that this new formulation of momentum does result in invariance of the conservation of momentum, consider the transformation of the x' component of relativistic momentum $\mathbf{p}'$, measured in the S' frame parallel to the motion $\mathbf{V} = V\mathbf{i}$ of S' relative to S. From Eqs. [19 . 20], this momentum is transformed to S coordinates by the relation

$$\begin{aligned} p'_x &= m_0\frac{dx'}{d\tau} = m_0\gamma\frac{dx - V\,dt}{d\tau} \\ &= m_0\gamma\frac{dx}{d\tau} - m_0\gamma V\frac{dt}{d\tau} \end{aligned} \qquad [19 \cdot 44]$$

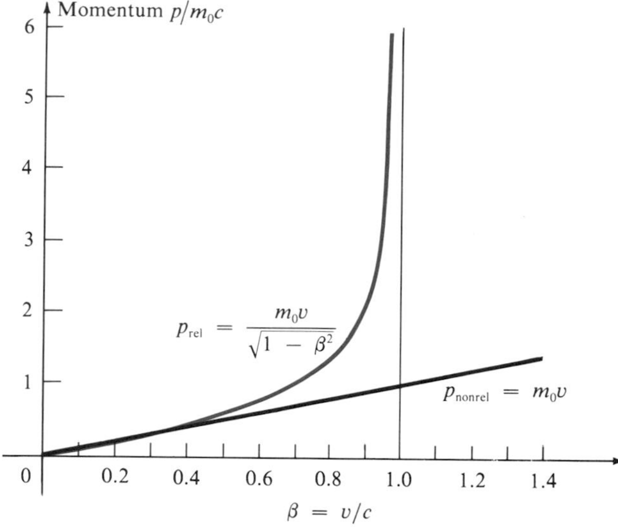

figure 19 • 12 *Comparison of relativistic and nonrelativistic momenta (in units of m_0c) as a function of $\beta = v/c$*

where $\gamma = 1/\sqrt{1 - V^2/c^2}$ is the system dilation factor for the S' frame. However, the particle dilation factor $dt/d\tau = \gamma_v$ and the observed x component of velocity in the S frame is $dx/dt = v_x$. Therefore $dx/d\tau = \gamma_v v_x$, and Eq. [19 . 44] becomes

$$p'_x = \gamma(m_0\gamma_v v_x - m_0 V\gamma_v) = \gamma(p_x - m_0 V\gamma_v) \qquad [19 \cdot 45]$$

Let us now consider a binary collision along the x' axis between two particles, of rest masses m_{01} and m_{02} in the S' frame, in which the total relativistic momentum is conserved and all momenta are in the x' direction:

$$p'_{1i} + p'_{2i} = p'_{1f} + p'_{2f} \qquad [19 \cdot 46]$$

When we transform these momenta to the S frame, using Eq. [19 . 45] and the particle dilation factors

$$\gamma_{u1} = \frac{1}{\sqrt{1 - u_1^2/c^2}} \qquad \gamma_{u2} = \frac{1}{\sqrt{1 - u_2^2/c^2}} \qquad \text{etc.}$$

and cancel common dilation factors γ, we obtain

$$(p_{1i} - m_{01}V\gamma_{u1}) + (p_{2i} - m_{02}V\gamma_{u2})$$
$$= (p_{1f} - m_{01}V\gamma_{v1}) + (p_{2f} - m_{02}V\gamma_{v2})$$

If we rearrange terms to collect factors of V, this equation becomes

$$p_{1i} + p_{2i} - p_{1f} - p_{2f}$$
$$= V(m_{01}\gamma_{u1} + m_{02}\gamma_{u2} - m_{01}\gamma_{v1} - m_{02}\gamma_{v2}) \qquad [19 \cdot 47]$$

Since the factor V occurs only on the right-hand side of Eq. [19 . 47], it follows that the two sides can be equal for all values of V only if each side equals some universal constant that holds whatever the speeds involved. We know that at very low collision speeds both sides vanish to zero. This, then, is the desired constant: both sides of Eq. [19 . 47] must vanish. The relativistic formulation of conservation of momentum therefore leads to a pair of conservation laws, one for conservation of relativistic momentum,

$$p_{1i} + p_{2i} = p_{1f} + p_{2f} \qquad [19 \cdot 48]$$

and the other for conservation of relativistic mass:

$$m_{01}\gamma_{u1} + m_{02}\gamma_{u2} = m_{01}\gamma_{v1} + m_{02}\gamma_{v2} \qquad [19 \cdot 49]$$

In the limit of small speeds, this second conservation law is simply the conservation of mass assumed by classical mechanics. The significance of this equation, however, is that we must redefine mass in relativistic terms, since our original definition of mass m in Sec. 5 . 2 was as a measure of inertia—that is, resistance to a *change in motion.* We must now define the relativistic mass m of a particle of rest mass m_0 and speed v as

$$m = \frac{m_0}{\sqrt{1 - v^2/c^2}} \qquad [19 \cdot 50]$$

The definition of relativistic momentum in Eq. [19 . 41] then becomes

$$\mathbf{p} = m_0\gamma_v\mathbf{v} = m\mathbf{v} \qquad [19 \cdot 51]$$

where m always refers to the relativistic mass of a particle of rest mass m_0. As Fig. 19 . 13 shows, the mass becomes infinite as the speed approaches the speed of light.

We have not discussed the transformation of momenta that are perpendicular to the relative motion $\mathbf{V}$ of the two reference frames. However, since $dy' = dy$ and $dz' = dz$, the transformation of any relativistic momentum $\mathbf{p}'$ to a relativistic momentum $\mathbf{p}$ has the component forms

$$p'_x = \gamma(p_x - m_0V\gamma_v) \qquad p'_y = p_y \qquad p'_z = p_z \qquad [19 \cdot 52]$$

figure 19 • 13 *Relativistic mass as a function of $\beta = v/c$*

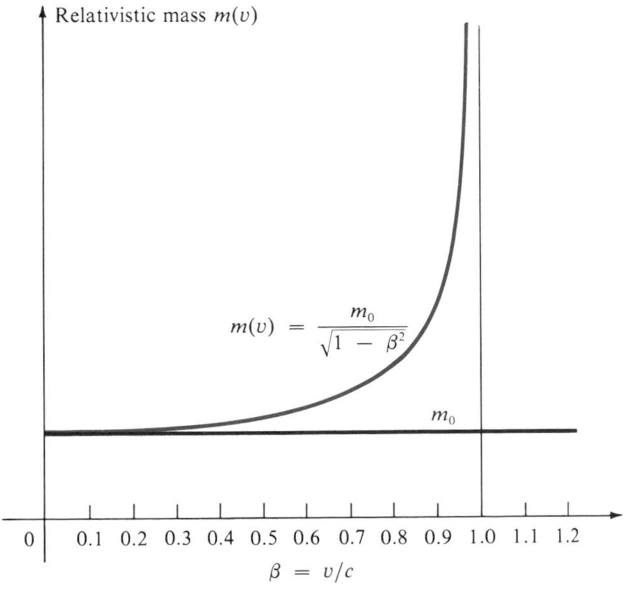

Note that these last two relations hold only for *relativistic momenta*, as defined in Eq. [19 . 51]. It would be incorrect, except in the classical limit, to say that $m_0 v_y = m_0 v'_y$ or $m_0 v_z = m_0 v'_z$. This is apparent if we consider a "system" consisting of one particle moving with speed v in the S frame and speed v' in some S' frame. Since v is not equal to v', Eq. [19 . 52] yields $\gamma'_v v'_y = \gamma_v v_y$, or

$$v'_y = \frac{\gamma_v}{\gamma'_v} v_y \neq v_y$$

example 19 • 6 A particle of rest mass m_0 collides with an identical stationary particle at a speed u_1 in the S frame. If the center of mass of the two-particle system is taken as the S' frame, what is its speed relative to the S frame (*a*) in the classical nonrelativistic case and (*b*) in the relativistic case?

solution (*a*) Recall from Sec. 10 . 1 that in center-of-mass coordinates the total momentum always equals zero. In terms of classical mechanics, this means that

$$m_0(u_1 - v_c) - m_0 v_c = 0 \qquad \text{or} \qquad v_c = \tfrac{1}{2} u_1$$

(*b*) If we retain this definition of the center-of-mass system, then the momenta measured in the S' frame also sum to zero:

$$p'_1 + p'_2 = 0$$

where these are the relativistic momenta, since m_0 refers to the S frame. However, when we transform to the S frame, where $u_2 = 0$, after canceling common factors $\gamma = 1/\sqrt{1 - v_c^2/c^2}$, we obtain

$$(m_0 u_1 \gamma_{u1} - m_0 v_c \gamma_{u1}) + (-m_0 v_c) = 0$$

If we now solve for v_c, we find

$$v_c = \frac{\gamma_{u1} u_1}{1 + \gamma_{u1}} = \frac{u_1}{\sqrt{1 - u_1^2/c^2} + 1} > \tfrac{1}{2} u_1$$

As a check, note that in the limit as u_1/c approaches zero, $v_c = \frac{1}{2}u_1$, and in the limit as u_1/c approaches unity, $v_c = u_1$. Because the relativistic mass of the approaching particle is increased by its motion, it literally carries more weight in determining the speed of the center of mass.

19 • 8 mass and energy

In any experiment which measures a moving mass, the ratio of momentum to speed, $m = p/v$, equals the *relativistic mass*, not the rest mass. This relativistic mass is not an illusion, but a fact of life to the observer who sees it traveling with velocity $\mathbf{v}$, since it is a measure of the particle's resistance to change in its motion. Of course, an observer moving with the particle would measure its mass as m_0. The relation between m and m_0 as a function of v/c is shown in Table 19 . 1. Note that the relativistic mass becomes infinite as

table 19 • 1 *Ratio of relativistic to nonrelativistic mass for various values of $\beta = v/c$*

v/c	m/m_0	v/c	m/m_0
0.01	1.000	0.9	2.294
0.1	1.005	0.99	7.09
0.6	1.250	0.999	22.4
0.8	1.667	0.999999	700.0

the particle approaches the speed of light, supporting our previous supposition that no material object can ever be accelerated to this speed.

Since we have redefined both mass and momentum in relativistic terms, we must now redefine force so that it is the time rate of change of *relativistic momentum*. Newton's second law of motion remains invariant in form, and for particle speeds much less than the speed of light, it reduces to the classical relation if we write:

$$\mathbf{F} = \frac{d\mathbf{p}}{dt} = m\frac{d\mathbf{v}}{dt} + \mathbf{v}\frac{dm}{dt} \qquad [19 \cdot 53]$$

As we saw in Sec. 5 . 3, this expression also takes into account the effect of changes in mass. If the mass remains constant, then dm/dt vanishes to zero. It is clear from Eq. [19 . 50], however, that an accelerating force as defined here does part of its work in increasing the relativistic mass m of the particle. The rate at which it does this work is, from Eq. [19 . 53],

$$\frac{dW}{dt} = \mathbf{F}\cdot\mathbf{v} = m\mathbf{v}\cdot\frac{d\mathbf{v}}{dt} + v^2\frac{dm}{dt} = \tfrac{1}{2}m\frac{dv^2}{dt} + v^2\frac{dm}{dt} \qquad [19 \cdot 54]$$

To see how this work affects the relativistic mass of the particle, let us express the right side of Eq. [19 . 54] entirely in terms of dm/dt. If we differentiate the formula for relativistic mass in Eq. [19 . 50] with respect to v^2, we obtain

$$\frac{dm}{dt} = \tfrac{1}{2}\gamma_v^3\frac{m_0}{c^2}\frac{dv^2}{dt} \qquad [19 \cdot 55]$$

and substitution of this expression for dv^2/dt into Eq. [19 . 54] yields

$$\frac{dW}{dt} = \left(\frac{c^2}{\gamma_v^2} + v^2\right)\frac{dm}{dt} = c^2\frac{dm}{dt} \qquad [19 \cdot 56]$$

If we identify the work done on the particle as an increase in its *relativistic kinetic energy,* then the solution to Eq. [19 . 56] is

$$W - W_0 = K = (m - m_0)c^2 \qquad [19 \cdot 57]$$

or

$$\blacktriangleright \quad K = m_0c^2(\gamma_v - 1) \qquad [19 \cdot 58]$$

where the constant of integration is determined by the obvious fact that $K = 0$ when the particle is at rest.

As a check on this result, we note that the classical nonrelativistic limit should be reached when v/c vanishes to zero. If we expand γ_v in a binomial series (see Appendix D),

$$\gamma_v = \frac{1}{\sqrt{1 - v^2/c^2}} = 1 + \frac{\frac{1}{2}v^2}{c^2} + \frac{\frac{3}{8}v^4}{c^4} + \cdots$$

then in the limit as v/c approaches zero, Eq. [19 . 58] becomes

$$\lim_{v/c \to 0} K = m_0 c^2 \left(1 + \frac{\frac{1}{2}v^2}{c^2} - 1\right) = \tfrac{1}{2} m_0 v^2 \qquad [19 \cdot 59]$$

which is the classical expression for kinetic energy. A more interesting point, however, is that a change in mass is the equivalent of a change in energy. And by any experiment the observer can devise, m is a palpable, tangible lump of mass.

This suggests that we can define the total relativistic energy E of a free particle as

$$\blacktriangleright \quad E = mc^2 \qquad [19 \cdot 60]$$

which implies that the particle also has a *rest energy* E_0:

$$\blacktriangleright \quad E_0 = m_0 c^2 \qquad [19 \cdot 61]$$

This result of relativity theory is the one most pregnant with consequences for the future of mankind, for it indicates that mass is but another form of energy, and therefore, like other energy forms, convertible. The quantity c^2 is merely a constant of proportionality. We can now see that Eq. [19 . 49] states that in a two-particle elastic collision, when the total momentum is conserved, the total relativistic energy must also be conserved, and vice-versa. Figure 19 . 14 compares total relativistic energy and classical kinetic energy for a free particle.

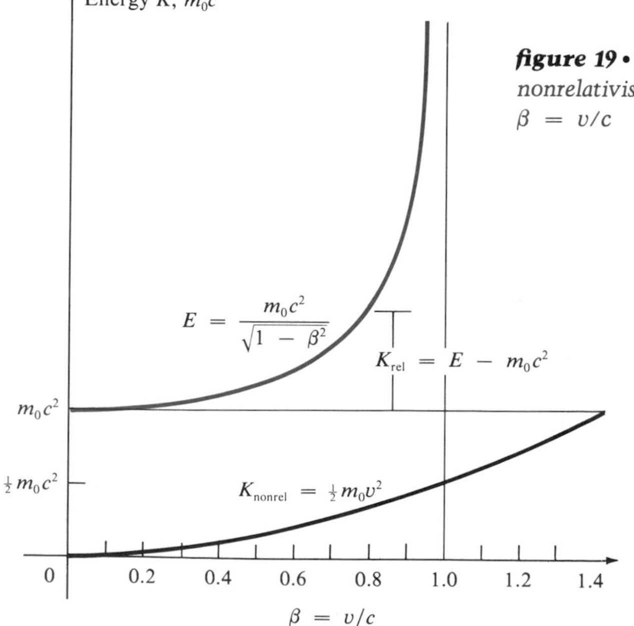

figure 19 • 14 *Relativistic energy E and nonrelativistic kinetic energy K plotted against* $\beta = v/c$

example 19 • 7 Two particles, each of rest mass m_0, are approaching each other with equal and opposite speeds $v = 0.7c$. If their collision results in coalescence into a single particle of mass M, then by symmetry that particle must be at rest. What is the rest mass M_0 of the combined particle?

solution The total energy of the two-particle system before collision is

$$E = 2mc^2 = 2\gamma_v m_0 c^2$$

where

$$\gamma_v = \frac{1}{\sqrt{1 - v^2/c^2}} = \frac{1}{\sqrt{0.51}} \doteq \sqrt{2}$$

Unlike the classical case, it is possible to prove that total relativistic energy is conserved in an isolated perfectly inelastic binary collision. After the collision, this energy is entirely equal to the rest energy of the combined particle:

$$M_0 c^2 = E_0 = 2\sqrt{2}\, m_0 c^2$$

Therefore,

$$M_0 = 2\sqrt{2}\, m_0$$

Note that the rest energy $M_0 c^2$ of the combined particle is greater than that of the two component particles by an amount

$$Q = (\sqrt{2} - 1) 2 m_0 c^2$$

This is the "threshold energy" that must be supplied by an external agent in the form of kinetic energy to make the reaction take place.

We can also express the total energy E of a particle in terms of its relativistic momentum. We first express its velocity as a function of p^2:

$$p^2 = m_0^2 \gamma_v^2 v^2 = m_0^2 \frac{1}{1 - v^2/c^2} v^2$$

Then, solving for v^2, we obtain

$$v^2 = \frac{p^2c^2}{m_0^2c^2 + p^2} \qquad \gamma_v^2 = \frac{m_0^2c^2 + p^2}{m_0^2c^2} = 1 + \frac{p^2}{m_0^2c^2} \tag{19 . 62}$$

We can write Eq. [19 . 60] as

$$E = mc^2 = \gamma_v m_0 c^2$$

and substitute the expression for γ_v^2 into the square of this equation:

► $$E^2 = \gamma_v^2 m_0^2 c^4 = m_0^2 c^4 + p^2 c^2 \tag{19 . 63}$$

This relation is extremely useful in considering collisions between particles traveling at very high speeds, as in nuclear reactions. It can be shown that a massless quantity of pure energy, such as light, can have momentum $p = E/c$ and can therefore exert pressure, as Descartes predicted, even though it has no rest mass.

example 19 • 8 Under certain conditions the subatomic particle known as the π-meson, or *pion*, completely disintegrates into a massless *neutrino* ν and a μ-meson, or *muon*. In the proper frame of the pion, find the neutrino's momentum p_ν and the kinetic energy K_μ of the muon. Assume rest masses $M_\pi > m_\mu$ to be known.

solution Since the total relativistic energy is conserved, we can use Eq. [19 . 63] to express the energies of the neutrino and muon. The total energy of the system is

$$M_\pi c^2 = (0 + p_\nu^2 c^2)^{1/2} + (m_\mu^2 c^4 + p_\mu^2 c^2)^{1/2}$$

and since momentum is conserved, $p_\nu = -p_\mu$. Therefore we can solve this equation for p_ν, the momentum of the neutrino:

$$p_\nu = \frac{(M_\pi^2 - m_\mu^2)c}{2M_\pi}$$

The kinetic energy K_μ of the muon is then

$$K_\mu = E_\mu - m_\mu c^2 = \sqrt{m_\mu^2 c^4 + p_\mu^2 c^2} - m_\mu c^2$$

and substituting for $p_\mu = p_\nu$ in the square root and collecting terms, we obtain

$$K_\mu = \frac{\frac{1}{2}(M_\pi - m_\mu)^2 c^2}{M_\pi}$$

The relativistic definitions of momentum, mass, and energy are, to the best of our knowledge, based on universal physical laws. They exist independently of the presence of light or its actual speed in a given medium. They hold in a pitch-black room just as well as in daylight and are in no way "caused" by light. Light just happens to be the wave which, in vacuum, propagates at the maximum relative speed allowed in the universe. It was this particular fact that led to the discovery of the relativistic laws of mechanics, just as the elliptic orbit of Mars was the particular fact that led to Newtonian mechanics. If the facts of nuclear physics had been available at the time, the laws of relativistic mechanics could just as well have been derived from them as from the speed of light.

19 • 9 relativistic center of mass

The problem of relativistic binary elastic collisions is an important one in nuclear physics and high-energy physics, which deals with nuclear and subnuclear particles moving at very high speeds. The relativistic center-of-mass system affords a useful approach to such problems. Suppose a particle of mass m_1 is approaching a stationary particle of mass m_2, where m_1 and m_2 are the rest masses measured in the stationary reference frame S. The approach speed of m_1 is measured as u_1 in the S frame.

We first set up the conservation equations for relativistic momentum and mass:

$$m_1\gamma_{u1}u_1 = m_1\gamma_{v1}v_1 + m_2\gamma_{v2}v_2 \qquad [19 \cdot 65]$$

and

$$m_1\gamma_{u1} + m_2 = m_1\gamma_{v1} + m_2\gamma_{v2} \qquad [19 . 66]$$

We can use the algebraic identities

$$\gamma_v = \frac{1}{\sqrt{1 - v^2/c^2}} \qquad \gamma_v v = \begin{cases} c\sqrt{\gamma_v^2 - 1} & v > 0 \\ -c\sqrt{\gamma_v^2 - 1} & v < 0 \end{cases}$$

to put Eq. [19 . 65] in the form

$$m_1\sqrt{\gamma_{u1}^2 - 1} = \pm m_1\sqrt{\gamma_{v1}^2 - 1} + m_2\sqrt{\gamma_{v2}^2 - 1} \qquad [19 . 67]$$

so that we have two equations, [19 . 66] and [19 . 67], to be solved for γ_{v1} and γ_{v2}. The plus-or-minus sign indicates our uncertainty about the sign of v_1, but there is only one physically consistent solution.

We can use the fact that a dilation factor is always greater than unity to establish limits on the solutions of Eq. [19 . 66]:

$$m_2 < m_2\gamma_{v2} = m_1\gamma_{u1} + m_2 - m_1\gamma_{v1}$$

$$m_1 < m_1\gamma_{v1} = m_1\gamma_{u1} + m_2 - m_2\gamma_{v2}$$

or, dividing these inequalities by m_2 and m_1 respectively,

$$1 < \gamma_{v1} < \gamma_{u1} \qquad 1 < \gamma_{v2} < \frac{m_1}{m_2}(\gamma_{u1} - 1) + 1 \qquad [19 . 68]$$

It is now feasible to solve for γ_{v1} and γ_{v2}, using a trial-and-error method with a programmable calculator, and then to find the velocities and momenta after collision. However, we can circumvent this cumbersome procedure by making use of the *relativistic center of mass* of a two-particle system moving with some velocity $\mathbf{v}_c$ relative to the S frame.

As we saw in Sec. 10 . 2, in center-of-mass coordinates the total momentum $\mathbf{p}'$ of the system is always zero:

$$\sum_i \mathbf{p}_i' = \mathbf{0}$$

This momentum is transformed back to the S frame by the relation

$$\sum_i \mathbf{p}_i' = \gamma\sum_i(\mathbf{p}_i - m_i\gamma_{vi}\mathbf{v}_c) = \mathbf{0}$$

where $\gamma = 1/\sqrt{1 - v_c^2/c^2}$ and $\gamma_{vi} = 1/\sqrt{1 - v_i^2/c^2}$. Therefore, in the S frame,

$$\sum_i \mathbf{p}_i = \left(\sum_i m_i\gamma_{vi}\right)\mathbf{v}_c = M\mathbf{v}_c \qquad [19 . 69]$$

where M is the total relativistic mass in the S frame. The magnitude of $\mathbf{v}_c$, then, is simply the ratio of the total momentum to the total relativistic mass in the S frame, and in the classical limit Eq. [19 . 69] reduces to Eq. [10 . 1].

Since the total momentum of our two-particle system is zero in center-of-mass coordinates, the initial and final momenta are

$$p_{1i}' = -p_{2i}' \qquad p_{1f}' = -p_{2f}' \qquad [19 . 70]$$

It can be shown that conservation of energy also requires that

$$p'_{1i} = -p'_{1f} \qquad p'_{2i} = -p'_{2f} \qquad [19 \cdot 71]$$

Therefore, if we know one momentum in center-of-mass coordinates, we know all four. The easiest one to find is

$$p'_{2i} = \gamma(0 - m_2 v_c) \qquad [19 \cdot 72]$$

The final momenta in the S frame can then immediately be computed from the transformation of Eq. [19 . 45], with $V = -v_c$:

$$p = \gamma(p' + m v_c \gamma'_v) \qquad [19 \cdot 73]$$

where $\gamma'_v = 1/\sqrt{1 - v'^2/c^2}$. The steps in the solution are as follows:

1 Find $v_c = (m_1\gamma_{u1} + m_2)^{-1} m_1 \gamma_{u1} u_1$.
2 Compute $p'_{2i} = -m_2\gamma v_c = m'_1 v'_1 = -m'_2 v'_2$.
3 Transform p'_{1f} and p'_{2f} to find p_{1f} and p_{2f} in the S system.
4 Verify the results against Eqs. [19 . 65] and [19 . 66].

As an example, consider the two-particle collision in Fig. 19 . 11, where $u_1 = 3v$, $m_1 = m$, and $m_2 = 2m$. For this computation let us take

$$3v = \tfrac{1}{2}\sqrt{3}\, c = 0.86602c$$

and find the final recoil momenta in S coordinates. First we compute $\gamma_{u1} = 2$ and $p_{1i} = m_1 u_1 \gamma_{u1} = 6mv$, so that Eq. [19 . 69] gives

$$(2m + 2m) v_c = 6mv,$$

or

$$v_c = \tfrac{3}{2}v \text{ and } \gamma = 1/\sqrt{1 - v_c^2/c^2} = 1.10940$$

Therefore, in step 2,

$$p'_{2i} = -\gamma(2mv_c) = -3.32820mv = p'_{1f} = -p'_{2f}$$

We now transform the final momenta to the S frame from Eq. [19 . 73]:

$$p_{1f} = \gamma(p'_{1f} + m\gamma'_{v1} v_c) \qquad p_{2f} = \gamma(p'_{2f} + 2m\gamma'_{v2} v_c) \qquad [19 \cdot 74]$$

where the dilation factors for $v^2/c^2 = \frac{1}{12}$ are, from Eq. [19 . 62]:

$$\gamma'_{v1} = \sqrt{1 + \frac{11.0769 m^2 v^2}{m^2 c^2}} = 1.38675$$

$$\gamma'_{v2} = \sqrt{1 + \frac{11.0769 m^2 v^2}{4m^2 c^2}} = 1.10940$$

Note as a check that $\gamma'_{v2} = \gamma$, since the velocity of the stationary particle in the S' frame will be $-v_c$. Substituting these values of γ' into Eq. [19 . 74] gives us the final momenta:

$$p_{1f} = 1.10940\,(-3.32820mv + 1.5 \times 1.38675mv)$$
$$= -1.38461mv$$

$$p_{2f} = 1.10940\,(3.32820mv + 3 \times 1.10940mv) = 7.38461mv$$

The sum of these two momenta is still $6mv$, so that momentum has been conserved (compare this result to the classical predictions of $-mv$ and $4mv$). It is also possible to verify that the initial and final relativistic mass of the system is $M = 4.00000$ and to show that the inequalities of [19 . 68] are satisfied:

$$1 < \gamma_{v1} = 1.46154 < 2 \qquad 1 < \gamma_{v2} = 1.07692 < 1.5$$

PROBLEMS

19 • 1 the Michelson-Morley experiment

19 • 1 If the ether were an elastic medium with shear modulus as great as that of steel, $S = 8 \times 10^{10}$ Pa, what would its density be?

19 • 2 Show that the law of reflection holds generally for sound waves in a wind if the windspeed is tangential to the reflecting surface. (*Hint:* Consider the path of reflected rays from the reference frame of an observer moving with respect to a stationary medium.)

19 • 3 What is the approximate formula for the difference Δt in the travel time of the two light beams at the telescope in Fig. 19 . 2? What is the time difference in seconds?

19 • 4 If the mirror M on the Michelson interferometer is movable, how far would it have to be moved to cause a shift from one dark fringe to the next?

****19 • 5*** An airplane has a sound reflector fixed at the end of one wingtip and another reflector on the tail, so that sound from a source near the center of the plane travels a distance l_t to the tail. When this sound is reflected back to the source, it meets sound pulses that are being reflected back at right angles to it from the wingtip reflector a distance l_w from the source. If the time delay Δt between two reflected pulses that were emitted simultaneously can be measured, set up a flowchart to compute the airspeed v of the airplane. (*Hint:* Begin by working in terms of $\gamma = (1 - v^2/c^2)^{-1/2}$, where c is the speed of sound.)

19 • 3 the Lorentz transformation

19 • 6 Find the Lorentz transformation equations (*a*) for $V = 0.8c$. (*b*) For $V = 0.1c$.

19 • 7 Derive the inverse transformation of Eqs. [19 . 20] by solving for x and t. Show that this is equivalent to transforming to a reference frame moving with velocity $\mathbf{V}' = -V\mathbf{i}$, in agreement with Einstein's first postulate.

19 • 8 A star is receding from the earth with a speed $v = 0.7c$. An observer on the ground measures two events taking place at the same location on the star, but 2 min apart. (*a*) Find the space-time distance of the two events. (*b*) Find the time between their occurrence as it would be measured by an observer moving with the star.

19 • 9 Show that a circle of unit radius traveling at speed V will appear to be an ellipse to a stationary observer. What is the ratio of the minor axis to the major axis of the ellipse?

19 • 10 Two events occur simultaneously in the S' frame at time $t' = 0$ and positions $x' = l$ and $-l$. What are the corresponding positions x and times t of these events in the S frame? Are the two events simultaneous in the S frame?

19 • 4 time dilation

19 • 11 Electrons are moving through a long vacuum tube at a constant speed of 2.4×10^8 m/s relative to the laboratory. The distance from a point O to the end of the tube is $l = 4$ m. (*a*) If an electron passes O at time $t = 0$, at what time t will it reach the end of the tube? (*b*) If one reading could be made on a single clock fixed in the proper frame of the electron, what would be the difference between times t and τ? (*c*) If the speed of the electron were 3×10^6 m/s, what would t and τ be?

19 • 12 The π-meson, or *pion,* is a short-lived electrically charged particle that occurs in cosmic radiation and can also be produced artificially in a high-energy accelerator. It is known from laboratory studies to have a proper mean life of 2.54×10^{-8} s before it converts spontaneously into other forms. In one experiment, pions in a long vacuum tube were found to travel an average distance of 3 m during this brief time. Show that their speed relative to the laboratory was 1.1×10^8 m/s.

19 • 13 A spaceship is equipped with a cesium atomic clock that is accurate to within 3 μs/day. If the spaceship is moving away from the earth at the escape velocity of $v = 11.2$ km/s, how slow is this clock relative to an identical clock on earth?

19 • 14 Assume that there are *two* clocks in a moving system, one at the origin O' and the other at a distance l' to the left of O'. (*a*) If the S' frame moves to the right with speed V, and O' crosses the stationary origin O at time $t'_0 = t_0 = 0$, at what time t will the second clock pass O, expressed in terms of the elapsed time $\Delta t'$ and the length l'? (*b*) What was the value of t for this clock at $t' = 0$? (*c*) Show that comparison of the elapsed times Δt and $\Delta t'$ still yield the same formula for the time dilation, despite the complications introduced into the Lorentz transformations by l'.

19 • 15 An astronaut goes on a two-week journey to the moon and back (770,000 km roundtrip). How many seconds of additional biological life has she gained as a result of her trip?

19 • 16 In the derivation of the Doppler effect in Example 19 . 3, we assumed that both observer and emitter were positioned on the x axis, parallel to the relative motion of the two reference frames. However, the relativistic Doppler effect also depends on the angle between the direction of motion x and the line of sight to the emitter. Show that for an observer at O and an emitter crossing the x axis at a distance y above the origin, there is a *transverse* Doppler effect, with $\nu'/\nu = \gamma$.

19 • 17 A homeward-bound astronaut is looking for Mars as a landmark, but the red ($\lambda = 7{,}000$ Å) planet appears blue ($\lambda = 5{,}000$ Å) to him, so he changes course. How rapidly was he approaching Mars before he made this error?

19 • 18 An observer in the S' frame, approaching a second observer in the S frame, emits signals with frequency ν'. Each time such a signal is received, the second observer returns the signals to the S' frame. What is the frequency of this return signal in the S' frame?

19 • 5 contraction and simultaneity

19 • 19 A spherical body of proper diameter $l_0 = 4$ m is moving past the earth at a speed of 1.8×10^8 m/s. Find its maximum dimensions parallel and perpendicular to the direction of motion as measured by an observer on the earth.

19 • 20 How does volume of the oblate spheroid seen by the stationary observer in Prob. 19 . 19 compare to that of the spherical body?

19 • 21 An aircraft of proper length $l_0 = 14$ m is flying parallel to the ground at a constant speed of 600 m/s ($\doteq$ 2200 km/h). (*a*) Show that observers on the ground should expect the aircraft to be 0.028 nm (1 nm $= 10^{-9}$ m) shorter than its proper length. (*b*) If the radical $\sqrt{1 - v^2/c^2}$ is evaluated in this problem by using the approximation formula $(1 \pm a)^m = 1 \pm ma$, is the error introduced large enough to affect the length difference 0.028 nm?

19 • 22 A cubical box of proper volume 27 m^3 is on a truck moving along a straight road at 50 m/s. If the speed of light in vacuum were only 100 m/s, what would be the apparent volume of the passing box to an observer standing by the side of the road?

19 • 23 A telescope of proper length l_0 is mounted on a moving flatcar traveling at a constant speed V on a straight horizontal track. An observer on the car finds that, relative to the surface of the flatcar, the angle of elevation of the telescope is θ'. (*a*) Find the length l of the telescope and its angle of elevation θ as seen by observers on the ground. (*b*) Graph l/l_0 and θ as functions of β for $\theta' = 45°$.

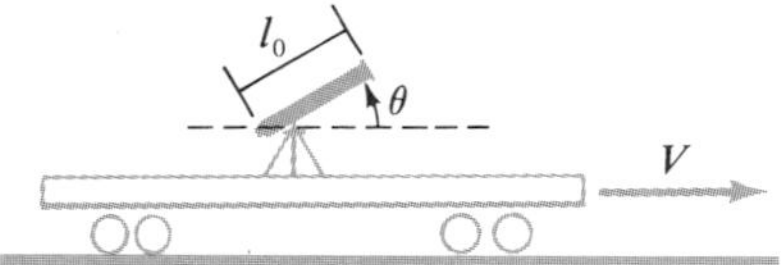

19 • 24 At two points x_1 and x_2 next to a railroad track flashbulbs are installed which flash simultaneously at time t and are recorded on film at points x_1' and x_2' on a passing train. (*a*) Show that a passenger on the train would conclude that the distance $\Delta x'$ is greater than the distance Δx between the lamps on the ground. (*b*) Show that if the rear and front lamps could be set to flash at times that differed from t by an amount $t \pm \Delta x'/2c$, the emitted signals would appear simultaneous to the passenger; that is, they would meet halfway between x_1' and x_2'.

19 • 25 At two points x_1' and x_2' on a train flashbulbs are installed to flash simultaneously when the train passes photographic recorders located at x_1 and x_2 beside the track. Show that an observer stationed on the ground midway between x_1 and x_2 will conclude that the rear lamp flashes first and that the time interval between the two flashes is $\Delta t = (V/c^2)\,\Delta x$, where V is the groundspeed of the train.

19 • 26 In Probs. 19 . 9 and 19 . 20 we saw that the Lorentz-FitzGerald contraction deforms a moving circle into an ellipse due to the shortening of the radius parallel to the direction of motion. However, this effect is predicated on simultaneous measurement of all points of the ellipse in the stationary frame. Suppose an observer in the S frame receives light pulses from various points on the moving circle and forms an image of the object from all the information *received* at a given instant of time t_1. In that case, the signals received must start at different points (x,y) and times t if they are all to reach the observer at the same instant. (*a*) Show that for points (x',y') of a moving system emitting light at time t, the Lorentz transformations predict an *apparent* position in the S frame of

$$x = \gamma\left[x' + \beta\gamma c t_1 - \beta\sqrt{y'^2 + (x' + \beta\gamma c t_1)^2}\right] \qquad y = y'$$

for an arrival time t_1 at the stationary observer. (*b*) Compute the apparent shape of a semicircle of radius

$$R = 1 \text{ light} \cdot \text{s} = 3 \times 10^8 \text{ m}$$

moving with speed $V = 0.9c$ whose center O' passes an observer at O at time $t_1 = 0$. [*Hint:* Take an evenly spaced collection of points (x',y'), compute (x,y), and connect the points by smooth curves.] (*c*) Repeat the computation for $t = 1$ s and graph your results in units of R.

19 • 6 relativistic transformation of velocities

19 • 27 Relative to the laboratory, one atomic particle P_1 is moving to the right with constant speed $0.94c$, and another atomic particle P_2 is moving to the left with constant speed $0.9c$. (*a*) Show that the velocity of P_1 relative to P_2 is $0.997c$, directed to the right. (*b*) What is the velocity of P_2 relative to P_1?

19 • 28 A spaceship is equipped with a radio transmitter that emits signals at intervals of 1 s. (*a*) Use the velocity-transformation equations to compute the speed of the signals relative to the earth if the spaceship is moving directly away from the earth with speed $V = 0.9c$. (*b*) Directly toward the earth with speed $V = 10^{-4}c$. (*c*) Why should we expect these computed values to be in agreement with Einstein's first postulate? (*d*) What is the time interval between successive signals received at the earth if the spaceship is receding with speed $V = 0.6c$? If it is approaching with speed $V = 0.6c$?

19 • 29 As seen from a reference frame S', a particle is moving along the y' axis with an instantaneous $v'_y = 0.9c$. The speed of frame S' relative to another frame S is $V = V_x = 0.9c$. Find the corresponding velocity $\mathbf{v}$ of the particle in frame S and the angle it makes with the x axis.

19 • 30 A particle that is accelerating can be thought of as passing through a succession of instantaneous proper reference frames, and the acceleration in a system is defined as dv/dt in that system. Show that the acceleration instantaneously parallel to the motion of a particle transforms as $a'_x = \gamma^3 a_x$, whereas the instantaneous acceleration perpendicular to its motion transforms as $a'_y = \gamma a_y$. Take the primed system as the one which is instantaneously moving with the particle (chosen as the x' direction).

19 • 7 relativistic momentum and mass

19 • 31 Solve Eq. [19 . 41] for v as a function of p and find v when $p = \frac{1}{2}mc$.

19 • 32 Express the speed v of a particle as a function of its velocity-dilation factor.

19 • 33 The rest mass of an electron is, to three significant figures, $m_e = 9.11 \times 10^{-31}$ kg. (*a*) Compute the relativistic momentum of an electron for speeds $v = 0.8c$ and $v = 0.001c$. (*b*) What is the relativistic mass m of the electron at each speed? (*c*) What would the momentum of an electron be if it were moving through water at a speed $v = 0.8c$, which is faster than the speed of light in water?

19 • 34 A cube of side length l has rest mass m_0. What is its density as seen by an observer moving with constant speed V parallel to one edge of the cube?

19 • 35 In a collision between two identical particles, one stationary and the other moving with an approach speed u, the center of mass of the two-particle system is seen by a stationary observer as moving with speed

$$v_c = \frac{u}{1 + \sqrt{1 - u^2/c^2}}$$

(*a*) Find the initial momenta of the two particles in the center-of-mass system. (*b*) If the collision is perfectly elastic, what are the recoil momenta of the two particles in the center-of-mass system? (*c*) Find the recoil momenta in the S frame and compare your result with the predictions of classical mechanics.

19 • 36 A stationary particle of rest mass m_0 is struck by an identical particle moving with speed u in a perfectly inelastic collision. (*a*) What is the combined relativistic mass of the two particles before collision? (*b*) After the collision? (*c*) What is the velocity of the two-particle system after collision?

19 • 37 A particle of rest mass m_0 is traveling with velocity $v_x\mathbf{i}$ in the S frame. (*a*) What is its velocity in an S' frame moving with velocity $V\mathbf{i}$ relative to the S frame? (*b*) Substitute this expression for v'_x in the momentum equation $p'_x = m'v'_x$ and show that this agrees with the transformation of Eq. [19 . 45].

19 • 8 mass and energy

19 • 38 A particle has kinetic energy $K = 9E_0$, where E_0 is its rest energy. (*a*) What is its speed? (*b*) Find the ratio of its relativistic kinetic energy to the classical value $K_0 = \frac{1}{2}m_0v^2$. (*c*) Find the ratio of its relativistic momentum to its classical momentum p_0.

19 • 39 Electrons have been accelerated to within one-millionth of the speed of light in vacuum—that is, to $v = (1 - 10^{-6})c$. For an electron of rest mass $m_0 = 9.11 \times 10^{-31}$ kg traveling at this speed, find (*a*) its rest energy, (*b*) its total energy, (*c*) its relativistic kinetic energy, (*d*) its relativistic momentum, and (*e*) the ratio of its relativistic and classical kinetic energies.

19 • 40 When the mass and kinetic energy of a particle are known, it is useful to have a ready means for judging whether the relativistic or the simpler Newtonian expressions should be used to compute speed and momentum. (*a*) Prove that

$$\frac{v^2}{c^2} = 1 - \frac{1}{(1 + K/E_0)^2} = \left(\frac{pc}{E}\right)^2$$

(*b*) Prove that if $K \ll E_0$, then $v^2/c^2 \doteq 2K/E_0$. (*c*) Compute and graph K/E_0 for $0 < \beta < 1$ at intervals of $\Delta\beta = 0.1$.

19 • 41 Show that relativistic momentum and kinetic energy are related by

$$p^2c^2 = 2m_0c^2K + K^2$$

and compare this formula with the results of Prob. 19 . 38.

19 • 42 In a typical nuclear reaction, a particle of mass m_1 and kinetic energy K_1 makes a direct and completely inelastic impact with an atomic nucleus of mass m_2 that is initially at rest ($K_2 = 0$). The resulting "compound" nucleus is highly unstable and it usually is very short-lived ($\Delta t \doteq 10^{-14}$ s). It spontaneously "explodes," usually into just two particles, of masses m_3 and m_4, that fly out from the point of disintegration with kinetic energies K_3 and K_4. The total relativistic energy, E_T, of the system is conserved; however, if the product masses are less than the reacting masses, the disintegration energy thus released is $\Delta K = \Delta m\, c^2$. In the famous Cockroft-Walton experiment of 1932, a hydrogen nucleus was captured by a lithium nucleus, and both disintegrated into two helium nuclei. The relativistic energy formula can be verified by comparing the measured values of K with those of the nuclear masses. Using $K_H = 8.01 \times 10^{-14}$ J and $K_{He} = 1.415 \times 10^{-12}$ J as experimental data and a table of nuclear masses, verify this formula. What is the percentage loss of mass?

19 • 43 In discussing world energy consumption, the unit used is 10^{21} J ($\doteq 10^{18}$ Btu), enough heat energy to raise Lake Huron from 20°C to 80°C. The projected consumption by 1984 is 0.35×10^{21} J/y, with an annual increase of 5 percent. To avoid drastic environmental change, consumption must ultimately be limited to 15×10^{21} J/y (world consumption from 1 A.D. to 1850 A.D. is estimated to have totaled less than 1×10^{21} J). Present natural reserves are estimated at 150×10^{21} J, and fissionable materials for atomic reactors are estimated at 5000×10^{21} J. In fusion processes it is impractical to try to reproduce the solar process; instead, two deuterium nuclei, or *deuterons,* are used to form helium. A deuteron consisting of one proton plus one neutron is an isotope of hydrogen. If the average energy release per deuteron is 8×10^{-13} J and deuterons make up 0.015 percent of the hydrogen in our 10^9 km^3 of sea water, how many years will the supply of energy last at 15×10^{21} J/y? (A gram of water contains 3×10^{22} molecules of H_2O.)

19 • 44 The electrically charged particle known as the K^+-meson has a mean life of about 10^{-2} μs and a mass 966 times the mass of an electron.

Experiment indicates that it disintegrates into a neutrino, which is a massless neutral particle, and an electrically charged muon with a mass 207 times that of an electron. With respect to the K^+-meson's proper frame, compute the momentum, kinetic energy, total energy, and speed (*a*) of the neutrino and (*b*) of the muon.

19 • 9 relativistic center of mass

***19 • 45** Construct a flowchart for the computation of recoil momenta, masses, and kinetic energies in a perfectly elastic binary collision of particles of rest masses m_1 and m_2, given approach velocities u_{1i}, u_{2i}. Check your results against the example computed in the text.

***19 • 46** Repeat Prob. 19 . 45 for a perfectly inelastic collision.

answers

19 • 1 $\rho = 9 \times 10^{-10}\,\mathrm{g/cm^3}$
19 • 3 $\Delta t \doteq (l/c)(v/c)^2 \doteq 4 \times 10^{-16}$ s
19 • 4 $\Delta l = \frac{1}{2}\lambda$
19 • 5 $\gamma = (1/2l_t)\left(l_w + \sqrt{l_w^2 + 2l_t c\,\Delta t}\right)$
19 • 6 (*a*) $x' = 1.667x - (4.000 \times 10^8\,\mathrm{m/s})t$, $t' = 1.667t - (4.444 \times 10^{-9}\,\mathrm{s/m})x$; (*b*) $x' = 1.005x - (3.015 \times 10^7\,\mathrm{m/s})t$, $t' = 1.005t - (3.350 \times 10^{-10}\,\mathrm{s/m})x$
19 • 8 (*a*) $\Delta s = 85.7\{c\}$ m; (*b*) $\Delta t' = 85.7$ s
19 • 9 $b/a = \sqrt{1 - V^2/c^2}$
19 • 10 $x = \pm\gamma l$, $t = \pm\gamma Vl/c^2$, $\Delta t = 2\gamma Vl/c^2$
19 • 11 (*a*) $t = 1.67 \times 10^{-8}$ s; (*b*) $t - \tau = 10^{-8}$ s; (*c*) $t = 400.00\,\mathrm{m}/c$, $\tau = 399.98\,\mathrm{m}/c$
19 • 12 $v = 1 \cdot 1 \times 10^8$ m/s
19 • 13 $\Delta t = 48\,\mu$s/day
19 • 14 (*a*) $t = l'\gamma/V$; (*b*) $t = -\gamma Vl'/c^2$; (*c*) $\Delta t = \gamma\,\Delta t'$
19 • 15 $\Delta t = 2.6\,\mu$s
19 • 17 $v = 0.324c$
19 • 18 $\nu = \nu'(1 + \beta)/(1 - \beta)$
19 • 19 $l_\parallel = 3.2$ m, $l_\perp = 4.0$ m
19 • 20 Smaller by the factor γ
19 • 21 (*b*) Yes
19 • 22 $V = 23.4\,\mathrm{m}^3$
19 • 23 (*a*) $l = l_0\sqrt{(1 - \beta^2)\cos^2\theta' + \sin^2\theta'}$, $\theta = \tan^{-1}\left(\theta'/\sqrt{1 - \beta^2}\right)$
19 • 27 (*b*) $v = 0.997c$ to the left
19 • 28 (*a*) $v = c$; (*b*) $v = c$; (*d*) $\Delta t = 2$ s, $\Delta t = 0.5$ s
19 • 29 $v = 0.9c\mathbf{i} + 0.39c\mathbf{j}$, $\theta = 23°\,23'$
19 • 31 $v = c/\sqrt{5}$, see Example 19 . 7
19 • 32 $v^2 = c^2(\gamma_v^2 - 1)/\gamma_v^2$
19 • 33 (*a*) $p = 3.64 \times 10^{-22}$ kg · m/s, $p = 2.73 \times 10^{-25}$ kg · m/s; (*b*) $m = 1.52 \times 10^{-30}$ kg, $m = 9.11 \times 10^{-31}$ kg; (*c*) same as *in vacuo*
19 • 34 $\rho = (m_0/l^3)/(1 - V^2/c^2)$
19 • 35 (*a*) $p'_{1i} = m_0V\gamma\gamma_u = -p'_{2i}$; (*b*) $-p'_{1f} = p'_{2f} = m_0V\gamma\gamma_u$; (*c*) $p_{1f} = 0$, $p_{2f} = m_0\gamma_u u$
19 • 36 (*a*) $m_{tot} = m(1 + \gamma_u)$; (*b*) $m_{tot} = m(1 + \gamma_u)$; (*c*) $v = \gamma_u u/(1 + \gamma_u)$
19 • 37 $v'_x\mathbf{i} = [(v_x - V)/(1 - v_xV/c^2)]\mathbf{i}$
19 • 38 (*a*) $v = 0.995c$; (*b*) $K/K_0 = 18.18$; (*c*) $p/p_0 = 10$
19 • 39 (*a*) $E_0 = 8 \times 10^{-14}$ J; (*b*) $E = 6 \times 10^{-11}$ J; (*c*) $K \doteq E$; (*d*) $p = 2 \times 10^{-19}$ kg · m/s; (*e*) $K/K_0 = 1000$
19 • 42 $\Delta m = -0.2$ percent
19 • 43 $\Delta t = 7.9 \times 10^9$ years
19 • 44 (*a*) $p_\nu = 1.26 \times 10^{-19}$ kg · m/s, $K_\nu = 3.78 \times 10^{-11}\,\mathrm{J} = E_\nu$, $v_\nu = c$; (*b*) $p_\mu = 1.26 \times 10^{-19}$ kg · m/s, $K_\mu = 2.44 \times 10^{-11}$ J, $E_\mu = 4.14 \times 10^{-11}$ J, $v_\mu = 0.91c$

CHAPTER TWENTY

temperature

Let the first instrument be that which may serve, (as likewise several others) to shew the changes of the Air, *in reference* to Heat *and* Cold, *and is commonly call'd a Thermometer:... Make the Ball of this Instrument of such a Capacity, and joyn thereto a Cane of such a bore, that by filling it to a certain mark in the Neck with Spirit of Wine, the simple cold of Snow or Ice* Externally Applyed, *may not be able to condense it below the 20* deg. *of the Cane; nor on the contrary, the greatest vigour of the Sun's Rays at* Midsummer, *to Rarifie it above 80 deg.... The next thing is to divide the Neck of the* Instrument *or Tube into Degrees exactly; therefore first, divide the whole* Tube *into Ten equal Parts with Compasses, marking each of them with a knob of white* Enamel.... *This done, and with the proof of* Sun *and* Ice, *the proportion of the Spirit of Wine found; the Mouth of the* Tube *must be closed with* Hermes *Seal at the flame of a Lamp, and the* Thermometer *is finish'd.*

Essayes of Natural Experiments Made in the Academie del Cimento, *translated by Richard Waller, 1684*

The concept of temperature and its measurement is the characteristic feature of the science of *thermodynamics,* which is concerned with physical phenomena involving heat. The main purpose of this chapter is to develop an operational definition of temperature and examine its consequences. This new variable usually refers to a number that expresses, on some definite scale, how "hot" or "cold" an object is. However, such descriptions are entirely subjective; when we say an object feels cold, we mean that heat, or thermal energy, is transferred from our skin to the object we have touched. Furthermore, the sensation is determined not only by the amount of energy transferred, but by the rate at which it is transferred. Therefore, in all processes involving heat energy, it is essential to have clear and definite ideas of what is meant by *temperature, heat,* and *heat flow,* as well as their interrelationships.

Temperature and heat are not the same thing. The amount of heat energy in a system depends on its size, whereas temperature refers to the intensity of this energy throughout the system, irrespective of its size. It takes a bigger fire to boil a gallon of water than to boil a cup of water, yet both have the same boiling temperature. Temperature differs fundamentally from the well-defined mechanical quantities discussed so far, because it is related to the average kinetic energy of a large aggregate of particles. Such a statistical relation cannot be made the basis of a constitutive definition. Therefore we shall develop several different operational definitions of temperature in this chapter.

20 • 1 thermometry

Let us approach the operational definition of temperature in two steps. First we must define what we mean by one temperature being less than, equal to, or greater than another temperature; then we can define a means of measurement—a *thermometer*—and a temperature *scale*. When our discussions involve the concept of heat, we shall, for the present, rely on intuition, experience, and Joule's discovery (see Sec. 8 . 6) that heat is just another form of energy.

A *thermodynamic system* consists of a finite, identifiable quantity of matter and energy. Such a system may have definite boundaries like the gas in a container, or its volume may be indefinite, like the sun's atmosphere, which is continually flowing outward through the solar system. Although the emphasis in thermodynamics is on heat, any type of energy is fair game, especially if it is ultimately converted to or from heat energy. In general we shall deal with systems of constant mass, although we do speak of mass or energy as "entering" or "leaving" a system in cases where such quantities can be unambiguously identified.

Two systems are said to be in *thermal contact* when heat energy can pass from one to the other. A pot on a stove is obviously in thermal contact with the cooking fire; less obviously, the earth is in thermal contact with the sun. We define "hotness" by saying that when two bodies are placed in thermal contact, heat energy is transferred from the hotter body to the colder. This leads directly to the first step in our definition of temperature:

> If no heat energy is transferred when two systems are placed in thermal contact, then they are at the same temperature and are said to be in *thermal equilibrium*. If heat is transferred from one system to the other, the system from which it flows is defined to be at a ***higher temperature***.

Clearly, if two finite bodies initially at different temperatures are kept in thermal contact long enough, heat energy will eventually cease to flow between them, and they will be in thermal equilibrium.

To "measure" the temperature of two finite bodies, we need only establish that one of them is also in thermal equilibrium with a third body which is taken as the standard of reference of a fixed value of temperature. This fact is embodied in the *zeroth law of thermodynamics:*

> If two systems are in thermal equilibrium with a third system, then they are in thermal equilibrium with each other.

As a standard of reference we can select any one of numerous temperature-dependent, or *thermometric,* properties, such as the volume or pressure of a gas, the length of a metal rod, or the electrical resistance of a wire, and use that property to indicate when some reference temperature is reached. Then, by the zeroth law, we can test whether other bodies are at the same temperature by placing them in contact with the standard and observing the net effect on its thermometric property when thermal equilibrium is attained. A device for doing this is called a *thermometer.*

In the sixteenth century there was a revival of interest in various ancient devices which depended on the expansion of air when heated. Their adaptation to the measurement of temperature is generally attributed to Galileo in about 1600. Two fixed points of reference were eventually selected: the *ice point,* defined as the temperature at which pure ice coexists in equilibrium with air-saturated water under a pressure of 1 atm, and the *steam point,* defined as the temperature at which steam is in equilibrium with pure water under 1 atm of pressure.*

After the fixed reference points are selected, the interval between them is divided into an arbitrary number of temperature intervals, or *degrees.* For example, we could observe the length l_i of a given iron rod at the ice point and its length l_s at the steam point, arbitrarily calling the corresponding temperatures $\tau = 0°$ and $\tau = 100°$, respectively. This is what is done in the *Celsius,* or *centigrade, system,* proposed in 1742 by the Swedish astronomer Anders Celsius, which measures temperature in *degrees centigrade* (°C). The temperature corresponding to some other length l is then, by definition,

$$\blacktriangleright \quad \tau = 100°\text{C}\,\frac{l - l_i}{l_s - l_i} \qquad [20 \cdot 1]$$

We shall use the symbol τ for centigrade temperature to distinguish it from the absolute Kelvin temperature T.

This method can also be applied to the observed lengths of a mercury column in a mercury-in-glass thermometer to define Celsius temperature on that instrument. Although an operational definition which embodies a *linear* dependence of the thermometric property on temperature is the most convenient, there are other types of thermometers which utilize a nonlinear dependence on temperature. The important point is that the dependence of the thermometric property be unique, well defined, and measurable. Pressure, the force per unit area which a substance exerts against the walls of its container, is such a property. As we saw in Secs. 14 . 2 and 15 . 1, a fluid transmits pressure equally in all directions. For a fluid under gravity, the

*The ice point was suggested by Hooke in 1664 and the steam point a few months later by Huygens.

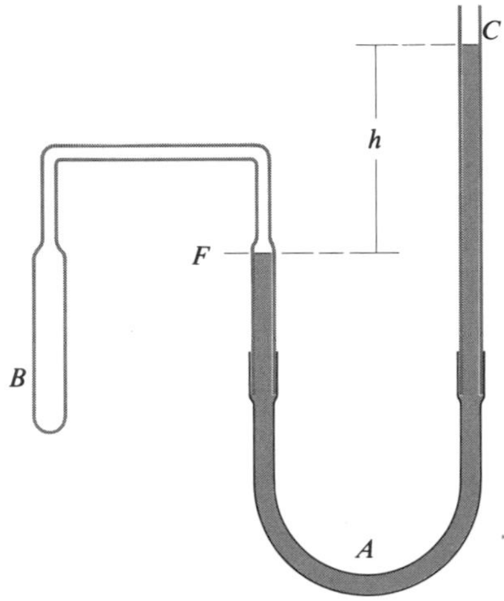

figure 20 • 1 *A simple constant-volume gas thermometer*

change in pressure with height h is given by $dP = -\rho g\, dh$, where ρ is the density of the fluid. For a liquid of constant density, the pressure on the bottom of a column of the liquid of height h is given by $P = \rho g h$.

The constant-volume gas thermometer, shown in Fig. 20 . 1, is based on the principle that a given mass of gas contained in a closed vessel of constant volume assumes a pressure which is entirely determined by the temperature. The bulb B, which contains the gas, is connected by a bent capillary tube to an open-tube manometer consisting of a flexible tube A and a glass tube C containing mercury. By raising or lowering tube C, the surface of the mercury in the other tube can always be brought to the *fiducial mark F*. This is done when the bulb B is first at the ice point, then at the steam point, and finally at the temperature τ to be determined. The pressure of the confined gas in each case is determined in the usual manner, by reading the difference in the heights of the two mercury columns and adding to this the atmospheric pressure in millimeters of mercury. The defining equation for Celsius temperature on a constant-volume gas scale is

$$\tau = 100°\text{C}\frac{P - P_i}{P_s - P_i} = 100°\text{C}\frac{h - h_i}{h_s - h_i} \qquad [20 \cdot 2]$$

This equation is similar to Eq. [20 . 1], except that the thermometric substance is now a gas and the thermometric property is its pressure.

Although the fixed-point temperatures are independent of the reference substance and its thermometric property, the properties of substances do not vary with temperature in exactly the same way, so that some particular substance and property must be taken as a standard. In 1842 the French physical scientist Henri Regnault selected the constant-volume gas thermometer, with hydrogen (the lightest gas) as the thermometric substance, as the ultimate standard of reference. However, there is a fundamental similarity in the behavior of gases at low pressures—so-called *ideal* gases—and today helium, nitrogen, and air are also used as thermometric substances in experimental work. One great advantage of helium is that it liquefies at a lower temperature than any other gas (−269°C) and can therefore be used for low-temperature measurements.

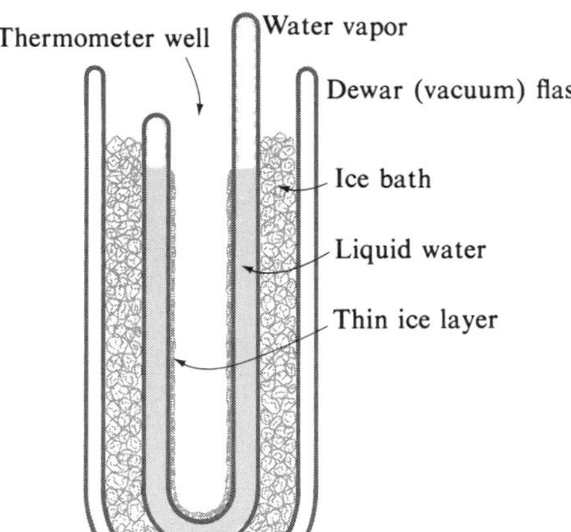

figure 20 • 2 *Schematic diagram of a triple-point cell*

In practice, neither the ice point nor the steam point can be reproduced experimentally with the accuracy necessary for present-day purposes. In 1954 the Tenth General Conference on Weights and Measures (Paris) adopted the "one-fixed-point" method of graduation proposed independently in 1665 by Boyle, Hooke, and Huygens, in which one point and the size of the degree is fixed. The point selected as being most accurately determinable is the *triple point of water*—the point at which ice, pure water, and water vapor coexist in equilibrium. The pressure of this point turns out to be 4.58 mm Hg, and its temperature is arbitrarily assigned the value 0.0100°C. If we added enough air to the triple-point cell of Fig. 20 . 2 to bring the pressure to 1 atm, the ice would melt and would not freeze again until the temperature was lowered to 0.000°C. These conditions, 0°C and 1 atm, are also known as *standard temperature and pressure.*

example 20 • 1 Suppose points C and F in Fig. 20 . 1 are at the same level, $h = 0$, in a room with the windows open. If the ice point is $h_i = 15$ mm and the steam point is $h_s = 45$ mm, where would you guess that the room is located?

solution From Eq. [20 . 2], the room temperature is

$$\tau = 100°\text{C}\frac{h - h_i}{h_s - h_i} = 100°\text{C}\frac{0 - 15}{45 - 15} = -50°\text{C}$$

Siberia seems a likely prospect.

example 20 • 2 D. G. Fahrenheit chose his lower fixed point to be 0°F, the temperature of "a mixture of water, of ice and of sal-ammoniac or even of sea-salt," and his upper point as 98°F, the temperature of "the blood of a healthy man." His ice and steam points therefore became 32°F and 212°F, respectively. Find the relation between Fahrenheit and Celsius temperatures.

solution Since both scales are linear, the corresponding temperature intervals are proportional:

$$\frac{\tau_{°F} - 32°F}{\tau_{°C} - 0°C} = \frac{212°F - 32°F}{100°C}$$

Therefore

$$\tau_{°F} = 32°F + \frac{9°F}{5°C}\tau_{°C}$$

20 • 2 the temperature scale

If we rearrange Eq. [20 . 2] for the constant-volume gas thermometer, we obtain the pressure equation

$$P = P_i(1 + \beta_0\tau) \qquad \beta_0 = \frac{0.01}{°C}\left(\frac{P_s}{P_i} - 1\right) \qquad [20 \cdot 3]$$

Ideal gases—which include most real gases at ordinary pressures and temperatures—give practically identical readings for $P(\tau)$. Therefore P_s/P_i and β_0 are constant for a constant-volume thermometer containing an ideal gas. A series of experiments with smaller and smaller pressures show that as the gas becomes more nearly ideal,

$$\lim_{P \to 0} \beta_0 = \frac{1}{273.15°C} \qquad [20 \cdot 4]$$

That is, a change of 1°C in temperature will result in a pressure change $\Delta P = P_i/273.15$. Provided the pressure is not too high and the temperature is well above that at which the gas liquefies, *all* real gases extrapolate to the *same* value of β_0 as the pressure becomes smaller.

Equation [20 . 3] is still somewhat clumsy. If we define a temperature scale such that the ice point is equal to $1/\beta_0 = 273.15°$, corresponding to $\tau = 0°C$, then we can substitute a new temperature scale $T = T_i + \tau$ in Eq. [20 . 3] to obtain

$$P = P_i(1 + \beta_0\tau) = P_i\left(1 + \frac{T - T_i}{T_i}\right) = P_i\frac{T}{T_i} \qquad [20 \cdot 5]$$

The result is an absolute temperature scale for the constant-volume ideal-gas thermometer known as the *Kelvin scale*, on which temperature is measured in *kelvins* (K) instead of degrees:

$$\blacktriangleright \quad T = (273.15\ \text{K})\frac{P}{P_i} \qquad [20 \cdot 6]$$

The kelvin is the same size as the degree on the Celsius scale, so that transformation from one scale to the other is just a matter of translating temperature coordinates by 273.15 K; ice melts at 0°C = 273.15 K and water boils at 100°C = 373.15 K. The zero point on the Kelvin scale, −273.15°C = 0 K, is known as *absolute zero.*

In the International System of units, the kelvin is one of the six basic units of measure. Since it is equal to an interval of 1 degree on either the centigrade or the absolute scale, the *difference* between two temperatures is, in kelvins,

$$100°\text{C} - 0°\text{C} = 373 \text{ K} - 273 \text{ K} = 100 \text{ K}$$

Note, however, that it would be incorrect to say that 1°C = 1 K, since 1°C = 274 K, 2°C = 275 K, ... For this reason it is particularly important to specify the scale in any temperature measurements.

The value of absolute zero, $T = 0$ K, can be approached, but it can never be attained; hence any temperature value is always greater than 0 K. According to Eq. [20 . 6], an ideal gas at absolute zero would exert no pressure on its surroundings, since $P = P_i T/T_i$ and it would therefore be reduced to a vanishingly small volume! If this extrapolation is more than a mathematical oddity, then we are also forced to conclude that it is not possible to attain a temperature below absolute zero.

> A system at absolute zero cannot give up any of the internal energy of its particles as heat.

This does not mean it has no internal energy, or that its particles are motionless; it does have internal energy, but it will not spontaneously transfer this energy to any other body with which it is placed in thermal contact.

Many attempts have been made to obtain temperatures close to absolute zero. In 1844 Michael Faraday reached a temperature of $\tau = -110°$C by rapid evaporation of ether and solid carbon dioxide. By 1926 temperatures as low as $T = 0.71$ K were achieved by solidifying helium by placing it in a brass tube in a helium bath and subjecting it to pressure. In 1963 Nicholas Kurti and his co-workers at Oxford used a method known as adiabatic demagnetization to obtain a temperature of 1.2×10^{-6} K. However, it appears that there is no way to reach absolute zero in a finite number of operations; experiments indicate that whatever method of cooling is used, the lower the temperature reached, the more difficult it is to go still lower. The quest for temperatures approaching absolute zero has furthered the development of *low-temperature physics*—an area in which new and often startling phenomena are continually being brought to light.

example 20 • 3 Another absolute temperature scale is the *Rankine scale*, which has degrees equal to those on the Fahrenheit scale. What is the relation between $T_{°\text{R}}$ and $\tau_{°\text{F}}$?

solution As in Example 20 . 2, we can transform $1/\beta_0$ to degrees Fahrenheit:

$$\frac{1}{\beta_0} = 273.15°\text{C} = 491.67°\text{F}$$

Since the ice point on the Fahrenheit scale is 32°F,

$$T_{°\text{R}} = (\tau_{°\text{F}} - 32°\text{F}) + 491.67°\text{F} = \tau_{°\text{F}} + 459.67°\text{F}$$

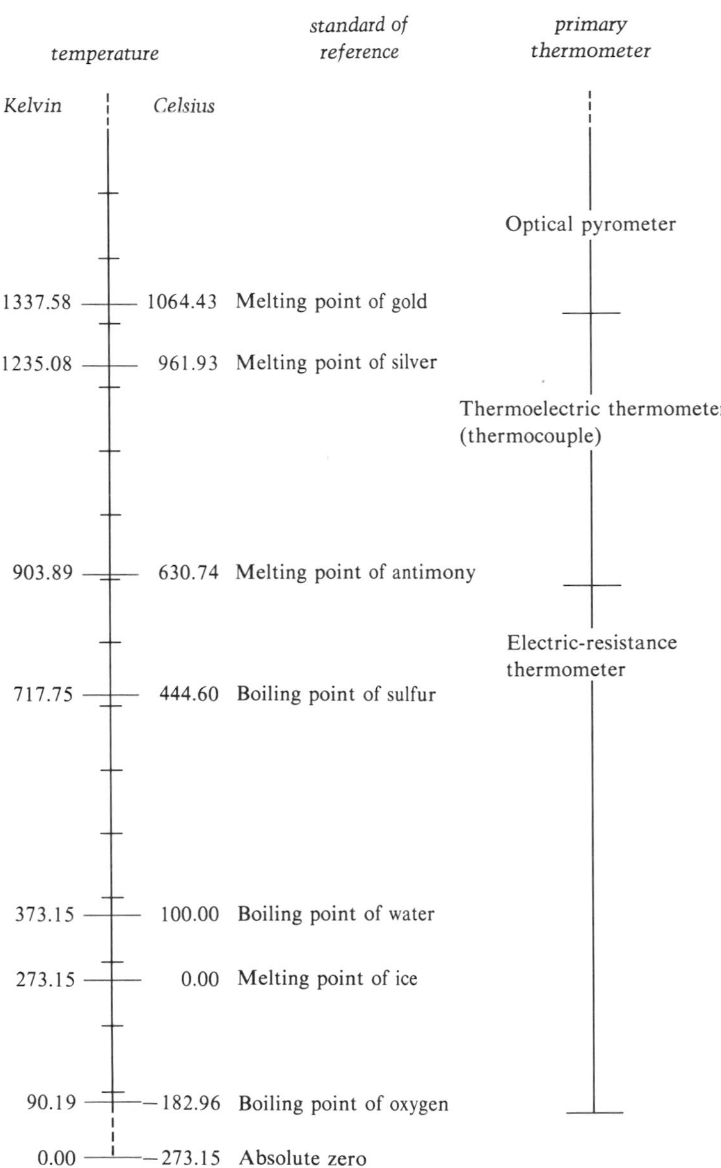

figure 20 • 3 *International temperature scale. All standards shown are at 1 atm*

Because of the experimental difficulties in using gas thermometers and the relatively low precision attainable in a single measurement, the General Conference on Weights and Measures adopted a practical working scale called the *international temperature scale*. This scale is defined by a series of fixed freezing and boiling points, the temperatures of which have been determined by gas-thermometer measurements, with suitable thermometers specified for interpolation between the fixed points and for extrapolation to higher temperatures. These relations are illustrated in Fig. 20 . 3.

The *electric-resistance thermometer* has a platinum wire wound in a helix or coil inside a sealed metal or porcelain tube. For reasons of economy,

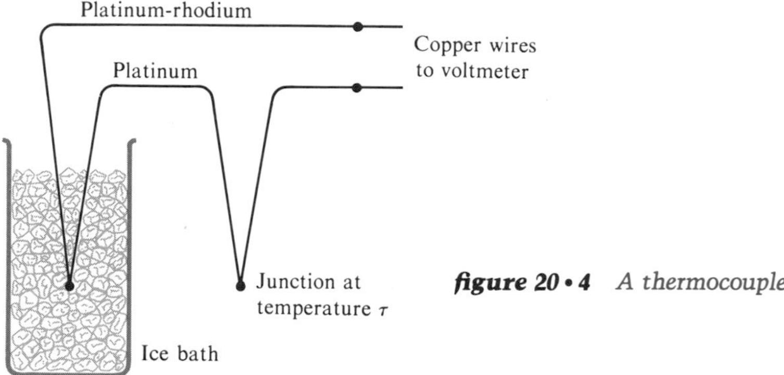

figure 20 • 4 *A thermocouple*

most industrial thermometers employ nickel or copper wire instead of platinum. As the temperature changes, the resistance of the wire to the passage of electrons (electric current) changes in a well-defined way. The *thermocouple*, shown in Fig. 20 . 4, depends on the experimental observation that when a pair of dissimilar conducting materials are joined together at one end and the junction is heated, a small but measurable electric current proportional to ΔT will flow from one to the other. The standard thermocouple is made of platinum joined to an alloy of platinum and rhodium. It is very useful at high temperature and for making precise measurements in small and/or inaccessible places, since only the hot junction need be in contact with the body whose temperature is to be measured, and the rest of the instrument can be some distance away.

The *optical pyrometer* depends on the fact that the color of the visible light radiated by any incandescent body depends on its temperature. When the filament of the lamp in Fig. 20 . 5 is made to radiate with a color identical to that of the body of interest, the temperature of the filament, and hence of the body, can be read directly from a suitably calibrated ammeter, an instrument for measuring the electric current passing through the filament.

Liquid-in-glass thermometers remain indispensable for use in the range $-190°C$ to about $600°C$ when facility of observation is of prime importance. As Table 20 . 1 shows, the variation between corresponding readings on several of the common temperature scales are negligible for most purposes. The wide variety of thermometers in use illustrates the general princi-

figure 20 • 5 *A simple optical pyrometer*

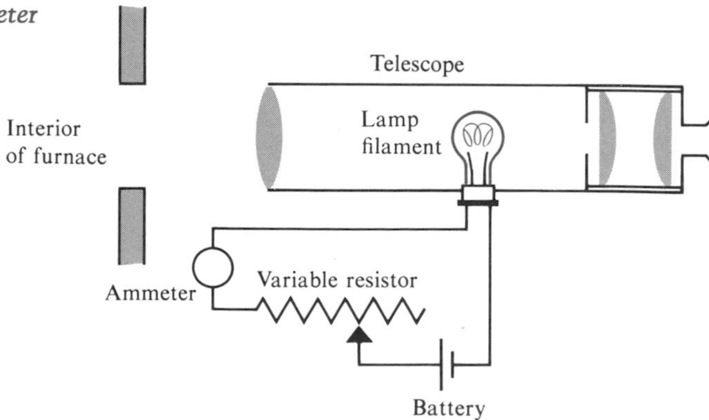

table 20 • 1 *Comparison of Celsius temperature scales*

constant volume hydrogen	constant volume air	platinum electric resistance	mercury in Jena 59^{III} glass
0.00	0.00	0.00	0.00
20.00	20.01	20.24	20.035
40.00	40.00	40.36	40.03
60.00	59.99	60.36	60.02
80.00	79.99	80.24	80.00
100.00	100.00	100.00	100.00

ple underlying all instrumentation: if a physical quantity x produces any unique and measurable effect y on system S, then this effect can be used to measure x. Linearity of the response y to the stimulus x is only a convenience, not a necessity.

20 • 3 equations of state

In the same way that we can uniquely specify a point in space by its three spatial coordinates, we can also uniquely specify a thermodynamic system by its physical properties, such as pressure, temperature, volume, or mass. These "thermodynamic coordinates" are known as *state variables;* taken together, they describe the *state* of a system. The values of these state variables are determined only by the present state of the system, not by its past history, just as the location of something on the earth's surface is a fact in itself, independent of how it arrived there. Thus a quart of water at 50°C and 1 atm pressure has values of temperature, pressure, volume, and mass which are independent of whether the water was condensed from steam or melted from ice.

State variables may be *extensive variables*, like volume and mass, which depend on the size or extent of the system. Or they may be *intensive variables*, like temperature and pressure, which are independent of the size of the system. Although many state variables may be defined, we shall mainly consider temperature, pressure, volume, and mass, usually for cases in which mass is constant.

We shall also assume that systems are in *thermodynamic equilibrium*—that is, that a system, if isolated from all thermal contact, will not change its properties with time. Whenever we do discuss processes in which a system is changing with time—as in the flow of heat in or out of the system—we shall assume the process to be *quasistatic*. That is, we shall consider a process to be taking place slowly enough that we can view the system as passing through a succession of intermediate equilibrium states—somewhat like cooking an egg by heating it one degree at a time. The reason for this requirement is fundamental: only such states of *aggregates* of matter can be uniquely specified, for it is only by their *average* properties that we can describe them.

The *equation of state* of a system relates its other state variables to its temperature. The simplest equation of state is that of the ideal gas, the study of which has been particularly important in the development of atomic theory. In Sec. 14 . 2 we studied Boyle's law, $PV =$ constant, which expresses the relation between the pressure and volume of a fixed mass of gas under *isothermal*, or constant-temperature, conditions. Now we turn to the question of how the volume and temperature of a fixed mass of gas are related in *isobaric processes*, in which the pressure is kept constant and the temperature is changed. The answer can then be combined with Boyle's law to yield the equation of state of an ideal gas.

Consider the simple device in Fig. 20 . 6, in which a definite mass of a gas is confined in the bulb by a globule of mercury, which moves forward or backward in the stem as the temperature rises and falls. The end of the stem is open so that a constant external pressure is provided by the atmosphere. Experiments show that the increase in the volume of the gas is approximately proportional to its original volume and the rise in temperature. If we agree always to take the volume V_i at the ice point as the "original volume," then we obtain *Charles' law:*

$$V = V_i(1 + \beta_i\tau) \qquad [20 \cdot 7]$$

where V is the volume at temperature τ and β_i is a proportionality constant termed the *coefficient of cubical expansion*, or simply the *expansion coefficient*.

In 1802 Louis Gay-Lussac carried out experiments with various gases, using the apparatus in Fig. 20 . 6, and concluded that all gases have the same expansion coefficient β_i when they are heated through the same range of temperatures. Like Boyle's law, Charles' law was later shown to be only approximately correct. However, for gases under moderate pressures and temperatures (nearly ideal gases), the errors are negligible.

Charles' law, Boyle's law, and the operational definition of temperature in Eq. [20 . 2] give three relations among three state variables describing an ideal gas: pressure, volume, and temperature. Since these relations are not independent of each other, we can combine them into a single equation of state. To combine Boyle's and Charles' laws, which hold precisely only for an ideal gas, let us assume a two-stage process in which we first heat a gas under constant pressure P_i from 0°C to a temperature τ (see Fig. 20 . 7). Its new volume, according to Eq. [20 . 7], is

$$V' = V_i(1 + \beta_i\tau) \qquad [20 \cdot 8]$$

We now keep the temperature constant at τ and change the pressure to

figure 20 • 6 *A simple device for studying the thermal expansion of a gas under isobaric, or constant-pressure, conditions. It may also be used as a constant-pressure gas thermometer.*

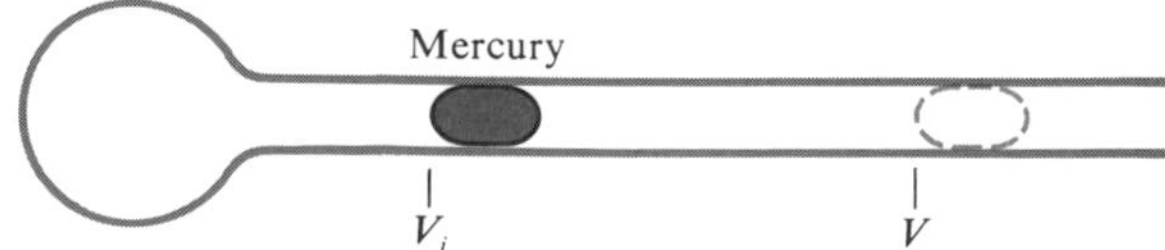

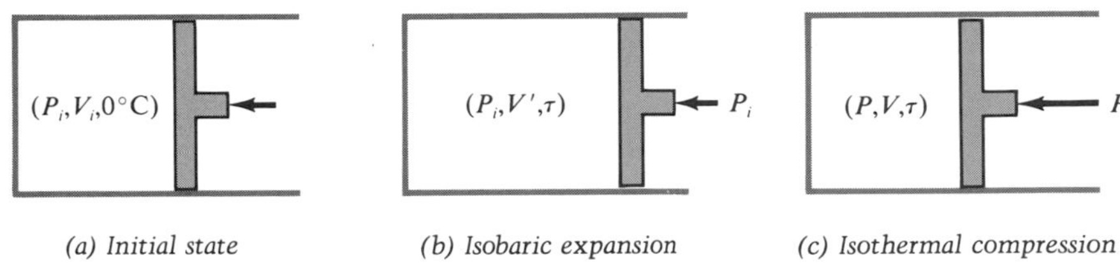

(a) Initial state (b) Isobaric expansion (c) Isothermal compression

figure 20 • 7 *A hypothetical experiment illustrating the derivation of the equation of state of an ideal gas*

some value P, so that the volume changes accordingly to V. Then, from Boyle's law,

$$PV = P_iV' = P_iV_i(1 + \beta_i\tau) \qquad [20 \cdot 9]$$

If we assume that the gas has no "memory," so that it does not matter how we go from one state to another, then Eq. [20 . 9] is the equation of state for an ideal gas in terms of P, V, and τ.

It only remains to evaluate β_i by requiring that this equation be consistent with our operational definition, Eqs. [20 . 3], of temperature on the constant-volume gas thermometer. Substituting for P_s and P_i from Eq. [20 . 9] for constant volume $V_s = V_i$ and $\tau_s = 100°C$, we obtain

$$\beta_i = \frac{0.01}{\mathrm{K}}\left(\frac{P_s}{P_i} - 1\right) = \beta_0 \qquad [20 \cdot 10]$$

Therefore we can use Eq. [20 . 5] to express the equation of state in terms T:

$$PV = P_iV_i(1 + \beta_0\tau) = \frac{P_iV_i}{T_i}T \qquad [20 \cdot 11]$$

Note that since the ratio PV/T is constant for any ideal gas, we can use *any* initial state (P_0,V_0,T_0) as the reference and write Eq. [20 . 11] as

$$\blacktriangleright \quad \frac{PV}{T} = \frac{P_0V_0}{T_0} = \text{constant} \qquad [20 \cdot 12]$$

example 20 • 4 Show that if volume is used as the thermometric property in the constant-pressure thermometer of Fig. 20 . 6, then for an ideal gas

$$\tau_V = 100°C\frac{V - V_i}{V_s - V_i}$$

gives precisely the same value of temperature as does Eq. [20 . 2].

solution We can substitute for the volumes, using Eq. [20 . 11] at constant pressure P_0. Then

$$V = \frac{V_iT}{T_i} = V_i(1 + \beta_0\tau) \qquad \text{and} \qquad \tau_V = 100°C\frac{\beta_0\tau}{100\beta_0} = \tau$$

Once again we see that the ideal-gas equation is consistent with the operational definition of temperature.

20 . 4 the ideal-gas equation

By international agreement, the carbon-12 atom, with a nucleus of six neutrons and six protons, has become the standard for atomic mass, taken as precisely 12 atomic mass units (u). The atomic weight of all other atoms and the molecular weight of compounds of two or more atoms is expressed relative to this standard. A *mole* of any substance is an amount of mass the numerical value of which exactly equals the numerical mass, in atomic mass units (u), of its constituent molecule. Hence, a mole, like a "dozen," contains a specific *number* of molecules, regardless of type, but the *weight* of the mole depends on the molecular or atomic weight of the substance. The kilogram-molecular weight, or weight of 1 kilogram-mole (kmole), of carbon-12 is exactly 12 kg.* The water atom (H_2O) has an atomic mass of 18 u; hence 1 kmole of water has a mass of 18 kg. A mole of carbon-12 and a mole of water, however, contain the same number of molecules. In the case of gases, experiment has shown that 1 kmole of any ideal gas at standard temperature and pressure (0°C and 1 atm) always occupies the *same volume*, $V_0 = 22.414\ \text{m}^3$. This is the kilogram-molecular volume.

This startling fact allows us to recast the equation of state for an ideal gas into a form which will include its mass. Let us take standard temperature and pressure as the reference conditions. Then, if n is the *number* of kilogram-moles in the system under the conditions $V_0 = 22.414\ \text{m}^3$, $P_0 = 1$ atm, and $T_0 = 273.15$ K, we have

$$PV/T = n\frac{P_0V_0}{T_0} = nR \qquad [20 \cdot 13]$$

where $R = P_0V_0/T_0$ is known as the *universal gas constant*, because Eq. [20 . 13] holds for any ideal gas. The value adopted by the International Council of Scientific Unions is

$$\begin{aligned} R &= (8.31441 \pm 0.00026) \times 10^3\ \text{J/K} \cdot \text{kmole} \\ &= 82.057\ \text{atm} \cdot \text{liters/K} \cdot \text{kmole} \end{aligned} \qquad [20 \cdot 14]$$

The *ideal-gas equation* is usually written as

► $$PV = nRT \qquad [20 \cdot 15]$$

although the *molar specific volume*,

$$v = \frac{V}{n}$$

is often employed to express the equation of state entirely in intensive variables:

► $$Pv = RT \qquad [20 \cdot 16]$$

The fact that the ideal-gas equation of state assumes a form as general as Eq. [20 . 15] suggests that the behavior of an ideal gas does not depend at all on the chemical properties or nature of the gaseous substance. It depends

*In the International System of Units (SI) a mole is the number of atoms in 12 g of Carbon-12—that is, a gram-mole.

only on the mechanical properties of the gas considered as a collection of point masses which do *not interact* with each other, except for elastic collisions. We shall see in Chapter 22 that, beginning with this hypothesis alone, it is possible to *derive* Eq. [20 . 15] theoretically, although we have derived it here from experimental results.

The analogy between state variables and the spatial coordinates of a point is more than metaphorical. A point in *PvT space*—one with coordinates (P,v,T)—can be said to represent a possible thermodynamic state of a system. For an ideal gas we can depict the *surface* of possible states in PvT space determined by $Pv = RT$ in either of two useful ways: by a three-dimensional surface like that in Fig. 20 . 8, or by a projection of the contours of the surface on one of the coordinate planes, as in Fig. 20 . 9. It is customary to take P as the ordinate in isothermal projections for convenience in

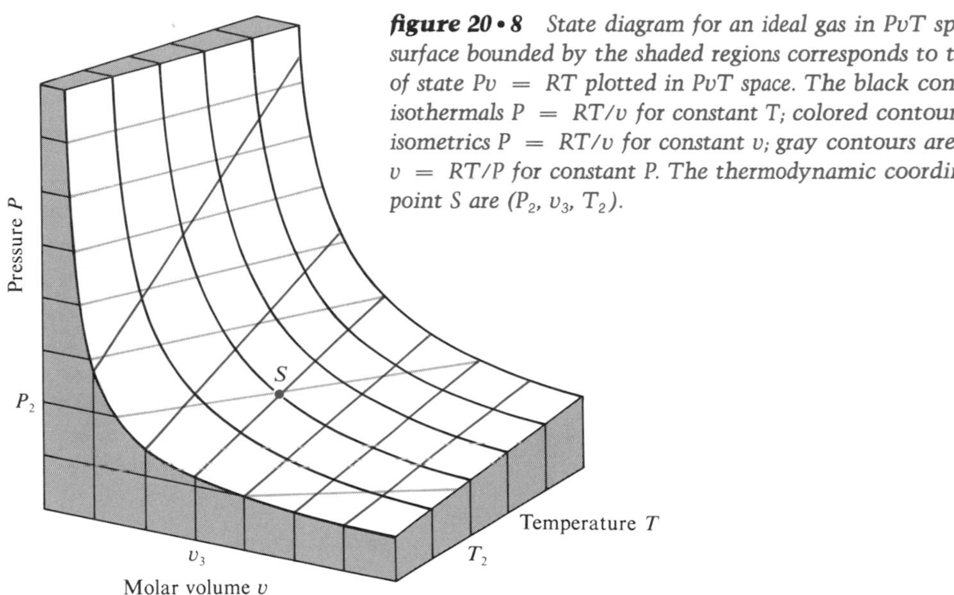

figure 20 • 8 *State diagram for an ideal gas in PvT space. The surface bounded by the shaded regions corresponds to the equation of state $Pv = RT$ plotted in PvT space. The black contours are isothermals $P = RT/v$ for constant T; colored contours are isometrics $P = RT/v$ for constant v; gray contours are isobars $v = RT/P$ for constant P. The thermodynamic coordinates of the point S are (P_2, v_3, T_2).*

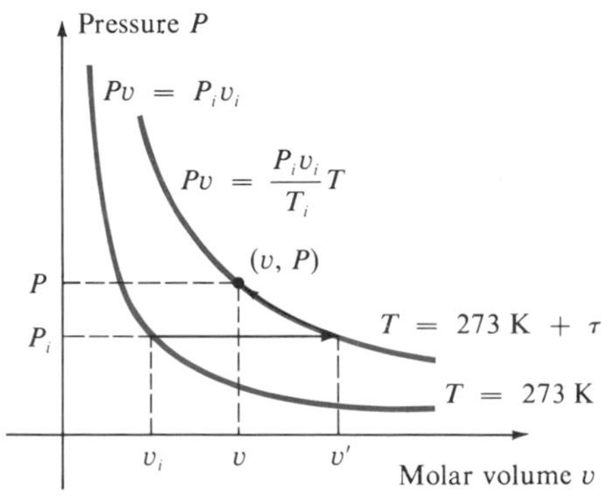

(a) Isothermal contours on Pv plane

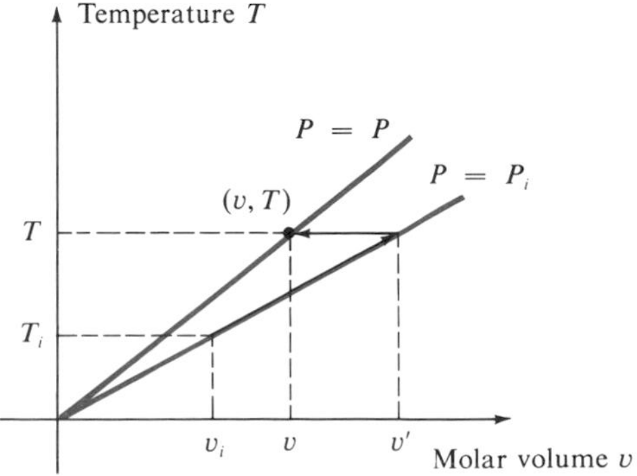

(b) Isobaric contours on vT plane

figure 20 • 9 *Isothermal and isobaric contours for the ideal-gas equation of state, $Pv = RT$*

graphing the work done in a volume expansion. The isobaric contours are a pencil of straight lines meeting at absolute zero, as already noted in Sec. 20 . 2. The solid arrows indicate the "path" of the two-step isobaric-isothermal process we used in deriving the equation of state, Eq. [20 . 10]. Such PvT diagrams are of great utility in thermodynamics and will repay careful study.

example 20 • 5 If the tires on your car have an inside diameter of 15 in and an outside diameter of 25 in, you can approximate the volume of one of the tires by a cylinder with a diameter of 5 in and a length of $\pi(15 + 5)$ in. The effective molecular weight of air (four-fifths nitrogen and one-fifth oxygen) is approximately 28.8 u. (*a*) How many kilograms of air does the tire contain at a gage pressure of 29.4 lbf/in^2 and a temperature of 273 K? (*b*) After driving on the highway for an hour, you find that the gage pressure is now 33 lbf/in^2. What is the temperature of the tire? Assume the ambient conditions to be standard temperature and pressure and the tire volume to be constant.

solution (*a*) The volume of the tire is

$$V = \tfrac{1}{4}\pi^2 \times 5 \times 20 \text{ in}^3 = 1234 \text{ in}^3 = 20.22 \text{ liters}$$

Its pressure is $P = (29.4 + 14.7)\ \text{lbf/in}^2 = 3$ atm, and since the temperature of the system is $T = 273$ K, its mass is

$$n = PV/RT = \frac{3 \text{ atm} \times 20.22 \text{ liters}}{(82.06 \text{ atm} \cdot \text{liters/kmole} \cdot \text{K}) \times 273 \text{ K}}$$

$$= 0.00271 \text{ kmole}$$

or a total mass of $m = 28.8n = 0.078$ kg of air.

(*b*) The temperature corresponding to an increase of pressure from $29.4 + 14.7 = 44.1\ \text{lbf/in}^2$ to $33 + 14.7 = 47.7\ \text{lbf/in}^2$ is

$$T = \frac{47.7}{44.1} \times 273 \text{ K} = 295.30 \text{ K} = 22.15°\text{C} \doteq 72°\text{F}$$

which is the temperature of the air inside the tire.

20 • 5 liquids and solids

Like gases, liquids and solids also undergo isobaric volume changes with temperature. However, since these changes are relatively small, we can make use of a Taylor's series expansion of $V(\tau)$ on the assumption that there is some continuous function $f(P,V,T) = 0$ which represents the equation of state of the substance, even if we do not happen to know it. Assuming that this equation can be solved for $V = V(P,T)$, for some fixed pressure P_0 we can express this series in *virial* form as an expansion in powers of τ about the ice point, 0°C:

$$V = V_i(1 + a\tau + b\tau^2 + c\tau^3 + \cdots) \qquad [20 \cdot 17]$$

where a, b, c, . . . are either constant or depend only on the pressure. These

table 20 • 2 Virial coefficients and coefficient of expansion for selected liquids under ordinary pressures

liquid	temperature range, °C	a, K^{-1}	b, K^{-2}	c, K^{-3}	β at 20°C, K^{-1}
alcohol, ethyl	27–46	1.012×10^{-3}	2.20×10^{-6}	—	1.12×10^{-3}
methyl	0–61	1.134×10^{-3}	1.363×10^{-6}	1.36×10^{-6}	1.199×10^{-3}
mercury	0–100	0.18182×10^{-3}	0.0078×10^{-6}	—	0.18186×10^{-3}
water	0–33	-0.06427×10^{-3}	8.505×10^{-6}	-6.79×10^{-8}	0.207×10^{-4}

virial coefficients are the ones which give the best fit to empirical data over a wide range of temperatures (see Table 20 . 2). The number of coefficients used depends on the accuracy desired, the temperature range involved, and the substance under investigation.

The volume-expansion coefficient of a substance at temperature τ and constant pressure is

$$\beta = \frac{1}{V}\frac{\partial V}{\partial \tau} = \frac{a + 2b\tau + 3c\tau^2 + \cdots}{1 + a\tau + b\tau^2 + c\tau^3 + \cdots} \qquad [20 \cdot 18]$$

where the partial-derivative operator ∂ shows that P is assumed constant in differentiating V with respect to τ. At the ice point, $\tau = 0°C$,

$$\beta = \beta_i = a \qquad [20 \cdot 19]$$

For many substances, b and c are so much smaller than a that the first virial coefficient is sufficiently accurate. In that case

$$V = V_i(1 + \beta_i \tau) \qquad [20 \cdot 20]$$

which corresponds to Charles' law, Eq. [20 . 7], although for solids and liquids, β_i depends on the nature of the specific substance involved. Equation [20 . 20] holds quite well for most solids and for liquid mercury below 100°C. For example, a mercury-in-glass thermometer graduated by dividing the increase in volume between 0°C and 100°C into 100 equal parts is not likely to differ from a hydrogen thermometer at any point in this interval by more than 0.2 K.

example 20 • 6 If the length of a mercury-in-glass thermometer is 15 cm and the cross section of the mercury column in its stem is 0.01 mm², how much mercury must the thermometer contain if it is to measure a range of 20 K? Neglect expansion of the glass with temperature.

solution A rise in temperature of $\Delta\tau = 20$ K must cause the volume V of mercury in the bulb to expand into the stem of the thermometer by an amount ΔV. According to Eq. [20 . 20], these quantities are related by the expansion coefficient for mercury, given in Table 20 . 2:

$$V = \frac{\Delta V}{\beta\,\Delta\tau} = \frac{15 \times 10^{-4}\ \text{cm}^3}{(0.18 \times 10^{-3}\ \text{K}^{-1}) \times 20\ \text{K}} = 0.42\ \text{cm}^3$$

table 20 • 3 *Linear expansion coefficients α of selected solids under ordinary pressures*

solid	*temperature range, °C*	α, K^{-1}
brass	25–100	1.9×10^{-5}
cobalt-iron-chromium	20–60	-1.1–1.7×10^{-5}
glass, Jena 59[III]	0–100	0.58×10^{-5}
glass, Pyrex	20–300	0.33×10^{-5}
	550–570	1.5×10^{-5}
ice	–250	-0.6×10^{-5}
	–50	$+4.5 \times 10^{-5}$
	–0	$+5.3 \times 10^{-5}$
platinum	0–100	0.90×10^{-5}
quartz, fused	20–1000	0.005×10^{-5}
// to axis	0–100	0.80×10^{-5}
⊥ to axis	0–100	1.4×10^{-5}
steel, stainless, annealed	20–100	1.0×10^{-5}
lead	20–100	2.92×10^{-5}

For solids we can also define a coefficient α for *linear expansion* in a given direction as

$$\alpha = \frac{1}{l}\frac{\partial l}{\partial \tau} \qquad [20 \cdot 21]$$

As with β, we can expand $l(\tau)$ in a virial representation about the ice point:

$$l = l_i(1 + a'\tau + b'\tau^2 + c'\tau^3 + \cdots) \qquad [20 \cdot 22]$$

Since a' is much greater for solids than the other coefficients (Table 20 . 3), yet much less than unity, we can use the approximation $\alpha \doteq a'$ and write

$$l = l_i(1 + \alpha\tau) \quad \text{or} \quad l = l_0[1 + \alpha(\tau - \tau_0)] \qquad [20 \cdot 23]$$

using any temperature τ_0 as a reference point.

If we know the linear expansion coefficients for three mutually perpendicular sides of a solid, then we can compute the volume coefficient. Suppose our specimen is a cube, with sides of length l_0, at some temperature τ_0. When the cube is heated to a temperature $\tau = \tau_0 + \Delta\tau$, it becomes a rectangular parallelepiped of volume

$$\begin{aligned} V &= l_0(1 + \alpha_x \Delta\tau)\, l_0(1 + \alpha_y \Delta\tau)\, l_0(1 + \alpha_z \Delta\tau) \\ &\doteq V_0[1 + (\alpha_x + \alpha_y + \alpha_z)\,\Delta\tau] \end{aligned} \qquad [20 \cdot 24]$$

where higher-order terms in $\Delta\tau$ are neglected, since linear expansion coefficients are always small. The volume expansion coefficient is therefore

$$\beta = \alpha_x + \alpha_y + \alpha_z \qquad [20 \cdot 25]$$

and if the solid is isotropic with respect to expansion, then

$$\alpha_x = \alpha_y = \alpha_z = \alpha \qquad \beta = 3\alpha \tag{20 · 26}$$

Since an isotropic solid expands evenly in all directions, if it contains a hole of radius r_0 before expansion, it will contain a hole of radius $r_0(1 + \alpha\,\Delta\tau)$ after expansion. That is, taking one material point as origin, all points move from $\mathbf{r}$ to $\mathbf{r}(1 + \alpha\,\Delta\tau)$ on expansion.

Although we have been discussing isobaric processes, there is no reason we could not consider pressure changes as well. For pressure changes which are small enough, it should be possible to express the change in volume under isothermal conditions by using the elasticity equation of Chapter 14:

$$\Delta V = -\kappa V\,\Delta P \tag{20 · 27}$$

where κ is the compressibility of the substance involved. Let us consider the volume changes in an isobaric-isothermal process, as we did for an ideal gas, but this time using Eq. [20 . 27] instead of Boyle's law.

If we let the reference volume V_0 expand with temperature to a volume V', and then let V' expand with pressure to a volume V, we obtain

$$V = V'(1 - \kappa\,\Delta P) = V_0(1 + \beta\,\Delta\tau)(1 - \kappa\,\Delta P)$$

and for small values of β, κ, and $\Delta\tau$ this becomes

$$V \doteq V_0(1 + \beta\,\Delta\tau - \kappa\,\Delta P) \tag{20 · 28}$$

This is, in effect, an *approximate equation of state* as defined for some particular substance over some specified ranges of temperature and pressure. If we substitute length l and longitudinal stress $-\Delta F/A$ for volume V and pressure P, Young's modulus Y for compressibility (where $\kappa = -1/Y$), and the expansion coefficient α for β, we have the equation of state for a *rod under tensile stress:*

$$l = l_0\left(1 + \alpha\,\Delta\tau + \frac{1}{Y}\frac{\Delta F}{A}\right) \tag{20 · 29}$$

The concept of the thermodynamic equation of state is quite universal and can, in principle, be applied to any system—even to those systems that include electric or magnetic fields and light energy among their state variables.

example 20 • 7 If water ($\kappa = 5 \times 10^{-10}\,\mathrm{m^2/N}$) is confined to a volume V_0 at 4°C and then cooled just to the freezing point, what is the increase in pressure required to keep the volume at V_0?

solution If the volume is unchanged, then from Eq. [20 . 28],

$$\Delta P \doteq \frac{\beta}{\kappa}\Delta\tau \doteq \frac{-3 \times 10^{-5}\,\mathrm{K^{-1}}}{5 \times 10^{-10}\,\mathrm{m^2/N}}(-4\,\mathrm{K})$$

or

$$\Delta P = 2.4 \times 10^5\,\mathrm{N/m^2} \doteq 2.4\,\mathrm{atm}$$

where we have used the coefficients from Table 20 . 2 to set

$$\beta = a + 2b \times 2\,\mathrm{K} \doteq -3.0 \times 10^{-5}\,\mathrm{K}^{-1}$$

Note that β is evaluated at $\tau = 2°\mathrm{C}$, the middle of the temperature interval, for greatest accuracy.

*20 . 6 *polynomial approximations*

In the real world, neat closed formulas are the exception rather than the rule. Often we must resort to polynomial approximations, such as the virial formulas of this chapter. Within their limitations they can be very useful, however, and it is important to know how to derive them quickly and simply from collections of data. In this section, we shall consider fitting a collection of $n + 1$ data points (x_i, y_i), for $i = 0, 1, 2, \ldots, n$, to a polynomial of nth degree by requiring that it pass through each data point.

This idea is not new; we have already applied it to linear interpolation between two points and to the derivation of Simpson's rule (Sec. 16 . 4). In the first case we approximated the curve between two points by a first-degree polynomial (a straight line), and in the second case we approximated a curve through three points by a second-degree polynomial (a parabola). In general, the requirement that the interpolating polynomial

$$P(x) = a_0 + a_1x + a_2x^2 + \cdots + a_ix^i + \cdots + a_nx^n = \sum_{i=0}^{n} a_ix^i \qquad [20 \cdot 30]$$

pass through $n + 1$ points means that a system of equations of the form

$$\begin{aligned} a_0 + a_1x_0 + a_2x_0^2 + \cdots + a_nx_0^n &= y_0 \\ a_0 + a_1x_1 + a_2x_1^2 + \cdots + a_nx_1^n &= y_1 \\ \cdots\cdots\cdots\cdots\cdots\cdots\cdots\cdots\cdots \\ a_0 + a_1x_n + a_2x_n^2 + \cdots + a_nx_n^n &= y_n \end{aligned} \qquad [20 \cdot 31]$$

must be solved for the appropriate values of the unknown coefficients a_0, $a_1, \cdots, a_n$. Given $n + 1$ independent data points (x_i, y_i), the resulting $n + 1$ equations of Eq. [20 . 31] are also independent and can, in principle, be solved by the method of determinants. For n so large that hand computation is impractical, (say, $n > 3$), it is usually possible to find a "canned" program to perform the computations on a computer, with the values of (x_i, y_i) merely given as input. Furthermore, at the price of adding more terms to $P(x)$ and more equations to [20 . 31], we are also at liberty to make additional requirements on the derivatives of $P(x)$. The important point, however, is this:

> The nth-degree polynomial $P(x)$ which fits $n + 1$ independent data points (x_i, y_i) is *unique*. Any other *form* of an nth-degree polynomial fitting the same data can be shown, upon rearrangement, to be identical to $P(x)$.

This means that if we can find any other more efficient way to approximate the data with an expression containing powers of x from x^0 to x^n, we shall have found the *unique interpolating polynomial* $P(x)$.

It is clear from the form of Eqs. [20 . 31] that, since the factors a and y occur only to the first power, all values of a will be a linear function of the ordinate y. Therefore the *Lagrange interpolating polynomial* is a possible form of $P(x)$, where

$$P(x) = A_0(x)y_0 + A_1(x)y_1 + \cdots + A_n(x)y_n \qquad [20 \cdot 32]$$

and each Lagrange coefficient $A(x)$ may itself be an nth-degree polynomial. Instead of vastly complicating matters, as it may seem at first glance, the added flexibility of this representation enables us immediately to formulate each $A_i(x)$ by inspection of the data.

Let us define a function $L_i(x)$ such that each $L_i(x)$ is a product of monomial factors $x - x_k$, with only the ith such factor missing:

$$L_i(x) = (x - x_0)(x - x_1)(x - x_2) \cdots (x - x_{i-1})(x - x_{i+1}) \cdots (x - x_n) \qquad [20 \cdot 33]$$

Then $L_i(x_i)$ is a nonzero constant. If we now set each coefficient of Eq. [20 . 32] equal to

$$A_i(x) = \frac{L_i(x)}{L_i(x_i)} = \begin{cases} 1 & \text{for } x = x_i \\ 0 & \text{for } x = x_0, x_1, \ldots, x_{i-1}, x_{i+1}, \ldots, x_n \end{cases} \qquad [20 \cdot 34]$$

then $P(x)$ in the form of Eq. [20 . 3] will *automatically* fit the data, since at any data point (x_i, y_i) all coefficients vanish except $A_i(x_i) = 1$; hence $P(x_i) = y_i$. In a sense, Eqs. [20 . 32] to [20 . 34] constitute an iterative procedure for evaluating $P(x)$ at some x. While this may not always be amenable to a hand computation, it is easy to translate this procedure into an algorithm, as shown in Fig. 20 . 10.

Further convenience is afforded if we take only *evenly spaced* data:

$$x_i = x_0 + i\,\Delta x \qquad \Delta x = \frac{x_n - x_0}{n}$$

If we substitute a normalized variable ν for x,

$$x = x_0 + \nu\,\Delta x \qquad 0 \le \nu \le n \qquad [20 \cdot 35]$$

then the $(\Delta x)^n$ factors in each $A_i(x)$ cancel, and we find the normalized Lagrange coefficients $\alpha_i(\nu)$:

$$A_i(x) = \alpha_i(\nu) = \frac{\nu(\nu - 1)(\nu - 2) \cdots (\nu - i + 1)(\nu - i - 1) \cdots (\nu - n)}{i(i - 1)(i - 2) \cdots (1)(-1) \cdots (i - n)} \qquad [20 \cdot 36]$$

The computation of $\alpha_i(\nu)$ is quite general and depends only on the *number* of evenly spaced data points and the noninteger number ν of intervals Δx corresponding to x:

$$\nu = n\frac{x - x_0}{x_n - x_0} \qquad [20 \cdot 37]$$

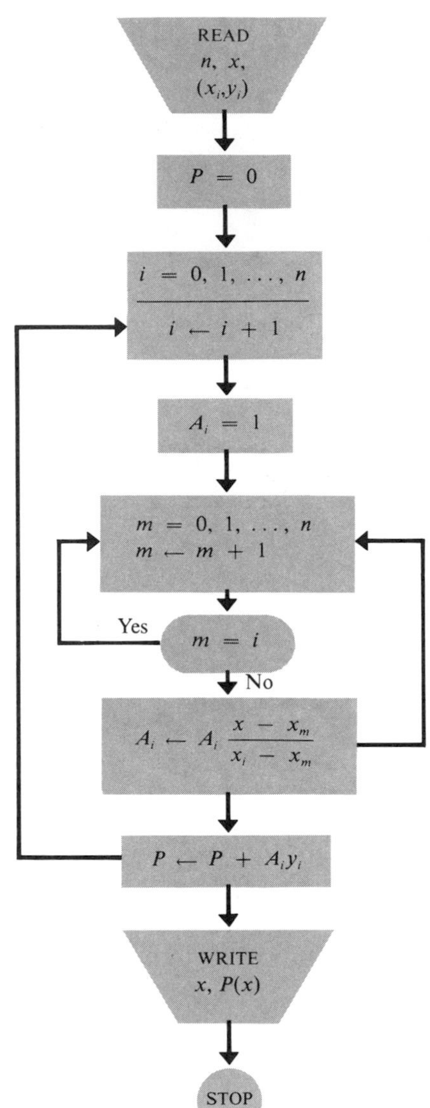

figure 20 • 10 *Algorithm for Lagrange interpolation. The inner loop, iterated on index m, evaluates the Lagrange coefficients by successive multiplications omitting the nth monomial fraction, according to Eq. [20 . 33]. The outer loop, iterated on index n, evaluates P(x) according to the Lagrange form, Eq. [20 . 32].*

The normalized Lagrange coefficients are independent of Δx and of the ordinate values y. Hence, once computed, they can be used with any set of $n + 1$ evenly spaced data points.*

*Tables of Lagrange coefficients up to $n = 7$ can be found in M. Abramowitz and I. A. Stegun, *Handbook of Mathematical Functions* (National Bureau of Standards, Washington, D.C., 1965).

example 20 • 8 Find the normalized Lagrange coefficients (a) for the case $n = 1$ and (b) for the case $n = 2$. Find $P(x)$ in each case.

solution (a) Each $L_i(x)$ can have only one monomial factor. Hence for $n = 1$ the coefficients are $\alpha_0(\nu) = 1 - \nu$ and $\alpha_1(\nu) = \nu$, and $P(x)$ is just the usual linear interpolation formula:

$$P(x) = (1 - \nu)y_0 + \nu y_1 = y_0 + \frac{x - x_0}{x_1 - x_0}(y_1 - y_0)$$

(*b*) For degree $n = 2$, we have $\alpha_0(\nu) = \frac{1}{2}(\nu - 1)(\nu - 2)$, $\alpha_1(\nu) = -\nu(\nu - 2)$, and $\alpha_2(\nu) = \frac{1}{2}\nu(\nu - 1)$. If we collect powers of ν, we find

$$P(x) = \tfrac{1}{2}[2y_0 - \nu(3y_0 - 4y_1 + y_2) + \nu^2(y_0 - 2y_1 + y_2)]$$

where $\nu = 2(x - x_0)/(x_2 - x_0)$. If we now substitute values $\nu =$ 0, 1, 2 into the equation for $P(x)$, we obtain $P(x) = y_0, y_1, y_2$, as required. Even for such a low degree, the Lagrange method is preferable to solution of Eqs. [20 . 31]. If $P(x)$ were integrated over $x_0 \leq x \leq x_2$, we would again obtain Simpson's rule.

With a little practice the linear expansion coefficients can be written down very easily. For example, for $n = 3$ and $0 \leq \nu \leq 3$,

$$\begin{aligned}
\alpha_0(\nu) &= -\tfrac{1}{6}(\nu - 1)(\nu - 2)(\nu - 3) \\
\alpha_1(\nu) &= \tfrac{1}{2}\nu(\nu - 2)(\nu - 3) \\
\alpha_2(\nu) &= -\tfrac{1}{2}\nu(\nu - 1)(\nu - 3) \\
\alpha_3(\nu) &= \tfrac{1}{6}\nu(\nu - 1)(\nu - 2)
\end{aligned} \qquad [20 \cdot 38]$$

Some values are tabulated in Table 20 . 4. Note that the table can be "folded" at $\nu = 1.5$ owing to the symmetry between the coefficients.

table 20 • 4 *Four-point Lagrangian interpolation coefficients $\alpha(\nu)$*

ν	α_0	α_1	α_2	α_3	
0.0	1.000	0.000	0.000	0.000	3.0
0.1	0.827	0.276	−0.131	0.029	2.9
0.2	0.672	0.504	−0.224	0.048	2.8
0.3	0.536	0.689	−0.284	0.060	2.7
0.4	0.416	0.832	−0.312	0.064	2.6
0.5	0.313	0.938	−0.313	0.063	2.5
0.6	0.224	1.008	−0.288	0.056	2.4
0.7	0.150	1.047	−0.242	0.046	2.3
0.8	0.088	1.056	−0.176	0.032	2.2
0.9	0.039	1.040	−0.0945	0.0165	2.1
1.0	0.000	1.000	0.000	0.000	2.0
1.1	−0.029	0.941	0.104	−0.017	1.9
1.2	−0.048	0.864	0.216	−0.032	1.8
1.3	−0.060	0.774	0.332	−0.046	1.7
1.4	−0.064	0.672	0.448	−0.056	1.6
1.5	−0.063	0.563	0.563	−0.063	1.5
	α_3	α_2	α_1	α_0	ν

As an example of the usefulness of Lagrange interpolating polynomials, recall that in Sec. 16 . 4 we found that the period of a real pendulum is proportional to the integral

$$K = \int_0^{\pi/2} (1 - k^2 \sin^2 \phi)^{-1/2} \, d\phi \qquad [20 \cdot 39]$$

known as the complete elliptic integral of the first kind. In this case the variable parameter k is equal to $\sin \frac{1}{2}\theta_m$, where θ_m is the angular amplitude of the oscillation. Although it is possible to evaluate this integral quite accurately for a given k by using Simpson's rule, it is tedious to do so every time a new k is encountered. Instead, we can construct a table of selected values from Simpson's rule and then use a Lagrange interpolating polynomial to approximate any desired K as a function of k. For example, for the selection of $\theta_m = 0°, 50°, 100°, 150°$,

$$K = 1.5708, 1.6490, 1.9356, 2.7681$$

and for any θ_m, where $\nu = \theta_m/50°$, the Lagrangian form of the interpolating polynomial for $K(k)$ is

$$P(\theta_m) = 1.5708\alpha_0\left(\frac{\theta_m}{50°}\right) + 1.6490\alpha_1\left(\frac{\theta_m}{50°}\right) + 1.9356\alpha_2\left(\frac{\theta_m}{50°}\right) + 2.7681\alpha_3\left(\frac{\theta_m}{50°}\right) \qquad [20 \cdot 40]$$

This polynomial predicts $K(\sin \frac{1}{2}\theta_m)$ with a maximum of 1.7 percent error over the range $0° \leq \theta_m \leq 150°$—much better accuracy than could be obtained by separate linear interpolations between each pair of successive values. For example, when $\theta_m = 125°$, then the true value is $K = 2.228$ and $P(\theta_m) = 2.263$, an error of 1.5 percent, while linear interpolation between 100° and 150° yields $K \doteq 2.352$, an error of 5.6 percent.

The unique interpolating polynomial has the advantage that it is continuous; hence it can easily be differentiated to give approximate values of y', y'', . . ., although the accuracy is reduced by each successive differentiation. It can be shown that the nth-order interpolating polynomial approximates the unknown but smoothly continuous function represented by the data with an error proportional to $(\Delta x)^{n+1}$. Therefore, for a given interval, as the number of available data points increases, the errors decrease as $(\Delta x)^{n+1} \sim (1/n)^{n+1}$.

In using such an approximation, we do not use all possible points, since a polynomial of high degree may well misbehave between the data points. It is best to use only some of the data points, leaving the rest as a check on the general suitability of the polynomial selected. The polynomial fit is not satisfactory in the case of functions which have very large higher-order derivatives, since an nth-degree polynomial has only n nonzero derivatives, and all higher derivatives vanish. An example of such an unpleasant function is $1/x$, whose nth derivative is proportional to $n!$, a number which grows very rapidly. Since we are not concerned here with the general theory of polynomial approximation, we shall rely on comparisons between the data and interpolating polynomials of different degree to serve as criteria of accuracy and suitability.

PROBLEMS

20 • 1–20 • 2 temperature measurement

20 • 1 In 1935 the French astronomer Delisle used a thermometric system in which the ice and steam point were assigned the respective values 150°D and 0°D. Show that, numerically, the Delisle temperature is

$$\tau_{°D} = 150 - \tfrac{3}{2}\tau = \tfrac{530}{3} - \tfrac{5}{6}\tau_{°F}$$

20 • 2 In a constant-volume gas thermometer like that of Fig. 20 . 2, the mercury in the tube C is 15 cm above the fiducial mark when the bulb is at the ice point and the atmospheric pressure is 750 mm Hg. Ignoring the expansion of the bulb and the dead-space correction, find the bulb temperature for which the mercury in tube C will be 5 cm below the fiducial mark.

20 • 3 The volume of a certain mass of carbon dioxide was found to be 100 ml at 0°C and 1 atm. When the temperature and pressure were raised to 100°C and 1.369 atm, the volume was found to be unchanged. Compute the mean pressure coefficient β_0 of carbon dioxide for the temperature range 0°C to 100°C.

20 • 4 A standard platinum electric-resistance thermometer is found to have resistances R of 16.50, 22.87, and 43.33 ohms at the ice, steam, and sulfur points, respectively. (*a*) Compute the constants A and B in the empirical equation $R(\tau) = R_0(1 + A\tau + B\tau^2)$. (*b*) When this thermometer is inserted in a certain liquid bath, the reading is 19.68 ohms. What is the temperature of the liquid? (*c*) If the thermometer were graduated by dividing resistance changes between the ice and steam points into 100 equal parts, what would be the temperature of the liquid on the resulting electric-resistance scale?

20 • 5 (*a*) Express the temperature 500 K in degrees Rankine and in degrees Celsius. (*b*) If an object warms through $\Delta T = 20$ K, what is this temperature change in degrees Rankine and in degrees Fahrenheit?

20 • 6 With a constant-volume gas thermometer, the ratio of the pressure P_s at the steam point to the pressure P_t at the triple point of water is determined for different masses of gas in the bulb. As the gas is made more and more nearly ideal by reducing its mass so that P_s and P_t both approach zero, it is found that the ratio P_s/P_t approaches the limiting value 1.36605. Find the temperature of the steam point on this scale.

20 • 7 The drop in absolute temperature $T_1 - T_2$ as the result of a single adiabatic demagnetization is roughly proportional to the initial temperature T_1. (*a*) Show that an infinite number n of successive adiabatic demagnetizations would be needed to reach the absolute zero. (*Hint:* Take n as a continuous variable.) (*b*) What is meant by the statement that the concept of an absolute zero can be entirely avoided by defining temperature in a manner different from that ordinarily used? Show that this statement is not in conflict with the assertion that the absolute zero of the Kelvin scale is unattainable. (*c*) In what sense is it correct to say that if a temperature of, say, 0.001 K is attained experimentally, this is still far above the zero of the Kelvin scale?

20 • 8 The first temperature scale proposed by Lord Kelvin, in 1848, was a logarithmic scale. The numerical values of any given temperature on the Kelvin logarithmic and absolute scales are related by the equation

$$\tau_{°L} = 100 \times \frac{\log T - \log 273.15}{\log 373.15 - \log 273.15}$$

where 273.15 K is the ice point. Find the temperature $\tau_{°L}$ on the logarithmic scale (*a*) when $T = \infty$. (*b*) When $T = 273.15$ K. (*c*) When $T = 0.001$ K. (*d*) When $T = 0$ K. (*e*) Show that a cooling from 3700 K (the melting point of tungsten) to 1336 K (the melting point of gold) and a cooling from 4.2 K (the boiling point of helium) to 0.005 K both represent a cooling through 2158°L.

20 • 9 An improperly calibrated thermometer reads 102°C in boiling water and −1°C in a water-ice mixture at 1 atm pressure. What does it read at the boiling point of sulfur?

20 • 3–20 • 4 ***equations of state and the ideal-gas law***

20 • 10 Explain why density is an *intensive* property of a system, whereas the ratio PV/T is an *extensive* quantity. Show that the average density of an inhomogeneous system can be measured indirectly in terms of extensive properties.

20 • 11 If the original volume in Eq. [20 . 13] were measured at the steam point instead of at the ice point, show that the coefficient of expansion of an ideal gas would be $2.68 \times 10^{-3}\ \mathrm{K}^{-1}$.

20 • 12 The gage pressure of an automobile tire was found to be 35 $\mathrm{lbf/in^2}$ on a cool day when the temperature was 16°C. Assuming that the volume change was negligible and that there was no leakage, compute the total pressure in the tire on a hot day when the temperature is 36°C.

20 • 13 If the molecular weight of a certain gas is 28, what volume will be occupied by 12 g of the gas at 35°C and a pressure of 2.0 atm?

20 • 14 What pressure is exerted by 0.5 g of argon (molecular weight 40) contained in a closed vessel of capacity 5 liters at 20°C?

20 • 15 A gas is enclosed in a tube placed in a dish of mercury as shown. If the mercury rises 30 cm when the gas temperature is −20°C and the pressure is 1 atm, at what temperature will it be at the same level as the mercury surface in the dish? Assume that the change in the density of the mercury and the change in the outside level are negligible.

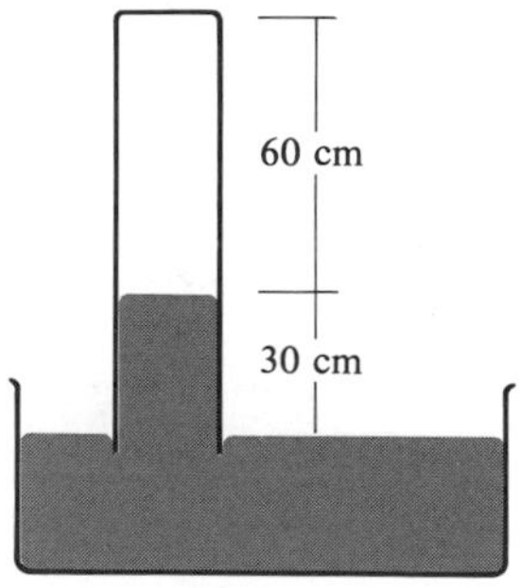

20 • 16 The density of dry air at 0°C and 1 atm pressure is approximately 1.29 kg/m³. (*a*) Use the general gas law and the definition of density to derive a single equation to compute the density at any other temperature and pressure, given an initial value. (*b*) What is the density of dry air at 16°C and a pressure of 740 mm Hg? (*c*) What volume will 28.9 g of dry air occupy under these conditions?

20 • 5 liquids and solids

20 • 17 From the data in Table 20 . 2, compute the *mean* value of β for water (*a*) in the range 0°C to 1°C. (*b*) In the range 0°C to 20°C. (*c*) Compute the true value of β at 0°C and at 10°C. (*d*) At what temperature is the specific volume of water a minimum?

20 • 18 The densities of a liquid or a solid at temperatures τ and 0°C are, respectively, ρ_τ and ρ_0. (*a*) Prove that

$$\rho_\tau = \rho_0(1 + \beta\tau)^{-1} \doteq \rho_0(1 - \beta\tau)$$

to the degree of approximation for which Eq. [20 . 20] is valid. (*b*) The experimentally determined values of β and ρ for mercury at 20°C are 1.82×10^{-4}/K and 13.546 g/cm³, respectively. Compute the density of mercury at 100°C and compare this value with the experimentally determined value, $\rho_{(100)} = 13.352$ g/cm³.

20 • 19 An ideal pendulum changes in length from l_0 to a slightly greater length $l_0 + \Delta l$ owing to thermal expansion. (*a*) Show that the number of vibrations lost in a day is $\Delta n = n\,\Delta l/2l_0$, where n is the number of vibrations completed in 24 h with length l_0. (*b*) At a certain temperature the pendulum keeps exact time, but how many vibrations would be lost in 24 h if the temperature were to increase by 10°C? Assume the linear expansivity of the material to be 1.0×10^{-5} K.

20 • 20 A thin, uniform metal hoop of mass m, diameter d, and expansion coefficient $\alpha = 2.6 \times 10^{-5}$/K hangs on a nail. (*a*) At 18°C, what is the change per degree of temperature in the moment of inertia I of the hoop about the nail as axis? (*b*) About the center of mass of the hoop as axis?

20 • 21 The bimetallic expansion strip in a thermostat consists of two metal strips of different linear expansion coefficients $\alpha_1 > \alpha_2$, riveted together to form a straight piece at room temperature τ. When the strip is heated to a temperature $\tau + \Delta\tau$, it bends into an arc as shown. In each strip there is a *neutral axis,* which coincides approximately with the centerline of the strip and along which there are no tensile or compressive stresses; hence each neutral axis expands as if the whole strip were free.
(*a*) If the expanded lengths of these neutral axes are l_1 and l_2, show that

$$\frac{l_1}{l_2} = 1 + (\alpha_1 - \alpha_2)\tau = \frac{r + \Delta r}{r}$$

(*b*) Explain the significance of the result in the event that $\alpha_1 = \alpha_2$.

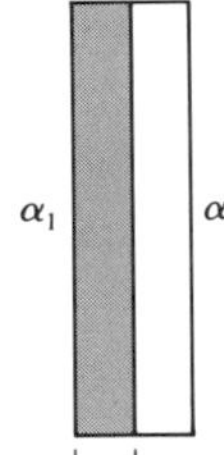

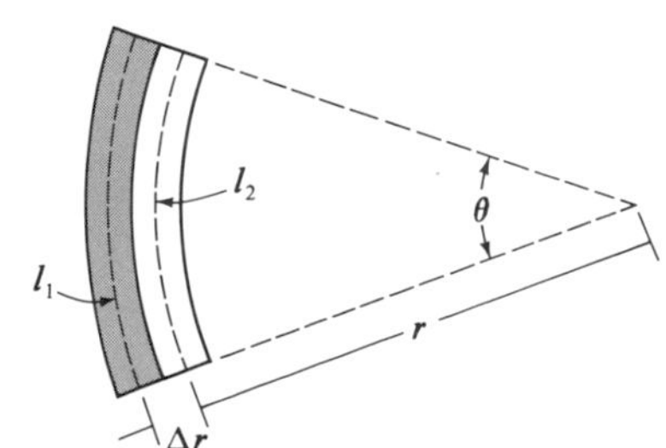

20 • 22 A wire 1 m long and 1.13 mm in diameter, for which $Y = 2 \times 10^{11}$ Pa and $\alpha = 1.3 \times 10^{-5}$/K, is attached at one end to a 20-kg block and at the other end to a rigid wall. The wire is slowly cooled from 20°C to −30°C before the block begins to move. What is the coefficient of static friction between the block and the horizontal surface on which it is resting? Assume that Y does not change appreciably during the cooling.

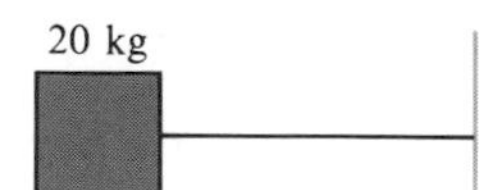

20 • 23 Bars of stainless steel ($Y = 2 \times 10^{11}$ Pa) and of aluminum ($\alpha = 2.5 \times 10^{-5}$/K) of the same length and thickness are placed end to end and wedged between rigid walls as shown. The temperature of the bars is now increased by 100 K. (*a*) Assuming that the common axis of the bars remains horizontal during the heating, find the pressure in each bar. (*b*) Which way and by what fractional amount does their interface move?

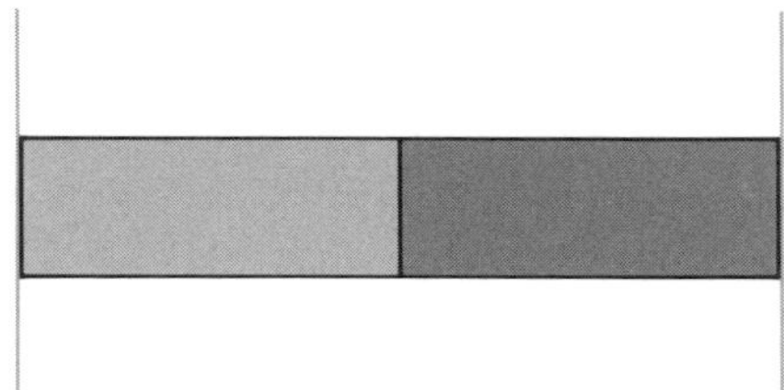

20 • 6 polynomial approximations

***20 • 24** Compute the five-point Lagrangian interpolation coefficients and print out their table (without folding, as a check for accuracy).

***20 • 25** Use the coefficients for the five-point Lagrangian interpolation in Prob. 20 . 24 to compute the unique fourth-order polynomial which approximates the sine function from $0 \leq \theta \leq 2\pi$. Compare the results with the true values and determine the percentage errors.

20 • 26 Differentiate the interpolating polynomial for $P(x)$ in Example 20 . 8 to find the approximation to the functional derivative $P'(x)$ in terms of the known data points y_0, y_1, and y_2.

***20 • 27** Rederive the interpolating polynomial for the elliptic integral K of Eq. [20 . 40], using $k^2 = \sin^2 \frac{1}{2}\theta_m$ as the independent variable over a range of values corresponding to $0° < \theta_m < 150°$. What is the maximum error in K as predicted by the interpolating polynomial? (Compare your results with a table of complete elliptic integrals of the first kind.)

20 • 28 Show that the numerical coefficient of each product α_i in a Lagrangian interpolation formula for $n + 1$ equally spaced points can be expressed in terms of the binomial coefficients

$$\binom{n}{i} = \frac{n!}{i!(n-i)!}$$

by the formula

$$\alpha_i(\nu) = \frac{(-1)^{n-i}}{n!}\binom{n}{i}\nu(\nu - 1)$$

$$\cdots(\nu - i + 1)(\nu - i - 1)\cdots(\nu - n)$$

20 • 29 Given that $y(0) = 1$, $y(60) = 0.5$, and $y(90) = 0$, find the Lagrange interpolating polynomial. Compare your results with $y = \cos x$ (x in degrees). Note that the points are not equally spaced.

answers

20 • 2 $\tau = -60.7°\text{C}$
20 • 3 $\beta_0 = 0.00369/\text{K}$
20 • 4 (*a*) $A = 3.92 \times 10^{-3}$ ohm/K, $B = -5.88 \times 10^{-7}$ ohm/K^2; (*b*) $\tau = 49.5°\text{C}$; (*c*) $\tau = 49.9°\text{C}$
20 • 5 (*a*) 500 K = 900°R = 227°C; (*b*) $\Delta T = 36°\text{R} = 36°\text{F}$
20 • 6 $\tau_s = 100.000°\text{C}$
20 • 7 (*a*) $T \doteq T_0 e^{-CN}$, where $C > 0$ and N = number of demagnetizations required to attain T, starting from T_0; (*b*) T could be defined in units of $N = -C^{-1} \ln (T/T_0)$
20 • 8 (*a*) $\tau_{°\text{L}} = \infty$; (*b*) $\tau_{°\text{L}} = 0°\text{L}$; (*c*) $\tau_{°\text{L}} = -4012°\text{L}$; (*d*) $\tau_{°\text{L}} = -\infty$
20 • 9 $\tau = 456.9°\text{C}$
20 • 12 $P = 53.2$ lbf/in^2
20 • 13 $V = 5.42$ liters
20 • 14 $P = 6090$ Pa
20 • 15 $\tau = 354°\text{C}$
20 • 16 (*a*) $\rho = (\rho_0 T_0/P_0)(P/T)$; (*b*) $\rho = 1.19$ kg/m^3; (*c*) $V = 24.4$ liters
20 • 17 (*a*) $\beta = -5.583 \times 10^{-5}/\text{K}$; (*b*) $\beta = -7.867 \times 10^{-5}/\text{K}$; (*c*) $\beta = -6.427 \times 10^{-5}/\text{K}$; (*d*) $\tau = 3.97°\text{C}$
20 • 18 $\rho_{(100)} = 13.348$ g/cm^3
20 • 19 2.2 vibrations are lost
20 • 20 (*a*) $\Delta I = 200\alpha$ percent/K; (*b*) $\Delta I = 200\alpha$ percent/K
20 • 22 $\mu_s = 0.665$
20 • 23 (*a*) $P = 18.15 \times 10^7$ Pa; (*b*) the iron expands 0.00926 percent
20 • 24 The numerical coefficients of $\alpha_0, \alpha_1, \ldots, \alpha_4$ are $\frac{1}{24}, -\frac{1}{6}, \frac{1}{4}, -\frac{1}{6}$, and $\frac{1}{24}$; some representative values are $\alpha_0(2.5) = 0.0234 = \alpha_4(1.5)$, $\alpha_1(2.5) = -0.156 = \alpha_3(1.5)$, and $\alpha_2(1.5) = 0.703$
20 • 26 $P'(x) = [4\nu(y_0 - 2y_1 + y_2) - (3y_0 - 4y_1 + y_2)]/(x_2 - x_0)$
20 • 29 $P(x) = (90 - x)(60 + \frac{1}{2}x)/5400$; errors less than 4 percent for $0 \leq x \leq 73°$ and less than 11 percent for $x \leq 89°$

CHAPTER TWENTY-ONE

heat

*It was formerly a common supposition, that the quantities of heat required to increase the heat of different bodies by the same number of degrees, were directly in proportion to the quantity of matter on each, and therefore, when the bodies were of equal size the quantities of heat were in proportion to their density. But very soon after I began to think on this subject (*anno *1760) I perceived that this opinion was a mistake, and that the quantities of heat which different kinds of matter must receive, to reduce them to an equilibrium with one another, or to raise their temperature by an equal number of degrees, are not in proportion to the quantity of matter in each, but in proportions widely different from this, and for which no general principle or reason can yet be assigned. . . .*

Joseph Black, Lecture on the Elements of Chemistry, *Edinburgh, 1803*

The concept of heat as some quality related to thermal phenomena has doubtless existed since remote antiquity. The fact that an object close to a fire becomes warmer must have suggested that something passes from the fire to the object, even though the nature of heat was not understood. The clarification started with the invention of the thermometer. Francis Bacon (1620), the Florentine Academicians (1657–1667), and Robert Boyle (1675) showed evidence of distinguishing between temperature and heat. Joseph Black, however, was the first to draw a sharp distinction between the two concepts. Black's clear formulation of heat as a measurable quantity and his discoveries of specific heats and heats of transition provided the foundation for a quantitative science of thermal phenomena.

In this chapter we shall consider heat in terms of its observable properties and examine the transfer of heat from one body to another, a phenomenon which plays a part in our subjective notions of thermal processes.

21 · 1 work

Before we can define heat, we must understand the work done on or by a thermodynamic system. Work is always the result of a force—whether it is mechanical, electrical, magnetic, etc., in origin—acting through a distance. It is apparent that work is being done on an object when its center of mass is accelerating or when it moves against some resistive force. But what about a tire that is being inflated? Its center of mass is not displaced appreciably, nor does it acquire any momentum. Yet work is obviously being done by the air pressure in expanding the volume of the tire against the restoring forces in the rubber. In general:

> Whenever one of the extensive state variables of a thermodynamic system of constant mass is changed, then work is done on or by the system.

In our subsequent discussions of work, heat, and energy, we shall assume that the center of mass is either stationary or moving *quasistatically*—that is, so slowly and gradually that we can ignore acceleration and assume the system to be in mechanical and thermal equilibrium at each instant. This is necessary because in the following chapters we shall be working with values of mass, momentum, and energy which are averaged over enormous numbers of particles. If these averages are to be meaningful, we must exclude cases in which significant fluctuations from the average could arise.

Suppose a fixed mass of any gas at pressure P is enclosed in a cylinder equipped with a piston of area A and smooth walls (Fig. 21 . 1). The gas will do an infinitesimal amount of work dW on the piston when it moves out a distance dx and the volume of the gas increases by an amount dV:

$$dW = F\,dx = \frac{F}{A}(A\,dx) = P\,dV \qquad [21 \cdot 1]$$

If the piston moves a finite distance, so that the volume changes from V_1 to V_2, the work done by the gas will obviously be

$$W = \int_{V_1}^{V_2} P\,dV \qquad [21 \cdot 2]$$

where P may vary during the process. In an expansion, where $V_2 > V_1$, the work done *by* the gas is positive, and the gas loses energy to the piston and surroundings. In a compression, $V_2 < V_1$, the work is negative, and work is done *on* the gas, so that it gains energy from its surroundings.

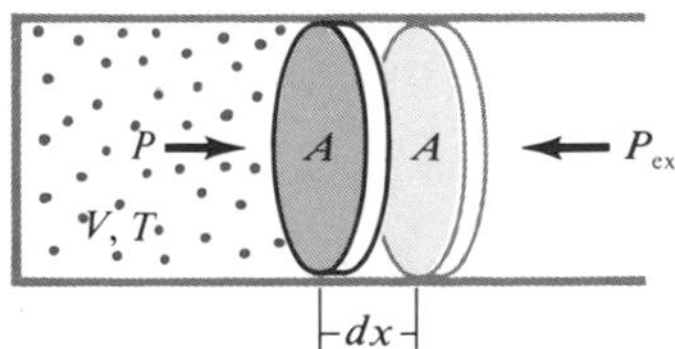

figure 21 • 1 *Work done by a gas in a volume expansion:* $dW = F\,dx = P\,dV$

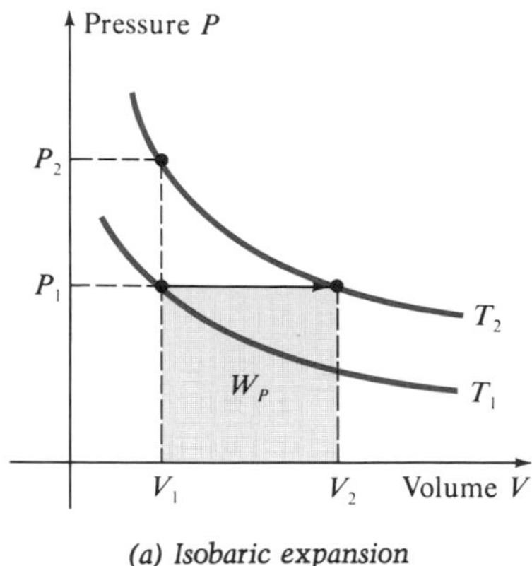

(a) Isobaric expansion

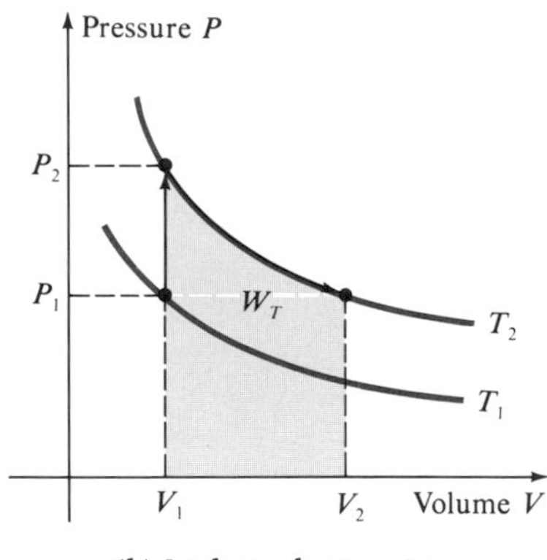

(b) Isothermal expansion

figure 21 • 2 *Work as a function of path in the PV plane*

To evaluate an integral such as Eq. [21 . 2] we need the equation of state. However, the equation of state of *any* system, such as $PV = nRT$ for ideal gases, is applicable only if the system is in a state of complete equilibrium. Therefore, if a system undergoes a change in volume, it must do so *quasistatically,* so that the equation of state holds throughout the process. When the piston moves as a result of a finite difference between the gas pressure and the external pressure on it, the piston will accelerate and so will the gas. There will also be pressure differences within the gas, so that during this process the gas will not be in mechanical or thermal equilibrium. We therefore assume that the piston moves by such small amounts that there is only an infinitesimal difference between P and the external pressure P_{ext}, and at any given instant the gas is infinitesimally close to a state of complete equilibrium. This allows us to set the pressure P in Eq. [21 . 2] equal to nRT/V, and if we know the behavior of T, we can then integrate over the change in volume as a function of temperature alone. Although a quasistatic process obviously is an idealization, there are many actual processes that approximate it very closely, even at speeds approaching the speed of sound.

Suppose that any ideal gas undergoes a quasistatic isobaric (constant-pressure) process, from a state (P_1,V_1,T_1) to a state (P_1,V_2,T_2), as shown in Fig. 21 . 2*a*. The work W_P done by the gas under constant pressure is

$$W_P = P_1\int_{V_1}^{V_2} dV = P_1(V_2 - V_1) = nR(T_2 - T_1) \qquad [21 \cdot 3]$$

as indicated by the shaded rectangle. Suppose now the gas arrives at the same state by means of an isometric (constant-volume) heating to temperature T_2 and pressure P_2, followed by an isothermal (constant-temperature) expansion to V_2. Then work is done by the gas only over the isothermal expansion, since the volume has remained constant during the isometric heating. In this case the work W_T done by the gas at constant temperature is

$$W_T = nRT_2\int_{V_1}^{V_2}\frac{dV}{V} = nRT_2 \ln\frac{V_2}{V_1}$$

$$= P_2V_1 \ln\frac{V_2}{V_1} = P_1V_2 \ln\frac{V_2}{V_1} \qquad [21 \cdot 4]$$

Graphically W_T equals the *entire* area under the T_2 isotherm between V_1 and V_2 (Fig. 21 . 2*b*). Therefore $W_T = \Delta W + W_P > W_P$.

> The work done in a change of state is not unique, but depends on the path taken by the system through a succession of equilibrium states.

That is, any curve between the endpoints of a process in the *PV* plane can represent the path of a quasistatic change of state, and in general the areas under the paths, which represent the work done in the process, are *not* equal. The "work content" of a system is therefore not properly a state variable, since it is not unique, but depends on the intermediate states through which the system passes between its initial state and its final state. This situation is distinctly different from a conservative force field such as gravity, where motion between two points always involves the same changes in energy regardless of the path. It suggests that heat transfer takes place somewhere in the process illustrated in Fig. 21 . 2.

example 21 • 1 The gas in Fig. 21 . 2 undergoes isometric-isothermal expansion to a state (P_1, V_2, T_2) and is then compressed isobarically back to state (P_1, V_1, T_1). What is the net work done by the gas on its surroundings?

solution This is an example of a cyclic process, in which the system returns to its initial state. During the compression the work W_P done by the gas at constant pressure P_1 is

$$\int_{V_2}^{V_1} P\,dV = P_1(V_1 - V_2) = -W_P < 0$$

That is, some outside agent does work W_P *on* the gas. Therefore the gas has done net work

$$\Delta W = W_T + (-W_P)$$

which is *the area bounded by the path* of the cyclic process in the *PV* plane.

example 21 • 2 Suppose 0.1 kmole of hydrogen is compressed from state $(P_1,\ 4\ \text{m}^3,\ 300\ \text{K})$ to state $(P_2,\ 2\ \text{m}^3,\ 300\ \text{K})$. (*a*) What is the net work done by the gas in an isobaric process followed by an isometric process? (*b*) In an isothermal process? (*c*) In an isometric process followed by an isobaric process?

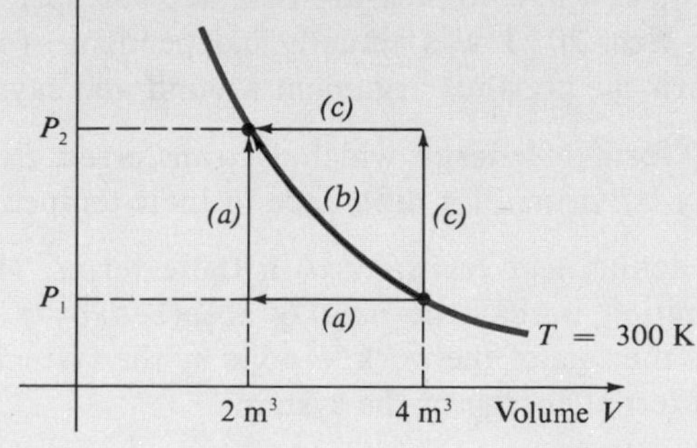

figure 21 • 3

solution Note that if $P_1 = nRT/V_1$, then $P_2 = 2P_1$ and

$$nRT = (0.1 \text{ kmole})(8300 \text{ J/kmole} \cdot \text{K})(300 \text{ K}) = 249{,}000 \text{ J}$$

Therefore the work done by the gas in each case is

$$(a) \quad W = P_1(V_2 - V_1) + 0 = nRT\left(\frac{V_2}{V_1} - 1\right) = -124{,}500 \text{ J}$$

$$(b) \quad W = nRT \ln \frac{V_2}{V_1} = 249{,}000 \text{ J} \times (-0.693) = -172{,}600 \text{ J}$$

$$(c) \quad W = 0 + P_2(V_2 - V_1) = 2P_1(V_2 - V_1) = -249{,}000 \text{ J}$$

example 21 • 3 As n moles of an ideal gas at pressure P_1 expands from V_1 to V_2, its temperature drops according to $T = T_1\sqrt{V_1/V}$ (the walls of its container are not insulated). How much work is done during expansion?

solution First we solve $PV = nRT$ and $T = T_1\sqrt{V_1/V}$ for $P(V)$. Since $P_1V_1 = nRT_1$,

$$P = P_1\left(\frac{V_1}{V}\right)^{3/2}$$

Therefore

$$W = \int_{V_1}^{V_2} P\,dV = P_1\int_{V_1}^{V_2}\left(\frac{V_1}{V}\right)^{3/2} dV$$

$$= 2P_1V_1\left(1 - \sqrt{\frac{V_1}{V_2}}\right)$$

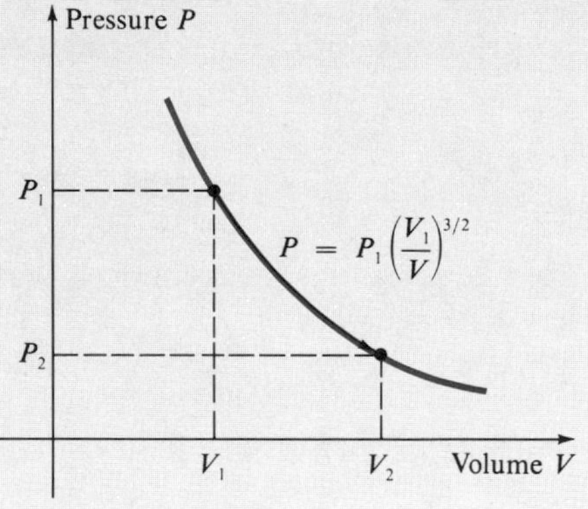

figure 21 • 4

21 • 2 *the first law of thermodynamics*

In our discussion of temperature in Chapter 20 we did not consider the direct measurement of heat. Now, however, we must develop a more rigorous concept of what heat is and how to measure it. Since our operational definition of temperature in Sec. 20 . 1 was actually independent of the definition of heat, we might turn the previous argument around and say:

> Heat is that form of energy which is transferred from one body to another solely by virtue of a difference in their temperatures.

We can, in fact, define *heat transferred* in these terms. However, a more fundamental definition is that the *heat* Q absorbed by a system in some physical process is the sum of the work W done by the system and the change $\Delta\mathcal{U}$ in the total internal energy of the system:

$$Q = \Delta\mathcal{U} + W \qquad [21 \cdot 5]$$

As we saw in Sec. 8 . 5, this is the *first law of thermodynamics:*

> The change in the internal energy of a system of constant mass is equal to the heat absorbed by the system minus the work done by the system on its external surroundings.

$$\Delta \mathcal{U} = Q - W \qquad [21 \cdot 6]$$

Hence we can view heat as that form of energy which is transferred from one body to another, but which is *not* attributable to work. If Q is negative, then an amount of heat $|Q|$ is given up by the system to its surroundings.

It is abundantly clear from experience that heat is "something," and not merely a bookkeeping device to maintain the principle of conservation of energy. The internal energy $\mathcal{U}$ of a system of constant mass in thermal equilibrium is the sum of the kinetic and potential energies of its particles. Countless experiments have shown that the *difference* $\Delta \mathcal{U} = Q - W$ between end states of the system depends only on the initial and final state variables, and not on the infinite number of different paths by which the final state can be reached. This has two immediate and important consequences.

1. The internal energy of a given system in a given equilibrium state depends only on the thermodynamic coordinates of that state; hence $\mathcal{U}$ itself must be a state variable.
2. Since W is not unique to a given equilibrium state, and is therefore not a state variable, $Q = \Delta \mathcal{U} + W$ is also not a state variable.

The internal energy of a system is sometimes called a *thermodynamic potential* because, like the potential-energy functions of classical mechanics, we can measure differences $\Delta \mathcal{U}$, but not the value of $\mathcal{U}$ itself. We are therefore free to assign an arbitrary value of energy to some *reference state* of the system and measure the differences in terms of this reference value.

Our definition of heat illustrates the fact that the unfolding of scientific knowledge is not a linear process; rather, it is an iterative process. We began in Chapter 8 with a subjective notion of heat and proceeded to a hypothesis, the first law of thermodynamics, defining its nature. We then used these concepts to develop an operational definition of temperature, which led us to the concept of absolute temperature and the equation of state. Now we shall use this knowledge to develop ways of measuring heat precisely, and then, by applying the first law of thermodynamics, show that this definition is perfectly consistent in so far as we are able to determine it.

In the eighteenth century the prevailing theory held that heat is an indestructible and imponderable elastic fluid, called *caloric*. The idea of a caloric fluid as a unique extensive state variable was very fruitful in its day, giving way to a better theory when its internal contradictions became insupportable. In fact, you will probably find yourself visualizing heat flow between bodies as analogous in some way to the flow of fluids. While this concept may be helpful at times, it is important to keep the fundamental distinctions in mind: heat is not a fluid, and a body does not "contain" heat any more than it "contains" work. Heat and work are both forms of energy *in the process of transference.*

Because of the early procedures for measuring heat, it was convenient to employ as the heat unit the quantity of heat that must enter or leave a unit

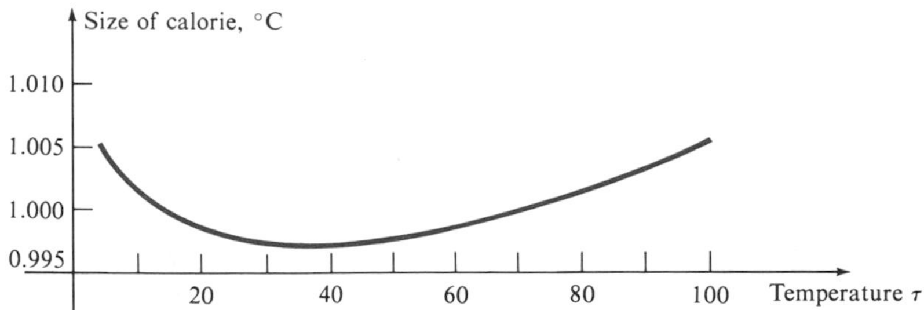

figure 21 • 5 *Relative values of the calorie at different temperatures, with the 15-degree calorie taken as unity*

mass of water in order to change its temperature one degree. This definition has been retained and has given us the familiar *calorie* (cal) and *British thermal unit* (Btu). However, increasingly accurate experiments have shown that the calorie so defined is not quite the same at all temperatures (see Fig. 21 . 5). The standard cgs unit is the 15-degree calorie, defined as the energy required to change the temperature of 1 g of water from 14.5 to 15.5°C. Also in use is the *mean calorie,* defined as $\frac{1}{100}$ of the energy required to change the temperature of 1 g of water from 0 to 100°C. The *kilogram-calorie* (kcal), or "large calorie," is defined as 1 kcal = 1000 cal, the heat required to warm 1 kg of water from 14.5 to 15.5°C. It is also used to measure the energy value of metabolized food. The *British thermal unit* (Btu) is the energy needed to raise 1 lbm of water through 1°F, usually from 59.5 to 60.5° F. Both definitions assume the pressure to be 1 atm.

Quantities of heat may be expressed in terms of any of the energy units—joules, foot-pounds, liter-atmospheres, and so on. In joules, the *mechanical equivalent of heat* is given by

$$1 \text{ cal} = 4.1855 \pm 0.004 \text{ J}$$

This number is known as *Joule's constant,* first measured by James Prescott Joule in 1843. From the conversion factors in Table 8 . 1, we also see that 1 Btu = 252.0 cal. At present, the most accurate measurements involve electrical quantities and devices and give the heat directly in joules. The mechanical equivalent of heat cannot be derived; it must be measured experimentally. Such measurements by Joule and others in the mid-nineteenth century were important in establishing the first law of thermodynamics.

21 . 3 specific heat capacity: latent heats

In 1760 Joseph Black arrived at the important conclusion that the quantity of heat required to raise the temperature of a substance a certain number of degrees depended very strongly on the nature of the substance, as well as on its amount. The heat given up by 1 g of water in cooling through one degree would raise very different masses of other substances through one degree—for example, about 30 g of mercury or 9 g of iron. Black explained these facts with the assumption that equal masses of different substances possess different capacities for heat. Thus the *heat capacity* of a system came to be defined as the quantity of heat absorbed or emitted by the system when its temperature changes by one degree. Symbolically, if ΔQ denotes the quantity of heat transferred when a system undergoes a temperature change ΔT, then the heat capacity of the system is, by definition,

$$C = \frac{\Delta Q}{\Delta T} \qquad [21 \cdot 7]$$

The heat capacity per unit mass of a system over the temperature interval ΔT is termed the *mean specific heat capacity*, or *mean specific heat* $\bar{c}$:

$$\bar{c} = \frac{\Delta Q}{M\,\Delta T} \qquad [21 \cdot 8]$$

where M is the total mass of the system. The specific heat is characteristic of the material of which the system is composed. For example, the mean specific heat of water in the temperature range 14.5 to 15.5°C is defined as equal to one 15-degree calorie per kelvin. Sometimes it is more convenient to speak of the heat capacity per mole of the material, or its *mean molar specific heat* $\overline{C}$:

$$\overline{C} = \frac{\Delta Q}{n\,\Delta T} = M_0\bar{c} \qquad [21 \cdot 9]$$

where $n = M/M_0$ is the number of moles and M_0 is the molecular weight of the substance in kilograms. Thus if a system of mass M consists of n moles of material, the amount of heat ΔQ required to raise it from a temperature T_1 to temperature T_2 is

$$\Delta Q = M\bar{c}(T_2 - T_1) = n\overline{C}(T_2 - T_1) \qquad [21 \cdot 10]$$

In this case ΔQ represents heat absorbed by the system. For a temperature decrease, $T_2 < T_1$, the negative quantity ΔQ represents heat lost in cooling.

As ΔQ and ΔT are made smaller and smaller, the mean specific heat approaches the *specific heat c;* that is, in the limit as ΔT approaches zero,

$$\begin{aligned} c(T) &= \lim_{\Delta T \to 0} \frac{\Delta Q}{M\,\Delta T} = \frac{1}{M}\frac{dQ}{dT} \\ C(T) &= \lim_{\Delta T \to 0} \frac{\Delta Q}{n\,\Delta T} = \frac{1}{n}\frac{dQ}{dT} \end{aligned} \qquad [21 \cdot 11]$$

For all substances specific heat capacity varies with temperature, although the variation is slight at moderate temperatures for water and most solids. For most liquids it is not negligible even at ordinary temperatures. At extremely low temperatures all specific heats are sharply reduced (see Table 21 . 1).

If a system consists of some substance of specific heat $c(T) = (n/M)C(T)$, the quantity of heat that must be transferred to it to increase its temperature from T_1 to T_2 is given by

$$Q = M\int_{T_1}^{T_2} c(T)\,dT = n\int_{T_1}^{T_2} C(T)\,dT \qquad [21 \cdot 12]$$

However, when the variation in $c(T)$ is small, it is usually permissible to set the specific heat c equal to the mean specific heat $\bar{c}$ over a suitable interval of moderate range and use Eq. [21 . 8].

table 21 • 1 *Specific heats of selected solids and liquids*

substance	temperature, °C	specific heat $c(\tau)$ or $\bar{c}$, kcal/kg · K*
solid		
aluminum	−250	0.0039
	0	0.2079
	16–100	0.212
copper	0–300	$0.0915 + 2.4 \times 10^{-5}\tau$
glass, thermometer	19–100	0.199
gold	−258	0.0018
	−209	0.0211
	18	0.0312
lead	0–300	$0.0295 + 2 \times 10^{-5}\tau$
mercury	−40	0.0337
quartz	12–100	0.188
silver	0–400	$0.0556 + 8 \times 10^{-6}\tau$
liquid		
alcohol, ethyl	−100	0.456
	25	0.581
ammonia	−60	1.047
benzene	20	0.406
mercury	−20	0.0335534
	20	0.033240
	100	0.032776

*The value listed is the specific heat $c(\tau)$ in entries where one temperature or a formula is given; otherwise it is the mean specific heat $\bar{c}$ for the indicated temperature range.

example 21 • 4 A mass M of water at 35°C is added to a mass $2M$ of water in an insulated container at 5°C. Ignoring any small changes in volume, find the final temperature of the mixture.

solution We can equate the heat lost by the smaller mass to the heat gained by the larger one:

$$|Q| = Mc(35°\text{C} - \tau) = 2Mc(\tau - 5°\text{C}) \qquad \text{or} \qquad \tau = 15°\text{C}$$

example 21 • 5 A 5-kg lead bar at 300°C is cooled by immersion in 1 kg of water at 25°C. Ignoring volume changes and assuming no heat loss to the external environment, find the final temperature of the water-lead mixture.

solution The specific heat capacity of lead is given in Table 21 . 1 as

$$c = (0.03 + 2 \times 10^{-5}\,\tau)(\text{kcal/kg} \cdot \text{K})$$

Therefore the heat it loses in cooling from 300°C to some temperature τ is

$$-Q = \int_{\tau}^{300^\circ\mathrm{C}} (0.15 + 10^{-4}\tau)\,d\tau$$

$$= (49.5 - 0.15\tau - 5 \times 10^{-5}\tau^2)\ \mathrm{kcal}$$

If we set this equal to $(\tau - 25)$ kcal, the heat gained by the water yields a quadratic in τ,

$$f(\tau) = 5 \times 10^{-5}\tau^2 + 1.15\tau - 74.5 = 0$$

which has only one physically meaningful solution, $\tau = 64.6^\circ$C. Had we not included the temperature dependence of c, we would have obtained 60.9°C as the final temperature, a large discrepancy.

example 21 • 6 Find the increase in internal energy of 1 kg of lead when its temperature increases from 20°C to 21°C at 1 atm $\doteq 10^5$ Pa. The density of lead is $\rho = 11.5\ \mathrm{g/cm^3}$.

solution From Table 21 . 1, the amount of heat which is absorbed by the lead is

$$Q = \bar{c}(1\ \mathrm{kg} \times 1\ \mathrm{K}) = 0.03\ \mathrm{kcal} = 0.03 \times 4186 \doteq 126\ \mathrm{J}$$

From Table 20 . 3, the linear expansion coefficient of lead at this temperature is $\alpha = 3 \times 10^{-5}/\mathrm{K}$, so the work done in expanding against atmospheric pressure is

$$W = P\,\Delta V = P\frac{3\alpha}{\rho}\Delta\tau$$

$$= 10^5\ \mathrm{Pa}\ \frac{9 \times 10^{-5}}{\mathrm{K}}\ \frac{1\ \mathrm{m^3}}{11{,}500\ \mathrm{kg}}\ 1\ \mathrm{K}$$

Therefore $W = 7.8 \times 10^{-4}\ \mathrm{J} \ll Q$, and the increase in the internal energy of the lead is

$$\Delta\mathcal{U} = Q - W \doteq Q$$

Until Black's time it was thought that the temperature rise of a substance in contact with a hot object was continuous. However, Black pointed out that while ice or snow is melting, if it is well stirred, it maintains a constant temperature; moreover, when water is boiled, no matter how long and violently, "we cannot make it in the least hotter than when it began to boil." To the relatively large quantities of added heat required to melt solids and to vaporize liquids Black gave the name *latent heats*. These are also the heats given off in freezing and condensation, respectively.

Each possible transition, or *change of phase,* with increasing temperature, whether from the solid to the liquid phase (fusion), from the liquid to vapor phase (vaporization), or from the solid directly to the vapor phase (sublimation), has a *latent heat* L_f, L_v, or L_s; this is the energy required to change a unit mass of substance from one phase to another (see Table 21 . 2). Figure 21 . 6 is the "cooling curve" for a typical pure crystalline substance; Fig. 21 . 7 is a typical "heating curve" of such a substance.

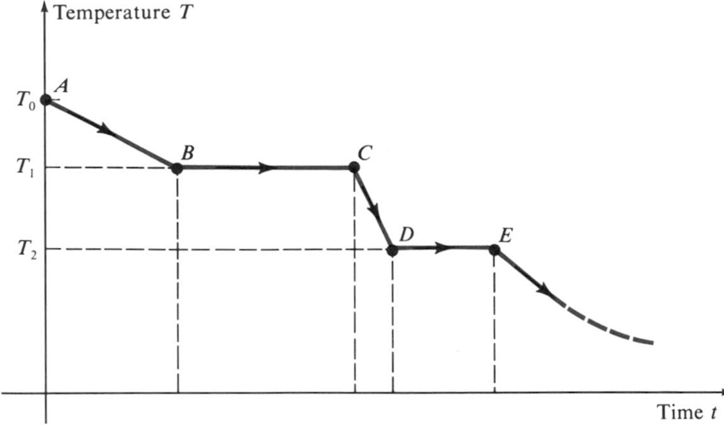

figure 21 • 6 *Cooling curve for a typical crystalline substance. As heat is removed, vapor cools (A → B) and then condenses to a liquid at constant temperature (B → C). The liquid cools (C → D) and then freezes to a solid (D → E), which then cools further.*

table 21 • 2 *Phase transitions and their temperatures and latent heats*

equal heats: 1.000 *J/g* = 0.2389 *cal/g* = 0.4300 *Btu/lbm*

substance	*boiling point* τ_v, °C	*heat of vaporization* L_v, J/g	*fusion point* τ_f, °C	*heat of fusion* L_f, J/g
alcohol, ethyl	78.5	885	−115	104
ammonia	−33.4	1369	—	—
ether, diethyl	34.6	351	—	—
gold	2660	1870	1064.43*	66.6
hydrogen	−252.9	447	—	—
lead	—	—	327	23
mercury	356.7	297	−39	11
nitrogen	−195.8	200	−210	25.6
oxygen	−182.960*	213	−219	13.9
platinum	4010	2680	1770	113
silver	—	—	961.93*	105
sulfur	444.600*	330	—	—
sulfur dioxide	−10.0	389	—	—
tungsten	—	—	3380	184
water (air-saturated)	100*	2257	0*	333.6

*Fixed points of the International Temperature Scale.

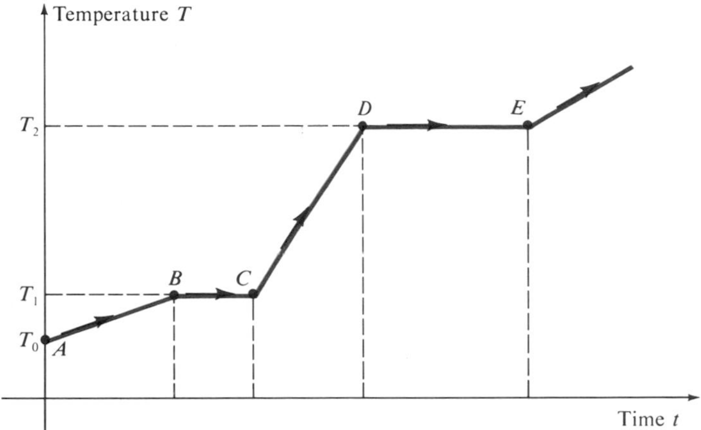

figure 21 • 7 *Heating curve for a typical crystalline substance. As heat is added, solid warms (A to B) and then melts (B to C) as the added energy breaks crystalline bonds. The liquid warms (C to D) and then boils (D to E) as added energy separates molecules to form a vapor, which then warms further.*

Latent heats, also known as heats of transformation or heats of transition, obviously represent the energies required to change the internal structure of the substance. When a solid melts, the latent heat represents the energy which must be added to the particles in the solid crystal in order to overcome the attractive forces that bind them together. This energy, supplied in the form of heat, "lifts" each particle out of the attractive potential well into which it was bound as part of the solid. It follows, then, that temperature changes are not caused by changes in the internal potential energy of a substance. This argument suggests that temperature, which is somehow related to the transfer of energy as heat, is therefore related to the internal kinetic energies of the particles of the system.

example 21 • 7 A 15-kg mass of metal at 400°C with a specific heat $c = 0.2$ kcal/kg · K, is cooled by being placed in thermal contact with 1 kg of ice at −10°C. The latent heats of water at 1 atm pressure are

$$L_v \doteq 539 \text{ kcal/kg at } 100°\text{C} \qquad L_f \doteq 79.7 \text{ kcal/kg at } 0°\text{C}$$

and the specific heats of ice and steam are

$$c_{\text{ice}} \doteq 0.5 \text{ kcal/kg} \cdot \text{K} \qquad c_{\text{steam}} = 0.5 \text{ kcal/kg} \cdot \text{K}$$

What is the final temperature of the metal?

solution Let us do this problem in three steps:

1 If we ignored any possible phase changes in the ice, then we could set

$$0.2 \times 15\,(400°\text{C} - \tau) = 0.5[\tau - (-10°\text{C})]$$

and find the equilibrium temperature, $\tau \doteq 340°\text{C}$. It is clear that the ice will be heated to 0°C and then melt to water, a process which requires

$$Q_1 = (0.5 \times 10 + 80)\ \text{kcal} = 85\ \text{kcal}$$

This would cool the metal to

$$\tau_1 = 400°\ \text{C} - \frac{85\ \text{kcal}}{15\ \text{kg} \times 0.2\ \text{kcal/kg} \cdot \text{K}} = (400 - \tfrac{85}{3})°\text{C}$$

2 If we next ignore phase changes in the water, we can set

$$0.2 \times 15(\tau_1 - \tau) = \tau - 0°$$

to find $\tau \doteq 260°$C. The water will reach the boiling point, absorbing an additional $Q_2 = 100$ kcal from the metal, which cools it to

$$\tau_2 = \tau_1 - \tfrac{100}{3}\ \text{K} = (400 - \tfrac{185}{3})°\text{C} > 100°\text{C}$$

The water then begins to vaporize. It is completely converted to steam at 100°C after absorbing $Q_3 = 540$ kcal from the metal, which reduces its temperature to

$$\tau_3 = \tau_2 - \tfrac{540}{3}\ \text{K} = (400 - 180 - \tfrac{185}{3})°\text{C} = 158.33°\text{C} > 100°\text{C}$$

3 Since the temperature of the metal is still above the boiling point, it will heat the steam above 100°C. (Had we found $\tau_3 < 100°$C, we would have realized that the metal was not hot enough to vaporize all the water.) To find the final temperature of the steam and the metal, τ_4, we set Q_4, the heat absorbed by the steam, equal to the heat lost by the metal, so that

$$Q_4 = 0.5(\tau_4 - 100°\ \text{C}) = 0.2 \times 15(\tau_3 - \tau_4)$$

$$\tau_4 = \tfrac{1}{7}(6\tau_3 + 100°\ \text{C}) = 150°\ \text{C}$$

Had we initially assumed steam to be formed, this would have required 5 + 80 + 100 + 540 = 725 kcal, a heat loss that is insufficient to cool the metal down to 100°C. Educated guesses are often useful time savers.

Calorimetry is concerned with the measurement of quantities of heat added to or subtracted from systems. Two general methods are used for such measurements: *thermometric calorimetry* and *change-of-phase calorimetry.* In the first method we measure temperatures, and the thermometer is the instrument of prime importance; in the second method fixed temperatures are employed and no thermometer is needed. Figure 21 . 8 shows an *adiabatic calorimeter,* so called because the outer vessel is kept at the same temperature as the test sample to prevent heat transfer between the sample and its surroundings. A given amount of heat energy is applied to the sample by an electric coil, and the rise in the temperature of the sample is measured.

One thermometric method, the *method of mixtures*, consists in mixing known masses of substances of different initial temperatures, observing the resulting temperature, and then writing an equation that includes all the heat quantities lost by the cooling bodies in one member and all the heat gained by the warming bodies in the other member. Another thermometric

figure 21 • 8 *An electrically heated adiabatic calorimeter*

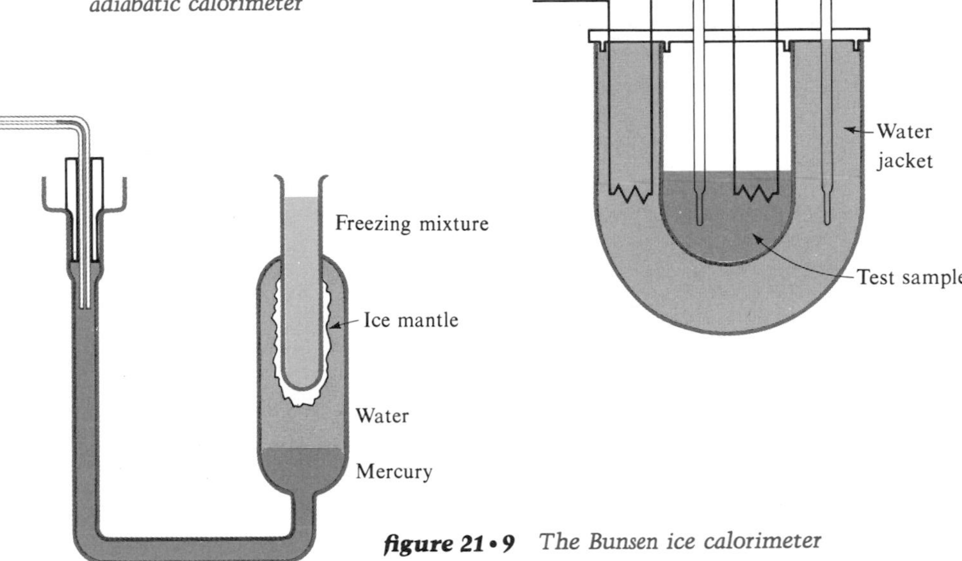

figure 21 • 9 *The Bunsen ice calorimeter*

method, known as the *method of cooling*, is used to determine specific heats. It consists in comparing the time required for a test sample to cool through a given temperature drop ΔT with the cooling time for a substance whose specific heat is known. A closed vessel is filled first with water of mass M_1 and specific heat c_1, and then with the substance of mass M_2 whose specific heat c_2 is to be determined. If the heat capacity of the cooling apparatus is C, then for a given temperature change ΔT, the rate of heat loss for each substance is

$$\frac{\Delta Q_1}{\Delta t_1} = \frac{(M_1c_1 + C)\,\Delta T}{\Delta t_1} \qquad \frac{\Delta Q_2}{\Delta t_2} = \frac{(M_2c_2 + C)\,\Delta T}{\Delta t_2} \qquad [21 \cdot 13]$$

Since the nature and area of the cooling surface is the same in both cases, the rate of loss must also be the same, $\Delta Q_1/\Delta t_1 = \Delta Q_2/\Delta t_2$. Therefore

$$\frac{M_1c_1 + C}{M_2c_2 + C} = \frac{\Delta t_1}{\Delta t_2} \qquad [21 \cdot 14]$$

where c_2 is the only unknown quantity.

Calorimetric methods that depend on change of phase consist essentially either in determining the mass of ice M_i that a heated body will melt while its temperature is falling to 0°C or in finding the mass of steam M_s that a cold body will condense while its temperature is rising to the boiling temperature of water. One such method is the ice calorimeter in Fig. 21 . 9, which depends on the fact that a volume change occurs during fusion. A mantle of ice is formed in the air-free water by inserting a freezing mixture in tube D, and the point to which the mercury rises in the graduated capillary tube is recorded. The heated specimen to be tested is then dropped into the tube, where it melts a certain amount of ice. The displacement of the mercury to the right because of contraction of the ice on fusion is proportional to the amount of ice melted.

21 · 4 specific heats of gases

We have tacitly assumed that the specific heats listed in Table 21 . 1 were measured at constant pressure (probably atmospheric). However, in an isobaric temperature rise some of the heat absorbed is used up as work when the heated substance expands, whereas in an isometric process *all* the heat absorbed goes into the internal energy $\Delta \mathcal{U}$ of the system, since no work $dW = P\,dV$ is done. Therefore it is reasonable to expect that different amounts of heat are required for a given temperature change, depending on the nature of the heating process. That is, the specific heat capacity depends on the path taken in the PV plane, and there are as many different specific heats as there are possible paths between two end states.

This plethora of specific heats may seem rather discouraging at first glance, but it does not pose a serious problem in practice. As we saw in Example 21 . 6, the work done in the thermal expansion of a solid or liquid is typically negligible compared to the heat absorbed, so that their specific heats c_P at constant pressure are approximately equal to c_V, their specific heats when they are confined to a constant volume during the heating process. The coefficients of expansion of gases, however, are relatively large, and to describe their thermal behavior we must distinguish between $c_P = (1/M)(dQ/dT)$ for constant pressure and $c_V = (1/M)(dQ/dT)$ for constant volume.

As might be expected, c_P and c_V are not unrelated. We can derive their relationship for an ideal gas by a heuristic argument based on the observation that the temperature of a substance depends only on its internal *kinetic energy*. Let us assume that for a given temperature change dT in two different processes—one isobaric and the other isometric—the change dK in internal kinetic energy is the same. However, according to Sec. 20 . 4, an ideal gas is one whose particles can be viewed as noninteracting point masses. It follows from this premise that there are no internal forces. Hence there is no internal potential energy, and the internal energy $\mathcal{U}$ of the ideal gas must be *entirely kinetic*. If we set $d\mathcal{U} = dK$, for a temperature change dT in an isobaric process, then

$$dQ_P = Mc_P\,dT = dK + P\,dV \qquad [21 \cdot 15]$$

and for the same temperature change in an isometric process,

$$dQ_V = Mc_V\,dT = dK = d\mathcal{U} \qquad [21 \cdot 16]$$

Thus for an ideal gas

$$\blacktriangleright \quad \mathcal{U} = Mc_VT = nC_VT \qquad [21 \cdot 17]$$

If we subtract Eq. [21 . 16] from Eq. [21 . 15] and differentiate the ideal-gas equation of state for an isobaric process, we obtain

$$M(c_P - c_V)\,dT = P\,dV = nR\,dT \qquad [21 \cdot 18]$$

Then, since the number of moles is $n = M/M_0$, the two specific heats are related by

$$c_P - c_V = \frac{R}{M_0} \qquad [21 \cdot 19]$$

and the two *molar* specific heats, $C_P = M_0 c_p$ and $C_V = M_0 c_V$, are related by R, the universal gas constant per mole:

$$C_P - C_V = R \qquad [21 \cdot 20]$$

Changes in internal energy can be predicted from C_P and C_V by applying the first law of thermodynamics. Since there are three state variables related by an equation of state (whatever its form), we can take T and V as the independent variables, with P and $\mathcal{U}$ assumed dependent on them. In that case, for small changes we can separate the increments due to T and V:

$$\Delta\mathcal{U} = Q - W = Mc\,\Delta T - P\,\Delta V \qquad [21 \cdot 21]$$

where c depends on the path taken. However, for a process at constant volume,

$$\Delta\mathcal{U} = Mc_V\,\Delta T \qquad [21 \cdot 22]$$

and for constant pressure,

$$\Delta\mathcal{U} = Mc_P\,\Delta T - P\,\Delta V \qquad [21 \cdot 23]$$

so that for quasistatic processes we can compute the change in internal energy during the process. Since $\mathcal{U}$ is a state variable, we can in principle use a combination of isometric and isobaric processes to determine $\Delta\mathcal{U}$ no matter how the final state is attained.

example 21 • 8 An ideal gas with a mass of 1 kmole expands from the state (P_1,V_1,T_1) to a state (P_2,V_2,T_2) by passing through a series of equilibrium states which lie along a straight line in the PV plane. (*a*) Compute the work done by the gas in the expansion and the change in internal energy of the gas. (*b*) Find $\overline{C}_{12}$, the mean specific heat for the process. (*c*) Find the specific heat $C_{12}(T)$.

solution (*a*) It is easy to show from elementary geometry that the area under the path of the process is given by

$$W = \tfrac{1}{2}(P_1 + P_2)(V_2 - V_1)$$

The change in $\mathcal{U}$ is entirely due to the change in temperature:

$$\Delta\mathcal{U} = C_V(T_2 - T_1)$$

(*b*) From the first law of thermodynamics, $Q = \Delta\mathcal{U} + W$ and

$$\frac{Q}{T_2 - T_1} = \overline{C}_{12} = C_V + \frac{\tfrac{1}{2}(P_1 + P_2)(V_2 - V_1)}{T_2 - T_1}$$

Replacing the temperatures from the ideal-gas law $T = PV/R$ then gives

$$\overline{C}_{12} = C_V + \tfrac{1}{2}R\left(1 + \frac{P_1V_2 - P_2V_1}{P_2V_2 - P_1V_1}\right)$$

Note that for an isometric process $V_1 = V_2$ and $\overline{C}_{12} = C_V$, while for an isobaric process $P_1 = P_2$ and $\overline{C}_{12} = C_P$, which is consistent.

(*c*) The specific heat can be found from the differential form of the first law of thermodynamics, $dQ = d\mathcal{U} + P\,dV$:

$$C_{12} = \frac{dQ}{dT} = \frac{d\mathcal{U}}{dT} + P\frac{dV}{dT} = C_V + P\frac{dV}{dT}$$

We can find dV/dT from the requirement that the ideal-gas law is obeyed, but that the process also follows a linear path in the PV plane. This gives two simultaneous equations relating the differentials of the state variables:

$$P\,dV + V\,dP = R\,dT \qquad dP = A\,dV$$

where $A = (P_2 - P_1)/(V_2 - V_1)$ is just the slope of the linear path in the PV plane. Eliminating dP gives

$$(P + AV)\frac{dV}{dT} = R \qquad \text{or} \qquad \frac{P\,dV}{dT} = \frac{PR}{P + AV}$$

and, substituting this in the equation for C_{12}, we obtain

$$C_{12} = C_V + \frac{PR}{P + AV}$$

If desired, P and V can be replaced by their expressions in terms of T along the path. Note that for an isobaric process $A = 0$ and $C_{12} = C_V + R = C_P$, and for an isometric process $A = \infty$ and $C_{12} = C_V$.

21 • 5 adiabatic processes

An *adiabatic process* is one in which no heat enters or leaves the system. Such a process can be realized in practice by enclosing the entire system in a structure whose walls are either impervious to heat or always kept in thermal equilibrium with the system, as in Fig. 21 . 8. It can also be achieved by having the process take place so *rapidly* that heat transfer is not appreciable even when the heat insulation is imperfect, as in a sound wave. To derive the relation between the pressures and volumes of an ideal gas in an adiabatic process, consider an adiabatic change from the state (P,V,T) to the state $(P + dP,\ V + dV,\ T + dT)$. Since $dQ = 0$ in an adiabatic process,

$$d\mathcal{U} + dW = 0 \qquad \text{or} \qquad d\mathcal{U} = -dW = -P\,dV \qquad [21 \cdot 24]$$

The minus sign indicates that if work is done by the gas, its internal energy is decreased. Since the internal energy of an ideal gas is all kinetic, we see from Eq. [21 . 16] that $d\mathcal{U} = dK = nC_V\,dT$, and therefore

$$nC_v\,dT = -P\,dV \qquad [21 \cdot 25]$$

In other words, if an ideal gas is prevented from absorbing heat from the external environment during a volume expansion, it will cool, because the work of expansion is drawn from its own internal energy. Conversely, a rapid adiabatic compression, as in an automobile engine, will increase the temperature of the gas in the cylinder. We can ignore any change in pressure, since it would only give rise to a second-order infinitesimal, $dP\,dV$, on the right side of Eq. [21 . 25].

Equation [21 . 25] is mathematically equivalent to the requirement that the process be adiabatic. If we add this equation to the equation of state of the ideal gas, $PV = nRT$, then we have two mathematical conditions on the three unknowns P, V, and T. We can, in principle, eliminate one of the variables, leaving a single relationship between the remaining two. To do this, let us first differentiate the equation of state. For an adiabatic process,

$$P\,dV + V\,dP = nR\,dT \qquad [21 \cdot 26]$$

and substituting $nR\,dT = -(RP/C_V)\,dV$ from Eq. [21 . 25] yields

$$P\,dV + V\,dP = -\frac{RP}{C_V}\,dV \qquad [21 \cdot 27]$$

If we now define the ratio of specific heats as

$$\gamma = \frac{C_P}{C_V} \qquad [21 \cdot 28]$$

then Eq. [21 . 23] shows that $R = (\gamma - 1)C_V$. Substituting this value of R into Eq. [21 . 27], we obtain

$$\frac{dP}{P} + \gamma\frac{dV}{V} = 0 \qquad [21 \cdot 29]$$

and integration gives, for constant γ,

$$\ln P + \gamma \ln V = \text{constant} \qquad \text{or} \qquad PV^{\gamma} = \text{constant} \qquad [21 \cdot 30]$$

Note that this equation does not replace the equation of state; it merely expresses the *additional adiabatic condition* of no heat transfer and is valid only for adiabatic processes. Moreover, it does not hold for every substance, but applies only to an ideal gas. At moderate temperatures the assumption of constant γ holds quite well for ideal gases (see Table 21 . 3), a fact we shall discuss in more detail in the next chapter. Figure 21 . 10 compares the curves $PV^{\gamma} = $ constant with the isothermals $PV = $ constant.

Direct determinations of c_V are difficult to make because of the low mass of the gas compared to that of its container. For this reason values of c_V are usually obtained from c_P and from the measurement of γ, the ratio of the specific heats as determined by measurements of the acoustic speed $v = \sqrt{\gamma P/\rho}$ (see Sec. 17 . 4). We can now derive this formula for the speed

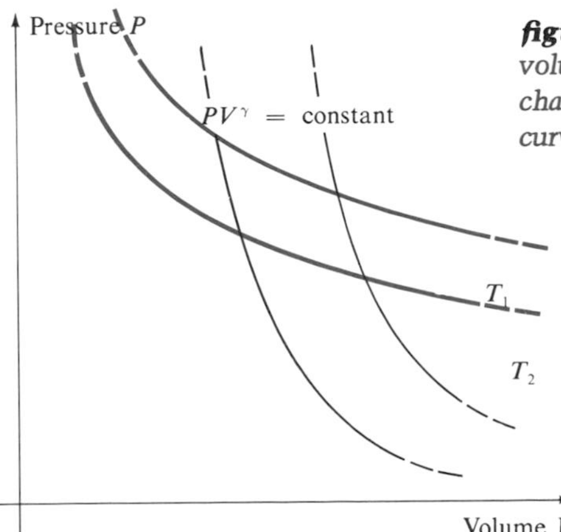

figure 21 • 10 *The relations between pressure and volume for a fixed mass of air undergoing an adiabatic change. The colored curves are isotherms and the black curves are "adiabats," PV^{γ} = constant.*

table 21 • 3 *Experimental values of the ratio γ for various gases*

gas	*formula*	*γ at 15°C, 1 atm*
helium	He	1.666
argon	A	1.668
hydrogen	H_2	1.410
nitrogen	N_2	1.404
oxygen	O_2	1.401
air	80% N_2, 20% O_2, approx.	1.403
carbon monoxide	CO	1.404
carbon dioxide	CO_2	1.302
sulfur dioxide	SO_2	1.29
ether vapor	$(C_2H_5)_2O$	1.024

of sound by a method similar to that used in Sec. 17 . 3. If small expansions and compressions of an ideal gas take place adiabatically, as in a sound wave, then, from Eq. [21 . 29], the *adiabatic bulk modulus* is

$$B_{ad} = -V\left(\frac{dP}{dV}\right)_{ad} = \gamma P \qquad [21 \cdot 31]$$

Since the density ρ of the gas can be found from the ideal-gas law,

$$\rho = \frac{M}{V} = \frac{nM_0}{V} = \frac{M_0 P}{RT} \qquad [21 \cdot 32]$$

the speed of the longitudinal compressional sound wave is obtained by substituting B_{ad} for B in Eq. [17 . 23]:

$$\sqrt{\frac{B_{ad}}{\rho}} = \sqrt{\frac{\gamma P}{\rho}} = \sqrt{\gamma\frac{RT}{M_0}} \qquad [21 \cdot 33]$$

where M_0 is the molecular weight of the gas. Measurements of acoustic speed can be made quite accurately and used to determine γ, since the other state variables are easily measured.

example 21 • 9 Determine the molar specific heats C_P and C_V for helium and for oxygen.

solution We can solve Eq. [21 . 28] for γ, where

$$C_P = \frac{\gamma R}{\gamma - 1} \qquad C_V = \frac{R}{\gamma - 1}$$

Then, from Table 21 . 3, the specific-heat ratio for helium is $\gamma = \frac{5}{3}$, and

$$C_P = \tfrac{5}{2}R \qquad C_V = \tfrac{3}{2}R$$

where $R = 1.986 \doteq 2$ kcal/kmole · K. The specific-heat ratio for oxygen is $\gamma = \frac{7}{5}$, so that

$$C_P = \tfrac{7}{2}R \qquad C_V = \tfrac{5}{2}R$$

example 21 • 10 An ideal gas is contained in a large vessel at room temperature T_0 and pressure P_1 slightly higher than the atmospheric pressure P_0. The vessel is opened just long enough for the pressure inside to drop to P_0, but not long enough to allow more than a negligible amount of the surrounding air to diffuse into the vessel. During this approximately adiabatic expansion, the temperature inside the vessel drops below T_0 (this is why spray cans feel cold). It warms up to T_0 again after the vessel is closed, but the final pressure is P_2, since some of the gas has been lost. Express the ratio γ in terms of the pressures.

solution If the number of moles of a gas is unchanged, then we can write the adiabatic condition in Eq. [21 . 30] in terms of the molar specific volume v:

$$Pv^\gamma = \text{constant}$$

If the initial molar volume is v_1 and the final volume is v_2, then

$$P_1 v_1^\gamma = P_0 v_2^\gamma$$

since P_0 is the pressure inside the vessel after closure. However, when the gas warms up to room temperature again, the ideal-gas equation of state gives

$$RT_0 = P_1 v_1 = P_2 v_2$$

Solving the preceding equations for v_2/v_1, we find

$$\left(\frac{P_1}{P_0}\right)^{1/\gamma} = \frac{P_1}{P_2} \qquad \text{or} \qquad \gamma = \frac{\ln (P_1/P_0)}{\ln (P_1/P_2)}$$

One of the earliest determinations of γ was made in this way by Clement and Desormes in 1819.

The basic energy exchanges as they take place in the *PV* plane for *n* moles of an ideal gas are summarized in Table 21 . 4.

table 21 • 4 *Energy exchanges for an ideal gas*

process	Q	W	$\Delta\mu$
isometric	$nC_V\,\Delta T$	0	$nC_V\,\Delta T$
isobaric	$nC_P\,\Delta T$	$P\,\Delta V$	$nC_V\,\Delta T$
isothermal	$nRT\ln\frac{V_f}{V_i}$	$nRT\ln\frac{V_f}{V_i}$	0
adiabatic	0	$-nC_V\,\Delta T$	$nC_V\,\Delta T$

21 · 6 heat conduction

Heat is transferred from one body to another, or from one point to another within a body, by three very different processes. These are *conduction,* in which the heat diffuses through a solid material or a stagnant fluid; *convection,* in which a moving liquid or gas absorbs heat at one place and gives it up at another place; and *radiation,* in which the transfer from one place to another is by means of electromagnetic waves, such as light waves. The theory of convection is enormously complex, involving advanced fluid mechanics, and the subject of radiation is appropriately deferred to our later study of light and quantum mechanics. Therefore we shall consider here only the transfer of heat by conduction.

When two parts of a body are maintained at different temperatures, experiment shows that there is a continuous gradation of temperature in the material between the two parts. The transfer of heat that occurs through the intervening material, without any evident motion of mass, is called *heat conduction.* The law of heat flow due to a temperature difference was first derived experimentally by determining the average time rate of heat flow through test bars of homogeneous solid materials. If one end face of the bar is kept at temperature T_1 and the other is kept at a higher temperature T_2, then at some time t_{ss} a *steady state* is reached, and the temperatures in the bar do not change with time (see Fig. 21 . 11). Measurements show that the rate of heat flow, Q/t, is proportional to the temperature difference $T_2 - T_1$ and the cross-sectional area A and is inversely proportional to the length l of the bar:

$$\frac{\Delta Q}{\Delta t} = \mathcal{K}A\frac{T_2 - T_1}{l} \qquad [21 \cdot 34]$$

where $\mathcal{K}$ is a factor of proportionality called the *thermal conductivity,* or *coefficient of thermal conduction,* of the material tested. The quantity $(T_2 - T_1)/l$ is the average temperature *gradient,* or average fall in temperature per unit distance between the end faces.

In the limiting case of a bar or slab of material of thickness dx, across which there is an infinitesimal temperature difference dT, the experimental equation [21 . 34] takes the form

$$\blacktriangleright \quad \frac{dQ}{dt} = -\mathcal{K}A\frac{dT}{dx} \qquad [21 \cdot 35]$$

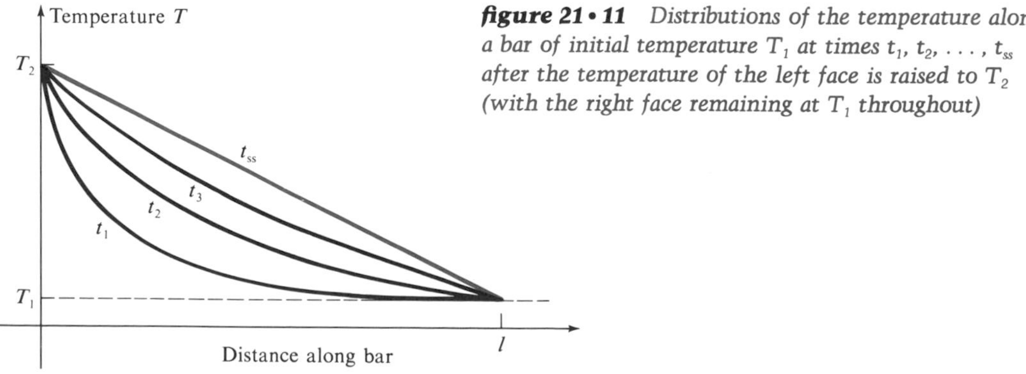

figure 21 • 11 *Distributions of the temperature along a bar of initial temperature T_1 at times $t_1, t_2, \ldots, t_{ss}$ after the temperature of the left face is raised to T_2 (with the right face remaining at T_1 throughout)*

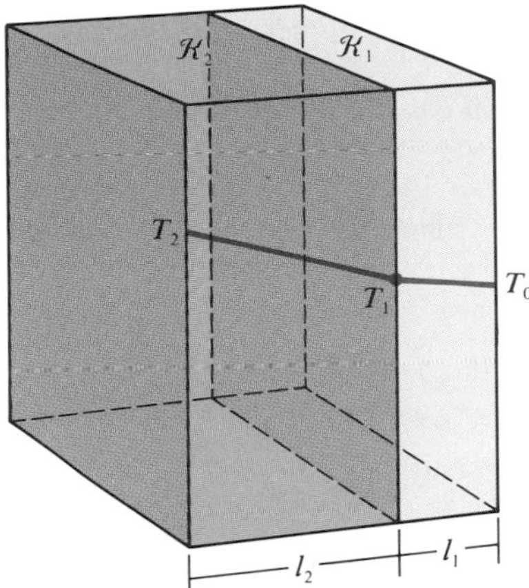

figure 21 • 12 *Temperature distribution in two heat conductors in series*

This is the fundamental law of heat conduction. Since heat flows in the direction of decreasing temperature—that is, the heat current dQ/dt is positive when the temperature gradient dT/dx is negative—the minus sign in the fundamental equation makes the thermal conductivity $\mathcal{K}$ of any substance a positive number.

Of all known substances metals have by far the highest conductivities, although the values vary widely for different metals. Nonmetallic solids such as natural fibers have great value as thermal insulators, because they conduct very poorly; liquids and gases also conduct poorly. Although $\mathcal{K}$ does vary with temperature, it can often be regarded as constant over a given range of temperatures. In a steady state there is just as much heat entering as leaving the system or any element of the system. The quantity dQ/dt is the time rate of transfer of heat energy dQ *through* a cross section of the conductor perpendicular to the x direction. It does not represent heat absorbed by the conductor. In the steady state, by definition, temperatures remain constant, and the conductor is neither absorbing nor losing heat with time, but is merely acting as a "heat pipe"; hence dQ/dt must have the same value at *every* cross section of the conductor.

Consider the composite wall in Fig. 21 . 12, which consists of two parallel slabs of material of the same cross-sectional area, but of different thicknesses and different thermal conductivities. The outer surfaces are at temperatures T_2 and T_0, and we want to know the temperature T_1 at the junction, as well as the steady-state heat current passing through the wall. Since the heat entering the interface must equal the heat leaving it, the heat current in the two slabs must be equal. Hence Eq. [21 . 34] gives two equations in the two unknowns T_1 and $\Delta Q/\Delta t$:

$$\frac{\Delta Q}{\Delta t} = \mathcal{K}_2 \frac{A(T_2 - T_1)}{l_2} = \mathcal{K}_1 \frac{A(T_1 - T_0)}{l_1} \qquad [21 \cdot 36]$$

which yields

$$T_1 = \frac{l_1 \mathcal{K}_2 T_2 + l_2 \mathcal{K}_1 T_0}{l_1 \mathcal{K}_2 + l_2 \mathcal{K}_1} \qquad [21 \cdot 37]$$

and

$$\frac{\Delta Q}{\Delta t} = \frac{A(T_2 - T_0)}{l_1/\mathcal{K}_1 + l_2/\mathcal{K}_2} \qquad [21 \cdot 38]$$

table 21 • 5 *Thermal conductivities of various substances*

equal conductivities: 1.0 cal/s · cm · K = 419 watts/m · K = 242 Btu/h · ft · °F

substance	*temperature, °C*	*thermal conductivity $\mathcal{K}$, cal/s · cm · K*
metals		
aluminum	−269	7.6
	18	0.048
bismuth	18	0.019
copper	−269	4.9
	18	0.92
	100	0.91
iconel (Ni alloy)	70	0.036
mercury	18	0.016
nickel	−160	0.13
	18	0.14
	100	0.14
silver	18	1.01
	100	0.99
steel, mild	18	0.11
tungsten	18	0.48
	1600	0.25
nonmetallic solids at ordinary temperatures		
concrete		2.2×10^{-3}
cotton, raw		0.1×10^{-3}
glass, window		2.0×10^{-3}
ice (0°C)		5.3×10^{-3}
oak, ∥ to grain		0.86×10^{-3}
⊥ to grain		0.50×10^{-3}
quartz, ∥ to axis		30.5×10^{-3}
⊥ to axis		16.5×10^{-3}
sawdust		0.14×10^{-3}
silica gel		0.05×10^{-3}
liquids at 20°C		
alcohol, ethyl		0.40×10^{-3}
methyl		0.48×10^{-3}
water		1.4×10^{-3}
gases at 0°C		
air		0.057×10^{-3}
hydrogen		0.42×10^{-3}
oxygen		0.057×10^{-3}

It can be shown that for a series of n slabs Eq. [21 . 38] generalizes to

$$\frac{\Delta Q}{\Delta t} = \frac{A(T_n - T_0)}{\sum_{i=1}^{n}(l_i/\mathcal{K}_i)} \qquad [21 \cdot 39]$$

The *effective* conductivity is therefore

$$\mathcal{K}_{\text{eff}} = \frac{\sum_{i=1}^{n} l_i}{\sum_{i=1}^{n}(l_i/\mathcal{K}_i)} \qquad [21 \cdot 40]$$

The temperature difference "drives" the heat current, analogous to the way a pressure difference causes a fluid to flow.

The thermal conductivities of some common substances at different temperatures are given in Table 21 . 5.

example 21 • 11 Often we have to deal with more complicated geometry, as when a substance of thermal conductivity $\mathcal{K}$ lies between two very long cylindrical pipes. The outer cylinder, of radius r_1, is kept at a constant temperature T_1 by a cooling stream of water, and the inner cylinder, of radius r_2, is maintained at a constant temperature T_2 by passing steam through it. Compute the steady-state heat current through the walls of the inner cylinder.

solution Consider a cylindrical shell of radius r and thickness dr, across which the temperature difference is dT. The inner surface of this shell is $2\pi rl$ and the temperature gradient is dT/dr. Hence the steady current through the shell is, from Eq. [21 . 35],

$$\frac{dQ}{dt} = -\mathcal{K}(2\pi rl)\frac{dT}{dr} = \text{constant}$$

Separating variables and integrating, we obtain

$$\frac{dQ}{dt}\int_{r_1}^{r_2}\frac{dr}{r} = 2\pi\mathcal{K}l\int_{T_2}^{T_1} dT$$

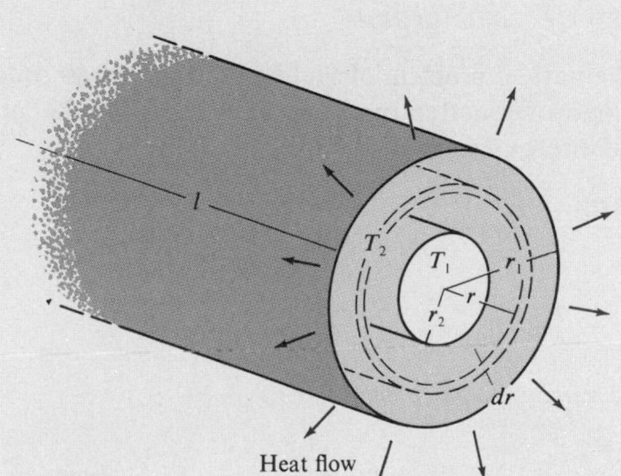

figure 21 • 13

Therefore the rate at which heat energy is lost through the walls of a cylindrical pipe is

$$\frac{dQ}{dt} = \frac{2\pi\mathcal{K}l(T_2 - T_1)}{\ln(r_1/r_2)}$$

example 21 • 12 A long copper rod of length l, radius r, density ρ, and specific heat c is heated to temperature T_0 and then placed in air of conductivity $\mathcal{K}$ to cool. Assume that beyond some distance $d \ll r$ from the rod the air may be considered at a constant temperature T_a. Find the time dependence of the temperature of rod as it cools.

solution The instantaneous heat flow through the thin cylindrical layer of air surrounding the rod is given by

$$\frac{dQ}{dt} = \frac{2\pi\mathcal{K}l\,\Delta T}{\ln(1 + d/r)}$$

where at any time t, $\Delta T = T(t) - T_a$. Since dQ represents the amount of heat *lost* by the rod in time dt, it must result in a decrease in the temperature of the rod. The amount of decrease depends on the mass M of the rod, $dQ = Mc\,dT$. If we substitute $M = \pi r^2 l\rho$ into the conduction equation and use the approximation $\ln(1 + x) \doteq x$ for $x \ll 1$, we have

$$\pi r^2 l\rho c\frac{dT}{dt} = \pi r^2 l\rho c\frac{d(\Delta T)}{dt} = -\frac{2\pi r l\mathcal{K}\,\Delta T}{d}$$

This equation has the solution

$$\Delta T = (T_0 - T_a)e^{-t/\tau}$$

where τ is the cooling time for the rod, $\tau = \frac{1}{2}\rho r d c/\mathcal{K}$. Therefore the temperature difference between the rod and the surrounding air decays exponentially. The only uncertainty in our approximate treatment is the effective value of d, but this affects only the time constant τ. The exponential decay is valid and is known as *Newton's law of cooling.*

*21 • 7 *roots of equations*

The general mathematical problem of finding the roots of an equation of the form $f(x) = 0$ arises very often in physics. For example, the integration of a one-dimensional energy equation in the form

$$\frac{dx}{dt} = \sqrt{\frac{2[E - U(x)]}{m}}$$

gives solutions

$$\int\frac{dx}{\sqrt{2(E - U)/m}} = \phi(x) = t$$

which are not, in general, readily soluble for $x = F(t)$. Hence to find x for a given value of $t = t_0$ it is necessary to solve $f(x) = \phi(x) - t_0 = 0$. The capacity to handle such problems will enable us to consider more realistic physical situations.

One rather obvious way to approach the problem would be to compute a table of values of $f(x)$. If we can find two successive x_i whose functional values are of opposite sign, $f(x_1)f(x_2) < 0$, we can interpolate linearly for $x_3 \doteq \alpha$ such that $f(\alpha) = 0$. This rather crude approximation can be refined by reiterating it over the subinterval, either $[x_1,x_3]$ or $[x_3,x_2]$, depending on the sign of $f(x_3)$, which contains the *true* root α. The procedure is then repeated until $f(x_i)$ is found sufficiently close to zero. This process is known as the *method of secants* or *regula falsi,* the "rule of false position." Although very reliable, it is relatively slow to approach the desired root; therefore we shall employ the more sophisticated *Newton-Raphson method.*

In the Newton-Raphson method, as in the method of secants, we make use of a straight-line approximation to the curve of $f(x)$, but we approximate the curve near some x_0 by a straight line passing through x_0 whose slope equals $f'(x_0)$. This approximation is equivalent to taking $\Delta f/\Delta x \doteq df/dx$; that is, we assume

$$\frac{\Delta f}{\Delta x} = \frac{f(x) - f(x_0)}{x - x_0} \doteq f'(x_0)$$

or

$$f(x) \doteq f(x_0) + f'(x_0)(x - x_0) \qquad [21 \cdot 41]$$

Hence our first approximation to the desired root is $x = x_1$ such that $f(x_1)$ is assumed to be zero,

$$0 = f(x_0) + f'(x_0)(x_1 - x_0)$$

which gives us

$$x_1 = x_0 - \frac{f(x_0)}{f'(x_0)} \qquad [21 \cdot 42]$$

Successively better approximations are generated as shown in Fig. 21 . 14, using the recursion formula

$$x_{i+1} = x_i - \frac{f(x_i)}{f'(x_i)} \qquad [21 \cdot 43]$$

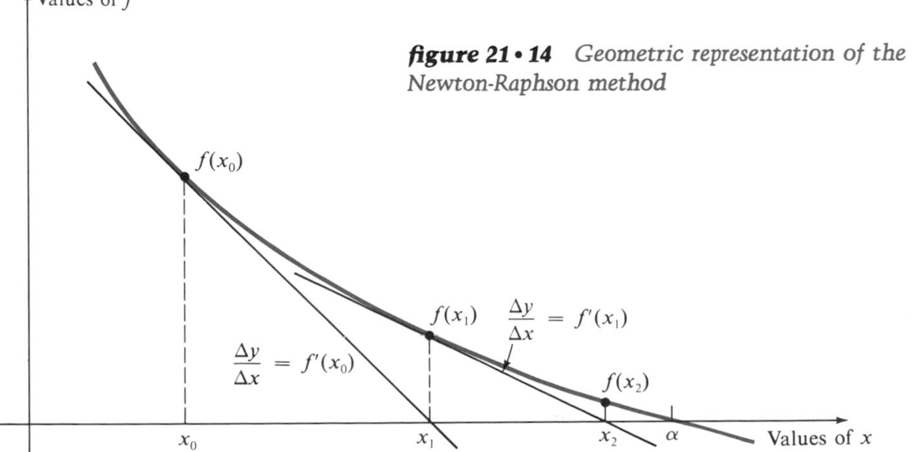

figure 21 • 14 *Geometric representation of the Newton-Raphson method*

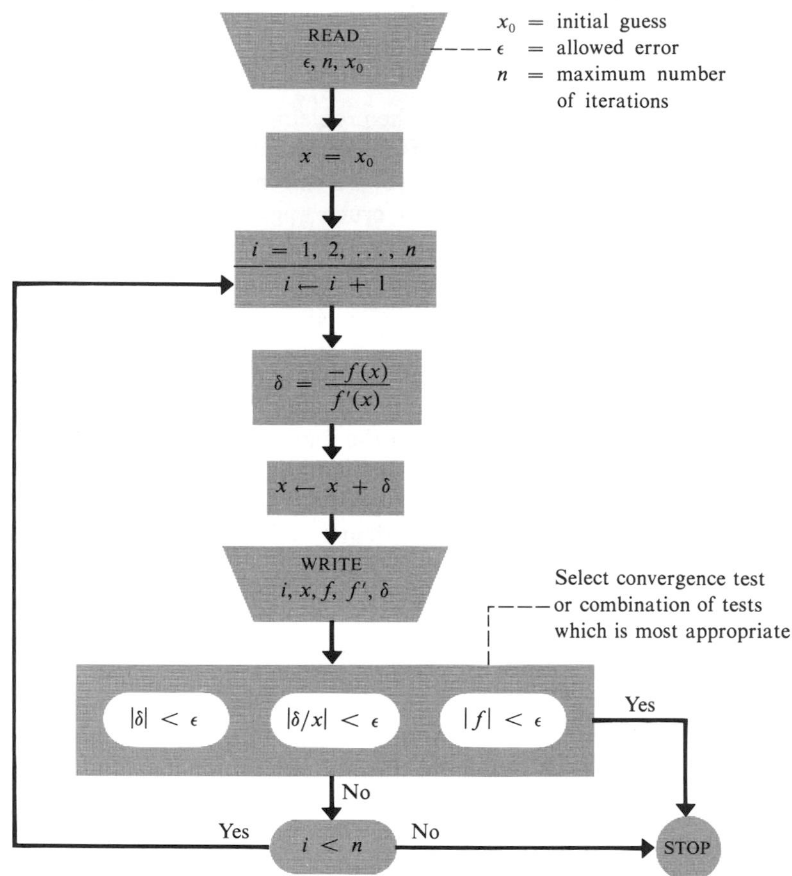

figure 21 • 15 *Algorithm for the Newton-Raphson method. Several different types of convergence tests on ε are indicated in the largest box. As a precaution against divergence, one should also set an upper bound n to the number of iterations.*

Provided the initial guess x_0 is not too far from the desired root and $f'(x_i) \neq 0$ for some x_i, the method converges very rapidly. The accuracy of x_i, the last value computed, is easily checked by evaluating $f(x_i)$. Figure 21 . 15 is the flowchart for this method.

The function on the right of Eq. [21 . 43] is called the *iteration function,*

$$F(x) = x - \frac{f(x)}{f'(x)} \qquad [21 \cdot 44]$$

because it gives the prescription for finding x_{i+1} from x_i. Mathematically, it can be shown that the sequence of values x_i converges to α, where $f(\alpha) = 0$, if $|F'(x_i)| < 1$ for every value of i.* This amounts to saying that

$$\left| \frac{F(x_{i+1}) - F(x_i)}{x_{i+1} - x_i} \right| = \left| \frac{x_{i+2} - x_{i+1}}{x_{i+1} - x_i} \right| \doteq |F'(x_i)| < 1 \qquad [21 \cdot 45]$$

*T. R. McCalla, *Introduction to Numerical Methods and FORTRAN Programming,* chap. 3 (John Wiley & Sons, New York, 1967).

which is approximately the condition for convergence of a sequence of differences: $\delta_i = (x_{i+1} - x_i)$ approaches zero as i approaches infinity. This implies that the success of the method in a given instance depends on making a reasonably good guess at the first value $x = x_1$ to begin the process. In general this is not hard to do, especially when a physical problem is involved and intuition provides a clue to the approximate value of the desired root. However, a poor guess can cause successive values x_i to diverge more and more *away* from α. This instability is usually signaled by increasingly erratic fluctuations of δ_i.

The Newton-Raphson method is a *second-order* iteration method, because the differences $\delta_i = x_{i+1} - x_i$ tend to decrease as the *square* of the preceding difference. That is, once convergence is under way, $\delta_{i+1} \doteq C\delta_i^2$. The method of secants is a first-order method in which $\delta_{i+1} \doteq C'\delta_i$, $\delta_{i+2} \doteq C'\delta_{i+1}$, etc.; hence it does not converge as rapidly to the desired root.

This method can also be used to find solutions in cases where analytic methods are well known, to improve either accuracy or efficiency. In Example 21 . 5 we had to find the final temperature of a water-lead mixture by solving the quadratic equation

$$f(\tau) = 5 \times 10^{-5}\tau^2 + 1.15\tau - 74.5 = 0 \qquad [21 \cdot 46]$$

Since the leading coefficient of the quadratic, 5×10^{-5}, is so small, it is necessary to compute the discriminant to at least five-place accuracy in order to obtain useful results by the conventional method. It is often more practical to solve by the Newton-Raphson method. We merely start from an inspired guess obtained by ignoring the quadratic leading term, solve for

$$\tau_0 = \frac{74.5}{1.15} \doteq 64.8°\text{C}$$

and then evaluate the iteration function at τ_0:

$$F(\tau_0) = \tau_1 = \tau_0 - \frac{f(\tau_0)}{f'(\tau_0)} = \tau_0 - \frac{5 \times 10^{-5}\tau_0^2}{10^{-4}\tau_0 + 1.15}$$

$$= 64.8 - \frac{0.210}{1.16} \doteq 64.6°\text{C}$$

which is the correct answer to three significant figures.

PROBLEMS

21 • 1 work

21 • 1 How much work is done against atmospheric pressure when a 15-g mass of nitrogen is heated from 0°C to 50°C under a constant pressure of 1 atm?

21 • 2 Two liters of an ideal gas has an initial pressure of 720 mm Hg. Calculate the work done in compressing it isothermally to half its original volume.

21 • 3 Compute the total work done *by n* moles of an ideal gas when it undergoes three consecutive quasistatic processes: (1) first it is heated at constant volume V_1 to pressure P_1 and temperature T_1; (2) it is then allowed to expand at constant temperature T_1 to pressure P_2 and volume $2V_1$; (3) finally it is returned to its initial state by compressing it under constant pressure.

21•4 What is the work done by 1 kmole of an ideal gas starting from a state (P,V) and going to state (P,V') by means of an exactly semicircular path in the PV plane (*a*) if the path passes through $P' > P$ as its maximum pressure? (*b*) If the path through $P'' < P$ as its minimum pressure?

21•5 Compute the work done by a brass block, 10 cm on a side, as it is heated from 20°C to 80°C.

***21•6** An observer records the following data on a complex gaseous chemical reaction:

V, m^3	1.0	1.1	1.2	1.25	1.33	1.50	1.62	1.75	1.80
P, atm	1.0	1.0	1.05	1.15	1.30	1.60	1.80	2.00	2.10

If the pressure and volume changes are due to changes in the number of reacting molecules and their temperature, how much work is done by the system during the reaction? Use Simpson's rule, first obtaining evenly spaced data by linear interpolation.

21•7 Suppose the cylinder of Fig. 21 . 1 initially contains 1 kmole of nitrogen gas at standard temperature and pressure, which forces the piston outward in such a way that $P = \exp(1 - V/22.4\ \mathrm{m}^3)$ atm as the gas expands. (*a*) How much work has been done by the gas when its volume has doubled? (*b*) During the process, did the gas gain or lose heat energy? (*Hint:* Find its final temperature.)

21•2 *first law of thermodynamics*

21•8 Show that the British thermal unit is equivalent to 252 cal, or 778 ft · lbf.

21•9 A 300-kg flywheel 2 m in diameter has an angular frequency of 1 Hz. (*a*) How much heat is produced in stopping the wheel by friction if the whole mass of the wheel is concentrated in the rim? (*b*) If the wheel is a homogenous solid disk?

21•10 An ideal gas absorbs 200 cal of heat while expanding from a volume of 2 liters to 3 liters under 1 atm of pressure. What is the change in its internal energy?

21•11 How much heat is absorbed by the gas in Example 21 . 1? Explain the physical process involved.

21•12 The internal energy of an ideal gas is directly proportional to its temperature. (*a*) How much heat is absorbed by 2 moles of nitrogen in expanding from 1 liter to 3 liters at 250 K? (*b*) If the container is thermally insulated so that no heat may flow through it, what happens to the temperature and internal energy of the gas when it is allowed to expand against an external pressure?

21 • 13 A system undergoes three separate processes, which can be represented as paths in the PV plane. Find Q, W, and $\Delta\mathcal{U}$ for each process: (*a*) Starting from (P_0,V_0), the system changes isometrically, absorbing 150 kcal. (*b*) It then changes isobarically, absorbing another 100 kcal and doing 83,700 J of work in the process. (*c*) When the system returns to its original state (P_0,V_0), it gives up 180 kcal of heat.

21 • 14 If the internal energy of 1 kmole of an ideal gas is given by $\mathcal{U} = C_V T$, where $C_V \doteq 5$ kcal/kmole, how much heat was absorbed or lost by the nitrogen gas during the process described in Prob. 21 . 7?

21 • 3 heat capacity: latent heats

21 • 15 In one of Joule's earlier experiments the work done by falling weights was expended in stirring water. After allowances for the frictional forces, the following data were obtained:

Total mass of weights, 57.8 lbm
Height of fall, 5.00 ft
Number of times weights were allowed to fall, 21
Rise in temperature of water, 0.563°F
Heat capacity of water and containing vessel, 13.9 Btu/°F

Calculate the value of the joule equivalent yielded by these experimental results.

21 • 16 A block of ice of mass M is at rest on a perfectly smooth surface. A bullet of mass $m \ll M$ is fired into one side of the block with speed v. If L_f is the heat of fusion of ice, what is the mass of ice melted?

21 • 17 Suppose the adiabatic calorimeter in Fig. 21 . 8 contains 400 g of water initially at room temperature and the power expended in the immersed heating coil is kept at an average value of 3000 W for 1000 s. (*a*) How much heat, expressed in calories, is generated by the coil? (*b*) What is the approximate rise in the temperature of the water?

21 • 18 In the manufacture of certain lead pipes, solid lead is forced through an annular die by applying a pressure of 1.38×10^8 Pa. The mean specific heat of lead in the temperature range involved here is 3.13×10^{-2} cal/g · K, and the density of the lead is 11.4 g/cm^3. How much will the temperature of the lead rise as it passes through the die?

21 • 19 A 50-g piece of ice is dropped into a 150-g brass calorimeter cup with a specific heat of 0.089 cal/g · K. If the cup contains 800 g of water at a temperature of 27°C, what will be the final temperature of the cup and its contents?

21 • 20 If 100 g of ice at −5°C is mixed with 40 g of water at 30°C and 10 g of steam at 100°C and 1 atm, what is the final temperature of the mixture?

21 • 21 A 50-kg block of ice falls 3 m. Assuming that the transformation of mechanical energy into heat was complete, determine how much ice was melted by the heat generated in the impact with the ground. The heat of fusion of ice is 79.7 cal/g.

21•22 The introduction of 5.0 g of a certain substance at 100°C into the Bunsen ice calorimeter in Fig. 21 . 9 causes the end of the mercury column to move 18.5 mm. If the mean inside diameter of the capillary tube is 1.1 mm and the reduction in volume of the ice upon fusion is 0.0905 cm^3/g, what is the specific heat of the substance?

21•4–21•5 *specific heats of gases*

21•23 The specific heat of water at temperature τ is given by

$$c = [4.2048 - (1.768 \times 10^{-3}/\text{K})\tau + (2.645 \times 10^{-5}/\text{K}^2)\tau^2]\ \text{J/g} \cdot \text{K}$$

(*a*) What error would result from assuming that the specific heat of water at 25°C is the same as its mean specific heat over the range 0 to 50°C? (*b*) How should the formula be altered to yield values of the molar specific heat of water at temperature τ in terms of 1 kcal/kmole · K as a unit of measure?

***21•24** Using the four-point Lagrange polynomial, interpolate between the following values of the specific heat of water:

c, cal	1.000	0.998	0.999	1.001
τ, °C	15	35	55	75

Compare your results with those obtained from the three-point formula in Prob. 21 . 23 for the range 15°C to 75°C.

21•25 Show that the heat per unit volume required to heat an ideal gas at constant pressure is inversely proportional to the absolute temperature of the gas.

21•26 Find the change in internal energy in the isobaric expansion of an ideal gas at pressure P by an amount ΔV. Show that owing to our initial assumption that the internal energy of an ideal gas of noninteracting point masses must be entirely kinetic, the result is consistent with the statement that $\mathcal{U} = nC_VT$.

21•27 The expansion of 1 kmole of ideal gas from V_1 to V_2 moves along the path $P = P_1(V_1/V)^3$ in the PV plane. (*a*) Find the work done. (*b*) What is the internal-energy change $\Delta\mathcal{U} = C_V\Delta T$? (*c*) Find the molar specific heat for this process.

21•28 Use Eq. [21 . 30] and the ideal-gas equation to show that the relations between the state variables in a quasistatic adiabatic process may be expressed (*a*) as $TV^{\gamma-1}$ = constant. (*b*) As $TP^{(1-\gamma)/\gamma}$ = constant.

21•29 Show that the work done by an ideal gas during an adiabatic expansion from the state (P_1,V_1) to the state (P_2,V_2) is given by

$$\Delta W = \int_{V_1}^{V_2} P\,dV = \frac{1}{\gamma - 1}(P_1V_1 - P_2V_2)$$

21•30 When 1 m^3 of nitrogen is compressed adiabatically to 0.2 m^3 and is then heated at constant volume to 400 K, its pressure changes from 1 atm to 10 atm. (*a*) How many kilogram-moles of nitrogen are present? (*b*) What is the change in its internal energy? (*c*) How much external work is done by the nitrogen in the process? (*d*) What is the mean molar specific heat for this process? (*Hint:* You may want to use the formula from Prob. 21 . 29.)

21 • 31 A quantity of hydrogen is contained in a cylinder fitted with a piston, both made of some material that is impermeable to heat. The initial pressure and volume of the gas are 2 atm and 1 m^3. (*a*) If the piston is allowed to move outward until the pressure is halved, by how much is the volume increased? (*b*) How much external work is done?

21 • 32 A given mass of air, initially at 27°C and 1 atm, is suddenly compressed to half its initial volume. Assuming that the compression is adiabatic, compute the temperature and pressure attained by the gas.

21 • 6 heat conduction

21 • 33 Compute the loss of heat through a furnace wall 20 in thick if the two faces are at 1390°F and the conductivity of the wall is $\mathcal{K} = 0.002$ cal/s · cm · K.

21 • 34 The heat of combustion of a certain grade of coal is 7200 kcal/kg and the furnace is only 50 percent efficient. How much coal must be burned each day to compensate for the heat loss through a windowpane of area 1.5 m^2, with glass 3.0 mm thick, when the outer surface is at −9°C and the inner surface is at 12°C?

21 • 35 A hollow cubical steel vessel 101 cm on a side with walls 0.5 cm thick is filled with ice at 0°C. It is then placed in a water bath maintained at a constant temperature of 100°C. Assuming the steady state to have been reached, find the mass of ice that melts in 10 s.

21 • 36 A composite wall consists of parallel slabs 4.5 and 3.5 cm thick with conductivities of 0.15 and 0.45 cal/s · cm · K, respectively. The temperatures of the two outer faces of the wall are 95°C and 5°C, respectively. (*a*) What is the temperature at the interface? (*b*) Find the temperature gradient in each slab. (*c*) What is the heat current through the wall? (*d*) Does it matter whether the temperature in Eq. [21 . 37] is measured on an absolute or a Celsius scale?

21 • 37 The conductivity $\mathcal{K}$ of a poor conducting material is sometimes measured by placing the material between two concentric spheres. If the radii and constant temperatures of the inner and outer spheres are R_2 and T_2 and R_1 and T_1, show that the steady-state heat flow through the material under test will be

$$\frac{\Delta Q}{\Delta t} = 4\pi \mathcal{K} R_1 R_2 \frac{T_2 - T_1}{R_1 - R_2}$$

21 • 38 Adhering to each surface of a windowpane 2.5 mm thick is a stagnant layer of air of effective thickness 2.5 mm and mean thermal conductivity 6.2×10^{-5} cal/cm · s · K. (*a*) If the room temperature is 24°C and the outside temperature is −4°C, what is the temperature gradient in each of the three conducting layers? (*b*) At what rate per square meter is heat conducted through the window? (*c*) By what factor would the heat loss be reduced if another pane of glass were added with a 2.5-mm dead-air space between the two panes?

21 • 39 Suppose n moles of helium at temperature T_0 is contained in a sphere of radius R, thickness $d \ll R$ and thermal conductivity $\mathcal{K}$. The sphere is immersed in a water bath at temperature $T_1 < T_0$. Assuming that the temperature T of the gas is always uniform, how does it vary with time?

21 • 7 *roots of equations*

***21 • 40** Construct an algorithm to find the roots of $f(x) = 0$ by evaluating $f(x)$ over $x_1 \leq x \leq x_2$ in intervals of Δx and interpolating between successive values of $f(x)$ of opposite sign. Include a provision to halve the interval *once* and repeat the process if no sign change is found. Also include a subroutine to subdivide into 10 equal parts any interval in which a root is found, reevaluate $f(x)$ 10 times on that interval, and then locate the root by interpolation between some $f(x)$ and $f(x + \Delta x/10)$. (This means that Δx can be chosen larger for the initial search or a broader range of x can be selected, which makes the program much more effective.)

21 • 41 Show that the iteration function for finding $x = \sqrt{y}$ is

$$F(x) = \tfrac{1}{2}\left(x + \frac{y}{x}\right)$$

This is the fundamental algorithm used in many computers when a square root is called for. Show that, for $x_0 = \frac{1}{2}y$ as an initial guess, if $y > \frac{4}{3}$, the process always converges. Try the algorithm by hand for $y = 100$, for example. Try it for $y = 2$.

21 • 42 The specific heat of platinum at temperatures τ in the range 0 to 1625°C is

$$c = 3.162 \times 10^{-2} + \frac{6.17 \times 10^{-6}}{\mathrm{K}}\tau + \frac{2.33 \times 10^{-10}}{\mathrm{K}^2}\tau^2 \quad \mathrm{cal/g \cdot K}$$

Find the temperature of a Bunsen flame if a 20-g platinum ball dropped from the flame into 347 g of water raises the water temperature 2°C.

***21 • 43** A high-altitude balloon of mass M and maximum volume V rises into the stratosphere from an altitude of 12 km (the *tropopause*). It contains helium gas of density $\rho_0 = 2M/V$, and the atmosphere outside the balloon has density $\rho_a = 5\rho_0$. Heat conduction through the skin of the balloon is negligible, and the balloon will rise into the stratosphere, expanding adiabatically against the atmospheric pressure. This pressure declines by a factor of e^{-y/y_0} with altitude y *above* the tropopause, where $y_0 = 6.37$ km, the scale height of the stratosphere (see Sec. 15 . 1). When the balloon reaches a height where it is in equilibrium with the buoyant force of the atmosphere, it will stop rising. (*a*) Show that maximum altitude is reached when

$$5x = x^{+3/5} + \tfrac{1}{2}$$

where $x = e^{-y/y_0}$. (*b*) Use the Newton-Raphson method to find x and the maximum altitude reached by the balloon. Show the reasoning behind your first guess for x_0 and subsequent iterations $x_1, x_2, \ldots$. (*c*) Find the iteration function $F(x)$.

answers

21•1 $W = 2.23 \times 10^9$ ergs
21•2 $W = 133$ J
21•3 $W_1 = 0, W_2 = nRT_1 \ln 2, W_3 = P_2V_1, W_{tot} = 0.193\, nRT_1$
21•4 (*a*) $W = P(V' - V) + \frac{1}{8}\pi(P' - P'')(V' - V)$; (*b*) $W = P(V' - V) - \frac{1}{8}\pi(P' - P'')(V' - V)$
21•5 $W = 0.346$ J
21•6 $W = 1.155$ atm · m^3
21•7 (*a*) $W = 14.2$ atm · m^3; (*b*) lost heat
21•9 (*a*) $Q = 5922$ J $= 1425$ cal; (*b*) $Q = 2961$ J $= 713$ cal
21•10 $\Delta\mathcal{U} = 740$ J
21•11 $\Delta Q = \Delta W, \Delta\mathcal{U} = 0$
21•12 (*a*) $Q = 1091$ cal; (*b*) T and $\mathcal{U}$ decrease
21•13 (*a*) $W = 0, \Delta\mathcal{U} = 150$ kcal; (*b*) $\Delta\mathcal{U} = 80$ kcal; (*c*) $\Delta\mathcal{U} = -230$ kcal, $Q = -180$ kcal, $W = 50$ kcal
21•14 $Q = -990$ kcal
21•15 $J_0 = 775$ ft · lbf/Btu $= 4.17$ J/cal
21•16 $\Delta M \doteq \frac{1}{2}mv^2(1 - m/M)/L_f$
21•17 (*a*) $Q = 0.717$ cal; (*b*) $\Delta\tau = 1.8$ K
21•18 $\Delta\tau = 94.4$ K
21•19 $\tau = 21°$C
21•20 $\tau = 49°$C
21•21 $\Delta m = 4.4$ g
21•22 $c = 0.0309$ cal/g · K
21•23 (*a*) 0.13 percent; (*b*) divide by 4.8155
21•24 Maximum error is 0.75 percent for $\tau = 75°$C
21•26 $\Delta\mathcal{U} = P\Delta V/(\gamma - 1) = nC_V\Delta T$
21•27 (*a*) $W = \frac{1}{2}P_1V_1(1 - V_1^2/V_2^2)$; (*b*) $\Delta\mathcal{U} = C_V(P_1V_1/R)(1 - V_1^2/V_2^2)$; (*c*) $C = C_V + \frac{1}{2}R$
21•30 (*a*) $n = 0.0609$ kmole; (*b*) $\Delta\mathcal{U} = 59.9$ kcal; (*c*) $W = 2.30 \times 10^5$ J $= 54.89$ kcal; (*d*) $C = 0.413$ kcal/kmole · K
21•31 (*a*) $\Delta V = 0.635$ m^3; (*b*) $W = 90{,}200$ J
21•32 $\tau = 123°$C, $P = 2.64$ atm
21•33 $dQ/dt = 1153$ W/m^2
21•34 $\Delta m = 50$ kg/day
21•35 $\Delta m = 169$ kg
21•36 (*a*) $\tau = 23.5°$C; (*b*) $d\tau/dx = 15.9$ K/cm, 5.29 K/cm; (*c*) $dQ/dt = 2.38$ cal/cm^2 · s; (*d*) no
21•38 (*a*) $d\tau/dx = 55.2$ K/cm in air, 1.6 K/cm in glass; (*b*) $dQ/dt = 9.15 \times 10^{-5}$ cal/s · cm · K; (*c*) 26.8 percent reduction
21•39 $T = T_1 + (T_0 - T_1)\exp(-4\pi\mathcal{K}R^2t/nC_Vd)$
21•42 $\tau = 1000°$C
21•43 (*b*) $x_0 = 0.125, x_1 = 0.170, x_2 = 0.169, y = 1.778, y_0 = 11.31$ km, or 23.31 km above the earth's surface; (*c*) $f(x) = 5x - x^{1.5} - \frac{1}{2}$, and $F(x) = (0.4x + \frac{1}{2}x^{0.4})/(5x^{0.4} - 0.6)$

CHAPTER TWENTY-TWO

kinetic theory: energy

Without the recognition of a causal connection between motion and heat, it is just as difficult to explain the production of heat as it is to give any account of the fact that motion disappears. . . . We prefer the assumption that heat proceeds from motion, to the assumption of a cause without effect and of an effect without a cause. . . .

Julius R. Mayer, in Annalen der Chemie und Pharmacie, *1842*

In the preceding two chapters we have focused on the macroscopic properties of thermodynamic systems; that is, we have discussed measurable state variables such as mass, volume, pressure, temperature, expansion coefficients, and specific heats in terms of the system as a whole. In this chapter we shall consider the microscopic mechanical properties of the particles that comprise thermodynamic systems. We shall begin, in fact, with the atomic-molecular theory of matter and proceed from there to derive a number of physical laws, ranging from the equation of state of an ideal gas to the specific heats of crystalline solids. The body of theory so derived is known as *kinetic theory.* In its more advanced aspects it is extremely powerful and exceedingly complex, and its applications range from gases to galaxies.

22 • 1 *the atomic theory*

By the seventeenth century Francis Bacon, Robert Boyle, and especially Newton were finding an atomic picture helpful in explaining such phenomena as the production of heat by friction and the elastic properties of gases. In the eighteenth century Daniel Bernoulli made the first significant attempt to explain quantitatively the pressure exerted by a gas conceived of as a collection of randomly moving particles. However, it was not until the beginning of the nineteenth century that the purely speculative views inherited from the Greek atomists were placed on the substantial foundation of experiment. This was accomplished by the English chemist John Dalton. Although Dalton was familiar with, and greatly influenced by, Newton's views on atomism, it was mainly his own studies of the physical properties of gases, made originally in connection with meteorological problems, that led him to construct a model of matter based on the following postulates:

1 Each chemical element consists of atoms that are identical in weight [in mass] and in all other respects.
2 Atoms of different elements have different weights [masses].
3 Atoms are indivisible and indestructible, thus preserving their identities in all chemical reactions.
4 When different elements combine to form a compound, the smallest possible particle [molecule] of the compound consists of a definite number of atoms of each element.

As is generally true in the initial development of a physical theory, Dalton did not proceed in a clearcut fashion from postulates to deductions and tests. Indeed, he often followed the reverse order. The study of thousands of experiments on the proportions by weight in which various substances combine chemically revealed several fundamental principles that were to find their most natural interpretation in this atomic hypothesis.

In 1799 Joseph Louis Proust stated the law of *definite proportions:*

> Compounds are substances containing two or more elements combined in definite, unvarying proportions by weight, independent of the way in which they are prepared.

Although this law was established experimentally and without reference to any particular hypothesis concerning the structure of matter, Dalton interpreted it to mean that the atoms of a given element are of constant mass and that a given compound consists of discrete molecules, each of which is always of the same atomic composition.

Dalton himself stated the law of *multiple proportions:*

> When two elements can combine to form more than one compound, the weights of one element that combine separately with a particular weight of the other are in the ratios of small integers.

This law, based on the atomic theory and only meager data, was suggested by the possibility that a new compound might be formed by adding an "extra" atom to a compound already containing atoms of the same species.

In a series of experimental papers published in Germany between 1792 and 1802, J. B. Richter stated the *law of equivalent proportions:*

> The weights of two elements that separately combine with identical weights of a third element are either the weights in which the two elements may combine together or are related to these weights in the ratio of small integers.

This shows the possibility of assigning to each element an *equivalent weight,* representing the relative proportion in which the element combines with other elements. If the element combines with another to form more than one compound, it has more than one equivalent weight, but these, by the law of multiple proportions, must be in the ratio of small integers. Evidently, then, it was possible to assign to each element an experimental number, which came to be called the *atomic weight.* When multiplied by some small integer, the atomic weight gives the mass by which the element enters into combination with other elements. Similarly, it became possible to identify the *molecular weight* of a compound as the sum of the atomic weights of its constituents, although errors were common owing to the lack of precise knowledge of atomic weights and compositions.*

In 1808 Gay-Lussac discovered the law of *combining volumes:*

> The volumes of gases participating in a chemical reaction are related as the ratios of whole numbers.

For example, two volumes of hydrogen gas can combine with one volume of oxygen gas to form two volumes of steam—corresponding to the currently well-known fact that a water molecule consists of two atoms of hydrogen and one of oxygen. Since the chemistry of the day, following the example of the great Lavoisier, emphasized analysis by weight, Gay-Lussac's discovery was not at all obvious. The law of combining volumes substantiated the idea that compounds were composed of definite and unvarying *numbers* of constituent atoms and suggested that these numbers were somehow related to the relative volumes involved in their formation. However, Dalton himself rejected Gay-Lussac's theory, partly on the basis of his own poorly done volumetric experiments and partly because in his atomic theory he had assumed that the molecules of a gas are in contact with each other, so that molecules of different sizes would require different volumes.

Dalton and Gay-Lussac were finally reconciled by Avogadro's hypothesis. Amedeo Avogadro suggested in 1811 that gas particles were so small compared to the total volume they occupied that their actual diameter was largely irrelevant. Furthermore, he postulated *Avogadro's law:*

> Equal volumes of [ideal] gas under identical conditions of temperature and pressure contain equal numbers of individual particles.

Hence the fact that two volumes of hydrogen combine with only one volume of oxygen actually does indicate that the water molecule is composed of hydrogen and oxygen atoms in the ratio 2:1, in agreement with Dalton's

*The existence of *isotopes,* atoms of different atomic masses but belonging to the same element—that is, having essentially the same chemical properties—means that the atomic weight of an element may best be defined as the weighted average of the masses of the atoms of the element as it is found in nature.

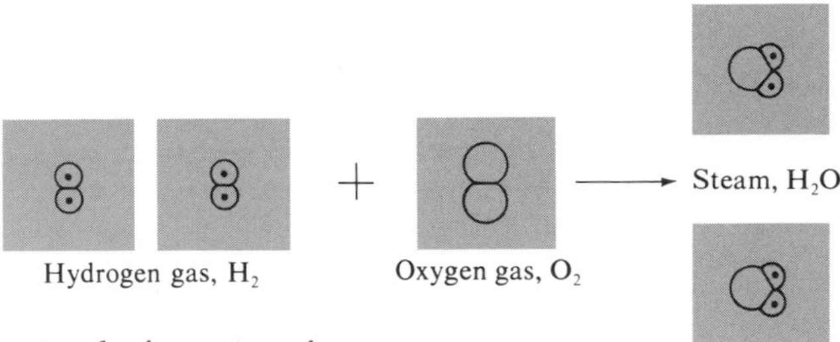

figure 22 • 1 *Volume diagram representing the formation of water*

hypotheses. But then how could one explain the fact that *two* volumes of steam are formed, and not one? There must have been some knowing smiles among Avogadro's contemporaries when he resolved this dilemma by further hypothesizing that some elementary gases are *diatomic*—that is, one particle or molecule of hydrogen gas, for example, consists of *two* hydrogen atoms. Thus if there are N molecules per unit volume, then two volumes, $2 \times 2N$ atoms, of hydrogen combine with one volume, $2N$ atoms, of oxygen to make two volumes, $2N$ molecules, of water. Figure 22 . 1 represents the formation of water according to Avogadro's hypothesis.

Avogadro's idea that for at least some of the elements the smallest units are not single atoms, but groups of atoms, ultimately proved correct. Formerly the term *molecule* had been reserved for the smallest group of atoms into which a compound can be divided without altering its chemical properties. The new conception, following Avogadro, is that molecules are the smallest units into which either compounds or elements can be divided without decomposition into chemically different constituents. The properties of single atoms of oxygen, for example, are distinctly different from those of diatomic oxygen. For any of the elements the molecule consists of one or more atoms of that element only, whereas for a compound the molecule consists of atoms of at least two different elements.

In light of Avogadro's hypothesis, it can be seen why molecular weights of compounds enjoy their peculiar chemical significance: it is because they contain *equal numbers of particles.* Thus a kilogram-mole of every ideal gas, occupies the same "molecular volume," $V_0 = 22.4 \text{ m}^3$ at standard temperature and pressure, and a mole of any substance in any phase whatever must represent a certain *number* of molecules, just as a "dozen" always represents 12. This number, known as *Avogadro's number,* has the experimentally determined value

$$N_A = (6.022045 \pm 0.000031) \times 10^{26} \text{ molecules/kmole} \qquad [22 \cdot 1]$$

For nearly half a century chemists ignored Avogadro's proposal, partly because they distrusted conceptions that were "too hypothetical," but also because it had not become clear that Avogadro's law could be used to remove uncertainties in the correct atomic weights of the elements and the correct formulas of compounds. The clarification came at last in 1858, in a remarkably lucid and convincing paper by the Italian chemist Stanislao Cannizzaro. Cannizzaro reasoned that if Avogadro's law is true, then the masses and densities of equal volumes of different gases are exactly proportional to their atomic masses, and hence to their molecular weights as determined from analysis of chemical reactions by weight. From available data Cannizzaro

table 22 • 1 *An experimental verification of Avogadro's law*

gas	*formula*	*density relative to oxygen as 32*	*molecular weight (chemical scale)*
hydrogen chloride	HCl	36.59	36.47
methane	CH_4	16.21	16.04
carbon monoxide	CO	28.17	28.01
carbon dioxide	CO_2	44.75	44.01
nitric oxide	NO	30.19	30.01
nitrous oxide	N_2O	44.45	44.02
water vapor	H_2O	18.17	18.02
			atomic weight (chemical scale)
hydrogen	H_2	2.012	1.008
helium	He	3.999	4.003
nitrogen	N_2	28.05	14.01
oxygen	O_2	32.00	16.00
bromine	Br_2	159.9	79.92

was able to construct tables like Table 22 . 1, which clearly corroborates Avogadro's law. That is, if equal volumes contain equal numbers of molecules, the relative densities of the gases must be proportional to the atomic weights of the gases. This set of numbers is, in fact, in very close agreement with molecular weights as determined by independent chemical analysis of the combining weights of the constituent atoms of each compound.

22 . 2 *ideal gases*

Many years before the establishment of the atomic-molecular theory of chemistry, the accumulating knowledge of such purely physical phenomena as the elasticity of gases and the effects of heat on matter had led to a revival of the ancient Greek conjecture that all matter is composed of discrete particles. To account for the elastic properties of air, two views were advanced. The first was a repulsion theory, according to which the pressure exerted by confined air was attributed to repellent forces existing between the particles, which were assumed to be at rest. However, this is contradicted by the observation that temperature decreases in an adiabatic expansion. If there were repulsive forces, they would add to the kinetic energy of the particles as they moved apart, much as in the release of a compressed spring.

Daniel Bernoulli formed the hypothesis that a gas consists of perfectly elastic particles in rapid motion of translation, and in this way made the first successful attempt to explain the physical properties of a gas on a simple mechanical basis. He explained the pressure on the walls of a vessel enclosing a gas as due to the impacts of the particles (Fig. 22 . 2) and deduced an

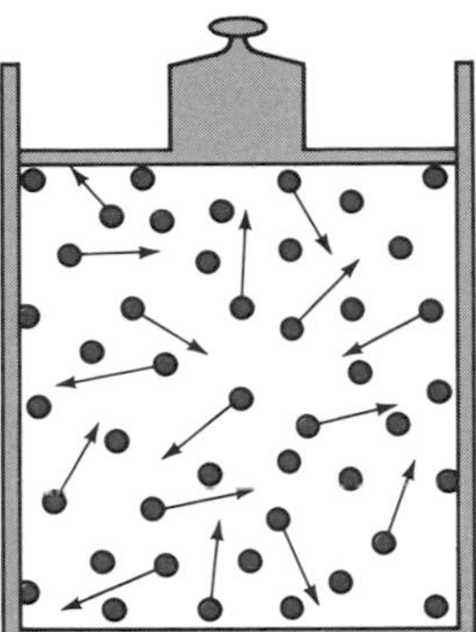

figure 22 • 2 *Bernoulli's model of a gas (1738). The arrows indicate random motion of the individual molecules.*

expression for the pressure in terms of their speeds. However, he did not specify accurately what is meant by the speeds of the gas particles, or by pressure, or by temperature—concepts that were still unclarified in his day. Therefore his formulation can be regarded as only the first rough quantitative sketch of what has since come to be the important branch of theoretical physics known as the *kinetic-molecular theory of gases.*

In 1848 James Joule, building on a sounder foundation of experimentation than had Bernoulli, and probably without even knowing of Bernoulli's work, did deduce an expression for the pressure of an ideal gas in terms of the number, the speed, and the mass of the molecules. The model he formulated was a chemically homogeneous gas, composed of a tremendous number of identical noninteracting molecules moving in completely random fashion through largely empty space. Although the particles could make perfectly elastic hard-sphere-type collisions, the effect, since the particles were identical, was as if they passed through each other. Macroscopically, the only significant interactions were the perfectly elastic collisions between the gas particles and the walls of their container.

Let us use Joule's model to compute the pressure which the moving particles exert on the container shown in Fig. 22 . 3. Consider the gas to be in a state of thermal equilibrium, with its pressure, density, and temperature constant and uniform throughout. Now visualize some particular molecule of mass m, which at the moment has a velocity $\mathbf{v}$, with rectangular components v_x, v_y, and v_z. When this molecule impinges on the top of the box, the normal

figure 22 • 3 *Joule's method of obtaining the kinetic-theory expression for pressure*

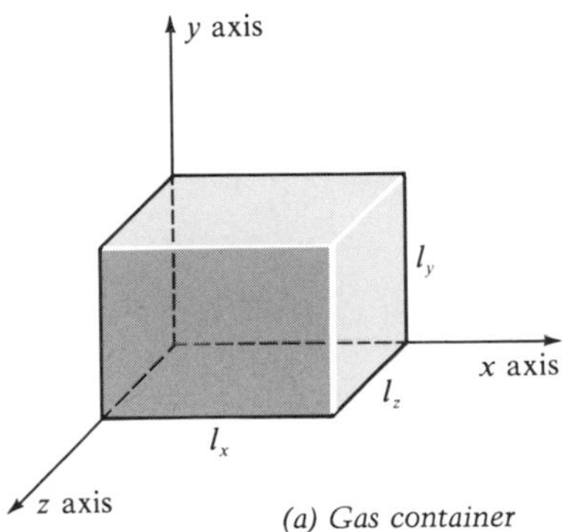

(a) Gas container

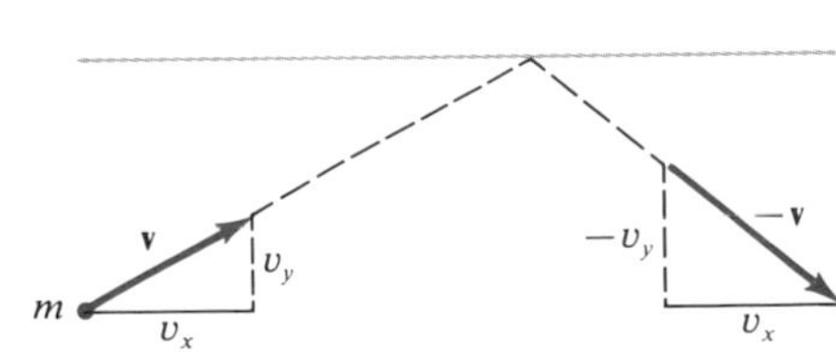

(b) Perfectly elastic impact between molecule and wall

force it exerts depends on the rate of change of its momentum in the y direction only. The net momentum change of the molecule in this collision is therefore

$$\Delta p_y = -mv_y - mv_y = -2mv_y \qquad [22 \cdot 2]$$

Since the speed of the molecule in the y direction is always the same, it will make one collision with the top of the box every round trip, or at intervals of

$$\Delta t = \frac{2l_y}{v_y} \qquad [22 \cdot 3]$$

where l_y is the height of the box. The average force that this wall exerts on the molecule in each collision is $F_y = \Delta p_y/\Delta t$. Therefore the reaction force that the molecule exerts on the wall is $F'_y = -F_y$, or

$$F'_y = -\frac{\Delta p_y}{\Delta t} = \frac{mv_y^2}{l_y} \qquad [22 \cdot 4]$$

The gas pressure on the top of the box is the sum of the forces exerted by all N molecules, divided by the area $l_x l_z$ of the top of the box:

$$P_y = \frac{m}{l_x l_y l_z} \sum_{i=1}^{N} v_{yi}^2 \qquad [22 \cdot 5]$$

If we denote the average of the *square* of the molecular speed by $\overline{v_y^2}$, then

$$\overline{v_y^2} = \frac{\sum_{i=1}^{N} v_{yi}^2}{N} \qquad [22 \cdot 6]$$

and for volume $l_x l_y l_z = V$ the pressure is

$$P_y = \frac{Nm}{V} \overline{v_y^2} \qquad [22 \cdot 7]$$

At this point the hypothesis of completely random motion, or "molecular chaos," is necessary if Eq. [22 . 7] is to agree with the experimental observation that the pressure is the same in all directions, $P_x = P_y = P_z$. If the molecules are moving at random, with no "preferred" direction, then

$$\overline{v_x^2} = \overline{v_y^2} = \overline{v_z^2} \qquad [22 \cdot 8]$$

and the average value of the squares of the molecular speeds is given by

$$\overline{v^2} = \frac{\sum_{i=1}^{N} (v_{xi}^2 + v_{yi}^2 + v_{zi}^2)}{N} = \overline{v_x^2} + \overline{v_y^2} + \overline{v_z^2} = 3\overline{v_y^2} \qquad [22 \cdot 9]$$

Substituting into Eq. [22 . 7], we obtain

$$\blacktriangleright \quad P = \frac{Nm}{3V} \overline{v^2} \qquad [22 \cdot 10]$$

where the quantity Nm/V may be recognized as the density:

$$\rho = \frac{Nm}{V} \qquad [22 \cdot 11]$$

For a stationary gas the total momentum must be zero. This is in agreement with the hypothesis of randomness, since, on the average, there must be as many molecules moving the $+x$ direction as in the $-x$ direction over any given range of speeds. Hence $\bar{v}_x = 0$, and the same argument holds for $\bar{v}_y = 0 = \bar{v}_z$.

example 22 • 1 Consider two particles with velocities $\mathbf{v}_1 = 2\mathbf{i}$ and $\mathbf{v}_2 = -3\mathbf{i}$. (*a*) What is their average squared speed $\overline{v^2}$? (*b*) Their average speed squared $\bar{v}^2$? (*c*) Their average velocity squared $\bar{\mathbf{v}} \cdot \bar{\mathbf{v}}$?

solution Note that these three values are not the same:

(*a*) $\overline{v^2} = \frac{1}{2}(v_1^2 + v_2^2) = \frac{1}{2}(4 + 9) = \frac{13}{2} = \frac{26}{4}$

(*b*) $\bar{v}^2 = [\frac{1}{2}(v_1 + v_2)]^2 = [\frac{1}{2}(2 + 3)]^2 = \frac{25}{4}$

(*c*) $\bar{\mathbf{v}} \cdot \bar{\mathbf{v}} = [\frac{1}{2}(\mathbf{v}_1 + \mathbf{v}_2)]^2 = [\frac{1}{2}(-\mathbf{i})]^2 = \frac{1}{4}$

Equation [22 . 10] affords a way to compute $\overline{v^2}$. This was the first molecular magnitude to be determined quantitatively, and the result was astonishing. By finding the *root-mean-square speed,*

$$v_{\mathrm{rms}} = \sqrt{\overline{v^2}} \qquad [22 \cdot 12]$$

we can estimate how fast the molecules are moving. This value is always somewhat greater than the arithmetic mean speed $\bar{v}$, because the squares of large numbers increase more rapidly than the numbers themselves. From Eqs. [22 . 10] and [22 . 11], $v_{\mathrm{rms}} = \sqrt{3P/\rho}$. Then if we use the equation of state in the form $PV = nRT$, for a gas of molecular weight M_0 we obtain

$$\blacktriangleright \quad v_{\mathrm{rms}} = \sqrt{\frac{3P}{\rho}} = \sqrt{\frac{3PV}{M}} = \sqrt{\frac{3RT}{M_0}} \qquad [22 \cdot 13]$$

which is a function of *temperature alone.* Of course, v_{rms} can be computed more easily from pressure and density data if they are available.

example 22 • 2 Calculate the density and root-mean-square speed of hydrogen molecules at 0°C and 1 atm.

solution From Table 22 . 1, the molecular weight of hydrogen is 2.016 kg/kmole. Therefore the density under standard conditions is

$$\rho = \frac{M}{V} = \frac{M_0}{V_0} = \frac{2.016\ \mathrm{kg/kmole}}{22.4\ \mathrm{m^3/kmole}} = 0.090\ \mathrm{kg/m^3}$$

and the root-mean-square speed of hydrogen molecules at 0°C is

$$v_{\mathrm{rms}} = \sqrt{\frac{3P}{\rho}} \doteq \sqrt{\frac{3 \times 10^5\ \mathrm{Pa}}{0.090\ \mathrm{kg/m^3}}} = 1826\ \mathrm{m/s}$$

or more than 4000 miles an hour!

22 . 3 temperature and heat

If we express the ideal-gas equation $PV = nRT$ in terms of the number of molecules $N = nN_A$ instead of the number of moles, we obtain

$$PV = N\frac{R}{N_A}T = NkT \qquad [22 . 14]$$

where $k = R/N_A$ is a new fundamental quantity known as the *Boltzmann constant,* or the universal gas constant *per molecule:**

$$\blacktriangleright \quad k = (1.380662 \pm 0.000044) \times 10^{-23}\ \text{J/mol} \cdot \text{K} \qquad [22 . 15]$$

In numerical computations it is usually satisfactory to approximate the value of k as $\frac{7}{5} \times 10^{-23}$.

We can also write Eq. [22 . 14] as $PV/N = kT$. Substitution of this expression into Eq. [22 . 10] then gives us the average kinetic energy per molecule of an ideal gas:

$$\blacktriangleright \quad \tfrac{3}{2}kT = \tfrac{1}{2}m\overline{v^2} = \overline{K}_{\text{tr}} \qquad [22 . 16]$$

where $\overline{K}_{\text{tr}}$ is, by definition, the average *translational* kinetic energy per molecule. Hence the root-mean-square speed is

$$\blacktriangleright \quad v_{\text{rms}} = \sqrt{\frac{3kT}{m}} \qquad [22 . 17]$$

In the next chapter we shall see that

$$\bar{v} = \sqrt{\frac{8kT}{\pi m}} = 0.921 v_{\text{rms}} \qquad [22 . 18]$$

Note that v_{rms} is comparable to the speed of sound in the gas (Eq. [21 . 33]), so that the speed with which a disturbance can be transmitted through a gas is closely related to the speed of the gas molecules themselves. This is consistent with our picture of an ideal gas as one in which the only interactions are collisions.

Equation [22 . 16] tells us that the average kinetic energy associated with the random translational motion of each molecule is a function of the temperature alone, independent of the pressure or the volume. It is still more specific, in that it shows the average translational energy of a molecule to be *directly proportional* to the temperature on the absolute thermodynamic scale. The kinetic theory thus explains how the temperature of a gas can be raised by compressing it—that is, doing work on the gas—or how a heated gas can itself do work when it expands against, say, a moving piston, becoming cooler in the process.

It is important to recognize that temperature is *statistical* in nature. Each individual molecule has its own instantaneous velocity and kinetic

*In the literature of physics and chemistry *molecule* is dimensionless and usually does not appear in formulas, while *mol* is the standard abbreviation for gram-molecular weight. In this book, however, molecular weights are identified as *mole* and *kmole,* so that the use of *mol* to denote molecule should cause no confusion.

energy, and their values may vary over an indefinitely wide range from particle to particle. The temperature, however, is an average property of the *system as a whole.*

The interpretation of temperature as a measure of kinetic energy, based on the kinetic model of an ideal gas, is related to the internal energy $\mathcal{U}$ of the gas, which must be entirely kinetic. Thus for n moles of an ideal monatomic gas,

$$\mathcal{U} = nN_A\overline{K}_{tr} = n(\tfrac{3}{2}N_Ak)T = \tfrac{3}{2}nRT \qquad [22 \cdot 19]$$

In an isometric process ($\Delta V = 0$) the heat absorbed by the gas goes entirely into kinetic internal energy. Therefore

$$dQ_V = nC_V\,dT = d\mathcal{U}$$

so that

$$C_V = \tfrac{3}{2}R \qquad C_P = C_V + R = \tfrac{5}{2}R \qquad [22 \cdot 20]$$

where $R = 8314$ J/kmole · K $= 1.986$ kcal/kmole · K. These values prove to be in excellent agreement with the measured values for helium and argon, monatomic gases whose lack of intramolecular structure makes them a close approximation to the ideal gas of point particles. Although Eqs. [22 . 20] only hold for monatomic gases, this is sufficient to verify the kinetic theory and our interpretation of temperature. For more complex molecules it becomes necessary to account for the vibrational and rotational energies of the atomic constituents with respect to the center of mass of the molecule itself.

It would be incorrect to infer from Eq. [22 . 16] that at absolute zero on the Kelvin or Rankine scale gas molecules have no translational kinetic energy and therefore come completely to rest. In the first place, these equations apply rigorously only to ideal gases. Moreover, all real gases condense to the liquid phase long before they get close to absolute zero. However, we are still left with the question of what does happen to the internal energy of a liquid or a solid as the temperature approaches absolute zero. To obtain an answer, we must turn to kinetic theory as it has been modified by *quantum mechanics,* which deals with the postulates and laws of mechanics that apply to systems of atomic dimensions. This modified kinetic theory shows that as the temperature of any substance approaches absolute zero, the kinetic energy of the particles approaches, not zero, but a constant value characteristic of the substance, which is appropriately termed its *zero-point energy.* Helium, for example, has a large zero-point energy—about 250 J/mole. Consequently, when liquid helium is cooled toward absolute zero, its atoms cannot get close enough together to form a solid unless the volume is reduced by subjecting the liquid to a very high pressure. Equation [22 . 16] does hold for an ideal gas—that is, for a gas that is behaving ideally and obeying the ideal-gas equation at moderate temperatures and pressures.

When two gases at different temperatures are placed in the same container, molecules of the two gases will collide and transfer momentum and kinetic energy back and forth. The result is a transfer of energy from the hot gas to the cool gas until both reach the same temperature (have the same average translational kinetic energy). It is this transferred kinetic energy that we have previously termed *heat,* and obviously heat transfer is also statistical in nature. It is possible for a rapidly moving particle in the cooler gas to collide with and transfer energy to a slowly moving particle in the hotter gas; however, the net effect of very many collisions is a transfer of kinetic energy from the hotter to the cooler gas. Since ordinary conditions of temperature and pressure involve such vast numbers of particles, any fluctuations from the average are relatively infinitesimal and are difficult, although not impossible, to detect. As a result, this average statistical behavior has become embodied in the second law of thermodynamics, discussed in Chapter 25. One form of this law states that heat will never flow spontaneously from a body of higher temperature to one of lower temperature.

example 22 • 3 Two volumes of helium at temperature 546°C are mixed with one volume of argon gas at 0°C. Both gases are monatomic and at 1 atm pressure. Find the equilibrium temperature of the mixture.

solution From the ideal-gas equation of state,

$$\frac{n_{\mathrm{He}}}{n_{\mathrm{Ar}}} = \frac{V_{\mathrm{He}}\, T_{\mathrm{Ar}}}{V_{\mathrm{Ar}}\, T_{\mathrm{He}}} = \frac{2 \times 273}{546 + 273} = \tfrac{2}{3}$$

Since both gases are monatomic, their molar specific heats are the same (see Table 22 . 2), and equating the heats transferred, we obtain

$$n_{\mathrm{He}} C_P (546°\mathrm{C} - \tau) = n_{\mathrm{Ar}} C_P (\tau - 0°\mathrm{C})$$

Therefore

$$\frac{n_{\mathrm{He}}}{n_{\mathrm{Ar}}} = \frac{\tau}{546°\mathrm{C} - \tau} = \tfrac{2}{3}$$

or

$$\tau = 218.4°\mathrm{C}$$

example 22 • 4 Helium has an atomic weight of 4.003 kg/kmole. (*a*) What is its molecular mass m? (*b*) What is the root-mean-square speed of helium molecules at 0°C? (*c*) At 273°C?

solution (*a*) The monatomic molecular mass is

$$m = \frac{M_0}{N_A} \doteq \frac{4}{6 \times 10^{26}}\ \mathrm{kg} = 0.67 \times 10^{-26}\ \mathrm{kg}$$

(*b*) At 0°C = 273 K the root-mean-square speed is

$$v_{\mathrm{rms}} = \sqrt{\frac{3kT}{m}} \doteq \sqrt{\frac{3 \times 273 \times 1.38 \times 10^{-23}}{0.67 \times 10^{-26}}} = 1300\ \mathrm{m/s}$$

(*c*) At 273°C = 2 × 273 K,

$$v_{\mathrm{rms}} = 1300\sqrt{2} = 1837\ \mathrm{m/s}$$

22 • 4 applications of the ideal-gas model

If two different gases are confined in separate vessels at the same temperature, the average kinetic energy per molecule is the same for both gases:

$$\tfrac{1}{2}m_1\overline{v_1^2} = \tfrac{1}{2}m_2\overline{v_2^2} = \tfrac{3}{2}kT \qquad [22 \cdot 21]$$

There is no reason to suppose that the same condition would not hold if the gases were confined in the same vessel; in fact, direct experiment shows that it does. It follows, then, that the molecules of each gas in a mixture of gases have the same average kinetic energy of translation $\overline{K}_{tr}$.

Equation [22 . 21] represents a particular case of a famous generalization known as the classical *law of equipartition of energy.* In 1802 Dalton discovered experimentally that if several gases that have no chemical action on each other are introduced into the same container, the pressure P exerted by the mixture is equal to the sum of the pressures P_i which would be observed if each gas alone occupied that container at the same temperature. This law, called *Dalton's law of partial pressures,* is exact only for ideal gases and can be derived directly from the kinetic model. The total pressure must be due to the sum of all the molecule-container impacts, regardless of the species of molecule:

$$P = \frac{1}{3V}\sum_i N_i m_i \overline{v_i^2} = \sum_i \frac{N_i kT}{V} = \sum_i P_i \qquad [22 \cdot 22]$$

where the index i refers to the range of different gases in the mixture. The collisions between molecules of different species ensure that each species considered separately comes to the same temperature $T = \overline{m_i v_i^2}/3k$ as the others.

The kinetic theory also explains a *law of effusion* that Thomas Graham had obtained experimentally in 1846. Graham confined two different gases, one at a time, in a vessel stoppered by a thin porous plug, and placed the vessel in an evacuated chamber. He postulated that if both gases were at the same temperature and pressure, they would escape through the fine pores of the plug with speeds that vary inversely as the square roots of the densities of the gases, and therefore of their molecular weights. On the assumption that the rate of effusion depends on the average frequency $\overline{1/\Delta t}$ with which the molecules strike the walls, for two gases at the same temperature and pressure Eq. [22 . 3] yields

$$\frac{\overline{1/\Delta t_1}}{\overline{1/\Delta t_2}} = \frac{\bar{v}_1}{\bar{v}_2} = \sqrt{\frac{\overline{v_1^2}}{\overline{v_2^2}}} = \sqrt{\frac{\rho_2}{\rho_1}} \qquad [22 \cdot 23]$$

Avogadro's law equates the ratio of the densities with the ratio of the molecular weights, so that

$$\frac{\bar{v}_1}{\bar{v}_2} = \sqrt{\frac{M_{02}}{M_{01}}} \qquad [22 \cdot 24]$$

These equations predict, for instance, that hydrogen molecules will effuse, on the average, nearly four times faster than oxygen molecules, for oxygen has a density and molecular weight nearly 16 times that of hydrogen at the same temperature and pressure. Indeed, it was the observation that hydrogen escaped from a cracked vessel faster than air passed in to replace it which led Graham to make his experimental studies of effusion (see Fig. 22 . 4).

In 1896 Lord Rayleigh extended the theory of effusion of gases to mixtures and showed that a mixture of two gases of different molecular weights can be partly separated by effusion into a vacuum through a porous plug. For any gas, the number ΔN of molecules that escape through the plug per unit time is proportional to both the number N of molecules in the vessel and to their arithmetic mean speed $\bar{v}$. Therefore the ratio of escaping molecules of the two gases in the mixture is

$$\frac{\Delta N_1}{\Delta N_2} = \frac{N_1 \bar{v}_1}{N_2 \bar{v}_2} \qquad [22 \cdot 25]$$

However, since both gases in a mixture must be at the same temperature, the ratio of their mean speeds is also proportional to the ratio of their molecular weights:

$$\blacktriangleright \quad \frac{\Delta N_1}{\Delta N_2} = \frac{N_1}{N_2}\sqrt{\frac{M_{02}}{M_{01}}} \qquad [22 \cdot 26]$$

This equation provides the basis of a widely used method for separating isotopes when they occur as a gaseous mixture at ordinary temperatures and has been applied to the large-scale separation of uranium isotopes for use in nuclear reactors.

Another highly interesting kinetic phenomenon known as ***Brownian movement*** was first reported in 1827 by the distinguished botanist Robert Brown, who found that "Extremely minute particles of solid matter, whether obtained from organic or inorganic substances, when suspended in pure water, or in some other aqueous fluids, exhibit motions for which I am unable to account and which, from their irregularity and seeming independence, resemble in a remarkable degree, the less rapid motions of some of the simplest animalcules of infusions." This type of irregular zigzag movement is typified by the dancing of dust particles in a beam of light (Fig. 22 . 5).

The cause of the Brownian movement was long in doubt, but with the development of the kinetic theory came the realization that the particles

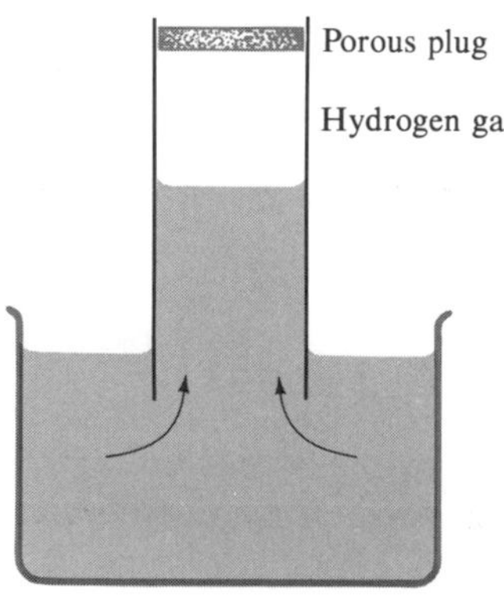

figure 22 • 4 *Graham's experimental study of effusion. A glass tube stoppered at one end by a porous plug is filled with hydrogen gas and placed with its open end under a liquid. The liquid rises in the tube, showing that more gas leaves the tube than enters it.*

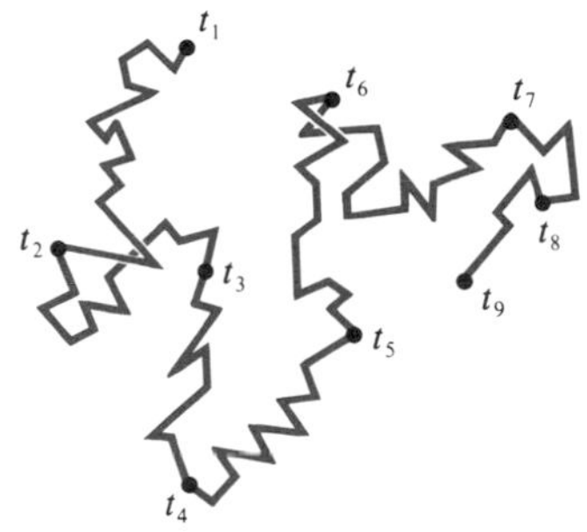

figure 22 • 5 *Typical successive positions of a particle undergoing Brownian movement, as observed at successive 30-s intervals*

move because they are bombarded unequally on different sides by the rapidly moving molecules of the fluid in which they are suspended. The Brownian movement never ceases. As Jean Perrin observed, "It can be seen in liquid occlusions in quartz, which have been sealed up for thousands of years. It is inherent and eternal." Here, then, is direct experimental evidence of the same sort of perpetual motion that we have assumed the molecules themselves to have.

This motion can be explained if we assume that the suspended particles behave like an ideal gas composed of massive molecules. Particles moving at speeds of, say, $v_{\text{rms}} \doteq 1$ cm/s at 273 K would have masses equal to $3kT/v^2_{\text{rms}}$, approximately 1.69×10^{10} times the mass of a helium molecule. This implies that, given measurements of v_{rms} for particles in Brownian motion, one could, by carefully weighing them, determine the Boltzmann constant, and hence Avogadro's number $N_A = R/k$. In fact, it was by careful and detailed measurements of various kinds of colloidal particles, studied under an ultramicroscope, that Perrin first obtained a value of $N_A = 6.85 \times 10^{26}$ mol/kmole. He also verified the equipartition law even for particles varying in mass by 60,000:1. The detailed theory of Brownian movement was worked out in 1904 by M. von Smoluchowski and in more final form in 1905 by Einstein.

22 • 5 equipartition of energy and specific heats

In deriving the difference in specific heats for an ideal gas, $C_P - C_V = R$, we have ignored the internal structure of the gas molecules. However, it is possible to loosen our assumptions about the nature of gas molecules and still obtain the same results. First, we must recognize that the internal energy of any substance can have several different sources:

1. K_{tr}, the translational kinetic energy of the molecules
2. K_{rot}, the rotational kinetic energy of molecules about their centers of mass
3. K_{vib}, the vibrational kinetic energy of the atoms of molecules—that part of the kinetic energy of the atoms which is due, not to their rotations, but to their periodic oscillations within the molecule
4. U_{vib}, the potential energy associated with the restoring forces in the vibrations of atoms within their molecules
5. U_{mol}, the potential energy due to the attractions or repulsions between the molecules of a system, which depend strongly on intermolecular distance

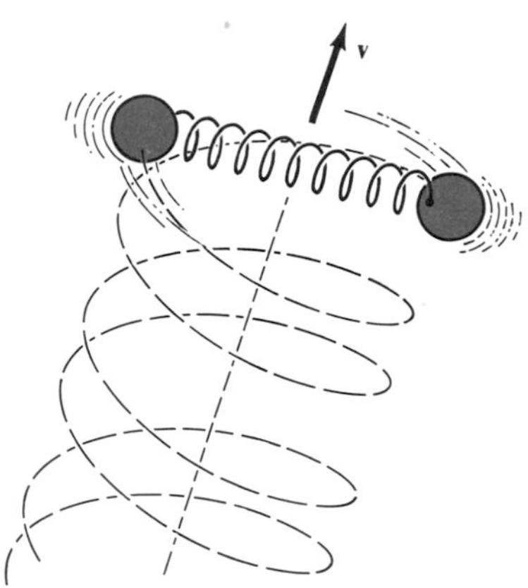

figure 22 • 6 *Combined translational, rotational, and vibrational motions of a diatomic molecule. The spring symbolizes the restoring force between the atoms which tends to keep them vibrating about some equilibrium separation.*

Figure 22 . 6 illustrates the combination of rotational and vibrational motions that a diatomic molecule might undergo as it translates with velocity $\mathbf{v}$.

Equal masses of different substances absorb different quantities of heat per degree of temperature rise for two reasons: First, they contain different numbers of atoms. Second, the heat energy absorbed is distributed, or "partitioned," in different ways among the various forms of energy listed above, and only translational kinetic energy K_{tr} is identifiable with a measurable temperature increase ΔT. If some of the absorbed energy goes into rotation or vibration, for example, the ratio $\Delta Q/\Delta T$ will be larger than if all of ΔQ were available for increasing the translational energy—that is, for increasing the value of ΔT. However, since this ratio is what we have termed specific heat, the specific heats of substances reflect the ways their internal energies are partitioned and are therefore important clues to their internal structure. Historically, some of the most important advances in the quantum mechanical theory of matter have come about through the study of specific heats.

In 1831 F. E. Neumann investigated the molar specific heats of substances whose molecules possess the same number of atoms and found them to be nearly the same for a given phase—for example, for the gaseous phase. This finding, known as *Neumann's law,* implies that the partition of internal energy is about the same for such similar substances. The law is not exact, nor could it be, in view of the differences in the attractive intermolecular forces that exist for different species of molecules.

Since the molecules of gases under ordinary pressures are not subject to appreciable mutual attraction, we might expect their potential-energy changes ΔU_{mol} to be relatively unimportant (compare the diatomic gases in Table 22 . 2). Changes in the internal energy of an ideal gas are therefore

$$\Delta \mathcal{U} = \Delta K_{tr} + \Delta K_{rot} + \Delta K_{vib} + \Delta U_{vib} \qquad [22 \cdot 27]$$

The last three types of energy can be ignored for monatomic gases. Although they become increasingly important with the complexity of the molecule, the atomic structure of such gases is significantly affected only at temperatures of 10,000 K or more. However, since these three energies are all internal to the molecule, ΔK_{rot}, ΔK_{vib}, and ΔU_{vib} have no effect on the *volume* of the system. Hence we can modify Eq. [21 . 15] by simply replacing dK by $d\mathcal{U}$:

$$\begin{aligned} dQ_P &= nC_P\,dT = d\mathcal{U} + P\,dV \\ dQ_V &= nC_V\,dT = d\mathcal{U} \end{aligned} \qquad [22 \cdot 28]$$

table 22 • 2 *Molar specific heats of various substances*

substance	*formula*	*molar heat at 15°C, 1 atm kcal/kmole · K*
gaseous		
A (monatomic)		
helium	He	4.97
argon	A	5.00
B (diatomic)		
hydrogen	H_2	6.83
nitrogen	N_2	6.93
oxygen	O_2	6.97
carbon monoxide	CO	6.94
nitric oxide	NO	6.99
hydrogen chloride	HCl	7.07
chlorine	Cl_2	8.15
C (polyatomic)		
carbon dioxide	CO_2	8.75
nitrous oxide	N_2O	8.82
ammonia	NH_3	8.91
methane	CH_4	8.48
ethane	C_2H_6	11.6
crystalline solids		
D (monatomic)		
sodium	Na	6.79
potassium	K	6.92
copper	Cu	5.85
mercury	Hg	6.67
E (diatomic)		
nickel monoxide	NiO	11.9
cupric oxide	CuO	11.3
mercuric oxide	HgO	11.2
F (triatomic)		
calcium chloride	$CaCl_2$	18.2
zinc chloride	$ZnCl_2$	18.6
barium chloride	$BaCl_2$	18.6

This means that the final result is still

$$C_P - C_V = R = 8314 \text{ J/kmole} \cdot \text{K}$$
$$= 1.986 \text{ kcal/kmole} \cdot \text{K} \qquad [22 \cdot 29]$$

for all ideal gases, regardless of the internal structure of their molecules.

In arguing that an ideal gas has negligible energy changes ΔU_{mol} we have reversed cause and effect. Actually, the distinctive physical difference between the behavior of an ideal gas and that of a real gas is the ΔU_{mol} term. One prominent feature of real gases is that they condense to liquids, so that this term becomes important for real gases at low temperatures and high pressures.

To understand the variations in specific heats shown in Table 22 . 2, as well as the variations due to temperature, it is necessary to introduce the important concept of *degrees of freedom* of a particle or a system. A particle regarded as a point may move in space in three, and only three, directions such that its motion along any one of these has no component along either of the other two. These three independent directions, obviously, are mutually perpendicular. A particle is therefore said to have three degrees of freedom. It cannot rotate because, as a point, it has no spatial extent. If a particle is constrained to move in a single plane, it is said to have two degrees of freedom. If it is still further limited to move in a straight line, it then has only one degree of freedom. In contrast, a *rigid body* thrown through the air has six degrees of freedom, of which three are translational and three are associated with the independent rotations of the body about three mutually perpendicular axes. If a ball were rolling on a table, it would have only five degrees of freedom—two of translation and three of rotation—since translational motion perpendicular to the table is impossible.

According to the classical law of equipartition, which we shall not prove here, each degree of freedom of a molecule of a gas has, on the average, an equal amount of kinetic energy $\frac{1}{2}kT$. We have already derived this result (see Eq. [22 . 16]) for the three translational degrees of freedom of a gas molecule, $\overline{K}_{\text{tr}} = \frac{3}{2}kT$, from kinetic theory. If we now let f represent the number of degrees of freedom of a molecule, then for an ideal gas of N molecules or $n = N/N_A$ moles, the law of equipartition yields an internal energy

$$\blacktriangleright \quad \mathcal{U} = Nf\tfrac{1}{2}kT = \tfrac{1}{2}nfRT \qquad [22 \cdot 30]$$

and from Eq. [22 . 28], specific heat is related to degrees of freedom by

$$nC_V \, dT = d\mathcal{U} = \tfrac{1}{2}nfR \, dT$$

Therefore

$$\blacktriangleright \quad C_V = \tfrac{1}{2}fR \qquad C_P = C_V + R = \tfrac{1}{2}(f + 2)R \qquad [22 \cdot 31]$$

The ratio of specific heats is a pure number given by

$$\blacktriangleright \quad \gamma = \frac{C_P}{C_V} = \frac{f + 2}{f} \qquad [22 \cdot 32]$$

This agrees with Eqs. [22 . 20] for a monatomic gas, for which $f = 3$. Now let us see how these formulas from kinetic theory fit the experimental facts more generally.

A glance at Table 22.2 shows that for monatomic gases $C_P \doteq 5$ kcal/kmole · K, whereas it is generally close to 7 kcal/kmole · K for diatomic molecules, and higher for more complex molecules. To understand the reason for this, note that a monatomic gas is like a point particle, with three degrees of freedom. Therefore $\mathcal{U} = \frac{3}{2}nRT$ and $C_P = \frac{5}{2}R \doteq 5$ kcal/kmole · K, as we could have predicted from Eq. [22 . 16]. However, to explain the specific heat of a diatomic molecule we need the law of equipartition, because such a molecule has rotational as well as translational degrees of freedom. At ordinary temperatures a diatomic molecule can be viewed as a rigid body, with the spring of Fig. 22 . 6 replaced by a rigid rod (a "dumbbell" molecule). However, its moment of inertia along the axis of the rod is negligible compared to its moments of inertia about axes normal to the rod. Hence we assign energies only to the rotational degrees of freedom about the two axes perpendicular to the rod and to each other. This gives a diatomic molecule five degrees of freedom, so that $C_P = \frac{7}{2}R \doteq$ 7 kcal/kmole · K, in agreement with Table 22 . 2. The values predicted by Eq. [22 . 32] are $\gamma = \frac{5}{3}$ for a monatomic gas, $\gamma = \frac{7}{5}$ for a diatomic gas, and $\gamma = \frac{8}{6}$ for a polyatomic gas of rigid molecules with all three degrees of rotational freedom—all in good agreement with Table 21 . 3. However, the last value in Table 21 . 3, $\gamma = 1.024$, would indicate a great many degrees of freedom for the ether molecule $(C_2H_5)_2$, so that it becomes necessary to consider the internal degrees of freedom of vibration of molecules.

Some diatomic gases, such as chlorine, and almost all polyatomic gases have specific heats distinctly larger than those calculated on the assumptions we have considered so far. These discrepancies tend to disappear if we assume that the molecules are not rigid, but that their atoms can vibrate under the influence of the attractive forces that bind them together and the repulsive forces that prevent them from coalescing. These internal forces act like the elastic force in a spring and vary in magnitude with the substance and the temperature. On the assumption of equipartition, such atomic vibrations within the molecule receive their share of the energy. Now, as in the case of the spring, the energy of these vibrations is at any moment part kinetic and part potential, and if the vibrations are of the simple harmonic variety, then the average amount of potential energy $\overline{U}_{\text{vib}}$ associated with each vibrational degree of freedom is, like the associated kinetic energy, $\frac{1}{2}kT$ per molecule. Therefore each mode of vibration within the molecule has two degrees of freedom and an average energy of

$$\overline{K}_{\text{vib}} + \overline{U}_{\text{vib}} = \tfrac{1}{2}kT + \tfrac{1}{2}kT = kT \qquad [22 \cdot 33]$$

A diatomic molecule with three translational, two rotational, and one vibrational degree of freedom appears to have $f = 3(\text{tr}) + 2(\text{rot}) + 2(\text{vib}) =$ 7 and so would have an average energy of $\frac{7}{2}kT$ per molecule, or $C_P =$ 9 kcal/kmole · K.

It turns out, then, that if the vibrational degrees of freedom are taken into account, the theoretical values are not lower than the experimental values. As a matter of fact, they are a little too high, although the agreement improves as the temperature increases. It would appear from this that some, but not all, of the molecules receive sufficient energy from their collisions with other molecules to set their atoms into vibration. This would also explain why many diatomic molecules, such as hydrogen, appear to have no

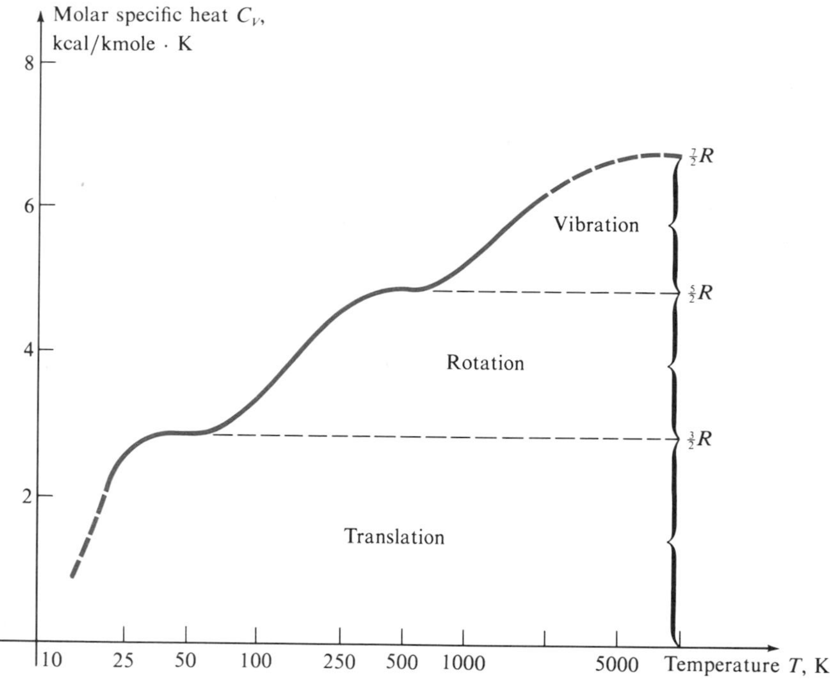

figure 22 • 7 *Variation with temperature of the molar specific heat C_V of ordinary diatomic hydrogen. Note the logarithmic temperature scale.*

vibrational degrees of freedom at ordinary temperatures; their molar specific heats C_V do not exceed 5 kcal/kmole · K simply because their atoms are too tightly bound to be set into vibration by impacts at ordinary temperatures. However, if the temperature of hydrogen is raised to about 400°C, then C_V increases steadily toward 7 kcal/kmole · K (see Fig. 22 . 7). Apparently more and more molecules are receiving sufficient energy from impacts to set their atoms into vibration.

Even more remarkable is the fact that when hydrogen is cooled below −25°C, the value of C_V decreases rapidly to 3 kcal/kmole · K. This means that it now behaves like a *monatomic* gas; the rotational degrees of freedom appear to have become inactive. We cannot assume that the two hydrogen atoms have come so close together that they act as a single molecule, for evidence from other sources shows that molecular dimensions change very little with cooling under ordinary pressures. The explanation is to be found in the quantum theory, discussed in a later chapter. For most other diatomic gases, the low-temperature approach to 3 kcal/kmole · K cannot be observed, but there is usually an increase in C_V at high temperatures.

example 22 • 5 An ideal diatomic gas expands from state (P_1,V_1) to state (P_1,V_2), and during this expansion its temperature drops according to $T = T_1\sqrt{V_1/V}$. (*a*) What is the change in the internal energy of the system? (*b*) How much heat is absorbed by the gas as it expands? (*c*) Define a specific heat for the expansion law $P = P_1\ (V_1/V)^{3/2}$.

solution (*a*) The change in internal energy is given by

$$\mathcal{U}_2 - \mathcal{U}_1 = \Delta\mathcal{U} = nC_V(T_2 - T_1) = nC_VT_1\left(\sqrt{\frac{V_1}{V_2}} - 1\right)$$

and since $C_V = \frac{5}{2}R$ for a diatomic gas,

$$\Delta\mathcal{U} = \tfrac{5}{2}P_1V_1\left(\sqrt{\frac{V_1}{V_2}} - 1\right) < 0$$

(*b*) In Example 21 . 3 we found that the work done during such an expansion is

$$W = 2P_1V_1\left(1 - \sqrt{\frac{V_1}{V_2}}\right)$$

Therefore the heat absorbed by the gas is

$$\Delta Q = W + \Delta\mathcal{U} = (2P_1V_1 - \tfrac{5}{2}P_1V_1)\left(1 - \sqrt{\frac{V_1}{V_2}}\right)$$

$$= -\tfrac{1}{2}P_1V_1\left(1 - \sqrt{\frac{V_1}{V_2}}\right) < 0$$

(*c*) If we let $V_2 \rightarrow V$, a variable, and we know that

$$\sqrt{\frac{V_1}{V}} = \frac{T}{T_1} \qquad \text{and} \qquad P_1V_1 = nRT_1$$

then $Q = -\frac{1}{2}nRT_1(1 - T/T_1)$. Therefore

$$nC = \frac{dQ}{dT} = \tfrac{1}{2}nR \qquad \text{or} \qquad C = \tfrac{1}{2}R$$

That is, in order to maintain the particular expansion law $P(V)$ stated, it is necessary to remove heat by cooling the container walls during the expansion, or else P would not decrease rapidly enough. However, since the internal energy of the gas decreases mainly by doing work on the external environment, C is much less than C_V, the case of simple isometric cooling without work.

22 • 6 *quantum theory of specific heats*

The idea that the average energy of a molecule must exceed a certain critical value before some degree of freedom becomes active offers a simple explanation of the variation of specific heats of gases with temperature. It presents the very great difficulty, however, of being contrary to the laws of classical mechanics. According to the classical theory, every degree of freedom should receive its share of energy at all temperatures, just as a pendulum may be set into vibration or a wheel into rotation, however minutely, by applying the slightest impulse. Theoretically, even the rotation of a monatomic molecule, or of a diatomic molecule about the line joining its nuclei, should not be ruled out, although the moment of inertia I in each of these cases is so small that the energy $\frac{1}{2}I\omega^2$ associated with these particular degrees of freedom can be disregarded as negligible.

Such considerations force us to extend the laws of physics if we are to interpret molecular and atomic phenomena satisfactorily. Evidently the energy associated with any vibration or rotation cannot have all values; it can have only certain special values. The difference between these values determines the amount of vibrational or rotational energy that a molecule can gain or lose through impact with other molecules in order to change its motion. This condition was first enunciated by Max Planck in 1900, in connection with his studies of the properties of thermal radiation. It was extended by Einstein in 1907 to vibrating ions and by Walter Nernst in 1911 to atomic vibration and rotating molecules. Applied to a diatomic molecule whose atoms have a natural vibration frequency ν, the condition is that the atoms will not be set into vibration until the molecules receive upon impact a quantity of energy equal to $h\nu$, where h is a universal constant known as *Planck's constant* and having the value 6.626×10^{-34} J/s. The quantity $h\nu$ is called a *quantum of energy*. Quantum theory shows that a diatomic molecule will not start rotating until it receives an energy of at least $h^2/8\pi^2 I$, where I is the moment of inertia about the axis of rotation.

The moment of inertia I of a hydrogen molecule about an axis at right angles to the line connecting its atoms is about 10^{-47} kg · m². Hence, as a simple calculation will show, the rotational quantum $h^2/8\pi^2 I$ is so large that rotation will not start until the temperature of hydrogen exceeds about −230°C; this is in accord with experimental findings (see Fig. 22 . 7). More massive diatomic molecules, such as oxygen, have larger moments of inertia, so that the energy $h^2/8\pi^2 I$ required to start them into rotation is smaller. We should therefore expect their rotational degrees of freedom to become active at much lower temperatures than for hydrogen, a conclusion that is again borne out by experiment.

example 22 • 6 If the vibrational degrees of freedom for hydrogen become fully activated at approximately 6000 K, what is the vibrational frequency of the hydrogen atoms?

solution At this temperature the average molecule has a vibrational energy equal to kT. Therefore, setting $h\nu = kT$, we obtain

$$\nu = \frac{kT}{h} \doteq 1.25 \times 10^{14} \text{ Hz}$$

This answer is quite close to experiment; the correct value is actually 1.2465×10^{14} Hz.

The specific-heat curves in Fig. 22 . 8 show that the change in C_V with temperature is gradual, although *individual* molecules can absorb energy only in discrete packets or quanta. This is because temperature is only an average statistical property. As the temperature of a gas increases, more and more molecules acquire energies significantly larger than the average and thus are able to cause quantum "jumps" when they collide with other mole-

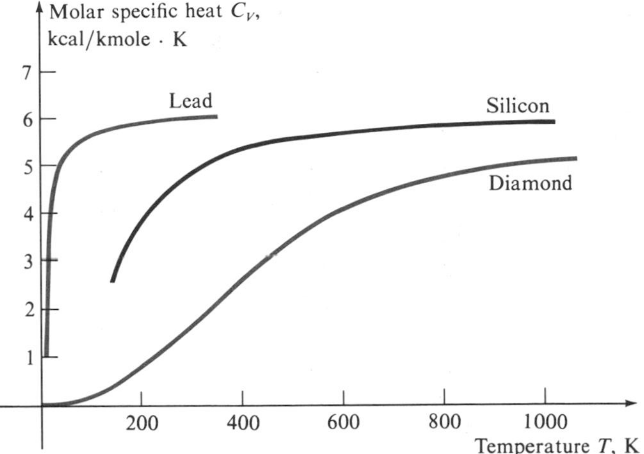

figure 22 • 8 *Temperature variation of the molar specific heat C_V at constant volume. The curves for nearly all solid elements lie between the curve for lead and that for diamond. All are of roughly the same form—that is, they can be made almost to coincide by choosing a suitable temperature scale for each.*

cules in the gas. Finally, when T is large enough, nearly all molecules have made the quantum jump in rotational energy or, at much higher temperatures, in vibrational energy.

In the case of solids at ordinary temperatures, all the internal energy is vibrational. Because the atoms or ions in a solid must be held in place between their neighbors, their only possible thermal motions are vibrations about their equilibrium positions, called *lattice vibrations*. Hence the atoms have only three degrees of freedom, all vibrational, corresponding to the three mutually perpendicular directions of vibration (see Fig. 14 . 14). There are as many molecules in a kilogram-molecular weight of a solid as there are molecules in a kilogram-mole of a gas. Therefore, as in the intramolecular vibration of gas molecules, each vibrational degree of freedom of an atom has an energy associated with it. The internal energy of a solid consisting of N single-atom molecules is

$$\mathcal{U} = 3N(2 \times \tfrac{1}{2}kT) = 3NkT = 3nRT \qquad [22 \cdot 34]$$

and its molar specific heat is

$$C_V = \frac{1}{n}\frac{d\mathcal{U}}{dT} = 3R \doteq 6 \text{ kcal/kmole} \cdot \text{K}$$

This value is, to a first approximation, equal to the empirical value of 6.4 kcal/kmole · K obtained experimentally by Dulong and Petit in 1818 and is close to the value approached by the experimental curves in Fig. 22 . 8 as T increases (see the monatomic solids in Table 22 . 2). In fact, lead and some 57 other elements are found to have approximately the Dulong-Petit value even at moderate temperatures. Since the coefficient of expansion of a solid or a liquid is generally very small, the work done by it during expansion is negligible, and $C_V \doteq C_P$.

The crystalline compounds in Table 22 . 2 also follow the Dulong-Petit principle. To see this it is necessary to recognize that the atoms of each molecule, such as Ni and O in nickel monoxide, occupy separate positions within the nickel monoxide crystal, and each atom has three degrees of vibrational freedom. The internal energy of one nickel monoxide molecule is therefore $6kT$, and $C_V = 6R \doteq 12$ kcal/kmole · K. In fact, the excellent agreement of this prediction with Table 22 . 2 is evidence that such crystals do consist of separate atoms. Otherwise we would expect a nickel monoxide molecule to have at most four degrees of vibrational freedom in a lattice, one of them internal to the molecule, resulting in $C_V = 4R$, which is not the case. The same reasoning can be applied to the triatomic molecules to predict $C_V = 9R$, in close agreement with the experimental values.

It was known very early that nonmetallic elements with low atomic weights and high fusion points, such as carbon, boron, and silicon, are notable exceptions to the Dulong-Petit law at ordinary temperatures. Extended research revealed that at low temperatures the specific heats of all solids examined approach zero, and that it is only at relatively high temperatures that they reach the Dulong-Petit value. As with gases, these variations with temperature have their explanation in quantum theory. If the binding forces acting in the directions of the three degrees of vibrational freedom are different, the atoms will have correspondingly different vibration frequencies in those directions. Hence the energy $h\nu$ imparted to an atom by an impact may be sufficient to start the vibration that characterizes one degree of freedom, but not of another. At low temperatures most collisions are too feeble to excite any of the degrees of vibrational freedom. Accordingly, the specific heat approaches zero with decreasing temperature, since energy transfer does not occur. The fact that elements having low atomic weights and high fusion points do not exhibit the Dulong-Petit value at moderate temperatures is also explained. The atoms of such an element are of small mass and are bound together by relatively large forces; their natural frequencies are therefore large, and a high temperature is needed to activate all degrees of freedom.

PROBLEMS

22 • 1–22 • 2 atomic theory and ideal gases

22 • 1 Numerous gaseous compounds of nitrogen and oxygen may be formed. Given the density of diatomic oxygen as 32 units, find the chemical formulas of the following compounds: (*a*) nitrogen dioxide, density 46; (*b*) nitrogen trioxide, density 76; (*c*) nitrogen pentoxide, density 108.

22 • 2 Calculate the volume occupied at standard temperature and pressure (*a*) by 2.016 g of hydrogen. (*b*) By 32.00 g of oxygen. (*c*) By 30.01 g of nitric oxide. Explain the connection between the results.

22 • 3 Compute *Loschmidt's constant*, the number of molecules per cubic centimeter of ideal gas at 0°C and 1 atm, and find the corresponding numbers of molecules (*a*) at 10^{-9} atm, a very high vacuum for commercial work. (*b*) At 10^{-13} atm, about the lowest pressure attainable by the most refined techniques.

22 • 4 (*a*) What is the density of nitrogen at standard temperature and pressure? What is the mass of a nitrogen molecule? (*b*) How many molecules are there in a gram-molecular volume? (*c*) What is this volume in liters?

22 • 5 (*a*) Prove that if the pressure exerted by a gas were the result of repulsions between the molecules, the repulsive force would have to vary inversely with the intermolecular distance for Boyle's law to hold. (*b*) Prove that, for such a force law, the effect of the distant parts of the gas on a given molecule would be greater than that of the neighboring parts, and consequently the pressure would depend not only on the density of the gas, but on the form and dimensions of the containing vessel.

22 • 6 A 2-mg spherical pellet is moving back and forth between two parallel walls that are 10 cm apart. The speed of the pellet is 2×10^3 cm/s and its impacts with the wall are completely elastic. (*a*) What average force is needed to keep the walls from moving under the influence of the impacts when the diameter of the pellet is negligibly small? (*b*) What average force is needed when the diameter is 4 mm?

22 • 7 Suppose the actual speeds of 10 molecules at a given instant are 1, 2, 3, . . . , 10 m/s, respectively. (*a*) Compute their root-mean-square speed v_{rms}. (*b*) What is their arithmetic-mean speed $\bar{v}$? (*c*) If each molecule is of mass *m*, what is the average kinetic energy? (*d*) Explain why the root-mean-square speed for gas molecules will always exceed the arithmetic-mean speed.

22 • 8 The density of nitrogen at 0°C and 1 atm is 1.25 kg/m^3. (*a*) Find the mean-square speed of nitrogen molecules at 0°C. (*b*) Calculate the Boltzmann constant R/N_A using Avogadro's constant and the molecular weight of nitrogen. (*c*) Natural nitrogen actually consists of two isotopes with atomic masses in the ratio 14.008/15.005 and with relative abundances of 99.63 and 0.37 percent. If this were taken into account, would it change your calculated results appreciably?

***22 • 9** Write a computer program to take as input data the different speeds v_i of $N = \Sigma_i N_i$ molecules, where the N_i molecules all have the same speed v_i. Then compute the root-mean-square speed v_{rms} and the average molecular speed $\bar{v}$. Test your program (*a*) with the results obtained in Prob. 22 . 7 and (*b*) for the case in which $v_i = 10i$ m/s and N_i is the largest integer,

$$N_i \leq \frac{100}{i^2 - 20i + 26}$$

where $i = 1, \ldots, 9$. The general formula for a weighted mean of any variable x_i is

$$\bar{x} = \frac{1}{N} \sum_i N_i x_i \qquad N = \sum_i N_i$$

The formula that gives the values of $N_i = f(i)$ is known as a *frequency-distribution function.*

22 • 3 temperature and heat

22 • 10 (*a*) Compute the average kinetic energy of translation of a molecule of any ideal gas at 300 K. (*b*) Find the total kinetic energy of translation of all the molecules in 1 liter of ideal gas at 0°C and 1 atm.

22 • 11 The temperature T of 1 kmole of hydrogen is raised one degree to $(T + 1)$ K. (*a*) Find the increase in the total kinetic energy of translation associated with the random motion of the molecules. (*b*) Find the fractional increase in the root-mean-square speed of the molecules.

22 • 12 If all the molecules of air in a massless container at room temperature, 22°C, were to suddenly move upward together, how high would they lift the container against gravity?

22 • 13 A 1-g mass of nitrous oxide was introduced into an evacuated spherical bulb of average inside diameter 10 cm, and the bulb was then sealed off. What pressure was exerted by the gas on the walls of the bulb when the temperature was 30°C?

22 • 14 Suppose 1 kmole of air has an average mass M and the average mass of an "air molecule" is m. (*a*) Prove that the isothermal barometric equation may be written in the alternative forms

$$P(y) = P_0 \exp\left(-\frac{Mgy}{RT}\right) = P_0 \exp\left(-\frac{mgy}{kT}\right)$$

where P_0 is the atmospheric pressure at sea level, $y_0 = 0$. (*b*) Prove that the ratio P/P_0 is equal to the ratio of the number densities of the molecules at the two levels. (*c*) What is the physical significance of the quantities Mgy, mgy, RT, and kT? (*d*) If the air is at 17°C and has an average molecular weight $M_0 = 29$, at what altitude will the pressure be $\frac{1}{2}P_0$?

***22 • 15** The temperature of the upper atmosphere varies considerably with altitude, time of day, and other factors. Given tabulated values of temperature T as a function of altitude y, draw a flow diagram for computing $P(y)$ by numerical quadrature. You may want to compare your results against those in the standard-atmosphere reference tables.

22 • 16 The differential steam calorimeter, devised by Joly in 1886, is used to measure molar specific heats $\mathcal{C}_V$. Two hollow copper bulbs hang in a closed chamber from the arms of a sensitive balance. One bulb contains n moles of gas and the other is evacuated. Steam is admitted to the chamber, and ΔM is the difference in the mass of steam condensing on the two bulbs. If T is the initial temperature in the chamber, and T_c and L_v are the condensation temperature and heat of vaporization of water, prove that

$$\mathcal{C}_V = L_v \frac{\Delta M}{T_c - T}$$

22 • 4 applications of the ideal-gas model

22 • 17 The root-mean-square speed of a hydrogen molecule at 0°C is 1.838 km/s. (*a*) Assuming the validity of the law of equipartition, calculate the v_{rms} for an oxygen molecule at this temperature. (*b*) When the temperature is 364 K, what is the arithmetic-mean speed of the hydrogen molecules?

22 • 18 A mixture of 1 g of nitrogen and 1 g of helium are introduced into an evacuated spherical bulb 20 cm in average inside diameter, and the bulb is then sealed off. What pressure does this mixture of gases exert on the walls of the bulb when the temperature is 25°C?

22 • 19 In 1 cm^3 of a certain mixture of nitrogen and carbon dioxide at 0°C there are 2×10^{20} molecules of nitrogen and 5×10^{19} molecules of carbon dioxide. (*a*) Compute the partial pressure due to each gas. (*b*) What is the total pressure?

22 • 20 A given mass of air is in the state (P_1,V_1,T_1), and after the oxygen has been removed the state of the remaining nitrogen is (P_2,V_2,T_2). Show that the ratio of the number of moles of nitrogen to the number of moles of oxygen in this sample is equal to

$$\frac{P_2T_1}{P_1T_2 - P_2T_1}$$

22 • 21 (*a*) Prove that the density of a homogenous mixture of inert gases is equal to the sum of the densities of the constituents. (*b*) Dry air at 0°C and 1 atm near ground level consists mainly of five constitutents having partial pressures and molecular weights (on the physical scale) as follows:

N_2 (0.78 atm, 28.02)
O_2 (0.21 atm, 32.01)
Ar (0.0094 atm, 39.95)
CO_2 (0.0003 atm, 44.02)
H_2 (0.0001 atm, 2.017)

Compute the density of this air.

22 • 22 Uranium hexafluoride, which is a gas at ordinary temperatures, consists mainly of two compounds, $^{235}U\,^{19}F_6$ and $^{238}U\,^{19}F_6$, of approximate molecular weights 349 and 352 and relative abundances of 0.72 and 99.28 percent. A very small fraction of a quantity of this gas is allowed to effuse through a porous plug into a vacuum. (*a*) Compute the ratio of the numbers of molecules of the two species effusing per unit time. (*b*) Find the relative abundance of $^{235}U\,^{19}F_6$ in the gaseous mixture after effusion. (*c*) How many successive effusions of progressively smaller portions of the gas would be needed to increase the abundance to 50 percent?

22 • 23 The separation of the two isotopes ^1H and ^2H (deuterium) of natural hydrogen by the effusion method is comparatively easy because of the large fractional difference in their molecular masses, which are in the ratio 2.016/4.029. The respective relative abundances of ^1H and ^2H are 99.985 and 0.015 percent. (*a*) In what ratio, $\Delta N_1/\Delta N_2$, will the two kinds of molecules *initially* effuse per second into a vacuum? (*b*) What are their initial relative abundances in the effused mixture? Why initial?

22 • 24 In a Brownian-movement experiment with smoke particles of mass 4.5×10^{-14} g suspended in air at 27°C, the average speed $\bar{v}$ of the particles was found to be 1.6 cm/s. Using these data and the law of equipartition of energy, calculate an approximate value for Avogadro's constant.

22 • 5–22 • 6 equipartition and specific heats

22 • 25 A smooth bar of hexagonal cross section has three close-fitting hexagonal nuts and three loose-fitting circular rings mounted on it. These fittings are all free to slide along the bar but cannot fall off the ends, which are flanged. (*a*) How many degrees of freedom does the system have if the bar remains stationary? (*b*) How many degrees of freedom does it have if the bar is free to move?

22 • 26 For liquids and solids $C_P \doteq C_V$, since their volumes change so little with temperature. (*a*) What is the value of C_V for water? (*b*) Assuming that the three atoms in a water molecule are not collinear and that there is vibrational energy associated with each pair of atoms at ordinary temperatures, what value would you predict for C_V? (*c*) How much potential energy per molecule must be attributed to intermolecular interactions between each water molecule and its nearest neighbors for the value in part (*b*) to agree with (*a*)?

22 • 27 At high altitudes and low pressures the diatomic constituents of the atmosphere dissociate into single atoms, with the fraction α of the diatomic molecules which dissociate increasing with altitude. Find the ratio γ of specific heats of these constituents as a function of α. (*Hint:* Consider the degrees of freedom in 1 mole of the mixture.)

22 • 28 Consider a rotating dumbbell molecule in which the atomic masses m_1 and m_2 are unequal and separated by distance r. (*a*) Show that $I = \mu r^2$, where μ is the reduced mass, given by

$$\frac{1}{\mu} = \frac{1}{m_1} + \frac{1}{m_2}$$

(*b*) Compute the minimum rotational frequency of the hydroxyl molecule OH, for which $r = 0.97$ Å. (*c*) Compute the minimum rotational frequency of the nitric oxide molecule NO, for which $r = 1.15$ Å.

22 • 29 For chloroform vapor at 100°C, $\gamma = 1.15$. (*a*) If the vapor is kept at constant volume and heated, what fraction of the heat goes into increasing the kinetic energy of translation of the molecules? (*b*) Where does the remaining heat go?

22 • 30 If we set $h^2/8\pi^2 I = kT$ for a rotating molecule, then T gives a measure of the temperature at which rotational degrees of freedom are becoming activated. Use the data in Prob. 22 . 28 to find the temperatures T for OH and NO molecules.

22 • 31 If the activation temperature for vibration of the O_2 molecule is 2260°C, find the vibration frequency.

22 • 32 A solid crystalline substance consists of molecules having A atoms each, and every degree of freedom of the molecule is considered vibrational. Find C_V and compare your answer with the experimental values in Table 22 . 2.

22 • 33 A monatomic crystalline solid begins to melt when the resultant vibrational amplitude of a molecule of mass m reaches the order of 20 percent of the intermolecular spacing a. Show that the vibration frequency can be determined from the fusion temperature T_f as

$$\nu = \frac{5}{a}\sqrt{\frac{6kT_f}{m}}$$

and compare this with the case of copper, for which $T_f = 1356$ K, $a = 3.6$ Å, and the activation temperature is 116 K.

answers

22•1 (*a*) NO_2; (*b*) N_2O_3; (*c*) N_2O_5
22•2 $V = 22.4$ liters in each case, by Avogadro's law
22•3 $N_L = 2.686 \times 10^{19}$ mol/cm^3
22•4 (*a*) $\rho = 1.25$ kg/m^3, $m = 4.65 \times 10^{-26}$ kg; (*b*) $N = N_A/1000$; (*c*) 22.4 liters
22•6 (*a*) $F = 800$ dynes; (*b*) $F = 833$ dynes
22•7 (*a*) $v_{rms} = 6.20$ m/s; (*b*) $\bar{v} = 5.50$ m/s; (*c*) $K = 19.25\{m_0\}$ J
22•8 (*a*) $\overline{v^2} = 24.3 \times 10^8$ cm^2/s^2; (*b*) $k = M_0P/N_A\rho T = 1.38 \times 10^{-23}$ J/mol · K; (*c*) no
22•9 (*b*) $\bar{v} = 5.16$ m/s
22•10 (*a*) $\overline{K}_{tr} = 6.21 \times 10^{-21}$ J/mol; (*b*) $K = 152$ J
22•11 (*a*) $\Delta K = 12{,}470$ J; (*b*) $\Delta v_{rms}/v_{rms} \doteq \frac{1}{2}T$
22•12 $\Delta y = 14.2$ km
22•13 $P = 1.08$ atm
22•14 (*c*) They are measures of molar potential energy, molecular potential energy, molar kinetic energy, and molecular kinetic energy, respectively; (*d*) $y = 5.88$ km
22•17 (*a*) $v_{rms} = 0.461$ km/s; (*b*) $\bar{v} = 1.96$ km/s
22•18 $P = 1.67$ atm
22•19 (*a*) $P = 7.44$ atm for N_2, 1.86 atm for CO_2; (*b*) $P_{tot} = 9.30$ atm
22•21 $\rho = 1.293$ kg/m^3
22•22 (*a*) $N_{235}/N_{238} = 0.00728$; (*b*) 0.723 percent; (*c*) 1151 effusions
22•23 (*a*) $\Delta N_1/\Delta N_2 = 9423$; (*b*) 99.9894 percent and 0.0106 percent
22•24 $N_A \doteq 5.5 \times 10^{26}$ mols/kmole
22•25 (*a*) $f = 9$; (*b*) $f = 15$
22•26 (*a*) $C_V = 18$ kcal/kmole; (*b*) $C_V = 12$ kcal/kmole; (*c*) $U = 3kT$
22•27 $\gamma = (7 + 3\alpha)/(5 + \alpha)$
22•28 (*b*) $\nu = 11.3 \times 10^{11}$ Hz; (*c*) $\nu = 1.02 \times 10^{11}$ Hz
22•29 (*a*) 22.5 percent of the heat
22•30 $T_{OH} = 27.4$ K, $T_{NO} = 2.47$ K
22•31 $\nu = 4.71 \times 10^{13}$ Hz
22•32 $C_V = 3AR$
22•33 $\omega = 1.43 \times 10^{13}$ rad/s, $2\pi kT/h = 1.52 \times 10^{13}$ rad/s

CHAPTER TWENTY-THREE

phases of matter

The ordinary gaseous and ordinary liquid states are, in short, only widely separated forms of the same condition of matter, and may be made to pass into one another by a series of gradations so gentle that the passage shall nowhere present any interruption or breach of continuity. From carbonic acid as a perfect gas to carbonic acid as a perfect liquid, the transition we have seen may be accomplished by a continuous process, and the gas and liquid are only distinct stages of a long series of continuous physical changes. Under certain conditions of temperature and pressure, carbonic acid finds itself, it is true, in what may be described as a state of instability, and suddenly passes, with the evolution of heat, and without the application of additional pressure or change of temperature, to the volume, which by the continuous process can only be reached through a long and circuitous route. In the abrupt change which here occurs, a marked difference is exhibited, while the process is going on, in the optical and other physical properties of the carbonic acid which has collapsed into the smaller volume, and of the carbonic acid not yet altered. There is no difficulty here, therefore, in distinguishing between the liquid and the gas. But in other cases the distinction cannot be made; and under many of the conditions I have described it would be vain to attempt to assign carbonic acid to the liquid rather than the gaseous state.

Thomas Andrews, On the Continuity of the Gaseous and Liquid States of Matter, Bakerian lecture, 1869

Although we could ignore the attractive and repulsive forces between molecules as a factor in the internal-energy changes of near-ideal gases, it is these forces that account for the structural changes in a substance as it passes through different phases—from a solid, to a liquid, to a gas. As we shall see, these intermolecular forces also account for the observed deviations of real gases from the ideal-gas law. In discussing the latent heats of transition from one phase to another, we did not consider the dynamics of phase transitions. However, when we include the forces between molecules, we can explain these transitions to some extent in terms of atomic-molecular concepts and kinetic theory (a complete explanation requires quantum theory).

Intermolecular forces actually consist of two different interactions between pairs of neighboring molecules. The first is a weak, relatively long-range attractive force known as the *van der Waals force,* and the second is a very strong, short-range repulsive force which prevents the molecules from coming too close together. Both component forces are electrical in origin and depend on the outer electronic structure of the given species of molecule. Experimental evidence for the existence of these components comes from the study of the *Joule-Thomson effect,* the cooling experienced by a compressed gas when released through a nozzle, and from the study of phase transitions themselves. In fact, the Joule-Thomson effect provided the means for liquefying gases.

The extent to which the ideal-gas laws fail to describe the behavior of real gases (real matter in the gaseous phase) was investigated first by Henri Regnault, and later by Émile Amagat. Boyle himself had noticed small departures from constancy of the product of pressure and volume, but he ignored the failure of his law to hold accurately because of its simplicity and general usefulness. Among the various subsequent attempts to formulate an equation of state to describe more than one phase of a given substance, one of the most successful is the van der Waals equation. This equation, despite some inadequacies, is relatively simple and gives a good qualitative description of the physical behavior of the liquid and gaseous phases of all substances. In this chapter we shall derive the van der Waals equation as another instance of the applicability of kinetic theory. Finally, we shall examine some empirical relations describing the dependence of phase transitions on state variables and energy transfer.

23 . 1 real gases

If Boyle's law did apply accurately to a real gas, then for a gas at temperature T_0 the product $Pv = RT_0$ (where v is the molar specific volume) should be a constant at all pressures. In fact, Regnault and Amagat found that when a gas is tested and Pv is plotted over a wide range of pressures and at different values of T_0, we obtain curves like those of Figs. 23 . 1 and 23 . 2 instead of the horizontal line predicted by Boyle's law. When the value of Pv decreases at first, as is generally the case, we see that the gas is *more* compressible than an ideal gas, suggesting the presence of attractive intermolecular forces which aid compression. However, when the pressure exceeds a certain value, depending on both the gas and its temperature, then Pv increases steadily with pressure, suggesting the presence of shorter-range repulsive intermolecular forces which resist compression.

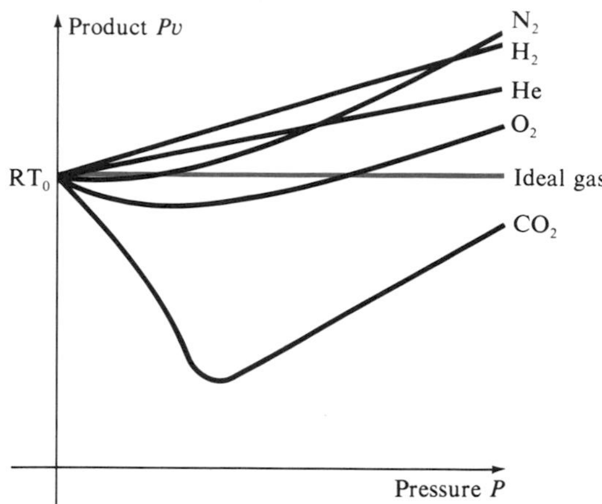

figure 23 • 1 *Variations of the product Pv with pressure for various gases, all kept at the same temperature T_0*

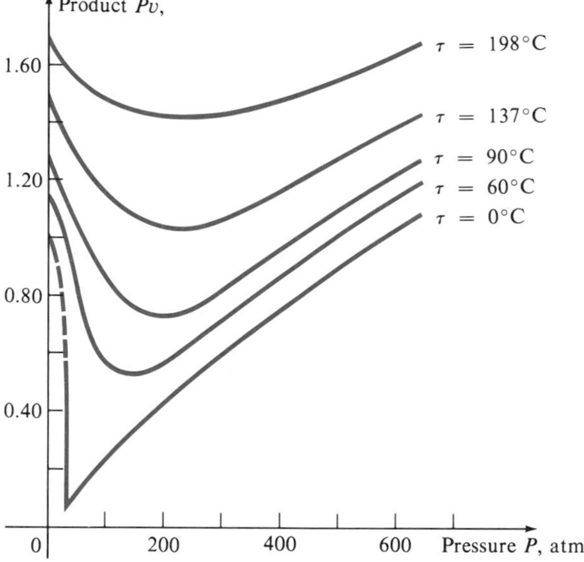

figure 23 • 2 *Variations of Pv with pressure for carbon dioxide, at various constant temperatures*

As the temperature increases, the marked drops in the curves disappear (Fig. 23 . 2). At sufficiently high temperatures, depending on the gas, the curves have only an upward slope. This is true for hydrogen and helium even at relatively low temperatures. Incidentally, Fig. 23 . 1 also illustrates the remarkable property of gases that makes them so valuable as thermometric substances: as the pressure approaches zero, the product Pv approaches the same value RT_0 for *all* gases at the same temperature.

In considering experimental results such as these, the question naturally arises of what happens to the properties of a gas when it approaches the liquid phase. When a liquid such as ethyl alcohol or ordinary (diethyl) ether is heated in a strong sealed tube, it vaporizes steadily as the temperature and pressure rise to a certain point. But then the interface between liquid and gas grows indistinct and completely disappears; the densities of the liquid and gas have become the same and the distinction between the two phases has vanished (see the chapter-opening quotation).

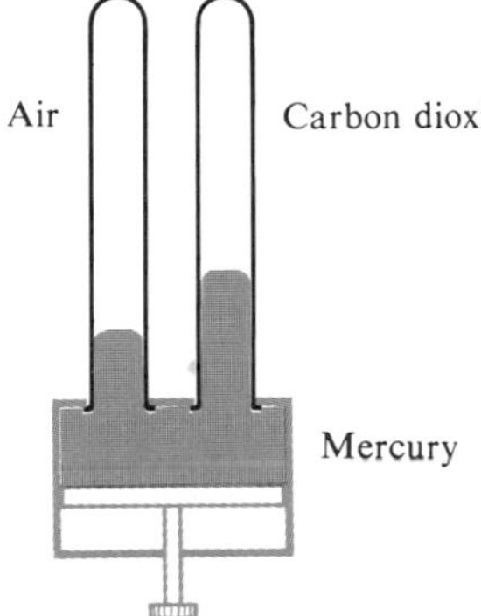

figure 23 • 3 *Schematic diagram of Andrews' apparatus. The gas is compressed isothermally by slowly forcing the mercury up into the tubes. One of the tubes contains air, which is used to measure the pressure applied to the carbon dioxide.*

Results typical of the kind Andrews obtained with the apparatus in Fig. 23 . 3 are shown in Fig. 23 . 4, which is a contour plot of the thermodynamic states of carbon dioxide in thermodynamic coordinates PvT. The curve for $\tau = 48.1°C$ is nearly a perfect hyperbola, $P \sim 1/v$, since carbon dioxide behaves almost like an ideal gas at this temperature. At lower temperatures the deviations from Boyle's law become greater, and the isotherms change in character, taking a form such as that at $\tau = 35.5°C$. The *slope* of this curve does not decrease monotonically with decreasing v, but has a local maximum (inflection point), while at higher pressures and smaller specific volumes it becomes very steep. Here the curves resemble the isotherm for a liquid, since a relatively large increase in pressure produces only a small reduction in volume. However, no sign of liquefaction of the carbon dioxide could be noted at this temperature or above it, no matter how much the pressure was increased.

At temperatures below 31.0°C there is a radical change in the shape of the isotherms. Let us trace the curve for 21.5°C along the path $ABDE$ shown in Fig. 23 . 5.

figure 23 • 4 *Isotherms for carbon dioxide*

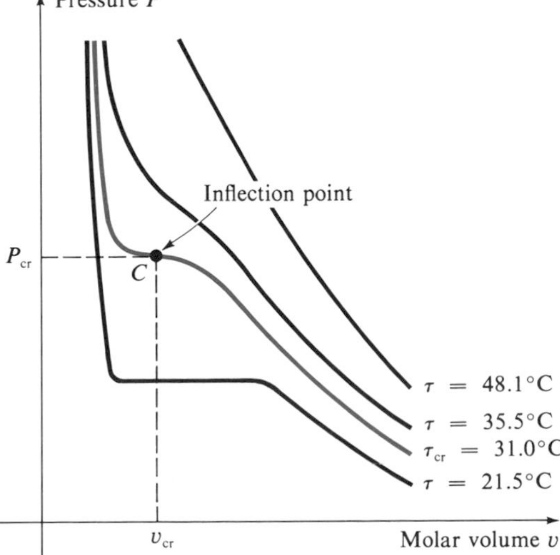

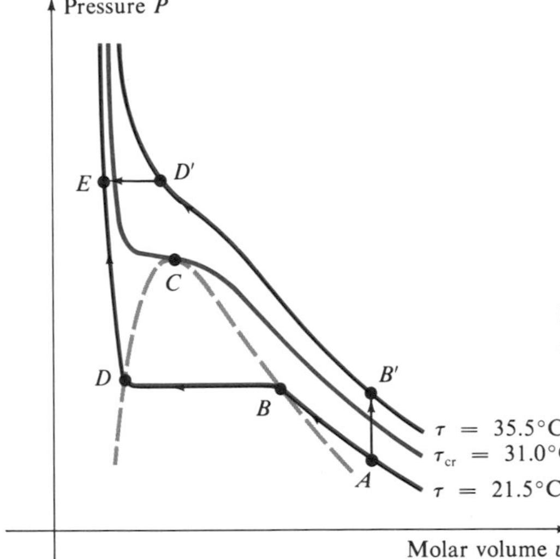

figure 23•5 *Two paths in the Pv plane for liquefaction of carbon dioxide: ABDE and AB′D′E*

As $A \longrightarrow B$, the pressure on the gas is increased and the volume diminishes rapidly.

As $B \longrightarrow D$, the gas is further compressed and some of it condenses and liquefies, so that considerable volume change takes place without any change of pressure; the gas and liquid coexist with a visible boundary between them

For $D \longrightarrow E$, the carbon dioxide is again homogeneous at D, but in the liquid phase; beyond this point the isotherm is almost vertical, since a liquid is not easily compressed

This behavior is again typical of a pure substance.

The dashed curve *BCD* in Fig. 23 . 5 delineates the region of two-phase liquid-vapor equilibrium states; the difference in specific volume between the two phases becomes smaller and smaller as the temperature increases. At last, when the contour $\tau_{cr} = 31.0°C$ is reached, the horizontal part of the isotherm disappears altogether, and *no* separation into liquid and gas can be effected, however much the pressure is increased. The temperature at which this occurs is called the *critical temperature* τ_{cr}. The contour corresponding to the critical temperature is the *critical isotherm;* and the point of inflection *C* on the critical isotherm, the point where the densities of the liquid and of the gas in equilibrium with it become equal, is called the *critical point.* The values of the state variables at *C* are called the *critical constants.*

Above the critical temperature τ_{cr} the gaseous and liquid phases cannot be separated. Indeed, it is possible to reduce a substance from any point *A* at which it would clearly be regarded as a gas to a point *E*, where it is in the dense, almost incompressible condition that characterizes a liquid, without the two phases being distinct at any time. All that is necessary is to vary the pressure, volume, and temperature in such a way as to avoid the region bounded by the dashed curve *BCD*. The curve *AB′D′E* represents such a path; the carbon dioxide gas is first heated to 35.5°C, then compressed isothermally to *D′*, and finally, simultaneously compressed and cooled back to 21.5°C, bringing it to a state *E* in the liquid phase without ever having passed through a distinct phase transition.

table 23 • 1 *Critical constants of selected substances*

substance	*critical temperature* τ_{cr}, °C	*critical pressure* P_{cr}, *atm*	*critical density* ρ_{cr}, g/cm^3
helium	−268	2.26	0.0693
hydrogen	−240	12.8	0.0310
nitrogen	−147	33.5	0.311
oxygen	−118	50.1	0.41
carbon dioxide	31.04	72.8	0.468
ammonia	132	111	0.235
uranium hexafluoride	230	45.5	—
water	374	218	0.326
sulfur	1040	116	—

It is customary, following a suggestion by Andrews, to give the name *vapor* to a substance in the gaseous state when it is below its critical temperature and to restrict the term *gas* to a substance that is gaseous above the critical temperature. A vapor that is not in contact with its liquid is said to be *unsaturated.* At *B* in Fig. 23 . 5, and anywhere on the isothermal-isobaric line between *B* and *D*, where vapor and liquid are in equilibrium, the vapor is said to be a *saturated vapor.* The constant pressure corresponding to the line *BC* is called the *vapor pressure* P_v of the substance for the temperature in question. For any pure substance, the vapor pressure is a function only of temperature, and not of volume; that is, in a closed vessel containing a liquid and its vapor in equilibrium at a fixed temperature, the pressure is independent of the relative amounts of liquid and vapor present.

Figure 23 . 6 illustrates a typical surface of thermodynamic states of a real substance with isothermal contours projected on the *Pv* plane. The

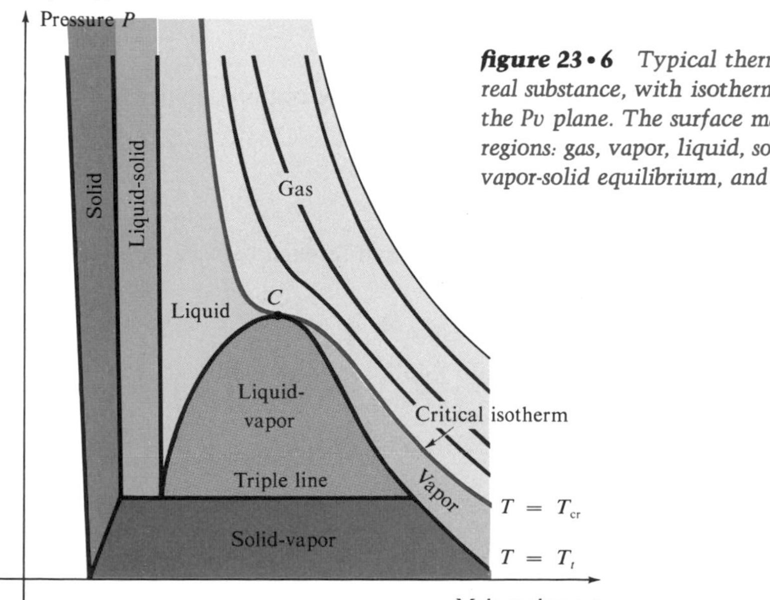

figure 23 • 6 *Typical thermodynamic states of a real substance, with isothermal contours projected on the Pv plane. The surface may be divided into seven regions: gas, vapor, liquid, solid, vapor-liquid equilibrium, vapor-solid equilibrium, and solid-liquid equilibrium.*

table 23 • 2 *Triple-point data*

equal pressures: 1.000 mm Hg = 133.3 Pa = 0.001316 atm

substance	*temperature, °C*	*pressure, mm Hg*
ammonia	−77.75	45.6
carbon dioxide	−56.60	3880
neon	−248.6	324
nitrogen	−209.97	95
oxygen	−218.8	1.1
water	+0.01	4.58

region of vapor-solid equilibrium occurs at temperatures lower than the *triple-point temperature.* For any substance there is obviously one isotherm whose horizontal portion represents the boundary between the solid-vapor and the liquid-vapor regions. Only for points on this line, called the *triple line,* can the vapor, liquid, and solid phases of a substance coexist in equilibrium, and at any point on the triple line, there is only one temperature and one pressure at which all three phases coexist in equilibrium. Such a point is called the *triple point.* As noted earlier, the triple point of water, $\tau_t = 0.0100°\text{C}$ and $P_t = 0.006$ atm has been adopted as the single fixed point for fundamental thermometry. Data for several other substances are listed in Table 23 . 2. The specific volume is not fixed, but depends on the location of the triple point on the triple line.

It is also possible to have temperatures and pressures at which some of the solid melts into liquid, but no vapor is formed, and the solid and liquid coexist in equilibrium (region 7 in Fig. 23 . 6). The phases of matter are more closely related than they may appear, and to describe them with precision it is necessary to specify their state variables and properties, such as compressibility.

example 23 • 1 Under isothermal conditions the compressibility κ of a substance is given by

$$\kappa = -\frac{dv/dP}{v}$$

Show that $d(Pv)/dP$ is proportional to the difference between the ideal and real compressibility of a gas.

solution Differentiating Pv with respect to P yields

$$\frac{d(Pv)}{dP} = v + P\frac{dv}{dP} = Pv\left(\frac{1}{P} + \frac{1}{v}\frac{dv}{dP}\right) = Pv\left(\frac{1}{P} - \kappa\right)$$

and since $\kappa_{\text{ideal}} = 1/P$,

$$\frac{d(Pv)}{dP} = Pv\,(\kappa_{\text{ideal}} - \kappa)$$

23 • 2 *intermolecular forces: the Joule-Thomson effect*

The first investigations of intermolecular forces in gases were made by Joseph Gay-Lussac, followed several decades later by Joule's free-expansion experiments. Joule employed the apparatus shown in Fig. 23 . 7, which consisted of a copper vessel A, containing air under a pressure of 22 atm, connected to another copper vessel B which had been evacuated and then immersed in water. When the stopcock was opened, the air would expand and fill both vessels. Since the air was expanding freely into a rigid container, no energy was being expended on external work by the system as a whole. However, any long-range attractive forces between molecules would slow the molecules down as they separated, and this reduction in their kinetic energy should be reflected by a drop in the temperature of the gas. Joule hoped to demonstrate the existence of intermolecular forces by measuring such temperature changes.

Even if there were no intermolecular forces, Joule expected to find some temperature change, since the gas remaining in vessel A did work on the gas drifting into vessel B. Once the gas molecules were inside vessel B, the kinetic energy of their drift velocity would soon become random as a result of collisions, and this increase in their kinetic energy should appear as a rise in temperature. However, Joule found that the drop in the temperature of vessel A was practically equal to the rise in the temperature of vessel B. He concluded that once the temperature of the two vessels equalized through conduction, there would be no observable temperature change for the gas as a whole, but this did not preclude the possibility of intermolecular forces that might be detected with a more sensitive experiment. In fact, the heat capacity of Joule's apparatus was so large compared with that of the gas that it would have taken a change of several degrees in the gas temperature to produce any noticeable effect on the water temperature.

The first successful attempt to show the existence of intermolecular forces in gases was the "porous-plug experiment" conducted by Thomson (Lord Kelvin) and Joule in 1852. Gas, kept under constant positive gage pressure by a pump, was allowed to seep steadily through a porous plug of cotton into a second enclosure, in which a constant lower pressure was main-

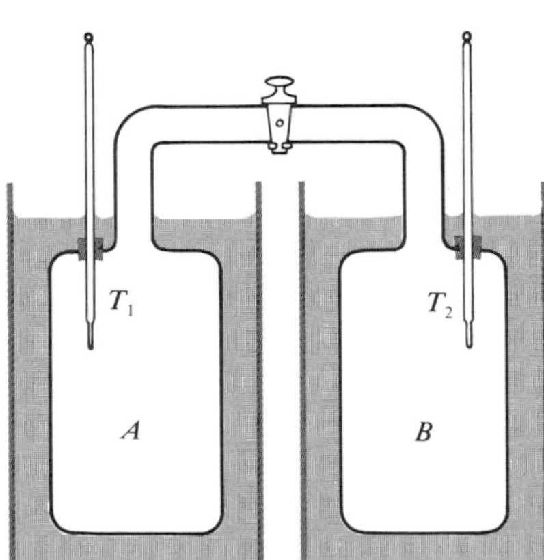

figure 23 • 7 *The Joule free-expansion experiment*

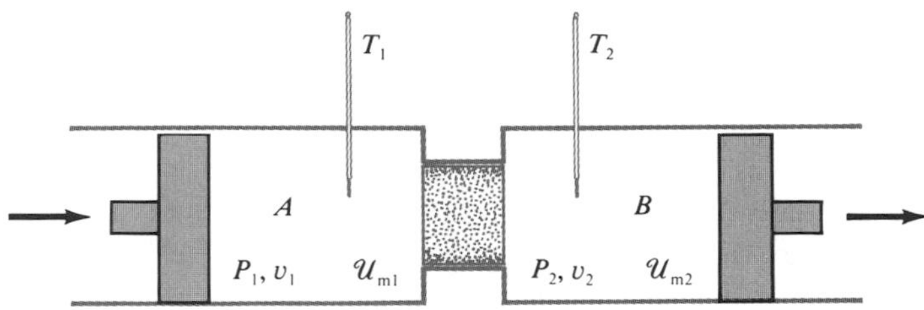

figure 23 • 8 *Schematic diagram of the porous-plug experiment. The cylinders and pistons are assumed to be impervious to heat.*

tained (Fig. 23 . 8). The pores of the plug were large enough in relation to the mean free path of the molecules that a flow, rather than an effusion, of the gas occurred, but they were small enough for the viscous resistance to reduce the kinetic energy of the flow to a negligible amount. Such a procedure is called a *throttling process;* it was made adiabatic by insulating the apparatus.

In one of their experiments, Joule and Thomson found that when air expanded from a pressure of 2.26 atm and a temperature of 15.0°C to a pressure of 1.00 atm, the temperature fell to 14.6°C. All subsequent porous-plug experiments have similarly shown that every gas except hydrogen and helium cools when expanded at ordinary temperatures and pressures. This is known as the *Joule-Thomson effect.* The drop in temperature is found to be nearly proportional to the difference between the pressures on the two sides of the plug and, over a considerable range of temperatures, inversely proportional to the square of the absolute temperature.

The ratio dT/dP, the change in temperature per unit difference in pressure on the two sides of the plug, is termed the *Joule-Thomson coefficient.* Since the pressure difference in the direction of expansion is always negative, a positive value of dT/dP indicates cooling upon throttling and a negative value indicates warming. For example, the Joule-Thomson coefficients for air and for helium, given initial conditions of 25°C and 1 atm, are 0.232 and −0.0624 K/atm, respectively.

Even hydrogen and helium are found to cool by their own expansion if their initial temperatures are within a certain range of values, called *Joule-Thomson inversion temperatures.* The inversion temperatures for any gas depend on the pressure, as shown for hydrogen in Fig. 23 . 9. Hydrogen will not cool on expansion at any pressure greater than 160 atm, and no cooling will occur if the initial temperature in chamber A exceeds 203 K = −71°C. Every point in the PT plane corresponds to a pair of initial thermodynamic coordinates (P,T), and the location of a point relative to the inversion curve tells us whether the temperature in chamber B will be less than, equal to, or greater than the temperature in chamber A after throttling. For helium the maximum inversion temperature is 43 K = −230°C, and for air it is 603 K = 330°C.

The Joule-Thomson effect confirmed the existence of intermolecular forces in gases. To see this let us relate the intermolecular forces to the Pv curves of Figs. 23 . 1 and 23 . 2 through the first law of thermodynamics.

Suppose 1 mole of a gas has an internal energy $\mathcal{U}_{m1}$ on entering the porous plug and an internal energy $\mathcal{U}_{m2}$ on leaving it, where the subscript denotes molar quantities. Then, since the throttling process is adiabatic, the work per mole done on the system equals the increase in internal energy:

$$-\Delta W = P_1 v_1 - P_2 v_2 = \mathcal{U}_{m1} - \mathcal{U}_{m2} \qquad [23 \cdot 1]$$

The left piston does work $P_1 v_1$ on each mole of gas in forcing it slowly through the plug, and the right piston absorbs work $P_2 v_2$ from the gas emerging on the right side of the plug.

If we set u_m equal to the molar potential energy of the intermolecular interaction and k_{tr} equal to the molar translational kinetic energy, then we can write Eq. [23 . 1] in the form

$$\Delta \mathcal{U}_m = \Delta k_{tr} + \Delta u_m = -\Delta(Pv) \qquad [23 \cdot 2]$$

When a gas expands freely into a vacuum under adiabatic conditions, then no work is done; hence $\Delta k_{tr} = -\Delta u_m$, and in the absence of intermolecular forces there would be no temperature change. The fact that temperature changes do occur in the Joule-Thomson process proves that such forces must exist in real gases.

As Fig. 23 . 1 shows, at low pressures and ordinary temperatures the product Pv increases as the pressure decreases for all gases except hydrogen and helium. Therefore, in the throttling process, $\Delta(Pv) > 0$, and from Eq. [23 . 2],

$$\Delta u_m < -\Delta k_{tr} = -\tfrac{3}{2} R \, \Delta T \qquad [23 \cdot 3]$$

If we assume for these gases and conditions that attractive intermolecular forces predominate, then as the molecules separate, Δu_m becomes positive and the temperature change ΔT must be negative. Cooling upon throttling should therefore be expected, and this is what is observed. In regions where

figure 23 • 9 *Inversion curve for hydrogen: a graph of $dT/dP = 0$ versus initial values of P and T before throttling*

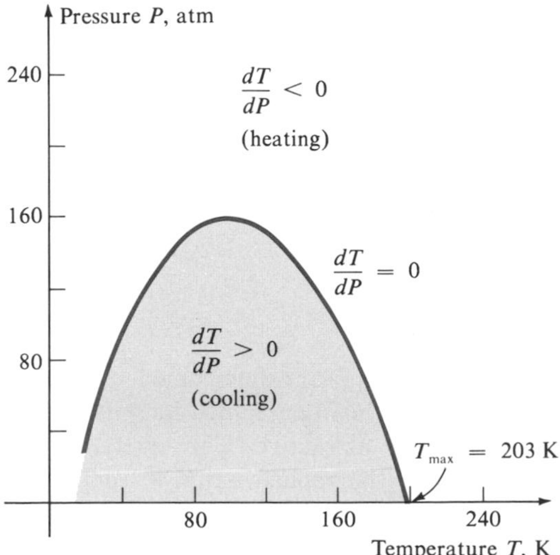

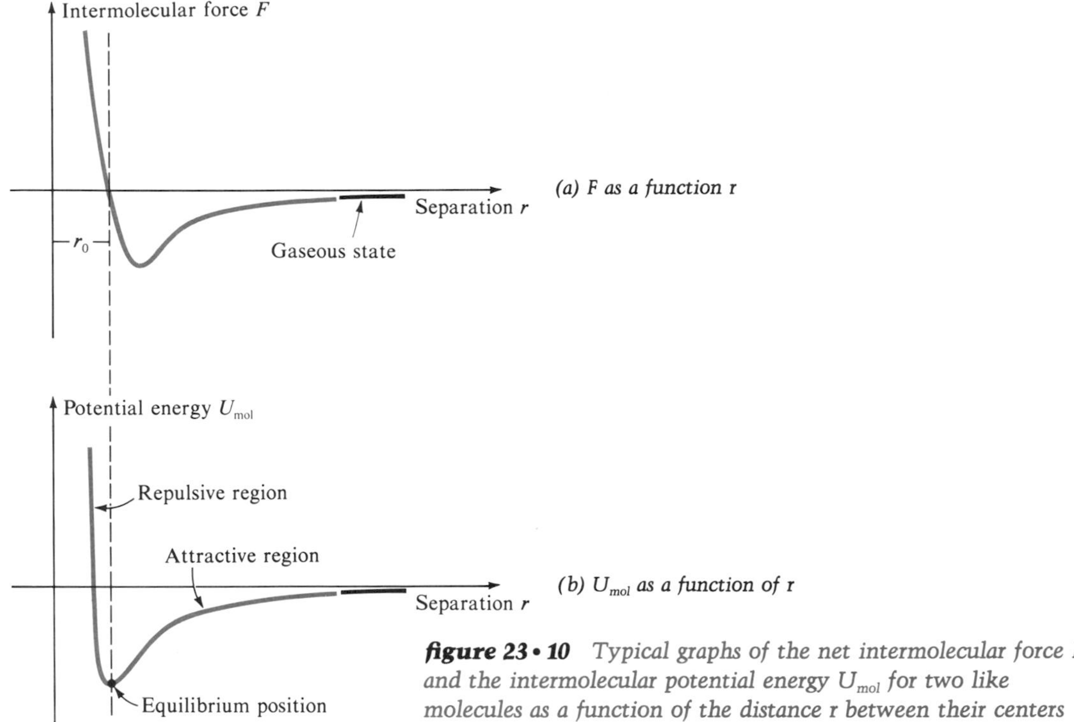

figure 23 • 10 *Typical graphs of the net intermolecular force F and the intermolecular potential energy U_{mol} for two like molecules as a function of the distance r between their centers*

the molecules are relatively far apart and the number of collisions per unit time relatively small, the predominant long-range attractive forces make the gases more compressible than an ideal gas.

If the product Pv decreases with decreasing pressure, as it does for all gases at high enough pressures and temperatures, then $\Delta(Pv)$ becomes negative; therefore, from Eq. [23 . 2], $\Delta k_{tr} > -\Delta u_m$. If we assume that here the dominant forces are the forces of repulsion that act during collisions, then $\Delta u_m < 0$, and hence $\Delta k_{tr} > 0$. Warming upon throttling should therefore be expected, and this is what is observed. At the higher temperatures and pressures, then, where collisions are frequent and molecules are close together, the strong repulsive forces predominate over the weaker attractive forces and make the gas less compressible than an ideal gas.

The potential energy of a gas molecule due to intermolecular force is a function of r, the center-to-center separation distance between two interacting molecules. This function can be expressed by a relation known as the *Lennard-Jones potential:*

$$U_{mol}(r) = U_0\left[\left(\frac{r_0}{r}\right)^{12} - 2\left(\frac{r_0}{r}\right)^6\right] \qquad [23 \cdot 4]$$

The equilibrium parameters, r_0 and $U_0 = U_{mol}(r_0)$, are determined by the structure of the individual molecules. As shown in Fig. 23 . 10, for $r > r_0$ the slope of U_{mol} is positive; hence the force is attractive. Conversely, for $r < r_0$ the force is repulsive, while at $r = r_0$ the repulsive and attractive components of the force exactly balance. Table 23 . 3 shows some typical values of U_0 and r_0.

table 23 • 3 *Intermolecular potential-energy parameters for gases*

gas	*formula*	*U_0, 10^{-23} J*	*r_0, angstroms (1 Å = 10^{-10} m)*
helium	He	14	2.9
hydrogen	H_2	43	3.3
neon	N	48	3.1
nitrogen	N_2	131	4.2
oxygen	O_2	162	3.9
argon	Ar	166	3.8

Now let us see how the presence of such forces can account for the Joule-Thomson effect. Each particle of the substance acts as if it were moving in a potential of the form shown in Fig. 23 . 11, at some average distance $\bar{r}$ from the origin. The rotational and vibrational forms of energy are *internal* to each molecule, and so are not affected by the intermolecular forces. Hence we can ignore them by considering the total energy E of the molecule as equal to its translational kinetic energy at a separation beyond the range of the attractive force. The distance (dashed lines) between the graph of U_{mol} and the horizontal line E represents the kinetic energy K_{tr} of the molecule. In a binary collision between molecules, d would be the distance of closest approach, or *collision diameter.* The values of d are nearly the same as r_0, although always somewhat smaller.

Note in Fig. 23 . 11 that when the average separation increases from r_1 to r_2 without any change in the total molecular energy, the repulsive forces cause an increase in the average kinetic energy, and hence in the temperature of the gas. However, if the average separation is r_3 and it increases to r_4, then the attractive forces decrease the average kinetic energy and the

figure 23 • 11 *Intermolecular potential $U_{\text{mol}} = U_0[(r_0/r)^{12} - 2(r_0/r)^6]$, showing regions giving rise to cooling (attraction) and heating (repulsion) in the Joule-Thomson process. The shaded area represents kinetic energy.*

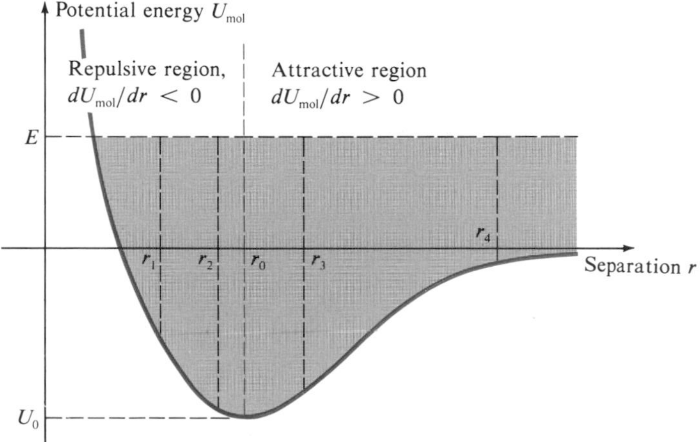

temperature decreases. If the separation is larger than r_4, then further increases in $\bar{r}$ cause only negligible changes in temperature.

For a gas consisting of N particles occupying a volume V, each particle can be said to occupy a sphere of average radius $\frac{1}{2}\bar{r}$, and volume

$$\tfrac{4}{3}\pi\,(\tfrac{1}{2}\bar{r})^3 = \frac{V}{N} = \frac{kT}{P} \qquad [23 \cdot 5]$$

where, prior to liquefaction, the ideal-gas law is a reasonable approximation. For low enough temperatures $\bar{r}$ is less than r_0, and we are in the heating portion, $dT/dP < 0$, of the PT plane (see Fig. 23 . 9). As T increases, $\bar{r}$ becomes greater than r_0, and we enter the cooling portion, $dT/dP > 0$, of the PT plane. Finally, when T becomes large enough, the attractive forces become negligible, and we have essentially a free expansion. At this stage the emergent gas is heated due to the work done against friction in forcing it through the porous plug, whereas previously this work had been overshadowed by the effects of the intermolecular forces. If we increase the pressure driving the expansion, and thus increase the density, we can force the system from the cooling region into the heating region by decreasing $\bar{r}$.

We can use Eq. [23 . 1] to derive a new thermodynamic state variable called the *molar enthalpy* h_m, which has the dimensions of energy. Note that the energy per mole conserved during the Joule-Thomson process is given by Eq. [23 . 1], rewritten in the form

$$\mathcal{U}_{m1} + P_1 v_1 = \mathcal{U}_{m2} + P_2 v_2 = h_m \qquad [23 \cdot 6]$$

where h_m is a constant. Molar enthalpy is a particularly useful variable in engineering applications. For an ideal gas,

$$\blacktriangleright \quad h_m = u_m + Pv = C_V T + RT = C_P T \qquad [23 \cdot 7]$$

example 23 • 2 Show that when water turns to steam, the first law of thermodynamics predicts that the latent heat of vaporization L_v is equal to the difference in molar enthalpies between the water and the steam:

$$L_v = h_{m,v} - h_{m,l}$$

solution The differential change in molar enthalpy is given by

$$dh_m = d\mathcal{U}_m + p\,dv + v\,dp$$

and from the first law of thermodynamics,

$$d\mathcal{U}_m = dq - p\,dv$$

where q is the heat Q per mole. Therefore

$$dh_m = dq - p\,dv + p\,dv + v\,dp$$

Since change of phase is isobaric,

$$dp = 0 \qquad dh_m = dq$$

Hence, for 1 mole of substance,

$$q = L_v = h_{m,v} - h_{m,l}$$

23 • 3 real-gas equations of state

Since the liquid and gaseous phases are continuous, it should be possible to obtain an equation of state relating the pressure, volume, and temperature of a substance whether it is in the liquid *or* the gaseous phase. The simple ideal-gas equation $PV = nRT$ obviously will not hold for the more general case, but must be modified to include two new factors: (1) the volume of the molecules themselves, which are not point particles, and (2) the intermolecular forces between the molecules. When this is done we shall have obtained the *van der Waals equation of state,* which is not only a significant improvement on the ideal-gas equation, but also accounts for many of the features of real gases, such as their critical constants and the gas-to-liquid transition region of PvT space.

For very high pressures the curves in Fig. 23 . 2 become a family of nearly parallel lines. The slope-intercept equation for any one of these lines will be $PV = b'P + c$, where the slope b' depends on the nature and mass of the gas considered and the intercept c depends on the temperature and mass. If we write this equation in the form

$$P(V - b') = c \qquad [23 \cdot 8]$$

it is apparent that P becomes infinite for $V = b'$. The *covolume* b' can be interpreted as the least volume into which the given mass of gas can be compressed. Thus $V - b'$ is the whole space in which the gas is enclosed, diminished by the least possible volume of the gas. If we consider $V - b'$ as the *effective* volume of the gas, then Boyle's law may be said to apply to all gases even at high pressures.

The assumption that the volume occupied by the molecules is negligible compared to the space between them is not completely valid for actual gases. Consequently, the number of collisions per unit time of any molecule with the other molecules or with the walls of the containing vessel is larger than that calculated from the ideal-gas theory. In collisions with the walls the effect is the same as if the molecules themselves were still negligibly small, but confined in a smaller space.

To determine b', the least possible volume of the gas, let us consider a pair of molecules in the gas with an average collision diameter d (Fig. 23 . 12). A sphere of volume $\frac{4}{3}\pi d^3$ is thus excluded from the total volume available for translational motion. If there are $\frac{1}{2}N$ such pairs, then the total covolume is

$$b' = \tfrac{1}{2}N\tfrac{4}{3}\pi d^3 = \tfrac{2}{3}N\pi d^3 \qquad [23 \cdot 9]$$

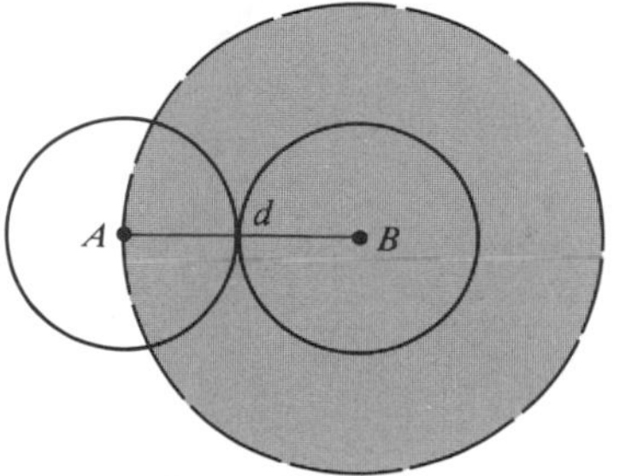

figure 23 • 12 *Two molecules with an average collision diameter d. The center of molecule A is excluded from the radius d about the center of molecule B.*

Substituting $c = nRT$ in Eq. [23 . 8] yields our first improvement on the real-gas equation, the *Clausius equation of state:*

$$P(V - b') = nRT \qquad [23 \cdot 10]$$

which can now be further modified in light of the Joule-Thomson effect.

Having established experimentally that appreciable attractive forces exist between gas molecules, we now wish to take them into account in obtaining the general equation of state applicable to real gases. In the interior of a fluid the molecules attract one another equally in all directions, so that the resultant effect is negligible. However, in the layers next to the walls of the containing vessel there is a resultant attraction directed inward toward the fluid which tends to reduce the volume it occupies, just as an increase in pressure would do. This suggests that the effect of intermolecular attraction may be taken into account by adding to the *externally* applied pressure P a quantity P' that represents the increased pressure in the fluid due to *internal* attraction. Equation [23 . 10] then becomes

$$(P + P')(V - b') = nRT \qquad [23 \cdot 11]$$

which leads to the van der Waals equation.

The van der Waals equation is the most celebrated of the numerous attempts that have been made to deduce a general equation of state holding for any substance throughout the liquid and gaseous phases. It was developed in 1873 by Johannes Diderik van der Waals, who assumed that the quantity P' is proportional both to the number of molecules striking a unit area of the wall in unit time and to the number of molecules attracting any given molecule. Since both of these factors are proportional to the molecular number density n_{mol} of the gas, P' will vary directly as $n_{mol}^2 = N_A^2 n^2/V^2$, where N_A is Avogadro's number. If we express Eq. [23 . 11] in terms of n, the number of moles, we obtain

$$\left(P + \frac{n^2 a}{V^2}\right)(V - nb) = nRT \qquad [23 \cdot 12]$$

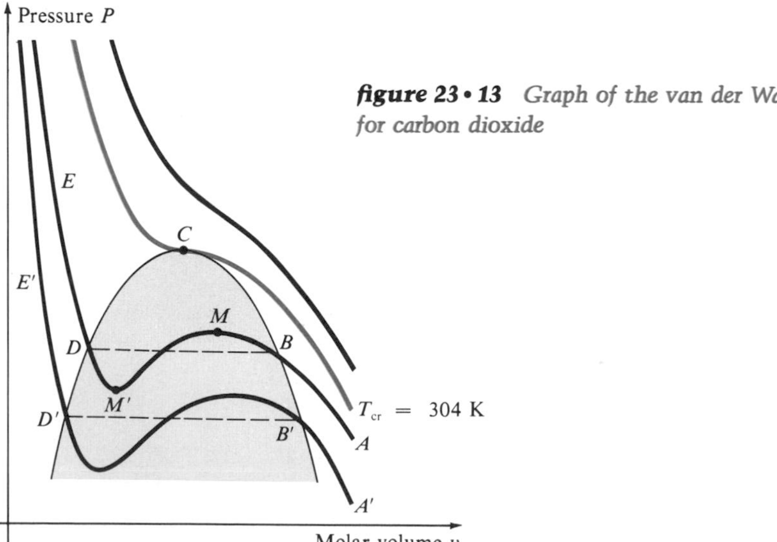

figure 23 • 13 *Graph of the van der Waals equation for carbon dioxide*

where $b = b'/n = \frac{2}{3}N_A\pi d^3$ is the molar covolume and a is an internal pressure constant that depends on the magnitude of the intermolecular attraction. This attraction varies widely from one substance to another and, as we might expect, is smallest for those substances with low critical temperatures. The covolume b varies less widely, indicating that the intermolecular distances d at which strong repulsive collision forces come into play do not differ greatly for different substances. In terms of the specific volume per mole, $v = V/n$,

$$\left(P + \frac{a}{v^2}\right)(v - b) = RT \qquad [23 \cdot 13]$$

which is the *van der Waals equation of state* of a real gas.

Although the van der Waals equation applies to both the gaseous and liquid phases, the isotherms plotted from it do not have the discontinuous horizontal segments which appear in the experimental curves of Figs. 23 . 4 and 23 . 6. As shown in Fig. 23 . 13, the van der Waals isotherms are continuous and cut the horizontal lines BD and $B'D'$ at three points. This is because the equation is of third degree in v, and in this region, for a given pressure and temperature, there are three roots of v which satisfy Eq. [23 . 13]. However, the mathematical behavior of the van der Waals equation inside the liquid-vapor region (shaded) has no physical significance. For example, between M and M' the shape of the curve $BMM'D$ would require

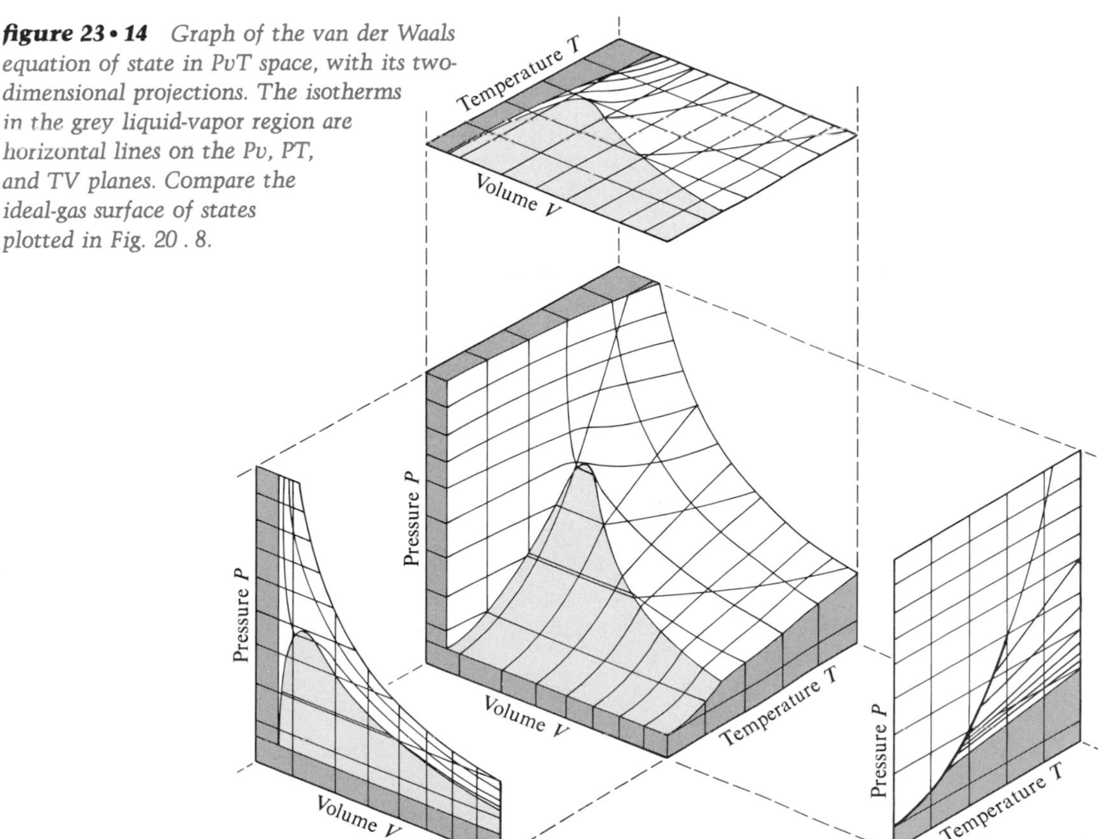

figure 23 • 14 *Graph of the van der Waals equation of state in PvT space, with its two-dimensional projections. The isotherms in the grey liquid-vapor region are horizontal lines on the Pv, PT, and TV planes. Compare the ideal-gas surface of states plotted in Fig. 20 . 8.*

that pressure and volume be directly proportional to each other. For temperatures above the critical temperature T_{cr} the van der Waals equation has only one real solution for $v(P,T)$.

In view of the simplicity of the van der Waals equation, the fact that it holds even fairly well near the critical point is remarkable. At high pressures it is quite inadequate, as van der Waals himself realized. However, this equation gives us a qualitative conception of the physical behavior of the liquid, vapor, and gaseous phases of all substances. More important, it has served to stimulate a tremendous amount of experimental and theoretical research on fluid properties. Equations that are improvements, in one respect or another, on the van der Waals equation have been proposed by Clausius (1880), Barthelot (1889), Dieterici (1890), Beattie and Bridgman (1927), and many others. One of the most useful and widely applicable forms of the equation of state is the so-called *virial equation,*

$$Pv = RT\left(1 + \frac{B}{v} + \frac{C}{v^2} + \cdots\right) \qquad [23 \cdot 14]$$

in which a series of correction terms accounts for the changes in Pv with increasing gas density. The *virial coefficients* $B, C, \ldots$ depend on the nature of the substance and on its temperature. Ordinarily C is negligible, as are higher-order terms; however, the coefficient B is directly related to the strength of the intermolecular force, and hence is a quantity of great interest.

example 23 • 3 What are the values of the critical constant (P_{cr}, v_{cr}, T_{cr}) as predicted by the van der Waals equation?

solution If we solve Eq. [23 . 13] for P_{cr}, we obtain

$$P_{cr} = \frac{RT_{cr}}{v_{cr} - b} - \frac{a}{v_{cr}^2}$$

Since the slope of the critical isotherm is horizontal at C,

$$\left(\frac{dP}{dv}\right)_{cr} = 0 \qquad \frac{RT_{cr}}{(v_{cr} - b)^2} = \frac{2a}{v_{cr}^3}$$

Furthermore, C is an inflection point, so that

$$\left(\frac{d^2P}{dv^2}\right)_{cr} = 0 \qquad \frac{2RT_{cr}}{(v_{cr} - b)^3} = \frac{6a}{v_{cr}^4}$$

Therefore, dividing this equation into the preceding one, we obtain

$$\tfrac{1}{2}(v_{cr} - b) = \tfrac{1}{3}v_{cr}$$

and the three critical values are

$$v_{cr} = 3b \qquad T_{cr} = \frac{8a}{27Rb} \qquad P_{cr} = \frac{a}{27b^2}$$

The values of a and b can then be determined from a table of critical constants (Table 23 . 1).

example 23 • 4 Find the collision diameter d for diatomic nitrogen, which has a gram-molecular weight $M_0 = 28$ g/mole.

solution From Table 23 . 1, the critical density of nitrogen is $\rho_{cr} = 0.31$ g/cm^3. Hence we can find $v_{cr} = M_0/\rho_{cr}$,

$$v_{cr} = 3b = 2N_A \pi d^3$$

and then compute d from the relation

$$d = \sqrt[3]{\frac{M_0}{2N_A \pi \rho_{cr}}} = \sqrt[3]{\frac{28 \text{ g/mole}}{2(6 \times 10^{23} \text{ mol/mole}) \times \pi(0.31 \text{ g/cm}^3)}}$$

$$= 2.9 \times 10^{-8} \text{ cm}$$

This value compares favorably with the correct value of 3.8×10^{-8} cm.

23 • 4 *vapor pressure*

If the molecules of a gas are in rapid motion, the same must be true of the molecules of a liquid, for near the critical point there is no fundamental distinction between the liquid and the gaseous phases. Indeed, at high temperatures the two phases become identical. Below the critical temperature a clearly delineated free surface may be regarded as the distinguishing feature of a liquid; that is, a liquid does not necessarily fill its container the way a gas does.

Direct experimental evidence of molecular motion in liquids is furnished by the Brownian movement, discussed in Sec. 22 . 4. However, little information is available about the exact nature of this motion. Certainly the molecules are crowded so close together that their paths between impacts are extremely minute, of the order of a collision diameter. There is evidence from x-ray studies, however, that some degree of order exists in the molecular arrangement, that the molecules tend to cluster in groups of slightly higher density than the average for the liquid, and that this tendency increases as the freezing temperature of the liquid is approached.

Owing to energy transfer in collisions, a molecule may gain sufficient kinetic energy to break loose from one group and join another. It may break loose at the surface of the liquid, but lack sufficient kinetic energy to overcome the powerful attractive forces of the other liquid molecules in or near the surface. Then it will simply rise to a certain distance and fall back into the liquid (see Fig. 23 . 15). But there also will be molecules at the surface

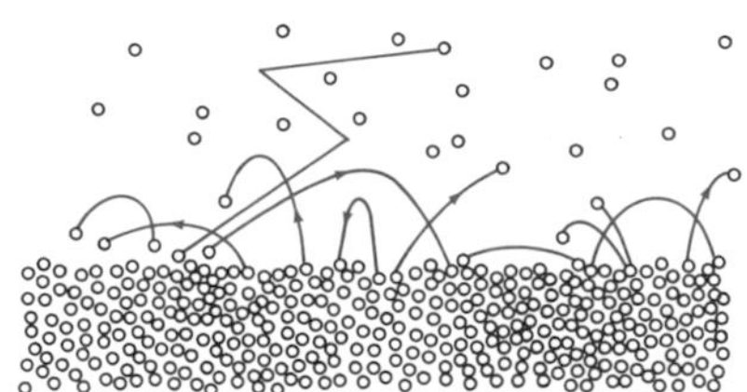

figure 23 • 15 *A liquid and its vapor, viewed at the molecular level*

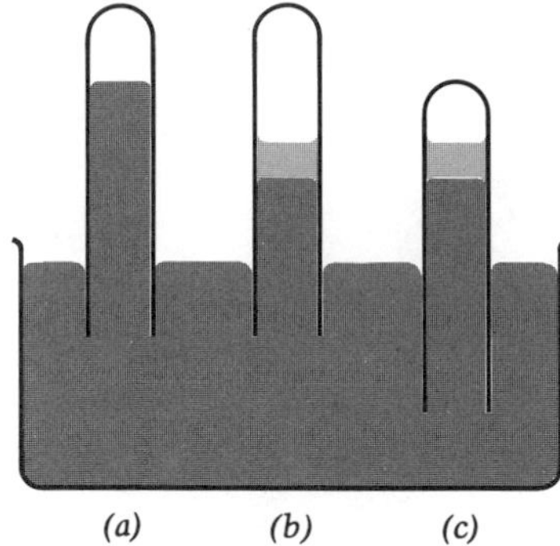

figure 23 • 16 *The vapor pressure Pv of a pure liquid. The space above the mercury column in (a) contains pure mercury vapor; in (b) a small quantity of some volatile liquid such as ether is introduced. When the tube is lowered (c), the volume occupied by the combined vapors decreases, but the height of the mercury column remains unchanged.*

that do have the kinetic energy necessary to escape and move off as independent vapor molecules into the space above; that is, *evaporation* occurs.

If the space above a liquid surface is enclosed, the vapor gradually fills it, becoming more and more dense as more molecules escape. However, many of the escaped molecules return to the surface and reenter the liquid. The number returning per unit time evidently increases as the density of the vapor increases. When this density has reached a certain limit, a condition of equilibrium is set up in which the average number returning will equal the average number escaping. This state is sometimes referred to as a *kinetic equilibrium,* to emphasize the idea that two opposing effects are taking place at the same rate, with the result that there is no net change. A vapor in kinetic equilibrium with its liquid is said to be *saturated.* If the vapor is not allowed to accumulate over the liquid, the system will remain unsaturated; equilibrium will not be reached, and the liquid will gradually evaporate.

If a vapor is in contact with its liquid in a closed vessel at constant temperature, all attempts to increase the density or the pressure must be futile. Suppose a few drops of some liquid, such as ether, are introduced into a barometer tube so as to fill the space above the mercury with ether and saturated ether vapor (Fig. 23 . 16). As soon as the density of this vapor is momentarily increased by lowering the tube farther into the mercury well, the equilibrium at the surface of the ether is destroyed, and more molecules begin to enter the surface per unit time than escape from it. In a short time enough ether condenses to restore the original density and pressure. These experimental facts may be summarized as follows:

> The vapor pressure P_v of a pure liquid is a function of temperature only.

This accounts for the horizontal segments of isotherms such as *ABDE* in Fig. 23 . 4.

When a closed vessel containing a liquid and its vapor is heated, obviously the vapor density will increase, for the number of molecules escaping per unit time must be greater at the higher temperature because of their higher mean kinetic energy of translation. Also, since the pressure exerted by the vapor is proportional to both its density and its mean molecular kinetic energy, it must rise rapidly as the temperature increases. If data are recorded at various temperatures, a curve can be constructed showing the vapor pressure P_v as a function of temperature. The curve for the water data in Table 23 . 4 is shown in Fig. 23 . 17. However, no general relation is

table 23 • 4 *Boiling temperature of water and density of steam corresponding to various external pressures*

equal pressures: 1.000 mm Hg = 133.3 Pa = 0.001316 atm

temperature, °C	*vapor pressure P_v, mm Hg*	*density ρ_v, g/cm³*
10.0*	1.95	2.16×10^{-6}
0.0	4.58	4.85×10^{-6}
10.0	9.20	9.41×10^{-6}
20.0	17.5	17.3×10^{-6}
30.0	31.7	30.4×10^{-6}
40.0	55.1	51.1×10^{-6}
65.0	187	161×10^{-6}
90.0	526	424×10^{-6}
99.0	733	579×10^{-6}
99.6	749	—
100	760.00	598×10^{-6}
101	787.5	618×10^{-6}
110	1,070	827×10^{-6}
150	3,570	$2{,}550 \times 10^{-6}$
200	11,700	$7{,}840 \times 10^{-6}$
374†	166,000	$326{,}000 \times 10^{-6}$

*Over water. †The critical temperature of water.

figure 23 • 17 *The pressure-temperature curve for saturated water vapor. Points to the right of this equilibrium curve represent PT conditions under which water in a closed vessel will be entirely vaporized.*

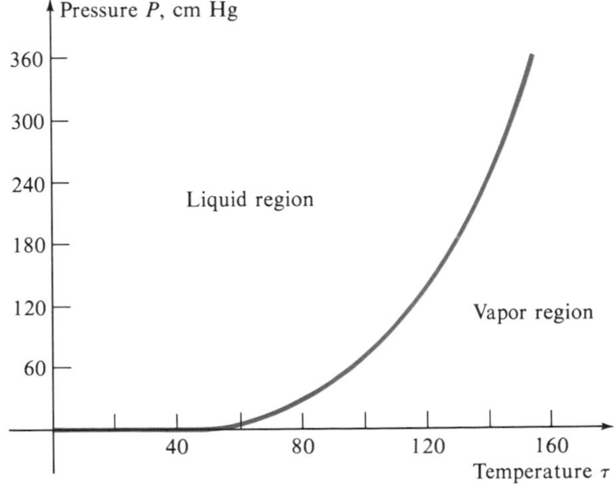

available for P_v as a function of temperature. The vapor pressure at a given temperature is affected very little by the presence of gases that do not combine chemically with the vapor. Evaporation occurs to some extent at all temperatures whenever the vapor above a liquid is not saturated. Under a constant external pressure an increase in temperature will result in more rapid evaporation, until finally a temperature is reached at which evaporation begins to take place, not just at the surface, but also within the body of the liquid. That is, there will be molecules of the liquid with enough kinetic energy to break away and form tiny bubbles of vapor beneath the surface. The temperature at which the pressure P_v of the saturated vapor becomes equal to the outside pressure is called the *boiling temperature.* A curve showing the vapor pressure P_v of a given liquid as a function of temperature provides boiling-temperature data for that liquid, with the pressures now signifying the external pressure on the liquid. However, the *boiling point* of a liquid, as distinguished from its boiling temperature, is defined as the temperature at which the vapor pressure P_v is equal to 1 atm.

23 · 5 the Clausius-Clapeyron equation

A typical surface of thermodynamic states was illustrated in Fig. 23 . 6. If we were to project this surface into three dimensions, it would look like the central portion of Fig. 23 . 18, with the *Pv* projection of Fig. 23 . 6 shown at the left and the *PT* projection of the surfaces of two-phase equilibrium shown as contours at the right. The vapor pressure of a pure liquid depends on temperature; however, it does not depend on specific volume as long as two-phase equilibrium is maintained, since any changes in pressure are immediately compensated by changes in the relative amounts of the two phases in the mixture. Consequently, the two-phase regions of the surfaces of states will project as contours in the *PT* plane, while the triple line projects as a single point—the triple point at which the three contours meet.

There is no general equation relating vapor pressure to temperature, although there are both theoretical and empirical relations that hold under certain limited conditions. However, it is possible to derive an exact differential equation for the *slope* of the curve *BC* which connects the vapor pressure of a liquid with temperature *T*. This expression is known as the *Clausius-Clapeyron equation:*

$$\frac{dP_v}{dT} = \frac{L_v}{T(v_v - v_l)} \qquad [23 \cdot 15]$$

where L_v is the latent heat of vaporization of the substance (see Sec. 21 . 3 and Table 21 . 2) and v_v and v_l are the specific volumes $v = 1/\rho$ of the saturated vapor and liquid at the temperature *T*. The derivation of this equation rests on the second law of thermodynamics, to be discussed in Chapter 25.

Solids also evaporate, changing from the solid phase directly to the vapor phase, as in the evaporation of camphor, of tungsten in an electric lamp bulb, or of snow in cold dry weather. This process is called *sublimation.* The low-temperature region in which this phenomenon occurs is the area below the vapor-pressure curve *AB* in Fig. 23 . 18. The vapor pressure P_s of a

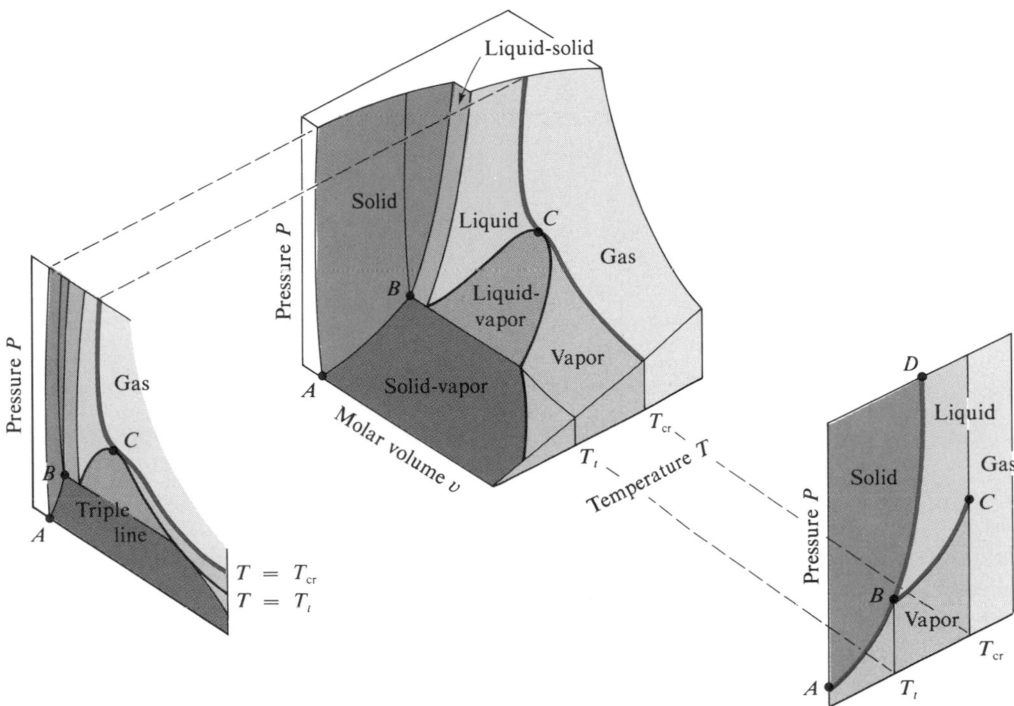

figure 23 • 18 Projections of the surface of thermodynamic states of a typical real substance in PvT space

pure solid is a function of temperature only, and it increases in much the same way as for a liquid. Some solids, such as iodine crystals, have appreciable vapor pressures even at room temperature. For solid carbon dioxide ("dry ice"), P_s reaches atmospheric pressure at $-78.5°C$, which is below the melting temperature of the carbon dioxide. Hence the solid sublimates, leaving no liquid behind, as ordinary ice would do. The temperature at which the vapor pressure P_s of a solid equals 1 atm is termed the ***sublimation point.***

For any substance the slope of the sublimation curve is given by the Clausius-Clapeyron equation in the form

$$\frac{dP_s}{dT} = \frac{L_s}{T(v_v - v_s)} \qquad [23 \cdot 16]$$

where L_s is the latent heat of sublimation. Heats of sublimation, unlike heats of vaporization, change very little with temperature. For instance, L_s for ice is 2800 J/g for all temperatures between -60 and $0°C$. If a substance at the triple point is compressed until no vapor is left, and the pressure on the resulting solid-liquid mixture is increased, the temperature will also have to be changed to preserve equilibrium between the solid and the liquid. These pressures and corresponding temperatures yield a third curve, *BD*, which starts at the triple point. This is the melting, or ***fusion,*** curve. Any given point on the fusion curve of a substance represents the ***fusion temperature***—the temperature of the solid and liquid phases when they coexist in equilibrium under an external pressure corresponding to the given point on the curve. At the fusion temperature the vapor pressures of the solid and the

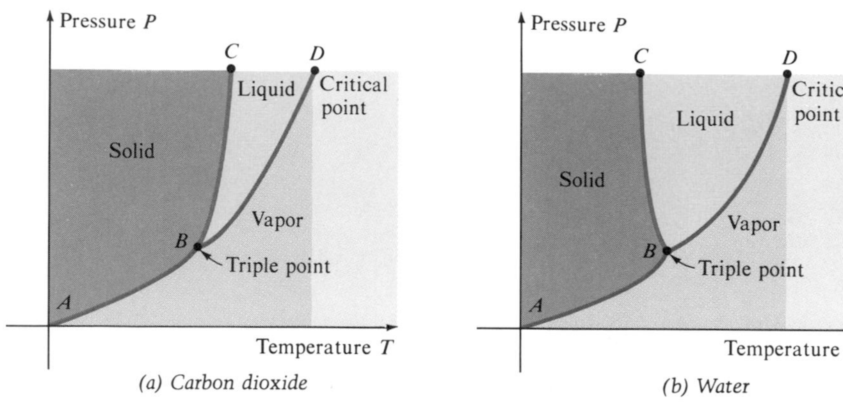

figure 23 • 19 *Triple-point diagrams for carbon dioxide, a substance which contracts on freezing, and water, a substance which expands on freezing*

liquid are equal; otherwise the phase with the higher vapor pressure would necessarily pass over gradually into the other. The melting, or *fusion, point* is defined as the temperature of the liquid-solid system when the external pressure is 1 atm.

The slope of the fusion curve is given by the Clausius-Clapeyron equation in the form

$$\frac{dP_f}{dT} = \frac{L_f}{T(v_l - v_s)} \qquad [23 \cdot 17]$$

where L_f is the latent heat of fusion of the substance. Since L_f and T are positive, the slope of the fusion curve is determined by the factor $v_l - v_s$. If expansion occurs on melting, as is true of most substances, then $v_l > v_s$ and the slope is positive, as in Fig. 23 . 19*a*; increasing the external pressure raises the fusion temperature of such a substance. If contraction occurs on melting, as is true of water and a few other substances, the slope is negative, as in Fig. 23 . 19*b*; increasing the external pressure lowers the fusion temperature of such a substance.

Definite fusion temperatures tend to be characteristic of solids that have well-defined crystal structures. Other solids, of which plastics, glass, and most alloys are examples, pass gradually through all stages of viscosity in melting or freezing. For these substances the temperature changes continuously and there is no definite point at which they may be said to melt or freeze. X-ray analyses show that many of these apparently amorphous solids actually consist of extremely small crystals, whereas others appear to be *undercooled liquids* of very high viscosity. For instance, glass may best be regarded as an undercooled liquid of such high viscosity that it is quite stable—although not completely stable, for it can devitrify and slowly build up a crystalline structure. Again, if quartz, which ordinarily occurs in the crystalline form, is melted and then allowed to cool, it undercools to form so-called "fused quartz." There is evidence of some orderly structural arrangement in these undercooled liquids, just as there is in ordinary liquids at higher temperatures.

example 23 • 5 Find the vapor pressure P_v as a function of temperature for a substance such that $v_v \gg v_l$, where the latent heat of vaporization L_v is constant and the vapor displays ideal-gas behavior.

solution We can express the specific volume of the vapor phase by the equation

$$v_v = \frac{v}{M_0} = \frac{RT}{P_v M_0}$$

where M_0 is the molecular weight. We can substitute this expression into the Clausius-Clapeyron equation, Eq. [23 . 15], in the form

$$\frac{dP_v}{dT} = \frac{L_v}{Tv_v}$$

We then obtain the differential equation

$$\frac{dP_v}{P_v} = \frac{M_0 L_v}{RT^2}\,dT$$

which has the solution

$$P_v = P_0 \exp\left[-\frac{M_0 L_v}{R}\left(\frac{1}{T} - \frac{1}{T_0}\right)\right]$$

example 23 • 6 What is the change in the melting point of ice at the triple point when $\Delta P_v = 1$ atm?

solution The temperature of ice at the triple point is $\tau_t \doteq 0°\text{C}$, and we can find from any handbook that at this point

$$v_{\text{ice}} - v_{\text{water}} = 0.091 \times 10^{-3}\ \text{m}^3/\text{kg}$$

$$L_f \doteq 80\ \text{kcal/kg}$$

$$P_v \doteq 0.006\ \text{atm}$$

If the pressure is increased to 1 atm, then the change in melting point is, from Eq. [23 . 17],

$$\Delta T = T(v_l - v_s)\frac{\Delta P}{L_f}$$

$$\doteq 273\ \text{K}(-9.1 \times 10^{-5}\ \text{m}^3/\text{kg})\frac{10^5\ \text{Pa}}{8(4.2 \times 10^4\ \text{J/kg})}$$

$$\doteq -0.0075°\text{C}$$

Since we defined the triple point of water to be 0.100°C, it would seem that the Clausius-Clapeyron equation is in error in predicting a temperature drop of only 0.0075°C. However, the discrepancy is due to the fact that under ordinary conditions ice is mainly in equilibrium with air, whereas the calculation above assumes it to be in equilibrium with water vapor.

PROBLEMS

23 • 1–23 • 2 real gases and intermolecular forces

23 • 1 A small amount of liquid is placed in a glass thermometer bulb, all the air is removed, and the stem of the bulb is sealed. The temperature is then gradually raised above the initial value T. (*a*) Describe the behavior of the contents if the volume of the bulb is much larger than the critical volume V_{cr} of the substance. (*b*) If the bulb volume is much smaller than V_{cr}. (*c*) If the bulb volume is approximately equal to V_{cr}.

23 • 2 When P is such that $d(Pv)/dP = 0$, as in Fig. 23 . 1, what does this imply about the relationship between the average separation $\bar{r}$ of molecules at that pressure and the equilibrium separation r_0?

23 • 3 A cylinder of volume V_1 contains a mass M of vapor. The vapor is maintained at a constant temperature and compressed until, at volume V_2, it becomes practically impossible to compress it any further. When the vapor is at some intermediate volume V, such that $V_1 > V > V_2$, what fraction of it has been condensed to the liquid phase?

23 • 4 Explain why the displacement of the minimum in the contours of Fig. 23 . 2 is toward the right as the temperature T increases. (*Hint:* See Prob. 23 . 2.)

23 • 5 If the total energy of a molecule in a Lennard-Jones potential well is $E = -0.3U_0$, find the minimum and maximum intermolecular separations r_1 and r_2 in units of equilibrium separation r_0.

23 • 6 Show that when ice turns to water, the first law of thermodynamics predicts that the latent heat of fusion per mole is equal to the difference in molar enthalpies between the ice and the water:

$$L_f = h_{ml} - h_{mv}$$

23 • 7 The latent heat of vaporization of water at 0°C and 1 atm is $L_v \doteq 539$ kcal/kg. (*a*) What percentage of L_v is used in doing work against the atmospheric pressure? (*b*) What percentage is used in increasing the internal energy of the water-steam substance?

23 • 8 The Lennard-Jones potential, Eq. [23 . 4], is most accurate for symmetrical molecules, particularly for the liquefied elementary gases. For liquid argon at a temperature of −186°C, the density is $\rho = 1.37$ g/cm^3 and $U_0 \doteq 1.7 \times 10^{-21}$ J. (*a*) Taking the mass of a single molecule as $m = 66.6 \times 10^{-27}$ kg, what is the average separation $\bar{r}$ between molecules? (*b*) If $\bar{r}$ is really $1.1r_0$ due to vibrational motions of a liquid molecule inside the potential well, what is the force on the molecule at $\bar{r}$? (*c*) Assuming that the average downward force on a molecule at the surface is due to the attractive force of its nearest neighbor below it, compute the interior pressure in the liquid.

****23 • 9*** The average intermolecular separation in liquid argon at −186°C does not occur at the minimum of the Lennard-Jones potential, but at $\bar{r} = 1.1r_0$, due to the motions of the molecules relative to each other.

(*a*) What are the limiting separations of the internal motions, $2\bar{r} \doteq r_{min} + r_{max}$? (*b*) What is the corresponding kinetic energy K_0 of the molecule at r_0? Compare this to the quantity kT and see if you can apply the law of equipartition to account for it.

23 • 3 the real-gas equation of state

23 • 10 Experiment shows that 1 mole of oxygen of volume 56.0 ml and temperature 0°C exerts a pressure of 440 atm. (*a*) Calculate this pressure by using the ideal-gas law. (*b*) By using the van der Waals equation (the van der Waals constants are $a = 1.36$ liter$^2 \cdot$ atm/mole^2 and $b = 0.0318$ liter/mole).

23 • 11 For oxygen at a temperature of 0°C the virial coefficients of Eq. [23 . 14] are

$$B = -1.92 \times 10^{-2} \text{ liter/mole}$$

$$C = 1.46 \times 10^{-3} \text{ liter}^2/\text{mole}^2$$

$$D = 2.14 \times 10^{-4} \text{ liter}^3/\text{mole}^3$$

Predict the pressure exerted by oxygen when its molar volume and temperature are 56.0 ml/mole and 0°C. How does this value compare with those obtained in Prob. 23 . 10?

23 • 12 By writing the van der Waals equation in powers of V and making use of the fact that the three roots of this cubic equation are *equal* at the critical point, show that the critical temperature, pressure, and volume per mole are as given in Eqs. [23 . 18].

23 • 13 (*a*) Show that if a fluid behaved like an ideal gas at the critical point, the ratio $RT_{cr}/P_{cr}v_{cr}$ would be equal to unity. (*b*) Show that if a substance were correctly described by the van der Waals equation, this ratio would be $RT_{cr}/P_{cr}v_{cr} = 8/3$, or 2.67. (Experimental values of this ratio for various common gases range from about 3 for hydrogen to about 4.5 for water.)

23 • 14 Use of the van der Waals equation is simplified if the pressure is expressed in atmospheres and the volume is expressed as the ratio of actual volume V to the volume $V_0 = 22.4n$ liters the substance would occupy if it were an ideal gas at 0°C at 1 atm. Consequently, tabulated values of the van der Waals constants a and b are often expressed in terms of these special units. (*a*) What is the resulting form of the van der Waals equation? (*b*) For these units n^2a/V_0^2 and nb/V_0 for air become 0.0026 atm and 0.0021. Given an initial volume of 1 m^3 of air at 0°C and 1 atm, what will its pressure be after it is compressed isothermally to 0.15 m^3? (*c*) What pressure at this smaller volume is predicted by Boyle's law?

23 • 15 If we express the actual pressure, volume, and temperature of a fluid as fractions of the critical values, we obtain the reduced pressure $P_r = P/P_{cr}$, the reduced volume $V_r = v/V_{cr}$, and the reduced temperature $T_r = T/T_{cr}$. (*a*) Express the van der Waals equation in terms of these quantities. (*b*) What is the remarkable characteristic of this *reduced van der Waals equation*, and how do you interpret it?

23 • 16 The van der Waals equation can be expanded into the form

$$PV = nRT + n\left(Pb - \frac{na}{V} + \frac{n^2ab}{V^2}\right)$$

(*a*) Show that for large volumes this reduces to

$$PV = RT\left[1 + \left(\frac{b}{V} - \frac{a}{RTV}\right)\right]$$

and that the virial coefficient B of Eq. [23 . 14] is therefore $B = b - a/RT$. Compare this result with the numerical values stated in Probs. 23 . 10 and 23 . 11.

***23 • 17** The van der Waals coefficients for carbon dioxide are $a = 3.61$ liter^2 · atm/mole^2 and $b = 0.0428$ liter/mole. (*a*) Expand the van der Waals equation as a cubic in v, the molar volume. (*b*) Use the Newton-Raphson method to find the possible values of v when $\tau = 21.5°C$ and $P = 50$ atm. (*Hint:* Take $v \gg v_{cr}$ as a first guess and check your final value by computing P from it.)

23 • 4 vapor pressure

23 • 18 (*a*) What are the forms of the isobars for ideal gases and for actual gases? (*b*) What are the forms of the isometric curves that connect P and T when V is kept constant, for ideal gases, for actual gases, and for a saturated vapor in contact with its liquid?

23 • 19 A boiler with a capacity of 425 liters contains 90 kg of water and steam at 200°C. Find the mass of the steam. (The density of water under these conditions is $\rho = 865$ kg/m^3.)

23 • 20 In a uniform barometer tube in which the mercury stands at 40 cm, the space above the liquid mercury is 40 cm long, and contains dry air and mercury vapor. A few drops of ether are then introduced at the bottom of the tube and rise to the surface of the mercury, where they evaporate. If the vapor pressures of ether and of mercury at the existing temperature are 0.03 and 0.0009 mmHg, respectively, to what point above the mercury level in the cistern will the mercury in the tube ultimately fall? Assume that the atmospheric pressure is 1.013×10^5 Pa.

23 • 21 If 150 cm^3 of oxygen is collected over water at a pressure of 740 mmHg and a temperature of 20°C, what volume would the dry oxygen occupy at 0°C and 1 atm?

23 • 22 Two thousand cubic centimeters of dry air at 15°C and 1 atm is passed through flasks that contain a known mass of carbon disulfide (CS_2) at 15°C, and the resulting mixture of air and carbon disulfide vapor is allowed to escape into the room at a pressure of 1 atm. When the flasks are reweighed, the reduction in mass is found to be 3.011 g. What is the vapor pressure of carbon disulfide at 15°C?

23 • 23 The molecular weights of water and air are 18.0 and 28.9, respectively. (*a*) Assuming that the ideal gas law is applicable to water vapor up to the point of saturation, prove that, under like conditions of temperature and pressure, the density of water vapor relative to the density of air is equal to the ratio of the molecular weights of water and air. (*b*) Using the density of air at 0°C and 1 atm, $\rho = 1.293\ \text{kg/m}^3$, and the vapor pressure of water at 0°C (Table 23 . 3), calculate the density of saturated water vapor at 0°C.

23 • 24 The vapor pressure of sulfur dioxide at 27°C is 4.08 atm and the density of liquid SO_2 at this temperature and pressure is $0.733\ \text{g/cm}^3$. A cylinder, provided with a tightly fitting piston, contains 1 mole of SO_2 at 27°C and 1 atm. The piston is slowly forced into the cylinder while the temperature is kept constant. (*a*) Assuming that the vapor behaves like an ideal gas, find the initial volume of the system. (*b*) To what value must the volume be reduced before condensation begins? (*c*) When the volume is reduced to 1 liter, how many grams of SO_2 have condensed?

23 • 5 *the Clausius-Clapeyron equation*

23 • 25 Describe the physical changes that will occur in a unit mass of water if its state on the *PT* projection changes (*a*) from *A* to *B*. (*b*) From *D* to *E*. (*c*) From *F* to *G*. (*d*) From *H* to *I*. (*e*) From *J* to *K*.

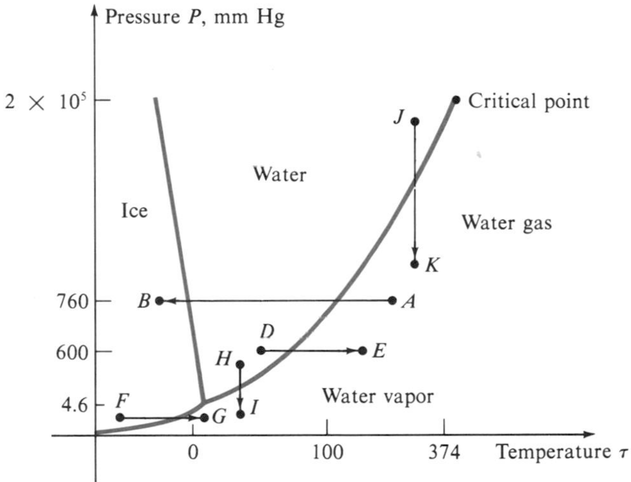

23 • 26 Faraday observed that when two pieces of ice are pressed together and then released, they are frozen together. (*a*) Explain this phenomenon, which Faraday termed *regelation*. (*b*) A metal bar of width *a*, with weights of mass *m* hung at each end, rests on a cake of ice of length *b*. The bar and the ice are at 0°C. Find the reduction in the temperature of the ice immediately under the bar.

23 • 27 How much work is done against external atmospheric pressure in evaporating 100 g of water at 98°C and 9.43×10^4 Pa? The density of steam under these conditions is $\rho = 0.56$ kg/m^3.

23 • 28 (*a*) Compute dP_v/dT for water at 100°C. (*b*) Assuming this rate to be constant for small changes in pressure, compute the boiling temperature of water when the atmospheric pressure is 10^5 Pa and also when it is 1.05×10^5 Pa. Compare these computed values with the experimental ones listed in Table 23 . 3.

23 • 29 (*a*) Calculate the slope of the fusion curve for water at 0°C. (*b*) What is the external pressure when the fusion temperature of ice is −0.75°C? (*c*) In calculating the slope of a sublimation curve, is it ever possible to neglect the specific volume of the solid in comparison with that of the saturated vapor? Is it ever possible to neglect one of the specific volumes in calculating the slope of a fusion curve? Of a vaporization curve?

23 • 30 (*a*) Explain why the heat of vaporization must decrease to zero as a substance approaches its critical temperature. (*b*) Assuming that $L_v = a - bT$, where a and b are positive constants, repeat the derivation of the differential equation in Example 23 . 4.

answers

23 • 1 (*a*) Liquid evaporates, vapor pressure increases; (*b*) substance remains liquid; (*c*) meniscus between liquid and vapor disappears
23 • 2 $\bar{r} = r_0$
23 • 3 $M_{liq}/M = (V_1 - V)/(V_1 - V_2)$
23 • 4 If T increases, then $\bar{r}$ increases, and $\bar{r} > r_0$; to force it back to r_0 where $d(Pv)/dP = 0$, it is necessary to increase P, shifting the minimum to the right
23 • 5 $r_1 = 0.904r_0$, $r_2 = 1.352r_0$
23 • 7 (*a*) 5.6 percent of L_v; (*b*) 94.4 percent of L_v
23 • 8 (*a*) $\bar{r} = 3.65$ Å; (*b*) $F = 2.34 \times 10^{-12}$ N; (*c*) $P = 176$ atm
23 • 9 (*a*) $r_{min} = 0.908r_0$, $r_{max} = 1.292r_0$; (*b*) $K_0 = 1.023 \times 10^{-21}$ J; (*c*) $K_0 = 0.852\ kT \doteq \frac{6}{7}kT$
23 • 10 (*a*) $P = 400$ atm; (*b*) $P = 492$ atm
23 • 11 $P = 446$ atm
23 • 14 (*a*) $(P + a/V'^2)(V' - b) = T/273$, where $V' = V/V_0$; (*b*) $P = 6.65$ atm; (*c*) $P = 6.67$ atm
23 • 15 (*b*) Independent of substance
23 • 17 (*a*) $Pv^3 - (Pb + RT)v^2 + av - ab = 0$; (*b*) $v = 0.341$ liters/mole

23 • 19 $M = 2.54$ kg
23 • 20 $h = 21.4$ cm
23 • 21 $V = 133$ cm^3
23 • 22 $P_v = 0.318$ atm
23 • 23 (*b*) $\rho = 4.85$ g/m^3
23 • 24 (*a*) $V = 24.6$ liters; (*b*) $V = 6.03$ liters; (*c*) $\Delta m = 54.2$ g (see Problem 23 . 3)
23 • 26 (*b*) $\Delta T = 2mgT(v_s - v_l)/abL_f$
23 • 27 $W = 16{,}800$ J
23 • 28 (*a*) $dP_v/dt = 36{,}100$ dyne/cm$^2 \cdot$ K; (*b*) $\tau_s = 99.6°$C, $100.6°$C
23 • 29 (*a*) $dP_f/dT = -1.343 \times 10^7$ Pa/K; (*b*) $P = 1.01 \times 10^7$ Pa
23 • 30 (*b*) $P_v = P_0(T_0/T)^\beta \times \exp\{-(M_0a/R)[(1/T) - (1/T_0)]\}$, where $\beta = M_0b/R$

CHAPTER TWENTY-FOUR

kinetic theory: transport properties

The distribution of the molecules according to their velocities is found to be of exactly the same mathematical form as the distributions of observations according to the magnitude of their errors, as described in the theory of errors of observation. The distribution of bullet holes in a target according to their distances from the point arrived at is found to be of the same form, provided a great many shots are fired by persons of the same degree of skill.

James Clerk Maxwell, Theory of Heat, *1891*

Many of the important macroscopic properties of gaseous systems are merely the net observable effects of the transport of mass, momentum, or kinetic energy through the system via the random motions of its particles. Properties such as diffusion, viscosity, and heat conduction can be attributed to these causes, and consequently, we can analyze them through the kinetic theory. To do so, however, we must be able to describe systems of large numbers of molecules moving along different paths with different velocities.

Since the velocities of the molecules are continually changing, any attempt to describe their individual behavior would be hopeless. However, it is just this disorder, or *randomness* of motion, combined with the enormous number of molecules in even the smallest volume of gas, that does enable us to describe the behavior of a gas in terms of the motion of its molecules. These are precisely the conditions under which we can apply the laws of probability.

The problem of how the speeds of the molecules are distributed in an ideal gas in a steady state was solved by the joint efforts of Maxwell and Boltzmann, and the law they developed is known as the *Maxwell-Boltzmann distribution of molecular speeds*. This distribution is closely related to the statistical theory of random distributions of quantities such as errors in repeated measurements, heights or weights of a population, yearly rainfall, and so on. Let us therefore begin with a discussion of the nature of experimental error and the reduction of experimental data under the assumption of randomness. Such a discussion of errors is very important in the analysis of experiments, and it will prove a useful context in which to develop those concepts of statistical theory which are relevant both here and in later chapters on quantum theory.

In this chapter we shall see how kinetic theory, applied via the statistical properties of the Maxwell-Boltzmann distribution, can be used to describe diffusion, viscosity, and heat conduction in a gaseous system in terms of the random transport of the mass, momentum, and kinetic energy, respectively, of its molecules. The agreement of the results with experiment provides important corroboration of both the kinetic theory and the Maxwell-Boltzmann distribution. Our most important goal, in a larger sense, will be to show that the macroscopic properties of matter are explainable through the atomic-molecular theory of matter.

24 • 1 *errors and statistics: least-squares fits*

There are two fundamental types of error in any measurement. Nonrandom, or *systematic, errors* are those due to the nature of a given process of measurement and generally result in similar deviations in all the data. For example, if your stopwatch runs fast, the time intervals you record will be larger than their true values, and the speeds determined from measured distances versus time will be correspondingly smaller. *Random errors* appear to be completely chance occurrences, in the sense that they involve such a complex set of variables that they must be treated as unpredictable in a given instance. In measuring time intervals for an object moving with constant speed, you will generally obtain slightly different readings due to the varying speed of your reflexes, differences in the position of your head as you read your watch, nonuniformities in the watch itself, and occasional errors of judgment. A well-designed experiment should eliminate systematic errors and minimize the amount of random error, although random errors can never be entirely eliminated.

There is a similar situation even in numerical computations performed by the computer. In this case the systematic error known as *truncation error* is due to the nature of the particular approximation selected, as in using the trapezoidal rule or Simpson's rule to perform a quadrature. The random error of *roundoff* is due to the computer's capacity to store only a finite number of digits in its register. Since it must either drop or round off digits beyond this number, on each operation a small error enters the computation. Such errors tend to accumulate, and for extensive or complicated computations they may become significant. Since there are so many of them, however, they can be treated in the aggregate as random errors.

To take the simplest type of measurement process, suppose we are measuring a rod of length l_0 by comparing it with an accurate meter stick (so that there are no systematic errors in the measurement process). In practice, we would make n measurements l_i, where $i = 1, 2, \ldots, n$, and take their average, or arithmetic mean $\bar{l}$:

$$\bar{l} = \frac{1}{n}\sum_{i=1}^{n} l_i \tag{24.1}$$

as our best estimate of l_0. This average gives us no information about the spread in the measurements or the random error involved in the process. To obtain a measure of the accuracy, we would compute the mean-square error, or *variance* $\overline{\epsilon^2}$:

$$\overline{\epsilon^2} = \frac{1}{n}\sum_{i=1}^{n}(l_i - \bar{l})^2 = \overline{(l - \bar{l})^2} \tag{24.2}$$

The root-mean-square error, or *standard deviation* $\sigma = \sqrt{\epsilon}$ is then the best estimate of the error to be expected on any individual measurement l_i. However, since $\bar{l}$ is the average of n measurements, probability theory shows that the *standard error of the mean* is $\sigma/\sqrt{n}$. Hence we can estimate l_0 as

$$l_0 = \bar{l} \pm \frac{\sigma}{\sqrt{n}} \tag{24.3}$$

The quantity $\sigma/\sqrt{n} = \Delta l$ represents the uncertainty of $\bar{l}$ as a measure of l_0. Statistical theory has also shown that for every set of n measurements l_i, 68.3 percent of the time l_0 will fall somewhere in the range $\bar{l} \pm \sigma/\sqrt{n}$.

The concepts of arithmetic mean and variance are fundamental in statistics, the branch of mathematics that deals with the analysis of large bodies of data. These notions arise naturally when a collection of n data points (x_i, y_i) is to be fitted to an approximating curve of form $y = f(x)$, where the criterion for "goodness of fit" is that all arbitrary parameters in $f(x)$ be chosen to minimize R^2, the total of the squared deviations of the predictions from actual data:

$$R^2 = \sum_{i=1}^{n}[y_i - f(x_i)]^2 \geq 0 \tag{24.4}$$

This is called the *least-squares method*. Under these circumstances there will be only one such curve $y = f(x)$, which, unlike the interpolating polynomial in Sec. 20.6, may not actually pass through a single data point. It will, however, always remain in the general vicinity of the ensemble of data points.

If we take the curve $f(x)$ to be a straight line,

$$f(x) = \alpha + \beta x \tag{24.5}$$

then after collecting terms in α, we have

$$R^2 = \sum_i [y_i - \alpha - \beta x_i]^2$$

$$= n\alpha^2 - 2\alpha \sum_i (y_i - \beta x_i) + \sum_i (y_i - \beta x_i)^2 \qquad [24 \cdot 6]$$

and completing the square in α gives

$$R^2 = n\left[\alpha - \frac{\sum_i (y_i - \beta x_i)}{n}\right]^2 + \sum_i (y_i - \beta x_i)^2 - \frac{\left[\sum_i (y_i - \beta x_i)\right]^2}{n} \geq 0 \qquad [24 \cdot 7]$$

The optimal choice of α is that which minimizes R^2 by making the first term in the equation vanish:

$$\alpha = \bar{y} - \beta\bar{x} \qquad [24 \cdot 8]$$

where $\bar{y} = \Sigma_i y_i/n$ and $\bar{x} = \Sigma_i x_i/n$. Thus Eq. [24 . 5] becomes

$$f(x) = \bar{y} + \beta(x - \bar{x}) \qquad [24 \cdot 9]$$

It is clear that, no matter what β is, the straight-line approximation must pass through the point $(\bar{x}, \bar{y})$ whose coordinates are the average values of x_i and y_i. In handling laboratory data it is often helpful to find $(\bar{x}, \bar{y})$ and estimate the slope β of the best straight-line fit by eye.

If we now substitute Eq. [24 . 9] into Eq. [24 . 4], we have

$$R^2 = \sum_i [y_i - \bar{y} - \beta(x_i - \bar{x})]^2 \qquad [24 \cdot 10]$$

To find the value of β for which R^2 is minimized, we set $dR^2/d\beta = 0$ and solve for β:

$$\frac{dR^2}{d\beta} = \sum_i \frac{d}{d\beta} [y_i - \bar{y} - \beta(x_i - \bar{x})]^2$$

$$= -2\sum_i (x_i - \bar{x})[y_i - \bar{y} - \beta(x_i - \bar{x})] = 0 \qquad [24 \cdot 11]$$

Since β is the same for every value of i, we can take it outside the summation:

$$\frac{dR^2}{d\beta} = 0 = -\sum_i (x_i - \bar{x})(y_i - \bar{y}) + \beta \sum_i (x_i - \bar{x})^2$$

Thus the value of β for the best straight-line fit to the data is

$$\beta = \frac{\sum_i (x_i - \bar{x})(y_i - \bar{y})}{\sum_i (x_i - \bar{x})^2} = \frac{\text{cov}\,(x,y)}{\sigma_x^2} \qquad [24 \cdot 12]$$

The denominator of β is $n\sigma_x^2$, where σ_x is the standard deviation of x, and the numerator is n cov (x,y), where cov (x,y) is the *covariance of x and y.*

We can now graph that straight line $f(x) = \alpha + \beta x$ which is the best fit to the data according to our least-squares criterion. Note that we do not exclude the possibility that some other type of functional dependence would provide a better fit. If the data points are regularly spaced and not too numerous, the unique interpolating polynomial passing through all or most of the data points might be preferable. However, this straight-line fit is very widely used, even on the most unlikely and variable data.

example 24 • 1 Show that $\overline{\epsilon^2} = \overline{x^2} - \bar{x}^2$.

solution By definition,

$$\overline{\epsilon^2} = \frac{1}{n}\sum_i (x_i - \bar{x})^2 = \frac{1}{n}\sum_i (x_i^2 - 2\bar{x}x_i + \bar{x}^2)$$

and separating the individual sums gives us

$$\overline{\epsilon^2} = \frac{1}{n}\sum_i x_i^2 - 2\bar{x}\frac{1}{n}\sum_i x_i + \frac{1}{n}\sum_i \bar{x}^2$$

then, since

$$n\overline{x^2} = \sum_i x_i^2 \qquad n\bar{x} = \sum_i x_i \qquad \sum_i \bar{x}^2 = n\bar{x}^2$$

this expression becomes

$$\overline{\epsilon^2} = \overline{x^2} - 2\bar{x}\bar{x} + \bar{x}^2 = \overline{x^2} - \bar{x}^2$$

example 24 • 2 Compute a straight-line fit to the following data:

x	0	1	2	3	4
y	2.28	2.48	2.86	3.16	3.32

solution First compute $\bar{x} = 2$, $\bar{y} = 2.82$, and the slope β. The results are

$$\beta = 0.276$$

$$\alpha = 2.82 - 0.276 \times 2 = 2.27$$

and

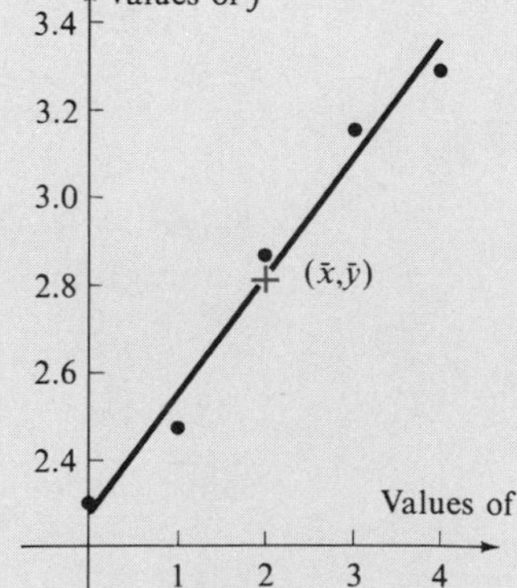

figure 24 • 1

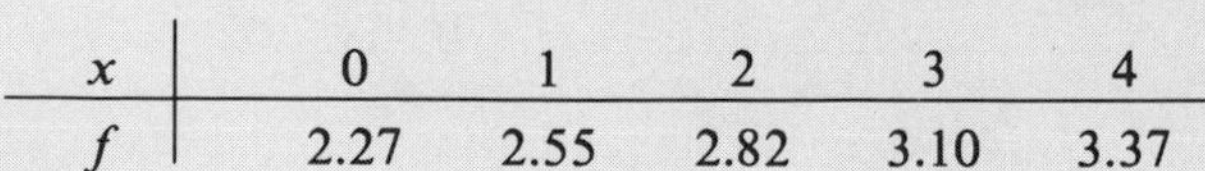

x	0	1	2	3	4
f	2.27	2.55	2.82	3.10	3.37

example 24 • 3 Find the best fit of the form $f(x) = ax^b$ to the following data:

x	1	2	3	4	5	6
y	2	20	85	220	485	870

solution Since $\log f = \log a + b \log x$, we can use a straight-line fit of $v = \log y$ to $u = \log x$, where $v \doteq \alpha + \beta u$, $\alpha = \log a$, and $\beta = b$. The data then become

u	0	0.301	0.477	0.602	0.699	0.778
v	0.301	1.301	1.929	2.343	2.686	2.940

Calculating α and β from Eqs. [24 . 8] and [24 . 12], we obtain

$$\alpha = 0.295 \qquad \beta = 3.408$$

so that $a = 10^{0.295} = 1.972$ and $b = 3.408$. It is easy to verify that $y \doteq f(x) = 1.972x^{3.408}$ fits the data to within 2 percent.

In dealing with raw data, one way to tell at a glance whether a power law gives a reasonable fit is to plot the data on log-log graph paper. If a straight line passes fairly through the points, then a computation like that above is in order. Dimensional consistency is preserved by assigning suitable dimensions to the constant of proportionality. Alternatively, $f(x)$ may be expressed in dimensionless form as

$$f(x) = \left(\frac{x}{0.819}\right)^{3.408}$$

24 • 2 *randomness and probability*

If we make repeated measurements of a length l_0, we obtain a number of individual measurements l_i. Suppose we now group these measurements into $i = 1, 2, 3, \ldots$ categories, as in Table 24 . 1, which shows the distribution of 200 values of l_i about the mean value $\bar{l} = 500$ mm. The *frequency of occurrence* is the number n_i in a given category, where the total number of measurements is $\Sigma_i n_i = n$. The same data are depicted in the histogram of Fig. 24 . 2, where each bar of width $\Delta x_i = 0.2$ mm has a height equal to the frequency of occurrence n_i. Probability theory tells us that as the number of data points becomes infinite and the categories Δx_i are made smaller and smaller, the polygonal curve formed by the top of the histogram will, in the limit as Δx_i approaches zero, become a smooth curve of the form

$$y = Y \exp\left[-\frac{(x - \mu)^2}{2\sigma^2}\right] \qquad [24 \cdot 13]$$

where the mean μ and the standard deviation σ are constant parameters determined by the nature of the particular data sample.

This type of curve is known as a *bell curve,* and as we shall see, in a statistical sense it represents a fit to the data. The exponent is always zero or negative, so that the curve has its maximum at the mean, where $x = \mu$ and $y = Y$. Since the exponent depends on the square of distance δ between x

table 24 • 1 *Hypothetical distribution of n = 200 measurements of x_0 = 500 mm*

k	n_k	*range of x, mm*	p_k
1	2	498.8–499.0	0.010
2	3	499.0–499.2	0.015
3	9	499.2–499.4	0.045
4	18	499.4–499.6	0.090
5	31	499.6–499.8	0.155
6	38	499.8–500.0	0.190
7	37	500.0–500.2	0.185
8	29	500.2–500.4	0.145
9	18	500.4–500.6	0.090
10	10	500.6–500.8	0.050
11	4	500.8–501.0	0.020
12	1	501.0–501.2	0.005
	200		1.000

and μ, it has the same sign for $x = \mu \pm \delta$. Hence the curve is symmetrical about $x = \mu$ and drops off to zero in the limit of $x = \pm\infty$. The standard deviation σ gives a measure of the spread, or *dispersion,* of the data relative to the height of the curve. That is, for a given value of $\delta = |x - \mu|$,

$$y = Y \exp\left(-\frac{\frac{1}{2}\delta^2}{\sigma^2}\right) \qquad [24 \cdot 14]$$

and the larger the value of σ, the smaller the exponent and the closer y is to Y. Conversely, if σ is very small, y will be an appreciable fraction of Y only at very small distances δ from the mean, and the curve will be very narrow, somewhat like a short pulse. In Fig. 24 . 2 the maximum value of y is $Y = 40$, and for $\mu = 500$ mm and $\delta = 0.4$ mm, the limiting curve is as

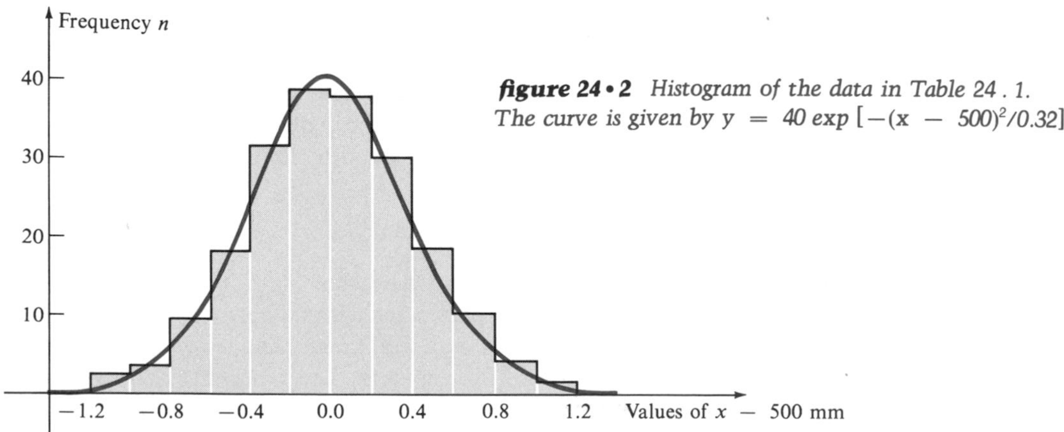

figure 24 • 2 *Histogram of the data in Table 24 . 1. The curve is given by y = 40 exp [−(x − 500)²/0.32]*

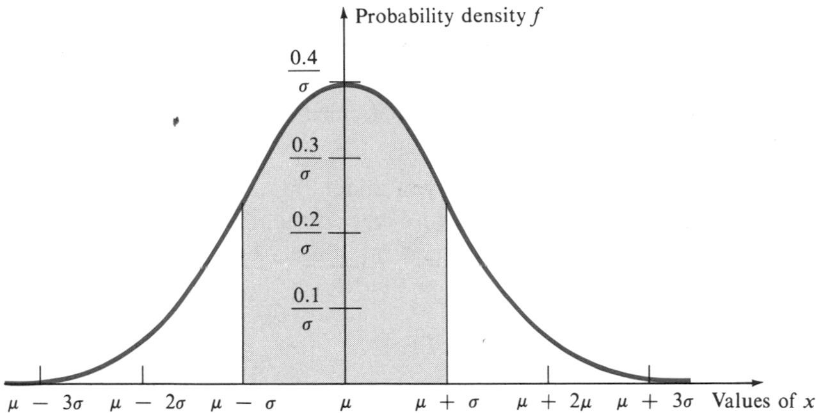

figure 24 • 3 *The Gaussian or normal distribution, from the probability-density function* $f(x) = (1/\sqrt{2\pi}\sigma) \exp[-(x - \mu)^2/2\sigma^2]$. *The area under the curve between the limits* $\mu - 3\sigma$ *and* $\mu + 3\sigma$ *is 0.997. The shaded area is 0.683. The total area under the curve is* $\int_{-\infty}^{\infty} f(x)\, dx = 1$.

shown. Although the curve extends to $x = \pm\infty$, for $\delta > 1.2$ mm the value of y is generally much less than 1 percent of its maximum.

We can predict the statistical distribution of future measurements by saying that there is a *probability* $P(i)$ that a future value will fall in a given category i. This probability can be estimated as

$$P(i) = \frac{n_i}{n} \qquad [24 \cdot 15]$$

In other words, "n_i times out of n" a measurement will fall in this category. For the data in Table 24 . 1, for example, 4 of the 200 measurements fall in category 11; therefore we would assign a probability of $P(11) = \frac{4}{200} = 0.02$, and for 100 additional measurements we would expect only $P(11) \times 100 = 2$ values to fall in this range.

In the limiting case of very large samples n and vanishingly narrow data groups $\Delta x_i \rightarrow dx_i$, the categories on either side of i will have nearly the same probabilities:

$$dP(i - 1) \doteq dP(i) \doteq dP(i + 1)$$

If we combine them into a single category i', then $dP(i') \doteq 3\, dP(i)$, indicating that, in the limit, probability is proportional to interval size; that is,

$$dP = \frac{dn}{n} = f(x)\, dx \qquad [24 \cdot 16]$$

The function $f(x) = dP/dx$ is known as the *probability density function,* defined by

$$\lim_{n \to \infty} \frac{dn/dx}{n} = f(x) = \frac{1}{\sqrt{2\pi}\sigma} \exp\left[-\frac{(x - \mu)^2}{2\sigma^2}\right] \qquad [24 \cdot 17]$$

This is the smooth bell-shaped curve of Fig. 24 . 3, which is known as the *Gaussian,* or *normal, distribution.*

It is clear from Eq. [24 . 16] that if we want to find the probability that a measurement will fall in the interval from x_1 to x_2, we simply integrate:

$$P(x_1 \le x \le x_2) = \int_{x_1}^{x_2} dP = \int_{x_1}^{x_2} f(x)\,dx \qquad [24 \cdot 18]$$

This integral is represented by the area under the curve from x_1 to x_2. To evaluate such integrals, we substitute the normalized random variable $u = (x - \mu)/\sigma$ into Eq. [24 . 17] and integrate from $-\infty$ to x to obtain the cumulative probability distribution $P(u)$:

$$\int_{-\infty}^{x} f(x)\,dx = \int_{-\infty}^{u} Z(u)\,du = P(u)$$

where the function $Z(u)$ is known as the *standardized normal distribution:**

$$Z(u) = \frac{1}{\sqrt{2\pi}} e^{-u^2/2} \qquad [24 \cdot 19]$$

Note that the total probability of measuring x anywhere between $-\infty$ and $+\infty$ is $P(\infty) = 1$, as it should be, since our distribution is normalized. Now, to predict a probability distribution of randomly selected values from Eq. [24 . 18], all we need is a method for estimating μ and σ.

When measurement data can be grouped into categories of frequency n_i and value x_i, then the arithmetic mean in Eq. [24 . 1] can be expressed as

$$\bar{x} = \frac{1}{n}\sum_i n_i x_i \qquad n = \sum_i n_i \qquad [24 \cdot 20]$$

and if we sum over infinitesimal categories, the summation becomes an integral. In this case the number in each interval from x to $x + dx$ is $dn = n\,dP$, and for a random variable x we have the integral

$$\lim_{n \to \infty} \bar{x} = \bar{x}_\infty = \frac{1}{n}\int_{-\infty}^{\infty} x(n\,dP) = \int_{-\infty}^{\infty} xf(x)\,dx \qquad [24 \cdot 21]$$

Then, since

$$\int_{-\infty}^{\infty} Z(u)\,du = 1$$

transforming the integral to the normalized form gives us

$$\bar{x}_\infty = \int_{-\infty}^{\infty} (\sigma u + \mu) Z(u)\,du$$

$$= \sigma\int_{-\infty}^{\infty} uZ(u)\,du + \mu\int_{-\infty}^{\infty} Z(u)\,du = \mu \qquad [24 \cdot 22]$$

where we have used the following facts: (1) The integrand $uZ(u)$ is an odd function of u and so integrates to zero and (2) the integral of Z over u is just 1. We can therefore use the mean of our data sample to *estimate* the true value of μ:

$$\bar{x} \doteq \bar{x}_\infty = \mu \qquad [24 \cdot 23]$$

*Some useful mathematical properties of $Z(u)$ are summarized in Appendix J.

The larger the data sample, the more closely the arithmetic mean $\bar{x}$ approximates μ.

It is evident from Fig. 24 . 3 that the height and breadth of the probability density function $f(x)$ depends on σ. As σ increases, the distribution becomes lower and broader. Fully 68.3 percent of the area under the curve occurs within $\pm\sigma$ of the mean value μ; 95.7 percent occurs within $\pm 2\sigma$, and 99.7 percent occurs within $\pm 3\sigma$. The probability of making a random measurement that is in error by more than $\pm 3\sigma$ is practically negligible. The variance of a continuous distribution is derived in the same way as the arithmetic mean. From Eq. [24 . 2],

$$\overline{\epsilon^2} = \frac{1}{n}\sum_i n_i\,(x_i - \bar{x})^2 \tag{24 . 24}$$

In the limit of very large data samples, $n_i/n \rightarrow dn/n = dP = f(x)\,dx$. Therefore

$$\lim_{n\to\infty} \overline{\epsilon^2} = \overline{e^2_\infty} = \frac{1}{n}\int_{-\infty}^{\infty} (x - \bar{x})^2\,dn$$
$$= \int_{-\infty}^{\infty} (x - \bar{x})^2 f(x)\,dx \tag{24 . 25}$$

and transforming to normalized form and integrating by parts, we have

$$\overline{\epsilon^2_\infty} = \int_{-\infty}^{\infty} (x - \mu)^2 f(x)\,dx = \sigma^2\int_{-\infty}^{\infty} u^2 Z(u)\,du = \sigma^2 \tag{24 . 26}$$

Thus we can use the mean-square error of our data sample to estimate the true value of σ:

$$\overline{\epsilon^2} = \overline{x^2} - \bar{x}^2 \doteq \overline{\epsilon^2_\infty} = \sigma^2 \tag{24 . 27}$$

With these estimates of μ and σ we can, in most cases, arrive at a good approximation to the true theoretical limiting distribution of measured values of x. Moreover, we can also estimate the amount of error inherent in the process of measurement itself. Figure 24 . 4 shows a flowchart for an algorithm to determine μ and σ from a set of n measurements of x.

Probability theory now tells us that in measuring a distance l, the true value l_0 that we wish to determine is best described by the relation

$$l_0 = \mu \pm \frac{\sigma}{\sqrt{n}} \tag{24 . 28}$$

In probabilistic terms, this equation states that if we perform $n = 1000$ measurements, then 683 times out of 1000 our estimate for μ will be within $\sigma/\sqrt{n}$ of l_0, 957 times out of 1000 it will be within $2\sigma/\sqrt{n}$ of l_0, and 997 times out of 1000 it will be within $3\sigma/\sqrt{n}$ of l_0. We still do not know the precise value of l_0, but the statistics determine the degree of *confidence* we can have in our predicted value $l_0 = \mu$. The uncertainty, $\Delta l = \sigma/\sqrt{n}$, must always be included in any precise specification of a constant. Uncertainties are usually expressed in parts per million (ppm) of μ, and most quantities can be measured within an accuracy of a few parts per million—roughly equivalent to measuring the length of a football field to within the thickness of this page.

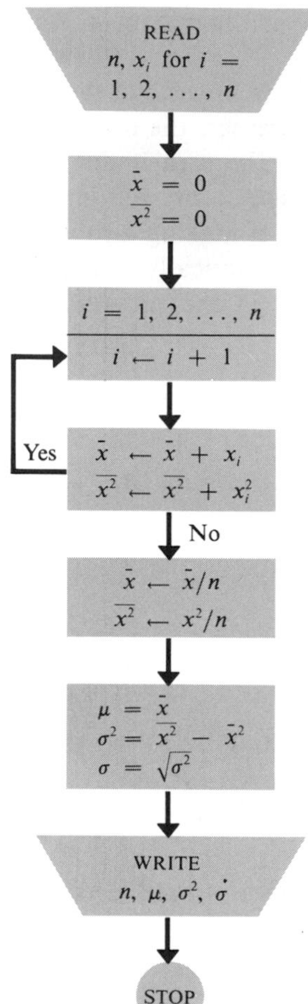

figure 24 • 4 *Algorithm for data reduction*

The fact that randomness has its own laws of behavior enables us to treat many problems statistically that would be otherwise hopeless. In fact, it is often necessary to create a mathematical model of randomness that can be studied with the aid of a computer. This requires a *random-number generator,* which is an algorithm for creating a completely random sequence of numbers. As might be expected, the creation of orderly rules for the simulation of complete disorder presents some subtle pitfalls, and much research has gone into the creation of random-number tables. Most computers provide a function which returns a random number on request.

Methods that simulate natural processes, such as diffusion or radioactive decay, by means of random numbers are called *Monte Carlo methods.* For example, to "roll dice," we would take pairs of digits from a random-number table (see Appendix J), discarding all 0's and digits greater than 6. The sequence

5446321018667390570638794125

when "cleaned" becomes

54, 46, 32, 11, 66, 35, 63*, 41, 25†

*Made the point. †Craps.

each digit occurring about three times, as might be expected for this many throws.

Because of the importance of statistical methods, most computer function libraries provide not only random-number generators, but also means to evaluate the functions $Z(u)$ and $P(u)$ for given values of u.

example 24 • 4 On the basis of the data in Table 24 . 1, predict the probability that x will fall in the range $499.6 \leq x \leq 500.2$ mm.

solution In this case we already know that $\mu = 500$ mm and $\sigma = 0.4$ mm; hence

$$(499.6 < x < 500.2) = \int_{499.6}^{500.2} f(x)\, dx$$

Since 499.6 is one standard deviation to the left of the mean and 500.2 is half a standard deviation to the right, we have

$$\int_{499.6}^{500.2} f(x)\, dx = \int_{-1}^{1/2} Z(u)\, du$$

$$= P(\tfrac{1}{2}) - P(-1) = 0.6915 - 0.1587 = 0.5328$$

Therefore $200 \times 0.5328 = 106.56$ measurements should fall in this range. We actually have 106 measurements in this range in Table 24 . 1.

The Gaussian distribution is often a useful approximation to other probability densities. In fact, the *central-limit theorem* of statistics states that if $y = \bar{x}_n$ is the arithmetic mean of n values of x selected from any population of values, then the *collection of mean values* y will become more and more randomly distributed as we take more average values, until it can be described by a Gaussian distribution. The standard deviation σ_y of these values decreases with $1/\sqrt{n}$. In other words, the more often an experiment is performed (neglecting systematic errors), the closer we can come statistically to the true value.

24 • 3 *the Maxwell-Boltzmann distribution*

Let us now use the statistical-probability concepts we have discussed to derive the distribution of molecular speeds in a stationary ideal gas. Since the velocities of the N molecules in a quantity of gas are perfectly random in all directions, the x component of their average velocity must vanish, $\bar{v}_x = 0$. If we apply the kinetic theory described in Secs. 22 . 2 and 22 . 3, we can show that the mean-square velocity is

$$\overline{v_x^2} = \tfrac{1}{3}\overline{v^2} = \frac{kT}{m} \qquad [24 \cdot 29]$$

where m is the mass of the molecule. Then, substituting into Eq. [24 . 27], we obtain the variance of the x component of velocity:

$$\sigma^2 = \overline{v_x^2} - \bar{v}_x^2 = \frac{kT}{m} \qquad [24 \cdot 30]$$

To represent the fraction dN_x/N of the molecules having velocities in the range from v_x to $v_x + dv_x$ by a Gaussian distribution, we simply substitute v_x for x and kT/m for σ^2 in Eq. [24 . 17]:

$$\frac{dN_x}{N} = \sqrt{\frac{m}{2\pi kT}} \exp\left(-\frac{\frac{1}{2}mv_x^2}{kT}\right) dv_x \qquad [24 \cdot 31]$$

The probability density function for v_x is shown in Fig. 24 . 5.

In nature even infinitesimal increments dv_x include enormous numbers of molecules, so that the differential notation is permissible. To find the actual number ΔN of molecules in some range of velocities from v_x to v'_x, we must evaluate the integral

$$\Delta N = \int dN = N\sqrt{\frac{m}{2\pi kT}} \int_{v_x}^{v'_x} \exp\left(-\frac{\frac{1}{2}mv_x^2}{kT}\right) dv_x \qquad [24 \cdot 32]$$

Note that this equation says nothing about the distribution of the other components of velocity, v_y and v_z. However, the molecules in the increment ΔN will contain all possible values of v_y and v_z.

Since the three degrees of translational freedom are independent, the fraction d^3N/N of the molecules with velocity components simultaneously in the ranges v_x to $v_x + dv_x$, v_y to $v_y + dv_y$, and v_z to $v_z + dv_z$ must be the product of the fraction $f(v_i)dv_i$ of molecules included in each of these ranges:

$$d^3N = Nf(v_x)f(v_y)f(v_z)\, dv_x\, dv_y\, dv_z \qquad [24 \cdot 33]$$

Therefore

$$d^3N = N\left(\sqrt{\frac{m}{2\pi kT}}\right)^3 \exp\left(-\frac{mv_x^2}{2kT}\right) \exp\left(-\frac{mv_y^2}{2kT}\right) \times \exp\left(-\frac{mv_z^2}{2kT}\right) dv_x\, dv_y\, dv_z$$

and since $v_x^2 + v_y^2 + v_z^2 = v^2$,

$$d^3N = N\left(\sqrt{\frac{m}{2\pi kT}}\right)^3 \exp\left(-\frac{mv^2}{2kT}\right) dv_x\, dv_y\, dv_z \qquad [24 \cdot 34]$$

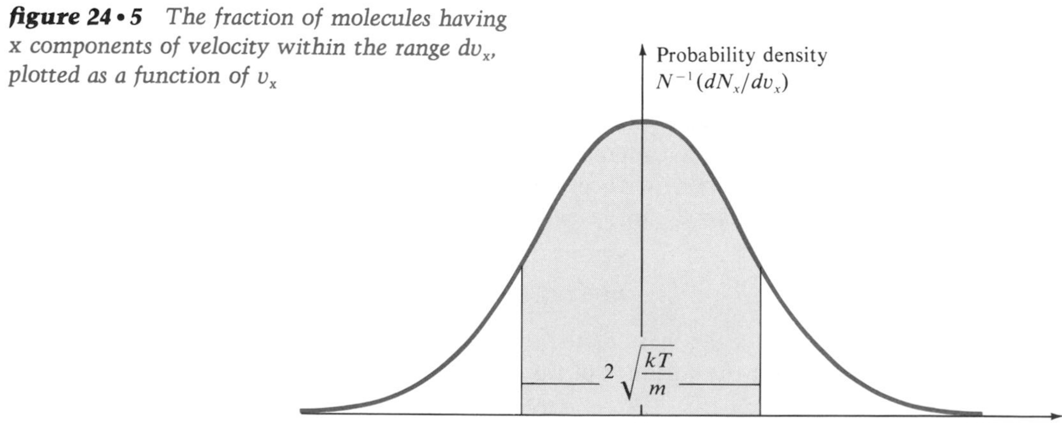

***figure 24 • 5** The fraction of molecules having x components of velocity within the range dv_x, plotted as a function of v_x*

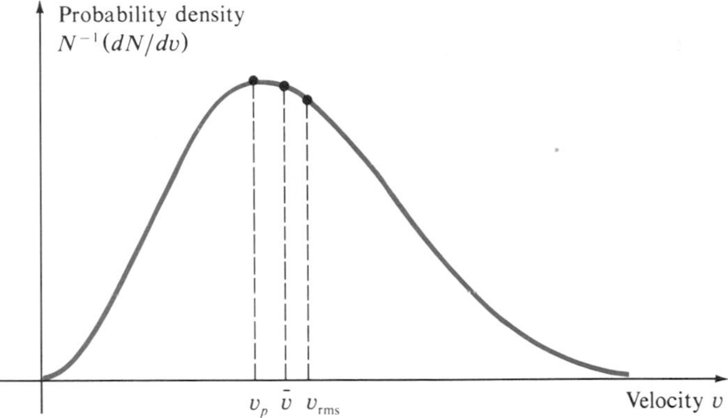

figure 24 • 6 *The Maxwell-Boltzmann distribution for a given temperature*

Then, to find the actual number ΔN of particles with velocity components simultaneously in the ranges v_x to v'_x, v_y to v'_y, and v_z to v'_z, we take the product of three separate and independent integrals:

$$\Delta N = N\left(\sqrt{\frac{m}{2\pi kT}}\right)^3 \int_{v_x}^{v'_x} \exp\left(-\frac{\frac{1}{2}mv_x^2}{kT}\right) dv_x \int_{v_y}^{v'_y} \exp\left(-\frac{\frac{1}{2}mv_y^2}{kT}\right) dv_y \times \int_{v_z}^{v'_z} \exp\left(-\frac{\frac{1}{2}mv_z^2}{kT}\right) dv_z \qquad [24 \cdot 35]$$

Since we were able to express Eq. [24 . 34] entirely in terms of v^2, independently of the direction of motion, we should be able to derive an expression for dN_v, the number of molecules with *speeds* in some range v to $v + dv$. This expression is known as the *Maxwell-Boltzmann distribution of molecular speeds* (see Prob. 24 . 20):

$$dN_v = 4\pi N\left(\sqrt{\frac{m}{2\pi kT}}\right)^3 v^2 \exp\left(-\frac{\frac{1}{2}mv^2}{kT}\right) dv \qquad [24 \cdot 36]$$

To find the number of molecules in some range from v to v', we simply integrate over speed. As we shall see, the presence of v^2 in the integrand causes no serious difficulty.

A graph of the probability density for speed, $(dN_v/N)/dv$, shows that the function is clearly asymmetrical and is defined only for $0 \leq v < \infty$ (see Fig. 24 . 6). However, the area under the curve is equal to unity. The abscissa value corresponding to the peak of the curve is the speed possessed at any moment by the largest number of molecules; it is therefore the *most probable speed* v_p. Because the curve is not symmetrical, the arithmetic-mean speed $\bar{v}$ is slightly greater than v_p, and v_{rms} comes still farther out on the abscissa. It is not difficult to show that

$$v_p : \bar{v} : v_{\text{rms}} = 0.8165 : 0.9213 : 1 \qquad [24 \cdot 37]$$

example 24•5 Prove that the area under the curve in Fig. 24 . 6 is equal to unity and find the arithmetic-mean speed $\bar{v}$.

solution First we set $\sigma = \sqrt{kT/m}$. Then, substituting the normalized variable $u = v/\sigma$, we obtain

$$\int_0^\infty \frac{dN}{N} = 4\pi\left(\sqrt{\frac{1}{2\pi\sigma^2}}\right)^3 \int_0^\infty v^2 e^{-v^2/2\sigma^2}\,dv = \frac{2}{\sqrt{2\pi}\sigma^3}\int_0^\infty \sigma^3 u^2 e^{-u^2/2}\,du$$

$$= 2\int_0^\infty u^2 Z(u)\,du = \int_{-\infty}^\infty u^2 Z(u)\,du = 1$$

To find $\bar{v} = \int_0^\infty v\,dN_v/N$ we express the integral as

$$\bar{v} = 4\pi\left(\sqrt{\frac{1}{2\pi\sigma^2}}\right)^3 \int_0^\infty v^3 e^{-v^2/2\sigma^2}\,dv = \frac{2\sigma}{\sqrt{2\pi}}\int_0^\infty u^3 e^{-u^2/2}\,du$$

The last integral is evaluated by parts. Since

$$\frac{dZ}{du} = -\frac{u}{\sqrt{2\pi}}e^{-u^2/2}$$

we obtain

$$\frac{2\sigma}{\sqrt{2\pi}}\int_0^\infty u^3 e^{-u^2/2}\,du = -2\sigma\int_0^\infty u^2\frac{dZ(u)}{du}\,du$$

$$= -2\sigma[u^2 Z(u)]_0^\infty + 4\sigma\int_0^\infty uZ(u)\,du$$

The term in brackets vanishes (see Appendix J), and the integral on the right becomes

$$4\sigma Z(0) = \frac{4\sigma}{\sqrt{2\pi}}$$

Therefore

$$\bar{v} = \frac{4\sigma}{\sqrt{2\pi}} = \sqrt{\frac{8kT}{\pi m}}$$

example 24•6 Find $\overline{v^2}$ for the Maxwell-Boltzmann speed distribution.

solution If each value of v^2 is weighted by dN_v, the number of particles in the speed range v to $v + dv$, then for $\sigma^2 = kT/m$,

$$N\overline{v^2} = \int v^2\,dN_v = \frac{2N}{\sqrt{2\pi}\sigma^3}\int_0^\infty v^4 e^{-v^2/2\sigma^2}\,dv$$

and successive integration by parts gives

$$v^2 = 2\sigma^2\int_0^\infty u^4 Z(u)\,du = 6\sigma^2\int_0^\infty u^2 Z(u)\,du = 3\sigma^2 = \frac{3kT}{m}$$

in agreement with Eq. [22 . 16].

example 24 • 7 Compute the relative increase in the number of ideal-gas molecules with kinetic energies greater than kT', $T' = 1350$ K, when the gas temperature is raised from $T_0 = 300$ K to $T_0' = 370$ K.

solution If we substitute the normalized variable $u = v/\sigma$ and integrate by parts, the probability of a molecular speed in the range $v_1 \leq v \leq \infty$ is

$$\int_{v_1}^{\infty} \frac{dN_v}{N} = 2\int_{u_1}^{\infty} u^2 Z(u)\, du = 2\int_{u_1}^{\infty} Z(u)\, du - 2[uZ(u)]_{u_1}^{\infty}$$

Therefore the probability that a molecule will have a speed $v \geq v_1$ is

$$\int_{v_1}^{\infty} \frac{dN_v}{N} = 2[P(\infty) - P(u_1)] + 2u_1 Z(u_1)$$

We next compute

$$u_1 = \frac{v_1}{\sigma} = \sqrt{\frac{mv_1^2}{kT}} = \sqrt{\frac{2T'}{T_0}} = 3.00$$

From Table J . 1 in Appendix J, we have

$$\int_{v_1}^{\infty} \frac{dN_v}{N} = 2(1 - 0.9987) + 6 \times 0.0044 = 0.0290$$

If the temperature is increased to $T_0' = 370$ K, then $u_1' = 3\sqrt{\frac{300}{370}} = 2.70$ and

$$\int_{v_1'}^{\infty} \frac{dN_v}{N} = 2\int_{2.70}^{\infty} u^2 Z(u)\, du = 2(1 - 0.9965) + 5.4 \times 0.0104$$

$$= 0.0632$$

Although the temperature was increased by only 23 percent, the number of particles with kinetic energies greater than kT' has increased by 118 percent.

Figure 24 . 7 shows the difference in speed distribution for an ideal gas at two different temperatures. The areas under the two curves are equal; that is, both curves refer to the same total number N of molecules. Notice that even at 773 K an appreciable fraction of the molecules have speeds greater than the most probable speed for 1273 K. These rather obvious differences are not nearly as dramatic as the effect of relatively small temperature changes.

In ordinary nonexplosive chemical reactions, it is the molecules in the "high-energy tail" of the Maxwell-Boltzmann speed distribution that have sufficient energy to form new compounds on collision. Reaction rates are so sensitive to temperature, however, that an increase of as little as 10 K may actually *double* the reaction speed by significantly increasing the population

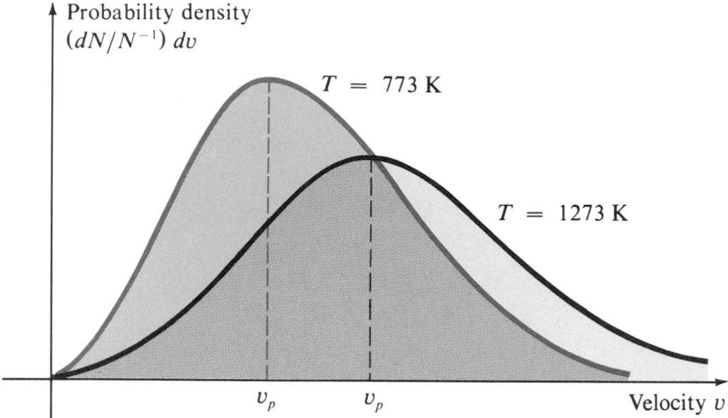

figure 24 • 7 *Maxwell-Boltzmann distributions for 773 K and 1273 K*

of particles in the high-energy tail. As we saw in Example 24 . 7, an increase of 70 K more than doubled the number of particles with energies greater than $1350k$ (reaction rates vary as the product of particle densities).

For very high activation temperatures, $T' > 10T$, we can ignore the $\int Z(u)\,du$ term, so that the relative increase in chemically active particles is

$$\frac{u_1'Z(u_1')}{u_1Z(u_1)} = \sqrt{\frac{T_0}{T_0'}}\exp\left(\frac{T'}{T_0} - \frac{T'}{T_0'}\right) \qquad [24 \cdot 38]$$

The reverse of this process is cooling by evaporation, in which the particles in the high-energy tail are removed from the system. Random collisions among the gas molecules soon repopulate the tail of the distribution, but the average kinetic energy of the system, and therefore the temperature, is lowered in the process.

Tests of the Maxwell-Boltzmann distributions were carried out from 1889 onward, but not until 1920 was the law confirmed by *direct* measurements of molecular speeds—a delay indicating the difficulty of the experimental techniques involved. For these measurements, Otto Stern developed the *molecular-beam method* illustrated in Fig. 24 . 8. Molecules evaporated

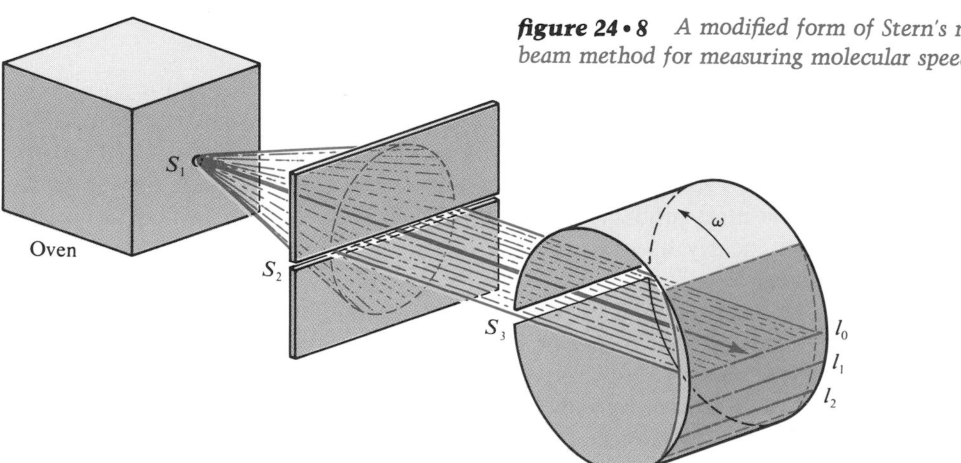

figure 24 • 8 *A modified form of Stern's molecular-beam method for measuring molecular speeds*

from a metal heated in a very hot oven were allowed to escape through a small hole S_1 and pass through two narrow slits S_2 and S_3. This last slit was in the side of a cylinder that could be rotated about its axis with constant angular speed ω. The apparatus was enclosed in an evacuated chamber.

With the cylinder at rest, a deposit of the metal gradually formed along a line l_0 on a curved glass plate attached to the inside wall. However, when the cylinder was rotating rapidly, the glass plate moved an appreciable distance while the molecules entering S_3 traversed the diameter d of the cylinder. Molecules having different speeds $v_1, v_2, \ldots$ formed deposits on the glass at different distances $l_1, l_2, \ldots$ from the center line l_0. It was therefore possible to compute $v = \frac{1}{2}\omega d^2/l$, and the relative thickness of the deposits plotted against $1/l$ yielded a curve that could be compared with the Maxwell-Boltzmann distribution curve appropriate to the temperature of the molecular beam. The similarity between Eq. [24 . 38] and Eq. [23 . 18] for the vapor pressure of an ideal gas implies a fundamental similarity in the temperature dependence of the distributions of molecular energies in a liquid and its vapor in equilibrium.

24 • 4 mean free path

As we have seen, kinetic theory indicates that the average speeds of gas molecules are of the order of 1 km/s. This result was very puzzling at first, in view of the fact that the time required for the odor, say, from an open bottle of ammonia, to travel a few meters across a room is a matter of minutes, rather than fractions of a second. Clausius saw that the explanation lay in the collisions of the molecules: a molecule of ammonia cannot move very far in one direction without striking a molecule of air, and the succession of unequal zigzag paths it traverses is such that a given molecule might take several hours to travel a few meters in any given direction. On this basis Clausius deduced a formula for the mean distance that a molecule travels between collisions. He called this distance the *mean free path,* and it has come to be an important concept in molecular and atomic theory.

Imagine that the "test molecule" in Fig. 24 . 9 is projected into a system of randomly located molecules which are *at rest.* The test molecule will

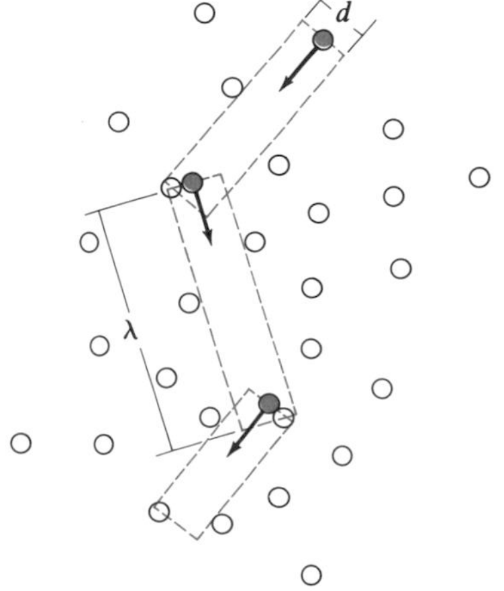

figure 24 • 9 *Method of computing the mean free path λ of a particle that has been projected into a group of stationary particles*

collide with any other molecule whose center lies within a distance d of the center of the test molecule. Its average collision speed is $\bar{v}$, the arithmetic-mean speed appropriate to the temperature of the system. In a time $t - t_0$ the test molecule travels a distance $\bar{v}t$. As the diameters of the molecules are assumed to be negligibly small compared with the distances between them, the number of collisions in this time interval is equal to the number of target particles whose centers lie in a cylinder of volume $\pi d^2\bar{v}t$. If n_{mol} is the *number density* of the molecules, then the number of collisions is

$$n_c = n_{\text{mol}}\pi d^2\bar{v}t \qquad \nu_c = \frac{n_c}{t} = n_{\text{mol}}\pi d^2\bar{v} \tag{24 . 39}$$

where ν_c is the *collision frequency* of the gas. The mean free path λ which the molecule traverses between collisions is found by dividing the actual distance traveled by n_c, the number of collisions:

$$\lambda = \frac{\bar{v}t}{n_c} = \frac{1}{n_{\text{mol}}\pi d^2} \tag{24 . 40}$$

Since it is implicit in our derivation that $\lambda \gg d$—that is, $n_{\text{mol}} \ll 1/\pi d^3$—our results are restricted to gases.

Obviously this simple calculation cannot be exact, for we have assumed all molecules except one to be at rest, so that the approach speed is $v_c = \bar{v}$. In an actual collision, however, both molecules are moving, and $0 < v_c < 2\bar{v}$, depending on the angle of approach. If we assume that the *average* angle of approach is 90 degrees, then as shown in Fig. 24 . 10, the average approach speed is

$$\bar{v}_c = \sqrt{2}\bar{v} \tag{24 . 41}$$

Substituting this value into Eqs. [24 . 39] gives us the expression for mean free path obtained by Maxwell:

$$\blacktriangleright\ \lambda = \frac{1}{\sqrt{2}n_{\text{mol}}\pi d^2} \tag{24 . 42}$$

The mean free path is inversely proportional to the number density of the molecules, and therefore to the density ρ of the gas. If both λ and n_{mol} are determined experimentally, Eq. [24 . 42] permits the calculation of molecular diameters d and collision cross sections πd^2.

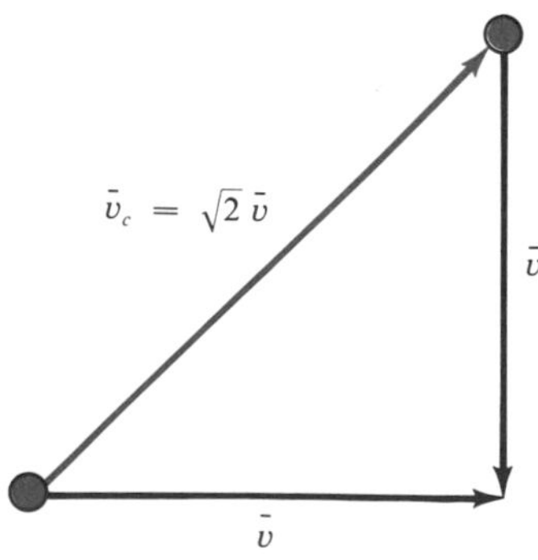

figure 24 • 10 *The average angle of approach for two identical particles in an ideal gas*

table 24 • 2 *Collision diameters calculated from the kinetic theory of gases*

gas	collision diameter d, 10^{-8} cm from viscosity η	from diffusivity D	from thermal conductivity $\mathcal{K}$
hydrogen	2.72	2.72	2.72
helium	2.18	—	2.20
nitrogen	3.78	3.84	3.78
oxygen	3.62	3.64	3.62
carbon dioxide	4.62	4.38	4.82

Methods are now available for making direct measurements of mean free paths, but originally these values had to be obtained indirectly from studies of the phenomena of viscosity, diffusion, and thermal conduction in gases. These phenomena provide three separate means for calculating λ and d, and the fairly good agreement of the values obtained was regarded as strong evidence for the validity and accuracy of the methods of kinetic theory (Table 24 . 2).

Current theory pictures molecules as consisting of positively charged and very compact nuclei surrounded by negatively charged "clouds" of electrons. It is the forces due to these electric charges that act between the molecules to divert them from their rectilinear paths. The deviations become larger as the two molecules approach each other; hence the molecular paths, although still nearly zigzag, are without any sharp corners. Obviously it is also improper to speak of the diameter of a molecule as though it were a rigid sphere. The quantity d in Eq. [24 . 42] is better called the *collision diameter,* defined as the distance of nearest approach of the *centers* of two molecules in a collision. If the molecules of a gas are not spherically symmetrical, the distance of nearest approach of two such molecules must depend on their momentary orientation with respect to each other. However, there is an *average* distance of approach that may be regarded as the collision diameter.

example 24 • 8 The smallest molecular radius is that of helium, approximately 10^{-10} m. Compute the mean free path of a helium molecule at 0°C and 1 atm.

solution Since 1 kmole of gas at 0°C and 1 atm occupies 22.4 m^3, the number density of the molecules is

$$n_{mol} = 6.02 \times 10^{26}\ \text{mol}/22.4\ \text{m}^3$$

Therefore, from Eq. [24 . 42],

$$\lambda = \frac{22.4\ \text{m}}{\sqrt{2}\,(6.02 \times 10^{26})\pi 10^{-20}} = 0.84 \times 10^{-6}\ \text{m}$$

24 • 5 momentum transport: viscosity

In Sec. 15 . 4 we saw that the shearing stress in a gas undergoing laminar flow is proportional to the velocity gradient in the gas. Moreover, in contrast to a liquid, the coefficient of viscosity for a gas increases with rising temperature and does not vary with pressure, except at very high and very low pressures. We are now in a position to investigate these peculiar properties of gases from the point of view of kinetic theory.

Suppose OO' in Fig. 24 . 11 represent the trace of an imaginary plane in a gas that is flowing from left to right with a velocity gradient such that v_x increases in successive upward layers from this plane. The vectors parallel to this plane represent the drift velocities of the layers of gas *relative* to the velocity of the layer OO'. Because the gas molecules in each layer are also in random thermal motion, some of the molecules in the layer above OO' will cross to the layer immediately below, and an equal number will pass upward to replace them. Thus the layer above OO' is continually losing the momentum associated with its flow, and the layer below OO' is continually gaining momentum. This will tend to decrease the relative motion of adjacent layers of the gas. As a result of this transport of momentum, there is a net tangential force on the plane OO' which causes the viscous drag on the fluid layer above it. Since the kinetic energy associated with the flow of the gas as a whole is being changed into the kinetic energy associated with the random motion of the molecules, the process is accompanied by a rise in temperature.

We can safely assume that "drift" speed u relative to the plane is much less than the average molecular speed $\bar{v}$. (It is important to note that this drift speed u is *not* the x component of $\bar{v}$, but represents the average fluid velocity in the x direction and varies with height y.) We can find the net momentum transport across plane OO' in three steps:

1. Find the *molecular flux density,* the number of molecules crossing the OO' plane per unit area per second.
2. Find the average vertical distance above or below the plane OO' at which the crossing molecule made its last collision.
3. From this determine the average drift speed $u(y)$ of the crossing molecules.

Combining these quantities, we can determine the *net* momentum transport across OO' and relate it to the coefficient of viscosity of the gas.

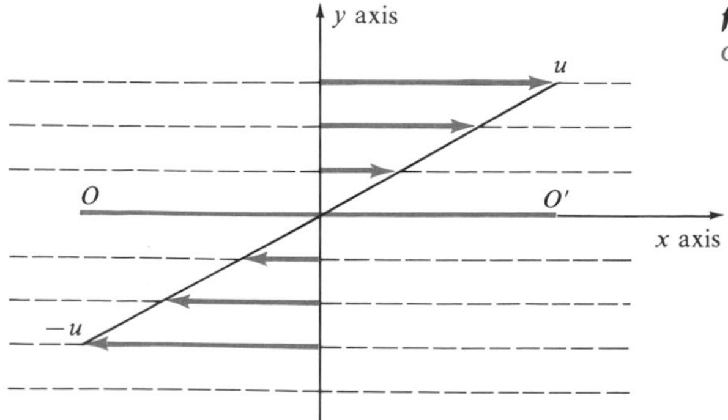

figure 24 • 11 *The kinetic-theory explanation of viscosity in a gas*

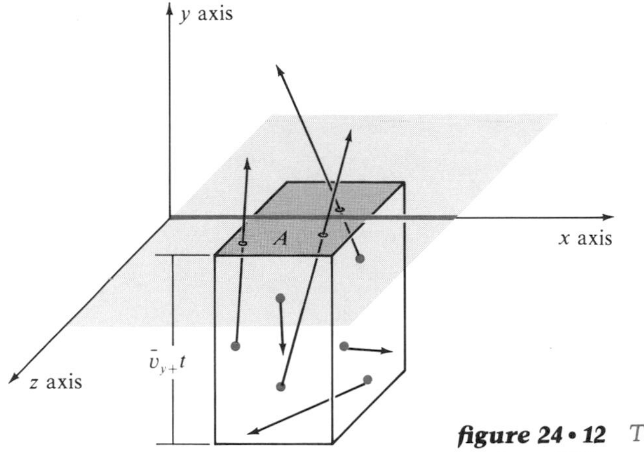

figure 24 • 12 *The particle flux across area A of plane OO′*

To determine the flux density we need to know the average velocity component $\bar{v}_{y+}$ of all $\frac{1}{2}N$ molecules traveling in the *positive y* direction (their other velocity components are irrelevant here). For $\sigma^2 = kT/m$ we can calculate

$$\tfrac{1}{2}N\bar{v}_{y+} = \frac{N}{\sqrt{2\pi}\sigma}\int_0^\infty v_y \exp\left(-\frac{v_y^2}{2\sigma^2}\right) dv_y = \frac{N\sigma}{\sqrt{2\pi}} \qquad [24 \cdot 43]$$

Recall from Example 24 . 6 that

$$\bar{v} = \sqrt{\frac{8kT}{\pi m}}$$

Therefore, integrating only over positive values of v_y, we find that the average velocity component is

$$\bar{v}_{y+} = \sqrt{\frac{2kT}{\pi m}} = \tfrac{1}{2}\bar{v} \qquad [24 \cdot 44]$$

Thus, as shown in Fig. 24 . 12, the *average* behavior is as if half of all particles contained in a box, whose upper surface of area A lies in the OO' plane and whose height is $\bar{v}_{y+}t$, cross the plane in time Δt. Hence the molecular flux density in the upward direction is

$$\blacktriangleright \quad \frac{\Delta N}{At} = \frac{\frac{1}{2}n_{\text{mol}}A\bar{v}_{y+}t}{At} = \tfrac{1}{2}n_{\text{mol}}\bar{v}_{y+} = \tfrac{1}{4}n_{\text{mol}}\bar{v} \qquad [24 \cdot 45]$$

This formula also holds generally for the number of gas molecules striking the wall of a container per unit area per second.

The second step, determining where the crossing molecule made its last collision, is a bit more difficult. On the average, this occurred at a distance of one mean-free-path length λ from A, but that could be in any direction whatsoever (see Fig. 24 . 13). Let us therefore assume that $A \ll \lambda^2$ and is located at the center of a hemisphere of radius λ. Since the distribution of velocities is random and the same in all directions, we can assume that incident molecules are coming in at all angles and are evenly distributed over

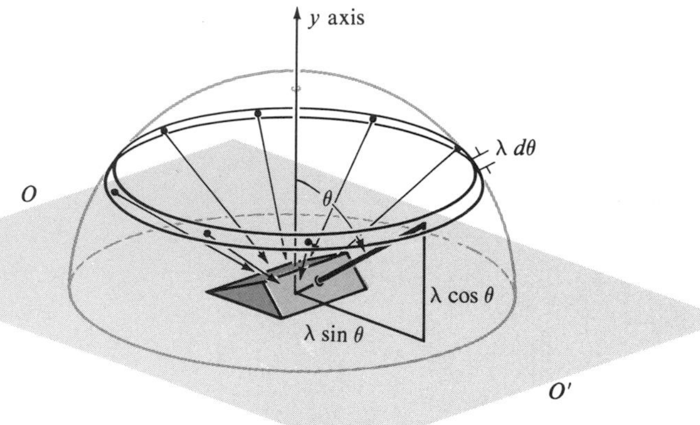

figure 24 • 13 *Hemispherical surface of particles incident upon A (area $A_\perp$ at angle θ) from a distance of one mean free path length λ*

the sphere. The number dN_θ crossing the plane OO' at angle θ is proportional to the surface area of the ring,

$$dS = 2\pi\lambda^2 \sin\theta \, d\theta \qquad [24 \cdot 46]$$

and to the *projected* area $A_\perp$ of A as seen from the ring. Thus

$$A_\perp = A\cos\theta \qquad [24 \cdot 47]$$

is the area projected perpendicular to a particle incident upon A from the ring dS. We now set

$$dN_\theta = CA_\perp \, dS = 2\pi A\lambda^2 C\cos\theta \sin\theta \, d\theta \qquad [24 \cdot 48]$$

where C is a constant of proportionality. The average height $\bar{y}$ from which particles arrive at area A is, for $y = \lambda\cos\theta$,

$$\bar{y} = \frac{\int_0^{\pi/2} y \, dN_\theta}{\int_0^{\pi/2} dN_\theta} = \frac{\int_0^{\pi/2} \lambda\cos^2\theta \sin\theta \, d\theta}{\int_0^{\pi/2} \cos\theta \sin\theta \, d\theta} = \tfrac{2}{3}\lambda \qquad [24 \cdot 49]$$

Therefore the particles arriving at A from this distance above and below the plane OO' cross with drift velocities

$$u_+ = u + \tfrac{2}{3}\lambda\frac{du}{dy} \qquad u_- = u - \tfrac{2}{3}\lambda\frac{du}{dy} \qquad [24 \cdot 50]$$

The molecular flux density is $\frac{1}{4}n_{\text{mol}}\bar{v}$ arriving at OO' from above and also from below. Hence the net momentum transfer into the fluid below OO' per unit area per second is

$$\frac{F}{A} = \frac{\Delta p/\Delta t}{A} = \begin{Bmatrix}\text{momentum gain} \\ \text{from above}\end{Bmatrix} + \begin{Bmatrix}\text{reaction to} \\ \text{momentum loss}\end{Bmatrix}$$

$$= \tfrac{1}{4}n_{\text{mol}}\bar{v}mu_+ - \tfrac{1}{4}n_{\text{mol}}\bar{v}mu_- = \tfrac{1}{3}n_{\text{mol}}m\bar{v}\lambda\frac{du}{dy} \qquad [24 \cdot 51]$$

which is the net tangential force per unit area exerted on the plane OO'. If we compare this to the viscosity equation $F/A = \eta\, du/dy$, we see that

$$\eta = \tfrac{1}{3} n_{\text{mol}} m \bar{v} \lambda = \tfrac{1}{3} \rho \bar{v} \lambda = \frac{2}{3\pi d^2} \sqrt{\frac{mkT}{\pi}} \qquad [24 \cdot 52]$$

which depends only on temperature.

example 24 • 9 A spaceship of volume $V = 27\ \text{m}^3$, with a cabin pressure of $P_0 = 1$ atm and a constant temperature such that $\bar{v} = 300$ m/s, is struck by a meteorite which opens a small hole of area $A = 1\ \text{cm}^2$, allowing the air to escape. How long will it take for the pressure to drop to $1/e \doteq 0.37$ atm? Assume that the temperature is maintained during the process.

solution The number N of air molecules in the ship changes at a rate

$$\frac{dN}{dt} = \tfrac{1}{4} n_{\text{mol}} \bar{v} A$$

due to the loss of air through the hole. Since $PV = NkT$ and $n_{\text{mol}} = N/V = P/kT$, the rate of pressure loss is

$$\frac{dP}{dt} = -\frac{A\bar{v}}{4V} P$$

and the pressure at any time t is given by

$$P = P_0 \exp\left(-\frac{A\bar{v}t}{4V}\right)$$

The pressure is therefore reduced to $1/e$ atm in a time

$$t = \frac{4V}{A\bar{v}} = \frac{4 \times 27\ \text{m}^3}{10^{-4}\ \text{m}^2 \times 300\ \text{m/s}} = 3600\ \text{s} = 1\ \text{h}$$

which is *independent* of the initial pressure. To make the air last longer we might reduce the temperature in the cabin, thus decreasing $\bar{v}$. An even better solution would be to make $A = 0$ by finding the hole and sealing it.

The coefficient of viscosity η for a given gas kept at constant temperature is independent of the pressure to which the gas is subjected. This fact led Maxwell to conclude that the torsional oscillations of a disk suspended in a gas should be equally damped by viscous resistance at any gas pressure. This hypothesis was so contrary to the general opinion of physicists at the time that its experimental confirmation by Meyer and Maxwell constituted one of the more brilliant successes of kinetic theory in its early history. Later experiments have shown that at pressures so low that the mean free path is comparable with the dimensions of the vessel enclosing the gas, viscosity is no longer independent of the pressure. The physical reason for this anomaly is that in a high vacuum a gas molecule makes many more impacts with the container walls than with other gas molecules, whereas at ordinary pressures the situation is reversed. Evidently one physical criterion of whether a gas is to be described by laws holding for low pressures or by laws holding for ordinary pressures is the length of the mean free path compared with the linear dimensions of the container. The "ordinary" case is $\lambda \ll l$, as in Example 24 . 8.

24 · 6 mass and energy transport

If the density of a gas at constant temperature is initially nonuniform, diffusion will take place from regions of higher density to those of lower density, ceasing only when the density becomes uniform. If two different gases that do not interact chemically are released in the same vessel, they will intermingle by the process of diffusion until the density is uniform throughout the vessel. Diffusion, like viscosity, is a *transport phenomenon.* However, there is an essential difference, in that diffusion depends on the transport of mass rather than momentum.

The general law of diffusion, applicable to both gases and liquids, was first stated by Fick in 1855. Fick saw that the time rate of flow of the diffusing fluid at any point in any direction y must depend on the *density gradient* $d\rho/dy$ at that point and in that direction. More specifically, for the case of steady diffusion in a direction perpendicular to an imaginary plane OO', Fick concluded that the *mass current,* or amount of mass diffusing per second, is proportional to both the area A of the plane OO' and to $-d\rho/dy$, the negative of the density gradient at OO':

$$\frac{dM}{dt} = -DA\frac{d\rho}{dy} \qquad [24 \cdot 53]$$

where D is a constant of proportionality termed the *diffusivity,* or *diffusion coefficient.* Experiment shows that D depends on the nature, common temperature, and total pressure of the mixing fluids, and to a slight extent on their relative proportions in the mixture. The following are approximate values of D for some gaseous mixtures at 0°C and 1 atm (in centimeters squared per second):

O_2, H_2	0.70
O_2, CO	0.18
O_2, N_2	0.19
H_2, water vapor	1.3

Let us now see how kinetic theory can be used to predict D.

If two interdiffusing gases consist of molecules of the same or nearly the same molecular mass and collision diameter—for example, N_2 and CO, O_2 and N_2, or the molecules of the isotopes of a given element—then we can readily calculate their diffusion coefficient from kinetic theory, using the results of Sec. 24 . 5. Consider the model of the diffusion process shown in Fig. 24 . 14. The number of molecules of gas crossing OO' into volume V_2 is $\frac{1}{4}n_{\mathrm{mol}+}\bar{v}$, where $n_{\mathrm{mol}+}$ is the average number density of the molecules crossing plane OO' just after their last collision. Similarly, the number of molecules per unit area crossing from volume V_2 into volume V_1 is $\frac{1}{4}n_{\mathrm{mol}-}\bar{v}$, assuming uniform temperature throughout.

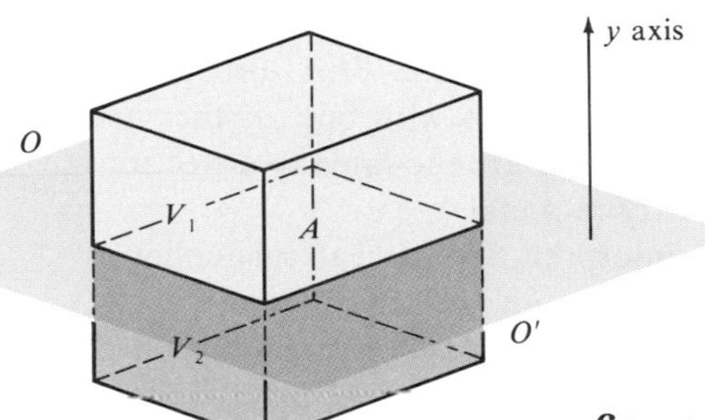

figure 24 • 14 *Model of the diffusion process*

If there is a constant density gradient dn_{mol}/dy, then the number density of the gas at height y is

$$n_{mol}(y) = n_{mol0} + y\frac{dn_{mol}}{dy} \qquad [24 \cdot 54]$$

The last collision takes place at an average of $\frac{2}{3}\lambda$ from the OO' plane, so that, from Eq. [24 . 47], we can express the average densities crossing OO' as

$$n_{mol+} = n_{mol}\,\tfrac{2}{3}\lambda = n_{mol0} + \tfrac{2}{3}\lambda\frac{dn_{mol}}{dy} \qquad [24 \cdot 55]$$

$$n_{mol-} = n_{mol}(-\tfrac{2}{3}\lambda) = n_{mol0} - \tfrac{2}{3}\lambda\frac{dn_{mol}}{dy} \qquad [24 \cdot 56]$$

The *net* molecular flux in the y direction is therefore

$$\frac{dN}{dt} = \tfrac{1}{4}\bar{v}(n_{mol-} - n_{mol+})A = -\tfrac{1}{3}\bar{v}\lambda A\frac{dn_{mol}}{dy} \qquad [24 \cdot 57]$$

Note that

$$\frac{dM}{dt} = m\frac{dN}{dt}$$

where m is the mass of the gas molecule and A is the total cross-sectional area of OO'. The mass density is thus $\rho = n_{mol}m$, and Eq. [24 . 57] becomes

$$\frac{dM}{dt} = -\tfrac{1}{3}\bar{v}\lambda A\frac{d\rho}{dy} \qquad [24 \cdot 58]$$

Comparing this to Fick's equation, Eq. [24 . 54], we obtain

$$D = \tfrac{1}{3}\bar{v}\lambda = \frac{\eta}{\rho} \qquad [24 \cdot 59]$$

These relations show that the diffusivity varies with the pressure even at ordinary pressures; indeed, for a fixed temperature, D is seen to be inversely proportional to the pressure. For a fixed pressure, D should vary as $T^{3/2}$; however, this prediction is found with actual gases to be inexact.

Heat conduction in *nonmetallic* solids and liquids consists in the transfer of kinetic energies through collisions of molecules or atoms in the high-temperature region with the slower-moving molecules or atoms in the low-temperature region. Because the thermally agitated atoms and molecules contain electrically charged particles, radiation between neighboring molecules and atoms may also be assumed to have a role in the energy transfer. In metals, heat conduction is due almost entirely to the free electrons, which are not attached to any particular atomic nucleus and can drift through the metal and transfer kinetic energy by their collisions with one another and with the massive ions in the crystal lattice. These are the same free electrons responsible for the conduction of electricity in metals.

It is in gases that a kinetic-theory interpretation of heat conduction is most simply obtained, for conduction depends on the transport of energy, just as viscosity in a gas depends on transport of momentum and diffusion on transport of mass. If a gas is confined between two parallel plates kept at different temperatures, the molecules hitting the hotter plate acquire additional kinetic energy which is then passed on to other molecules in collisions and eventually reaches the cooler plate. When a steady state is finally reached, the heat current and the temperature gradient remain constant throughout the gas. If the gas is considered to be composed of layers parallel to the confining planes, then molecules pass from layer to layer, the result being a net transport of kinetic energy from the hotter layers to the cooler ones. In a monatomic gas this transported kinetic energy is solely translational, whereas in a diatomic or polyatomic gas it includes rotational and vibrational energies. It was in light of such a picture that Maxwell, in 1860, carried out the first kinetic-theory calculation of heat conduction.

Consider a gas whose molecules have f degrees of freedom and an average energy $E_{mol} = \frac{1}{2}fkT$. If we assume a constant temperature gradient dT/dy and take the average drift speed as $\bar{v}_+ = \bar{v}_- = \bar{v}$, then substitution of the average molecular energy E_{mol} for momentum in Eq. [24 . 51] yields the net *heat-flux density* in the y direction:

$$q = \frac{\Delta \mathcal{U}}{A\,\Delta t} = -\left(\tfrac{1}{4}n_{mol}\bar{v}_+E_{mol+} - \tfrac{1}{4}n_{mol}\bar{v}_-E_{mol-}\right)$$

$$= \tfrac{1}{4}n_{mol}\bar{v}\tfrac{1}{2}fk(T_- - T_+) \qquad [24 \cdot 60]$$

where q is the heat crossing the plane OO' per unit area per second. Since, on the average, a molecule crossing OO' makes its last collision at $\frac{2}{3}\lambda$ from plane OO', we have

$$q = \tfrac{1}{4}n_{mol}\bar{v}\tfrac{1}{2}fk\left[\left(T_0 - \tfrac{2}{3}\lambda\frac{dT}{dy}\right) - \left(T_0 + \tfrac{2}{3}\lambda\frac{dT}{dy}\right)\right]$$

$$= -\tfrac{1}{6}n_{mol}\bar{v}fk\lambda\frac{dT}{dy} = -\mathcal{K}\frac{dT}{dy} \qquad [24 \cdot 61]$$

The thermal conductivity $\mathcal{K}$ of the gas is therefore

$$\mathcal{K} = \tfrac{1}{6}n_{mol}\bar{v}fk\lambda \qquad [24 \cdot 62]$$

and substitution of $1/\sqrt{2}\pi\, d^2$ for $n_{mol}\lambda$ gives us

$$\mathcal{K} = \frac{\bar{v}fk}{6\sqrt{2}\pi\, d^2} \qquad [24 \cdot 63]$$

In other words, the conductivity of a gas is independent of the pressure and depends only on temperature T, through $\bar{v}$, and on the number of degrees of freedom of each molecule of the gas. This has been demonstrated by experiment and confirms that kinetic theory can be applied to energy transport, as well as to the transport of mass and momentum.

table 24 • 3 *Summary of formulas*

arithmetic mean

$\bar{x} = \frac{1}{n}\sum_i n_i x_i \qquad n = \sum_i n_i$	discrete categories with frequency n_i
$\bar{x} = \int_{-\infty}^{\infty} x f(x)\, dx$	continuous distribution, probability density $f(x)$

mean-squared error (variance)

Formula	Description
$\overline{\epsilon^2} = \frac{1}{n}\sum_i n_i(x_i - \bar{x})^2 = \overline{x^2} - \bar{x}^2$	discrete categories with frequency n_i
$\overline{\epsilon^2} = \int_{-\infty}^{\infty} (x - \bar{x})^2 f(x)\, dx$	continuous distribution, probability density $f(x)$

normal distribution

$$f(x) = (2\pi\sigma^2)^{-1/2} \exp\left[-\frac{\frac{1}{2}(x - \mu)^2}{\sigma^2}\right]$$

$$\bar{x} = \mu$$

$$\overline{\epsilon^2} = \sigma^2$$

*standardized distribution functions**

$$Z(u) = (2\pi)^{-1/2} \exp\left(-\tfrac{1}{2}u^2\right)$$

$$P(u) = \int_{-\infty}^{u} Z(u)\, du$$

straight-line least-squares fit to data (x_i, y_i)

$$y \doteq f(x) = \bar{y} + \beta(x - \bar{x})$$

$$\beta = \frac{\sum_i (x_i - \bar{x})(y_i - \bar{y})}{\sum_i (x_i - \bar{x})^2} = \frac{\overline{xy} - \bar{x}\bar{y}}{\overline{x^2} - \bar{x}^2}$$

Maxwell-Boltzmann distributions, $\sigma^2 = kT/m$

Formula	Description
$dN_x = Nf(v_x)\, dv_x = N(2\pi\sigma^2)^{-1/2} \exp\left(-\frac{\frac{1}{2}v_x^2}{\sigma^2}\right)$	number of molecules with velocity components in the range $[v_x, v_x + dv_x]$
$d^3N = Nf(v_x)f(v_y)f(v_z)\, dv_x\, dv_y\, dv_z$	number of molecules with velocity components $[v_x, v_x + dv_x]$, $[v_y, v_y + dv_y]$, $[v_z, v_z + dv_z]$
$dN_v = 4\pi N(2\pi\sigma^2)^{-3/2} v^2 \exp\left[-\tfrac{1}{2}v^2/\sigma^2\right] dv$	number of molecules with speeds in the range $[v, v + dv]$

transport properties

Formula	Description	Formula	Description
$\lambda = (\sqrt{2}\, n\pi d^2)^{-1}$	mean free path	$\eta = \frac{1}{3} nm\bar{v}\, \lambda$	viscosity
$D = \frac{1}{3}\bar{v}\, \lambda$	diffusivity	$\mathcal{K} = \frac{1}{6} n\bar{v}\, fk\, \lambda$	thermal conductivity

*See Appendix J.

PROBLEMS

24 • 1 errors and statistics

24 • 1 In discussing random errors, why do we not use the average deviation from the mean, $\overline{x - \bar{x}}$, instead of ϵ?

24 • 2 We can think of the overbar used to denote averages as an "averaging operator." (*a*) Prove that the averaging operator is a *linear* operator; that is,

$$\overline{[af(x) + bg(x)]} = a\bar{f} + b\bar{g}$$

where a and b are constants. (*b*) Use this property to show that $\overline{\epsilon^2} = \overline{x^2} - \bar{x}^2$, without referring to the original summation notation.

24 • 3 Use the linear property of the averaging operator in Prob. 24 . 2 to show that the slope of the linear least-squares fit is given by

$$\beta = \frac{\overline{xy} - \bar{x}\bar{y}}{\overline{x^2} - \bar{x}^2}$$

and that if $y = A + Bx$ is truly a linear function of x, then $\alpha = A$ and $\beta = B$.

***24 • 4** Construct the algorithm for fitting a straight line to n data points and draw the corresponding flowchart.

***24 • 5** Compute a straight-line fit to the following data and sketch the results:

x	0.1	0.34	0.51	0.6	0.95	1.01	1.17	1.17	1.19	1.35
y	7	8	11	11	17	20	30	28	29	33

Compute the average squared error $R^2/10$ per data point.

24 • 6 Fit a straight line to the vapor-pressure data of Table 23 . 3, plotting $\ln P$ against $1/T$ (omit the critical temperature $T = 647$ K). Show how α and β are related to the parameters of P_v in Example 23 . 5 and compare a with the conditions at 100°C, using β to determine the best empirical fit to L_v of this (rather wide) range. Use your formula to predict P_v at the critical temperature and compare your result with the known value in Table 23 . 3.

24 • 2 randomness and probability

24 • 7 Verify that the normalization constant in Eq. [24 . 17] must be $1/\sqrt{2\pi\sigma^2}$ in order that $P(-\infty < x < \infty) = 1$.

24 • 8 For $\mu = 52$ and $\sigma = 7$, find $P(40 \leq x \leq 60)$, where x is a random variable.

24 • 9 Suppose you were to flip a coin eight times, counting 1 for heads and 0 for tails, and repeated the experiment many times. (*a*) What would be your average score? (*b*) What is the probability of getting eight heads in a row? (*c*) Compare this answer with $P(7.5 \leq \text{score} < \infty)$ computed from a Gaussian distribution with $\mu = 4$ and $\sigma = \sqrt{2}$.

24 • 10 The Gaussian distribution of Prob. 24 . 9 is a fair approximation to the probability density distribution of heads and tails in eight flips of the coin. (*a*) What is the probability of getting seven heads and one tail? (*b*) Compare this with the Gaussian prediction of $P(6.5 < \text{score} < 7.5)$. (*c*) What does the Gaussian distribution predict as the probability of obtaining four heads?

***24 • 11** It is possible to compute the area under a curve by Monte Carlo methods. If a function $f(x)$ and its argument x are transformed and scaled so that $0 \leq x \leq 1$ and $0 \leq f \leq 1$, then the computer generates pairs of random numbers (x, y) between 0 and 1. If $y \leq f(x)$ it scores one point, but no points for $y > f(x)$. After N = several hundred such computations, the ratio of points to tries is approximately $\int_0^1 f(x)\, dx$. Write an algorithm to perform such a computation. Using four-digit numbers for accuracy, compute $\int_0^1 \sin(\pi x)\, dx$. Compare your results for different values of $N = 100$, 1000, 10,000 with the known answer, $2/\pi$.

24 • 3 the Maxwell-Boltzmann distribution

24 • 12 Compute the *fraction* of molecules in a gas whose velocity components fall in the range $0 \leq v_x \leq \sigma$, $-\sigma \leq v_y \leq 0$, and $-\sigma \leq v_z \leq \sigma$, where $\sigma^2 = kT/m$.

24 • 13 For oxygen (O_2) gas at 1 atm and 22°C, compute the number of molecules per cubic centimeter whose three velocity components are each within ±1 of 300 m/s.

24 • 14 Show that the *most probable* speed in a Maxwell-Boltzmann distribution is $v_p = \sqrt{2kT/m}$.

24 • 15 A mixture of two gases at temperature T with molecular masses m and m' consists of a fraction in the ratio f by *weight* of the first gas to the second. Derive a formula for the arithmetic mean speed of the *mixture*.

24 • 16 Find the fraction of molecules in a gas whose kinetic energies are less than half the average energy.

24 • 17 Stern's experimental apparatus for measuring molecular speeds is shown in Fig. 24 . 8. (*a*) Derive the formula $v = \omega d^2/2l$ for the speeds of the molecules in the molecular beam. (*b*) In one such experiment a cylinder 10 cm in diameter was spun with a frequency of 240 Hz. If the beam temperature was 831°C and the duration of the experiment was 22 h, at what mean distance from the centerline would molecules traveling at 470 m/s have struck the glass plate?

24 • 18 For chemically interacting gaseous particles at room temperature with an activation temperature $T' = 3000$ K, find the relative increase in chemically active particles for a temperature change of $\Delta T_0 = 10$°K. Show that $\exp[T' \Delta T_0/T_0^2]$ is a good approximation to this figure.

24 • 19 If all molecules with speeds greater than 3σ are removed from a gas, what is the ratio of the temperature T' of the remaining molecules to the original temperature T_0 after equilibrium has been reestablished? (*Hint:* Define T' in terms of the average kinetic energy of the remaining molecules.)

24 • 20 To derive Eq. [24 . 36] it is helpful to imagine a "velocity space" in which the coordinates v_x, v_y, and v_z are exactly analogous to geometric xyz space. The state of motion of each molecule of the gas can then be represented by a "point" with coordinates (v_x, v_y, v_z). As shown below, the radial "distance" of such a point from the origin is just the speed

$$v = \sqrt{v_x^2 + v_y^2 + v_z^2}$$

(*a*) Find the "volume" in velocity space of a thin spherical shell of radius v and thickness dv. (*b*) Assume that the density of points in velocity space is given by

$$\rho_v = \frac{d^3N}{dv_x\, dv_y\, dv_z}$$

and relate this ratio to the number dN_v of molecules whose velocity coordinates fall within the spherical shell in part (*a*). (*c*) Determine the distribution function for the number of particles in a speed interval from v to $v + dv$.

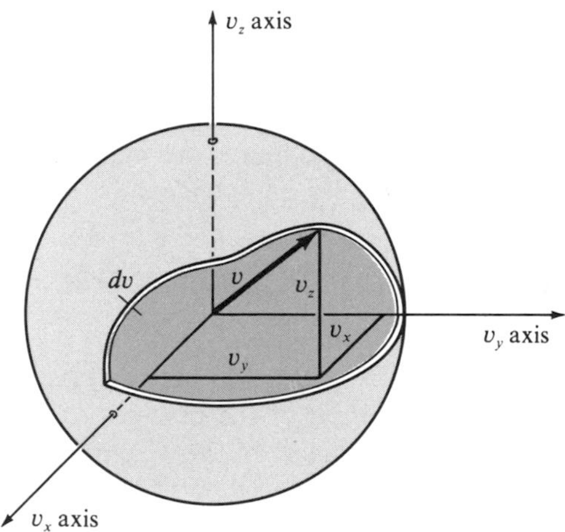

24 • 4 mean free path

24 • 21 Nitrogen molecules have a mean free path of $\lambda = 6.5 \times 10^{-6}$ cm at 17°C and 1 atm pressure. What is their mean free path at 17°C and 2×10^{-4} mm Hg?

24 • 22 The mean free path of oxygen molecules at 1 atm is about 10^{-5} cm. What is the approximate mean free path at a pressure of 10^{-6} mm Hg?

24 • 23 For nitrogen at 0°C and 1 atm, the mean free path of the molecules is 8×10^{-6} cm and their number density is 2.7×10^{19} mol/cm^3. (*a*) Assuming the molecules to be hard spheres, calculate the molecular diameter. (*b*) What is the actual volume occupied by the molecules comprising 1 m^3 of N_2? (*c*) Calculate the collision frequency of the gas molecules.

24 • 24 A neutron is moving through a system of similar spherical particles that are at rest and distributed at random. Assuming the diameter of the neutron to be negligibly small compared with the diameter d of the particles, show that the mean free path of the neutron is $4/\pi n d^2$.

24 • 5 viscosity

24 • 25 Using the value of the collision diameter of helium molecules in Table 24 . 2, find the total number of molecular collisions per second per cubic meter in helium (*a*) at 27°C and 1.0 atm. (*b*) At 27°C and 0.10 atm. (*c*) At 227°C and 1.0 atm. (*Hint:* Remember that each collision requires a *pair* of molecules.)

24 • 26 Two trains are moving eastward on parallel tracks at speeds of 64 and 16 km/h, respectively. It takes 30 s for one train to pass the other completely, and during this time 2 metric tons of bricks are thrown from each train to the other. What average force, in addition to that necessary because of the friction of the track, is required to keep each train moving at its original speed?

24 • 27 The coefficient of viscosity of hydrogen at 0°C is 0.867×10^{-5} dekapoise. (*a*) Calculate the mean free path in hydrogen at 0°C and 1 atm. (*b*) What is the collision frequency of hydrogen molecules at 0°C and 1 atm? (*c*) The collision diameter d of a hydrogen molecule?

24 • 28 If gas is leaking out of a small hole of area A in a container of molecules m at temperature T, show that the reaction force on the container must be $\frac{1}{8} nm(\bar{v})^2$.

24 • 29 A liquid is in equilibrium with its vapor at some vapor pressure P and temperature T. No other gas is present. In order to maintain equilibrium, how many molecules of mass m must be evaporating from the liquid per unit area per second?

24 • 30 A completely evacuated container of volume V is exposed to the atmosphere at pressure P_0, and a very small hole of area A is then drilled in the container, allowing the container to fill with air of average molecular weight $M_0 = 28.9$. (*a*) Find the pressure P inside the container as a function of time, assuming that $\bar{v}$ is the same inside and outside the container. (*b*) If $V = 1\ \text{m}^3$ and the area of the hole is $A = 1\ \text{cm}^2$, how long will it take, under standard temperature and pressure conditions, for the pressure inside the container to reach 0.9 atm? (*Hint:* You must also account for the escape of air from the container.)

24 • 31 Suppose N gas molecules of mass m are at temperature T inside an oven of volume V and have a Maxwell-Boltzmann speed distribution dN_v. (*a*) What is the flux density of molecules striking the walls of the oven at speeds in the range v to $v + dv$? (*b*) A very small hole of area A is drilled in the side of the oven which does not appreciably disturb the gas within. How much energy per second is carried away by the particles escaping through the hole at speeds v to $v + dv$? (*c*) Show that the average kinetic energy of the escaping molecules is given by

$$\overline{K} = \frac{1}{2}m \frac{\int_0^\infty v^3\, dN_v}{\int_0^\infty v\, dN_v}$$

(*d*) Show that the escaping molecules have a root-mean-square speed 15.5

percent greater than the molecules inside the oven by applying the following formula:

$$\int_0^\infty v^{2n+1} \exp\left(-\frac{\frac{1}{2}v^2}{\sigma^2}\right) dv = 2^n n!\,(\sigma^2)^{n+1}$$

where n is an integer.

24 • 32 An experimental test of the Maxwell-Boltzmann distribution law carried out in 1947 was based on the principle that a beam of particles is deflected downward by the earth's gravitational field. In an evacuated chamber, a narrow beam of cesium atoms (molecular weight 133) emerges horizontally from a source of temperature 177°C and travels a distance of 150 cm before striking a vertical screen. At what distance below the original level will the deposits of atoms on the screen be heaviest? (*Hint:* The distribution function of particles in a beam is different from that in a gas which is stationary; see the results of Prob. 24 . 31.)

24 • 6 mass and energy transport

24 • 33 Prove that the diffusivity for a mixture of gases having molecules of similar masses and similar collision diameters is (*a*) inversely proportional to the pressure when the temperature T is constant. (*b*) Proportional to $T^{1/2}$ when the density is constant. (*c*) Proportional to $T^{3/2}$ when the pressure is constant.

24 • 34 A vertical tube containing gas of molecular mass m at temperature T has pressure distributed along it according to the law of atmospheres:

$$P = P_0 e^{-y/h} \qquad h = \frac{kT}{mg}$$

When the tube is laid on its side, derive an expression for the mass current per unit area as a function of horizontal distance from the base of the tube. (*Note:* These are initial values; density gradients vanish with time in a closed container.)

24 • 35 Compute the thermal conductivity for argon and for oxygen at 1 atm and 0°C and compare your results with the experimental values of 0.016 and 0.024 J/m · s · K. Take d = 3.6 Å for both gases.

24 • 36 Show that kinetic theory predicts that $\mathcal{K} M_0/\eta C_V = 1$.

answers

24 • 1 The average deviation is zero by definition
24 • 5 $\alpha = 0.729$, $\beta = 22.0$, $R^2/10 = 9.76$
24 • 6 $\alpha = 20.30$, $\beta = 5112$, $P = P_0 \exp[-5112(1/T - 1/289)]$, $P_0 = 13.53$ mm Hg, $L_v = 564$ kcal/kg, error less than 4.7 percent
24 • 8 $P(40 \leq x \leq 60) = 0.83$
24 • 9 (*a*) Average score = 4; (*b*) P(8 consecutive heads) $= 2^{-8} = 0.0039$; (*c*) $P(7.5 \leq \text{score}) = 0.0066$
24 • 10 (*a*) P(7 heads, 1 tail) $= 8 \times 2^{-8} = 0.03125$; (*b*) $P(6.5 \leq x \leq 7.5) = 0.0218$; (*c*) $P(3.5 \leq 4 \leq 4.5 \text{ heads}) = 0.276 \doteq 70 \times 2^{-8} = 0.273$
24 • 12 $P = 0.34 \times 0.34 \times 0.68 = 0.079$
24 • 13 $4.1 \times 10^{-9} N = 1.03 \times 10^{11}$ mol/cm^3
24 • 15 $\bar{v} = \sqrt{8kT/\pi m m'}\,(m^{3/2} + fm'^{3/2})/(m + fm')$
24 • 16 31.8 percent

24 • 17 (*b*) $l = 1.60$ cm
24 • 18 Increase = 38 percent
24 • 19 $T'/T_0 = 0.918$
24 • 20 (*a*) $dV_v = 4\pi v^2\,dv$;
(*b*) $dN_v = 4\pi\rho_v v^2\,dv$; (*c*) see Eq. [24 . 36]
24 • 21 $\lambda = 24.7$ cm
24 • 22 $\lambda = 76$ m
24 • 23 (*a*) $d = 3.23$ Å;
(*b*) $V_{mol} = 4.76 \times 10^{-4}\ \text{cm}^3$;
(*c*) $\nu_c = 4.0 \times 10^9$ collisions/s
24 • 25 (*a*) $N_c = 5.7 \times 10^{34}$ collisions/s · m³;
(*b*) $N_c = 5.7 \times 10^{32}$ collisions/s · m³;
(*c*) $N_c = 2.6 \times 10^{34}$ collisions/s · m³
24 • 26 $F = \pm 889$ N, east
24 • 27 (*a*) $\lambda = 1.72 \times 10^{-7}$ cm; (*b*) $\nu = 9.91 \times 10^9$ collisions/s; (*c*) $d = 2.63$ Å
24 • 29 $dN/dt = P_v/\sqrt{2\pi mkT}$
24 • 30 (*a*) $P = P_0[1 - \exp(-t/\tau)]$, where $\tau = 4V/A\bar{v}$; (*b*) $t = 206$ s
24 • 31 (*a*) $dn/dt = \frac{1}{4}v\,dN_v/V$; (*b*) $d\mathcal{U}/dt = \frac{1}{8}(mA/V)v^3\,dN_v$; (*d*) $v_{rms} = \sqrt{4kT/m}$
24 • 32 $\Delta y = 0.30$ mm
24 • 34 $dM/dt = \frac{1}{3}m\bar{v}/\pi hd^2 = 0.169\sqrt{m^3/kT}\,g/d^2$
24 • 35 $\mathcal{K}_{Ar} = 0.0064$ J/kmole · s · K, $\mathcal{K}_{O} = 0.012$ J/kmole · s · K

CHAPTER TWENTY-FIVE

entropy

If we think of that quantity which with reference to a single body I have called its entropy, as formed in a consistent way, with consideration of all the circumstances, for the whole universe, and if we use in connection with it the other simpler concept of energy, we can express the fundamental laws of the mechanical theory of heat [thermodynamics] in the following simple form.

1 *The energy of the universe is constant.*
2 *The entropy of the universe tends toward a maximum.*

Clausius, in Annalen der Physik und Chemie, *1865*

The convertibility of heat and work expressed by the first law of thermodynamics places heat on the same footing as all forms of mechanical energy. It assures, for example, that the heat generated when a rotating flywheel is stopped by friction, or when a falling object is stopped by the ground, is accounted for by a corresponding loss of kinetic energy. However, since there is nothing in the first law that specifies the direction of these changes, the reverse of these processes should also be true. A flywheel should therefore start rotating spontaneously if it acquires enough heat from the bearings as they cool, and an object on the ground should spontaneously jump upward while the ground beneath it becomes cooler. Neither of these processes violates the first law, yet they have never been observed to happen. Such considerations indicate that when heat is one of the forms of energy in question, the first law of thermodynamics alone is not a sufficient condition for determining the direction of a thermal process.

Such reversals of ordinary experience, although theoretically possible, are so improbable that we may, in effect, decree them impossible, and in so doing, proclaim a new law, the *second law of thermodynamics.* This law states that physical changes in the universe always move in the direction of increasing disorder, or increasing *entropy.* The concept that all motion in the universe tends to become increasingly random is statistical in nature and has no counterpart in classical Newtonian mechanics. However, this law has important consequences for all the natural sciences. Its philosophical implications have also greatly exercised scholars of every shade of opinion during the century since its enunciation. In this chapter we shall consider the physical concepts which underlie the second law of thermodynamics.

25 · 1 engines

The problems raised by the development of the steam engine primarily involved the conversion of heat energy into work. In 1712 Thomas Newcomen built a successful piston-and-cylinder steam engine for pumping water from mines. It was through studies of a model of the Newcomen engine owned by Glasgow University that James Watt, in 1763, started the series of discoveries and improvements that made the steam engine an effective source of power. The successful development of the reciprocating steam engine was mainly responsible for the drastic social and economic changes of the first industrial revolution, as well as for the inception of the science of thermodynamics.

In 1824, when Watt's steam engine was already in wide use, and many years before the first law of thermodynamics was firmly established, the French engineer N. L. Sadi Carnot published a remarkable memoir that laid the basis for thermodynamics and the solution of one of the most fundamental problems in the entire range of physical science.* Carnot pointed out that the *working substance* of a heat engine, such as the steam of Watt's engines, performs work by absorbing heat from a high-temperature source and deivering some of it to a receiver, or *heat sink,* of lower temperature. Realizing that any actual heat engine involves some processes that are not amenable to exact mathematical analysis, Carnot began with an *ideal* heat engine, which has a cycle of operations consisting of processes that are thermodynamically reversible. He assumed a heat source and sink extensive enough to be unaffected themselves by the heat transfer, so that the temperatures of these "heat reservoirs" remain constant during the process.

A thermodynamically reversible process has two salient features:

1 It is devoid of all dissipative effects, such as viscosity in the steam or other working substance and friction in the piston or other moving parts.
2 It is performed quasistatically.

Any process having these two features is called *thermodynamically reversible,* or simply *reversible.* Since no dissipative effects are involved, all the external work done by the working substance in the process of passing from

*Carnot, son of the man whose genius preserved the First Republic of revolutionary France, died at 36 and has only this one publication to his credit. Yet he is regarded as one of the most original and profound scientific thinkers.

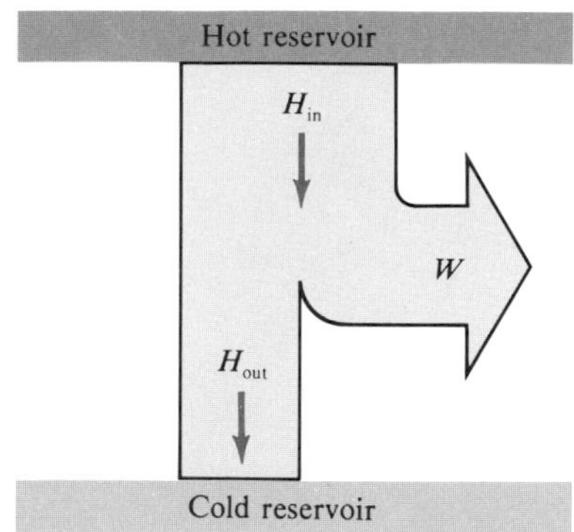

figure 25 • 1 *Schematic diagram showing the work* $W = H_{in} - H_{out} = \eta H_{in}$ *per cycle absorbed from a heat reservoir by an engine*

an initial to a final state will be returned to the working substance if the system is brought back to its initial state by precisely the same process in reverse order. Since the process is performed quasistatically, the working substance and its surroundings pass only through thermodynamic equilibrium states, and these can be traversed just as well in one direction as in the other.

Experience shows that no actual physical process entirely meets the two stringent conditions for thermodynamic reversibility; all processes are more or less irreversible. However, an actual process carried out under controlled conditions can often be made approximately reversible—as when a gas is expanded or compressed very slowly and the pressure and temperature differences between the gas and its surroundings are kept very small throughout the process.

The *efficiency* of any operation, defined in general as the ratio of the useful energy output to the total energy input, was originally formulated to provide a criterion of the performance of machines. For a reversible heat engine, which does external work continuously by performing the same cycle of operations over and over again, a useful concept is the *thermal efficiency* η:

$$\eta = \frac{W}{H_{in}} \qquad [25 \cdot 1]$$

where W is the net external work done by the steam or other working substance during one cycle and H_{in} is the amount of heat the working substance absorbs per cycle from the high-temperature source or "hot" reservoir (Fig. 25 . 1). In this chapter H_{in} is used instead of Q_{in} so that we can indicate the absorption or rejection of heat explicitly by plus or minus signs.

If the working substance "rejects" an amount of heat H_{out} to the low-temperature cold reservoir, then the *net* heat absorbed per cycle is $\Delta Q = H_{in} - H_{out}$. In any complete reversible cycle the working substance returns to its original state. Hence the net change in internal energy of the working substance is, for a reversible process,

$$\Delta \mathcal{U} = \Delta Q - W = 0 \qquad [25 \cdot 2]$$

$$\eta = \frac{W}{H_{in}} = \frac{\Delta Q}{H_{in}} = 1 - \frac{H_{out}}{H_{in}} \qquad [25 \cdot 3]$$

As we shall see later, every heat engine, even a reversible engine, rejects some heat to the cold reservoir, so that thermal efficiency is always less than unity. For example, if the working substance of an ideal reversible heat engine absorbs 10 kcal/cycle from the hot reservoir and rejects 8 kcal/cycle to the cold reservoir, then $\eta = 1 - \frac{8}{10} = 0.2$, or 20 percent.

25 • 2 the ideal-gas Carnot engine

Carnot sought to determine whether some particular kind of ideal engine has a higher thermal efficiency than any other conceivable engine operating under similar conditions. He reasoned that the efficiency of a reversible engine would be a maximum if the process was as smoothly quasistatic as possible. Therefore H_{in} should be absorbed from a single reservoir kept at a constant high temperature T_1, while H_{out} should be rejected to a single cold reservoir kept at a constant temperature T_2. Furthermore, transitions *between* these two temperatures should be adiabatic, so that no heat transfer is involved. The resulting reversible *Carnot cycle,* shown in Fig. 25 . 2, consists of two isothermal and two adiabatic processes. Any engine employing a Carnot cycle is called a *Carnot engine.*

To see what happens during the Carnot cycle *ABCD* in Fig. 25 . 2, let us assume that the cylinder in Fig. 25 . 3 contains volume V_A of n moles of an ideal gas which has previously been in contact with the cold reservoir at temperature T_2. The energy equations for each part of the Carnot cycle are then as follows:

$A \rightarrow B$: With the end of the cylinder against the insulating end block, the gas is compressed adiabatically, so that $\Delta Q = 0$, until its temperature rises to T_1. At this point its volume has been reduced to V_B. From the ideal-gas equation and the adiabatic equation $PV^\gamma = $ constant, we have

$$T_1 V_B^{\gamma - 1} = T_2 V_A^{\gamma - 1} \qquad [25 \cdot 4]$$

figure 25 • 2 *The ideal Carnot cycle with an ideal gas as the working substance*

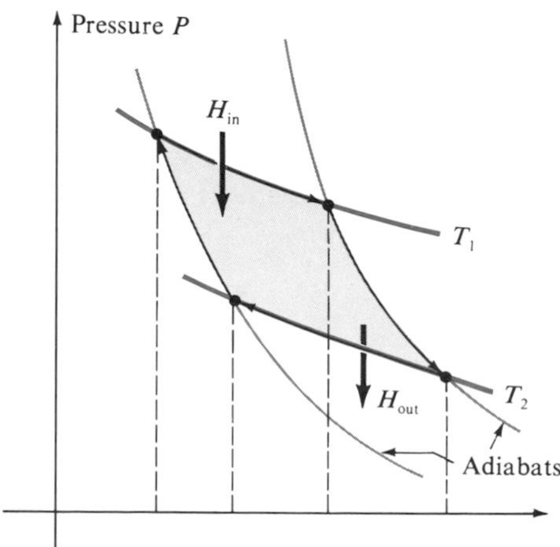

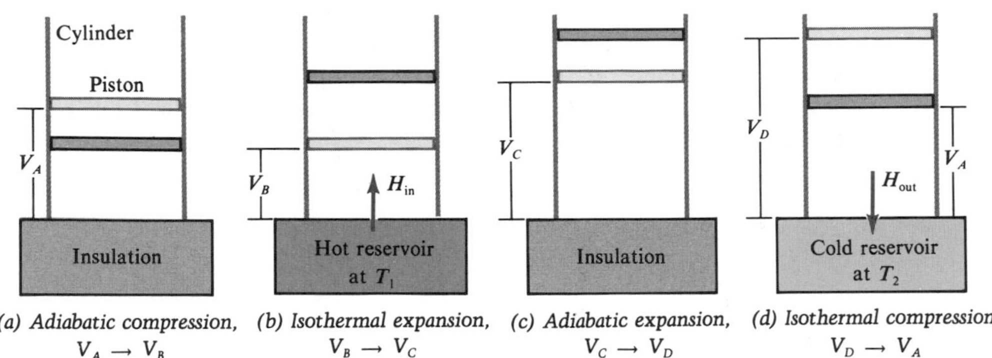

(a) Adiabatic compression, $V_A \rightarrow V_B$ *(b) Isothermal expansion,* $V_B \rightarrow V_C$ *(c) Adiabatic expansion,* $V_C \rightarrow V_D$ *(d) Isothermal compression,* $V_D \rightarrow V_A$

figure 25 • 3 *Physical processes in the ideal Carnot cycle*

in an adiabatic compression. Since $\mathcal{U} = nC_VT$ for an ideal gas, the work done by the gas during this process is

$$\Delta W = -\Delta\mathcal{U} = -nC_V(T_1 - T_2) < 0 \qquad [25 \cdot 5]$$

$B \rightarrow C$: With the cylinder now in thermal contact with the hot reservoir at T_1, the gas is allowed to expand isothermally to some arbitrary volume V_C, doing work W_1 on the piston while it absorbs heat Q_1 from the heat source. Since the internal energy of an ideal gas does not change during any isothermal process, $Q_1 = W_1$ and the heat absorbed by the system is

$$\begin{aligned} H_{in} = Q_1 = W_1 &= \int_{V_B}^{V_C} P\,dV = nRT_1 \int_{V_B}^{V_C} \frac{dV}{V} \\ &= nRT_1 \ln \frac{V_C}{V_B} > 0 \end{aligned} \qquad [25 \cdot 6]$$

$C \rightarrow D$: With the cylinder against the nonconducting end block, the gas is now allowed to expand adiabatically to volume V_D, doing external work with no gain or loss of heat, until its temperature falls to T_2. At this point,

$$\begin{aligned} T_1V_C^{\gamma-1} &= T_2V_D^{\gamma-1} \\ \Delta W' = -\Delta\mathcal{U} &= nC_V(T_1 - T_2) > 0 \end{aligned} \qquad [25 \cdot 7]$$

$D \rightarrow A$: With the cylinder against the cold reservoir at T_2, the gas is now compressed isothermally until it reaches its initial state V_A and T_2. The piston does work W_2 on the gas, and the gas rejects heat H_{out} to the cold reservoir. Thus, as in Eq. [25 . 6],

$$\begin{aligned} -H_{out} = Q_2 = W_2 &= nRT_2 \ln \frac{V_A}{V_D} \\ &= -nRT_2 \ln \frac{V_D}{V_A} < 0 \end{aligned} \qquad [25 \cdot 8]$$

where Q_1 and Q_2 conventionally refer to heat *absorbed* by the working substance.

The cycle is now complete: the working substance has been returned to its initial state, whereas the hot reservoir has lost an amount of heat $H_{in} = Q_1$ and the cold reservoir has gained a smaller quantity $H_{out} = -Q_2$. The work done on the adiabatic paths cancels, for $\Delta W + \Delta W' = 0$; hence the net work done is $W = W_1 + W_2 = Q_1 + Q_2$. The thermal efficiency of the Carnot cycle is

$$\eta_c = 1 + \frac{Q_2}{Q_1} = 1 - \frac{H_{out}}{H_{in}} \tag{25 · 9}$$

In subsequent cycles the same fixed mass of working substance is used over and over again. The adiabatic temperature-volume relations of Eqs. [25 . 4] and [25 . 7] show that

$$\frac{V_A}{V_B} = \frac{V_D}{V_C} = \left(\frac{T_1}{T_2}\right)^{1/(\gamma - 1)} \tag{25 · 10}$$

or

$$\frac{V_A}{V_D} = \frac{V_B}{V_C} \qquad \ln\frac{V_A}{V_D} = -\ln\frac{V_C}{V_B}$$

Substituting from Eqs. [25 . 6], [25 . 8], and [25 . 10] for Q_2 and Q_1 in Eq. [25 . 9], we have

$$\eta_c = 1 + \frac{Q_2}{Q_1} = 1 + \frac{T_2 \ln (V_A/V_D)}{T_1 \ln (V_C/V_B)} = 1 - \frac{T_2}{T_1} \tag{25 · 11}$$

or

$$\eta_c = \frac{T_1 - T_2}{T_1} \tag{25 · 12}$$

This last equation shows that the thermal efficiency of a Carnot engine using an ideal gas is a function solely of the absolute temperatures of the heat source and heat sink. Note that if T_2 could be reduced to 0 K, the Carnot efficiency would be unity; however, it cannot be greater than unity. Equation [25 . 12] also implies that we could replace our original operational definition of the temperature scale in terms of the expansion of an ideal gas (Sec. 22 . 2) by a definition in terms of the efficiency of an ideal-gas Carnot engine. Kelvin, in a paper written in 1851, specified that the ratio T_1/T_2 be *defined* by the ratio of heats absorbed and rejected by a Carnot engine operating between two constant-temperature reservoirs.

$$\frac{T_2}{T_1} = \frac{H_{out}}{H_{in}} = \frac{-Q_2}{Q_1} \tag{25 · 13}$$

In fact, as we shall see, this scale is *independent of the working substance.*

To complete the definition of the thermodynamic temperature scale, we must establish its zero point and specify one or more fixed points. Today the one-fixed-point method is used, with the point chosen by international agreement as the triple point of pure water. The temperature of this triple point is assigned the *exact* value 273.16 K, so that any temperature T on the Kelvin thermodynamic scale is given by

$$T = 273.16\,\text{K}\,\frac{H}{H_t} \tag{25 · 14}$$

where H_t is the heat transferred isothermally between the Carnot engine and a reservoir kept at the temperature of the triple point of water. In principle, therefore, any unknown temperature T can be determined by operating a Carnot engine between one reservoir at the unknown temperature and another at the triple-point temperature and measuring heat transferred to or from the unknown reservoir. The thermodynamic scale thus utilizes *energy transferred as heat* as the thermometric property.

If H_{out} in Eq. [25 . 13] has its smallest possible value, $H_{out} = 0$, then $T_2 = 0$ K. Therefore the working substance of a Carnot engine could undergo an isothermal process without heat transfer *only at absolute zero;* in other words, an isothermal process and an adiabatic process would be identical at the absolute zero. Here we have a new *definition* of absolute zero on the Kelvin scale that is independent of any particular substance. And as is characteristic in thermodynamics, the definition is expressed solely in terms of macroscopic quantities, no model of the structure of matter being required. Furthermore, a *negative* thermodynamic temperature is impossible.

example 25 • 1 Compute $\int_A^B P\,dV = \Delta W$ for process $A \rightarrow B$ in Fig. 25 . 2.

solution For an adiabatic process, $PV^\gamma = P_A V_A^\gamma =$ constant. Therefore

$$\Delta W = P_A \int \left(\frac{V_A}{V}\right)^\gamma dV = P_A V_A^\gamma \frac{V_A^{1-\gamma} - V_B^{1-\gamma}}{\gamma - 1}$$

Since $P_B = P_A (V_A/V_B)^\gamma$,

$$\Delta W = \frac{P_A V_A - P_B V_B}{\gamma - 1}$$

and since $\gamma - 1 = R/C_V$,

$$\Delta W = nC_V(T_2 - T_1) = -\Delta \mathcal{U} < 0$$

This is consistent, of course, with Eq. [25 . 5].

25 • 3 the second law of thermodynamics

The criterion that determines which thermodynamic processes, among the multitude of those allowed by the first law of thermodynamics, actually will occur is implicit in Carnot's treatment of engines. This criterion, first formulated by Clausius and later stated independently by Kelvin, was called the second law of thermodynamics. Clausius stated the law in two forms:

1 Heat cannot, of itself, pass from a colder to a hotter body.*
2 It is impossible for a self-acting device unaided by an external agency to convey heat spontaneously from one body to another at a higher temperature.

*Recall that we used the direction of heat flow to define the concepts "hotter" and "colder" (Sec. 20 . 1). Clausius obviously used the operational definition of temperature to define these concepts.

And Kelvin added:

3 It is impossible, by means of an inanimate material agency, to derive mechanical work from any portion of matter by cooling it below the temperature of the coldest of surrounding objects.

He also noted that if this axiom, as he called it, were not valid, it would be possible for an engine to do external work simply by cooling the earth or sea, thus drawing on their vast stores of internal energy.

If it were not for the second law, heat from the condenser of a steam engine could be collected, used to raise steam to boiler temperature without doing work, and used again as a source of energy for the engine. Any such self-acting engine would have a kind of perpetual motion which, because its existence is denied by the second law of thermodynamics, came to be called *perpetual motion of the second kind.** The law does not exclude the transfer of heat from a colder to a hotter body; this occurs in refrigerators. But it does assert that this transfer cannot occur unless some outside agent does work or there is some other additional effect that is compensation for this "unnatural" result.

If a heat engine is operated in reverse, so that external work W is done in each cycle *on* the working substance, instead of by it, the result is a refrigerator. The working substance now *absorbs* heat $H_{in} = Q_2 > 0$ from a low-temperature reservoir, or refrigerator compartment, and will *reject* a larger quantity $H_{out} = -Q_1 > 0$ to the high-temperature reservoir, such as a room. This is the effect, as Carnot pointed out, if the Carnot engine cycle of Fig. 25 . 2 is performed counterclockwise—that is, $A \to D \to C \to B \to A$. This *reversed* reversible cycle is a Carnot refrigerator, or *heat pump*. The quantities Q_1, A_2, and $W = Q_1 + Q_2$ are numerically equal to the same quantities in the Carnot engine cycle, but are of opposite sign. Refrigerator performance is expressed in terms of a *coefficient of performance* K_r, defined as $K_r = -Q_2/W$. For the Carnot refrigerator cycle $|Q_1/Q_2| = T_1/T_2$; therefore

$$K_r = \frac{T_2}{T_1 - T_2} \qquad [25 \cdot 15]$$

(Note that K_r is *not* the reciprocal of η.)

Equation [25 . 12] for the thermal efficiency of a Carnot engine was based on the assumption that the working substance is an ideal gas. However, Carnot concluded that the thermal efficiency is the same for *all* Carnot engines working between any two given temperatures, irrespective of working substance. Consider two Carnot engines C and C' that are operating between the same two heat reservoirs, with the working substance of engine C an ideal gas and that of engine C' any substance other than an ideal gas. The initial pressures are adjusted so that both engines do the same amount of

*A perpetual-motion machine of the *first* kind is one which violates the first law of thermodynamics. The historic failure of all attempts to create such devices formed the introduction to Hermann Helmholtz' great paper of 1847 on the first law of thermodynamics. The *Encyclopaedia Britannica's* excellent article "Perpetual Motion" remarks of one John Keeley, a swindler who claimed to have invented a perpetual-motion machine: "In the course of his long career Keeley may have broken a number of laws, but he left the first and second laws of thermodynamics . . . inviolate."

external work W per cycle. In Fig. 25 . 4 the two engines are coupled mechanically so that engine C drives engine C' *backward* as a Carnot refrigerator. The result is a self-acting device, since all the work needed to operate C' is supplied by C. The quantities $Q_1 = H_{in}$, $Q_2 = -H_{out}$, and W for the heat engine are opposite in sign to the corresponding quantities $Q_1' = H'_{out}$, $Q_2' = H'_{in}$, and W' for the refrigerator.

For the sake of discussion, let us assume that the thermal efficiency of C exceeds that of C'. Then for $\eta_c > \eta_c'$, we have

$$\frac{W}{H_{in}} > \frac{-W'}{H'_{out}} = \frac{W}{H'_{out}} \quad \text{or} \quad H_{in} < H'_{out} \qquad [25 \cdot 16]$$

Therefore the net heat transferred to the hot reservoir by the entire two-cycle engine system is

$$H_{out} - H_{in} > 0 \qquad [25 \cdot 17]$$

Since both cycles are reversible, the first law of thermodynamics states that

$$W = H_{in} - H_{out} = H'_{out} - H'_{in} = -W' \qquad [25 \cdot 18]$$

and substituting this into Eq. [25 . 17] yields

$$\Delta Q = H'_{out} - H_{in} = H'_{in} - H_{out} > 0 \qquad [25 \cdot 19]$$

The net amount of heat transferred *out* of the cold reservoir is positive. This means that the assumption $\eta_c > \eta_c'$ implies the existence of a self-acting device capable of transferring a quantity of heat ΔQ from a cold reservoir to a hot reservoir. Since this contradicts the second form of the second law, the assumption $\eta_c > \eta_c'$ is therefore *false.*

If the functions of the two Carnot engines are now interchanged—that is, the process is run "backwards"—then we can repeat the argument above, just reversing the inequality, to show that $\eta_c < \eta_c'$ is also a false assumption. In fact, there is only one possible conclusion that does not violate the second law:

> All Carnot engines have the same thermal efficiency, irrespective of the nature of the working substance.

figure 25 • 4 *The equivalence of all Carnot engines operating between the same temperatures. The second law of thermodynamics would be violated if η_c were not equal to η_c'.*

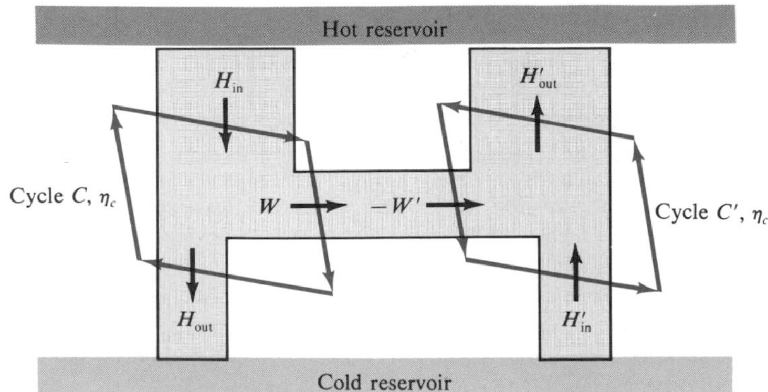

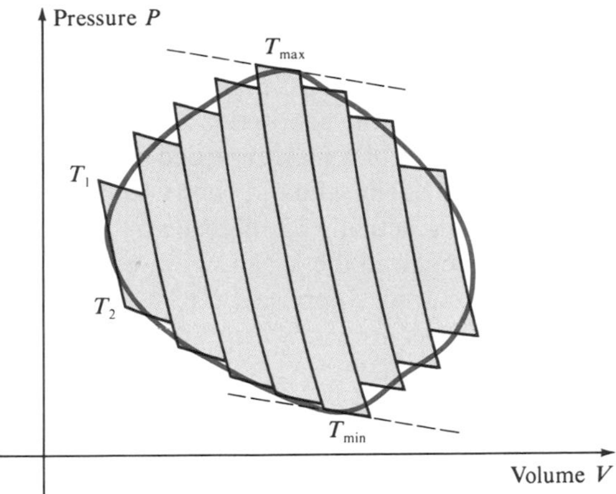

figure 25 • 5 *Approximation of a reversible cycle by an ensemble of Carnot cycles*

It follows that any Carnot engine has the same efficiency as the ideal-gas Carnot engine, justifying Kelvin's assertion that any Carnot engine can be used as a *thermometer,* since $H_{in}/H_{out} = T_1/T_2$, independent of the properties of any particular thermometric substance. Note also that the efficiency of a Carnot engine is completely independent of the *size* of the engine.

Any *real* reversible engine would have a cycle differing from the Carnot cycle, in that there would have to be a series of external reservoirs, each differing in temperature from the next in the series by an infinitesimal amount. The heat transfers would occur when the system was brought into contact with each reservoir in turn, undergoing an infinitely slow (quasi-static) expansion or compression. Any reversible cycle can be approximated, however, by resolving it into Carnot subcycles in the following way. Draw a number of adiabats through it as in Fig. 25 . 5. Then join each adjacent pair of adiabats with the two isotherms corresponding to the observed temperatures at the top and bottom of the strip. Each strip then constitutes a Carnot cycle, representing a small Carnot engine.

The efficiency of this ensemble of small Carnot engines all taken together will be approximately the same as that of the real engine. However, the least efficient Carnot engine is that operating between T_1 and T_2, and the most efficient is that operating between T_{max} and T_{min}. The actual efficiency of the assemblage, and hence of the real engine, is then somewhere between $(T_1 - T_2)/T_1 = \eta'$ and $(T_{max} - T_{min})/T_{max} = \eta$. And as we take more and more small engines, further subdividing the area within *ABCD*, η' approaches zero, so that

$$0 < \eta_{real} < \eta_c(T_{max}, T_{min}) \qquad [25 \cdot 20]$$

The efficiency of any real reversible engine operating between two temperatures T_{max} and T_{min} is always less than that of a Carnot engine operating between the same two temperatures. An *irreversible* engine is always less efficient than the most closely corresponding reversible engine:

> The efficiency of a real engine operating between two temperatures is always less than that of a Carnot engine operating between the same two temperatures.

Carnot used the caloric theory, in which heat was assumed to be an indestructible material substance, in his development. Consequently, when extensive experimentation increasingly supported the conception of heat as a form of energy, all of Carnot's conclusions were for a time regarded as being in conflict with this theory. Clausius and Kelvin showed, however, that Carnot's argument leading to the second law actually involved no assumption concerning the *nature* of heat. They recognized the broad significance of the second law and extended it far beyond Carnot's application of it to heat engines, to show that the first and second laws are completely independent and necessary postulates of thermodynamics.

example 25 • 2 In 1852 Kelvin showed how a refrigerator could be used to heat a building in the winter by pumping heat into the building from a colder outdoor source, such as the ground. If a heat pump is designed to operate between $-18°C$ and $22°C$ with an electric motor supplying 3000 kcal/h, how much heat does the cold reservoir supply to the building? Assume the pump to be a reversible Carnot engine.

solution When we solve the two equations

$$\frac{H_{in}}{H_{out}} = \frac{T_2}{T_1} \qquad H_{in} = H_{out} - W$$

for H_{in}, taking all quantities as greater than zero, we obtain

$$H_{in} = W\frac{T_2}{T_1 - T_2} \doteq 19{,}000 \text{ kcal/h}$$

According to a principle first stated by Walter Nernst in 1906:

It is impossible to reduce the temperature of a system to absolute zero in a finite number of operations.

This generalization, now firmly supported by experiment, cannot be derived from the first two laws of thermodynamics. It is an independent postulate, the *third law of thermodynamics,* and it implies that as the reservoir temperatures approach zero, the efficiencies can be expected to decrease.

25 • 4 entropy

In Sec. 22 . 2 we saw that $\Delta\mathcal{U}$, the change in the internal energy of a system, has the same value for all conceivable paths, irreversible or reversible, by which the system can pass from one equilibrium state to another. We shall now see that there is another important state variable whose value, like the internal energy, depends on the state of the system, and not on how the system reached that state from any previous state. For this property Clausius selected the symbol S and coined the term *entropy* "so as to be as similar as possible to the word *energy,* since both these quantities . . . are so nearly related to each other in their physical significance that a certain similarity in their names seemed to be advantageous." Like energy, the entropy of a system in any given state provides a measure of its capacity to do useful work.

To investigate the nature of this new physical quantity Clausius began by considering a Carnot cycle. From Eq. [25 . 13],

$$\frac{Q_1}{T_1} + \frac{Q_2}{T_2} = 0 \qquad [25 \cdot 21]$$

That is, for any two isothermal processes carried out between the same two adiabats, the algebraic sum of the quantities Q/T from each process is zero.

Now imagine any reversible engine operating between T_{max} and T_{min} as in Fig. 25 . 5, but this time subdivide the cycle into an *infinite* number of Carnot cycles of *infinitesimal* width. The resulting infinitesimal isothermal segments between adiabats will, in the limit, approach the path of the reversible cycle. Let dQ_i (positive or negative) be the heat transferred and let T_i be the temperature in the ith infinitesimal isothermal segment. No heat is transferred in the adiabatic processes. For the isothermal processes of any *reversible* cycle, if we sum the ratios dQ_i/T_i for all segments of the cycle, we obtain the integral

$$\oint \frac{dQ}{T} = 0 \qquad [25 \cdot 22]$$

where the symbol $\oint$ signifies integration around the complete cycle. From the derivation of Eq. [25 . 22] it is apparent that this equation holds for any reversible cycle irrespective of the working substance. As in the case of the Carnot engine, however, note that $\Sigma Q \neq 0$.

Suppose next that we divide any reversible cycle C into two parts, as in Fig. 25 . 6. If we carry the integration from state A to state B along path C_1 and from B to A along path C_2, adding the infinitesimals dQ/T as we move along each path, we have

$$\oint_C \frac{dQ}{T} = \int_{C_1:A\to B} \frac{dQ}{T} + \int_{C_2:B\to A} \frac{dQ}{T}$$

$$= \int_{C_1:A\to B} \frac{dQ}{T} - \int_{C_2:A\to B} \frac{dQ}{T} = 0$$

or

$$\int_{C_1:A\to B} \frac{dQ}{T} = \int_{C_2:A\to B} \frac{dQ}{T} \qquad [25 \cdot 23]$$

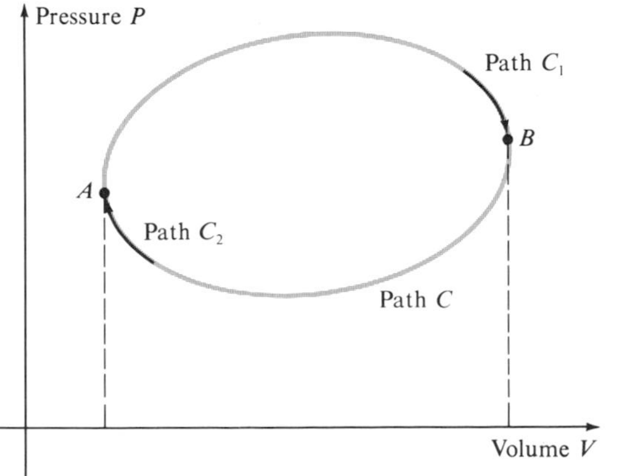

figure 25 • 6 *Entropy as a state function;* $S_B - S_A = \int_A^B dQ/T$, *independent of path. Path C is a reversible cycle starting at A and returning to A, divided into the two partial cycles* C_1 *and* C_2.

This relation holds for any reversible path between two equilibrium states, so that $\int_A^B dQ/T$ has the same value for *all reversible paths* joining any given equilibrium states A and B. Therefore:

There exists a function of the state of a system, called the *entropy* S, whose value differs from state A to state B by an amount

$$S_B - S_A = \Delta S = \int_A^B \frac{dQ}{T} \qquad [25 \cdot 24]$$

The entropy must be a state function, since it is independent of path; hence it depends only on the end states of the system. The heat dQ absorbed by the system, although not itself a state function, is in fact proportional to the change in the entropy state function:

$$dQ = T\,dS \qquad [25 \cdot 25]$$

Note that we are concerned only with *differences* in entropy, as in the case of internal energy, which leaves us free to choose the initial value of S arbitrarily.

The cyclic integral in Eq. [25 . 22] means that the net entropy change in any reversible cycle is zero. A system returns to its initial state in a complete cycle, and its entropy returns to the value for that state. For an infinitesimal segment of a reversible path the entropy change is infinitesimal. Hence we can write the first law of thermodynamics for *reversible* processes as

$$dQ = T\,dS = d\mathcal{U} + P\,dV \qquad [25 \cdot 26]$$

In the special case of an ideal gas, $d\mathcal{U} = nC_V\,dT$, and substituting for $P = nRT/V$, we can integrate Eq. [25 . 26] between the corresponding end states (P_1,V_1,T_1) and (P_2,V_2,T_2) to obtain $S_2 - S_1 = \Delta S$:

$$dS = nC_V \frac{dT}{T} + nR \frac{dV}{V}$$

$$\Delta S = nC_V \ln \frac{T_2}{T_1} + nR \ln \frac{V_2}{V_1} \qquad [25 \cdot 27]$$

In other words, although heat is not a state variable, if we know the path of a reversible process and the values of the system entropy and temperature at every point along the path, we can compute the heat absorbed in the process by integrating the first law as in Eq. [25 . 27]. There are tables of entropy values for various substances, and if necessary, such computations of heat absorption can be made in numerical form.

example 25 • 3 Prove that the following expressions give the entropy changes of n moles of a substance for various *reversible* paths connecting initial and final equilibrium states:

(*a*) $\Delta S = n\int_{T_1}^{T_2} C_P \frac{dT}{T}$ isobaric, $P_1 = P_2$

(*b*) $\Delta S = n\int_{T_1}^{T_2} C_V \frac{dT}{T}$ isometric, $V_1 = V_2$

(c) $\Delta S = 0$ adiabatic, $Q = 0$

(d) $\Delta S = \frac{Q}{T}$ isothermal, T = constant

solution In the first three cases the process determines substitutions in Eq. [25 . 27] for dQ equal to (a) $nC_P\,dT$, (b) $nC_V\,dT$, and (c) zero. In (d)

$$\Delta S = \int \frac{dQ}{T} = \frac{1}{T}\int dQ = \frac{Q}{T}$$

example 25 • 4 Compute the entropy imparted to 1 kmole of an ideal gas when it undergoes a reversible expansion from state (4 atm, V_1, 15°C) to state (2 atm, V_2, 15°C).

solution From Eq. [25 . 27],

$$\Delta S = nR \ln \frac{V_2}{V_1} = nR \ln \frac{P_1}{P_2}$$

$$\doteq 1 \text{ kmole} \times 2\text{kcal/K} \cdot \text{kmole} \times \ln \tfrac{4}{2} \doteq 1.39 \text{ kcal/K}$$

This value is positive, which means that the entropy has increased as a result of the expansion. What would be the entropy change of the gas if it were returned reversibly from the second to the first state?

25 • 5 *irreversible processes*

The efficiency of an irreversible engine must be less than that of an ideal Carnot engine operating between the same two temperatures. An irreversible process cannot be represented by a continuous path on the surface of thermodynamic states, since it does not proceed through a sequence of equilibrium states; it proceeds through nonequilibrium states over at least part of its cycle. Nevertheless, we can approximate the net change in entropy in an irreversible process by the entropy changes in a series of infinitesimal Carnot cycles in the sense depicted in Fig. 25 . 5. Consider a portion of an irreversible cycle with maximum and minimum temperatures T_1 and T_2 approximated by an infinitesimal Carnot cycle operating between T_{1i} and T_{2i}. The actual irreversible heat transfers are denoted by $đQ_{1i}$ (positive) and $đQ_{2i}$ (negative). Since the irreversible cycle is less efficient than the reversible cycle,

$$1 + \frac{đQ_{2i}}{đQ_{1i}} < 1 - \frac{T_{2i}}{T_{1i}} \qquad [25 \cdot 28]$$

which can be rearranged as

$$\frac{đQ_{1i}}{T_{1i}} + \frac{đQ_{2i}}{T_{2i}} < 0 \qquad [25 \cdot 29]$$

Summing $đQ/T$ over all such infinitesimal cycles yields

$$\oint \frac{đQ}{T} < 0 \qquad [25 \cdot 30]$$

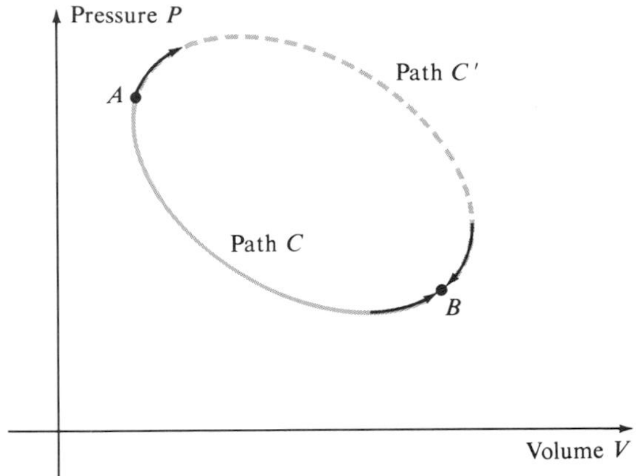

figure 25 • 7 *An irreversible process C′ and a reversible process C*

for the working substance in an *irreversible cycle.* This relation, the *Clausius inequality,* can now be used to prove that entropy always increases in an irreversible process between two end states.

Recall that entropy is a state function; hence $\Delta S = 0$ over a complete cycle whether the cyclic process is reversible or not. However, we cannot set ΔS equal to $\oint (dQ/T)$ for an irreversible process, since the integral is not equal to zero. We can resolve this problem by joining the two end states of any real irreversible change of state by a reversible process. In Fig. 25 . 7 the dotted line C' depicts an irreversible process between state A and state B, and path C depicts a reversible process. If we reverse process C, then the two processes together constitute an irreversible cycle:

$$\oint \frac{đQ}{T} = \int_{C':A\to B} \frac{đQ}{T} - \int_{C:A\to B} \frac{dQ}{T} < 0 \qquad [25 \cdot 31]$$

If the system is *isolated* during C', then $đQ = 0$ and the first integral on the right vanishes. Hence for any irreversible process,

$$0 < \int_A^B \frac{dQ}{T} = S_B - S_A \qquad [25 \cdot 32]$$

In other words, to compute the entropy change between two equilibrium end states, we must compute it for an arbitrary *reversible* path connecting the two states. Furthermore:

> The entropy of an isolated system must increase whenever it undergoes an irreversible process.

In fact, we can *define* as irreversible those processes for which the total entropy of the "universe" increases. It is often convenient in thermodynamics to replace the phrase "system plus the surroundings with which it interacts" by the single word *universe.* For instance, a Carnot engine is a universe consisting of a working substance, or an isolated system, and two heat reservoirs. Obviously, *universe* is used here in a special sense to refer to a discrete and observable portion of nature whose thermodynamic properties we are investigating.

Analyses of all kinds of reversible processes support the following conclusion:

> When all the processes performed in a universe are reversible, the entropy of that universe remains unchanged.

For instance, when a gas undergoes a reversible isothermal expansion, the interacting bodies are the gas and a surrounding heat reservoir, both at the same constant temperature T. The gas absorbs heat H_{in} and the reservoir loses the same amount. Therefore

$$\Delta S + \Delta S' = \frac{H_{\text{in}}}{T} - \frac{H_{\text{in}}}{T} = 0$$

For an isothermal compression

$$\Delta S + \Delta S' = \frac{-H_{\text{out}}}{T} + \frac{H_{\text{out}}}{T} = 0$$

Any adiabatic process is *isentropic,* since no heat is transferred; hence $\Delta S = \Delta S' = 0$. Therefore:

> The total entropy change of a Carnot engine or refrigerator and its reservoirs is zero.

Of more general importance is the question of how the total entropy changes when the processes are irreversible. Every actual process is more or less irreversible, and this of course includes every spontaneous process occurring in nature. All thermodynamic analyses support the same general conclusion:

> When any irreversible process occurs in a universe, the entropy of that universe increases.

Some parts of the universe may decrease in entropy as a result of an irreversible process, but these reductions will be more than balanced by entropy increases in the remaining parts. Consider the following examples, all of which involve irreversible processes in particular universes:

1 An ideal gas undergoes a free expansion in an insulated container (Fig. 23 . 7). Although the entropy of the external environment is unchanged, the process is irreversible; hence the entropy of the gas must have increased.

2 The part of a body to which a frictional force is applied absorbs heat, and therefore its entropy increases. Since the heat has come from external work and not from other parts of the body, the entropy of the whole body has increased.

3 Steady-state heat conduction takes place through a metal bar from a hot reservoir to a cold one. The total *rate* of change in entropy of the two reservoirs is

$$\frac{dS}{dt} = \frac{dQ}{dt}\left(-\frac{1}{T_1} + \frac{1}{T_2}\right) > 0$$

However, the conductor suffers no change in entropy, since its state variables remain constant.

4 A pail of water is stirred violently and allowed to come to rest. As the circulatory motion of the water in the pail decays through dissipation, the resultant heat increases the entropy of the water. As the heat is conducted out through the pail, the water returns eventually to its original state and its *net* change in entropy is equal to zero. However, the external reservoir has absorbed this mechanical energy as heat, thus increasing the entropy of the reservoir.

Since no system is ever in perfect thermal isolation, we can extend this principle to include the cosmos:*

The entropy of the entire physical universe is always increasing.

As an example, consider the entropy change when n moles of an ideal gas undergo a free expansion into an insulated, evacuated vessel as in the Joule experiment in Fig. 23 . 7. Free expansion is an irreversible process, because the gas does not pass through a succession of equilibrium states while rushing into the evacuated vessel. The freely expanding gas is an isolated or closed system, for the gas does not interact with its surroundings; that is, $\Delta Q = 0$ and $W = 0$. If, in addition, the gas is ideal, then its internal energy, and hence its temperature T, is the same in the initial and final equilibrium states. To compute ΔS we just substitute $T_2/T_1 = 1$ and $V_2/V_1 = 2$ into Eq. [25 . 27]:

$$\Delta S = nR \ln 2 \qquad [25 \cdot 33]$$

In a free expansion an ideal gas does no external work and neither absorbs nor rejects heat. Instead it increases in entropy and loses some capacity to do work, since work must be done on it to compress it back to its original state.

Increase in entropy is accompanied by loss of work capacity.

Notice that if we had mistakenly treated the free expansion as a reversible process, our conclusion would have been that $S_2 - S_1 = Q/T = 0$, or that $S_2 = S_1$, in contradiction to Eq. [25 . 27].

*See Clausius' statement of the second law in the introductory quotation.

example 25 • 5 Two conducting bodies at temperatures T_1 and $T_2 < T_1$ are placed in thermal contact while insulated from the rest of the universe. Show that the total entropy of the two bodies has increased by the time they reach thermal equilibrium at temperature T.

solution Since the heat lost by the hotter body equals the heat gained by the colder body,

$$C_1 \int_T^{T_1} dT = C_2 \int_{T_2}^{T} dT$$

and since $T_2 < T < T_1$,

$$C_1 \int_T^{T_1} \frac{dT}{T} < C_2 \int_{T_2}^{T} \frac{dT}{T}$$

Therefore, reversing the limits on the first integral, we have

$$0 < C_1\int_{T_1}^{T}\frac{dT}{T} + C_2\int_{T_2}^{T}\frac{dT}{T} = \Delta S$$

where ΔS is, by definition, the total change in entropy of the two bodies. Although the entropy of the hotter body decreases, that of the colder body increases still more, so that the total entropy increases.

25 • 6 *statistical interpretation of entropy*

The thermodynamic state of a system is defined by a small number of macroscopic properties of the system, such as pressure, volume, and temperature. In statistical mechanics and kinetic theory, a *microstate* is defined by all the individual positions and velocities of all the molecules of the system at a given instant. These positions and velocities are constantly changing, owing to motions and collisions, even while the thermodynamic state of the system remains unchanged. Thus to any given thermodynamic state, or *macrostate,* of a system there corresponds an immense number $\mathcal{W}$ of different possible microstates. According to classical mechanics, $\mathcal{W}$ would be infinite for any given macrostate. However, modern statistical mechanics, which utilizes the quantum theory, shows that this number, while very large, is finite. The number $\mathcal{W}$, is known as the *thermodynamic probability* of the given macrostate.

Thermodynamic theory asserts that any process occurring in an isolated system takes the system from a macrostate of lower entropy to one of higher entropy. In statistical mechanics such a process in an isolated system is interpreted as the passage of the system from a macrostate of lower thermodynamic probability to one of higher thermodynamic probability—that is, to a state having a larger number $\mathcal{W}$ of allowable microstates. This implies a greater degree of randomness or disorder, since each of the $\mathcal{W}$ microstates is equally likely. The parallel between entropy and thermodynamic probability suggests that some way might be found to relate these two quantities mathematically.

Such a relationship was gradually clarified by Boltzmann, starting in about 1877. Consider a universe consisting of two independent parts in states of equilibrium. Let S_1 and S_2 be the entropies of the two parts, and $\mathcal{W}_1$ and $\mathcal{W}_2$ their thermodynamic probabilities. The probability of several independent events occurring together is the product of the probabilities of each of the events separately.* Hence the thermodynamic probability of this universe is $\mathcal{W} = \mathcal{W}_1\mathcal{W}_2$. Since the entropy of this universe is $S = S_1 + S_2$, it follows that S and $\mathcal{W}$ cannot be simply proportional. However,

$$\ln \mathcal{W} = \ln \mathcal{W}_1\mathcal{W}_2 = \ln \mathcal{W}_1 + \ln \mathcal{W}_2$$

It is plausible, therefore, that S is proportional to the logarithm $\ln \mathcal{W}$. This is the celebrated *Boltzmann entropy principle.* Max Planck stated in 1900 that

*For example, the probability of getting two heads in a row by tossing a coin is $P = \frac{1}{2} \times \frac{1}{2} = \frac{1}{4}$.

the constant of proportionality was just the Boltzmann constant, with which we are already familiar:

$$S = k \ln \mathcal{W} \qquad [25 \cdot 34]$$

Equation [25 . 34] provides a statistical interpretation of the second law of thermodynamics.

example 25 • 6 An ideal gas is allowed to expand freely from a vessel of volume V_1 into an identical connecting vessel in Joule's apparatus (see Fig. 23 . 7). What is the probability P that the gas will spontaneously return to its originally smaller volume V_1 and entropy $S_1 = k \ln \mathcal{W}_1$ instead of remaining in the two vessels of combined volume V_2, where its entropy is $S_2 > S_1$? (For an ideal gas in Joule's apparatus, $T =$ constant.)

solution From Eq. [25 . 34], the change in entropy is

$$\Delta S = S_1 - S_2 = k(\ln \mathcal{W}_1 - \ln \mathcal{W}_2) = k \ln \left(\frac{\mathcal{W}_1}{\mathcal{W}_2}\right)$$

$$\frac{\mathcal{W}_1}{\mathcal{W}_2} = \exp \frac{\Delta S}{k}$$

As we saw in Example 25 . 4,

$$\frac{\Delta S}{k} = N \ln \frac{V_1}{V_2}$$

and for a typical case, $N = N_A = 6 \times 10^{26}$ and $V_1/V_2 = 1/2$, we find

$$\frac{\mathcal{W}_1}{\mathcal{W}_2} \doteq \exp(-4 \times 10^{26})$$

Since each microstate is equally likely, $P = \mathcal{W}_1/\mathcal{W}_2$, a probability so infinitesimal that it can be regarded as zero.

Thermodynamic probability may be regarded as a measure of the degree of *disorder* existing in the positions and motions of the molecules comprising a system. Therefore entropy, in view of the Boltzmann entropy principle, also is a measure of such disorder. For example, when a crystalline solid melts, its entropy increases while its particles change from the highly ordered arrangement and confined motions of the crystals to the more disorderly arrangements and motions of the liquid phase. The overwhelming tendency for natural processes to proceed in the direction of increased entropy means that disorder is always far more probable than order.

Since any universe or closed system that gains entropy loses some of its capacity for doing work, energy in nature is continually being degraded—converted to disordered, random motions unavailable for doing work. In a heat engine only part of H_{in} is converted into work; the remainder, H_{out}, is absorbed by the cold reservoir and there converted into internal energy of the reservoir. This internal energy can be extracted in the form of heat and part of it used to do work *only* if another reservoir of still lower temperature is available. Eventually, though, we would reach a point in this and all natural processes when there will no longer be any available reservoirs or bodies of still lower temperature; then all temperature differences will disappear, no energy will be available for doing work, and gross mechanical motion will no longer be possible.* The philosophers of the nineteenth century called this the *Wärmetod,* or "heat death"—a sort of thermodynamic *Götterdämmerung.*

Such a conclusion involves the questionable assumption that the cosmological universe constitutes a single universe or closed system in the thermodynamic sense. Moreover, it must be remembered that thermodynamics ignores the mechanism of a process and makes no assumptions about the structure of matter. The state of maximum entropy predicted for any isolated system by the second law must be interpreted as the *most probable state.* Thus the second law, with its implication of constantly increasing unavailable energy in nature, applies to the *average-probability* condition to be met over exceedingly long periods of time. From this point of view the "heat death" need not be regarded as inevitable.†

*There the wicked cease from troubling, and there the weary be at rest. *The Old Testament,* Book of Job 3:17.

†Hope springs eternal in the human breast. . . . *Essay on Man,* Alexander Pope, 1733.

PROBLEMS

25 • 1–25 • 2 engines and the Carnot cycle

25 • 1 An ideal gas absorbs 200 cal of heat and at the same time undergoes a reversible expansion from 2 to 3 liters under a constant pressure of 1 atm. Compute the change in the internal energy of the gas.

25 • 2 A cylinder contains 1.5 moles of helium at 0°C and 1 atm. In a reversible process the gas is (1) heated under constant pressure until its volume has doubled, (2) cooled at constant volume to 0°C, and then (3) compressed isothermally to a pressure of 1 atm. (*a*) Draw a *PV* diagram of the cycle, with the initial point of each process labeled with the values of P, V, and T. (*b*) Find the *net* work done by the gas, the *net* change in internal energy, and the *net* heat absorbed.

25 • 3 For each of the three processes in Prob. 25 . 2 find the work done by the gas, the change in its internal energy, and the heat absorbed. Your answers must be consistent with the answers to Prob. 25 . 2.

25 • 4 The working substance of a Carnot engine operating between 227°C and 27°C acquires heat from the hot reservoir at the rate of 2 kW. (*a*) Find the efficiency of the engine, its mechanical power, and the rate at which heat is transferred to the cold reservoir. (*b*) Which would cause a greater increase in the efficiency of this or any other Carnot engine: increasing T_1 by an amount ΔT and not changing T_2, or decreasing T_2 by the same amount ΔT and not changing T_1?

25 • 5 As we saw in Chapter 21, specific heats may be temperature-dependent. Show that Eq. [25 . 12] applies even if C_V changes during the Carnot engine cycle. In other words, show that the efficiency of the cycle depends on the thermodynamic process, not on the working substance.

25 • 6 A Carnot engine having 58 g of dry air as the working substance absorbs 500 J/cycle of heat at 127°C and rejects heat at −1°C. The air is of molecular weight 29 and $\gamma = C_P/C_V = 1.4$. (*a*) Find the thermal efficiency, the heat rejected, and the net work done by the engine. (*b*) Find the change in internal energy of the air in each of the four processes of the cycle and in the whole cycle.

25 • 7 For a certain small reciprocating steam engine, the boiler and exhaust temperatures are 180°C and 102°C, the heat intake is 10 kcal/cycle, and the overall efficiency is $\eta = \frac{1}{2}\eta_c$. How fast must the engine run in order to deliver 40 kW of power?

25 • 8 The following data were obtained for a certain reciprocating steam engine:

Operating speed, 600 cycles/min
Temperature of steam entering engine, 370°F
Exhaust temperature 215°F
Intake of heat to engine, 40 Btu/cycle
Engine power output, 79 hp
Shaft power output (useful work), 71 hp
Efficiency of steam-generating unit, 75 percent
Heat of combustion of coal, 12,000 Btu/lbm

(*a*) Find the efficiency of the Carnot cycle. (*b*) Find the actual thermal efficiency of the engine. (*c*) Find the mechanical efficiency of the engine. (*d*) What is the overall engine efficiency? (*e*) What is the overall efficiency of the whole steam power plant? (*f*) Find the rate of coal consumption.

25 • 9 Consider the Carnot cycle in Prob. 25 . 6. (*a*) Find the ratio of the volumes of air at the end and the beginning of the adiabatic expansion. (*b*) At the end and the beginning of the isothermal expansion. (*c*) How do these two ratios compare with those for the adiabatic compression and the isothermal compression? (*d*) Taking the pressure of the air at the start of the adiabatic compression to be 1 atm, compute the pressure and volume at the starts of the remaining three processes.

25 • 3 the second law of thermodynamics

25 • 10 For a Carnot refrigerator cycle with ideal gas as the refrigerant, describe the successive processes and write one or more equations applicable to each process.

25 • 11 A Carnot refrigerator operates between 30°F and 100°F at the rate of 500 cycles/min, and the refrigerant absorbs 10,000 Btu/min from the cold region. (*a*) Show that the external agent does work on the refrigerant at the rate of 34 hp, or 2200 ft · lbf/cycle. (*b*) Show that the refrigerant rejects 11,000 Btu/min to the warmer region. (*c*) What is the advantage, if any, of placing the condenser of a household refrigerator as close as possible to the floor?

25 • 12 A refrigerating unit working between $-14°C$ and $40°C$ and driven by a 0.2-kW motor is found to convert water that is initially at $16°C$ into ice of final temperature $-10°C$ under a constant pressure of 1 atm and at the rate of 96 kg/day. Show that this unit has an overall coefficient of performance K_r equal to 49 percent of that of a Carnot refrigerator cycle working between the same two temperatures.

25 • 13 For the coupled Carnot engines of Fig. 25 . 4, imagine that engine C' drives engine C backward as a refrigerator and thus prove that the inequality $\eta_c < \eta_c'$ is false.

25 • 14 The following data are given for an electric refrigerator:

Actual K_r, 2.4
Interior temperature, $-9°C$
Temperature of refrigerator radiator, $40°C$
Room temperature, $35°C$
Area of refrigerator walls, 3.2 m^2
Wall thickness, 4.0 cm
Thermal conductivity of the walls, 2.0×10^{-5} kcal/sec · m · K

(*a*) Find the power needed to operate this refrigerator. (*b*) What power would be needed if the coefficient of performance were that of a Carnot refrigerator operating between the same two temperatures?

25 • 15 The vapor-absorption type of refrigerator differs basically from the vapor-compression type in that the energy required to absorb heat H_{in} from the cold interior and reject heat H_{out} to a room at temperature T_1 is obtained, not by having an external agent do work W on the refrigerant, but by having the refrigerant absorb heat Q from a source of higher temperature, $T_1 + \Delta T$, such as a gas flame. (*a*) Show that for this type of refrigerator,

$$K_r = \frac{H_{\text{in}}}{Q} = \frac{H_{\text{out}}}{Q} - 1$$

(*b*) Show that this refrigeration process does not violate the second law of thermodynamics.

25 • 4–25 • 5 entropy and irreversibility

25 • 16 Consider the Carnot engine in Prob. 25 . 4. (*a*) What is the entropy change per unit time in the hot reservoir? (*b*) In the cold reservoir? (*c*) What is the net change in the engine and both reservoirs?

25 • 17 A mole of ideal gas changes from state A to state C as shown. (*a*) Find Q, $\Delta\mathcal{U}$, and ΔS of the gas if the change of state occurs reversibly along path ABC. (*b*) Along path $AB'C$. (*c*) What is the difference in heat absorbed for the two paths?

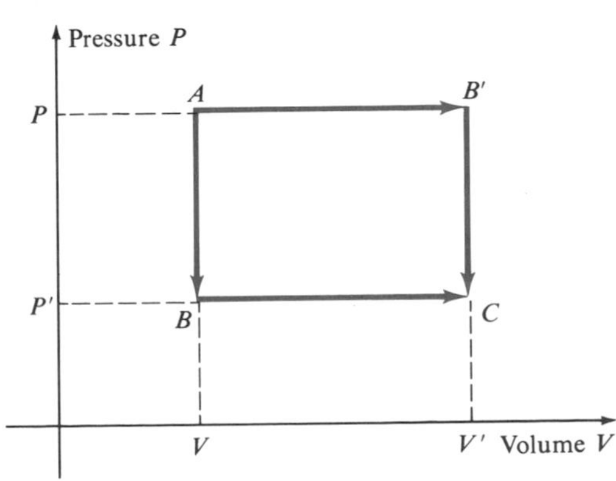

25 • 18 Show that the Carnot cycle diagram simply becomes a square of area $Q_1 + Q_2$ if plotted against entropy and temperature, rather than pressure and volume, as the state variables.

25 • 19 Prove that setting $\Delta S = 0$ in Eq. [25 . 27] is equivalent to the condition for an adiabatic change of state of an ideal gas, $TV^{\gamma - 1} = \text{constant}$.

25 • 20 In a given time an amount of heat Q passes in the form of solar radiation from the sun to a parabolic mirror on the earth, which focuses it on a small object. Prove that the temperature of the object cannot exceed the temperature of the sun.

25 • 21 The ends of a uniform copper rod 10 cm long with a cross-sectional area of 2 cm^2 are in thermal contact with two reservoirs of constant temperatures 100°C and 0°C. The rest of the rod is enclosed in a thermal insulator. (*a*) Assuming that the steady state has been reached, what is the entropy change during 5 min in the rod? (*b*) In the reservoirs? (*c*) In the universe consisting of the rod and reservoirs?

25 • 22 Under a constant pressure of 1 atm, 1 kg of ice at 0°C is converted reversibly to steam at 100°C. (*a*) What is the entropy change of the ice-water-steam system? (*b*) Find the increase in the total entropy of the universe if the surrounding reservoirs that furnish the heat for the process are at 120°C. (*c*) If the value zero is arbitrarily assigned to the entropy of water when it is in the liquid phase at 1 atm and 0°C, what is the entropy of 1 kg of liquid water at 1 atm and 100°C?

25 • 23 Experiments show that the molar specific heats C_P for many gases under a constant pressure of 1 atm are given by the empirical equation

$$C_P = a + bT + cT^2$$

where a, b, and c are constants that must be determined experimentally for each gas. For nitrogen gas C_P is obtained in kcal/kmole · K by setting $a = 6.52$, $b = 1.25 \times 10^{-3}/\text{K}$, and $c = 1.0 \times 10^{-9}/\text{K}^2$. (*a*) What is the energy required to heat 1 kmole of nitrogen under atmospheric pressure from 27°C to 127°C? (*b*) What is the change in entropy under these conditions?

25 • 24 Consider the system described in Prob. 25 . 22. (*a*) What would be the entropy change if the steam were converted reversibly back to ice by thermal contact with a reservoir at −20°C? (*b*) Show that the total entropy of the universe would increase in this case.

25 • 25 At 27°C the equilibrium vapor pressure of water is 3.55×10^3 Pa and the heat of vaporization is 2400 J/g. Find the entropy change of the water-vapor system in the phase change of 1 kg of liquid water at 3.55×10^3 Pa and 27°C to vapor of the same pressure and temperature.

25 • 26 In an isobaric process 1 kg of water at 1 atm and 27°C is converted to vapor at 27°C. Given that

c_P (water) = 1.0 kcal/kg · K
L_v (1 atm, 100°C) = 539 kcal/kg
c_P (water vapor) = 0.5 kcal/kg · K

find the entropy change $\Delta S'$ of the liquid-vapor system. [*Hint:* The vaporization of water under 1 atm of pressure can be carried out *reversibly* only at 100°C. Therefore $\Delta S'$ must be computed for a series of reversible processes connecting the initial and final equilibrium states.]

25 • 27 Suppose 1 g of ice at 1 atm and −13°C were to melt into water at the same temperature and pressure. (*a*) Show that this process would be irreversible and hence that a method similar to the one suggested in Prob. 25 . 26 must be used to compute the entropy change. (*b*) Given that c_p (ice) = 0.49 kcal/kg and L_f (1 atm, 273°K) = 80 kcal/kg, find the resulting entropy change of the ice-water system.

25 • 28 Suppose the valve between the two vessels in the Joule free-expansion apparatus is closed and each vessel contains 2 moles of an ideal gas at 1 atm and 20°C. The gases are of different molecular weights and will not react chemically when mixed. (*a*) Show that when the valve is opened each gas will undergo a free expansion. (*b*) Find the total increase in entropy as a result of the expansions. (*c*) What will be the pressure and temperature of the mixture in the final equilibrium state? (*d*) Discuss the meaning of the statement that the two gases when mixed have a higher degree of disorder than when they are separated.

25 • 29 Two copper slabs have equal masses m and constant specific heats c, but different initial temperatures T_h and T_c, with $T_h > T_c$. The slabs are brought into contact while thermally insulated from all other bodies, whereupon heat flows irreversibly until their temperatures reach the equilibrium value T. (*a*) Find the total entropy change of the slabs. (*b*) Show that the total entropy has increased as a result of this irreversible process. (*Hint:* The arithmetic mean of two numbers is always greater than their geometric mean.) (*c*) Given that m = 100 g and $T_h = 5\,T_c$, find the total entropy change.

***25 • 30** Entropy tables for copper give the following experimentally determined values:

T, K	20	30	40	50	60	70	80
S, J/kmole · K	152	541	1294	2385	3724	5222	6806

(*a*) Compute the average molar specific heat $\overline{C}$ at T = 25, 35, 45, 55, 65, and 75 K. (*b*) Determine the best fit of the form $\overline{C} = aT^b$ to the data (see Sec. 24 . 1).

25 • 6 statistical interpretation of entropy

25 • 31 A simple pendulum consisting of a 0.5-g bob suspended from a fixed point by a thread of negligible mass is initially at rest in still air of uniform temperature 27°C. There is a chance that the pendulum will acquire mechanical energy at the expense of the internal energy of the air and thus spontaneously begin to oscillate, with the particle initially swinging up to some height y above its rest position. (*a*) How does the probability $\mathcal{W}(y)$ of this state of motion, assuming y to be only 1 mm, compare with the probability $\mathcal{W}(0)$ that the pendulum will remain in the state of rest? (*b*) State the main assumptions involved in your computation.

25 • 32 If we assume that the entropy of liquid helium at absolute zero is also zero, what does this imply about the allowed microstates of the helium? Recast your answer into a new definition of absolute zero in terms of the microstates of helium.

25 • 33 A deck of cards is a "system" of 52 parts having 13 possible values and 4 possible suits. Let the microstate of a deck of cards be specified by the suits and values and the *order* in which they appear, with the suits ordered by hearts, diamonds, clubs, and spades and the values ordered by 2 to 10, J, Q, K, and A. (*a*) What is the number of microstates of the deck when it is ordered first by suit and then by value within each suit? (*b*) When it is ordered by values without regard to suit? (*c*) When it is ordered by suit without regard to value? (*d*) When it is completely random (disordered)? (*e*) What is the entropy of the deck in each of the above situations? (*Hint:* Stirling's approximation is $\ln(n!) \doteq 0.91894 + (n + \frac{1}{2})\ln n - n$ for $n \geq 5$.)

answers

25 • 1 $\Delta\mathcal{U} = 740$ J
25 • 2 (*b*) $Q_{net} = W = 1044$ J, $\Delta\mathcal{U} = 0$
25 • 3 (Process 1) $Q = 8511$ J, $W = 3404$ J, $\Delta\mathcal{U} = 5107$ J; (process 2) $Q = \Delta\mathcal{U} = -5107$ J, $W = 0$; (process 3) $Q = W = -2360$ J, $\Delta\mathcal{U} = 0$
25 • 4 (*a*) $\eta = 40$ percent, $dW/dt = 800$ W, $dQ/dt = 1200$ W
25 • 6 (*a*) $\eta = 32$ percent, $H_{out} = 340$ J/cycle, $W = 160$ J/cycle;
(*b*) $\Delta\mathcal{U} = 0, -5321, 0, +5321, 0$ J, $\Delta\mathcal{U}_{net} = 0$
25 • 7 666 cycles/min = 11.1 Hz
25 • 8 (*a*) $\eta_c = 19$ percent; (*b*) $\eta = 14$ percent; (*c*) $71/79 = 90$ percent; (*d*) 13 percent; (*e*) 10 percent; (*f*) 160 lbm/h
25 • 9 (*a*) $V_D/V_C = 2.6$; (*b*) $V_C/V_B = 1.1$
25 • 14 (*a*) $dW/dt = 122.8$ W;
(*b*) $dW/dt = 50.8$ W
25 • 16 $\Delta S/\Delta t = 4$ W/K;
(*b*) $\Delta S/\Delta t = -4$ W/K;
(*c*) $\Delta S/\Delta t = 0$
25 • 17 (*a*) $\Delta S = C_V \ln(P'/P) + C_P \ln(V'/V)$, $Q = (C_V/R)(P'V - PV) + (C_P/R)(P'V' - P'V)$, $\Delta\mathcal{U} = (C_V/R)(P'V' - PV)$;
(*b*) ΔS and $\Delta\mathcal{U}$ same as in part (*a*), $Q = (C_P/R)(PV' - PV) + (C_V/R)(P'V' - PV)$;
(*c*) $\Delta Q = (P - P')(V' - V)$
25 • 21 (*a*) $\Delta S = 0$; (*b*) $\Delta S = 5.4$ cal/K;
(*c*) $\Delta S = 5.4$ cal/K
25 • 22 (*a*) $\Delta S = 8577$ J/K; (*b*) $\Delta S = 922$ J/K; (*c*) $S = 1306$ J/K
25 • 23 (*a*) $Q = 696$ kcal;
(*b*) $\Delta S = 2.00$ kcal/K

25 • 24 (a) $\Delta S = -8577$ J/K;
(b) $\Delta S = 3323$ J/K
25 • 25 $\Delta S = 8000$ J/K
25 • 26 $\Delta S' = 6504$ J/K
25 • 27 (b) $\Delta S = 0.268$ cal/K
25 • 28 (b) $\Delta S = 23.0$ J/K;
(c) $P = 1$ atm, $\tau = 20°C$
25 • 29 (a) $\Delta S = mc \ln [(T_h + T_c)^2/4T_hT_c]$;
(c) $\Delta S = 59.8c$
25 • 30 (a) $\overline{C}_V = 973, 2636, 4910, 7365, 9737, 11{,}880$ J/kmole · K;
(b) $\overline{C}_V = 0.736T^{2.276}$
25 • 31 (a) $\mathcal{W}(y)/\mathcal{W}(0) = \exp(-10^{15})$
25 • 33 (a) $\mathcal{W} = 1$;
(b) $\mathcal{W} = (4!)^{13}$;
(c) $\mathcal{W} = (13!)^4$; (d) $\mathcal{W} = 52!$; (e) $S = 0$, $41.3k$, $90.2k$, $156.4k$

appendices

APPENDIX A

glossary of symbols and abbreviations

table A • 1 *Greek alphabet*

Α	α	alpha	Ι	ι	iota	Ρ	ρ	rho
Β	β	beta	Κ	κ	kappa	Σ	σ	sigma
Γ	γ	gamma	Λ	λ	lambda	Τ	τ	tau
Δ	δ	delta	Μ	μ	mu	Υ	υ	upsilon
Ε	ϵ	epsilon	Ν	ν	nu	Φ	ϕ	phi
Ζ	ζ	zeta	Ξ	ξ	xi	Χ	χ	chi
Η	η	eta	Ο	o	omicron	Ψ	ψ	psi
Θ	θ	theta	Π	π	pi	Ω	ω	omega

table A • 2 *Symbols used in text*

The number following the definition indicates the page on which the symbol is first used. Abbreviations of units are listed in Table A . 3.

α	angular acceleration, 265
α	direction angle between vector and x axis, 26
α	linear expansion coefficient, 570
α	y intercept of linear least-squares fit, 676
$\alpha_n(\nu)$	normalized Lagrange coefficients, 573
β	direction angle between vector and y axis, 26
β	ratio of speed to the speed of light, 524
β	coefficient of cubical expansion, 569
β	slope of linear least-squares fit, 679
γ	direction angle between vector and z axis, 26
γ	ratio of specific heats, 462
γ	damping coefficient, 435
γ	system dilation factor, 524
γ_v	particle dilation factor, 537
Δ	"change in" or "difference in," 25
$\Delta\mathbf{r}$	finite vector displacement, 25
δ	logarithmic decrement, 437
δ	very small but finite increment, 452
∂	partial differential, 186
ϵ	error, 5
$\overline{\epsilon^2}$	average squared error, 5

η	mechanical efficiency, 192
η	viscosity, 395
η	thermal efficiency, 710
η_c	thermal efficiency of a Carnot engine, 713
$\hat{\theta}$	unit vector in θ direction, 259
θ	angular distance, 8
θ	polar-coordinate angle, 15
θ'	angle of refraction, 503
θ_i	angle of incidence, 502
θ_r	angle of reflection, 502
κ	compressibility, 364
λ	latitude, 310
λ	wavelength, 455
λ	scaled length variable, 133
λ	mean free path, 692
μ	specific damping coefficient, 436
μ	average (mean) of Gaussian distribution, 679
μ_k	coefficient of kinetic friction, 125
μ_s	coefficient of static friction, 125
ν	frequency in cycles/sec (Hz), 415
ν	angular frequency in revolutions/sec (Hz), 259
ν	scaled velocity variable, 133
π	ratio of circumference to diameter of a circle (3.14159265), 12
ρ	density, 255
ρ_l	linear density of a string, 353
Σ	summation, 101
σ	Poisson ratio, 371
σ	standard deviation, 5
σ	surface mass density, 242
τ	torque, 272
τ	scaled time variable, 133
τ	proper time, 527
τ	Celsius (centigrade) temperature, 556
$\tau_{°F}$	Fahrenheit temperature, 559
ϕ	spherical-polar-coordinate angle, 16
ϕ	scattering angle, 211
ϕ	shear angle, 366
ϕ	phase constant, 417
$\mathbf{\Omega}$	angular velocity of precession, 315
ω	angular velocity, 265
ω	angular speed in rad/sec, 256
ω	angular frequency in rad/sec, 415
ω_d	damped angular frequency of oscillation, 436
ω_n	natural angular frequency of oscillation, 418
ω_s	spin angular velocity of a spinning rigid body, 315
A	area, 5
A	amplitude of oscillation, 416
$\mathbf{a}$	acceleration vector, 51
a	internal pressure constant, 659

B	bulk modulus, 363
b	molar covolume, 659
C	molar specific heat, 589
C_D	drag coefficient, 403
C_P	molar specific heat at constant pressure, 597
C_V	molar specific heat at constant volume, 596
c	speed of light, 520
c	specific heat, 589
D	diffusion coefficient, 699
d	differential notation (as in ds, $d\tau$, etc.), 46
d	collision diameter, 692
$đQ$	irreversible heat transfer, 721
E	total mechanical energy, 176
E	total relativistic energy, 542
E_V	energy density, 470
e	base of natural logarithms (2.7182818), 217
e	coefficient of restitution, 210
$\mathbf{F}$	force vector, 104
$\mathbf{F}'$	reaction force, 110
$\mathbf{F}_c$	Coriolis force, 311
$\mathbf{F}_f$	frictional force vector, 125
$\mathbf{F}_g$	gravitational force vector, 106
$\mathbf{f}$	gravitational field strength, 339
f	number of degrees of freedom, 632
$f(x)$	Gaussian probability density of x, 681
G	universal gravitational constant, 330
g	acceleration due to gravity, 6
g'	numerical value of g_s in mks units, 106
g''	numerical value of g_s in fps units, 106
g_s	standard acceleration due to gravity, 95
H_{in}	heat absorbed by engine or refrigerator, 710
H_{out}	heat ejected by engine or refrigerator, 710
h	Planck's constant, 636
h	molar specific enthalpy, 656
h	height, 75
h	scale height of atmosphere, 386
I	moment of inertia, 293
I	intensity of a wave, 470
$\mathbf{i}$	unit vector in the positive x direction, 25
$\mathbf{j}$	unit vector in the positive y direction, 25
$\mathcal{K}$	thermal conductivity, 602
K	kinetic energy, 155
K_r	refrigerator coefficient of performance, 715
K_{rot}	rotational kinetic energy, 629
K_{tr}	translational kinetic energy, 629
K_{vib}	vibrational kinetic energy, 629
$\mathbf{k}$	unit vector in the positive z direction, 25
k	radius of gyration, 298
k	restoring force constant, 132
k	Boltzmann's constant, 624

k_{tr}	molar translational kinetic energy, 653
k_p	Kepler's planetary constant, 328
k_s	Kepler's solar constant, 329
k_s	stiffness coefficient, 359
L	angular momentum, 299
L	length dimension, 6
L	length scale, 133
L_f	latent heat of fusion, 591
L_s	latent heat of sublimation, 591
L_v	latent heat of vaporization, 591
l	length, 5
l	scale of a conic section, 333
M	mass dimension, 399
M	mass of a system, 228
M_0	molecular weight of a kmole in kilograms, 463
m	mass of a particle, 98
$\mathbf{N}$	reaction force vector normal to a surface, 123
N	number of molecules, 686
N_A	Avogadro's number, 619
N_{Re}	Reynolds number, 401
$\hat{\mathbf{n}}$	unit vector normal to a surface, 211
n	number of moles, 566
n_c	number of collisions, 692
n_{mol}	number density of molecules, 692
O	origin of coordinates, 15
O'	origin of moving coordinate system, 79
$\mathcal{P}$	angular impulse, 302
$\mathbf{P}$	impulse vector, 202
P	power, 153
P	pressure, 363
$P(i)$	probability of an observation falling in the ith category, 681
$P(x)$	unique interpolating polynomial, 572
$P(u)$	cumulative normal distribution integral, 682
P_s	vapor pressure of a solid, 664
P_v	vapor pressure of a liquid, 649
$\mathbf{p}$	linear momentum vector, 98
Q	heat absorbed by a system, 191
q	heat flux density, 700
R	radius, 6
R	radius of rotation, 259
R	universal gas constant, 566
R^2	total squared error between data and prediction, 676
R_e	radius of the earth, 330
$\mathbf{r}$	radial position vector, 25
$\hat{\mathbf{r}}$	unit vector in the direction of **r**, 259
$\mathbf{r}_c$	position vector of the center of mass, 228
r	radial coordinate, or radial distance, 15
r_a	apogée radius, 336
r_p	perigée radius, 336
S	shear modulus, 367
S	entropy, 720

S	fixed reference frame, 79
S'	moving reference frame, 79
s	arc length, 8
s	path length, 44
T	tension, 118
T	time dimension, 6
T	time scale, 133
T	period of oscillation, 183
T	absolute temperature (in kelvins), 556
$\hat{\mathbf{t}}$	unit vector tangential to a surface, 211
t	time, 6
$\mathcal{U}$	total internal energy of a system, 191
U	potential energy, 172
U_g	gravitational potential energy, 174
U_m	intermolecular potential energy, 629
U_{mol}	Lennard–Jones intermolecular potential energy per pair of interacting molecules, 654
U_s	spring or elastic potential energy, 175
U_{vib}	vibrational potential energy, 629
u	speed before binary collision, 95
u	normalized random variable, 682
u	molar internal energy, 653
u	drift velocity, 696
u_m	molar potential energy, 653
$\mathbf{V}$	constant velocity of a moving frame of reference relative to a fixed observer, 79
V	velocity scale, 133
V	volume, 6
V_0	molecular volume of an ideal gas, 566
$\mathbf{v}$	instantaneous velocity, 48
$\mathbf{v}_c$	velocity of center of mass, 228
v	instantaneous speed, 45
v	specific volume, 364
v	molar specific volume, 566
$\bar{v}_c$	average speed of approach on collision, 692
v_{rms}	root-mean-square speed, 623
$\mathcal{W}$	thermodynamic probability, 724
W	work, 144
(x,y)	two-dimensional cartesian coordinates, 15
(x,y,z)	three-dimensional cartesian coordinates, 16
Y	Young's modulus, 360
$Z(u)$	standardized normal distribution, 682

General forms:

$\{q\}$	numerical value, or measure of quantity q, 5
$[q]$	unit of quantity q, 4
$\bar{a}$	average value of a, 5
$\overline{\epsilon^2}$	average value of (ϵ^2), 5
$x \to a$	the value of x approaches the value a, 45
$\Delta t \to dt$	the increment Δt becomes the infinitesimally small differential dt, 45

$\lim_{x \to 0} f(x)$	the limiting value of the expression $f(x)$ as x approaches zero, 45
$[t_1,t_2]$	the interval or range of values $t_1 \leq t \leq t_2$, 57
$\int$	integral, 58
$\oint$	integral over a closed path, 171
Σ	summation, 101
$\frac{\partial f}{\partial x}$	partial derivative of f with respect to x, treating any other variables as constants, 186
$\doteq$	approximately equals, 131
$\sim$	is proportional to, 103
$x \approx 10^2$	x is of the order of magnitude 100 (i.e., $10^{1.5} < x < 10^{2.5}$), 402
$\mathbf{A} \cdot \mathbf{B}$	scalar dot product of vectors **A** and **B**, 30
$\mathbf{A} \times \mathbf{B}$	vector cross product of vectors **A** and **B**, 32
$\|\mathbf{A}\|$ or A	magnitude of the vector **A**, 21
$\hat{\mathbf{A}}$	unit vector in the direction of **A**, 21
$\mathbf{A}_\perp$	component of vector **A** perpendicular to some selected direction, 28
$\mathbf{A}_\parallel$	component of vector **A** parallel to some selected direction, 32
$\overline{AB}$	line segment from point A to point B, 22
$\overrightarrow{AB}$	vector (directed line segment) from point A to point B, 22
A_x, A_y, A_z	cartesian components of vector **A**, 26
v_r, v_θ	polar components of vector **v**, 333

table A • 3 *Abbreviations for units*

See Appendix B for SI units and prefixes, and see Appendix C for conversion tables.

Ω	ohm (SI derived unit of electric resistance)
A	ampere (SI base unit of electric current)
Å	angstrom (unit of length)
A · h	ampere hour (unit of electric charge)
A · s	ampere second (unit of electric charge)
atm	standard atmosphere (unit of pressure)
AU	astronomical unit (unit of length)
b	barn (unit of area)
bar	bar (unit of pressure)
bbl	barrel (unit of volume)
Bq	becquerel (SI derived unit of activity: 1 Bq = 1 event/s)
Btu	British thermal unit (unit of energy)
C	coulomb (SI derived unit of electric charge)
°C	degree Celsius or centigrade (unit of temperature *change:* 1°C = 1 K)
°C	degree Celsius or centigrade (unit of temperature *value:* 0°C = 273.15 K)
cal	gram-calorie or calorie (unit of energy)
cd	candela (SI base unit of luminous intensity)
cgs	centimeter-gram-second system of units
Ci	curie (unit of radioactivity: 1 Ci = 3.7×10^{10} events/s)

cm	centimeter (cgs unit of length)
cwt	short hundredweight (unit of force: 1 cwt = 100 lbf)
d	day (unit of time)
dyne	dyne (cgs unit of force)
erg	erg (cgs unit of energy)
eV	electronvolt (unit of energy)
F	farad (SI derived unit of capacitance)
°F	degree Fahrenheit (unit of temperature *change:* 1°F = $\frac{5}{9}$ K = 0.5556 K)
°F	degree Fahrenheit (unit of temperature *value:* 32°F = 0°C = 273.15 K)
fdy	faraday (unit of electric charge)
fm	fermi or femtometer (unit of length)
fps	foot-pound-second system of units
ft	foot (fps unit of length)
ft · lbf	foot-pound (fps unit of energy)
g	gram (cgs unit of mass)
Gal	gal or galileo (unit of acceleration: 1 Gal = 1 cm/s^2 = 10^{-2} m/s^2)
gal	gallon (unit of volume)
gmole	gram-mole or mole (SI base unit of amount of substance)
Gy	gray (SI derived unit of absorbed energy dose)
H	henry (SI derived unit of inductance)
h	hour (unit of time)
ha	hectare (unit of area)
hhd	hogshead (unit of volume)
hp	horsepower (unit of power)
Hz	hertz (SI derived unit of frequency: 1 Hz = 1 cycle/s)
in	inch (unit of length)
J	joule (SI derived unit of energy)
K	kelvin (SI base unit of temperature)
kcal	kilogram-calorie, large calorie, or Calorie (unit of energy)
kg	kilogram (SI base unit of mass)
kmole	kilogram-mole (unit of amount of substance: 1 kmole = 10^3 moles)
knot	knot (unit of speed)
L	lambert (unit of surface brightness: 1 L = 3183 cd/m^2)
l	liter or litre (in this text abbreviated only with prefixes)
lbf	pound-force (fps unit of force)
lbm	pound-mass (fps unit of mass)
liter	liter or litre (unit of volume: 1 liter = 1 dm^3 = 10^{-3} m^3)
lm	lumen (SI derived unit of luminous flux)
lx	lux (SI derived unit of illuminance)
m	meter or metre (SI base unit of length)
mi	mile (unit of length)
mil	mil (unit of length: 1 mil = 10^{-3} in)
min	minute (unit of time)
mol	molecule (NOTE: this abbreviation stands for "mole" in many other books)
mole	gram-mole (SI base unit of amount of substance)
mph	miles per hour = mi/h (unit of speed)

N	newton (SI derived unit of force)
nmi	nautical mile (unit of length)
oz	ounce (unit of *force:* 1 oz $= \frac{1}{16}$ lbf)
oz	ounce (unit of *mass:* 1 oz $= \frac{1}{16}$ lbm)
oz	fluid ounce (unit of *volume:* 1 oz $= \frac{1}{128}$ gal)
P	poise (cgs unit of viscosity: 1 P = 1 g/cm · s = 0.1 kg/m · s)
Pa	pascal (SI derived unit of pressure)
pc	parsec (unit of length)
pdl	poundal (unit of force: 1 pdl = 0.031081 lbf = 0.13825 N)
psi	pound per square inch = lbf/in^2 (unit of pressure)
pt	pint (unit of volume)
qt	quart (unit of volume)
R	roentgen or röntgen (unit of radiation exposure: 1 R $= 2.58 \times 10^{-4}$ C/kg)
rad	radian (SI derived unit of plane angle)
rod	rod (unit of length)
rpm	revolution per minute (unit of frequency: 1 rpm $= \frac{1}{60}$ Hz = 0.01667 Hz)
rps	revolution per second (unit of frequency: 1 rps = 1 Hz)
S	siemens (SI derived unit of conductance)
s	second (SI base unit of time)
sr	steradian (SI derived unit of solid angle)
T	tesla (SI derived unit of magnetic flux density)
t	tonne or metric ton (unit of mass)
ton	ton (unit of mass)
Torr	torr or torricelli (unit of pressure)
u	atomic mass unit (unit of mass)
V	volt (SI derived unit of electric potential)
W	watt (SI derived unit of power)
Wb	weber (SI derived unit of magnetic flux)
yd	yard (unit of length)
yr	year (unit of time)
°	degree (unit of temperature: *see* °C and °F above)
°	degree (unit of plane angle)
′	minute (unit of plane angle)
″	second (unit of plane angle)

APPENDIX B

the international system of units (SI)

The International System of Units (SI) is based upon the older metric system of units. The SI system has been developed in a series of international conferences and agreements to provide a logical and coherent set of units for all measurements in science, industry, and commerce. Seven SI base units are defined operationally (see Table B . 1).

The other SI units (SI derived units) are defined in terms of the base units. For example, the SI derived unit of speed is the m/s, and the SI derived unit of density is the kg/m^3. Table B . 2 lists the SI derived units that have special names.

table B • 1 *The SI base units*

base quantity	*SI base unit*	*operational definition*
length	meter or metre (m)	The length equal to 1,650,763.73 wavelengths in vacuum of the orange-red radiation corresponding to the transition between the levels $2p_{10}$ and $5d_5$ of the krypton-86 atom
mass	kilogram (kg)	The mass of the international prototype of the kilogram
time	second (s)	The duration of 9,192,631,770 periods of the radiation corresponding to the transition between the two hyperfine levels of the ground state of the cesium-133 atom
electric current	ampere (A)	That constant current which, if maintained in two straight parallel conductors of infinite length, of negligible circular cross-section, and placed 1 m apart in vacuum, would produce between these conductors a force equal to 2×10^{-7} N per meter of length
thermodynamic temperature	kelvin (K)	The fraction 1/273.16 of the thermodynamic temperature of the triple point of water
amount of substance	mole or gram-mole (mole)	The amount of substance of a system which contains as many elementary entities as there are atoms in exactly 0.012 kg of carbon-12
luminous intensity	candela (cd)	The luminous intensity, in the perpendicular direction, of a surface of 1/600,000 m^2 of a black body at the temperature of freezing platinum under a pressure of 101,325 N/m^2

table B•2 *The SI derived units*

quantity	*SI derived unit*	*definition*
plane angle	radian (rad)	angle at center of circle that subtends an arc of length equal to radius of circle (a dimensionless unit)
solid angle	steradian (sr)	solid angle at center of sphere of radius r that subtends an area of r^2 on the surface of the sphere (dimensionless)
frequency	hertz (Hz)	1 Hz = 1 cycle/s
force	newton (N)	$1\ \text{N} = 1\ \text{m} \cdot \text{kg/s}^2 = 1\ \text{J/m}$
pressure	pascal (Pa)	$1\ \text{Pa} = 1\ \text{N/m}^2 = 1\ \text{kg/m} \cdot \text{s}^2$
energy, work, quantity of heat	joule (J)	$1\ \text{J} = 1\ \text{N} \cdot \text{m} = 1\ \text{m}^2 \cdot \text{kg/s}^2$
power, radiant flux	watt (W)	$1\ \text{W} = 1\ \text{J/s} = 1\ \text{m}^2 \cdot \text{kg/s}^3$
quantity of electricity, electric charge	coulomb (C)	$1\ \text{C} = 1\ \text{A} \cdot \text{s}$
electric potential, potential difference, electromotive force	volt (V)	$1\ \text{V} = 1\ \text{J/C} = 1\ \text{W/A} = 1\text{m}^2 \cdot \text{kg/s}^3 \cdot \text{A}$
capacitance	farad (F)	$1\ \text{F} = 1\ \text{C/V} = 1\ \text{s}^4 \cdot \text{A}^2/\text{m}^2 \cdot \text{kg}$
electric resistance	ohm (Ω)	$1\ \Omega = 1\ \text{V/A} = 1\ \text{m}^2 \cdot \text{kg/s}^3 \cdot \text{A}^2$
conductance	siemens (S)	$1\ \text{S} = 1\ \text{A/V} = 1\ \text{s}^3 \cdot \text{A}^2/\text{m}^2 \cdot \text{kg}$
magnetic flux	weber (Wb)	$1\ \text{Wb} = 1\ \text{V} \cdot \text{s} = 1\ \text{m}^2 \cdot \text{kg/s}^2 \cdot \text{A}$
magnetic flux density	tesla (T)	$1\ \text{T} = 1\ \text{Wb/m}^2 = 1\ \text{kg/s}^2 \cdot \text{A}$
inductance	henry (H)	$1\ \text{H} = 1\ \text{Wb/A} = 1\ \text{m}^2 \cdot \text{kg/s}^2 \cdot \text{A}^2$
luminous flux	lumen (lm)	$1\ \text{lm} = 1\ \text{cd} \cdot \text{sr}$
illuminance	lux (lx)	$1\ \text{lx} = 1\ \text{lm/m}^2 = 1\ \text{cd} \cdot \text{sr/m}^2$
activity	becquerel (Bq)	1 Bq = 1 event/s
absorbed dose	gray (Gy)	$1\ \text{Gy} = 1\ \text{J/kg} = 1\ \text{m}^2/\text{s}^2$

The prefixes listed in Table B . 3 can be combined with any SI unit to obtain larger or smaller units suitable for the measurements at hand. Each prefix indicates multiplication of the SI unit by a specified power of 10. For example, $1\ \text{mm} = 10^{-3}\ \text{m}$. Note that $1\ \text{mm}^3 = (10^{-3}\ \text{m})^3 = 10^{-9}\ \text{m}^3$.

Appendix C provides conversion tables for use in translating between the SI units and other common units.

table B • 3 *The SI prefixes*

prefix	*multiple*	*prefix*	*multiple*
deci- (d)	10^{-1}	*deca- (da)	10^{1}
centi- (c)	10^{-2}	hecto- (h)	10^{2}
milli- (m)	10^{-3}	kilo- (k)	10^{3}
micro- (μ)	10^{-6}	mega- (M)	10^{6}
nano- (n)	10^{-9}	giga- (G)	10^{9}
pico- (p)	10^{-12}	tera- (T)	10^{12}
femto- (f)	10^{-15}	peta- (P)	10^{15}
atto- (a)	10^{-18}	exa- (E)	10^{18}

*This prefix is often written as "deka-," but the officially recommended SI form is "deca-."

APPENDIX C

conversion factors

The following tables can be used to determine conversion factors between different units. The columns of any row in a table can be read as if they were consecutive equalities between units. For example, the first two rows of Table C . 1 can be read as

$$1 \text{ m} = 3.281 \text{ ft} = 39.37 \text{ in} = 6.214 \times 10^{-4} \text{ mi}$$

$$0.3048 \text{ m} = 1.000 \text{ ft} = 12.00 \text{ in} = 1.894 \times 10^{-4} \text{ mi}$$

Consider a quantity q with measure $\{q\}$, dimension Q, and unit $[Q]$. To convert this quantity to a different unit $[Q]^*$, we set up the equality

$$q = \{q\}[Q] = \{q\}^*[Q]^* \qquad [\text{C} \cdot 1]$$

where $\{q\}^*$ is the measure of q in units of $[Q]^*$. We can determine $\{q\}^*$ as

$$\{q\}^* = \{q\}\frac{[Q]}{[Q]^*} = \{q\}c \qquad [\text{C} \cdot 2]$$

The conversion factor $c = [Q]/[Q]^*$ is the ratio of two physical units having the same dimension Q, so this ratio is actually dimensionless. For example, from the second line of Table C . 2,

$$c = \frac{1 \text{ ft}^2}{1 \text{ in}^2} = \frac{144 \text{ in}^2}{1 \text{ in}^2} = 144$$

The entries in the tables represent these dimensionless conversion ratios.

In order to avoid errors in conversion, we actually multiply the quantity to be converted by a factor of unity,

$$q = \{q\}[Q] \times \frac{c[Q]^*}{[Q]} = \{q\}c[Q]^* \qquad [\text{C} \cdot 3]$$

thus cancelling the original unit. For example, to convert $A = 7 \text{ ft}^2$ to square inches, we would write

$$A = 7 \text{ ft}^2 \times \frac{144 \text{ in}^2}{1 \text{ ft}^2} = 1008 \text{ in}^2$$

If multiple units are involved, this method can be applied on a unit-by-unit basis.

For simplicity, we have not tabulated units that differ from standard units only by factors of powers of ten. In most cases, the conversion factors are given to four significant figures. In some cases, the product of reciprocal conversion factors may not be exactly unity due to round-off errors; however the given conversion factors are correct to at least four places.

table C • 1 *Length conversion factors*

meters (m)	*feet (ft)*	*inches (in)*	*miles (mi)*
1	3.281	39.37	6.214×10^{-4}
0.3048	1	12	1.894×10^{-4}
0.02540	0.08333	1	1.578×10^{-5}
1609	5280	63,360	1

equal lengths:

1 m = 100 cm
1000 m = 1 km
10^{-10} = 1 angstrom (Å)

9.464×10^{12} km = 1 light-year
1.495×10^{8} km = 1 astronomical unit
3.084×10^{13} km = 1 parsec (pc)

3 ft = 1 yd
16.5 ft = 1 rod

1 nautical mile = 1.152 mi = 6080 ft
1 in = 1000 mils = 6 picas = 72 points

table C • 2 *Area conversion factors*

square meters (m^2)	*square feet (ft^2)*	*square inches (in^2)*	*square miles (mi^2)*
1	10.76	1550	3.861×10^{-7}
0.0929	1	144	3.587×10^{-8}
6.452×10^{-4}	6.944×10^{-3}	1	2.491×10^{-10}
2.590×10^{6}	2.788×10^{7}	4.015×10^{9}	1

equal areas:

1 mi^2 = 640 acres   1 acre = 43,560 ft^2 = 4840 yd^2
1 barn = 10^{-28} m^2
1 circular inch = area of a circle one inch in diameter = 0.7854 in^2
1 circular mil = 10^{-6} circular inches = 0.7854 mil^2 = 0.7854×10^{-6} in^2
1 km^2 = 10^6 m^2 1 m^2 = 10^4 cm^2
1 hectare (ha) = 100 ares = 10^4 m^2

table C • 3 *Volume conversion factors*

cubic meters (m^3)	*cubic feet (ft^3)*	*cubic inches (in^3)*
1	35.31	61,023
0.02832	1	1728
1.639×10^{-5}	5.787×10^{-4}	1

equal volumes:

1 m^3 = 10^3 liters = 10^6 cm^3 = 10^9 mm^3
1 U.S. quart (qt) = 0.9463 liter
1 U.S. gallon (gal) = 3.785 liters = 231.0 in^3 = 0.8327 British gal
1 U.S. gal = 4 U.S. qt = 8 U.S. pints = 32 U.S. gills
= 128 U.S. fluid ounces
1 U.S. hogshead = 2 U.S. barrels = 63 U.S. gal
1 oil barrel = 42 U.S. gal
1 ft^3 = 7.481 U.S. gal = 6.229 British gal = 28.32 liters
1 in^3 = 0.5541 U.S. fluid ounce = 0.01639 liter   1 stere = 1 m^3

table C • 4 *Angular conversion factors*

radians (rad)	*degrees (°)*	*minutes (′)*	*seconds (″)*	*revolutions (rev)*
1	57.2958	3437.75	2.063×10^5	0.1592
0.0174533	1	60	3600	2.778×10^{-3}
2.909×10^{-4}	0.01667	1	60	4.630×10^{-5}
4.848×10^{-6}	2.778×10^{-4}	0.01667	1	7.716×10^{-7}
6.28319	360	21,600	1.296×10^6	1

Note: A circle subtends a plane angle of 2π radians (rad).
A sphere subtends a solid angle of 4π steradians (sr).

table C • 5 *Time conversion factors*

seconds (s)	*minutes (min)*	*hours (h)*	*days (d)*	*years (y)*
1	0.01667	2.778×10^{-4}	1.157×10^{-5}	3.169×10^{-8}
60	1	0.01667	6.944×10^{-4}	1.901×10^{-6}
3600	60	1	0.04167	1.141×10^{-4}
86,400	1440	24	1	2.738×10^{-3}
3.156×10^{7}	5.259×10^{5}	8766	365.242	1

equal times:

1 sidereal year = 365.256 mean solar days = 8766.14 mean solar hours
1 sidereal day = 23.93 mean solar hours = 86,164 mean solar seconds
1 sidereal year = 1.000038 solar tropical years
1 sidereal day = 0.9973 mean solar day
1 sidereal year = 366.32 sidereal days

Note: The year given in this table is the solar tropical year, which is the time between two successive northward crossings of the celestial equator by the sun. The day listed here is the mean solar tropical day, which is the mean time between two successive passages of the sun over the north-south meridian from some fixed observation point. The mean solar tropical hour, minute, and second as given here are defined as fractions of the mean solar tropical day. In contrast, the sidereal year is defined as the time between two passages of the sun across the same spot in the background of fixed stars. The sidereal day is the time between two successive passages of a fixed star over the north-south meridian from some fixed observation point. The mean solar and sidereal measurements differ because of the precession of the equinoxes. The mean solar time units are used for most measurements except those in astronomy that deal with observation of the fixed stars.

table C • 6 *Speed conversion factors*

meters/second (m/s)	*feet/second (ft/s)*	*kilometers/hour (km/h)*	*miles/hour (mi/h)*
1	3.281	3.600	2.237
0.3048	1	1.097	0.6818
0.2778	0.9113	1	0.6214
0.4470	1.467	1.609	1

equal speeds:

1 knot = 1 nautical mile per hour = 1.1516 mi/h = 0.5148 m/s

table C • 7 *Mass conversion factors*

kilograms (kg)	*pound-masses (lbm)*	*slugs*
1	2.205	0.06852
0.4536	1	0.03108
14.59	32.17	1

equal masses:

1 kg = 10^3 g = 10^6 mg = 10^9 μg 1 ounce-mass = 28.35 g
1 tonne (metric ton) = 10^3 kg 1 ton-mass = 2000 lbm
1 lbm = 16 ounce-masses
1 atomic mass unit (u) = 1.660×10^{-27} kg
= 0.9312×10^9 electronvolts (eV)

table C • 8 *Force conversion factors*

newtons (N)	*kilogram-forces (kgf)*	*pound-forces (lbf)*
1	0.1020	0.2248
9.807	1	2.205
4.448	0.4536	1

equal forces:

1 N = 10^5 dynes 1 kgf = 10^3 gram-forces (gf)
1 lbf = 16 ounce-forces = 32.17 poundals (pdl)

table C • 9 *Pressure conversion factors*

pascals (Pa)	*atmosphere (atm)*	*lbf/ft²*	*lbf/in² (psi)*	*kgf/cm²*	*cm of Hg (at 0°C)*
1	9.869×10^{-6}	0.02089	1.450×10^{-4}	1.020×10^{-5}	7.501×10^{-4}
1.013×10^5	1	2116	14.70	1.033	76.00
47.88	4.725×10^{-4}	1	6.944×10^{-3}	4.882×10^{-4}	0.03591
6895	0.06805	144	1	0.07031	5.172
9.807×10^4	0.9678	2048	14.22	1	73.56
1333	0.01316	27.85	0.1934	0.01360	1

equal pressures:

1 Pa = 10 dynes/cm² = 1 N/m² = 1 J/m³
1 bar = 1000 millibars = 10^6 baryes = 10^6 dyn/cm² = 10^5 Pa = 0.9869 atm
1 atm = 1.013 bars = 1033 cm of H_2O (4°C) = 33.90 ft of H_2O (4°C)
= 1.058 tons/ft² = 29.92 in of Hg (0°C)
1 torr = 1 mm of Hg at 0°C

table C • 10 *Energy conversion factors*

joules (J)	*gram-calories (cal)*	*kilowatt-hours (kW · h)*	*foot-pounds (ft · lbf)*	*British thermal units (Btu)*	*horsepower hours (hp · h)*	*electron-volts (eV)*
1	0.2389	2.778×10^{-7}	0.7376	9.480×10^{-4}	3.725×10^{-7}	6.242×10^{18}
4.186	1	1.163×10^{-6}	3.087	3.969×10^{-3}	1.559×10^{-6}	2.613×10^{19}
3.600×10^{6}	8.600×10^{5}	1	2.655×10^{6}	3413	1.341	2.247×10^{25}
1.356	0.3239	3.766×10^{-7}	1	1.285×10^{-3}	5.051×10^{-7}	8.463×10^{18}
1055	252.0	2.930×10^{-4}	778.0	1	3.929×10^{-4}	6.584×10^{21}
2.685×10^{6}	6.413×10^{5}	0.7457	1.980×10^{6}	2545	1	1.676×10^{25}
1.602×10^{-19}	3.827×10^{-20}	4.451×10^{-26}	1.182×10^{-19}	1.519×10^{-22}	5.968×10^{-26}	1

relativistic mass-energy equivalents ($E = mc^2$):

1 kg = 5.60953×10^{29} MeV = 8.98755×10^{16} J
1 electron-mass = m_e = 0.511004 MeV = 8.18725×10^{-14} J
1 proton-mass = m_p = 938.259 MeV = 1.50327×10^{-10} J
1 neutron-mass = m_n = 939.553 MeV = 1.50534×10^{-10} J

equal energies:

1 J = 10^{7} erg = 3.485×10^{-4} ft^3 · atm = 23.73 ft · poundals = 1 W · s
1 erg = 1 dyne · cm = 1 g · cm^2/s^2 = 10^{-7} J

thermal temperature-energy equivalents ($E = kT$):

1 K = 1.38063×10^{-23} J = 8.6171×10^{-5} eV
1 eV = 11,604.8 K

power equivalents:

1 watt (W) = 1 J/s 1 hp = 550 ft · lbf/s
1 hp = 745.7 W = 178.1 cal/s = 2545 Btu/h

APPENDIX D

formulas and approximations from alegebra and geometry

quadratic equations

If $ax^2 + bx + c = 0$ for $a \neq 0$, then

$$x = \frac{-b \pm \sqrt{b^2 - 4ac}}{2a} \tag{D·1}$$

If a, b, and c are real, then

for $b^2 - 4ac > 0$, the roots are real;
for $b^2 - 4ac = 0$, the roots are real and equal;
for $b^2 - 4ac < 0$, the roots are imaginary conjugates.

binomial series

If n is a positive integer, then

$$\begin{aligned}(x + y)^n = x^n + nx^{n-1}y &+ \frac{n(n-1)}{2}x^{n-2}y^2 \\ &+ \frac{n(n-1)(n-2)}{2 \times 3}x^{n-3}y^3 + \cdots \\ &+ nxy^{n-1} + y^n\end{aligned}$$

or

$$(x + y)^n = \sum_{k=0}^{n} \binom{n}{k} x^{n-k}y^k \tag{D·2}$$

where

$$\binom{n}{k} = \frac{n(n-1)\cdots(n-k+1)}{1 \times 2 \times 3 \times \cdots \times k}$$

If n is not a positive integer and $y \ll x$, we can use the first few terms of the resulting infinite (nonterminating) series to generate useful approximations, including the following:

$$\begin{aligned}(1 + y)^n &= 1 + ny + \tfrac{1}{2}n(n-1)y^2 + \cdots \\ &\doteq 1 + ny\end{aligned} \tag{D·3}$$

$$(1 + y)^{-1} = \frac{1}{1 + y} = 1 - y + y^2 - \cdots$$
$$\doteq 1 - y \qquad [D \cdot 4]$$

$$(1 + y)^{1/2} = \sqrt{1 + y} = 1 + \tfrac{1}{2}y - \tfrac{1}{8}y^2 + \cdots$$
$$\doteq 1 + \tfrac{1}{2}y \qquad [D \cdot 5]$$

$$(1 + y)^{-1/2} = \frac{1}{\sqrt{1 + y}} = 1 - \tfrac{1}{2}y + \tfrac{1}{8}y^2 - \cdots$$
$$\doteq 1 - \tfrac{1}{2}y \qquad [D \cdot 6]$$

If $x \doteq y$ such that $(y - x)/x = \epsilon \ll 1$, then

$$\sqrt{xy} = x(1 + \epsilon)^{1/2} \doteq \tfrac{1}{2}(x + y) \qquad [D \cdot 7]$$

In other words, the geometric and arithmetic means are approximately equal.

geometric progressions

Consider the geometric progression a, ax, ax^2, ax^3, ... The sum of the first $n + 1$ terms in this progression is

$$a + ax + \cdots + ax^n = a\sum_{k=0}^{n} x^k = a\frac{1 - x^{n+1}}{1 - x} \qquad [D \cdot 8]$$

In the case that $x < 1$ and $n \to \infty$,

$$a\sum_{k=0}^{\infty} x^k = a\frac{1}{1 - x} \qquad [D \cdot 9]$$

geometry

For a right circular cylinder of radius r and altitude h, the volume V and lateral area A are given by

$$V = \pi r^2 h \qquad A = 2\pi r h \qquad [D \cdot 10]$$

For a right circular cone of base radius r, altitude h, and slant height $l = \sqrt{r^2 + h^2}$, the volume V and lateral area A are given by

$$V = \tfrac{1}{3}\pi r^2 h \qquad A = \pi r l \qquad [D \cdot 11]$$

For a circle of radius r, the area A and circumference c are given by

$$A = \pi r^2 \qquad c = 2\pi r \qquad [D \cdot 12]$$

For a sphere of radius r, the volume V and surface area A are given by

$$V = \tfrac{4}{3}\pi r^3 \qquad A = 4\pi r^2 \qquad [D \cdot 13]$$

For an ellipse of semimajor axes a and b, the area A is given by

$$A = \pi ab \qquad [D \cdot 14]$$

summation notation

By definition, the summation notation

$$\sum_{i=m}^{n} x_i = x_m + x_{m+1} + x_{m+2} + \cdots$$
$$+ x_{n-2} + x_{n-1} + x_n \qquad [D \cdot 15]$$

Thus the summation notation is merely a shorthand symbol for the sum of all subscripted values of x from x_m through x_n. For example, if $x_n = n^2$, then

$$\sum_{i=1}^{5} x_i = 1 + 4 + 9 + 16 + 25 = 55$$

The individual terms of the sum can be functions $f_i(x)$ or tabulated values of x_i. Depending on the context, the summation notation may be written in various ways:

$$\sum_{i=m}^{n} x_i = \sum_{i=m}^{i=n} x_i = \sum_{i} x_i = \sum x_i$$

In the latter two forms, the limiting values of the summation must be indicated by the context.

The summation operator Σ is a linear operator. Given two constants a and b and two families of functions (or subscripted variables) $f_i(x)$ and $g_i(x)$,

$$\sum_{i=m}^{n} (af_i + bg_i) = a \sum_{i=m}^{n} f_i + b \sum_{i=m}^{n} g_i \qquad [D \cdot 16]$$

Thus, for $f_i \equiv 1$,

$$\sum_{i=1}^{n} a = na \qquad [D \cdot 17]$$

APPENDIX E

trigonometry and vector algebra

trigonometric functions

The six classic functions of trigonometry are defined with respect to a right triangle and the ratios of its sides. For the triangle *ABC* shown in Fig. E . 1, these ratios are

$$\begin{aligned}
\text{sine}\,\theta &= \sin\theta = \frac{a}{c} = \cos\phi = \text{cosine}\,\phi \\
\text{tangent}\,\theta &= \tan\theta = \frac{a}{b} = \cot\phi = \text{cotangent}\,\phi \\
\text{secant}\,\theta &= \sec\theta = \frac{c}{b} = \csc\phi = \text{cosecant}\,\phi \\
\text{cosine}\,\theta &= \cos\theta = \frac{b}{c} = \sin\phi = \text{sine}\,\phi \\
\text{cotangent}\,\theta &= \cot\theta = \frac{b}{a} = \tan\phi = \text{tangent}\,\phi \\
\text{cosecant}\,\theta &= \csc\theta = \frac{c}{a} = \sec\phi = \text{secant}\,\phi
\end{aligned} \qquad [\text{E} \cdot 1]$$

Note that the functions of θ are the cofunctions of its complement ϕ, and vice versa. (The cotangent is sometimes abbreviated as ctn or cotan, and the cosecant is sometimes abbreviated as cosec.)

The concept of the trigonometric function is generalized to include all possible angles from 0° to 360° (0 to 2π rad) by defining the functions in the *xy* plane with the angle measured positive counterclockwise and negative clockwise. Thus, in polar coordinates, a point (x,y) defines the functions

$$\sin\theta = \frac{y}{r} \qquad \cos\theta = \frac{x}{r} \qquad \tan\theta = \frac{y}{x} \qquad \text{etc.} \qquad [\text{E} \cdot 2]$$

where $r = \sqrt{x^2 + y^2}$, by Pythagoras' theorem. Thus the sign of the function

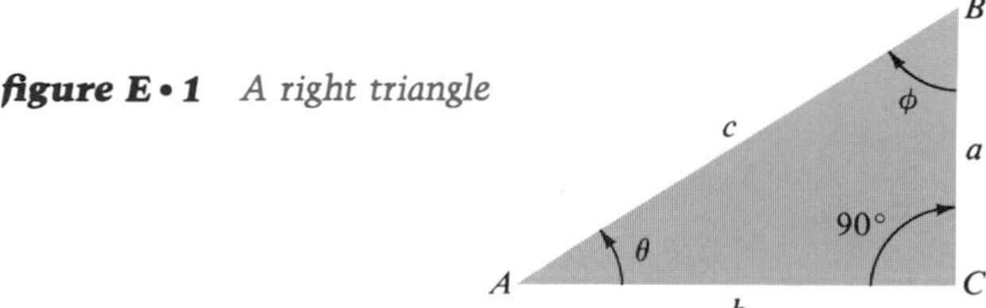

figure E • 1 *A right triangle*

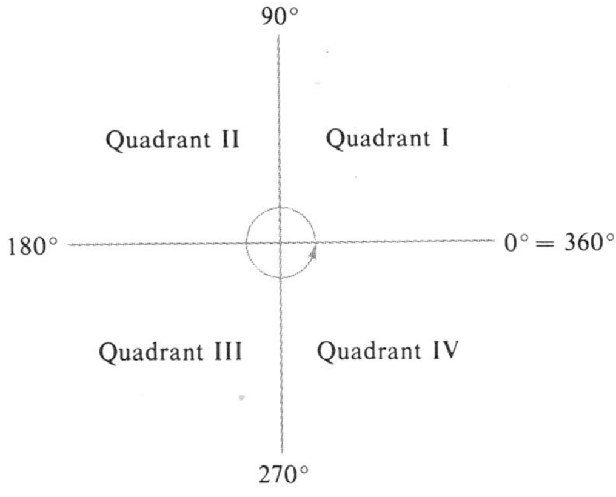

figure E • 2 *The four quadrants of the trigonometric functions*

will depend on the quadrant of the xy plane in which the point $P(x,y) = P(r,\theta)$ is located. The quadrants are customarily identified as the first (I), second (II), third (III), and fourth (IV) quadrants, as shown in Fig. E . 2.

The following discussion focuses on the most commonly used trigonometric functions: the sine, cosine, tangent, and cotangent.

relationships among trigonometric functions

As a result of their interrelated definitions and Pythagoras' theorem relating the sides of a right triangle, the trigonometric functions satisfy the following identities:

$$\sin\theta = \frac{1}{\csc\theta} \qquad \cos\theta = \frac{1}{\sec\theta} \qquad \tan\theta = \frac{1}{\cot\theta} = \frac{\sin\theta}{\cos\theta}$$
$$\csc\theta = \frac{1}{\sin\theta} \qquad \sec\theta = \frac{1}{\cos\theta} \qquad \cot\theta = \frac{1}{\tan\theta} = \frac{\cos\theta}{\sin\theta} \qquad [E \cdot 3]$$

$$\sin^2\theta + \cos^2\theta = 1 \qquad 1 + \tan^2\theta = \sec^2\theta$$
$$1 + \cot^2\theta = \csc^2\theta \qquad [E \cdot 4]$$

Therefore,

$$\sin\theta = \pm\sqrt{1 - \cos^2\theta} = \pm\frac{\tan\theta}{\sqrt{1 + \tan^2\theta}} = \pm\frac{1}{\sqrt{1 + \cot^2\theta}} \qquad [E \cdot 5]$$

$$\cos\theta = \pm\sqrt{1 - \sin^2\theta} = \pm\frac{1}{\sqrt{1 + \tan^2\theta}} = \pm\frac{\cot\theta}{\sqrt{1 + \cot^2\theta}} \qquad [E \cdot 6]$$

$$\tan\theta = \pm\frac{\sin\theta}{\sqrt{1 - \sin^2\theta}} = \pm\frac{\sqrt{1 - \cos^2\theta}}{\cos\theta} \qquad [E \cdot 7]$$

Table E . 1 summarizes some useful special values of the trigonometric functions.

table E • 1 *Special values of trigonometric functions*

θ, degrees	0°	30°	45°	60°	90°	180°	270°
θ, radians	0	$\frac{1}{6}\pi$	$\frac{1}{4}\pi$	$\frac{1}{3}\pi$	$\frac{1}{2}\pi$	π	$\frac{3}{2}\pi$
$\sin\theta$	0	$\frac{1}{2}$	$\frac{1}{2}\sqrt{2}$	$\frac{1}{2}\sqrt{3}$	1	0	-1
$\cos\theta$	1	$\frac{1}{2}\sqrt{3}$	$\frac{1}{2}\sqrt{2}$	$\frac{1}{2}$	0	-1	0
$\tan\theta$	0	$\frac{1}{3}\sqrt{3}$	1	$\sqrt{3}$	$\pm\infty$	0	$\pm\infty$
$\cot\theta$	$\pm\infty$	$\sqrt{3}$	1	$\frac{1}{3}\sqrt{3}$	0	$\pm\infty$	0
$\sec\theta$	1	$\frac{2}{3}\sqrt{3}$	$\sqrt{2}$	2	$\pm\infty$	-1	$\pm\infty$
$\csc\theta$	$\pm\infty$	2	$\sqrt{2}$	$\frac{2}{3}\sqrt{3}$	1	$\pm\infty$	-1

Furthermore, we may find the trigonometric functions of sums and differences of angles:

$$\sin(\alpha \pm \beta) = \sin\alpha\cos\beta \pm \cos\alpha\sin\beta \qquad [E \cdot 8]$$

$$\cos(\alpha \pm \beta) = \cos\alpha\cos\beta \mp \sin\alpha\sin\beta \qquad [E \cdot 9]$$

$$\tan(\alpha \pm \beta) = \frac{\tan\alpha \pm \tan\beta}{1 \mp \tan\alpha\tan\beta} \qquad [E \cdot 10]$$

where the sign in the righthand expression is the same as that in the lefthand expression if $\pm$ appears and is the opposite of that in the lefthand expression if $\mp$ appears.

We can also find the sums and differences of certain trigonometric functions in terms of sums and differences of their angles:

$$\sin\alpha + \sin\beta = 2\sin\tfrac{1}{2}(\alpha + \beta) \times \cos\tfrac{1}{2}(\alpha - \beta) \qquad [E \cdot 11]$$

$$\sin\alpha - \sin\beta = 2\cos\tfrac{1}{2}(\alpha + \beta) \times \sin\tfrac{1}{2}(\alpha - \beta) \qquad [E \cdot 12]$$

$$\cos\alpha + \cos\beta = 2\cos\tfrac{1}{2}(\alpha + \beta) \times \cos\tfrac{1}{2}(\alpha - \beta) \qquad [E \cdot 13]$$

$$\cos\alpha - \cos\beta = -2\sin\tfrac{1}{2}(\alpha + \beta) \times \sin\tfrac{1}{2}(\alpha - \beta) \qquad [E \cdot 14]$$

or, conversely,

$$\sin\alpha\sin\beta = \tfrac{1}{2}\cos(\alpha - \beta) - \tfrac{1}{2}\cos(\alpha + \beta) \qquad [E \cdot 15]$$

$$\cos\alpha\cos\beta = \tfrac{1}{2}\cos(\alpha - \beta) + \tfrac{1}{2}\cos(\alpha + \beta) \qquad [E \cdot 16]$$

$$\sin\alpha\cos\beta = \tfrac{1}{2}\sin(\alpha + \beta) + \tfrac{1}{2}\sin(\alpha - \beta) \qquad [E \cdot 17]$$

The addition formulas for angles yield the following widely-used double-angle and half-angle relationships:

$$\sin 2\theta = 2\sin\theta\cos\theta \qquad [E \cdot 18]$$

$$\cos 2\theta = \cos^2\theta - \sin^2\theta = 2\cos^2\theta - 1 = 1 - 2\sin^2\theta \qquad [E \cdot 19]$$

$$\sin\tfrac{1}{2}\theta = \pm\sqrt{\frac{1 - \cos\theta}{2}} \qquad \cos\tfrac{1}{2}\theta = \pm\sqrt{\frac{1 + \cos\theta}{2}} \qquad [E \cdot 20]$$

inverse trigonometric functions

Let $f(\theta)$ be a trigonometric function of the angle θ. By definition, the inverse trigonometric function $\text{arc} f(x) = f^{-1}(x)$ is that angle θ that satisfies the equation

$$x = f(\theta) \qquad \text{hence} \qquad \theta = f^{-1}(x) = \text{arc} f(x) \qquad [E \cdot 21]$$

Thus, if n is an integer and the angle θ is expressed in radians,

$$\theta = (-1)^n \sin^{-1} x + n\pi \qquad \text{where } x = \sin\theta \qquad [E \cdot 22]$$

$$\theta = \pm\cos^{-1} x + 2n\pi \qquad \text{where } x = \cos\theta \qquad [E \cdot 22]$$

$$\theta = \tan^{-1} x + n\pi \qquad \text{where } x = \tan\theta \qquad [E \cdot 24]$$

Since the inverse trigonometric functions are multivalued, we always (unless otherwise specified) take the *principal value,* which falls in the following ranges:

$$-\tfrac{1}{2}\pi \leq \sin^{-1} x \leq \tfrac{1}{2}\pi \qquad 0 \leq \cos^{-1} x \leq \pi$$

$$-\tfrac{1}{2}\pi \leq \tan^{-1} x \leq \tfrac{1}{2}\pi \qquad 0 \leq \cot^{-1} x \leq \pi$$

Where necessary for clarity, the inverse trigonometric functions may be abbreviated as arcsin, arccos, arctan, and arccot rather than using the $\sin^{-1}$ form of notation.

Because of the identities that hold between the trigonometric functions, we can also relate the inverse trigonometric functions to one another for their principal values:

$$\theta = \sin^{-1} x = \cos^{-1}\sqrt{1 - x^2} = \tan^{-1}\frac{x}{\sqrt{1 - x^2}} \qquad [E \cdot 25]$$

$$\theta = \cos^{-1} x = \sin^{-1}\sqrt{1 - x^2} = \tan^{-1}\frac{\sqrt{1 - x^2}}{x} \qquad [E \cdot 26]$$

$$\theta = \tan^{-1} x = \sin^{-1}\frac{x}{\sqrt{1 + x^2}} = \cos^{-1}\frac{1}{\sqrt{1 + x^2}} \qquad [E \cdot 27]$$

These relationships are particularly useful in programming for those computer languages that supply only the arctan function in their libraries, making it necessary to program computation of the arcsin and arccos functions (see Appendix I).

vector algebra

The following relationships are useful and are directly related to vector usage. For any triangle with sides a, b, and c and opposite angles A, B, and C,

$$\text{area} = \tfrac{1}{2}ab\sin C \qquad [E \cdot 28]$$

$$\frac{a}{\sin A} = \frac{b}{\sin B} = \frac{c}{\sin C} \qquad [E \cdot 29]$$

$$c^2 = a^2 + b^2 - 2ab\cos C \qquad [E \cdot 30]$$

Eq. [E . 29] is called the law of sines, and Eq. [E . 30] is called the law of cosines.

Most of the following vector relationships are derived in Chapter 2 and are merely listed here. The unit vectors in the x, y, and z directions are **i**, **j**, and **k**, respectively. The vector **A** is represented by its components as

$$\mathbf{A} = A_x\mathbf{i} + A_y\mathbf{j} + A_z\mathbf{k} \tag{E . 31}$$

Its magnitude is

$$|\mathbf{A}| = A = \sqrt{A_x^2 + A_y^2 + A_z^2} \tag{E . 32}$$

and it points in the direction of the unit vector

$$\hat{\mathbf{A}} = \frac{1}{A}\mathbf{A} \tag{E . 33}$$

The product of a scalar c and a vector **A** is the vector sum of the product of c with the separate vector components:

$$a\mathbf{A} = cA_x\mathbf{i} + cA_y\mathbf{j} + cA_z\mathbf{k} \tag{E . 34}$$

which is distributive:

$$(c + d)(\mathbf{A} + \mathbf{B}) = c\mathbf{A} + c\mathbf{B} + d\mathbf{A} + d\mathbf{B} \tag{E . 35}$$

Two or more vectors are summed by summing their components separately:

$$\sum_m \mathbf{A}_m = \left(\sum_m A_{mx}\right)\mathbf{i} + \left(\sum_m A_{my}\right)\mathbf{j} + \left(\sum_m A_{mz}\right)\mathbf{k} \tag{E . 36}$$

Three vector products are defined in Chapter 2. The *scalar dot product* of vectors **A** and **B** is the product of the vector magnitudes and the cosine of the angle θ between them:

$$\mathbf{A} \cdot \mathbf{B} = AB\cos\theta \tag{E . 37}$$

The *vector cross product* of vectors **A** and **B** is a vector with magnitude equal to the products of the vector magnitudes and the sine of the angle θ between them:

$$\mathbf{A} \times \mathbf{B} = (AB\sin\theta)\hat{\mathbf{n}} \tag{E . 38}$$

where the unit vector $\hat{\mathbf{n}}$ points in the direction of travel of a right-handed screw rotating from **A** to **B** through the smallest angle, and hence $\hat{\mathbf{n}}$ is normal to the plane of **A** and **B**. The *box product* of vectors **A**, **B**, and **C** is the volume of the parallelepiped whose sides are formed by **A**, **B**, and **C**; the box product is written as $\mathbf{A} \cdot \mathbf{B} \times \mathbf{C}$.

If these vector products are evaluated by components, their formulas are

$$\mathbf{A} \cdot \mathbf{B} = A_xB_x + A_yB_y + A_zB_z \tag{E . 39}$$

$$\begin{aligned}\mathbf{A} \times \mathbf{B} &= \begin{vmatrix} \mathbf{i} & \mathbf{j} & \mathbf{k} \\ A_x & A_y & A_z \\ B_x & B_y & B_z \end{vmatrix} \\ &= (A_yB_z - A_zB_y)\mathbf{i} - (A_xB_z - A_zB_x)\mathbf{j} \\ &\qquad + (A_xB_y - A_yB_x)\mathbf{k}\end{aligned} \tag{E . 40}$$

and therefore

$$\mathbf{A \cdot B \times C} = \begin{vmatrix} A_x & A_y & A_z \\ B_x & B_y & B_z \\ C_x & C_y & C_z \end{vmatrix}$$
$$= A_x(B_yC_z - B_zC_y) - A_y(B_xC_z - B_zC_x) + A_z(B_xC_y - B_yC_x) \qquad [E \cdot 41]$$

The dot and cross products of the unit vectors **i**, **j**, and **k** are

$$\mathbf{i \cdot i} = \mathbf{j \cdot j} = \mathbf{k \cdot k} = 1 \qquad \mathbf{i \cdot j} = \mathbf{j \cdot k} = \mathbf{k \cdot i} = 0 \qquad [E \cdot 42]$$

$$\mathbf{i \times i} = \mathbf{j \times j} = \mathbf{k \times k} = \mathbf{0}$$
$$\mathbf{i \times j} = \mathbf{k} \qquad \mathbf{j \times k} = \mathbf{i} \qquad \mathbf{k \times i} = \mathbf{j} \qquad [E \cdot 43]$$

Both scalar and vector products are distributive:

$$\mathbf{A \cdot (B + C)} = \mathbf{A \cdot B + A \cdot C}$$
$$\mathbf{A \times (B + C)} = \mathbf{A \times B + A \times C} \qquad [E \cdot 44]$$

However, the vector cross product is anticommutative:

$$\mathbf{A \cdot B} = \mathbf{B \cdot A} \qquad \mathbf{A \times B} = -\mathbf{B \times A} \qquad [E \cdot 45]$$

As a result, the box product obeys the identity

$$\mathbf{A \cdot (B \times C)} = \mathbf{(A \times B) \cdot C} = \mathbf{B \cdot (C \times A)}$$
$$= -\mathbf{B \cdot (A \times C)} \qquad [E \cdot 46]$$

In the *vector triple product*, the order of multiplication is fixed:

$$\mathbf{A \times (B \times C)} = \mathbf{(A \cdot C)B} - \mathbf{(A \cdot B)C} \neq \mathbf{(A \times B) \times C} \qquad [E \cdot 47]$$

APPENDIX F

differential, integral, and vector calculus

differential calculus

If $y = f(x)$, then the first derivative of y with respect to x is

$$\frac{dy}{dx} = \lim_{\Delta x \to 0} \frac{\Delta y}{\Delta x} = \lim_{\Delta x \to 0} \frac{f(x + \Delta x) - f(x)}{\Delta x} = f'(x) \qquad [F \cdot 1]$$

The second derivative is

$$\frac{d^2y}{dx^2} = \frac{d}{dx}\left(\frac{dy}{dx}\right) = \frac{df'(x)}{dx} = f''(x) \qquad [F \cdot 2]$$

Similarly, the nth derivative is

$$\frac{d^ny}{dx^n} = \frac{d}{dx}\left(\frac{d^{n-1}y}{dx^{n-1}}\right) = \frac{df^{(n-1)}(x)}{dx} = f^{(n)}(x) \qquad [F \cdot 3]$$

A function given in the form $x = g(y)$ may imply a relationship $y = f(x)$ that cannot be expressed explicitly. However,

$$\frac{dy}{dx} = \frac{1}{\dfrac{dx}{dy}} \quad \text{or} \quad f'(x) = \frac{1}{g'(y)} \quad \text{where } x = g(y) \qquad [F \cdot 4]$$

For example,

$$x = y^3 + y^2 \qquad \frac{dx}{dy} = 3y^2 + 2y \qquad \frac{dy}{dx} = \frac{1}{3y^2 + 2y}$$

For a function of a function, we have the *chain rule:* if $y = F(u)$ and $u = f(x)$, then

$$\frac{dy}{dx} = \frac{dy}{du}\frac{du}{dx} = F'(u) \times f'(x) \qquad [F \cdot 5]$$

If a is a constant, then

$$\frac{da}{dx} = 0 \quad \text{and} \quad \frac{d(ay)}{dx} = a\frac{dy}{dx} \qquad [F \cdot 6]$$

If $u = u(x)$ and $v = v(x)$, then

$$\frac{d}{dx}(u + v) = \frac{du}{dx} + \frac{dv}{dx}$$

$$\frac{d}{dx}(uv) = u\frac{dv}{dx} + v\frac{du}{dx} \qquad \frac{d}{dx}\left(\frac{u}{v}\right) = \frac{v\dfrac{du}{dx} - u\dfrac{dv}{dx}}{v^2} \qquad [F \cdot 7]$$

If n is any nonzero number, then

$$\frac{d}{dx}(x^n) = nx^{n-1} \qquad [\text{F} \cdot 8]$$

The derivatives of the trigonometric functions are the following:

$$\begin{aligned} \frac{d}{dx}(\sin x) &= \cos x & \frac{d}{dx}(\cot x) &= -\csc^2 x \\ \frac{d}{dx}(\cos x) &= -\sin x & \frac{d}{dx}(\sec x) &= \sec x \tan x \\ \frac{d}{dx}(\tan x) &= \sec^2 x & \frac{d}{dx}(\csc x) &= -\csc x \cot x \end{aligned} \qquad [\text{F} \cdot 9]$$

The exponential function is defined as $\exp x = e^x$, where $e = 2.71828183\ldots$ is the base of the natural logarithms: $e^{(\ln x)} = x$. The derivatives of these functions are

$$\frac{d}{dx}(e^x) = \frac{d}{dx}(\exp x) = e^x \qquad \frac{d}{dx}(\ln x) = \frac{1}{x} \qquad [\text{F} \cdot 10]$$

For a constant a,

$$\frac{d}{dx}(a^x) = a^x \ln a \qquad [\text{F} \cdot 11]$$

The derivatives of the inverse trigonometric functions (assuming principal values for the functions—see Appendix E) are the following:

$$\begin{aligned} \frac{d}{dx}(\sin^{-1} x) &= (1 - x^2)^{-1/2} = -\frac{d}{dx}(\cos^{-1} x) \\ \frac{d}{dx}(\tan^{-1} x) &= \frac{1}{1 + x^2} = -\frac{d}{dx}(\cot^{-1} x) \end{aligned} \qquad [\text{F} \cdot 12]$$

series

If a function $f(x)$ has finite derivatives of all other orders in the neighborhood of a point x_0, then it can be expanded in a power series about x_0:

$$f(x) = \sum_{n=0}^{\infty} a_n(x - x_0)^n \qquad [\text{F} \cdot 13]$$

where the constant coefficients a_n are determined from the nth-order derivatives of the function evaluated at $x = x_0$:

$$a_n = \frac{1}{n!} f^{(n)}(x_0) \qquad [\text{F} \cdot 14]$$

The series defined by Eqs. [F . 13] and [F . 14] is called the *Taylor series* of the function $f(x)$, and it converges to the function if the remainder term

$$R_N(x) = f(x) - \sum_{n=0}^{N} a_n(x - x_0)^n = \frac{(x - x_0)^{N+1}}{(N + 1)!} f^{(N+1)}(\xi) \qquad [\text{F} \cdot 15]$$

(where ξ lies between x and x_0) is such that $R_N(x)$ vanishes as N approaches infinity.

The following series can be obtained by expansion of the functions in Taylor series:

$$e^x = 1 + x + \frac{x^2}{2!} + \frac{x^3}{3!} + \cdots \qquad \text{[F}\cdot\text{16]}$$

$$e = 1 + 1 + \tfrac{1}{2} + \tfrac{1}{6} + \tfrac{1}{24} + \cdots \qquad \text{[F}\cdot\text{17]}$$

$$\ln x = (x - 1) - \tfrac{1}{2}(x - 1)^2 + \tfrac{1}{3}(x - 1)^3 - \cdots \quad \text{for } 0 < x \leq 2 \qquad \text{[F}\cdot\text{18]}$$

$$\ln(1 + x) = x - \tfrac{1}{2}x^2 + \tfrac{1}{3}x^3 - \tfrac{1}{4}x^4 + \cdots \quad \text{for } -1 \leq x \leq 1 \qquad \text{[F}\cdot\text{19]}$$

$$\ln\left(\frac{1 + x}{1 - x}\right) = 2(x + \tfrac{1}{3}x^3 + \tfrac{1}{5}x^5 + \cdots) \quad \text{for } x^2 < 1 \qquad \text{[F}\cdot\text{20]}$$

$$\ln\left(\frac{x + 1}{x - 1}\right) = 2\left(\frac{1}{x} + \frac{1}{3x^3} + \frac{1}{5x^5} + \cdots\right) \quad \text{for } x^2 > 1 \qquad \text{[F}\cdot\text{21]}$$

$$\sin x = x - \frac{x^3}{3!} + \frac{x^5}{5!} - \frac{x^7}{7!} + \cdots \quad \text{for } x \text{ in radians} \qquad \text{[F}\cdot\text{22]}$$

$$\cos x = 1 - \frac{x^2}{2!} + \frac{x^4}{4!} - \frac{x^6}{6!} + \cdots \quad \text{for } x \text{ in radians} \qquad \text{[F}\cdot\text{23]}$$

$$\tan x = x + \frac{x^3}{3} + \frac{2x^5}{15} + \frac{17x^7}{315} + \cdots \quad \text{for } x \text{ in radians and } x^2 < \tfrac{1}{4}\pi^2 \qquad \text{[F}\cdot\text{24]}$$

In the limit of small x, the final three equations yield

$$\sin x \doteq x \qquad \cos x \doteq 1 - \tfrac{1}{2}x^2 \qquad \tan x \doteq x \qquad \text{for } x \ll 1 \text{ rad} \qquad \text{[F}\cdot\text{25]}$$

Thus,

$$\frac{d}{dx}(\sin x) \doteq 1 \qquad \frac{d}{dx}(\cos x) \doteq -x \qquad \frac{d}{dx}(\tan x) \doteq 1 \quad \text{for } x \ll 1 \text{ rad} \qquad \text{[F}\cdot\text{26]}$$

The inverse trigonometric functions can also be expanded in Taylor series:

$$\sin^{-1} x = x + \tfrac{1}{2} \times \tfrac{1}{3}x^3 + \tfrac{1}{2} \times \tfrac{3}{4} \times \tfrac{1}{5}x^5 + \tfrac{1}{2} \times \tfrac{3}{4} \times \tfrac{5}{6} \times \tfrac{1}{7}x^7 + \cdots \quad \text{for } x^2 < 1 \qquad \text{[F}\cdot\text{27]}$$

$$\tan^{-1} x = x - \tfrac{1}{3}x^3 + \tfrac{1}{5}x^5 - \tfrac{1}{7}x^7 + \cdots \quad \text{for } x^2 < 1 \qquad \text{[F}\cdot\text{28]}$$

$$\tan^{-1} x = \tfrac{1}{2}\pi - \frac{1}{x} + \frac{1}{3x^3} - \frac{1}{5x^5} + \cdots \qquad \text{for } x^2 > 1 \qquad [F \cdot 29]$$

Stirling's approximation is a useful expression based upon a series expansion:

$$\ln (N!) \doteq 0.91894 + (N + \tfrac{1}{2}) \ln N - N \qquad [F \cdot 30]$$

This approximation produces an error of less than 1.7 percent for $N > 5$.

integrals

Many integrals can be obtained by inspection of the differentiation formulas given earlier in this appendix, since $F(x) = \int f(x)\, dx$ is by definition a function such that $dF/dx = f(x)$. Thus,

$$\int x^n\, dx = \frac{x^{n+1}}{n+1} \qquad \int \frac{dx}{x} = \ln x$$

$$\int \sin x\, dx = -\cos x \qquad \int e^x\, dx = e^x \qquad \text{etc.}$$

Table F . 1 is a table of indefinite integrals for a wide range of functions. In this table, an arbitrary constant of integration should be understood on the righthand side of each equation—for example, $\int dx = x + C$. The following general formulas are useful in applying the indefinite integrals from Table F . 1.

$$\int_a^b f(x)\, dx = F(x)\Big|_a^b = F(b) - F(a) \qquad [F \cdot 31]$$

where $F(x) = \int f(x)\, dx$ is seen by definition to be a function such that $dF/dx = f(x)$.

$$\int_a^b [f_1(x) + f_2(x) + \cdots + f_n(x)]\, dx$$

$$= \int_a^b f_1(x)\, dx + \int_a^b f_2(x)\, dx + \cdots + \int_a^b f_n(x)\, dx \qquad [F \cdot 32]$$

$$\int_a^b f(x)\, dx = -\int_b^a f(x)\, dx \qquad [F \cdot 33]$$

$$\int_a^b cf(x)\, dx = c\int_a^b f(x)\, dx \qquad \text{where } c \text{ is a constant} \qquad [F \cdot 34]$$

$$\int_a^b f(x)\, dx = \int_a^c f(x)\, dx + \int_c^b f(x)\, dx \qquad \text{for any value } c \qquad [F \cdot 35]$$

$$\int u\, dv = uv - \int v\, du \qquad \text{(integration by parts)} \qquad [F \cdot 36]$$

$$\int_a^b u \frac{dv}{dx}\, dx = uv\Big|_a^b - \int_a^b v \frac{du}{dx}\, dx \qquad \text{(integration by parts)} \qquad [F \cdot 37]$$

$$\int_a^b f(x)\, dx = \int_{u(a)}^{u(b)} f(x(u)) \frac{dx}{du}\, du \qquad \text{(substitution)} \qquad [F \cdot 38]$$

table F • 1 *Indefinite integrals and definite integrals*

---

1 $\int (ax + b)^n \, dx = \frac{(ax + b)^{n+1}}{a(n + 1)}$ for $n \neq -1$

2 $\int \frac{dx}{ax + b} = \frac{1}{a} \ln (ax + b)$

3 $\int \frac{dx}{x(ax + b)} = \frac{1}{b} \ln \left(\frac{x}{ax + b}\right)$

4 $\int \frac{dx}{x^2 + a^2} = \frac{1}{a} \tan^{-1} \left(\frac{x}{a}\right)$

5 $$\int \frac{dx}{x^2 - a^2} = \begin{cases} \frac{1}{2a} \ln \left(\frac{a - x}{a + x}\right) & \text{for } x^2 < a^2 \\ \frac{1}{2a} \ln \left(\frac{x - a}{x + a}\right) & \text{for } x^2 > a^2 \end{cases}$$

6 $\int \sqrt{ax + b} \, dx = \frac{2}{3a} \sqrt{(ax + b)^3}$

7 $\int \frac{dx}{\sqrt{ax + b}} = \frac{2}{a} \sqrt{ax + b}$

8 $\int \frac{dx}{\sqrt{a^2 \pm x^2}} = \ln \left(x + \sqrt{x^2 \pm a^2}\right)$

9 $\int \frac{dx}{\sqrt{2ax - x^2}} = \cos^{-1} \left(\frac{a - x}{a}\right)$ for $a > 0$

10 $\int \frac{dx}{x\sqrt{x^2 - a^2}} = \frac{1}{a} \cos^{-1} \left(\frac{a}{x}\right)$ for $a > 0$

11 $\int \frac{dx}{x\sqrt{a^2 \pm x^2}} = -\frac{1}{a} \ln \left(\frac{a + \sqrt{a^2 \pm x^2}}{x}\right)$ for $a > 0$

12 $\int \sqrt{x^2 \pm a^2} \, dx = \frac{1}{2}\left[x\sqrt{x^2 \pm a^2} \pm a^2 \ln \left(x + \sqrt{x^2 \pm a^2}\right)\right]$

13 $\int \sqrt{a^2 - x^2} \, dx = \frac{1}{2} x\sqrt{a^2 - x^2} + \frac{1}{2} a^2 \sin^{-1} \left(\frac{x}{a}\right)$

14 $\int \frac{\sqrt{x^2 - a^2}}{x} \, dx = \sqrt{x^2 - a^2} - a \tan^{-1} \sqrt{\frac{x^2 - a^2}{a^2}}$

15 $\int \frac{\sqrt{x^2 + a^2}}{x} \, dx = \sqrt{x^2 + a^2} + a \ln \left(\frac{\sqrt{x^2 + a^2} - a}{x}\right)$

In integrals 16–19, $P = ax^2 + bx + c$, and $D = b^2 - 4ac$

16 $\int \frac{dx}{P} = \frac{1}{\sqrt{D}} \ln \left(\frac{2ax + b - D}{2ax + b + D}\right)$ for $D > 0$

17 $\int \frac{dx}{P} = \frac{2}{\sqrt{-D}} \tan^{-1} \left(\frac{2ax + b}{\sqrt{-D}}\right)$ for $D < 0$

18 $\displaystyle\int \frac{dx}{\sqrt{P}} = -\frac{1}{\sqrt{-a}} \sin^{-1}\left(\frac{2ax + b}{\sqrt{D}}\right)$ for $a < 0$ and $D > 0$

19 $\displaystyle\int \frac{dx}{\sqrt{P}} = \frac{1}{\sqrt{a}} \ln\left(2ax + b + 2\sqrt{aP}\right)$ for $a > 0$

20 $\displaystyle\int \frac{dx}{(x^2 + a)\sqrt{x^2 + 2a}} = \frac{1}{a} \tan^{-1}\left(\frac{x}{\sqrt{x^2 + 2a}}\right)$

21 $\displaystyle\int x^3\sqrt{a^2 - x^2}\, dx = -\tfrac{1}{3}x^2(a^2 - x^2)^{3/2} - \tfrac{2}{15}(a^2 - x^2)^{3/2}$

22 $\int \tan x\, dx = -\ln(\cos x)$

23 $\int \cot x\, dx = \ln(\sin x)$

24 $\int \sec x\, dx = \ln(\sec x + \tan x)$

25 $\int \csc x\, dx = \ln(\csc x - \cot x)$

26 $\displaystyle\int \sin^2 ax\, dx = \tfrac{1}{2}x - \frac{1}{4a}\sin 2ax$

27 $\displaystyle\int \cos^2 ax\, dx = \tfrac{1}{2}x + \frac{1}{4a}\sin 2ax$

28 $\displaystyle\int \sin^3 ax\, dx = -\frac{1}{a}\cos ax + \frac{1}{3a}\cos^3 ax$

29 $\displaystyle\int \cos^3 ax\, dx = \frac{1}{a}\sin ax - \frac{1}{3a}\sin^3 ax$

30 $\displaystyle\int x \sin ax\, dx = \frac{1}{a^2}\sin ax - \frac{x}{a}\cos ax$

31 $\displaystyle\int x \cos ax\, dx = \frac{1}{a^2}\cos ax + \frac{x}{a}\sin ax$

32 $\displaystyle\int \frac{\sin ax}{x}\, dx = ax - \frac{a^3x^3}{3 \times 3!} + \frac{a^5x^5}{5 \times 5!} - \cdots$

33 $\displaystyle\int \frac{\cos ax}{x}\, dx = \ln ax - \frac{a^2x^2}{2 \times 2!} + \frac{a^4x^4}{4 \times 4!} - \cdots$

34 $\displaystyle\int a^x\, dx = \frac{a^x}{\ln a}$

35 $\displaystyle\int xe^{ax}\, dx = \frac{e^{ax}}{a^2}(ax - 1)$

36 $\displaystyle\int xb^{ax}\, dx = \frac{xb^{ax}}{a \ln b} - \frac{b^{ax}}{a^2(\ln b)^2}$

37 $\displaystyle\int x^2e^{ax}\, dx = \frac{e^{ax}}{a^3}(a^2x^2 - 2ax + 2)$

38 $\displaystyle\int \frac{e^{ax}}{x}\, dx = \ln x + ax + \frac{a^2x^2}{2 \times 2!} + \frac{a^3x^3}{3 \times 3!} + \cdots$

39 $\displaystyle\int \frac{dx}{b + ce^{ax}} = \frac{1}{ab}[ax - \ln(b + ce^{ax})]$

40 $\displaystyle\int \frac{dx}{be^{ax} - ce^{-ax}} = \frac{1}{a\sqrt{bc}} \tan^{-1}\left(\sqrt{\frac{b}{c}}\, e^{ax}\right)$ for $b > 0$ and $c > 0$

41 $$\int e^{ax}\sin bx\,dx = \frac{e^{ax}}{a^2+b^2}(a\sin bx - b\cos bx)$$

42 $$\int e^{ax}\cos bx\,dx = \frac{e^{ax}}{a^2+b^2}(a\cos bx + b\sin bx)$$

43 $$\int \ln ax\,dx = x\ln ax - x$$

44 $$\int x^n \ln ax\,dx = x^{n+1}\left(\frac{\ln ax}{n+1} - \frac{1}{(n+1)^2}\right)$$

45 $$\int \frac{(\ln ax)^n}{x}\,dx = \frac{(\ln ax)^{n+1}}{n+1}$$

46 $$\int \sin^{-1} ax\,dx = x\sin^{-1} ax + \frac{1}{a}\sqrt{1-a^2x^2}$$

47 $$\int \cos^{-1} ax\,dx = x\cos^{-1} ax - \frac{1}{a}\sqrt{1-a^2x^2}$$

48 $$\int \tan^{-1} ax\,dx = x\tan^{-1} ax - \frac{1}{2a}\ln(1+a^2x^2)$$

49 $$\int \cot^{-1} ax\,dx = x\cot^{-1} ax + \frac{1}{2a}\ln(1+a^2x^2)$$

50 $$\int \sec^{-1} ax\,dx = x\sec^{-1} ax - \frac{1}{a}\ln\left(ax + \sqrt{a^2x^2-1}\right)$$

51 $$\int \csc^{-1} ax\,dx = x\csc^{-1} ax + \frac{1}{a}\ln\left(ax + \sqrt{a^2x^2-1}\right)$$

definite integrals:

52 $$\int_0^\pi \sin^2 nx\,dx = \int_0^\pi \cos^2 nx\,dx = \tfrac{1}{2}\pi \qquad \text{for } n = \text{integer}$$

53 $$\int_0^{\pi/2} \sin^n x\,dx = \int_0^{\pi/2} \cos^n x\,dx$$
$$= \begin{cases} \dfrac{2\times 4\times\cdots\times(n-1)}{3\times 5\times\cdots\times n} & \text{for } n = \text{odd integer} \\[2ex] \left(\dfrac{1\times 3\times\cdots\times(n-1)}{2\times 4\times\cdots\times n}\right)\dfrac{\pi}{2} & \text{for } n = \text{even integer} > 0 \end{cases}$$

54 $$\int_0^{\pi/2} \sin^m x\cos^n x\,dx = \begin{cases} \dfrac{2\times 4\times\cdots\times(n-1)}{(m+1)(m+3)\cdots(m+n)} & \text{for } n = \text{odd integer} > 1 \\[2ex] \dfrac{2\times 4\times\cdots\times(m-1)}{(n+1)(n+3)\cdots(n+m)} & \text{for } m = \text{odd integer} > 1 \end{cases}$$

55 $$\int_0^\pi \sin ax\sin bx\,dx = \int_0^\pi \cos ax\cos bx\,dx = 0 \qquad \text{for } a \neq b$$

56 $$\int_0^\pi \sin ax\cos ax\,dx = \int_0^{\pi/a} \sin ax\cos ax\,dx = 0$$

57 $$\int_0^\pi \sin ax\cos bx\,dx = \begin{cases} \dfrac{2a}{a^2-b^2} & \text{for } a-b = \text{odd integer} \\[2ex] 0 & \text{for } a-b = \text{even integer} \end{cases}$$

58 $$\int_0^\infty \frac{\sin ax}{x}\,dx = \int_0^\infty \frac{\sin^2 x}{x^2}\,dx = \tfrac{1}{2}\pi$$

59 $$\int_0^\infty \sin x^2\,dx = \int_0^\infty \cos x^2\,dx = \tfrac{1}{2}\sqrt{\tfrac{1}{2}\pi}$$

60 $$\int_0^\infty \exp(-a^2x^2)\,dx = \frac{\sqrt{\pi}}{2a} \qquad \text{for } a > 0$$

61 $$\int_0^\infty x^n e^{-ax}\,dx = \frac{n!}{a^{n+1}} \qquad \text{for } n = \text{integer} > 0 \text{ and } a > 0$$

62 $$\int_0^\infty x^{2n} \exp(-ax^2)\,dx = \frac{1 \times 3 \times 5 \times \cdots \times (2n-1)}{2^{n+1}a^n}\sqrt{\frac{\pi}{a}}$$

$$\int_0^\infty x^{2n+1} \exp(-ax^2)\,dx = \frac{n!}{2a^{n+1}}$$

63 $$a\int_0^\infty e^{-ax}\sin bx\,dx = b\int_0^\infty e^{-ax}\cos bx\,dx = \frac{ab}{a^2+b^2} \qquad \text{for } a > 0$$

64 $$\int_0^\infty \exp\left(-x^2 - \frac{a^2}{x^2}\right)dx = \tfrac{1}{2}\sqrt{\pi}\,e^{-2a} \qquad \text{for } a > 0$$

65 $$\int_0^\infty \exp(-a^2x^2)\cos bx\,dx = \frac{\sqrt{\pi}}{2a}\exp\left(\frac{-b^2}{4a^2}\right) \qquad \text{for } a > 0$$

66 $$\int_0^\infty \frac{e^x}{1+e^x}\,dx = -\int_0^1 \frac{\ln x}{1+x}\,dx = \tfrac{1}{12}\pi^2$$

67 $$\int_0^\infty \frac{\cos ax}{1+x^2}\,dx = \tfrac{1}{2}\pi e^{-|a|}$$

68 $$\int_1^\infty \frac{e^{-ax}}{x}\,dx = -\gamma - \ln a + a - \frac{a^2}{2 \times 2!} + \frac{a^3}{3 \times 3!} - \cdots$$

for $a > 0$ where $\gamma = 0.5772\cdots$

69 $$\int_0^{\pi/2} \frac{dx}{\sqrt{1 - k^2\sin^2 x}} =$$

$$\tfrac{1}{2}\pi\left[1 + \left(\frac{1}{2}\right)^2 k^2 + \left(\frac{1 \times 3}{2 \times 4}\right)^2 k^4 + \left(\frac{1 \times 3 \times 5}{2 \times 4 \times 6}\right)^2 k^6 + \cdots\right]$$

for $k^2 < 1$

70 $$\int_0^{\pi/2} \sqrt{1 - k^2\sin^2 x}\,dx =$$

$$\tfrac{1}{2}\pi\left[1 - \left(\frac{1}{2}\right)^2 k^2 - \left(\frac{1 \times 3}{2 \times 4}\right)^2 \frac{k^4}{3} - \left(\frac{1 \times 3 \times 5}{2 \times 4 \times 6}\right)^2 \frac{k^6}{5} - \cdots\right]$$

for $k^2 < 1$

71 $$\int_0^\infty \frac{x^n}{e^x - 1}\,dx = n!\zeta(n+1) \qquad \text{where } \zeta(n) = \text{Riemann zeta function}$$

$$= \sum_{k=1}^{\infty} k^{-n}$$

Note: Arguments of all trigonometric functions are assumed to be expressed in radians. Many more integrals may be found in any of the standard mathematics handbooks, including R. S. Burington, *Handbook of Mathematical Tables and Formulas,* 5th ed (New York: McGraw-Hill, 1973); *Handbook of Tables for Mathematics,* 4th ed (Cleveland: Chemical Rubber Co., 1970); M. Abramowitz and I. A. Stegun, *Handbook of Mathematical Functions* (Washington, D.C.: National Bureau of Standards, 1965); I. S. Gradshteyn and I. M. Ryzhik, *Tables of Integrals, Series and Products,* 4th ed (New York: Academic Press, 1965).

vector calculus

The following expressions summarize the differentiation of vectors.

$$\frac{d}{dt}\mathbf{A}(t) = \frac{d}{dt}[A_x(t)\mathbf{i} + A_y(t)\mathbf{j} + A_z(t)\mathbf{k}]$$

$$= \frac{dA_x}{dt}\mathbf{i} + \frac{dA_y}{dt}\mathbf{j} + \frac{dA_z}{dt}\mathbf{k} \qquad [\mathrm{F}\cdot 39]$$

$$\frac{d}{dt}(\mathbf{A}\cdot\mathbf{B}) = \frac{d\mathbf{A}}{dt}\cdot\mathbf{B} + \mathbf{A}\cdot\frac{d\mathbf{B}}{dt}$$

$$\frac{d}{dt}(\mathbf{A}\times\mathbf{B}) = \frac{d\mathbf{A}}{dt}\times\mathbf{B} + \mathbf{A}\times\frac{d\mathbf{B}}{dt} \qquad [\mathrm{F}\cdot 40]$$

$$\frac{d}{dt}(\tfrac{1}{2}A^2) = \frac{d}{dt}(\tfrac{1}{2}\mathbf{A}\cdot\mathbf{A}) = \mathbf{A}\cdot\frac{d\mathbf{A}}{dt} = A\frac{dA}{dt}$$

[F · 41]

The *del operator* ∇ is defined as follows:

$$\nabla = \frac{\partial}{\partial x}\mathbf{i} + \frac{\partial}{\partial y}\mathbf{j} + \frac{\partial}{\partial z}\mathbf{k} \qquad [\mathrm{F}\cdot 42]$$

The *gradient* of a function $f(x,y,z)$ is obtained by application of the del operator:

$$\mathbf{grad}\, f(x,y,z) = \nabla f = \frac{\partial f}{\partial x}\mathbf{i} + \frac{\partial f}{\partial y}\mathbf{j} + \frac{\partial f}{\partial z}\mathbf{k} \qquad [\mathrm{F}\cdot 43]$$

The *divergence* of a vector function is obtained as a dot product with ∇:

$$\mathrm{div}\,\mathbf{A}(x,y,z) = \nabla\cdot\mathbf{A} = \frac{\partial A_x}{\partial x} + \frac{\partial A_y}{\partial y} + \frac{\partial A_z}{\partial z} \qquad [\mathrm{F}\cdot 44]$$

The *curl* of a vector function is obtained as the cross product with ∇:

$$\mathbf{curl}\,\mathbf{A}(x,y,z) = \nabla\times\mathbf{A} = \begin{vmatrix} \mathbf{i} & \mathbf{j} & \mathbf{k} \\ \dfrac{\partial}{\partial x} & \dfrac{\partial}{\partial y} & \dfrac{\partial}{\partial z} \\ A_x & A_y & A_z \end{vmatrix} \qquad [\mathrm{F}\cdot 45]$$

The Laplacian operator ∇^2 represents the divergence of the gradient of a function:

$$\mathrm{div}\,\mathbf{grad}\, f = \nabla^2 f(x,y,z) = \frac{\partial^2 f}{\partial x^2} + \frac{\partial^2 f}{\partial y^2} + \frac{\partial^2 f}{\partial z^2} \qquad [\mathrm{F}\cdot 46]$$

Also note the following relationships:

$$\mathbf{curl}\,\mathbf{grad}\, f(x,y,z) \equiv \mathbf{0} \qquad \mathrm{div}\,\mathbf{curl}\,\mathbf{A}(x,y,z) \equiv 0 \qquad [\mathrm{F}\cdot 47]$$

APPENDIX G

periodic table of the elements

notes

(1) Atomic weights in atomic mass units (u) based on an atomic weight of 12.00000 u for carbon-12. Values in parentheses represent atomic weights of the most stable isotopes.
(2) Density values given for 300 K, except where asterisks indicate gaseous elements at 273 K and 1 atm and given in kg/m^3.
(3) Specific heats at 300 K. The specific heats listed for N, O, H, Cl, Br, I, and Se are for diatomic molecules.
(4) Electrical conductivities at 20°C.
(5) Melting points at 1 atm except where otherwise noted.
(6) Where more than one crystalline form exists, the properties are indicated for the α form; the properties of the white form of phosphorus are given. Values in parentheses indicate best estimates.

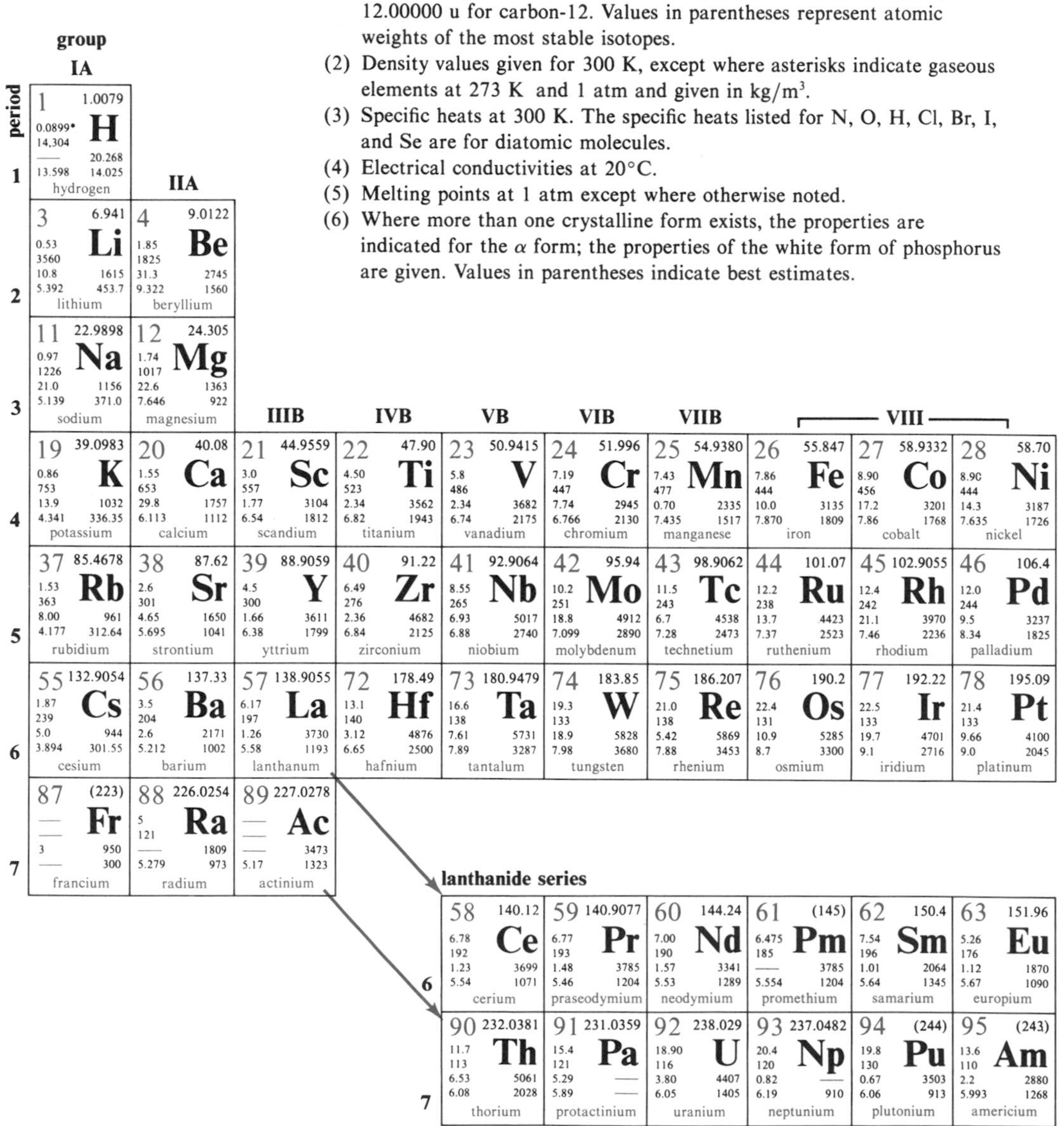

period	group	no.	symbol	atomic weight	left 1	left 2	left 3	left 4	right 1	right 2	name
1	IA	1	H	1.0079	0.0899*	14,304	—	13.598	20.268	14.025	hydrogen
2	IA	3	Li	6.941	0.53	3560	10.8	5.392	1615	453.7	lithium
2	IIA	4	Be	9.0122	1.85	1825	31.3	9.322	2745	1560	beryllium
3	IA	11	Na	22.9898	0.97	1226	21.0	5.139	1156	371.0	sodium
3	IIA	12	Mg	24.305	1.74	1017	22.6	7.646	1363	922	magnesium
4	IA	19	K	39.0983	0.86	753	13.9	4.341	1032	336.35	potassium
4	IIA	20	Ca	40.08	1.55	653	29.8	6.113	1757	1112	calcium
4	IIIB	21	Sc	44.9559	3.0	557	1.77	6.54	3104	1812	scandium
4	IVB	22	Ti	47.90	4.50	523	2.34	6.82	3562	1943	titanium
4	VB	23	V	50.9415	5.8	486	2.34	6.74	3682	2175	vanadium
4	VIB	24	Cr	51.996	7.19	447	7.74	6.766	2945	2130	chromium
4	VIIB	25	Mn	54.9380	7.43	477	0.70	7.435	2335	1517	manganese
4	VIII	26	Fe	55.847	7.86	444	10.0	7.870	3135	1809	iron
4	VIII	27	Co	58.9332	8.90	456	17.2	7.86	3201	1768	cobalt
4	VIII	28	Ni	58.70	8.90	444	14.3	7.635	3187	1726	nickel
5	IA	37	Rb	85.4678	1.53	363	8.00	4.177	961	312.64	rubidium
5	IIA	38	Sr	87.62	2.6	301	4.65	5.695	1650	1041	strontium
5	IIIB	39	Y	88.9059	4.5	300	1.66	6.38	3611	1799	yttrium
5	IVB	40	Zr	91.22	6.49	276	2.36	6.84	4682	2125	zirconium
5	VB	41	Nb	92.9064	8.55	265	6.93	6.88	5017	2740	niobium
5	VIB	42	Mo	95.94	10.2	251	18.8	7.099	4912	2890	molybdenum
5	VIIB	43	Tc	98.9062	11.5	243	6.7	7.28	4538	2473	technetium
5	VIII	44	Ru	101.07	12.2	238	13.7	7.37	4423	2523	ruthenium
5	VIII	45	Rh	102.9055	12.4	242	21.1	7.46	3970	2236	rhodium
5	VIII	46	Pd	106.4	12.0	244	9.5	8.34	3237	1825	palladium
6	IA	55	Cs	132.9054	1.87	239	5.0	3.894	944	301.55	cesium
6	IIA	56	Ba	137.33	3.5	204	2.6	5.212	2171	1002	barium
6	IIIB	57	La	138.9055	6.17	197	1.26	5.58	3730	1193	lanthanum
6	IVB	72	Hf	178.49	13.1	140	3.12	6.65	4876	2500	hafnium
6	VB	73	Ta	180.9479	16.6	138	7.61	7.89	5731	3287	tantalum
6	VIB	74	W	183.85	19.3	133	18.9	7.98	5828	3680	tungsten
6	VIIB	75	Re	186.207	21.0	138	5.42	7.88	5869	3453	rhenium
6	VIII	76	Os	190.2	22.4	131	10.9	8.7	5285	3300	osmium
6	VIII	77	Ir	192.22	22.5	133	19.7	9.1	4701	2716	iridium
6	VIII	78	Pt	195.09	21.4	133	9.66	9.0	4100	2045	platinum
7	IA	87	Fr	(223)	—	—	3	—	950	300	francium
7	IIA	88	Ra	226.0254	5	121	—	5.279	1809	973	radium
7	IIIB	89	Ac	227.0278	—	—	—	5.17	3473	1323	actinium

lanthanide series

period	no.	symbol	atomic weight	left 1	left 2	left 3	left 4	right 1	right 2	name
6	58	Ce	140.12	6.78	192	1.23	5.54	3699	1071	cerium
6	59	Pr	140.9077	6.77	193	1.48	5.46	3785	1204	praseodymium
6	60	Nd	144.24	7.00	190	1.57	5.53	3341	1289	neodymium
6	61	Pm	(145)	6.475	185	—	5.554	3785	1204	promethium
6	62	Sm	150.4	7.54	196	1.01	5.64	2064	1345	samarium
6	63	Eu	151.96	5.26	176	1.12	5.67	1870	1090	europium

actinide series

period	no.	symbol	atomic weight	left 1	left 2	left 3	left 4	right 1	right 2	name
7	90	Th	232.0381	11.7	113	6.53	6.08	5061	2028	thorium
7	91	Pa	231.0359	15.4	121	5.29	5.89	—	—	protactinium
7	92	U	238.029	18.90	116	3.80	6.05	4407	1405	uranium
7	93	Np	237.0482	20.4	120	0.82	6.19	—	910	neptunium
7	94	Pu	(244)	19.8	130	0.67	6.06	3503	913	plutonium
7	95	Am	(243)	13.6	110	2.2	5.993	2880	1268	americium

SOURCE: Adapted with permission from *Periodic Table of the Elements* (Skokie IL: Sargent-Welch Scientific Company, 1980).

Key

30	65.38
7.14	**Zn**
388	
16.9	1180
9.394	692.73
zinc	

- atomic number
- atomic weight[(1)]
- density[(2)] (g/cm^3)
- specific heat[(3)] ($J/kg \cdot K$)
- conductivity[(4)] ($\mu\Omega^{-1} \cdot m^{-1}$)
- first ionization potential (V)
- boiling point (K)
- melting point[(5)] (K)

Group	Atomic number	Symbol	Name	Atomic weight	Density	Specific heat	Conductivity	First ionization potential	Boiling point	Melting point
VIII	2	**He**	helium	4.0026	0.1787*	5193	—	24.587	4.215	0.95
IIIA	5	**B**	boron	10.81	2.34	1025	$\approx 10^{-10}$	8.298	4275	2300
IVA	6	**C**	carbon	12.011	2.25	712	0.0727	11.260	(4470)	(4100)
VA	7	**N**	nitrogen	14.0067	1.251*	1042	—	14.534	77.35	63.14
VIA	8	**O**	oxygen	15.9994	1.429*	917	—	13.618	90.18	50.35
VIIA	9	**F**	fluorine	18.9984	1.696*	825	—	17.422	84.95	53.48
VIII	10	**Ne**	neon	20.179	0.901*	1027	—	21.564	27.096	24.553
IIIA	13	**Al**	aluminum	26.9815	2.70	900	37.7	5.986	2793	933.25
IVA	14	**Si**	silicon	28.0855	2.33	710	10.0	8.151	3540	1685
VA	15	**P**	phosphorus	30.9738	1.82	770	$\approx 10^{-15}$	10.486	550	317.30
VIA	16	**S**	sulfur	32.06	2.07	710	$\approx 10^{-21}$	10.360	717.75	388.36
VIIA	17	**Cl**	chlorine	35.453	3.17*	477	—	12.967	239.1	172.16
VIII	18	**Ar**	argon	39.948	1.784*	519	—	15.759	87.30	83.81
IB	29	**Cu**	copper	63.546	8.96	385	59.6	7.726	2836	1357.6
IIB	30	**Zn**	zinc	65.38	7.14	388	16.9	9.394	1180	692.73
IIIA	31	**Ga**	gallium	69.72	5.91	373	6.78	5.999	2478	302.90
IVA	32	**Ge**	germanium	72.59	5.32	322	$\approx 10^{-6}$	7.899	3107	1210.4
VA	33	**As**	arsenic	74.922	5.72	329	3.45	9.81	876(subl.)	1081(28 atm)
VIA	34	**Se**	selenium	78.96	4.80	321	8.33	9.752	958	494
VIIA	35	**Br**	bromine	79.904	3.12	473	—	11.814	332.25	265.90
VIII	36	**Kr**	krypton	83.80	3.74*	248	—	13.999	119.80	115.78
IB	47	**Ag**	silver	107.868	10.5	236	62.9	7.576	2436	1234
IIB	48	**Cd**	cadmium	112.41	8.65	232	13.8	8.993	1040	594.18
IIIA	49	**In**	indium	114.82	7.31	234	11.6	5.786	2346	429.76
IVA	50	**Sn**	tin	118.69	7.30	217	9.07	7.344	2876	505.06
VA	51	**Sb**	antimony	121.75	6.68	207	2.43	8.641	1860	904
VIA	52	**Te**	tellurium	127.60	6.24	201	$\approx 10^{-4}$	9.009	1261	722.65
VIIA	53	**I**	iodine	126.9045	4.92	214	$\approx 10^{-13}$	10.451	458.4	386.7
VIII	54	**Xe**	xenon	131.30	5.89*	158	—	12.130	165.03	161.36
IB	79	**Au**	gold	196.9665	19.3	129	45.2	9.225	3130	1337.58
IIB	80	**Hg**	mercury	200.59	13.53	139	1.04	10.437	630	234.28
IIIA	81	**Tl**	thallium	204.37	11.85	129	6.17	6.108	1746	577
IVA	82	**Pb**	lead	207.2	11.4	127	4.81	7.416	2023	600.6
VA	83	**Bi**	bismuth	208.9804	9.8	124	0.867	7.289	1837	544.52
VIA	84	**Po**	polonium	(209)	9.4	126	2.19	8.42	1235	527
VIIA	85	**At**	astatine	(210)	—	—	—	—	610	575
VIII	86	**Rn**	radon	(222)	9.91*	94	—	10.748	211	202
	64	**Gd**	gadolinium	157.25	7.89	230	0.74	6.15	3539	1585
	65	**Tb**	terbium	158.9254	8.27	183	0.89	5.86	3496	1630
	66	**Dy**	dysprosium	162.50	8.54	173	1.08	5.94	2835	1682
	67	**Ho**	holmium	164.9304	8.80	165	1.24	6.018	2968	1743
	68	**Er**	erbium	167.26	9.05	168	1.17	6.101	3136	1795
	69	**Tm**	thulium	168.9342	9.33	160	1.50	6.184	2220	1818
	70	**Yb**	ytterbium	173.04	6.98	145	3.51	6.254	1467	1097
	71	**Lu**	lutetium	174.967	9.84	155	1.85	5.43	3668	1936
	96	**Cm**	curium	(247)	13.511	—	—	6.02	—	1340
	97	**Bk**	berkelium	(247)	—	—	—	6.23	—	—
	98	**Cf**	californium	(251)	—	—	—	6.30	—	900
	99	**Es**	einsteinium	(252)	—	—	—	6.42	—	—
	100	**Fm**	fermium	(257)	—	—	—	6.50	—	—
	101	**Md**	mendelevium	(258)	—	—	—	6.58	—	—
	102	**No**	nobelium	(259)	—	—	—	6.65	—	—
	103	**Lr**	lawrencium	(260)	—	—	—	—	—	—

APPENDIX H

flowcharting conventions

The concept that is common to all computer programming is the algorithm, the step-by-step procedure or method whereby the computation is performed. An algorithm is conveniently portrayed with a diagram called a *flowchart*. Figure H . 1 shows the symbols used in this book to represent the various components of algorithms.

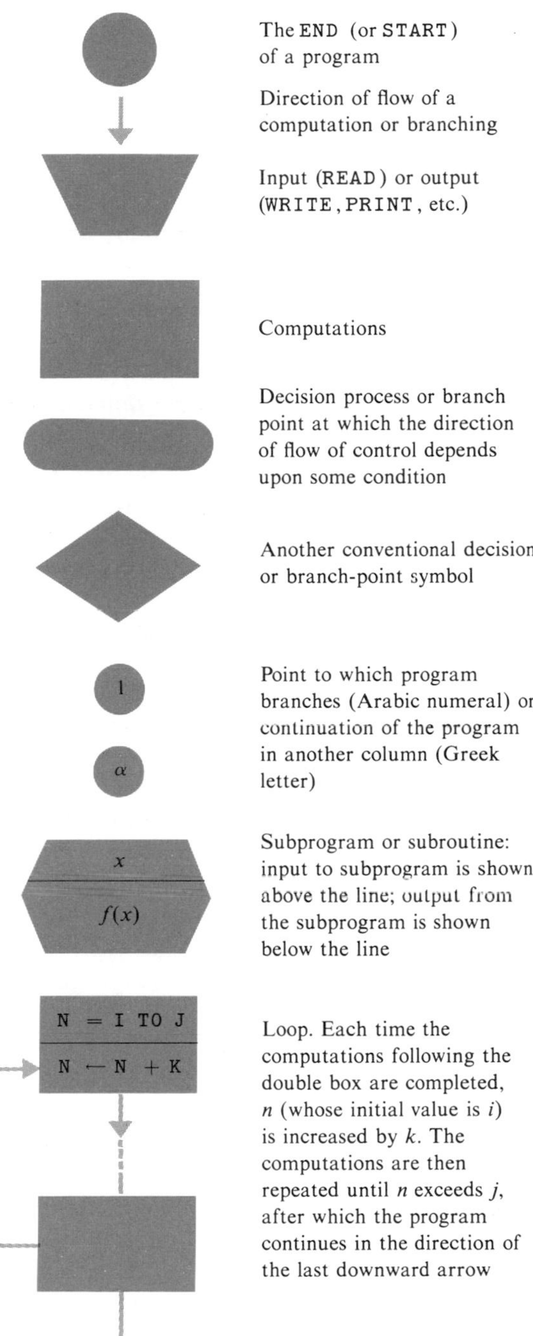

figure H • 1 *Flowcharting conventions*

APPENDIX I

the BASIC computer language

The only computer "language" used in the text is the flowchart (see Appendix H). However, for actual computation, the procedure *algorithm* embodied in the flowchart must be presented to the computer as a set of instructions called a *program*. The functional form of the program within the computer is a string of binary digits known as the machine-language program. A simple language of symbols and abbreviations (*mnemonics*) can be used to represent the string of digits in a form more readily understood by the programmer. A program called an *assembler* automatically converts the symbolic input into machine language and converts machine language into symbolic output. However, only a skilled programmer with extensive knowledge can efficiently program in such a "first-order" machine language.

To make computers more accessible to potential users, special "higher-order" languages have been developed which look very much like ordinary English or like algebraic notation. However, each such language does have a set of rules of usage (its grammar and syntax) which must be followed precisely and completely. A master program or *processor* called a *compiler* or *interpreter* acts as the interface between the language and the computer.

A wide variety of higher-order languages have been developed for various purposes: BASIC, FORTRAN, APL, PASCAL, ALGOL, and many others. The typical computer is equipped with compilers for a variety of languages. The most popular and widely used language for scientific purposes is FORTRAN. However, for any problem likely to be encountered in undergraduate physics, the BASIC language should be quite adequate, and it is relatively simple to learn and to use.* Furthermore, some dialect of BASIC generally is available on microcomputers and on computers providing a *time-shared* environment, in which many users employ the computer simultaneously. Each time-sharing user interacts with the computer as if the computer were responding swiftly and exclusively to the input from that user, so that programs can be corrected and altered while remaining connected (*on-line*) to the computer. Major scientific applications generally are handled through *batch processing,* in which the program is submitted to the computer operators on punched cards and the output is returned hours (or days) later in printed form. Batch processing generally is more economical and efficient for skilled users and for programs requiring lengthy and repetitive computations. However, time sharing is more fun and more practical for the unskilled user working with relatively simple programs.

*John G. Kemeny and Thomas E. Kurtz, *BASIC Programming,* 3rd ed (New York: Wiley, 1980). This is the classic text on the BASIC language by the original authors of the language.

The BASIC processor is a special program which must be "called" in a fashion unique to the particular computer system being used. The compiler not only can process the instructions in a BASIC program but can also be commanded to edit the program and perform other operations upon the program itself. Different versions or dialects of BASIC are found in different computer systems, but they are generally quite similar; experience with one dialect makes it easy to learn another.*

The algorithm (as illustrated in a flowchart) is the universal and fundamental concept of computer processing. Any flowchart in this text can be translated into a BASIC program. The following discussion provides only a handy outline of the generally accepted fundamentals of BASIC. If you wish to program in BASIC, you should consult the manual appropriate to the computer you will use and ask at your computer center about available references and short courses in BASIC programming.

format of the program

A BASIC program is a collection of one-line statements, each uniquely identified by a line number (an integer from 1 to 9999 on smaller computers or up to 99999 on larger computers). Each statement serves a specific purpose in the structure and operation of the algorithm except a line that begins with **REM** ("remark"). A line preceded by **REM** has no effect when the program is executed, but it remains in the listing (printout) of the program as a helpful comment (a blank **REM** line may be used as spacing to improve the legibility of the program).

```
100 REM THIS PROGRAM CALCULATES ORBITS. T IS
110 REM TIME, D = DT IS THE TIME-STEP, AND R
120 REM AND P REPRESENT THE POSITION OF THE
130 REM PARTICLE IN POLAR COORDINATES.
140 REM
```

You should never neglect the **REM** documentation. Always start your program by noting its identifying name, purpose, and author. Define each variable in a **REM** statement for later reference, and use **REM** at any point in the program where the insertion of a comment might prove useful. It is good practice to begin with line number 100 and to increase the number by 10 for each consecutive line (making it easy to insert additional lines in later modification of the program). No matter where the lines are written in the program, the statements are processed by the compiler in the numerical order of their line numbers.

It is most common to use exclusively uppercase letters in the BASIC language, and we recommend this practice even for a dialect that permits the use of lowercase letters.

In the following discussion we have roughly subdivided the features of

**ANSI Minimal BASIC X3.60-1978* (American National Standards Institute, 1430 Broadway, New York, NY 10018) outlines generally accepted principles of the language. *CONDUIT BASIC Guide* (CONDUIT, P.O. Box 388, Iowa City, IA 52240), edited by Janet Frederick, is an excellent comparative anatomy of the major BASIC dialects and a guide to good BASIC programming practice. It is not a textbook but is readable and recommended.

the BASIC language into nine topics: constants and variables, operations and expressions, functions, control statements, recursive statements (loops), subroutines, input and output, matrix operations, and system commands.

constants and variables

Any positive or negative number (a *numerical constant*) may be expressed in BASIC as an integer, a decimal, or an exponential. Thus `104` and `104.00` and `1.04E2` represent the same number. In the exponential form the number following `E` signifies the power of ten. In most dialects powers of ten from -38 through $+38$ are permitted. Negative exponents may be used also. Here are some examples of equivalent forms for numerical constants:

```
-0.032     -3.2E-2
73.8123E-16     .0738123E-13
62.1    6.21E1     0.621E2
.00123     0.00123     1.23E-3
123456E7     1.23456E12
-573     -5.73E2     -5730E-1
```

All types of numerical constants may occur in a single line. Some computer systems permit only six significant digits, exclusive of the exponent; other systems allow more. However, there must be at least one significant digit; thus 100 may be written as `1E2` but not as `E2`.

A collection of symbols (delimited by quotes) can be treated as a constant called a *literal constant* or *string*. The computer recognizes `"TOM LONDON"` as a single literal constant (a *ten*-letter string, with the space counting as a "letter"). The following statement stores this ten-letter string at an "address" in the computer's memory labeled `A$`:

```
150 LET A$ = "TOM LONDON"
```

Most systems can accept strings of length at least up to the maximum length of a line. Obviously, mathematical operations cannot be performed on strings. (Except within a string, the BASIC compiler ignores spaces.)

Every computer has a memory where values may be stored. The name of a variable is equivalent to the "address" of the location in memory where its current value (a constant) is stored by the supervisory program which controls the operation of the entire computer. Thus the statement

```
100 LET X = 1.3
```

means that the number 1.3 is stored in the memory location designated as `X`. Whenever the program calls for `X`, the computer retrieves the number stored at that address as the value of `X` (in this case, the value 1.3). The distinction between the name (address) of a variable and its value is fundamental. Because of this distinction, we can write the *assignment* statement

```
180 LET X = X + 1
```

This statement causes the computer to retrieve the value 1.3 from location `X`, add 1 to it, and store the result (the value 2.3) in location `X` (now replacing the old value 1.3).

The following rules govern the naming of BASIC variables.

1 A *simple variable* is named by a single letter (from A to Z) or by a single letter followed by a digit (from 0 through 9): A, B, . . . , A1, A2, . . . , Z9. Use of the letter O and the digit 0 is not recommended because they are easily confused.

2 A *subscripted variable* such as x_i, y_{mn}, or z_{3k} can be represented in BASIC as X(I), Y(M,N), or Z(3,K). No more than two subscripts may be used, and each subscript must be either a constant or a variable of positive value. Many dialects of BASIC also allow the use of an expression as a subscript, so that a_{2n+1} can be represented as A(2*N+1). The subscript should be an integer, although some dialects automatically truncate or round off any noninteger subscripts. In many dialects subscripted variables can be named just like simple variables, but some dialects permit only a single-letter name for a subscripted variable. Some dialects allow variables with three or four subscripts.

Whenever a subscripted variable is to be used in a program, space must be reserved for it in memory by means of a *dimension statement*. For example, if you wish to use the arrays of quantities $x_1, x_2, \ldots, x_{20}$ and $a_{11}, a_{12}, \ldots, a_{77}$, you must first use the dimension statement

```
100 DIM X(20), A(7,7)
```

Most dialects automatically allow a dimension of 10 for each subscript if no dimension statement is used. However, it is best to dimension all arrays of subscripted variables through a dimension statement inserted before the first executable instruction in your program. This practice not only makes efficient use of memory locations (and avoids the danger of trying to use a subscript value for which no space has been reserved), but also provides helpful documentation by its presence. In general, you should be able to dimension your arrays so as to have a total of no more than 1500 numerical variables (simple and subscripted) in your program. Most dialects provide for a lowest subscript value of zero, but it is best to begin your array entries with subscript 1. Remember never to use the same name for both a subscripted and a simple variable in the same program.

3 A *string variable* has a literal constant or string as its value. One of the 26 allowed names A$, B$, . . . , Z$ must be used to name a string variable. It is also possible to designate a single list of literal constants by an array; for example,

```
100 DIM L$(15)
```

Such arrays may have no more than one subscript.

operations and expressions

Seven operations on constants or variables can be specified in BASIC: five computational operations, one grouping operation, and the assignment operation. Using the *operators* (the symbols specifying the operations), we can instruct the computer to manipulate the values of variables, perform operations with constants, or form any desired algebraic expression. The operators must be stated explicitly. Table I . 1 lists the *computational operators.*

table I • 1 *BASIC computational operators*

operation	*operator*	*priority*	*BASIC expression*	*algebraic or numerical equivalent*
exponentiation*	↑	1	`X↑2.7`	$x^{2.7}$
			`2↑Y`	2^y
			`U↑V`	u^v
multiplication	*	2	`X*Y`	xy
			`3E-3*7E13`	$3 \times 10^{-3} \times 7 \times 10^{13}$
division	/	2	`X/Y`	x/y
			`X/1E-3`	$x/10^{-3} = 1000x$
addition	+	3	`X+Y`	$x + y$
			`3E3+7E4`	$3 \times 10^3 + 7 \times 10^4 = 73{,}000$
			`3+X`	$3 + x$
subtraction	−	3	`X-Y`	$x - y$
			`3E3-7E4`	$3 \times 10^3 - 7 \times 10^4 = -67{,}000$
			`3-X`	$3 - x$

*In some BASIC dialects, the exponentiation operator is ∧ or **.

The BASIC compiler treats the computational operators according to a hierarchy of priority, proceeding from priority 1 to priority 3 as listed in Table I . 1. The compiler first performs the leftmost exponentiation specified in the line, then the next exponentiation to the right, and so on until it has performed all exponentiations. It next performs the leftmost multiplication or division, proceeding through the line toward the right until it has completed all multiplications and divisions. Finally, it performs the additions and subtractions in sequence from left to right. In each case, the computation is performed on the constants or variables (or results of earlier operations) immediately preceding and following the operator. Thus

`A*B↑C` represents $a(b^c)$ `A/B*C` represents $\frac{a}{b} \times c$

The sequence of operations can be altered by use of the *grouping operator,* a pair of parentheses that gives highest priority to the computation specified within the parentheses. Thus

`(A*B)↑C` represents $(ab)^c$ `A/(B*C)` represents $\frac{a}{b \times c}$

Nested parentheses may be used. Highest priority is given to the computations specified within the innermost set of parentheses. Parentheses should be used liberally for improved legibility and to avoid careless errors.

The computational and grouping operations can be used to specify any algebraic or mathematical expression. A BASIC *expression* consists of one or more operations performed on two or more constants or variables. In writing a BASIC expression, be sure that the specified priority of operations corresponds to the sequence of computations you wish carried out. Also beware of trying to raise zero or a negative number to a power (negative powers of positive numbers are legitimate).

The *assignment operation* replaces the value of a variable by the value of a constant, variable, or expression. The operator is LET . . . = . . . , as in

```
LET A = B
```

where B is the value to be assigned to the variable A. (In some dialects the keyword LET is optional.) The assignment operator has the lowest priority in the statement. Only one assignment operator may be used in a statement.

Consider the sequence in which the compiler carries out the following statement:

```
190 LET P = ((X↑Y/Z*W)↑U−Q/R+T)*S
```

The compiler first evaluates the expression to the right of the = operator, obtaining from the appropriate memory location the current numerical value of each variable. In the following representation we use *a*, *b*, . . . to indicate the intermediate values calculated as the evaluation progresses:

$a = x^y$ expression now equivalent to	((*a*/Z*W) ↑ U − Q/R + T) * S
$b = a/z$	((*b* *W) ↑ U − Q/R + T) * S
$c = bw$	(*c*↑U − Q/R + T) * S
$d = c^u$	(*d* − Q/R + T) * S
$e = q/r$	(*d* − *e* + T) * S
$f = d - e$	(*f* + T) * S
$g = f + t$	*g* * S
$h = g \times s$	*h*

The expression has now been given the value we have called *h*, where

$$h = \left[\left(\frac{wx^y}{z}\right)^u - \frac{q}{r} + t\right]s$$

Finally, the assignment operation is performed, replacing the old value of P with the value *h* calculated for the expression. Note that the value of any expression is calculated by using the current values for any variables in the expression.

Because the BASIC compiler ignores spaces (except within strings), the legibility of an expression can often be improved by inserting spaces. For example, the statement just discussed can be written as

```
190 LET P = ((X↑Y/Z*W)↑U−Q/R+T)*S
```

functions

Certain commonly used mathematical functions are built into the BASIC language (see Table I . 2). The argument X of such a *library function* may be a constant, a variable, an expression, or another function. For example, the following statements would assign a random integer from 0 to 99 as the value of X:

```
100 RANDOMIZE
110 LET X = INT(100*RND(Y))
```

The following is another valid statement using nested functions:

```
120 LET Z3 = 3*SQR(5 + SIN(X)/LOG(10−Y))
```

table I • 2 *BASIC library functions*

function	*value of function*
ABS(X)	$\lvert x \rvert$ = absolute value of x
SGN(X)	−1 if x is negative, or 0 if x is zero, or 1 if x is positive
SQR(X)	$\sqrt{x}$ = square root of x, which must be nonnegative
INT(X)	greatest integer less than or equal to x
SIN(X)	$\sin x$ = sine of x (with x assumed to be expressed in radians)
COS(X)	$\cos x$ = cosine of x (assumed to be expressed in radians)
TAN(X)	$\tan x$ = tangent of x (assumed to be expressed in radians)
ATN(X)	$\tan^{-1} x$ = arctangent of x (principal value* given in radians)
EXP(X)	e^x, where e is the base of the natural logarithms
LOG(X)	$\ln x$ = natural logarithm of x, which must be nonnegative
RND(X)	a randomly selected number between 0 and 1 (this function always returns the same sequence of random numbers unless it is preceded in the program by a RANDOM or RANDOMIZE statement)

*See Appendix E.

Note that SQR(X) is equivalent to X↑.5, but the functional form is preferred for clarity and economy. Also note that the variable Y in the function RND(X) is a dummy variable that has no effect on the output of the function. Any constant or variable could be used as the argument when this function is "called" without affecting the value of the function output.

The user can create up to 26 *defined functions* by inserting a DEF ("definition") statement in the program prior to the first use of the function. Only the names FNA, FNB, . . . , FNZ are allowed as names of defined functions, and only one DEF statement is permitted for each name within a program. The statement

```
130 DEF FNA(R) = 4*3.1416*R↑3/3
```

defines the function FNA. The function may be "called" in any subsequent statement by using FNA(X) in an expression, where X can be any constant, variable, or expression. The value of FNA(X) will be evaluated as 4*3.1416*X↑3/3 or $\frac{4}{3}\pi x^3$. That is, upon each use of the function it is evaluated by using the argument given in the calling statement to replace the dummy variable given as the argument in the definition statement. The use of R as a dummy variable within the DEF statement has no effect on any other possible uses of R as a variable within the rest of the program. The statement

```
140 LET V = FNA(R1) + FNA(12)
```

will assign to v a value equal to the sum of the volumes of two spheres of radii r_1 and 12. Of course, some value of r_1 must have been assigned in a preceding statement.

Two or more arguments may be used in defining a function, but values must be given for each defined argument when the function is called. If variables other than the arguments are used in a function definition, the current values of those variables will be used when the function is called. A function with no argument may be defined if all variables used in the definition have current values when the function is used. Consider the following function definitions:

```
100 DEF FNA(X,Y)  = X*COS(Y)
110 DEF FNB  = X↑2 + Y↑2
120 DEF FNC(Z)  = B*Z↑3
130 DEF FND(X,Y)  = 3*FNA(X,Y)/FNB
```

Suppose that the following statement appears later in the program:

```
250 LET H = FNC(13.5) - FND(T,-2*SIN(1.2))
```

When this statement is executed, `FNC(13.5)` will be evaluated by using the constant `13.5` to replace the dummy variable `Z` in the defining expression for `FNC`, yielding a value of `B*13.5↑3`, using the current value of `B`. Similarly, `FND(T, -2*SIN(1.2))` will be evaluated as

```
3*FNA(T, -2*SIN(1.2))/FNB     or
3*(T*COS( -2*SIN(1.2)))/(X↑2 +Y↑2)
```

using the current values of `T`, `X`, and `Y`. Note that the dummy variables `X` and `Y` are replaced by other constants or variables in evaluating `FND` and `FNA`, but that current values of variables `X` and `Y` are used in evaluating `FNB`.

control statements

The BASIC compiler executes the statements of a program in numerical sequence except where one of three possible *control statements* in the program alters that sequence.

The `GOTO` (or `GO TO`) statement is an *unconditional transfer* to some particular point in the program. For example,

```
210 GOTO 300
```

causes the compiler to move immediately to statement 300 (skipping any statements between 210 and 300) and then continue in numerical sequence from there. Some dialects do not permit transfers to nonexecutable statements such as `REM`, `DIM`, and `DEF` statements, so it is best always to write the transfer to an executable statement.

Most dialects also permit a computed `GOTO` operation based on the value of some variable `X` by using a statement such as

```
220 ON X GOTO 110, 300, 450
```

which will transfer control to statement 110 if the value of `X` is 1, to statement 300 if `X` = 2, and to statement 450 if `X` = 3. If `X` has any value other than 1, 2, or 3, execution will halt with an error report to the user (although some dialects automatically utilize `INT(X)` instead of `X`). Any variable or expression can be used in place of `X` in such a statement, but care must be taken to see that it can have only the desired values.

It is possible to transfer control to a given line from any of several

table I • 3 *BASIC relational operators*

operator	*comparison*
=	is equal to (or is identical to, if strings are being compared)*
<>	is not equal to (or is not identical to, if strings are being compared)*
>	is greater than
<	is less than
>=	is greater than or equal to
<=	is less than or equal to

*Only the first two relational operators may be applied to strings or string variables. Otherwise, any constants, variables, and/or expressions may be used as the arguments of the operators.

different GOTO statements within a single program. Control may be transferred forward or backward in a program.

The IF-THEN statement is a *conditional transfer* of control that permits alteration of the sequence of computations depending on the current values of variables and/or expressions. The ***relational operators*** listed in Table I . 3 are used only in IF-THEN statements. The form of the statement is

000 IF (*relational expression*) THEN (*statement number*)

The compiler evaluates the relational expression and determines its truth or falsity. If the expression is true, the compiler transfers control to the specified statement. If the expression is false, the compiler simply moves on to the next statement in numerical sequence. Each of the following statements would result in a transfer of control to statement 300 *only if* the specified relational expression is true.

```
180 IF X < Y THEN 300
190 IF X + Y > 0 THEN 300
200 IF 3+X*SIN(Y) < =Q↑2.5 THEN 300
205 IF A$ <> B$ THEN 300
210 IF A$ = "STOP" THEN 300
```

If none of the specified relations is true, then control moves on to the next statement in sequence, with no effect from this string of conditional control statements.

The END statement is a control statement that orders the compiler to terminate execution of the program. The END statement will cause termination of the program even if the control is being shifted past that statement by a GOTO statement, so no statement following the END statement can be executed and only one END statement can appear in a program. It is wise therefore to form the habit of assigning the highest possible number to the END statement:

```
9999 END
```

A STOP statement is a form of unconditional transfer that orders the compiler to skip forward through the program until it reaches the END statement; it can be used to indicate points in a branching program where termination is desired. A statement such as GOTO 9999 can be used instead of the STOP statement. There is no limit to the possible number of STOP statements in a program. It is possible to shift control past a STOP by means of GOTO or IF-THEN statements.

iteration, recursion, and looping

In many algorithms, a certain sequence of computations must be repeated (a *recursion* or *interation*). Such a *loop* can be created in a BASIC program by using the FOR-NEXT statements:

```
240 FOR N  = X TO Y STEP Z
     . . .
290 NEXT N
```

Statement 240 causes the value of N to be set equal to X before control moves on to the next statement. When statement 290 is reached, the value of N is compared to the value Y. If N is between X and Y, then the value of N is changed to N + Z and control is returned to the next statement after statement 240. If N is equal to Y, or no longer falls between X and Y, then control moves on to the next statement after statement 290. Thus the loop of statements between lines 240 and 290 is repeated some predetermined number of times. Any constants, variables, or expressions may be used for X, Y, and Z in the FOR statement, but these values may not be altered by any statements within the loop. The values must be chosen so that the loop will terminate after a finite number of iterations. If the STEP Z portion of the FOR statement is omitted, Z will be automatically set to 1. In general, it is best to design the loop so that X, Y, and Z are integers. Any variable name may be used for N. It is possible to choose negative values of X, Y, or Z so long as the loop terminates (finite loop).

For example, the following statements will compute the sum $x_1 + x_2 + \cdots + x_{25}$:

```
320 LET S  = 0
330 FOR I  = 1 TO 25
340        LET S  = S  + X(I)
350 NEXT I
```

The loop will be iterated until all 25 terms have been added, and control will then move on to the next statement after line 350 with the value of S set equal to the sum.

A conditional loop can be created by use of an IF-THEN statement within the loop. For example, if the addition is to be terminated whenever the sum exceeds 100, the following statement can be inserted within the loop:

```
345 IF S  > 100 THEN 360
```

In this case the value of I upon termination of the loop would indicate the number of terms that have been summed.

It is possible to nest loops within loops, but each loop must be wholly contained within the next surrounding loop. The number of nestings allowed varies from dialect to dialect but is usually around 10. The following statements could be used to find a double sum:

```
360 LET S2 = 0
370 FOR I = 1 TO 7
380        FOR J = 1 TO 5
390               LET S2 = S2 + Y(I,J)
400        NEXT J
410 NEXT I
```

Indentations are useful for legibility and to ensure that each loop is wholly contained within the next outer loop (recall that the compiler ignores spaces except in strings).

subroutines

A *subroutine* is a subprogram within a larger program. A subroutine is generally used when some procedure must be employed at several different points within a program. Rather than repeat the same sequence of statements at each point, we can use the **GOSUB-RETURN** statements to create a subroutine. Suppose that the statements of the subroutine appear as lines 1000 through 1080 in the program. We then add

```
1090 RETURN
```

If the **RETURN** statement is encountered during the normal sequential processing of statements, it is ignored. However, we can insert a statement such as

```
330 GOSUB 1000
```

at any point in the program (except within the subroutine itself) to transfer control to line 1000. In this case the **RETURN** statement will cause control to be transferred back to the next line after the **GOSUB** statement.

The **GOSUB** statement is similar to the **GOTO** statement, but the number of the line following the **GOSUB** statement is stored for a later **RETURN**. Each subroutine may in turn call upon some other subroutine, although no more than 10 **GOSUB** statements should be used before the next **RETURN** is encountered.

input and output

The typical program is written for repeated use with different sets of information or data. In a BASIC program, data may be entered into the computer by any number of **DATA** lines containing numerical or string constants:

```
1000 DATA 1.7,-6.234E6,"RIDE",13,"?"
2000 DATA 2,4,8,16,32
```

These statements may be located anywhere in the program, but it is good practice to keep them grouped together with a special set of line numbers for ready recognition and replacement on reuse of the program.

The data from the **DATA** statements are brought into the computations by **READ** statements, which take the data in order of appearance and assign these values to appropriate variables. Suppose that lines 1000 and 2000 as shown are the only **DATA** statements in a program. The first **READ** statement in the program is

```
160 READ A,B,C$,D
```

This causes the compiler to locate the first **DATA** statement in the program (line 1000) and to assign the first four constants found there as the values of the four variables listed in the **READ** statement. The next **READ** statement in the program is

```
210 READ E$,U,V,W
```

The value for **E$** is set to **"?"**, the remaining constant in line 1000. The values for the other three variables are taken from the next **DATA** statement in the program (line 2000), so that U = 2, V = 4, and W = 8.

A **RESTORE** statement can be used at any point in the program to instruct the computer to begin again with the first **DATA** statement for constants required in subsequent **READ** statements:

```
220 RESTORE
```

In most dialects the statement

```
220 RESTORE 2000
```

will cause the next **READ** statement to begin taking data from line 2000.

When the user is interacting directly with the computer through a terminal, the **INPUT** statement can be used to allow entry of data during execution. In most dialects

```
130 INPUT A,B,F$
```

will cause the computer to halt execution and display a query symbol (? or > in most dialects) on the terminal. The user can then enter three constants (for example, 6,2E7,"GO") which will be stored as the values of **A**, **B**, and **F$**, respectively. Execution of the program then resumes with these values in the memory. Obviously the program should be designed to print out instructions to the user on the terminal before pausing for the input. The **INPUT** statement is a very useful device for creating a programmer-machine dialog.

The **PRINT** statement is used to output the results of the program on the terminal during execution. In a typical dialect the statement

```
580 PRINT A, B, A*B, "RESULT IS", F$
```

would print out the *values* of the variables and constant in five *zones* of about 15 spaces each across the page or terminal:

space no.	1	16	31	46	61
	↓	↓	↓	↓	↓
output	6	2E7	12E7	RESULT IS	GO

The spacing of the zones varies with the dialect and the output device. If the **PRINT** statement lists more than five values to be output, the remainder will appear in consecutive zones on succeeding lines of output until all have been printed. A blank zone can easily be created by use of a literal constant consisting of one or more spaces (" ").

The zoning feature can be suppressed by using a semicolon instead of a comma between items in the `PRINT` statement. Thus, if `A = 6`, then

```
590 PRINT 17,"A = ";A
```

yields the output

```
17          A = 6
```

The semicolon after a string or string variable causes the next entry to be printed with no intervening spaces. The semicolon after a numerical constant, variable, or expression may cause insertion of one to three spaces between entries, depending on the system.

Note that the output entries in the `PRINT` statement may be constants, variables, expressions, or strings. The expressions may even include library or defined functions.

The `TAB` feature in the `PRINT` statement allows placement of the first character of an output entry in any desired column (space) across the line. Thus

```
600 PRINT TAB(6); A
```

will cause the value of `A` to be printed, beginning in column `6`. More than one `TAB` may be used in a single `PRINT` statement. The statement

```
610 PRINT TAB(N):"*"
```

will print an asterisk in the *N*th column. This technique can be employed to plot useful (though imprecise) graphs. The argument of `TAB` should be an integer to avoid roundoff and truncation errors.

`PRINT` statements should be used liberally to provide headings for columns of data or instructions for the user working at a terminal. A blank `PRINT` statement such as

```
620 PRINT
```

will cause the output of a blank line, improving the legibility of the output. The terminal is automatically positioned at the next line after each `PRINT` statement unless this feature is suppressed by a semicolon at the end of the `PRINT` statement.

matrix operations

Most computers and some microcomputers provide a set of `MAT` statements which simplify operations on arrays (matrices) of doubly subscripted variables. Table I.4 shows these matrix operations for the arrays `A`, `B`, and `C` of variables a_{ij}, b_{ij}, and c_{ij}, where the arrays have the same dimensions. In general, the matrix `A` may not appear on both sides of the `MAT` statement (thus `MAT A = A + B` is not allowed), and only one operation is allowed after the equal sign.

A matrix `X(I,J)` can be filled with data by the statement

```
870 MAT READ X
```

table I • 4 *BASIC matrix statements*

statement	*meaning*
`MAT A = B`	sets $a_{ij} = b_{ij}$ for every value of i and j
`MAT A = B + C`	sets $a_{ij} = b_{ij} + c_{ij}$ for every value of i and j
`MAT A = B - C`	sets $a_{ij} = b_{ij} - c_{ij}$ for every value of i and j
`MAT A = (M)*B`	sets $a_{ij} = mb_{ij}$ for every value of i and j (scalar multiplication)
`MAT A = B*C`	sets $a_{ij} = \sum_k a_{ik}b_{kj}$ for every value of i and j, where the sum is taken over all possible values of k (matrix multiplication)

which reads in the values of $x_{11}, x_{12}, \ldots, x_{21}, x_{22}, \ldots$ row by row, using sequential data items from the next available **DATA** statement(s). If the data for the matrix are to be entered by the user during execution,

```
870 MAT INPUT X
```

solicits this information at the terminal and accepts it row by row for the memory locations that were reserved for **X** in a **DIM** statement. Other available input operations are

```
880 MAT A = ZER
890 MAT B = CON
900 MAT C = IDN
```

where line 880 sets each a_{ij} equal to zero, line 890 sets each b_{ij} equal to 1, and line 900 sets a_{ij} equal to zero for $i \neq j$ and equal to 1 for $i = j$.

The statement

```
910 MAT PRINT A;
```

outputs the matrix **A**, one row at a time, in packed (nonzoned) format. Only one matrix at a time should be output in this fashion to avoid confusion. If zoned format is desired, the semicolon at the end of the **MAT PRINT** line should be replaced by a comma, or simply deleted on some systems.

The **MAT** statements are very useful in the solution of simultaneous linear equations, using the three **MAT** library functions: the statement

```
850 MAT A = TRN(B)
```

sets each a_{ij} to the value of b_{ji}, so that **A** becomes the transpose of **B**;

```
860 LET X = DET(B)
```

sets the variable **X** equal to the value of the determinant of matrix **B**; and

```
870 MAT C = INV(A)
```

sets matrix **C** equal to the inverse matrix of matrix **A**—that is, **C** is the unique matrix such that the matrix product

```
MAT I = C*A
```

is the *identity matrix* **I** with each entry equal to zero except for those on the main diagonal, which have the value 1 (see the IDN operation above). When any matrix is multiplied by **I**, the product is equal to the original matrix; that is, A = A*I = I*A (although matrix multiplication is not in general commutative).

A set of simultaneous linear equations in $x_1, \ldots, x_n$ consists of n equations of the form

$$\sum_{j=i}^{n} a_{ij}x_j = b_i \qquad \text{for } i = 1, \ldots, n$$

This set of equations can be represented and solved as follows:

```
100 DIM A(N,N), X(N,1), B(N,1), C(N,N)
110 MAT READ A
120 MAT READ B
130 MAT C = INV(A)
140 MAT X = C*B
150 MAT PRINT X;
```

This procedure is the equivalent of the matrix equations

$$\mathbf{AX} = \mathbf{B} \qquad \mathbf{CAX} = \mathbf{IX} = \mathbf{X} = \mathbf{CB}$$

which is one of the standard techniques of matrix algebra.

system commands and debugging

Line numbers of consecutive statements are normally increased by 10 so that additional lines can later be inserted if necessary. For example, if the programmer forgot to specify the value of a constant, the necessary statement can be inserted between lines 230 and 240 at any point before or after execution by adding the statement

```
235 LET A = 3.5
```

This statement will be inserted in proper sequence in the program before the next execution. If a line is in error, it can be replaced by typing a new statement with the same line number. On a terminal a minor error during typing of a statement can be deleted by immediately using the backspace (←) key, prior to the carriage return (or transmit) key. For example, typing

```
620 LET A = SIM←N(3.2←1416/2(←)
```

is equivalent to typing

```
620 LET A = SIN(3.1416/2)
```

with each use of the ← key causing deletion of the preceding character and a return to that space in the line for the next character. Two successive strikes of the ← key will delete the *two* characters immediately preceding them, and so on. The deletion symbols vary with the dialect, and they may or may not be printed explicitly on the terminal when used.

The process of locating and correcting errors in a program is known as *debugging*. When the computer is unable to carry out an instruction in the program, it will halt execution and print out an error report with some information about the nature of the problem and its location in the program. Consult your system manual for the listing of error reports as they appear on your system. When working at a terminal you can fix the error in the program by using editing commands (see below) and then resume execution of the program.

System commands are not a part of the program, and they have no line numbers. When a program is ready to be executed, the system command

```
RUN
```

will cause computing to begin. After the program has been completed without error reports, try a few simple and obvious computations to test the results for any errors in algorithm or logic. When you have learned to set up files on your system, the commands

```
NEW GARBLE
SAVE
```

will name the program **GARBLE** and store it for future use. To access the program the next time you call BASIC, give the command

```
OLD GARBLE
```

The computer will make the program available as if you have just typed it in. You can rename the program **JUMBLE** after accessing it by using the command

```
RENAME JUMBLE
```

The command

```
CATALOG
```

will cause the computer to print out a list of the names of the programs in your file. After accessing a program you can delete it from storage with the command

```
UNSAVE
```

These system commands may vary slightly in some computer systems.

After a program has been debugged the line numbers typically are irregular. The system command

```
EDIT RESEQUENCE
```

will renumber all lines, beginning with 100 and incrementing by 10. All **IF-THEN**, **GOTO**, and **GOSUB** statements are automatically renumbered during this process. Most systems allow resequencing of selected portions of a program by selected increments. An **EDIT DELETE** command can be used to remove a set of statements:

```
EDIT DELETE 50, 80, 230-250
```

will cause deletion of lines 50, 80, and 230 through 250 inclusive. Conversely,

```
EDIT EXTRACT 50, 80, 230-250
```

will delete all lines *except* those specified. These command statements may have varying forms in different dialects and systems; some systems do not require the word EDIT, but most have an extensive repertoire of editing commands.

You can cause a printout of the complete program at any time with the system command

```
LIST
```

It is a good idea to do this before running the program and, if possible, to save a printed copy of the listing for future reference. Various forms of the LIST command can also be used to print out selected portions of the program.

Most BASIC compilers terminate user-computer interactions upon the command

```
BYE
```

example

The preceding sections of this appendix contain all the information you need to write a BASIC program for any algorithm discussed in this text. For example, the following program was used to perform the Hooke's law force computation in Sec. 6 . 6:

```
100 REM HOOKE'S LAW CALCULATION. T = TIME,
110 REM D = DT, X = EXTENSION, V = VELOCITY,
120 REM T1 = TIME FOR WHICH RESULT IS
130 REM PRINTED, X1 = EXACT ANSWER,
140 REM P = PERCENT ERROR
150 LET T = 0
160 LET D = .01
170 LET X = 1
180 LET V = 0
190 LET T1 = 19*D
200 LET T = T + D
210 LET V1 = V - X*D
220 LET X = X + V*D
230 LET V = V1
240 IF T < T1 THEN 200
250 LET X1 = COS(T)
260 LET P = 100*(X - X1)/X1
270 PRINT T;X;V;X1;P
280 LET T1 = T1 + 20*D
290 IF T < 7 THEN 200
9999 END
```

APPENDIX J

statistical tables

Table J . 1 shows values of the standardized normal distribution function

$$Z(u) = \frac{\exp(-u^2/2)}{\sqrt{2\pi}} \qquad [J \cdot 1]$$

and its cumulative probability function

$$P(u) = \int_{-\infty}^{u} Z(u)\, du \qquad [J \cdot 2]$$

for three standard deviations ($0 \leq u \leq 3.00$). The following are some useful relationships involving these functions:

$$P(\infty) = \int_{-\infty}^{\infty} Z(u)\, du = 1 \qquad [J \cdot 3]$$

$$\int_{-u}^{u} Z(u)\, du = 2\int_{0}^{u} Z(u)\, du \qquad \text{since } Z(u) = Z(-u) \qquad [J \cdot 4]$$

$$P(u) = \int_{-\infty}^{u} Z(u)\, du = 0.5 + \int_{0}^{u} Z(u)\, du$$

$$= 1 - \int_{u}^{\infty} Z(u)\, du = 1 - P(-u) \qquad [J \cdot 5]$$

$$\frac{dZ}{du} = -uZ \qquad \int_{0}^{u} uZ\, du = -Z(u) + Z(0)$$

$$\int_{0}^{\infty} uZ\, du = \frac{1}{\sqrt{2\pi}} \qquad [J \cdot 6]$$

$$\lim_{u \to \infty} u^n Z(u) = 0 \qquad \text{for } n < \infty \qquad [J \cdot 7]$$

$$\int_{-u}^{u} u^n Z\, du = 0 \qquad \text{for } n = \text{odd integer} \qquad [J \cdot 8]$$

Integrating by parts, we obtain

$$\int_{-\infty}^{\infty} u^2 Z\, du = -\int_{-\infty}^{\infty} u \frac{dZ}{du}\, du = -[uZ(u)]_{-\infty}^{\infty} + \int_{-\infty}^{\infty} Z\, du = 1 \qquad [J \cdot 9]$$

Similarly,

$$\int_{-\infty}^{\infty} u^4 Z\, du = 3\int_{-\infty}^{\infty} u^2 Z\, du = 3 \qquad [J \cdot 10]$$

$$\int u^{2n+1} Z\, du = -(2n)!!Z(u)\sum_{k=0}^{n} \frac{u^{2k}}{(2k)!!} \qquad \text{for } n = \text{integer} \qquad [J \cdot 11]$$

where $(2n)!! = 2 \times 4 \times 6 \times \cdots \times 2n$, and $0!! = 1$.

$$\int u^{2n} Z\, du = (2n-1)!! \int Z\, du - Z(u) \sum_{k=1}^{n} \frac{(2n-1)!!}{(2k-1)!!} u^{2k-1} \quad \text{for } n = \text{integer} \qquad [\text{J}\cdot 12]$$

where $(2n-1)!! = 1 \times 3 \times 5 \times \cdots \times (2n-1)$.

$$J_0 = \int_{-\infty}^{\infty} \exp(-\alpha x^2 - \beta x)\, dx = \sqrt{\frac{\pi}{\alpha}} \exp\left(\frac{\beta^2}{4\alpha}\right) \quad \text{for } \alpha \text{ real and positive, } \beta \text{ complex} \qquad [\text{J}\cdot 13]$$

If $\quad J_n = \int_{-\infty}^{\infty} x^n \exp(-\alpha x^2 - \beta x)\, dx \quad$ then

$$J_{n+1} = -\frac{\partial J_n}{\partial \beta} \quad \text{for } \beta \text{ complex} \qquad [\text{J}\cdot 14]$$

table J • 1 *Standardized normal distribution function* $Z(u)$ *and its cumulative probability function* $P(u)$ *for 3 standard deviations*

u	$P(u)$	$Z(u)$	u	$P(u)$	$Z(u)$	u	$P(u)$	$Z(u)$
0.00	0.5000	0.3989	1.00	0.8413	0.2420	2.00	0.9772	0.0540
0.01	0.5040	0.3989	1.01	0.8438	0.2396	2.01	0.9778	0.0529
0.02	0.5080	0.3989	1.02	0.8461	0.2371	2.02	0.9783	0.0519
0.03	0.5120	0.3988	1.03	0.8485	0.2347	2.03	0.9788	0.0508
0.04	0.5160	0.3986	1.04	0.8508	0.2323	2.04	0.9793	0.0498
0.05	0.5199	0.3984	1.05	0.8531	0.2299	2.05	0.9798	0.0488
0.06	0.5239	0.3982	1.06	0.8554	0.2275	2.06	0.9803	0.0478
0.07	0.5279	0.3980	1.07	0.8577	0.2251	2.07	0.9808	0.0468
0.08	0.5319	0.3977	1.08	0.8599	0.2227	2.08	0.9812	0.0459
0.09	0.5359	0.3973	1.09	0.8621	0.2203	2.09	0.9817	0.0449
0.10	0.5398	0.3970	1.10	0.8643	0.2179	2.10	0.9821	0.0440
0.11	0.5438	0.3965	1.11	0.8665	0.2155	2.11	0.9826	0.0431
0.12	0.5478	0.3961	1.12	0.8686	0.2131	2.12	0.9830	0.0422
0.13	0.5517	0.3956	1.13	0.8708	0.2107	2.13	0.9834	0.0413
0.14	0.5557	0.3951	1.14	0.8729	0.2083	2.14	0.9838	0.0404
0.15	0.5596	0.3945	1.15	0.8749	0.2059	2.15	0.9842	0.0395
0.16	0.5636	0.3939	1.16	0.8770	0.2036	2.16	0.9846	0.0387
0.17	0.5675	0.3932	1.17	0.8790	0.2012	2.17	0.9850	0.0379
0.18	0.5714	0.3925	1.18	0.8810	0.1989	2.18	0.9854	0.0371
0.19	0.5753	0.3918	1.19	0.8830	0.1965	2.19	0.9857	0.0363
0.20	0.5793	0.3910	1.20	0.8849	0.1942	2.20	0.9861	0.0355
0.21	0.5832	0.3902	1.21	0.8869	0.1919	2.21	0.9864	0.0347
0.22	0.5871	0.3894	1.22	0.8888	0.1895	2.22	0.9868	0.0339
0.23	0.5910	0.3885	1.23	0.8907	0.1872	2.23	0.9871	0.0332
0.24	0.5948	0.3876	1.24	0.8925	0.1849	2.24	0.9875	0.0325
0.25	0.5987	0.3867	1.25	0.8944	0.1826	2.25	0.9878	0.0317
0.26	0.6026	0.3857	1.26	0.8962	0.1804	2.26	0.9881	0.0310
0.27	0.6064	0.3847	1.27	0.8980	0.1781	2.27	0.9884	0.0303
0.28	0.6103	0.3836	1.28	0.8997	0.1758	2.28	0.9887	0.0297

0.29	0.6141	0.3825	1.29	0.9015	0.1736	2.29	0.9890	0.0290
0.30	0.6179	0.3814	1.30	0.9032	0.1714	2.30	0.9893	0.0283
0.31	0.6217	0.3802	1.31	0.9049	0.1691	2.31	0.9896	0.0277
0.32	0.6255	0.3790	1.32	0.9066	0.1669	2.32	0.9898	0.0270
0.33	0.6293	0.3778	1.33	0.9082	0.1647	2.33	0.9901	0.0264
0.34	0.6331	0.3765	1.34	0.9099	0.1626	2.34	0.9904	0.0258
0.35	0.6368	0.3752	1.35	0.9115	0.1604	2.35	0.9906	0.0252
0.36	0.6406	0.3739	1.36	0.9131	0.1582	2.36	0.9909	0.0246
0.37	0.6443	0.3725	1.37	0.9147	0.1561	2.37	0.9911	0.0241
0.38	0.6480	0.3712	1.38	0.9162	0.1539	2.38	0.9913	0.0235
0.39	0.6517	0.3697	1.39	0.9177	0.1518	2.39	0.9916	0.0229
0.40	0.6554	0.3683	1.40	0.9192	0.1497	2.40	0.9918	0.0224
0.41	0.6591	0.3668	1.41	0.9207	0.1476	2.41	0.9920	0.0219
0.42	0.6628	0.3653	1.42	0.9222	0.1456	2.42	0.9922	0.0213
0.43	0.6664	0.3637	1.43	0.9236	0.1435	2.43	0.9926	0.0208
0.44	0.6700	0.3621	1.44	0.9251	0.1415	2.44	0.9927	0.0203
0.45	0.6736	0.3605	1.45	0.9265	0.1394	2.45	0.9929	0.0198
0.46	0.6772	0.3589	1.46	0.9279	0.1374	2.46	0.9931	0.0194
0.47	0.6808	0.3572	1.47	0.9292	0.1354	2.47	0.9932	0.0189
0.48	0.6844	0.3555	1.48	0.9306	0.1334	2.48	0.9934	0.0184
0.49	0.6879	0.3538	1.49	0.9319	0.1315	2.49	0.9936	0.0180
0.50	0.6915	0.3521	1.50	0.9332	0.1295	2.50	0.9938	0.0175
0.51	0.6950	0.3503	1.51	0.9345	0.1276	2.51	0.9940	0.0171
0.52	0.6985	0.3485	1.52	0.9357	0.1257	2.52	0.9941	0.0167
0.53	0.7019	0.3467	1.53	0.9370	0.1238	2.53	0.9943	0.0163
0.54	0.7054	0.3448	1.54	0.9382	0.1219	2.54	0.9945	0.0158
0.55	0.7088	0.3429	1.55	0.9394	0.1200	2.55	0.9946	0.0154
0.56	0.7123	0.3410	1.56	0.9406	0.1182	2.56	0.9948	0.0151
0.57	0.7157	0.3391	1.57	0.9418	0.1163	2.57	0.9949	0.0147
0.58	0.7190	0.3372	1.58	0.9429	0.1145	2.58	0.9951	0.0143
0.59	0.7224	0.3352	1.59	0.9441	0.1127	2.59	0.9952	0.0139
0.60	0.7257	0.3332	1.60	0.9452	0.1109	2.60	0.9953	0.0136
0.61	0.7291	0.3312	1.61	0.9463	0.1092	2.61	0.9955	0.0132
0.62	0.7324	0.3292	1.62	0.9474	0.1074	2.62	0.9956	0.0129
0.63	0.7357	0.3271	1.63	0.9484	0.1057	2.63	0.9957	0.0126
0.64	0.7389	0.3251	1.64	0.9495	0.1040	2.64	0.9959	0.0122
0.65	0.7422	0.3230	1.65	0.9505	0.1023	2.65	0.9960	0.0119
0.66	0.7454	0.3209	1.66	0.9515	0.1006	2.66	0.9961	0.0116
0.67	0.7486	0.3187	1.67	0.9525	0.0989	2.67	0.9962	0.0113
0.68	0.7517	0.3166	1.68	0.9535	0.0973	2.68	0.9963	0.0110
0.69	0.7549	0.3144	1.69	0.9545	0.0957	2.69	0.9964	0.0107
0.70	0.7580	0.3123	1.70	0.9554	0.0940	2.70	0.9965	0.0104
0.71	0.7611	0.3101	1.71	0.9564	0.0925	2.71	0.9966	0.0101
0.72	0.7642	0.3079	1.72	0.9573	0.0909	2.72	0.9967	0.0099
0.73	0.7673	0.3056	1.73	0.9582	0.0893	2.73	0.9968	0.0096
0.74	0.7704	0.3034	1.74	0.9591	0.0878	2.74	0.9969	0.0093
0.75	0.7734	0.3011	1.75	0.9599	0.0863	2.75	0.9970	0.0091
0.76	0.7764	0.2989	1.76	0.9608	0.0848	2.76	0.9971	0.0088
0.77	0.7794	0.2966	1.77	0.9616	0.0833	2.77	0.9972	0.0086
0.78	0.7823	0.2943	1.78	0.9625	0.0818	2.78	0.9973	0.0084

0.79	0.7852	0.2920	1.79	0.9633	0.0804	2.79	0.9974	0.0081
0.80	0.7881	0.2897	1.80	0.9641	0.0790	2.80	0.9974	0.0079
0.81	0.7910	0.2874	1.81	0.9649	0.0775	2.81	0.9975	0.0077
0.82	0.7939	0.2850	1.82	0.9656	0.0761	2.82	0.9976	0.0075
0.83	0.7967	0.2827	1.83	0.9664	0.0748	2.83	0.9977	0.0073
0.84	0.7995	0.2803	1.84	0.9671	0.0734	2.84	0.9977	0.0071
0.85	0.8023	0.2780	1.85	0.9678	0.0721	2.85	0.9978	0.0069
0.86	0.8051	0.2756	1.86	0.9686	0.0707	2.86	0.9979	0.0067
0.87	0.8078	0.2732	1.87	0.9693	0.0694	2.87	0.9979	0.0065
0.88	0.8106	0.2709	1.88	0.9699	0.0681	2.88	0.9980	0.0063
0.89	0.8133	0.2685	1.89	0.9706	0.0669	2.89	0.9981	0.0061
0.90	0.8159	0.2661	1.90	0.9713	0.0656	2.90	0.9981	0.0059
0.91	0.8186	0.2637	1.91	0.9719	0.0644	2.91	0.9982	0.0058
0.92	0.8212	0.2613	1.92	0.9726	0.0632	2.92	0.9982	0.0056
0.93	0.8238	0.2589	1.93	0.9732	0.0620	2.93	0.9983	0.0055
0.94	0.8264	0.2565	1.94	0.9738	0.0608	2.94	0.9984	0.0053
0.95	0.8289	0.2541	1.95	0.9744	0.0596	2.95	0.9984	0.0051
0.96	0.8315	0.2516	1.96	0.9750	0.0584	2.96	0.9985	0.0050
0.97	0.8340	0.2492	1.97	0.9756	0.0573	2.97	0.9985	0.0048
0.98	0.8365	0.2468	1.98	0.9761	0.0562	2.98	0.9986	0.0047
0.99	0.8389	0.2444	1.99	0.9767	0.0551	2.99	0.9986	0.0046
						3.00	0.9987	0.0044

table J • 2 *A table of random digits*

14 92 47 16 07	74 63 80 19 39	77 04 41 82 77	68 22 90 93 10	63 25 88 50 64
60 10 25 65 87	29 84 21 20 31	00 79 49 06 38	24 86 27 24 54	19 38 08 08 95
33 39 33 01 09	42 00 54 79 77	26 45 31 53 33	45 39 08 98 62	70 35 13 69 53
92 59 27 82 32	19 32 82 96 45	69 83 69 11 69	95 18 81 50 46	45 67 99 45 00
27 42 17 33 56	53 39 72 69 16	52 01 52 68 54	42 83 36 42 57	49 42 53 94 40
63 91 98 39 77	72 41 65 54 50	47 65 20 10 13	17 56 05 41 89	73 52 33 36 81
97 59 57 83 00	93 50 22 58 47	19 52 01 68 96	69 28 21 69 17	04 19 42 57 97
17 11 64 26 49	89 25 26 10 52	78 33 21 51 02	47 70 72 83 69	60 40 01 59 08
34 03 78 45 73	49 91 45 96 88	18 58 45 34 98	94 90 51 49 14	23 26 15 81 33
08 81 31 44 53	70 43 74 90 93	69 88 09 99 44	40 48 20 42 54	62 84 26 80 89
33 86 71 71 21	77 86 11 59 89	35 66 32 30 46	74 16 37 51 08	07 99 97 08 21
95 67 26 34 40	79 34 76 76 88	56 11 31 13 83	51 70 76 73 90	92 52 09 10 99
60 63 30 80 53	46 74 33 59 51	06 52 09 02 32	30 39 05 46 91	35 98 62 24 80
56 61 92 28 00	20 40 16 36 02	73 04 25 08 42	69 24 74 98 46	50 84 02 45 73
54 94 72 92 93	71 19 09 59 08	67 84 53 76 43	68 53 10 25 67	33 52 37 90 01
31 14 38 74 32	86 31 65 87 39	82 98 06 28 19	96 94 67 15 78	84 37 64 95 99
26 01 19 06 92	39 46 74 63 09	15 97 66 67 63	10 48 57 05 68	00 20 45 86 88
71 87 71 17 86	01 60 45 74 81	87 00 76 47 31	38 08 98 82 60	43 69 92 54 38
15 87 65 77 25	10 24 25 04 92	84 28 65 27 05	23 77 28 08 93	13 45 05 08 11
89 93 70 05 05	33 43 90 79 06	85 50 49 01 22	70 68 65 09 41	76 77 10 25 89
67 11 54 10 50	65 45 30 45 10	35 51 88 88 43	15 80 18 61 63	49 91 34 68 49
14 59 66 28 02	33 00 97 40 95	34 54 97 66 34	64 27 06 74 91	83 94 15 51 54
70 25 82 69 48	60 86 34 76 66	58 51 29 14 04	97 35 78 57 10	03 41 45 65 91

83 36 98 11 79	32 24 93 47 81	00 00 66 59 51	40 25 47 11 02	60 41 64 32 96
83 81 14 93 58	75 17 13 47 68	70 07 07 65 50	14 25 29 73 78	79 50 09 71 23
26 52 82 25 50	36 69 22 52 62	61 41 95 39 86	73 89 88 02 37	71 38 73 25 32
07 66 41 07 52	95 02 11 70 30	76 78 48 68 61	12 78 04 43 79	31 26 11 84 24
37 95 47 29 29	29 52 40 91 19	13 44 42 26 45	78 34 57 98 76	37 58 27 55 49
66 49 51 12 33	81 10 16 93 28	84 83 87 63 95	35 99 70 72 59	66 25 44 74 51
72 50 62 85 24	39 59 54 93 56	92 73 34 29 15	47 77 47 54 62	64 40 96 91 16
89 54 12 83 45	22 88 77 92 69	88 76 57 24 64	41 58 65 22 41	50 50 23 98 90
81 46 50 04 49	09 93 11 26 67	30 27 86 39 63	84 19 22 52 66	30 86 26 19 96
55 62 06 82 69	73 24 61 25 54	01 20 06 32 34	41 27 72 04 77	71 89 90 53 47
36 73 55 83 41	35 28 75 11 12	47 32 28 13 54	51 08 07 27 71	53 65 34 21 08
44 56 18 12 04	68 17 59 30 37	47 96 77 40 85	05 90 58 46 17	47 67 11 84 56
83 96 78 48 10	51 61 28 15 01	10 44 04 85 53	67 91 97 36 78	83 14 94 70 96
49 99 00 20 51	64 57 57 03 23	07 86 35 92 20	01 74 69 42 13	82 97 74 23 84
65 33 21 64 88	04 43 33 79 90	93 51 71 83 95	72 84 89 70 25	38 89 44 66 11
42 30 90 40 63	55 29 17 21 65	96 92 15 33 50	34 17 33 99 08	11 65 48 71 45
84 71 54 18 08	12 08 57 25 25	91 17 10 97 79	07 22 37 55 77	20 93 49 20 47
09 71 37 02 70	28 07 71 56 34	36 52 59 12 04	48 44 35 05 61	92 39 46 06 36
51 85 27 07 82	93 49 84 46 69	79 64 70 68 21	04 02 00 01 11	24 73 70 55 21
31 46 02 22 22	18 80 40 06 75	57 56 29 79 23	45 97 47 51 58	94 91 84 01 44
14 32 80 50 24	04 33 30 41 35	53 14 37 92 07	85 86 30 49 62	89 54 96 33 08
84 00 29 80 73	32 99 80 57 49	45 53 82 61 64	68 67 29 74 86	32 54 19 94 19
53 48 83 72 53	44 30 25 39 06	28 24 38 08 53	87 74 69 25 30	35 75 46 97 37
42 65 21 94 38	90 78 28 64 82	84 57 53 55 09	84 07 45 23 50	53 12 33 75 36
97 58 83 26 81	74 25 83 50 37	81 84 75 66 86	29 56 61 50 20	20 67 79 62 58
23 79 03 04 32	93 47 09 97 68	41 62 43 03 53	60 42 81 75 54	50 80 10 60 13
07 19 29 91 11	58 89 33 18 56	07 16 37 96 63	05 86 72 70 43	71 74 75 60 66

APPENDIX K

Fourier series

In Chapter 16 we note that simple waves may combine to produce waves of any shape whatever, depending upon the relationships between their amplitudes and frequencies. In fact, Jean Fourier showed in 1822 that any periodic function can be represented by a linear combination of simple sinusoids whose frequencies are integer multiples of the frequency of the periodic function. This statement, now known as *Fourier's theorem,* may be rephrased as follows:

> Any function $f(\theta)$ which is defined and periodic over intervals $\Delta\theta = 2\pi$ with no more than a finite number of discontinuities over each interval can be represented as an infinite series of trigonometric functions:

$$f(\theta) = a_0 + \sum_{n=1}^{\infty}(a_n \cos n\theta + b_n \sin n\theta) \qquad [K \cdot 1]$$

If we accept Fourier's theorem as given, we can then evaluate the coefficients by suitable operations on both sides of Eq. [K . 1]. The Fourier coefficients are obtained by integration, making use of the following rules (which may be derived either through substitution or through integration by parts):

$$\int_0^{2\pi} \sin^2 mx\, dx = \int_0^{2\pi} \cos^2 mx\, dx = \pi \qquad \text{for any } m \neq 0 \qquad [K \cdot 2]$$

whereas

$$\int_0^{2\pi} \sin mx \cos mx\, dx = 0 \qquad \text{for any } m \qquad [K \cdot 3]$$

and

$$\int_0^{2\pi} \sin mx \sin nx\, dx = \int_0^{2\pi} \cos mx \cos nx\, dx = 0 \qquad \text{for any } m \neq n \qquad [K \cdot 4]$$

where m and n are integers. Thus if we multiply $f(\theta)$ by some $\cos m\theta$ and integrate over one period, all terms on the righthand side of Eq. [K . 1] become zero except

$$a_m = \frac{1}{\pi}\int_0^{2\pi} f(\theta) \cos m\theta\, d\theta \qquad \text{for } m \neq 0 \qquad [K \cdot 5]$$

Carrying out the same procedure with $\sin m\theta$, we find

$$b_m = \frac{1}{\pi}\int_0^{2\pi} f(\theta) \sin m\theta\, d\theta \qquad \text{for } m \neq 0 \qquad [K \cdot 6]$$

And because the average value of a trigonometric function is zero,

$$a_0 = \frac{1}{2\pi}\int_0^{2\pi} f(\theta)\,d\theta \qquad [\text{K}\cdot 7]$$

These quadratures may be performed analytically or numerically.

example K • 1 Compute the Fourier coefficients of a "square wave" whose equation is

$$f(\theta) = \begin{cases} 1 & \text{for } 0 \leq \theta < \pi \\ -1 & \text{for } \pi \leq \theta < 2\pi \end{cases}$$

and which repeats periodically every $\Delta\theta = \pm 2\pi$.

solution The cosine coefficients are identically zero:

$$a_m = \int_0^{\pi} \cos m\theta\,d\theta - \int_{\pi}^{2\pi} \cos m\theta\,d\theta = 0$$

However,

$$b_m = \int_0^{\pi} \sin m\theta\,d\theta - \int_{\pi}^{2\pi} \sin m\theta\,d\theta = -\frac{1}{m\pi}\left(\cos m\theta\Big|_0^{\pi} - \cos m\theta\Big|_{\pi}^{2\pi}\right)$$

$$= -\frac{2}{m\pi}[(-1)^m - 1]$$

Thus

$$b_m = \begin{cases} \dfrac{4}{m\pi} & \text{for } m \text{ odd} \\ 0 & \text{for } m \text{ even} \end{cases}$$

Therefore

$$f(\theta) = \frac{4}{\pi}\left(\sin\theta + \frac{1}{3}\sin 3\theta + \frac{1}{5}\sin 5\theta + \cdots\right)$$

$$= \frac{4}{\pi}\sum_{n=1}^{\infty} \frac{\sin(2n-1)\theta}{(2n-1)}$$

The contribution from each term in the Fourier series diminishes with increasing m. Fig. K . 1 illustrates the convergence of the Fourier series in Example K . 1 toward the desired discontinuous function.

In theory it is possible not only to add simple waves to produce any complex wave form, but conversely to perform a Fourier analysis of any complex form $f(\theta)$ to determine its simple Fourier components a_n and b_n.

Note in Fig. K . 1 that the Fourier-series approximation seems to oscillate about the desired function, with the oscillations growing smaller and smaller as we add more terms to the approximation. We state without proof that this is a fundamental property of any Fourier series:

> Given a finite number of terms in a Fourier series (starting with the lowest-order terms), the coefficients a_n and b_n are exactly those that

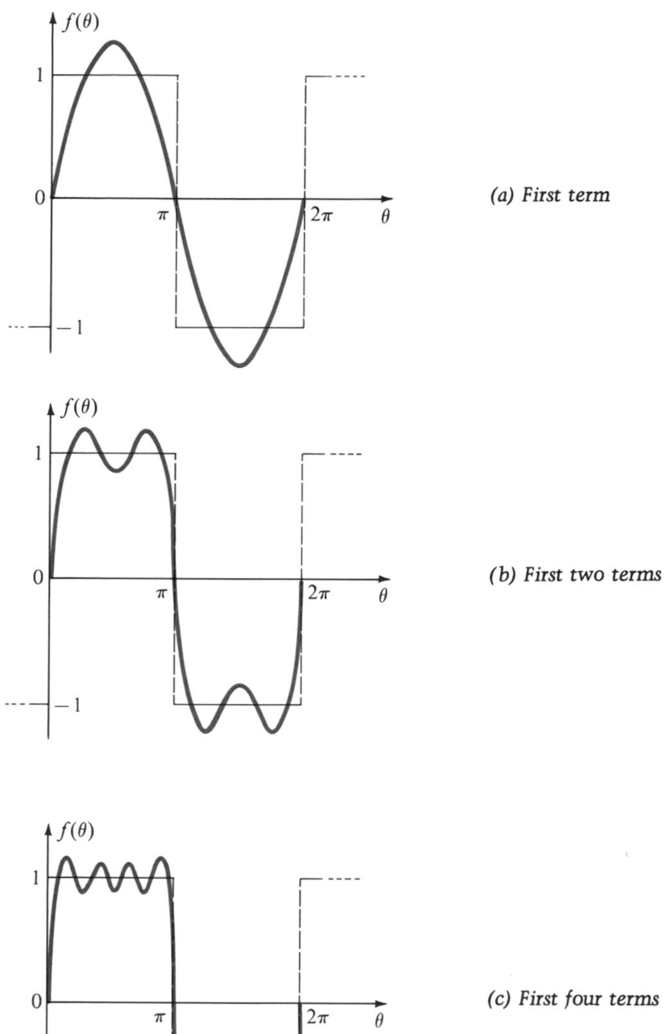

figure K • 1 *Fourier-series approximation of a square wave*

minimize the average squared distance between the curve $f(\theta)$ and the Fourier-series approximation to the curve as represented by the truncated Fourier series.

APPENDIX L

fundamental physical constants

table L • 1 *Fundamental physical constants*

symbol	*value*	*name*
α	$(7.29735 \pm 0.00001) \times 10^{-3}$	Fine structure constant
ϵ_0	8.85419×10^{-12} F/m	Permittivity of vacuum
$1/4\pi\epsilon_0$	8.98755×10^{9} m/F	
μ_0	1.25664×10^{-6} H/m	Permeability of vacuum
$\mu_0/4\pi$	1.00000×10^{-7} m/H	
μ_B	$(9.27408 \pm 0.00004) \times 10^{-24}$ J/T	Bohr magneton
σ	$(5.67032 \pm 0.00071) \times 10^{-8}$ W/m$^2 \cdot$ K^4	Stefan-Boltzmann constant
a_0	5.29177×10^{-11} m	Bohr radius
c	2.99793×10^{8} m/s	Speed of light in vacuum
c_1	$(3.74183 \pm 0.00002) \times 10^{-16}$ W $\cdot$ m^2	First radiation constant
c_2	$(1.43879 \pm 0.00005) \times 10^{-2}$ m $\cdot$ K	Second radiation constant
e	1.60219×10^{-19} C	Elementary charge
e/m_e	1.75880×10^{11} C/kg	
F	$(9.64846 \pm 0.00003) \times 10^{7}$ C/kmole	Faraday constant
G	$(6.6720 \pm 0.0041) \times 10^{-11}$ N $\cdot$ m^2/kg^2	Gravitational constant
h	$(6.62618 \pm 0.00004) \times 10^{-34}$ J $\cdot$ s	Planck constant
h	$(4.13570 \pm 0.00001) \times 10^{-15}$ eV $\cdot$ s	Planck constant
$\hbar = h/2\pi$	$(1.05459 \pm 0.00001) \times 10^{-34}$ J $\cdot$ s	
k	$(1.38066 \pm 0.00004) \times 10^{-23}$ J/K	Boltzmann constant
k	$(8.61735 \pm 0.00028) \times 10^{-5}$ eV/K	Boltzmann constant
m_e	$(9.10953 \pm 0.00005) \times 10^{-31}$ kg	Electron rest mass
m_e	$(5.11003 \pm 0.00001) \times 10^{5}$ eV	Electron rest mass
m_n	$(1.67495 \pm 0.00001) \times 10^{-27}$ kg	Neutron rest mass
m_n	$(9.39573 \pm 0.00003) \times 10^{8}$ eV	Neutron rest mass
m_p	$(.167265 \pm 0.00001) \times 10^{-27}$ kg	Proton rest mass
m_p	$(9.38280 \pm 0.00003) \times 10^{8}$ eV	Proton rest mass

***table L • 1** (continued)*

symbol	*value*	*name*
m_p/m_e	1.83615×10^3	
m_μ	$(1.88357 \pm 0.00001) \times 10^{-28}$ kg	Muon rest mass
m_μ	1.05659×10^8 eV	Muon rest mass
N_A	$(6.02205 \pm 0.00003) \times 10^{26}$ kmole^{-1}	Avogardro constant
R	$(8.31441 \pm 0.00026) \times 10^3$ J/kmole · K	Universal gas constant
R	1.98648 kcal/kmole · K	Universal gas constant
R_∞	1.09737×10^7 m^{-1}	Rydberg constant

Note: The constants are given to six significant digits; experimental uncertainties are indicated where they affect the digits listed. For more accurate data, refer to the text (see index) or to any of the standard reference works. Constants from both Volume I and Volume II are listed here.

Source: International Union of Pure and Applied Physics, *Symbols, Units and Nomenclature in Physics* (The Netherlands: IUPAC, Document U.I.P. 20, 1978).

table L • 2 *Useful physical constants*

symbol	*value*	*name*
ρ_0	1.2929 kg/m^2	Density of dry air (at 0°C, 1 atm)
c_a	331.4 m/s	Speed of sound in dry air (at 0°C, 1 atm)
g_s	9.80665 m/s^2	Standard acceleration due to gravity
L_f	79.7 kcal/kg	Heat of fusion of water (at 0°C, 1 atm)
L_v	539.6 kcal/kg	Heat of vaporization of water (at 100°C, 1 atm)
M_{air}	28.98 kg/kmole	Average molecular weight of air (at 0°C, 1 atm)
n_0	$(2.68675 \pm 0.00009) \times 10^{25}$ m^{-3}	Loschmidt constant (number density of molecules in an ideal gas at 0°C, 1 atm)
T_0	273.15 K	Freezing point of water (at 1 atm)
V_0	(22.4138 ± 0.0007) m^3/kmole	Molar volume of an ideal gas (at 0°C, 1 atm)

subject index

name index

Note: Numbers in italic refer to chapter-opening quotations.